Fundamentals of Error-Correcting Codes

Fundamentals of Error-Correcting Codes is an in-depth introduction to coding theory from both an engineering and mathematical viewpoint. As well as covering classical topics, much coverage is included of recent techniques that until now could only be found in specialist journals and book publications. Numerous exercises and examples and an accessible writing style make this a lucid and effective introduction to coding theory for advanced undergraduate and graduate students, researchers and engineers, whether approaching the subject from a mathematical, engineering, or computer science background.

Professor W. Cary Huffman graduated with a PhD in mathematics from the California Institute of Technology in 1974. He taught at Dartmouth College and Union College until he joined the Department of Mathematics and Statistics at Loyola in 1978, serving as chair of the department from 1986 through 1992. He is an author of approximately 40 research papers in finite group theory, combinatorics, and coding theory, which have appeared in journals such as the *Journal of Algebra*, *IEEE Transactions on Information Theory*, and the *Journal of Combinatorial Theory*.

Professor Vera Pless was an undergraduate at the University of Chicago and received her PhD from Northwestern in 1957. After ten years at the Air Force Cambridge Research Laboratory, she spent a few years at MIT's project MAC. She joined the University of Illinois-Chicago's Department of Mathematics, Statistics, and Computer Science as a full professor in 1975 and has been there ever since. She is a University of Illinois Scholar and has published over 100 papers.

Fundamentals of
Error-Correcting Codes

W. Cary Huffman
Loyola University of Chicago

and

Vera Pless
University of Illinois at Chicago

CAMBRIDGE
UNIVERSITY PRESS

CAMBRIDGE UNIVERSITY PRESS
Cambridge, New York, Melbourne, Madrid, Cape Town, Singapore,
São Paulo, Delhi, Dubai, Tokyo

Cambridge University Press
The Edinburgh Building, Cambridge CB2 8RU, UK

Published in the United States of America by Cambridge University Press, New York

www.cambridge.org
Information on this title: www.cambridge.org/9780521131704

First published 2003
This digitally printed version 2010

A catalogue record for this publication is available from the British Library

Library of Congress Cataloguing in Publication data

Huffman, W. C. (William Cary)
Fundamentals of error-correcting codes / W. Cary Huffman, Vera Pless.
    p.   cm.
Includes bibliographical references and index.
ISBN 0 521 78280 5
1. Error-correcting codes (Information theory) I. Pless, Vera. II. Title.
QA268 .H84 2003
005.7′2 – dc21 2002067236

ISBN 978-0-521-78280-7 Hardback
ISBN 978-0-521-13170-4 Paperback

Contents

3   Finite fields

4   Cyclic codes

14 Convolutional codes

15 Soft decision and iterative decoding

Preface

Coding theory originated with the 1948 publication of the paper "A mathematical theory of communication" by Claude Shannon. For the past half century, coding theory has grown into a discipline intersecting mathematics and engineering with applications to almost every area of communication such as satellite and cellular telephone transmission, compact disc recording, and data storage.

During the 50th anniversary year of Shannon's seminal paper, the two volume *Handbook of Coding Theory*, edited by the authors of the current text, was published by Elsevier Science. That Handbook, with contributions from 33 authors, covers a wide range of topics at the frontiers of research. As editors of the Handbook, we felt it would be appropriate to produce a textbook that could serve in part as a bridge to the Handbook. This textbook is intended to be an in-depth introduction to coding theory from both a mathematical and engineering viewpoint suitable either for the classroom or for individual study. Several of the topics are classical, while others cover current subjects that appear only in specialized books and journal publications. We hope that the presentation in this book, with its numerous examples and exercises, will serve as a lucid introduction that will enable readers to pursue some of the many themes of coding theory.

Fundamentals of Error-Correcting Codes is a largely self-contained textbook suitable for advanced undergraduate students and graduate students at any level. A prerequisite for this book is a course in linear algebra. A course in abstract algebra is recommended, but not essential. This textbook could be used for at least three semesters. A wide variety of examples illustrate both theory and computation. Over 850 exercises are interspersed at points in the text where they are most appropriate to attempt. Most of the theory is accompanied by detailed proofs, with some proofs left to the exercises. Because of the number of examples and exercises that directly illustrate the theory, the instructor can easily choose either to emphasize or deemphasize proofs.

In this preface we briefly describe the contents of the 15 chapters and give a suggested outline for the first semester. We also propose blocks of material that can be combined in a variety of ways to make up subsequent courses. Chapter 1 is basic with the introduction of linear codes, generator and parity check matrices, dual codes, weight and distance, encoding and decoding, and the Sphere Packing Bound. The Hamming codes, Golay codes, binary Reed–Muller codes, and the hexacode are introduced. Shannon's Theorem for the binary symmetric channel is discussed. Chapter 1 is certainly essential for the understanding of the remainder of the book.

Chapter 2 covers the main upper and lower bounds on the size of linear and nonlinear codes. These include the Plotkin, Johnson, Singleton, Elias, Linear Programming, Griesmer,

Gilbert, and Varshamov Bounds. Asymptotic versions of most of these are included. MDS codes and lexicodes are introduced.

Chapter 3 is an introduction to constructions and properties of finite fields, with a few proofs omitted. A quick treatment of this chapter is possible if the students are familiar with constructing finite fields, irreducible polynomials, factoring polynomials over finite fields, and Galois theory of finite fields. Much of Chapter 3 is immediately used in the study of cyclic codes in Chapter 4. Even with a background in finite fields, cyclotomic cosets (Section 3.7) may be new to the student.

Chapter 4 gives the basic theory of cyclic codes. Our presentation interrelates the concepts of idempotent generator, generator polynomial, zeros of a code, and defining sets. Multipliers are used to explore equivalence of cyclic codes. Meggitt decoding of cyclic codes is presented as are extended cyclic and affine-invariant codes.

Chapter 5 looks at the special families of BCH and Reed–Solomon cyclic codes as well as generalized Reed–Solomon codes. Four decoding algorithms for these codes are presented. Burst errors and the technique of concatenation for handling burst errors are introduced with an application of these ideas to the use of Reed–Solomon codes in the encoding and decoding of compact disc recorders.

Continuing with the theory of cyclic codes, Chapter 6 presents the theory of duadic codes, which include the family of quadratic residue codes. Because the complete theory of quadratic residue codes is only slightly simpler than the theory of duadic codes, the authors have chosen to present the more general codes and then apply the theory of these codes to quadratic residue codes. Idempotents of binary and ternary quadratic residue codes are explicitly computed. As a prelude to Chapter 8, projective planes are introduced as examples of combinatorial designs held by codewords of a fixed weight in a code.

Chapter 7 expands on the concept of weight distribution defined in Chapter 1. Six equivalent forms of the MacWilliams equations, including the Pless power moments, that relate the weight distributions of a code and its dual, are formulated. MDS codes, introduced in Chapter 2, and coset weight distributions, introduced in Chapter 1, are revisited in more depth. A proof of a theorem of MacWilliams on weight preserving transformations is given in Section 7.9.

Chapter 8 delineates the basic theory of block designs particularly as they arise from the supports of codewords of fixed weight in certain codes. The important theorem of Assmus–Mattson is proved. The theory of projective planes in connection with codes, first introduced in Chapter 6, is examined in depth, including a discussion of the nonexistence of the projective plane of order 10.

Chapter 9 consolidates much of the extensive literature on self-dual codes. The Gleason–Pierce–Ward Theorem is proved showing why binary, ternary, and quaternary self-dual codes are the most interesting self-dual codes to study. Gleason polynomials are introduced and applied to the determination of bounds on the minimum weight of self-dual codes. Techniques for classifying self-dual codes are presented. Formally self-dual codes and additive codes over $\mathbb{F}_4$, used in correcting errors in quantum computers, share many properties of self-dual codes; they are introduced in this chapter.

The Golay codes and the hexacode are the subject of Chapter 10. Existence and uniqueness of these codes are proved. The Pless symmetry codes, which generalize the ternary Golay codes, are defined and some of their properties are given. The connection between codes and lattices is developed in the final section of the chapter.

The theory of the covering radius of a code, first introduced in Chapter 1, is the topic of Chapter 11. The covering radii of BCH codes, Reed–Muller codes, self-dual codes, and subcodes are examined. The length function, a basic tool in finding bounds on the covering radius, is presented along with many of its properties.

Chapter 12 examines linear codes over the ring $\mathbb{Z}_4$ of integers modulo 4. The theory of these codes is compared and contrasted with the theory of linear codes over fields. Cyclic, quadratic residue, and self-dual linear codes over $\mathbb{Z}_4$ are defined and analyzed. The nonlinear binary Kerdock and Preparata codes are presented as the Gray image of certain linear codes over $\mathbb{Z}_4$, an amazing connection that explains many of the remarkable properties of these nonlinear codes. To study these codes, Galois rings are defined, analogously to extension fields of the binary field.

Chapter 13 presents a brief introduction to algebraic geometry which is sufficient for a basic understanding of algebraic geometry codes. Goppa codes, generalized Reed–Solomon codes, and generalized Reed–Muller codes can be realized as algebraic geometry codes. A family of algebraic geometry codes has been shown to exceed the Gilbert–Varshamov Bound, a result that many believed was not possible.

Until Chapter 14, the codes considered were block codes where encoding depended only upon the current message. In Chapter 14 we look at binary convolutional codes where each codeword depends not only on the current message but on some messages in the past as well. These codes are studied as linear codes over the infinite field of binary rational functions. State and trellis diagrams are developed for the Viterbi Algorithm, one of the main decoding algorithms for convolutional codes. Their generator matrices and free distance are examined.

Chapter 15 concludes the textbook with a look at soft decision and iterative decoding. Until this point, we had only examined hard decision decoding. We begin with a more detailed look at communication channels, particularly those subject to additive white Gaussian noise. A soft decision Viterbi decoding algorithm is developed for convolutional codes. Low density parity check codes and turbo codes are defined and a number of decoders for these codes are examined. The text concludes with a brief history of the application of codes to deep space exploration.

The following chapters and sections of this book are recommended as an introductory one-semester course in coding theory:
- Chapter 1 (except Section 1.7),
- Sections 2.1, 2.3.4, 2.4, 2.7–2.9,
- Chapter 3 (except Section 3.8),
- Chapter 4 (except Sections 4.6 and 4.7),
- Chapter 5 (except Sections 5.4.3, 5.4.4, 5.5, and 5.6), and
- Sections 7.1–7.3.

If it is unlikely that a subsequent course in coding theory will be taught, the material in Chapter 7 can be replaced by the last two sections of Chapter 5. This material will show how a compact disc is encoded and decoded, presenting a nice real-world application that students can relate to.

For subsequent semesters of coding theory, we suggest a combination of some of the following blocks of material. With each block we have included sections that will hopefully make the blocks self-contained under the assumption that the first course given above has been completed. Certainly other blocks are possible. A semester can be made up of more than one block. Later we give individual chapters or sections that stand alone and can be used in conjunction with each other or with some of these blocks. The sections and chapters are listed in the order they should be covered.

- Sections 1.7, 8.1–8.4, 9.1–9.7, and Chapter 10. Sections 8.1–8.4 of this block present the essential material relating block designs to codes with particular emphasis on designs arising from self-dual codes. The material from Chapter 9 gives an in-depth study of self-dual codes with connections to designs. Chapter 10 studies the Golay codes and hexacode in great detail, again using designs to help in the analysis. Section 2.11 can be added to this block as the binary Golay codes are lexicodes.
- Sections 1.7, 7.4–7.10, Chapters 8, 9, and 10, and Section 2.11. This is an extension of the above block with more on designs from codes and codes from designs. It also looks at weight distributions in more depth, part of which is required in Section 9.12. Codes closely related to self-dual codes are also examined. This block may require an entire semester.
- Sections 4.6, 5.4.3, 5.4.4, 5.5, 5.6, and Chapters 14 and 15. This block covers most of the decoding algorithms described in the text but not studied in the first course, including both hard and soft decision decoding. It also introduces the important classes of convolutional and turbo codes that are used in many applications particularly in deep space communication. This would be an excellent block for engineering students or others interested in applications.
- Sections 2.2, 2.3, 2.5, 2.6, 2.10, and Chapter 13. This block finishes the nonasymptotic bounds not covered in the first course and presents the asymptotic versions of these bounds. The algebraic geometry codes and Goppa codes are important for, among other reasons, their relationship to the bounds on families of codes.
- Section 1.7 and Chapters 6 and 12. This block studies two families of codes extensively: duadic codes, which include quadratic residue codes, and linear codes over $\mathbb{Z}_4$. There is some overlap between the two chapters to warrant studying them together. When presenting Section 12.5.1, ideas from Section 9.6 should be discussed. Similarly it is helpful to examine Section 10.6 before presenting Section 12.5.3.

The following mini-blocks and chapters could be used in conjunction with one another or with the above blocks to construct a one-semester course.

- Section 1.7 and Chapter 6. Chapter 6 can stand alone after Section 1.7 is covered.
- Sections 1.7, 8.1–8.4, Chapter 10, and Section 2.11. This mini-block gives an in-depth study of the Golay codes and hexacode with the prerequisite material on designs covered first.
- Section 1.7 and Chapter 12. After Section 1.7 is covered, Chapter 12 can be used alone with the exception of Sections 12.4 and 12.5. Section 12.4 can either be omitted or supplemented with material from Section 6.6. Section 12.5 can either be skipped or supplemented with material from Sections 9.6 and 10.6.
- Chapter 11. This chapter can stand alone.
- Chapter 14. This chapter can stand alone.

The authors would like to thank a number of people for their advice and suggestions for this book. Philippe Gaborit tested portions of the text in its earliest form in a coding theory course he taught at the University of Illinois at Chicago resulting in many helpful insights. Philippe also provided some of the data used in the tables in Chapter 6. Judy Walker's monograph [343] on algebraic geometry codes was invaluable when we wrote Chapter 13; Judy kindly read this chapter and offered many helpful suggestions. Ian Blake and Frank Kschischang read and critiqued Chapters 14 and 15 providing valuable direction. Bob McEliece provided data for some of the figures in Chapter 15. The authors also wish to thank the staff and associates of Cambridge University Press for their valuable assistance with production of this book. In particular we thank editorial manager Dr. Philip Meyler, copy editor Dr. Lesley J. Thomas, and production editor Ms. Lucille Murby. Finally, the authors would like to thank their students in coding theory courses whose questions and comments helped refine the text. In particular Jon Lark Kim at the University of Illinois at Chicago and Robyn Canning at Loyola University of Chicago were most helpful.

We have taken great care to read and reread the text, check the examples, and work the exercises in an attempt to eliminate errors. As with all texts, errors are still likely to exist. The authors welcome corrections to any that the readers find. We can be reached at our e-mail addresses below.

W. Cary Huffman
wch@math.luc.edu

Vera Pless
pless@math.uic.edu

February 1, 2003

1 Basic concepts of linear codes

In 1948 Claude Shannon published a landmark paper "A mathematical theory of communication" [306] that signified the beginning of both information theory and coding theory. Given a communication channel which may corrupt information sent over it, Shannon identified a number called the capacity of the channel and proved that arbitrarily reliable communication is possible at any rate below the channel capacity. For example, when transmitting images of planets from deep space, it is impractical to retransmit the images. Hence if portions of the data giving the images are altered, due to noise arising in the transmission, the data may prove useless. Shannon's results guarantee that the data can be encoded before transmission so that the altered data can be decoded to the specified degree of accuracy. Examples of other communication channels include magnetic storage devices, compact discs, and any kind of electronic communication device such as cellular telephones.

The common feature of communication channels is that information is emanating from a source and is sent over the channel to a receiver at the other end. For instance in deep space communication, the message source is the satellite, the channel is outer space together with the hardware that sends and receives the data, and the receiver is the ground station on Earth. (Of course, messages travel from Earth to the satellite as well.) For the compact disc, the message is the voice, music, or data to be placed on the disc, the channel is the disc itself, and the receiver is the listener. The channel is "noisy" in the sense that what is received is not always the same as what was sent. Thus if binary data is being transmitted over the channel, when a 0 is sent, it is hopefully received as a 0 but sometimes will be received as a 1 (or as unrecognizable). Noise in deep space communications can be caused, for example, by thermal disturbance. Noise in a compact disc can be caused by fingerprints or scratches on the disc. The fundamental problem in coding theory is to determine what message was sent on the basis of what is received.

A communication channel is illustrated in Figure 1.1. At the source, a message, denoted $\mathbf{x}$ in the figure, is to be sent. If no modification is made to the message and it is transmitted directly over the channel, any noise would distort the message so that it is not recoverable. The basic idea is to embellish the message by adding some redundancy to it so that hopefully the received message is the original message that was sent. The redundancy is added by the encoder and the embellished message, called a codeword $\mathbf{c}$ in the figure, is sent over the channel where noise in the form of an error vector $\mathbf{e}$ distorts the codeword producing a received vector $\mathbf{y}$.[1] The received vector is then sent to be decoded where the errors are

---

[1] Generally our codeword symbols will come from a field $\mathbb{F}_q$, with q elements, and our messages and codewords will be vectors in vector spaces $\mathbb{F}_q^k$ and $\mathbb{F}_q^n$, respectively; if $\mathbf{c}$ entered the channel and $\mathbf{y}$ exited the channel, the difference $\mathbf{y} - \mathbf{c}$ is what we have termed the error $\mathbf{e}$ in Figure 1.1.

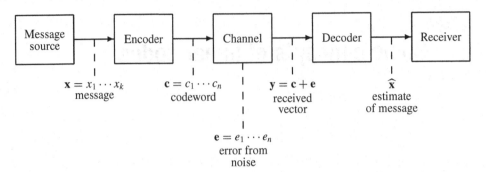

Figure 1.1 Communication channel.

removed, the redundancy is then stripped off, and an estimate $\widehat{\mathbf{x}}$ of the original message is produced. Hopefully $\widehat{\mathbf{x}} = \mathbf{x}$. (There is a one-to-one correspondence between codewords and messages. Thus we will often take the point of view that the job of the decoder is to obtain an estimate $\widehat{\mathbf{y}}$ of $\mathbf{y}$ and hope that $\widehat{\mathbf{y}} = \mathbf{c}$.) Shannon's Theorem guarantees that our hopes will be fulfilled a certain percentage of the time. With the right encoding based on the characteristics of the channel, this percentage can be made as high as we desire, although not 100%.

The proof of Shannon's Theorem is probabilistic and nonconstructive. In other words, no specific codes were produced in the proof that give the desired accuracy for a given channel. Shannon's Theorem only guarantees their existence. The goal of research in coding theory is to produce codes that fulfill the conditions of Shannon's Theorem. In the pages that follow, we will present many codes that have been developed since the publication of Shannon's work. We will describe the properties of these codes and on occasion connect these codes to other branches of mathematics. Once the code is chosen for application, encoding is usually rather straightforward. On the other hand, decoding efficiently can be a much more difficult task; at various points in this book we will examine techniques for decoding the codes we construct.

1.1   Three fields

Among all types of codes, linear codes are studied the most. Because of their algebraic structure, they are easier to describe, encode, and decode than nonlinear codes. The code alphabet for linear codes is a finite field, although sometimes other algebraic structures (such as the integers modulo 4) can be used to define codes that are also called "linear."

In this chapter we will study linear codes whose alphabet is a field $\mathbb{F}_q$, also denoted GF(q), with q elements. In Chapter 3, we will give the structure and properties of finite fields. Although we will present our general results over arbitrary fields, we will often specialize to fields with two, three, or four elements.

A field is an algebraic structure consisting of a set together with two operations, usually called addition (denoted by $+$) and multiplication (denoted by $\cdot$ but often omitted), which satisfy certain axioms. Three of the fields that are very common in the study of linear codes are the *binary* field with two elements, the *ternary* field with three elements, and the *quaternary* field with four elements. One can work with these fields by knowing their addition and multiplication tables, which we present in the next three examples.

Example 1.1.1 The binary field $\mathbb{F}_2$ with two elements $\{0, 1\}$ has the following addition and multiplication tables:

+	0	1
0	0	1
1	1	0

·	0	1
0	0	0
1	0	1

This is also the ring of integers modulo 2. ∎

Example 1.1.2 The ternary field $\mathbb{F}_3$ with three elements $\{0, 1, 2\}$ has addition and multiplication tables given by addition and multiplication modulo 3:

+	0	1	2
0	0	1	2
1	1	2	0
2	2	0	1

·	0	1	2
0	0	0	0
1	0	1	2
2	0	2	1

∎

Example 1.1.3 The quaternary field $\mathbb{F}_4$ with four elements $\{0, 1, \omega, \overline{\omega}\}$ is more complicated. It has the following addition and multiplication tables; $\mathbb{F}_4$ is not the ring of integers modulo 4:

+	0	1	ω	$\overline{\omega}$
0	0	1	ω	$\overline{\omega}$
1	1	0	$\overline{\omega}$	ω
ω	ω	$\overline{\omega}$	0	1
$\overline{\omega}$	$\overline{\omega}$	ω	1	0

·	0	1	ω	$\overline{\omega}$
0	0	0	0	0
1	0	1	ω	$\overline{\omega}$
ω	0	ω	$\overline{\omega}$	1
$\overline{\omega}$	0	$\overline{\omega}$	1	ω

Some fundamental equations are observed in these tables. For instance, one notices that $x + x = 0$ for all $x \in \mathbb{F}_4$. Also $\overline{\omega} = \omega^2 = 1 + \omega$ and $\omega^3 = \overline{\omega}^3 = 1$. ∎

1.2 Linear codes, generator and parity check matrices

Let $\mathbb{F}_q^n$ denote the vector space of all n-tuples over the finite field $\mathbb{F}_q$. An (n, M) *code* $\mathcal{C}$ over $\mathbb{F}_q$ is a subset of $\mathbb{F}_q^n$ of size M. We usually write the vectors $(a_1, a_2, \ldots, a_n)$ in $\mathbb{F}_q^n$ in the form $a_1 a_2 \cdots a_n$ and call the vectors in $\mathcal{C}$ *codewords*. Codewords are sometimes specified in other ways. The classic example is the polynomial representation used for codewords in cyclic codes; this will be described in Chapter 4. The field $\mathbb{F}_2$ of Example 1.1.1 has had a very special place in the history of coding theory, and codes over $\mathbb{F}_2$ are called *binary codes*. Similarly codes over $\mathbb{F}_3$ are termed *ternary codes*, while codes over $\mathbb{F}_4$ are called *quaternary codes*. The term "quaternary" has also been used to refer to codes over the ring $\mathbb{Z}_4$ of integers modulo 4; see Chapter 12.

Without imposing further structure on a code its usefulness is somewhat limited. The most useful additional structure to impose is that of linearity. To that end, if C is a k-dimensional subspace of $\mathbb{F}_q^n$, then C will be called an $[n, k]$ *linear code* over $\mathbb{F}_q$. The linear code C has q^k codewords. The two most common ways to present a linear code are with either a generator matrix or a parity check matrix. A *generator matrix* for an $[n, k]$ code C is any $k \times n$ matrix G whose rows form a basis for C. In general there are many generator matrices for a code. For any set of k independent columns of a generator matrix G, the corresponding set of coordinates forms an *information set* for C. The remaining $r = n - k$ coordinates are termed a *redundancy set* and r is called the *redundancy* of C. If the first k coordinates form an information set, the code has a unique generator matrix of the form $[I_k \mid A]$ where I_k is the $k \times k$ identity matrix. Such a generator matrix is in *standard form*. Because a linear code is a subspace of a vector space, it is the kernel of some linear transformation. In particular, there is an $(n - k) \times n$ matrix H, called a *parity check matrix* for the $[n, k]$ code C, defined by

$$C = \left\{ \mathbf{x} \in \mathbb{F}_q^n \mid H\mathbf{x}^{\mathsf{T}} = \mathbf{0} \right\}. \tag{1.1}$$

Note that the rows of H will also be independent. In general, there are also several possible parity check matrices for C. The next theorem gives one of them when C has a generator matrix in standard form. In this theorem A^{T} is the transpose of A.

Theorem 1.2.1 *If $G = [I_k \mid A]$ is a generator matrix for the $[n, k]$ code C in standard form, then $H = [-A^{\mathsf{T}} \mid I_{n-k}]$ is a parity check matrix for C.*

Proof: We clearly have $HG^{\mathsf{T}} = -A^{\mathsf{T}} + A^{\mathsf{T}} = O$. Thus C is contained in the kernel of the linear transformation $\mathbf{x} \mapsto H\mathbf{x}^{\mathsf{T}}$. As H has rank $n - k$, this linear transformation has kernel of dimension k, which is also the dimension of C. The result follows. $\square$

Exercise 1 Prior to the statement of Theorem 1.2.1, it was noted that the rows of the $(n - k) \times n$ parity check matrix H satisfying (1.1) are independent. Why is that so? Hint: The map $\mathbf{x} \mapsto H\mathbf{x}^{\mathsf{T}}$ is a linear transformation from $\mathbb{F}_q^n$ to $\mathbb{F}_q^{n-k}$ with kernel C. From linear algebra, what is the rank of H? ◆

Example 1.2.2 The simplest way to encode information in order to recover it in the presence of noise is to repeat each message symbol a fixed number of times. Suppose that our information is binary with symbols from the field $\mathbb{F}_2$, and we repeat each symbol n times. If for instance $n = 7$, then whenever we want to send a 0 we send 0000000, and whenever we want to send a 1 we send 1111111. If at most three errors are made in transmission and if we decode by "majority vote," then we can correctly determine the information symbol, 0 or 1. In general, our code C is the $[n, 1]$ binary linear code consisting of the two codewords $\mathbf{0} = 00 \cdots 0$ and $\mathbf{1} = 11 \cdots 1$ and is called the *binary repetition code* of length n. This code can correct up to $e = \lfloor (n - 1)/2 \rfloor$ errors: if at most e errors are made in a received vector, then the majority of coordinates will be correct, and hence the original sent codeword can be recovered. If more than e errors are made, these errors cannot be corrected. However, this code can detect $n - 1$ errors, as received vectors with between 1 and $n - 1$ errors will definitely not be codewords. A generator matrix for the repetition code is

$$G = [1 \mid 1 \quad \cdots \quad 1],$$

which is of course in standard form. The corresponding parity check matrix from Theorem 1.2.1 is

$$H = \begin{bmatrix} 1 & \\ 1 & \\ \vdots & I_{n-1} \\ 1 & \end{bmatrix}.$$

The first coordinate is an information set and the last $n - 1$ coordinates form a redundancy set. ∎

Exercise 2 How many information sets are there for the $[n, 1]$ repetition code of Example 1.2.2? ◆

Example 1.2.3 The matrix $G = [I_4 \mid A]$, where

$$G = \begin{bmatrix} 1 & 0 & 0 & 0 & 0 & 1 & 1 \\ 0 & 1 & 0 & 0 & 1 & 0 & 1 \\ 0 & 0 & 1 & 0 & 1 & 1 & 0 \\ 0 & 0 & 0 & 1 & 1 & 1 & 1 \end{bmatrix}$$

is a generator matrix in standard form for a $[7, 4]$ binary code that we denote by $\mathcal{H}_3$. By Theorem 1.2.1 a parity check matrix for $\mathcal{H}_3$ is

$$H = [A^{\mathrm{T}} \mid I_3] = \begin{bmatrix} 0 & 1 & 1 & 1 & 1 & 0 & 0 \\ 1 & 0 & 1 & 1 & 0 & 1 & 0 \\ 1 & 1 & 0 & 1 & 0 & 0 & 1 \end{bmatrix}.$$

This code is called the $[7, 4]$ *Hamming code*. ∎

Exercise 3 Find at least four information sets in the $[7, 4]$ code $\mathcal{H}_3$ from Example 1.2.3. Find at least one set of four coordinates that do not form an information set. ◆

Often in this text we will refer to a *subcode* of a code $\mathcal{C}$. If $\mathcal{C}$ is not linear (or not known to be linear), a subcode of $\mathcal{C}$ is any subset of $\mathcal{C}$. If $\mathcal{C}$ is linear, a subcode will be a subset of $\mathcal{C}$ which must also be linear; in this case a subcode of $\mathcal{C}$ is a subspace of $\mathcal{C}$.

1.3   Dual codes

The generator matrix G of an $[n, k]$ code $\mathcal{C}$ is simply a matrix whose rows are independent and span the code. The rows of the parity check matrix H are independent; hence H is the generator matrix of some code, called the *dual* or *orthogonal* of $\mathcal{C}$ and denoted $\mathcal{C}^{\perp}$. Notice that $\mathcal{C}^{\perp}$ is an $[n, n - k]$ code. An alternate way to define the dual code is by using inner products.

Recall that the ordinary inner product of vectors $\mathbf{x} = x_1 \cdots x_n$, $\mathbf{y} = y_1 \cdots y_n$ in $\mathbb{F}_q^n$ is

$$\mathbf{x} \cdot \mathbf{y} = \sum_{i=1}^{n} x_i y_i.$$

Therefore from (1.1), we see that $C^\perp$ can also be defined by

$$C^\perp = \{\mathbf{x} \in \mathbb{F}_q^n \mid \mathbf{x} \cdot \mathbf{c} = 0 \text{ for all } \mathbf{c} \in C\}. \tag{1.2}$$

It is a simple exercise to show that if G and H are generator and parity check matrices, respectively, for C, then H and G are generator and parity check matrices, respectively, for $C^\perp$.

Exercise 4 Prove that if G and H are generator and parity check matrices, respectively, for C, then H and G are generator and parity check matrices, respectively, for $C^\perp$. ◆

Example 1.3.1 Generator and parity check matrices for the $[n, 1]$ repetition code C are given in Example 1.2.2. The dual code $C^\perp$ is the $[n, n-1]$ code with generator matrix H and thus consists of all binary n-tuples $a_1 a_2 \cdots a_{n-1} b$, where $b = a_1 + a_2 + \cdots + a_{n-1}$ (addition in $\mathbb{F}_2$). The nth coordinate b is an overall parity check for the first $n-1$ coordinates chosen, therefore, so that the sum of all the coordinates equals 0. This makes it easy to see that G is indeed a parity check matrix for $C^\perp$. The code $C^\perp$ has the property that a single transmission error can be detected (since the sum of the coordinates will not be 0) but not corrected (since changing any one of the received coordinates will give a vector whose sum of coordinates will be 0). ■

A code C is *self-orthogonal* provided $C \subseteq C^\perp$ and *self-dual* provided $C = C^\perp$. The length n of a self-dual code is even and the dimension is $n/2$.

Exercise 5 Prove that a self-dual code has even length n and dimension $n/2$. ◆

Example 1.3.2 One generator matrix for the $[7, 4]$ Hamming code $\mathcal{H}_3$ is presented in Example 1.2.3. Let $\widehat{\mathcal{H}}_3$ be the code of length 8 and dimension 4 obtained from $\mathcal{H}_3$ by adding an overall parity check coordinate to each vector of G and thus to each codeword of $\mathcal{H}_3$. Then

$$\widehat{G} = \begin{bmatrix} 1 & 0 & 0 & 0 & 0 & 1 & 1 & 1 \\ 0 & 1 & 0 & 0 & 1 & 0 & 1 & 1 \\ 0 & 0 & 1 & 0 & 1 & 1 & 0 & 1 \\ 0 & 0 & 0 & 1 & 1 & 1 & 1 & 0 \end{bmatrix}$$

is a generator matrix for $\widehat{\mathcal{H}}_3$. It is easy to verify that $\widehat{\mathcal{H}}_3$ is a self-dual code. ■

Example 1.3.3 The $[4, 2]$ ternary code $\mathcal{H}_{3,2}$, often called the *tetracode*, has generator matrix G, in standard form, given by

$$G = \begin{bmatrix} 1 & 0 & 1 & 1 \\ 0 & 1 & 1 & -1 \end{bmatrix}.$$

This code is also self-dual. ■

Exercise 6 Prove that $\widehat{\mathcal{H}}_3$ from Example 1.3.2 and $\mathcal{H}_{3,2}$ from Example 1.3.3 are self-dual codes. ◆

Exercise 7 Find all the information sets of the tetracode given in Example 1.3.3. ◆

When studying quaternary codes over the field $\mathbb{F}_4$ (Example 1.1.3), it is often useful to consider another inner product, called the *Hermitian inner product*, given by

$$\langle \mathbf{x}, \mathbf{y} \rangle = \mathbf{x} \cdot \overline{\mathbf{y}} = \sum_{i=1}^{n} x_i \overline{y_i},$$

where $^-$, called *conjugation*, is given by $\overline{0} = 0$, $\overline{1} = 1$, and $\overline{\omega} = \omega$. Using this inner product, we can define the *Hermitian dual* of a quaternary code $\mathcal{C}$ to be, analogous to (1.2),

$$\mathcal{C}^{\perp_H} = \{\mathbf{x} \in \mathbb{F}_q^n \mid \langle \mathbf{x}, \mathbf{c} \rangle = 0 \text{ for all } \mathbf{c} \in \mathcal{C}\}.$$

Define the *conjugate* of $\mathcal{C}$ to be

$$\overline{\mathcal{C}} = \{\overline{\mathbf{c}} \mid \mathbf{c} \in \mathcal{C}\},$$

where $\overline{\mathbf{c}} = \overline{c_1} \, \overline{c_2} \cdots \overline{c_n}$ when $\mathbf{c} = c_1 c_2 \cdots c_n$. Notice that $\mathcal{C}^{\perp_H} = \overline{\mathcal{C}}^{\perp}$. We also have Hermitian self-orthogonality and Hermitian self-duality: namely, $\mathcal{C}$ is *Hermitian self-orthogonal* if $\mathcal{C} \subseteq \mathcal{C}^{\perp_H}$ and *Hermitian self-dual* if $\mathcal{C} = \mathcal{C}^{\perp_H}$.

Exercise 8 Prove that if $\mathcal{C}$ is a code over $\mathbb{F}_4$, then $\mathcal{C}^{\perp_H} = \overline{\mathcal{C}}^{\perp}$. ◆

Example 1.3.4 The [6, 3] quaternary code $\mathcal{G}_6$ has generator matrix G_6 in standard form given by

$$G_6 = \begin{bmatrix} 1 & 0 & 0 & 1 & \omega & \omega \\ 0 & 1 & 0 & \omega & 1 & \omega \\ 0 & 0 & 1 & \omega & \omega & 1 \end{bmatrix}.$$

This code is often called the *hexacode*. It is Hermitian self-dual. ∎

Exercise 9 Verify the following properties of the Hermitian inner product on $\mathbb{F}_4^n$:
(a) $\langle \mathbf{x}, \mathbf{x} \rangle \in \{0, 1\}$ for all $\mathbf{x} \in \mathbb{F}_4^n$.
(b) $\langle \mathbf{x}, \mathbf{y} + \mathbf{z} \rangle = \langle \mathbf{x}, \mathbf{y} \rangle + \langle \mathbf{x}, \mathbf{z} \rangle$ for all $\mathbf{x}, \mathbf{y}, \mathbf{z} \in \mathbb{F}_4^n$.
(c) $\langle \mathbf{x} + \mathbf{y}, \mathbf{z} \rangle = \langle \mathbf{x}, \mathbf{z} \rangle + \langle \mathbf{y}, \mathbf{z} \rangle$ for all $\mathbf{x}, \mathbf{y}, \mathbf{z} \in \mathbb{F}_4^n$.
(d) $\overline{\langle \mathbf{x}, \mathbf{y} \rangle} = \langle \mathbf{y}, \mathbf{x} \rangle$ for all $\mathbf{x}, \mathbf{y} \in \mathbb{F}_4^n$.
(e) $\langle \alpha \mathbf{x}, \mathbf{y} \rangle = \alpha \langle \mathbf{x}, \mathbf{y} \rangle$ for all $\mathbf{x}, \mathbf{y} \in \mathbb{F}_4^n$.
(f) $\langle \mathbf{x}, \alpha \mathbf{y} \rangle = \overline{\alpha} \langle \mathbf{x}, \mathbf{y} \rangle$ for all $\mathbf{x}, \mathbf{y} \in \mathbb{F}_4^n$. ◆

Exercise 10 Prove that the hexacode $\mathcal{G}_6$ from Example 1.3.4 is Hermitian self-dual. ◆

1.4 Weights and distances

An important invariant of a code is the minimum distance between codewords. The *(Hamming) distance* $d(\mathbf{x}, \mathbf{y})$ between two vectors $\mathbf{x}, \mathbf{y} \in \mathbb{F}_q^n$ is defined to be the number of coordinates in which **x** and **y** differ. The proofs of the following properties of distance are left as an exercise.

Theorem 1.4.1 *The distance function* d(**x**, **y**) *satisfies the following four properties*:
(i)   (*non-negativity*) d(**x**, **y**) ≥ 0 *for all* **x**, **y** $\in \mathbb{F}_q^n$.
(ii)   d(**x**, **y**) $= 0$ *if and only if* **x** $=$ **y**.
(iii)   (*symmetry*) d(**x**, **y**) $=$ d(**y**, **x**) *for all* **x**, **y** $\in \mathbb{F}_q^n$.
(iv)   (*triangle inequality*) d(**x**, **z**) $\leq$ d(**x**, **y**) $+$ d(**y**, **z**) *for all* **x**, **y**, **z** $\in \mathbb{F}_q^n$.

This theorem makes the distance function a metric on the vector space $\mathbb{F}_q^n$.

Exercise 11   Prove Theorem 1.4.1.   ◆

The (*minimum*) *distance* of a code C is the smallest distance between distinct codewords and is important in determining the error-correcting capability of C; as we see later, the higher the minimum distance, the more errors the code can correct. The (*Hamming*) *weight* wt(**x**) of a vector **x** $\in \mathbb{F}_q^n$ is the number of nonzero coordinates in **x**. The proof of the following relationship between distance and weight is also left as an exercise.

Theorem 1.4.2 *If* **x**, **y** $\in \mathbb{F}_q^n$, *then* d(**x**, **y**) $=$ wt(**x** $-$ **y**). *If* C *is a linear code, the minimum distance* d *is the same as the minimum weight of the nonzero codewords of* C.

As a result of this theorem, for linear codes, the minimum distance is also called the *minimum weight* of the code. If the minimum weight d of an $[n, k]$ code is known, then we refer to the code as an $[n, k, d]$ code.

Exercise 12   Prove Theorem 1.4.2.   ◆

When dealing with codes over $\mathbb{F}_2$, $\mathbb{F}_3$, or $\mathbb{F}_4$, there are some elementary results about codeword weights that prove to be useful. We collect them here and leave the proof to the reader.

Theorem 1.4.3 *The following hold*:
(i)   *If* **x**, **y** $\in \mathbb{F}_2^n$, *then*

$$\text{wt}(\mathbf{x} + \mathbf{y}) = \text{wt}(\mathbf{x}) + \text{wt}(\mathbf{y}) - 2\text{wt}(\mathbf{x} \cap \mathbf{y}),$$

where **x** $\cap$ **y** *is the vector in* $\mathbb{F}_2^n$, *which has* 1s *precisely in those positions where both* **x** *and* **y** *have* 1s.
(ii)   *If* **x**, **y** $\in \mathbb{F}_2^n$, *then* wt(**x** $\cap$ **y**) $\equiv$ **x** $\cdot$ **y** (mod 2).
(iii)   *If* **x** $\in \mathbb{F}_2^n$, *then* wt(**x**) $\equiv$ **x** $\cdot$ **x** (mod 2).
(iv)   *If* **x** $\in \mathbb{F}_3^n$, *then* wt(**x**) $\equiv$ **x** $\cdot$ **x** (mod 3).
(v)   *If* **x** $\in \mathbb{F}_4^n$, *then* wt(**x**) $\equiv \langle$**x**,**x**$\rangle$ (mod 2).

Exercise 13   Prove Theorem 1.4.3.   ◆

Let A_i, also denoted $A_i(C)$, be the number of codewords of weight i in C. The list A_i for $0 \leq i \leq n$ is called the *weight distribution* or *weight spectrum* of C. A great deal of research is devoted to the computation of the weight distribution of specific codes or families of codes.

Example 1.4.4 Let C be the binary code with generator matrix

$$G = \begin{bmatrix} 1 & 1 & 0 & 0 & 0 & 0 \\ 0 & 0 & 1 & 1 & 0 & 0 \\ 0 & 0 & 0 & 0 & 1 & 1 \end{bmatrix}.$$

The weight distribution of C is $A_0 = A_6 = 1$ and $A_2 = A_4 = 3$. Notice that only the nonzero A_i are usually listed. ∎

Exercise 14 Find the weight distribution of the ternary code with generator matrix

$$G = \begin{bmatrix} 1 & 1 & 0 & 0 & 0 & 0 \\ 0 & 0 & 1 & 1 & 0 & 0 \\ 0 & 0 & 0 & 0 & 1 & 1 \end{bmatrix}.$$

Compare your result to Example 1.4.4. ♦

Certain elementary facts about the weight distribution are gathered in the following theorem. Deeper results on the weight distribution of codes will be presented in Chapter 7.

Theorem 1.4.5 Let C be an $[n, k, d]$ code over $\mathbb{F}_q$. Then:
(i) $A_0(C) + A_1(C) + \cdots + A_n(C) = q^k$.
(ii) $A_0(C) = 1$ and $A_1(C) = A_2(C) = \cdots = A_{d-1}(C) = 0$.
(iii) If C is a binary code containing the codeword $\mathbf{1} = 11 \cdots 1$, then $A_i(C) = A_{n-i}(C)$ for $0 \le i \le n$.
(iv) If C is a binary self-orthogonal code, then each codeword has even weight, and $C^\perp$ contains the codeword $\mathbf{1} = 11 \cdots 1$.
(v)   If C is a ternary self-orthogonal code, then the weight of each codeword is divisible by three.
(vi) If C is a quaternary Hermitian self-orthogonal code, then the weight of each codeword is even.

Exercise 15 Prove Theorem 1.4.5. ♦

Theorem 1.4.5(iv) states that all codewords in a binary self-orthogonal code C have even weight. If we look at the subset of codewords of C that have weights divisible by four, we surprisingly get a subcode of C; that is, the subset of codewords of weights divisible by four form a subspace of C. This is not necessarily the case for non-self-orthogonal codes.

Theorem 1.4.6 Let C be an $[n, k]$ self-orthogonal binary code. Let C_0 be the set of codewords in C whose weights are divisible by four. Then either:
(i) $C = C_0$, or
(ii) C_0 is an $[n, k-1]$ subcode of C and $C = C_0 \cup C_1$, where $C_1 = \mathbf{x} + C_0$ for any codeword $\mathbf{x}$ whose weight is even but not divisible by four. Furthermore C_1 consists of all codewords of C whose weights are not divisible by four.

Proof: By Theorem 1.4.5(iv) all codewords have even weight. Therefore either (i) holds or there exists a codeword $\mathbf{x}$ of even weight but not of weight a multiple of four. Assume the latter. Let $\mathbf{y}$ be another codeword whose weight is even but not a multiple of four. Then by Theorem 1.4.3(i), $\text{wt}(\mathbf{x} + \mathbf{y}) = \text{wt}(\mathbf{x}) + \text{wt}(\mathbf{y}) - 2\text{wt}(\mathbf{x} \cap \mathbf{y}) \equiv 2 + 2 - 2\text{wt}(\mathbf{x} \cap \mathbf{y})$ (mod 4). But by Theorem 1.4.3(ii), $\text{wt}(\mathbf{x} \cap \mathbf{y}) \equiv \mathbf{x} \cdot \mathbf{y}$ (mod 2). Hence $\text{wt}(\mathbf{x} + \mathbf{y})$ is divisible by four. Therefore $\mathbf{x} + \mathbf{y} \in C_0$. This shows that $\mathbf{y} \in \mathbf{x} + C_0$ and $C = C_0 \cup (\mathbf{x} + C_0)$. That C_0 is a subcode of C and that $C_1 = \mathbf{x} + C_0$ consists of all codewords of C whose weights are not divisible by four follow from a similar argument. $\qquad\square$

There is an analogous result to Theorem 1.4.6 where you consider the subset of codewords of a binary code whose weights are even. In this case the self-orthogonality requirement is unnecessary; we leave its proof to the exercises.

Theorem 1.4.7 *Let C be an $[n, k]$ binary code. Let C_e be the set of codewords in C whose weights are even. Then either*:
(i) $C = C_e$, *or*
(ii) *C_e is an $[n, k - 1]$ subcode of C and $C = C_e \cup C_o$, where $C_o = \mathbf{x} + C_e$ for any codeword $\mathbf{x}$ whose weight is odd. Furthermore C_o consists of all codewords of C whose weights are odd.*

Exercise 16   Prove Theorem 1.4.7. $\qquad\blacklozenge$

Exercise 17   Let C be the $[6, 3]$ binary code with generator matrix

$$G = \begin{bmatrix} 1 & 1 & 0 & 0 & 0 & 0 \\ 0 & 1 & 1 & 0 & 0 & 0 \\ 1 & 1 & 1 & 1 & 1 & 1 \end{bmatrix}.$$

(a) Prove that C is not self-orthogonal.
(b) Find the weight distribution of C.
(c) Show that the codewords whose weights are divisible by four do not form a subcode of C. $\qquad\blacklozenge$

The next result gives a way to tell when Theorem 1.4.6(i) is satisfied.

Theorem 1.4.8 *Let C be a binary linear code.*
(i) *If C is self-orthogonal and has a generator matrix each of whose rows has weight divisible by four, then every codeword of C has weight divisible by four.*
(ii) *If every codeword of C has weight divisible by four, then C is self-orthogonal.*

Proof: For (i), let $\mathbf{x}$ and $\mathbf{y}$ be rows of the generator matrix. By Theorem 1.4.3(i), $\text{wt}(\mathbf{x} + \mathbf{y}) = \text{wt}(\mathbf{x}) + \text{wt}(\mathbf{y}) - 2\text{wt}(\mathbf{x} \cap \mathbf{y}) \equiv 0 + 0 - 2\text{wt}(\mathbf{x} \cap \mathbf{y}) \equiv 0$ (mod 4). Now proceed by induction as every codeword is a sum of rows of the generator matrix. For (ii), let $\mathbf{x}, \mathbf{y} \in C$. By Theorem 1.4.3(i) and (ii), $2(\mathbf{x} \cdot \mathbf{y}) \equiv 2\text{wt}(\mathbf{x} \cap \mathbf{y}) \equiv 2\text{wt}(\mathbf{x} \cap \mathbf{y}) - \text{wt}(\mathbf{x}) - \text{wt}(\mathbf{y}) \equiv -\text{wt}(\mathbf{x} + \mathbf{y}) \equiv 0$ (mod 4). Thus $\mathbf{x} \cdot \mathbf{y} \equiv 0$ (mod 2). $\qquad\square$

It is natural to ask if Theorem 1.4.8(ii) can be generalized to codes whose codewords have weights that are divisible by numbers other than four. We say that a code C (over any field) is *divisible* provided all codewords have weights divisible by an integer $\Delta > 1$. The code is said to be *divisible by* Δ; Δ is called *a divisor* of C, and the largest such divisor is called *the divisor* of C. Thus Theorem 1.4.8(ii) says that binary codes divisible by $\Delta = 4$ are self-orthogonal. This is not true when considering binary codes divisible by $\Delta = 2$, as the next example illustrates. Binary codes divisible by $\Delta = 2$ are called *even*.

Example 1.4.9 The dual of the $[n, 1]$ binary repetition code C of Example 1.2.2 consists of all the even weight vectors of length n. (See also Example 1.3.1.) If $n > 2$, this code is not self-orthogonal. ■

When considering codes over $\mathbb{F}_3$ and $\mathbb{F}_4$, the divisible codes with divisors three and two, respectively, are self-orthogonal as the next theorem shows. This theorem includes the converse of Theorem 1.4.5(v) and (vi). Part (ii) is found in [217].

Theorem 1.4.10 *Let C be a code over $\mathbb{F}_q$, with $q = 3$ or 4.*
(i) *When $q = 3$, every codeword of C has weight divisible by three if and only if C is self-orthogonal.*
(ii) *When $q = 4$, every codeword of C has weight divisible by two if and only if C is Hermitian self-orthogonal.*

Proof: In (i), if C is self-orthogonal, the codewords have weights divisible by three by Theorem 1.4.5(v). For the converse let $\mathbf{x}, \mathbf{y} \in C$. We need to show that $\mathbf{x} \cdot \mathbf{y} = 0$. We can view the codewords $\mathbf{x}$ and $\mathbf{y}$ having the following parameters:

$$
\begin{array}{ccccc}
\mathbf{x}: & \star & 0 & = & \neq & 0 \\
\mathbf{y}: & 0 & \star & = & \neq & 0 \\
& a & b & c & d & e
\end{array}
$$

where there are a coordinates where $\mathbf{x}$ is nonzero and $\mathbf{y}$ is zero, b coordinates where $\mathbf{y}$ is nonzero and $\mathbf{x}$ is zero, c coordinates where both agree and are nonzero, d coordinates when both disagree and are nonzero, and e coordinates where both are zero. So $\mathrm{wt}(\mathbf{x} + \mathbf{y}) = a + b + c$ and $\mathrm{wt}(\mathbf{x} - \mathbf{y}) = a + b + d$. But $\mathbf{x} \pm \mathbf{y} \in C$ and hence $a + b + c \equiv a + b + d \equiv 0$ (mod 3). In particular $c \equiv d$ (mod 3). Therefore $\mathbf{x} \cdot \mathbf{y} = c + 2d \equiv 0$ (mod 3), proving (i).

In (ii), if C is Hermitian self-orthogonal, the codewords have even weights by Theorem 1.4.5(vi). For the converse let $\mathbf{x} \in C$. If $\mathbf{x}$ has a 0s, b 1s, c ωs, and d $\overline{\omega}$s, then $b + c + d$ is even as $\mathrm{wt}(\mathbf{x}) = b + c + d$. However, $\langle \mathbf{x}, \mathbf{x} \rangle$ also equals $b + c + d$ (as an element of $\mathbb{F}_4$). Therefore $\langle \mathbf{x}, \mathbf{x} \rangle = 0$ for all $\mathbf{x} \in C$. Now let $\mathbf{x}, \mathbf{y} \in C$. So both $\mathbf{x} + \mathbf{y}$ and $\omega\mathbf{x} + \mathbf{y}$ are in C. Using Exercise 9 we have $0 = \langle \mathbf{x} + \mathbf{y}, \mathbf{x} + \mathbf{y} \rangle = \langle \mathbf{x}, \mathbf{x} \rangle + \langle \mathbf{x}, \mathbf{y} \rangle + \langle \mathbf{y}, \mathbf{x} \rangle + \langle \mathbf{y}, \mathbf{y} \rangle = \langle \mathbf{x}, \mathbf{y} \rangle + \langle \mathbf{y}, \mathbf{x} \rangle$. Also $0 = \langle \omega\mathbf{x} + \mathbf{y}, \omega\mathbf{x} + \mathbf{y} \rangle = \langle \mathbf{x}, \mathbf{x} \rangle + \omega\langle \mathbf{x}, \mathbf{y} \rangle + \overline{\omega}\langle \mathbf{y}, \mathbf{x} \rangle + \langle \mathbf{y}, \mathbf{y} \rangle = \omega\langle \mathbf{x}, \mathbf{y} \rangle + \overline{\omega}\langle \mathbf{y}, \mathbf{x} \rangle$. Combining these $\langle \mathbf{x}, \mathbf{y} \rangle$ must be 0, proving (ii). □

The converse of Theorem 1.4.5(iv) is in general not true. The best that can be said in this case is contained in the following theorem, whose proof we leave as an exercise.

Theorem 1.4.11 *Let C be a binary code with a generator matrix each of whose rows has even weight. Then every codeword of C has even weight.*

Exercise 18   Prove Theorem 1.4.11.   ◆

Binary codes for which all codewords have weight divisible by four are called *doubly-even*.[2] By Theorem 1.4.8, doubly-even codes are self-orthogonal. A self-orthogonal code must be even by Theorem 1.4.5(iv); one which is not doubly-even is called *singly-even*.

Exercise 19   Find the minimum weights and weight distributions of the codes $\mathcal{H}_3$ in Example 1.2.3, $\mathcal{H}_3^\perp$, $\widehat{\mathcal{H}}_3$ in Example 1.3.2, the tetracode in Example 1.3.3, and the hexacode in Example 1.3.4. Which of the binary codes listed are self-orthogonal? Which are doubly-even? Which are singly-even?   ◆

There is a generalization of the concepts of even and odd weight binary vectors to vectors over arbitrary fields, which is useful in the study of many types of codes. A vector $\mathbf{x} = x_1 x_2 \cdots x_n$ in $\mathbb{F}_q^n$ is *even-like* provided that

$$\sum_{i=1}^{n} x_i = 0$$

and is *odd-like* otherwise. A binary vector is even-like if and only if it has even weight; so the concept of even-like vectors is indeed a generalization of even weight binary vectors. The even-like vectors in a code form a subcode of a code over $\mathbb{F}_q$ as did the even weight vectors in a binary code. Except in the binary case, even-like vectors need not have even weight. The vectors $(1, 1, 1)$ in $\mathbb{F}_3^3$ and $(1, \omega, \overline{\omega})$ in $\mathbb{F}_4^3$ are examples. We say that a code is *even-like* if it has only even-like codewords; a code is *odd-like* if it is not even-like.

Theorem 1.4.12 *Let C be an $[n, k]$ code over $\mathbb{F}_q$. Let C_e be the set of even-like codewords in C. Then either:*
(i) *$C = C_e$, or*
(ii) *C_e is an $[n, k - 1]$ subcode of C.*

Exercise 20   Prove Theorem 1.4.12.   ◆

There is an elementary relationship between the weight of a codeword and a parity check matrix for a linear code. This is presented in the following theorem whose proof is left as an exercise.

Theorem 1.4.13 *Let C be a linear code with parity check matrix H. If $\mathbf{c} \in C$, the columns of H corresponding to the nonzero coordinates of $\mathbf{c}$ are linearly dependent. Conversely, if a linear dependence relation with nonzero coefficients exists among w columns of H, then there is a codeword in C of weight w whose nonzero coordinates correspond to these columns.*

One way to find the minimum weight d of a linear code is to examine all the nonzero codewords. The following corollary shows how to use the parity check matrix to find d.

---

[2] Some authors reserve the term "doubly-even" for self-dual codes for which all codewords have weight divisible by four.

Corollary 1.4.14 *A linear code has minimum weight d if and only if its parity check matrix has a set of d linearly dependent columns but no set of d − 1 linearly dependent columns.*

Exercise 21 Prove Theorem 1.4.13 and Corollary 1.4.14.   ◆

The minimum weight is also characterized in the following theorem.

Theorem 1.4.15 *If C is an* $[n, k, d]$ *code, then every* $n - d + 1$ *coordinate position contains an information set. Furthermore, d is the largest number with this property.*

Proof: Let G be a generator matrix for C, and consider any set X of s coordinate positions. To make the argument easier, we assume X is the set of the last s positions. (After we develop the notion of equivalent codes, the reader will see that this argument is in fact general.) Suppose X does not contain an information set. Let $G = [A \mid B]$, where A is $k \times (n - s)$ and B is $k \times s$. Then the column rank of B, and hence the row rank of B, is less than k. Hence there exists a nontrivial linear combination of the rows of B which equals $\mathbf{0}$, and hence a codeword $\mathbf{c}$ which is $\mathbf{0}$ in the last s positions. Since the rows of G are linearly independent, $\mathbf{c} \neq \mathbf{0}$ and hence $d \leq n - s$, equivalently, $s \leq n - d$. The theorem now follows.   □

Exercise 22 Find the number of information sets for the $[7, 4]$ Hamming code $\mathcal{H}_3$ given in Example 1.2.3. Do the same for the extended Hamming code $\widehat{\mathcal{H}}_3$ from Example 1.3.2.   ◆

1.5   New codes from old

As we will see throughout this book, many interesting and important codes will arise by modifying or combining existing codes. We will discuss five ways to do this.

1.5.1   Puncturing codes

Let C be an $[n, k, d]$ code over $\mathbb{F}_q$. We can *puncture* C by deleting the same coordinate i in each codeword. The resulting code is still linear, a fact that we leave as an exercise; its length is $n - 1$, and we often denote the punctured code by C^*. If G is a generator matrix for C, then a generator matrix for C^* is obtained from G by deleting column i (and omitting a zero or duplicate row that may occur). What are the dimension and minimum weight of C^*? Because C contains q^k codewords, the only way that C^* could contain fewer codewords is if two codewords of C agree in all but coordinate i. In that case C has minimum distance $d = 1$ and a codeword of weight 1 whose nonzero entry is in coordinate i. The minimum distance decreases by 1 only if a minimum weight codeword of C has a nonzero ith coordinate. Summarizing, we have the following theorem.

Theorem 1.5.1 *Let C be an* $[n, k, d]$ *code over* $\mathbb{F}_q$, *and let* C^* *be the code C punctured on the i th coordinate.*

(i) If $d > 1$, C^* is an $[n - 1, k, d^*]$ code where $d^* = d - 1$ if C has a minimum weight codeword with a nonzero ith coordinate and $d^* = d$ otherwise.

(ii) When $d = 1$, C^* is an $[n - 1, k, 1]$ code if C has no codeword of weight 1 whose nonzero entry is in coordinate i; otherwise, if $k > 1$, C^* is an $[n - 1, k - 1, d^*]$ code with $d^* \geq 1$.

Exercise 23 Prove directly from the definition that a punctured linear code is also linear.   ◆

Example 1.5.2 Let C be the $[5, 2, 2]$ binary code with generator matrix

$$G = \begin{bmatrix} 1 & 1 & 0 & 0 & 0 \\ 0 & 0 & 1 & 1 & 1 \end{bmatrix}.$$

Let C_1^* and C_5^* be the code C punctured on coordinates 1 and 5, respectively. They have generator matrices

$$G_1^* = \begin{bmatrix} 1 & 0 & 0 & 0 \\ 0 & 1 & 1 & 1 \end{bmatrix} \quad \text{and} \quad G_5^* = \begin{bmatrix} 1 & 1 & 0 & 0 \\ 0 & 0 & 1 & 1 \end{bmatrix}.$$

So C_1^* is a $[4, 2, 1]$ code, while C_5^* is a $[4, 2, 2]$ code.   ■

Example 1.5.3 Let D be the $[4, 2, 1]$ binary code with generator matrix

$$G = \begin{bmatrix} 1 & 0 & 0 & 0 \\ 0 & 1 & 1 & 1 \end{bmatrix}.$$

Let D_1^* and D_4^* be the code D punctured on coordinates 1 and 4, respectively. They have generator matrices

$$D_1^* = \begin{bmatrix} 1 & 1 & 1 \end{bmatrix} \quad \text{and} \quad D_4^* = \begin{bmatrix} 1 & 0 & 0 \\ 0 & 1 & 1 \end{bmatrix}.$$

So D_1^* is a $[3, 1, 3]$ code and D_4^* is a $[3, 2, 1]$ code.   ■

Notice that the code D of Example 1.5.3 is the code C_1^* of Example 1.5.2. Obviously D_4^* could have been obtained from C directly by puncturing on coordinates $\{1, 5\}$. In general a code C can be punctured on the coordinate set T by deleting components indexed by the set T in all codewords of C. If T has size t, the resulting code, which we will often denote C^T, is an $[n - t, k^*, d^*]$ code with $k^* \geq k - t$ and $d^* \geq d - t$ by Theorem 1.5.1 and induction.

1.5.2 Extending codes

We can create longer codes by adding a coordinate. There are many possible ways to extend a code but the most common is to choose the extension so that the new code has only even-like vectors (as defined in Section 1.4). If C is an $[n, k, d]$ code over $\mathbb{F}_q$, define the *extended* code $\widehat{C}$ to be the code

$$\widehat{C} = \{x_1 x_2 \cdots x_{n+1} \in \mathbb{F}_q^{n+1} \mid x_1 x_2 \cdots x_n \in C \text{ with } x_1 + x_2 + \cdots + x_{n+1} = 0\}.$$

We leave it as an exercise to show that $\widehat{C}$ is linear. In fact $\widehat{C}$ is an $[n+1, k, \widehat{d}]$ code, where $\widehat{d} = d$ or $d + 1$. Let G and H be generator and parity check matrices, respectively, for C. Then a generator matrix $\widehat{G}$ for $\widehat{C}$ can be obtained from G by adding an extra column to G so that the sum of the coordinates of each row of $\widehat{G}$ is 0. A parity check matrix for $\widehat{C}$ is the matrix

$$
\widehat{H} = \left[
\begin{array}{ccc|c}
1 & \cdots & 1 & 1 \\
\hline
 & & & 0 \\
 & H & & \vdots \\
 & & & 0
\end{array}
\right]. \tag{1.3}
$$

This construction is also referred to as *adding an overall parity check*. The $[8, 4, 4]$ binary code $\widehat{\mathcal{H}}_3$ in Example 1.3.2 obtained from the $[7, 4, 3]$ Hamming code $\mathcal{H}_3$ by adding an overall parity check is called the *extended Hamming code*.

Exercise 24 Prove directly from the definition that an extended linear code is also linear. ♦

Exercise 25 Suppose we extend the $[n, k]$ linear code C over the field $\mathbb{F}_q$ to the code $\widetilde{C}$ where

$$
\widetilde{C} = \{x_1 x_2 \cdots x_{n+1} \in \mathbb{F}_q^{n+1} \mid x_1 x_2 \cdots x_n \in C \text{ with } x_1^2 + x_2^2 + \cdots + x_{n+1}^2 = 0\}.
$$

Under what conditions is $\widetilde{C}$ linear? ♦

Exercise 26 Prove that $\widehat{H}$ in (1.3) is the parity check matrix for an extended code $\widehat{C}$, where C has parity check matrix H. ♦

If C is an $[n, k, d]$ binary code, then the extended code $\widehat{C}$ contains only even weight vectors and is an $[n+1, k, \widehat{d}]$ code, where $\widehat{d}$ equals d if d is even and equals $d + 1$ if d is odd. This is consistent with the results obtained by extending $\mathcal{H}_3$. In the nonbinary case, however, whether or not $\widehat{d}$ is d or $d + 1$ is not so straightforward. For an $[n, k, d]$ code C over $\mathbb{F}_q$, call the minimum weight of the even-like codewords, respectively the odd-like codewords, the *minimum even-like weight*, respectively the *minimum odd-like weight*, of the code. Denote the minimum even-like weight by d_e and the minimum odd-like weight by d_o. So $d = \min\{d_e, d_o\}$. If $d_e \leq d_o$, then $\widehat{C}$ has minimum weight $\widehat{d} = d_e$. If $d_o < d_e$, then $\widehat{d} = d_o + 1$.

Example 1.5.4 Recall that the tetracode $\mathcal{H}_{3,2}$ from Example 1.3.3 is a $[4, 2, 3]$ code over $\mathbb{F}_3$ with generator matrix G and parity check matrix H given by

$$
G = \begin{bmatrix} 1 & 0 & 1 & 1 \\ 0 & 1 & 1 & -1 \end{bmatrix} \quad \text{and} \quad H = \begin{bmatrix} -1 & -1 & 1 & 0 \\ -1 & 1 & 0 & 1 \end{bmatrix}.
$$

The codeword $(1, 0, 1, 1)$ extends to $(1, 0, 1, 1, 0)$ and the codeword $(0, 1, 1, -1)$ extends to $(0, 1, 1, -1, -1)$. Hence $d = d_e = d_o = 3$ and $\widehat{d} = 3$. The generator and parity check matrices for $\widehat{\mathcal{H}}_{3,2}$ are

$$\widehat{G} = \begin{bmatrix} 1 & 0 & 1 & 1 & 0 \\ 0 & 1 & 1 & -1 & -1 \end{bmatrix} \quad \text{and} \quad \widehat{H} = \begin{bmatrix} 1 & 1 & 1 & 1 & 1 \\ -1 & -1 & 1 & 0 & 0 \\ -1 & 1 & 0 & 1 & 0 \end{bmatrix}. \qquad \blacksquare$$

If we extend a code and then puncture the new coordinate, we obtain the original code. However, performing the operations in the other order will in general result in a different code.

Example 1.5.5 If we puncture the binary code $\mathcal{C}$ with generator matrix

$$G = \begin{bmatrix} 1 & 1 & 0 & 0 & 1 \\ 0 & 0 & 1 & 1 & 0 \end{bmatrix}$$

on its last coordinate and then extend (on the right), the resulting code has generator matrix

$$G = \begin{bmatrix} 1 & 1 & 0 & 0 & 0 \\ 0 & 0 & 1 & 1 & 0 \end{bmatrix}. \qquad \blacksquare$$

In this example, our last step was to extend a binary code with only even weight vectors. The extended coordinate was always 0. In general, that is precisely what happens when you extend a code that has only even-like codewords.

Exercise 27 Do the following.

(a) Let $\mathcal{C} = \mathcal{H}_{3,2}$ be the [4, 2, 3] tetracode over $\mathbb{F}_3$ defined in Example 1.3.3 with generator matrix

$$G = \begin{bmatrix} 1 & 0 & 1 & 1 \\ 0 & 1 & 1 & -1 \end{bmatrix}.$$

Give the generator matrix of the code obtained from $\mathcal{C}$ by puncturing on the right-most coordinate and then extending on the right. Also determine the minimum weight of the resulting code.

(b) Let $\mathcal{C}$ be a code over $\mathbb{F}_q$. Let $\mathcal{C}_1$ be the code obtained from $\mathcal{C}$ by puncturing on the right-most coordinate and then extending this punctured code on the right. Prove that $\mathcal{C} = \mathcal{C}_1$ if and only if $\mathcal{C}$ is an even-like code.

(c) With $\mathcal{C}_1$ defined as in (b), prove that if $\mathcal{C}$ is self-orthogonal and contains the all-one codeword $\mathbf{1}$, then $\mathcal{C} = \mathcal{C}_1$.

(d) With $\mathcal{C}_1$ defined as in (b), prove that $\mathcal{C} = \mathcal{C}_1$ if and only if the all-one vector $\mathbf{1}$ is in $\mathcal{C}^{\perp}$. ◆

1.5.3 Shortening codes

Let $\mathcal{C}$ be an $[n, k, d]$ code over $\mathbb{F}_q$ and let T be any set of t coordinates. Consider the set $\mathcal{C}(T)$ of codewords which are $\mathbf{0}$ on T; this set is a subcode of $\mathcal{C}$. Puncturing $\mathcal{C}(T)$ on T gives a code over $\mathbb{F}_q$ of length $n - t$ called the code *shortened* on T and denoted $\mathcal{C}_T$.

Example 1.5.6 Let C be the $[6, 3, 2]$ binary code with generator matrix

$$G = \begin{bmatrix} 1 & 0 & 0 & 1 & 1 & 1 \\ 0 & 1 & 0 & 1 & 1 & 1 \\ 0 & 0 & 1 & 1 & 1 & 1 \end{bmatrix}.$$

$C^\perp$ is also a $[6, 3, 2]$ code with generator matrix

$$G^\perp = \begin{bmatrix} 1 & 1 & 1 & 1 & 0 & 0 \\ 1 & 1 & 1 & 0 & 1 & 0 \\ 1 & 1 & 1 & 0 & 0 & 1 \end{bmatrix}.$$

If the coordinates are labeled $1, 2, \ldots, 6$, let $T = \{5, 6\}$. Generator matrices for the shortened code C_T and punctured code C^T are

$$G_T = \begin{bmatrix} 1 & 0 & 1 & 0 \\ 0 & 1 & 1 & 0 \end{bmatrix} \quad \text{and} \quad G^T = \begin{bmatrix} 1 & 0 & 0 & 1 \\ 0 & 1 & 0 & 1 \\ 0 & 0 & 1 & 1 \end{bmatrix}.$$

Shortening and puncturing the dual code gives the codes $(C^\perp)_T$ and $(C^\perp)^T$, which have generator matrices

$$(G^\perp)_T = \begin{bmatrix} 1 & 1 & 1 & 1 \end{bmatrix} \quad \text{and} \quad (G^\perp)^T = \begin{bmatrix} 1 & 1 & 1 & 1 \\ 1 & 1 & 1 & 0 \end{bmatrix}.$$

From the generator matrices G_T and G^T, we find that the duals of C_T and C^T have generator matrices

$$(G_T)^\perp = \begin{bmatrix} 1 & 1 & 1 & 0 \\ 0 & 0 & 0 & 1 \end{bmatrix} \quad \text{and} \quad (G^T)^\perp = \begin{bmatrix} 1 & 1 & 1 & 1 \end{bmatrix}.$$

Notice that these matrices show that $(C^\perp)_T = (C^T)^\perp$ and $(C^\perp)^T = (C_T)^\perp$. ∎

The conclusions observed in the previous example hold in general.

Theorem 1.5.7 *Let C be an $[n, k, d]$ code over $\mathbb{F}_q$. Let T be a set of t coordinates. Then:*
(i) $(C^\perp)_T = (C^T)^\perp$ *and* $(C^\perp)^T = (C_T)^\perp$, *and*
(ii) *if $t < d$, then C^T and $(C^\perp)_T$ have dimensions k and $n - t - k$, respectively;*
(iii) *if $t = d$ and T is the set of coordinates where a minimum weight codeword is nonzero, then C^T and $(C^\perp)_T$ have dimensions $k - 1$ and $n - d - k + 1$, respectively.*

Proof: Let $\mathbf{c}$ be a codeword of $C^\perp$ which is $\mathbf{0}$ on T and $\mathbf{c}^*$ the codeword with the coordinates in T removed. So $\mathbf{c}^* \in (C^\perp)_T$. If $\mathbf{x} \in C$, then $0 = \mathbf{x} \cdot \mathbf{c} = \mathbf{x}^* \cdot \mathbf{c}^*$, where $\mathbf{x}^*$ is the codeword $\mathbf{x}$ punctured on T. Thus $(C^\perp)_T \subseteq (C^T)^\perp$. Any vector $\mathbf{c} \in (C^T)^\perp$ can be extended to a vector $\widehat{\mathbf{c}}$ by inserting 0s in the positions of T. If $\mathbf{x} \in C$, puncture $\mathbf{x}$ on T to obtain $\mathbf{x}^*$. As $0 = \mathbf{x}^* \cdot \mathbf{c} = \mathbf{x} \cdot \widehat{\mathbf{c}}$, $\mathbf{c} \in (C^\perp)_T$. Thus $(C^\perp)_T = (C^T)^\perp$. Replacing C by $C^\perp$ gives $(C^\perp)^T = (C_T)^\perp$, completing (i).

Assume $t < d$. Then $n - d + 1 \leq n - t$, implying any $n - t$ coordinates of C contain an information set by Theorem 1.4.15. Therefore C^T must be k-dimensional and hence $(C^\perp)_T = (C^T)^\perp$ has dimension $n - t - k$ by (i); this proves (ii).

As in (ii), (iii) is completed if we show that C^T has dimension $k - 1$. If $S \subset T$ with S of size $d - 1$, C^S has dimension k by part (ii). Clearly C^S has minimum distance 1 and C^T is obtained by puncturing C^S on the nonzero coordinate of a weight 1 codeword in C^S. By Theorem 1.5.1(ii) C^T has dimension $k - 1$. $\qquad\square$

Exercise 28 Let C be the binary repetition code of length n as described in Example 1.2.2. Describe $(C^\perp)_T$ and $(C_T)^\perp$ for any T. $\qquad\blacklozenge$

Exercise 29 Let C be the code of length 6 in Example 1.4.4. Give generator matrices for $(C^\perp)_T$ and $(C_T)^\perp$ when $T = \{1, 2\}$ and $T = \{1, 3\}$. $\qquad\blacklozenge$

1.5.4   Direct sums

For $i \in \{1, 2\}$ let C_i be an $[n_i, k_i, d_i]$ code, both over the same finite field $\mathbb{F}_q$. Then their *direct sum* is the $[n_1 + n_2, k_1 + k_2, \min\{d_1, d_2\}]$ code

$$C_1 \oplus C_2 = \{(\mathbf{c}_1, \mathbf{c}_2) \mid \mathbf{c}_1 \in C_1, \mathbf{c}_2 \in C_2\}.$$

If C_i has generator matrix G_i and parity check matrix H_i, then

$$G_1 \oplus G_2 = \begin{bmatrix} G_1 & O \\ O & G_2 \end{bmatrix} \quad \text{and} \quad H_1 \oplus H_2 = \begin{bmatrix} H_1 & O \\ O & H_2 \end{bmatrix} \qquad (1.4)$$

are a generator matrix and parity check matrix for $C_1 \oplus C_2$.

Exercise 30 Let C_i have generator matrix G_i and parity check matrix H_i for $i \in \{1, 2\}$. Prove that the generator and parity check matrices for $C_1 \oplus C_2$ are as given in (1.4). $\qquad\blacklozenge$

Exercise 31 Let C be the binary code with generator matrix

$$G = \begin{bmatrix} 1 & 1 & 0 & 0 & 1 & 1 & 0 \\ 1 & 0 & 1 & 0 & 1 & 0 & 1 \\ 1 & 0 & 0 & 1 & 1 & 1 & 0 \\ 1 & 0 & 1 & 0 & 1 & 1 & 0 \\ 1 & 0 & 0 & 1 & 0 & 1 & 1 \end{bmatrix}.$$

Give another generator matrix for C that shows that C is a direct sum of two binary codes. $\qquad\blacklozenge$

Example 1.5.8 The $[6, 3, 2]$ binary code C of Example 1.4.4 is the direct sum $\mathcal{D} \oplus \mathcal{D} \oplus \mathcal{D}$ of the $[2, 1, 2]$ code $\mathcal{D} = \{00, 11\}$. $\qquad\blacksquare$

Since the minimum distance of the direct sum of two codes does not exceed the minimum distance of either of the codes, the direct sum of two codes is generally of little use in applications and is primarily of theoretical interest.

1.5.5   The $(\mathbf{u} \mid \mathbf{u} + \mathbf{v})$ construction

Two codes of the same length can be combined to form a third code of twice the length in a way similar to the direct sum construction. Let C_i be an $[n, k_i, d_i]$ code for $i \in \{1, 2\}$, both over the same finite field $\mathbb{F}_q$. The $(\mathbf{u} \mid \mathbf{u} + \mathbf{v})$ *construction* produces the $[2n, k_1 + k_2, \min\{2d_1, d_2\}]$ code

$$\mathcal{C} = \{(\mathbf{u}, \mathbf{u} + \mathbf{v}) \mid \mathbf{u} \in \mathcal{C}_1, \mathbf{v} \in \mathcal{C}_2\}.$$

If $\mathcal{C}_i$ has generator matrix G_i and parity check matrix H_i, then generator and parity check matrices for $\mathcal{C}$ are

$$\begin{bmatrix} G_1 & G_1 \\ O & G_2 \end{bmatrix} \quad \text{and} \quad \begin{bmatrix} H_1 & O \\ -H_2 & H_2 \end{bmatrix}. \tag{1.5}$$

Exercise 32 Prove that generator and parity check matrices for the code obtained in the $(\mathbf{u} \mid \mathbf{u} + \mathbf{v})$ construction from the codes $\mathcal{C}_i$ are as given in (1.5). ♦

Example 1.5.9 Consider the $[8, 4, 4]$ binary code $\mathcal{C}$ with generator matrix

$$G = \begin{bmatrix} 1 & 0 & 1 & 0 & 1 & 0 & 1 & 0 \\ 0 & 1 & 0 & 1 & 0 & 1 & 0 & 1 \\ 0 & 0 & 1 & 1 & 0 & 0 & 1 & 1 \\ 0 & 0 & 0 & 0 & 1 & 1 & 1 & 1 \end{bmatrix}.$$

Then $\mathcal{C}$ can be produced from the $[4, 3, 2]$ code $\mathcal{C}_1$ and the $[4, 1, 4]$ code $\mathcal{C}_2$ with generator matrices

$$G_1 = \begin{bmatrix} 1 & 0 & 1 & 0 \\ 0 & 1 & 0 & 1 \\ 0 & 0 & 1 & 1 \end{bmatrix} \quad \text{and} \quad G_2 = [1 \ \ 1 \ \ 1 \ \ 1],$$

respectively, using the $(\mathbf{u} \mid \mathbf{u} + \mathbf{v})$ construction. Notice that the code $\mathcal{C}_1$ is also constructed using the $(\mathbf{u} \mid \mathbf{u} + \mathbf{v})$ construction from the $[2, 2, 1]$ code $\mathcal{C}_3$ and the $[2, 1, 2]$ code $\mathcal{C}_4$ with generator matrices

$$G_3 = \begin{bmatrix} 1 & 0 \\ 0 & 1 \end{bmatrix} \quad \text{and} \quad G_4 = [1 \ \ 1],$$

respectively. ∎

Unlike the direct sum construction of the previous section, the $(\mathbf{u} \mid \mathbf{u} + \mathbf{v})$ construction can produce codes that are important for reasons other than theoretical. For example, the family of Reed–Muller codes can be constructed in this manner as we see in Section 1.10. The code in the previous example is one of these codes.

Exercise 33 Prove that the $(\mathbf{u} \mid \mathbf{u} + \mathbf{v})$ construction using $[n, k_i, d_i]$ codes $\mathcal{C}_i$ produces a code of dimension $k = k_1 + k_2$ and minimum weight $d = \min\{2d_1, d_2\}$. ♦

1.6   Permutation equivalent codes

In this section and the next, we ask when two codes are "essentially the same." We term this concept "equivalence." Often we are interested in properties of codes, such as weight distribution, which remain unchanged when passing from one code to another that is essentially the same. Here we focus on the simplest form of equivalence, called permutation equivalence, and generalize this concept in the next section.

One way to view codes as "essentially the same" is to consider them "the same" if they are isomorphic as vector spaces. However, in that case the concept of weight, which we will see is crucial to the study and use of codes, is lost: codewords of one weight may be sent to codewords of a different weight by the isomorphism. A theorem of MacWilliams [212], which we will examine in Section 7.9, states that a vector space isomorphism of two binary codes of length n that preserves the weight of codewords (that is, send codewords of one weight to codewords of the same weight) can be extended to an isomorphism of $\mathbb{F}_2^n$ that is a permutation of coordinates. Clearly any permutation of coordinates that sends one code to another preserves the weight of codewords, regardless of the field. This leads to the following natural definition of permutation equivalent codes.

Two linear codes C_1 and C_2 are *permutation equivalent* provided there is a permutation of coordinates which sends C_1 to C_2. This permutation can be described using a *permutation matrix*, which is a square matrix with exactly one 1 in each row and column and 0s elsewhere. Thus C_1 and C_2 are permutation equivalent provided there is a permutation matrix P such that G_1 is a generator matrix of C_1 if and only if $G_1 P$ is a generator matrix of C_2. The effect of applying P to a generator matrix is to rearrange the columns of the generator matrix. If P is a permutation sending C_1 to C_2, we will write $C_1 P = C_2$, where $C_1 P = \{\mathbf{y} \mid \mathbf{y} = \mathbf{x} P$ for $\mathbf{x} \in C_1\}$.

Exercise 34 Prove that if G_1 and G_2 are generator matrices for a code C of length n and P is an $n \times n$ permutation matrix, then $G_1 P$ and $G_2 P$ are generator matrices for CP. ◆

Exercise 35 Suppose C_1 and C_2 are permutation equivalent codes where $C_1 P = C_2$ for some permutation matrix P. Prove that:
(a) $C_1^\perp P = C_2^\perp$, and
(b) if C_1 is self-dual, so is C_2. ◆

Example 1.6.1 Let C_1, C_2, and C_3 be binary codes with generator matrices

$$G_1 = \begin{bmatrix} 1 & 1 & 0 & 0 & 0 & 0 \\ 0 & 0 & 1 & 1 & 0 & 0 \\ 0 & 0 & 0 & 0 & 1 & 1 \end{bmatrix}, \quad G_2 = \begin{bmatrix} 1 & 0 & 0 & 0 & 0 & 1 \\ 0 & 0 & 1 & 1 & 0 & 0 \\ 0 & 1 & 0 & 0 & 1 & 0 \end{bmatrix}, \quad \text{and}$$

$$G_3 = \begin{bmatrix} 1 & 1 & 0 & 0 & 0 & 0 \\ 1 & 0 & 1 & 0 & 0 & 0 \\ 1 & 1 & 1 & 1 & 1 & 1 \end{bmatrix},$$

respectively. All three codes have weight distribution $A_0 = A_6 = 1$ and $A_2 = A_4 = 3$. (See Example 1.4.4 and Exercise 17.) The permutation switching columns 2 and 6 sends G_1 to G_2, showing that C_1 and C_2 are permutation equivalent. Both C_1 and C_2 are self-dual, consistent with (a) of Exercise 35. C_3 is not self-dual. Therefore C_1 and C_3 are not permutation equivalent by part (b) of Exercise 35. ∎

The next theorem shows that any code is permutation equivalent to one with generator matrix in standard form.

Theorem 1.6.2 *Let C be a linear code.*
(i) *C is permutation equivalent to a code which has generator matrix in standard form.*
(ii) *If $\mathcal{I}$ and $\mathcal{R}$ are information and redundancy positions, respectively, for C, then $\mathcal{R}$ and $\mathcal{I}$ are information and redundancy positions, respectively, for the dual code $C^{\perp}$.*

Proof: For (i), apply elementary row operations to any generator matrix of C. This will produce a new generator matrix of C which has columns the same as those in I_k, but possibly in a different order. Now choose a permutation of the columns of the new generator matrix so that these columns are moved to the order that produces $[I_k \mid A]$. The code generated by $[I_k \mid A]$ is equivalent to C.

If $\mathcal{I}$ is an information set for C, then by row reducing a generator matrix for C, we obtain columns in the information positions that are the columns of I_k in some order. As above, choose a permutation matrix P to move the columns so that CP has generator matrix $[I_k \mid A]$; P has moved $\mathcal{I}$ to the first k coordinate positions. By Theorem 1.2.1, $(CP)^{\perp}$ has the last $n - k$ coordinates as information positions. By Exercise 35, $(CP)^{\perp} = C^{\perp}P$, implying that $\mathcal{R}$ is a set of information positions for $C^{\perp}$, proving (ii). $\qquad\square$

It is often more convenient to use permutations (in cycle form) rather than permutation matrices to express equivalence. Let Sym_n be the set of all permutations of the set of n coordinates. If $\sigma \in \text{Sym}_n$ and $\mathbf{x} = x_1 x_2 \cdots x_n$, define

$$\mathbf{x}\sigma = y_1 y_2 \cdots y_n, \quad \text{where } y_j = x_{j\sigma^{-1}} \text{ for } 1 \le j \le n.$$

So $\mathbf{x}\sigma = \mathbf{x}P$, where $P = [p_{i,j}]$ is the permutation matrix given by

$$p_{i,j} = \begin{cases} 1 & \text{if } j = i\sigma, \\ 0 & \text{otherwise.} \end{cases} \tag{1.6}$$

This is illustrated in the next example.

Example 1.6.3 Let $n = 3$, $\mathbf{x} = x_1 x_2 x_3$, and $\sigma = (1, 2, 3)$. Then $1\sigma^{-1} = 3$, $2\sigma^{-1} = 1$, and $3\sigma^{-1} = 2$. So $\mathbf{x}\sigma = x_3 x_1 x_2$. Let

$$P = \begin{bmatrix} 0 & 1 & 0 \\ 0 & 0 & 1 \\ 1 & 0 & 0 \end{bmatrix}.$$

Then $\mathbf{x}P$ also equals $x_3 x_1 x_2$. ∎

Exercise 36 If $\sigma, \tau \in \text{Sym}_n$, show that $\mathbf{x}(\sigma \tau) = (\mathbf{x}\sigma)\tau$. ◆

Exercise 37 Let S be the set of all codes over $\mathbb{F}_q$ of length n. Let $C_1, C_2 \in S$. Define $C_1 \sim C_2$ to mean that there exists an $n \times n$ permutation matrix P such that $C_1 P = C_2$. Prove that $\sim$ is an equivalence relation on S. Recall that $\sim$ is an equivalence relation on a

set $\mathcal{S}$ if the following three conditions are fulfilled:
(i)   (reflexive) $\mathcal{C} \sim \mathcal{C}$ for all $\mathcal{C} \in \mathcal{S}$,
(ii) (symmetric) if $\mathcal{C}_1 \sim \mathcal{C}_2$, then $\mathcal{C}_2 \sim \mathcal{C}_1$, and
(iii) (transitive) if $\mathcal{C}_1 \sim \mathcal{C}_2$ and $\mathcal{C}_2 \sim \mathcal{C}_3$, then $\mathcal{C}_1 \sim \mathcal{C}_3$.   ◆

The set of coordinate permutations that map a code $\mathcal{C}$ to itself forms a *group*, that is, a set with an associative binary operation which has an identity and where all elements have inverses, called the *permutation automorphism group* of $\mathcal{C}$. This group is denoted by PAut($\mathcal{C}$). So if $\mathcal{C}$ is a code of length n, then PAut($\mathcal{C}$) is a subgroup of the *symmetric group* Sym$_n$.

Exercise 38 Show that if $\mathcal{C}$ is the $[n, 1]$ binary repetition code of Example 1.2.2, then PAut($\mathcal{C}$) $=$ Sym$_n$.   ◆

Exercise 39 Show that $(1, 2)(5, 6)$, $(1, 2, 3)(5, 6, 7)$, and $(1, 2, 4, 5, 7, 3, 6)$ are automorphisms of the $[7, 4]$ binary code $\mathcal{H}_3$ given in Example 1.2.3. These three permutations generate a group of order 168 called the *projective special linear group* PSL$_2$(7). This is in fact the permutation automorphism group of $\mathcal{H}_3$.   ◆

Knowledge of the permutation automorphism group of a code can give important theoretical and practical information about the code. While these groups for some codes have been determined, they are in general difficult to find. The following result shows the relationship between the permutation automorphism group of a code and that of its dual; it also establishes the connection between automorphism groups of permutation equivalent codes. Its proof is left to the reader.

Theorem 1.6.4 *Let $\mathcal{C}, \mathcal{C}_1,$ and $\mathcal{C}_2$ be codes over $\mathbb{F}_q$. Then:*
(i)   PAut($\mathcal{C}$) $=$ PAut($\mathcal{C}^\perp$),
(ii) *if $q = 4$,* PAut($\mathcal{C}$) $=$ PAut($\mathcal{C}^{\perp_H}$), *and*
(iii) *if $\mathcal{C}_1 P = \mathcal{C}_2$ for a permutation matrix P, then P^{-1}*PAut($\mathcal{C}_1$)$P =$ PAut($\mathcal{C}_2$).   ◆

Exercise 40 Prove Theorem 1.6.4.   ◆

One can prove that if two codes are permutation equivalent, so are their extensions; see Exercise 41. This is not necessarily the case for punctured codes.

Exercise 41 Prove that if $\mathcal{C}_1$ and $\mathcal{C}_2$ are permutation equivalent codes, then so are $\widehat{\mathcal{C}}_1$ and $\widehat{\mathcal{C}}_2$.   ◆

Example 1.6.5 Let $\mathcal{C}$ be the binary code with generator matrix

$$G = \begin{bmatrix} 1 & 1 & 0 & 0 & 0 \\ 0 & 0 & 1 & 1 & 1 \end{bmatrix}.$$

Let $\mathcal{C}_1^*$ and $\mathcal{C}_5^*$ be $\mathcal{C}$ punctured on coordinate 1 and 5, respectively. Then $\mathcal{C}_5^*$ has only even weight vectors, while $\mathcal{C}_1^*$ has odd weight codewords. Thus although $\mathcal{C}$ is certainly permutation equivalent to itself, $\mathcal{C}_1^*$ and $\mathcal{C}_5^*$ are not permutation equivalent.   ■

In some instances, the group PAut(C) is *transitive* as a permutation group; thus for every ordered pair (i, j) of coordinates, there is a permutation in PAut(C) which sends coordinate i to coordinate j. When PAut(C) is transitive, we have information about the structure of its punctured codes. When PAut($\widehat{C}$) is transitive, we have information about the minimum weight of C.

Theorem 1.6.6 *Let C be an $[n, k, d]$ code.*

(i) *Suppose that PAut(C) is transitive. Then the n codes obtained from C by puncturing C on a coordinate are permutation equivalent.*

(ii) *Suppose that PAut($\widehat{C}$) is transitive. Then the minimum weight d of C is its minimum odd-like weight d_o. Furthermore, every minimum weight codeword of C is odd-like.*

Proof: The proof of assertion (i) is left to the reader in Exercise 42. Now assume that PAut($\widehat{C}$) is transitive. Applying (i) to $\widehat{C}$ we conclude that puncturing $\widehat{C}$ on any coordinate gives a code permutation equivalent to C. Let $\mathbf{c}$ be a minimum weight vector of C and assume that $\mathbf{c}$ is even-like. Then wt($\widehat{\mathbf{c}}$) $= d$, where $\widehat{\mathbf{c}} \in \widehat{C}$ is the extended vector. Puncturing $\widehat{C}$ on a coordinate where $\mathbf{c}$ is nonzero gives a vector of weight $d - 1$ in a code permutation equivalent to C, a contradiction.   □

Exercise 42 Prove Theorem 1.6.6(i).   ♦

Exercise 43 Let C be the code of Example 1.4.4.

(a) Is PAut(C) transitive?

(b) Find generator matrices for all six codes punctured on one point. Which of these punctured codes are equivalent?

(c) Find generator matrices for all 15 codes punctured on two points. Which of these punctured codes are equivalent?   ♦

Exercise 44 Let $C = C_1 \oplus C_2$, where C_1 and C_2 are of length n_1 and n_2, respectively. Prove that

PAut(C_1) × PAut(C_2) ⊆ PAut(C),

where PAut(C_1) × PAut(C_2) is the direct product of the groups PAut(C_1) (acting on the first n_1 coordinates of C) and PAut(C_2) (acting on the last n_2 coordinates of C).   ♦

For binary codes, the notion of permutation equivalence is the most general form of equivalence. However, for codes over other fields, other forms of equivalence are possible.

1.7   More general equivalence of codes

When considering codes over fields other than $\mathbb{F}_2$, equivalence takes a more general form. For these codes there are other maps which preserve the weight of codewords. These maps include those which rescale coordinates and those which are induced from field automorphisms (a topic we study more extensively in Chapter 3). We take up these maps one at a time.

First, recall that a *monomial matrix* is a square matrix with exactly one nonzero entry in each row and column. A monomial matrix M can be written either in the form DP or the form PD_1, where D and D_1 are diagonal matrices and P is a permutation matrix.

Example 1.7.1 The monomial matrix

$$M = \begin{bmatrix} 0 & a & 0 \\ 0 & 0 & b \\ c & 0 & 0 \end{bmatrix}$$

equals

$$DP = \begin{bmatrix} a & 0 & 0 \\ 0 & b & 0 \\ 0 & 0 & c \end{bmatrix} \begin{bmatrix} 0 & 1 & 0 \\ 0 & 0 & 1 \\ 1 & 0 & 0 \end{bmatrix} = PD_1 = \begin{bmatrix} 0 & 1 & 0 \\ 0 & 0 & 1 \\ 1 & 0 & 0 \end{bmatrix} \begin{bmatrix} c & 0 & 0 \\ 0 & a & 0 \\ 0 & 0 & b \end{bmatrix}.$$ ∎

We will generally choose the form $M = DP$ for representing monomial matrices; D is called the *diagonal part* of M and P is the *permutation part*. This notation allows a more compact form using (1.6), as we now illustrate.

Example 1.7.2 The monomial matrix $M = DP$ of Example 1.7.1 can be written

$\mathrm{diag}(a, b, c)(1, 2, 3),$

where $\mathrm{diag}(a, b, c)$ is the diagonal matrix D and $(1, 2, 3)$ is the permutation matrix P written in cycle form. ∎

We will apply monomial maps $M = DP$ on the right of row vectors $\mathbf{x}$ in the manner of the next example.

Example 1.7.3 Let $M = \mathrm{diag}(a, b, c)(1, 2, 3)$ be the monomial map of Example 1.7.2 and $\mathbf{x} = x_1x_2x_3 = (x_1, x_2, x_3)$. Then

$\mathbf{x}M = \mathbf{x}DP = (ax_1, bx_2, cx_3)P = (cx_3, ax_1, bx_2).$ ∎

This example illustrates the more general principle of how to apply $M = DP$ to a vector $\mathbf{x}$ where σ is the permutation (in cycle form) associated to P. For all i:
- first, multiply the ith component of $\mathbf{x}$ by the ith diagonal entry of D, and
- second, move this product to coordinate position $i\sigma$.

With this concept of monomial maps, we now are ready to define monomial equivalence. Let C_1 and C_2 be codes of the same length over a field $\mathbb{F}_q$, and let G_1 be a generator matrix for C_1. Then C_1 and C_2 are *monomially equivalent* provided there is a monomial matrix M so that G_1M is a generator matrix of C_2. More simply, C_1 and C_2 are monomially equivalent if there is a monomial map M such that $C_2 = C_1M$. Monomial equivalence and permutation equivalence are precisely the same for binary codes.

Exercise 45 Let $\mathcal{S}$ be the set of all codes over $\mathbb{F}_q$ of length n. Let $\mathcal{C}_1, \mathcal{C}_2 \in \mathcal{S}$. Define $\mathcal{C}_1 \sim \mathcal{C}_2$ to mean that there exists an $n \times n$ monomial matrix M such that $\mathcal{C}_1 M = \mathcal{C}_2$. Prove that $\sim$ is an equivalence relation on $\mathcal{S}$. (The definition of "equivalence relation" is given in Exercise 37.) ◆

There is one more type of map that we need to consider: that arising from automorphisms of the field $\mathbb{F}_q$, called Galois automorphisms. We will apply this in conjunction with monomial maps. (We will apply field automorphisms on the right of field elements since we are applying matrices on the right of vectors.) If γ is a field automorphism of $\mathbb{F}_q$ and $M = DP$ is a monomial map with entries in $\mathbb{F}_q$, then applying the map $M\gamma$ to a vector $\mathbf{x}$ is described by the following process, where again σ is the permutation associated to the matrix P. For all i:

- first, multiply the ith component of $\mathbf{x}$ by the ith diagonal entry of D,
- second, move this product to coordinate position $i\sigma$, and
- third, apply γ to this component.

Example 1.7.4 The field $\mathbb{F}_4$ has automorphism γ given by $x\gamma = x^2$. If $M = DP = \mathrm{diag}(a, b, c)(1, 2, 3)$ is the monomial map of Example 1.7.2 and $\mathbf{x} = x_1 x_2 x_3 = (x_1, x_2, x_3) \in \mathbb{F}_4^3$, then

$$\mathbf{x}M\gamma = (ax_1, bx_2, cx_3)P\gamma = (cx_3, ax_1, bx_2)\gamma = ((cx_3)^2, (ax_1)^2, (bx_2)^2).$$

For instance,

$$(1, \omega, 0)\mathrm{diag}(\omega, \overline{\omega}, 1)(1, 2, 3)\gamma = (0, \overline{\omega}, 1).$$ ∎

We say that two codes $\mathcal{C}_1$ and $\mathcal{C}_2$ of the same length over $\mathbb{F}_q$ are *equivalent* provided there is a monomial matrix M and an automorphism γ of the field such that $\mathcal{C}_2 = \mathcal{C}_1 M\gamma$. This is the most general notion of equivalence that we will consider. Thus we have three notions of when codes are the "same": permutation equivalence, monomial equivalence, and equivalence. All three are the same if the codes are binary; monomial equivalence and equivalence are the same if the field considered has a prime number of elements. The fact that these are the appropriate maps to consider for equivalence is a consequence of a theorem by MacWilliams [212] regarding weight preserving maps discussed in Section 7.9.

Two equivalent codes have the same weight distribution. However, two codes with the same weight distribution need not be equivalent as Example 1.6.1 shows. Exercise 35 shows that if $\mathcal{C}_1$ and $\mathcal{C}_2$ are permutation equivalent codes, then so are their duals under the same map. However, if $\mathcal{C}_1 M = \mathcal{C}_2$, it is not necessarily the case that $\mathcal{C}_1^\perp M = \mathcal{C}_2^\perp$.

Example 1.7.5 Let $\mathcal{C}_1$ and $\mathcal{C}_2$ be $[2, 1, 2]$ codes over $\mathbb{F}_4$ with generator matrices $[1 \ 1]$ and $[1 \ \omega]$, respectively. Then the duals $\mathcal{C}_1^\perp$ and $\mathcal{C}_2^\perp$ under the ordinary inner product have generator matrices $[1 \ 1]$ and $[1 \ \overline{\omega}]$, respectively. Notice that $\mathcal{C}_1 \mathrm{diag}(1, \omega) = \mathcal{C}_2$, but $\mathcal{C}_1^\perp \mathrm{diag}(1, \omega) \neq \mathcal{C}_2^\perp$. ∎

In the above example, $\mathcal{C}_1$ is self-dual but $\mathcal{C}_2$ is not. Thus equivalence may not preserve self-duality. However, the following theorem is valid, and its proof is left as an exercise.

Theorem 1.7.6 *Let C be a code over $\mathbb{F}_q$. The following hold:*

(i)   *If M is a monomial matrix with entries only from $\{0, -1, 1\}$, then C is self-dual if and only if CM is self-dual.*

(ii)   *If $q = 3$ and C is equivalent to C_1, then C is self-dual if and only if C_1 is self-dual.*

(iii)   *If $q = 4$ and C is equivalent to C_1, then C is Hermitian self-dual if and only if C_1 is Hermitian self-dual.*

As there are three versions of equivalence, there are three possible automorphism groups. Let C be a code over $\mathbb{F}_q$. We defined the permutation automorphism group PAut(C) of C in the last section. The set of monomial matrices that map C to itself forms the group MAut(C) called the *monomial automorphism group* of C. Finally, the set of maps of the form $M\gamma$, where M is a monomial matrix and γ is a field automorphism, that map C to itself forms the group ΓAut(C), called *automorphism group* of C.[3] That MAut(C) and ΓAut(C) are groups is left as an exercise. In the binary case all three groups are identical. If q is a prime, MAut(C) = ΓAut(C). In general, PAut(C) $\subseteq$ MAut(C) $\subseteq$ ΓAut(C).

Exercise 46   For $1 \le i \le 3$ let D_i be diagonal matrices, P_i permutation matrices, and γ_i automorphisms of $\mathbb{F}_q$.

(a)   You can write $(D_1 P_1 \gamma_1)(D_2 P_2 \gamma_2)$ in the form $D_3 P_3 \gamma_3$. Find D_3, P_3, and γ_3 in terms of D_1, D_2, P_1, P_2, γ_1, and γ_2.

(b)   You can write $(D_1 P_1 \gamma_1)^{-1}$ in the form $D_2 P_2 \gamma_2$. Find D_2, P_2, and γ_2 in terms of D_1, P_1, and γ_1.   ◆

Exercise 47   Let $\mathcal{S}$ be the set of all codes over $\mathbb{F}_q$ of length n. Let $C_1, C_2 \in \mathcal{S}$. Define $C_1 \sim C_2$ to mean that there exists an $n \times n$ monomial matrix M and an automorphism γ of $\mathbb{F}_q$ such that $C_1 M \gamma = C_2$. Prove that $\sim$ is an equivalence relation on $\mathcal{S}$. (The definition of "equivalence relation" is given in Exercise 37.) You may find Exercise 46 helpful.   ◆

Exercise 48   Prove that MAut(C) and ΓAut(C) are groups. (Hint: Use Exercise 46.)   ◆

Example 1.7.7   Let C be the tetracode with generator matrix as in Example 1.3.3. Labeling the coordinates by $\{1, 2, 3, 4\}$, PAut(C) is the group of order 3 generated by the permutation $(1, 3, 4)$. MAut(C) = ΓAut(C) is a group of order 48 generated by diag$(1, 1, 1, -1)(1, 2, 3, 4)$ and diag$(1, 1, 1, -1)(1, 2)$.   ∎

Exercise 49   Let C be the tetracode of Examples 1.3.3 and 1.7.7.

(a)   Verify that the maps listed in Example 1.7.7 are indeed automorphisms of the tetracode.

(b)   Write the generator $(1, 3, 4)$ of PAut(C) as a product of the two generators given for MAut(C) in Example 1.7.7.

(c)   (Hard) Prove that the groups PAut(C) and ΓAut(C) are as claimed in Example 1.7.7.   ◆

---

[3]   The notation for automorphism groups is not uniform in the literature. For example, G(C) or Aut(C) are sometimes used for one of the automorphism groups of C. As a result of this, we avoid both of these notations.

Example 1.7.8 Let G_6' be the generator matrix of a [6, 3] code $\mathcal{G}_6'$ over $\mathbb{F}_4$, where

$$G_6' = \begin{bmatrix} 1 & \omega & 1 & 0 & 0 & \omega \\ 0 & 1 & \omega & 1 & 0 & \omega \\ 0 & 0 & 1 & \omega & 1 & \omega \end{bmatrix}.$$

Label the columns $\{1, 2, 3, 4, 5, 6\}$. If this generator matrix is row reduced, we obtain the matrix

$$\begin{bmatrix} 1 & 0 & 0 & 1 & \omega & \omega \\ 0 & 1 & 0 & \omega & \omega & 1 \\ 0 & 0 & 1 & \omega & 1 & \omega \end{bmatrix}.$$

Swapping columns 5 and 6 gives the generator matrix $\mathcal{G}_6$ of Example 1.3.4; thus $\mathcal{G}_6'(5, 6) = \mathcal{G}_6$. (The codes are equivalent and both are called the hexacode.) Using group theoretic arguments, one can verify the following information about the three automorphism groups of $\mathcal{G}_6'$; however, one can also use algebraic systems such as Magma or Gap to carry out this verification: $\mathrm{PAut}(\mathcal{G}_6')$ is a group of order 60 generated by the permutations $(1, 2, 6)(3, 5, 4)$ and $(1, 2, 3, 4, 5)$. $\mathrm{MAut}(\mathcal{G}_6')$ is a group of order $3 \cdot 360$ generated by the monomial map $\mathrm{diag}(\omega, 1, 1, \omega, \overline{\omega}, \overline{\omega})(1, 2, 6)$ and the permutation $(1, 2, 3, 4, 5)$.[4] Finally, $\Gamma\mathrm{Aut}(\mathcal{G}_6')$ is a group twice as big as $\mathrm{MAut}(\mathcal{G}_6')$ generated by $\mathrm{MAut}(\mathcal{G}_6')$ and $\mathrm{diag}(1, \overline{\omega}, \omega, \omega, \overline{\omega}, 1)(1, 6)\gamma$, where γ is the automorphism of $\mathbb{F}_4$ given by $x\gamma = x^2$. ∎

Exercise 50   Let $\mathcal{G}_6'$ be the hexacode of Example 1.7.8.
(a) Verify that $(1, 2, 6)(3, 5, 4)$ and $(1, 2, 3, 4, 5)$ are elements of $\mathrm{PAut}(\mathcal{G}_6')$.
(b) Verify that $\mathrm{diag}(\omega, 1, 1, \omega, \overline{\omega}, \overline{\omega})(1, 2, 6)$ is an element of $\mathrm{MAut}(\mathcal{G}_6')$.
(c) Verify that $\mathrm{diag}(1, \overline{\omega}, \omega, \omega, \overline{\omega}, 1)(1, 6)\gamma$ is an element of $\Gamma\mathrm{Aut}(\mathcal{G}_6')$.   ◆

Recall that $\mathrm{PAut}(\mathcal{C}^\perp) = \mathrm{PAut}(\mathcal{C})$, by Theorem 1.6.4. One can find $\mathrm{MAut}(\mathcal{C}^\perp)$ and $\Gamma\mathrm{Aut}(\mathcal{C}^\perp)$ from $\mathrm{MAut}(\mathcal{C})$ and $\Gamma\mathrm{Aut}(\mathcal{C})$ although the statement is not so simple.

Theorem 1.7.9 *Let $\mathcal{C}$ be a code over $\mathbb{F}_q$. Then:*
(i) $\mathrm{MAut}(\mathcal{C}^\perp) = \{D^{-1}P \mid DP \in \mathrm{MAut}(\mathcal{C})\}$, *and*
(ii) $\Gamma\mathrm{Aut}(\mathcal{C}^\perp) = \{D^{-1}P\gamma \mid DP\gamma \in \Gamma\mathrm{Aut}(\mathcal{C})\}$.

In the case of codes over $\mathbb{F}_4$, $\mathrm{PAut}(\mathcal{C}^{\perp_H}) = \mathrm{PAut}(\mathcal{C})$ by Theorem 1.6.4; the following extends this in the nicest possible fashion.

Theorem 1.7.10 *Let $\mathcal{C}$ be a code over $\mathbb{F}_4$. Then:*
(i) $\mathrm{MAut}(\mathcal{C}^{\perp_H}) = \mathrm{MAut}(\mathcal{C})$, *and*
(ii) $\Gamma\mathrm{Aut}(\mathcal{C}^{\perp_H}) = \Gamma\mathrm{Aut}(\mathcal{C})$.

The third part of Theorem 1.6.4 is generalized as follows.

---

[4] This group is isomorphic to the nonsplitting central extension of the cyclic group of order 3 by the alternating group on six points.

Theorem 1.7.11 *Let C_1 and C_2 be codes over $\mathbb{F}_q$. Let P be a permutation matrix, M a monomial matrix, and γ an automorphism of $\mathbb{F}_q$.*
(i)   *If $C_1 P = C_2$, then $P^{-1}\mathrm{PAut}(C_1)P = \mathrm{PAut}(C_2)$.*
(ii) *If $C_1 M = C_2$, then $M^{-1}\mathrm{MAut}(C_1)M = \mathrm{MAut}(C_2)$.*
(iii) *If $C_1 M\gamma = C_2$, then $(M\gamma)^{-1}\Gamma\mathrm{Aut}(C_1)M\gamma = \Gamma\mathrm{Aut}(C_2)$.*

Exercise 51   Prove Theorems 1.7.9, 1.7.10, and 1.7.11.   ◆

Exercise 52   Using Theorems 1.6.4 and 1.7.11 give generators of $\mathrm{PAut}(\mathcal{G}_6)$, $\mathrm{MAut}(\mathcal{G}_6)$, and $\Gamma\mathrm{Aut}(\mathcal{G}_6)$ from the information given in Example 1.7.8.   ◆

As with $\mathrm{PAut}(C)$, we can speak of transitivity of the automorphism groups $\mathrm{MAut}(C)$ or $\Gamma\mathrm{Aut}(C)$. To do this we consider only the permutation parts of the maps in these groups. Specifically, define $\mathrm{MAut}_{\mathrm{Pr}}(C)$ to be the set $\{P \mid DP \in \mathrm{MAut}(C)\}$ and $\Gamma\mathrm{Aut}_{\mathrm{Pr}}(C)$ to be $\{P \mid DP\gamma \in \Gamma\mathrm{Aut}(C)\}$. (The subscript Pr stands for projection. The groups $\mathrm{MAut}(C)$ and $\Gamma\mathrm{Aut}(C)$ are semi-direct products; the groups $\mathrm{MAut}_{\mathrm{Pr}}(C)$ and $\Gamma\mathrm{Aut}_{\mathrm{Pr}}(C)$ are obtained from $\mathrm{MAut}(C)$ and $\Gamma\mathrm{Aut}(C)$ by projecting onto the permutation part of the semi-direct product.) For instance, in Example 1.7.7, $\mathrm{MAut}_{\mathrm{Pr}}(C) = \Gamma\mathrm{Aut}_{\mathrm{Pr}}(C) = \mathrm{Sym}_4$; in Example 1.7.8, $\mathrm{MAut}_{\mathrm{Pr}}(C)$ is the alternating group on six points and $\Gamma\mathrm{Aut}_{\mathrm{Pr}}(C) = \mathrm{Sym}_6$. We leave the proof of the following theorem as an exercise.

Theorem 1.7.12 *Let C be a linear code over $\mathbb{F}_q$. Then:*
(i)   *$\mathrm{MAut}_{\mathrm{Pr}}(C)$ and $\Gamma\mathrm{Aut}_{\mathrm{Pr}}(C)$ are subgroups of the symmetric group Sym_n, and*
(ii) *$\mathrm{PAut}(C) \subseteq \mathrm{MAut}_{\mathrm{Pr}}(C) \subseteq \Gamma\mathrm{Aut}_{\mathrm{Pr}}(C)$.*

Exercise 53   Prove Theorem 1.7.12. (Hint: Use Exercise 46.)   ◆

We now say that $\mathrm{MAut}(C)$ ($\Gamma\mathrm{Aut}(C)$, respectively) is *transitive* as a permutation group if $\mathrm{MAut}_{\mathrm{Pr}}(C)$ ($\Gamma\mathrm{Aut}_{\mathrm{Pr}}(C)$, respectively) is transitive. The following is a generalization of Theorem 1.6.6.

Theorem 1.7.13 *Let C be an $[n, k, d]$ code.*
(i)   *Suppose that $\mathrm{MAut}(C)$ is transitive. Then the n codes obtained from C by puncturing C on a coordinate are monomially equivalent.*
(ii) *Suppose that $\Gamma\mathrm{Aut}(C)$ is transitive. Then the n codes obtained from C by puncturing C on a coordinate are equivalent.*
(iii) *Suppose that either $\mathrm{MAut}(\widehat{C})$ or $\Gamma\mathrm{Aut}(\widehat{C})$ is transitive. Then the minimum weight d of C is its minimum odd-like weight d_o. Furthermore, every minimum weight codeword of C is odd-like.*

Exercise 54   Prove Theorem 1.7.13.   ◆

1.8   Hamming codes

We now generalize the binary code $\mathcal{H}_3$ of Example 1.2.3. The parity check matrix obtained in that example was

$$H = [A^T \mid I_3] = \begin{bmatrix} 0 & 1 & 1 & 1 & 1 & 0 & 0 \\ 1 & 0 & 1 & 1 & 0 & 1 & 0 \\ 1 & 1 & 0 & 1 & 0 & 0 & 1 \end{bmatrix}.$$

Notice that the columns of this parity check matrix are all the distinct nonzero binary columns of length 3. So $\mathcal{H}_3$ is equivalent to the code with parity check matrix

$$H' = \begin{bmatrix} 0 & 0 & 0 & 1 & 1 & 1 & 1 \\ 0 & 1 & 1 & 0 & 0 & 1 & 1 \\ 1 & 0 & 1 & 0 & 1 & 0 & 1 \end{bmatrix}$$

whose columns are the numbers 1 through 7 written as binary numerals (with leading 0s as necessary to have a 3-tuple) in their natural order.

This form generalizes easily. Let $n = 2^r - 1$, with $r \geq 2$. Then the $r \times (2^r - 1)$ matrix H_r whose columns, in order, are the numbers $1, 2, \ldots, 2^r - 1$ written as binary numerals, is the parity check matrix of an $[n = 2^r - 1, k = n - r]$ binary code. Any rearrangement of columns of H_r gives an equivalent code, and hence any one of these equivalent codes will be called the *binary Hamming code of length* $n = 2^r - 1$ and denoted by either $\mathcal{H}_r$ or $\mathcal{H}_{2,r}$. It is customary when naming a code, such as the Hamming code, the tetracode, or the hexacode, to identify equivalent codes. We will follow this practice with these and other codes as well.

Since the columns of H_r are distinct and nonzero, the minimum distance is at least 3 by Corollary 1.4.14. Since the columns corresponding to the numbers 1, 2, and 3 are linearly dependent, the minimum distance equals 3, by the same corollary. Thus $\mathcal{H}_r$ is a binary $[2^r - 1, 2^r - 1 - r, 3]$ code. In the following sense, these codes are unique.

Theorem 1.8.1 *Any* $[2^r - 1, 2^r - 1 - r, 3]$ *binary code is equivalent to the binary Hamming code* $\mathcal{H}_r$.

Exercise 55   Prove Theorem 1.8.1.   ◆

Exercise 56   Prove that every $[8, 4, 4]$ binary code is equivalent to the extended Hamming code $\widehat{\mathcal{H}}_3$. (So for that reason, we say that the $[8, 4, 4]$ binary code is unique.)   ◆

Similarly, Hamming codes $\mathcal{H}_{q,r}$ can be defined over an arbitrary finite field $\mathbb{F}_q$. For $r \geq 2$, $\mathcal{H}_{q,r}$ has parity check matrix $H_{q,r}$ defined by choosing for its columns a nonzero vector from each 1-dimensional subspace of $\mathbb{F}_q^r$. (Alternately, these columns are the points of the projective geometry PG$(r - 1, q)$.) There are $(q^r - 1)/(q - 1)$ 1-dimensional subspaces. Therefore $\mathcal{H}_{q,r}$ has length $n = (q^r - 1)/(q - 1)$, dimension $n - r$, and redundancy r. As no two columns are multiples of each other, $\mathcal{H}_{q,r}$ has minimum weight at least 3. Adding two nonzero vectors from two different 1-dimensional subspaces gives a nonzero vector from yet a third 1-dimensional space; hence $\mathcal{H}_{q,r}$ has minimum weight 3. When $q = 2$, $\mathcal{H}_{2,r}$ is precisely the code $\mathcal{H}_r$.

Suppose you begin with one particular order of the 1-dimensional subspaces and one particular choice for representatives for those subspaces to form the parity check matrix $H_{q,r}$ for $\mathcal{H}_{q,r}$. If you choose a different parity check matrix $H'_{q,r}$ by choosing a different order for the list of subspaces and choosing different representatives from these subspaces, $H'_{q,r}$ can be obtained from $H_{q,r}$ by rescaling and reordering the columns – precisely what is accomplished by multiplying $H_{q,r}$ on the right by some monomial matrix. So any code you get in the above manner is monomially equivalent to any other code obtained in the same manner. Again $\mathcal{H}_{q,r}$ will therefore refer to any code in the equivalence class. As in the binary case, these codes are unique, up to equivalence.

Theorem 1.8.2 *Any* $[(q^r − 1)/(q − 1), (q^r − 1)/(q − 1) − r, 3]$ *code over* $\mathbb{F}_q$ *is monomially equivalent to the Hamming code* $\mathcal{H}_{q,r}$.

Exercise 57   Prove Theorem 1.8.2.   ◆

Exercise 58   Prove that the tetracode of Example 1.3.3 was appropriately denoted in that example as $\mathcal{H}_{3,2}$. In other words, show that the tetracode is indeed a Hamming code.   ◆

The duals of the Hamming codes are called *simplex codes*. They are $[(q^r − 1)/(q − 1), r]$ codes whose codeword weights have a rather interesting property. The simplex code $\mathcal{H}_3^{\perp}$ has only nonzero codewords of weight 4 (see Example 1.2.3). The tetracode, being a self-dual Hamming code, is a simplex code; its nonzero codewords all have weight 3. In general, we have the following, which will be proved as part of Theorem 2.7.5.

Theorem 1.8.3 *The nonzero codewords of the* $[(q^r − 1)/(q − 1), r]$ *simplex code over* $\mathbb{F}_q$ *all have weights* $q^{r−1}$.

We now give a construction of the binary simplex codes and prove Theorem 1.8.3 in this case. These codes are produced by a modification of the $(\mathbf{u} \mid \mathbf{u} + \mathbf{v})$ construction of Section 1.5.5.

Let G_2 be the matrix

$$G_2 = \begin{bmatrix} 0 & 1 & 1 \\ 1 & 0 & 1 \end{bmatrix}.$$

For $r \geq 3$, define G_r inductively by

$$G_r = \left[\begin{array}{c|c|c} 0 \cdots 0 & 1 & 1 \cdots 1 \\ \hline & 0 & \\ G_{r-1} & \vdots & G_{r-1} \\ & 0 & \end{array} \right].$$

We claim the code $\mathcal{S}_r$ generated by G_r is the dual of $\mathcal{H}_r$. Clearly, G_r has one more row than G_{r-1} and, as G_2 has 2 rows, G_r has r rows. Let G_r have n_r columns. So $n_2 = 2^2 − 1$ and $n_r = 2n_{r-1} + 1$; by induction $n_r = 2^r − 1$. The columns of G_2 are nonzero and distinct; clearly by construction, the columns of G_r are nonzero and distinct if the columns of G_{r-1} are also nonzero and distinct. So by induction G_r has $2^r - 1$ distinct nonzero columns of length r. But there are only $2^r - 1$ possible distinct nonzero r-tuples; these are the binary expansions of $1, 2, \ldots, 2^r - 1$. (In fact, the columns are in this order.) So $\mathcal{S}_r = \mathcal{H}_r^{\perp}$.

The nonzero codewords of $\mathcal{S}_2$ have weight 2. Assume the nonzero codewords of $\mathcal{S}_{r-1}$ have weight 2^{r-2}. Then the nonzero codewords of the subcode generated by the last $r - 1$ rows of G_r have the form $(\mathbf{a}, 0, \mathbf{b})$, where $\mathbf{a}, \mathbf{b} \in \mathcal{S}_{r-1}$. So these codewords have weight $2 \cdot 2^{r-2} = 2^{r-1}$. Also the top row of G_r has weight $1 + 2^{r-1} - 1 = 2^{r-1}$. The remaining nonzero codewords of $\mathcal{S}_r$ have the form $(\mathbf{a}, 1, \mathbf{b} + \mathbf{1})$, where $\mathbf{a}, \mathbf{b} \in \mathcal{S}_{r-1}$. As $\mathrm{wt}(\mathbf{b} + \mathbf{1}) = 2^{r-2} - 1$, $\mathrm{wt}(\mathbf{a}, 1, \mathbf{b} + \mathbf{1}) = 2^{r-2} + 1 + 2^{r-2} - 1 = 2^{r-1}$. Thus by induction $\mathcal{S}_r$ has all nonzero codewords of weight 2^{r-1}.

1.9 The Golay codes

In this section we define four codes that are called the Golay codes. The first two are binary and the last two are ternary. In the binary case, the shorter of the codes is obtained from the longer by puncturing and the longer from the shorter by extending. The same holds in the ternary case if the generator matrix is chosen in the right form. (See Exercise 61.) Although the hexacode $\mathcal{G}_6$ of Example 1.3.4 and the punctured code $\mathcal{G}_6^*$ are technically not Golay codes, they have so many properties similar to the binary and ternary Golay codes, they are often referred to as the Golay codes over $\mathbb{F}_4$. These codes have had an exceptional place in the history of coding theory. The binary code of length 23 and the ternary code of length 11 were first described by M. J. E. Golay in 1949 [102].

1.9.1 The binary Golay codes

We let $\mathcal{G}_{24}$ be the $[24, 12]$ code with generator matrix $G_{24} = [I_{12} \mid A]$ in standard form, where

$$A = \begin{bmatrix} 0 & 1 & 1 & 1 & 1 & 1 & 1 & 1 & 1 & 1 & 1 & 1 \\ 1 & 1 & 1 & 0 & 1 & 1 & 1 & 0 & 0 & 0 & 1 & 0 \\ 1 & 1 & 0 & 1 & 1 & 1 & 0 & 0 & 0 & 1 & 0 & 1 \\ 1 & 0 & 1 & 1 & 1 & 0 & 0 & 0 & 1 & 0 & 1 & 1 \\ 1 & 1 & 1 & 1 & 0 & 0 & 0 & 1 & 0 & 1 & 1 & 0 \\ 1 & 1 & 1 & 0 & 0 & 0 & 1 & 0 & 1 & 1 & 0 & 1 \\ 1 & 1 & 0 & 0 & 0 & 1 & 0 & 1 & 1 & 0 & 1 & 1 \\ 1 & 0 & 0 & 0 & 1 & 0 & 1 & 1 & 0 & 1 & 1 & 1 \\ 1 & 0 & 0 & 1 & 0 & 1 & 1 & 0 & 1 & 1 & 1 & 0 \\ 1 & 0 & 1 & 0 & 1 & 1 & 0 & 1 & 1 & 1 & 0 & 0 \\ 1 & 1 & 0 & 1 & 1 & 0 & 1 & 1 & 1 & 0 & 0 & 0 \\ 1 & 0 & 1 & 1 & 0 & 1 & 1 & 1 & 0 & 0 & 0 & 1 \end{bmatrix}.$$

Notice how A is constructed. The matrix A is an example of a bordered reverse circulant matrix. Label the columns of A by $\infty, 0, 1, 2, \ldots, 10$. The first row contains 0 in column ∞

and 1 elsewhere. To obtain the second row, a 1 is placed in column ∞ and a 1 is placed in columns 0, 1, 3, 4, 5, and 9; these numbers are precisely the squares of the integers modulo 11. That is $0^2 = 0$, $1^2 \equiv 10^2 \equiv 1 \pmod{11}$, $2^2 \equiv 9^2 \equiv 4 \pmod{11}$, etc. The third row of A is obtained by putting a 1 in column ∞ and then shifting the components in the second row one place to the left and wrapping the entry in column 0 around to column 10. The fourth row is obtained from the third in the same manner, as are the remaining rows.

We give some elementary properties of $\mathcal{G}_{24}$. For ease of notation, let A_1 be the 11×11 reverse circulant matrix obtained from A by deleting row one and column ∞. Note first that the rows of $\mathcal{G}_{24}$ have weights 8 and 12. In particular the inner product of any row of $\mathcal{G}_{24}$ with itself is 0. The inner product of row one with any other row is also 0 as each row of A_1 has weight 6. To find the inner product of any row below the first with any other row below the first, by the circulant nature of A_1, we can shift both rows so that one of them is row two. (For example, the inner product of row four with row seven is the same as the inner product of row two with row five.) The inner product of row two with any row below it is 0 by direct inspection. Therefore $\mathcal{G}_{24}$ is self-dual with all rows in the generator matrix of weight divisible by four. By Theorem 1.4.8(i), all codewords of $\mathcal{G}_{24}$ have weights divisible by four.

Thus $\mathcal{G}_{24}$ is a $[24, 12, d]$ self-dual code, with $d = 4$ or 8. Suppose $d = 4$. Notice that $A^{\mathrm{T}} = A$. As $\mathcal{G}_{24}$ is self-dual, by Theorem 1.2.1, $[A^{\mathrm{T}} \mid I_{12}] = [A \mid I_{12}]$ is also a generator matrix. Hence if $(\mathbf{a}, \mathbf{b})$ is a codeword of $\mathcal{G}_{24}$, where $\mathbf{a}, \mathbf{b} \in \mathbb{F}_2^{12}$, so is $(\mathbf{b}, \mathbf{a})$. Then if $\mathbf{c} = (\mathbf{a}, \mathbf{b})$ is a codeword of $\mathcal{G}_{24}$ of weight 4, we may assume $\mathrm{wt}(\mathbf{a}) \leq \mathrm{wt}(\mathbf{b})$. If $\mathrm{wt}(\mathbf{a}) = 0$, $\mathbf{a} = \mathbf{0}$ and as $\mathcal{G}_{24}$ is in standard form, $\mathbf{b} = \mathbf{0}$, which is a contradiction. If $\mathrm{wt}(\mathbf{a}) = 1$, then $\mathbf{c}$ is one of the rows of $\mathcal{G}_{24}$, which is also a contradiction. Finally, if $\mathrm{wt}(\mathbf{a}) = 2$, then $\mathbf{c}$ is the sum of two rows of $\mathcal{G}_{24}$. The same shifting argument as earlier shows that the weight of $\mathbf{c}$ is the same as the weight of a codeword that is the sum of row two of $\mathcal{G}_{24}$ and another row. By inspection, none of these sums contributes exactly two to the weight of the right 12 components. So $d = 8$.

If we puncture in any of the coordinates, we obtain a $[23, 12, 7]$ binary code $\mathcal{G}_{23}$. It turns out, as we will see later, that all these punctured codes are equivalent. By Exercise 59, adding an overall parity check to one of these punctured codes (in the same position which had been punctured) gives exactly the same $\mathcal{G}_{24}$ back. In the future, any code equivalent to $\mathcal{G}_{23}$ will be called the *binary Golay code* and any code equivalent to $\mathcal{G}_{24}$ will be called the *extended binary Golay code*. The codes $\mathcal{G}_{23}$ and $\mathcal{G}_{24}$ have amazing properties and a variety of constructions, as we will see throughout this book.

Exercise 59 Prove that if $\mathcal{G}_{24}$ is punctured in any coordinate and the resulting code is extended in the same position, exactly the same code $\mathcal{G}_{24}$ is obtained. Hint: See Exercise 27. ♦

1.9.2   The ternary Golay codes

The ternary code $\mathcal{G}_{12}$ is the $[12, 6]$ code over $\mathbb{F}_3$ with generator matrix $G_{12} = [I_6 \mid A]$ in standard form, where

$$A = \begin{bmatrix} 0 & 1 & 1 & 1 & 1 & 1 \\ 1 & 0 & 1 & -1 & -1 & 1 \\ 1 & 1 & 0 & 1 & -1 & -1 \\ 1 & -1 & 1 & 0 & 1 & -1 \\ 1 & -1 & -1 & 1 & 0 & 1 \\ 1 & 1 & -1 & -1 & 1 & 0 \end{bmatrix}.$$

In a fashion analogous to that of Section 1.9.1, we can show that $\mathcal{G}_{12}$ is a $[12, 6, 6]$ self-dual code. The code $\mathcal{G}_{11}$ is a $[11, 6, 5]$ code obtained from $\mathcal{G}_{12}$ by puncturing. Again, equivalent codes are obtained regardless of the coordinate. However, adding an overall parity check to $\mathcal{G}_{11}$ in the same coordinate may not give the same $\mathcal{G}_{12}$ back; it will give either a $[12, 6, 6]$ code or a $[12, 6, 5]$ code depending upon the coordinate; see Exercise 61.

Exercise 60 Prove that $\mathcal{G}_{12}$ is a $[12, 6, 6]$ self-dual ternary code. ◆

Exercise 61 Number the columns of the matrix A used to generate $\mathcal{G}_{12}$ by $\infty, 0, 1, 2, 3, 4$. Let $\mathcal{G}'_{12}$ be obtained from $\mathcal{G}_{12}$ by scaling column ∞ by -1.
(a) Show how to give the entries in row two of A using squares and non-squares of integers modulo 5.
(b) Why is $\mathcal{G}'_{12}$ a $[12, 6, 6]$ self-dual code? Hint: Use Exercise 60.
(c) Show that puncturing $\mathcal{G}'_{12}$ in any coordinate and adding back an overall parity check in that same position gives the same code $\mathcal{G}'_{12}$.
(d) Show that if $\mathcal{G}_{12}$ is punctured in coordinate ∞ and this code is then extended in the same position, the resulting code is $\mathcal{G}'_{12}$.
(e) Show that if $\mathcal{G}_{12}$ is punctured in any coordinate other than ∞ and this code is then extended in the same position, the resulting code is a $[12, 6, 5]$ code. ◆

In Exercise 61, we scaled the first column of A by -1 to obtain a $[12, 6, 6]$ self-dual code $\mathcal{G}'_{12}$ equivalent to $\mathcal{G}_{12}$. By that exercise, if we puncture $\mathcal{G}'_{12}$ in any coordinate and then extend in the same coordinate, we get $\mathcal{G}'_{12}$ back. In Chapter 10 we will see that these punctured codes are all equivalent to each other and to $\mathcal{G}_{11}$. As a result any $[11, 6, 5]$ code equivalent to one obtained by puncturing $\mathcal{G}'_{12}$ in any coordinate will be called the *ternary Golay code*; any $[12, 6, 6]$ code equivalent to $\mathcal{G}'_{12}$ (or $\mathcal{G}_{12}$) will be called the *extended ternary Golay code*.

1.10 Reed–Muller codes

In this section, we introduce the binary Reed–Muller codes. Nonbinary generalized Reed–Muller codes will be examined in Section 13.2.3. The binary codes were first constructed and explored by Muller [241] in 1954, and a majority logic decoding algorithm for them was described by Reed [293] also in 1954. Although their minimum distance is relatively small, they are of practical importance because of the ease with which they can be implemented and decoded. They are of mathematical interest because of their connection with finite affine and projective geometries; see [4, 5]. These codes can be defined in several different ways. Here we choose a recursive definition based on the $(\mathbf{u} \mid \mathbf{u} + \mathbf{v})$ construction.

Let m be a positive integer and r a nonnegative integer with $r \le m$. The binary codes we construct will have length 2^m. For each length there will be $m + 1$ linear codes, denoted $\mathcal{R}(r, m)$ and called the rth order Reed–Muller, or RM, code of length 2^m. The codes $\mathcal{R}(0, m)$ and $\mathcal{R}(m, m)$ are trivial codes: the 0th order RM code $\mathcal{R}(0, m)$ is the binary repetition code of length 2^m with basis $\{\mathbf{1}\}$, and the mth order RM code $\mathcal{R}(m, m)$ is the entire space $\mathbb{F}_2^{2^m}$. For $1 \le r < m$, define

$$\mathcal{R}(r, m) = \{(\mathbf{u}, \mathbf{u} + \mathbf{v}) \mid \mathbf{u} \in \mathcal{R}(r, m - 1), \mathbf{v} \in \mathcal{R}(r - 1, m - 1)\}. \tag{1.7}$$

Let $G(0, m) = [11 \cdots 1]$ and $G(m, m) = I_{2^m}$. From the above description, these are generator matrices for $\mathcal{R}(0, m)$ and $\mathcal{R}(m, m)$, respectively. For $1 \le r < m$, using (1.5), a generator matrix $G(r, m)$ for $\mathcal{R}(r, m)$ is

$$G(r, m) = \begin{bmatrix} G(r, m - 1) & G(r, m - 1) \\ 0 & G(r - 1, m - 1) \end{bmatrix}.$$

We illustrate this construction by producing the generator matrices for $\mathcal{R}(r, m)$ with $1 \le r < m \le 3$:

$$G(1, 2) = \begin{bmatrix} 1 & 0 & 1 & 0 \\ 0 & 1 & 0 & 1 \\ 0 & 0 & 1 & 1 \end{bmatrix}, \quad G(1, 3) = \begin{bmatrix} 1 & 0 & 1 & 0 & 1 & 0 & 1 & 0 \\ 0 & 1 & 0 & 1 & 0 & 1 & 0 & 1 \\ 0 & 0 & 1 & 1 & 0 & 0 & 1 & 1 \\ 0 & 0 & 0 & 0 & 1 & 1 & 1 & 1 \end{bmatrix}, \quad \text{and}$$

$$G(2, 3) = \begin{bmatrix} 1 & 0 & 0 & 0 & 1 & 0 & 0 & 0 \\ 0 & 1 & 0 & 0 & 0 & 1 & 0 & 0 \\ 0 & 0 & 1 & 0 & 0 & 0 & 1 & 0 \\ 0 & 0 & 0 & 1 & 0 & 0 & 0 & 1 \\ 0 & 0 & 0 & 0 & 1 & 0 & 1 & 0 \\ 0 & 0 & 0 & 0 & 0 & 1 & 0 & 1 \\ 0 & 0 & 0 & 0 & 0 & 0 & 1 & 1 \end{bmatrix}.$$

From these matrices, notice that $\mathcal{R}(1, 2)$ and $\mathcal{R}(2, 3)$ are both the set of all even weight vectors in $\mathbb{F}_2^4$ and $\mathbb{F}_2^8$, respectively. Notice also that $\mathcal{R}(1, 3)$ is an $[8, 4, 4]$ self-dual code, which must be $\widehat{\mathcal{H}}_3$ by Exercise 56.

The dimension, minimum weight, and duals of the binary Reed–Muller codes can be computed directly from their definitions.

Theorem 1.10.1 *Let r be an integer with $0 \le r \le m$. Then the following hold:*
(i) $\mathcal{R}(i, m) \subseteq \mathcal{R}(j, m)$, *if* $0 \le i \le j \le m$.
(ii) *The dimension of* $\mathcal{R}(r, m)$ *equals*

$$\binom{m}{0} + \binom{m}{1} + \cdots + \binom{m}{r}.$$

(iii) *The minimum weight of* $\mathcal{R}(r, m)$ *equals* 2^{m-r}.
(iv) $\mathcal{R}(m, m)^{\perp} = \{\mathbf{0}\}$, *and if* $0 \le r < m$, *then* $\mathcal{R}(r, m)^{\perp} = \mathcal{R}(m - r - 1, m)$.

Proof: Part (i) is certainly true if $m = 1$ by direct computation and if $j = m$ as $\mathcal{R}(m, m)$ is the full space $\mathbb{F}_2^{2^m}$. Assume inductively that $\mathcal{R}(k, m - 1) \subseteq \mathcal{R}(\ell, m - 1)$ for all $0 \leq k \leq \ell < m$. Let $0 < i \leq j < m$. Then:

$$\mathcal{R}(i, m) = \{(\mathbf{u}, \mathbf{u} + \mathbf{v}) \mid \mathbf{u} \in \mathcal{R}(i, m - 1), \mathbf{v} \in \mathcal{R}(i - 1, m - 1)\}$$
$$\subseteq \{(\mathbf{u}, \mathbf{u} + \mathbf{v}) \mid \mathbf{u} \in \mathcal{R}(j, m - 1), \mathbf{v} \in \mathcal{R}(j - 1, m - 1)\}$$
$$= \mathcal{R}(j, m).$$

So (i) follows by induction if $0 < i$. If $i = 0$, we only need to show that the all-one vector of length 2^m is in $\mathcal{R}(j, m)$ for $j < m$. Inductively assume the all-one vector of length 2^{m-1} is in $\mathcal{R}(j, m - 1)$. Then by definition (1.7), we see that the all-one vector of length 2^m is in $\mathcal{R}(j, m)$ as one choice for $\mathbf{u}$ is $\mathbf{1}$ and one choice for $\mathbf{v}$ is $\mathbf{0}$.

For (ii) the result is true for $r = m$ as $\mathcal{R}(m, m) = \mathbb{F}_2^{2^m}$ and

$$\binom{m}{0} + \binom{m}{1} + \cdots + \binom{m}{m} = 2^m.$$

It is also true for $m = 1$ by inspection. Now assume that $\mathcal{R}(i, m - 1)$ has dimension

$$\binom{m-1}{0} + \binom{m-1}{1} + \cdots + \binom{m-1}{i} \quad \text{for all } 0 \leq i < m.$$

By the discussion in Section 1.5.5 (and Exercise 33), $\mathcal{R}(r, m)$ has dimension the sum of the dimensions of $\mathcal{R}(r, m - 1)$ and $\mathcal{R}(r - 1, m - 1)$, that is,

$$\binom{m-1}{0} + \binom{m-1}{1} + \cdots + \binom{m-1}{r} + \binom{m-1}{0} + \binom{m-1}{1} + \cdots + \binom{m-1}{r-1}.$$

The result follows by the elementary properties of binomial coefficients:

$$\binom{m-1}{0} = \binom{m}{0} \quad \text{and} \quad \binom{m-1}{i-1} + \binom{m-1}{i} = \binom{m}{i}.$$

Part (iii) is again valid for $m = 1$ by inspection and for both $r = 0$ and $r = m$ as $\mathcal{R}(0, m)$ is the binary repetition code of length 2^m and $\mathcal{R}(m, m) = \mathbb{F}_2^{2^m}$. Assume that $\mathcal{R}(i, m - 1)$ has minimum weight 2^{m-1-i} for all $0 \leq i < m$. If $0 < r < m$, then by definition (1.7) and the discussion in Section 1.5.5 (and Exercise 33), $\mathcal{R}(r, m)$ has minimum weight $\min\{2 \cdot 2^{m-1-r}, 2^{m-1-(r-1)}\} = 2^{m-r}$.

To prove (iv), we first note that $\mathcal{R}(m, m)^\perp$ is $\{\mathbf{0}\}$ since $\mathcal{R}(m, m) = \mathbb{F}_2^{2^m}$. So if we define $\mathcal{R}(-1, m) = \{\mathbf{0}\}$, then $\mathcal{R}(-1, m)^\perp = \mathcal{R}(m - (-1) - 1, m)$ for all $m > 0$. By direct computation, $\mathcal{R}(r, m)^\perp = \mathcal{R}(m - r - 1, m)$ for all r with $-1 \leq r \leq m = 1$. Assume inductively that if $-1 \leq i \leq m - 1$, then $\mathcal{R}(i, m - 1)^\perp = \mathcal{R}((m - 1) - i - 1, m - 1)$. Let $0 \leq r < m$. To prove $\mathcal{R}(r, m)^\perp = \mathcal{R}(m - r - 1, m)$, it suffices to show that $\mathcal{R}(m - r - 1, m) \subseteq \mathcal{R}(r, m)^\perp$ as $\dim \mathcal{R}(r, m) + \dim \mathcal{R}(m - r - 1, m) = 2^m$ by (ii). Notice that with the definition of $\mathcal{R}(-1, m)$, (1.7) extends to the case $r = 0$. Let $\mathbf{x} = (\mathbf{a}, \mathbf{a} + \mathbf{b}) \in \mathcal{R}(m - r - 1, m)$ where $\mathbf{a} \in \mathcal{R}(m - r - 1, m - 1)$ and $\mathbf{b} \in \mathcal{R}(m - r - 2, m - 1)$, and let $\mathbf{y} = (\mathbf{u}, \mathbf{u} + \mathbf{v}) \in \mathcal{R}(r, m)$ where $\mathbf{u} \in \mathcal{R}(r, m - 1)$ and $\mathbf{v} \in \mathcal{R}(r - 1, m - 1)$. Then $\mathbf{x} \cdot \mathbf{y} = 2\mathbf{a} \cdot \mathbf{u} + \mathbf{a} \cdot \mathbf{v} + \mathbf{b} \cdot \mathbf{u} + \mathbf{b} \cdot \mathbf{v} = \mathbf{a} \cdot \mathbf{v} + \mathbf{b} \cdot \mathbf{u} + \mathbf{b} \cdot \mathbf{v}$. Each term is 0 as follows. As $\mathbf{a} \in \mathcal{R}(m - r - 1, m - 1) = \mathcal{R}(r - 1, m - 1)^\perp$, $\mathbf{a} \cdot \mathbf{v} = 0$. As $\mathbf{b} \in \mathcal{R}(m - r - 2, m - 1) = \mathcal{R}(r, m - 1)^\perp$, $\mathbf{b} \cdot \mathbf{u} = 0$ and

$\mathbf{b} \cdot \mathbf{v} = 0$ using $\mathcal{R}(r - 1, m - 1) \subseteq \mathcal{R}(r, m - 1)$ from (i). We conclude that $\mathcal{R}(m - r - 1, m) \subseteq \mathcal{R}(r, m)^{\perp}$, completing (iv). $\square$

We make a few observations based on this theorem. First, since $\mathcal{R}(0, m)$ is the length 2^m repetition code, $\mathcal{R}(m - 1, m) = \mathcal{R}(0, m)^{\perp}$ is the code of all even weight vectors in $\mathbb{F}_2^{2^m}$. We had previously observed this about $\mathcal{R}(1, 2)$ and $\mathcal{R}(2, 3)$. Second, if m is odd and $r = (m - 1)/2$ we see from parts (iii) and (iv) that $\mathcal{R}(r, m) = \mathcal{R}((m - 1)/2, m)$ is self-dual with minimum weight $2^{(m-1)/2}$. Again we had observed this about $\mathcal{R}(1, 3)$. In the exercises, you will also verify the general result that puncturing $\mathcal{R}(1, m)$ and then taking the subcode of even weight vectors produces the simplex code $\mathcal{S}_m$ of length $2^m - 1$.

Exercise 62 In this exercise we produce another generator matrix $G''(1, m)$ for $\mathcal{R}(1, m)$. Define

$$G''(1, 1) = \begin{bmatrix} 1 & 1 \\ 0 & 1 \end{bmatrix}.$$

For $m \geq 2$, recursively define

$$G'(1, m) = \begin{bmatrix} G''(1, m - 1) & G''(1, m - 1) \\ 00 \cdots 0 & 11 \cdots 1 \end{bmatrix},$$

and define $G''(1, m)$ to be the matrix obtained from $G'(1, m)$ by removing the bottom row and placing it as row two in the matrix, moving the rows below down.

(a) Show that $G''(1, 1)$ is a generator matrix for $\mathcal{R}(1, 1)$.

(b) Find the matrices $G'(1, 2)$, $G''(1, 2)$, $G'(1, 3)$, and $G''(1, 3)$.

(c) What do you notice about the columns of the matrices obtained from $G''(1, 2)$ and $G''(1, 3)$ by deleting the first row and the first column?

(d) Show using induction, part (a), and the definition (1.7) that $G''(1, m)$ is a generator matrix for $\mathcal{R}(1, m)$.

(e) Formulate a generalization of part (c) that applies to the matrix obtained from $G''(1, m)$ by deleting the first row and the first column. Prove your generalization is correct.

(f) Show that the code generated by the matrix obtained from $G''(1, m)$ by deleting the first row and the first column is the simplex code $\mathcal{S}_m$.

(g) Show that the code $\mathcal{R}(m - 2, m)$ is the extended binary Hamming code $\widehat{\mathcal{H}}_m$.

Notice that this problem shows that the extended binary Hamming codes and their duals are Reed–Muller codes.   ◆

1.11   Encoding, decoding, and Shannon's Theorem

Since the inception of coding theory, codes have been used in many diverse ways; in addition to providing reliability in communication channels and computers, they give high fidelity on compact disc recordings, and they have also permitted successful transmission of pictures from outer space. New uses constantly appear. As a primary application of codes is to store or transmit data, we introduce the process of encoding and decoding a message.

1.11.1   Encoding

Let C be an $[n, k]$ linear code over the field $\mathbb{F}_q$ with generator matrix G. This code has q^k codewords which will be in one-to-one correspondence with q^k messages. The simplest way to view these messages is as k-tuples $\mathbf{x}$ in $\mathbb{F}_q^k$. The most common way to encode the message $\mathbf{x}$ is as the codeword $\mathbf{c} = \mathbf{x}G$. If G is in standard form, the first k coordinates of the codeword $\mathbf{c}$ are the information symbols $\mathbf{x}$; the remaining $n - k$ symbols are the parity check symbols, that is, the redundancy added to $\mathbf{x}$ in order to help recover $\mathbf{x}$ if errors occur. The generator matrix G may not be in standard form. If, however, there exist column indices $i_1, i_2, \ldots, i_k$ such that the $k \times k$ matrix consisting of these k columns of G is the $k \times k$ identity matrix, then the message is found in the k coordinates $i_1, i_2, \ldots, i_k$ of the codeword scrambled but otherwise unchanged; that is, the message symbol x_j is in component i_j of the codeword. If this occurs, we say that the encoder is *systematic*. If G is replaced by another generator matrix, the encoding of $\mathbf{x}$ will, of course, be different. By row reduction, one could always choose a generator matrix so that the encoder is systematic. Furthermore, if we are willing to replace the code with a permutation equivalent one, by Theorem 1.6.2, we can choose a code with generator matrix in standard form, and therefore the first k bits of the codeword make up the message.

The method just described shows how to encode a message $\mathbf{x}$ using the generator matrix of the code C. There is a second way to encode using the parity check matrix H. This is easiest to do when G is in standard form $[I_k \mid A]$. In this case $H = [-A^T \mid I_{n-k}]$ by Theorem 1.2.1. Suppose that $\mathbf{x} = x_1 \cdots x_k$ is to be encoded as the codeword $\mathbf{c} = c_1 \cdots c_n$. As G is in standard form, $c_1 \cdots c_k = x_1 \cdots x_k$. So we need to determine the $n - k$ parity check symbols (redundancy symbols) $c_{k+1} \cdots c_n$. As $\mathbf{0} = H\mathbf{c}^T = [-A^T \mid I_{n-k}]\mathbf{c}^T$, $A^T\mathbf{x}^T = [c_{k+1} \cdots c_n]^T$. One can generalize this when G is a systematic encoder.

Example 1.11.1 Let C be the $[6, 3, 3]$ binary code with generator and parity check matrices

$$
G = \begin{bmatrix} 1 & 0 & 0 & 1 & 0 & 1 \\ 0 & 1 & 0 & 1 & 1 & 0 \\ 0 & 0 & 1 & 0 & 1 & 1 \end{bmatrix} \quad \text{and} \quad H = \begin{bmatrix} 1 & 1 & 0 & 1 & 0 & 0 \\ 0 & 1 & 1 & 0 & 1 & 0 \\ 1 & 0 & 1 & 0 & 0 & 1 \end{bmatrix},
$$

respectively. Suppose we desire to encode the message $\mathbf{x} = x_1 x_2 x_3$ to obtain the codeword $\mathbf{c} = c_1 c_2 \cdots c_6$. Using G to encode yields

$$
\mathbf{c} = \mathbf{x}G = (x_1, x_2, x_3, x_1 + x_2, x_2 + x_3, x_1 + x_3). \tag{1.8}
$$

Using H to encode, $\mathbf{0} = H\mathbf{c}^T$ leads to the system

$$
0 = c_1 + c_2 + c_4,
$$
$$
0 = c_2 + c_3 + c_5,
$$
$$
0 = c_1 + c_3 + c_6.
$$

As G is in standard form, $c_1 c_2 c_3 = x_1 x_2 x_3$, and solving this system clearly gives the same codeword as in (1.8). ∎

Exercise 63 Let C be the Hamming code $\mathcal{H}_3$ of Example 1.2.3, with parity check matrix

$$H = \begin{bmatrix} 0 & 1 & 1 & 1 & 1 & 0 & 0 \\ 1 & 0 & 1 & 1 & 0 & 1 & 0 \\ 1 & 1 & 0 & 1 & 0 & 0 & 1 \end{bmatrix}.$$

(a) Construct the generator matrix for C and use it to encode the message 0110.
(b) Use your generator matrix to encode $x_1 x_2 x_3 x_4$.
(c) Use H to encode the messages 0110 and $x_1 x_2 x_3 x_4$. ◆

Since there is a one-to-one correspondence between messages and codewords, one often works only with the encoded messages (the codewords) at both the sending and receiving end. In that case, at the decoding end in Figure 1.1, we are satisfied with an estimate $\hat{\mathbf{c}}$ obtained by the decoder from $\mathbf{y}$, hoping that this is the codeword $\mathbf{c}$ that was transmitted. However, if we are interested in the actual message, a question arises as to how to recover the message from a codeword. If the codeword $\mathbf{c} = \mathbf{x}G$, and G is in standard form, the message is the first k components of $\mathbf{c}$; if the encoding is systematic, it is easy to recover the message by looking at the coordinates of G containing the identity matrix. What can be done otherwise? Because G has independent rows, there is an $n \times k$ matrix K such that $GK = I_k$; K is called a *right inverse* for G and is not necessarily unique. As $\mathbf{c} = \mathbf{x}G$, $\mathbf{c}K = \mathbf{x}GK = \mathbf{x}$.

Exercise 64 Let G be a $k \times n$ generator matrix for a binary code C.
(a) Suppose $G = [I_k \ A]$. Show that

$$\begin{bmatrix} I_k \\ O \end{bmatrix},$$

where O is the $(n - k) \times k$ zero matrix, is a right inverse of G.
(b) Find a 7×3 right inverse K of G, where

$$G = \begin{bmatrix} 1 & 0 & 1 & 1 & 0 & 1 & 1 \\ 1 & 1 & 0 & 1 & 0 & 1 & 0 \\ 0 & 0 & 1 & 1 & 1 & 1 & 0 \end{bmatrix}.$$

Hint: One way K can be found is by using four zero rows and the three rows of I_3.
(c) Find a 7×4 right inverse K of G, where

$$G = \begin{bmatrix} 1 & 1 & 0 & 1 & 0 & 0 & 0 \\ 0 & 1 & 1 & 0 & 1 & 0 & 0 \\ 0 & 0 & 1 & 1 & 0 & 1 & 0 \\ 0 & 0 & 0 & 1 & 1 & 0 & 1 \end{bmatrix}.$$

Remark: In Chapter 4, we will see that G generates a cyclic code and the structure of G is typical of the structure of generator matrices of such codes.
(d) What is the message $\mathbf{x}$ if $\mathbf{x}G = 1000110$, where G is given in part (c)? ◆

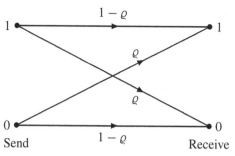

Figure 1.2 Binary symmetric channel.

1.11.2 Decoding and Shannon's Theorem

The process of decoding, that is, determining which codeword (and thus which message $\mathbf{x}$) was sent when a vector $\mathbf{y}$ is received, is more complex. Finding efficient (fast) decoding algorithms is a major area of research in coding theory because of their practical applications. In general, encoding is easy and decoding is hard, if the code has a reasonably large size.

In order to set the stage for decoding, we begin with one possible mathematical model of a channel that transmits binary data. This model is called the *binary symmetric channel* (or *BSC*) with *crossover probability* ϱ and is illustrated in Figure 1.2. If 0 or 1 is sent, the probability it is received without error is $1 - \varrho$; if a 0 (respectively 1) is sent, the probability that a 1 (respectively 0) is received is ϱ. In most practical situations ϱ is very small. This is an example of a *discrete memoryless channel* (or *DMC*), a channel in which inputs and outputs are discrete and the probability of error in one bit is independent of previous bits. We will assume that it is more likely that a bit is received correctly than in error; so $\varrho < 1/2$.[5]

If E_1 and E_2 are events, let $\text{prob}(E_1)$ denote the probability that E_1 occurs and $\text{prob}(E_1 \mid E_2)$ the probability that E_1 occurs given that E_2 occurs. Assume that $\mathbf{c} \in \mathbb{F}_2^n$ is sent and $\mathbf{y} \in \mathbb{F}_2^n$ is received and decoded as $\widehat{\mathbf{c}} \in \mathbb{F}_2^n$. So $\text{prob}(\mathbf{c} \mid \mathbf{y})$ is the probability that the codeword $\mathbf{c}$ is sent given that $\mathbf{y}$ is received, and $\text{prob}(\mathbf{y} \mid \mathbf{c})$ is the probability that $\mathbf{y}$ is received given that the codeword $\mathbf{c}$ is sent. These probabilities can be computed from the statistics associated with the channel. The probabilities are related by Bayes' Rule

$$\text{prob}(\mathbf{c} \mid \mathbf{y}) = \frac{\text{prob}(\mathbf{y} \mid \mathbf{c})\text{prob}(\mathbf{c})}{\text{prob}(\mathbf{y})},$$

where $\text{prob}(\mathbf{c})$ is the probability that $\mathbf{c}$ is sent and $\text{prob}(\mathbf{y})$ is the probability that $\mathbf{y}$ is received. There are two natural means by which a decoder can make a choice based on these two probabilities. First, the decoder could choose $\widehat{\mathbf{c}} = \mathbf{c}$ for the codeword $\mathbf{c}$ with $\text{prob}(\mathbf{c} \mid \mathbf{y})$ maximum; such a decoder is called a *maximum a posteriori probability* (or *MAP*) *decoder*.

---

[5] While ϱ is usually very small, if $\varrho > 1/2$, the probability that a bit is received in error is higher than the probability that it is received correctly. So one strategy is to interchange 0 and 1 immediately at the receiving end. This converts the BSC with crossover probability ϱ to a BSC with crossover probability $1 - \varrho < 1/2$. This of course does not help if $\varrho = 1/2$; in this case communication is not possible – see Exercise 77.

Symbolically, a MAP decoder makes the decision

$$\widehat{\mathbf{c}} = \arg \max_{\mathbf{c} \in C} \text{prob}(\mathbf{c} \mid \mathbf{y}).$$

Here $\arg \max_{\mathbf{c} \in C} \text{prob}(\mathbf{c} \mid \mathbf{y})$ is the argument $\mathbf{c}$ of the probability function $\text{prob}(\mathbf{c} \mid \mathbf{y})$ that maximizes this probability. Alternately, the decoder could choose $\widehat{\mathbf{c}} = \mathbf{c}$ for the codeword $\mathbf{c}$ with $\text{prob}(\mathbf{y} \mid \mathbf{c})$ maximum; such a decoder is called a *maximum likelihood* (or *ML*) *decoder*. Symbolically, a ML decoder makes the decision

$$\widehat{\mathbf{c}} = \arg \max_{\mathbf{c} \in C} \text{prob}(\mathbf{y} \mid \mathbf{c}). \tag{1.9}$$

Consider ML decoding over a BSC. If $\mathbf{y} = y_1 \cdots y_n$ and $\mathbf{c} = c_1 \cdots c_n$,

$$\text{prob}(\mathbf{y} \mid \mathbf{c}) = \prod_{i=1}^{n} \text{prob}(y_i \mid c_i),$$

since we assumed that bit errors are independent. By Figure 1.2, $\text{prob}(y_i \mid c_i) = \varrho$ if $y_i \neq c_i$ and $\text{prob}(y_i \mid c_i) = 1 - \varrho$ if $y_i = c_i$. Therefore

$$\text{prob}(\mathbf{y} \mid \mathbf{c}) = \varrho^{d(\mathbf{y}, \mathbf{c})}(1 - \varrho)^{n - d(\mathbf{y}, \mathbf{c})} = (1 - \varrho)^n \left(\frac{\varrho}{1 - \varrho} \right)^{d(\mathbf{y}, \mathbf{c})}. \tag{1.10}$$

Since $0 < \varrho < 1/2$, $0 < \varrho/(1 - \varrho) < 1$. Therefore maximizing $\text{prob}(\mathbf{y} \mid \mathbf{c})$ is equivalent to minimizing $d(\mathbf{y}, \mathbf{c})$, that is, finding the codeword $\mathbf{c}$ closest to the received vector $\mathbf{y}$ in Hamming distance; this is called *nearest neighbor decoding*. Hence on a BSC, maximum likelihood and nearest neighbor decoding are the same.

Let $\mathbf{e} = \mathbf{y} - \mathbf{c}$ so that $\mathbf{y} = \mathbf{c} + \mathbf{e}$. The effect of noise in the communication channel is to add an *error vector* $\mathbf{e}$ to the codeword $\mathbf{c}$, and the goal of decoding is to determine $\mathbf{e}$. Nearest neighbor decoding is equivalent to finding a vector $\mathbf{e}$ of smallest weight such that $\mathbf{y} - \mathbf{e}$ is in the code. This error vector need not be unique since there may be more than one codeword closest to $\mathbf{y}$; in other words, (1.9) may not have a unique solution. When we have a decoder capable of finding all codewords nearest to the received vector $\mathbf{y}$, then we have a *complete decoder*.

To examine vectors closest to a given codeword, the concept of spheres about codewords proves useful. The *sphere* of radius r centered at a vector $\mathbf{u}$ in $\mathbb{F}_q^n$ is defined to be the set

$$S_r(\mathbf{u}) = \left\{ \mathbf{v} \in \mathbb{F}_q^n \mid d(\mathbf{u}, \mathbf{v}) \leq r \right\}$$

of all vectors whose distance from $\mathbf{u}$ is less than or equal to r. The number of vectors in $S_r(\mathbf{u})$ equals

$$\sum_{i=0}^{r} \binom{n}{i} (q - 1)^i. \tag{1.11}$$

These spheres are pairwise disjoint provided their radius is chosen small enough.

Theorem 1.11.2 *If d is the minimum distance of a code C (linear or nonlinear) and $t = \lfloor (d - 1)/2 \rfloor$, then spheres of radius t about distinct codewords are disjoint.*

Proof: If $z \in S_t(c_1) \cap S_t(c_2)$, where c_1 and c_2 are codewords, then by the triangle inequality (Theorem 1.4.1(iv)),

$$d(c_1, c_2) \leq d(c_1, z) + d(z, c_2) \leq 2t < d,$$

implying that $c_1 = c_2$. $\square$

Corollary 1.11.3 *With the notation of the previous theorem, if a codeword c is sent and* y *is received where t or fewer errors have occurred, then* c *is the unique codeword closest to* y. *In particular, nearest neighbor decoding uniquely and correctly decodes any received vector in which at most t errors have occurred in transmission.*

Exercise 65   Prove that the number of vectors in $S_r(u)$ is given by (1.11).   ◆

For purposes of decoding as many errors as possible, this corollary implies that for given n and k, we wish to find a code with as high a minimum weight d as possible. Alternately, given n and d, one wishes to send as many messages as possible; thus we want to find a code with the largest number of codewords, or, in the linear case, the highest dimension. We may relax these requirements somewhat if we can find a code with an efficient decoding algorithm.

Since the minimum distance of C is d, there exist two distinct codewords such that the spheres of radius $t + 1$ about them are not disjoint. Therefore if more than t errors occur, nearest neighbor decoding may yield more than one nearest codeword. Thus C is a *t-error-correcting code* but not a $(t + 1)$-error-correcting code. The *packing radius* of a code is the largest radius of spheres centered at codewords so that the spheres are pairwise disjoint. This discussion shows the following two facts about the packing radius.

Theorem 1.11.4 *Let C be an $[n, k, d]$ code over $\mathbb{F}_q$. The following hold:*
(i)   *The packing radius of C equals $t = \lfloor (d - 1)/2 \rfloor$.*
(ii) *The packing radius t of C is characterized by the property that nearest neighbor decoding always decodes correctly a received vector in which t or fewer errors have occurred but will not always decode correctly a received vector in which $t + 1$ errors have occurred.*

The decoding problem now becomes one of finding an efficient algorithm that will correct up to t errors. One of the most obvious decoding algorithms is to examine all codewords until one is found with distance t or less from the received vector. But obviously this is a realistic decoding algorithm only for codes with a small number of codewords. Another obvious algorithm is to make a table consisting of a nearest codeword for each of the q^n vectors in $\mathbb{F}_q^n$ and then look up a received vector in the table in order to decode it. This is impractical if q^n is very large.

For an $[n, k, d]$ linear code C over $\mathbb{F}_q$, we can, however, devise an algorithm using a table with q^{n-k} rather than q^n entries where one can find the nearest codeword by looking up one of these q^{n-k} entries. This general decoding algorithm for linear codes is called syndrome decoding. Because our code C is an elementary abelian subgroup of the additive group of $\mathbb{F}_q^n$, its distinct cosets $x + C$ partition $\mathbb{F}_q^n$ into q^{n-k} sets of size q^k. Two vectors x and y belong to the same coset if and only if $y - x \in C$. The *weight of a coset* is the smallest weight of a vector in the coset, and any vector of this smallest weight in the coset is called a *coset leader*. The zero vector is the unique coset leader of the code C. More generally, every coset of weight at most $t = \lfloor (d-1)/2 \rfloor$ has a unique coset leader.

Exercise 66 Do the following:
(a) Prove that if C is an $[n, k, d]$ code over $\mathbb{F}_q$, every coset of weight at most $t = \lfloor (d-1)/2 \rfloor$ has a unique coset leader.
(b) Find a nonzero binary code of length 4 and minimum weight d in which all cosets have unique coset leaders and some coset has weight greater than $t = \lfloor (d-1)/2 \rfloor$.   ◆

Choose a parity check matrix H for C. The *syndrome* of a vector $\mathbf{x}$ in $\mathbb{F}_q^n$ with respect to the parity check matrix H is the vector in $\mathbb{F}_q^{n-k}$ defined by

$$\mathrm{syn}(\mathbf{x}) = H\mathbf{x}^{\mathrm{T}}.$$

The code C consists of all vectors whose syndrome equals $\mathbf{0}$. As H has rank $n - k$, every vector in $\mathbb{F}_q^{n-k}$ is a syndrome. If $\mathbf{x}_1, \mathbf{x}_2 \in \mathbb{F}_q^n$ are in the same coset of C, then $\mathbf{x}_1 - \mathbf{x}_2 = \mathbf{c} \in C$. Therefore $\mathrm{syn}(\mathbf{x}_1) = H(\mathbf{x}_2 + \mathbf{c})^{\mathrm{T}} = H\mathbf{x}_2^{\mathrm{T}} + H\mathbf{c}^{\mathrm{T}} = H\mathbf{x}_2^{\mathrm{T}} = \mathrm{syn}(\mathbf{x}_2)$. Hence $\mathbf{x}_1$ and $\mathbf{x}_2$ have the same syndrome. On the other hand, if $\mathrm{syn}(\mathbf{x}_1) = \mathrm{syn}(\mathbf{x}_2)$, then $H(\mathbf{x}_2 - \mathbf{x}_1)^{\mathrm{T}} = \mathbf{0}$ and so $\mathbf{x}_2 - \mathbf{x}_1 \in C$. Thus we have the following theorem.

Theorem 1.11.5 *Two vectors belong to the same coset if and only if they have the same syndrome.*

Hence there exists a one-to-one correspondence between cosets of C and syndromes. We denote by $C_{\mathbf{s}}$ the coset of C consisting of all vectors in $\mathbb{F}_q^n$ with syndrome $\mathbf{s}$.

Suppose a codeword sent over a communication channel is received as a vector $\mathbf{y}$. Since in nearest neighbor decoding we seek a vector $\mathbf{e}$ of smallest weight such that $\mathbf{y} - \mathbf{e} \in C$, nearest neighbor decoding is equivalent to finding a vector $\mathbf{e}$ of smallest weight in the coset containing $\mathbf{y}$, that is, a coset leader of the coset containing $\mathbf{y}$. *The Syndrome Decoding Algorithm* is the following implementation of nearest neighbor decoding. We begin with a fixed parity check matrix H.

I. For each syndrome $\mathbf{s} \in \mathbb{F}_q^{n-k}$, choose a coset leader $\mathbf{e}_{\mathbf{s}}$ of the coset $C_{\mathbf{s}}$. Create a table pairing the syndrome with the coset leader.

This process can be somewhat involved, but this is a one-time preprocessing task that is carried out before received vectors are analyzed. One method of computing this table will be described shortly. After producing the table, received vectors can be decoded.

II. After receiving a vector $\mathbf{y}$, compute its syndrome $\mathbf{s}$ using the parity check matrix H.

III. $\mathbf{y}$ is then decoded as the codeword $\mathbf{y} - \mathbf{e}_{\mathbf{s}}$.

Syndrome decoding requires a table with only q^{n-k} entries, which may be a vast improvement over a table of q^n vectors showing which codeword is closest to each of these. However, there is a cost for shortening the table: before looking in the table of syndromes, one must perform a matrix-vector multiplication in order to determine the syndrome of the received vector. Then the table is used to look up the syndrome and find the coset leader.

How do we construct the table of syndromes as described in Step I? We briefly discuss this for binary codes; one can extend this easily to nonbinary codes. Given the t-error-correcting code C of length n with parity check matrix H, we can construct the syndromes as follows. The coset of weight 0 has coset leader $\mathbf{0}$. Consider the n cosets of weight 1.

Choose an n-tuple with a 1 in position i and 0s elsewhere; the coset leader is the n-tuple and the associated syndrome is column i of H. For the $\binom{n}{2}$ cosets of weight 2, choose an n-tuple with two 1s in positions i and j, with $i < j$, and the rest 0s; the coset leader is the n-tuple and the associated syndrome is the sum of columns i and j of H. Continue in this manner through the cosets of weight t. We could choose to stop here. If we do, we can decode any received vector with t or fewer errors, but if the received vector has more than t errors, it will be either incorrectly decoded (if the syndrome of the received vector is in the table) or not decoded at all (if the syndrome of the received vector is not in the table). If we decide to go on and compute syndromes of weights w greater than t, we continue in the same fashion with the added feature that we must check for possible repetition of syndromes. This repetition will occur if the n-tuple of weight w is not a coset leader or it is a coset leader with the same syndrome as another leader of weight w, in which cases we move on to the next n-tuple. We continue until we have 2^{n-k} syndromes. The table produced will allow us to perform nearest neighbor decoding.

Syndrome decoding is particularly simple for the binary Hamming codes $\mathcal{H}_r$ with parameters $[n = 2^r - 1, 2^r - 1 - r, 3]$. We do not have to create the table for syndromes and corresponding coset leaders. This is because the coset leaders are unique and are the 2^r vectors of weight at most 1. Let H_r be the parity check matrix whose columns are the binary numerals for the numbers $1, 2, \ldots, 2^r - 1$. Since the syndrome of the binary n-tuple of weight 1 whose unique 1 is in position i is the r-tuple representing the binary numeral for i, the syndrome immediately gives the coset leader and no table is required for syndrome decoding. Thus *Syndrome Decoding for Binary Hamming Codes* takes the form:

I. After receiving a vector $\mathbf{y}$, compute its syndrome $\mathbf{s}$ using the parity check matrix H_r.

II. If $\mathbf{s} = \mathbf{0}$, then $\mathbf{y}$ is in the code and $\mathbf{y}$ is decoded as $\mathbf{y}$; otherwise, $\mathbf{s}$ is the binary numeral for some positive integer i and $\mathbf{y}$ is decoded as the codeword obtained from $\mathbf{y}$ by adding 1 to its ith bit.

The above procedure is easily modified for Hamming codes over other fields. This is explored in the exercises.

Exercise 67 Construct the parity check matrix of the binary Hamming code $\mathcal{H}_4$ of length 15 where the columns are the binary numbers $1, 2, \ldots, 15$ in that order. Using this parity check matrix decode the following vectors, and then check that your decoded vectors are actually codewords.
(a) 001000001100100,
(b) 101001110101100,
(c) 000100100011000. ♦

Exercise 68 Construct a table of all syndromes of the ternary tetracode of Example 1.3.3 using the generator matrix of that example to construct the parity check matrix. Find a coset leader for each of the syndromes. Use your parity check matrix to decode the following vectors, and then check that your decoded vectors are actually codewords.
(a) $(1, 1, 1, 1)$,
(b) $(1, -1, 0, -1)$,
(c) $(0, 1, 0, 1)$. ♦

Exercise 69 Let C be the $[6, 3, 3]$ binary code with generator matrix G and parity check matrix H given by

$$G = \begin{bmatrix} 1 & 0 & 0 & 0 & 1 & 1 \\ 0 & 1 & 0 & 1 & 0 & 1 \\ 0 & 0 & 1 & 1 & 1 & 0 \end{bmatrix} \quad \text{and} \quad H = \begin{bmatrix} 0 & 1 & 1 & 1 & 0 & 0 \\ 1 & 0 & 1 & 0 & 1 & 0 \\ 1 & 1 & 0 & 0 & 0 & 1 \end{bmatrix}.$$

(a) Construct a table of coset leaders and associated syndromes for the eight cosets of C.

(b) One of the cosets in part (a) has weight 2. This coset has three coset leaders. Which coset is it and what are its coset leaders?

(c) Using part (a), decode the following received vectors:
  (i) 110110,
  (ii) 110111,
  (iii) 110001.

(d) For one of the received vectors in part (c) there is ambiguity as to what codeword it should be decoded to. List the other nearest neighbors possible for this received vector. ◆

Exercise 70 Let $\widehat{\mathcal{H}}_3$ be the extended Hamming code with parity check matrix

$$\widehat{H}_3 = \begin{bmatrix} 1 & 1 & 1 & 1 & 1 & 1 & 1 & 1 \\ 0 & 0 & 0 & 0 & 1 & 1 & 1 & 1 \\ 0 & 0 & 1 & 1 & 0 & 0 & 1 & 1 \\ 0 & 1 & 0 & 1 & 0 & 1 & 0 & 1 \end{bmatrix}.$$

Number the coordinates $0, 1, 2, \ldots, 7$. Notice that if we delete the top row of $\widehat{H}_3$, we have the coordinate numbers in binary. We can decode $\widehat{\mathcal{H}}_3$ without a table of syndromes and coset leaders using the following algorithm. If $\mathbf{y}$ is received, compute $\mathrm{syn}(\mathbf{y})$ using the parity check matrix $\widehat{H}_3$. If $\mathrm{syn}(\mathbf{y}) = (0, 0, 0, 0)^{\mathrm{T}}$, then $\mathbf{y}$ has no errors. If $\mathrm{syn}(\mathbf{y}) = (1, a, b, c)^{\mathrm{T}}$, then there is a single error in the coordinate position abc (written in binary). If $\mathrm{syn}(\mathbf{y}) = (0, a, b, c)^{\mathrm{T}}$ with $(a, b, c) \neq (0, 0, 0)$, then there are two errors in coordinate position 0 and in the coordinate position abc (written in binary).

(a) Decode the following vectors using this algorithm:
  (i) 10110101,
  (ii) 11010010,
  (iii) 10011100.

(b) Verify that this procedure provides a nearest neighbor decoding algorithm for $\widehat{\mathcal{H}}_3$. To do this, the following must be verified. All weight 0 and weight 1 errors can be corrected, accounting for nine of the 16 syndromes. All weight 2 errors cannot necessarily be corrected but all weight 2 errors lead to one of the seven syndromes remaining. ◆

A received vector may contain both *errors* (where a transmitted symbol is read as a different symbol) and *erasures* (where a transmitted symbol is unreadable). These are fundamentally different in that the locations of errors are unknown, whereas the locations of erasures are known. Suppose $\mathbf{c} \in C$ is sent, and the received vector $\mathbf{y}$ contains ν errors and ϵ erasures. One could certainly not guarantee that $\mathbf{y}$ can be corrected if $\epsilon \geq d$ because there may be a codeword other than $\mathbf{c}$ closer to $\mathbf{y}$. So assume that $\epsilon < d$. Puncture C in the

ϵ positions where the erasures occurred in **y** to obtain an $[n - \epsilon, k^*, d^*]$ code C^*. Note that $k^* = k$ by Theorem 1.5.7(ii), and $d^* \geq d - \epsilon$. Puncture **c** and **y** similarly to obtain $\mathbf{c}^*$ and $\mathbf{y}^*$; these can be viewed as sent and received vectors using the code C^* with $\mathbf{y}^*$ containing ν errors but no erasures. If $2\nu < d - \epsilon \leq d^*$, $\mathbf{c}^*$ can be recovered from $\mathbf{y}^*$ by Corollary 1.11.3. There is a unique codeword $\mathbf{c} \in C$ which when punctured produces $\mathbf{c}^*$; otherwise if puncturing both **c** and $\mathbf{c}'$ yields $\mathbf{c}^*$, then $\mathrm{wt}(\mathbf{c} - \mathbf{c}') \leq \epsilon < d$, a contradiction unless $\mathbf{c} = \mathbf{c}'$. The following theorem summarizes this discussion and extends Corollary 1.11.3.

Theorem 1.11.6 *Let C be an $[n, k, d]$ code. If a codeword **c** is sent and **y** is received where ν errors and ϵ erasures have occurred, then **c** is the unique codeword in C closest to **y** provided $2\nu + \epsilon < d$.*

Exercise 71 Let $\widehat{\mathcal{H}}_3$ be the extended Hamming code with parity check matrix

$$\widehat{H}_3 = \begin{bmatrix} 1 & 1 & 1 & 1 & 1 & 1 & 1 & 1 \\ 0 & 0 & 0 & 0 & 1 & 1 & 1 & 1 \\ 0 & 0 & 1 & 1 & 0 & 0 & 1 & 1 \\ 0 & 1 & 0 & 1 & 0 & 1 & 0 & 1 \end{bmatrix}.$$

Correct the received vector $101 \star 0111$, where $\star$ is an erasure. ◆

In Exercises 70 and 71 we explored the decoding of the $[8, 4, 4]$ extended Hamming code $\widehat{\mathcal{H}}_3$. In Exercise 70, we had the reader verify that there are eight cosets of weight 1 and seven of weight 2. Each of these cosets is a nonlinear code and so it is appropriate to discuss the weight distribution of these cosets and to tabulate the results. In general, the *complete coset weight distribution* of a linear code is the weight distribution of each coset of the code. The next example gives the complete coset weight distribution of $\widehat{\mathcal{H}}_3$. As every $[8, 4, 4]$ code is equivalent to $\widehat{\mathcal{H}}_3$, by Exercise 56, this is the complete coset weight distribution of any $[8, 4, 4]$ binary code.

Example 1.11.7 The complete coset weight distribution of the $[8, 4, 4]$ extended binary Hamming code $\widehat{\mathcal{H}}_3$ is given in the following table:

Coset weight	Number of vectors of given weight									Number of cosets
	0	1	2	3	4	5	6	7	8	
0	1	0	0	0	14	0	0	0	1	1
1	0	1	0	7	0	7	0	1	0	8
2	0	0	4	0	8	0	4	0	0	7

Note that the first line is the weight distribution of $\widehat{\mathcal{H}}_3$. The second line is the weight distribution of each coset of weight one. This code has the special property that all cosets of a given weight have the same weight distribution. This is not the case for codes in general. In Exercise 73 we ask the reader to verify some of the information in the table. Notice that this code has the all-one vector **1** and hence the table is symmetric about the middle weight. Notice also that an even weight coset has only even weight vectors, and an odd weight coset has only odd weight vectors. These observations hold in general; see Exercise 72. The information in this table helps explain the decoding of $\widehat{\mathcal{H}}_3$. We see that all the cosets of weight 2 have four coset leaders. This implies that when we decode a received vector in which two errors had been made, we actually have four equally likely codewords that could have been sent. ∎

Exercise 72 Let C be a binary code of length n. Prove the following.
(a) If C is an even code, then an even weight coset of C has only even weight vectors, and an odd weight coset has only odd weight vectors.
(b) If C contains the all-one vector $\mathbf{1}$, then in a fixed coset, the number of vectors of weight i is the same as the number of vectors of weight $n - i$, for $0 \leq i \leq n$.   ◆

Exercise 73 Consider the complete coset weight distribution of $\widehat{\mathcal{H}}_3$ given in Example 1.11.7. The results of Exercise 72 will be useful.
(a) Prove that the weight distribution of the cosets of weight 1 are as claimed.
(b) (Harder) Prove that the weight distribution of the cosets of weight 2 are as claimed.   ◆

We conclude this section with a discussion of Shannon's Theorem in the framework of the decoding we have developed. Assume that the communication channel is a BSC with crossover probability ϱ on which syndrome decoding is used. The *word error rate* P_{err} for this channel and decoding scheme is the probability that the decoder makes an error, averaged over all codewords of C; for simplicity we assume that each codeword of C is equally likely to be sent. A decoder error occurs when $\widehat{\mathbf{c}} = \arg\max_{\mathbf{c} \in C} \text{prob}(\mathbf{y} \mid \mathbf{c})$ is not the originally transmitted word $\mathbf{c}$ when $\mathbf{y}$ is received. The syndrome decoder makes a correct decision if $\mathbf{y} - \mathbf{c}$ is a chosen coset leader. This probability is

$$\varrho^{\text{wt}(\mathbf{y}-\mathbf{c})}(1 - \varrho)^{n - \text{wt}(\mathbf{y}-\mathbf{c})}$$

by (1.10). Therefore the probability that the syndrome decoder makes a correct decision is $\sum_{i=0}^{n} \alpha_i \varrho^i (1 - \varrho)^{n-i}$, where α_i is the number of cosets weight i. Thus

$$P_{\text{err}} = 1 - \sum_{i=0}^{n} \alpha_i \varrho^i (1 - \varrho)^{n-i}. \tag{1.12}$$

Example 1.11.8 Suppose binary messages of length k are sent unencoded over a BSC with crossover probability ϱ. This in effect is the same as using the $[k, k]$ code $\mathbb{F}_2^k$. This code has a unique coset, the code itself, and its leader is the zero codeword of weight 0. Hence (1.12) shows that the probability of decoder error is

$$P_{\text{err}} = 1 - \varrho^0 (1 - \varrho)^k = 1 - (1 - \varrho)^k.$$

This is precisely what we expect as the probability of no decoding error is the probability $(1 - \varrho)^k$ that the k bits are received without error. ∎

Example 1.11.9 We compare sending $2^4 = 16$ binary messages unencoded to encoding using the $[7, 4]$ binary Hamming code $\mathcal{H}_3$. Assume communication is over a BSC with crossover probability ϱ. By Example 1.11.8, $P_{\text{err}} = 1 - (1 - \varrho)^4$ for the unencoded data. $\mathcal{H}_3$ has one coset of weight 0 and seven cosets of weight 1. Hence $P_{\text{err}} = 1 - (1 - \varrho)^7 - 7\varrho(1 - \varrho)^6$ by (1.12). For example if $\varrho = 0.01$, P_{err} without coding is $0.039\,403\,99$. Using $\mathcal{H}_3$, it is $0.002\,031\,04\ldots$. ∎

Exercise 74 Assume communication is over a BSC with crossover probability ϱ.
(a) Using Example 1.11.7, compute P_{err} for the extended Hamming code $\widehat{\mathcal{H}}_3$.
(b) Prove that the values of P_{err} for both $\mathcal{H}_3$, found in Example 1.11.9, and $\widehat{\mathcal{H}}_3$ are equal.
(c) Which code $\mathcal{H}_3$ or $\widehat{\mathcal{H}}_3$ would be better to use when communicating over a BSC? Why?   ◆

Exercise 75 Assume communication is over a BSC with crossover probability ϱ using the [23, 12, 7] binary Golay code $\mathcal{G}_{23}$.
(a) In Exercises 78 and 80 you will see that for $\mathcal{G}_{23}$ there are $\binom{23}{i}$ cosets of weight i for $0 \le i \le 3$ and no others. Compute P_{err} for this code.
(b) Compare P_{err} for sending 2^{12} binary messages unencoded to encoding with $\mathcal{G}_{23}$ when $\varrho = 0.01$.   ◆

Exercise 76 Assume communication is over a BSC with crossover probability ϱ using the [24, 12, 8] extended binary Golay code $\mathcal{G}_{24}$.
(a) In Example 8.3.2 you will see that for $\mathcal{G}_{24}$ there are 1, 24, 276, 2024, and 1771 cosets of weights 0, 1, 2, 3, and 4, respectively. Compute P_{err} for this code.
(b) Prove that the values of P_{err} for both $\mathcal{G}_{23}$, found in Exercise 75, and $\mathcal{G}_{24}$ are equal.
(c) Which code $\mathcal{G}_{23}$ or $\mathcal{G}_{24}$ would be better to use when communicating over a BSC? Why?   ◆

For a BSC with crossover probability ϱ, the *capacity* of the channel is

$$C(\varrho) = 1 + \varrho \log_2 \varrho + (1 - \varrho) \log_2(1 - \varrho).$$

The capacity $C(\varrho) = 1 - H_2(\varrho)$, where $H_2(\varrho)$ is the Hilbert entropy function that we define in Section 2.10.3. For binary symmetric channels, Shannon's Theorem is as follows.[6]

Theorem 1.11.10 (Shannon) *Let $\delta > 0$ and $R < C(\varrho)$. Then for large enough n, there exists an $[n, k]$ binary linear code C with $k/n \ge R$ such that $P_{err} < \delta$ when C is used for communication over a BSC with crossover probability ϱ. Furthermore no such code exists if $R > C(\varrho)$.*

Shannon's Theorem remains valid for nonbinary codes and other channels provided the channel capacity is defined appropriately. The fraction k/n is called the *rate*, or *information rate*, of an $[n, k]$ code and gives a measure of how much information is being transmitted; we discuss this more extensively in Section 2.10.

Exercise 77 Do the following.
(a) Graph the channel capacity as a function of ϱ for $0 < \varrho < 1$.
(b) In your graph, what is the region in which arbitrarily reliable communication can occur according to Shannon's Theorem?
(c) What is the channel capacity when $\varrho = 1/2$? What does Shannon's Theorem say about communication when $\varrho = 1/2$? (See Footnote 5 earlier in this section.)   ◆

---

[6] Shannon's original theorem was stated for nonlinear codes but was later shown to be valid for linear codes as well.

1.12   Sphere Packing Bound, covering radius, and perfect codes

The minimum distance d is a simple measure of the goodness of a code. For a given length and number of codewords, a fundamental problem in coding theory is to produce a code with the largest possible d. Alternatively, given n and d, determine the maximum number $A_q(n, d)$ of codewords in a code over $\mathbb{F}_q$ of length n and minimum distance at least d. The number $A_2(n, d)$ is also denoted by $A(n, d)$. The same question can be asked for linear codes. Namely, what is the maximum number $B_q(n, d)$ ($B(n, d)$ in the binary case) of codewords in a linear code over $\mathbb{F}_q$ of length n and minimum weight at least d? Clearly, $B_q(n, d) \leq A_q(n, d)$. For modest values of n and d, $A(n, d)$ and $B(n, d)$ have been determined and tabulated; see Chapter 2.

The fact that the spheres of radius t about codewords are pairwise disjoint immediately implies the following elementary inequality, commonly referred to as the *Sphere Packing Bound* or the *Hamming Bound*.

Theorem 1.12.1 (Sphere Packing Bound)

$$B_q(n, d) \leq A_q(n, d) \leq \frac{q^n}{\displaystyle\sum_{i=0}^{t} \binom{n}{i}(q-1)^i},$$

where $t = \lfloor (d-1)/2 \rfloor$.

Proof: Let $\mathcal{C}$ be a (possibly nonlinear) code over $\mathbb{F}_q$ of length n and minimum distance d. Suppose that $\mathcal{C}$ contains M codewords. By Theorem 1.11.2, the spheres of radius t about distinct codewords are disjoint. As there are $\alpha = \sum_{i=0}^{t} \binom{n}{i}(q-1)^i$ total vectors in any one of these spheres by (1.11) and the spheres are disjoint, $M\alpha$ cannot exceed the number q^n of vectors in $\mathbb{F}_q^n$. The result is now clear. $\square$

From the proof of the Sphere Packing Bound, we see that when we get equality in the bound, we actually fill the space $\mathbb{F}_q^n$ with disjoint spheres of radius t. In other words, every vector in $\mathbb{F}_q^n$ is contained in precisely one sphere of radius t centered about a codeword. When we have a code for which this is true, the code is called *perfect*.

Example 1.12.2 Recall that the Hamming code $\mathcal{H}_{q,r}$ over $\mathbb{F}_q$ is an $[n, k, 3]$ code, where $n = (q^r - 1)/(q - 1)$ and $k = n - r$. Then $t = 1$ and

$$\frac{q^n}{\displaystyle\sum_{i=0}^{t} \binom{n}{i}(q-1)^i} = \frac{q^n}{1 + n(q-1)} = \frac{q^n}{q^r} = q^k.$$

Thus $\mathcal{H}_{q,r}$ is perfect.   ∎

Exercise 78 Prove that the $[23, 12, 7]$ binary and the $[11, 6, 5]$ ternary Golay codes are perfect.   ♦

Exercise 79 Show that the following codes are perfect:

(a) the codes $C = \mathbb{F}_q^n$,

(b) the codes consisting of exactly one codeword (the zero vector in the case of linear codes),

(c) the binary repetition codes of odd length, and

(d) the binary codes of odd length consisting of a vector $\mathbf{c}$ and the complementary vector $\bar{\mathbf{c}}$ with 0s and 1s interchanged.

These codes are called *trivial perfect codes*.　　　　　　　　　　　　　◆

Exercise 80 Prove that a perfect t-error-correcting linear code of length n has precisely $\binom{n}{i}$ cosets of weight i for $0 \le i \le t$ and no other cosets. Hint: How many weight i vectors in $\mathbb{F}_q^n$ are there? Could distinct vectors of weights i and j with $i \le t$ and $j \le t$ be in the same coset? Use the equality in the Sphere Packing Bound.　　　　　　　◆

So the Hamming codes are perfect, as are two of the Golay codes, as shown in Exercise 78. Furthermore, Theorem 1.8.2 shows that all linear codes of the same length, dimension, and minimum weight as a Hamming code are equivalent. Any of these codes can be called the Hamming code. There are also some trivial perfect codes as described in Exercise 79. Thus we have part of the proof of the following theorem.

Theorem 1.12.3

(i) *There exist perfect single error-correcting codes over $\mathbb{F}_q$ which are not linear and all such codes have parameters corresponding to those of the Hamming codes, namely, length $n = (q^r - 1)/(q - 1)$ with q^{n-r} codewords and minimum distance 3. The only perfect single error-correcting linear codes over $\mathbb{F}_q$ are the Hamming codes.*

(ii) *The only nontrivial perfect multiple error-correcting codes have the same length, number of codewords, and minimum distance as either the [23, 12, 7] binary Golay code or the [11, 6, 5] ternary Golay code.*

(iii) *Any binary (respectively, ternary) possibly nonlinear code with 2^{12} (respectively, 3^6) vectors containing the $\mathbf{0}$ vector with length 23 (respectively, 11) and minimum distance 7 (respectively, 5) is equivalent to the [23, 12, 7] binary (respectively, [11, 6, 5] ternary) Golay code.*

The classification of the perfect codes as summarized in this theorem was a significant and difficult piece of mathematics, in which a number of authors contributed. We will prove part (iii) in Chapter 10. The rest of the proof can be found in [137, Section 5]. A portion of part (ii) is proved in Exercise 81.

Exercise 81 The purpose of this exercise is to prove part of Theorem 1.12.3(ii). Let C be an $[n, k, 7]$ perfect binary code.

(a) Using equality in the Sphere Packing Bound, prove that

$$(n + 1)[(n + 1)^2 - 3(n + 1) + 8] = 3 \cdot 2^{n-k+1}.$$

(b) Prove that $n + 1$ is either 2^b or $3 \cdot 2^b$ where, in either case, $b \le n - k + 1$.

(c) Prove that $b < 4$.

(d) Prove that $n = 23$ or $n = 7$.

(e) Name two codes that are perfect $[n, k, 7]$ codes, one with $n = 7$ and the other with $n = 23$. ◆

One can obtain nonlinear perfect codes by taking a coset of a linear perfect code; see Exercise 82. Theorem 1.12.3 shows that all multiple error-correcting nonlinear codes are cosets of the binary Golay code of length 23 or the ternary Golay code of length 11. On the other hand, there are nonlinear single error-correcting codes which are not cosets of Hamming codes; these were first constructed by Vasil'ev [338].

Exercise 82   Prove that a coset of a linear perfect code is also a perfect code. ◆

Let C be an $[n, k, d]$ code over $\mathbb{F}_q$ and let $t = \lfloor (d - 1)/2 \rfloor$. When you do not have a perfect code, in order to fill the space $\mathbb{F}_q^n$ with spheres centered at codewords, the spheres must have radius larger than t. Of course when you increase the sphere size, not all spheres will be pairwise disjoint. We define the *covering radius* $\rho = \rho(C)$ to be the smallest integer s such that $\mathbb{F}_q^n$ is the union of the spheres of radius s centered at the codewords of C. Equivalently,

$$\rho(C) = \max_{\mathbf{x} \in \mathbb{F}_q^n} \min_{\mathbf{c} \in C} d(\mathbf{x}, \mathbf{c}).$$

Obviously, $t \leq \rho(C)$ and $t = \rho(C)$ if and only if C is perfect. By Theorem 1.11.4, the packing radius of a code is the largest radius of spheres centered at codewords so that the spheres are disjoint. So a code is perfect if and only if its covering radius equals its packing radius. If the code is not perfect, its covering radius is larger than its packing radius.

For a nonlinear code C, the covering radius $\rho(C)$ is defined in the same way to be

$$\rho(C) = \max_{\mathbf{x} \in \mathbb{F}_q^n} \min_{\mathbf{c} \in C} d(\mathbf{x}, \mathbf{c}).$$

Again if d is the minimum distance of C and $t = \lfloor (d - 1)/2 \rfloor$, then $t \leq \rho(C)$ and $t = \rho(C)$ if and only if C is perfect. The theorems that we prove later in this section are only for linear codes.

If C is a code with packing radius t and covering radius $t + 1$, C is called *quasi-perfect*. There are many known linear and nonlinear quasi-perfect codes (e.g. certain double error-correcting BCH codes and some punctured Preparata codes). However, unlike perfect codes, there is no general classification.

Example 1.12.4   By Exercise 56, the binary $[8, 4, 4]$ code is shown to be unique, in the sense that all such codes are equivalent to $\widehat{\mathcal{H}}_3$. In Example 1.11.7, we give the complete coset weight distribution of this code. Since there are no cosets of weight greater than 2, the covering radius, $\rho(\widehat{\mathcal{H}}_3)$, is 2. Since the packing radius is $t = \lfloor (4 - 1)/2 \rfloor = 1$, this code is quasi-perfect. Both the covering and packing radius of the nonextended Hamming code $\mathcal{H}_3$ equal 1. This is an illustration of the fact that extending a binary code will not increase its packing radius (error-correcting capability) but will increase its covering radius. See Theorem 1.12.6(iv) below. ∎

Recall that the weight of a coset of a code C is the smallest weight of a vector in the coset. The definition of the covering radius implies the following characterization of the covering radius of a linear code in terms of coset weights and in terms of syndromes.

Theorem 1.12.5 *Let C be a linear code with parity check matrix H. Then:*
(i) *$\rho(C)$ is the weight of the coset of largest weight;*
(ii) *$\rho(C)$ is the smallest number s such that every nonzero syndrome is a combination of s or fewer columns of H, and some syndrome requires s columns.*

Exercise 83 Prove Theorem 1.12.5.   ♦

We conclude this chapter by collecting some elementary facts about the covering radius of codes and coset leaders, particularly involving codes arising in Section 1.5. More on covering radius can be found in Chapter 11.

Theorem 1.12.6 *Let C be an $[n, k]$ code over $\mathbb{F}_q$. Let $\widehat{C}$ be the extension of C, and let C^* be a code obtained from C by puncturing on some coordinate. The following hold:*
(i) *If $C = C_1 \oplus C_2$, then $\rho(C) = \rho(C_1) + \rho(C_2)$.*
(ii) *$\rho(C^*) = \rho(C)$ or $\rho(C^*) = \rho(C) - 1$.*
(iii) *$\rho(\widehat{C}) = \rho(C)$ or $\rho(\widehat{C}) = \rho(C) + 1$.*
(iv) *If $q = 2$, then $\rho(\widehat{C}) = \rho(C) + 1$.*
(v) *Assume that $\mathbf{x}$ is a coset leader of C. If $\mathbf{x}' \in \mathbb{F}_q^n$ all of whose nonzero components agree with the same components of $\mathbf{x}$, then $\mathbf{x}'$ is also a coset leader of C. In particular, if there is a coset of weight s, there is also a coset of any weight less than s.*

Proof: The proofs of the first three assertions are left as exercises.

For (iv), let $\mathbf{x} = x_1 \cdots x_n$ be a coset leader of C. Let $\mathbf{x}' = x_1 \cdots x_n'$. By part (iii), it suffices to show that $\mathbf{x}'$ is a coset leader of $\widehat{C}$. Let $\mathbf{c} = c_1 \cdots c_n \in C$, and let $\widehat{\mathbf{c}} = c_1 \cdots c_n c_{n+1}$ be its extension. If $\mathbf{c}$ has even weight, then $\text{wt}(\widehat{\mathbf{c}} + \mathbf{x}') = \text{wt}(\mathbf{c} + \mathbf{x}) + 1 \geq \text{wt}(\mathbf{x}) + 1$. Assume $\mathbf{c}$ has odd weight. Then $\text{wt}(\widehat{\mathbf{c}} + \mathbf{x}') = \text{wt}(\mathbf{c} + \mathbf{x})$. If $\mathbf{x}$ has even (odd) weight, then $\mathbf{c} + \mathbf{x}$ has odd (even) weight by Theorem 1.4.3, and so $\text{wt}(\mathbf{c} + \mathbf{x}) > \text{wt}(\mathbf{x})$ as $\mathbf{x}$ is a coset leader. Thus in all cases, $\text{wt}(\widehat{\mathbf{c}} + \mathbf{x}') \geq \text{wt}(\mathbf{x}) + 1 = \text{wt}(\mathbf{x}')$ and so $\mathbf{x}'$ is a coset leader of $\widehat{C}$.

To prove (v), it suffices, by induction, to verify the result when $\mathbf{x} = x_1 \cdots x_n$ is a coset leader and $\mathbf{x}' = x_1' \cdots x_n'$, where $x_j = x_j'$ for all $j \neq i$ and $x_i \neq x_i' = 0$. Notice that $\text{wt}(\mathbf{x}) = \text{wt}(\mathbf{x}') + 1$. Suppose that $\mathbf{x}'$ is not a coset leader. Then there is a codeword $\mathbf{c} \in C$ such that $\mathbf{x}' + \mathbf{c}$ is a coset leader and hence

$$\text{wt}(\mathbf{x}' + \mathbf{c}) \leq \text{wt}(\mathbf{x}') - 1 = \text{wt}(\mathbf{x}) - 2. \tag{1.13}$$

But as $\mathbf{x}$ and $\mathbf{x}'$ disagree in only one coordinate, $\text{wt}(\mathbf{x} + \mathbf{c}) \leq \text{wt}(\mathbf{x}' + \mathbf{c}) + 1$. Using (1.13), this implies that $\text{wt}(\mathbf{x} + \mathbf{c}) \leq \text{wt}(\mathbf{x}) - 1$, a contradiction as $\mathbf{x}$ is a coset leader.   □

Exercise 84 Prove parts (i), (ii), and (iii) of Theorem 1.12.6.   ♦

The next example illustrates that it is possible to extend or puncture a code and leave the covering radius unchanged. Compare this to Theorem 1.12.6(ii) and (iii).

Example 1.12.7 Let C be the ternary code with generator matrix $[1 \; 1 \; -1]$. Computing the covering radius, we see that $\rho(C) = \rho(\widehat{C}) = 2$. If $D = \widehat{C}$ and we puncture D on the last coordinate to obtain $D^* = C$, we have $\rho(D) = \rho(D^*)$. ∎

In the binary case by Theorem 1.12.6(iv), whenever we extend a code, we increase the covering radius by 1. But when we puncture a binary code we may not reduce the covering radius.

Example 1.12.8 Let C be the binary code with generator matrix

$$\begin{bmatrix} 1 & 0 & 1 & 1 \\ 0 & 1 & 1 & 1 \end{bmatrix},$$

and let C^* be obtained from C by puncturing on the last coordinate. It is easy to see that $\rho(C) = \rho(C^*) = 1$. Also if D is the extension of C^*, $\rho(D) = 2$, consistent with Theorem 1.12.6. ∎

2 Bounds on the size of codes

In this chapter, we present several bounds on the number of codewords in a linear or nonlinear code given the length n and minimum distance d of the code. In Section 1.12 we proved the Sphere Packing (or Hamming) Bound, which gives an upper bound on the size of a code. This chapter is devoted to developing several other upper bounds along with two lower bounds. There are fewer lower bounds presented, as lower bounds are often tied to particular constructions of codes. For example, if a code with a given length n and minimum distance d is produced, its size becomes a lower bound on the code size. In this chapter we will speak about codes that *meet* a given bound. If the bound is a lower bound on the size of a code in terms of its length and minimum distance, then a code C *meets* the lower bound if the size of C is at least the size given by the lower bound. If the bound is an upper bound on the size of a code in terms of its length and minimum distance, then C *meets* the upper bound if its size equals the size given by the upper bound.

We present the upper bounds first after we take a closer look at the concepts previously developed.

2.1 $A_q(n, d)$ **and** $B_q(n, d)$

In this section, we will consider both linear and nonlinear codes. An (n, M, d) code C over $\mathbb{F}_q$ is a code of length n with M codewords whose minimum distance is d. The code C can be either linear or nonlinear; if it is linear, it is an $[n, k, d]$ code, where $k = \log_q M$ and d is the minimum weight of C; see Theorem 1.4.2.

We stated the Sphere Packing Bound using the notation $B_q(n, d)$ and $A_q(n, d)$, where $B_q(n, d)$, respectively $A_q(n, d)$, is the largest number of codewords in a linear, respectively arbitrary (linear or nonlinear), code over $\mathbb{F}_q$ of length n and minimum distance at least d. A code of length n over $\mathbb{F}_q$ and minimum distance at least d will be called *optimal* if it has $A_q(n, d)$ codewords (or $B_q(n, d)$ codewords in the case that C is linear). There are other perspectives on optimizing a code. For example, one could ask to find the largest d, given n and M, such that there is a code over $\mathbb{F}_q$ of length n with M codewords and minimum distance d. Or, find the smallest n, given M and d such that there is a code over $\mathbb{F}_q$ of length n with M codewords and minimum distance d. We choose to focus on $A_q(n, d)$ and $B_q(n, d)$ here.

We begin with some rather simple properties of $A_q(n, d)$ and $B_q(n, d)$. First we have the following obvious facts:

Table 2.1 *Upper and lower bounds on* $A_2(n, d)$ *for*
$6 \le n \le 24$

n	$d = 4$	$d = 6$	$d = 8$	$d = 10$
6	4	2	1	1
7	8	2	1	1
8	16	2	2	1
9	20	4	2	1
10	40	6	2	2
11	72	12	2	2
12	144	24	4	2
13	256	32	4	2
14	512	64	8	2
15	1024	128	16	4
16	2048	256	32	4
17	2720–3276	256–340	36–37	6
18	5312–6552	512–680	64–72	10
19	10496–13104	1024–1288	128–144	20
20	20480–26208	2048–2372	256–279	40
21	36864–43689	2560–4096	512	42–48
22	73728–87378	4096–6941	1024	50–88
23	147456–173491	8192–13774	2048	76–150
24	294912–344308	16384–24106	4096	128–280

Theorem 2.1.1 $B_q(n, d) \le A_q(n, d)$ *and* $B_q(n, d)$ *is a nonnegative integer power of* q.

So $B_q(n, d)$ is a lower bound for $A_q(n, d)$ and $A_q(n, d)$ is an upper bound for $B_q(n, d)$. The Sphere Packing Bound is an upper bound on $A_q(n, d)$ and hence on $B_q(n, d)$.

Tables which lead to information about the values of $A_q(n, d)$ or $B_q(n, d)$ have been computed and are regularly updated. These tables are for small values of q and moderate to large values of n. The most comprehensive table is compiled by A. E. Brouwer [32], which gives upper and lower bounds on the minimum distance d of an $[n, k]$ linear code over $\mathbb{F}_q$. A less extensive table giving bounds for $A_2(n, d)$ is kept by S. Litsyn [205].

To illustrate, we reproduce a table due to many authors and recently updated by Agrell, Vardy, Zeger, and Litsyn in [2, 205]. Most of the upper bounds in Table 2.1 are obtained from the bounds presented in this chapter together with the Sphere Packing Bound; *ad hoc* methods in certain cases produce the remaining values. Notice that Table 2.1 contains only even values of d, a consequence of the following result.

Theorem 2.1.2 *Let* $d > 1$. *Then*:
(i) $A_q(n, d) \le A_q(n - 1, d - 1)$ *and* $B_q(n, d) \le B_q(n - 1, d - 1)$, *and*
(ii) *if d is even,* $A_2(n, d) = A_2(n - 1, d - 1)$ *and* $B_2(n, d) = B_2(n - 1, d - 1)$.
Furthermore:
(iii) *if d is even and* $M = A_2(n, d)$, *then there exists a binary (n, M, d) code such that all codewords have even weight and the distance between all pairs of codewords is also even.*

Proof: Let C be a code (linear or nonlinear) with M codewords and minimum distance d. Puncturing on any coordinate gives a code C^* also with M codewords; otherwise if C^* has fewer codewords, there would exist two codewords of C which differ in one position implying $d = 1$. This proves (i); to complete (ii), we only need to show that $A_2(n, d) \geq A_2(n - 1, d - 1)$ (or $B_2(n, d) \geq B_2(n - 1, d - 1)$) when C is linear. To that end let C be a binary code with M codewords, length $n - 1$, and minimum distance $d - 1$. Extend C by adding an overall parity check to obtain a code $\widehat{C}$ of length n and minimum distance d, since $d - 1$ is odd. Because $\widehat{C}$ has M codewords, $A_2(n, d) \geq A_2(n - 1, d - 1)$ (or $B_2(n, d) \geq B_2(n - 1, d - 1)$). For (iii), if C is a binary (n, M, d) code with d even, the punctured code C^* as previously stated is an $(n - 1, M, d - 1)$ code. Extending C^* produces an (n, M, d) code $\widehat{C^*}$ since $d - 1$ is odd; furthermore this code has only even weight codewords. Since $d(\mathbf{x}, \mathbf{y}) = \text{wt}(\mathbf{x} + \mathbf{y}) = \text{wt}(\mathbf{x}) + \text{wt}(\mathbf{y}) - 2\text{wt}(\mathbf{x} \cap \mathbf{y})$, the distance between codewords is even. $\qquad \square$

Exercise 85 In the proof of Theorem 2.1.2, we claim that if C is a binary code of length $n - 1$ and minimum weight $d - 1$, where $d - 1$ is odd, then the extended code $\widehat{C}$ of length n has minimum distance d. In Section 1.5.2 we stated that this is true if C is linear, where it is obvious since the minimum distance is the minimum weight. Prove that it is also true when C is nonlinear. $\qquad \blacklozenge$

Theorem 2.1.2(ii) shows that any table of values of $A_2(n, d)$ or $B_2(n, d)$ only needs to be compiled for d either always odd or d always even. Despite the fact that $A_2(n, d) = A_2(n - 1, d - 1)$ when d is even, we want to emphasize that a given bound for $A_2(n, d)$ may not be the same bound as for $A_2(n - 1, d - 1)$. So since these values are equal, we can always choose the smaller upper bound, respectively larger lower bound, as a common upper bound, respectively lower bound, for both $A_2(n, d)$ and $A_2(n - 1, d - 1)$.

Example 2.1.3 By using $n = 7$ and $d = 4$ (that is, $t = 1$) in the Sphere Packing Bound, we find that $A_2(7, 4) \leq 16$. On the other hand, using $n = 6$ and $d = 3$ (still, $t = 1$) the Sphere Packing Bound yields $64/7$ implying that $A_2(6, 3) \leq 9$. So by Theorem 2.1.2(ii) an upper bound for both $A_2(7, 4)$ and $A_2(6, 3)$ is 9. $\qquad \blacksquare$

Exercise 86 Let C be a code (possibly nonlinear) over $\mathbb{F}_q$ with minimum distance d. Fix a codeword $\mathbf{c}$ in C. Let $C_1 = \{\mathbf{x} - \mathbf{c} \mid \mathbf{x} \in C\}$. Prove that C_1 contains the zero vector $\mathbf{0}$, has the same number of codewords as C, and also has minimum distance d. Prove also that $C = C_1$ if C is linear. $\qquad \blacklozenge$

Exercise 87 By Example 2.1.3, $A_2(7, 4) \leq 9$.
(a) Prove that $B_2(7, 4) \leq 8$.
(b) Find a binary $[7, 3, 4]$ code thus verifying $B_2(7, 4) = 2^3 = 8$. In our terminology, this code is optimal.
(c) Show that $A_2(7, 4)$ is either 8 or 9. (Table 2.1 shows that it is actually 8, a fact we will verify in the next section.) $\qquad \blacklozenge$

Exercise 88 By Table 2.1, $A_2(13, 10) = 2$.
(a) By computing the Sphere Packing Bound using $(n, d) = (13, 10)$ and $(12, 9)$, find the best sphere packing upper bound for $A_2(13, 10)$.

(b) Using part (a), give an upper bound on $B_2(13, 10)$.

(c) Prove that $B_2(13, 10) = A_2(13, 10)$ is exactly 2 by carrying out the following.

    (i) Construct a [13, 1, 10] linear code.

    (ii) Show that no binary code of length 13 and minimum distance 10 or more can contain three codewords. (Hint: By Exercise 86, you may assume that such a code contains the vector **0**.)   ◆

Exercise 89 This exercise verifies the entry for $n = 16$ and $d = 4$ in Table 2.1.

(a) Use the Sphere Packing Bound to get an upper bound on $A_2(15, 3)$. What (linear) code meets this bound? Is this code perfect?

(b) Use two pairs of numbers (n, d) and the Sphere Packing Bound to get an upper bound on $A_2(16, 4)$. What (linear) code meets this bound? Is this code perfect?

(c) Justify the value of $A_2(16, 4)$ given in Table 2.1.   ◆

In examining Table 2.1, notice that $d = 2$ is not considered. The values for $A_2(n, 2)$ and $B_2(n, 2)$ can be determined for all n.

Theorem 2.1.4 $A_2(n, 2) = B_2(n, 2) = 2^{n-1}$.

Proof: By Theorem 2.1.2(ii), $A_2(n, 2) = A_2(n - 1, 1)$. But clearly $A_2(n - 1, 1) \leq 2^{n-1}$, and the entire space $\mathbb{F}_2^{n-1}$ is a code of length $n - 1$ and minimum distance 1, implying $A_2(n - 1, 1) = 2^{n-1}$. By Theorem 2.1.1 as $\mathbb{F}_2^{n-1}$ is linear, $2^{n-1} = B_2(n - 1, 1) = B_2(n, 2)$.   □

There is another set of table values that can easily be found as a result of the next theorem.

Theorem 2.1.5 $A_q(n, n) = B_q(n, n) = q$.

Proof: The linear code of size q consisting of all multiples of the all-one vector of length n (that is, the repetition code over $\mathbb{F}_q$) has minimum distance n. So by Theorem 2.1.1, $A_q(n, n) \geq B_q(n, n) \geq q$. If $A_q(n, n) > q$, there exists a code with more than q codewords and minimum distance n. Hence at least two of the codewords agree on some coordinate; but then these two codewords are less than distance n apart, a contradiction. So $A_q(n, n) = B_q(n, n) = q$.   □

In tables such as Table 2.1, it is often the case that when one bound is found for a particular n and d, this bound can be used to find bounds for "nearby" n and d. For instance, once you have an upper bound for $A_q(n - 1, d)$ or $B_q(n - 1, d)$, there is an upper bound on $A_q(n, d)$ or $B_q(n, d)$.

Theorem 2.1.6 $A_q(n, d) \leq q A_q(n - 1, d)$ *and* $B_q(n, d) \leq q B_q(n - 1, d)$.

Proof: Let $\mathcal{C}$ be a (possibly nonlinear) code over $\mathbb{F}_q$ of length n and minimum distance at least d with $M = A_q(n, d)$ codewords. Let $\mathcal{C}(\alpha)$ be the subcode of $\mathcal{C}$ in which every codeword has α in coordinate n. Then, for some α, $\mathcal{C}(\alpha)$ contains at least M/q codewords. Puncturing this code on coordinate n produces a code of length $n - 1$ and minimum distance d. Therefore $M/q \leq A_q(n - 1, d)$ giving $A_q(n, d) \leq q A_q(n - 1, d)$. We leave the second inequality as an exercise.   □

Exercise 90 Let C be an $[n, k, d]$ linear code over $\mathbb{F}_q$.

(a) Prove that if i is a fixed coordinate, either all codewords of C have 0 in that coordinate position or the subset consisting of all codewords which have a 0 in coordinate position i is an $[n, k-1, d]$ linear subcode of C.

(b) Prove that $B_q(n, d) \leq q B_q(n-1, d)$. ♦

Exercise 91 Verify the following values for $A_2(n, d)$.

(a) Show that $A_2(8, 6) = 2$ by direct computation. (That is, show that there is a binary code with two codewords of length 8 that are distance 6 apart; then show that no code with three such codewords can exist. Use Exercise 86.)

(b) Show that $A_2(9, 6) \leq 4$ using part (a) and Theorem 2.1.6. Construct a code meeting this bound.

(c) What are $B_2(8, 6)$ and $B_2(9, 6)$? Why? ♦

Exercise 92 Assume that $A_2(13, 6) = 32$, as indicated in Table 2.1.

(a) Show that $A_2(14, 6) \leq 64$, $A_2(15, 6) \leq 128$, and $A_2(16, 6) \leq 256$.

(b) Show that if you can verify that $A_2(16, 6) = 256$, then there is equality in the other bounds in (a).

See also Exercise 108. ♦

Exercise 93 Show that $B_2(13, 6) \leq 32$ by assuming that a $[13, 6, 6]$ binary code exists. Obtain a contradiction by attempting to construct a generator matrix for this code in standard form. ♦

Exercise 94 Verify that $A_2(24, 8) = 4096$ consistent with Table 2.1. ♦

Before proceeding to the other bounds, we observe that the covering radius of a code C with $A_q(n, d)$ codewords is at most $d - 1$. For if a code C with $A_q(n, d)$ codewords has covering radius d or higher, there is a vector $\mathbf{x}$ in $\mathbb{F}_q^n$ at distance d or more from every codeword of C; hence $C \cup \{\mathbf{x}\}$ has one more codeword and minimum distance at least d. The same observation holds for linear codes with $B_q(n, d)$ codewords; such codes have covering radius $d - 1$ or less, a fact left to the exercises. For future reference, we state these in the following theorem.

Theorem 2.1.7 *Let C be either a code over $\mathbb{F}_q$ with $A_q(n, d)$ codewords or a linear code over $\mathbb{F}_q$ with $B_q(n, d)$ codewords. Then C has covering radius $d - 1$ or less.*

Exercise 95 Prove that if C is a linear code over $\mathbb{F}_q$ with $B_q(n, d)$ codewords, then C has covering radius at most $d - 1$. ♦

There are two types of bounds that we consider in this chapter. Until Section 2.10 the bounds we consider in the chapter are valid for arbitrary values of n and d. Most of these have asymptotic versions which hold for families of codes having lengths that go to infinity. These asymptotic bounds are considered in Section 2.10.

There is a common technique used in many of the proofs of the upper bounds that we examine. A code will be chosen and its codewords will become the rows of a matrix. There will be some expression, related to the bound we are seeking, which must itself be bounded. We will often look at the number of times a particular entry occurs in a particular column of the matrix of codewords. From there we will be able to bound the expression and that will lead directly to our desired upper bound.

2.2   The Plotkin Upper Bound

The purpose of having several upper bounds is that one may be smaller than another for a given value of n and d. In general, one would like an upper bound as tight (small) as possible so that there is hope that codes meeting this bound actually exist. The Plotkin Bound [285] is an upper bound which often improves the Sphere Packing Bound on $A_q(n, d)$; however, it is only valid when d is sufficiently close to n.

Theorem 2.2.1 (Plotkin Bound) *Let* C *be an* (n, M, d) *code over* $\mathbb{F}_q$ *such that* $rn < d$ *where* $r = 1 - q^{-1}$. *Then*

$$M \le \left\lfloor \frac{d}{d - rn} \right\rfloor.$$

In particular,

$$A_q(n, d) \le \left\lfloor \frac{d}{d - rn} \right\rfloor, \tag{2.1}$$

provided $rn < d$. *In the binary case,*

$$A_2(n, d) \le 2 \left\lfloor \frac{d}{2d - n} \right\rfloor \tag{2.2}$$

if $n < 2d$.

Proof: Let

$$S = \sum_{\mathbf{x} \in C} \sum_{\mathbf{y} \in C} d(\mathbf{x}, \mathbf{y}).$$

If $\mathbf{x} \ne \mathbf{y}$ for $\mathbf{x}, \mathbf{y} \in C$, then $d \le d(\mathbf{x}, \mathbf{y})$ implying that

$$M(M - 1)d \le S. \tag{2.3}$$

Let $\mathcal{M}$ be the $M \times n$ matrix whose rows are the codewords of C. For $1 \le i \le n$, let $n_{i,\alpha}$ be the number of times $\alpha \in \mathbb{F}_q$ occurs in column i of $\mathcal{M}$. As $\sum_{\alpha \in \mathbb{F}_q} n_{i,\alpha} = M$ for $1 \le i \le n$, we have

$$S = \sum_{i=1}^{n} \sum_{\alpha \in \mathbb{F}_q} n_{i,\alpha}(M - n_{i,\alpha}) = nM^2 - \sum_{i=1}^{n} \sum_{\alpha \in \mathbb{F}_q} n_{i,\alpha}^2. \tag{2.4}$$

By the Cauchy–Schwartz inequality,

$$\left(\sum_{\alpha \in \mathbb{F}_q} n_{i,\alpha} \right)^2 \le q \sum_{\alpha \in \mathbb{F}_q} n_{i,\alpha}^2.$$

Using this, we obtain

$$S \leq nM^2 - \sum_{i=1}^{n} q^{-1} \left(\sum_{\alpha \in \mathbb{F}_q} n_{i,\alpha} \right)^2 = nrM^2. \tag{2.5}$$

Combining (2.3) and (2.5) we obtain $M \leq \lfloor d/(d - rn) \rfloor$ since M is an integer, which gives bound (2.1).

In the binary case, this can be slightly improved. We still have

$$M \leq \frac{d}{d - n/2} = \frac{2d}{2d - n},$$

using (2.3) and (2.5). If M is even, we can round the expression $2d/(2d - n)$ down to the nearest even integer, which gives (2.2). When M is odd, we do not use Cauchy–Schwartz. Instead, from (2.4), we observe that

$$S = \sum_{i=1}^{n} [n_{i,0}(M - n_{i,0}) + n_{i,1}(M - n_{i,1})] = \sum_{i=1}^{n} 2n_{i,0}n_{i,1} \tag{2.6}$$

because $n_{i,0} + n_{i,1} = M$. But the right-hand side of (2.6) is maximized when $\{n_{i,0}, n_{i,1}\} = \{(M - 1)/2, (M + 1)/2\}$; thus using (2.3)

$$M(M - 1)d \leq n\frac{1}{2}(M - 1)(M + 1).$$

Simplifying,

$$M \leq \frac{n}{2d - n} = \frac{2d}{2d - n} - 1,$$

which proves (2.2) in the case that M is odd.   □

The Plotkin Bound has rather limited scope as it is only valid when $n < 2d$ in the binary case. However, we can examine what happens for "nearby" values, namely $n = 2d$ and $n = 2d + 1$.

Corollary 2.2.2 *The following bounds hold:*
(i) *If d is even, $A_2(2d, d) \leq 4d$.*
(ii) *If d is odd, $A_2(2d, d) \leq 2d + 2$.*
(iii) *If d is odd, $A_2(2d + 1, d) \leq 4d + 4$.*

Proof: By Theorem 2.1.6, $A_2(2d, d) \leq 2A_2(2d - 1, d)$. But by the Plotkin Bound, $A_2(2d - 1, d) \leq 2d$, giving (i), regardless of the parity of d. If d is odd, we obtain a better bound. By Theorem 2.1.2(ii), $A_2(2d, d) = A_2(2d + 1, d + 1)$ if d is odd. Applying the Plotkin Bound, $A_2(2d + 1, d + 1) \leq 2d + 2$, producing bound (ii). Finally if d is odd, $A_2(2d + 1, d) = A_2(2d + 2, d + 1)$ by Theorem 2.1.2(ii). Since $A_2(2d + 2, d + 1) \leq 4(d + 1)$ by (i), we have (iii).   □

Example 2.2.3 The Sphere Packing Bound for $A_2(17, 9)$ is $65\,536/1607$, yielding $A_2(18, 10) = A_2(17, 9) \leq 40$. However, by the Plotkin Bound, $A_2(18, 10) \leq 10$. There is a code meeting this bound as indicated by Table 2.1.   ■

Example 2.2.4 The Sphere Packing Bound for $A_2(14, 7)$ is $8192/235$, yielding $A_2(15, 8) = A_2(14, 7) \leq 34$. However, by Corollary 2.2.2(ii), $A_2(14, 7) \leq 16$. Again this bound is attained as indicated by Table 2.1. ∎

Exercise 96 (a) Find the best Sphere Packing Bound for $A_2(n, d)$ by choosing the smaller of the Sphere Packing Bounds for $A_2(n, d)$ and $A_2(n - 1, d - 1)$ for the following values of (n, d). (Note that d is even in each case and so $A_2(n, d) = A_2(n - 1, d - 1)$.) (b) From the Plotkin Bound and Corollary 2.2.2, where applicable, compute the best bound. (c) For each (n, d), which bound (a) or (b) is the better bound? (d) What is the true value of $A_2(n, d)$ according to Table 2.1?
(i) $(n, d) = (7, 4)$ (compare to Example 2.1.3), $(8, 4)$.
(ii) $(n, d) = (9, 6)$ (compare to Exercise 91), $(10, 6)$, $(11, 6)$.
(iii) $(n, d) = (14, 8)$, $(15, 8)$, $(16, 8)$.
(iv) $(n, d) = (16, 10)$, $(17, 10)$, $(18, 10)$, $(19, 10)$. $(20, 10)$. ◆

2.3 The Johnson Upper Bounds

In this section we present a series of bounds due to Johnson [159]. In connection with these bounds, we introduce the concept of constant weight codes. Bounds on these constant weight codes will be used in Section 2.3.3 to produce upper bounds on $A_q(n, d)$. A (nonlinear) (n, M, d) code C over $\mathbb{F}_q$ is a *constant weight code* provided every codeword has the same weight w. For example, the codewords of fixed weight in a linear code form a constant weight code. If $\mathbf{x}$ and $\mathbf{y}$ are distinct codewords of weight w, then $d(\mathbf{x}, \mathbf{y}) \leq 2w$. Therefore we have the following simple observation.

Theorem 2.3.1 *If C is a constant weight (n, M, d) code with codewords of weight w and if $M > 1$, then $d \leq 2w$.*

Define $A_q(n, d, w)$ to be the maximum number of codewords in a constant weight (n, M) code over $\mathbb{F}_q$ of length n and minimum distance at least d whose codewords have weight w. Obviously $A_q(n, d, w) \leq A_q(n, d)$.

Example 2.3.2 It turns out that there are 759 weight 8 codewords in the [24, 12, 8] extended binary Golay code. These codewords form a $(24, 759, 8)$ constant weight code with codewords of weight 8; thus $759 \leq A_2(24, 8, 8)$. ∎

We have the following bounds on $A_q(n, d, w)$.

Theorem 2.3.3
(i) $A_q(n, d, w) = 1$ if $d > 2w$.
(ii) $A_q(n, 2w, w) \leq \lfloor (n(q - 1)/w) \rfloor$.
(iii) $A_2(n, 2w, w) = \lfloor n/w \rfloor$.
(iv) $A_2(n, 2e - 1, w) = A_2(n, 2e, w)$.

Proof: Part (i) is a restatement of Theorem 2.3.1. In an $(n, M, 2w)$ constant weight code $\mathcal{C}$ over $\mathbb{F}_q$ with codewords of weight w, no two codewords can have the same nonzero entries in the same coordinate. Thus if $\mathcal{M}$ is the $M \times n$ matrix whose rows are the codewords of $\mathcal{C}$, each column of $\mathcal{M}$ can have at most $q - 1$ nonzero entries. So $\mathcal{M}$ has at most $n(q - 1)$ nonzero entries. However, each row of $\mathcal{M}$ has w nonzero entries and so $Mw \leq n(q - 1)$. This gives (ii). For (iii), let $\mathcal{C} = \{\mathbf{c}_1, \ldots, \mathbf{c}_M\}$, where $M = \lfloor n/w \rfloor$ and $\mathbf{c}_i$ is the vector of length n consisting of $(i - 1)w$ 0s followed by w 1s followed by $n - iw$ 0s, noting that $n - Mw \geq 0$. Clearly $\mathcal{C}$ is a constant weight binary $(n, \lfloor n/w \rfloor, 2w)$ code. The existence of this code and part (ii) with $q = 2$ give (iii). Part (iv) is left for Exercise 97. □

Exercise 97   Show that two binary codewords of the same weight must have even distance between them. Then use this to show that $A_2(n, 2e - 1, w) = A_2(n, 2e, w)$.   ◆

2.3.1   The Restricted Johnson Bound

We consider two bounds on $A_q(n, d, w)$, the first of which we call the Restricted Johnson Bound.

Theorem 2.3.4 (Restricted Johnson Bound for $A_q(n, d, w)$)

$$A_q(n, d, w) \leq \left\lfloor \frac{nd(q - 1)}{qw^2 - 2(q - 1)nw + nd(q - 1)} \right\rfloor$$

provided $qw^2 - 2(q - 1)nw + nd(q - 1) > 0$, *and*

$$A_2(n, d, w) \leq \left\lfloor \frac{nd}{2w^2 - 2nw + nd} \right\rfloor$$

provided $2w^2 - 2nw + nd > 0$.

Proof: The second bound is a special case of the first. The proof of the first uses the same ideas as in the proof of the Plotkin Bound. Let $\mathcal{C}$ be an (n, M, d) constant weight code with codewords of weight w. Let $\mathcal{M}$ be the $M \times n$ matrix whose rows are the codewords of $\mathcal{C}$. Let

$$S = \sum_{\mathbf{x} \in \mathcal{C}} \sum_{\mathbf{y} \in \mathcal{C}} d(\mathbf{x}, \mathbf{y}).$$

If $\mathbf{x} \neq \mathbf{y}$ for $\mathbf{x}, \mathbf{y} \in \mathcal{C}$, then $d \leq d(\mathbf{x}, \mathbf{y})$ implying that

$$M(M - 1)d \leq S. \tag{2.7}$$

For $1 \leq i \leq n$, let $n_{i,\alpha}$ be the number of times $\alpha \in \mathbb{F}_q$ occurs in column i of $\mathcal{M}$. So

$$S = \sum_{i=1}^{n} \sum_{\alpha \in \mathbb{F}_q} n_{i,\alpha}(M - n_{i,\alpha}) = \sum_{i=1}^{n} \left(Mn_{i,0} - n_{i,0}^2 \right) + \sum_{i=1}^{n} \sum_{\alpha \in \mathbb{F}_q^*} \left(Mn_{i,\alpha} - n_{i,\alpha}^2 \right), \tag{2.8}$$

where $\mathbb{F}_q^*$ denotes the nonzero elements of $\mathbb{F}_q$. We analyze each of the last two terms separately. First,

$$\sum_{i=1}^{n} n_{i,0} = (n - w)M$$

because the left-hand side counts the number of 0s in the matrix $\mathcal{M}$ and each of the M rows of $\mathcal{M}$ has $n - w$ 0s. Second, by the Cauchy–Schwartz inequality,

$$\left(\sum_{i=1}^{n} n_{i,0} \right)^2 \le n \sum_{i=1}^{n} n_{i,0}^2.$$

Combining these we see that the first summation on the right-hand side of (2.8) satisfies

$$\sum_{i=1}^{n} \left(M n_{i,0} - n_{i,0}^2 \right) \le (n - w)M^2 - \frac{1}{n} \left(\sum_{i=1}^{n} n_{i,0} \right)^2$$

$$= (n - w)M^2 - \frac{(n - w)^2 M^2}{n}. \tag{2.9}$$

A similar argument is used on the second summation of the right-hand side of (2.8). This time

$$\sum_{i=1}^{n} \sum_{\alpha \in \mathbb{F}_q^*} n_{i,\alpha} = wM$$

because the left-hand side counts the number of nonzero elements in the matrix $\mathcal{M}$ and each of the M rows of $\mathcal{M}$ has w nonzero components. By the Cauchy–Schwartz inequality,

$$\left(\sum_{i=1}^{n} \sum_{\alpha \in \mathbb{F}_q^*} n_{i,\alpha} \right)^2 \le n(q - 1) \sum_{i=1}^{n} \sum_{\alpha \in \mathbb{F}_q^*} n_{i,\alpha}^2.$$

This yields

$$\sum_{i=1}^{n} \sum_{\alpha \in \mathbb{F}_q^*} \left(M n_{i,\alpha} - n_{i,\alpha}^2 \right) \le wM^2 - \frac{1}{n(q - 1)} \left(\sum_{i=1}^{n} \sum_{\alpha \in \mathbb{F}_q^*} n_{i,\alpha} \right)^2$$

$$= wM^2 - \frac{1}{n(q - 1)} (wM)^2. \tag{2.10}$$

Combining (2.7), (2.8), (2.9), and (2.10), we obtain:

$$M(M - 1)d \le (n - w)M^2 - \frac{(n - w)^2 M^2}{n} + wM^2 - \frac{1}{n(q - 1)} (wM)^2,$$

which simplifies to

$$(M - 1)d \le M \left[\frac{2(q - 1)nw - qw^2}{n(q - 1)} \right].$$

Solving this inequality for M, we get

$$M \le \frac{nd(q-1)}{qw^2 - 2(q-1)nw + nd(q-1)},$$

provided the denominator is positive. This produces our bound. $\square$

Example 2.3.5 By the Restricted Johnson Bound $A_2(7,4,4) \le 7$. The subcode $\mathcal{C}$ of $\mathcal{H}_3$ consisting of the even weight codewords is a $[7,3,4]$ code with exactly seven codewords of weight 4. Therefore these seven vectors form a $(7,7,4)$ constant weight code with codewords of weight 4. Thus $A_2(7,4,4) = 7$.   ■

Exercise 98 Verify all claims in Example 2.3.5.   ◆

Exercise 99 Using the Restricted Johnson Bound, show that $A_2(10,6,4) \le 5$. Also construct a $(10,5,6)$ constant weight binary code with codewords of weight 4.   ◆

2.3.2   The Unrestricted Johnson Bound

The bound in the previous subsection is "restricted" in the sense that $qw^2 - 2(q-1)nw + nd(q-1) > 0$ is necessary. There is another bound on $A_q(n,d,w)$, also due to Johnson, which has no such restriction.

Theorem 2.3.6 (Unrestricted Johnson Bound for $A_q(n,d,w)$)
(i)   If $2w < d$, then $A_q(n,d,w) = 1$.
(ii) If $2w \ge d$ and $d \in \{2e-1, 2e\}$, then, setting $q^* = q-1$,

$$A_q(n,d,w) \le \left\lfloor \frac{nq^*}{w} \left\lfloor \frac{(n-1)q^*}{w-1} \left\lfloor \cdots \left\lfloor \frac{(n-w+e)q^*}{e} \right\rfloor \cdots \right\rfloor \right\rfloor \right\rfloor.$$

(iii) If $w < e$, then $A_2(n, 2e-1, w) = A_2(n, 2e, w) = 1$.
(iv) If $w \ge e$, then

$$A_2(n, 2e-1, w) = A_2(n, 2e, w) \le \left\lfloor \frac{n}{w} \left\lfloor \frac{n-1}{w-1} \left\lfloor \cdots \left\lfloor \frac{n-w+e}{e} \right\rfloor \cdots \right\rfloor \right\rfloor \right\rfloor.$$

Proof: Part (i) is clear from Theorem 2.3.1. For part (ii), let $\mathcal{C}$ be an (n, M, d) constant weight code over $\mathbb{F}_q$ with codewords of weight w where $M = A_q(n, d, w)$. Let $\mathcal{M}$ be the $M \times n$ matrix of the codewords of $\mathcal{C}$. Let $\mathbb{F}_q^*$ be the nonzero elements of $\mathbb{F}_q$. For $1 \le i \le n$ and $\alpha \in \mathbb{F}_q^*$, let $\mathcal{C}_i(\alpha)$ be the codewords in $\mathcal{C}$ which have α in column i. Suppose that $\mathcal{C}_i(\alpha)$ has $m_{i,\alpha}$ codewords. The expression $\sum_{i=1}^{n} m_{i,\alpha}$ counts the number of times α occurs in the matrix $\mathcal{M}$. Therefore $\sum_{\alpha \in \mathbb{F}_q^*} \sum_{i=1}^{n} m_{i,\alpha}$ counts the number of nonzero entries in $\mathcal{M}$. Since $\mathcal{C}$ is a constant weight code,

$$\sum_{\alpha \in \mathbb{F}_q^*} \sum_{i=1}^{n} m_{i,\alpha} = wM.$$

But if you puncture $C_i(\alpha)$ on coordinate i, you obtain an $(n-1, m_{i,\alpha}, d)$ code with codewords of weight $w-1$. Thus $m_{i,\alpha} \le A_q(n-1, d, w-1)$, yielding

$$wM = \sum_{\alpha \in \mathbb{F}_q^*} \sum_{i=1}^{n} m_{i,\alpha} \le q^* n A_q(n-1, d, w-1).$$

Therefore,

$$A_q(n, d, w) \le \left\lfloor \frac{nq^*}{w} A_q(n-1, d, w-1) \right\rfloor. \tag{2.11}$$

By induction, repeatedly using (2.11),

$$A_q(n, d, w) \le \left\lfloor \frac{nq^*}{w} \left\lfloor \frac{(n-1)q^*}{w-1} \left\lfloor \cdots \left\lfloor \frac{(n-i+1)q^*}{w-i+1} A_q(n-i, d, w-i) \right\rfloor \cdots \right\rfloor \right\rfloor \right\rfloor$$

for any i. If $d = 2e - 1$, let $i = w - e + 1$; then $A_q(n-i, d, w-i) = A_q(n-w+e-1, 2e-1, e-1) = 1$ by Theorem 2.3.3(i) and part (ii) holds in this case. If $d = 2e$, let $i = w - e$; then $A_q(n-i, d, w-i) = A_q(n-w+e, 2e, e) \le \lfloor ((n-w+e)q^*)/e \rfloor$ by Theorem 2.3.3(ii) and part (ii) again holds in this case.

Parts (iii) and (iv) follow from (i) and (ii) with $d = 2e - 1$ using Theorem 2.3.3(iv). □

Example 2.3.7 In Example 2.3.2 we showed that $A_2(24, 8, 8) \ge 759$. The Restricted Johnson Bound cannot be used to obtain an upper bound on $A_2(24, 8, 8)$, but the Unrestricted Johnson Bound can. By this bound,

$$A_2(24, 8, 8) \le \left\lfloor \frac{24}{8} \left\lfloor \frac{23}{7} \left\lfloor \frac{22}{6} \left\lfloor \frac{21}{5} \left\lfloor \frac{20}{4} \right\rfloor \right\rfloor \right\rfloor \right\rfloor \right\rfloor = 759.$$

Thus $A_2(24, 8, 8) = 759$. ■

Exercise 100 Do the following:
(a) Prove that $A_2(n, d, w) = A_2(n, d, n-w)$. Hint: If C is a binary constant weight code of length n with all codewords of weight w, what is the code $\mathbf{1} + C$?
(b) Prove that

$$A_2(n, d, w) \le \left\lfloor \frac{n}{n-w} A_2(n-1, d, w) \right\rfloor.$$

Hint: Use (a) and (2.11).
(c) Show directly that $A_2(7, 4, 6) = 1$.
(d) Show using parts (b) and (c) that $A_2(8, 4, 6) \le 4$. Construct a binary constant weight code of length 8 with four codewords of weight 6 and all with distance at least 4 apart, thus showing that $A_2(8, 4, 6) = 4$.
(e) Use parts (b) and (d) to prove that $A_2(9, 4, 6) \le 12$.
(f) What are the bounds on $A_2(9, 4, 6)$ and $A_2(9, 4, 3)$ using the Unrestricted Johnson Bound? Note that $A_2(9, 4, 6) = A_2(9, 4, 3)$ by part (a).
(g) Show that $A_2(9, 4, 6) = 12$. Hint: By part (a), $A_2(9, 4, 6) = A_2(9, 4, 3)$. A binary $(9, 12, 4)$ code with all codewords of weight 3 exists where, for each coordinate, there are exactly four codewords with a 1 in that coordinate. ◆

Exercise 101 Do the following:

(a) Use the techniques given in Exercise 100 to prove that $A_2(8, 4, 5) \leq 8$.

(b) Show that $A_2(8, 4, 5) = 8$. Hint: By Exercise 100, $A_2(8, 4, 5) = A_2(8, 4, 3)$. A binary $(8, 8, 4)$ code with all codewords of weight 3 exists where, for each coordinate, there are exactly three codewords with a 1 in that coordinate.   ◆

2.3.3   The Johnson Bound for $A_q(n, d)$

The bounds on $A_q(n, d, w)$ can be used to give upper bounds on $A_q(n, d)$ also due to Johnson [159]. As can be seen from the proof, these bounds strengthen the Sphere Packing Bound. The idea of the proof is to count not only the vectors in $\mathbb{F}_q^n$ that are within distance $t = \lfloor (d - 1)/2 \rfloor$ of all codewords (that is, the disjoint spheres of radius t centered at codewords) but also the vectors at distance $t + 1$ from codewords that are not within these spheres. To accomplish this we need the following notation. If C is a code of length n over $\mathbb{F}_q$ and $\mathbf{x} \in \mathbb{F}_q^n$, let $d(C, \mathbf{x})$ denote the distance from $\mathbf{x}$ to C. So $d(C, \mathbf{x}) = \min\{d(\mathbf{c}, \mathbf{x}) \mid \mathbf{c} \in C\}$.

Theorem 2.3.8 (Johnson Bound for $A_q(n, d)$) *Let* $t = \lfloor (d - 1)/2 \rfloor$.

(i) *If d is odd, then*

$$A_q(n, d) \leq \frac{q^n}{\displaystyle\sum_{i=0}^{t} \binom{n}{i}(q - 1)^i + \dfrac{\dbinom{n}{t+1}(q - 1)^{t+1} - \dbinom{d}{t}A_q(n, d, d)}{A_q(n, d, t + 1)}}.$$

(ii) *If d is even, then*

$$A_q(n, d) \leq \frac{q^n}{\displaystyle\sum_{i=0}^{t} \binom{n}{i}(q - 1)^i + \dfrac{\dbinom{n}{t+1}(q - 1)^{t+1}}{A_q(n, d, t + 1)}}.$$

(iii) *If d is odd, then*

$$A_2(n, d) \leq \frac{2^n}{\displaystyle\sum_{i=0}^{t} \binom{n}{i} + \dfrac{\dbinom{n}{t+1} - \dbinom{d}{t}A_2(n, d, d)}{\left\lfloor \dfrac{n}{t + 1} \right\rfloor}}.$$

(iv) *If d is even, then*

$$A_2(n, d) \leq \frac{2^n}{\displaystyle\sum_{i=0}^{t} \binom{n}{i} + \dfrac{\dbinom{n}{t+1}}{\left\lfloor \dfrac{n}{t + 1} \right\rfloor}}. \tag{2.12}$$

(v) *If d is odd, then*

$$A_2(n,d) \le \cfrac{2^n}{\sum_{i=0}^{t} \binom{n}{i} + \cfrac{\binom{n}{t}\left(\frac{n-t}{t+1} - \left\lfloor \frac{n-t}{t+1} \right\rfloor\right)}{\left\lfloor \frac{n}{t+1} \right\rfloor}}. \tag{2.13}$$

Proof: Let C be an (n, M, d) code over $\mathbb{F}_q$. Notice that t is the packing radius of C; $d = 2t + 1$ if d is odd and $d = 2t + 2$ if d is even. So the spheres of radius t centered at codewords are disjoint. The vectors in these spheres are precisely the vectors in $\mathbb{F}_q^n$ that are distance t or less from C. We will count these vectors together with those vectors at distance $t + 1$ from C and use this count to obtain our bounds. To that end let $\mathcal{N}$ be the vectors at distance $t + 1$ from C; so $\mathcal{N} = \{\mathbf{x} \in \mathbb{F}_q^n \mid d(C, \mathbf{x}) = t + 1\}$. Let $|\mathcal{N}|$ denote the size of $\mathcal{N}$. Therefore,

$$M \sum_{i=0}^{t} \binom{n}{i}(q-1)^i + |\mathcal{N}| \le q^n, \tag{2.14}$$

as the summation on the left-hand side counts the vectors in the spheres of radius t centered at codewords; see (1.11). Our bounds will emerge after we obtain a lower bound on $|\mathcal{N}|$.

Let $\mathcal{X} = \{(\mathbf{c}, \mathbf{x}) \in C \times \mathcal{N} \mid d(\mathbf{c}, \mathbf{x}) = t + 1\}$. To get the lower bound on $|\mathcal{N}|$ we obtain lower and upper estimates for $|\mathcal{X}|$.

We first obtain a lower estimate on $|\mathcal{X}|$. Let $\mathcal{X}_\mathbf{c} = \{\mathbf{x} \in \mathcal{N} \mid (\mathbf{c}, \mathbf{x}) \in \mathcal{X}\}$. Then

$$|\mathcal{X}| = \sum_{\mathbf{c} \in C} |\mathcal{X}_\mathbf{c}|. \tag{2.15}$$

Fix $\mathbf{c} \in C$. Let $\mathbf{x} \in \mathbb{F}_q^n$ be a vector at distance $t + 1$ from $\mathbf{c}$ so that $\text{wt}(\mathbf{c} - \mathbf{x}) = t + 1$. There are exactly

$$\binom{n}{t+1}(q-1)^{t+1}$$

such vectors $\mathbf{x}$ because they are obtained by freely changing any $t + 1$ coordinates of $\mathbf{c}$. Some of these lie in $\mathcal{X}_\mathbf{c}$ and some do not. Because $\text{wt}(\mathbf{c} - \mathbf{x}) = t + 1$, $d(C, \mathbf{x}) \le t + 1$. Let $\mathbf{c}' \in C$ with $\mathbf{c}' \ne \mathbf{c}$. Then by the triangle inequality of Theorem 1.4.1, $d \le \text{wt}(\mathbf{c}' - \mathbf{c}) = \text{wt}(\mathbf{c}' - \mathbf{x} - (\mathbf{c} - \mathbf{x})) \le \text{wt}(\mathbf{c}' - \mathbf{x}) + \text{wt}(\mathbf{c} - \mathbf{x}) = \text{wt}(\mathbf{c}' - \mathbf{x}) + t + 1$, implying

$$d - t - 1 \le \text{wt}(\mathbf{c}' - \mathbf{x}). \tag{2.16}$$

If $d = 2t + 2$, $\text{wt}(\mathbf{c}' - \mathbf{x}) \ge t + 1$ yielding $d(C, \mathbf{x}) = t + 1$ since $\mathbf{c}' \in C$ was arbitrary and we saw previously that $d(C, \mathbf{x}) \le t + 1$. Therefore all such $\mathbf{x}$ lie in $\mathcal{X}_\mathbf{c}$ giving

$$|\mathcal{X}_\mathbf{c}| = \binom{n}{t+1}(q-1)^{t+1};$$

hence

$$|\mathcal{X}| = M \binom{n}{t+1}(q-1)^{t+1} \qquad \text{if } d = 2t + 2. \tag{2.17}$$

If $d = 2t + 1$, $\mathrm{wt}(\mathbf{c}' - \mathbf{x}) \geq t$ by (2.16) yielding $t \leq d(\mathcal{C}, \mathbf{x}) \leq t + 1$. As we only want to count the $\mathbf{x}$ where $d(\mathcal{C}, \mathbf{x}) = t + 1$, we will throw away those with $d(\mathcal{C}, \mathbf{x}) = t$. Such $\mathbf{x}$ must simultaneously be at distance t from some codeword $\mathbf{c}' \in \mathcal{C}$ and at distance $t + 1$ from $\mathbf{c}$. Hence the distance from $\mathbf{c}'$ to $\mathbf{c}$ is at most $2t + 1 = d$, by the triangle inequality; this distance must also be at least d as that is the minimum distance of $\mathcal{C}$. Therefore we have $\mathrm{wt}(\mathbf{c}' - \mathbf{c}) = 2t + 1$. How many $\mathbf{c}'$ are possible? As the set $\{\mathbf{c}' - \mathbf{c} \mid \mathbf{c}' \in \mathcal{C}\}$ forms a constant weight code of length n and minimum distance d, whose codewords have weight d, there are at most $A_q(n, d, d)$ such $\mathbf{c}'$. For each $\mathbf{c}'$, how many $\mathbf{x}$ are there with $\mathrm{wt}(\mathbf{x} - \mathbf{c}) = t + 1$ and $t = \mathrm{wt}(\mathbf{c}' - \mathbf{x}) = \mathrm{wt}((\mathbf{c}' - \mathbf{c}) - (\mathbf{x} - \mathbf{c}))$? Since $\mathrm{wt}(\mathbf{c}' - \mathbf{c}) = 2t + 1$, $\mathbf{x} - \mathbf{c}$ is obtained from $\mathbf{c}' - \mathbf{c}$ by arbitrarily choosing t of its $2t + 1$ nonzero components and making them zero. This can be done in $\binom{d}{t}$ ways. Therefore

$$\binom{n}{t+1}(q-1)^{t+1} - \binom{d}{t}A_q(n, d, d) \leq |\mathcal{X}_\mathbf{c}|$$

showing by (2.15) that

$$M\left[\binom{n}{t+1}(q-1)^{t+1} - \binom{d}{t}A_q(n, d, d)\right] \leq |\mathcal{X}| \qquad \text{if } d = 2t + 1. \tag{2.18}$$

We are now ready to obtain our upper estimate on $|\mathcal{X}|$. Fix $\mathbf{x} \in \mathcal{N}$. How many $\mathbf{c} \in \mathcal{C}$ are there with $d(\mathbf{c}, \mathbf{x}) = t + 1$? The set $\{\mathbf{c} - \mathbf{x} \mid \mathbf{c} \in \mathcal{C}$ with $d(\mathbf{c}, \mathbf{x}) = t + 1\}$ is a constant weight code of length n with words of weight $t + 1$ and minimum distance d because $(\mathbf{c}' - \mathbf{x}) - (\mathbf{c} - \mathbf{x}) = \mathbf{c}' - \mathbf{c}$. Thus for each $\mathbf{x} \in \mathcal{N}$ there are at most $A_q(n, d, t + 1)$ choices for $\mathbf{c}$ with $d(\mathbf{c}, \mathbf{x}) = t + 1$. Hence $|\mathcal{X}| \leq |\mathcal{N}|A_q(n, d, t + 1)$ or

$$\frac{|\mathcal{X}|}{A_q(n, d, t + 1)} \leq |\mathcal{N}|. \tag{2.19}$$

We obtain bound (i) by combining (2.14), (2.18), and (2.19) and bound (ii) by combining (2.14), (2.17), and (2.19). Bounds (iii) and (iv) follow from (i) and (ii) by observing that $A_2(n, 2t + 1, t + 1) = A_2(n, 2t + 2, t + 1) = \lfloor n/(t + 1) \rfloor$ by Theorem 2.3.3. Finally, bound (v) follows from (iii) and the observation (with details left as Exercise 102) that

$$\binom{d}{t}A_2(n, d, d) = \binom{d}{t}A_2(n, 2t + 1, 2t + 1)$$

$$= \binom{d}{t}A_2(n, 2t + 2, 2t + 1) \leq \binom{n}{t}\left\lfloor\frac{n - t}{t + 1}\right\rfloor$$

by Theorem 2.3.3(iv) and the Unrestricted Johnson Bound. $\qquad\square$

Exercise 102 Show that

$$\binom{d}{t}A_2(n, 2t + 2, 2t + 1) \leq \binom{n}{t}\left\lfloor\frac{n - t}{t + 1}\right\rfloor,$$

using the Unrestricted Johnson Bound. $\qquad\blacklozenge$

Example 2.3.9 Using (2.12) we compute that an upper bound for $A_2(16, 6)$ is 263. Recall that $A_2(16, 6) = A_2(15, 5)$. Using (2.13), we discover that $A_2(15, 5)$, and hence $A_2(16, 6)$, is bounded above by 256. In the next subsection, we present a code that meets this bound. ∎

Exercise 103 Compute the best possible upper bounds for $A_2(9, 4) = A_2(8, 3)$ and $A_2(13, 6) = A_2(12, 5)$ using the Johnson Bound. Compare these values to those in Table 2.1. ◆

2.3.4 The Nordstrom–Robinson code

The existence of the Nordstrom–Robinson code shows that the upper bound on $A_2(16, 6)$ discovered in Example 2.3.9 is met, and hence that $A_2(16, 6) = 256$.

The Nordstrom–Robinson code was discovered by Nordstrom and Robinson [247] and later independently by Semakov and Zinov'ev [303]. This code can be defined in several ways; one of the easiest is the following and is due to Goethals [100] and Semakov and Zinov'ev [304].

Let C be the [24, 12, 8] extended binary Golay code chosen to contain the weight 8 codeword $\mathbf{c} = 11 \cdots 100 \cdots 0$. Let T be the set consisting of the first eight coordinates. Let $C(T)$ be the subcode of C which is zero on T, and let C_T be C shortened on T. Let $C^{\overline{T}}$ be C punctured on the positions of $\overline{T} = \{9, 10, \ldots, 24\}$. By Corollary 1.4.14, as C is self-dual, the first seven coordinate positions of C are linearly independent. Thus as $\mathbf{c} \in C$ and C is self-dual, $C^{\overline{T}}$ is the [8, 7, 2] binary code consisting of all even weight vectors of length 8. Exercise 104 shows that the dimension of C_T is 5. Hence C_T is a [16, 5, 8] code. (In fact C_T is equivalent to $\mathcal{R}(1, 4)$ as Exercise 121 shows.) For $1 \le i \le 7$, let $\mathbf{c}_i \in C$ be a codeword of C with zeros in the first eight coordinates except coordinate i and coordinate 8; such codewords are present in C because $C^{\overline{T}}$ is all length 8 even weight vectors. Let $\mathbf{c}_0 = \mathbf{0}$. For $0 \le j \le 7$, let C_j be the coset $\mathbf{c}_j + C(T)$ of $C(T)$ in the extended Golay code C. These cosets are distinct, as you can verify in Exercise 105. Let $\mathcal{N}$ be the union of the eight cosets $C_0, \ldots, C_7$. The *Nordstrom–Robinson code* $\mathcal{N}_{16}$ is the code obtained by puncturing $\mathcal{N}$ on T. Thus $\mathcal{N}_{16}$ is the union of $C_0^*, \ldots, C_7^*$, where C_j^* is C_j punctured on T. Figure 2.1 gives a picture of the construction. Clearly, $\mathcal{N}_{16}$ is a (16, 256) code, as Exercise 106 shows. Let $\mathbf{a}, \mathbf{b} \in \mathcal{N}$ be distinct. Then $d(\mathbf{a}, \mathbf{b}) \ge 8$, as C has minimum distance 8. Since $\mathbf{a}$ and $\mathbf{b}$

	T	$\mathcal{N}_{16}$
$C_0\{$	0 0 0 0 0 0 0 0	32 codewords of C_0^*
$C_1\{$	1 0 0 0 0 0 0 1	32 codewords of C_1^*
$C_2\{$	0 1 0 0 0 0 0 1	32 codewords of C_2^*
$\vdots$	$\vdots$	$\vdots$
$C_7\{$	0 0 0 0 0 0 1 1	32 codewords of C_7^*

Figure 2.1 The Nordstrom–Robinson code inside the extended Golay code.

disagree on at most two of the first eight coordinates, the codewords of $\mathcal{N}_{16}$ obtained from **a** and **b** by puncturing on T are distance 6 or more apart. Thus $\mathcal{N}_{16}$ has minimum distance at least 6 showing that $A_2(16, 6) = 256$. In particular, $\mathcal{N}_{16}$ is optimal.

Exercise 104 Show that $\mathcal{C}_T$ has dimension 5. Hint: $\mathcal{C} = \mathcal{C}^{\perp}$; apply Theorem 1.5.7(iii).
♦

Exercise 105 Show that the cosets $\mathcal{C}_j$ for $0 \le j \le 7$ are distinct.   ♦

Exercise 106 Show that $\mathcal{N}_{16}$ has 256 codewords as claimed.   ♦

We compute the weight distribution $A_i(\mathcal{N}_{16})$ of $\mathcal{N}_{16}$. Clearly

$$\sum_{j=0}^{7} A_i(\mathcal{C}_j^*) = A_i(\mathcal{N}_{16}). \tag{2.20}$$

By Theorem 1.4.5(iv), $\mathcal{C}$ contains the all-one codeword **1**. Hence as $\mathbf{c} + \mathbf{1}$ has 0s in the first eight coordinates and 1s in the last 16, $\mathcal{C}_0^*$ contains the all-one vector of length 16. By Exercise 107 $A_{16-i}(\mathcal{C}_j^*) = A_i(\mathcal{C}_j^*)$ for $0 \le i \le 16$ and $0 \le j \le 7$. As $\mathcal{C}_j^*$ is obtained from $\mathcal{C}$ by deleting eight coordinates on which the codewords have even weight, $A_i(\mathcal{C}_j^*) = 0$ if i is odd. By construction $\mathcal{N}_{16}$ contains **0**. As $\mathcal{N}_{16}$ has minimum distance 6, we deduce that $A_i(\mathcal{N}_{16}) = 0$ for $1 \le i \le 5$ and therefore that $A_i(\mathcal{C}_j^*) = 0$ for $1 \le i \le 5$ and $11 \le i \le 15$. As $\mathcal{C}_0^* = \mathcal{C}_T$ and the weights of codewords in $\mathcal{C}(T)$ are multiples of 4, so are the weights of vectors in $\mathcal{C}_0^*$. Since $\mathcal{C}_0^*$ has 32 codewords, $A_0(\mathcal{C}_0^*) = A_{16}(\mathcal{C}_0^*) = 1$ and $A_8(\mathcal{C}_0^*) = 30$, the other $A_i(\mathcal{C}_0^*)$ being 0. For $1 \le j \le 7$, the codewords in $\mathcal{C}_j$ have weights a multiple of 4; since each codeword has two 1s in the first eight coordinates, the vectors in $\mathcal{C}_j^*$ have weights that are congruent to 2 modulo 4. Therefore the only possible weights of vectors in $\mathcal{C}_j$ are 6 and 10, and since $A_6(\mathcal{C}_j^*) = A_{10}(\mathcal{C}_j^*)$, these both must be 16. Therefore by (2.20), $A_0(\mathcal{N}_{16}) = A_{16}(\mathcal{N}_{16}) = 1$, $A_6(\mathcal{N}_{16}) = A_{10}(\mathcal{N}_{16}) = 7 \cdot 16 = 112$, and $A_8(\mathcal{N}_{16}) = 30$, the other $A_i(\mathcal{N}_{16})$ being 0.

Exercise 107 Prove that $A_{16-i}(\mathcal{C}_j^*) = A_i(\mathcal{C}_j^*)$ for $0 \le i \le 16$ and $0 \le j \le 7$.   ♦

It turns out that $\mathcal{N}_{16}$ is unique [317] in the following sense. If $\mathcal{C}$ is any binary $(16, 256, 6)$ code, and **c** is a codeword of $\mathcal{C}$, then the code $\mathbf{c} + \mathcal{C} = \{\mathbf{c} + \mathbf{x} \mid \mathbf{x} \in \mathcal{C}\}$ is also a $(16, 256, 6)$ code containing the zero vector (see Exercise 86) and this code is equivalent to $\mathcal{N}_{16}$.

Exercise 108 From $\mathcal{N}_{16}$, produce $(15, 128, 6)$, $(14, 64, 6)$, and $(13, 32, 6)$ codes. Note that these codes are optimal; see Table 2.1 and Exercise 92.   ♦

2.3.5 Nearly perfect binary codes

We explore the case when bound (2.13) is met. This bound strengthens the Sphere Packing Bound and the two bounds in fact agree precisely when $(t + 1) \mid (n - t)$. Recall that codes that meet the Sphere Packing Bound are called perfect. An $(n, M, 2t + 1)$ binary code with $M = A_2(n, 2t + 1)$ which attains the Johnson Bound (2.13) is called *nearly perfect*.

A natural problem is to classify the nearly perfect codes. As just observed, the Johnson Bound strengthens the Sphere Packing Bound, and so perfect codes are nearly perfect (and $(t + 1) \mid (n - t)$); see Exercise 109. Nearly perfect codes were first examined by Semakov, Zinov'ev, and Zaitsev [305] and independently by Goethals and Snover [101]. The next two examples, found in [305, Theorem 1], give parameters for other nearly perfect codes.

Exercise 109 As stated in the text, because the Johnson Bound strengthens the Sphere Packing Bound, perfect codes are nearly perfect. Fill in the details showing why this is true.   ♦

Example 2.3.10 Let C be an $(n, M, 3)$ nearly perfect code. So $t = 1$. If n is odd, $(t + 1) \mid (n - t)$ and so the Sphere Packing Bound and (2.13) agree. Thus C is a perfect single error-correcting code and must have the parameters of $\mathcal{H}_{2,r}$ by Theorem 1.12.3. (We do not actually need Theorem 1.12.3 because the Sphere Packing Bound gives $M = 2^n/(1 + n)$; so $n = 2^r - 1$ for some r and $M = 2^{2^r - 1 - r}$.) If n is even, equality in (2.13) produces

$$M = \frac{2^n}{n + 2}.$$

Hence M is an integer if and only if $n = 2^r - 2$ for some integer r. Therefore the only possible sets of parameters for nearly perfect $(n, M, 3)$ codes that are not perfect are $(n, M, 3) = (2^r - 2, 2^{2^r - 2 - r}, 3)$ for $r \geq 3$. For example, the code obtained by puncturing the subcode of even weight codewords in $\mathcal{H}_{2,r}$ is a linear code having these parameters. (See Exercise 110.)   ∎

Exercise 110 Prove that the code obtained by puncturing the subcode of even weight codewords in $\mathcal{H}_{2,r}$ is a linear code having parameters $(n, M, 3) = (2^r - 2, 2^{2^r - 2 - r}, 3)$.   ♦

Example 2.3.11 Let C be an $(n, M, 5)$ nearly perfect code. So $t = 2$. If $n \equiv 2 \pmod{3}$, $(t + 1) \mid (n - t)$ and again the Sphere Packing Bound and (2.13) agree. Thus C is a perfect double error-correcting code which does not exist by Theorem 1.12.3. If $n \equiv 1 \pmod{3}$, equality in (2.13) yields

$$M = \frac{2^{n+1}}{(n + 2)(n + 1)}.$$

So for M to be an integer, both $n + 1$ and $n + 2$ must be powers of 2, which is impossible for $n \geq 1$. Finally, consider the case $n \equiv 0 \pmod{3}$. Equality in (2.13) gives

$$M = \frac{2^{n+1}}{(n + 1)^2}.$$

So $n = 2^m - 1$ for some m, and as $3 \mid n$, m must be even. Thus C is a $(2^m - 1, 2^{2^m - 2m}, 5)$ code, a code that has the same parameters as the punctured Preparata code $\mathcal{P}(m)^*$, which we will describe in Chapter 12.   ∎

These two examples provide the initial steps in the classification of the nearly perfect codes, a work begun by Semakov, Zinov'ev, and Zaitsev [305] and completed by Lindström [199, 200]; one can also define nearly perfect nonbinary codes. These authors show that all of the nearly perfect binary codes are either perfect, or have parameters of either the codes in Example 2.3.10 or the punctured Preparata codes in Example 2.3.11, and that all nearly perfect nonbinary codes must be perfect.

2.4 The Singleton Upper Bound and MDS codes

The next upper bound for $A_q(n, d)$ and $B_q(n, d)$, called the Singleton Bound, is much simpler to prove than the previous upper bounds. It is a rather weak bound in general but does lead to the class of codes called MDS codes; this class contains the very important family of codes known as Reed–Solomon codes, which are generally very useful in many applications. They correct burst errors and provide the high fidelity in CD players.

Theorem 2.4.1 (Singleton Bound [312]) *For $d \leq n$,*

$$A_q(n, d) \leq q^{n-d+1}.$$

Furthermore if an $[n, k, d]$ linear code over $\mathbb{F}_q$ exists, then $k \leq n - d + 1$.

Proof: The second statement follows from the first by Theorem 2.1.1. Recall that $A_q(n, n) = q$ by Theorem 2.1.5 yielding the bound when $d = n$. Now assume that $d < n$. By Theorem 2.1.6 $A_q(n, d) \leq q A_q(n - 1, d)$. Inductively we have that $A_q(n, d) \leq q^{n-d} A_q(d, d)$. Since $A_q(d, d) = q$, $A_q(n, d) \leq q^{n-d+1}$. $\qquad\square$

Example 2.4.2 The hexacode of Example 1.3.4 is a $[6, 3, 4]$ linear code over $\mathbb{F}_4$. In this code, $k = 3 = 6 - 4 + 1 = n - d + 1$ and the Singleton Bound is met. So $A_4(6, 4) = 4^3$. $\qquad\blacksquare$

Exercise 111 Prove using either the parity check matrix or the standard form of the generator matrix for an $[n, k, d]$ linear code that $d \leq n - k + 1$, hence verifying directly the linear version of the Singleton Bound. $\qquad\blacklozenge$

A code for which equality holds in the Singleton Bound is called *maximum distance separable*, abbreviated *MDS*. No code of length n and minimum distance d has more codewords than an MDS code with parameters n and d; equivalently, no code of length n with M codewords has a larger minimum distance than an MDS code with parameters n and M. We briefly discuss some results on linear MDS codes.

Theorem 2.4.3 *Let C be an $[n, k]$ code over $\mathbb{F}_q$ with $k \geq 1$. Then the following are equivalent:*
(i) *C is MDS.*
(ii) *Every set of k coordinates is an information set for C.*
(iii) *$C^\perp$ is MDS.*
(iv) *Every set of $n - k$ coordinates is an information set for $C^\perp$.*

Proof: The first two statements are equivalent by Theorem 1.4.15 as an $[n, k]$ code is MDS if and only if $k = n - d + 1$. Similarly the last two are equivalent. Finally, (ii) and (iv) are equivalent by Theorem 1.6.2. $\qquad\square$

We say that C is a *trivial* MDS code over $\mathbb{F}_q$ if and only if $C = \mathbb{F}_q^n$ or C is monomially equivalent to the code generated by **1** or its dual. By examining the generator matrix in standard form, it is straightforward to verify the following result about binary codes.

Theorem 2.4.4 *Let C be an $[n, k, d]$ binary code.*
(i) *If C is MDS, then C is trivial.*
(ii) *If $3 \leq d$ and $5 \leq k$, then $k \leq n - d - 1$.*

Exercise 112 Prove Theorem 2.4.4.   ◆

We will discuss other aspects of MDS codes in Section 7.4. Trivial MDS codes are arbitrarily long. Examples of nontrivial MDS codes are Reed–Solomon codes over $\mathbb{F}_q$ of length $n = q - 1$ and extensions of these codes of lengths q and $q + 1$. Reed–Solomon codes and their generalizations will be examined in Chapter 5. The weight distribution of an MDS code over $\mathbb{F}_q$ is determined by its parameters n, k, and q (see Theorem 7.4.1). If an MDS code is nontrivial, its length is bounded as a function of q and k (see Corollary 7.4.4).

2.5   The Elias Upper Bound

The ideas of the proof of the Plotkin Bound can be used to find a bound that applies to a larger range of minimum distances. This extension was discovered in 1960 by Elias but he did not publish it. Unfortunately, the Elias Bound is sometimes rather weak; see Exercises 114 and 115. However, the Elias Bound is important because the asymptotic form of this bound, derived later in this chapter, is superior to all of the asymptotic bounds we discuss except the MRRW Bounds. Before stating the Elias Bound, we need two lemmas.

Lemma 2.5.1 *Let C be an (n, K, d) code over $\mathbb{F}_q$ such that all codewords have weights at most w, where $w \leq rn$ with $r = 1 - q^{-1}$. Then*

$$d \leq \frac{Kw}{K - 1} \left(2 - \frac{w}{rn} \right).$$

Proof: As in the proof of the Plotkin Bound, let $\mathcal{M}$ be the $K \times n$ matrix whose rows are the codewords of C. For $1 \leq i \leq n$, let $n_{i,\alpha}$ be the number of times $\alpha \in \mathbb{F}_q$ occurs in column i of $\mathcal{M}$. Clearly,

$$\sum_{\alpha \in \mathbb{F}_q} n_{i,\alpha} = K \qquad \text{for } 1 \leq i \leq n. \tag{2.21}$$

Also, if $T = \sum_{i=1}^{n} n_{i,0}$, then

$$T = \sum_{i=1}^{n} n_{i,0} \geq K(n - w) \tag{2.22}$$

as every row of $\mathcal{M}$ has at least $n - w$ zeros. By the Cauchy–Schwartz inequality and (2.21)

$$\sum_{\alpha \in \mathbb{F}_q^*} n_{i,\alpha}^2 \geq \frac{1}{q-1} \left(\sum_{\alpha \in \mathbb{F}_q^*} n_{i,\alpha} \right)^2 = \frac{1}{q-1}(K - n_{i,0})^2 \qquad \text{and} \qquad (2.23)$$

$$\sum_{i=1}^{n} n_{i,0}^2 \geq \frac{1}{n} \left(\sum_{i=1}^{n} n_{i,0} \right)^2 = \frac{1}{n} T^2. \qquad (2.24)$$

Again, exactly as in the proof of the Plotkin Bound, using (2.3) and (2.4),

$$K(K-1)d \leq \sum_{\mathbf{x} \in C} \sum_{\mathbf{y} \in C} d(\mathbf{x}, \mathbf{y}) = \sum_{i=1}^{n} \sum_{\alpha \in \mathbb{F}_q} n_{i,\alpha}(K - n_{i,\alpha}). \qquad (2.25)$$

Using first (2.21), then (2.23), and finally (2.24),

$$\sum_{i=1}^{n} \sum_{\alpha \in \mathbb{F}_q} n_{i,\alpha}(K - n_{i,\alpha}) = nK^2 - \sum_{i=1}^{n} \left(n_{i,0}^2 + \sum_{\alpha \in \mathbb{F}_q^*} n_{i,\alpha}^2 \right)$$

$$\leq nK^2 - \frac{1}{q-1} \sum_{i=1}^{n} \left(qn_{i,0}^2 + K^2 - 2Kn_{i,0} \right)$$

$$\leq nK^2 - \frac{1}{q-1} \left(\frac{q}{n} T^2 + nK^2 - 2KT \right). \qquad (2.26)$$

Since $w \leq rn$, $qw \leq (q-1)n$ implying $n \leq qn - qw$ and hence

$$K \leq \frac{q}{n} K(n - w). \qquad (2.27)$$

Also as $w \leq rn$, by (2.22)

$$K \leq \frac{q}{n} T. \qquad (2.28)$$

Adding (2.27) and (2.28) gives $2K \leq qn^{-1}(T + K(n - w))$. Multiplying both sides by $T - K(n - w)$, which is nonnegative by (2.22), produces

$$2K[T - K(n - w)] \leq \frac{q}{n}[T^2 - K^2(n - w)^2]$$

and hence

$$\frac{q}{n} K^2(n - w)^2 - 2K^2(n - w) \leq \frac{q}{n} T^2 - 2KT.$$

Substituting this into (2.26) and using (2.25) yields

$$K(K-1)d \leq nK^2 - \frac{1}{q-1} \left[\frac{q}{n} K^2(n-w)^2 + nK^2 - 2K^2(n-w) \right].$$

Simplifying the right-hand side produces

$$K(K-1)d \leq K^2 w \left(2 - \frac{q}{q-1} \frac{w}{n} \right) = K^2 w \left(2 - \frac{w}{rn} \right).$$

Solving for d verifies the lemma. $\qquad\qquad\qquad\qquad\qquad\qquad\qquad\square$

By (1.11) the number of vectors in a sphere of radius a in $\mathbb{F}_q^n$ centered at some vector in $\mathbb{F}_q^n$, denoted $V_q(n, a)$, is

$$V_q(n, a) = \sum_{i=0}^{a} \binom{n}{i}(q - 1)^i. \tag{2.29}$$

Lemma 2.5.2 *Suppose C is an (n, M, d) code over $\mathbb{F}_q$. Then there is an (n, M, d) code C' over $\mathbb{F}_q$ with an (n, K, d) subcode $\mathcal{A}$ containing only codewords of weight at most w such that $K \geq MV_q(n, w)/q^n$.*

Proof: Let $S_w(\mathbf{0})$ be the sphere in $\mathbb{F}_q^n$ of radius w centered at $\mathbf{0}$. Let $\mathbf{x} \in \mathbb{F}_q^n$ be chosen so that $|S_w(\mathbf{0}) \cap (\mathbf{x} + C)|$ is maximal. Then

$$|S_w(\mathbf{0}) \cap (\mathbf{x} + C)| \geq \frac{1}{q^n} \sum_{\mathbf{y} \in \mathbb{F}_q^n} |S_w(\mathbf{0}) \cap (\mathbf{y} + C)|$$

$$= \frac{1}{q^n} \sum_{\mathbf{y} \in \mathbb{F}_q^n} \sum_{\mathbf{b} \in S_w(\mathbf{0})} \sum_{\mathbf{c} \in C} |\{\mathbf{b}\} \cap \{\mathbf{y} + \mathbf{c}\}|$$

$$= \frac{1}{q^n} \sum_{\mathbf{b} \in S_w(\mathbf{0})} \sum_{\mathbf{c} \in C} 1 = \frac{1}{q^n}|S_w(\mathbf{0})||C| = \frac{1}{q^n}V_q(n, w)M.$$

The result follows by letting $C' = \mathbf{x} + C$ and $\mathcal{A} = S_w(\mathbf{0}) \cap C'$. □

Theorem 2.5.3 (Elias Bound) *Let $r = 1 - q^{-1}$. Suppose that $w \leq rn$ and $w^2 - 2rnw + rnd > 0$. Then*

$$A_q(n, d) \leq \frac{rnd}{w^2 - 2rnw + rnd} \cdot \frac{q^n}{V_q(n, w)}.$$

Proof: Let $M = A_q(n, d)$. By Lemma 2.5.2 there is an (n, K, d) code over $\mathbb{F}_q$ containing only codewords of weight at most w such that

$$MV_q(n, w)/q^n \leq K. \tag{2.30}$$

As $w \leq rn$, Lemma 2.5.1 implies that $d \leq Kw(2 - w/(rn))/(K - 1)$. Solving for K and using $w^2 - 2rnw + rnd > 0$ yields

$$K \leq \frac{rnd}{w^2 - 2rnw + rnd}. \tag{2.31}$$

Putting (2.30) and (2.31) together gives the bound. □

Example 2.5.4 By Theorem 2.1.2, $A_2(13, 5) = A_2(14, 6)$. By the Sphere Packing Bound, $A_2(13, 5) \leq 2048/23$ implying $A_2(13, 5) \leq 89$, and $A_2(14, 6) \leq 8192/53$ implying $A_2(14, 6) \leq 154$. The Johnson Bound yields $A_2(13, 5) \leq 8192/105$, showing $A_2(13, 5) \leq 78$; and $A_2(14, 6) \leq 16\,384/197$, showing $A_2(14, 6) \leq 83$. The following table gives the upper bounds on $A_2(13, 5)$ and $A_2(14, 6)$ using the Elias Bound. Note that each $w \leq n/2$

such that $w^2 - nw + (nd/2) > 0$ for $(n, d) = (13, 5)$ and $(n, d) = (14, 6)$ must be tried.

w	$A_2(13, 5)$	$A_2(14, 6)$
0	8192	16 384
1	927	1581
2	275	360
3	281	162
4		233

Thus the best upper bound from the Elias Bound for $A_2(13, 5) = A_2(14, 6)$ is 162, while the best bound from the Sphere Packing and Johnson Bounds is 78. By Table 2.1, $A_2(13, 5) = A_2(14, 6) = 64$; see also Exercises 92 and 108. ∎

Exercise 113   Verify that the entries in the table of Example 2.5.4 are correct and that all possible values of w have been examined. ◆

Exercise 114   By Theorem 2.1.2, $A_2(9, 5) = A_2(10, 6)$.
(a) Compute upper bounds on both $A_2(9, 5)$ and $A_2(10, 6)$ using the Sphere Packing Bound, the Plotkin Bound, and the Elias Bound. When computing the Elias Bound make sure all possible values of w have been checked.
(b) What is the best upper bound for $A_2(9, 5) = A_2(10, 6)$?
(c) Find a binary code of length 10 and minimum distance 6 meeting the bound in part (b). Hint: This can be constructed using the zero vector with the remaining codewords having weight 6. (Note: This verifies the entry in Table 2.1.) ◆

Exercise 115   Prove that when $w = rn$ the condition $w^2 - 2rnw + rnd > 0$ becomes $rn < d$ and the Elias Bound is weaker than the Plotkin Bound in this case. ◆

2.6   The Linear Programming Upper Bound

The next upper bound that we present is the linear programming bound which uses results of Delsarte [61, 62, 63]. In general, this is the most powerful of the bounds we have presented but, as its name signifies, does require the use of linear programming. In order to present this bound, we introduce two concepts. First, we generalize the notion of weight distribution of a code. The (*Hamming*) *distance distribution* or *inner distribution* of a code C of length n is the list $B_i = B_i(C)$ for $0 \leq i \leq n$, where

$$B_i(C) = \frac{1}{|C|} \sum_{c \in C} |\{v \in C \mid d(v, c) = i\}|.$$

By Exercise 117, the distance distribution and weight distribution of a linear code are identical. In particular, if C is linear, the distance distribution is a list of nonnegative integers; if C is nonlinear, the B_i are nonnegative but need not be integers. We will see the distance distribution again in Chapter 12. Second, we define the *Krawtchouk polynomial* $K_k^{n,q}(x)$

of degree k to be

$$K_k^{n,q}(x) = \sum_{j=0}^{k}(-1)^j(q-1)^{k-j}\binom{x}{j}\binom{n-x}{k-j} \qquad \text{for } 0 \le k \le n.$$

In 1957, Lloyd [206], in his work on perfect codes, was the first to use the Krawtchouk polynomials in connection with coding theory. We will see related applications of Krawtchouk polynomials in Chapters 7 and 12.

Exercise 116 Let $q = 2$. Then $K_k^{n,2}(x) = \sum_{j=0}^{k}(-1)^j\binom{x}{j}\binom{n-x}{k-j}$.
(a) Prove that if w is an integer with $0 \le w \le n$, then

$$K_k^{n,2}(w) = \sum_{j=0}^{n}(-1)^j\binom{w}{j}\binom{n-w}{k-j} = \sum_{j=0}^{w}(-1)^j\binom{w}{j}\binom{n-w}{k-j}.$$

Hint: Observe when some of the binomial coefficients are 0.
(b) Prove that $K_k^{n,2}(w) = (-1)^w K_{n-k}^{n,2}(w)$. Hint: By part (a),

$$K_k^{n,2}(w) = \sum_{j=0}^{w}(-1)^j\binom{w}{j}\binom{n-w}{k-j} \qquad \text{and hence}$$

$$K_{n-k}^{n,2}(w) = \sum_{j=0}^{w}(-1)^j\binom{w}{j}\binom{n-w}{n-k-j}.$$

In one of the summations replace j by $w - j$ and use $\binom{r}{s} = \binom{r}{r-s}$.   ◆

Exercise 117 Let B_i, $0 \le i \le n$, be the distance distribution of an (n, M) code C over $\mathbb{F}_q$.
(a) Prove that $\sum_{i=0}^{n} B_i = M$.
(b) Prove that $B_0 = 1$.
(c) Prove that if $q = 2$, $B_n \le 1$.
(d) Prove that the distance distribution and weight distribution of C are identical if C is linear.   ◆

As with the other upper bounds presented so far, the Linear Programming Bound applies to codes that may be nonlinear. In fact, the alphabet over which the code is defined is not required to be a finite field; the Linear Programming Bound depends only on the code parameters, including the alphabet size, but not the specific alphabet. This is an advantage as the preliminary lemmas needed to derive the bound can most easily be proved if we use $\mathbb{Z}_q$, the integers modulo q, as the alphabet. To facilitate this, define $\alpha: \mathbb{F}_q \to \mathbb{Z}_q$ to be any bijection from $\mathbb{F}_q$ to $\mathbb{Z}_q$ with $\alpha(0) = 0$. Of course if q is a prime, we can choose α to be the identity. If $\mathbf{c} = (c_1, c_2, \ldots, c_n) \in \mathbb{F}_q^n$, define $\alpha(\mathbf{c}) = (\alpha(c_1), \alpha(c_2), \ldots, \alpha(c_n)) \in \mathbb{Z}_q^n$. If C is an (n, M) code over $\mathbb{F}_q$ of length n, define $\alpha(C) = \{\alpha(\mathbf{c}) \mid \mathbf{c} \in C\}$. As $\alpha(0) = 0$, if $\mathbf{c}$ and $\mathbf{v}$ are in $\mathbb{F}_q^n$, then $\mathrm{wt}(\mathbf{c}) = \mathrm{wt}(\alpha(\mathbf{c}))$ and $d(\mathbf{c}, \mathbf{v}) = d(\alpha(\mathbf{c}), \alpha(\mathbf{v}))$ implying that the weight and distance distributions of C are identical to the weight and distance distributions, respectively, of $\alpha(C)$. In particular, when replacing $\mathbb{Z}_q$ with any alphabet with q elements, this discussion shows the following.

Theorem 2.6.1 *There exists an (n, M) code C over $\mathbb{F}_q$ if and only if there exists an (n, M) code over any alphabet with q elements having the same distance distribution as C.*

This theorem shows that any of the bounds on the size of a (possibly) nonlinear code over $\mathbb{F}_q$ that we have derived apply to codes over any alphabet with q elements. As we will see in Chapter 12, codes over $\mathbb{Z}_q$ have been studied extensively.

In $\mathbb{Z}_q^n$ define the ordinary inner product as done over fields, namely, $\mathbf{u} \cdot \mathbf{v} = u_1 v_1 + \cdots + u_n v_n$. As with fields, let $\mathbb{Z}_q^* = \mathbb{Z}_q \setminus \{0\}$.

Lemma 2.6.2 *Let $\xi = e^{2\pi i/q}$ in the complex numbers $\mathbb{C}$, where $i = \sqrt{-1}$. Let $\mathbf{u} \in \mathbb{Z}_q^n$ with* $\mathrm{wt}(\mathbf{u}) = w$. *Then*

$$\sum_{\substack{\mathbf{v} \in \mathbb{Z}_q^n \\ \mathrm{wt}(\mathbf{v})=k}} \xi^{\mathbf{u} \cdot \mathbf{v}} = K_k^{n,q}(w).$$

Proof: Rearrange coordinates so that $\mathbf{u} = u_1 u_2 \cdots u_w 0 \cdots 0$, where $u_m \neq 0$ for $1 \leq m \leq w$. Let $\mathcal{A} = \{a_1, a_2, \ldots, a_k\}$ be a set of k coordinates satisfying

$$1 \leq a_1 < a_2 < \cdots < a_j \leq w < a_{j+1} < \cdots < a_k.$$

Let $\mathcal{S} = \{\mathbf{v} \in \mathbb{Z}_q^n \mid \mathrm{wt}(\mathbf{v}) = k$ with nonzero coordinates exactly in $\mathcal{A}\}$. Then

$$\sum_{\substack{\mathbf{v} \in \mathbb{Z}_q^n \\ \mathrm{wt}(\mathbf{v})=k}} \xi^{\mathbf{u} \cdot \mathbf{v}} = \sum_{j=0}^{k} \sum_{\mathcal{A}} \sum_{\mathbf{v} \in \mathcal{S}} \xi^{\mathbf{u} \cdot \mathbf{v}}. \tag{2.32}$$

The lemma follows if we show that the inner sum $\sum_{\mathbf{v} \in \mathcal{S}} \xi^{\mathbf{u} \cdot \mathbf{v}}$ always equals $(-1)^j (q-1)^{k-j}$ as there are $\binom{w}{j}\binom{n-w}{k-j}$ choices for $\mathcal{A}$.

Before we do this, notice that

$$\sum_{v=1}^{q-1} \xi^{uv} = -1 \qquad \text{if } u \in \mathbb{Z}_q \text{ with } u \neq 0. \tag{2.33}$$

This is because $\sum_{v=0}^{q-1} \xi^{uv} = ((\xi^u)^q - 1)/(\xi^u - 1) = 0$ as $\xi^q = 1$. Examining the desired inner sum of (2.32), we obtain

$$\sum_{\mathbf{v} \in \mathcal{S}} \xi^{\mathbf{u} \cdot \mathbf{v}} = \sum_{v_{a_1} \in \mathbb{Z}_q^*} \sum_{v_{a_2} \in \mathbb{Z}_q^*} \cdots \sum_{v_{a_k} \in \mathbb{Z}_q^*} \xi^{u_{a_1} v_{a_1}} \xi^{u_{a_2} v_{a_2}} \cdots \xi^{u_{a_k} v_{a_k}}$$

$$= (q-1)^{k-j} \sum_{v_{a_1} \in \mathbb{Z}_q^*} \sum_{v_{a_2} \in \mathbb{Z}_q^*} \cdots \sum_{v_{a_j} \in \mathbb{Z}_q^*} \xi^{u_{a_1} v_{a_1}} \xi^{u_{a_2} v_{a_2}} \cdots \xi^{u_{a_j} v_{a_j}}. \tag{2.34}$$

The last equality follows because $u_{a_{j+1}} = u_{a_{j+2}} = \cdots = u_{a_k} = 0$. But using (2.33),

$$\sum_{v_{a_1} \in \mathbb{Z}_q^*} \sum_{v_{a_2} \in \mathbb{Z}_q^*} \cdots \sum_{v_{a_j} \in \mathbb{Z}_q^*} \xi^{u_{a_1} v_{a_1}} \xi^{u_{a_2} v_{a_2}} \cdots \xi^{u_{a_j} v_{a_j}} = \prod_{m=1}^{j} \sum_{v=0}^{q-1} \xi^{u_{a_m} v} = (-1)^j.$$

Combining this with (2.34), we have $\sum_{\mathbf{v} \in \mathcal{S}} \xi^{\mathbf{u} \cdot \mathbf{v}} = (-1)^j (q-1)^{k-j}$ as required. $\qquad \square$

In Theorem 7.2.3 we will show that if C is a linear code of length n over $\mathbb{F}_q$ with weight distribution A_w for $0 \le w \le n$, then $A_k^{\perp} = 1/|C| \sum_{w=0}^{n} A_w K_k^{n,q}(w)$ for $0 \le k \le n$ is the weight distribution of $C^{\perp}$. As C is linear $A_w = B_w$ by Exercise 117. In particular for linear codes $\sum_{w=0}^{n} B_w K_k^{n,q}(w) \ge 0$. The next lemma, which is the basis of the Linear Programming Bound, shows that this inequality holds where B_w is the distance distribution of a (possibly) nonlinear code.

Lemma 2.6.3 *Let* $B_w, 0 \le w \le n$, *be the distance distribution of a code over* $\mathbb{F}_q$. *Then*

$$\sum_{w=0}^{n} B_w K_k^{n,q}(w) \ge 0 \tag{2.35}$$

for $0 \le k \le n$.

Proof: By Theorem 2.6.1 we only need to verify these inequalities for an (n, M) code C over $\mathbb{Z}_q$. By definition of the distance distribution and Lemma 2.6.2,

$$M \sum_{w=0}^{n} B_w K_k^{n,q}(w) = \sum_{w=0}^{n} \sum_{\substack{(\mathbf{x},\mathbf{y}) \in C^2 \\ d(\mathbf{x},\mathbf{y})=w}} K_k^{n,q}(w) = \sum_{w=0}^{n} \sum_{\substack{(\mathbf{x},\mathbf{y}) \in C^2 \\ d(\mathbf{x},\mathbf{y})=w}} \sum_{\substack{\mathbf{v} \in \mathbb{Z}_q^n \\ \mathrm{wt}(\mathbf{v})=k}} \xi^{(\mathbf{x}-\mathbf{y})\cdot\mathbf{v}}$$

$$= \sum_{\substack{(\mathbf{x},\mathbf{y}) \in C^2 \\ \mathrm{wt}(\mathbf{v})=k}} \sum_{\substack{\mathbf{v} \in \mathbb{Z}_q^n}} \xi^{\mathbf{x}\cdot\mathbf{v}} \xi^{-\mathbf{y}\cdot\mathbf{v}} = \sum_{\substack{\mathbf{v} \in \mathbb{Z}_q^n \\ \mathrm{wt}(\mathbf{v})=k}} \sum_{\mathbf{x} \in C} \xi^{\mathbf{x}\cdot\mathbf{v}} \sum_{\mathbf{y} \in C} \xi^{-\mathbf{y}\cdot\mathbf{v}}$$

$$= \sum_{\substack{\mathbf{v} \in \mathbb{Z}_q^n \\ \mathrm{wt}(\mathbf{v})=k}} \sum_{\mathbf{x} \in C} \xi^{\mathbf{x}\cdot\mathbf{v}} \sum_{\mathbf{x} \in C} \xi^{-\mathbf{x}\cdot\mathbf{v}} = \sum_{\substack{\mathbf{v} \in \mathbb{Z}_q^n \\ \mathrm{wt}(\mathbf{v})=k}} \left| \sum_{\mathbf{x} \in C} \xi^{\mathbf{x}\cdot\mathbf{v}} \right|^2 \ge 0,$$

proving the result. □

If C is an (n, M, d) code over $\mathbb{F}_q$ with distance distribution $B_w, 0 \le w \le n$, then $M = \sum_{w=0}^{n} B_w$ and $B_0 = 1$ by Exercise 117. Also $B_1 = B_2 = \cdots = B_{d-1} = 0$. By Lemma 2.6.3, we also have $\sum_{w=0}^{n} B_w K_k^{n,q}(w) \ge 0$ for $0 \le k \le n$. However, this inequality is merely $\sum_{w=0}^{n} B_w \ge 0$ when $k = 0$ as $K_0^{n,q}(w) = 1$, which is clearly already true. If $q = 2$, again by Exercise 117, $B_n \le 1$, and furthermore if d is even, by Theorem 2.1.2, we may also assume that $B_w = 0$ when w is odd. By Exercise 116, when w is even, $K_k^{n,2}(w) = K_{n-k}^{n,2}(w)$. Thus the kth inequality in (2.35) is the same as the $(n - k)$th as the only (possibly) nonzero B_ws are when w is even. This discussion yields our bound.

Theorem 2.6.4 (Linear Programming Bound) *The following hold:*
(i) *When* $q \ge 2$, $A_q(n, d) \le \max\{\sum_{w=0}^{n} B_w\}$, *where the maximum is taken over all* B_w *subject to the following conditions:*
   (a) $B_0 = 1$ *and* $B_w = 0$ *for* $1 \le w \le d - 1$,
   (b) $B_w \ge 0$ *for* $d \le w \le n$, *and*
   (c) $\sum_{w=0}^{n} B_w K_k^{n,q}(w) \ge 0$ *for* $1 \le k \le n$.
(ii) *When* d *is even and* $q = 2$, $A_2(n, d) \le \max\{\sum_{w=0}^{n} B_w\}$, *where the maximum is taken over all* B_w *subject to the following conditions:*
   (a) $B_0 = 1$ *and* $B_w = 0$ *for* $1 \le w \le d - 1$ *and all odd* w,

(b) $B_w \geq 0$ for $d \leq w \leq n$ and $B_n \leq 1$, and
(c) $\sum_{w=0}^{n} B_w K_k^{n,2}(w) \geq 0$ for $1 \leq k \leq \lfloor n/2 \rfloor$.

Solving the inequalities of this theorem is accomplished by linear programming, hence the name. At times other inequalities can be added to the list which add more constraints to the linear program and reduce the size of $\sum_{w=0}^{n} B_w$. In specific cases other variations to the Linear Programming Bound can be performed to achieve a smaller upper bound. Many of the upper bounds in Table 2.1 come from the Linear Programming Bound and these variations.

Example 2.6.5 We apply the Linear Programming Bound to obtain an upper bound on $A_2(8, 3)$. By Theorem 2.1.2, $A_2(8, 3) = A_2(9, 4)$. Hence we apply the Linear Programming Bound (ii) with $q = 2$, $n = 9$, and $d = 4$. Thus we are trying to find a solution to $\max\{1 + B_4 + B_6 + B_8\}$ where

$$
\begin{aligned}
9 + B_4 - 3B_6 - 7B_8 &\geq 0 \\
36 - 4B_4 + 20B_8 &\geq 0 \\
84 - 4B_4 + 8B_6 - 28B_8 &\geq 0 \\
126 + 6B_4 - 6B_6 + 14B_8 &\geq 0
\end{aligned}
\tag{2.36}
$$

with $B_4 \geq 0$, $B_6 \geq 0$, and $B_8 \geq 0$. The unique solution to this linear program is $B_4 = 18$, $B_6 = 24/5$, and $B_8 = 9/5$. Hence $\max\{1 + B_4 + B_6 + B_8\} = 1 + (123/5)$, implying $A_2(9, 4) \leq 25$.

We can add two more inequalities to (2.36). Let C be a $(9, M, 4)$ code, and let $\mathbf{x} \in C$. Define $C_{\mathbf{x}} = \mathbf{x} + C$; note that $C_{\mathbf{x}}$ is also a $(9, M, 4)$ code. The number of codewords in C at distance 8 from $\mathbf{x}$ is the same as the number of vectors of weight 8 in $C_{\mathbf{x}}$. As there is clearly no more than one vector of weight 8 in a $(9, M, 4)$ code, we have

$$B_8 \leq 1. \tag{2.37}$$

Also the number of codewords at distance 6 from $\mathbf{x}$ is the same as the number of vectors of weight 6 in $C_{\mathbf{x}}$; this number is at most $A_2(9, 4, 6)$, which is 12 by Exercise 100. Furthermore, if there is a codeword of weight 8 in $C_{\mathbf{x}}$, then every vector in $C_{\mathbf{x}}$ of weight 6 has a 1 in the coordinate where the weight 8 vector is 0. This means that the number of weight 6 vectors in $C_{\mathbf{x}}$ is the number of weight 5 vectors in the code obtained by puncturing $C_{\mathbf{x}}$ on this coordinate. That number is at most $A_2(8, 4, 5)$, which is 8 by Exercise 101. Putting these together shows that

$$B_6 + 4B_8 \leq 12. \tag{2.38}$$

Including inequalities (2.37) and (2.38) with (2.36) gives a linear program which when solved yields the unique solution $B_4 = 14$, $B_6 = 16/3$, and $B_8 = 1$ implying $\max\{1 + B_4 + B_6 + B_8\} = 64/3$. Thus $A_2(9, 4) \leq 21$. By further modifying the linear program (see [218, Chapter 17]) it can be shown that $A_2(9, 4) \leq 20$. In fact $A_2(9, 4) = 20$; see Table 2.1. See also Exercise 103. ∎

Exercise 118 Find an upper bound on $A_2(9, 3) = A_2(10, 4)$ as follows. (It may be helpful to use a computer algebra program that can perform linear programming.)

(a) Give the Sphere Packing, Johnson, and Elias Bounds for $A_2(9, 3)$ and $A_2(10, 4)$.
(b) Apply the Linear Programming Bound (ii) with $n = 10$, $q = 2$, and $d = 4$.
(c) Prove that $B_8 + 5B_{10} \leq 5$.
(d) Combine the inequalities from the Linear Programming Bound (ii) with the inequality in (c) to obtain an upper bound for $A_2(10, 4)$. Is it an improvement over the bound found in (b)?
(e) What is the best bound on $A_2(9, 3) = A_2(10, 4)$ from parts (a), (b), and (d)? How does it compare to the value in Table 2.1?
(f) Assuming the value of $A_2(9, 4)$ in Table 2.1 is correct, what bound do you get on $A_2(9, 3) = A_2(10, 4)$ from Theorem 2.1.6?   ◆

Exercise 119 Do the following to obtain bounds on $A_2(13, 5) = A_2(14, 6)$. (It may be helpful to use a computer algebra program that can perform linear programming.)
(a) Give the Sphere Packing, Johnson, and Elias Bounds for $A_2(13, 5)$ and $A_2(14, 6)$.
(b) Apply the Linear Programming Bound (ii) with $n = 14$, $q = 2$, and $d = 6$ to obtain an upper bound for $A_2(14, 6)$.
(c) What is the best bound on $A_2(13, 5) = A_2(14, 6)$ from parts (a) and (b)? How does it compare to the value in Table 2.1?
(d) Assuming the value of $A_2(13, 6)$ in Table 2.1 is correct, what bound do you get on $A_2(13, 5) = A_2(14, 6)$ from Theorem 2.1.6?   ◆

2.7   The Griesmer Upper Bound

The final upper bound we discuss is a generalization of the Singleton Bound known as the Griesmer Bound. We place it last because, unlike our other upper bounds, this one applies only to linear codes.

To prove this bound we first discuss the generally useful idea of a residual code due to H. J. Helgert and R. D. Stinaff [120]. Let C be an $[n, k]$ code and let $\mathbf{c}$ be a codeword of weight w. Let the set of coordinates on which $\mathbf{c}$ is nonzero be $\mathcal{I}$. Then the *residual code of C with respect to* $\mathbf{c}$, denoted Res(C, $\mathbf{c}$), is the code of length $n - w$ punctured on all the coordinates of $\mathcal{I}$. The next result gives a lower bound for the minimum distance of residual codes [131].

Theorem 2.7.1 Let C be an $[n, k, d]$ code over $\mathbb{F}_q$ and let $\mathbf{c}$ be a codeword of weight $w < (q/(q - 1))d$. Then Res(C, $\mathbf{c}$) is an $[n - w, k - 1, d']$ code, where $d' \geq d - w + \lceil w/q \rceil$.

Proof: By replacing C by a monomially equivalent code, we may assume that $\mathbf{c} = 11 \cdots 100 \cdots 0$. Since puncturing $\mathbf{c}$ on its nonzero coordinates gives the zero vector, Res(C, $\mathbf{c}$) has dimension less than k. Assume that the dimension is strictly less than $k - 1$. Then there exists a nonzero codeword $\mathbf{x} = x_1 \cdots x_n \in C$ which is not a multiple of $\mathbf{c}$ with $x_{w+1} \cdots x_n = \mathbf{0}$. There exists $\alpha \in \mathbb{F}_q$ such that at least w/q coordinates of $x_1 \cdots x_w$ equal α. Therefore

$$d \leq \mathrm{wt}(\mathbf{x} - \alpha\mathbf{c}) \leq w - \frac{w}{q} = \frac{w(q - 1)}{q},$$

contradicting our assumption on w. Hence Res(C, $\mathbf{c}$) has dimension $k - 1$.

We now establish the lower bound for d'. Let $x_{w+1} \cdots x_n$ be any nonzero codeword in $\text{Res}(\mathcal{C}, \mathbf{c})$, and let $\mathbf{x} = x_1 \cdots x_w x_{w+1} \cdots x_n$ be a corresponding codeword in $\mathcal{C}$. There exists $\alpha \in \mathbb{F}_q$ such that at least w/q coordinates of $x_1 \cdots x_w$ equal α. So

$$d \leq \text{wt}(\mathbf{x} - \alpha\mathbf{c}) \leq w - \frac{w}{q} + \text{wt}(x_{w+1} \cdots x_n).$$

Thus every nonzero codeword of $\text{Res}(\mathcal{C}, \mathbf{c})$ has weight at least $d - w + \lceil w/q \rceil$. $\square$

Applying Theorem 2.7.1 to a codeword of minimum weight we obtain the following.

Corollary 2.7.2 *If $\mathcal{C}$ is an $[n, k, d]$ code over $\mathbb{F}_q$ and $\mathbf{c} \in \mathcal{C}$ has weight d, then $\text{Res}(\mathcal{C}, \mathbf{c})$ is an $[n - d, k - 1, d']$ code, where $d' \geq \lceil d/q \rceil$.*

Recall that the Nordstrom–Robinson code defined in Section 2.3.4 is a nonlinear binary code of length 16 and minimum distance 6, with $256 = 2^8$ codewords, and, as we described, its existence together with the Johnson Bound (see Example 2.3.9) implies that $A_2(16, 6) = 2^8$. It is natural to ask whether $B_2(16, 6)$ also equals 2^8. In the next example we illustrate how residual codes can be used to show that no $[16, 8, 6]$ binary linear code exists, thus implying that $B_2(16, 6) \leq 2^7$.

Example 2.7.3 Let $\mathcal{C}$ be a $[16, 8, 6]$ binary linear code. Let $\mathcal{C}_1$ be the residual code of $\mathcal{C}$ with respect to a weight 6 vector. By Corollary 2.7.2, $\mathcal{C}_1$ is a $[10, 7, d']$ code with $3 \leq d'$; by the Singleton Bound $d' \leq 4$. If $d' = 4$, $\mathcal{C}_1$ is a nontrivial binary MDS code, which is impossible by Theorem 2.4.4. So $d' = 3$. Notice that we have now reduced the problem to showing the nonexistence of a $[10, 7, 3]$ code. But the nonexistence of this code follows from the Sphere Packing Bound as

$$2^7 > \frac{2^{10}}{\binom{10}{0} + \binom{10}{1}}.$$ ■

Exercise 120 In Exercise 93, we showed that $B_2(13, 6) \leq 2^5$. Show using residual codes that $B_2(13, 6) \leq 2^4$. Also construct a code that meets this bound.   ◆

Exercise 121 Do the following:
(a) Use the residual code to prove that a $[16, 5, 8]$ binary code contains the all-one codeword $\mathbf{1}$.
(b) Prove that a $[16, 5, 8]$ binary code has weight distribution $A_0 = A_{16} = 1$ and $A_8 = 30$.
(c) Prove that all $[16, 5, 8]$ binary codes are equivalent.
(d) Prove that $\mathcal{R}(1, 4)$ is a $[16, 5, 8]$ binary code.   ◆

Theorem 2.7.4 (Griesmer Bound [112]) *Let $\mathcal{C}$ be an $[n, k, d]$ code over $\mathbb{F}_q$ with $k \geq 1$. Then*

$$n \geq \sum_{i=0}^{k-1} \left\lceil \frac{d}{q^i} \right\rceil.$$

Proof: The proof is by induction on k. If $k = 1$ the conclusion clearly holds. Now assume that $k > 1$ and let $\mathbf{c} \in C$ be a codeword of weight d. By Corollary 2.7.2, $\mathrm{Res}(C, \mathbf{c})$ is an $[n - d, k - 1, d']$ code, where $d' \geq \lceil d/q \rceil$. Applying the inductive assumption to $\mathrm{Res}(C, \mathbf{c})$, we have $n - d \geq \sum_{i=0}^{k-2} \lceil d/q^{i+1} \rceil$ and the result follows.   □

Since $\lceil d/q^0 \rceil = d$ and $\lceil d/q^i \rceil \geq 1$ for $i = 1, \ldots, k - 1$, the Griesmer Bound implies the linear case of the Singleton Bound.

The Griesmer Bound gives a lower bound on the length of a code over $\mathbb{F}_q$ with a prescribed dimension k and minimum distance d. The Griesmer Bound does provide an upper bound on $B_q(n, d)$ because, given n and d, there is a largest k for which the Griesmer Bound holds. Then $B_q(n, d) \leq q^k$.

Given k, d, and q there need not exist an $[n, k, d]$ code over $\mathbb{F}_q$ which meets the Griesmer Bound; that is, no code may exist where there is equality in the Griesmer Bound. For example, by the Griesmer Bound, a binary code of dimension $k = 12$ and minimum distance $d = 7$ has length $n \geq 22$. Thus the $[23, 12, 7]$ binary Golay code does not meet the Griesmer Bound. But a $[22, 12, 7]$ binary code does not exist because the Johnson Bound (2.13) gives $A_2(22, 7) \leq \lfloor 2^{22}/2025 \rfloor = 2071$, implying $B_2(22, 7) \leq 2^{11}$.

It is natural to try to construct codes that meet the Griesmer Bound. We saw that the $[23, 12, 7]$ binary Golay code does not meet the Griesmer Bound. Neither does the $[24, 12, 8]$ extended binary Golay code, but both the $[12, 6, 6]$ and $[11, 6, 5]$ ternary Golay codes do. (See Exercise 122.) In the next theorem, we show that the $[(q^r - 1)/(q - 1), r]$ simplex code meets the Griesmer Bound; we also show that all its nonzero codewords have weight q^{r-1}, a fact we verified in Section 1.8 when $q = 2$.

Theorem 2.7.5 *Every nonzero codeword of the r-dimensional simplex code over $\mathbb{F}_q$ has weight q^{r-1}. The simplex codes meet the Griesmer Bound.*

Proof: Let G be a generator matrix for the r-dimensional simplex code C over $\mathbb{F}_q$. The matrix G is formed by choosing for its columns a nonzero vector from each 1-dimensional subspace of $\mathbb{F}_q^r$. Because $C = \{\mathbf{x}G \mid \mathbf{x} \in \mathbb{F}_q^r\}$, if $\mathbf{x} \neq \mathbf{0}$, then $\mathrm{wt}(\mathbf{x}G) = n - s$, where s is the number of columns $\mathbf{y}$ of G such that $\mathbf{x} \cdot \mathbf{y}^T = 0$. The set of vectors of $\mathbb{F}_q^r$ orthogonal to $\mathbf{x}$ is an $(r - 1)$-dimensional subspace of $\mathbb{F}_q^r$ and thus exactly $(q^{r-1} - 1)/(q - 1)$ columns $\mathbf{y}$ of G satisfy $\mathbf{x} \cdot \mathbf{y}^T = 0$. Thus $\mathrm{wt}(\mathbf{x}G) = (q^r - 1)/(q - 1) - (q^{r-1} - 1)/(q - 1) = q^{r-1}$ proving that each nonzero codeword has weight q^{r-1}.

In particular, the minimum distance is q^{r-1}. Since

$$\sum_{i=0}^{r-1} \left\lceil \frac{q^{r-1}}{q^i} \right\rceil = \sum_{i=0}^{r-1} q^i = (q^r - 1)/(q - 1),$$

the simplex codes meet the Griesmer Bound.   □

Exercise 122 Prove that the $[11, 6, 5]$ and $[12, 6, 6]$ ternary Golay codes meet the Griesmer Bound, but that the $[24, 12, 8]$ extended binary Golay code does not.   ♦

Solomon and Stiffler [319] and Belov [15] each construct a family of codes containing the simplex codes which meet the Griesmer Bound. Helleseth [122] has shown that in many cases there are no other binary codes meeting the bound. For nonbinary fields, the situation is much more complex. Projective geometries have also been used to construct codes meeting the Griesmer Bound (see [114]).

In general, an $[n, k, d]$ code may not have a basis of minimum weight vectors. However, in the binary case, if the code meets the Griesmer Bound, it has such a generator matrix, as the following result of van Tilborg [328] shows.

Theorem 2.7.6 *Let C be an $[n, k, d]$ binary code that meets the Griesmer Bound. Then C has a basis of minimum weight codewords.*

Proof: We proceed by induction on k. The result is clearly true if $k = 1$. Assume that $\mathbf{c}$ is a codeword of weight d. By permuting coordinates, we may assume that C has a generator matrix

$$G = \left[\begin{array}{c|c} 1 \cdots 1 & 0 \cdots 0 \\ \hline G_0 & G_1 \end{array} \right],$$

where the first row is $\mathbf{c}$ and G_1 is a generator matrix of the $[n - d, k - 1, d']$ residual code $C_1 = \mathrm{Res}(C, \mathbf{c})$; $d' \geq d_1 = \lceil d/2 \rceil$ by Corollary 2.7.2. As C meets the Griesmer Bound,

$$n - d = \sum_{i=1}^{k-1} \left\lceil \frac{d}{2^i} \right\rceil = \sum_{i=0}^{k-2} \left\lceil \frac{d_1}{2^i} \right\rceil, \tag{2.39}$$

by Exercise 123. Suppose that $d' > d_1$. Then $n - d < \sum_{i=0}^{k-2} \lceil d'/2^i \rceil$ by (2.39) and C_1 violates the Griesmer Bound. Therefore $d' = d_1$ and C_1 is an $[n - d, k - 1, d_1]$ code meeting the Griesmer Bound. By induction, we may assume the rows of G_1 have weight d_1. For $i \geq 2$, let $\mathbf{r}_i = (\mathbf{s}_{i-1}, \mathbf{t}_{i-1})$ be row i of G, where $\mathbf{s}_{i-1}$ is row $i - 1$ of G_0 and $\mathbf{t}_{i-1}$ is row $i - 1$ of G_1. By Exercise 124, one of $\mathbf{r}_i$ or $\mathbf{c} + \mathbf{r}_i$ has weight d. Hence C has a basis of weight d codewords. $\square$

Exercise 123 Prove that for $i \geq 1$,

$$\left\lceil \frac{d}{2^i} \right\rceil = \left\lceil \frac{d_1}{2^{i-1}} \right\rceil,$$

where $d_1 = \lceil d/2 \rceil$.   ◆

Exercise 124 In the notation of the proof of Theorem 2.7.6 show that one of $\mathbf{r}_i$ or $\mathbf{c} + \mathbf{r}_i$ has weight d.   ◆

Exercise 125 Prove that if d is even, a binary code meeting the Griesmer Bound has only even weight codewords. Do not use Theorem 2.7.9.   ◆

This result has been generalized by Dodunekov and Manev [70]. Let

$$g(k, d) = \sum_{i=0}^{k-1} \left\lceil \frac{d}{2^i} \right\rceil \tag{2.40}$$

be the summation in the binary Griesmer Bound. The Griesmer Bound says that for an $[n, k, d]$ binary code to exist, $n \geq g(k, d)$. So $n - g(k, d)$ is a measure of how close the length of the code is to one that meets the Griesmer Bound. It also turns out to be a measure of how much larger than minimum weight the weights of your basis vectors may need to be.

Theorem 2.7.7 *Let C be an $[n, k, d]$ binary code with $h = n - g(k, d)$. Then C has a basis of codewords of weight at most $d + h$.*

Proof: We proceed by induction on h. The case $h = 0$ is covered by Theorem 2.7.6. For fixed h, proceed by induction on k. In the case $k = 1$, there certainly is a basis of one codeword of weight d. When $k > 1$ assume that c is a codeword of weight d. By permuting coordinates, we may assume that C has a generator matrix

$$G = \left[\begin{array}{c|c} 1 \cdots 1 & 0 \cdots 0 \\ \hline G_0 & G_1 \end{array}\right],$$

where the first row is c and G_1 is a generator matrix of the $[n - d, k - 1, d']$ residual code $C_1 = \text{Res}(C, c)$; $d' \geq d_1 = \lceil d/2 \rceil$ by Corollary 2.7.2. Let $d' = d_1 + \epsilon$. Using Exercise 126,

$$g(k - 1, d') = \sum_{i=0}^{k-2} \left\lceil \frac{d'}{2^i} \right\rceil = \sum_{i=0}^{k-2} \left\lceil \frac{d_1 + \epsilon}{2^i} \right\rceil$$

$$\geq \sum_{i=0}^{k-2} \left\lceil \frac{d_1}{2^i} \right\rceil + \sum_{i=0}^{k-2} \left\lfloor \frac{\epsilon}{2^i} \right\rfloor = g(k - 1, d_1) + \sum_{i=0}^{k-2} \left\lfloor \frac{\epsilon}{2^i} \right\rfloor.$$

As part of the proof of Theorem 2.7.6, we in effect showed that $g(k - 1, d_1) = g(k - 1, \lceil d/2 \rceil) = g(k, d) - d$ (see Exercise 127). As $h = n - g(k, d)$,

$$g(k - 1, d') \geq g(k - 1, d_1) + \sum_{i=0}^{k-2} \left\lfloor \frac{\epsilon}{2^i} \right\rfloor = n - d + \left(\sum_{i=0}^{k-2} \left\lfloor \frac{\epsilon}{2^i} \right\rfloor - h \right). \tag{2.41}$$

Therefore as C_1 exists, $n - d \geq g(k - 1, d')$. Let $n - d - g(k - 1, d') = h_1 \geq 0$. By (2.41), $h_1 \leq h - \sum_{i=0}^{k-2} \lfloor \epsilon/2^i \rfloor$. By induction, we may assume the rows of G_1 have weight at most $d' + h_1$. For $j \geq 2$, let $r_j = (s_{j-1}, t_{j-1})$ be row j of G, where s_{j-1} is row $j - 1$ of G_0 and t_{j-1} is row $j - 1$ of G_1. So

$$\text{wt}(t_{j-1}) \leq d' + h_1 \leq d_1 + \epsilon + h - \sum_{i=0}^{k-2} \left\lfloor \frac{\epsilon}{2^i} \right\rfloor \leq d_1 + h = \left\lceil \frac{d}{2} \right\rceil + h,$$

since $\epsilon \leq \sum_{i=0}^{k-2} \lfloor \epsilon/2^i \rfloor$. By Exercise 128 one of r_j or $c + r_j$ has weight between d and $d + h$. Hence C has a basis of codewords of weights at most $d + h$. $\qquad \square$

Exercise 126 Let x and y be nonnegative real numbers. Show that $\lceil x + y \rceil \geq \lceil x \rceil + \lfloor y \rfloor$. $\blacklozenge$

Exercise 127 Show that $g(k - 1, \lceil d/2 \rceil) = g(k, d) - d$. $\blacklozenge$

Exercise 128 In the notation of the proof of Theorem 2.7.7 show that one of r_j or $c + r_j$ has weight between d and $d + h$. $\blacklozenge$

Exercise 129

(a) Compute $g(5, 4)$ from (2.40).

(b) What are the smallest weights that Theorem 2.7.7 guarantees can be used in a basis of a $[10, 5, 4]$ binary code?

(c) Show that a $[10, 5, 4]$ binary code with only even weight codewords has a basis of weight 4 codewords.

(d) Construct a $[10, 5, 4]$ binary code with only even weight codewords.   ◆

Both Theorems 2.7.6 and 2.7.7 can be generalized in the obvious way to codes over $\mathbb{F}_q$; see [68]. In particular, codes over $\mathbb{F}_q$ that meet the Griesmer Bound have a basis of minimum weight codewords. This is not true of codes in general but is true for at least one code with the same parameters, as the following theorem of Simonis [308] shows. This result may prove to be useful in showing the nonexistence of linear codes with given parameters $[n, k, d]$.

Theorem 2.7.8 *Suppose that there exists an $[n, k, d]$ code C over $\mathbb{F}_q$. Then there exists an $[n, k, d]$ code C' with a basis of codewords of weight d.*

Proof: Let s be the maximum number of independent codewords $\{c_1, \ldots, c_s\}$ in C of weight d. Note that $s \geq 1$ as C has minimum weight d. We are done if $s = k$. So assume $s < k$. The theorem will follow by induction if we show that we can create from C an $[n, k, d]$ code C_1 with at least $s + 1$ independent codewords of weight d. Let $S = \text{span}\{c_1, \ldots, c_s\}$. By the maximality of s, every vector in $C \setminus S$ has weight greater than d. Let e_1 be a minimum weight vector in $C \setminus S$ with $\text{wt}(e_1) = d_1 > d$. Complete $\{c_1, \ldots, c_s, e_1\}$ to a basis $\{c_1, \ldots, c_s, e_1, e_2, \ldots, e_{k-s}\}$ of C. Choose $d_1 - d$ nonzero coordinates of e_1 and create e_1' to be the same as e_1 except on these $d_1 - d$ coordinates, where it is 0. So $\text{wt}(e_1') = d$. Let $C_1 = \text{span}\{c_1, \ldots, c_s, e_1', e_2, \ldots, e_{k-s}\}$. We show C_1 has minimum weight d and dimension k. The vectors in C_1 fall into two disjoint sets: $S = \text{span}\{c_1, \ldots, c_s\}$ and $C_1 \setminus S$. The nonzero codewords in S have weight d or more as $S \subset C$. The codewords in $C_1 \setminus S$ are obtained from those in $C \setminus S$ by modifying $d_1 - d$ coordinates; therefore as $C \setminus S$ has minimum weight d_1, $C_1 \setminus S$ has minimum weight at least d. So C_1 has minimum weight d. Suppose that C_1 has dimension less than k. Then by our construction, e_1' must be in $\text{span}\{c_1, \ldots, c_s, e_2, \ldots, e_{k-s}\} \subset C$. By maximality of s, e_1' must in fact be in S. So $e_1 - e_1' \in C \setminus S$ as $e_1 \notin S$. By construction $\text{wt}(e_1 - e_1') = d_1 - d$; on the other hand as $e_1 - e_1' \in C \setminus S$, $\text{wt}(e_1 - e_1') \geq d_1$, since d_1 is the minimum weight of $C \setminus S$. This contradiction shows that C_1 is an $[n, k, d]$ code with at least $s + 1$ independent codewords of weight d.   □

Exercise 130 Let C be the binary $[9, 4]$ code with generator matrix

$$\begin{bmatrix} 1 & 0 & 0 & 0 & 1 & 1 & 1 & 0 & 0 \\ 0 & 1 & 0 & 0 & 1 & 1 & 0 & 1 & 0 \\ 0 & 0 & 1 & 0 & 1 & 0 & 1 & 1 & 1 \\ 0 & 0 & 0 & 1 & 0 & 1 & 1 & 1 & 1 \end{bmatrix}.$$

(a) Find the weight distribution of C and show that the minimum weight of C is 4.

(b) Apply the technique of Theorem 2.7.8 to construct a $[9, 4, 4]$ code with a basis of weight 4 vectors.

(c) Choose any three independent weight 4 vectors in C and any weight 5 vector in C. Modify the latter vector by changing one of its 1s to 0. Show that these four weight 4 vectors always generate a $[9, 4, 4]$ code. ◆

Exercise 125 shows that a binary code meeting the Griesmer Bound has only even weight codewords if d is even. The next theorem extends this result. The binary case is due to Dodunekov and Manev [69] and the nonbinary case is due to Ward [346].

Theorem 2.7.9 *Let C be a linear code over $\mathbb{F}_p$, where p is a prime, which meets the Griesmer Bound. Assume that $p^i \mid d$. Then p^i divides the weights of all codewords of C; that is, p^i is a divisor of C.*

2.8   The Gilbert Lower Bound

We now turn to lower bounds on $A_q(n, d)$ and $B_q(n, d)$. The Gilbert Bound is a lower bound on $B_q(n, d)$ and hence a lower bound on $A_q(n, d)$.

Theorem 2.8.1 (Gilbert Bound [98])

$$B_q(n, d) \geq \frac{q^n}{\displaystyle\sum_{i=0}^{d-1} \binom{n}{i}(q-1)^i}.$$

Proof: Let C be a linear code over $\mathbb{F}_q$ with $B_q(n, d)$ codewords. By Theorem 2.1.7 the covering radius of C is at most $d - 1$. Hence the spheres of radius $d - 1$ about the code-words cover $\mathbb{F}_q^n$. By (2.29) a sphere of radius $d - 1$ centered at a codeword contains $\alpha = \sum_{i=0}^{d-1} \binom{n}{i}(q-1)^i$ vectors. As the $B_q(n, d)$ spheres centered at codewords must fill the space $\mathbb{F}_q^n$, $B_q(n, d)\alpha \geq q^n$ giving the bound. □

The Gilbert Bound can be also stated as

$$B_q(n, d) \geq q^{n - \log_q \sum_{i=0}^{d-1} \binom{n}{i}(q-1)^i}.$$

We present this formulation so that it can be compared to the Varshamov Bound given in the next section.

The proof of Theorem 2.8.1 suggests a nonconstructive "greedy" algorithm for producing a linear code with minimum distance at least d which meets the Gilbert Bound:

(a) Begin with any nonzero vector $\mathbf{c}_1$ of weight at least d.

(b) While the covering radius of the linear code C_i generated by $\{\mathbf{c}_1, \ldots, \mathbf{c}_i\}$ is at least d, choose any vector $\mathbf{c}_{i+1}$ in a coset of C_i of weight at least d.

No matter how this algorithm is carried out the resulting linear code has at least

$$q^{n - \log_q \sum_{i=0}^{d-1} \binom{n}{i}(q-1)^i}$$

codewords.

For (possibly) nonlinear codes the greedy algorithm is even easier.

(a) Start with any vector in $\mathbb{F}_q^n$.

(b) Continue to choose a vector whose distance is at least d to all previously chosen vectors as long as there are such vectors.

The result is again a code with minimum distance at least d (and covering radius at most $d - 1$) which meets the Gilbert Bound.

Exercise 131 Show that the two greedy constructions described after the proof of the Gilbert Bound indeed yield codes with at least

$$q^{n-\log_q \sum_{i=0}^{d-1} \binom{n}{i}(q-1)^i}$$

codewords.   ◆

Exercise 132 Show that any (n, M, d) code with covering radius $d - 1$ or less meets the Gilbert Bound.   ◆

2.9   The Varshamov Lower Bound

The Varshamov Bound is similar to the Gilbert Bound, and, in fact, asymptotically they are the same. The proof of the Varshamov Bound uses a lemma in which we show that if a code's parameters satisfies a certain inequality, then using a different greedy algorithm we can attach another column to the parity check matrix and increase the length, and therefore dimension, without decreasing the minimum distance.

Lemma 2.9.1 *Let n, k, and d be integers with $2 \le d \le n$ and $1 \le k \le n$, and let q be a prime power. If*

$$\sum_{i=0}^{d-2} \binom{n-1}{i}(q-1)^i < q^{n-k}, \tag{2.42}$$

then there exists an $(n - k) \times n$ matrix H over $\mathbb{F}_q$ such that every set of $d - 1$ columns of H is linearly independent.

Proof: We define a greedy algorithm for finding the columns $\mathbf{h}_1, \ldots, \mathbf{h}_n$ of H. From the set of all q^{n-k} column vectors of length $n - k$ over $\mathbb{F}_q$, choose:

(1) $\mathbf{h}_1$ to be any nonzero vector;

(2) $\mathbf{h}_2$ to be any vector that is not a multiple of $\mathbf{h}_1$;

$\vdots$

(j) $\mathbf{h}_j$ to be any vector that is not a linear combination of $d - 2$ (or fewer) of the vectors $\mathbf{h}_1, \ldots, \mathbf{h}_{j-1}$;

$\vdots$

(n) $\mathbf{h}_n$ to be any vector that is not a linear combination of $d - 2$ (or fewer) of the vectors $\mathbf{h}_1, \ldots, \mathbf{h}_{n-1}$.

If we can carry out this algorithm to completion, then $\mathbf{h}_1, \ldots, \mathbf{h}_n$ are the columns of an $(n - k) \times n$ matrix no $d - 1$ of which are linearly dependent. By Corollary 1.4.14, this matrix is the parity check matrix for a linear code with minimum weight at least d. We show that the construction can indeed be completed. Let j be an integer with $1 \leq j \leq n - 1$ and assume that vectors $\mathbf{h}_1, \ldots, \mathbf{h}_j$ have been found. Since $j \leq n - 1$, the number of different linear combinations of $d - 2$ or fewer of $\mathbf{h}_1, \ldots, \mathbf{h}_j$ is:

$$\sum_{i=0}^{d-2} \binom{j}{i} (q - 1)^i \leq \sum_{i=0}^{d-2} \binom{n-1}{i} (q - 1)^i.$$

Hence if (2.42) holds, then there is some vector $\mathbf{h}_{j+1}$ which is not a linear combination of $d - 2$ (or fewer) of $\mathbf{h}_1, \ldots, \mathbf{h}_j$. Thus the fact that $\mathbf{h}_1, \mathbf{h}_2, \ldots, \mathbf{h}_n$ can be chosen follows by induction on j. □

The matrix H in Lemma 2.9.1 is the parity check matrix of a code of length n over $\mathbb{F}_q$ that has dimension at least k and minimum distance at least d. Since the minimum distance of a subcode of a code is at least the minimum distance of the code, we have the following corollary.

Corollary 2.9.2 *Let n, k, and d be integers with $2 \leq d \leq n$ and $1 \leq k \leq n$. Then there exists an $[n, k]$ linear code over $\mathbb{F}_q$ with minimum distance at least d, provided*

$$1 + \sum_{i=0}^{d-2} \binom{n-1}{i} (q - 1)^i \leq q^{n-k}. \tag{2.43}$$

Theorem 2.9.3 (Varshamov Bound [337])

$$B_q(n, d) \geq q^{n - \left\lceil \log_q \left(1 + \sum_{i=0}^{d-2} \binom{n-1}{i} (q-1)^i \right) \right\rceil}.$$

Proof: Let L be the left-hand side of (2.43). By Corollary 2.9.2, there exists an $[n, k]$ code over $\mathbb{F}_q$ with minimum weight at least d provided $\log_q(L) \leq n - k$, or equivalently $k \leq n - \log_q(L)$. The largest integer k satisfying this inequality is $n - \lceil \log_q(L) \rceil$. Thus

$$B_q(n, d) \geq q^{n - \lceil \log_q(L) \rceil},$$

giving the theorem. □

2.10 Asymptotic bounds

In this section we will study some of the bounds from previous sections as the code lengths go to infinity. The resulting bounds are called *asymptotic bounds*. Before beginning this exploration, we need to define two terms. For a (possibly) nonlinear code over $\mathbb{F}_q$ with M codewords the *information rate*, or simply *rate*, of the code is defined to be $n^{-1} \log_q M$. Notice that if the code were actually an $[n, k, d]$ linear code, it would contain $M = q^k$ codewords and $n^{-1} \log_q M = k/n$; so for an $[n, k, d]$ linear code, the ratio k/n is the rate of the code consistent with the definition of "rate" in Section 1.11.2. In the linear case the rate of a code is a measure of the number of information coordinates relative to the total number of coordinates. The higher the rate, the higher the proportion of coordinates in a codeword actually contain information rather than redundancy. If a linear or nonlinear code of length n has minimum distance d, the ratio d/n is called the *relative distance* of the code; the relative distance is a measure of the error-correcting capability of the code relative to its length. Our asymptotic bounds will be either an upper or lower bound on the largest possible rate for a family of (possibly nonlinear) codes over $\mathbb{F}_q$ of lengths going to infinity with relative distances approaching δ. The function which determines this rate is

$$\alpha_q(\delta) = \limsup_{n \to \infty} n^{-1} \log_q A_q(n, \delta n).$$

The exact value of $\alpha_q(\delta)$ is unknown and hence we want upper and lower bounds on this function. An upper bound on $\alpha_q(\delta)$ would indicate that all families with relative distances approaching δ have rates, in the limit, at most this upper bound. A lower bound on $\alpha_q(\delta)$ indicates that there exists a family of codes of lengths approaching infinity and relative distances approaching δ whose rates are at least this bound. A number of upper and lower bounds exist; we investigate six upper and one lower bound arising from the nonasymptotic bounds already presented in this chapter and in Section 1.12, beginning with the upper bounds.

2.10.1 Asymptotic Singleton Bound

Our first asymptotic upper bound on $\alpha_q(\delta)$ is a simple consequence of the Singleton Bound; we leave its proof as an exercise.

Theorem 2.10.1 (Asymptotic Singleton Bound) *If $0 \le \delta \le 1$, then $\alpha_q(\delta) \le 1 - \delta$.*

Exercise 133 Prove Theorem 2.10.1. ◆

2.10.2 Asymptotic Plotkin Bound

The Plotkin Bound can be used to give an improved (smaller) upper bound on $\alpha_q(\delta)$ compared to the Asymptotic Singleton Bound.

Theorem 2.10.2 (Asymptotic Plotkin Bound) *Let $r = 1 - q^{-1}$. Then*

$$\alpha_q(\delta) = 0 \qquad \text{if } r \le \delta \le 1, \quad \text{and}$$
$$\alpha_q(\delta) \le 1 - \delta/r \qquad \text{if } 0 \le \delta \le r.$$

Proof: Note that the two formulas agree when $\delta = r$. First, assume that $r < \delta \le 1$. By the Plotkin Bound (2.1), as $rn < \delta n$, $A_q(n, \delta n) \le \delta n/(\delta n - rn)$ implying that $0 \le A_q(n, \delta n) \le \delta/(\delta - r)$, independent of n. Thus $\alpha_q(\delta) = 0$ follows immediately.

Now assume that $0 \le \delta \le r$. Suppose that $\mathcal{C}$ is an $(n, M, \delta n)$ code with $M = A_q(n, \delta n)$. We can shorten $\mathcal{C}$ in a manner analogous to that given in Section 1.5.3 as follows. Let $n' = \lfloor (\delta n - 1)/r \rfloor$; $n' < n$ as $\delta \le r$. Fix an $(n - n')$-tuple of elements from $\mathbb{F}_q$. For at least one choice of this $(n - n')$-tuple, there is a subset of at least $M/q^{n-n'}$ codewords of $\mathcal{C}$ whose right-most $n - n'$ coordinates equal this $(n - n')$-tuple. For this subset of $\mathcal{C}$, puncture the right-most $n - n'$ coordinates to form the code C' of length n' with $M' \geq M/q^{n-n'}$ codewords, noting that distinct codewords in C remain distinct in C'; the minimum distance of C' is at least that of C by our construction. This $(n', M', \delta n)$ code satisfies $rn' < \delta n$ by our choice of n'. Applying the Plotkin Bound to C' gives

$$\frac{M}{q^{n-n'}} \leq M' \leq \frac{\delta n}{\delta n - rn'} \leq \delta n,$$

as $\delta n - rn' \geq 1$. Therefore $A_q(n, \delta n) = M \leq q^{n-n'}\delta n$ and so

$$\alpha_q(\delta) \leq \limsup_{n\to\infty} n^{-1}\log_q(q^{n-n'}\delta n)$$

$$= \limsup_{n\to\infty} \left(1 - \frac{n'}{n} + \frac{\log_q \delta}{n} + \frac{\log_q n}{n}\right)$$

$$= \lim_{n\to\infty} 1 - \frac{n'}{n} = 1 - \frac{\delta}{r}.$$

This completes the proof. $\square$

Exercise 134 Draw the graphs of the inequalities given by the Asymptotic Singleton and Asymptotic Plotkin Bounds when $q = 2$, where the horizontal axis is the relative distance δ and the vertical axis is the rate $R = \alpha_q(\delta)$. Why is the Asymptotic Plotkin Bound stronger than the Asymptotic Singleton Bound? Repeat this for the cases $q = 3$ and $q = 4$. $\blacklozenge$

2.10.3 Asymptotic Hamming Bound

By the Asymptotic Plotkin Bound, when bounding $\alpha_q(\delta)$ we can assume that $0 \leq \delta < r = 1 - q^{-1}$ as otherwise $\alpha_q(\delta) = 0$. There is an asymptotic bound derived from the Hamming Bound (Sphere Packing Bound) that is superior to the Asymptotic Plotkin Bound on an interval of values for δ. In order to derive this bound, we define the *Hilbert entropy function* on $0 \leq x \leq r$ by

$$H_q(x) = \begin{cases} 0 & \text{if } x = 0, \\ x\log_q(q - 1) - x\log_q x - (1 - x)\log_q(1 - x) & \text{if } 0 < x \leq r. \end{cases}$$

We will need to estimate factorials; this can be done with Stirling's Formula [115, Chapter 9], one version of which is

$$n^{n+1/2}e^{-n+7/8} \leq n! \leq n^{n+1/2}e^{-n+1}.$$

The results of Exercises 135 and 136 will also be needed.

Exercise 135 Let $0 < \delta \leq 1 - q^{-1}$.
(a) Show that $n\log_q n - \lfloor \delta n \rfloor \log_q \lfloor \delta n \rfloor - (n - \lfloor \delta n \rfloor)\log_q(n - \lfloor \delta n \rfloor) \leq -(\delta n - 1)\times \log_q(\delta - \frac{1}{n}) + \log_q n - n(1 - \delta)\log_q(1 - \delta)$ when $\delta n > 2$. Hint: $(\delta - (1/n))n = \delta n - 1 \leq \lfloor \delta n \rfloor \leq \delta n$.
(b) Show that $n\log_q n - \lfloor \delta n \rfloor \log_q \lfloor \delta n \rfloor - (n - \lfloor \delta n \rfloor)\log_q(n - \lfloor \delta n \rfloor) \geq -\log_q n - \delta n \log_q \delta - (n - \delta n + 1)\log_q(1 - \delta + (1/n))$ when $\delta n \geq 1$. Hint: $n - \lfloor \delta n \rfloor \leq n - (\delta - (1/n))n$ and $\lfloor \delta n \rfloor \leq \delta n$.

(c) Show that $\lim_{n\to\infty} n^{-1}(n \log_q n - \lfloor \delta n \rfloor \log_q \lfloor \delta n \rfloor - (n - \lfloor \delta n \rfloor) \log_q (n - \lfloor \delta n \rfloor)) = -\delta \log_q \delta - (1 - \delta) \log_q (1 - \delta)$. ◆

Exercise 136 Let $0 < i \le \delta n$ where $0 < \delta \le 1 - q^{-1}$ and $q \ge 2$. Prove that $\binom{n}{i-1} \times (q - 1)^{i-1} < \binom{n}{i}(q - 1)^i$. ◆

Recall that a sphere of radius a in $\mathbb{F}_q^n$ centered at some vector in $\mathbb{F}_q^n$ contains

$$V_q(n, a) = \sum_{i=0}^{a} \binom{n}{i}(q - 1)^i \tag{2.44}$$

vectors. The following lemma gives a relationship between $V_q(n, a)$ and the entropy function.

Lemma 2.10.3 Let $0 < \delta \le 1 - q^{-1}$ where $q \ge 2$. Then

$$\lim_{n\to\infty} n^{-1} \log_q V_q(n, \lfloor \delta n \rfloor) = H_q(\delta).$$

Proof: In (2.44) with $a = \lfloor \delta n \rfloor$, the largest of the $1 + \lfloor \delta n \rfloor$ terms is the one with $i = \lfloor \delta n \rfloor$ by Exercise 136. Thus

$$\binom{n}{\lfloor \delta n \rfloor}(q - 1)^{\lfloor \delta n \rfloor} \le V_q(n, \lfloor \delta n \rfloor) \le (1 + \lfloor \delta n \rfloor)\binom{n}{\lfloor \delta n \rfloor}(q - 1)^{\lfloor \delta n \rfloor}.$$

Taking logarithms and dividing by n gives

$$A + n^{-1}\lfloor \delta n \rfloor \log_q(q - 1) \le n^{-1} \log_q V_q(n, \lfloor \delta n \rfloor)$$
$$\le A + n^{-1}\lfloor \delta n \rfloor \log_q(q - 1) + n^{-1} \log_q(1 + \lfloor \delta n \rfloor)$$

where $A = n^{-1} \log_q \binom{n}{\lfloor \delta n \rfloor}$. Therefore,

$$\lim_{n\to\infty} n^{-1} \log_q V_q(n, \lfloor \delta n \rfloor) = \lim_{n\to\infty} A + \delta \log_q(q - 1) \tag{2.45}$$

as $\lim_{n\to\infty} n^{-1} \log_q(1 + \lfloor \delta n \rfloor) = 0$.

As $\binom{n}{\lfloor \delta n \rfloor} = n!/(\lfloor \delta n \rfloor!(n - \lfloor \delta n \rfloor)!)$, by Stirling's Formula,

$$\frac{n^{n+1/2}e^{-n+7/8}}{\lfloor \delta n \rfloor^{\lfloor \delta n \rfloor+1/2}e^{-\lfloor \delta n \rfloor+1}(n - \lfloor \delta n \rfloor)^{n-\lfloor \delta n \rfloor+1/2}e^{-n+\lfloor \delta n \rfloor+1}} \le \binom{n}{\lfloor \delta n \rfloor}$$

and

$$\binom{n}{\lfloor \delta n \rfloor} \le \frac{n^{n+1/2}e^{-n+1}}{\lfloor \delta n \rfloor^{\lfloor \delta n \rfloor+1/2}e^{-\lfloor \delta n \rfloor+7/8}(n - \lfloor \delta n \rfloor)^{n-\lfloor \delta n \rfloor+1/2}e^{-n+\lfloor \delta n \rfloor+7/8}};$$

hence

$$Be^{-9/8} \le \binom{n}{\lfloor \delta n \rfloor} \le Be^{-3/4},$$

where

$$B = \frac{n^{n+1/2}}{\lfloor \delta n \rfloor^{\lfloor \delta n \rfloor+1/2}(n - \lfloor \delta n \rfloor)^{n-\lfloor \delta n \rfloor+1/2}}.$$

Since

$$\lim_{n\to\infty} n^{-1}\log_q(Be^k) = \lim_{n\to\infty} n^{-1}(\log_q B + k\log_q e) = \lim_{n\to\infty} n^{-1}\log_q B,$$

we conclude that

$$
\begin{aligned}
\lim_{n\to\infty} A &= \lim_{n\to\infty} n^{-1}\log_q B\\
&= \lim_{n\to\infty} n^{-1}[n\log_q n - \lfloor \delta n\rfloor \log_q \lfloor \delta n\rfloor - (n-\lfloor \delta n\rfloor)\log_q(n-\lfloor \delta n\rfloor)]\\
&\quad + \lim_{n\to\infty} n^{-1}(1/2)[\log_q n - \log_q\lfloor \delta n\rfloor - \log_q(n-\lfloor \delta n\rfloor)]\\
&= -\delta\log_q \delta - (1-\delta)\log_q(1-\delta) + 0
\end{aligned}
$$

by Exercise 135. Plugging into (2.45), we obtain

$$\lim_{n\to\infty} n^{-1}\log_q V_q(n,\lfloor \delta n\rfloor) = -\delta\log_q \delta - (1-\delta)\log_q(1-\delta) + \delta\log_q(q-1),$$

which is $H_q(\delta)$, proving the result.   □

Theorem 2.10.4 (Asymptotic Hamming Bound) *Let* $0 < \delta \le 1 - q^{-1}$, *where* $q \ge 2$. *Then* $\alpha_q(\delta) \le 1 - H_q(\delta/2)$.

Proof: Note first that $A_q(n,\delta n) = A_q(n,\lceil \delta n\rceil) \le q^n/V_q(n,\lfloor(\lceil \delta n\rceil - 1)/2\rfloor)$ by the Sphere Packing Bound. If $n \ge N$, $\lfloor(\lceil \delta n\rceil - 1)/2\rfloor \ge \lfloor \delta n/2\rfloor - 1 \ge \lfloor(\delta - (2/N))n/2\rfloor$. Thus $A_q(n,\delta n) \le q^n/V_q(n,\lfloor(\delta - (2/N))n/2\rfloor)$, implying that

$$
\begin{aligned}
\alpha_q(\delta) &= \limsup_{n\to\infty} n^{-1}\log_q A_q(n,\delta n)\\
&\le \limsup_{n\to\infty} 1 - n^{-1}\log_q V_q\left(n,\left\lfloor\frac{1}{2}\left(\delta - \frac{2}{N}\right)n\right\rfloor\right)\\
&= 1 - H_q\left(\frac{1}{2}\left(\delta - \frac{2}{N}\right)\right)
\end{aligned}
$$

by Lemma 2.10.3. But as n goes to infinity, we may let N get as large as we please showing that $\alpha_q(\delta) = 1 - H_q(\delta/2)$ since H_q is continuous.   □

Exercise 137 Continuing with Exercise 134, add to the graphs drawn for $q = 2$, $q = 3$, and $q = 4$ the graph of the inequality from the Asymptotic Hamming Bound. (A computer graphing tool may be helpful.) In each case, for what values of δ is the Asymptotic Hamming Bound stronger than the Asymptotic Singleton Bound and the Asymptotic Plotkin Bound?   ◆

2.10.4 Asymptotic Elias Bound

The Asymptotic Elias Bound surpasses the Asymptotic Singleton, Plotkin, and Hamming Bounds. As we know $\alpha_q(\delta) = 0$ for $1 - q^{-1} = r \le \delta \le 1$, we only examine the case $0 < \delta < r$.

Theorem 2.10.5 (Asymptotic Elias Bound) *Let* $0 < \delta < r = 1 - q^{-1}$, *where* $q \geq 2$. *Then* $\alpha_q(\delta) \leq 1 - H_q(r - \sqrt{r(r-\delta)}\,)$.

Proof: Choose x so that $0 < x < r - \sqrt{r(r-\delta)}$. Then $x^2 - 2rx + r\delta > 0$. Let $w = \lfloor xn \rfloor$ and $d = \lfloor \delta n \rfloor$. By Exercise 138, for n sufficiently large, $w^2 - 2rnw + rnd > 0$. As $x < r$, $w < rn$. Thus by the Elias Bound, for n large enough,

$$A_q(n, \delta n) = A_q(n, d) \leq \frac{rnd}{w^2 - 2rnw + rnd} \cdot \frac{q^n}{V_q(n, w)}.$$

So

$$n^{-1} \log_q A_q(n, \delta n) \leq n^{-1} \log_q \left(\frac{rnd}{w^2 - 2rnw + rnd} \cdot \frac{q^n}{V_q(n, w)} \right)$$

$$= n^{-1} \log_q \left(\frac{r\dfrac{d}{n}}{\left(\dfrac{w}{n}\right)^2 - 2r\dfrac{w}{n} + r\dfrac{d}{n}} \right) + 1 - n^{-1} V_q(n, \lfloor xn \rfloor).$$

Observing that $\lim_{n \to \infty} d/n = \delta$ and $\lim_{n \to \infty} w/n = x$, by taking the limit of the above and using Lemma 2.10.3,

$$\alpha_q(\delta) \leq 1 - H_q(x).$$

Since this is valid for all x with $0 < x < r - \sqrt{r(r-\delta)}$ and $H_q(x)$ is continuous, $\alpha_q(\delta) \leq 1 - H_q(r - \sqrt{r(r-\delta)}\,)$. $\qquad\qquad\Box$

Exercise 138 Assume $x^2 - 2rx + r\delta > 0$, $w = \lfloor xn \rfloor$, and $d = \lfloor \delta n \rfloor$ where δ, x, and n are positive. Show that for n large enough, $w^2 - 2rnw + rnd > 0$. Hint: Obtain a lower bound on $(w^2 - 2rnw + rnd)/n^2$ by using $xn - 1 < w \leq xn$ and $\delta n - 1 < d$. ◆

Exercise 139 Continuing with Exercise 137, add to the graphs drawn for $q = 2$, $q = 3$, and $q = 4$ the graph of the inequality from the Asymptotic Elias Bound. (A computer graphing tool may be helpful.) In each case, for what values of δ is the Asymptotic Elias Bound stronger than the Asymptotic Singleton Bound, the Asymptotic Plotkin Bound, and the Asymptotic Hamming Bound? ◆

2.10.5 The MRRW Bounds

There are two asymptotic versions of the Linear Programming Bound that are generally the best upper bounds on $\alpha_q(\delta)$. These bounds were discovered by McEliece, Rodemich, Rumsey, and Welch [236]. As a result they are called the MRRW Bounds. The first of these was originally developed for binary codes but has been generalized to codes over any field; see Levenshtein [194, Theorem 6.19]. The second holds only for binary codes. The First MRRW Bound, when considered only for binary codes, is a consequence of the Second MRRW Bound as we see in Exercise 141. The proofs of these bounds, other than what is shown in Exercise 141, is beyond the scope of this text. Again $\alpha_q(\delta) = 0$ for $1 - q^{-1} = r \leq \delta \leq 1$; the MRRW Bounds apply only when $0 < \delta < r$.

Theorem 2.10.6 (The First MRRW Bound) *Let $0 < \delta < r = 1 - q^{-1}$. Then*

$$\alpha_q(\delta) \leq H_q \left(\frac{1}{q} [q - 1 - (q - 2)\delta - 2\sqrt{(q - 1)\delta(1 - \delta)}] \right).$$

In particular if $q = 2$, then when $0 < \delta < 1/2$,

$$\alpha_2(\delta) \leq H_2 \left(\frac{1}{2} - \sqrt{\delta(1 - \delta)} \right).$$

Theorem 2.10.7 (The Second MRRW Bound) *Let $0 < \delta < 1/2$. Then*

$$\alpha_2(\delta) \leq \min_{0 \leq u \leq 1 - 2\delta} \{1 + g(u^2) - g(u^2 + 2\delta u + 2\delta)\}$$

where $g(x) = H_2((1 - \sqrt{1 - x})/2)$.

The Second MRRW Bound is better than the First MRRW Bound, for $q = 2$, when $\delta < 0.272$. Amazingly, if $q = 2$, the bounds agree when $0.273 \leq \delta \leq 0.5$. Exercise 142 shows that the Second MRRW Bound is strictly smaller than the Asymptotic Elias Bound when $q = 2$.

Exercise 140 Continuing with Exercise 139, add to the graphs drawn for $q = 2, q = 3$, and $q = 4$ the graph of the inequality from the First MRRW Bound. (A computer graphing tool may be helpful.) In each case, for what values of δ is the First MRRW Bound stronger than the Asymptotic Singleton Bound, the Asymptotic Plotkin Bound, the Asymptotic Hamming Bound, and the Asymptotic Elias Bound? ◆

Exercise 141 By the Second MRRW Bound, $\alpha_2(\delta) \leq 1 + g((1 - 2\delta)^2) - g((1 - 2\delta)^2 + 2\delta(1 - 2\delta) + 2\delta)$. Verify that this is the First MRRW Bound when $q = 2$. ◆

Exercise 142 This exercise shows that Second MRRW Bound is strictly smaller than the Asymptotic Elias Bound when $q = 2$.
(a) By the Second MRRW Bound, $\alpha_2(\delta) \leq 1 + g(0) - g(2\delta)$. Verify that this is the Asymptotic Elias Bound when $q = 2$.
(b) Verify that the derivative of $1 + g(u^2) - g(u^2 + 2\delta u + 2\delta)$ is negative at $u = 0$.
(c) How do parts (a) and (b) show that the Second MRRW Bound is strictly smaller than the Asymptotic Elias Bound when $q = 2$? ◆

2.10.6 Asymptotic Gilbert–Varshamov Bound

We now turn to the only asymptotic lower bound we will present. This bound is the asymptotic version of both the Gilbert and the Varshamov Bounds. We will give this asymptotic bound using the Gilbert Bound and leave as an exercise the verification that asymptotically the Varshamov Bound gives the same result.

Theorem 2.10.8 (Asymptotic Gilbert–Varshamov Bound) *If $0 < \delta \leq 1 - q^{-1}$ where $q \geq 2$, then $\alpha_q(\delta) \geq 1 - H_q(\delta)$.*

Proof: By the Gilbert Bound $A_q(n, \delta n) = A_q(n, \lceil \delta n \rceil) \geq q^n / V_q(n, \lceil \delta n \rceil - 1)$. Since $\lceil \delta n \rceil - 1 \leq \lfloor \delta n \rfloor$, $A_q(n, \delta n) \geq q^n / V_q(n, \lfloor \delta n \rfloor)$. Thus

$$\alpha_q(\delta) = \limsup_{n \to \infty} n^{-1} \log_q A_q(n, \delta n)$$

$$\geq \limsup_{n \to \infty} 1 - n^{-1} \log_q V_q(n, \lfloor \delta n \rfloor) = 1 - H_q(\delta)$$

by Lemma 2.10.3. □

Exercise 143 Verify that, for $q \geq 2$, the asymptotic version of the Varshamov Bound produces the lower bound $\alpha_q(\delta) \geq 1 - H_q(\delta)$ when $0 < \delta \leq 1 - q^{-1}$. ◆

The Asymptotic Gilbert–Varshamov Bound was discovered in 1952. This bound guarantees (theoretically) the existence of a family of codes of increasing length whose relative distances approach δ while their rates approach $1 - H_q(\delta)$. In the next section and in Chapter 13 we will produce specific families of codes, namely lexicodes and Goppa codes, which meet this bound. For 30 years no one was able to produce any family that exceeded this bound and many thought that the Asymptotic Gilbert–Varshamov Bound in fact gave the true value of $\alpha_q(\delta)$. However, in 1982, Tsfasman, Vlăduţ, and Zink [333] demonstrated that a certain family of codes of increasing length exceeds the Asymptotic Gilbert–Varshamov Bound. This family of codes comes from a collection of codes called algebraic geometry codes, described in Chapter 13, that generalize Goppa codes. There is, however, a restriction on $\mathbb{F}_q$ for which this construction works: q must be a square with $q \geq 49$. In particular, no family of binary codes is currently known that surpasses the Asymptotic Gilbert–Varshamov Bound. In Figure 2.2 we give five upper bounds and one lower bound on $\alpha_q(\delta)$, for $q = 2$, discussed in this section and Section 1.12; see Exercise 140. In this figure, the actual value of $\alpha_2(\delta)$ is 0 to the right of the dashed line. Families of binary codes meeting or exceeding the Asymptotic Gilbert–Varshamov Bound lie in the dotted region of the figure.

Exercise 144 Continuing with Exercise 140, add to the graphs drawn for $q = 3$ and $q = 4$ the graph of the inequality from the Asymptotic Gilbert–Varshamov Bound as done in Figure 2.2 for $q = 2$. ◆

2.11 Lexicodes

It is interesting that there is a class of binary linear codes whose construction is the greedy construction for nonlinear codes described in Section 2.8, except that the order in which the vectors are chosen is determined ahead of time. These codes are called lexicodes [39, 57, 192], and we will show that they indeed are linear. The construction implies that the lexicodes meet the Gilbert Bound, a fact we leave as an exercise. This implies that we can choose a family of lexicodes of increasing lengths which meet the Asymptotic Gilbert–Varshamov Bound.

The algorithm for constructing lexicodes of length n and minimum distance d proceeds as follows.

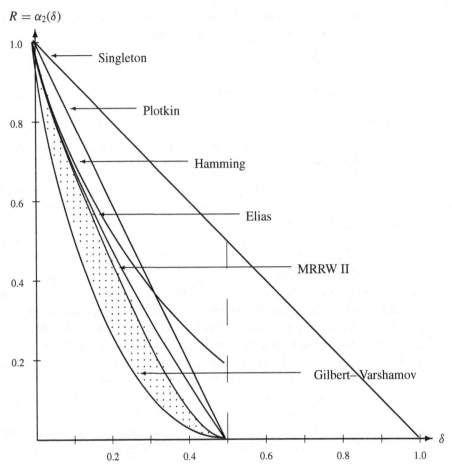

Figure 2.2 Asymptotic Bounds with $q = 2$.

I. Order all n-tuples in lexicographic order:

$$0\cdots000$$
$$0\cdots001$$
$$0\cdots010$$
$$0\cdots011$$
$$0\cdots100$$
$$\vdots$$

II. Construct the class $\mathcal{L}$ of vectors of length n as follows:
   (a) Put the zero vector $\mathbf{0}$ in $\mathcal{L}$.
   (b) Look for the first vector $\mathbf{x}$ of weight d in the lexicographic ordering. Put $\mathbf{x}$ in $\mathcal{L}$.
   (c) Look for the next vector in the lexicographic ordering whose distance from each vector in $\mathcal{L}$ is d or more and add this to $\mathcal{L}$.
   (d) Repeat (c) until there are no more vectors in the lexicographic list to look at.

The set $\mathcal{L}$ is actually a linear code, called the *lexicode* of length n and minimum distance d. In fact if we halt the process earlier, but at just the right spots, we have linear subcodes of $\mathcal{L}$. To prove this, we need a preliminary result due to Levenshtein [192]. If $\mathbf{u}$ and $\mathbf{v}$ are binary vectors of length n, we say that $\mathbf{u} < \mathbf{v}$ provided that $\mathbf{u}$ comes before $\mathbf{v}$ in the lexicographic order.

Lemma 2.11.1 *If* $\mathbf{u}$, $\mathbf{v}$, *and* $\mathbf{w}$ *are binary vectors of length* n *with* $\mathbf{u} < \mathbf{v} + \mathbf{w}$ *and* $\mathbf{v} < \mathbf{u} + \mathbf{w}$, *then* $\mathbf{u} + \mathbf{v} < \mathbf{w}$.

Proof: If $\mathbf{u} < \mathbf{v} + \mathbf{w}$, then $\mathbf{u}$ and $\mathbf{v} + \mathbf{w}$ have the same leading entries after which $\mathbf{u}$ has a 0 and $\mathbf{v} + \mathbf{w}$ has a 1. We can represent this as follows:

$$\mathbf{u} = \mathbf{a}0\cdots,$$
$$\mathbf{v} + \mathbf{w} = \mathbf{a}1\cdots.$$

Similarly as $\mathbf{v} < \mathbf{u} + \mathbf{w}$, we have

$$\mathbf{v} = \mathbf{b}0\cdots,$$
$$\mathbf{u} + \mathbf{w} = \mathbf{b}1\cdots.$$

However, we do not know that the length i of $\mathbf{a}$ and the length j of $\mathbf{b}$ are the same. Assume they are different, and by symmetry, that $i < j$. Then we have

$$\mathbf{v} = \mathbf{b}'x\cdots,$$
$$\mathbf{u} + \mathbf{w} = \mathbf{b}'x\cdots,$$

where $\mathbf{b}'$ is the first i entries of $\mathbf{b}$. Computing $\mathbf{w}$ in two ways, we obtain

$$\mathbf{w} = \mathbf{u} + (\mathbf{u} + \mathbf{w}) = (\mathbf{a} + \mathbf{b}')x\cdots,$$
$$\mathbf{w} = \mathbf{v} + (\mathbf{v} + \mathbf{w}) = (\mathbf{a} + \mathbf{b}')(1 + x)\cdots,$$

a contradiction. So $\mathbf{a}$ and $\mathbf{b}$ are the same length, giving

$$\mathbf{u} + \mathbf{v} = (\mathbf{a} + \mathbf{b})0\cdots,$$
$$\mathbf{w} = (\mathbf{a} + \mathbf{b})1\cdots,$$

showing $\mathbf{u} + \mathbf{v} < \mathbf{w}$. $\square$

Theorem 2.11.2 *Label the vectors in the lexicode in the order in which they are generated so that* $\mathbf{c}_0$ *is the zero vector.*
(i)   *$\mathcal{L}$ is a linear code and the vectors $\mathbf{c}_{2^i}$ are a basis of $\mathcal{L}$.*
(ii) *After $\mathbf{c}_{2^i}$ is chosen, the next $2^i - 1$ vectors generated are $\mathbf{c}_1 + \mathbf{c}_{2^i}$, $\mathbf{c}_2 + \mathbf{c}_{2^i}, \ldots,$* $\mathbf{c}_{2^i-1} + \mathbf{c}_{2^i}$.
(iii) *Let $\mathcal{L}_i = \{\mathbf{c}_0, \mathbf{c}_1, \ldots, \mathbf{c}_{2^i-1}\}$. Then $\mathcal{L}_i$ is an $[n, i, d]$ linear code.*

Proof: If we prove (ii), we have (i) and (iii). The proof of (ii) is by induction on i. Clearly it is true for $i = 1$. Assume the first 2^i vectors generated are as claimed. Then $\mathcal{L}_i$ is linear with basis $\{\mathbf{c}_1, \mathbf{c}_2, \ldots, \mathbf{c}_{2^i-1}\}$. We show that $\mathcal{L}_{i+1}$ is linear with the next 2^i vectors generated in order by adding $\mathbf{c}_{2^i}$ to each of the previously chosen vectors in $\mathcal{L}_i$. If not, there is an $r < 2^i$ such that $\mathbf{c}_{2^i+r} \neq \mathbf{c}_{2^i} + \mathbf{c}_r$. Choose r to be the smallest such value. As $\mathrm{d}(\mathbf{c}_{2^i} + \mathbf{c}_r, \mathbf{c}_{2^i} + \mathbf{c}_j) = \mathrm{d}(\mathbf{c}_r, \mathbf{c}_j) \geq d$ for $j < r$, $\mathbf{c}_{2^i} + \mathbf{c}_r$ was a possible vector to be chosen. Since it was not, it must have come too late in the lexicographic order; so

$$\mathbf{c}_{2^i+r} < \mathbf{c}_{2^i} + \mathbf{c}_r. \tag{2.46}$$

As $\mathrm{d}(\mathbf{c}_r + \mathbf{c}_{2^i+r}, \mathbf{c}_j) = \mathrm{d}(\mathbf{c}_{2^i+r}, \mathbf{c}_r + \mathbf{c}_j) \geq d$ for $j < 2^i$ by linearity of $\mathcal{L}_i$, $\mathbf{c}_r + \mathbf{c}_{2^i+r}$ could have been chosen to be in the code instead of $\mathbf{c}_{2^i}$ (which it cannot equal). So it must be that

$$\mathbf{c}_{2^i} < \mathbf{c}_r + \mathbf{c}_{2^i+r}. \tag{2.47}$$

If $j < r$, then $\mathbf{c}_{2^i+j} = \mathbf{c}_{2^i} + \mathbf{c}_j$ by the assumption on r. Hence $\mathrm{d}(\mathbf{c}_{2^i+r} + \mathbf{c}_{2^i}, \mathbf{c}_j) = \mathrm{d}(\mathbf{c}_{2^i+r}, \mathbf{c}_{2^i+j}) \geq d$ for $j < r$. So $\mathbf{c}_{2^i+r} + \mathbf{c}_{2^i}$ could have been chosen to be a codeword instead of $\mathbf{c}_r$. The fact that it was not implies that

$$\mathbf{c}_r < \mathbf{c}_{2^i+r} + \mathbf{c}_{2^i}. \tag{2.48}$$

But then (2.47) and (2.48) with Lemma 2.11.1 imply $\mathbf{c}_{2^i} + \mathbf{c}_r < \mathbf{c}_{2^i+r}$ contradicting (2.46). □

The codes $\mathcal{L}_i$ satisfy the inclusions $\mathcal{L}_1 \subset \mathcal{L}_2 \subset \cdots \subset \mathcal{L}_k = \mathcal{L}$, where k is the dimension of $\mathcal{L}$. In general this dimension is not known before the construction. If $i < k$, the left-most coordinates are always 0 (exactly how many is also unknown). If we puncture $\mathcal{L}_i$ on these zero coordinates, we actually get a lexicode of smaller length.

Exercise 145 Do the following:
(a) Construct the codes $\mathcal{L}_i$ of length 5 and minimum distance 2.
(b) Verify that these codes are linear and the vectors are generated in the order described by Theorem 2.11.2.
(c) Repeat (a) and (b) for length 5 and minimum distance 3. ◆

Exercise 146 Find an ordering of $\mathbb{F}_2^5$ so that the greedy algorithm does not produce a linear code. ◆

Exercise 147 Prove that the covering radius of $\mathcal{L}$ is $d - 1$ or less. Also prove that the lexicodes meet the Gilbert Bound. Hint: See Exercise 132. ◆

The lexicode $\mathcal{L}$ is the largest of the $\mathcal{L}_i$ constructed in Theorem 2.11.2. We can give a parity check matrix for $\mathcal{L}$ provided $d \geq 3$, which is reminiscent of the parity check matrix constructed in the proof of the Varshamov Bound. If $\mathcal{C}$ is a lexicode of length n with $d \geq 3$, construct its parity check matrix $H = [\mathbf{h}_n \cdots \mathbf{h}_1]$ as follows (where $\mathbf{h}_i$ is a column vector). Regard the columns $\mathbf{h}_i$ as binary numbers where $1 \leftrightarrow (\cdots 001)^\mathrm{T}$, $2 \leftrightarrow (\cdots 010)^\mathrm{T}$, etc. Let $\mathbf{h}_1$ be the column corresponding to 1 and $\mathbf{h}_2$ the column corresponding to 2. Once $\mathbf{h}_{i-1}, \ldots, \mathbf{h}_1$ are chosen, choose $\mathbf{h}_i$ to be the column corresponding to the smallest number which is not a linear combination of $d - 2$ or fewer of $\mathbf{h}_{i-1}, \ldots, \mathbf{h}_1$. Note that the length of the columns does not have to be determined ahead of time. Whenever $\mathbf{h}_i$ corresponds to a number that is a power of 2, the length of the columns increases and zeros are placed on the tops of the columns $\mathbf{h}_j$ for $j < i$.

Example 2.11.3 If $n = 7$ and $d = 3$, the parity check matrix for the lexicode $\mathcal{L}$ is

$$H = \begin{bmatrix} 1 & 1 & 1 & 1 & 0 & 0 & 0 \\ 1 & 1 & 0 & 0 & 1 & 1 & 0 \\ 1 & 0 & 1 & 0 & 1 & 0 & 1 \end{bmatrix}.$$

So we recognize this lexicode as the $[7, 4, 3]$ Hamming code.   ■

As this example illustrates, the Hamming code $\mathcal{H}_3$ is a lexicode. In fact, all binary Hamming and Golay codes are lexicodes [39].

Exercise 148 Compute the parity check matrix for the lexicode of length 5 and minimum distance 3. Check that it yields the code produced in Exercise 145(c).   ◆

Exercise 149 Compute the generator and parity check matrices for the lexicodes of length 6 and minimum distance 2, 3, 4, and 5.   ◆

3 Finite fields

For deeper analysis and construction of linear codes we need to make use of the basic theory of finite fields. In this chapter we review that theory. We will omit many of the proofs; for those readers interested in the proofs and other properties of finite fields, we refer you to [18, 196, 233].

3.1 Introduction

A *field* is a set $\mathbb{F}$ together with two operations: $+$, called addition, and $\cdot$, called multiplication, which satisfy the following axioms. The set $\mathbb{F}$ is an abelian group under $+$ with additive identity called *zero* and denoted 0; the set $\mathbb{F}^*$ of all nonzero elements of $\mathbb{F}$ is also an abelian group under multiplication with multiplicative identity called *one* and denoted 1; and multiplication distributes over addition. We will usually omit the symbol for multiplication and write ab for the product $a \cdot b$. The field is *finite* if $\mathbb{F}$ has a finite number of elements; the number of elements in $\mathbb{F}$ is called the *order* of $\mathbb{F}$. In Section 1.1 we gave three fields denoted $\mathbb{F}_2$, $\mathbb{F}_3$, and $\mathbb{F}_4$ of orders 2, 3, and 4, respectively. In general, we will denote a field with q elements by $\mathbb{F}_q$; another common notation is $\mathrm{GF}(q)$ and read "the Galois field with q elements."

If p is a prime, the integers modulo p form a field, which is then denoted $\mathbb{F}_p$. This is not true if p is not a prime. These are the simplest examples of finite fields. As we will see momentarily, every finite field contains some $\mathbb{F}_p$ as a subfield.

Exercise 150 Prove that the integers modulo n do not form a field if n is not prime. ◆

The finiteness of $\mathbb{F}_q$ implies that there exists a smallest positive integer p such that $1 + \cdots + 1$ (p 1s) is 0. The integer p is a prime, as verified in Exercise 151, and is called the *characteristic* of $\mathbb{F}_q$. If a is a positive integer, we will denote the sum of a 1s in the field by a. Also if we wish to write the sum of a αs where α is in the field, we write this as either $a\alpha$ or $a \cdot \alpha$. Notice that $p\alpha = 0$ for all $\alpha \in \mathbb{F}_q$. The set of p distinct elements $\{0, 1, 2, \ldots, (p-1)\}$ of $\mathbb{F}_q$ is isomorphic to the field $\mathbb{F}_p$ of integers modulo p. As a field isomorphic to $\mathbb{F}_p$ is contained in $\mathbb{F}_q$, we will simplify terminology and say that $\mathbb{F}_p$ is a subfield of $\mathbb{F}_q$; this subfield $\mathbb{F}_p$ is called the *prime subfield* of $\mathbb{F}_q$. The fact that $\mathbb{F}_p$ is a subfield of $\mathbb{F}_q$ gives us crucial information about q; specifically, by Exercise 151, the field $\mathbb{F}_q$ is also a finite dimensional vector space over $\mathbb{F}_p$, say of dimension m. Therefore $q = p^m$ as this is the number of vectors in a vector space of dimension m over $\mathbb{F}_p$.

Although it is not obvious, all finite fields with the same number of elements are isomorphic. Thus our notation $\mathbb{F}_q$ is not ambiguous; $\mathbb{F}_q$ will be any representation of a field with q elements. As we did with the prime subfield of $\mathbb{F}_q$, if we say that $\mathbb{F}_r$ is a subfield of $\mathbb{F}_q$, we actually mean that $\mathbb{F}_q$ contains a subfield with r elements. If $\mathbb{F}_q$ has a subfield with r elements, that subfield is unique. Hence there is no ambiguity when we say that $\mathbb{F}_r$ is a subfield of $\mathbb{F}_q$. It is important to note that although all finite fields of order q are isomorphic, one field may have many different representations. The exact form that we use for the field may be crucial in its application to coding theory. We summarize the results we have just given in a theorem; all but the last part are proved in Exercise 151.

Theorem 3.1.1 *Let $\mathbb{F}_q$ be a finite field with q elements. Then:*
(i) $q = p^m$ *for some prime p,*
(ii) $\mathbb{F}_q$ *contains the subfield $\mathbb{F}_p$,*
(iii) $\mathbb{F}_q$ *is a vector space over $\mathbb{F}_p$ of dimension m,*
(iv) $p\alpha = 0$ *for all $\alpha \in \mathbb{F}_q$, and*
(v) $\mathbb{F}_q$ *is unique up to isomorphism.*

Exercise 151 Prove the following:
(a) If a and b are in a field $\mathbb{F}$ with $ab = 0$, then either $a = 0$ or $b = 0$.
(b) If $\mathbb{F}$ is a finite field, then the characteristic of $\mathbb{F}$ is a prime p and $\{0, 1, 2, \ldots, (p-1)\}$ is a subfield of $\mathbb{F}$.
(c) If $\mathbb{F}$ is a field of characteristic p, then $p\alpha = 0$ for all $\alpha \in \mathbb{F}$.
(d) A finite field $\mathbb{F}$ of characteristic p is a finite dimensional vector space over its prime subfield and contains p^m elements, where m is the dimension of this vector space. ◆

Exercise 152 Let $\mathbb{F}_q$ have characteristic p. Prove that $(\alpha + \beta)^p = \alpha^p + \beta^p$ for all $\alpha, \beta \in \mathbb{F}_q$. ◆

Throughout this chapter we will let p denote a prime number and $q = p^m$, where m is a positive integer.

3.2   Polynomials and the Euclidean Algorithm

Let x be an indeterminate. The set of polynomials in x with coefficients in $\mathbb{F}_q$ is denoted by $\mathbb{F}_q[x]$. This set forms a commutative ring with unity under ordinary polynomial addition and multiplication. A *commutative ring with unity* satisfies the same axioms as a field except the nonzero elements do not necessarily have multiplicative inverses. In fact $\mathbb{F}_q[x]$ is an integral domain as well; recall that an *integral domain* is a commutative ring with unity such that the product of any two nonzero elements in the ring is also nonzero. The ring $\mathbb{F}_q[x]$ plays a key role not only in the construction of finite fields but also in the construction of certain families of codes. So a typical polynomial in $\mathbb{F}_q[x]$ is $f(x) = \sum_{i=0}^{n} a_i x^i$. As usual a_i is the *coefficient* of the *term $a_i x^i$* of *degree i*. The *degree of a polynomial* $f(x)$ is the highest degree of any term with a nonzero coefficient and is denoted $\deg f(x)$; the zero polynomial does not have a degree. The coefficient of the highest degree term is called the *leading* *coefficient.* We will usually write polynomials with terms in either increasing or decreasing degree order; e.g. $a_0 + a_1 x + a_2 x^2 + \cdots + a_n x^n$ or $a_n x^n + a_{n-1} x^{n-1} + \cdots + a_1 x + a_0$.

Exercise 153 Prove that $\mathbb{F}_q[x]$ is a commutative ring with unity and an integral domain as well. ◆

Exercise 154 Multiply $(x^3 + x^2 + 1)(x^3 + x + 1)(x + 1)$ in the ring $\mathbb{F}_2[x]$. Do the same multiplication in $\mathbb{F}_3[x]$. ◆

Exercise 155 Show that the degree of the product of two polynomials is the sum of the degrees of the polynomials. ◆

A polynomial is *monic* provided its leading coefficient is 1. Let $f(x)$ and $g(x)$ be polynomials in $\mathbb{F}_q[x]$. We say that $f(x)$ *divides* $g(x)$, denoted $f(x) \mid g(x)$, if there exists a polynomial $h(x) \in \mathbb{F}_q[x]$ such that $g(x) = f(x)h(x)$. The polynomial $f(x)$ is called a *divisor* or *factor* of $g(x)$. The *greatest common divisor* of $f(x)$ and $g(x)$, at least one of which is nonzero, is the monic polynomial in $\mathbb{F}_q[x]$ of largest degree dividing both $f(x)$ and $g(x)$. The greatest common divisor of two polynomials is uniquely determined and is denoted by $\gcd(f(x), g(x))$. The polynomials $f(x)$ and $g(x)$ are *relatively prime* if $\gcd(f(x), g(x)) = 1$.

Many properties of the ring $\mathbb{F}_q[x]$ are analogous to properties of the integers. One can divide two polynomials and obtain a quotient and remainder just as one can do with integers. In part (i) of the next theorem, we state this fact, which is usually called the Division Algorithm.

Theorem 3.2.1 *Let $f(x)$ and $g(x)$ be in $\mathbb{F}_q[x]$ with $g(x)$ nonzero.*
(i) *(Division Algorithm) There exist unique polynomials $h(x), r(x) \in \mathbb{F}_q[x]$ such that*

$$f(x) = g(x)h(x) + r(x), \qquad \text{where } \deg r(x) < \deg g(x) \text{ or } r(x) = 0.$$

(ii) *If $f(x) = g(x)h(x) + r(x)$, then $\gcd(f(x), g(x)) = \gcd(g(x), r(x))$.*

We can use the Division Algorithm recursively together with part (ii) of the previous theorem to find the gcd of two polynomials. This process is known as the Euclidean Algorithm. The Euclidean Algorithm for polynomials is analogous to the Euclidean Algorithm for integers. We state it in the next theorem and then illustrate it with an example.

Theorem 3.2.2 (Euclidean Algorithm) *Let $f(x)$ and $g(x)$ be polynomials in $\mathbb{F}_q[x]$ with $g(x)$ nonzero.*
(i) *Perform the following sequence of steps until $r_n(x) = 0$ for some n:*

$$f(x) = g(x)h_1(x) + r_1(x), \qquad \text{where } \deg r_1(x) < \deg g(x),$$
$$g(x) = r_1(x)h_2(x) + r_2(x), \qquad \text{where } \deg r_2(x) < \deg r_1(x),$$
$$r_1(x) = r_2(x)h_3(x) + r_3(x), \qquad \text{where } \deg r_3(x) < \deg r_2(x),$$
$$\vdots$$
$$r_{n-3}(x) = r_{n-2}(x)h_{n-1}(x) + r_{n-1}(x), \qquad \text{where } \deg r_{n-1}(x) < \deg r_{n-2}(x),$$
$$r_{n-2}(x) = r_{n-1}(x)h_n(x) + r_n(x), \qquad \text{where } r_n(x) = 0.$$

Then $\gcd(f(x), g(x)) = cr_{n-1}(x)$, *where* $c \in \mathbb{F}_q$ *is chosen so that* $cr_{n-1}(x)$ *is monic.*

(ii) *There exist polynomials* $a(x), b(x) \in \mathbb{F}_q[x]$ *such that*

$$a(x)f(x) + b(x)g(x) = \gcd(f(x), g(x)).$$

The sequence of steps in (i) eventually terminates because at each stage the degree of the remainder decreases by at least 1. By repeatedly applying Theorem 3.2.1(ii), we have that $cr_{n-1}(x) = \gcd(r_{n-2}(x), r_{n-3}(x)) = \gcd(r_{n-3}(x), r_{n-4}(x)) = \cdots = \gcd(f(x), g(x))$. This explains why (i) produces the desired gcd.

Technically, the Euclidean Algorithm is only (i) of this theorem. However, (ii) is a natural consequence of (i), and seems to possess no name of its own in the literature. As we use (ii) so often, we include both in the term "Euclidean Algorithm." Part (ii) is obtained by beginning with the next to last equation $r_{n-3}(x) = r_{n-2}(x)h_{n-1}(x) + r_{n-1}(x)$ in the sequence in (i) and solving for $r_{n-1}(x)$ in terms of $r_{n-2}(x)$ and $r_{n-3}(x)$. Using the previous equation, solve for $r_{n-2}(x)$ and substitute into the equation for $r_{n-1}(x)$ to obtain $r_{n-1}(x)$ as a combination of $r_{n-3}(x)$ and $r_{n-4}(x)$. Continue up through the sequence until we obtain $r_{n-1}(x)$ as a combination of $f(x)$ and $g(x)$. We illustrate all of this in the next example.

Example 3.2.3 We compute $\gcd(x^5 + x^4 + x^2 + 1, x^3 + x^2 + x)$ in the ring $\mathbb{F}_2[x]$ using the Euclidean Algorithm. Part (i) of the algorithm produces the following sequence.

$$x^5 + x^4 + x^2 + 1 = (x^3 + x^2 + x)(x^2 + 1) + x + 1$$
$$x^3 + x^2 + x = (x + 1)(x^2 + 1) + 1$$
$$x + 1 = 1(x + 1) + 0.$$

Thus $1 = \gcd(x + 1, 1) = \gcd(x^3 + x^2 + x, x + 1) = \gcd(x^5 + x^4 + x^2 + 1, x^3 + x^2 + x)$. Now we find $a(x)$ and $b(x)$ such that $a(x)(x^5 + x^4 + x^2 + 1) + b(x)(x^3 + x^2 + x) = 1$ by reversing the above steps. We begin with the last equation in our sequence with a nonzero remainder, which we first solve for. This yields

$$1 = (x^3 + x^2 + x) - (x + 1)(x^2 + 1). \tag{3.1}$$

Now $x + 1$ is the remainder in the first equation in our sequence; solving this for $x + 1$ and plugging into (3.1) produces

$$1 = (x^3 + x^2 + x) - [(x^5 + x^4 + x^2 + 1) - (x^3 + x^2 + x)(x^2 + 1)](x^2 + 1)$$
$$= (x^2 + 1)(x^5 + x^4 + x^2 + 1) + x^4(x^3 + x^2 + x).$$

So $a(x) = x^2 + 1$ and $b(x) = x^4$. ∎

Exercise 156 Find $\gcd(x^6 + x^5 + x^4 + x^3 + x + 1, x^5 + x^3 + x^2 + x)$ in $\mathbb{F}_2[x]$. Find $a(x)$ and $b(x)$ such that $\gcd(x^6 + x^5 + x^4 + x^3 + x + 1, x^5 + x^3 + x^2 + x) = a(x)(x^6 + x^5 + x^4 + x^3 + x + 1) + b(x)(x^5 + x^3 + x^2 + x)$. ◆

Exercise 157 Find $\gcd(x^5 - x^4 + x + 1, x^3 + x)$ in $\mathbb{F}_3[x]$. Find $a(x)$ and $b(x)$ such that $\gcd(x^5 - x^4 + x + 1, x^3 + x) = a(x)(x^5 - x^4 + x + 1) + b(x)(x^3 + x)$. ◆

Exercise 158 Let $f(x)$ and $g(x)$ be polynomials in $\mathbb{F}_q[x]$.
(a) If $k(x)$ is a divisor of $f(x)$ and $g(x)$, prove that $k(x)$ is a divisor of $a(x)f(x) + b(x)g(x)$ for any $a(x), b(x) \in \mathbb{F}_q[x]$.
(b) If $k(x)$ is a divisor of $f(x)$ and $g(x)$, prove that $k(x)$ is a divisor of $\gcd(f(x), g(x))$. ◆

Exercise 159 Let $f(x)$ be a polynomial over $\mathbb{F}_q$ of degree n.
(a) Prove that if $\alpha \in \mathbb{F}_q$ is a root of $f(x)$, then $x - \alpha$ is a factor of $f(x)$.
(b) Prove that $f(x)$ has at most n roots in any field containing $\mathbb{F}_q$. ◆

3.3 Primitive elements

When working with a finite field, one needs to be able to add and multiply as simply as possible. In Theorem 3.1.1(iii), we stated that $\mathbb{F}_q$ is a vector space over $\mathbb{F}_p$ of dimension m. So a simple way to add field elements is to write them as m-tuples over $\mathbb{F}_p$ and add them using ordinary vector addition, as we will see in Section 3.4. Unfortunately, multiplying such m-tuples is far from simple. We need another way to write the field elements so that multiplication is easy; then we need a way to connect this form of the field elements to the m-tuple form. The following theorem will assist us with this. Recall that the set $\mathbb{F}_q^*$ of nonzero elements in $\mathbb{F}_q$ is a group.

Theorem 3.3.1 *We have the following:*
(i) *The group $\mathbb{F}_q^*$ is cyclic of order $q - 1$ under the multiplication of $\mathbb{F}_q$.*
(ii) *If γ is a generator of this cyclic group, then*

$$\mathbb{F}_q = \{0, 1 = \gamma^0, \gamma, \gamma^2, \ldots, \gamma^{q-2}\},$$

and $\gamma^i = 1$ if and only if $(q - 1) \mid i$.

Proof: From the Fundamental Theorem of Finite Abelian Groups [130], $\mathbb{F}_q^*$ is a direct product of cyclic groups of orders $m_1, m_2, \ldots, m_a$, where $m_i \mid m_{i+1}$ for $1 \leq i < a$ and $m_1 m_2 \cdots m_a = q - 1$. In particular $\alpha^{m_a} = 1$ for all $\alpha \in \mathbb{F}_q^*$. Thus the polynomial $x^{m_a} - 1$ has at least $q - 1$ roots, a contradiction to Exercise 159 unless $a = 1$ and $m_a = q - 1$. Thus $\mathbb{F}_q^*$ is cyclic giving (i). Part (ii) follows from the properties of cyclic groups. □

Each generator γ of $\mathbb{F}_q^*$ is called a *primitive element* of $\mathbb{F}_q$. When the nonzero elements of a finite field are expressed as powers of γ, the multiplication in the field is easily carried out according to the rule $\gamma^i \gamma^j = \gamma^{i+j} = \gamma^s$, where $0 \leq s \leq q - 2$ and $i + j \equiv s \pmod{q - 1}$.

Exercise 160 Find all primitive elements in the fields $\mathbb{F}_2$, $\mathbb{F}_3$, and $\mathbb{F}_4$ of Examples 1.1.1, 1.1.2, and 1.1.3. Pick one of the primitive elements γ of $\mathbb{F}_4$ and rewrite the addition and multiplication tables of $\mathbb{F}_4$ using the elements $\{0, 1 = \gamma^0, \gamma, \gamma^2\}$. ◆

Let γ be a primitive element of $\mathbb{F}_q$. Then $\gamma^{q-1} = 1$ by definition. Hence $(\gamma^i)^{q-1} = 1$ for $0 \leq i \leq q - 2$ showing that the elements of $\mathbb{F}_q^*$ are roots of $x^{q-1} - 1 \in \mathbb{F}_p[x]$ and hence of $x^q - x$. As 0 is a root of $x^q - x$, by Exercise 159 we now see that the elements of $\mathbb{F}_q$ are precisely the roots of $x^q - x$ giving this important theorem.

Theorem 3.3.2 *The elements of $\mathbb{F}_q$ are precisely the roots of $x^q - x$.*

In Theorem 3.1.1(v) we claim that the field with $q = p^m$ elements is unique. Theorem 3.3.2 shows that a field with q elements is the smallest field containing $\mathbb{F}_p$ and all the roots of $x^q - x$. Such a field is termed a splitting field of the polynomial $x^q - x$ over $\mathbb{F}_p$, that is, the smallest extension field of $\mathbb{F}_p$ containing all the roots of the polynomial. In general splitting fields of a fixed polynomial over a given field are isomorphic; by carefully defining a map between the roots of the polynomial in one field and the roots in the other, the map can be shown to be an isomorphism of the splitting fields.

Exercise 161 Using the table for $\mathbb{F}_4$ in Example 1.1.3, verify that all the elements of $\mathbb{F}_4$ are roots of $x^4 - x = 0$. ◆

In analyzing the field structure, it will be useful to know the number of primitive elements in $\mathbb{F}_q$ and how to find them all once one primitive element has been found. Since $\mathbb{F}_q^*$ is cyclic, we recall a few facts about finite cyclic groups. In any finite cyclic group $\mathcal{G}$ of order n with generator g, the generators of $\mathcal{G}$ are precisely the elements g^i where $\gcd(i, n) = 1$. We let $\phi(n)$ be the number of integers i with $1 \leq i \leq n$ such that $\gcd(i, n) = 1$; ϕ is called the *Euler totient* or the *Euler ϕ-function*. So there are $\phi(n)$ generators of $\mathcal{G}$. The *order* of an element $\alpha \in \mathcal{G}$ is the smallest positive integer i such that $\alpha^i = 1$. An element of $\mathcal{G}$ has order d if and only if $d \mid n$. Furthermore g^i has order $d = n/\gcd(i, n)$, and there are $\phi(d)$ elements of order d. When speaking of field elements $\alpha \in \mathbb{F}_q^*$, the order of α is its order in the multiplicative group $\mathbb{F}_q^*$. In particular, primitive elements of $\mathbb{F}_q$ are those of order $q - 1$. The next theorem follows from this discussion.

Theorem 3.3.3 *Let γ be a primitive element of $\mathbb{F}_q$.*
(i) *There are $\phi(q - 1)$ primitive elements in $\mathbb{F}_q$; these are the elements γ^i where $\gcd(i, q - 1) = 1$.*
(ii) *For any d where $d \mid (q - 1)$, there are $\phi(d)$ elements in $\mathbb{F}_q$ of order d; these are the elements $\gamma^{(q-1)i/d}$ where $\gcd(i, d) = 1$.*

An element $\xi \in \mathbb{F}_q$ is an *n*th *root of unity* provided $\xi^n = 1$, and is a *primitive nth root of unity* if in addition $\xi^s \neq 1$ for $0 < s < n$. A primitive element γ of $\mathbb{F}_q$ is therefore a primitive $(q - 1)$st root of unity. It follows from Theorem 3.3.1 that the field $\mathbb{F}_q$ contains a primitive *n*th root of unity if and only if $n \mid (q - 1)$, in which case $\gamma^{(q-1)/n}$ is a primitive *n*th root of unity.

Exercise 162 (a) Find a primitive element γ in the field $\mathbb{F}_q$ given below. (b) Then write every nonzero element in $\mathbb{F}_q$ as a power of γ. (c) What is the order of each element and which are primitive? (d) Verify that there are precisely $\phi(d)$ elements of order d for every $d \mid (q - 1)$.
(i) $\mathbb{F}_5$,
(ii) $\mathbb{F}_7$,
(iii) $\mathbb{F}_{13}$. ◆

Exercise 163 Let γ be a primitive element of $\mathbb{F}_q$, where q is odd.
(a) Show that the equation $x^2 = 1$ has only two solutions, 1 and -1.
(b) Show that $\gamma^{(q-1)/2} = -1$. ♦

Exercise 164 If $q \neq 2$, show that

$$\sum_{\alpha \in \mathbb{F}_q} \alpha = 0.$$ ♦

Exercise 165 What is the smallest field of characteristic 2 that contains a:
(a) primitive nineth root of unity?
(b) primitive 11th root of unity?
What is the smallest field of characteristic 3 that contains a:
(c) primitive seventh root of unity?
(d) primitive 11th root of unity? ♦

3.4   Constructing finite fields

We are now ready to link the additive structure of a finite field arising from the vector space interpretation with the multiplicative structure arising from the powers of a primitive element and actually construct finite fields.

A nonconstant polynomial $f(x) \in \mathbb{F}_q[x]$ is *irreducible over* $\mathbb{F}_q$ provided it does not factor into a product of two polynomials in $\mathbb{F}_q[x]$ of smaller degree. The irreducible polynomials in $\mathbb{F}_q[x]$ are like the prime numbers in the ring of integers. For example, every integer greater than 1 is a unique product of positive primes. The same result holds in $\mathbb{F}_q[x]$, making $\mathbb{F}_q[x]$ a unique factorization domain.

Theorem 3.4.1 *Let $f(x)$ be a nonconstant polynomial. Then*

$$f(x) = p_1(x)^{a_1} p_2(x)^{a_2} \cdots p_k(x)^{a_k},$$

where each $p_i(x)$ is irreducible, the $p_i(x)$s are unique up to scalar multiplication, and the a_is are unique.

Not only is $\mathbb{F}_q[x]$ a unique factorization domain, it is also a principal ideal domain. An *ideal* $\mathcal{I}$ in a commutative ring $\mathcal{R}$ is a nonempty subset of the ring that is closed under subtraction such that the product of an element in $\mathcal{I}$ with an element in $\mathcal{R}$ is always in $\mathcal{I}$. The ideal $\mathcal{I}$ is *principal* provided there is an $a \in \mathcal{R}$ such that $\mathcal{I} = \{ra \mid r \in \mathcal{R}\}$; this ideal will be denoted (a). A *principal ideal domain* is an integral domain in which each ideal is principal. Exercises 153 and 166 show that $\mathbb{F}_q[x]$ is a principal ideal domain. (The fact that $\mathbb{F}_q[x]$ is a unique factorization domain actually follows from the fact that it is a principal ideal domain.)

Exercise 166 Show using the Division Algorithm that every ideal of $\mathbb{F}_q[x]$ is a principal ideal. ♦

To construct a field of characteristic p, we begin with a polynomial $f(x) \in \mathbb{F}_p[x]$ which is irreducible over $\mathbb{F}_p$. Suppose that $f(x)$ has degree m. By using the Euclidean Algorithm it can be proved that the *residue class ring*

$$\mathbb{F}_p[x]/(f(x))$$

is actually a field and hence a finite field $\mathbb{F}_q$ with $q = p^m$ elements; see Exercise 167. Every element of the residue class ring is a coset $g(x) + (f(x))$, where $g(x)$ is uniquely determined of degree at most $m - 1$. We can compress the notation by writing the coset as a vector in $\mathbb{F}_p^m$ with the correspondence

$$g_{m-1}x^{m-1} + g_{m-2}x^{m-2} + \cdots + g_1x + g_0 + (f(x)) \leftrightarrow g_{m-1}g_{m-2}\cdots g_1g_0. \tag{3.2}$$

This vector notation allows you to add in the field using ordinary vector addition.

Exercise 167   Let $f(x)$ be an irreducible polynomial of degree m in $\mathbb{F}_p[x]$. Prove that

$$\mathbb{F}_p[x]/(f(x))$$

is a finite field with p^m elements. ♦

Example 3.4.2   The polynomial $f(x) = x^3 + x + 1$ is irreducible over $\mathbb{F}_2$; if it were reducible, it would have a factor of degree 1 and hence a root in $\mathbb{F}_2$, which it does not. So $\mathbb{F}_8 = \mathbb{F}_2[x]/(f(x))$ and, using the correspondence (3.2), the elements of $\mathbb{F}_8$ are given by

Cosets	Vectors
$0 + (f(x))$	000
$1 + (f(x))$	001
$x + (f(x))$	010
$x + 1 + (f(x))$	011
$x^2 + (f(x))$	100
$x^2 + 1 + (f(x))$	101
$x^2 + x + (f(x))$	110
$x^2 + x + 1 + (f(x))$	111

As an illustration of addition, adding $x + (f(x))$ to $x^2 + x + 1 + (f(x))$ yields $x^2 + 1 + (f(x))$, which corresponds to adding 010 to 111 and obtaining 101 in $\mathbb{F}_2^3$. ∎

How do you multiply? To multiply $g_1(x) + (f(x))$ times $g_2(x) + (f(x))$, first use the Division Algorithm to write

$$g_1(x)g_2(x) = f(x)h(x) + r(x), \tag{3.3}$$

where $\deg r(x) \leq m - 1$ or $r(x) = 0$. Then $(g_1(x) + (f(x)))(g_2(x) + (f(x))) = r(x) + (f(x))$. The notation is rather cumbersome and can be simplified if we replace x by α and let $f(\alpha) = 0$; we justify this shortly. From (3.3), $g_1(\alpha)g_2(\alpha) = r(\alpha)$ and we extend our correspondence (3.2) to

$$g_{m-1}g_{m-2}\cdots g_1g_0 \leftrightarrow g_{m-1}\alpha^{m-1} + g_{m-2}\alpha^{m-2} + \cdots + g_1\alpha + g_0. \tag{3.4}$$

So to multiply in $\mathbb{F}_q$, we simply multiply polynomials in α in the ordinary way and use the equation $f(\alpha) = 0$ to reduce powers of α higher than $m - 1$ to polynomials in α of degree less than m. Notice that the subset $\{0\alpha^{m-1} + 0\alpha^{m-2} + \cdots + 0\alpha + a_0 \mid a_0 \in \mathbb{F}_p\} = \{a_0 \mid a_0 \in \mathbb{F}_p\}$ is the prime subfield of $\mathbb{F}_q$.

We continue with our example of $\mathbb{F}_8$ adding this new correspondence. Notice that the group $\mathbb{F}_8^*$ is cyclic of order 7 and hence all nonidentity elements of $\mathbb{F}_8^*$ are primitive. In particular α is primitive, and we include powers of α in our table below.

Example 3.4.3 Continuing with Example 3.4.2, using correspondence (3.4), we obtain

Vectors	Polynomials in α	Powers of α
000	0	0
001	1	$1 = \alpha^0$
010	α	α
011	$\alpha + 1$	α^3
100	α^2	α^2
101	$\alpha^2 + 1$	α^6
110	$\alpha^2 + \alpha$	α^4
111	$\alpha^2 + \alpha + 1$	α^5

The column "powers of α" is obtained by using $f(\alpha) = \alpha^3 + \alpha + 1 = 0$, which implies that $\alpha^3 = \alpha + 1$. So $\alpha^4 = \alpha\alpha^3 = \alpha(\alpha + 1) = \alpha^2 + \alpha$, $\alpha^5 = \alpha\alpha^4 = \alpha(\alpha^2 + \alpha) = \alpha^3 + \alpha^2 = \alpha^2 + \alpha + 1$, etc. ■

Exercise 168 In the field $\mathbb{F}_8$ given in Example 3.4.3, simplify

$$(\alpha^2 + \alpha^6 - \alpha + 1)(\alpha^3 + \alpha)/(\alpha^4 + \alpha).$$

Hint: Use the vector form of the elements to do additions and subtractions and the powers of α to do multiplications and divisions. ◆

We describe this construction by saying that $\mathbb{F}_q$ is obtained from $\mathbb{F}_p$ by "adjoining" a root α of $f(x)$ to $\mathbb{F}_p$. This root α is formally given by $\alpha = x + (f(x))$ in the residue class ring $\mathbb{F}_p[x]/(f(x))$; therefore $g(x) + (f(x)) = g(\alpha)$ and $f(\alpha) = f(x + (f(x))) = f(x) + (f(x)) = 0 + (f(x))$.

In Example 3.4.3, we were fortunate that α was a primitive element of $\mathbb{F}_8$. In general, this will not be the case. We say that an irreducible polynomial over $\mathbb{F}_p$ of degree m is *primitive* provided that it has a root that is a primitive element of $\mathbb{F}_q = \mathbb{F}_{p^m}$. Ideally we would like to start with a primitive polynomial to construct our field, but that is not a requirement (see Exercise 174). It is worth noting that the irreducible polynomial we begin with can be multiplied by a constant to make it monic as that has no effect on the ideal generated by the polynomial or the residue class ring.

It is not obvious, but either by using the theory of splitting fields or by counting the number of irreducible polynomials over a finite field, one can show that irreducible polynomials of any degree exist. In particular we have the following result.

Theorem 3.4.4 *For any prime p and any positive integer m, there exists a finite field, unique up to isomorphism, with $q = p^m$ elements.*

Since constructing finite fields requires irreducible polynomials, we note that tables of irreducible and primitive polynomials over $\mathbb{F}_2$ can be found in [256].

Remark: In the construction of $\mathbb{F}_q$ by adjoining a root of an irreducible polynomial $f(x)$ to $\mathbb{F}_p$, the field $\mathbb{F}_p$ can be replaced by any finite field $\mathbb{F}_r$, where r is a power of p and $f(x)$ by an irreducible polynomial of degree m in $\mathbb{F}_r[x]$ for some positive integer m. The field constructed contains $\mathbb{F}_r$ as a subfield and is of order r^m.

Exercise 169
(a) Find all irreducible polynomials of degrees 1, 2, 3, and 4 over $\mathbb{F}_2$.
(b) Compute the product of all irreducible polynomials of degrees 1 and 2 in $\mathbb{F}_2[x]$.
(c) Compute the product of all irreducible polynomials of degrees 1 and 3 in $\mathbb{F}_2[x]$.
(d) Compute the product of all irreducible polynomials of degrees 1, 2, and 4 in $\mathbb{F}_2[x]$.
(e) Make a conjecture based on the results of (b), (c), and (d).
(f) In part (a), you found two irreducible polynomials of degree 3. The roots of these polynomials lie in $\mathbb{F}_8$. Using the table in Example 3.4.3 find the roots of these two polynomials as powers of α. ◆

Exercise 170 Find all monic irreducible polynomials of degrees 1 and 2 over $\mathbb{F}_3$. Then compute their product in $\mathbb{F}_3[x]$. Does this result confirm your conjecture of Exercise 169(e)? If not, modify your conjecture. ◆

Exercise 171 Find all monic irreducible polynomials of degrees 1 and 2 over $\mathbb{F}_4$. Then compute their product in $\mathbb{F}_4[x]$. Does this result confirm your conjecture of Exercise 169(e) or your modified conjecture in Exercise 170? If not, modify your conjecture again. ◆

Exercise 172 In Exercise 169, you found an irreducible polynomial of degree 3 different from the one used to construct $\mathbb{F}_8$ in Examples 3.4.2 and 3.4.3. Let β be a root of this second polynomial and construct the field $\mathbb{F}_8$ by adjoining β to $\mathbb{F}_2$ and giving a table with each vector in $\mathbb{F}_2^3$ associated to 0 and the powers of β. ◆

Exercise 173 By Exercise 169, the polynomial $f(x) = x^4 + x + 1 \in \mathbb{F}_2[x]$ is irreducible over $\mathbb{F}_2$. Let α be a root of $f(x)$.
(a) Construct the field $\mathbb{F}_{16}$ by adjoining α to $\mathbb{F}_2$ and giving a table with each vector in $\mathbb{F}_2^4$ associated to 0 and the powers of α.
(b) Which powers of α are primitive elements of $\mathbb{F}_{16}$?
(c) Find the roots of the irreducible polynomials of degrees 1, 2, and 4 from Exercise 169(a). ◆

Exercise 174 Let $f(x) = x^2 + x + 1 \in \mathbb{F}_5[x]$.
(a) Prove that $f(x)$ is irreducible over $\mathbb{F}_5$.
(b) By part (a) $\mathbb{F}_{25} = \mathbb{F}_5[x]/(f(x))$. Let α be a root of $f(x)$. Show that α is not primitive.
(c) Find a primitive element in $\mathbb{F}_{25}$ of the form $a\alpha + b$, where $a, b \in \mathbb{F}_5$. ◆

Exercise 175 By Exercise 169, $x^2 + x + 1$, $x^3 + x + 1$, and $x^4 + x + 1$ are irreducible over $\mathbb{F}_2$. Is $x^5 + x + 1$ irreducible over $\mathbb{F}_2$?   ♦

Exercise 176 Define a function $\tau : \mathbb{F}_q \to \mathbb{F}_q$ by $\tau(0) = 0$ and $\tau(\alpha) = \alpha^{-1}$ for $\alpha \in \mathbb{F}_q^*$.
(a) Show that $\tau(ab) = \tau(a)\tau(b)$ for all $a, b \in \mathbb{F}_q$.
(b) Show that if $q = 2, 3$, or 4, then $\tau(a + b) = \tau(a) + \tau(b)$.
(c) Show that if $\tau(a + b) = \tau(a) + \tau(b)$ for all $a, b \in \mathbb{F}_q$, then $q = 2, 3$, or 4. Hint: Let $\alpha \in \mathbb{F}_q$ with $\alpha + \alpha^2 \neq 0$. Then set $a = \alpha$ and $b = \alpha^2$.   ♦

3.5   Subfields

In order to understand the structure of a finite field $\mathbb{F}_q$, we must find the subfields contained in $\mathbb{F}_q$.

Recall that $\mathbb{F}_q$ has a primitive element of order $q - 1 = p^m - 1$. If $\mathbb{F}_s$ is a subfield of $\mathbb{F}_q$, then $\mathbb{F}_s$ has a primitive element of order $s - 1$ where $(s - 1) \mid (q - 1)$. Because the identity element 1 is the same for both $\mathbb{F}_q$ and $\mathbb{F}_s$, $\mathbb{F}_s$ has characteristic p implying that $s = p^r$. So it is necessary to have $(p^r - 1) \mid (p^m - 1)$. The following lemma shows when that can happen.

Lemma 3.5.1 *Let $a > 1$ be an integer. Then $(a^r - 1) \mid (a^m - 1)$ if and only if $r \mid m$.*

Proof: If $r \mid m$, then $m = rh$ and $a^m - 1 = (a^r - 1) \sum_{i=0}^{h-1} a^{ir}$. Conversely, by the Division Algorithm, let $m = rh + u$, with $0 \leq u < r$. Then

$$\frac{a^m - 1}{a^r - 1} = a^u \frac{a^{rh} - 1}{a^r - 1} + \frac{a^u - 1}{a^r - 1}.$$

As $r \mid rh$, by the above, $(a^{rh} - 1)/(a^r - 1)$ is an integer; also $(a^u - 1)/(a^r - 1)$ is strictly less than 1. Thus for $(a^m - 1)/(a^r - 1)$ to be an integer, $(a^u - 1)/(a^r - 1)$ must be 0 and so $u = 0$ yielding $r \mid m$.   □

The same argument used in the proof of Lemma 3.5.1 shows that $(x^{s-1} - 1) \mid (x^{q-1} - 1)$ if and only if $(s - 1) \mid (q - 1)$. Thus $(x^s - x) \mid (x^q - x)$ if and only if $(s - 1) \mid (q - 1)$. So if $s = p^r$, $(x^s - x) \mid (x^q - x)$ if and only if $r \mid m$ by Lemma 3.5.1. So we have the following lemma.

Lemma 3.5.2 *Let $s = p^r$ and $q = p^m$. Then $(x^s - x) \mid (x^q - x)$ if and only if $r \mid m$.*

In particular if $r \mid m$, all of the roots of $x^{p^r} - x$ are in $\mathbb{F}_q$. Exercise 177 shows that the roots of $x^{p^r} - x$ in $\mathbb{F}_q$ form a subfield of $\mathbb{F}_q$. Since any subfield of order p^r must consist of the roots of $x^{p^r} - x$ in $\mathbb{F}_q$, this subfield must be unique. The following theorem summarizes this discussion and completely characterizes the subfield structure of $\mathbb{F}_q$.

Theorem 3.5.3 *When $q = p^m$,*
(i)   *$\mathbb{F}_q$ has a subfield of order $s = p^r$ if and only if $r \mid m$,*
(ii) *the elements of the subfield $\mathbb{F}_s$ are exactly the elements of $\mathbb{F}_q$ that are roots of $x^s - x$, and*
(iii) *for each $r \mid m$ there is only one subfield $\mathbb{F}_{p^r}$ of $\mathbb{F}_q$.*

Corollary 3.5.4 *If γ is a primitive element of $\mathbb{F}_q$ and $\mathbb{F}_s$ is a subfield of $\mathbb{F}_q$, then the elements of $\mathbb{F}_s$ are $\{0, 1, \alpha, \ldots, \alpha^{s-2}\}$, where $\alpha = \gamma^{(q-1)/(s-1)}$.*

Exercise 177 If $q = p^m$ and $r \mid m$, prove that the roots of $x^{p^r} - x$ in $\mathbb{F}_q$ form a subfield of $\mathbb{F}_q$. Hint: See Exercise 152. ◆

We can picture the subfield structure very nicely using a lattice as the next example shows.

Example 3.5.5 The lattice of subfields of $\mathbb{F}_{2^{24}}$ is:

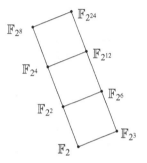

If $\mathbb{F}$ is a subfield of $\mathbb{E}$, we also say that $\mathbb{E}$ is an *extension* field of $\mathbb{F}$. In the lattice of Example 3.5.5 we connect two fields $\mathbb{F}$ and $\mathbb{E}$ if $\mathbb{F} \subset \mathbb{E}$ with no proper subfields between; the extension field $\mathbb{E}$ is placed above $\mathbb{F}$. From this lattice one can find the intersection of two subfields as well as the smallest subfield containing two other subfields.

Exercise 178 Draw the lattice of subfields of $\mathbb{F}_{2^{30}}$. ◆

Corollary 3.5.6 *The prime subfield of $\mathbb{F}_q$ consists of those elements α in $\mathbb{F}_q$ that satisfy the equation $x^p = x$.*

Exercise 179 Prove Corollary 3.5.6. ◆

3.6 Field automorphisms

The field automorphisms of $\mathbb{F}_q$ form a group under function composition. We can describe this group completely.

Recall that an *automorphism* σ of $\mathbb{F}_q$ is a bijection $\sigma : \mathbb{F}_q \to \mathbb{F}_q$ such that $\sigma(\alpha + \beta) = \sigma(\alpha) + \sigma(\beta)$ and $\sigma(\alpha\beta) = \sigma(\alpha)\sigma(\beta)$ for all $\alpha, \beta \in \mathbb{F}_q$.[1] Define $\sigma_p : \mathbb{F}_q \to \mathbb{F}_q$ by

$$\sigma_p(\alpha) = \alpha^p \quad \text{for all } \alpha \in \mathbb{F}_q.$$

---

[1] In Section 1.7, where we used field automorphisms to define equivalence, the field automorphism σ acted on the right $x \to (x)\sigma$, because the monomial maps act most naturally on the right. Here we have σ act on the left by $x \to \sigma(x)$, because it is probably more natural to the reader. They are interchangeable because two field automorphisms σ and τ commute by Theorem 3.6.1, implying that the right action $(x)(\sigma\tau) = ((x)\sigma)\tau$ agrees with the left action $(\sigma\tau)(x) = (\tau\sigma)(x) = \tau(\sigma(x))$.

Obviously $\sigma_p(\alpha\beta) = \sigma_p(\alpha)\sigma_p(\beta)$, and $\sigma_p(\alpha + \beta) = \sigma_p(\alpha) + \sigma_p(\beta)$ follows from Exercise 152. As σ_p has kernel $\{0\}$, σ_p is an automorphism of $\mathbb{F}_q$, called the *Frobenius automorphism*. Analogously, define $\sigma_{p^r}(\alpha) = \alpha^{p^r}$.

The automorphism group of $\mathbb{F}_q$, denoted $\mathrm{Gal}(\mathbb{F}_q)$, is called the *Galois group* of $\mathbb{F}_q$. We have the following theorem characterizing this group. Part (ii) of this theorem follows from Corollary 3.5.6.

Theorem 3.6.1
(i) $\mathrm{Gal}(\mathbb{F}_q)$ *is cyclic of order m and is generated by the Frobenius automorphism σ_p.*
(ii) *The prime subfield of $\mathbb{F}_q$ is precisely the set of elements in $\mathbb{F}_q$ such that $\sigma_p(\alpha) = \alpha$.*
(iii) *The subfield $\mathbb{F}_q$ of $\mathbb{F}_{q^t}$ is precisely the set of elements in $\mathbb{F}_{q^t}$ such that $\sigma_q(\alpha) = \alpha$.*

We use σ_p to denote the Frobenius automorphism of any field of characteristic p. If $\mathbb{E}$ and $\mathbb{F}$ are fields of characteristic p with $\mathbb{E}$ an extension field of $\mathbb{F}$, then the Frobenius automorphism of $\mathbb{E}$ when restricted to $\mathbb{F}$ is the Frobenius automorphism of $\mathbb{F}$. An element $\alpha \in \mathbb{F}$ is *fixed* by an automorphism σ of $\mathbb{F}$ provided $\sigma(\alpha) = \alpha$. Let $r \mid m$. Then σ_p^r generates a cyclic subgroup of $\mathrm{Gal}(\mathbb{F}_q)$ of order m/r. By Exercises 180 and 181, the elements of $\mathbb{F}_q$ fixed by this subgroup are precisely the elements of the subfield $\mathbb{F}_{p^r}$. We let $\mathrm{Gal}(\mathbb{F}_q : \mathbb{F}_{p^r})$ denote automorphisms of $\mathbb{F}_q$ which fix $\mathbb{F}_{p^r}$. Our discussion shows that $\sigma_p^r \in \mathrm{Gal}(\mathbb{F}_q : \mathbb{F}_{p^r})$. The following theorem strengthens this.

Theorem 3.6.2 $\mathrm{Gal}(\mathbb{F}_q : \mathbb{F}_{p^r})$ *is the cyclic group generated by σ_p^r.*

Exercise 180   Let σ be an automorphism of a field $\mathbb{F}$. Prove that the elements of $\mathbb{F}$ fixed by σ form a subfield of $\mathbb{F}$.   ◆

Exercise 181   Prove that if $r \mid m$, then the elements in $\mathbb{F}_q$ fixed by σ_p^r are exactly the elements of the subfield $\mathbb{F}_{p^r}$.   ◆

Exercise 182   Prove Theorem 3.6.2.   ◆

3.7   Cyclotomic cosets and minimal polynomials

Let $\mathbb{E}$ be a finite extension field of $\mathbb{F}_q$. Then $\mathbb{E}$ is a vector space over $\mathbb{F}_q$ and so $\mathbb{E} = \mathbb{F}_{q^t}$ for some positive integer t. By Theorem 3.3.2, each element α of $\mathbb{E}$ is a root of the polynomial $x^{q^t} - x$. Thus there is a monic polynomial $M_\alpha(x)$ in $\mathbb{F}_q[x]$ of smallest degree which has α as a root; this polynomial is called the *minimal polynomial of α over $\mathbb{F}_q$*. In the following theorem we collect some elementary facts about minimal polynomials.

Theorem 3.7.1 *Let $\mathbb{F}_{q^t}$ be an extension field of $\mathbb{F}_q$ and let α be an element of $\mathbb{F}_{q^t}$ with minimal polynomial $M_\alpha(x)$ in $\mathbb{F}_q[x]$. The following hold:*
(i) $M_\alpha(x)$ *is irreducible over $\mathbb{F}_q$.*
(ii) *If $g(x)$ is any polynomial in $\mathbb{F}_q[x]$ satisfying $g(\alpha) = 0$, then $M_\alpha(x) \mid g(x)$.*

(iii) $M_\alpha(x)$ *is unique; that is, there is only one monic polynomial in* $\mathbb{F}_q[x]$ *of smallest degree which has* α *as a root.*

Exercise 183 Prove Theorem 3.7.1.   ♦

If we begin with an irreducible polynomial $f(x)$ over $\mathbb{F}_q$ of degree r, we can adjoin a root of $f(x)$ to $\mathbb{F}_q$ and obtain the field $\mathbb{F}_{q^r}$. Amazingly, all the roots of $f(x)$ lie in $\mathbb{F}_{q^r}$.

Theorem 3.7.2 *Let* $f(x)$ *be a monic irreducible polynomial over* $\mathbb{F}_q$ *of degree* r. *Then:*
(i)   *all the roots of* $f(x)$ *lie in* $\mathbb{F}_{q^r}$ *and in any field containing* $\mathbb{F}_q$ *along with one root of* $f(x)$,
(ii) $f(x) = \prod_{i=1}^{r}(x - \alpha_i)$, *where* $\alpha_i \in \mathbb{F}_{q^r}$ *for* $1 \le i \le r$, *and*
(iii) $f(x) \mid x^{q^r} - x$.

Proof: Let α be a root of $f(x)$ which we adjoin to $\mathbb{F}_q$ to form a field $\mathbb{E}_\alpha$ with q^r elements. If β is another root of $f(x)$, not in $\mathbb{E}_\alpha$, it is a root of some irreducible factor, over $\mathbb{E}_\alpha$, of $f(x)$. Adjoining β to $\mathbb{E}_\alpha$ forms an extension field $\mathbb{E}$ of $\mathbb{E}_\alpha$. However, inside $\mathbb{E}$, there is a subfield $\mathbb{E}_\beta$ obtained by adjoining β to $\mathbb{F}_q$. $\mathbb{E}_\beta$ must have q^r elements as $f(x)$ is irreducible of degree r over $\mathbb{F}_q$. Since $\mathbb{E}_\alpha$ and $\mathbb{E}_\beta$ are subfields of $\mathbb{E}$ of the same size, by Theorem 3.5.3(iii), $\mathbb{E}_\alpha = \mathbb{E}_\beta$ proving that all roots of $f(x)$ lie in $\mathbb{F}_{q^r}$. Since any field containing $\mathbb{F}_q$ and one root of $f(x)$ contains $\mathbb{F}_{q^r}$, part (i) follows. Part (ii) now follows from Exercise 159. Part (iii) follows from part (ii) and the fact that $x^{q^r} - x = \prod_{\alpha \in \mathbb{F}_{q^r}}(x - \alpha)$ by Theorem 3.3.2.   □

In particular this theorem holds for minimal polynomials $M_\alpha(x)$ over $\mathbb{F}_q$ as such polynomials are monic irreducible.

Theorem 3.7.3 *Let* $\mathbb{F}_{q^t}$ *be an extension field of* $\mathbb{F}_q$ *and let* α *be an element of* $\mathbb{F}_{q^t}$ *with minimal polynomial* $M_\alpha(x)$ *in* $\mathbb{F}_q[x]$. *The following hold:*
(i) $M_\alpha(x) \mid x^{q^t} - x$.
(ii) $M_\alpha(x)$ *has distinct roots all lying in* $\mathbb{F}_{q^t}$.
(iii) *The degree of* $M_\alpha(x)$ *divides* t.
(iv) $x^{q^t} - x = \prod_\alpha M_\alpha(x)$, *where* α *runs through some subset of* $\mathbb{F}_{q^t}$ *which enumerates the minimal polynomials of all elements of* $\mathbb{F}_{q^t}$ *exactly once.*
(v) $x^{q^t} - x = \prod_f f(x)$, *where* f *runs through all monic irreducible polynomials over* $\mathbb{F}_q$ *whose degree divides* t.

Proof: Part (i) follows from Theorem 3.7.1(ii), since $\alpha^{q^t} - \alpha = 0$ by Theorem 3.3.2. Since the roots of $x^{q^t} - x$ are the q^t elements of $\mathbb{F}_{q^t}$, $x^{q^t} - x$ has distinct roots, and so (i) and Theorem 3.7.2(i) imply (ii). By Theorem 3.4.1 $x^{q^t} - x = \prod_{i=1}^{n} p_i(x)$, where $p_i(x)$ is irreducible over $\mathbb{F}_q$. As $x^{q^t} - x$ has distinct roots, the factors $p_i(x)$ are distinct. By scaling them, we may assume that each is monic as $x^{q^t} - x$ is monic. So $p_i(x) = M_\alpha(x)$ for any $\alpha \in \mathbb{F}_{q^t}$ with $p_i(\alpha) = 0$. Thus (iv) holds. But if $M_\alpha(x)$ has degree r, adjoining α to $\mathbb{F}_q$ gives the subfield $\mathbb{F}_{q^r} = \mathbb{F}_{p^{mr}}$ of $\mathbb{F}_{q^t} = \mathbb{F}_{p^{mt}}$ implying $mr \mid mt$ by Theorem 3.5.3 and hence (iii). Part (v) follows from (iv) if we show that every monic irreducible polynomial over $\mathbb{F}_q$ of degree r dividing t is a factor of $x^{q^t} - x$. But $f(x) \mid (x^{q^r} - x)$ by Theorem 3.7.2(iii). Since $mr \mid mt$, $(x^{q^r} - x) \mid (x^{q^t} - x)$ by Lemma 3.5.2.   □

Two elements of $\mathbb{F}_{q^t}$ which have the same minimal polynomial in $\mathbb{F}_q[x]$ are called *conjugate over* $\mathbb{F}_q$. It will be important to find all the conjugates of $\alpha \in \mathbb{F}_q$, that is, all the roots of $M_\alpha(x)$. We know by Theorem 3.7.3(ii) that the roots of $M_\alpha(x)$ are distinct and lie in $\mathbb{F}_{q^t}$. We can find these roots with the assistance of the following theorem.

Theorem 3.7.4 *Let $f(x)$ be a polynomial in $\mathbb{F}_q[x]$ and let α be a root of $f(x)$ in some extension field $\mathbb{F}_{q^t}$. Then:*
(i) $f(x^q) = f(x)^q$, *and*
(ii) α^q *is also a root of $f(x)$ in $\mathbb{F}_q$.*

Proof: Let $f(x) = \sum_{i=0}^{n} a_i x^i$. Since $q = p^m$, where p is the characteristic of $\mathbb{F}_q$, $f(x)^q = \sum_{i=0}^{n} a_i^q x^{iq}$ by applying Exercise 152 repeatedly. However, $a_i^q = a_i$, because $a_i \in \mathbb{F}_q$ and elements of $\mathbb{F}_q$ are roots of $x^q - x$ by Theorem 3.3.2. Thus (i) holds. In particular, $f(\alpha^q) = f(\alpha)^q = 0$, implying (ii). $\square$

Repeatedly applying this theorem we see that $\alpha, \alpha^q, \alpha^{q^2}, \ldots$ are all roots of $M_\alpha(x)$. Where does this sequence stop? It will stop after r terms, where $\alpha^{q^r} = \alpha$. Suppose now that γ is a primitive element of $\mathbb{F}_{q^t}$. Then $\alpha = \gamma^s$ for some s. Hence $\alpha^{q^r} = \alpha$ if and only if $\gamma^{sq^r - s} = 1$. By Theorem 3.3.1(ii), $sq^r \equiv s \pmod{q^t - 1}$. Based on this, we define the q-cyclotomic coset of s modulo $q^t - 1$ to be the set

$$C_s = \{s, sq, \ldots, sq^{r-1}\} \pmod{q^t - 1},$$

where r is the smallest positive integer such that $sq^r \equiv s \pmod{q^t - 1}$. The sets C_s partition the set $\{0, 1, 2, \ldots, q^t - 2\}$ of integers into disjoint sets. When listing the cyclotomic cosets, it is usual to list C_s only once, where s is the smallest element of the coset.

Example 3.7.5 The 2-cyclotomic cosets modulo 15 are $C_0 = \{0\}$, $C_1 = \{1, 2, 4, 8\}$, $C_3 = \{3, 6, 12, 9\}$, $C_5 = \{5, 10\}$, and $C_7 = \{7, 14, 13, 11\}$. ∎

Exercise 184 Compute the 2-cyclotomic cosets modulo:
(a) 7,
(b) 31,
(c) 63. ◆

Exercise 185 Compute the 3-cyclotomic cosets modulo:
(a) 8,
(b) 26. ◆

Exercise 186 Compute the 4-cyclotomic cosets modulo:
(a) 15,
(b) 63.
Compare your answers to those of Example 3.7.5 and Exercise 184. ◆

We now know that the roots of $M_\alpha(x) = M_{\gamma^s}(x)$ include $\{\gamma^i \mid i \in C_s\}$. In fact these are all of the roots. So if we know the size of C_s, we know the degree of $M_{\gamma^s}(x)$, as these are the same.

Theorem 3.7.6 *If γ is a primitive element of $\mathbb{F}_{q^t}$, then the minimal polynomial of γ^s over $\mathbb{F}_q$ is*

$$M_{\gamma^s}(x) = \prod_{i \in C_s}(x - \gamma^i).$$

Proof: This theorem is claiming that, when expanded, $f(x) = \prod_{i \in C_s}(x - \gamma^i) = \sum_j f_j x^j$ is a polynomial in $\mathbb{F}_q[x]$, not merely $\mathbb{F}_{q^t}[x]$. Let $g(x) = f(x)^q$. Then $g(x) = \prod_{i \in C_s}(x^q - \gamma^{qi})$ by Exercise 152. As C_s is a q-cyclotomic coset, qi runs through C_s as i does. Thus $g(x) = f(x^q) = \sum_j f_j x^{qj}$. But $g(x) = (\sum_j f_j x^j)^q = \sum_j f_j^q x^{qj}$, again by Exercise 152. Equating coefficients, we have $f_j^q = f_j$ and hence by Theorem 3.6.1(iii), $f(x) \in \mathbb{F}_q[x]$. $\square$

Exercise 187 Prove that the size r of a q-cyclotomic coset modulo $q^t - 1$ satisfies $r \mid t$. ◆

Example 3.7.7 The field $\mathbb{F}_8$ was constructed in Examples 3.4.2 and 3.4.3. In the table below we give the minimal polynomial over $\mathbb{F}_2$ of each element of $\mathbb{F}_8$ and the associated 2-cyclotomic coset modulo 7.

Roots	Minimal polynomial	2-cyclotomic coset
0	x	
1	$x + 1$	$\{0\}$
$\alpha, \alpha^2, \alpha^4$	$x^3 + x + 1$	$\{1, 2, 4\}$
$\alpha^3, \alpha^5, \alpha^6$	$x^3 + x^2 + 1$	$\{3, 5, 6\}$

Notice that $x^8 - x = x(x + 1)(x^3 + x + 1)(x^3 + x^2 + 1)$ is the factorization of $x^8 - x$ into irreducible polynomials in $\mathbb{F}_2[x]$ consistent with Theorem 3.7.3(iv). The polynomials $x^3 + x + 1$ and $x^3 + x^2 + 1$ are primitive polynomials of degree 3. ∎

Example 3.7.8 In Exercise 173, you are asked to construct $\mathbb{F}_{16}$ using the irreducible polynomial $x^4 + x + 1$ over $\mathbb{F}_2$. With α as a root of this polynomial, we give the minimal polynomial over $\mathbb{F}_2$ of each element of $\mathbb{F}_{16}$ and the associated 2-cyclotomic coset modulo 15 in the table below.

Roots	Minimal polynomial	2-cyclotomic coset
0	x	
1	$x + 1$	$\{0\}$
$\alpha, \alpha^2, \alpha^4, \alpha^8$	$x^4 + x + 1$	$\{1, 2, 4, 8\}$
$\alpha^3, \alpha^6, \alpha^9, \alpha^{12}$	$x^4 + x^3 + x^2 + x + 1$	$\{3, 6, 9, 12\}$
α^5, α^{10}	$x^2 + x + 1$	$\{5, 10\}$
$\alpha^7, \alpha^{11}, \alpha^{13}, \alpha^{14}$	$x^4 + x^3 + 1$	$\{7, 11, 13, 14\}$

The factorization of $x^{15} - 1$ into irreducible polynomials in $\mathbb{F}_2[x]$ is

$$(x + 1)(x^4 + x + 1)(x^4 + x^3 + x^2 + x + 1)(x^2 + x + 1)(x^4 + x^3 + 1),$$

again consistent with Theorem 3.7.3(iv). ∎

Exercise 188 Referring to Example 3.7.8:

(a) verify that the table is correct,

(b) find the elements of $\mathbb{F}_{16}$ that make up the subfields $\mathbb{F}_2$ and $\mathbb{F}_4$, and

(c) find which irreducible polynomials of degree 4 are primitive.   ♦

Exercise 189 The irreducible polynomials of degree 2 over $\mathbb{F}_3$ were found in Exercise 170.

(a) Which of the irreducible polynomials of degree 2 are primitive?

(b) Let α be a root of one of these primitive polynomials. Construct the field $\mathbb{F}_9$ by adjoining α to $\mathbb{F}_3$ and giving a table with each vector in $\mathbb{F}_3^2$ associated to 0 and the powers of α.

(c) In a table as in Examples 3.7.7 and 3.7.8, give the minimal polynomial over $\mathbb{F}_3$ of each element of $\mathbb{F}_9$ and the associated 3-cyclotomic coset modulo 8.

(d) Verify that the product of all the polynomials in your table is $x^9 - x$.   ♦

Exercise 190 Without factoring $x^{63} - 1$, how many irreducible factors does it have over $\mathbb{F}_2$ and what are their degrees? Answer the same question about $x^{63} - 1$ over $\mathbb{F}_4$. See Exercises 184 and 186.   ♦

Exercise 191 Without factoring $x^{26} - 1$, how many irreducible factors does it have over $\mathbb{F}_3$ and what are their degrees? See Exercise 185.   ♦

Exercise 192 Let $f(x) = f_0 + f_1 x + \cdots + f_a x^a$ be a polynomial of degree a in $\mathbb{F}_q[x]$. The *reciprocal polynomial* of $f(x)$ is the polynomial

$$f^*(x) = x^a f(x^{-1}) = f_a + f_{a-1} x + \cdots + f_0 x^a.$$

(We will study reciprocal polynomials further in Chapter 4.)

(a) Give the reciprocal polynomial of each of the polynomials in the table of Example 3.7.7.

(b) Give the reciprocal polynomial of each of the polynomials in the table of Example 3.7.8.

(c) What do you notice about the roots and the irreducibility of the reciprocal polynomials that you found in parts (a) and (b)?

(d) How can you use what you have learned about reciprocal polynomials in parts (a), (b), and (c) to help find irreducible factors of $x^q - x$ over $\mathbb{F}_p$ where $q = p^m$ with p a prime?   ♦

3.8   Trace and subfield subcodes

Suppose that we have a code $\mathcal{C}$ over a field $\mathbb{F}_{q^t}$. It is possible that some of the codewords have all their components in the subfield $\mathbb{F}_q$. Can much be said about the code consisting of such codewords? Using the trace function, a surprising amount can be said.

Let $\mathcal{C}$ be an $[n, k]$ code over $\mathbb{F}_{q^t}$. The *subfield subcode* $\mathcal{C}|_{\mathbb{F}_q}$ of $\mathcal{C}$ with respect to $\mathbb{F}_q$ is the set of codewords in $\mathcal{C}$ each of whose components is in $\mathbb{F}_q$. Because $\mathcal{C}$ is linear over $\mathbb{F}_{q^t}$, $\mathcal{C}|_{\mathbb{F}_q}$ is a linear code over $\mathbb{F}_q$.

We first describe how to find a parity check matrix for $\mathcal{C}|_{\mathbb{F}_q}$ beginning with a parity check matrix H of $\mathcal{C}$. Because $\mathbb{F}_{q^t}$ is a vector space of dimension t over $\mathbb{F}_q$, we can choose a basis $\{b_1, b_2, \ldots, b_t\} \subset \mathbb{F}_{q^t}$ of $\mathbb{F}_{q^t}$ over $\mathbb{F}_q$. Each element $z \in \mathbb{F}_{q^t}$ can be uniquely written as $z = z_1 b_1 + \cdots + z_t b_t$, where $z_i \in \mathbb{F}_q$ for $1 \leq i \leq t$. Associate to z the $t \times 1$ column vector $\tilde{z} = [z_1 \cdots z_t]^T$. Create $\tilde{H}$ from H by replacing each entry h by $\tilde{h}$. Because H is an $(n-k) \times n$ matrix with entries in $\mathbb{F}_{q^t}$, $\tilde{H}$ is a $t(n-k) \times n$ matrix over $\mathbb{F}_q$. The rows of $\tilde{H}$ may be dependent. So a parity check matrix for $\mathcal{C}|_{\mathbb{F}_q}$ is obtained from $\tilde{H}$ by deleting dependent rows; the details of this are left as an exercise. Denote this parity check matrix by $H|_{\mathbb{F}_q}$.

Exercise 193 Prove that by deleting dependent rows from $\tilde{H}$, a parity check matrix for $\mathcal{C}|_{\mathbb{F}_q}$ is obtained. ◆

Example 3.8.1 A parity check matrix for the $[6, 3, 4]$ hexacode $\mathcal{G}_6$ over $\mathbb{F}_4$ given in Example 1.3.4 is

$$H = \begin{bmatrix} 1 & \omega & \omega & 1 & 0 & 0 \\ \omega & 1 & \omega & 0 & 1 & 0 \\ \omega & \omega & 1 & 0 & 0 & 1 \end{bmatrix}.$$

The set $\{1, \omega\}$ is a basis of $\mathbb{F}_4$ over $\mathbb{F}_2$. So

$$\tilde{0} = \begin{bmatrix} 0 \\ 0 \end{bmatrix}, \quad \tilde{1} = \begin{bmatrix} 1 \\ 0 \end{bmatrix}, \quad \tilde{\omega} = \begin{bmatrix} 0 \\ 1 \end{bmatrix}, \quad \text{and} \quad \tilde{\overline{\omega}} = \begin{bmatrix} 1 \\ 1 \end{bmatrix}.$$

Thus

$$\tilde{H} = \begin{bmatrix} 1 & 0 & 0 & 1 & 0 & 0 \\ 0 & 1 & 1 & 0 & 0 & 0 \\ 0 & 1 & 0 & 0 & 1 & 0 \\ 1 & 0 & 1 & 0 & 0 & 0 \\ 0 & 0 & 1 & 0 & 0 & 1 \\ 1 & 1 & 0 & 0 & 0 & 0 \end{bmatrix} \quad \text{and} \quad H|_{\mathbb{F}_2} = \begin{bmatrix} 1 & 0 & 0 & 1 & 0 & 0 \\ 0 & 1 & 1 & 0 & 0 & 0 \\ 0 & 1 & 0 & 0 & 1 & 0 \\ 1 & 0 & 1 & 0 & 0 & 0 \\ 0 & 0 & 1 & 0 & 0 & 1 \end{bmatrix}.$$

So we see that $\mathcal{G}_6|_{\mathbb{F}_2}$ is the $[6, 1, 6]$ binary repetition code. ∎

Exercise 194 The $[6, 3, 4]$ code $\mathcal{C}$ over $\mathbb{F}_4$ with generator matrix G given by

$$G = \begin{bmatrix} 1 & 0 & 0 & 1 & 1 & 1 \\ 0 & 1 & 0 & 1 & \omega & \overline{\omega} \\ 0 & 0 & 1 & 1 & \overline{\omega} & \omega \end{bmatrix}$$

is the hexacode (equivalent to but not equal to the code of Example 3.8.1). Using the basis $\{1, \omega\}$ of $\mathbb{F}_4$ over $\mathbb{F}_2$, find a parity check matrix for $\mathcal{C}|_{\mathbb{F}_2}$. ◆

Example 3.8.2 In Examples 3.4.2 and 3.4.3, we constructed $\mathbb{F}_8$ by adjoining a primitive element α to $\mathbb{F}_2$ where $\alpha^3 = \alpha + 1$. Consider the three codes $\mathcal{C}_1, \mathcal{C}_2$, and $\mathcal{C}_3$ of length 7 over

$\mathbb{F}_8$ given by the following parity check matrices, respectively:

$$H_1 = \begin{bmatrix} 1 & \alpha & \alpha^2 & \alpha^3 & \alpha^4 & \alpha^5 & \alpha^6 \end{bmatrix},$$

$$H_2 = \begin{bmatrix} 1 & \alpha & \alpha^2 & \alpha^3 & \alpha^4 & \alpha^5 & \alpha^6 \\ 1 & \alpha^2 & \alpha^4 & \alpha^6 & \alpha & \alpha^3 & \alpha^5 \end{bmatrix}, \quad \text{and}$$

$$H_3 = \begin{bmatrix} 1 & \alpha & \alpha^2 & \alpha^3 & \alpha^4 & \alpha^5 & \alpha^6 \\ 1 & \alpha^3 & \alpha^6 & \alpha^2 & \alpha^5 & \alpha & \alpha^4 \end{bmatrix}.$$

The code $\mathcal{C}_1$ is a $[7, 6, 2]$ MDS code, while $\mathcal{C}_2$ and $\mathcal{C}_3$ are $[7, 5, 3]$ MDS codes. Choosing $\{1, \alpha, \alpha^2\}$ as a basis of $\mathbb{F}_8$ over $\mathbb{F}_2$, we obtain the following parity check matrices for $\mathcal{C}_1|_{\mathbb{F}_2}$, $\mathcal{C}_2|_{\mathbb{F}_2}$, and $\mathcal{C}_3|_{\mathbb{F}_2}$:

$$H_1|_{\mathbb{F}_2} = H_2|_{\mathbb{F}_2} = \begin{bmatrix} 1 & 0 & 0 & 1 & 0 & 1 & 1 \\ 0 & 1 & 0 & 1 & 1 & 1 & 0 \\ 0 & 0 & 1 & 0 & 1 & 1 & 1 \end{bmatrix} \quad \text{and}$$

$$H_3|_{\mathbb{F}_2} = \begin{bmatrix} 1 & 0 & 0 & 1 & 0 & 1 & 1 \\ 0 & 1 & 0 & 1 & 1 & 1 & 0 \\ 0 & 0 & 1 & 0 & 1 & 1 & 1 \\ 1 & 1 & 1 & 0 & 1 & 0 & 0 \\ 0 & 1 & 0 & 0 & 1 & 1 & 1 \\ 0 & 0 & 1 & 1 & 1 & 0 & 1 \end{bmatrix}.$$

So $\mathcal{C}_1|_{\mathbb{F}_2} = \mathcal{C}_2|_{\mathbb{F}_2}$ are both representations of the $[7, 4, 3]$ binary Hamming code $\mathcal{H}_3$, whereas $\mathcal{C}_3|_{\mathbb{F}_2}$ is the $[7, 1, 7]$ binary repetition code. ■

Exercise 195 Verify all the claims in Example 3.8.2. ◆

This example illustrates that there is no elementary relationship between the dimension of a code and the dimension of its subfield subcode. The following theorem does exhibit a lower bound on the dimension of a subfield subcode; its proof is left as an exercise.

Theorem 3.8.3 Let C be an $[n, k]$ code over $\mathbb{F}_{q^t}$. Then:
(i) $C|_{\mathbb{F}_q}$ is an $[n, k_q]$ code over $\mathbb{F}_q$, where $k_q \geq n - t(n - k)$, and
(ii) if the entries of a monomial matrix $M \in \Gamma\mathrm{Aut}(C)$ belong to $\mathbb{F}_q$, then $M \in \Gamma\mathrm{Aut}(C|_{\mathbb{F}_q})$.

Exercise 196 Prove Theorem 3.8.3. ◆

An upper bound on the dimension of a subfield subcode is given by the following theorem.

Theorem 3.8.4 Let C be an $[n, k]$ code over $\mathbb{F}_{q^t}$. Then $C|_{\mathbb{F}_q}$ is an $[n, k_q]$ code over $\mathbb{F}_q$, where $k_q \leq k$. If C has a basis of codewords in $\mathbb{F}_q^n$, then this is also a basis of $C|_{\mathbb{F}_q}$ and $C|_{\mathbb{F}_q}$ is k-dimensional.

Proof: Let G be a generator matrix of $C|_{\mathbb{F}_q}$. Then G has k_q independent rows; thus the rank of G is k_q when considered as a matrix with entries in $\mathbb{F}_q$ or in $\mathbb{F}_{q^t}$. Hence the rows of G remain independent in $\mathbb{F}_{q^t}^n$, implying all parts of the theorem. □

Another natural way to construct a code over $\mathbb{F}_q$ from a code over $\mathbb{F}_{q^t}$ is to use the *trace function* $\mathrm{Tr}_t : \mathbb{F}_{q^t} \to \mathbb{F}_q$ defined by

$$\mathrm{Tr}_t(\alpha) = \sum_{i=0}^{t-1} \alpha^{q^i} = \sum_{i=0}^{t-1} \sigma_q^i(\alpha), \qquad \text{for all } \alpha \in \mathbb{F}_{q^t},$$

where $\sigma_q = \sigma_p^m$ and $\sigma_q(\alpha) = \alpha^q$. Furthermore, $\sigma_q^t(\alpha) = \alpha^{q^t} = \alpha$ as every element of $\mathbb{F}_{q^t}$ is a root of $x^{q^t} - x$ by Theorem 3.3.2. So σ_q^t is the identity and therefore $\sigma_q(\mathrm{Tr}_t(\alpha)) = \mathrm{Tr}_t(\alpha)$. Thus $\mathrm{Tr}_t(\alpha)$ is a root of $x^q - x$ and hence is in $\mathbb{F}_q$ as required by the definition of trace.

Exercise 197 Fill in the missing steps that show that $\sigma_q(\mathrm{Tr}_t(\alpha)) = \mathrm{Tr}_t(\alpha)$. ◆

Exercise 198 Using the notation of Examples 3.4.2 and 3.4.3, produce a table of values of $\mathrm{Tr}_3(\beta)$ for all $\beta \in \mathbb{F}_8$ where $\mathrm{Tr}_3 : \mathbb{F}_8 \to \mathbb{F}_2$. ◆

As indicated in the following lemma, the trace function is a nontrivial linear functional on the vector space $\mathbb{F}_{q^t}$ over $\mathbb{F}_q$.

Lemma 3.8.5 *The following properties hold for* $\mathrm{Tr}_t : \mathbb{F}_{q^t} \to \mathbb{F}_q$:
(i) Tr_t *is not identically zero,*
(ii) $\mathrm{Tr}_t(\alpha + \beta) = \mathrm{Tr}_t(\alpha) + \mathrm{Tr}_t(\beta)$, *for all* $\alpha, \beta \in \mathbb{F}_{q^t}$, *and*
(iii) $\mathrm{Tr}_t(a\alpha) = a\mathrm{Tr}_t(\alpha)$, *for all* $\alpha \in \mathbb{F}_{q^t}$ *and all* $a \in \mathbb{F}_q$.

Proof: Part (i) is clear because $\mathrm{Tr}_t(x)$ is a polynomial in x of degree q^{t-1} and hence has at most q^{t-1} roots in $\mathbb{F}_q^t$ by Exercise 159. Parts (ii) and (iii) follow from the facts that $\sigma_q(\alpha + \beta) = \sigma_q(\alpha) + \sigma_q(\beta)$ and $\sigma_q(a\alpha) = a\sigma_q(\alpha)$ for all $\alpha, \beta \in \mathbb{F}_{q^m}$ and all $a \in \mathbb{F}_q$ since σ_q is a field automorphism that fixes $\mathbb{F}_q$. □

The *trace of a vector* $\mathbf{c} = (c_1, c_2, \ldots, c_n) \in \mathbb{F}_{q^t}^n$ is defined by $\mathrm{Tr}_t(\mathbf{c}) = (\mathrm{Tr}_t(c_1), \mathrm{Tr}_t(c_2), \ldots, \mathrm{Tr}_t(c_n))$. The *trace of a linear code* $\mathcal{C}$ of length n over $\mathbb{F}_{q^t}$ is defined by

$$\mathrm{Tr}_t(\mathcal{C}) = \{\mathrm{Tr}_t(\mathbf{c}) \mid \mathbf{c} \in \mathcal{C}\}.$$

The trace of $\mathcal{C}$ is a linear code of length n over $\mathbb{F}_q$. The following theorem of Delsarte [64] exhibits a dual relation between subfield subcodes and trace codes.

Theorem 3.8.6 (Delsarte) *Let* $\mathcal{C}$ *be a linear code of length n over* $\mathbb{F}_{q^t}$. *Then*

$$(\mathcal{C}|_{\mathbb{F}_q})^{\perp} = \mathrm{Tr}_t(\mathcal{C}^{\perp}).$$

Proof: We first show that $(\mathcal{C}|_{\mathbb{F}_q})^{\perp} \supseteq \mathrm{Tr}_t(\mathcal{C}^{\perp})$. Let $\mathbf{c} = (c_1, c_2, \ldots, c_n)$ be in $\mathcal{C}^{\perp}$ and let $\mathbf{b} = (b_1, \ldots, b_n)$ be in $\mathcal{C}|_{\mathbb{F}_q}$. Then by Lemma 3.8.5(ii) and (iii),

$$\mathrm{Tr}_t(\mathbf{c}) \cdot \mathbf{b} = \sum_{i=1}^{n} \mathrm{Tr}_t(c_i)b_i = \mathrm{Tr}_t\left(\sum_{i=1}^{n} c_i b_i\right) = \mathrm{Tr}_t(0) = 0$$

as $\mathbf{c} \in \mathcal{C}^{\perp}$ and $\mathbf{b} \in \mathcal{C}|_{\mathbb{F}_q} \subseteq \mathcal{C}$. Thus $\mathrm{Tr}_t(\mathbf{c}) \in (\mathcal{C}|_{\mathbb{F}_q})^{\perp}$.

We now show that $(\mathcal{C}|_{\mathbb{F}_q})^{\perp} \subseteq \mathrm{Tr}_t(\mathcal{C}^{\perp})$ by showing that $(\mathrm{Tr}_t(\mathcal{C}^{\perp}))^{\perp} \subseteq \mathcal{C}|_{\mathbb{F}_q}$. Let $\mathbf{a} = (a_1, \ldots, a_n) \in (\mathrm{Tr}_t(\mathcal{C}^{\perp}))^{\perp}$. Since $a_i \in \mathbb{F}_q$ for $1 \leq i \leq n$, we need only show that $\mathbf{a} \in \mathcal{C}$, and for this it suffices to show that $\mathbf{a} \cdot \mathbf{b} = 0$ for all $\mathbf{b} \in \mathcal{C}^\perp$. Let $\mathbf{b} = (b_1, \ldots, b_n)$ be a vector in $\mathcal{C}^\perp$. Then $\beta \mathbf{b} \in \mathcal{C}^\perp$ for all $\beta \in \mathbb{F}_{q^t}$, and

$$0 = \mathbf{a} \cdot \text{Tr}_t(\beta \mathbf{b}) = \sum_{i=1}^{n} a_i \text{Tr}_t(\beta b_i) = \text{Tr}_t\left(\beta \sum_{i=1}^{n} a_i b_i\right) = \text{Tr}_t(\beta x),$$

where $x = \mathbf{a} \cdot \mathbf{b}$ by Lemma 3.8.5(ii) and (iii). If $x \neq 0$, then we contradict (i) of Lemma 3.8.5. Hence $x = 0$ and the theorem follows. $\qquad\square$

Example 3.8.7 Delsarte's theorem can be used to compute the subfield subcodes of the codes in Example 3.8.2. For instance, $\mathcal{C}_1^\perp$ has generator matrix

$$H_1 = [1 \quad \alpha \quad \alpha^2 \quad \alpha^3 \quad \alpha^4 \quad \alpha^5 \quad \alpha^6].$$

The seven nonzero vectors of $\mathcal{C}_1^\perp$ are $\alpha^i(1, \alpha, \alpha^2, \ldots, \alpha^6)$ for $0 \leq i \leq 6$. Using Exercise 198, we have $\text{Tr}_3(\alpha^i) = 1$ for $i \in \{0, 3, 5, 6\}$ and $\text{Tr}_3(\alpha^i) = 0$ for $i \in \{1, 2, 4\}$. Thus the nonzero vectors of $(\mathcal{C}_1|_{\mathbb{F}_2})^\perp = \text{Tr}_3(\mathcal{C}_1^\perp)$ are the seven cyclic shifts of 1001011, namely 1001011, 1100101, 1110010, 0111001, 1011100, 0101110, and 0010111. A generator matrix for $(\mathcal{C}_1|_{\mathbb{F}_2})^\perp$ is obtained by taking three linearly independent cyclic shifts of 1001011. Such a matrix is the parity check matrix of $\mathcal{C}_1|_{\mathbb{F}_2}$ given in Example 3.8.2. $\qquad\blacksquare$

Exercise 199 Verify all the claims made in Example 3.8.7. $\qquad\blacklozenge$

Exercise 200 Use Delsarte's Theorem to find the subfield subcodes of $\mathcal{C}_2$ and $\mathcal{C}_3$ from Example 3.8.2 in an analogous manner to that given in Example 3.8.7. $\qquad\blacklozenge$

If $\mathcal{C}$ is a code of length n over $\mathbb{F}_{q^t}$ which has a basis of vectors in $\mathbb{F}_q^n$, the minimum weight vectors in $\mathcal{C}$ and $\mathcal{C}|_{\mathbb{F}_q}$ are the same up to scalar multiplication as we now see. In particular the minimum weight vectors in the two codes have the same set of supports, where the *support of a vector* $\mathbf{c}$, denoted $\text{supp}(\mathbf{c})$, is the set of coordinates where $\mathbf{c}$ is nonzero.

Theorem 3.8.8 *Let $\mathcal{C}$ be an $[n, k, d]$ code over $\mathbb{F}_{q^t}$. Assume that $\mathcal{C}$ has a basis of codewords in $\mathbb{F}_q^n$. Then every vector in $\mathcal{C}$ of weight d is a multiple of a vector of weight d in $\mathcal{C}|_{\mathbb{F}_q}$.*

Proof: Let $\{\mathbf{b}_1, \ldots, \mathbf{b}_k\}$ be a basis of $\mathcal{C}$ with $\mathbf{b}_i \in \mathbb{F}_q^n$ for $1 \leq i \leq k$. Then $\{\mathbf{b}_1, \ldots, \mathbf{b}_k\}$ is also a basis of $\mathcal{C}|_{\mathbb{F}_q}$, by Theorem 3.8.4. Let $\mathbf{c} = c_1 \cdots c_n = \sum_{i=1}^{k} \alpha_i \mathbf{b}_i$, with $\alpha_i \in \mathbb{F}_{q^t}$. Thus $\mathbf{c} \in \mathcal{C}$, and by Lemma 3.8.5, $\text{Tr}_t(\mathbf{c}) = \sum_{i=1}^{k} \text{Tr}_t(\alpha_i) \mathbf{b}_i \in \mathcal{C}|_{\mathbb{F}_q} \subseteq \mathcal{C}$. Also $\text{wt}(\text{Tr}_t(\mathbf{c})) \leq \text{wt}(\mathbf{c})$ as $\text{Tr}_t(0) = 0$. Assume now that $\mathbf{c}$ has weight d. Then either $\text{Tr}_t(\mathbf{c}) = \mathbf{0}$ or $\text{Tr}_t(\mathbf{c})$ has weight d with $\text{supp}(\mathbf{c}) = \text{supp}(\text{Tr}_t(\mathbf{c}))$. By replacing $\mathbf{c}$ by $\alpha \mathbf{c}$ for some $\alpha \in \mathbb{F}_{q^t}$, we may assume that $\text{Tr}_t(\mathbf{c}) \neq \mathbf{0}$ using Lemma 3.8.5(i). Choose i so that $c_i \neq 0$. As $\text{supp}(\mathbf{c}) = \text{supp}(\text{Tr}_t(\mathbf{c}))$, $\text{Tr}_t(c_i) \neq 0$ and $\text{wt}(c_i \text{Tr}_t(\mathbf{c}) - \text{Tr}_t(c_i)\mathbf{c}) < d$. Thus $c_i \text{Tr}_t(\mathbf{c}) - \text{Tr}_t(c_i)\mathbf{c} = \mathbf{0}$ and $\mathbf{c}$ is a multiple of $\text{Tr}_t(\mathbf{c})$. $\qquad\square$

4 Cyclic codes

We now turn to the study of an extremely important class of codes known as cyclic codes. Many families of codes including the Golay codes, the binary Hamming codes, and codes equivalent to the Reed–Muller codes are either cyclic or extended cyclic codes. The study of cyclic codes began with two 1957 and 1959 AFCRL reports by E. Prange. The 1961 book by W. W. Peterson [255] compiled extensive results about cyclic codes and laid the framework for much of the present-day theory; it also stimulated much of the subsequent research on cyclic codes. In 1972 this book was expanded and published jointly by Peterson and E. J. Weldon [256].

In studying cyclic codes of length n, it is convenient to label the coordinate positions as $0, 1, \ldots, n - 1$ and think of these as the integers modulo n. A linear code C of length n over $\mathbb{F}_q$ is *cyclic* provided that for each vector $\mathbf{c} = c_0 \cdots c_{n-2}c_{n-1}$ in C the vector $c_{n-1}c_0 \cdots c_{n-2}$, obtained from $\mathbf{c}$ by the *cyclic shift* of coordinates $i \mapsto i + 1 \pmod{n}$, is also in C. So a cyclic code contains all n cyclic shifts of any codeword. Hence it is convenient to think of the coordinate positions cyclically where, once you reach $n - 1$, you begin again with coordinate 0. When we speak of consecutive coordinates, we will always mean consecutive in that cyclical sense.

When examining cyclic codes over $\mathbb{F}_q$, we will most often represent the codewords in polynomial form. There is a bijective correspondence between the vectors $\mathbf{c} = c_0 c_1 \cdots c_{n-1}$ in $\mathbb{F}_q^n$ and the polynomials $c(x) = c_0 + c_1 x + \cdots + c_{n-1} x^{n-1}$ in $\mathbb{F}_q[x]$ of degree at most $n - 1$. We order the terms of our polynomials from smallest to largest degree. We allow ourselves the latitude of using the vector notation $\mathbf{c}$ and the polynomial notation $c(x)$ interchangeably. Notice that if $c(x) = c_0 + c_1 x + \cdots + c_{n-1} x^{n-1}$, then $xc(x) = c_{n-1}x^n + c_0 x + c_1 x^2 + \cdots + c_{n-2}x^{n-1}$, which would represent the codeword $\mathbf{c}$ cyclically shifted one to the right if x^n were set equal to 1. More formally, the fact that a cyclic code C is invariant under a cyclic shift implies that if $c(x)$ is in C, then so is $xc(x)$ provided we multiply modulo $x^n - 1$. This suggests that the proper context for studying cyclic codes is the residue class ring

$$\mathcal{R}_n = \mathbb{F}_q[x]/(x^n - 1).$$

Under the correspondence of vectors with polynomials as given above, cyclic codes are ideals of $\mathcal{R}_n$ and ideals of $\mathcal{R}_n$ are cyclic codes. *Thus the study of cyclic codes in $\mathbb{F}_q^n$ is equivalent to the study of ideals in $\mathcal{R}_n$.* The study of ideals in $\mathcal{R}_n$ hinges on factoring $x^n - 1$, a topic we now explore.

4.1   Factoring $x^n - 1$

We are interested in finding the irreducible factors of $x^n - 1$ over $\mathbb{F}_q$. Two possibilities arise: either $x^n - 1$ has repeated irreducible factors or it does not. The study of cyclic codes has primarily focused on the latter case. By Exercise 201, $x^n - 1$ has no repeated factors if and only if q and n are relatively prime, an assumption we make throughout this chapter.

Exercise 201 For $f(x) = a_0 + a_1 x + \cdots + a_n x^n \in \mathbb{F}_q[x]$, define the *formal derivative* of $f(x)$ to be the polynomial $f'(x) = a_1 + 2a_2 x + 3a_3 x^2 + \cdots + na_n x^{n-1} \in \mathbb{F}_q[x]$. From this definition, show that the following rules hold:
(a) $(f + g)'(x) = f'(x) + g'(x)$.
(b) $(fg)'(x) = f'(x)g(x) + f(x)g'(x)$.
(c) $(f(x)^m)' = m(f(x))^{m-1} f'(x)$ for all positive integers m.
(d) If $f(x) = f_1(x)^{a_1} f_2(x)^{a_2} \cdots f_n(x)^{a_n}$, where $a_1, \ldots, a_n$ are positive integers and $f_1(x), \ldots, f_n(x)$ are distinct and irreducible over $\mathbb{F}_q$, then

$$\frac{f(x)}{\gcd(f(x), f'(x))} = f_1(x) \cdots f_n(x).$$

(e) Show that $f(x)$ has no repeated irreducible factors if and only if $f(x)$ and $f'(x)$ are relatively prime.
(f) Show that $x^n - 1$ has no repeated irreducible factors if and only if q and n are relatively prime.   ◆

To help factor $x^n - 1$ over $\mathbb{F}_q$, it is useful to find an extension field $\mathbb{F}_{q^t}$ of $\mathbb{F}_q$ that contains all of its roots. In other words, $\mathbb{F}_{q^t}$ must contain a primitive nth root of unity, which occurs precisely when $n \mid (q^t - 1)$ by Theorem 3.3.3. Define the *order* $\mathrm{ord}_n(q)$ of q modulo n to be the smallest positive integer a such that $q^a \equiv 1 \pmod{n}$. Notice that if $t = \mathrm{ord}_n(q)$, then $\mathbb{F}_{q^t}$ contains a primitive nth root of unity α, but no smaller extension field of $\mathbb{F}_q$ contains such a primitive root. As the α^i are distinct for $0 \le i < n$ and $(\alpha^i)^n = 1$, $\mathbb{F}_{q^t}$ contains all the roots of $x^n - 1$. Consequently, $\mathbb{F}_{q^t}$ is called a *splitting field of $x^n - 1$ over $\mathbb{F}_q$*. So the irreducible factors of $x^n - 1$ over $\mathbb{F}_q$ must be the product of the distinct minimal polynomials of the nth roots of unity in $\mathbb{F}_{q^t}$. Suppose that γ is a primitive element of $\mathbb{F}_{q^t}$. Then $\alpha = \gamma^d$ is a primitive nth root of unity where $d = (q^t - 1)/n$. The roots of $M_{\alpha^s}(x)$ are $\{\gamma^{ds}, \gamma^{dsq}, \gamma^{dsq^2}, \ldots, \gamma^{dsq^{r-1}}\} = \{\alpha^s, \alpha^{sq}, \alpha^{sq^2}, \ldots, \alpha^{sq^{r-1}}\}$, where r is the smallest positive integer such that $dsq^r \equiv ds \pmod{q^t - 1}$ by Theorem 3.7.6. But $dsq^r \equiv ds \pmod{q^t - 1}$ if and only if $sq^r \equiv s \pmod{n}$.

This leads us to extend the notion of q-cyclotomic cosets first developed in Section 3.7. Let s be an integer with $0 \le s < n$. The *q-cyclotomic coset of s modulo n* is the set

$$C_s = \{s, sq, \ldots, sq^{r-1}\} \pmod{n},$$

where r is the smallest positive integer such that $sq^r \equiv s \pmod{n}$. It follows that C_s is the orbit of the permutation $i \mapsto iq \pmod{n}$ that contains s. The distinct q-cyclotomic cosets modulo n partition the set of integers $\{0, 1, 2, \ldots, n - 1\}$. As before we normally denote a cyclotomic coset in this partition by choosing s to be the smallest integer contained in the cyclotomic coset. In Section 3.7 we had studied the more restricted case where $n = q^t - 1$. Notice that $\mathrm{ord}_n(q)$ is the size of the q-cyclotomic coset C_1 modulo n. This discussion gives the following theorem.

Theorem 4.1.1 *Let n be a positive integer relatively prime to q. Let $t = \mathrm{ord}_n(q)$. Let α be a primitive nth root of unity in $\mathbb{F}_{q^t}$.*

(i) *For each integer s with $0 \le s < n$, the minimal polynomial of α^s over $\mathbb{F}_q$ is*

$$M_{\alpha^s}(x) = \prod_{i \in C_s} (x - \alpha^i),$$

where C_s is the q-cyclotomic coset of s modulo n.

(ii) *The conjugates of α^s are the elements α^i with $i \in C_s$.*

(iii) *Furthermore,*

$$x^n - 1 = \prod_s M_{\alpha^s}(x)$$

is the factorization of $x^n - 1$ into irreducible factors over $\mathbb{F}_q$, where s runs through a set of representatives of the q-cyclotomic cosets modulo n.

Example 4.1.2 The 2-cyclotomic cosets modulo 9 are $C_0 = \{0\}$, $C_1 = \{1, 2, 4, 8, 7, 5\}$, and $C_3 = \{3, 6\}$. So $\mathrm{ord}_9(2) = 6$ and the primitive ninth roots of unity lie in $\mathbb{F}_{64}$ but in no smaller extension field of $\mathbb{F}_2$. The irreducible factors of $x^9 - 1$ over $\mathbb{F}_2$ have degrees 1, 6, and 2. These are the polynomials $M_1(x) = x + 1$, $M_\alpha(x)$, and $M_{\alpha^3}(x)$, where α is a primitive ninth root of unity in $\mathbb{F}_{64}$. The only irreducible polynomial of degree 2 over $\mathbb{F}_2$ is $x^2 + x + 1$, which must therefore be $M_{\alpha^3}(x)$. (Notice also that α^3 is a primitive third root of unity, which must lie in the subfield $\mathbb{F}_4$ of $\mathbb{F}_{64}$.) Hence the factorization of $x^9 - 1$ is $x^9 - 1 = (x + 1)(x^2 + x + 1)(x^6 + x^3 + 1)$ and $M_\alpha(x) = x^6 + x^3 + 1$. ∎

Example 4.1.3 The 3-cyclotomic cosets modulo 13 are $C_0 = \{0\}$, $C_1 = \{1, 3, 9\}$, $C_2 = \{2, 6, 5\}$, $C_4 = \{4, 12, 10\}$, and $C_7 = \{7, 8, 11\}$. So $\mathrm{ord}_{13}(3) = 3$ and the primitive 13th roots of unity lie in $\mathbb{F}_{27}$ but in no smaller extension field of $\mathbb{F}_3$. The irreducible factors of $x^{13} - 1$ over $\mathbb{F}_3$ have degrees 1, 3, 3, 3, and 3. These are the polynomials $M_1(x) = x - 1$, $M_\alpha(x)$, $M_{\alpha^2}(x)$, $M_{\alpha^4}(x)$, and $M_{\alpha^7}(x)$, where α is a primitive 13th root of unity in $\mathbb{F}_{27}$. ∎

In these examples we notice that the size of each q-cyclotomic coset is a divisor of $\mathrm{ord}_n(q)$. This holds in general.

Theorem 4.1.4 *The size of each q-cyclotomic coset is a divisor of $\mathrm{ord}_n(q)$. Furthermore, the size of C_1 is $\mathrm{ord}_n(q)$.*

Proof: Let $t = \mathrm{ord}_n(q)$ and let m be the size of C_s. Then $M_{\alpha^s}(x)$ has degree m where α is a primitive nth root of unity. So $m \mid t$ by Theorem 3.7.3. The fact that the size of C_1 is $\mathrm{ord}_n(q)$ follows directly from the definitions of q-cyclotomic cosets and $\mathrm{ord}_n(q)$ as mentioned prior to Theorem 4.1.1. □

Exercise 202 Let $q = 2$.

(a) Find the q-cyclotomic cosets modulo n where n is:
   (i) 23,
   (ii) 45.

(b) Find $\mathrm{ord}_n(q)$ for the two values of n given in part (a).

(c) What are the degrees of the irreducible factors of $x^n - 1$ over $\mathbb{F}_q$ for the two values of n given in part (a)?   ◆

Exercise 203 Repeat Exercise 202 with $q = 3$ and $n = 28$ and $n = 41$.   ◆

Exercise 204 Factor $x^{13} - 1$ over $\mathbb{F}_3$.   ◆

We conclude this section by noting that there are efficient computer algorithms for factoring polynomials over a finite field, among them the algorithm of Berlekamp, MacWilliams, and Sloane [18]. Many of the algebraic manipulation software packages can factor $x^n - 1$ over a finite field for reasonably sized integers n. There are also extensive tables in [256] listing all irreducible polynomials over $\mathbb{F}_2$ for $n \leq 34$.

4.2 Basic theory of cyclic codes

We noted earlier that cyclic codes over $\mathbb{F}_q$ are precisely the ideals of

$$\mathcal{R}_n = \mathbb{F}_q[x]/(x^n - 1).$$

Exercises 153 and 166 show that $\mathbb{F}_q[x]$ is a principal ideal domain. It is straightforward then to show that the ideals of $\mathcal{R}_n$ are also principal, and hence cyclic codes are the principal ideals of $\mathcal{R}_n$. When writing a codeword of a cyclic code as $c(x)$, we technically mean the coset $c(x) + (x^n - 1)$ in $\mathcal{R}_n$. However, such notation is too cumbersome, and we will write $c(x)$ even when working in $\mathcal{R}_n$. Thus we think of the elements of $\mathcal{R}_n$ as the polynomials in $\mathbb{F}_q[x]$ of degree less than n with multiplication being carried out modulo $x^n - 1$. So when working in $\mathcal{R}_n$, to multiply two polynomials, we multiply them as we would in $\mathbb{F}_q[x]$ and then replace any term of the form ax^{ni+j}, where $0 \leq j < n$, by ax^j. We see immediately that when writing a polynomial as both an element in $\mathbb{F}_q[x]$ and an element in $\mathcal{R}_n$, confusion can easily arise. The reader should be aware of which ring is being considered.

As stated earlier, throughout our study of cyclic codes, we make the basic assumption that the characteristic p of $\mathbb{F}_q$ does not divide the length n of the cyclic codes being considered, or equivalently, that $\gcd(n, q) = 1$. The primary reason for this assumption is that $x^n - 1$ has distinct roots in an extension field of $\mathbb{F}_q$ by Exercise 201, and this enables us to describe its roots (and as we shall see, cyclic codes) by q-cyclotomic cosets modulo n. The assumption that p does not divide n also implies that $\mathcal{R}_n$ is semi-simple and thus that the Wedderburn Structure Theorems apply; we shall not invoke these structure theorems preferring rather to derive the needed consequences for the particular case of $\mathcal{R}_n$. The theory of cyclic codes with $\gcd(n, q) \neq 1$ is discussed in [49, 201], but, to date, these "repeated root" cyclic codes do not seem to be of much interest.

To distinguish the principal ideal $(g(x))$ of $\mathbb{F}_q[x]$ from that ideal in $\mathcal{R}_n$, we use the notation $\langle g(x)\rangle$ for the principal ideal of $\mathcal{R}_n$ generated by $g(x)$. We now show that there is a bijective correspondence between the cyclic codes in $\mathcal{R}_n$ and the monic polynomial divisors of $x^n - 1$.

Theorem 4.2.1 *Let C be a nonzero cyclic code in $\mathcal{R}_n$. There exists a polynomial $g(x) \in C$ with the following properties:*
(i)   *$g(x)$ is the unique monic polynomial of minimum degree in C,*
(ii) *$C = \langle g(x)\rangle$, and*
(iii) *$g(x) \mid (x^n - 1)$.*
Let $k = n - \deg g(x)$, and let $g(x) = \sum_{i=0}^{n-k} g_i x^i$, where $g_{n-k} = 1$. Then:
(iv) *the dimension of C is k and $\{g(x), xg(x), \ldots, x^{k-1}g(x)\}$ is a basis for C,*
(v)   *every element of C is uniquely expressible as a product $g(x)f(x)$, where $f(x) = 0$ or $\deg f(x) < k$,*
(vi)

$$
G = \begin{bmatrix}
g_0 & g_1 & g_2 & \cdots & g_{n-k} & & & 0 \\
0 & g_0 & g_1 & \cdots & g_{n-k-1} & g_{n-k} & & \\
\cdots & \cdots & \cdots & & \cdots & & \cdots & \\
0 & & g_0 & & & \cdots & & g_{n-k}
\end{bmatrix}
$$

$$
\leftrightarrow \begin{bmatrix}
g(x) & & & \\
 & xg(x) & & \\
 & & \cdots & \\
 & & & x^{k-1}g(x)
\end{bmatrix}
$$

is a generator matrix for C, and
(vii) *if α is a primitive nth root of unity in some extension field of $\mathbb{F}_q$, then*

$$
g(x) = \prod_s M_{\alpha^s}(x),
$$

where the product is over a subset of representatives of the q-cyclotomic cosets modulo n.

Proof: Let $g(x)$ be a monic polynomial of minimum degree in C. Since C is nonzero, such a polynomial exists. If $c(x) \in C$, then by the Division Algorithm in $\mathbb{F}_q[x]$, $c(x) = g(x)h(x) + r(x)$, where either $r(x) = 0$ or $\deg r(x) < \deg g(x)$. As C is an ideal in $\mathcal{R}_n$, $r(x) \in C$ and the minimality of the degree of $g(x)$ implies $r(x) = 0$. This gives (i) and (ii). By the Division Algorithm $x^n - 1 = g(x)h(x) + r(x)$, where again $r(x) = 0$ or $\deg r(x) < \deg g(x)$ in $\mathbb{F}_q[x]$. As $x^n - 1$ corresponds to the zero codeword in C and C is an ideal in $\mathcal{R}_n$, $r(x) \in C$, a contradiction unless $r(x) = 0$, proving (iii).

Suppose that $\deg g(x) = n - k$. By parts (ii) and (iii), if $c(x) \in C$ with $c(x) = 0$ or $\deg c(x) < n$, then $c(x) = g(x)f(x)$ in $\mathbb{F}_q[x]$. If $c(x) = 0$, then $f(x) = 0$. If $c(x) \neq 0$, $\deg c(x) < n$ implies that $\deg f(x) < k$, by Exercise 155. Therefore

$$C = \{g(x)f(x) \mid f(x) = 0 \text{ or } \deg f(x) < k\}.$$

So C has dimension at most k and

$$\{g(x), xg(x), \ldots, x^{k-1}g(x)\}$$

spans C. Since these k polynomials are of different degrees, they are independent in $\mathbb{F}_q[x]$. Since they are of degree at most $n - 1$, they remain independent in $\mathcal{R}_n$, yielding (iv) and (v). The codewords in this basis, written as n-tuples, give G in part (vi). Part (vii) follows from Theorem 4.1.1. □

We remark that part (ii) shows that $\mathcal{R}_n$ is a principal ideal ring. Also parts (i) through (vi) of Theorem 4.2.1 hold even if $\gcd(n, q) \neq 1$. However, part (vii) requires that $\gcd(n, q) = 1$. A parity check matrix for a cyclic code will be given in Theorem 4.2.7.

Corollary 4.2.2 *Let C be a nonzero cyclic code in $\mathcal{R}_n$. The following are equivalent:*
(i) *$g(x)$ is the monic polynomial of minimum degree in C.*
(ii) *$C = \langle g(x) \rangle$, $g(x)$ is monic, and $g(x) \mid (x^n - 1)$.*

Proof: That (i) implies (ii) was shown in the proof of Theorem 4.2.1. Assume (ii). Let $g_1(x)$ be the monic polynomial of minimum degree in C. By the proof of Theorem 4.2.1(i) and (ii), $g_1(x) \mid g(x)$ in $\mathbb{F}_q[x]$ and $C = \langle g_1(x) \rangle$. As $g_1(x) \in C = \langle g(x) \rangle$, $g_1(x) \equiv g(x)a(x)$ (mod $x^n - 1$) implying $g_1(x) = g(x)a(x) + (x^n - 1)b(x)$ in $\mathbb{F}_q[x]$. Since $g(x) \mid (x^n - 1)$, $g(x) \mid g(x)a(x) + (x^n - 1)b(x)$ or $g(x) \mid g_1(x)$. As both $g_1(x)$ and $g(x)$ are monic and divide one another in $\mathbb{F}_q[x]$, they are equal. □

Theorem 4.2.1 shows that there is a monic polynomial $g(x)$ dividing $x^n - 1$ and generating C. Corollary 4.2.2 shows that the monic polynomial dividing $x^n - 1$ which generates C is unique. This polynomial is called *the generator polynomial* of the cyclic code C. By the corollary, this polynomial is both the monic polynomial in C of minimum degree and the monic polynomial dividing $x^n - 1$ which generates C. So there is a one-to-one correspondence between the nonzero cyclic codes and the divisors of $x^n - 1$, not equal to $x^n - 1$. In order to have a bijective correspondence between all the cyclic codes in $\mathcal{R}_n$ and all the monic divisors of $x^n - 1$, we define the generator polynomial of the zero cyclic code $\{0\}$ to be $x^n - 1$. (Note that $x^n - 1$ equals 0 in $\mathcal{R}_n$.) This bijective correspondence leads to the following corollary.

Corollary 4.2.3 *The number of cyclic codes in $\mathcal{R}_n$ equals 2^m, where m is the number of q-cyclotomic cosets modulo n. Moreover, the dimensions of cyclic codes in $\mathcal{R}_n$ are all possible sums of the sizes of the q-cyclotomic cosets modulo n.*

Example 4.2.4 In Example 4.1.2, we showed that, over $\mathbb{F}_2$, $x^9 - 1 = (1 + x)(1 + x + x^2)(1 + x^3 + x^6)$, and so there are eight binary cyclic codes C_i of length 9 with generator polynomial $g_i(x)$ given in the following table.

i	dim	$g_i(x)$
0	0	$1 + x^9$
1	1	$(1 + x + x^2)(1 + x^3 + x^6) = 1 + x + x^2 + \cdots + x^8$
2	2	$(1 + x)(1 + x^3 + x^6) = 1 + x + x^3 + x^4 + x^6 + x^7$
3	3	$1 + x^3 + x^6$
4	6	$(1 + x)(1 + x + x^2) = 1 + x^3$
5	7	$1 + x + x^2$
6	8	$1 + x$
7	9	1

∎

The following corollary to Theorem 4.2.1 shows the relationship between the generator polynomials of two cyclic codes when one code is a subcode of the other. Its proof is left as an exercise.

Corollary 4.2.5 *Let C_1 and C_2 be cyclic codes over $\mathbb{F}_q$ with generator polynomials $g_1(x)$ and $g_2(x)$, respectively. Then $C_1 \subseteq C_2$ if and only if $g_2(x) \mid g_1(x)$.*

Exercise 205 Prove Corollary 4.2.5.   ♦

Exercise 206 Find all pairs of codes C_i and C_j from Example 4.2.4 where $C_i \subseteq C_j$.   ♦

Exercise 207 Over $\mathbb{F}_2$, $(1 + x) \mid (x^n - 1)$. Let C be the binary cyclic code $\langle 1 + x \rangle$ of length n. Let C_1 be any binary cyclic code of length n with generator polynomial $g_1(x)$.
(a) What is the dimension of C?
(b) Prove that C is the set of all vectors in $\mathbb{F}_2^n$ with even weight.
(c) If C_1 has only even weight codewords, what is the relationship between $1 + x$ and $g_1(x)$?
(d) If C_1 has some odd weight codewords, what is the relationship between $1 + x$ and $g_1(x)$?   ♦

Not surprisingly, the dual of a cyclic code is also cyclic, a fact whose proof we also leave as an exercise.

Theorem 4.2.6 *The dual code of a cyclic code is cyclic.*

In Section 4.4, we will develop the tools to prove the following theorem about the generator polynomial and generator matrix of the dual of a cyclic code; see Theorem 4.4.9. The generator matrix of the dual is of course a parity check matrix of the original cyclic code.

Theorem 4.2.7 *Let C be an $[n, k]$ cyclic code with generator polynomial $g(x)$. Let $h(x) = (x^n - 1)/g(x) = \sum_{i=0}^{k} h_i x^i$. Then the generator polynomial of $C^{\perp}$ is $g^{\perp}(x) = x^k h(x^{-1})/h(0)$. Furthermore, a generator matrix for $C^{\perp}$, and hence a parity check matrix for C, is*

$$\begin{bmatrix} h_k & h_{k-1} & h_{k-2} & \cdots & h_0 & & 0 \\ 0 & h_k & h_{k-1} & \cdots & h_1 & h_0 & \\ & \cdots & \cdots & \cdots & \cdots & \cdots & \\ 0 & & h_k & & & \cdots & & h_0 \end{bmatrix}. \tag{4.1}$$

Example 4.2.8 The cyclic codes $\mathbb{F}_q^n$ and $\{\mathbf{0}\}$ are duals of one another. The repetition code of length n over $\mathbb{F}_q$ is a cyclic code whose dual is the cyclic code of even-like vectors of $\mathbb{F}_q^n$. ∎

Exercise 208 Prove Theorem 4.2.6. ♦

Exercise 209 Based on dimension only, for $0 \le i \le 7$ find $C_i^\perp$ for the cyclic codes C_i in Example 4.2.4. ♦

It is also not surprising that a subfield subcode of a cyclic code is cyclic.

Theorem 4.2.9 *Let C be a cyclic code over $\mathbb{F}_{q^t}$. Then $C|_{\mathbb{F}_q}$ is also cyclic.*

Exercise 210 Prove Theorem 4.2.9. ♦

Exercise 211 Verify that the three codes C_1, C_2, and C_3 of length 7 over $\mathbb{F}_8$ of Example 3.8.2 are cyclic. Verify that $C_i|_{\mathbb{F}_2}$ are all cyclic as well. ♦

Cyclic codes are easier to decode than other codes because of their additional structure. We will examine decoding algorithms for general cyclic codes in Section 4.6 and specific families in Section 5.4. We now examine three ways to encode cyclic codes. Let C be a cyclic code of length n over $\mathbb{F}_q$ with generator polynomial $g(x)$ of degree $n - k$; so C has dimension k.

The first encoding is based on the natural encoding procedure described in Section 1.11. Let G be the generator matrix obtained from the shifts of $g(x)$ in Theorem 4.2.1. We encode the message $\mathbf{m} \in \mathbb{F}_q^k$ as the codeword $\mathbf{c} = \mathbf{m}G$. We leave it as Exercise 212 to show that if $m(x)$ and $c(x)$ are the polynomials in $\mathbb{F}_q[x]$ associated to $\mathbf{m}$ and $\mathbf{c}$, then $c(x) = m(x)g(x)$. However, this encoding is not systematic.

Exercise 212 Let C be a cyclic code of length n over $\mathbb{F}_q$ with generator polynomial $g(x)$. Let G be the generator matrix obtained from the shifts of $g(x)$ in Theorem 4.2.1. Prove that the encoding of the message $\mathbf{m} \in \mathbb{F}_q^k$ as the codeword $\mathbf{c} = \mathbf{m}G$ is the same as forming the product $c(x) = m(x)g(x)$ in $\mathbb{F}_q[x]$, where $m(x)$ and $c(x)$ are the polynomials in $\mathbb{F}_q[x]$ associated to $\mathbf{m}$ and $\mathbf{c}$. ♦

The second encoding procedure is systematic. The polynomial $m(x)$ associated to the message $\mathbf{m}$ is of degree at most $k - 1$ (or is the zero polynomial). The polynomial $x^{n-k}m(x)$ has degree at most $n - 1$ and has its first $n - k$ coefficients equal to 0; thus the message is contained in the coefficients of $x^{n-k}, x^{n-k+1}, \ldots, x^{n-1}$. By the Division Algorithm,

$$x^{n-k}m(x) = g(x)a(x) + r(x), \quad \text{where } \deg r(x) < n - k \text{ or } r(x) = 0.$$

Let $c(x) = x^{n-k}m(x) - r(x)$; as $c(x)$ is a multiple of $g(x)$, $c(x) \in \mathcal{C}$. Also $c(x)$ differs from $x^{n-k}m(x)$ in the coefficients of $1, x, \ldots, x^{n-k-1}$ as deg $r(x) < n - k$. So $c(x)$ contains the message **m** in the coefficients of the terms of degree at least $n - k$.

The third encoding procedure, also systematic, for $\mathcal{C} = \langle g(x) \rangle$ uses the generator polynomial $g^{\perp}(x)$ of the dual code $\mathcal{C}^{\perp}$ as given in Theorem 4.2.7. As $\mathcal{C}$ is an $[n, k]$ code, if $\mathbf{c} = (c_0, c_1, \ldots, c_{n-1}) \in \mathcal{C}$, once $c_0, c_1, \ldots, c_{k-1}$ are known, then the remaining components $c_k, \ldots, c_{n-1}$ are determined from $H\mathbf{c}^{\mathsf{T}} = \mathbf{0}$, where H is the parity check matrix (4.1). We can scale the rows of H so that its rows are shifts of the monic polynomial $g^{\perp}(x) = h'_0 + h'_1 x + \cdots + h'_{k-1}x^{k-1} + x^k$. To encode $\mathcal{C}$, choose k information bits $c_0, c_1, \ldots, c_{k-1}$; then

$$c_i = -\sum_{j=0}^{k-1} h'_j c_{i-k+j},\tag{4.2}$$

where the computation c_i is performed in the order $i = k, k+1, \ldots, n-1$.

Exercise 213 Let $\mathcal{C}$ be a binary cyclic code of length 15 with generator polynomial $g(x) = (1 + x + x^4)(1 + x + x^2 + x^3 + x^4)$.
(a) Encode the message $m(x) = 1 + x^2 + x^5$ using the first encoding procedure (the non-systematic encoding) described in this section.
(b) Encode the message $m(x) = 1 + x^2 + x^5$ using the second encoding procedure (the first systematic encoding) described in this section.
(c) Encode the message $m(x) = 1 + x^2 + x^5$ using the third encoding procedure (the second systematic encoding) described in this section.   ◆

Exercise 214 Show that in any cyclic code of dimension k, any set of k consecutive coordinates forms an information set.   ◆

Each of the encoding schemes can be implemented using linear shift-registers. We illustrate this for binary codes with the last scheme using a linear feedback shift-register. For more details on implementations of the other encoding schemes we refer the reader to [18, 21, 233, 256]. The main components of a linear feedback shift-register are *delay elements* (also called *flip-flops*) and *binary adders* shown in Figure 4.1. The shift-register is run by an external clock which generates a timing signal, or *clock cycle*, every t_0 seconds, where t_0 can be very small. Generally, a delay element stores one bit (a 0 or a 1) for one clock cycle, after which the bit is pushed out and replaced by another bit. A *linear shift-register* is a series of delay elements; a bit enters at one end of the shift-register and moves to the next delay element with each new clock cycle. A *linear feedback shift-register* is a linear shift-register in which the output is fed back into the shift-register as part of the input. The

Delay element   Binary adder

Figure 4.1 Delay element and 3-input binary adder.

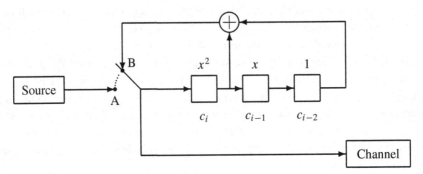

Figure 4.2 Encoder for C, where $C^\perp = \langle 1 + x^2 + x^3 \rangle$.

adder takes all its input signals and adds them in binary; this process is considered to occur instantaneously.

Example 4.2.10 In Figure 4.2, we construct a linear feedback shift-register for encoding the [7, 3] binary cyclic code C with generator polynomial $g(x) = 1 + x + x^2 + x^4$. Then $g^\perp(x) = 1 + x^2 + x^3$, and so the parity check equations from (4.2) are:

$$c_0 + c_2 = c_3,$$
$$c_1 + c_3 = c_4,$$
$$c_2 + c_4 = c_5,$$
$$c_3 + c_5 = c_6.$$

The shift-register has three flip-flops. Suppose we input $c_0 = 1$, $c_1 = 0$, and $c_2 = 0$ into the shift-register. Initially, before the first clock cycle, the shift-register has three unknown quantities, which we can denote by $\star\star\star$. The switch is set at position A for three clock cycles, in which case the digits 1, 0, 0 from the source are moved into the shift-register from left to right, as indicated in Table 4.1. These three bits also move to the transmission channel. Between clock cycles 3 and 4, the switch is set to position B, which enables the

Table 4.1 *Shift-register for C, where $C^\perp = \langle 1 + x^2 + x^3 \rangle$ with input 100*

Clock cycle	Switch	Source	Channel	Register
1	A	1	1	1$\star\star$
2	A	0	0	01$\star$
3	A	0	0	001
4	B		1	100
5	B		1	110
6	B		1	111
7	B		0	011

feedback to take place. The switch remains in this position for 4 clock cycles; during this time no further input arrives from the source. During cycle 3, the register reads 001, which corresponds to $c_0c_1c_2 = 100$. Then at cycle 4, $c_0 = 1$ and $c_2 = 0$ from the shift-register pass through the binary adder and are added to give $c_3 = 1$; the result both passes to the channel and is fed back into the shift-register from the left. Note that the shift-register has merely executed the first of the above parity check equations. The shift-register now contains 100, which corresponds to $c_1c_2c_3 = 001$. At clock cycle 5, the shift-register performs the binary addition $c_1 + c_3 = 0 + 1 = 1 = c_4$, which satisfies the second of the parity check equations. Clock cycles 6 and 7 produce c_5 and c_6 as indicated in the table; these satisfy the last two parity check equations. The shift-register has sent $c_0c_1 \cdots c_6 = 1001110$ to the channel. (Notice also that the codeword is the first entry of the column "Register" in Table 4.1.) Then the switch is reset to position A, and the shift-register is ready to receive input for the next codeword. ∎

Exercise 215 Give the contents of the shift-register in Figure 4.2 in the form of a table similar to Table 4.1 for the following input sequences. Also give the codeword produced.
(a) $c_0c_1c_2 = 011$,
(b) $c_0c_1c_2 = 101$. ◆

Notice that we labeled the three delay elements in Figure 4.2 with x^2, x, and 1. The vertical wires entering the binary adder came after the delay elements x^2 and 1. These are the nonzero terms in $g^\perp(x)$ of degree less than 3. In general if $\deg g^\perp(x) = k$, the k delay elements are labeled $x^{k-1}, x^{k-2}, \ldots, 1$ from left to right. The delay elements with wires to a binary adder are precisely those with labels coming after the nonzero terms in $g^\perp(x)$ of degree less than k. Examining (4.2) shows why this works.

Exercise 216 Do the following:
(a) Let C be the [7, 4] binary cyclic code with generator polynomial $g(x) = 1 + x + x^3$. Find $g^\perp(x)$.
(b) Draw a linear feedback shift-register to encode C.
(c) Give the contents of the shift-register from part (b) in the form of a table similar to Table 4.1 for the input sequence $c_0c_1c_2c_3 = 1001$. Also give the codeword produced. ◆

Exercise 217 Do the following:
(a) Let C be the [9, 7] binary cyclic code with generator polynomial $g(x) = 1 + x + x^2$ shown in Example 4.2.4. Find $g^\perp(x)$.
(b) Draw a linear feedback shift-register to encode C.
(c) Give the contents of the shift-register from part (b) in the form of a table similar to Table 4.1 for the input sequence $c_0c_1 \cdots c_6 = 1011011$. Also give the codeword produced. ◆

The idea of a cyclic code has been generalized in the following way. If a code C has the property that there exists an integer s such that the shift of a codeword by s positions is again a codeword, C is called a *quasi-cyclic* code. Cyclic codes are quasi-cyclic codes with $s = 1$. Quasi-cyclic codes with $s = 2$ are sometimes monomially equivalent to *double circulant codes*; a double circulant code has generator matrix $[I \ A]$, where A is a circulant matrix. We note that the term "double circulant code" is sometimes used when the generator matrix has other "cyclic-like" structures such as

$$\left[I \ \left| \begin{array}{cccc} 0 & 1 \cdots 1 \\ 1 & \\ \vdots & B \\ 1 & \end{array} \right. \right],$$

where B is a circulant matrix; such a code may be called a "bordered double circulant code." See Section 9.8 where we examine more extensively the construction of codes using circulant matrices.

4.3   Idempotents and multipliers

Besides the generator polynomial, there are many polynomials that can be used to generate a cyclic code. A general result about which polynomials generate a given cyclic code will be presented in Theorem 4.4.4. There is another very specific polynomial, called an idempotent generator, which can be used to generate a cyclic code.

An element e of a ring satisfying $e^2 = e$ is called an *idempotent*. As stated earlier without proof, the ring $\mathcal{R}_n$ is semi-simple when $\gcd(n, q) = 1$. Therefore it follows from the Wedderburn Structure Theorems that each cyclic code in $\mathcal{R}_n$ contains a unique idempotent which generates the ideal. This idempotent is called the *generating idempotent* of the cyclic code. In the next theorem we prove this fact directly and in the process show how to determine the generating idempotent of a cyclic code. Recall that a *unity* in a ring is a (nonzero) multiplicative identity in the ring, which may or may not exist; however, if it exists, it is unique.

Example 4.3.1 The generating idempotent for the zero cyclic code $\{0\}$ is 0, while that for the cyclic code $\mathcal{R}_n$ is 1. ∎

Theorem 4.3.2 *Let C be a cyclic code in $\mathcal{R}_n$. Then:*
(i) *there exists a unique idempotent $e(x) \in C$ such that $C = \langle e(x) \rangle$, and*
(ii) *if $e(x)$ is a nonzero idempotent in C, then $C = \langle e(x) \rangle$ if and only if $e(x)$ is a unity of C.*

Proof: If C is the zero code, then the idempotent is the zero polynomial and (i) is clear and (ii) does not apply.

So we assume that C is nonzero. We prove (ii) first. Suppose that $e(x)$ is a unity in C. Then $\langle e(x) \rangle \subseteq C$ as C is an ideal. If $c(x) \in C$, then $c(x)e(x) = c(x)$ in C, implying that $\langle e(x) \rangle = C$. Conversely, suppose that $e(x)$ is a nonzero idempotent such that $C = \langle e(x) \rangle$. Then every element $c(x) \in C$ can be written in the form $c(x) = f(x)e(x)$. But $c(x)e(x) = f(x)(e(x))^2 = f(x)e(x) = c(x)$ implying $e(x)$ is a unity for C.

As C is nonzero, by (ii) if $e_1(x)$ and $e_2(x)$ are generating idempotents, then both are unities and $e_1(x) = e_2(x)e_1(x) = e_2(x)$. So we only need to show that a generating idempotent exists. If $g(x)$ is the generator polynomial for C, then $g(x) \mid (x^n - 1)$ by Theorem 4.2.1. Let $h(x) = (x^n - 1)/g(x)$. Then $\gcd(g(x), h(x)) = 1$ in $\mathbb{F}_q[x]$ as $x^n - 1$ has distinct roots. By the Euclidean Algorithm there exist polynomials $a(x), b(x) \in \mathbb{F}_q[x]$ so that $a(x)g(x) + b(x)h(x) = 1$. Let $e(x) \equiv a(x)g(x) \pmod{x^n - 1}$; that is, $e(x)$ is the coset representative of $a(x)g(x) + (x^n - 1)$ in $\mathcal{R}_n$. Then in $\mathcal{R}_n$, $e(x)^2 \equiv (a(x)g(x))(1 - b(x)h(x)) \equiv a(x)g(x) \equiv e(x) \pmod{x^n - 1}$ as $g(x)h(x) = x^n - 1$. Also if $c(x) \in C$, $c(x) = f(x)g(x)$ implying $c(x)e(x) \equiv f(x)g(x)(1 - b(x)h(x)) \equiv f(x)g(x) \equiv c(x) \pmod{x^n - 1}$; so $e(x)$ is a unity in C, and (i) follows from (ii). $\qquad\square$

The proof shows that one way to find the generating idempotent $e(x)$ for a cyclic code C from the generator polynomial $g(x)$ is to solve $1 = a(x)g(x) + b(x)h(x)$ for $a(x)$ using the Euclidean Algorithm, where $h(x) = (x^n - 1)/g(x)$. Then reducing $a(x)g(x)$ modulo $x^n - 1$ produces $e(x)$. We can produce $g(x)$ if we know $e(x)$ as the following theorem shows.

Theorem 4.3.3 *Let C be a cyclic code over $\mathbb{F}_q$ with generating idempotent $e(x)$. Then the generator polynomial of C is $g(x) = \gcd(e(x), x^n - 1)$ computed in $\mathbb{F}_q[x]$.*

Proof: Let $d(x) = \gcd(e(x), x^n - 1)$ in $\mathbb{F}_q[x]$, and let $g(x)$ be the generator polynomial for C. As $d(x) \mid e(x)$, $e(x) = d(x)k(x)$ implying that every element of $C = \langle e(x) \rangle$ is also a multiple of $d(x)$; thus $C \subseteq \langle d(x) \rangle$. By Theorem 4.2.1, in $\mathbb{F}_q[x]$ $g(x) \mid (x^n - 1)$ and $g(x) \mid e(x)$ as $e(x) \in C$. So by Exercise 158, $g(x) \mid d(x)$ implying $d(x) \in C$. Thus $\langle d(x) \rangle \subseteq C$, and so $C = \langle d(x) \rangle$. Since $d(x)$ is a monic divisor of $x^n - 1$ generating C, $d(x) = g(x)$ by Corollary 4.2.2. $\qquad\square$

Example 4.3.4 The following table gives all the cyclic codes C_i of length 7 over $\mathbb{F}_2$ together with their generator polynomials $g_i(x)$ and their generating idempotents $e_i(x)$.

i	dim	$g_i(x)$	$e_i(x)$
0	0	$1 + x^7$	0
1	1	$1 + x + x^2 + \cdots + x^6$	$1 + x + x^2 + \cdots + x^6$
2	3	$1 + x^2 + x^3 + x^4$	$1 + x^3 + x^5 + x^6$
3	3	$1 + x + x^2 + x^4$	$1 + x + x^2 + x^4$
4	4	$1 + x + x^3$	$x + x^2 + x^4$
5	4	$1 + x^2 + x^3$	$x^3 + x^5 + x^6$
6	6	$1 + x$	$x + x^2 + \cdots + x^6$
7	7	1	1

The two codes of dimension 4 are $[7, 4, 3]$ Hamming codes. ∎

Example 4.3.5 The following table gives all the cyclic codes C_i of length 11 over $\mathbb{F}_3$ together with their generator polynomials $g_i(x)$ and their generating idempotents $e_i(x)$.

i	dim	$g_i(x)$	$e_i(x)$
0	0	$x^{11} - 1$	0
1	1	$1 + x + x^2 + \cdots + x^{10}$	$-1 - x - x^2 - \cdots - x^{10}$
2	5	$1 - x - x^2 - x^3 + x^4 + x^6$	$1 + x + x^3 + x^4 + x^5 + x^9$
3	5	$1 + x^2 - x^3 - x^4 - x^5 + x^6$	$1 + x^2 + x^6 + x^7 + x^8 + x^{10}$
4	6	$-1 + x^2 - x^3 + x^4 + x^5$	$-x^2 - x^6 - x^7 - x^8 - x^{10}$
5	6	$-1 - x + x^2 - x^3 + x^5$	$-x - x^3 - x^4 - x^5 - x^9$
6	10	$-1 + x$	$-1 + x + x^2 + \cdots + x^{10}$
7	11	1	1

The two codes of dimension 6 are [11, 6, 5] ternary Golay codes.   ∎

Notice that Theorem 1.8.1 shows that the only [7, 4, 3] binary code is the Hamming code. In Section 10.4.1 we will show that the only [11, 6, 5] ternary code is the Golay code. By Examples 4.3.4 and 4.3.5 these two codes have cyclic representations.

Exercise 218 Verify the entries in the table in Example 4.3.4.   ◆

Exercise 219 Verify the entries in the table in Example 4.3.5.   ◆

Exercise 220 Find the generator polynomials and generating idempotents of all cyclic codes over $\mathbb{F}_3$ of length 8 and dimensions 3 and 5.   ◆

Exercise 221 Let $j(x) = 1 + x + x^2 + \cdots + x^{n-1}$ in $\mathcal{R}_n$ and $\overline{j}(x) = (1/n)j(x)$.
(a) Prove that $j(x)^2 = nj(x)$ in $\mathcal{R}_n$.
(b) Prove that $\overline{j}(x)$ is an idempotent in $\mathcal{R}_n$.
(c) Prove that $\overline{j}(x)$ is the generating idempotent of the repetition code of length n over $\mathbb{F}_q$.
(d) Prove that if $c(x)$ is in $\mathcal{R}_n$, then $c(x)j(x) = c(1)j(x)$ in $\mathcal{R}_n$.
(e) Prove that if $c(x)$ is in $\mathcal{R}_n$, then $c(x)\overline{j}(x) = 0$ in $\mathcal{R}_n$ if $c(x)$ corresponds to an even-like vector in $\mathbb{F}_q^n$ and $c(x)\overline{j}(x)$ is a nonzero multiple of $\overline{j}(x)$ in $\mathcal{R}_n$ if $c(x)$ corresponds to an odd-like vector in $\mathbb{F}_q^n$.   ◆

The next theorem shows that, just as for the generator polynomial, the generating idempotent and its first $k - 1$ cyclic shifts form a basis of a cyclic code.

Theorem 4.3.6 *Let C be an $[n, k]$ cyclic code with generating idempotent $e(x) = \sum_{i=0}^{n-1} e_i x^i$. Then the $k \times n$ matrix*

$$\begin{bmatrix} e_0 & e_1 & e_2 & \cdots & e_{n-2} & e_{n-1} \\ e_{n-1} & e_0 & e_1 & \cdots & e_{n-3} & e_{n-2} \\ & & & \vdots & & \\ e_{n-k+1} & e_{n-k+2} & e_{n-k+3} & \cdots & e_{n-k-1} & e_{n-k} \end{bmatrix}$$

is a generator matrix for C.

Proof: This is equivalent to saying that $\{e(x), xe(x), \ldots, x^{k-1}e(x)\}$ is a basis of C. Therefore it suffices to show that if $a(x) \in \mathbb{F}_q[x]$ has degree less than k such that $a(x)e(x) = 0$, then $a(x) = 0$. Let $g(x)$ be the generator polynomial for C. If $a(x)e(x) = 0$, then $0 = a(x)e(x)g(x) = a(x)g(x)$ as $e(x)$ is the unity of C by Theorem 4.3.2, contradicting Theorem 4.2.1(v) unless $a(x) = 0$. $\qquad\square$

If C_1 and C_2 are codes of length n over $\mathbb{F}_q$, then $C_1 + C_2 = \{\mathbf{c}_1 + \mathbf{c}_2 \mid \mathbf{c}_1 \in C_1 \text{ and } \mathbf{c}_2 \in C_2\}$ is the *sum* of C_1 and C_2. Both the intersection and the sum of two cyclic codes are cyclic, and their generator polynomials and generating idempotents are determined in the next theorem.

Theorem 4.3.7 *Let C_i be a cyclic code of length n over $\mathbb{F}_q$ with generator polynomial $g_i(x)$ and generating idempotent $e_i(x)$ for $i = 1$ and 2. Then:*

(i) *$C_1 \cap C_2$ has generator polynomial $\mathrm{lcm}(g_1(x), g_2(x))$ and generating idempotent $e_1(x)e_2(x)$, and*

(ii) *$C_1 + C_2$ has generator polynomial $\gcd(g_1(x), g_2(x))$ and generating idempotent $e_1(x) + e_2(x) - e_1(x)e_2(x)$.*

Proof: We prove (ii) and leave the proof of (i) as an exercise. We also leave it as an exercise to show that the sum of two cyclic codes is cyclic. Let $g(x) = \gcd(g_1(x), g_2(x))$. It follows from the Euclidean Algorithm that $g(x) = g_1(x)a(x) + g_2(x)b(x)$ for some $a(x)$ and $b(x)$ in $\mathbb{F}_q[x]$. So $g(x) \in C_1 + C_2$. Since $C_1 + C_2$ is cyclic, $\langle g(x) \rangle \subseteq C_1 + C_2$. On the other hand $g(x) \mid g_1(x)$, which shows that $C_1 \subseteq \langle g(x) \rangle$ by Corollary 4.2.5; similarly $C_2 \subseteq \langle g(x) \rangle$ implying $C_1 + C_2 \subseteq \langle g(x) \rangle$. So $C_1 + C_2 = \langle g(x) \rangle$. Since $g(x) \mid (x^n - 1)$ as $g(x) \mid g_1(x)$ and $g(x)$ is monic, $g(x)$ is the generator polynomial for $C_1 + C_2$ by Corollary 4.2.2. If $c(x) = c_1(x) + c_2(x)$ where $c_i(x) \in C_i$ for $i = 1$ and 2, then $c(x)(e_1(x) + e_2(x) - e_1(x)e_2(x)) = c_1(x) + c_1(x)e_2(x) - c_1(x)e_2(x) + c_2(x)e_1(x) + c_2(x) - c_2(x)e_1(x) = c(x)$. Thus (ii) follows by Theorem 4.3.2 since $e_1(x) + e_2(x) - e_1(x)e_2(x) \in C_1 + C_2$. $\qquad\square$

Exercise 222 Prove part (i) of Theorem 4.3.7. Also prove that if $e_1(x)$ and $e_2(x)$ are idempotents, so are $e_1(x)e_2(x)$, $e_1(x) + e_2(x) - e_1(x)e_2(x)$, and $1 - e_1(x)$. ◆

Exercise 223 Show that the sum of two cyclic codes is cyclic as claimed in Theorem 4.3.7. ◆

Exercise 224 Let C_i be a cyclic code of length n over $\mathbb{F}_q$ for $i = 1$ and 2. Let α be a primitive nth root of unity in some extension field of $\mathbb{F}_q$. Suppose C_i has generator polynomial $g_i(x)$, where

$$g_i(x) = \prod_{s \in K_i} M_{\alpha^s}(x)$$

is the factorization of $g_i(x)$ into minimal polynomials over $\mathbb{F}_q$ with K_i a subset of the representatives of the q-cyclotomic cosets modulo n. Assume that the representative of a coset is the smallest element in the coset. What are the subsets of representatives of q-cyclotomic cosets that will produce the generator polynomials for the codes $C_1 + C_2$ and $C_1 \cap C_2$? ◆

Exercise 225 Find the generator polynomials and the generating idempotents of the following codes from Example 4.3.4: $C_1 + C_6, C_2 + C_3, C_2 + C_4, C_2 + C_5, C_3 + C_4, C_3 + C_5,$ $C_1 \cap C_6, C_2 \cap C_3, C_2 \cap C_4, C_2 \cap C_5, C_3 \cap C_4,$ and $C_3 \cap C_5$. ♦

Exercise 226 Which pairs of codes in Exercise 220 sum to the code $\mathbb{F}_3^8$? Which pairs of codes in that example have intersection $\{0\}$? ♦

Exercise 227 If C_i is a cyclic code with generator polynomial $g_i(x)$ and generating idempotent $e_i(x)$ for $1 \le i \le 3$, what are the generator polynomial and generating idempotent of $C_1 + C_2 + C_3$? ♦

We are now ready to describe a special set of idempotents, called primitive idempotents, that, once known, will produce all the idempotents in $\mathcal{R}_n$ and therefore all the cyclic codes. Let $x^n - 1 = f_1(x) \cdots f_s(x)$, where $f_i(x)$ is irreducible over $\mathbb{F}_q$ for $1 \le i \le s$. The $f_i(x)$ are distinct as $x^n - 1$ has distinct roots. Let $\widehat{f_i}(x) = (x^n - 1)/f_i(x)$. In the next theorem we show that the ideals $\langle \widehat{f_i}(x) \rangle$ of $\mathcal{R}_n$ are the minimal ideals of $\mathcal{R}_n$. Recall that an ideal $\mathcal{I}$ in a ring $\mathcal{R}$ is a *minimal ideal* provided there is no proper ideal between $\{0\}$ and $\mathcal{I}$. We denote the generating idempotent of $\langle \widehat{f_i}(x) \rangle$ by $\widehat{e_i}(x)$. The idempotents $\widehat{e_1}(x), \ldots, \widehat{e_s}(x)$ are called the *primitive idempotents* of $\mathcal{R}_n$.

Theorem 4.3.8 *The following hold in $\mathcal{R}_n$.*
(i)   *The ideals $\langle \widehat{f_i}(x) \rangle$ for $1 \le i \le s$ are all the minimal ideals of $\mathcal{R}_n$.*
(ii)   *$\mathcal{R}_n$ is the vector space direct sum of $\langle \widehat{f_i}(x) \rangle$ for $1 \le i \le s$.*
(iii) *If $i \ne j$, then $\widehat{e_i}(x)\widehat{e_j}(x) = 0$ in $\mathcal{R}_n$.*
(iv)   *$\sum_{i=1}^{s} \widehat{e_i}(x) = 1$ in $\mathcal{R}_n$.*
(v)   *The only idempotents in $\langle \widehat{f_i}(x) \rangle$ are 0 and $\widehat{e_i}(x)$.*
(vi)   *If $e(x)$ is a nonzero idempotent in $\mathcal{R}_n$, then there is a subset T of $\{1, 2, \ldots, s\}$ such that $e(x) = \sum_{i \in T} \widehat{e_i}(x)$ and $\langle e(x) \rangle = \sum_{i \in T} \langle \widehat{f_i}(x) \rangle$.*

Proof: Suppose that $\langle \widehat{f_i}(x) \rangle$ is not a minimal ideal of $\mathcal{R}_n$. By Corollary 4.2.5, there would be a generator polynomial $g(x)$ of a nonzero ideal properly contained in $\langle \widehat{f_i}(x) \rangle$ such that $\widehat{f_i}(x) \mid g(x)$ with $g(x) \ne \widehat{f_i}(x)$. As $f_i(x)$ is irreducible and $g(x) \mid (x^n - 1)$, this is impossible. So $\langle \widehat{f_i}(x) \rangle$ is a minimal ideal of $\mathcal{R}_n$, completing part of (i).

As $\{\widehat{f_i}(x) \mid 1 \le i \le s\}$ has no common irreducible factor of $x^n - 1$ and each polynomial in the set divides $x^n - 1$, $\gcd(\widehat{f_1}(x), \ldots, \widehat{f_s}(x)) = 1$. Applying the Euclidean Algorithm inductively,

$$1 = \sum_{i=1}^{s} a_i(x)\widehat{f_i}(x) \tag{4.3}$$

for some $a_i(x) \in \mathbb{F}_q[x]$. So 1 is in the sum of the ideals $\langle \widehat{f_i}(x) \rangle$, which is itself an ideal of $\mathcal{R}_n$. In any ring, the only ideal containing the identity of the ring is the ring itself. This proves that $\mathcal{R}_n$ is the vector space sum of the ideals $\langle \widehat{f_i}(x) \rangle$. To prove it is a direct sum, we must show that $\langle \widehat{f_i}(x) \rangle \cap \sum_{j \ne i} \langle \widehat{f_j}(x) \rangle = \{0\}$ for $1 \le i \le s$. As $f_i(x) \mid \widehat{f_j}(x)$ for $j \ne i$, $f_j(x) \nmid \widehat{f_j}(x)$, and the irreducible factors of $x^n - 1$ are distinct, we conclude that $f_i(x) = \gcd\{\widehat{f_j}(x) \mid 1 \le j \le s, j \ne i\}$. Applying induction to the results of

Theorem 4.3.7(ii) shows that $\langle f_i(x)\rangle = \sum_{j\neq i}\langle \widehat{f}_j(x)\rangle$. So $\langle \widehat{f}_i(x)\rangle \cap \sum_{j\neq i}\langle \widehat{f}_j(x)\rangle = \langle \widehat{f}_i(x)\rangle \cap \langle f_i(x)\rangle = \langle \mathrm{lcm}(\widehat{f}_i(x), f_i(x))\rangle = \langle x^n - 1\rangle = \{0\}$ by Theorem 4.3.7 completing (ii). Let $\mathcal{M} = \langle m(x)\rangle$ be any minimal ideal of $\mathcal{R}_n$. As

$$0 \neq m(x) = m(x)\cdot 1 = \sum_{i=1}^{s} m(x)a_i(x)\widehat{f}_i(x)$$

by (4.3), there is an i such that $m(x)a_i(x)\widehat{f}_i(x) \neq 0$. Hence $\mathcal{M} \cap \langle \widehat{f}_i(x)\rangle \neq \{0\}$ as $m(x)a_i(x)\widehat{f}_i(x) \in \mathcal{M} \cap \langle \widehat{f}_i(x)\rangle$, and therefore $\mathcal{M} = \langle \widehat{f}_i(x)\rangle$ by minimality of $\mathcal{M}$ and $\langle \widehat{f}_i(x)\rangle$. This completes the proof of (i).

If $i \neq j$, $\widehat{e}_i(x)\widehat{e}_j(x) \in \langle \widehat{f}_i(x)\rangle \cap \langle \widehat{f}_j(x)\rangle = \{0\}$ by (ii), yielding (iii). By using (iii) and applying induction to Theorem 4.3.7(ii), $\sum_{i=1}^{s}\widehat{e}_i(x)$ is the generating idempotent of $\sum_{i=1}^{s}\langle \widehat{f}_i(x)\rangle = \mathcal{R}_n$ by part (ii). The generating idempotent of $\mathcal{R}_n$ is 1, verifying (iv).

If $e(x)$ is a nonzero idempotent in $\langle \widehat{f}_i(x)\rangle$, then $\langle e(x)\rangle$ is an ideal contained in $\langle \widehat{f}_i(x)\rangle$. By minimality as $e(x)$ is nonzero, $\langle \widehat{f}_i(x)\rangle = \langle e(x)\rangle$, implying by Theorem 4.3.2 that $e(x) = \widehat{e}_i(x)$ as both are the unique unity of $\langle \widehat{f}_i(x)\rangle$. Thus (v) holds.

For (vi), note that $e(x)\widehat{e}_i(x)$ is an idempotent in $\langle \widehat{f}_i(x)\rangle$. Thus either $e(x)\widehat{e}_i(x)$ is 0 or $\widehat{e}_i(x)$ by (v). Let $T = \{i \mid e(x)\widehat{e}_i(x) \neq 0\}$. Then by (iv), $e(x) = e(x)\cdot 1 = e(x)\sum_{i=1}^{s}\widehat{e}_i(x) = \sum_{i=1}^{s}e(x)\widehat{e}_i(x) = \sum_{i\in T}\widehat{e}_i(x)$. Furthermore, $\langle e(x)\rangle = \langle \sum_{i\in T}\widehat{e}_i(x)\rangle = \sum_{i\in T}\langle \widehat{e}_i(x)\rangle$ by Theorem 4.3.7(ii) and induction. $\qquad\square$

We remark that the minimal ideals in this theorem are extension fields of $\mathbb{F}_q$. Theorem 4.4.19 will also characterize these minimal ideals using the trace map.

Theorem 4.3.9 *Let $\mathcal{M}$ be a minimal ideal of $\mathcal{R}_n$. Then $\mathcal{M}$ is an extension field of $\mathbb{F}_q$.*

Proof: We only need to show that every nonzero element in $\mathcal{M}$ has a multiplicative inverse in $\mathcal{M}$. Let $a(x) \in \mathcal{M}$ with $a(x)$ not zero. Then $\langle a(x)\rangle$ is a nonzero ideal of $\mathcal{R}_n$ contained in $\mathcal{M}$, and hence $\langle a(x)\rangle = \mathcal{M}$. So if $e(x)$ is the unity of $\mathcal{M}$, there is an element $b(x)$ in $\mathcal{R}_n$ with $a(x)b(x) = e(x)$. Now $c(x) = b(x)e(x) \in \mathcal{M}$ as $e(x) \in \mathcal{M}$. Hence $a(x)c(x) = e(x)^2 = e(x)$. $\qquad\square$

Exercise 228 What fields arise as the minimal ideals in $\mathcal{R}_7$ and $\mathcal{R}_{15}$ over $\mathbb{F}_2$?   ◆

Theorem 4.3.8 shows that every idempotent is a sum of primitive idempotents and that cyclic codes are sums of minimal cyclic codes. An interesting consequence, found in [280], of this characterization of cyclic codes is that the dimension of a sum of cyclic codes satisfies the same formula as that of the inclusion–exclusion principle, a fact that fails in general.

Theorem 4.3.10 *Let $\mathcal{C}_i$ be a cyclic code of length n over $\mathbb{F}_q$ for $1 \leq i \leq a$. Then:*

$$\dim(\mathcal{C}_1 + \mathcal{C}_2 + \cdots + \mathcal{C}_a) = \sum_i \dim(\mathcal{C}_i) - \sum_{i<j}\dim(\mathcal{C}_i \cap \mathcal{C}_j)$$

$$+ \sum_{i<j<k}\dim(\mathcal{C}_i \cap \mathcal{C}_j \cap \mathcal{C}_k) - \cdots$$

$$+ (-1)^{a-1}\dim(\mathcal{C}_1 \cap \mathcal{C}_2 \cap \cdots \cap \mathcal{C}_a).$$

Proof: Let $\{\widehat{e}_i(x) \mid 1 \leq i \leq s\}$ be the primitive idempotents of $\mathcal{R}_n$. By Theorem 4.3.8, the minimal ideals of $\mathcal{R}_n$ are $\langle \widehat{e}_i(x)\rangle$. Fix a basis B_i of $\langle \widehat{e}_i(x)\rangle$ for $1 \leq i \leq s$. Also by

Theorem 4.3.8, each C_i is a direct sum of $\{\langle \widehat{e}_j(x) \rangle \mid j \in S_i\}$ for some subset S_i of $\{1, 2, \dots, s\}$. Thus a basis of $C_{i_1} + \dots + C_{i_b}$ is $B_{i_1} \cup \dots \cup B_{i_b}$, and this basis contains $|B_{i_1} \cup \dots \cup B_{i_b}| = \dim(C_{i_1} + \dots + C_{i_b})$ elements, where $|B|$ is the number of (distinct) elements in B. A basis of $C_{i_1} \cap \dots \cap C_{i_b}$ is $B_{i_1} \cap \dots \cap B_{i_b}$, and this basis contains $|B_{i_1} \cap \dots \cap B_{i_b}| = \dim(C_{i_1} \cap \dots \cap C_{i_b})$ elements. Since $\dim(C_1 + C_2 + \dots + C_a) = |B_1 \cup B_2 \cup \dots \cup B_a|$, we can apply the inclusion–exclusion principle to obtain the result. $\square$

Example 4.3.11 Theorem 4.3.10 does not work in general for noncyclic codes. For example, for $1 \leq i \leq 3$, let C_i be a binary code of length 2 with generator matrix G_i, where

$$G_1 = [1 \ \ 0], \qquad G_2 = [0 \ \ 1], \qquad \text{and} \qquad G_3 = [1 \ \ 1].$$

Then $\dim(C_i) = 1$ for $1 \leq i \leq 3$, $\dim(C_i \cap C_j) = 0$ for $i \neq j$, and $\dim(C_1 \cap C_2 \cap C_3) = 0$. But $\dim(C_1 + C_2 + C_3) = 2$, which does not equal $1 + 1 + 1 - 0 - 0 - 0 + 0$.   ∎

Exercise 229 Prove that if C_1 and C_2 are linear codes of length n over $\mathbb{F}_q$, then $\dim(C_1 + C_2) = \dim(C_1) + \dim(C_2) - \dim(C_1 \cap C_2)$.   ◆

We turn now to a particular permutation which maps idempotents of $\mathcal{R}_n$ to idempotents of $\mathcal{R}_n$. Let a be an integer such that $\gcd(a, n) = 1$. The function μ_a defined on $\{0, 1, \dots, n-1\}$ by $i\mu_a \equiv ia \pmod{n}$ is a permutation of the coordinate positions $\{0, 1, \dots, n-1\}$ of a cyclic code of length n and is called a *multiplier*. Because cyclic codes of length n are represented as ideals in $\mathcal{R}_n$, for $a > 0$ it is convenient to regard μ_a as acting on $\mathcal{R}_n$ by

$$f(x)\mu_a \equiv f(x^a) \pmod{x^n - 1}. \tag{4.4}$$

This equation is consistent with the original definition of μ_a because $x^i \mu_a = x^{ia} = x^{ia+jn}$ in $\mathcal{R}_n$ for an integer j such that $0 \leq ia + jn < n$ since $x^n = 1$ in $\mathcal{R}_n$. In other words $x^i \mu_a = x^{ia \bmod n}$. If $a < 0$, we can attach meaning to $f(x^a)$ in $\mathcal{R}_n$ by defining $x^i \mu_a = x^{ia \bmod n}$, where, of course, $0 \leq ia \bmod n < n$. With this interpretation, (4.4) is consistent with the original definition of μ_a when $a < 0$. We leave the proof of the following as an exercise.

Theorem 4.3.12 *Let $f(x)$ and $g(x)$ be elements of $\mathcal{R}_n$. Suppose $e(x)$ is an idempotent of $\mathcal{R}_n$. Let a be relatively prime to n. Then:*
(i)   *if $b \equiv a \pmod{n}$, then $\mu_b = \mu_a$,*
(ii) *$(f(x) + g(x))\mu_a = f(x)\mu_a + g(x)\mu_a$,*
(iii) *$(f(x)g(x))\mu_a = (f(x)\mu_a)(g(x)\mu_a)$,*
(iv) *μ_a is an automorphism of $\mathcal{R}_n$,*
(v)   *$e(x)\mu_a$ is an idempotent of $\mathcal{R}_n$, and*
(vi) *μ_q leaves invariant each q-cyclotomic coset modulo n and has order equal to $\mathrm{ord}_n(q)$.*

Exercise 230 Prove that if $\gcd(a, n) = 1$, then the map μ_a is indeed a permutation of $\{0, 1, \dots, n-1\}$ as claimed in the text. What happens if $\gcd(a, n) \neq 1$?   ◆

Exercise 231 Prove Theorem 4.3.12. ◆

Theorem 4.3.13 *Let C be a cyclic code of length n over $\mathbb{F}_q$ with generating idempotent $e(x)$. Let a be an integer with $\gcd(a, n) = 1$. Then:*
(i) *$C\mu_a = \langle e(x)\mu_a \rangle$ and $e(x)\mu_a$ is the generating idempotent of the cyclic code $C\mu_a$, and*
(ii) *$e(x)\mu_q = e(x)$ and $\mu_q \in \mathrm{PAut}(C)$.*

Proof: Using Theorem 4.3.12(iii), $C\mu_a = \{(e(x)f(x))\mu_a \mid f(x) \in \mathcal{R}_n\} = \{e(x)\mu_a \times f(x)\mu_a \mid f(x)\mu_a \in \mathcal{R}_n\} = \{e(x)\mu_a h(x) \mid h(x) \in \mathcal{R}_n\} = \langle e(x)\mu_a \rangle$ as μ_a is an automorphism of $\mathcal{R}_n$ by Theorem 4.3.12(iv). Hence $C\mu_a$ is cyclic and has generating idempotent $e(x)\mu_a$ by Theorem 4.3.12(v), proving (i).

If we show that $e(x)\mu_q = e(x)$, then by part (i), $C\mu_q = C$ and so $\mu_q \in \mathrm{PAut}(C)$. By Theorem 4.3.8(vi), $e(x) = \sum_{i \in T} \widehat{e}_i(x)$ for some set T. By Theorem 4.3.12(ii), $e(x)\mu_q = e(x)$ if $\widehat{e}_i(x)\mu_q = \widehat{e}_i(x)$ for all i. But $\widehat{e}_i(x)\mu_q = \widehat{e}_i(x^q) = (\widehat{e}_i(x))^q$ by Theorem 3.7.4, the latter certainly being a nonzero element of $\langle \widehat{e}_i(x) \rangle$. But by Theorem 4.3.12(v), $\widehat{e}_i(x)\mu_q$ is also an idempotent of $\langle \widehat{e}_i(x) \rangle$. Hence $\widehat{e}_i(x)\mu_q = \widehat{e}_i(x)$ by Theorem 4.3.8(v). □

Exercise 232 Consider the cyclic codes of length 11 over $\mathbb{F}_3$ as given in Example 4.3.5.
(a) Find the image of each generating idempotent, and hence each cyclic code, under μ_2.
(b) Verify that μ_3 fixes each idempotent.
(c) Write the image of each generator polynomial under μ_3 as an element of $\mathcal{R}_{11}$. Do generator polynomials get mapped to generator polynomials? ◆

Exercise 233 Show that any two codes of the same dimension in Examples 4.3.4 and 4.3.5 are permutation equivalent. ◆

Note that arbitrary permutations in general do not map idempotents to idempotents, nor do they even map cyclic codes to cyclic codes.

Corollary 4.3.14 *Let C be a cyclic code of length n over $\mathbb{F}_q$. Let $\mathcal{A}$ be the group of order n generated by the cyclic shift $i \mapsto i + 1 \pmod{n}$. Let $\mathcal{B}$ be the group of order $\mathrm{ord}_n(q)$ generated by the multiplier μ_q. Then the group $\mathcal{G}$ of order $n \cdot \mathrm{ord}_n(q)$ generated by $\mathcal{A}$ and $\mathcal{B}$ is a subgroup of $\mathrm{PAut}(C)$.*

Proof: The corollary follows from the structure of the normalizer of $\mathcal{A}$ in the symmetric group Sym_n and Theorem 4.3.13(ii). In fact, $\mathcal{G}$ is the semidirect product of $\mathcal{A}$ extended by $\mathcal{B}$. □

Exercise 234 In the notation of Corollary 4.3.14, what is the order of the subgroup $\mathcal{G}$ of $\mathrm{PAut}(C)$ for the following values of n and q?
(a) $n = 15, q = 2$.
(b) $n = 17, q = 2$.
(c) $n = 23, q = 2$.
(d) $n = 15, q = 4$.
(e) $n = 25, q = 3$. ◆

Corollary 4.3.15 *Let C be a cyclic code of length n over $\mathbb{F}_q$ with generating idempotent $e(x) = \sum_{i=0}^{n-1} e_i x^i$. Then:*
(i)   *$e_i = e_j$ if i and j are in the same q-cyclotomic coset modulo n,*
(ii) *if $q = 2$,*

$$e(x) = \sum_{j \in J} \sum_{i \in C_j} x^i,$$

where J is some subset of representatives of 2-cyclotomic cosets modulo n, and
(iii) *if $q = 2$, every element of $\mathcal{R}_n$ of the form*

$$\sum_{j \in J} \sum_{i \in C_j} x^i,$$

where J is some subset of representatives of 2-cyclotomic cosets modulo n, is an idempotent of $\mathcal{R}_n$.

Proof: By Theorem 4.3.13(ii), $e(x)\mu_q = e(x)$. Thus $e(x)\mu_q = \sum_{i=0}^{n-1} e_i x^{iq} \equiv \sum_{i=0}^{n-1} e_i x^i \equiv \sum_{i=0}^{n-1} e_{iq} x^{iq} \pmod{x^n - 1}$, where subscripts are read modulo n. Hence (i) holds, and (ii) is a special case of (i). Part (iii) follows as $e(x)^2 = \sum_{j \in J} \sum_{i \in C_j} x^{2i} = \sum_{j \in J} \sum_{i \in C_j} x^i = e(x)$ by Exercise 152 and the fact that $2C_j \equiv C_j \pmod{n}$.   □

Since any idempotent is a generating idempotent of some code, the preceding corollary shows that each idempotent in $\mathcal{R}_n$ has the form

$$e(x) = \sum_j a_j \sum_{i \in C_j} x^i, \tag{4.5}$$

where the outer sum is over a system of representatives of the q-cyclotomic cosets modulo n and each a_j is in $\mathbb{F}_q$. For $q = 2$, but not for arbitrary q, all such expressions are idempotents. (Compare Examples 4.3.4 and 4.3.5.)

We can also give the general form for the idempotents in $\mathcal{R}_n$ over $\mathbb{F}_4$. We can construct a set S of representatives of all the distinct 4-cyclotomic cosets modulo n as follows. The set $S = K \cup L_1 \cup L_2$, where K, L_1, and L_2 are pairwise disjoint. K consists of distinct representatives k, where $C_k = C_{2k}$. L_1 and L_2 are chosen so that if $k \in L_1 \cup L_2$, $C_k \neq C_{2k}$; furthermore $L_2 = \{2k \mid k \in L_1\}$. Squaring $e(x)$ in (4.5), we obtain

$$e(x)^2 = \sum_{j \in S} a_j^2 \sum_{i \in C_j} x^{2i},$$

as $\mathcal{R}_n$ has characteristic 2. But if $j \in K$, then $\sum_{i \in C_j} x^{2i} = \sum_{i \in C_j} x^i$; if $j \in L_1$, then $\sum_{i \in C_j} x^{2i} = \sum_{i \in C_{2j}} x^i$ and $2j \in L_2$; and if $2j \in L_2$, then $\sum_{i \in C_{2j}} x^{2i} = \sum_{i \in C_j} x^i$ and $j \in L_1$ as i and $4i$ are in the same 4-cyclotomic coset. Therefore $e(x)$ is an idempotent if and only if $a_j^2 = a_j$ for all $j \in K$ and $a_{2j} = a_j^2$ for all $j \in L_1$. In particular $e(x)$ is an idempotent in $\mathcal{R}_n$ if and only if

$$e(x) = \sum_{j \in K} a_j \sum_{i \in C_j} x^i + \sum_{j \in L_1} \left(a_j \sum_{i \in C_j} x^i + a_j^2 \sum_{i \in C_j} x^{2i} \right), \tag{4.6}$$

where $a_j \in \{0, 1\}$ if $j \in K$. Recall that in $\mathbb{F}_4$, $\bar{\ }$ is called conjugation, and is given by $\bar{0} = 0$, $\bar{1} = 1$, and $\bar{\bar{\omega}} = \omega$; alternately, $\bar{a} = a^2$. If $e(x)$ is the generating idempotent of C, then we leave it as an exercise to show that $\overline{C}$ is a cyclic code with generating idempotent $\overline{e(x)} = \sum_j \overline{a}_j \sum_{i \in C_j} x^i$. Furthermore, by examining (4.6), we see that $\overline{e(x)} = e(x)\mu_2$. By Theorem 4.3.13, $\overline{C} = C\mu_2$. We summarize these results.

Theorem 4.3.16 *Let C be a cyclic code over $\mathbb{F}_4$ with generating idempotent $e(x)$. Then $e(x)$ has the form given in (4.6). Also $\overline{C} = C\mu_2$ is cyclic with generating idempotent $e(x)\mu_2$.*

Exercise 235 Show that if $e(x)$ is the generating idempotent of a cyclic code C over $\mathbb{F}_4$, then $\overline{C}$ is a cyclic code with generating idempotent $\overline{e(x)} = \sum_j \overline{a}_j \sum_{i \in C_j} x^i$. Show also that $\overline{e(x)} = e(x)\mu_2$. ♦

Exercise 236 Do the following:
(a) List the 4-cyclotomic cosets modulo 21.
(b) Construct a set $S = K \cup L_1 \cup L_2$ of distinct 4-cyclotomic coset representatives modulo 21 which can be used to construct idempotents in $\mathcal{R}_{21}$ over $\mathbb{F}_4$ as in the discussion prior to Theorem 4.3.16.
(c) Give the general form of such an idempotent.
(d) How many of these idempotents are there?
(e) Write down four of these idempotents. ♦

Theorem 4.3.13 shows that μ_a maps cyclic codes to cyclic codes with the generating idempotent mapped to the generating idempotent; however, the generator polynomial may not be mapped to the generator polynomial of the image code. In fact, the automorphism μ_q maps the generator polynomial to its qth power. See Exercise 232.

A multiplier takes a cyclic code into an equivalent cyclic code. The following theorem, a special case of a theorem of Pálfy (see [150]), implies that, in certain instances, two cyclic codes are permutation equivalent if and only if a multiplier takes one to the other. This is a very powerful result when it applies.

Theorem 4.3.17 *Let C_1 and C_2 be cyclic codes of length n over $\mathbb{F}_q$. Assume that $\gcd(n, \phi(n)) = 1$, where ϕ is the Euler ϕ-function. Then C_1 and C_2 are permutation equivalent if and only if there is a multiplier that maps C_1 to C_2.*

Since multipliers send generating idempotents to generating idempotents, we have the following corollary.

Corollary 4.3.18 *Let C_1 and C_2 be cyclic codes of length n over $\mathbb{F}_q$. Assume that $\gcd(n, \phi(n)) = 1$, where ϕ is the Euler ϕ-function. Then C_1 and C_2 are permutation equivalent if and only if there is a multiplier that maps the idempotent of C_1 to the idempotent of C_2.*

4.4   Zeros of a cyclic code

Recall from Section 4.1 and, in particular Theorem 4.1.1, that if $t = \text{ord}_n(q)$, then $\mathbb{F}_{q^t}$ is a splitting field of $x^n - 1$; so $\mathbb{F}_{q^t}$ contains a primitive nth root of unity α, and $x^n - 1 = \prod_{i=0}^{n-1}(x - \alpha^i)$ is the factorization of $x^n - 1$ into linear factors over $\mathbb{F}_{q^t}$. Furthermore $x^n - 1 = \prod_s M_{\alpha^s}(x)$ is the factorization of $x^n - 1$ into irreducible factors over $\mathbb{F}_q$, where s runs through a set of representatives of the q-cyclotomic cosets modulo n.

Let C be a cyclic code in $\mathcal{R}_n$ with generator polynomial $g(x)$. By Theorems 4.1.1(i) and 4.2.1(vii), $g(x) = \prod_s M_{\alpha^s}(x) = \prod_s \prod_{i \in C_s}(x - \alpha^i)$, where s runs through some subset of representatives of the q-cyclotomic cosets C_s modulo n. Let $T = \bigcup_s C_s$ be the union of these q-cyclotomic cosets. The roots of unity $\mathcal{Z} = \{\alpha^i \mid i \in T\}$ are called the *zeros* of the cyclic code C and $\{\alpha^i \mid i \notin T\}$ are the *nonzeros* of C. The set T is called the *defining set* of C. (Note that if you change the primitive nth root of unity, you change T; so T is computed relative to a fixed primitive root. This will be discussed further in Section 4.5.) It follows that $c(x)$ belongs to C if and only if $c(\alpha^i) = 0$ for each $i \in T$ by Theorem 4.2.1. Notice that T, and hence either the set of zeros or the set of nonzeros, completely determines the generator polynomial $g(x)$. By Theorem 4.2.1, the dimension of C is $n - |T|$ as $|T|$ is the degree of $g(x)$.

Example 4.4.1 In Example 4.3.4 a table giving the dimension, generator polynomials $g_i(x)$, and generating idempotents $e_i(x)$ of all the cyclic codes C_i of length 7 over $\mathbb{F}_2$ was given. We add to that table the defining sets of each code relative to the primitive root α given in Example 3.4.3.

i	dim	$g_i(x)$	$e_i(x)$	Defining set
0	0	$1 + x^7$	0	$\{0, 1, 2, 3, 4, 5, 6\}$
1	1	$1 + x + x^2 + \cdots + x^6$	$1 + x + x^2 + \cdots + x^6$	$\{1, 2, 3, 4, 5, 6\}$
2	3	$1 + x^2 + x^3 + x^4$	$1 + x^3 + x^5 + x^6$	$\{0, 1, 2, 4\}$
3	3	$1 + x + x^2 + x^4$	$1 + x + x^2 + x^4$	$\{0, 3, 5, 6\}$
4	4	$1 + x + x^3$	$x + x^2 + x^4$	$\{1, 2, 4\}$
5	4	$1 + x^2 + x^3$	$x^3 + x^5 + x^6$	$\{3, 5, 6\}$
6	6	$1 + x$	$x + x^2 + \cdots + x^6$	$\{0\}$
7	7	1	1	$\emptyset$ ∎

Exercise 237 What would be the defining sets of each of the codes in Example 4.4.1 if the primitive root $\beta = \alpha^3$ were used to determine the defining set rather than α? ♦

Our next theorem gives basic properties of cyclic codes in terms of their defining sets, summarizing the above discussion.

Theorem 4.4.2 *Let α be a primitive nth root of unity in some extension field of $\mathbb{F}_q$. Let C be a cyclic code of length n over $\mathbb{F}_q$ with defining set T and generator polynomial $g(x)$. The following hold.*
(i) *T is a union of q-cyclotomic cosets modulo n.*
(ii) *$g(x) = \prod_{i \in T}(x - \alpha^i)$.*
(iii) *$c(x) \in \mathcal{R}_n$ is in C if and only if $c(\alpha^i) = 0$ for all $i \in T$.*
(iv) *The dimension of C is $n - |T|$.*

Exercise 238 Let C be a cyclic code over $\mathbb{F}_q$ with defining set T and generator polynomial $g(x)$. Let C_e be the subcode of all even-like vectors in C.
(a) Prove that C_e is cyclic and has defining set $T \cup \{0\}$.
(b) Prove that $C = C_e$ if and only if $0 \in T$ if and only if $g(1) = 0$.

(c) Prove that if $C \neq C_e$, then the generator polynomial of C_e is $(x - 1)g(x)$.
(d) Prove that if C is binary, then C contains the all-one vector if and only if $0 \notin T$.   ◆

Exercise 239 Let C_i be cyclic codes of length n over $\mathbb{F}_q$ with defining sets T_i for $i = 1, 2$.
(a) Prove that $C_1 \cap C_2$ has defining set $T_1 \cup T_2$.
(b) Prove that $C_1 + C_2$ has defining set $T_1 \cap T_2$.
(c) Prove that $C_1 \subseteq C_2$ if and only if $T_2 \subseteq T_1$.
Note: This exercise shows that the lattice of cyclic codes of length n over $\mathbb{F}_q$, where the join of two codes is the sum of the codes and the meet of two codes is their intersection, is isomorphic to the "upside-down" version of the lattice of subsets of $\mathcal{N} = \{0, 1, \ldots, n - 1\}$ that are unions of q-cyclotomic cosets modulo n, where the join of two such subsets is the set union of the subsets and the meet of two subsets is the set intersection of the subsets.   ◆

The zeros of a cyclic code can be used to obtain a parity check matrix (possibly with dependent rows) as explained in the next theorem. The construction presented in this theorem is analogous to that of the subfield subcode construction in Section 3.8.

Theorem 4.4.3 *Let C be an $[n, k]$ cyclic code over $\mathbb{F}_q$ with zeros $\mathcal{Z}$ in a splitting field $\mathbb{F}_{q^t}$ of $x^n - 1$ over $\mathbb{F}_q$. Let $\alpha \in \mathbb{F}_{q^t}$ be a primitive nth root of unity in $\mathbb{F}_{q^t}$, and let $\mathcal{Z} = \{\alpha^j \mid j \in C_{i_1} \cup \cdots \cup C_{i_w}\}$, where $C_{i_1}, \ldots, C_{i_w}$ are distinct q-cyclotomic cosets modulo n. Let L be the $w \times n$ matrix over $\mathbb{F}_{q^t}$ defined by*

$$
L = \begin{bmatrix}
1 & \alpha^{i_1} & \alpha^{2i_1} & \cdots & \alpha^{(n-1)i_1} \\
1 & \alpha^{i_2} & \alpha^{2i_2} & \cdots & \alpha^{(n-1)i_2} \\
& & \vdots & & \\
1 & \alpha^{i_w} & \alpha^{2i_w} & \cdots & \alpha^{(n-1)i_w}
\end{bmatrix}.
$$

Then $\mathbf{c}$ is in C if and only if $L\mathbf{c}^T = \mathbf{0}$. Choosing a basis of $\mathbb{F}_{q^t}$ over $\mathbb{F}_q$, we may represent each element of $\mathbb{F}_{q^t}$ as a $t \times 1$ column vector over $\mathbb{F}_q$. Replacing each entry of L by its corresponding column vector, we obtain a $tw \times n$ matrix H over $\mathbb{F}_q$ which has the property that $\mathbf{c} \in C$ if and only if $H\mathbf{c}^T = \mathbf{0}$. In particular, $k \geq n - tw$.

Proof: We have $c(x) \in C$ if and only if $c(\alpha^j) = 0$ for all $j \in C_{i_1} \cup \cdots \cup C_{i_w}$, which by Theorem 3.7.4 is equivalent to $c(\alpha^{i_j}) = 0$ for $1 \leq j \leq w$. Clearly, this is equivalent to $L\mathbf{c}^T = \mathbf{0}$, which is a system of homogeneous linear equations with coefficients that are powers of α. Expanding each of these powers of α in the chosen basis of $\mathbb{F}_{q^t}$ over $\mathbb{F}_q$ yields the equivalent system $H\mathbf{c}^T = \mathbf{0}$. As the rows of H may be dependent, $k \geq n - tw$.   □

If C' is the code over $\mathbb{F}_{q^t}$ with parity check matrix L in this theorem, then the code C is actually the subfield subcode $C'|_{\mathbb{F}_q}$.

Exercise 240 Show that the matrix L in Theorem 4.4.3 has rank w. Note: The matrix L is related to a Vandermonde matrix. See Lemma 4.5.1.   ◆

For a cyclic code C in $\mathcal{R}_n$, there are in general many polynomials $v(x)$ in $\mathcal{R}_n$ such that $C = \langle v(x) \rangle$. However, by Theorem 4.2.1 and its corollary, there is exactly one such polynomial, namely the monic polynomial in C of minimal degree, which also divides $x^n - 1$ and which we call the generator polynomial of C. In the next theorem we characterize all polynomials $v(x)$ which generate C.

Theorem 4.4.4 *Let C be a cyclic code of length n over $\mathbb{F}_q$ with generator polynomial $g(x)$. Let $v(x)$ be a polynomial in $\mathcal{R}_n$.*
(i) *$C = \langle v(x) \rangle$ if and only if $\gcd(v(x), x^n - 1) = g(x)$.*
(ii) *$v(x)$ generates C if and only if the nth roots of unity which are zeros of $v(x)$ are precisely the zeros of C.*

Proof: First assume that $\gcd(v(x), x^n - 1) = g(x)$. As $g(x) \mid v(x)$, multiples of $v(x)$ are multiples of $g(x)$ in $\mathcal{R}_n$ and so $\langle v(x) \rangle \subseteq C$. By the Euclidean Algorithm there exist polynomials $a(x)$ and $b(x)$ in $\mathbb{F}_q[x]$ such that $g(x) = a(x)v(x) + b(x)(x^n - 1)$. Hence $g(x) = a(x)v(x)$ in $\mathcal{R}_n$ and so multiples of $g(x)$ are multiples of $v(x)$ in $\mathcal{R}_n$ implying $\langle v(x) \rangle \supseteq C$. Thus $C = \langle v(x) \rangle$.

For the converse, assume that $C = \langle v(x) \rangle$. Let $d(x) = \gcd(v(x), x^n - 1)$. As $g(x) \mid v(x)$ and $g(x) \mid (x^n - 1)$ by Theorem 4.2.1, $g(x) \mid d(x)$ by Exercise 158. As $g(x) \in C = \langle v(x) \rangle$, there exists a polynomial $a(x)$ such that $g(x) = a(x)v(x)$ in $\mathcal{R}_n$. So there exists a polynomial $b(x)$ such that $g(x) = a(x)v(x) + b(x)(x^n - 1)$ in $\mathbb{F}_q[x]$. Thus $d(x) \mid g(x)$ by Exercise 158. Hence as both $d(x)$ and $g(x)$ are monic and divide each other, $d(x) = g(x)$ and (i) holds.

As the only roots of both $g(x)$ and $x^n - 1$ are nth roots of unity, $g(x) = \gcd(v(x), x^n - 1)$ if and only if the nth roots of unity which are zeros of $v(x)$ are precisely the zeros of $g(x)$; the latter are the zeros of C. □

Corollary 4.4.5 *Let C be a cyclic code of length n over $\mathbb{F}_q$ with zeros $\{\alpha^i \mid i \in T\}$ for some primitive nth root of unity α where T is the defining set of C. Let a be an integer such that $\gcd(a, n) = 1$ and let a^{-1} be the multiplicative inverse of a in the integers modulo n. Then $\{\alpha^{a^{-1}i} \mid i \in T\}$ are the zeros of the cyclic code $C\mu_a$ and $a^{-1}T \bmod n$ is the defining set for $C\mu_a$.*

Proof: Let $e(x)$ be the generating idempotent of C. By Theorem 4.3.13, the generating idempotent of the cyclic code $C\mu_a$ is $e(x)\mu_a$. By Theorem 4.4.4, the zeros of C and $C\mu_a$ are the nth roots of unity which are also roots of $e(x)$ and $e'(x) = e(x)\mu_a$, respectively. As $e'(x) = e(x)\mu_a \equiv e(x^a) \pmod{x^n - 1}$, $e'(x) = e(x^a) + b(x)(x^n - 1)$ in $\mathbb{F}_q[x]$. The corollary now follows from the fact that the nth root of unity α^j is a root of $e'(x)$ if and only if α^{aj} is a root of $e(x)$. □

Theorem 4.3.13 implies that the image of one special vector, the generating idempotent, of a cyclic code under a multiplier determines the image code. As described in Corollary 4.4.5 a multiplier maps the defining set, and hence the zeros, of a cyclic code to the defining set, and hence the zeros, of the image code. Such an assertion is not true for a general permutation.

Exercise 241 An *equivalence class* of codes is the set of all codes that are equivalent to one another. Give a defining set for a representative of each equivalence class of the binary cyclic codes of length 15. Example 3.7.8, Theorem 4.3.17, and Corollary 4.4.5 will be useful.   ◆

Exercise 242 Continuing with Exercise 241, do the following:

(a) List the 2-cyclotomic cosets modulo 31.

(b) List the defining sets for all $[31, 26]$ binary cyclic codes. Give a defining set for a representative of each equivalence class of the $[31, 26]$ binary cyclic codes. Hint: Use Theorem 4.3.17 and Corollary 4.4.5.

(c) Repeat part (b) for $[31, 5]$ binary cyclic codes. (Take advantage of your work in part (b).)

(d) List the 15 defining sets for all $[31, 21]$ binary cyclic codes, and give a defining set for a representative of each equivalence class of these codes.

(e) List the 20 defining sets for all $[31, 16]$ binary cyclic codes, and give a defining set for a representative of each equivalence class of these codes.   ◆

If C is a code of length n over $\mathbb{F}_q$, then a *complement* of C is a code C^C such that $C + C^C = \mathbb{F}_q^n$ and $C \cap C^C = \{0\}$. In general, the complement is not unique. However, Exercise 243 shows that if C is a cyclic code, there is a unique complement of C that is also cyclic. We call this code *the cyclic complement* of C. In the following theorem we give the generator polynomial and generating idempotent of the cyclic complement.

Exercise 243 Prove that a cyclic code has a unique complement that is also cyclic.   ◆

Theorem 4.4.6 *Let C be a cyclic code of length n over $\mathbb{F}_q$ with generator polynomial $g(x)$, generating idempotent $e(x)$, and defining set T. Let C^C be the cyclic complement of C. The following hold.*

(i) *$h(x) = (x^n - 1)/g(x)$ is the generator polynomial for C^C and $1 - e(x)$ is its generating idempotent.*

(ii) *C^C is the sum of the minimal ideals of $\mathcal{R}_n$ not contained in C.*

(iii) *If $\mathcal{N} = \{0, 1, \ldots, n - 1\}$, then $\mathcal{N} \setminus T$ is the defining set of C^C.*

Exercise 244 Prove Theorem 4.4.6.   ◆

The dual $C^\perp$ of a cyclic code C is also cyclic as Theorem 4.2.6 shows. The generator polynomial and generating idempotent for $C^\perp$ can be obtained from the generator polynomial and generating idempotent of C. To find these, we reintroduce the concept of the reciprocal polynomial encountered in Exercise 192. Let $f(x) = f_0 + f_1 x + \cdots + f_a x^a$ be a polynomial of degree a in $\mathbb{F}_q[x]$. The *reciprocal polynomial* of $f(x)$ is the polynomial

$$f^*(x) = x^a f(x^{-1}) = x^a (f(x)\mu_{-1}) = f_a + f_{a-1} x + \cdots + f_0 x^a.$$

So $f^*(x)$ has coefficients the reverse of those of $f(x)$. Furthermore, $f(x)$ is *reversible* provided $f(x) = f^*(x)$.

Exercise 245 Show that a monic irreducible reversible polynomial of degree greater than 1 cannot be a primitive polynomial except for the polynomial $1 + x + x^2$ over $\mathbb{F}_2$.   ◆

We have the following basic properties of reciprocal polynomials. Their proofs are left as an exercise.

Lemma 4.4.7 *Let $f(x) \in \mathbb{F}_q[x]$.*
(i) *If $\beta_1, \ldots, \beta_r$ are the nonzero roots of f in some extension field of $\mathbb{F}_q$, then $\beta_1^{-1}, \ldots,$ β_r^{-1} are the nonzero roots of f^* in that extension field.*
(ii) *If $f(x)$ is irreducible over $\mathbb{F}_q$, so is $f^*(x)$.*
(iii) *If $f(x)$ is a primitive polynomial, so is $f^*(x)$.*

Exercise 246 Prove Lemma 4.4.7. ◆

Exercise 247 In Example 3.7.8, the factorization of $x^{15} - 1$ into irreducible polynomials over $\mathbb{F}_2$ was found. Find the reciprocal polynomial of each of these irreducible polynomials. How does this confirm Lemma 4.4.7? ◆

Exercise 248 Prove that if $f_1(x)$ and $f_2(x)$ are reversible polynomials in $\mathbb{F}_q[x]$, so is $f_1(x)f_2(x)$. What about $f_1(x) + f_2(x)$? ◆

The connection between dual codes and reciprocal polynomials is clear from the following lemma.

Lemma 4.4.8 *Let $\mathbf{a} = a_0 a_1 \cdots a_{n-1}$ and $\mathbf{b} = b_0 b_1 \cdots b_{n-1}$ be vectors in $\mathbb{F}_q^n$ with associated polynomials $a(x)$ and $b(x)$. Then $\mathbf{a}$ is orthogonal to $\mathbf{b}$ and all its shifts if and only if $a(x)b^*(x) = 0$ in $\mathcal{R}_n$.*

Proof: Let $\mathbf{b}^{(i)} = b_i b_{i+1} \cdots b_{n+i-1}$ be the ith cyclic shift of $\mathbf{b}$, where the subscripts are read modulo n. Then

$$\mathbf{a} \cdot \mathbf{b}^{(i)} = 0 \text{ if and only if } \sum_{j=0}^{n-1} a_j b_{j+i} = 0. \tag{4.7}$$

But $a(x)b^*(x) = 0$ in $\mathcal{R}_n$ if and only if $a(x)(x^{n-1-\deg b(x)})b^*(x) = 0$ in $\mathcal{R}_n$. But $a(x)(x^{n-1-\deg b(x)})b^*(x) = \sum_{i=0}^{n-1}(\sum_{j=0}^{n-1} a_j b_{j+i} x^{n-1-i})$. Thus $a(x)b^*(x) = 0$ in $\mathcal{R}_n$ if and only if (4.7) holds for $0 \le i \le n - 1$. $\square$

We now give the generator polynomial and generating idempotent of the dual of a cyclic code. The proof is left as an exercise.

Theorem 4.4.9 *Let C be an $[n, k]$ cyclic code over $\mathbb{F}_q$ with generator polynomial $g(x)$, generating idempotent $e(x)$, and defining set T. Let $h(x) = (x^n - 1)/g(x)$. The following hold.*
(i) *$C^\perp$ is a cyclic code and $C^\perp = C^C \mu_{-1}$.*
(ii) *$C^\perp$ has generating idempotent $1 - e(x)\mu_{-1}$ and generator polynomial*

$$\frac{x^k}{h(0)} h(x^{-1}).$$

(iii) *If $\beta_1, \ldots, \beta_k$ are the zeros of C, then $\beta_1^{-1}, \ldots, \beta_k^{-1}$ are the nonzeros of $C^\perp$.*
(iv) *If $\mathcal{N} = \{0, 1, \ldots, n - 1\}$, then $\mathcal{N} \setminus (-1)T \bmod n$ is the defining set of $C^\perp$.*
(v) *Precisely one of C and $C^\perp$ is odd-like and the other is even-like.*

The polynomial $h(x) = (x^n - 1)/g(x)$ in this theorem is called the *check polynomial* of C. The generator polynomial of $C^\perp$ in part (ii) of the theorem is the reciprocal polynomial of $h(x)$ rescaled to be monic.

Exercise 249 Prove Theorem 4.4.9.   ◆

Exercise 250 Let C be a cyclic code with cyclic complement C^c. Prove that if C is MDS so is C^c.   ◆

The following corollary determines, from the generator polynomial, when a cyclic code is self-orthogonal.

Corollary 4.4.10 *Let C be a cyclic code over $\mathbb{F}_q$ of length n with generator polynomial $g(x)$ and check polynomial $h(x) = (x^n - 1)/g(x)$. Then C is self-orthogonal if and only if $h^*(x) \mid g(x)$.*

Exercise 251 Prove Corollary 4.4.10.   ◆

Exercise 252 Using Corollary 4.4.10 and Examples 3.7.8, 4.2.4, and 4.3.4 give the generator polynomials of the self-orthogonal binary cyclic codes of lengths 7, 9, and 15.   ◆

In the next theorem we show how to decide when a cyclic code is self-orthogonal. In particular, this characterization shows that all self-orthogonal cyclic codes are even-like. In this theorem we use the observation that if C is a q-cyclotomic coset modulo n, either $C\mu_{-1} = C$ or $C\mu_{-1} = C'$ for some different q-cyclotomic coset C', in which case $C'\mu_{-1} = C$ as μ_{-1}^2 is the identity.

Theorem 4.4.11 *Let C be a self-orthogonal cyclic code over $\mathbb{F}_q$ of length n with defining set T. Let $C_1, \ldots, C_k, D_1, \ldots, D_\ell, E_1, \ldots, E_\ell$ be all the distinct q-cyclotomic cosets modulo n partitioned so that $C_i = C_i\mu_{-1}$ for $1 \le i \le k$ and $D_i = E_i\mu_{-1}$ for $1 \le i \le \ell$. The following hold.*
(i) *$C_i \subseteq T$ for $1 \le i \le k$ and at least one of D_i or E_i is contained in T for $1 \le i \le \ell$.*
(ii) *C is even-like.*
(iii) *$C \cap C\mu_{-1} = \{0\}$.*
Conversely, if C is a cyclic code with defining set T that satisfies (i), then C is self-orthogonal.

Proof: Let $\mathcal{N} = \{0, 1, \ldots, n - 1\}$. Let $T^\perp$ be the defining set of $C^\perp$. By Theorem 4.4.9, $T^\perp = \mathcal{N} \setminus (-1)T \bmod n$. As $C \subseteq C^\perp, \mathcal{N} \setminus (-1)T \bmod n \subseteq T$ by Exercise 239. If $C_i \not\subseteq T$, then $C_i \not\subseteq (-1)T \bmod n$ because $C_i = C_i\mu_{-1}$ implying that $C_i \subseteq \mathcal{N} \setminus (-1)T \bmod n \subseteq T$, a contradiction. If $D_i \not\subseteq T$, then $E_i \not\subseteq (-1)T \bmod n$ because $E_i = D_i\mu_{-1}$ implying that $E_i \subseteq \mathcal{N} \setminus (-1)T \bmod n \subseteq T$, proving (i). Part (ii) follows from part (i) and Exercise 238 as $C_i = \{0\}$ for some i. By Corollary 4.4.5, $C\mu_{-1}$ has defining set $(-1)T \bmod n$. By (i) $T \cup (-1)T \bmod n = \mathcal{N}$ yielding (iii) using Exercise 239.

For the converse, assume T satisfies (i). We only need to show that $T^\perp \subseteq T$, where $T^\perp = \mathcal{N} \setminus (-1)T \bmod n$ by Exercise 239. As $C_i \subseteq T$ for $1 \le i \le k, C_i \subseteq (-1)T \bmod n$ implying $C_i \not\subseteq T^\perp$. Hence $T^\perp$ is a union of some D_is and E_is. If $D_i \subseteq \mathcal{N} \setminus (-1)T \bmod n$, then $D_i \not\subseteq (-1)T \bmod n$ and so $E_i \not\subseteq T$. By (i) $D_i \subseteq T$. Similarly if $E_i \subseteq \mathcal{N} \setminus (-1)T \bmod n$, then $E_i \subseteq T$. Hence $T^\perp \subseteq T$ implying C is self-orthogonal.   □

Exercise 253 Continuing with Exercise 242, do the following:
(a) Show that all $[31, 5]$ binary cyclic codes are self-orthogonal.
(b) Show that there are two inequivalent $[31, 15]$ self-orthogonal binary cyclic codes, and give defining sets for a code in each equivalence class.   ◆

Corollary 4.4.12 *Let* $\mathcal{D} = \mathcal{C} + \langle \mathbf{1} \rangle$ *be a cyclic code of length n over* $\mathbb{F}_q$, *where* $\mathcal{C}$ *is self-orthogonal. Then* $\mathcal{D} \cap \mathcal{D}\mu_{-1} = \langle \mathbf{1} \rangle$.

Exercise 254 Prove Corollary 4.4.12. ◆

Corollary 4.4.13 *Let* $p_1(x), \ldots, p_k(x), q_1(x), \ldots, q_\ell(x), r_1(x), \ldots, r_\ell(x)$ *be the monic irreducible factors of* $x^n - 1$ *over* $\mathbb{F}_q[x]$ *arranged as follows. For* $1 \leq i \leq k$, $p_i^*(x) = a_i p_i(x)$ *for some* $a_i \in \mathbb{F}_q$, *and for* $1 \leq i \leq \ell$, $r_i^*(x) = b_i q_i(x)$ *for some* $b_i \in \mathbb{F}_q$. *Let* $\mathcal{C}$ *be a cyclic code of length n over* $\mathbb{F}_q$ *with generator polynomial* $g(x)$. *Then* $\mathcal{C}$ *is self-orthogonal if and only if* $g(x)$ *has factors* $p_1(x) \cdots p_k(x)$ *and at least one of* $q_i(x)$ *or* $r_i(x)$ *for* $1 \leq i \leq \ell$.

Exercise 255 Prove Corollary 4.4.13. ◆

Exercise 256 Using Corollary 4.4.13 and Examples 3.7.8, 4.2.4, and 4.3.4 give the generator polynomials of the self-orthogonal binary cyclic codes of lengths 7, 9, and 15. Compare your answers to those of Exercise 252. ◆

Exercise 257 Let $j(x) = 1 + x + x^2 + \cdots + x^{n-1}$ in $\mathcal{R}_n$ and $\overline{j}(x) = (1/n)j(x)$. In Exercise 221 we gave properties of $j(x)$ and $\overline{j}(x)$. Let $\mathcal{C}$ be a cyclic code over $\mathbb{F}_q$ with generating idempotent $i(x)$. Let $\mathcal{C}_e$ be the subcode of all even-like vectors in $\mathcal{C}$. In Exercise 238 we found the generator polynomial of $\mathcal{C}_e$.
(a) Prove that $1 - \overline{j}(x)$ is the generating idempotent of the $[n, n-1]$ cyclic code over $\mathbb{F}_q$ consisting of all even-like vectors in $\mathcal{R}_n$.
(b) Prove that $i(1) = 0$ if $\mathcal{C} = \mathcal{C}_e$ and $i(1) = 1$ if $\mathcal{C} \neq \mathcal{C}_e$.
(c) Prove that if $\mathcal{C} \neq \mathcal{C}_e$, then $i(x) - \overline{j}(x)$ is the generating idempotent of $\mathcal{C}_e$. ◆

We illustrate Theorems 4.3.13, 4.3.17, 4.4.6, and 4.4.9 by returning to Examples 4.3.4 and 4.3.5.

Example 4.4.14 In Examples 4.3.4 and 4.3.5, the following codes are cyclic complementary pairs: $\mathcal{C}_1$ and $\mathcal{C}_6$, $\mathcal{C}_2$ and $\mathcal{C}_5$, and $\mathcal{C}_3$ and $\mathcal{C}_4$. In both examples, the following are dual pairs: $\mathcal{C}_1$ and $\mathcal{C}_6$, $\mathcal{C}_2$ and $\mathcal{C}_4$, and $\mathcal{C}_3$ and $\mathcal{C}_5$. In Example 4.3.4, $\mathcal{C}_2$ and $\mathcal{C}_3$ are equivalent under μ_3, as are $\mathcal{C}_4$ and $\mathcal{C}_5$. In Example 4.3.5, the same pairs are equivalent under μ_2. In both examples, the permutation automorphism group for each of $\mathcal{C}_1$, $\mathcal{C}_6$, and $\mathcal{C}_7$ is the full symmetric group. (In general, the permutation automorphism group of the repetition code of length n, and hence its dual, is the symmetric group on n letters.) In Example 4.3.4, the group of order 3 generated by μ_2 is a subgroup of the automorphism group of the remaining four codes; in Example 4.3.5, the group of order 5 generated by μ_3 is a subgroup of the automorphism group of the remaining four codes. ■

Exercise 258 Verify all the claims in Example 4.4.14. ◆

Exercise 259 Let C be a cyclic code of length n over $\mathbb{F}_q$ with generator polynomial $g(x)$. What conditions on $g(x)$ must be satisfied for the dual of C to equal the cyclic complement of C? ◆

Exercise 260 Identify all binary cyclic codes of lengths 7, 9, and 15 whose duals equal their cyclic complements. (Examples 3.7.8, 4.2.4, and 4.3.4 will be useful.) ◆

In Theorem 4.4.9 we found the generating idempotent of the dual of any code. The multiplier μ_{-1} was key in that theorem. We can also find the generating idempotent of the Hermitian dual of a cyclic code over $\mathbb{F}_4$. Here μ_{-2} will play the role of μ_{-1}. Recall from Exercise 8 that if C is a code over $\mathbb{F}_4$, then $C^{\perp_H} = \overline{C}^{\perp}$.

Theorem 4.4.15 *Let C be a cyclic code of length n over $\mathbb{F}_4$ with generating idempotent $e(x)$ and defining set T. The following hold.*
(i) *$C^{\perp_H}$ is a cyclic code and $C^{\perp_H} = C^c \mu_{-2}$, where C^c is the cyclic complement of C.*
(ii) *$C^{\perp_H}$ has generating idempotent $1 - e(x)\mu_{-2}$.*
(iii) *If $\mathcal{N} = \{0, 1, \ldots, n-1\}$, then $\mathcal{N} \setminus (-2)T \bmod n$ is the defining set of $C^{\perp_H}$.*
(iv) *Precisely one of C and $C^{\perp_H}$ is odd-like and the other is even-like.*

Proof: We leave the fact that $C^{\perp_H}$ is a cyclic code as an exercise. Exercise 8 shows that $C^{\perp_H} = \overline{C}^{\perp}$. By Theorem 4.4.9 $\overline{C}^{\perp} = \overline{C}^c \mu_{-1}$. Theorem 4.3.16 shows that $\overline{C}^c = \overline{C^c} = C^c \mu_2$, and (i) follows since $\mu_2 \mu_{-1} = \mu_{-2}$. By Theorem 4.4.6 $C^{\perp_H} = C^c \mu_{-2}$ has generating idempotent $(1 - e(x))\mu_{-2} = 1 - e(x)\mu_{-2}$, giving (ii). $C\mu_{-2}$ has defining set $(-2)^{-1}T \bmod n$ by Corollary 4.4.5. However, $(-2)^{-1}T = (-2)T$ modulo n because $\mu_{-2}^2 = \mu_4$ and μ_4 fixes all 4-cyclotomic cosets. By Theorem 4.4.6(iii) and Corollary 4.4.5, $C^{\perp_H} = C^c \mu_{-2}$ has defining set $(-2)^{-1}(\mathcal{N} \setminus T) = \mathcal{N} \setminus (-2)T \bmod n$, giving (iii). Part (iv) follows from (iii) and Exercise 238 as precisely one of $\mathcal{N}$ and $\mathcal{N} \setminus (-2)T \bmod n$ contains 0. ☐

Exercise 261 Prove that if C be a cyclic code over $\mathbb{F}_4$, then $C^{\perp_H}$ is also a cyclic code. ◆

Exercise 262 Using Theorem 4.3.16 find generating idempotents of all the cyclic codes C of length 9 over $\mathbb{F}_4$, their ordinary duals $C^{\perp}$, and their Hermitian duals $C^{\perp_H}$. ◆

We can obtain a result analogous to Theorem 4.4.11 for Hermitian self-orthogonal cyclic codes over $\mathbb{F}_4$. Again we simply replace μ_{-1} by μ_{-2} and apply Theorem 4.4.15. Notice that if C is a 4-cyclotomic coset modulo n, then either $C\mu_{-2} = C$ or $C\mu_{-2} = C'$ for a different 4-cyclotomic coset C', in which case $C'\mu_{-2} = C$ as $\mu_{-2}^2 = \mu_4$ and μ_4 fixes all 4-cyclotomic cosets.

Theorem 4.4.16 *Let C be a Hermitian self-orthogonal cyclic code over $\mathbb{F}_4$ of length n with defining set T. Let $C_1, \ldots, C_k, D_1, \ldots, D_\ell, E_1, \ldots, E_\ell$ be all the distinct 4-cyclotomic cosets modulo n partitioned so that $C_i = C_i\mu_{-2}$ for $1 \le i \le k$ and $D_i = E_i\mu_{-2}$ for $1 \le i \le \ell$. The following hold:*
(i) *$C_i \subseteq T$ for $1 \le i \le k$ and at least one of D_i or E_i is contained in T for $1 \le i \le \ell$.*
(ii) *C is even-like.*
(iii) *$C \cap C\mu_{-2} = \{\mathbf{0}\}$.*

Conversely, if C is a cyclic code with defining set T that satisfies (i), then C is Hermitian self-orthogonal.

Exercise 263 Prove Theorem 4.4.16.   ◆

Corollary 4.4.17 *Let $\mathcal{D} = \mathcal{C} + \langle \mathbf{1} \rangle$ be a cyclic code of length n over $\mathbb{F}_4$ such that $\mathcal{C}$ is Hermitian self-orthogonal. Then $\mathcal{D} \cap \mathcal{D}\mu_{-2} = \langle \mathbf{1} \rangle$.*

Exercise 264 Prove Corollary 4.4.17.   ◆

The next theorem shows the rather remarkable fact that a binary self-orthogonal cyclic code must be doubly-even.

Theorem 4.4.18 *A self-orthogonal binary cyclic code is doubly-even.*

Proof: Let $\mathcal{C}$ be an $[n, k]$ self-orthogonal binary cyclic code with defining set T. By Theorem 1.4.5(iv), $\mathcal{C}$ has only even weight codewords and hence $0 \in T$ by Exercise 238. Suppose that $\mathcal{C}$ is not doubly-even. Then the subcode $\mathcal{C}_0$ of $\mathcal{C}$ consisting of codewords of weights divisible by 4 has dimension $k - 1$ by Theorem 1.4.6. Clearly, $\mathcal{C}_0$ is cyclic as the cyclic shift of a vector is a vector of the same weight. By Theorem 4.4.2 and Corollary 4.2.5 (or Exercise 239), the defining set of $\mathcal{C}_0$ is $T \cup \{a\}$ for some $a \notin T$. But then $\{a\}$ must be a 2-cyclotomic coset modulo n, which implies that $2a \equiv a \pmod{n}$. Hence $a = 0$ as n is odd, which is impossible as $0 \in T$.   □

In Theorem 4.3.8, the minimal cyclic codes in $\mathcal{R}_n$ are shown to be those with generator polynomials $g(x)$ where $(x^n - 1)/g(x)$ is irreducible over $\mathbb{F}_q$. So minimal cyclic codes are sometimes called *irreducible cyclic codes*. These minimal cyclic codes can be described using the trace function.

Theorem 4.4.19 *Let $g(x)$ be an irreducible factor of $x^n - 1$ over $\mathbb{F}_q$. Suppose $g(x)$ has degree s, and let $\gamma \in \mathbb{F}_{q^s}$ be a root of $g(x)$. Let $\mathrm{Tr}_s : \mathbb{F}_{q^s} \to \mathbb{F}_q$ be the trace map from $\mathbb{F}_{q^s}$ to $\mathbb{F}_q$. Then*

$$
C_\gamma = \left\{ \sum_{i=0}^{n-1} \mathrm{Tr}_s(\xi \gamma^i) x^i \mid \xi \in \mathbb{F}_{q^s} \right\}
$$

is the $[n, s]$ irreducible cyclic code with nonzeros $\{\gamma^{-q^i} \mid 0 \le i < s\}$.

Proof: By Lemma 3.8.5, C_γ is a nonzero linear code over $\mathbb{F}_q$. If $c_\xi(x) = \sum_{i=0}^{n-1} \mathrm{Tr}_s(\xi \gamma^i) x^i$, then $c_{\xi \gamma^{-1}}(x) = c_\xi(x)x$ in $\mathcal{R}_n$ implying that C_γ is cyclic. Let $g(x) = \sum_{i=0}^{n-1} g_i x^i$. By Lemma 3.8.5, as $g_i \in \mathbb{F}_q$ and $g(\gamma) = 0$,

$$
\sum_{i=0}^{n-1} g_i \mathrm{Tr}_s(\xi \gamma^i) = \mathrm{Tr}_s \left(\xi \sum_{i=0}^{n-1} g_i \gamma^i \right) = \mathrm{Tr}_s(0) = 0.
$$

Hence $\langle g(x) \rangle \subseteq C_\gamma^\perp$. By Theorem 4.4.9, $C_\gamma^\perp$ is a cyclic code not equal to $\mathcal{R}_n$ as $C_\gamma \neq \{\mathbf{0}\}$. As $g(x)$ is irreducible over $\mathbb{F}_q$, there can be no proper cyclic codes between $\langle g(x) \rangle$ and $\mathcal{R}_n$. So $\langle g(x) \rangle = C_\gamma^\perp$. The result follows from Theorem 4.4.9(iii).   □

4.5   Minimum distance of cyclic codes

With any code, it is important to be able to determine the minimum distance in order to determine its error-correcting capability. It is therefore helpful to have bounds on the minimum distance, particularly lower bounds. There are several known lower bounds for the minimum distance of a cyclic code. The oldest of these is the Bose–Ray-Chaudhuri–Hocquenghem Bound [28, 132], usually called the BCH Bound, which is fundamental to the definition of the BCH codes presented in Chapter 5. Improvements of this bound have been obtained by Hartmann and Tzeng [117], and later van Lint and Wilson [203]. The BCH Bound, the Hartmann–Tzeng Bound, and a bounding technique of van Lint and Wilson are presented here. The BCH and Hartmann–Tzeng Bounds depend on the zeros of the code and especially on the ability to find strings of "consecutive" zeros.

Before proceeding with the BCH Bound, we state a lemma, used in the proof of the BCH Bound and useful elsewhere as well, about the determinant of a Vandermonde matrix. Let $\alpha_1, \ldots, \alpha_s$ be elements in a field $\mathbb{F}$. The $s \times s$ matrix $V = [v_{i,j}]$, where $v_{i,j} = \alpha_j^{i-1}$ is called a *Vandermonde matrix*. Note that the transpose of this matrix is also called a Vandermonde matrix.

Lemma 4.5.1 $\det V = \prod_{1 \le i < j \le s}(\alpha_j - \alpha_i)$. *In particular, V is nonsingular if the elements $\alpha_1, \ldots, \alpha_s$ are distinct.*

In this section we will assume that $\mathcal{C}$ is a cyclic code of length n over $\mathbb{F}_q$ and that α is a primitive nth root of unity in $\mathbb{F}_{q^t}$, where $t = \mathrm{ord}_n(q)$. Recall that T is a defining set for $\mathcal{C}$ provided the zeros of $\mathcal{C}$ are $\{\alpha^i \mid i \in T\}$. So T must be a union of q-cyclotomic cosets modulo n. We say that T contains a set of s *consecutive* elements $\mathcal{S}$ provided there is a set $\{b, b+1, \ldots, b+s-1\}$ of s consecutive integers such that

$$\{b, b+1, \ldots, b+s-1\} \bmod n = \mathcal{S} \subseteq T.$$

Example 4.5.2 Consider the binary cyclic code $\mathcal{C}$ of length 7 with defining set $T = \{0, 3, 6, 5\}$. Then T has a set of three consecutive elements $\mathcal{S} = \{5, 6, 0\}$.   ∎

Theorem 4.5.3 (BCH Bound) *Let $\mathcal{C}$ be a cyclic code of length n over $\mathbb{F}_q$ with defining set T. Suppose $\mathcal{C}$ has minimum weight d. Assume T contains $\delta - 1$ consecutive elements for some integer δ. Then $d \ge \delta$.*

Proof: By assumption, $\mathcal{C}$ has zeros that include $\alpha^b, \alpha^{b+1}, \ldots, \alpha^{b+\delta-2}$. Let $c(x)$ be a nonzero codeword in $\mathcal{C}$ of weight w, and let

$$c(x) = \sum_{j=1}^{w} c_{i_j} x^{i_j}.$$

Assume to the contrary that $w < \delta$. As $c(\alpha^i) = 0$ for $b \le i \le b + \delta - 2$, $M\mathbf{u}^{\mathsf{T}} = \mathbf{0}$, where

$$M = \begin{bmatrix} \alpha^{i_1 b} & \alpha^{i_2 b} & \cdots & \alpha^{i_w b} \\ \alpha^{i_1(b+1)} & \alpha^{i_2(b+1)} & \cdots & \alpha^{i_w(b+1)} \\ & & \vdots & \\ \alpha^{i_1(b+w-1)} & \alpha^{i_2(b+w-1)} & \cdots & \alpha^{i_w(b+w-1)} \end{bmatrix}$$

and $\mathbf{u} = c_{i_1} c_{i_2} \cdots c_{i_w}$. Since $\mathbf{u} \neq \mathbf{0}$, M is a singular matrix and hence $\det M = 0$. But $\det M = \alpha^{(i_1 + i_2 + \cdots + i_w)b} \det V$, where V is the Vandermonde matrix

$$V = \begin{bmatrix} 1 & 1 & \cdots & 1 \\ \alpha^{i_1} & \alpha^{i_2} & \cdots & \alpha^{i_w} \\ & & \vdots & \\ \alpha^{i_1(w-1)} & \alpha^{i_2(w-1)} & \cdots & \alpha^{i_w(w-1)} \end{bmatrix}.$$

Since the α^{i_j} are distinct, $\det V \neq 0$ by Lemma 4.5.1, contradicting $\det M = 0$. $\square$

Exercise 265 Find a generator polynomial and generator matrix of a triple error-correcting [15, 5] binary cyclic code. ◆

The BCH Bound asserts that you want to find the longest set of consecutive elements in the defining set. However, the defining set depends on the primitive element chosen. Let $\beta = \alpha^a$, where $\gcd(a, n) = 1$; so β is also a primitive nth root of unity. Therefore, if a^{-1} is the multiplicative inverse of a modulo n, the minimal polynomials $M_{\alpha^s}(x)$ and $M_{\beta^{a^{-1}s}}(x)$ are equal. So the code with defining set T, relative to the primitive element α, is the same as the code with defining set $a^{-1}T \bmod n$, relative to the primitive element β. Thus when applying the BCH Bound, or any of our other lower bounds, a higher lower bound may be obtained if you apply a multiplier to the defining set. Alternately, the two codes C and $C\mu_a$ are equivalent and have defining sets T and $a^{-1}T$ (by Corollary 4.4.5) with respect to the same primitive element α; hence they have the same minimum weight and so either defining set can be used to produce the best bound.

Example 4.5.4 Let C be the $[31, 25, d]$ binary cyclic code with defining set $T = \{0, 3, 6, 12, 24, 17\}$. Applying the BCH Bound to C, we see that $d \geq 2$, as the longest consecutive set in T is size 1. However, multiplying T by $3^{-1} \equiv 21 \pmod{31}$, we have $3^{-1}T \bmod 31 = \{0, 1, 2, 4, 8, 16\}$. Replacing α by α^3 or C by $C\mu_3$ and applying the BCH Bound, we obtain $d \geq 4$. In fact C is the even weight subcode of the Hamming code $\mathcal{H}_5$ and $d = 4$. ∎

Example 4.5.5 Let C be the $[23, 12, d]$ binary cyclic code with defining set $T = \{1, 2, 3, 4, 6, 8, 9, 12, 13, 16, 18\}$. The BCH Bound implies that $d \geq 5$ as T has four consecutive elements. Notice that $T = C_1$, the 2-cyclotomic coset modulo 23 containing 1. Modulo 23, there are only two other 2-cyclotomic cosets: $C_0 = \{0\}$ and $C_5 = \{5, 7, 10, 11, 14, 15, 17, 19, 20, 21, 22\}$. Let C_e be the subcode of C of even weight codewords; C_e is cyclic with defining set $C_0 \cup C_1$ by Exercise 238. By Theorem 4.4.11, C_e is self-orthogonal. Hence C_e is doubly-even by Theorem 4.4.18. Therefore its minimum weight is at least 8; it must be exactly 8 by the Sphere Packing Bound. So C_e has nonzero codewords of weights 8, 12, 16, and 20 only. As C contains the all-one codeword by Exercise 238, C_e cannot contain a codeword of weight 20 as adding such a codeword to $\mathbf{1}$ produces a weight 3 codeword, contradicting $d \geq 5$. Therefore C_e has nonzero codewords of weights 8, 12, and 16 only. Since $C = C_e \cup \mathbf{1} + C_e$, C has nonzero codewords of weights 7, 8, 11, 12, 15, 16, and 23 only. In particular $d = 7$. By Theorem 1.12.3, this code must be the [23, 12, 7] binary Golay code.   ∎

Hartmann and Tzeng [117] showed that if there are several consecutive sets of $\delta - 1$ elements in the defining set that are spaced properly, then the BCH Bound can be improved. To state the Hartmann–Tzeng Bound, which we do not prove, we develop the following notation. If A and B are subsets of the integers modulo n, then $A + B = \{a + b \bmod n \mid a \in A, b \in B\}$.

Theorem 4.5.6 (Hartmann–Tzeng Bound) *Let C be a cyclic code of length n over $\mathbb{F}_q$ with defining set T. Let A be a set of $\delta - 1$ consecutive elements of T and $B = \{jb \bmod n \mid 0 \leq j \leq s\}$, where $\gcd(b, n) < \delta$. If $A + B \subseteq T$, then the minimum weight d of C satisfies $d \geq \delta + s$.*

Clearly the BCH Bound is the Hartmann–Tzeng Bound with $s = 0$.

Example 4.5.7 Let C be the binary cyclic code of length 17 with defining set $T = \{1, 2, 4, 8, 9, 13, 15, 16\}$. There are two consecutive elements in T and so the BCH Bound gives $d \geq 3$. The Hartmann–Tzeng Bound improves this. Let $A = \{1, 2\}$ and $B = \{0, 7, 14\}$. So $\delta = 3$, $b = 7$, and $s = 2$; also $\gcd(7, 17) = 1 < \delta$. So C has minimum weight $d \geq 5$. Note that C is a [17, 9] code. By the Griesmer Bound, there is no [17, 9, 7] code. Hence $d = 5$ or 6. In fact, $d = 5$; see [203].   ∎

Example 4.5.8 Let C be the binary cyclic code of length 31 with defining set $T = \{1, 2, 4, 5, 8, 9, 10, 16, 18, 20\}$ and minimum weight d. The BCH Bound shows that $d \geq 4$ as the consecutive elements $\{8, 9, 10\}$ are in T. If $d = 4$, then the minimum weight vectors are in the subcode C_e of all even-like vectors in C. By Exercise 238, C_e is cyclic with defining set $T_e = \{0, 1, 2, 4, 5, 8, 9, 10, 16, 18, 20\}$. Applying the Hartmann–Tzeng Bound with $A = \{0, 1, 2\}$ and $B = \{0, 8\}$ (since $\gcd(8, 31) = 1 < \delta = 4$), the minimum weight of C_e is at least 5. Hence the minimum weight of C is at least 5, which is in fact the true minimum weight by [203].   ∎

Example 4.5.9 Let C be the binary cyclic code of length 31 with defining set $T = \{1, 2, 3, 4, 5, 6, 8, 9, 10, 11, 12, 13, 16, 17, 18, 20, 21, 22, 24, 26\}$ and minimum weight d. Applying the Hartmann–Tzeng Bound with $A = \{1, 2, 3, 4, 5, 6\}$ and $B = \{0, 7\}$ (since $\gcd(7, 31) = 1 < \delta = 7$), we obtain $d \geq 8$. Suppose that $d = 8$. Then the cyclic subcode C_e of even-like codewords has defining set $T_e = T \cup \{0\}$ by Exercise 238. But $17T_e$ contains the nine consecutive elements $\{29, 30, 0, 1, 2, 3, 4, 5, 6\}$. Hence, the code with defining set $17T_e$ has minimum weight at least 10 by the BCH Bound, implying that $d = 8$ is impossible.

Thus $d \geq 9$. We reconsider this code in Example 4.5.14 where we eliminate $d = 9$ and $d = 10$; the actual value of d is 11.   ∎

Exercise 266 Let C be the code of Example 4.5.9. Eliminate the possibility that $d = 10$ by showing that Theorem 4.4.18 applies to C_e.   ◆

Exercise 267 Let C_i be the 2-cyclotomic coset modulo n containing i. Apply the BCH Bound and the Hartmann–Tzeng Bound to cyclic codes of given length and defining set T. For which of the codes is one of the bounds improved if you multiply the defining set by a, where $\gcd(a, n) = 1$?
(a) $n = 15$ and $T = C_5 \cup C_7$,
(b) $n = 15$ and $T = C_3 \cup C_5$,
(c) $n = 39$ and $T = C_3 \cup C_{13}$,
(d) $n = 45$ and $T = C_1 \cup C_3 \cup C_5 \cup C_9 \cup C_{15}$,
(e) $n = 51$ and $T = C_1 \cup C_9$.   ◆

Exercise 268 Find the dimensions of the codes in Examples 4.5.8 and 4.5.9. For each code also give upper bounds on the minimum distance from the Griesmer Bound.   ◆

Generalizations of the Hartmann–Tzeng Bound discovered by Roos can be found in [296, 297].

In [203] van Lint and Wilson give techniques that can be used to produce lower bounds on the minimum weight of a cyclic code; we present one of these, which we will refer to as the van Lint–Wilson Bounding Technique. The van Lint–Wilson Bounding Technique can be used to prove both the BCH and the Hartmann–Tzeng Bounds. In order to use the van Lint–Wilson Bounding Technique, a sequence of subsets of the integers modulo n, related to the defining set of a code, must be constructed. Let $\mathcal{N} = \{0, 1, \ldots, n - 1\}$. Let $S \subseteq \mathcal{N}$. A sequence $I_0, I_1, I_2, \ldots$ of subsets of $\mathcal{N}$ is an *independent sequence with respect to S* provided
1. $I_0 = \emptyset$, and
2. if $i > 0$, either $I_i = I_j \cup \{a\}$ for some $0 \leq j < i$ such that $I_j \subseteq S$ and $a \in \mathcal{N} \setminus S$, or $I_i = \{b\} + I_j$ for some $0 \leq j < i$ and $b \neq 0$.

A subset I of $\mathcal{N}$ is *independent with respect to S* provided that I is a set in an independent sequence with respect to S. If $S = \mathcal{N}$, then only the empty set is independent with respect to S. Recall that α is a primitive nth root of unity in $\mathbb{F}_{q^t}$, where $t = \text{ord}_n(q)$.

Theorem 4.5.10 (van Lint–Wilson Bounding Technique) *Suppose $f(x)$ is in $\mathbb{F}_q[x]$ and has degree at most n. Let I be any subset of $\mathcal{N}$ that is independent with respect to $S = \{i \in \mathcal{N} \mid f(\alpha^i) = 0\}$. Then the weight of $f(x)$ is at least the size of I.*

Proof: Let $f(x) = c_1 x^{i_1} + c_2 x^{i_2} + \cdots + c_w x^{i_w}$, where $c_j \neq 0$ for $1 \leq j \leq w$, and let I be any independent set with respect to S. Let $I_0, I_1, I_2, \ldots$ be an independent sequence with respect to S such that $I = I_i$ for some $i \geq 0$. If $J \subseteq \mathcal{N}$, let $V(J)$ be the set of vectors in $\mathbb{F}_{q^t}^w$ defined by

$$V(J) = \{(\alpha^{ki_1}, \alpha^{ki_2}, \ldots, \alpha^{ki_w}) \mid k \in J\}.$$

If we prove that $V(I)$ is linearly independent in $\mathbb{F}_{q^t}^w$, then $w \geq |V(I)| = |I|$. Hence it suffices to show that $V(I_i)$ is linearly independent for all $i \geq 0$, which we do by induction on i. As $I_0 = \emptyset$, $V(I_0)$ is the empty set and hence is linearly independent. Assume that $V(I_j)$ is linearly independent for all $0 \leq j < i$. I_i is formed in one of two ways. First suppose $I_i = I_j \cup \{a\}$ for some $0 \leq j < i$, where $I_j \subseteq S$ and $a \in \mathcal{N} \setminus S$. By the inductive assumption, $V(I_j)$ is linearly independent. As $0 = f(\alpha^k) = c_1\alpha^{ki_1} + c_2\alpha^{ki_2} + \cdots + c_w\alpha^{ki_w}$ for $k \in I_j$, the vector $\mathbf{c} = (c_1, c_2, \ldots, c_w)$ is orthogonal to $V(I_j)$. If $V(I_i)$ is linearly dependent, then $(\alpha^{ai_1}, \alpha^{ai_2}, \ldots, \alpha^{ai_w})$ is in the span of $V(I_j)$ and hence is orthogonal to $(c_1, c_2, \ldots, c_w)$. So $0 = c_1\alpha^{ai_1} + c_2\alpha^{ai_2} + \cdots + c_w\alpha^{ai_w} = f(\alpha^a)$, a contradiction as $a \notin S$. Now suppose that $I_i = \{b\} + I_j$ for some $0 \leq j < i$ where $b \neq 0$. Then $V(I_i) = \{(\alpha^{(b+k)i_1}, \alpha^{(b+k)i_2}, \ldots, \alpha^{(b+k)i_w}) \mid k \in I_j\} = V(I_j)D$, where D is the nonsingular diagonal matrix $\text{diag}(\alpha^{bi_1}, \ldots, \alpha^{bi_w})$. As $V(I_j)$ is independent, so is $V(I_i)$. $\square$

The BCH Bound is a corollary of the van Lint–Wilson Bounding Technique.

Corollary 4.5.11 *Let $\mathcal{C}$ be a cyclic code of length n over $\mathbb{F}_q$. Suppose that $f(x)$ is a nonzero codeword such that $f(\alpha^b) = f(\alpha^{b+1}) = \cdots = f(\alpha^{b+w-1}) = 0$ but $f(\alpha^{b+w}) \neq 0$. Then $\text{wt}(f(x)) \geq w + 1$.*

Proof: Let $S = \{i \in \mathcal{N} \mid f(\alpha^i) = 0\}$. So $\{b, b+1, \ldots, b+w-1\} \subseteq S$ but $b + w \notin S$. Inductively define a sequence $I_0, I_1, \ldots, I_{2w+1}$ as follows: Let $I_0 = \emptyset$. Let $I_{2i+1} = I_{2i} \cup \{b + w\}$ for $0 \leq i \leq w$ and $I_{2i} = \{-1\} + I_{2i-1}$ for $1 \leq i \leq w$. Inductively $I_{2i} = \{b + w - i, b + w - i + 1, \ldots, b + w - 1\} \subseteq S$ for $0 \leq i \leq w$. Therefore $I_0, I_1, \ldots, I_{2w+1}$ is an independent sequence with respect to S. Since $I_{2w+1} = \{b, b+1, \ldots, b+w\}$, the van Lint–Wilson Bounding Technique implies that $\text{wt}(f(x)) \geq |I_{2w+1}| = w + 1$. $\square$

Exercise 269 Why does Corollary 4.5.11 prove that the BCH Bound holds? ◆

Example 4.5.12 Let $\mathcal{C}$ be the $[17, 9, d]$ binary cyclic code with defining set $T = \{1, 2, 4, 8, 9, 13, 15, 16\}$. In Example 4.5.7, we saw that the BCH Bound implies that $d \geq 3$, and the Hartmann–Tzeng Bound improves this to $d \geq 5$ using $A = \{1, 2\}$ and $B = \{0, 7, 14\}$. The van Lint–Wilson Bounding Technique gives the same bound using the following argument. Let $f(x) \in \mathcal{C}$ be a nonzero codeword of weight less than 5. T is the 2-cyclotomic coset C_1. If $f(\alpha^i) = 0$ for some $i \in C_3$, then $f(x)$ is a nonzero codeword in the cyclic code with defining set $C_1 \cup C_3$, which is the repetition code, and hence $\text{wt}(f(x)) = 17$, a contradiction. Letting $S = \{i \in \mathcal{N} \mid f(\alpha^i) = 0\}$, we assume that S has no elements of $C_3 = \{3, 5, 6, 7, 10, 11, 12, 14\}$. Then the following sequence of subsets of $\mathcal{N}$ is independent with respect to S:

$I_0 = \emptyset,$

$I_1 = I_0 \cup \{6\} = \{6\},$

$I_2 = \{-7\} + I_1 = \{16\} \subseteq S,$

$I_3 = I_2 \cup \{6\} = \{6, 16\},$

$I_4 = \{-7\} + I_3 = \{9, 16\} \subseteq S,$

$I_5 = I_4 \cup \{6\} = \{6, 9, 16\},$

$I_6 = \{-7\} + I_5 = \{2, 9, 16\} \subseteq S,$

$I_7 = I_6 \cup \{3\} = \{2, 3, 9, 16\},$

$I_8 = \{-1\} + I_7 = \{1, 2, 8, 15\} \subseteq S,$

$I_9 = I_8 \cup \{3\} = \{1, 2, 3, 8, 15\}.$

Since $\mathrm{wt}(f(x)) \geq |I_9| = 5$, the van Lint–Wilson Bounding Technique shows that $d \geq 5$. In fact, $d = 5$ by [203]. ∎

   This example shows the difficulty in applying the van Lint–Wilson Bounding Technique. In this example, the Hartmann–Tzeng and van Lint–Wilson Bounding Technique give the same bound. In this example, in order to apply the van Lint–Wilson Bounding Technique, we construct an independent sequence whose sets are very closely related to the sets A and B used in the Hartmann–Tzeng Bound. This construction mimics that used in the proof of Corollary 4.5.11 where we showed that the BCH Bound follows from the van Lint–Wilson Bounding Technique. If you generalize the construction in this example, you can show that the Hartmann–Tzeng Bound also follows from the van Lint–Wilson Bounding Technique.

Exercise 270 Prove that the Hartmann–Tzeng Bound follows from the van Lint–Wilson Bounding Technique. Hint: See Example 4.5.12. ◆

Exercise 271 Let C be the $[21, 13, d]$ binary cyclic code with defining set $T = \{3, 6, 7, 9, 12, 14, 15, 18\}$. Find a lower bound on the minimum weight of C using the BCH Bound, the Hartmann–Tzeng Bound, and the van Lint–Wilson Bounding Technique. Also using the van Lint–Wilson Bounding Technique, show that an odd weight codeword has weight at least 7. Hint: For the latter, note that α^0 is not a root of an odd weight codeword. ◆

Exercise 272 Let C be the $[41, 21, d]$ binary cyclic code with defining set $T = \{3, 6, 7, 11, 12, 13, 14, 15, 17, 19, 22, 24, 26, 27, 28, 29, 30, 34, 35, 38\}$.
(a) Find the 2-cyclotomic cosets modulo 41.
(b) Let $f(x)$ be a nonzero codeword and let $S = \{i \in \mathcal{N} \mid f(\alpha^i) = 0\}$. Show that if $1 \in S$, then $f(x)$ is either $\mathbf{0}$ or $1 + x + \cdots + x^{40}$.
(c) Assume that $1 \notin S$. Show that either $S = T$ or $S = T \cup \{0\}$.
(d) Now assume that $f(x)$ has weight 8 or less. Show, by applying the rules for constructing independent sequences, that the following sets are part of an independent sequence

with respect to S: $\{28\}$, $\{12, 17\}$, $\{19, 22, 27\}$, $\{3, 6, 11, 30\}$, $\{11, 12, 14, 19, 38\}$, $\{6, 12, 26, 27, 29, 34\}$, $\{6, 7, 13, 27, 28, 30, 35\}$, and $\{1, 6, 7, 13, 27, 28, 30, 35\}$.

(e) From part (d), show that $\text{wt}(f(x)) \geq 8$.

(f) If $\text{wt}(f(x)) = 8$, show that $\{0, 14, 15, 17, 22, 29, 34, 35\}$ and $\{0, 1, 14, 15, 17, 22, 29, 34, 35\}$ are independent with respect to S.

(g) Show that $d \geq 9$. (In fact, $d = 9$ by [203].)   ◆

We conclude this section with the binary version of a result of McEliece [231] that shows what powers of 2 are divisors of a binary cyclic code; we give only a partial proof as the full proof is very difficult. Recall that Δ is a divisor of a code provided every codeword has weight a multiple of Δ.

Theorem 4.5.13 (McEliece) *Let C be a binary cyclic code with defining set T. Let $a \geq 2$ be the smallest number of elements in $\mathcal{N} \setminus T$, with repetitions allowed, that sum to 0. Then 2^{a-1} is a divisor of C but 2^a is not.*

Proof: We prove only the case $a = 2$ and part of the case $a = 3$.

If $a = 2$, then $0 \in T$; otherwise $a = 1$. Thus C has only even weight vectors by Exercise 238 proving that $2^{a-1} = 2$ is a divisor of C. By definition of a, there is an element $b \in \mathcal{N} \setminus T$ such that $-b \mod n \in \mathcal{N} \setminus T$. So $b \in \mathcal{N} \setminus (-1)T \mod n$; as $0 \in T$, $b \neq 0$. By Theorem 4.4.9 $\mathcal{N} \setminus (-1)T \mod n$ is the defining set for $C^{\perp}$. If $2^a = 4$ is a divisor of C, by Theorem 1.4.8, C is self-orthogonal. Hence $C \subseteq C^{\perp}$ and by Corollary 4.2.5 (or Exercise 239), $\mathcal{N} \setminus (-1)T \mod n \subseteq T$. This is a contradiction as $b \in \mathcal{N} \setminus (-1)T \mod n$ and $b \in \mathcal{N} \setminus T$.

If $a = 3$, we only prove that $2^{a-1} = 4$ is a divisor of C. We show that $\mathcal{N} \setminus (-1)T \mod n \subseteq T$. Suppose not. Then there exists $b \in \mathcal{N} \setminus (-1)T \mod n$ with $b \notin T$. So $-b \mod n \in \mathcal{N} \setminus T$ and $b \in \mathcal{N} \setminus T$. Since $b + (-b) = 0$, $a \neq 3$, and this is a contradiction. So $\mathcal{N} \setminus (-1)T \mod n \subseteq T$. Therefore $C \subseteq C^{\perp}$ by Theorem 4.4.9 and Corollary 4.2.5 (or Exercise 239). Thus C is a self-orthogonal binary cyclic code which must be doubly-even by Theorem 4.4.18.   □

Example 4.5.14 Let C be the $[31, 11, d]$ binary cyclic code with defining set $T = \{1, 2, 3, 4, 5, 6, 8, 9, 10, 11, 12, 13, 16, 17, 18, 20, 21, 22, 24, 26\}$ that we considered in Example 4.5.9. There we showed that $d \geq 9$. The subcode C_e of even weight vectors in C is a cyclic code with defining set $T_e = \{0\} \cup T$ by Exercise 238. Notice that $\mathcal{N} \setminus T_e = \{7, 14, 15, 19, 23, 25, 27, 28, 29, 30\}$ and that the sum of any two elements in this set is not 0 modulo 31, but $7 + 25 + 30 \equiv 0 \pmod{31}$. So $a = 3$ in McEliece's Theorem applied to C_e. Thus all vectors in C_e, and hence all even weight vectors in C, have weights a multiple of 4. In particular, C has no codewords of weight 10. Furthermore $1 + x + \cdots + x^{30} = (x^n - 1)/(x - 1) \in C$ as $0 \notin T$. Thus if C has a vector of weight w, it has a vector of weight $31 - w$. Specifically, if C has a codeword of weight 9, it has a codeword of weight 22, a contradiction. Thus we have $d \geq 11$. The Griesmer Bound gives $d \leq 12$. By [203], $d = 11$.   ■

Exercise 273 Let C be the $[21, 9, d]$ binary cyclic code with defining set $T = \{0, 1, 2, 3, 4, 6, 7, 8, 11, 12, 14, 16\}$. Show that $d \geq 8$.   ◆

4.6   Meggitt decoding of cyclic codes

In this section we present a technique for decoding cyclic codes called Meggitt decoding [238, 239]. There are several variations of Meggitt decoding; we will present two of them. Meggitt decoding is a special case of permutation decoding, which will be explored further in Section 10.2. Permutation decoding itself is a special case of error trapping.

Let C be an $[n, k, d]$ cyclic code over $\mathbb{F}_q$ with generator polynomial $g(x)$ of degree $n - k$; C will correct $t = \lfloor (d - 1)/2 \rfloor$ errors. Suppose that $c(x) \in C$ is transmitted and $y(x) = c(x) + e(x)$ is received, where $e(x) = e_0 + e_1 x + \cdots + e_{n-1} x^{n-1}$ is the error vector with $\mathrm{wt}(e(x)) \leq t$. The Meggitt decoder stores syndromes of error patterns with coordinate $n - 1$ in error. The two versions of the Meggitt Decoding Algorithm that we present can briefly be described as follows. In the first version, by shifting $y(x)$ at most n times, the decoder finds the error vector $e(x)$ from the list and corrects the errors. In the second version, by shifting $y(x)$ until an error appears in coordinate $n - 1$, the decoder finds the error in that coordinate, corrects only that error, and then corrects errors in coordinates $n - 2, n - 3, \ldots, 1, 0$ in that order by further shifting. As you can see, Meggitt decoding takes advantage of the cyclic nature of the code.

For any vector $v(x) \in \mathbb{F}_q[x]$, let $R_{g(x)}(v(x))$ be the unique remainder when $v(x)$ is divided by $g(x)$ according to the Division Algorithm; that is, $R_{g(x)}(v(x)) = r(x)$, where $v(x) = g(x)f(x) + r(x)$ with $r(x) = 0$ or $\deg r(x) < n - k$. The function $R_{g(x)}$ satisfies the following properties; the proofs are left as an exercise.

Theorem 4.6.1 *With the preceding notation the following hold*:
(i) $R_{g(x)}(av(x) + bv'(x)) = a R_{g(x)}(v(x)) + b R_{g(x)}(v'(x))$ *for all* $v(x), v'(x) \in \mathbb{F}_q[x]$ *and all* $a, b \in \mathbb{F}_q$.
(ii) $R_{g(x)}(v(x) + a(x)(x^n - 1)) = R_{g(x)}(v(x))$.
(iii) $R_{g(x)}(v(x)) = 0$ *if and only if* $v(x) \bmod (x^n - 1) \in C$.
(iv)   *If* $c(x) \in C$, *then* $R_{g(x)}(c(x) + e(x)) = R_{g(x)}(e(x))$.
(v)   *If* $R_{g(x)}(e(x)) = R_{g(x)}(e'(x))$, *where* $e(x)$ *and* $e'(x)$ *each have weight at most* t, *then* $e(x) = e'(x)$.
(vi) $R_{g(x)}(v(x)) = v(x)$ *if* $\deg v(x) < n - k$.

Exercise 274 Prove Theorem 4.6.1.   ◆

Part (ii) of this theorem shows that we can apply $R_{g(x)}$ to either elements of $\mathcal{R}_n$ or elements of $\mathbb{F}_q[x]$ without ambiguity. We now need a theorem due to Meggitt that will simplify our computations with $R_{g(x)}$.

Theorem 4.6.2 *Let* $g(x)$ *be a monic divisor of* $x^n - 1$ *of degree* $n - k$. *If* $R_{g(x)}(v(x)) = s(x)$, *then*

$$R_{g(x)}(xv(x) \bmod (x^n - 1)) = R_{g(x)}(xs(x)) = xs(x) - g(x)s_{n-k-1},$$

where s_{n-k-1} *is the coefficient of* x^{n-k-1} *in* $s(x)$.

Proof: By definition $v(x) = g(x)f(x) + s(x)$, where $s(x) = \sum_{i=0}^{n-k-1} s_i x^i$. So $xv(x) = xg(x)f(x) + xs(x) = xg(x)f(x) + g(x)f_1(x) + s'(x)$, where $s'(x) = R_{g(x)}(xs(x))$. Also $xv(x) \bmod (x^n - 1) = xv(x) - (x^n - 1)v_{n-1}$. Thus $xv(x) \bmod (x^n - 1) = xg(x)f(x) + g(x)f_1(x) + s'(x) - (x^n - 1)v_{n-1} = (xf(x) + f_1(x) - h(x)v_{n-1})g(x) + s'(x)$, where $g(x)h(x) = x^n - 1$. Therefore $R_{g(x)}(xv(x) \bmod (x^n - 1)) = s'(x) = R_{g(x)}(xs(x))$ because deg $s'(x) < n - k$. As $g(x)$ is monic of degree $n - k$ and $xs(x) = \sum_{i=0}^{n-k-1} s_i x^{i+1}$, the remainder when $xs(x)$ is divided by $g(x)$ is $xs(x) - g(x)s_{n-k-1}$. $\qquad\square$

We now describe our first version of the *Meggitt Decoding Algorithm* and use an example to illustrate each step. Define the *syndrome polynomial* $S(v(x))$ of any $v(x)$ to be

$$S(v(x)) = R_{g(x)}(x^{n-k}v(x)).$$

By Theorem 4.6.1(iii), if $v(x) \in \mathcal{R}_n$, then $S(v(x)) = 0$ if and only if $v(x) \in \mathcal{C}$.

Step I:
This is a one-time precomputation. Find all the syndrome polynomials $S(e(x))$ of error patterns $e(x) = \sum_{i=0}^{n-1} e_i x^i$ such that wt$(e(x)) \le t$ and $e_{n-1} \ne 0$.

Example 4.6.3 Let $\mathcal{C}$ be the $[15, 7, 5]$ binary cyclic code with defining set $T = \{1, 2, 3, 4, 6, 8, 9, 12\}$. Let α be a 15th root of unity in $\mathbb{F}_{16}$. Then $g(x) = 1 + x^4 + x^6 + x^7 + x^8$ is the generator polynomial of $\mathcal{C}$ and the syndrome polynomial of $e(x)$ is $S(e(x)) = R_{g(x)}(x^8 e(x))$. Step I produces the following syndrome polynomials:

$e(x)$	$S(e(x))$	$e(x)$	$S(e(x))$
x^{14}	x^7	$x^6 + x^{14}$	$x^3 + x^5 + x^6$
$x^{13} + x^{14}$	$x^6 + x^7$	$x^5 + x^{14}$	$x^2 + x^4 + x^5 + x^6 + x^7$
$x^{12} + x^{14}$	$x^5 + x^7$	$x^4 + x^{14}$	$x + x^3 + x^4 + x^5 + x^7$
$x^{11} + x^{14}$	$x^4 + x^7$	$x^3 + x^{14}$	$1 + x^2 + x^3 + x^4 + x^7$
$x^{10} + x^{14}$	$x^3 + x^7$	$x^2 + x^{14}$	$x + x^2 + x^5 + x^6$
$x^9 + x^{14}$	$x^2 + x^7$	$x + x^{14}$	$1 + x + x^4 + x^5 + x^6 + x^7$
$x^8 + x^{14}$	$x + x^7$	$1 + x^{14}$	$1 + x^4 + x^6$
$x^7 + x^{14}$	$1 + x^7$		

The computations of these syndrome polynomials were aided by Theorems 4.6.1 and 4.6.2. For example, in computing the syndrome polynomial of $x^{12} + x^{14}$, we have $S(x^{12} + x^{14}) = R_{g(x)}(x^8(x^{12} + x^{14})) = R_{g(x)}(x^5 + x^7) = x^5 + x^7$ using Theorem 4.6.1(vi). In computing the syndrome polynomial for $1 + x^{14}$, first observe that $R_{g(x)}(x^8) = 1 + x^4 + x^6 + x^7$; then $S(1 + x^{14}) = R_{g(x)}(x^8(1 + x^{14})) = R_{g(x)}(x^8) + R_{g(x)}(x^7) = 1 + x^4 + x^6$. We see by Theorem 4.6.2 that $R_{g(x)}(x^9) = R_{g(x)}(xx^8) = R_{g(x)}(x(1 + x^4 + x^6 + x^7)) = R_{g(x)}(x + x^5 + x^7) + R_{g(x)}(x^8) = x + x^5 + x^7 + 1 + x^4 + x^6 + x^7 = 1 + x + x^4 + x^5 + x^6$. Therefore in computing the syndrome polynomial for $x + x^{14}$, we have $S(x + x^{14}) = R_{g(x)}(x^8(x + x^{14})) = R_{g(x)}(x^9) + R_{g(x)}(x^7) = 1 + x + x^4 + x^5 + x^6 + x^7$. The others follow similarly. $\qquad\blacksquare$

Exercise 275 Verify the syndrome polynomials found in Example 4.6.3. ♦

Step II:
Suppose that $y(x)$ is the received vector. Compute the syndrome polynomial $S(y(x)) = R_{g(x)}(x^{n-k}y(x))$. By Theorem 4.6.1(iv), $S(y(x)) = S(e(x))$, where $y(x) = c(x) + e(x)$ with $c(x) \in C$.

Example 4.6.4 Continuing with Example 4.6.3, suppose that $y(x) = 1 + x^4 + x^7 + x^9 + x^{10} + x^{12}$ is received. Then $S(y(x)) = x + x^2 + x^6 + x^7$. ∎

Exercise 276 Verify that $S(y(x)) = x + x^2 + x^6 + x^7$ in Example 4.6.4. ♦

Step III:
If $S(y(x))$ is in the list computed in Step I, then you know the error polynomial $e(x)$ and this can be subtracted from $y(x)$ to obtain the codeword $c(x)$. If $S(y(x))$ is not in the list, go on to Step IV.

Example 4.6.5 $S(y(x))$ from Example 4.6.4 is not in the list of syndrome polynomials given in Example 4.6.3. ∎

Step IV:
Compute the syndrome polynomial of $xy(x)$, $x^2y(x)$, $\ldots$ in succession until the syndrome polynomial is in the list from Step I. If $S(x^i y(x))$ is in this list and is associated with the error polynomial $e'(x)$, then the received vector is decoded as $y(x) - x^{n-i}e'(x)$.

The computation in Step IV is most easily carried out using Theorem 4.6.2. As $R_{g(x)}(x^{n-k}y(x)) = S(y(x)) = \sum_{i=0}^{n-k-1} s_i x^i$,

$$S(xy(x)) = R_{g(x)}(x^{n-k}xy(x)) = R_{g(x)}(x(x^{n-k}y(x))) = R_{g(x)}(xS(y(x)))$$
$$= xS(y(x)) - s_{n-k-1}g(x). \tag{4.8}$$

We proceed in the same fashion to get the syndrome of $x^2y(x)$ from that of $xy(x)$.

Example 4.6.6 Continuing with Example 4.6.4, we have, using (4.8) and $S(y(x)) = x + x^2 + x^6 + x^7$, that $S(xy(x)) = x(x + x^2 + x^6 + x^7) - 1 \cdot g(x) = 1 + x^2 + x^3 + x^4 + x^6$, which is not in the list in Example 4.6.3. Using (4.8), $S(x^2y(x)) = x(1 + x^2 + x^3 + x^4 + x^6) - 0 \cdot g(x) = x + x^3 + x^4 + x^5 + x^7$, which corresponds to the error $x^4 + x^{14}$ implying that $y(x)$ is decoded as $y(x) - (x^2 + x^{12}) = 1 + x^2 + x^4 + x^7 + x^9 + x^{10}$. Note that this is the codeword $(1 + x^2)g(x)$. ∎

The Meggitt decoder can be implemented with shift-register circuits, and a second version of the *Meggitt Decoding Algorithm* is often employed in order to simplify this circuit. In this second implementation, the circuitry only corrects coordinate $n - 1$ of the vector in the shift-register. The vector in the shift-register starts out as the received vector $y(x)$; if there is an error in position $n - 1$, it is corrected. Then the shift $xy(x)$ is moved into the shift-register. If there is an error in coordinate $n - 1$, it is corrected; this in effect corrects coordinate $n - 2$ of $y(x)$. This continues until $y(x)$ and its $n - 1$ shifts have been moved into the shift-register and have been examined. At each stage only the error at the end of the shift-register is corrected. This process corrects the original received vector $y(x)$ since a correction in coordinate $n-1$ of $x^i y(x)$ produces a correction in coordinate $n-1-i$ of $y(x)$. This version of Meggitt decoding allows the use of shift-registers whose internal stages do not need to be directly accessed and modified and allows for fewer wires in the circuit. For those interested in the circuit designs, consult [21]. We illustrate this in the binary case by continuing with Example 4.6.6.

Example 4.6.7 After computing the syndrome polynomials of $y(x)$, $xy(x)$, and $x^2 y(x)$, we conclude there is an error in coordinate 14 of $x^2 y(x)$, that is, coordinate 12 of $y(x)$. Thus we modify $x^2 y(x)$ in coordinate 14 to obtain $x^2 y'(x) = x^2 + x^6 + x^9 + x^{11} + x^{12}$ (where $y'(x) = 1 + x^4 + x^7 + x^9 + x^{10}$ is $y(x)$ corrected in coordinate 12). This changes the syndrome as well. Fortunately, the change is simple to deal with as follows. $S(x^2 y'(x)) = R_{g(x)}(x^8(x^2 y'(x))) = R_{g(x)}(x^8(x^2 y(x))) - R_{g(x)}(x^8 x^{14}) = S(x^2 y(x)) - R_{g(x)}(x^{22})$ because $x^2 y'(x) = x^2 y(x) - x^{14}$ since we changed only coordinate 14 of $x^2 y(x)$. But $S(x^2 y(x)) - R_{g(x)}(x^{22}) = x + x^3 + x^4 + x^5 + x^7 - x^7 = x + x^3 + x^4 + x^5$ from Example 4.6.6 as $R_{g(x)}(x^{22}) = R_{g(x)}(x^7) = x^7$. Thus to get the new syndrome for $x^2 y'(x)$, we take $S(x^2 y(x))$ and subtract x^7. This holds in general: namely, to obtain the new syndrome polynomial after an error has been corrected, take the old syndrome and subtract x^7. This simple modification in the syndrome polynomial is precisely why the definition of syndrome $S(v(x))$ as $R_{g(x)}(x^{n-k} v(x))$ was given; had we used the more natural $R_{g(x)}(v(x))$ as the definition of the syndrome polynomial of $v(x)$, we would have had a more complicated modification of the syndrome at this juncture. So we compute and obtain the following table:

Syndrome of	Syndrome	Syndrome of	Syndrome
$x^3 y'(x)$	$x^2 + x^4 + x^5 + x^6$	$x^8 y'(x)$	x^3
$x^4 y'(x)$	$x^3 + x^5 + x^6 + x^7$	$x^9 y'(x)$	x^4
$x^5 y'(x)$	1	$x^{10} y'(x)$	x^5
$x^6 v'(x)$	x	$x^{11} y'(x)$	x^6
$x^7 y'(x)$	x^2	$x^{12} y'(x)$	x^7

We see that none of our syndromes is in the list from Step I until that of $x^{12} y'(x)$. Hence we change coordinate 14 of $x^{12} y'(x) = x + x^4 + x^6 + x^7 + x^{12}$ to obtain $x^{12} y''(x) = x + x^4 + x^6 + x^7 + x^{12} + x^{14}$ (where $y''(x) = 1 + x^2 + x^4 + x^7 + x^9 + x^{10}$ is $y'(x)$ changed in coordinate 2). As above, the syndrome of $x^{12} y''(x)$ is $x^7 - x^7 = 0$. We could stop here, if our circuit is designed to check for the zero syndrome polynomial. Otherwise we compute the syndromes of $x^{13} y''(x)$ and $x^{14} y''(x)$, which are both 0 and not on the list, indicating that the procedure should halt as we have considered all cyclic shifts of our received vector. So the correct vector is $y''(x)$ and in fact turns out to be the vector output of the shift-register circuit used in this version of Meggitt decoding. ∎

Exercise 277 Verify the results in Example 4.6.7. ◆

In the binary case where we correct only the high-order errors one at a time, it is unnecessary to store the error polynomial corresponding to each syndrome. Obviously the speed of our decoding algorithm depends on the size of the list and the speed of the pattern recognition of the circuit. There are further variations of Meggitt decoding for which we again refer the reader to [21].

Exercise 278 With the same code as in Example 4.6.3, find the codeword sent if two or fewer errors have occurred and $y(x) = 1 + x + x^6 + x^9 + x^{11} + x^{12} + x^{13}$ is received. Do this in two ways:
(a) Carry out Steps I–IV (that is, the first version of the Meggitt Decoding Algorithm).
(b) Carry out Meggitt decoding by correcting only the high-order bits and shifting as in Example 4.6.7 (that is, the second version of the Meggitt Decoding Algorithm).   ◆

4.7   Affine-invariant codes

In this section we will look at extending certain cyclic codes and examine an important class of codes called affine-invariant codes. Reed–Muller codes and some BCH codes, defined in Chapter 5, are affine-invariant.

We will present a new setting for "primitive" cyclic codes that will assist us in the description of affine-invariant codes. A *primitive cyclic code over* $\mathbb{F}_q$ is a cyclic code of length $n = q^t - 1$ for some t.[1] To proceed, we need some notation.

Let $\mathcal{I}$ denote the field of order q^t, which is then an extension field of $\mathbb{F}_q$. The set $\mathcal{I}$ will be the index set of our extended cyclic codes of length q^t. Let $\mathcal{I}^*$ be the nonzero elements of $\mathcal{I}$, and suppose α is a primitive nth root of unity in $\mathcal{I}$ (and hence a primitive element of $\mathcal{I}$). The set $\mathcal{I}^*$ will be the index set of our primitive cyclic codes of length $n = q^t - 1$. With X an indeterminate, let

$$\mathbb{F}_q[\mathcal{I}] = \left\{ a = \sum_{g \in \mathcal{I}} a_g X^g \mid a_g \in \mathbb{F}_q \text{ for all } g \in \mathcal{I} \right\}.$$

The set $\mathbb{F}_q[\mathcal{I}]$ is actually an algebra under the operations

$$c \sum_{g \in \mathcal{I}} a_g X^g + d \sum_{g \in \mathcal{I}} b_g X^g = \sum_{g \in \mathcal{I}} (ca_g + db_g) X^g$$

for $c, d \in \mathbb{F}_q$, and

$$\sum_{g \in \mathcal{I}} a_g X^g \sum_{g \in \mathcal{I}} b_g X^g = \sum_{g \in \mathcal{I}} \left(\sum_{h \in \mathcal{I}} a_h b_{g-h} \right) X^g.$$

The zero and unity of $\mathbb{F}_q[\mathcal{I}]$ are $\sum_{g \in \mathcal{I}} 0 X^g$ and X^0, respectively. This is the group algebra of the additive group of $\mathcal{I}$ over $\mathbb{F}_q$. Let

$$\mathbb{F}_q[\mathcal{I}^*] = \left\{ a = \sum_{g \in \mathcal{I}^*} a_g X^g \mid a_g \in \mathbb{F}_q \text{ for all } g \in \mathcal{I}^* \right\}.$$

---

[1] Cyclic codes of length $n = q^t - 1$ over an extension field of $\mathbb{F}_q$ are also called primitive, but we will only study the more restricted case.

$\mathbb{F}_q[\mathcal{I}^*]$ is a subspace of $\mathbb{F}_q[\mathcal{I}]$ but not a subalgebra. So elements of $\mathbb{F}_q[\mathcal{I}^*]$ are of the form

$$\sum_{i=0}^{n-1} a_{\alpha^i} X^{\alpha^i},$$

while elements of $\mathbb{F}_q[\mathcal{I}]$ are of the form

$$a_0 X^0 + \sum_{i=0}^{n-1} a_{\alpha^i} X^{\alpha^i}.$$

The vector space $\mathbb{F}_q[\mathcal{I}^*]$ will be the new setting for primitive cyclic codes, and the algebra $\mathbb{F}_q[\mathcal{I}]$ will be the setting for the extended cyclic codes. So in fact both codes are contained in $\mathbb{F}_q[\mathcal{I}]$, which makes the discussion of affine-invariant codes more tractable.

Suppose that $\mathcal{C}$ is a cyclic code over $\mathbb{F}_q$ of length $n = q^t - 1$. The coordinates of $\mathcal{C}$ have been denoted $\{0, 1, \ldots, n-1\}$. In $\mathcal{R}_n$, the ith component c_i of a codeword $\mathbf{c} = c_0 c_1 \cdots c_{n-1}$, with associated polynomial $c(x)$, is the coefficient of the term $c_i x^i$ in $c(x)$; the component c_i is kept in position x^i. Now we associate $\mathbf{c}$ with an element $C(X) \in \mathbb{F}_q[\mathcal{I}^*]$ as follows:

$$\mathbf{c} \leftrightarrow C(X) = \sum_{i=0}^{n-1} C_{\alpha^i} X^{\alpha^i} = \sum_{g \in \mathcal{I}^*} C_g X^g, \tag{4.9}$$

where $C_{\alpha^i} = c_i$. Thus the ith component of $\mathbf{c}$ is the coefficient of the term $C_{\alpha^i} X^{\alpha^i}$ in $C(X)$; the component c_i is kept in position X^{α^i}.

Example 4.7.1 Consider the element $c(x) = 1 + x + x^3$ in $\mathcal{R}_7$ over $\mathbb{F}_2$. So $n = 7 = 2^3 - 1$. Let α be a primitive element of $\mathbb{F}_8$. Then $c_0 = C_{\alpha^0} = 1$, $c_1 = C_{\alpha^1} = 1$, and $c_3 = C_{\alpha^3} = 1$, with the other $c_i = C_{\alpha^i} = 0$. So

$$c(x) = 1 + x + x^3 \leftrightarrow C(X) = X + X^\alpha + X^{\alpha^3}.$$

∎

We now need to examine the cyclic shift $xc(x)$ under the correspondence (4.9). We have

$$xc(x) = c_{n-1} + \sum_{i=1}^{n-1} c_{i-1} x^i \leftrightarrow \sum_{i=0}^{n-1} C_{\alpha^{i-1}} X^{\alpha^i} = \sum_{i=0}^{n-1} C_{\alpha^i} X^{\alpha \alpha^i}.$$

Example 4.7.2 We continue with Example 4.7.1. Namely,

$$xc(x) = x + x^2 + x^4 \leftrightarrow X^\alpha + X^{\alpha^2} + X^{\alpha^4} = X^{\alpha 1} + X^{\alpha \alpha} + X^{\alpha \alpha^3}.$$

∎

In our new notation a primitive cyclic code over $\mathbb{F}_q$ of length $n = q^t - 1$ is any subset $\mathcal{C}$ of $\mathbb{F}_q[\mathcal{I}^*]$ such that

$$\sum_{i=0}^{n-1} C_{\alpha^i} X^{\alpha^i} = \sum_{g \in \mathcal{I}^*} C_g X^g \in \mathcal{C} \qquad \text{if and only if}$$

$$\sum_{i=0}^{n-1} C_{\alpha^i} X^{\alpha \alpha^i} = \sum_{g \in \mathcal{I}^*} C_g X^{\alpha g} \in \mathcal{C}. \tag{4.10}$$

The coordinates of our cyclic codes are indexed by $\mathcal{I}^*$. We are now ready to extend our cyclic codes. We use the element $0 \in \mathcal{I}$ to index the extended coordinate. The extended codeword of $C(X) = \sum_{g \in \mathcal{I}^*} C_g X^g$ is $\widehat{C}(X) = \sum_{g \in \mathcal{I}} C_g X^g$ such that $\sum_{g \in \mathcal{I}} C_g = 0$.

Example 4.7.3 We continue with Examples 4.7.1 and 4.7.2. If $C(X) = X + X^\alpha + X^{\alpha^3}$ then $\widehat{C}(X) = X^0 + X + X^\alpha + X^{\alpha^3} = 1 + X + X^\alpha + X^{\alpha^3}$. ∎

From (4.10), together with the observation that $X^{\alpha 0} = X^0 = 1$, in this terminology an extended cyclic code is a subspace $\widehat{C}$ of $\mathbb{F}_q[\mathcal{I}]$ such that

$$\sum_{g \in \mathcal{I}} C_g X^g \in \widehat{C} \quad \text{if and only if} \quad \sum_{g \in \mathcal{I}} C_g X^{\alpha g} \in \widehat{C} \text{ and } \sum_{g \in \mathcal{I}} C_g = 0.$$

With this new notation we want to see where the concepts of zeros and defining sets come in. This can be done with the assistance of a function ϕ_s. Let $\widehat{\mathcal{N}} = \{s \mid 0 \le s \le n\}$. For $s \in \widehat{\mathcal{N}}$ define $\phi_s : \mathbb{F}_q[\mathcal{I}] \to \mathcal{I}$ by

$$\phi_s\left(\sum_{g \in \mathcal{I}} C_g X^g\right) = \sum_{g \in \mathcal{I}} C_g g^s,$$

where by convention $0^0 = 1$ in $\mathcal{I}$. Thus $\phi_0(\widehat{C}(X)) = \sum_{g \in \mathcal{I}} C_g$ implying that $\widehat{C}(X)$ is the extended codeword of $C(X)$ if and only if $\phi_0(\widehat{C}(X)) = 0$. In particular, if $\widehat{C}$ is extended cyclic, then $\phi_0(\widehat{C}(X)) = 0$ for all $\widehat{C}(X) \in \widehat{C}$. What is $\phi_s(\widehat{C}(X))$ when $1 \le s \le n-1$? As $0^s = 0$ in $\mathcal{I}$,

$$\phi_s(\widehat{C}(X)) = \sum_{i=0}^{n-1} C_{\alpha^i}(\alpha^i)^s = \sum_{i=0}^{n-1} C_{\alpha^i}(\alpha^s)^i = \sum_{i=0}^{n-1} c_i(\alpha^s)^i = c(\alpha^s), \tag{4.11}$$

where $c(x)$ is the polynomial in $\mathcal{R}_n$ associated to $C(X)$ in $\mathbb{F}_q[\mathcal{I}^*]$. Suppose our original code C defined in $\mathcal{R}_n$ had defining set T relative to the nth root of unity α. Then (4.11) shows that if $1 \le s \le n-1$, $s \in T$ if and only if $\phi_s(\widehat{C}(X)) = 0$ for all $\widehat{C}(X) \in \widehat{C}$. Finally, what is $\phi_n(\widehat{C}(X))$? Equation (4.11) works in this case as well, implying that $\alpha^n = 1$ is a zero of C if and only if $\phi_n(\widehat{C}(X)) = 0$ for all $\widehat{C}(X) \in \widehat{C}$. But $\alpha^0 = \alpha^n = 1$. Hence we have, by Exercise 238, that $0 \in T$ if and only if $\phi_n(\widehat{C}(X)) = 0$ for all $\widehat{C}(X) \in \widehat{C}$. We can now describe an extended cyclic code in terms of a defining set as follows: a code $\widehat{C}$ of length q^t is an *extended cyclic code with defining set* $\widehat{T}$ provided $\widehat{T} \subseteq \widehat{\mathcal{N}}$ is a union of q-cyclotomic cosets modulo $n = q^t - 1$ with $0 \in \widehat{T}$ and

$$\widehat{C} = \{\widehat{C}(X) \in \mathbb{F}_q[\mathcal{I}] \mid \phi_s(\widehat{C}(X)) = 0 \text{ for all } s \in \widehat{T}\}. \tag{4.12}$$

We make several remarks.

- 0 and n are distinct in $\widehat{T}$ and each is its own q-cyclotomic coset.
- $0 \in \widehat{T}$ because we need all codewords in $\widehat{C}$ to be even-like.
- If $n \in \widehat{T}$, then $\phi_n(\widehat{C}(X)) = \sum_{g \in \mathcal{I}^*} C_g = 0$. As the extended codeword is even-like, since $0 \in \widehat{T}$, $\sum_{g \in \mathcal{I}} C_g = 0$. Thus the extended coordinate is always 0, a condition that makes the extension trivial.
- The defining set T of C and the defining set $\widehat{T}$ are closely related: $\widehat{T}$ is obtained by taking T, changing 0 to n if $0 \in T$, and adding 0.

- $s \in \widehat{T}$ with $1 \le s \le n$ if and only if α^s is a zero of $\mathcal{C}$.
- By employing the natural ordering $0, \alpha^n, \alpha^1, \ldots, \alpha^{n-1}$ on $\mathcal{I}$, we can make the extended cyclic nature of these codes apparent. The coordinate labeled 0 is the parity check coordinate, and puncturing the code $\widehat{\mathcal{C}}$ on this coordinate gives the code $\mathcal{C}$ that admits the standard cyclic shift $\alpha^i \to \alpha\alpha^i$ as an automorphism.
- To check that a code $\mathcal{C}$ in $\mathcal{R}_n$ is cyclic with defining set T, one only needs to verify that $c(x) \in \mathcal{C}$ if and only if $c(\alpha^s) = 0$ for all $s \in T$. Parallel to that, to check that a code $\widehat{\mathcal{C}}$ in $\mathbb{F}_q[\mathcal{I}]$ is extended cyclic with defining set $\widehat{T}$ one only needs to verify that $\widehat{C}(X) \in \widehat{\mathcal{C}}$ if and only if $\phi_s(\widehat{C}(X)) = 0$ for all $s \in \widehat{T}$.

Example 4.7.4 Let $\mathcal{C}$ be the $[7, 4]$ binary cyclic code with defining set $T = \{1, 2, 4\}$ and generator polynomial $g(x) = \prod_{i \in T}(x - \alpha^i) = 1 + x + x^3$, where α is a primitive element of $\mathbb{F}_8$ satisfying $\alpha^3 = 1 + \alpha$. See Examples 3.4.3 and 4.3.4. The extended generator is $1 + X + X^\alpha + X^{\alpha^3}$. Then a generator matrix for $\widehat{\mathcal{C}}$ is

$$
\begin{array}{cccccccc}
1 & X & X^\alpha & X^{\alpha^2} & X^{\alpha^3} & X^{\alpha^4} & X^{\alpha^5} & X^{\alpha^6} \\
\left[\begin{array}{cccccccc}
1 & 1 & 1 & 0 & 1 & 0 & 0 & 0 \\
1 & 0 & 1 & 1 & 0 & 1 & 0 & 0 \\
1 & 0 & 0 & 1 & 1 & 0 & 1 & 0 \\
1 & 0 & 0 & 0 & 1 & 1 & 0 & 1
\end{array}\right].
\end{array}
$$

Notice that this is the extended Hamming code $\widehat{\mathcal{H}}_3$. It has defining set $\widehat{T} = \{0, 1, 2, 4\}$. ∎

In Section 1.6, we describe how permutation automorphisms act on codes. From the discussion in that section, a permutation σ of $\mathcal{I}$ acts on $\widehat{\mathcal{C}}$ as follows:

$$
\left(\sum_{g \in \mathcal{I}} C_g X^g\right)\sigma = \sum_{g \in \mathcal{I}} C_g X^{g\sigma}.
$$

We define the *affine group* $\mathrm{GA}_1(\mathcal{I})$ by $\mathrm{GA}_1(\mathcal{I}) = \{\sigma_{a,b} \mid a \in \mathcal{I}^*, b \in \mathcal{I}\}$, where $g\sigma_{a,b} = ag + b$. Notice that the maps $\sigma_{a,0}$ are merely the cyclic shifts on the coordinates $\{\alpha^n, \alpha^1, \ldots, \alpha^{n-1}\}$ each fixing the coordinate 0. The group $\mathrm{GA}_1(\mathcal{I})$ has order $(n + 1)n = q^t(q^t - 1)$. An *affine-invariant code* is an extended cyclic code $\widehat{\mathcal{C}}$ over $\mathbb{F}_q$ such that $\mathrm{GA}_1(\mathcal{I}) \subseteq \mathrm{PAut}(\widehat{\mathcal{C}})$. Amazingly, we can easily decide which extended cyclic codes are affine-invariant by examining their defining sets. In order to do this we introduce a partial ordering $\preceq$ on $\widehat{\mathcal{N}}$. Suppose that $q = p^m$, where p is a prime. Then $\widehat{\mathcal{N}} = \{0, 1, \ldots, n\}$, where $n = q^t - 1 = p^{mt} - 1$. So every element $s \in \widehat{\mathcal{N}}$ can be written in its p-adic expansion

$$
s = \sum_{i=0}^{mt-1} s_i p^i, \qquad \text{where } 0 \le s_i < p \text{ for } 0 \le i < mt.
$$

We say that $r \preceq s$ provided $r_i \le s_i$ for all $0 \le i < mt$, where $r = \sum_{i=0}^{mt-1} r_i p^i$ is the p-adic expansion of r. Notice that if $r \preceq s$, then in particular $r \le s$. We also need a result called Lucas' Theorem [209], a proof of which can be found in [18].

Theorem 4.7.5 (Lucas) *Let $r = \sum_{i=0}^{mt-1} r_i p^i$ and $s = \sum_{i=0}^{mt-1} s_i p^i$ be the p-adic expansions of r and s. Then*

$$\binom{s}{r} \equiv \prod_{i=0}^{mt-1} \binom{s_i}{r_i} \pmod{p}.$$

We are now ready to determine the affine-invariant codes from their defining sets, a result due to Kasami, Lin, and Peterson [162].

Theorem 4.7.6 *Let $\widehat{C}$ be an extended cyclic code of length q^t with defining set $\widehat{T}$. The code $\widehat{C}$ is affine-invariant if and only if whenever $s \in \widehat{T}$ then $r \in \widehat{T}$ for all $r \in \widehat{N}$ with $r \preceq s$.*

Proof: Suppose that $\widehat{C}(X) = \sum_{g \in \mathcal{I}} C_g X^g \in \widehat{C}$. Let $s \in \widehat{N}$ and $a, b \in \mathcal{I}$ with $a \neq 0$. So $(\widehat{C}(X))\sigma_{a,b} = \sum_{g \in \mathcal{I}} C_g X^{ag+b}$. Therefore,

$$\phi_s((\widehat{C}(X))\sigma_{a,b}) = \sum_{g \in \mathcal{I}} C_g (ag + b)^s = \sum_{g \in \mathcal{I}} C_g \sum_{r=0}^{s} \binom{s}{r} (ag)^r b^{s-r}.$$

By Lucas' Theorem, $\binom{s}{r}$ is nonzero modulo p if and only if $r_i \leq s_i$ for all $0 \leq i < mt$ where $r = \sum_{i=0}^{mt-1} r_i p^i$ and $s = \sum_{i=0}^{mt-1} s_i p^i$ are the p-adic expansions of r and s. Hence

$$\phi_s((\widehat{C}(X))\sigma_{a,b}) = \sum_{g \in \mathcal{I}} C_g \sum_{r \preceq s} \binom{s}{r} (ag)^r b^{s-r} = \sum_{r \preceq s} \binom{s}{r} a^r b^{s-r} \sum_{g \in \mathcal{I}} C_g g^r.$$

Therefore,

$$\phi_s((\widehat{C}(X))\sigma_{a,b}) = \sum_{r \preceq s} \binom{s}{r} a^r b^{s-r} \phi_r(\widehat{C}(X)). \tag{4.13}$$

Let s be an arbitrary element of $\widehat{T}$ and assume that if $r \preceq s$, then $r \in \widehat{T}$. By (4.12), $\phi_r(\widehat{C}(X)) = 0$ as $r \in \widehat{T}$, and therefore by (4.13), $\phi_s(\widehat{C}(X)\sigma_{a,b}) = 0$. As s was an arbitrary element of $\widehat{T}$, by (4.12), $\widehat{C}$ is affine-invariant.

Conversely, assume that $\widehat{C}$ is affine-invariant. Assume that $s \in \widehat{T}$ and that $r \preceq s$. We need to show that $r \in \widehat{T}$, that is $\phi_r(\widehat{C}(X)) = 0$ by (4.12). As $\widehat{C}$ is affine-invariant, $\phi_s(\widehat{C}(X)\sigma_{a,b}) = 0$ for all $a \in \mathcal{I}^*$ and $b \in \mathcal{I}$. In particular this holds for $a = 1$; letting $a = 1$ in (4.13) yields

$$0 = \sum_{r \preceq s} \binom{s}{r} \phi_r(\widehat{C}(X)) b^{s-r}$$

for all $b \in \mathcal{I}$. But the right-hand side of this equation is a polynomial in b of degree at most $s < q^t$ with all q^t possible $b \in \mathcal{I}$ as roots. Hence this must be the zero polynomial. So $\binom{s}{r}\phi_r(\widehat{C}(X)) = 0$ in $\mathcal{I}$ for all $r \preceq s$. However, by Lucas' Theorem again, $\binom{s}{r} \neq 0 \pmod{p}$ and thus these binomial coefficients are nonzero in $\mathcal{I}$. Hence $\phi_r(\widehat{C}(X)) = 0$ implying that $r \in \widehat{T}$. $\qquad \square$

Example 4.7.7 Suppose that $\widehat{C}$ is an affine-invariant code with defining set $\widehat{T}$ such that $n \in \widehat{T}$. By Exercise 279, $r \preceq n$ for all $r \in \widehat{N}$. Thus $\widehat{T} = \widehat{N}$ and $\widehat{C}$ is the zero code. This makes sense because if $n \in \widehat{T}$, then the extended component of any codeword is always

0. Since the code is affine-invariant and the group $GA_1(\mathcal{I})$ is transitive (actually doubly-transitive), every component of any codeword must be zero. Thus $\widehat{\mathcal{C}}$ is the zero code.   ■

Exercise 279   Prove that $r \preceq n$ for all $r \in \widehat{\mathcal{N}}$.   ◆

Example 4.7.8   The following table gives the defining sets for the binary extended cyclic codes of length 8. The ones that are affine-invariant are marked.

Defining set	Affine-invariant
$\{0\}$	yes
$\{0, 1, 2, 4\}$	yes
$\{0, 1, 2, 3, 4, 5, 6\}$	yes
$\{0, 1, 2, 4, 7\}$	no
$\{0, 1, 2, 3, 4, 5, 6, 7\}$	yes
$\{0, 3, 5, 6\}$	no
$\{0, 3, 5, 6, 7\}$	no
$\{0, 7\}$	no

Notice the rather interesting phenomenon that the extended cyclic codes with defining sets $\{0, 1, 2, 4\}$ and $\{0, 3, 5, 6\}$ are equivalent, yet one is affine-invariant while the other is not. The equivalence of these codes follows from Exercise 41 together with the fact that the punctured codes of length 7 are cyclic with defining sets $\{1, 2, 4\}$ and $\{3, 5, 6\}$ and are equivalent under the multiplier μ_3. These two extended codes have isomorphic automorphism groups. The automorphism group of the affine-invariant one contains $GA_1(\mathbb{F}_8)$ while the other contains a subgroup isomorphic, but not equal, to $GA_1(\mathbb{F}_8)$. The code with defining set $\{0\}$ is the $[8, 7, 2]$ code consisting of the even weight vectors in $\mathbb{F}_2^8$. The code with defining set $\{0, 1, 2, 3, 4, 5, 6\}$ is the repetition code, and the code with defining set $\{0, 1, 2, 3, 4, 5, 6, 7\}$ is the zero code. Finally, the code with defining set $\{0, 1, 2, 4\}$ is one particular form of the extended Hamming code $\widehat{\mathcal{H}}_3$.   ■

Exercise 280   Verify all the claims in Example 4.7.8.   ◆

Exercise 281   Give the defining sets of all binary affine-invariant codes of length 16.   ◆

Exercise 282   Give the defining sets of all affine-invariant codes over $\mathbb{F}_4$ of length 16.   ◆

Exercise 283   Give the defining sets of all affine-invariant codes over $\mathbb{F}_3$ of length 9.   ◆

Corollary 4.7.9   *If $\mathcal{C}$ is a primitive cyclic code such that $\widehat{\mathcal{C}}$ is a nonzero affine-invariant code, then the minimum weight of $\mathcal{C}$ is its minimum odd-like weight. In particular, the minimum weight of a binary primitive cyclic code, whose extension is affine-invariant, is odd.*

Proof: As $GA_1(\mathcal{I})$ is transitive, the result follows from Theorem 1.6.6.   □

5   BCH and Reed–Solomon codes

In this chapter we examine one of the many important families of cyclic codes known as BCH codes together with a subfamily of these codes called Reed–Solomon codes.

The binary BCH codes were discovered around 1960 by Hocquenghem [132] and independently by Bose and Ray-Chaudhuri [28, 29], and were generalized to all finite fields by Gorenstein and Zierler [109]. At about the same time as BCH codes appeared in the literature, Reed and Solomon [294] published their work on the codes that now bear their names. These codes, which can be described as special BCH codes, were actually first constructed by Bush [42] in 1952 in the context of orthogonal arrays. Because of their burst error-correction capabilities, Reed–Solomon codes are used to improve the reliability of compact discs, digital audio tapes, and other data storage systems.

5.1   BCH codes

BCH codes are cyclic codes designed to take advantage of the BCH Bound. We would like to construct a cyclic code C of length n over $\mathbb{F}_q$ with simultaneously high minimum weight and high dimension. Having high minimum weight, by the BCH Bound, can be accomplished by choosing a defining set T for C with a large number of consecutive elements. Since the dimension of C is $n - |T|$ by Theorem 4.4.2, we would like $|T|$ to be as small as possible. So if we would like C to have minimum distance at least δ, we can choose a defining set as small as possible that is a union of q-cyclotomic cosets with $\delta - 1$ consecutive elements.

Let δ be an integer with $2 \leq \delta \leq n$. A *BCH code* C over $\mathbb{F}_q$ of length n and *designed distance* δ is a cyclic code with defining set

$$T = C_b \cup C_{b+1} \cup \cdots \cup C_{b+\delta-2}, \tag{5.1}$$

where C_i is the q-cyclotomic coset modulo n containing i. By the BCH Bound this code has minimum distance at least δ.

Theorem 5.1.1 *A BCH code of designed distance δ has minimum weight at least δ.*

Proof: The defining set (5.1) contains $\delta - 1$ consecutive elements. The result follows by the BCH Bound. $\qquad\square$

Varying the value of b produces a variety of codes with possibly different minimum distances and dimensions. When $b = 1$, C is called a *narrow-sense* BCH code. As with any cyclic code, if $n = q^t - 1$, then C is called a *primitive* BCH code.

BCH codes are *nested* in the following sense.

Theorem 5.1.2 *For $i = 1$ and 2, let C_i be the BCH code over $\mathbb{F}_q$ with defining set $T_i = C_b \cup C_{b+1} \cup \cdots \cup C_{b+\delta_i-2}$, where $\delta_1 < \delta_2$. Then $C_2 \subseteq C_1$.*

Exercise 284 Prove Theorem 5.1.2.   ♦

Example 5.1.3 We construct several BCH codes over $\mathbb{F}_3$ of length 13. The 3-cyclotomic cosets modulo 13 are

$$C_0 = \{0\}, \quad C_1 = \{1, 3, 9\}, \quad C_2 = \{2, 5, 6\}, \quad C_4 = \{4, 10, 12\}, \quad C_7 = \{7, 8, 11\}.$$

As $\mathrm{ord}_{13}(3) = 3$, $x^{13} - 1$ has its roots in $\mathbb{F}_{3^3}$. There is a primitive element α in $\mathbb{F}_{3^3}$ which satisfies $\alpha^3 + 2\alpha + 1 = 0$. Then $\beta = \alpha^2$ is a primitive 13th root of unity in $\mathbb{F}_{3^3}$. Using β, the narrow-sense BCH code C_1 of designed distance 2 has defining set C_1 and generator polynomial $g_1(x) = 2 + x + x^2 + x^3$. By Theorem 5.1.1, the minimum distance is at least 2. However, $C_1\mu_2$, which is equivalent to C_1, is the (non-narrow-sense) BCH code with defining set $2^{-1}C_1 = C_7 = C_8$ by Corollary 4.4.5. This code has designed distance 3 and generator polynomial $g_7(x) = 2 + 2x + x^3$. Thus C_1 is a $[13, 10, 3]$ BCH code. The even-like subcode $C_{1,e}$ of C_1 is the BCH code with defining set $C_0 \cup C_1$. $C_{1,e}$ has designed distance 3 and minimum distance 3 as $(x - 1)g_1(x) = 1 + x + x^4$ is even-like of weight 3. Note that the even-like subcode of $C_1\mu_2$ is equivalent to $C_{1,e}$ but is not BCH. The narrow-sense BCH code C_2 of designed distance 3 has defining set $C_1 \cup C_2$. As this defining set also equals $C_1 \cup C_2 \cup C_3$, C_2 also has designed distance 4. Its generator polynomial is $g_{1,2}(x) = 1 + 2x + x^2 + 2x^3 + 2x^4 + 2x^5 + x^6$, and $(1 + x)g_{1,2}(x) = 1 + x^4 + x^5 + x^7$ has weight 4. Thus C_2 is a $[13, 7, 4]$ BCH code. Finally, the narrow-sense BCH code C_3 of designed distance 5 has defining set $C_1 \cup C_2 \cup C_3 \cup C_4$; this code is also the narrow-sense BCH code of designed distance 7. C_3 has generator polynomial $2 + 2x^2 + 2x^3 + x^5 + 2x^7 + x^8 + x^9$, which has weight 7; thus C_3 is a $[13, 4, 7]$ BCH code. Notice that $C_3 \subset C_2 \subset C_1$ by the nesting property of Theorem 5.1.2. ∎

Exercise 285 This exercise continues Example 5.1.3. The minimal polynomials of β, β^2, β^4, and β^7 are $g_1(x) = 2 + x + x^2 + x^3$, $g_2(x) = 2 + x^2 + x^3$, $g_4(x) = 2 + 2x + 2x^2 + x^3$, and $g_7(x) = 2 + 2x + x^3$, respectively. Find generator polynomials, designed distances, and minimum distances of all BCH codes over $\mathbb{F}_3$ of length 13. Note that computations will be reduced if multipliers are used to find some equivalences between BCH codes.   ♦

The next theorem shows that many Hamming codes are narrow-sense BCH codes.

Theorem 5.1.4 *Let $n = (q^r - 1)/(q - 1)$ where $\gcd(r, q - 1) = 1$. Let C be the narrow-sense BCH code with defining set $T = C_1$. Then C is the Hamming code $\mathcal{H}_{q,r}$.*

Proof: Let γ be a primitive element of $\mathbb{F}_{q^r}$. The code C is generated by $M_\alpha(x)$, where $\alpha = \gamma^{q-1}$ is a primitive nth root of unity. An easy calculation shows that $n = (q - 1)\sum_{i=1}^{r-1} iq^{r-1-i} + r$. So $\gcd(n, q - 1) = \gcd(r, q - 1) = 1$. By Theorem 3.5.3, the nonzero elements of $\mathbb{F}_q$ in $\mathbb{F}_{q^r}$ are powers of γ where the power is a divisor of $n = (q^r - 1)/(q - 1)$. As $\gcd(n, q - 1) = 1$, the only element of $\mathbb{F}_q$ that is a power of α is the identity. Therefore if we write the elements of $\mathbb{F}_{q^r}$ as r-tuples in $\mathbb{F}_q^r$, none of the

r-tuples corresponding to $\alpha^0, \alpha^1, \ldots, \alpha^{n-1}$ are multiples of one another using only elements of $\mathbb{F}_q$. This implies that these elements are the distinct points of $\mathrm{PG}(r-1, q)$. The $r \times n$ matrix H, whose columns are the r-tuples corresponding to $\alpha^0, \alpha^1, \ldots, \alpha^{n-1}$, is the parity check matrix $H_{q,r}$ of $\mathcal{H}_{q,r}$ as given in Section 1.8. $\qquad\square$

Corollary 5.1.5 *Every binary Hamming code is a primitive narrow-sense BCH code.*

Exercise 286 Verify the claim made in the proof of Theorem 5.1.4 that $n = (q - 1) \times \sum_{i=1}^{r-1} iq^{r-1-i} + r$, where $n = (q^r - 1)/(q - 1)$. $\qquad\blacklozenge$

Exercise 287 Prove Corollary 5.1.5. $\qquad\blacklozenge$

But not every Hamming code is equivalent to a BCH code. Indeed as the next example shows, some Hamming codes are not equivalent to any cyclic code.

Example 5.1.6 In Example 1.3.3, a generator matrix for the [4, 2, 3] ternary Hamming code $\mathcal{C} = \mathcal{H}_{3,2}$, also called the tetracode, was presented. In Example 1.7.7, its monomial automorphism group was given; namely $\mathrm{MAut}(\mathcal{C}) = \Gamma\mathrm{Aut}(\mathcal{C})$ is a group of order 48 generated by $\mathrm{diag}(1, 1, 1, -1)(1, 2, 3, 4)$ and $\mathrm{diag}(1, 1, 1, -1)(1, 2)$, where $\mathrm{MAut}_{\mathrm{Pr}}(\mathcal{C}) = \mathrm{Sym}_4$. Using this fact a straightforward argument, which we leave to Exercise 288, shows that a monomial map M, such that $\mathcal{C}M$ is cyclic, does not exist. $\qquad\blacksquare$

Exercise 288 Verify that in Example 5.1.6 a monomial map M, such that $\mathcal{C}M$ is cyclic, does not exist. $\qquad\blacklozenge$

The Hamming codes of Theorem 5.1.4 have designed distance $\delta = 2$, yet their actual minimum distance is 3. In the binary case this can be explained as follows. These Hamming codes are the narrow-sense BCH codes of designed distance $\delta = 2$ with defining set $T = C_1$. But in the binary case, $C_1 = C_2$ and so T is also the defining set of the narrow-sense BCH code of designed distance $\delta = 3$. This same argument can be used with every narrow-sense binary BCH code in that the designed distance can always be assumed to be odd. In the next theorem we give a lower bound on the dimension of a BCH code in terms of δ and $\mathrm{ord}_n(q)$. Of course, the exact dimension is determined by the size of the defining set.

Theorem 5.1.7 *Let $\mathcal{C}$ be an $[n, k]$ BCH code over $\mathbb{F}_q$ of designed distance δ. The following hold:*

(i) *$k \geq n - \mathrm{ord}_n(q)(\delta - 1)$.*

(ii) *If $q = 2$ and $\mathcal{C}$ is a narrow-sense BCH code, then δ can be assumed to be odd; furthermore if $\delta = 2w + 1$, then $k \geq n - \mathrm{ord}_n(q)w$.*

Proof: By Theorem 4.1.4 each q-cyclotomic coset has size a divisor of $\mathrm{ord}_n(q)$. The defining set for a BCH code of designed distance δ is the union of at most $\delta - 1$ q-cyclotomic cosets each of size at most $\mathrm{ord}_n(q)$. Hence the dimension of the code is at least $n - \mathrm{ord}_n(q)(\delta - 1)$, giving (i). If the code is narrow-sense and binary, then $\{1, 2, \ldots, \delta - 1\} \subseteq T$. Suppose that δ is even. Then $\delta \in C_{\delta/2} \subseteq T$, implying that T contains the set $\{1, 2, \ldots, \delta\}$ of δ consecutive elements. Hence we can increase the designed distance by 1 whenever the designed distance is assumed to be even. So we may assume that δ is odd. If $\delta = 2w + 1$, then

$$T = \mathcal{C}_1 \cup \mathcal{C}_2 \cup \cdots \cup \mathcal{C}_{2w} = \mathcal{C}_1 \cup \mathcal{C}_3 \cup \cdots \cup \mathcal{C}_{2w-1},$$

as $\mathcal{C}_{2i} = \mathcal{C}_i$. Hence T is the union of at most w q-cyclotomic cosets of size at most $\mathrm{ord}_n(q)$, yielding $k \geq n - \mathrm{ord}_n(q)w$. $\qquad\square$

As we see in the proof of Theorem 5.1.7, it is possible for more than one value of δ to be used to construct the same BCH code. The binary Golay code provides a further example.

Example 5.1.8 In Example 4.5.5, we saw that the [23, 12, 7] binary Golay code is a cyclic code with defining set $T = \mathcal{C}_1$. Thus it is a narrow-sense BCH code with designed distance $\delta = 2$. As $\mathcal{C}_1 = \mathcal{C}_2 = \mathcal{C}_3 = \mathcal{C}_4$, it is also a BCH code with designed distance any of 3, 4, or 5. $\qquad\blacksquare$

Because of examples such as this, we call the largest designed distance δ' defining a BCH code $\mathcal{C}$ the *Bose distance* of $\mathcal{C}$. Thus we have $d \geq \delta' \geq \delta$, where d is the actual minimum distance of $\mathcal{C}$ by Theorem 5.1.1. For the Golay code of Example 5.1.8, the Bose distance is 5; notice that the true minimum distance is 7, which is still greater than the Bose distance. As we saw in Chapter 4, there are techniques to produce lower bounds on the minimum distance which, when applied to BCH codes, may produce lower bounds above the Bose distance.

Determining the actual minimum distance of specific BCH codes is an important, but difficult, problem. Section 3 of [50] discusses this problem extensively. Tables of minimum distances for very long codes have been produced. For example, Table 2 of [50] contains a list of primitive narrow-sense binary BCH codes whose minimum distance is the Bose distance. As an illustration of the lengths of codes involved, the primitive narrow-sense binary BCH code of length $n = 2^{4199} - 1$ and designed distance 49 actually has minimum distance 49, as does every such code of length $n = 2^{4199k} - 1$.

For narrow-sense BCH codes, the BCH Bound has been very good in general. In fact, that is how the minimum weights of many of the BCH codes of high dimension have been determined: find a vector in the code whose weight is the Bose designed distance. There has been a great deal of effort to find the true minimum distance of all primitive narrow-sense binary BCH codes of a fixed length. This has been done completely for lengths 3, 7, 31, 63, 127, and 255; for lengths up to 127 see Figure 9.1 of [218] and Table 2 of [50] and for length 255 see [10]. The true minimum distance of all but six codes of length 511 has been found in [10] and [46]; see also [50]. Of the 51 codes of length 511 whose minimum distance is known, 46 have minimum distance equal to their Bose distance, four have minimum distance equal to two more than their Bose distance, and one has minimum distance four more than its Bose distance (the [511, 103, 127] code of designed distance 123). The following conjecture has been formulated by Charpin.

Conjecture *Every binary primitive narrow-sense BCH code of Bose designed distance δ has minimum distance no more than $\delta + 4$.*

If the code is not primitive, some codes fail to satisfy this conjecture. For example, the binary narrow-sense BCH code of length 33 and designed distance 5 has minimum distance 10 (see [203]).

As we see, finding the minimum distance for a specific code or a family of codes, such as BCH codes, has been an important area of research. In this connection, it would be very useful to have improved algorithms for accomplishing this. There is also interest in determining as much as one can about the weight distribution or automorphism group of a specific code or a family of codes. For example, in Theorem 5.1.9 we will show that extended narrow-sense primitive BCH codes are affine-invariant; the full automorphism groups of these codes are known [16].

By Theorem 4.2.6 and Exercises 238 and 243, the dual, even-like subcode, and cyclic complement of a cyclic code C are all cyclic. If C is a BCH code, are its dual, even-like subcode, and cyclic complement also BCH? In general the answer is no to each of these, although in certain cases some of these are BCH. For example, suppose that C is a narrow-sense BCH code of length n and designed distance δ. Then as $1 + \delta - 2 \le n - 1, 0$ is not in the defining set T of C, and hence, by Exercise 238, C is odd-like. Thus $T = C_1 \cup C_2 \cup \cdots \cup C_{\delta-1}$ while the defining set of the even-like subcode C_e is $T \cup \{0\} = C_0 \cup C_1 \cup \cdots \cup C_{\delta-1}$; hence C_e is a BCH code of designed distance $\delta + 1$.

Exercise 289 Let $2 \le \delta \le 15$.
(a) Give the defining set of all binary BCH codes of designed distance δ and length 15.
(b) What is the Bose distance of each of the codes in part (a)?
(c) What is the defining set of each of the duals of the codes in part (a)? Which of these duals are BCH?
(d) What is the defining set of each of the even-like subcodes of the codes in part (a)? Which of these even-like subcodes are BCH?
(e) What is the defining set of each of the cyclic complements of the codes in part (a)? Which of these cyclic complements are BCH?
(f) Find the defining sets of all binary cyclic codes of length 15 that are not BCH.
(g) Find the minimum weight of the BCH codes of length 15.   ◆

Let C be a primitive narrow-sense BCH code of length $n = q^t - 1$ over $\mathbb{F}_q$ with designed distance δ. The defining set T is $C_1 \cup C_2 \cup \cdots \cup C_{\delta-1}$. The extended BCH code $\widehat{C}$ has defining set $\widehat{T} = \{0\} \cup T$. The reader is asked to show in Exercise 290 that if $s \in \widehat{T}$ and $r \preceq s$, then $r \in \widehat{T}$ where $\preceq$ is the partial order on $\mathcal{N} = \{0, 1, \ldots, n\}$ of Section 4.7. By Theorem 4.7.6, $\widehat{C}$ is affine-invariant, and by Corollary 4.7.9 the minimum weight of C is its minimum odd-like weight. So we have the following theorem.

Theorem 5.1.9 *Let C be a primitive narrow-sense BCH code of length $n = q^t - 1$ over $\mathbb{F}_q$ with designed distance δ. Then $\widehat{C}$ is affine-invariant, and the minimum weight of C is its minimum odd-like weight.*

Exercise 290 Let $\widehat{C}$ be an extended BCH code of length q^t over $\mathbb{F}_q$ with defining set $\widehat{T} = \{0\} \cup C_1 \cup C_2 \cup \cdots \cup C_{\delta-1}$. Let $q = p^m$, where p is a prime, and let $n = q^t - 1 = p^{mt} - 1$. Show, by carrying out the following, that if $s \in \widehat{T}$ and $r \preceq s$, then $r \in \widehat{T}$.
(a) Prove that if $r \preceq s$, then $r' \preceq s'$ where $r' \equiv qr \pmod{n}$ and $s' \equiv qs \pmod{n}$.
(b) Prove that if there exists an $s \in \widehat{T}$ and an $r \preceq s$ but $r \notin \widehat{T}$, then there is an $s'' \in \widehat{T}$ with $s'' \le \delta - 1$ and an $r'' \preceq s''$ such that $r'' \notin \widehat{T}$.
(c) Prove that r'' and s'' from part (b) do not exist.   ◆

When examining a family of codes, it is natural to ask if this family is asymptotically "good" or "bad" in the following sense. We say that a family of codes is *asymptotically good* provided that there exists an infinite subset of $[n_i, k_i, d_i]$ codes from this family with $\lim_{i \to \infty} n_i = \infty$ such that both $\liminf_{i \to \infty} k_i / n_i > 0$ and $\liminf_{i \to \infty} d_i / n_i > 0$. For example, codes that meet the Asymptotic Gilbert–Varshamov Bound are asymptotically good. The family is *asymptotically bad* if no asymptotically good subfamily exists. Recall from Section 2.10 that for an $[n, k, d]$ code, the ratio k/n is called the rate of the code and the ratio d/n is called the relative distance of the code. The former measures the number of information coordinates relative to the total number of coordinates, and the latter measures the error-correcting capability of the code. Ideally, we would like the rate and relative distance to both be high, in order to be able to send a large number of messages while correcting a large number of errors. But these are conflicting goals. So in a family of good codes, we want an infinite subfamily where both the code rates and relative distances are bounded away from 0; hence in this subfamily neither rate nor relative distance are low. In general, the rates and relative distances for any class of codes is difficult or impossible to determine. Unfortunately, primitive BCH codes are known to be bad [198].

Theorem 5.1.10 *The family of primitive BCH codes over $\mathbb{F}_q$ is asymptotically bad.*

Note that this negative result does not say that individual codes, particularly those of modest length, are not excellent codes. BCH codes, or codes constructed from them, are often the codes closest to optimal that are known [32].

As a corollary of Theorem 5.1.10, we see that the primitive narrow-sense BCH codes are asymptotically bad. These codes also have extensions that are affine-invariant by Theorem 5.1.9. The fact that primitive narrow-sense BCH codes are asymptotically bad also, is a corollary of the following result of Kasami [161].

Theorem 5.1.11 *Any family of cyclic codes whose extensions are affine-invariant is asymptotically bad.*

It is natural to ask whether or not there is any asymptotically good family of codes. The answer is yes, as the Asymptotic Gilbert–Varshamov Bound implies. As we saw, the proof of this bound is nonconstructive; it shows that a family of good codes exists but does not give a construction of such a family. In Section 2.11 we saw that the lexicodes meet the Asymptotic Gilbert–Varshamov Bound and hence are asymptotically good. We will examine another family of codes in Chapter 13 that meets the Asymptotic Gilbert–Varshamov Bound.

5.2 Reed–Solomon codes

In this section we will define Reed–Solomon codes as a subfamily of BCH codes. We will also give another equivalent definition for the narrow-sense Reed–Solomon codes that will allow us to generalize these important codes.

A *Reed–Solomon code*, abbreviated *RS code*, $\mathcal{C}$ over $\mathbb{F}_q$ is a BCH code of length $n = q - 1$. Thus $\text{ord}_n(q) = 1$ implying that all irreducible factors of $x^n - 1$ are of degree 1 and all

q-cyclotomic cosets modulo n have size 1. In fact, the roots of $x^n - 1$ are exactly the nonzero elements of $\mathbb{F}_q$, and a primitive nth root of unity is a primitive element of $\mathbb{F}_q$. So if C has designed distance δ, the defining set of C has size $\delta - 1$ and is $T = \{b, b+1, \ldots, b+\delta-2\}$ for some integer b. By Theorem 5.1.1 and the Singleton Bound, the dimension k and minimum distance d of C satisfy $k = n - \delta + 1 \geq n - d + 1 \geq k$. Thus both inequalities are equalities implying $d = \delta$ and $k = n - d + 1$. In particular, C is MDS. We summarize this information in the following theorem.

Theorem 5.2.1 *Let C be an RS code over $\mathbb{F}_q$ of length $n = q - 1$ and designed distance δ. Then:*

(i) *C has defining set $T = \{b, b+1, \ldots, b+\delta-2\}$ for some integer b,*

(ii) *C has minimum distance $d = \delta$ and dimension $k = n - d + 1$, and*

(iii) *C is MDS.*

Recall that in general the dual and cyclic complement of a BCH code are not BCH codes; that is not the case with RS codes. Suppose that T is the defining set for an RS code C of length n and designed distance δ. Then T is a set of $\delta - 1$ consecutive elements from $\mathcal{N} = \{0, 1, \ldots, n-1\}$. By Theorem 4.4.6, the defining set of the cyclic complement C^C of C is $\mathcal{N} \setminus T$, which is a set of $n - \delta + 1$ consecutive elements implying that C^C is RS. Similarly, as $(-1)T \bmod n$ is also a set of $\delta - 1$ consecutive elements from $\mathcal{N}$, we have using Theorem 4.4.9 that the defining set $\mathcal{N} \setminus (-1)T \bmod n$ of $C^\perp$ is a consecutive set of $n - \delta + 1$ elements also. Therefore $C^\perp$ is an RS code.

Example 5.2.2 A primitive element of $\mathbb{F}_{13}$ is 2. Let C be the narrow-sense Reed–Solomon code of designed distance 5 over $\mathbb{F}_{13}$. It is a code of length 12 with defining set $\{1, 2, 3, 4\}$ and generator polynomial $(x - 2)(x - 2^2)(x - 2^3)(x - 2^4) = 10 + 2x + 7x^2 + 9x^3 + x^4$. By Theorem 5.2.1, C has minimum distance 5 and C is a $[12, 8, 5]$ MDS code. $C^\perp$ is the $[12, 4, 9]$ Reed–Solomon code with defining set $\{0, 1, 2, 3, 4, 5, 6, 7\}$ and generator polynomial $(x - 2^0)(x - 2^1)(x - 2^2) \cdots (x - 2^7) = 3 + 12x + x^2 + 5x^3 + 11x^4 + 4x^5 + 10x^6 + 5x^7 + x^8$. The cyclic complement of C is the $[12, 4, 9]$ Reed–Solomon code with defining set $\{5, 6, 7, 8, 9, 10, 11, 0\}$ and generator polynomial $(x - 2^0)(x - 2^5)(x - 2^6) \cdots (x - 2^{11}) = 9 + 6x + 12x^2 + 10x^3 + 8x^4 + 6x^5 + 9x^6 + 4x^7 + x^8$. ∎

We now present an alternative formulation of narrow-sense Reed–Solomon codes, which is the original formulation of Reed and Solomon. This alternative formulation of narrow-sense RS codes is of particular importance because it is the basis for the definitions of generalized Reed–Solomon codes, Goppa codes, and algebraic geometry codes, as we will see in Chapter 13. For $k \geq 0$, let $\mathcal{P}_k$ denote the set of polynomials of degree less than k, including the zero polynomial, in $\mathbb{F}_q[x]$. Note that $\mathcal{P}_0$ is precisely the zero polynomial.

Theorem 5.2.3 *Let α be a primitive element of $\mathbb{F}_q$ and let k be an integer with $0 \leq k \leq n = q - 1$. Then*

$$C = \{(f(1), f(\alpha), f(\alpha^2), \ldots, f(\alpha^{q-2})) \mid f \in \mathcal{P}_k\}$$

is the narrow-sense $[n, k, n - k + 1]$ RS code over $\mathbb{F}_q$.

Proof: Clearly C is a linear code over $\mathbb{F}_q$ as $\mathcal{P}_k$ is a linear subspace over $\mathbb{F}_q$ of $\mathbb{F}_q[x]$. As $\mathcal{P}_k$ is k-dimensional, C will also be k-dimensional if we can show that if f and f_1 are distinct elements of $\mathcal{P}_k$, then the corresponding elements in C are distinct. If the latter are equal, then their difference is $\{0\}$ implying that $f - f_1$, which is a nonzero polynomial of degree at most $k - 1$, has $n \geq k$ roots, contradicting Exercise 159. Thus C is k-dimensional.

Let $\mathcal{D}$ be the narrow-sense $[n, k, n - k + 1]$ RS code over $\mathbb{F}_q$. So $\mathcal{D}$ has defining set $T = \{1, 2, \ldots, n - k\}$. We show that $C = \mathcal{D}$; it suffices to prove that $C \subseteq \mathcal{D}$ as both codes are k-dimensional. Let $c(x) = \sum_{j=0}^{n-1} c_j x^j \in C$. Then there exists some $f(x) = \sum_{m=0}^{k-1} f_m x^m \in \mathcal{P}_k$ such that $c_j = f(\alpha^j)$ for $0 \leq j < n$. To show that $c(x) \in \mathcal{D}$ we need to show that $c(\alpha^i) = 0$ for $i \in T$ by Theorem 4.4.2. If $i \in T$, then

$$c(\alpha^i) = \sum_{j=0}^{n-1} c_j \alpha^{ij} = \sum_{j=0}^{n-1} \left(\sum_{m=0}^{k-1} f_m \alpha^{jm} \right) \alpha^{ij}$$

$$= \sum_{m=0}^{k-1} f_m \sum_{j=0}^{n-1} \alpha^{(i+m)j} = \sum_{m=0}^{k-1} f_m \frac{\alpha^{(i+m)n} - 1}{\alpha^{i+m} - 1}.$$

But $\alpha^{(i+m)n} = 1$ and $\alpha^{i+m} \neq 1$ as $1 \leq i + m \leq n - 1 = q - 2$ and α is a primitive nth root of unity. Therefore $c(\alpha^i) = 0$ for $i \in T$ implying that $C \subseteq \mathcal{D}$. Hence $C = \mathcal{D}$.   □

Exercise 291   Let ev : $\mathcal{P}_k \to \mathbb{F}_q^n$ be given by

$$\mathrm{ev}(f) = (f(1), f(\alpha), f(\alpha^2), \ldots, f(\alpha^{q-2})),$$

where α is a primitive element of $\mathbb{F}_q$ and $n = q - 1$. Prove that the *evaluation map* ev is a nonsingular linear transformation.   ◆

Note that the narrow-sense RS code defined in this alternate sense with $k = 0$ is precisely the zero code.

This alternate formulation of narrow-sense RS codes gives an alternate encoding scheme as well. Suppose that $f_0, f_1, \ldots, f_{k-1}$ are k information symbols and $f(x) = f_0 + f_1 x + \cdots + f_{k-1} x^{k-1}$, then

$$(f_0, f_1, \ldots, f_{k-1}) \xrightarrow{\text{encode}} (f(1), f(\alpha), \ldots, f(\alpha^{q-2})). \tag{5.2}$$

Notice that this encoding scheme is not systematic. There is a decoding scheme for RS codes that is unusual in the sense that it finds the information symbols directly, under the assumption that they have been encoded using (5.2). It is not syndrome decoding but is an instance of a decoding scheme based on majority logic and was the original decoding developed by Reed and Solomon [294]. Currently other schemes are used for decoding and for that reason we do not present the original decoding here.

5.3   Generalized Reed–Solomon codes

The construction of the narrow-sense RS codes in Theorem 5.2.3 can be generalized to (possibly noncyclic) codes. Let n be any integer with $1 \leq n \leq q$. Choose $\gamma = (\gamma_0, \ldots, \gamma_{n-1})$

to be an n-tuple of *distinct* elements of $\mathbb{F}_q$, and $\mathbf{v} = (v_0, \ldots, v_{n-1})$ to be an n-tuple of *nonzero*, but not necessarily distinct, elements of $\mathbb{F}_q$. Let k be an integer with $1 \leq k \leq n$. Then the codes

$$\mathrm{GRS}_k(\boldsymbol{\gamma}, \mathbf{v}) = \{(v_0 f(\gamma_0), v_1 f(\gamma_1), \ldots, v_{n-1} f(\gamma_{n-1})) \mid f \in \mathcal{P}_k\}$$

are the *generalized Reed–Solomon* or *GRS* codes. Because no v_i is 0, by repeating the proof of Theorem 5.2.3, we see that $\mathrm{GRS}_k(\boldsymbol{\gamma}, \mathbf{v})$ is k-dimensional. Because a nonzero polynomial $f \in \mathcal{P}_k$ has at most $k - 1$ zeros, $\mathrm{GRS}_k(\boldsymbol{\gamma}, \mathbf{v})$ has minimum distance at least $n - (k - 1) = n - k + 1$. By the Singleton Bound, it has minimum distance at most $n - k + 1$; hence, $\mathrm{GRS}_k(\boldsymbol{\gamma}, \mathbf{v})$ has minimum distance exactly $n - k + 1$. Thus GRS codes are also MDS, as were RS codes. It is obvious that if $\mathbf{w}$ is another n-tuple of nonzero elements of $\mathbb{F}_q$, then $\mathrm{GRS}_k(\boldsymbol{\gamma}, \mathbf{v})$ is monomially equivalent to $\mathrm{GRS}_k(\boldsymbol{\gamma}, \mathbf{w})$. The narrow-sense RS codes are GRS codes with $n = q - 1$, $\gamma_i = \alpha^i$, where α is a primitive nth root of unity, and $v_i = 1$ for $0 \leq i \leq n - 1$. We summarize this information in the following theorem.

Theorem 5.3.1 *With the notation as above:*
(i) $\mathrm{GRS}_k(\boldsymbol{\gamma}, \mathbf{v})$ *is an* $[n, k, n - k + 1]$ *MDS code,*
(ii) $\mathrm{GRS}_k(\boldsymbol{\gamma}, \mathbf{v})$ *is monomially equivalent to* $\mathrm{GRS}_k(\boldsymbol{\gamma}, \mathbf{w})$*, and*
(iii) *narrow-sense RS codes are GRS codes with* $n = q - 1$*,* $\gamma_i = \alpha^i$*, and* $v_i = 1$ *for* $0 \leq i \leq n - 1$.

Narrow-sense $[q - 1, k, q - k]$ Reed–Solomon codes over $\mathbb{F}_q$ can be extended to MDS codes as follows. Let $\mathcal{C} = \{(f(1), f(\alpha), f(\alpha^2), \ldots, f(\alpha^{q-2})) \mid f \in \mathcal{P}_k\}$ be such a code. Exercise 292 shows that if $f \in \mathcal{P}_k$, where $k < q$, then $\sum_{\beta \in \mathbb{F}_q} f(\beta) = 0$. So

$$\widehat{\mathcal{C}} = \{(f(1), f(\alpha), f(\alpha^2), \ldots, f(\alpha^{q-2}), f(0)) \mid f \in \mathcal{P}_k\}$$

is the extension of $\mathcal{C}$. Notice that this is also a GRS code with $n = q$, $\gamma_i = \alpha^i$ for $0 \leq i \leq n - 2$, $\gamma_{n-1} = 0$, and $v_i = 1$ for $0 \leq i \leq n - 1$. Therefore $\widehat{\mathcal{C}}$ is a $[q, k, q - k + 1]$ MDS code. In other words, when extending the narrow sense RS codes by adding an overall parity check, the minimum weight increases, an assertion that can be guaranteed in general for codes over arbitrary fields only if the minimum weight vectors are all odd-like. This results in the following theorem.

Theorem 5.3.2 *The* $[q, k, q - k + 1]$ *extended narrow-sense Reed–Solomon code over* $\mathbb{F}_q$ *is generalized Reed–Solomon and MDS.*

Exercise 292 Prove that if $f \in \mathcal{P}_k$, where $k < q$, then $\sum_{\beta \in \mathbb{F}_q} f(\beta) = 0$. See Exercise 164.   ♦

We now show that the dual of a GRS code is also GRS.

Theorem 5.3.3 *Let* $\boldsymbol{\gamma} = (\gamma_0, \ldots, \gamma_{n-1})$ *be an* n-tuple of distinct elements of $\mathbb{F}_q$ *and let* $\mathbf{v} = (v_0, \ldots, v_{n-1})$ *be an* n-tuple of nonzero elements of $\mathbb{F}_q$*. Then there exists an* n-tuple $\mathbf{w} = (w_0, \ldots, w_{n-1})$ *of nonzero elements of* $\mathbb{F}_q$ *such that* $\mathrm{GRS}_k(\boldsymbol{\gamma}, \mathbf{v})^{\perp} = \mathrm{GRS}_{n-k}(\boldsymbol{\gamma}, \mathbf{w})$ *for all* k *with* $0 \leq k \leq n - 1$*. Furthermore, the vector* $\mathbf{w}$ *is any nonzero codeword in the*

1-*dimensional code* $\text{GRS}_{n-1}(\gamma, \mathbf{v})^\perp$ *and satisfies*

$$\sum_{i=0}^{n-1} w_i v_i h(\gamma_i) = 0 \tag{5.3}$$

for any polynomial $h \in \mathcal{P}_{n-1}$.

Proof: Let $\mathcal{C} = \text{GRS}_k(\gamma, \mathbf{v})$. We first consider the case $k = n - 1$. Since the dual of an MDS code is also MDS by Theorem 2.4.3, $\mathcal{C}^\perp$ is an $[n, 1, n]$ code with a basis vector $\mathbf{w} = (w_0, w_1, \ldots, w_{n-1})$ having no zero components. But the code $\text{GRS}_1(\gamma, \mathbf{w})$ is precisely all multiples of $\mathbf{w}$, implying that $\mathcal{C}^\perp = \text{GRS}_1(\gamma, \mathbf{w})$, verifying the result when $k = n - 1$. This also shows that if h is any polynomial in $\mathcal{P}_{n-1}$, then (5.3) holds because $(v_0 h(\gamma_0), \ldots, v_{n-1} h(\gamma_{n-1})) \in \text{GRS}_{n-1}(\gamma, \mathbf{v}) = \text{GRS}_1(\gamma, \mathbf{w})^\perp$. Now let $0 \le k \le n - 1$. When $f \in \mathcal{P}_k$ and $g \in \mathcal{P}_{n-k}$, $h = fg \in \mathcal{P}_{n-1}$. Thus, by (5.3), $\sum_{i=0}^{n-1} w_i g(\gamma_i) v_i f(\gamma_i) = \sum_{i=0}^{n-1} w_i v_i h(\gamma_i) = 0$. Therefore $\text{GRS}_k(\gamma, \mathbf{v})^\perp \subseteq \text{GRS}_{n-k}(\gamma, \mathbf{w})$. Since the dimension of $\text{GRS}_k(\gamma, \mathbf{v})^\perp$ is $n - k$, the theorem follows. $\qquad\square$

A generator matrix of $\text{GRS}_k(\gamma, \mathbf{v})$ is

$$G = \begin{bmatrix} v_0 & v_1 & \cdots & v_{n-1} \\ v_0\gamma_0 & v_1\gamma_1 & \cdots & v_{n-1}\gamma_{n-1} \\ v_0\gamma_0^2 & v_1\gamma_1^2 & \cdots & v_{n-1}\gamma_{n-1}^2 \\ & & \vdots & \\ v_0\gamma_0^{k-1} & v_1\gamma_1^{k-1} & \cdots & v_{n-1}\gamma_{n-1}^{k-1} \end{bmatrix}. \tag{5.4}$$

By Theorem 5.3.3, a parity check matrix of $\text{GRS}_k(\gamma, \mathbf{v})$ is the generator matrix of $\text{GRS}_{n-k}(\gamma, \mathbf{w})$, where $\mathbf{w}$ is given in Theorem 5.3.3. Therefore a parity check matrix for $\text{GRS}_k(\gamma, \mathbf{v})$ is

$$H = \begin{bmatrix} w_0 & w_1 & \cdots & w_{n-1} \\ w_0\gamma_0 & w_1\gamma_1 & \cdots & w_{n-1}\gamma_{n-1} \\ w_0\gamma_0^2 & w_1\gamma_1^2 & \cdots & w_{n-1}\gamma_{n-1}^2 \\ & & \vdots & \\ w_0\gamma_0^{n-k-1} & w_1\gamma_1^{n-k-1} & \cdots & w_{n-1}\gamma_{n-1}^{n-k-1} \end{bmatrix}.$$

Exercise 293 Prove that the matrix given in (5.4) is a generator matrix of $\text{GRS}_k(\gamma, \mathbf{v})$. ◆

We know from Theorem 5.3.1 that $\mathcal{C} = \text{GRS}_k(\gamma, \mathbf{v})$ is MDS. We want to describe an extension of $\mathcal{C}$, denoted $\check{\mathcal{C}}$, that is also MDS. Let v be a nonzero element of $\mathbb{F}_q$. The generator matrix of $\check{\mathcal{C}}$ is $\check{G} = [G \; \mathbf{u}^\mathsf{T}]$, where $\mathbf{u} = (0, 0, \ldots, 0, v)$. This extended code will generally not be even-like. Choose $w \in \mathbb{F}_q$ so that

$$\sum_{i=0}^{n-1} v_i w_i \gamma_i^{n-1} + vw = 0.$$

Such an element w exists as $v \neq 0$. Using (5.3) and the definition of w, we leave it to the reader in Exercise 294 to verify that $\check{C}$ has parity check matrix

$$
\check{H} = \begin{bmatrix}
w_0 & w_1 & \cdots & w_{n-1} & 0 \\
w_0\gamma_0 & w_1\gamma_1 & \cdots & w_{n-1}\gamma_{n-1} & 0 \\
w_0\gamma_0^2 & w_1\gamma_1^2 & \cdots & w_{n-1}\gamma_{n-1}^2 & 0 \\
& & \vdots & & \\
w_0\gamma_0^{n-k} & w_1\gamma_1^{n-k} & \cdots & w_{n-1}\gamma_{n-1}^{n-k} & w
\end{bmatrix}.
$$

Notice that if $w = 0$, $\sum_{i=0}^{n-1} w_i v_i h(\gamma_i) = 0$ for all $h \in \mathcal{P}_n$, implying that $\mathbf{v}$ is a nonzero vector in $\mathbb{F}_q^n$ orthogonal to all of $\mathbb{F}_q^n$, which is a contradiction. So $w \neq 0$.

Exercise 294 Verify that $\check{H}$ is a parity check matrix for $\check{C}$.   ◆

We now verify that $\check{C}$ is MDS. Consider the $(n - k + 1) \times (n - k + 1)$ submatrix M of $\check{H}$ formed by any $n - k + 1$ columns of $\check{H}$. If the right-most column of $\check{H}$ is not among the $n - k + 1$ chosen, then $M = VD$, where V is a Vandermonde matrix and D is a diagonal matrix. The entries of V are powers of $n - k + 1$ of the (distinct) γ_is; the diagonal entries of D are all chosen from $\{w_0, \ldots, w_{n-1}\}$. As the γ_is are distinct and the w_is are nonzero, the determinants of V and D are both nonzero, using Lemma 4.5.1. Therefore M is nonsingular in this case. Suppose the right-most column of $\check{H}$ is among the $n - k + 1$ chosen. By Theorem 2.4.3 any $n - k$ columns of H are independent (and hence so are the corresponding $n - k$ columns of $\check{H}$) as C is MDS. This implies that the right-most column of $\check{H}$ must be independent of any $n - k$ other columns of $\check{H}$. So all of our chosen columns are independent. Thus by Corollary 1.4.14, $\check{C}$ has minimum weight at least $n - k + 2$. By the Singleton Bound, the minimum weight is at most $n - k + 2$ implying that $\check{C}$ is MDS.

In summary this discussion and Theorem 5.3.1(i) proves the following theorem.

Theorem 5.3.4 *For $1 \leq k \leq n \leq q$, the GRS code $\mathrm{GRS}_k(\boldsymbol{\gamma}, \mathbf{v})$ is an MDS code, and it can be extended to an MDS code of length $n + 1$.*

Recall that a $[q - 1, k, q - k]$ narrow-sense RS code over $\mathbb{F}_q$ can be extended by adding an overall parity check; the resulting $[q, k, q - k + 1]$ code is a GRS code which is MDS by Theorem 5.3.2. This code itself can be extended to an MDS code by Theorem 5.3.4. Thus a narrow-sense RS code of length $q - 1$ can be extended twice to an MDS code of length $q + 1$.

So, in general, GRS codes C and their extensions $\check{C}$ are MDS. There are MDS codes that are not equivalent to such codes. However, no MDS code with parameters other than those arising from GRS codes or their extensions is presently known [298].

5.4   Decoding BCH codes

In this section we present three algorithms for nearest neighbor decoding of BCH codes. The first method is known as Peterson–Gorenstein–Zierler decoding. It was originally developed for binary BCH codes by Peterson [254] in 1960 and generalized shortly thereafter by Gorenstein and Zierler to nonbinary BCH codes [109]. We will describe this decoding method as a four step procedure. The second step of this procedure is the most complicated and time consuming. The second method, known as Berlekamp–Massey decoding, presents a more efficient alternate approach to carrying out step two of the Peterson–Gorenstein–Zierler Algorithm. This decoding method was developed by Berlekamp in 1967 [18]. Massey [224] recognized that Berlekamp's method provided a way to construct the shortest linear feedback shift-register capable of generating a specified sequence of digits. The third decoding algorithm, discovered by Sugiyama, Kasahara, Hirasawa, and Namekawa in 1975 [324], is also an alternate method to execute the second step of the Peterson–Gorenstein–Zierler Algorithm. Known as the Sugiyama Algorithm, it is a simple, yet powerful, application of the Euclidean Algorithm for polynomials.

In this section we also present the main ideas in a list decoding algorithm which can be applied to decoding generalized Reed–Solomon codes. This algorithm, known as the Sudan–Guruswami Algorithm, will accomplish decoding beyond the packing radius, that is, the bound obtained from the minimum distance of the code. When decoding beyond the packing radius, one must expect more than one nearest codeword to the received vector by Theorem 1.11.4. The Sudan–Guruswami Algorithm produces a complete list of all codewords within a certain distance of the received vector. While we present this algorithm applied to generalized Reed–Solomon codes, it can be used to decode BCH codes, Goppa codes, and algebraic geometry codes with some modifications.

5.4.1 The Peterson–Gorenstein–Zierler Decoding Algorithm

Let C be a BCH code over $\mathbb{F}_q$ of length n and designed distance δ. As the minimum distance of C is at least δ, C can correct at least $t = \lfloor (\delta - 1)/2 \rfloor$ errors. The Peterson–Gorenstein–Zierler Decoding Algorithm will correct up to t errors. While the algorithm will apply to any BCH code, the proofs are simplified if we assume that C is narrow-sense. Therefore the defining set T of C will be assumed to contain $\{1, 2, \ldots, \delta - 1\}$, with α the primitive nth root of unity in the extension field $\mathbb{F}_{q^m}$ of $\mathbb{F}_q$, where $m = \mathrm{ord}_n(q)$, used to determine this defining set. The algorithm requires four steps, which we describe in order and later summarize.

Suppose that $y(x)$ is received, where we assume that $y(x)$ differs from a codeword $c(x)$ in at most t coordinates. Therefore $y(x) = c(x) + e(x)$ where $c(x) \in C$ and $e(x)$ is the *error vector* which has weight $v \leq t$. Suppose that the errors occur in the unknown coordinates $k_1, k_2, \ldots, k_v$. Therefore

$$e(x) = e_{k_1} x^{k_1} + e_{k_2} x^{k_2} + \cdots + e_{k_v} x^{k_v}. \tag{5.5}$$

Once we determine $e(x)$, which amounts to finding the error locations k_j and the error magnitudes e_{k_j}, we can decode the received vector as $c(x) = y(x) - e(x)$. Recall by Theorem 4.4.2 that $c(x) \in C$ if and only if $c(\alpha^i) = 0$ for all $i \in T$. In particular $y(\alpha^i) = c(\alpha^i) + e(\alpha^i) = e(\alpha^i)$ for all $1 \leq i \leq 2t$, since $2t \leq \delta - 1$. For $1 \leq i \leq 2t$ we define the *syndrome* S_i of $y(x)$ to be the element $S_i = y(\alpha^i)$ in $\mathbb{F}_{q^m}$. (Exercise 295 will explore the connection between this notion of syndrome and that developed in Section 1.11.)

The first step in the algorithm is to compute the syndromes $S_i = y(\alpha^i)$ for $1 \le i \le 2t$ from the received vector. This process is aided by the following theorem proved in Exercise 296. In the theorem we allow S_i to be defined as $y(\alpha^i)$ even when $i > 2t$; these may not be legitimate syndromes as $c(\alpha^i)$ may not be 0 in those cases.

Theorem 5.4.1 $S_{iq} = S_i^q$ for all $i \ge 1$.

Exercise 295 Let H be the $t \times n$ matrix

$$H = \begin{bmatrix} 1 & \alpha & \alpha^2 & \cdots & \alpha^{n-1} \\ 1 & \alpha^2 & \alpha^4 & \cdots & \alpha^{(n-1)2} \\ & & \vdots & & \\ 1 & \alpha^t & \alpha^{2t} & \cdots & \alpha^{(n-1)t} \end{bmatrix}.$$

If $y(x) = y_0 + y_1 x + \cdots + y_{n-1} x^{n-1}$, let $\mathbf{y} = (y_0, y_1, \ldots, y_{n-1})$. Finally, let $\mathbf{S} = (S_1, S_2, \ldots, S_t)$, where $S_i = y(\alpha^i)$.
(a) Show that $H\mathbf{y}^{\mathrm{T}} = \mathbf{S}^{\mathrm{T}}$.
(b) Use Theorem 4.4.3 and part (a) to explain the connection between the notion of syndrome given in this section and the notion of syndrome given in Section 1.11. ◆

Exercise 296 Prove Theorem 5.4.1. ◆

The syndromes lead to a system of equations involving the unknown error locations and the unknown error magnitudes. Notice that from (5.5) the syndromes satisfy

$$S_i = y(\alpha^i) = \sum_{j=1}^{\nu} e_{k_j}(\alpha^i)^{k_j} = \sum_{j=1}^{\nu} e_{k_j}(\alpha^{k_j})^i, \tag{5.6}$$

for $1 \le i \le 2t$. To simplify the notation, for $1 \le j \le \nu$, let $E_j = e_{k_j}$ denote the *error magnitude at coordinate* k_j and $X_j = \alpha^{k_j}$ denote the *error location number corresponding to the error location* k_j. By Theorem 3.3.1, if $\alpha^i = \alpha^k$ for i and k between 0 and $n - 1$, then $i = k$. Thus knowing X_j uniquely determines the error location k_j. With this notation (5.6) becomes

$$S_i = \sum_{j=1}^{\nu} E_j X_j^i, \quad \text{for } 1 \le i \le 2t, \tag{5.7}$$

which in turn leads to the system of equations:

$$\begin{aligned} S_1 &= E_1 X_1 + E_2 X_2 + \cdots + E_\nu X_\nu, \\ S_2 &= E_1 X_1^2 + E_2 X_2^2 + \cdots + E_\nu X_\nu^2, \\ S_3 &= E_1 X_1^3 + E_2 X_2^3 + \cdots + E_\nu X_\nu^3, \end{aligned} \tag{5.8}$$

$$\vdots$$

$$S_{2t} = E_1 X_1^{2t} + E_2 X_2^{2t} + \cdots + E_\nu X_\nu^{2t}.$$

This system is nonlinear in the X_js with unknown coefficients E_j. The strategy is to use (5.7) to set up a linear system, involving new variables $\sigma_1, \sigma_2, \ldots, \sigma_\nu$, that will lead directly to the error location numbers. Once these are known, we return to the system (5.8), which is then a linear system in the E_js and solve for the error magnitudes.

To this end, define the *error locator polynomial* to be

$$\sigma(x) = (1 - xX_1)(1 - xX_2)\cdots(1 - xX_\nu) = 1 + \sum_{i=1}^{\nu}\sigma_i x^i.$$

The roots of $\sigma(x)$ are the inverses of the error location numbers and thus

$$\sigma\left(X_j^{-1}\right) = 1 + \sigma_1 X_j^{-1} + \sigma_2 X_j^{-2} + \cdots + \sigma_\nu X_j^{-\nu} = 0$$

for $1 \le j \le \nu$. Multiplying by $E_j X_j^{i+\nu}$ produces

$$E_j X_j^{i+\nu} + \sigma_1 E_j X_j^{i+\nu-1} + \cdots + \sigma_\nu E_j X_j^{i} = 0$$

for any i. Summing this over j for $1 \le j \le \nu$ yields

$$\sum_{j=1}^{\nu} E_j X_j^{i+\nu} + \sigma_1 \sum_{j=1}^{\nu} E_j X_j^{i+\nu-1} + \cdots + \sigma_\nu \sum_{j=1}^{\nu} E_j X_j^{i} = 0. \tag{5.9}$$

As long as $1 \le i$ and $i + \nu \le 2t$, these summations are the syndromes obtained in (5.7). Because $\nu \le t$, (5.9) becomes

$$S_{i+\nu} + \sigma_1 S_{i+\nu-1} + \sigma_2 S_{i+\nu-2} + \cdots + \sigma_\nu S_i = 0$$

or

$$\sigma_1 S_{i+\nu-1} + \sigma_2 S_{i+\nu-2} + \cdots + \sigma_\nu S_i = -S_{i+\nu} \tag{5.10}$$

valid for $1 \le i \le \nu$. Thus we can find the σ_ks if we solve the matrix equation

$$\begin{bmatrix} S_1 & S_2 & S_3 & \cdots & S_{\nu-1} & S_\nu \\ S_2 & S_3 & S_4 & \cdots & S_\nu & S_{\nu+1} \\ S_3 & S_4 & S_5 & \cdots & S_{\nu+1} & S_{\nu+2} \\ & & & \vdots & & \\ S_\nu & S_{\nu+1} & S_{\nu+2} & \cdots & S_{2\nu-2} & S_{2\nu-1} \end{bmatrix} \begin{bmatrix} \sigma_\nu \\ \sigma_{\nu-1} \\ \sigma_{\nu-2} \\ \vdots \\ \sigma_1 \end{bmatrix} = \begin{bmatrix} -S_{\nu+1} \\ -S_{\nu+2} \\ -S_{\nu+3} \\ \vdots \\ -S_{2\nu} \end{bmatrix} \tag{5.11}$$

that arises from (5.10). The second step of our algorithm is to solve (5.11) for $\sigma_1, \ldots, \sigma_\nu$.

Once this second step has been completed, $\sigma(x)$ has been determined. However, determining $\sigma(x)$ is complicated by the fact that we do not know ν, and hence we do not know the size of the system involved. We are searching for the solution which has the smallest value of ν, and this is aided by the following lemma.

Lemma 5.4.2 *Let $\mu \le t$ and let*

$$M_\mu = \begin{bmatrix} S_1 & S_2 & \cdots & S_\mu \\ S_2 & S_3 & \cdots & S_{\mu+1} \\ & & \vdots & \\ S_\mu & S_{\mu+1} & \cdots & S_{2\mu-1} \end{bmatrix}.$$

Then M_μ is nonsingular if $\mu = \nu$ and singular if $\mu > \nu$, where ν is the number of errors that have occurred.

Proof: If $\mu > \nu$, let $X_{\nu+1} = X_{\nu+2} = \cdots = X_\mu = 0$ and $E_{\nu+1} = E_{\nu+2} = \cdots = E_\mu = 0$.
Exercise 297 shows that if A_μ and B_μ are given by

$$
A_\mu = \begin{bmatrix} 1 & 1 & \cdots & 1 \\ X_1 & X_2 & \cdots & X_\mu \\ & & \vdots & \\ X_1^{\mu-1} & X_2^{\mu-1} & \cdots & X_\mu^{\mu-1} \end{bmatrix} \quad \text{and} \quad B_\mu = \begin{bmatrix} E_1 X_1 & 0 & \cdots & 0 \\ 0 & E_2 X_2 & \cdots & 0 \\ & & \vdots & \\ 0 & 0 & \cdots & E_\mu X_\mu \end{bmatrix},
$$

then $M_\mu = A_\mu B_\mu A_\mu^T$. Therefore $\det M_\mu = \det A_\mu \det B_\mu \det A_\mu$. If $\mu > \nu$, $\det B_\mu = 0$ as
B_μ is a diagonal matrix with 0 on the diagonal. If $\mu = \nu$, then $\det B_\mu \neq 0$ as B_μ is a
diagonal matrix with only nonzero entries on the diagonal. Also $\det A_\mu \neq 0$ by Lemma 4.5.1
because A_μ is a Vandermonde matrix with $X_1, \ldots, X_\mu$ distinct. Hence M_μ is nonsingular if
$\mu = \nu$. □

Exercise 297 Do the following, where the notation is given in Lemma 5.4.2:
(a) Show that if $\mu > \nu$, $S_i = \sum_{j=1}^\mu E_j X_j^i$ for $1 \leq i \leq 2t$.
(b) Show that $M_\mu = A_\mu B_\mu A_\mu^T$. ◆

To execute the second step of our algorithm, we attempt to guess the number ν of
errors. Call our guess μ and begin with $\mu = t$, which is the largest that ν could be. The
coefficient matrix of the linear system (5.11) that we are attempting to solve is $M_\mu = M_t$ in
Lemma 5.4.2. If M_μ is singular, we reduce our guess μ to $\mu = t - 1$ and decide whether or
not $M_\mu = M_{t-1}$ is singular. As long as we obtain a singular matrix, we continue to reduce
our guess μ of the number of errors by one until some M_μ is nonsingular. With $\nu = \mu$,
solve (5.11) and thereby determine $\sigma(x)$.

The third step is then to find the roots of $\sigma(x)$ and invert them to determine the error
location numbers. This is usually done by exhaustive search checking $\sigma(\alpha^i)$ for $0 \leq i < n$.
The fourth step is to plug these numbers into (5.8) and solve this linear system for the
error magnitudes E_j. In fact we only need to consider the first ν equations in (5.8) for the
following reason. The coefficient matrix of the first ν equations has determinant

$$
\det \begin{bmatrix} X_1 & X_2 & \cdots & X_\nu \\ X_1^2 & X_2^2 & \cdots & X_\nu^2 \\ & & \vdots & \\ X_1^\nu & X_2^\nu & \cdots & X_\nu^\nu \end{bmatrix} = X_1 X_2 \cdots X_\nu \det \begin{bmatrix} 1 & 1 & \cdots & 1 \\ X_1 & X_2 & \cdots & X_\nu \\ & & \vdots & \\ X_1^{\nu-1} & X_2^{\nu-1} & \cdots & X_\nu^{\nu-1} \end{bmatrix}.
$$

The right-hand side determinant is the determinant of a Vandermonde matrix; the latter is
nonzero as the X_js are distinct.

The *Peterson–Gorenstein–Zierler Decoding Algorithm* for BCH codes is therefore the
following:
I.   Compute the syndromes $S_i = y(\alpha^i)$ for $1 \leq i \leq 2t$.
II. In the order $\mu = t, \mu = t - 1, \ldots$ decide if M_μ is singular, stopping at the first value
     of μ where M_μ is nonsingular. Set $\nu = \mu$ and solve (5.11) to determine $\sigma(x)$.
III. Find the roots of $\sigma(x)$ by computing $\sigma(\alpha^i)$ for $0 \leq i < n$. Invert the roots to get the
     error location numbers X_j.
IV. Solve the first ν equations of (5.8) to obtain the error magnitudes E_j.

We now discuss why this algorithm actually works. We are assuming that a codeword has been transmitted and a vector received that differs from the transmitted codeword in $\nu \le t$ coordinates. Thus there is only one correct set of error location numbers and one correct set of error magnitudes. These lead to a unique error locator polynomial. Step II must determine correctly the value of ν since, by Lemma 5.4.2, ν is the largest value less than or equal to t such that M_ν is nonsingular. Once we know the number of errors, we solve (5.11) to obtain a possible solution for the unknown coefficients of the error locator polynomial. Because the matrix of the linear system used is nonsingular and our correct set of coefficients of the error locator polynomial must also be a solution, these must agree. Thus Step II correctly determines the error locator polynomial and hence Step III correctly determines the error location numbers. Once those are computed, the first ν equations in (5.8) have a unique solution for the error magnitudes that Step IV computes. Because the correct set of error magnitudes also is a solution, it must be the one computed.

What happens if a received vector is more than distance t from every codeword? In that case just about anything could happen. For example, the supposed error locator polynomial $\sigma(x)$ found in Step II may fail in Step III to have deg $\sigma(x)$ distinct roots that are all nth roots of unity. For instance, the roots might be repeated or they might lie in an extension field of $\mathbb{F}_q$ but not be nth roots of unity. If this were to occur at Step III, the decoder should declare that more than t errors have been made. Another problem could occur in Step IV. Suppose an error locator polynomial has been found whose roots are all distinct nth roots of unity and the number of these roots agrees with the degree of the error locator polynomial. Step IV fails if the error magnitudes do not lie in $\mathbb{F}_q$. This is certainly possible since the entries in the coefficient matrix and the syndromes in (5.8) generally lie in an extension field of $\mathbb{F}_q$. Again were this to occur, the decoder should declare that more than t errors have been made.

We make several remarks about this algorithm before presenting some examples.

- After the errors are found and the received vector is corrected, the resulting vector should be checked to make sure it is in the code. (This can be accomplished, for example, either by dividing the corrected vector $c(x)$ by the generator polynomial to verify that the generator polynomial is a factor of the corrected vector, or by computing $c(\alpha^i)$ and verifying that these values are 0 for all i in the defining set.) If it is not, and all steps have been performed correctly, more than t errors occurred.
- If the BCH code is binary, all error magnitudes must be 1. Hence Step IV can be skipped, provided the corrected vector is verified to be in the code, as just remarked.
- If all the syndromes are 0 in Step I, the received vector is in fact a codeword and the received vector should be considered to be the transmitted vector.
- As with all nearest neighbor decoders, the decoder will make a decoding error if the received vector is more than distance t from the transmitted codeword but less than or equal to distance t from some other codeword. The decoder will give the latter codeword as the nearest one, precisely as it is designed to do.
- If the BCH code is not narrow-sense, the algorithm still works as presented.
- In addition to the number of errors, ν is the length of the shortest linear feedback shift-register capable of generating the sequence $S_1, S_2, \ldots$; see (5.10).

Table 5.1 $\mathbb{F}_{16}$ *with primitive element* α, *where*
$\alpha^4 = 1 + \alpha$

0000	0	1000	α^3	1011	α^7	1110	α^{11}
0001	1	0011	α^4	0101	α^8	1111	α^{12}
0010	α	0110	α^5	1010	α^9	1101	α^{13}
0100	α^2	1100	α^6	0111	α^{10}	1001	α^{14}

Example 5.4.3 Let C be the $[15, 7]$ narrow-sense binary BCH code of designed distance $\delta = 5$, which has defining set $T = \{1, 2, 3, 4, 6, 8, 9, 12\}$. Using the primitive 15th root of unity α from Table 5.1, the generator polynomial of C is

$$g(x) = 1 + x^4 + x^6 + x^7 + x^8.$$

Suppose that C is used to transmit a codeword and $y(x) = 1 + x + x^5 + x^6 + x^9 + x^{10}$ is received. Using Table 5.1 and Theorem 5.4.1, Step I produces

$$S_1 = 1 + \alpha + \alpha^5 + \alpha^6 + \alpha^9 + \alpha^{10} = \alpha^2,$$
$$S_2 = S_1^2 = \alpha^4,$$
$$S_3 = 1 + \alpha^3 + \alpha^{15} + \alpha^{18} + \alpha^{27} + \alpha^{30} = \alpha^{11},$$
$$S_4 = S_2^2 = \alpha^8.$$

For Step II, we note that

$$M_2 = \begin{bmatrix} S_1 & S_2 \\ S_2 & S_3 \end{bmatrix} = \begin{bmatrix} \alpha^2 & \alpha^4 \\ \alpha^4 & \alpha^{11} \end{bmatrix}$$

is nonsingular with inverse

$$M_2^{-1} = \begin{bmatrix} \alpha^8 & \alpha \\ \alpha & \alpha^{14} \end{bmatrix}.$$

Thus $\nu = 2$ errors have been made, and we must solve

$$\begin{bmatrix} S_1 & S_2 \\ S_2 & S_3 \end{bmatrix} \begin{bmatrix} \sigma_2 \\ \sigma_1 \end{bmatrix} = \begin{bmatrix} -S_3 \\ -S_4 \end{bmatrix} \quad \text{or} \quad \begin{bmatrix} \alpha^2 & \alpha^4 \\ \alpha^4 & \alpha^{11} \end{bmatrix} \begin{bmatrix} \sigma_2 \\ \sigma_1 \end{bmatrix} = \begin{bmatrix} \alpha^{11} \\ \alpha^8 \end{bmatrix}.$$

The solution is $[\sigma_2 \ \sigma_1]^T = M_2^{-1}[\alpha^{11} \ \alpha^8]^T = [\alpha^{14} \ \alpha^2]^T$. Thus Step II produces the error locator polynomial $\sigma(x) = 1 + \alpha^2 x + \alpha^{14} x^2$. Step III yields the roots α^{11} and α^5 of $\sigma(x)$ and hence the error location numbers $X_1 = \alpha^4$ and $X_2 = \alpha^{10}$. As the code is binary, we skip Step IV. So the error vector is $e(x) = x^4 + x^{10}$, and the transmitted codeword is $c(x) = 1 + x + x^4 + x^5 + x^6 + x^9$, which is $(1 + x)g(x)$. ∎

Example 5.4.4 Let C be the code of Example 5.4.3. Suppose that $y(x) = 1 + x^2 + x^8$ is received. Then Step I produces $S_1 = S_2 = S_4 = 0$ and $S_3 = \alpha^{10}$. In Step II, the matrix

$$M_2 = \begin{bmatrix} S_1 & S_2 \\ S_2 & S_3 \end{bmatrix} = \begin{bmatrix} 0 & 0 \\ 0 & \alpha^{10} \end{bmatrix}$$

is singular, as is $M_1 = [S_1] = [0]$. Since the syndromes are not all 0 and we cannot complete the algorithm, we must conclude that more than two errors were made. ∎

Example 5.4.5 Let C be the binary $[15, 5]$ narrow-sense BCH code of designed distance $\delta = 7$, which has defining set $T = \{1, 2, 3, 4, 5, 6, 8, 9, 10, 12\}$. Using the primitive 15th root of unity α in Table 5.1, the generator polynomial of C is

$$g(x) = 1 + x + x^2 + x^4 + x^5 + x^8 + x^{10}.$$

Suppose that using C, $y(x) = x + x^4 + x^5 + x^7 + x^9 + x^{12}$ is received. Step I produces $S_1 = \alpha^{14}$, $S_2 = \alpha^{13}$, $S_3 = \alpha^{14}$, $S_4 = \alpha^{11}$, $S_5 = 1$, and $S_6 = \alpha^{13}$. The matrix M_3 is singular, and we have

$$M_2 = \begin{bmatrix} S_1 & S_2 \\ S_2 & S_3 \end{bmatrix} = \begin{bmatrix} \alpha^{14} & \alpha^{13} \\ \alpha^{13} & \alpha^{14} \end{bmatrix} \quad \text{and} \quad M_2^{-1} = \begin{bmatrix} \alpha^{10} & \alpha^{9} \\ \alpha^{9} & \alpha^{10} \end{bmatrix}.$$

Then $[\sigma_2 \ \sigma_1]^{\mathrm{T}} = M_2^{-1}[\alpha^{14} \ \alpha^{11}]^{\mathrm{T}} = [\alpha^6 \ \alpha^{14}]^{\mathrm{T}}$. Thus Step II produces the error locator polynomial $\sigma(x) = 1 + \alpha^{14}x + \alpha^6 x^2$. Step III yields the roots α^5 and α^4 of $\sigma(x)$ and hence the error location numbers $X_1 = \alpha^{10}$ and $X_2 = \alpha^{11}$. Skipping Step IV, the error vector is $e(x) = x^{10} + x^{11}$, and the transmitted codeword is $c(x) = x + x^4 + x^5 + x^7 + x^9 + x^{10} + x^{11} + x^{12}$, which is $(x + x^2)g(x)$.   ∎

Example 5.4.6 Let C be the $[15, 9]$ narrow-sense RS code over $\mathbb{F}_{16}$ of designed distance $\delta = 7$, which has defining set $T = \{1, 2, 3, 4, 5, 6\}$. Using the primitive 15th root of unity α in Table 5.1, the generator polynomial of C is

$$g(x) = (\alpha + x)(\alpha^2 + x) \cdots (\alpha^6 + x)$$
$$= \alpha^6 + \alpha^9 x + \alpha^6 x^2 + \alpha^4 x^3 + \alpha^{14} x^4 + \alpha^{10} x^5 + x^6.$$

Suppose that a codeword of C is received as

$$y(x) = \alpha^7 + \alpha^{10} x^2 + x^3 + \alpha^2 x^4 + \alpha^5 x^5 + \alpha^4 x^6 + \alpha^4 x^7 + \alpha^7 x^{11}.$$

Step I produces $S_1 = \alpha^5$, $S_2 = \alpha^7$, $S_3 = \alpha^{10}$, $S_4 = \alpha^5$, $S_5 = \alpha^7$, and $S_6 = \alpha^3$. The matrix M_3 is nonsingular and we have to solve

$$\begin{bmatrix} S_1 & S_2 & S_3 \\ S_2 & S_3 & S_4 \\ S_3 & S_4 & S_5 \end{bmatrix} \begin{bmatrix} \sigma_3 \\ \sigma_2 \\ \sigma_1 \end{bmatrix} = \begin{bmatrix} -S_4 \\ -S_5 \\ -S_6 \end{bmatrix} \quad \text{or} \quad \begin{bmatrix} \alpha^5 & \alpha^7 & \alpha^{10} \\ \alpha^7 & \alpha^{10} & \alpha^5 \\ \alpha^{10} & \alpha^5 & \alpha^7 \end{bmatrix} \begin{bmatrix} \sigma_3 \\ \sigma_2 \\ \sigma_1 \end{bmatrix} = \begin{bmatrix} \alpha^5 \\ \alpha^7 \\ \alpha^3 \end{bmatrix}.$$

The solution is $\sigma_1 = \alpha^5$, $\sigma_2 = \alpha^6$, and $\sigma_3 = \alpha^4$. Thus Step II produces the error locator polynomial $\sigma(x) = 1 + \alpha^5 x + \alpha^6 x^2 + \alpha^4 x^3$. Step III yields the roots α^{13}, α^9, and α^4 of $\sigma(x)$ and hence the error location numbers $X_1 = \alpha^2$, $X_2 = \alpha^6$, and $X_3 = \alpha^{11}$. For Step IV, solve the first three equations of (5.8) or

$$\alpha^5 = E_1 \alpha^2 + E_2 \alpha^6 + E_3 \alpha^{11},$$
$$\alpha^7 = E_1 \alpha^4 + E_2 \alpha^{12} + E_3 \alpha^7,$$
$$\alpha^{10} = E_1 \alpha^6 + E_2 \alpha^3 + E_3 \alpha^3.$$

The solution is $E_1 = 1$, $E_2 = \alpha^3$, and $E_3 = \alpha^7$. Thus the error vector is $e(x) = x^2 + \alpha^3 x^6 + \alpha^7 x^{11}$, and the transmitted codeword is $c(x) = \alpha^7 + \alpha^5 x^2 + x^3 + \alpha^2 x^4 + \alpha^5 x^5 + \alpha^7 x^6 + \alpha^4 x^7$, which is $(\alpha + \alpha^4 x)g(x)$.   ∎

Exercise 298 Verify the calculations in Examples 5.4.3, 5.4.4, 5.4.5, and 5.4.6.   ◆

Exercise 299 The following vectors were received using the BCH code $\mathcal{C}$ of Example 5.4.3. Correct these received vectors:

(a) $y(x) = 1 + x + x^4 + x^5 + x^6 + x^7 + x^{10} + x^{11} + x^{13}$,

(b) $y(x) = x + x^4 + x^7 + x^8 + x^{11} + x^{12} + x^{13}$,

(c) $y(x) = 1 + x + x^5$.   ◆

Exercise 300 The following vectors were received using the BCH code $\mathcal{C}$ of Example 5.4.5. Correct these received vectors:

(a) $y(x) = 1 + x^5 + x^6 + x^7 + x^8 + x^{12} + x^{13}$,

(b) $y(x) = 1 + x + x^2 + x^4 + x^7 + x^8 + x^9 + x^{13}$,

(c) $y(x) = 1 + x + x^2 + x^6 + x^9 + x^{10} + x^{12} + x^{14}$.   ◆

Exercise 301 The following vectors were received using the RS code $\mathcal{C}$ of Example 5.4.6. Correct these received vectors:

(a) $y(x) = \alpha^3 x^3 + \alpha^9 x^4 + \alpha^{14} x^5 + \alpha^{13} x^6 + \alpha^6 x^7 + \alpha^{14} x^8 + \alpha^4 x^9 + \alpha^{12} x^{10} + \alpha^2 x^{11}$,

(b) $y(x) = \alpha^6 + \alpha^9 x + \alpha^4 x^3 + \alpha^{14} x^4 + \alpha^6 x^6 + \alpha^{10} x^7 + \alpha^8 x^8 + \alpha^3 x^9 + \alpha^{14} x^{10} + \alpha^4 x^{11}$.   ◆

How can this algorithm be improved? As stated, this algorithm is quite efficient if the error-correcting capability of the code is rather small. It is not unreasonable to work, even by hand, with 3×3 matrices over finite fields. With computer algebra packages, larger size matrices can be handled. But when the size of these matrices becomes quite large (i.e. when the error-correcting capability of the code is very large), Step II becomes very time consuming. The Berlekamp–Massey Algorithm introduced in the next subsection uses an iterative approach to compute the error locator polynomial in a more efficient manner when t is large. There is another method due to Sugiyama, Kasahava, Hirasawa, and Namekawa [324] that uses the Euclidean Algorithm to find the error locator polynomial; this algorithm is quite comparable in efficiency with the Berlekamp–Massey Algorithm and is described in Section 5.4.3. Step III can also be quite time consuming if the code is long. Little seems to have been done to improve this step although there is a circuit design using Chien search that is often used; see [18] for a description. Step IV can be accomplished using a technique due to Forney [86]; see [21].

5.4.2   The Berlekamp–Massey Decoding Algorithm

The Berlekamp–Massey Decoding Algorithm is a modification of the second step of Peterson–Gorenstein–Zierler decoding. The verification that it works is quite technical and is omitted; readers interested should consult [18, 21, 22]. Although the algorithm applies to BCH codes, it is simplified if the codes are binary, and we will present only that case.

We will adopt the same notation as in the previous subsection. In Step II of the Peterson–Gorenstein–Zierler Algorithm, the error locator polynomial is computed by solving a system of v linear equations in v unknowns, where v is the number of errors made. If v is large, this step is time consuming. For binary codes, the Berlekamp–Massey Algorithm builds the error locator polynomial by requiring that its coefficients satisfy a set of equations called the Newton identities rather than (5.10). These identities hold over general fields provided all error magnitudes are 1, which is precisely the case when the field is $\mathbb{F}_2$. The equations (5.10) are sometimes called generalized Newton identities. The *Newton identities* are:

$$S_1 + \sigma_1 = 0,$$
$$S_2 + \sigma_1 S_1 + 2\sigma_2 = 0,$$
$$S_3 + \sigma_1 S_2 + \sigma_2 S_1 + 3\sigma_3 = 0,$$
$$\vdots$$
$$S_\nu + \sigma_1 S_{\nu-1} + \cdots + \sigma_{\nu-1} S_1 + \nu\sigma_\nu = 0,$$

and for $j > \nu$:

$$S_j + \sigma_1 S_{j-1} + \cdots + \sigma_\nu S_{j-\nu} = 0.$$

A proof that these identities hold is found in [210] or [50, Theorem 3.3]. It turns out that we only need to look at the first, third, fifth, ... of these. For convenience we number these Newton identities (noting that $i\sigma_i = \sigma_i$ when i is odd):

(1) $S_1 + \sigma_1 = 0,$
(2) $S_3 + \sigma_1 S_2 + \sigma_2 S_1 + \sigma_3 = 0,$
(3) $S_5 + \sigma_1 S_4 + \sigma_2 S_3 + \sigma_3 S_2 + \sigma_4 S_1 + \sigma_5 = 0,$
$$\vdots$$
(μ) $S_{2\mu-1} + \sigma_1 S_{2\mu-2} + \sigma_2 S_{2\mu-3} + \cdots + \sigma_{2\mu-2} S_1 + \sigma_{2\mu-1} = 0,$
$$\vdots$$

Define a sequence of polynomials $\sigma^{(\mu)}(x)$ of degree d_μ indexed by μ as follows:

$$\sigma^{(\mu)}(x) = 1 + \sigma_1^{(\mu)} x + \sigma_2^{(\mu)} x^2 + \cdots + \sigma_{d_\mu}^{(\mu)} x^{d_\mu}.$$

The polynomial $\sigma^{(\mu)}(x)$ is calculated to be the minimum degree polynomial whose coefficients $\sigma_1^{(\mu)}, \sigma_2^{(\mu)}, \sigma_3^{(\mu)}, \ldots$ satisfy all of the first μ numbered Newton identities. Associated to each polynomial is its *discrepancy* Δ_μ, which measures how far $\sigma^{(\mu)}(x)$ is from satisfying the $\mu + 1$st identity:

$$\Delta_\mu = S_{2\mu+1} + \sigma_1^{(\mu)} S_{2\mu} + \sigma_2^{(\mu)} S_{2\mu-1} + \cdots + \sigma_{2\mu}^{(\mu)} S_1 + \sigma_{2\mu+1}^{(\mu)}.$$

We start with two initial polynomials, $\sigma^{(-1/2)}(x) = 1$ and $\sigma^{(0)}(x) = 1$, and then generate $\sigma^{(\mu)}(x)$ inductively in a manner that depends on the discrepancy. The discrepancy $\Delta_{-1/2} = 1$ by convention; the remaining discrepancies are calculated. Plugging the coefficients of $\sigma^{(0)}(x)$ into identity (1), we obtain S_1 (as $\sigma_1^{(0)} = 0$) and so the discrepancy of $\sigma^{(0)}(x)$ is $\Delta_0 = S_1$.

We proceed with the first few polynomials to illustrate roughly the ideas involved. Noting the discrepancy $\Delta_0 = S_1$, if $\sigma^{(0)}(x)$ had an additional term $S_1 x$, the coefficients of this polynomial $\sigma^{(0)}(x) + S_1 x = 1 + S_1 x$ would satisfy identity (1) since $S_1 + S_1 = 0$. So $\sigma^{(1)}(x) = 1 + S_1 x$. Plugging the coefficients of $\sigma^{(1)}(x)$ into (2), we have $\Delta_1 = S_3 + \sigma_1^{(1)} S_2 = S_3 + S_1 S_2$. If $\Delta_1 = 0$, then $\sigma^{(1)}(x)$ satisfies (2) also. If $\Delta_1 \neq 0$ and if $S_1 \neq 0$, then setting

$\sigma^{(2)}(x) = \sigma^{(1)}(x) + (S_3 + S_1 S_2)S_1^{-1}x^2 = \sigma^{(1)}(x) + \Delta_1 \Delta_0^{-1}x^2$, we see that this polynomial satisfies (1) and (2). If $\Delta_1 \neq 0$ but $S_1 = 0$, then $\sigma^{(1)}(x) = 1$, $\Delta_1 = S_3$, and the lowest degree polynomial that will satisfy (1) and (2) is $\sigma^{(2)}(x) = \sigma^{(1)}(x) + S_3 x^3 = \sigma^{(1)}(x) + \Delta_1 x^3$. The choices get more complicated as the process continues but, remarkably, the Berlekamp–Massey Algorithm reduces each stage down to one of two choices.

The *Berlekamp–Massey Algorithm* for computing an error locator polynomial for binary BCH codes is the following iterative algorithm that begins with $\mu = 0$ and terminates when $\sigma^{(t)}(x)$ is computed:

I. If $\Delta_\mu = 0$, then

$$\sigma^{(\mu+1)}(x) = \sigma^{(\mu)}(x).$$

II. If $\Delta_\mu \neq 0$, find a value $-(1/2) \leq \rho < \mu$ such that $\Delta_\rho \neq 0$ and $2\rho - d_\rho$ is as large as possible. Then

$$\sigma^{(\mu+1)}(x) = \sigma^{(\mu)}(x) + \Delta_\mu \Delta_\rho^{-1} x^{2(\mu-\rho)}\sigma^{(\rho)}(x).$$

The error locator polynomial is $\sigma(x) = \sigma^{(t)}(x)$; if this polynomial has degree greater than t, more than t errors have been made, and the decoder should declare the received vector is uncorrectable. Once the error locator polynomial is determined, one of course proceeds as in the Peterson–Gorenstein–Zierler Algorithm to complete the decoding. We now reexamine Examples 5.4.3, 5.4.4, and 5.4.5 using the Berlekamp–Massey Algorithm. It is helpful in keeping track of the computations to fill out the following table.

μ	$\sigma^{(\mu)}(x)$	Δ_μ	d_μ	$2\mu - d_\mu$
$-1/2$	1	1	0	-1
0	1	S_1	0	0
1				
$\vdots$				
t				

Example 5.4.7 We recompute $\sigma(x)$ from Example 5.4.3, using Table 5.1. In that example $t = 2$, and the syndromes are $S_1 = \alpha^2$, $S_2 = \alpha^4$, $S_3 = \alpha^{11}$, and $S_4 = \alpha^8$. We obtain the table

μ	$\sigma^{(\mu)}(x)$	Δ_μ	d_μ	$2\mu - d_\mu$
$-1/2$	1	1	0	-1
0	1	α^2	0	0
1	$1 + \alpha^2 x$	α	1	1
2	$1 + \alpha^2 x + \alpha^{14}x^2$			

We explain how this table was obtained. Observe that $\Delta_0 = S_1 = \alpha^2$ and so II is used in computing $\sigma^{(1)}(x)$. We must choose $\rho < 0$ and the only choice is $\rho = -1/2$. So

$$\sigma^{(1)}(x) = \sigma^{(0)}(x) + \Delta_0 \Delta_{-1/2}^{-1} x^{2(0+1/2)}\sigma^{(-1/2)}(x) = 1 + \alpha^2 x.$$

After computing $\Delta_1 = S_3 + \sigma_1^{(1)}S_2 + \sigma_2^{(1)}S_1 + \sigma_3^{(1)} = \alpha^{11} + \alpha^2\alpha^4 = \alpha$, we again use II to find $\sigma^{(2)}(x)$. We must find $\rho < 1$ with $\Delta_\rho \neq 0$ and $2\rho - d_\rho$ as large as possible.

Thus $\rho = 0$ and

$$\sigma^{(2)}(x) = \sigma^{(1)}(x) + \Delta_1 \Delta_0^{-1} x^{2(1-0)} \sigma^{(0)}(x) = 1 + \alpha^2 x + \alpha^{14} x^2.$$

As $t = 2$, $\sigma(x) = \sigma^{(2)}(x) = 1 + \alpha^2 x + \alpha^{14} x^2$, which agrees with the result of Example 5.4.3. ∎

Example 5.4.8 We recompute $\sigma(x)$ from Example 5.4.4 where $t = 2$, $S_1 = S_2 = S_4 = 0$, and $S_3 = \alpha^{10}$. We obtain the table

μ	$\sigma^{(\mu)}(x)$	Δ_μ	d_μ	$2\mu - d_\mu$
$-1/2$	1	1	0	-1
0	1	0	0	0
1	1	α^{10}	0	2
2	$1 + \alpha^{10} x^3$			

Since $\Delta_0 = S_1 = 0$, I is used to compute $\sigma^{(1)}(x)$, yielding $\sigma^{(1)}(x) = \sigma^{(0)}(x) = 1$. Then $\Delta_1 = S_3 + \sigma_1^{(0)} S_2 + \sigma_2^{(0)} S_1 + \sigma_3^{(0)} = \alpha^{10}$. So we use II to find $\sigma^{(2)}(x)$. We must find $\rho < 1$ with $\Delta_\rho \neq 0$ and $2\rho - d_\rho$ as large as possible. Thus $\rho = -1/2$ and

$$\sigma^{(2)}(x) = \sigma^{(1)}(x) + \Delta_1 \Delta_{-1/2}^{-1} x^{2(1+1/2)} \sigma^{(-1/2)}(x) = 1 + \alpha^{10} x^3.$$

So $\sigma(x) = \sigma^{(2)}(x) = 1 + \alpha^{10} x^3$, which has degree greater than $t = 2$; hence the received vector is uncorrectable, which agrees with Example 5.4.4. ∎

Example 5.4.9 We recompute $\sigma(x)$ from Example 5.4.5 where $t = 3$ and the syndromes are $S_1 = \alpha^{14}$, $S_2 = \alpha^{13}$, $S_3 = \alpha^{14}$, $S_4 = \alpha^{11}$, $S_5 = 1$, and $S_6 = \alpha^{13}$. We obtain

μ	$\sigma^{(\mu)}(x)$	Δ_μ	d_μ	$2\mu - d_\mu$
$-1/2$	1	1	0	-1
0	1	α^{14}	0	0
1	$1 + \alpha^{14} x$	α^5	1	1
2	$1 + \alpha^{14} x + \alpha^6 x^2$	0	2	2
3	$1 + \alpha^{14} x + \alpha^6 x^2$			

As $\Delta_0 = S_1 = \alpha^{14}$, II is used to compute $\sigma^{(1)}(x)$. As $\rho < 0$, $\rho = -1/2$ yielding

$$\sigma^{(1)}(x) = \sigma^{(0)}(x) + \Delta_0 \Delta_{-1/2}^{-1} x^{2(0+1/2)} \sigma^{(-1/2)}(x) = 1 + \alpha^{14} x.$$

Since $\Delta_1 = S_3 + \sigma_1^{(1)} S_2 + \sigma_2^{(1)} S_1 + \sigma_3^{(1)} = \alpha^{14} + \alpha^{14} \alpha^{13} = \alpha^5$, we again use II to find $\sigma^{(2)}(x)$. We choose $\rho < 1$ with $\Delta_\rho \neq 0$ and $2\rho - d_\rho$ as large as possible. Thus $\rho = 0$ and

$$\sigma^{(2)}(x) = \sigma^{(1)}(x) + \Delta_1 \Delta_0^{-1} x^{2(1-0)} \sigma^{(0)}(x) = 1 + \alpha^{14} x + \alpha^6 x^2.$$

Now $\Delta_2 = S_5 + \sigma_1^{(2)} S_4 + \sigma_2^{(2)} S_3 + \sigma_3^{(2)} S_2 + \sigma_4^{(2)} S_1 + \sigma_5^{(2)} = 1 + \alpha^{14} \alpha^{11} + \alpha^6 \alpha^{14} = 0$. So to compute $\sigma^{(3)}(x)$, use I to obtain $\sigma^{(3)}(x) = \sigma^{(2)}(x)$; thus $\sigma(x) = 1 + \alpha^{14} x + \alpha^6 x^2$, agreeing with Example 5.4.5. ∎

Exercise 302 Recompute the error locator polynomials from Exercise 299 using the Berlekamp–Massey Algorithm. ◆

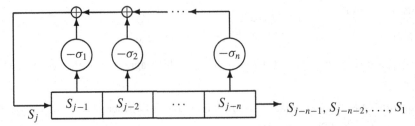

Figure 5.1 Linear feedback shift-register.

Exercise 303 Recompute the error locator polynomials from Exercise 300 using the Berlekamp–Massey Algorithm.   ♦

As stated earlier, this decoding algorithm for BCH codes over arbitrary fields was first developed by Berlekamp in the first edition of [18]. Shortly after, Massey [224] showed that this decoding algorithm actually gives the shortest length recurrence relation which generates the (finite or infinite) sequence $S_1, S_2, \ldots$ whether or not this sequence comes from syndromes. This is the same as the minimum length n of a linear feedback shift-register that generates the entire sequence when $S_1, \ldots, S_n$ is the initial contents of the shift-register. In this context, the algorithm produces a sequence of polynomials $\sigma^{(i)}(x)$ associated with a shift-register which generates $S_1, \ldots, S_i$. The discrepancy Δ_i of $\sigma^{(i)}(x)$ measures how close the shift-register also comes to generating S_{i+1}. If the discrepancy is 0, then the shift-register also generates S_{i+1}. Otherwise, the degree of the polynomial must be increased with a new longer associated shift-register. In the end, the algorithm produces a polynomial $\sigma(x) = 1 + \sum_{i=1}^{n} \sigma_i x^i$, called the connection polynomial, leading to the shift-register of Figure 5.1.

5.4.3   The Sugiyama Decoding Algorithm

The Sugiyama Algorithm is another method to find the error locator polynomial, and thus presents another alternative to complete Step II of the Peterson–Gorenstein–Zierler Algorithm. This algorithm, developed in [324], applies to a class of codes called Goppa codes that include BCH codes as a subclass. This algorithm is a relatively simple, but clever, application of the Euclidean Algorithm. We will only study the algorithm as applied to BCH codes.

Recall that the error locator polynomial $\sigma(x)$ is defined as $\prod_{j=1}^{v}(1 - xX_j)$. The *error evaluator polynomial* $\omega(x)$ is defined as

$$\omega(x) = \sum_{j=1}^{v} E_j X_j \prod_{\substack{i=1 \\ i \neq j}}^{v}(1 - xX_i) = \sum_{j=1}^{v} E_j X_j \frac{\sigma(x)}{1 - xX_j}. \tag{5.12}$$

Note that $\deg(\sigma(x)) = v$ and $\deg(\omega(x)) \leq v - 1$. We define the polynomial $S(x)$ of degree at most $2t - 1$ by

$$S(x) = \sum_{i=0}^{2t-1} S_{i+1} x^i,$$

where S_i for $1 \leq i \leq 2t$ are the syndromes of the received vector.

Expanding the right-hand side of (5.12) in a formal power series and using (5.7) together with the definition of $S(x)$, we obtain

$$\omega(x) = \sigma(x) \sum_{j=1}^{\nu} E_j X_j \frac{1}{1 - xX_j} = \sigma(x) \sum_{j=1}^{\nu} E_j X_j \sum_{i=0}^{\infty} (xX_j)^i$$

$$= \sigma(x) \sum_{i=0}^{\infty} \left(\sum_{j=1}^{\nu} E_j X_j^{i+1} \right) x^i \equiv \sigma(x) \sum_{i=0}^{2t-1} \left(\sum_{j=1}^{\nu} E_j X_j^{i+1} \right) x^i \pmod{x^{2t}}$$

$$\equiv \sigma(x) S(x) \pmod{x^{2t}}.$$

Therefore we have what is termed the *key equation*

$$\omega(x) \equiv \sigma(x) S(x) \pmod{x^{2t}}.$$

Exercise 304   You may wonder if using power series to derive the key equation is legitimate. Give a non-power series derivation. Hint:

$$\sigma(x) \frac{1}{1 - xX_j} = \sigma(x) \frac{1 - x^{2t} X_j^{2t}}{1 - xX_j} + x^{2t} \prod_{\substack{i=1 \\ i \neq j}}^{\nu} (1 - xX_i). \qquad \blacklozenge$$

The following observation about $\sigma(x)$ and $\omega(x)$ will be important later.

Lemma 5.4.10   *The polynomials $\sigma(x)$ and $\omega(x)$ are relatively prime.*

Proof: The roots of $\sigma(x)$ are precisely X_j^{-1} for $1 \leq j \leq \nu$. But

$$\omega\left(X_j^{-1}\right) = E_j X_j \prod_{\substack{i=1 \\ i \neq j}}^{\nu} \left(1 - X_j^{-1} X_i\right) \neq 0,$$

proving the lemma. $\square$

The Sugiyama Algorithm uses the Euclidean Algorithm to solve the key equation. The *Sugiyama Algorithm* is as follows.

I.   Suppose that $f(x) = x^{2t}$ and $s(x) = S(x)$. Set $r_{-1}(x) = f(x)$, $r_0(x) = s(x)$, $b_{-1}(x) = 0$, and $b_0(x) = 1$.

II. Repeat the following two computations finding $h_i(x)$, $r_i(x)$, and $b_i(x)$ inductively for $i = 1, 2, \ldots, I$, until I satisfies $\deg r_{I-1}(x) \geq t$ and $\deg r_I(x) < t$:

$$r_{i-2}(x) = r_{i-1}(x) h_i(x) + r_i(x), \qquad \text{where } \deg r_i(x) < \deg r_{i-1}(x),$$
$$b_i(x) = b_{i-2}(x) - h_i(x) b_{i-1}(x).$$

III. $\sigma(x)$ is some nonzero scalar multiple of $b_I(x)$.

Note that I from Step II is well-defined as $\deg r_i(x)$ is a strictly decreasing sequence with $\deg r_{-1}(x) > t$. In order to prove that the Sugiyama Algorithm works, we need the following lemma.

Lemma 5.4.11   *Using the notation of the Sugiyama Algorithm, let $a_{-1}(x) = 1$, $a_0(x) = 0$, and $a_i(x) = a_{i-2}(x) - h_i(x) a_{i-1}(x)$ for $i \geq 1$. The following hold.*

(i) $a_i(x)f(x) + b_i(x)s(x) = r_i(x)$ for $i \geq -1$.

(ii) $b_i(x)r_{i-1}(x) - b_{i-1}(x)r_i(x) = (-1)^i f(x)$ for $i \geq 0$.

(iii) $a_i(x)b_{i-1}(x) - a_{i-1}(x)b_i(x) = (-1)^{i+1}$ for $i \geq 0$.

(iv) $\deg b_i(x) + \deg r_{i-1}(x) = \deg f(x)$ for $i \geq 0$.

Proof: All of these are proved by induction. For (i), the cases $i = -1$ and $i = 0$ follow directly from the initial values set in Step I of the Sugiyama Algorithm and the values $a_{-1}(x) = 1$ and $a_0(x) = 0$. Assuming (i) holds with i replaced by $i - 1$ and $i - 2$, we have

$$a_i(x)f(x) + b_i(x)s(x) = [a_{i-2}(x) - h_i(x)a_{i-1}(x)]f(x)$$
$$+ [b_{i-2}(x) - h_i(x)b_{i-1}(x)]s(x)$$
$$= a_{i-2}(x)f(x) + b_{i-2}(x)s(x)$$
$$- h_i(x)[a_{i-1}(x)f(x) + b_{i-1}(x)s(x)]$$
$$= r_{i-2}(x) - h_i(x)r_{i-1}(x) = r_i(x),$$

completing (i).

Again when $i = 0$, (ii) follows from Step I of the Sugiyama Algorithm. Assume (ii) holds with i replaced by $i - 1$. Then

$$b_i(x)r_{i-1}(x) - b_{i-1}(x)r_i(x) = [b_{i-2}(x) - h_i(x)b_{i-1}(x)]r_{i-1}(x) - b_{i-1}(x)r_i(x)$$
$$= b_{i-2}(x)r_{i-1}(x) - b_{i-1}(x)[h_i(x)r_{i-1}(x) + r_i(x)]$$
$$= b_{i-2}(x)r_{i-1}(x) - b_{i-1}(x)r_{i-2}(x)$$
$$= -(-1)^{i-1}f(x) = (-1)^i f(x),$$

verifying (ii).

When $i = 0$, (iii) follows from Step I of the Sugiyama Algorithm, $a_{-1}(x) = 1$, and $a_0(x) = 0$. Assume (iii) holds with i replaced by $i - 1$. Then

$$a_i(x)b_{i-1}(x) - a_{i-1}(x)b_i(x) = [a_{i-2}(x) - h_i(x)a_{i-1}(x)]b_{i-1}(x)$$
$$- a_{i-1}(x)[b_{i-2}(x) - h_i(x)b_{i-1}(x)]$$
$$= -[a_{i-1}(x)b_{i-2}(x) - a_{i-2}(x)b_{i-1}(x)]$$
$$= -(-1)^i = (-1)^{i+1},$$

proving (iii).

When $i = 0$, (iv) follows again from Step I of the Sugiyama Algorithm. Assume (iv) holds with i replaced by $i - 1$, that is, $\deg b_{i-1}(x) + \deg r_{i-2}(x) = \deg f(x)$. In Step II of the Sugiyama Algorithm, we have $\deg r_i(x) < \deg r_{i-2}(x)$. So $\deg(b_{i-1}(x)r_i(x)) = \deg b_{i-1}(x) + \deg r_i(x) < \deg f(x)$ implying (iv) for case i using part (ii).   □

We now verify that the Sugiyama Algorithm works. By Lemma 5.4.11(i) we have

$$a_I(x)x^{2t} + b_I(x)S(x) = r_I(x). \tag{5.13}$$

From the key equation, we also know that

$$a(x)x^{2t} + \sigma(x)S(x) = \omega(x) \tag{5.14}$$

for some polynomial $a(x)$. Multiply (5.13) by $\sigma(x)$ and (5.14) by $b_I(x)$ to obtain

$$a_I(x)\sigma(x)x^{2t} + b_I(x)\sigma(x)S(x) = r_I(x)\sigma(x) \quad \text{and} \tag{5.15}$$
$$a(x)b_I(x)x^{2t} + \sigma(x)b_I(x)S(x) = \omega(x)b_I(x). \tag{5.16}$$

Modulo x^{2t} these imply that

$$r_I(x)\sigma(x) \equiv \omega(x)b_I(x) \,(\text{mod } x^{2t}). \tag{5.17}$$

As $\deg \sigma(x) \le t$, by the choice of I, $\deg(r_I(x)\sigma(x)) = \deg r_I(x) + \deg \sigma(x) < t + t = 2t$. By Lemma 5.4.11(iv), the choice of I, and the fact that $\deg \omega(x) < t$, $\deg(\omega(x) \times b_I(x)) = \deg \omega(x) + \deg b_I(x) < t + \deg b_I(x) = t + (\deg x^{2t} - \deg r_{I-1}(x)) \le 3t - t = 2t$. Therefore (5.17) implies that $r_I(x)\sigma(x) = \omega(x)b_I(x)$. This, together with (5.15) and (5.16), shows that

$$a_I(x)\sigma(x) = a(x)b_I(x). \tag{5.18}$$

However, Lemma 5.4.11(iii) implies that $a_I(x)$ and $b_I(x)$ are relatively prime and hence $a(x) = \lambda(x)a_I(x)$ by (5.18). Substituting this into (5.18) produces

$$\sigma(x) = \lambda(x)b_I(x). \tag{5.19}$$

Plugging these into (5.14) we obtain $\lambda(x)a_I(x)x^{2t} + \lambda(x)b_I(x)S(x) = \omega(x)$. Thus (5.13) implies that

$$\omega(x) = \lambda(x)r_I(x). \tag{5.20}$$

By Lemma 5.4.10, (5.19), and (5.20), $\lambda(x)$ must be a nonzero constant, verifying Step III of the Sugiyama Algorithm.

Since we are only interested in the roots of $\sigma(x)$, it suffices to find the roots of $b_I(x)$ produced in Step II; this gives the desired error location numbers.

Example 5.4.12 We obtain a scalar multiple of $\sigma(x)$ from Example 5.4.3, using the Sugiyama Algorithm and Table 5.1. In that example $t = 2$, and the syndromes are $S_1 = \alpha^2$, $S_2 = \alpha^4$, $S_3 = \alpha^{11}$, and $S_4 = \alpha^8$. The following table summarizes the results.

i	$r_i(x)$	$h_i(x)$	$b_i(x)$
-1	x^4		0
0	$\alpha^8 x^3 + \alpha^{11}x^2 + \alpha^4 x + \alpha^2$		1
1	$\alpha x^2 + \alpha^4 x + \alpha^{12}$	$\alpha^7 x + \alpha^{10}$	$\alpha^7 x + \alpha^{10}$
2	α^2	$\alpha^7 x$	$\alpha^{14}x^2 + \alpha^2 x + 1$

The first index I where $\deg r_I(x) < t = 2$ is $I = 2$. Hence $\sigma(x)$ is a multiple of $b_2(x) = \alpha^{14}x^2 + \alpha^2 + 1$; in fact from Example 5.4.3, $b_2(x) = \sigma(x)$. ∎

Example 5.4.13 Using the Sugiyama Algorithm we examine Example 5.4.4 where $t = 2$, $S_1 = S_2 = S_4 = 0$, and $S_3 = \alpha^{10}$. The following table summarizes the computations.

i	$r_i(x)$	$h_i(x)$	$b_i(x)$
-1	x^4		0
0	$\alpha^{10}x^2$		1
1	0	α^5x^2	α^5x^2

The first index I where $\deg r_I(x) < t = 2$ is $I = 1$. But in this case $b_1(x) = \alpha^5x^2$, which has 0 for its roots indicating that more than two errors were made, in agreement with Example 5.4.4. Note also that $r_1(x) = 0$ implies by (5.20) that $\omega(x) = 0$, which is obviously impossible as $\sigma(x)$ and $\omega(x)$ are supposed to be relatively prime by Lemma 5.4.10. ∎

Example 5.4.14 We obtain a scalar multiple of $\sigma(x)$ from Example 5.4.5 using the Sugiyama Algorithm. Here $t = 3$ and $S_1 = \alpha^{14}$, $S_2 = \alpha^{13}$, $S_3 = \alpha^{14}$, $S_4 = \alpha^{11}$, $S_5 = 1$, and $S_6 = \alpha^{13}$. The following table summarizes the results.

i	$r_i(x)$	$h_i(x)$	$b_i(x)$
-1	x^6		0
0	$\alpha^{13}x^5 + x^4 + \alpha^{11}x^3 + \alpha^{14}x^2 + \alpha^{13}x + \alpha^{14}$		1
1	$\alpha^{11}x^4 + \alpha^4x^3 + \alpha^{14}x^2 + \alpha^5x + \alpha^3$	$\alpha^2x + \alpha^4$	$\alpha^2x + \alpha^4$
2	α^{12}	$\alpha^2x + \alpha^2$	$\alpha^4x^2 + \alpha^{12}x + \alpha^{13}$

The first index I where $\deg r_I(x) < t = 3$ is $I = 2$. Hence $\sigma(x)$ is a multiple of $b_2(x) = \alpha^4x^2 + \alpha^{12}x + \alpha^{13}$. From Example 5.4.5, $\sigma(x) = \alpha^2b_2(x)$. ∎

Exercise 305 Verify the calculations in Examples 5.4.12, 5.4.13, and 5.4.14. ◆

Exercise 306 Using the Sugiyama Algorithm, find a scalar multiple of the error locator polynomial from Example 5.4.6. ◆

Exercise 307 Using the Sugiyama Algorithm, find scalar multiples of the error locator polynomials from Exercise 299. ◆

Exercise 308 Using the Sugiyama Algorithm, find scalar multiples of the error locator polynomials from Exercise 300. ◆

Exercise 309 Using the Sugiyama Algorithm, find scalar multiples of the error locator polynomials from Exercise 301. ◆

We remark that the Sugiyama Algorithm applies with other choices for $f(x)$ and $s(x)$, with an appropriate modification of the condition under which the algorithm stops in Step II. Such a modification works for decoding Goppa codes; see [232].

It is worth noting that the Peterson–Gorenstein–Zierler, the Berlekamp–Massey, or the Sugiyama Algorithm can be used to decode any cyclic code up to the BCH Bound. Let $\mathcal{C}$ be a cyclic code with defining set T and suppose that T contains δ consecutive elements $\{b, b+1, \ldots, b+\delta-2\}$. Let $\mathcal{B}$ be the BCH code with defining set $C_b \cup C_{b+1} \cup \cdots \cup C_{b+\delta-2}$, which is a subset of T. By Exercise 239, $\mathcal{C} \subseteq \mathcal{B}$. Let $t = \lfloor (\delta - 1)/2 \rfloor$. Suppose that a codeword $c(x) \in \mathcal{C}$ is transmitted and $y(x)$ is received where t or fewer errors have been made. Then $c(x) \in \mathcal{B}$ and any of the decoding algorithms applied to $\mathcal{B}$ will correct $y(x)$ and produce $c(x)$. Thus these algorithms will correct a received word in any cyclic code provided that if v errors are made, $2v + 1$ does not exceed the BCH Bound of the code. Of course this number of errors may be less than the actual number of errors that $\mathcal{C}$ is capable of correcting.

5.4.4   The Sudan–Guruswami Decoding Algorithm

In a 1997 paper Madhu Sudan [323] developed a procedure for decoding $[n, k, d]$ Reed–Solomon codes that is capable of correcting some e errors where $e > \lfloor (d - 1)/2 \rfloor$. This method was extended by Guruswami and Sudan [113] to remove certain restrictions in the original Sudan Algorithm. To be able to correct e errors where $e > \lfloor (d - 1)/2 \rfloor$, the algorithm produces a list of all possible codewords within Hamming distance e of any received vector; such an algorithm is called a *list-decoding algorithm*. The Sudan–Guruswami Algorithm applies to generalized Reed–Solomon codes as well as certain BCH and algebraic geometry codes. In this section we present this algorithm for generalized Reed–Solomon codes and refer the interested reader to [113] for the other codes. The Sudan–Guruswami Algorithm has itself been generalized by Kötter and Vardy [179] to apply to soft decision decoding.

To prepare for the algorithm we need some preliminary notation involving polynomials in two variables. Suppose x and y are independent indeterminates and $p(x, y) = \sum_i \sum_j p_{i,j} x^i y^j$ is a polynomial in $\mathbb{F}_q[x, y]$, the ring of all polynomials in the two variables x and y. Let w_x and w_y be nonnegative real numbers. The (w_x, w_y)-*weighted degree* of $p(x, y)$ is defined to be

$$\max\{w_x i + w_y j \mid p_{i,j} \neq 0\}.$$

Notice that the $(1, 1)$-weighted degree of $p(x, y)$ is merely the degree of $p(x, y)$. For positive integers s and δ, let $N_s(\delta)$ denote the number of monomials $x^i y^j$ whose $(1, s)$-weighted degree is δ or less. We say that the point $(\alpha, \beta) \in \mathbb{F}_q^2$ *lies on* or *is a root of* $p(x, y)$ provided $p(\alpha, \beta) = 0$. We will need the multiplicity of this root. To motivate the definition of multiplicity, recall that if $f(x) \in \mathbb{F}_q[x]$ and α is a root of $f(x)$, then its multiplicity as a root of $f(x)$ is the number m where $f(x) = (x - \alpha)^m g(x)$ for some $g(x) \in \mathbb{F}_q[x]$ with $g(\alpha) \neq 0$. When working with two variables we cannot generalize this notion directly. Notice, however, that $f(x + \alpha) = x^m h(x)$, where $h(x) = g(x + \alpha)$; also $h(0) \neq 0$. In particular $f(x + \alpha)$ contains a monomial of degree m but none of smaller degree. This concept can be generalized. The root (α, β) of the polynomial $p(x, y)$ has *multiplicity* m provided the shifted polynomial $p(x + \alpha, y + \beta)$ contains a monomial of degree m but no monomial of lower degree.

Exercise 310 Let $p(x, y) = 1 + x + y - x^2 - y^2 - 2x^2 y + xy^2 - y^3 + x^4 - 2x^3 y - x^2 y^2 + 2xy^3 \in \mathbb{F}_5[x]$. Show that $(1, 2) \in \mathbb{F}_5^2$ is a root of $p(x, y)$ with multiplicity 3.   ◆

Recall that an $[n, k]$ generalized Reed–Solomon code over $\mathbb{F}_q$ is defined by

$$\mathrm{GRS}_k(\boldsymbol{\gamma}, \mathbf{v}) = \{(v_0 f(\gamma_0), v_1 f(\gamma_1), \dots, v_{n-1} f(\gamma_{n-1})) \mid f \in \mathcal{P}_k\},$$

where $\boldsymbol{\gamma} = (\gamma_0, \gamma_1, \dots, \gamma_{n-1})$ is an n-tuple of distinct elements of $\mathbb{F}_q$, $\mathbf{v} = (v_0, v_1, \dots, v_{n-1})$ is an n-tuple of nonzero elements of $\mathbb{F}_q$, and $\mathcal{P}_k$ is the set of polynomials in $\mathbb{F}_q[x]$ of degree $k - 1$ or less including the zero polynomial. Suppose that $\mathbf{c} = c_0 c_1 \cdots c_{n-1} \in \mathrm{GRS}_k(\boldsymbol{\gamma}, \mathbf{v})$ is sent and $\mathbf{y}' = y_0' y_1' \cdots y_{n-1}' = \mathbf{c} + \mathbf{e}$ is received. Then there is a unique $f \in \mathcal{P}_k$ such that $c_i = v_i f(\gamma_i)$ for $0 \le i \le n - 1$. We can find $\mathbf{c}$ if we can determine the polynomial f. Let $\mathcal{A} = \{(\gamma_0, y_0), (\gamma_1, y_1), \dots, (\gamma_{n-1}, y_{n-1})\}$ where $y_i = y_i'/v_i$. Suppose for a moment that no errors occurred in the transmission of $\mathbf{c}$. Then $y_i = c_i/v_i = f(\gamma_i)$ for $0 \le i \le n - 1$. In particular, all points of $\mathcal{A}$ lie on the polynomial $p(x, y) = y - f(x)$. Now suppose that errors occur. Varying slightly our terminology from earlier, define an *error locator polynomial* $\Lambda(x, y)$ to be any polynomial in $\mathbb{F}_q[x, y]$ such that $\Lambda(\gamma_i, y_i) = 0$ for all i such that $y_i \ne c_i/v_i$. Since $y_i - f(\gamma_i) = 0$ if $y_i = c_i/v_i$, all points of $\mathcal{A}$ lie on the polynomial $p(x, y) = \Lambda(x, y)(y - f(x))$. The basic idea of the Sudan–Guruswami Algorithm is to find a polynomial $p(x, y) \in \mathbb{F}_q[x, y]$ where each element of $\mathcal{A}$ is a root with a certain multiplicity and then find the factors of that polynomial of the form $y - f(x)$. Further restrictions on $p(x, y)$ are imposed to guarantee the error-correcting capability of the algorithm.

The *Sudan–Guruswami Decoding Algorithm* for the $[n, k, n - k + 1]$ code $\mathrm{GRS}_k(\boldsymbol{\gamma}, \mathbf{v})$ is:

I. Fix a positive integer m. Pick δ to be the smallest positive integer to satisfy

$$\frac{nm(m + 1)}{2} < N_{k-1}(\delta). \tag{5.21}$$

Recall that $N_{k-1}(\delta)$ is the number of monomials $x^i y^j$ whose $(1, k - 1)$-weighted degree is δ or less. Set

$$t = \left\lfloor \frac{\delta}{m} \right\rfloor + 1.$$

II. Construct a nonzero polynomial $p(x, y) \in \mathbb{F}_q[x, y]$ such that each element of $\mathcal{A}$ is a root of $p(x, y)$ of multiplicity at least m and $p(x, y)$ has $(1, k - 1)$-weighted degree at most δ.

III. Find all factors of $p(x, y)$ of the form $y - f(x)$ where $f(x) \in \mathcal{P}_k$ and $f(\gamma_i) = y_i$ for at least t γ_is. For each such f produce the corresponding codeword in $\mathrm{GRS}_k(\boldsymbol{\gamma}, \mathbf{v})$.

We must verify that this algorithm works and give a bound on the number of errors that it will correct. The following three lemmas are needed.

Lemma 5.4.15 *Let* $(\alpha, \beta) \in \mathbb{F}_q^2$ *be a root of* $p(x, y) \in \mathbb{F}_q[x, y]$ *of multiplicity m or more. If* $f(x)$ *is a polynomial in* $\mathbb{F}_q[x]$ *such that* $f(\alpha) = \beta$, *then* $g(x) = p(x, f(x)) \in \mathbb{F}_q[x]$ *is divisible by* $(x - \alpha)^m$.

Proof: Let $f_1(x) = f(x + \alpha) - \beta$. By our hypothesis, $f_1(0) = 0$, and so $f_1(x) = x f_2(x)$ for some polynomial $f_2(x) \in \mathbb{F}_q[x]$. Define $g_1(x) = p(x + \alpha, f_1(x) + \beta)$. Since (α, β) is a root of $p(x, y)$ of multiplicity m or more, $p(x + \alpha, y + \beta)$ has no monomial of degree less than m. Setting $y = f_1(x) = x f_2(x)$ in $p(x + \alpha, y + \beta)$ shows that $g_1(x)$ is divisible by x^m, which implies that $g_1(x - \alpha)$ is divisible by $(x - \alpha)^m$. However,

$$g_1(x - \alpha) = p(x, f_1(x - \alpha) + \beta) = p(x, f(x)) = g(x),$$

showing that $g(x)$ is divisible by $(x - \alpha)^m$.   □

Lemma 5.4.16 *Fix positive integers $m, t,$ and δ such that $mt > \delta$. Let $p(x, y) \in \mathbb{F}_q[x, y]$ be a polynomial such that (γ_i, y_i) is a root of $p(x, y)$ of multiplicity at least m for $0 \leq i \leq n - 1$. Furthermore, assume that $p(x, y)$ has $(1, k - 1)$-weighted degree at most δ. Let $f(x) \in \mathcal{P}_k$ where $y_i = f(\gamma_i)$ for at least t values of i with $0 \leq i \leq n - 1$. Then $y - f(x)$ divides $p(x, y)$.*

Proof: Let $g(x) = p(x, f(x))$. As $p(x, y)$ has $(1, k - 1)$-weighted degree at most δ, $g(x)$ is either the zero polynomial or a nonzero polynomial of degree at most δ. Assume $g(x)$ is nonzero. Let $S = \{i \mid 0 \leq i \leq n - 1, \ f(\gamma_i) = y_i\}$. By Lemma 5.4.15, $(x - \gamma_i)^m$ divides $g(x)$ for $i \in S$. As the γ_is are distinct, $h(x) = \prod_{i \in S}(x - \gamma_i)^m$ divides $g(x)$. Since $h(x)$ has degree at least $mt > \delta$ and δ is the maximum degree of $g(x)$, we have a contradiction if $g(x)$ is nonzero. Thus $g(x)$ is the zero polynomial, which implies that $y = f(x)$ is a root of $p(x, y)$ viewed as a polynomial in y over the field of rational functions in x. By the Division Algorithm, $y - f(x)$ is a factor of this polynomial.   □

Lemma 5.4.17 *Let $p(x, y) = \sum_j \sum_\ell p_{j,\ell} x^j y^\ell \in \mathbb{F}_q[x, y]$. Suppose that $(\alpha, \beta) \in \mathbb{F}_q^2$ and that $p'(x, y) = \sum_a \sum_b p'_{a,b} x^a y^b = p(x + \alpha, y + \beta)$. Then*

$$p'_{a,b} = \sum_{j \geq a} \sum_{\ell \geq b} \binom{j}{a} \binom{\ell}{b} \alpha^{j-a} \beta^{\ell-b} p_{j,\ell}.$$

Proof: We have

$$p'(x, y) = \sum_j \sum_\ell p_{j,\ell}(x + \alpha)^j (y + \beta)^\ell$$

$$= \sum_j \sum_\ell p_{j,\ell} \sum_{a=0}^{j} \binom{j}{a} x^a \alpha^{j-a} \sum_{b=0}^{\ell} \binom{\ell}{b} y^b \beta^{\ell-b}.$$

Clearly, the coefficient $p'_{a,b}$ of $x^a y^b$ is as claimed.   □

We are now in a position to verify the Sudan–Guruswami Algorithm and give the error bound for which the algorithm is valid.

Theorem 5.4.18 *The Sudan–Guruswami Decoding Algorithm applied to the $[n, k, n - k + 1]$ code $\mathrm{GRS}_k(\gamma, v)$ will produce all codewords within Hamming distance e or less of a received vector where $e = n - \lfloor \delta/m \rfloor - 1$.*

Proof: We first must verify that the polynomial $p(x, y)$ from Step II actually exists. For $p(x, y)$ to exist, $p(x + \gamma_i, y + y_i)$ must have no terms of degree less than m for $0 \leq i \leq n - 1$. By Lemma 5.4.17, this is accomplished if for each i with $0 \leq i \leq n - 1$,

$$\sum_{j \geq a} \sum_{\ell \geq b} \binom{j}{a} \binom{\ell}{b} \gamma_i^{j-a} y_i^{\ell-b} p_{j,\ell} = 0 \quad \text{for all } a \geq 0, \ b \geq 0 \text{ with } a + b < m. \tag{5.22}$$

For each i, there are $(m(m+1))/2$ equations in (5.22) since the set $\{(a,b) \in \mathbb{Z}^2 \mid a \geq 0, \ b \geq 0, \ a+b < m\}$ has size $(m(m+1))/2$; hence there are a total of $(nm(m+1))/2$ homogeneous linear equations in the unknown coefficients $p_{j,\ell}$. Since we wish to produce a nontrivial polynomial of $(1, k-1)$-weighted degree at most δ, there are a total of $N_{k-1}(\delta)$ unknown coefficients $p_{j,\ell}$ in this system of $(nm(m+1))/2$ homogeneous linear equations. As there are fewer equations than unknowns by (5.21), a nontrivial solution exists and Step II can be completed. By our choice of t in Step I, $mt > \delta$. If $f(x) \in \mathcal{P}_k$ is a polynomial with $f(\gamma_i) = y_i$ for at least t values of i, by Lemma 5.4.16, $y - f(x)$ is a factor of $p(x,y)$. Thus Step III of the algorithm will produce all codewords at distance $e = n - t = n - \lfloor \delta/m \rfloor - 1$ or less from the received vector.   □

As Step I requires computation of $N_{k-1}(\delta)$, the next lemma proves useful.

Lemma 5.4.19 *Let s and δ be positive integers. Then*

$$N_s(\delta) = \left(\delta + 1 - \frac{s}{2}\left\lfloor \frac{\delta}{s} \right\rfloor\right)\left(\left\lfloor \frac{\delta}{s} \right\rfloor + 1\right) \geq \frac{\delta(\delta+2)}{2s}.$$

Proof: By definition,

$$N_s(\delta) = \sum_{i=0}^{\lfloor \frac{\delta}{s} \rfloor} \sum_{j=0}^{\delta - is} 1 = \sum_{i=0}^{\lfloor \frac{\delta}{s} \rfloor}(\delta + 1 - is)$$

$$= (\delta + 1)\left(\left\lfloor \frac{\delta}{s} \right\rfloor + 1\right) - \frac{s}{2}\left\lfloor \frac{\delta}{s} \right\rfloor\left(\left\lfloor \frac{\delta}{s} \right\rfloor + 1\right)$$

$$\geq \left(\left\lfloor \frac{\delta}{s} \right\rfloor + 1\right)\left(\delta + 1 - \frac{\delta}{2}\right) \geq \frac{\delta}{s} \cdot \frac{\delta + 2}{2}.$$

The result follows.   □

Example 5.4.20 Let $\mathcal{C}$ be a $[15, 6, 10]$ Reed–Solomon code over $\mathbb{F}_{16}$. The Peterson–Gorenstein–Zierler Decoding Algorithm can correct up to four errors. If we choose $m = 2$ in the Sudan–Guruswami Decoding Algorithm, then $(nm(m+1))/2 = 45$ and the smallest value of δ for which $45 < N_5(\delta)$ is $\delta = 18$, in which case $N_5(18) = 46$ by Lemma 5.4.19. Then $t = \lfloor \delta/m \rfloor + 1 = 10$ and by Theorem 5.4.18, the Sudan–Guruswami Algorithm can correct $15 - 10 = 5$ errors.   ∎

Exercise 311 Let $\mathcal{C}$ be the code of Example 5.4.20. Choose $m = 6$ in the Sudan–Guruswami Algorithm. Show that the smallest value of δ for which $(nm(m+1))/2 = 315 < N_5(\delta)$ is $\delta = 53$. Verify that the Sudan–Guruswami Algorithm can correct six errors with these parameters.   ◆

Exercise 312 Let $\mathcal{C}$ be a $[31, 8, 24]$ Reed–Solomon code over $\mathbb{F}_{32}$. The Peterson–Gorenstein–Zierler Algorithm can correct up to 11 errors.
(a) Choose $m = 1$ in the Sudan–Guruswami Algorithm. Find the smallest value of δ for which $(nm(m+1))/2 = 31 < N_7(\delta)$. Using $m = 1$, how many errors can the Sudan–Guruswami Algorithm correct?

(b) Choose $m = 2$ in the Sudan–Guruswami Algorithm. Find the smallest value of δ for which $(nm(m + 1))/2 = 93 < N_7(\delta)$. Using $m = 2$, how many errors can the Sudan–Guruswami Algorithm correct?

(c) Choose $m = 3$ in the Sudan–Guruswami Algorithm. Find the smallest value of δ for which $(nm(m + 1))/2 = 186 < N_7(\delta)$. Using $m = 3$, how many errors can the Sudan–Guruswami Algorithm correct?   ♦

As can be seen in Example 5.4.20 and Exercises 311 and 312, the error-correcting capability of the Sudan–Guruswami Decoding Algorithm can grow if m is increased. The tradeoff for higher error-correcting capability is an increase in the $(1, k - 1)$-weighted degree of $p(x, y)$, which of course increases the complexity of the algorithm. The following corollary gives an idea of how the error-correcting capability varies with m.

Corollary 5.4.21 *The Sudan–Guruswami Decoding Algorithm applied to the $[n, k, n - k + 1]$ code $\mathrm{GRS}_k(\gamma, \mathbf{v})$ will produce all codewords within Hamming distance e or less of a received vector where $e \geq n - 1 - n\sqrt{R(m + 1)/m}$ and $R = k/n$.*

Proof: As δ is chosen to be the smallest positive integer such that (5.21) holds,

$$N_{k-1}(\delta - 1) \leq \frac{nm(m + 1)}{2}.$$

By Lemma 5.4.19, $N_{k-1}(\delta - 1) \geq (\delta - 1)(\delta + 1)/(2(k - 1))$. Hence

$$\frac{\delta^2 - 1}{2(k - 1)} \leq \frac{nm(m + 1)}{2}.$$

If $\delta^2 < k$, then $\delta^2/(2k) < 1/2 \leq nm(m + 1)/2$. If $\delta^2 \geq k$, then $\delta^2/(2k) \leq (\delta^2 - 1)/(2(k - 1))$ by Exercise 313. In either case,

$$\frac{\delta^2}{2k} \leq \frac{nm(m + 1)}{2},$$

implying

$$\frac{\delta}{m} \leq n\sqrt{\frac{k}{n} \cdot \frac{m + 1}{m}},$$

which produces the desired result from Theorem 5.4.18.   □

Exercise 313 Do the following:

(a) Show that if $\delta^2 \geq k$, then $\delta^2/(2k) \leq (\delta^2 - 1)/(2(k - 1))$.

(b) Show that if $\delta^2/(2k) \leq nm(m + 1)/2$, then $\delta/m \leq n\sqrt{(k/n)((m + 1)/m)}$.   ♦

In this corollary, $R = k/n$ is the information rate. If m is large, we see that the fraction e/n of errors that the Sudan–Guruswami Decoding Algorithm can correct is approximately $1 - \sqrt{R}$. The fraction of errors that the Peterson–Gorenstein–Zierler Decoding Algorithm can correct is approximately $(1 - R)/2$. Exercise 314 explores the relationship between these two functions.

Exercise 314 Do the following:

(a) Verify that the fraction of errors that the Peterson–Gorenstein–Zierler Decoding Algorithm can correct in a GRS code is approximately $(1 - R)/2$.

(b) Plot the two functions $y = 1 - \sqrt{R}$ and $y = (1 - R)/2$ for $0 \le R \le 1$ on the same graph. What do these graphs show about the comparative error-correcting capability of the Peterson–Gorenstein–Zierler and the Sudan–Guruswami Decoding Algorithms? ◆

To carry out the Sudan–Guruswami Decoding Algorithm we must have a method to compute the polynomial $p(x, y)$ of Step II and then find the factors of $p(x, y)$ of the form $y - f(x)$ in Step III. (Finding $p(x, y)$ can certainly be accomplished by solving the $nm(m + 1)/2$ equations from (5.22), but as the values in Exercise 312 indicate, the number of equations and unknowns gets rather large rather quickly.) A variety of methods have been introduced to carry out Steps II and III. We will not examine these methods here, but the interested reader can consult [11, 178, 245, 248, 300, 354].

5.5   Burst errors, concatenated codes, and interleaving

Reed–Solomon codes, used in concatenated form, are very useful in correcting burst errors. As the term implies, a *burst error* occurs when several consecutive components of a codeword may be in error; such a burst often extends over several consecutive codewords which are received in sequence.

Before giving the actual details, we illustrate the process. Suppose that C is an $[n, k]$ binary code being used to transmit information. Each message from $\mathbb{F}_2^k$ is encoded to a codeword from $\mathbb{F}_2^n$ using C. The message is transmitted then as a sequence of n binary digits. In reality, several codewords are sent one after the other, which then appear to the receiver as a very long string of binary digits. Along the way these digits may have been changed. A random individual symbol may have been distorted so that one cannot recognize it as either 0 or 1, in which case the received symbol is considered erased. Or a random symbol could be changed into another symbol and the received symbol is in error. As we discussed in Section 1.11, more erasures than errors can be corrected because error locations are unknown, whereas erasure locations are known. Sometimes several consecutive symbols, a burst, may have been erased or are in error. The receiver then breaks up the string into codewords of length n and decodes each string, if possible. However, the presence of burst errors can make decoding problematic as the codes we have developed are designed to correct random errors. However, we can modify our codes to also handle bursts. An example where the use of coding has made a significant impact is in compact disc recording; a scratch across the disc can destroy several consecutive bits of information. The ability to correct burst errors has changed the entire audio industry. We will take this up in the next section.

Burst errors are often handled using concatenated codes, which are sometimes then interleaved. Concatenated codes were introduced by Forney in [87]. We give a simple version of concatenated codes; the more general theory can be found, for example, in [75].

Let $\mathcal{A}$ be an $[n, k, d]$ code over $\mathbb{F}_q$. Let $Q = q^k$ and define $\psi : \mathbb{F}_Q \to \mathcal{A}$ to be a one-to-one $\mathbb{F}_q$-linear map; that is $\psi(x + y) = \psi(x) + \psi(y)$ for all x and y in $\mathbb{F}_Q$, and $\psi(\alpha x) = \alpha \psi(x)$ for all $x \in \mathbb{F}_Q$ and $\alpha \in \mathbb{F}_q$, noting that $\mathbb{F}_Q$ is an extension field of $\mathbb{F}_q$. Let $\mathcal{B}$ be an $[N, K, D]$ code over $\mathbb{F}_Q$. The *concatenation* of $\mathcal{A}$ and $\mathcal{B}$ is the code

$$\mathcal{C} = \{\psi(b_1, b_2, \ldots, b_N) \mid (b_1, b_2, \ldots, b_N) \in \mathcal{B}\},$$

where $\psi(b_1, b_2, \ldots, b_N) = (\psi(b_1), \psi(b_2), \ldots, \psi(b_N))$. $\mathcal{C}$ is called a *concatenated code* with *inner code* $\mathcal{A}$ and *outer code* $\mathcal{B}$. In effect, a codeword in $\mathcal{C}$ is obtained by taking a codeword in $\mathcal{B}$ and replacing each component by a codeword of $\mathcal{A}$ determined by the image of that component under ψ. The code $\mathcal{C}$ is then a code of length nN over $\mathbb{F}_q$. The following theorem gives further information about $\mathcal{C}$.

Theorem 5.5.1 *Let $\mathcal{A}$, $\mathcal{B}$, and $\mathcal{C}$ be as above. Then $\mathcal{C}$ is a linear $[nN, kK]$ code over $\mathbb{F}_q$ whose minimum distance is at least dD.*

Exercise 315 Prove Theorem 5.5.1. ◆

Example 5.5.2 Let $\mathcal{B}$ be the $[6, 3, 4]$ hexacode over $\mathbb{F}_4$ with generator matrix

$$G = \begin{bmatrix} 1 & 0 & 0 & 1 & \omega & \omega \\ 0 & 1 & 0 & \omega & 1 & \omega \\ 0 & 0 & 1 & \omega & \omega & 1 \end{bmatrix}.$$

Let $\mathcal{A}$ be the $[2, 2, 1]$ binary code $\mathbb{F}_2^2$ and define $\psi : \mathbb{F}_4 \to \mathcal{A}$ by the following:

$$\psi(0) = 00, \quad \psi(1) = 10, \quad \psi(\omega) = 01, \quad \psi(\overline{\omega}) = 11.$$

This is $\mathbb{F}_2$-linear, as can be easily verified. The concatenated code $\mathcal{C}$ with inner code $\mathcal{A}$ and outer code $\mathcal{B}$ has generator matrix

$$\begin{bmatrix} 1 & 0 & 0 & 0 & 0 & 0 & 1 & 0 & 0 & 1 & 0 & 1 \\ 0 & 1 & 0 & 0 & 0 & 0 & 0 & 1 & 1 & 1 & 1 & 1 \\ 0 & 0 & 1 & 0 & 0 & 0 & 0 & 1 & 1 & 0 & 0 & 1 \\ 0 & 0 & 0 & 1 & 0 & 0 & 1 & 1 & 0 & 1 & 1 & 1 \\ 0 & 0 & 0 & 0 & 1 & 0 & 0 & 1 & 0 & 1 & 1 & 0 \\ 0 & 0 & 0 & 0 & 0 & 1 & 1 & 1 & 1 & 1 & 0 & 1 \end{bmatrix}.$$

Rows two, four, and six of this matrix are obtained after multiplying the rows of G by ω. $\mathcal{C}$ is a $[12, 6, 4]$ code. ∎

Exercise 316 Let $\mathcal{B}$ be the hexacode given in Example 5.5.2, and let $\mathcal{A}$ be the $[3, 2, 2]$ even binary code. Define $\psi : \mathbb{F}_4 \to \mathcal{A}$ by the following:

$$\psi(0) = 000, \quad \psi(1) = 101, \quad \psi(\omega) = 011, \quad \psi(\overline{\omega}) = 110.$$

(a) Verify that ψ is $\mathbb{F}_2$-linear.
(b) Give a generator matrix for the concatenated code $\mathcal{C}$ with inner code $\mathcal{A}$ and outer code $\mathcal{B}$.
(c) Show that $\mathcal{C}$ is an $[18, 6, 8]$ code. ◆

We now briefly discuss how codes, such as Reed–Solomon codes, can be used to correct burst errors. Let C be an $[n, k]$ code over $\mathbb{F}_q$. A b-*burst* is a vector in $\mathbb{F}_q^n$ whose nonzero coordinates are confined to b consecutive positions, the first and last of which are nonzero. The code C is b-*burst error-correcting* provided there do not exist distinct codewords $\mathbf{c}_1$ and $\mathbf{c}_2$, and a b'-burst $\mathbf{u}_1$ and a b''-burst $\mathbf{u}_2$ with $b' \leq b$ and $b'' \leq b$ such that $\mathbf{c}_1 + \mathbf{u}_1 = \mathbf{c}_2 + \mathbf{u}_2$. If C is a linear, b-burst error-correcting code then no b'-burst is a codeword for any b' with $1 \leq b' \leq 2b$.

Now let $Q = 2^m$, and let $\mathcal{B}$ be an $[N, K, D]$ code over $\mathbb{F}_Q$. (A good choice for $\mathcal{B}$ is a Reed–Solomon code or a shortened Reed–Solomon code as such a code will be MDS, hence maximizing D given N and K. See Exercise 317.) Let $\mathcal{A}$ be the $[m, m, 1]$ binary code $\mathbb{F}_2^m$. Choosing a basis $\mathbf{e}_1, \mathbf{e}_2, \ldots, \mathbf{e}_m$ of $\mathbb{F}_Q = \mathbb{F}_{2^m}$ over $\mathbb{F}_2$, we define $\psi : \mathbb{F}_Q \to \mathcal{A}$ by $\psi(a_1 \mathbf{e}_1 + \cdots + a_m \mathbf{e}_m) = a_1 \cdots a_m$. The map ψ is one-to-one and $\mathbb{F}_2$-linear; see Exercise 319. If we refer to elements of $\mathbb{F}_{2^m}$ as *bytes* and elements of $\mathbb{F}_2$ as *bits*, each component byte of a codeword in $\mathcal{B}$ is replaced by the associated vector from $\mathbb{F}_2^m$ of m bits to form the corresponding codeword in C. The concatenated code C with inner code $\mathcal{A}$ and outer code $\mathcal{B}$ is an $[n, k, d]$ binary code with $n = mN, k = mK$, and $d \geq D$. This process is illustrated by the concatenated code constructed in Example 5.5.2. There the basis of $\mathbb{F}_4$ is $\mathbf{e}_1 = 1$ and $\mathbf{e}_2 = \omega$, and the bytes (elements of $\mathbb{F}_4$) each correspond to two bits determined by the map ψ.

Exercise 317 Let C be an $[n, k, n - k + 1]$ MDS code over $\mathbb{F}_q$. Let C_1 be the code obtained from C by shortening on some coordinate. Show that C_1 is an $[n - 1, k - 1, n - k + 1]$ code; that is, show that C_1 is also MDS. ◆

Exercise 318 In compact disc recording two shortened Reed–Solomon codes over $\mathbb{F}_{256}$ are used. Beginning with a Reed–Solomon code of length 255, explain how to obtain $[32, 28, 5]$ and $[28, 24, 5]$ shortened Reed–Solomon codes. ◆

Exercise 319 Let $\mathcal{A}$ be the $[m, m, 1]$ binary code $\mathbb{F}_2^m$. Let $\mathbf{e}_1, \mathbf{e}_2, \ldots, \mathbf{e}_m$ be a basis of $\mathbb{F}_{2^m}$ over $\mathbb{F}_2$. Define $\psi : \mathbb{F}_{2^m} \to \mathcal{A}$ by $\psi(a_1 \mathbf{e}_1 + \cdots + a_m \mathbf{e}_m) = a_1 \cdots a_m$. Prove that ψ is one-to-one and $\mathbb{F}_2$-linear. ◆

Let C be the binary concatenated $[mN, mk]$ code, as above, with inner code $\mathcal{A}$ and outer code $\mathcal{B}$. Let $\mathbf{c} = \psi(\mathbf{b})$ be a codeword in C where $\mathbf{b}$ is a codeword in $\mathcal{B}$. Let $\mathbf{u}$ be a b-burst in $\mathbb{F}_2^{mN}$. We can break the burst into N strings of m bits each and use ψ^{-1} to map the burst into a vector of N bytes. More formally, let $\mathbf{u} = \mathbf{u}_1 \cdots \mathbf{u}_N$, where $\mathbf{u}_i \in \mathbb{F}_2^m$ for $1 \leq i \leq N$. Map $\mathbf{u}$ into the vector $\mathbf{u}' = \psi^{-1}(\mathbf{u}_1) \cdots \psi^{-1}(\mathbf{u}_N)$ in $\mathbb{F}_Q^N$. As $\mathbf{u}$ is a b-burst, then the number of nonzero bytes of $\mathbf{u}'$ is, roughly speaking, at most b/m. For instance, if $\mathbf{u}$ is a $(3m + 1)$-burst, then at most four of the bytes of $\mathbf{u}'$ can be nonzero; see Exercise 320. So burst error-correction is accomplished as follows. Break the received codeword into N bit strings of length m, apply ψ^{-1} to each bit string to produce a vector in $\mathbb{F}_Q^N$, and correct that vector using $\mathcal{B}$. More on this is left to Exercise 320.

Exercise 320 Let $\mathcal{A} = \mathbb{F}_2^m$ be the $[m, m, 1]$ binary code of length m and let $\mathcal{B}$ be an $[N, K, D]$ code over $\mathbb{F}_{2^m}$. Let ψ be a one-to-one $\mathbb{F}_2$-linear map from $\mathbb{F}_{2^m}$ onto $\mathbb{F}_2^m$. Let C be the concatenated code with inner code $\mathcal{A}$ and outer code $\mathcal{B}$.

(a) Let $\mathbf{u}$ be a b-burst of length mN in $\mathbb{F}_2^{mN}$ associated to $\mathbf{u}'$, a vector in $\mathbb{F}_Q^N$ where $Q = 2^m$. Let $b \leq am + 1$. Show that $\text{wt}(\mathbf{u}') \leq a + 1$.

(b) Show that $\mathcal{C}$ corrects bursts of length $b \leq am + 1$, where $a = \lfloor (D-1)/2 \rfloor - 1$.

(c) Let $m = 5$ and let $\mathcal{B}$ be a $[31, 7, 25]$ Reed–Solomon code over $\mathbb{F}_{32}$. What is the maximum length burst that the $[156, 35]$ binary concatenated code $\mathcal{C}$ can correct? ◆

There is another technique, called interleaving, that will improve the burst error-correcting capability of a code. Let $\mathcal{C}$ be an $[n, k]$ code over $\mathbb{F}_q$ that can correct a burst of length b. Define $I(\mathcal{C}, t)$ to be a set of vectors in $\mathbb{F}_q^{nt}$ constructed as follows. For any set of t codewords $\mathbf{c}_1, \ldots, \mathbf{c}_t$ from $\mathcal{C}$, with $\mathbf{c}_i = c_{i1}c_{i2} \cdots c_{in}$, form the matrix

$$
M = \begin{bmatrix} c_{11} & c_{12} & \cdots & c_{1n} \\ c_{21} & c_{22} & \cdots & c_{2n} \\ & & \vdots & \\ c_{t1} & c_{t2} & \cdots & c_{tn} \end{bmatrix}
$$

whose rows are the codewords $\mathbf{c}_1, \ldots, \mathbf{c}_t$. The codewords of $I(\mathcal{C}, t)$ are the vectors

$$c_{11}c_{21} \cdots c_{t1}c_{12}c_{22} \cdots c_{t2} \cdots c_{1n}c_{2n} \cdots c_{tn}$$

of length nt obtained from M by reading down consecutive columns. The code $I(\mathcal{C}, t)$ is $\mathcal{C}$ *interleaved to depth t*.

Theorem 5.5.3 *If $\mathcal{C}$ is an $[n, k]$ code over $\mathbb{F}_q$ that can correct any burst of length b, then $I(\mathcal{C}, t)$ is an $[nt, kt]$ code over $\mathbb{F}_q$ that can correct any burst of length bt.*

Exercise 321 Prove Theorem 5.5.3. ◆

Example 5.5.4 The $[7, 4, 3]$ binary Hamming code $\mathcal{H}_3$ can correct only bursts of length 1. However, interleaving $\mathcal{H}_3$ to depth 4 produces a $[28, 16]$ binary code $I(\mathcal{H}_3, 4)$ that can correct bursts of length 4. Note, however, that the minimum distance of $I(\mathcal{H}_3, 4)$ is 3 and hence this code can correct single errors, but not all double errors. It can correct up to four errors as long as they are confined to four consecutive components. See Exercise 322. ∎

Exercise 322 Prove that the minimum distance of $I(\mathcal{H}_3, 4)$ is 3. ◆

5.6 Coding for the compact disc

In this section we give an overview of the encoding and decoding used for the compact disc (CD) recorder. The compact disc digital audio system standard currently in use was developed by N. V. Philips of The Netherlands and Sony Corporation of Japan in an agreement signed in 1979. Readers interested in further information on coding for CDs should consult [47, 99, 119, 133, 154, 202, 253, 286, 335, 341].

A compact disc is an aluminized disc, 120 mm in diameter, which is coated with a clear plastic coating. On each disc is one spiral track, approximately 5 km in length (see Exercise 324), which is optically scanned by an AlGaAs laser, with wavelength approximately 0.8 μm, operating at a constant speed of about 1.25 m/s. The speed of rotation of the disc varies from approximately 8 rev/s for the inner portion of the track to 3.5 rev/s for the outer portion. Along the track are depressions, called *pits*, and flat segments between pits, called *lands*. The width of the track is 0.6 μm and the depth of a pit is 0.12 μm. The laser light is reflected with differing intensities between pits and lands because of interference. The data carried by these pits and lands is subject to error due to such problems as stray particles on the disc or embedded in the disc, air bubbles in the plastic coating, fingerprints, or scratches. These errors tend to be burst errors; fortunately, there is a very efficient encoding and decoding system involving both shortened Reed–Solomon codes and interleaving.

5.6.1   Encoding

We first describe how audio data is encoded and placed on a CD. Sound-waves are first converted from analog to digital using sampling. The amplitude of a waveform is sampled at a given point in time and assigned a binary string of length 16. As before we will call a binary digit 0 or 1 from $\mathbb{F}_2$ a bit. Because the sound is to be reproduced in stereo, there are actually two samples taken at once, one for the left channel and one for the right. Waveform sampling takes place at the rate of 44 100 pairs of samples per second (44.1 kHz). (The sampling rate of 44.1 kHz was chosen to be compatible with a standard already existing for video recording.) Thus each sample produces two binary vectors from $\mathbb{F}_2^{16}$, one for each channel. Each vector from $\mathbb{F}_2^{16}$ is cut in half and is used to represent an element of the field $\mathbb{F}_{2^8}$, which as before we call a byte. Each sample then produces four bytes of data. For every second of sound recording, $44\,100 \cdot 32 = 1\,411\,200$ bits or $44\,100 \cdot 4 = 176\,400$ bytes are generated. We are now ready to encode the bytes. This requires the use of two shortened Reed–Solomon codes, $\mathcal{C}_1$ and $\mathcal{C}_2$, and two forms of interleaving. This combination is called a *cross-interleaved Reed–Solomon code* or *CIRC*. The purpose of the cross-interleaving, which is a variation of interleaving, is to break up long burst errors.

Step I: Encoding using $\mathcal{C}_1$ and interleaving
The bytes are encoded in the following manner. Six samples of four bytes each are grouped together to form a *frame* consisting of 24 bytes. We can view a frame as $L_1 R_1 L_2 R_2 \cdots L_6 R_6$, where L_i is two bytes for the left channel from the ith sample of the frame, and R_i is two bytes for the right channel from the ith sample of the frame. Before any encoding is done, the bytes are permuted in two ways. First, the odd-numbered samples $L_1 R_1$, $L_3 R_3$, and $L_5 R_5$ are grouped with the even-numbered samples $\widetilde{L}_2 \widetilde{R}_2$, $\widetilde{L}_4 \widetilde{R}_4$, and $\widetilde{L}_6 \widetilde{R}_6$ taken from two frames later. So we are now looking at a new frame of 24 bytes:

$$L_1 R_1 \widetilde{L}_2 \widetilde{R}_2 L_3 R_3 \widetilde{L}_4 \widetilde{R}_4 L_5 R_5 \widetilde{L}_6 \widetilde{R}_6.$$

Thus samples that originally were consecutive in time are now two frames apart. Second, these new frames are rearranged internally into 24 bytes by separating the odd-numbered samples from the even-numbered samples to form

$$L_1 L_3 L_5 R_1 R_3 R_5 \widetilde{L}_2 \widetilde{L}_4 \widetilde{L}_6 \widetilde{R}_2 \widetilde{R}_4 \widetilde{R}_6.$$

$c_{1,1}$	$c_{2,1}$	$c_{3,1}$	$c_{4,1}$	$c_{5,1}$	$c_{6,1}$	$c_{7,1}$	$c_{8,1}$	$c_{9,1}$	$c_{10,1}$	$c_{11,1}$	$c_{12,1}$	$c_{13,1}$	$\cdots$
0	0	0	0	$c_{1,2}$	$c_{2,2}$	$c_{3,2}$	$c_{4,2}$	$c_{5,2}$	$c_{6,2}$	$c_{7,2}$	$c_{8,2}$	$c_{9,2}$	$\cdots$
0	0	0	0	0	0	0	0	$c_{1,3}$	$c_{2,3}$	$c_{3,3}$	$c_{4,3}$	$c_{5,3}$	$\cdots$
0	0	0	0	0	0	0	0	0	0	0	0	$c_{1,4}$	$\cdots$

$$\vdots$$

Figure 5.2 4-frame delay interleaving.

This separates samples as far apart as possible within the new frame. These permutations of bytes allow for error concealment as we discuss later. This 24-byte message consisting of a vector in $\mathbb{F}_{256}^{24}$ is encoded using a systematic encoder for a $[28, 24, 5]$ shortened Reed–Solomon code, which we denote $\mathcal{C}_1$. This encoder produces four bytes of redundancy, that is, two pairs P_1 and P_2 each with two bytes of parity which are then placed in the middle of the above to form

$$L_1 L_3 L_5 R_1 R_3 R_5 P_1 P_2 \widetilde{L}_2 \widetilde{L}_4 \widetilde{L}_6 \widetilde{R}_2 \widetilde{R}_4 \widetilde{R}_6,$$

further separating the odd-numbered samples from the even-numbered samples.

Thus from $\mathcal{C}_1$ we produce a string of 28-byte codewords which we interleave to a depth of 28 using 4-frame delay interleaving as we now describe. Begin with codewords $c_1, c_2, c_3, \ldots$ from $\mathcal{C}_1$ in the order they are generated. Form an array with 28 rows and a large number of columns in the following fashion. Row 1 consists of the first byte of c_1 in column 1, the first byte of c_2 in column 2, the first byte of c_3 in column 3, etc. Row 2 begins with four bytes equal to 0 followed by the second byte of c_1 in column 5, the second byte of c_2 in column 6, the second byte of c_3 in column 7, etc. Row 3 begins with eight bytes equal to 0 followed by the third byte of c_1 in column 9, the third byte of c_2 in column 10, the third byte of c_3 in column 11, etc. Continue in this manner filling out all 28 rows. If $c_i = c_{i,1} c_{i,2} \cdots c_{i,28}$, the resulting array begins as in Figure 5.2, and thus the original codewords are found going diagonally down this array with slope $-1/4$. This array will be as long as necessary to accommodate all the encoded frames of data. All the rows except row 28 will need to be padded with zeros so that the array is rectangular; see Exercise 323.

Exercise 323 Suppose a CD is used to record 72 minutes of sound. How many frames of data does this represent? How long is the array obtained by interleaving the codewords in $\mathcal{C}_1$ corresponding to all these frames to a depth of 28 using 4-frame delay interleaving? ◆

Step II: Encoding using $\mathcal{C}_2$ and interleaving
Each column of the array is a vector in $\mathbb{F}_{256}^{28}$ which is then encoded using a $[32, 28, 5]$ shortened Reed–Solomon code $\mathcal{C}_2$. Thus we now have a list of codewords, which are generated in the order of the columns, each consisting of 32 bytes. The codewords are regrouped with the odd-numbered symbols of one codeword grouped with the even-numbered symbols of the next codeword. This regrouping is another form of interleaving which further breaks up short bursts that may still be present after the 4-frame delay interleaving. The regrouped bytes are written consecutively in one long stream. We now re-divide this long string into

segments of 32 bytes, with 16 bytes from one C_2 codeword and 16 bytes from another C_2 codeword because of the above regrouping. At the end of each of these segments a 33rd byte is added which contains control and display information.[1] Thus each frame of six samples eventually leads to 33 bytes of data. A schematic of the encoding using C_1 and C_2 can be found in [154].

Step III: Imprinting and EFM

Each byte of data must now be imprinted onto the disc. First, the bytes are converted to strings of bits using EFM described shortly. Each bit is of length 0.3 μm when imprinted along the track. Each land-to-pit or pit-to-land transition is imprinted with a single 1, while the track along the pit or land is imprinted with a string of 0s whose number corresponds to the length of the pit or land. For example, a pit of length 2.1 μm followed by a land of length 1.2 μm corresponds to the string 10000001000. For technical reasons each land or pit must be between 0.9 and 3.3 μm in length. Therefore each pair of 1s is separated by at least two 0s and at most ten 0s. Thus the 256 possible bytes must be converted to bit strings in such a way that this criterion is satisfied. Were it not for this condition, one could convert the bytes to elements of $\mathbb{F}_2^8$; it turns out that the smallest string length such that there are at least 256 different strings where each 1 is separated by at least two 0s but no more than ten 0s is length 14. In fact there are 267 binary strings of length 14 satisfying this condition; 11 of these are not used. This conversion from bytes to strings of length 14 is called *EFM* or *eight-to-fourteen modulation*. Note, however, that bytes must be encoded in succession, and so two consecutive 14-bit strings may fail to satisfy our conditions on minimum and maximum numbers of 0s between 1s. For example 10010000000100 and 00000000010001 are both allowable strings but if they follow one after the other, we obtain

10010000000100000000000010001,

which has 11 consecutive 0s. To overcome this problem three additional bits called *merge bits* are added to the end of each 14-bit string. In our example, if we add 001 to the end of the first string, we have

10010000000100001000000000010001,

which satisfies our criterion. So our frame of six samples leads to 33 bytes each of which is converted to 17 bits. Finally, at the end of these $33 \cdot 17$ bits, 24 synchronization bits plus three merging bits are added; thus each frame of six samples leads to 588 bits.

Exercise 324 Do the following; see the related Exercise 323:
(a) How many bits are on the track of a CD with 72 minutes of sound?
(b) How long must the track be if each bit is 0.3 μm in length?   ◆

---

[1] The control and display bytes include information for the listener such as playing time, composer, and title of the piece, as well as technical information required by the CD player.

5.6.2   Decoding

We are now ready to see how decoding and error-correction is performed by the CD player.[2] The process reverses the encoding.

Step I: Decoding with C_2

First, the synchronization bits, control and display bits, and merging bits are removed. Then the remaining binary strings are converted from EFM form into byte form, a process called *demodulation*, using table look-up; we now have our data as a stream of bytes. Next we undo the scrambling done in the encoding process. The stream is divided into segments of 32 bytes. Each of these 32-byte segments contains odd-numbered bytes from one codeword (with possible errors, of course) and even-numbered bytes from the next. The bytes in the segments are regrouped to restore the positions in order and are passed on to the decoder for C_2. Note that if a short burst error had occurred on the disc, the burst may be split up into shorter bursts by the regrouping. As C_2 is a $[32, 28, 5]$ code over $\mathbb{F}_{256}$, it can correct two errors. However, it is only used to correct single errors or detect the presence of multiple errors, including all errors of size two or three and some of larger size. The sphere of radius 1 centered at some codeword c_1 does not contain a vector that differs from another codeword c_2 in at most three positions as c_1 and c_2 are distance at least five from one another. Therefore, if a single error has occurred, C_2 can correct that error; if two or three errors have occurred, C_2 can detect the presence of those errors (but will not be used to correct them).

   What is the probability that C_2 will fail to detect four or more errors when we use C_2 to correct only single errors? Such a situation would arise if errors are made in one codeword so that the resulting vector lies in a sphere of radius 1 about another codeword. Assuming all vectors are equally likely, the probability of this occurring is approximately the ratio of the total number of vectors inside spheres of radius 1 centered at codewords to the total number of vectors in $\mathbb{F}_{256}^{32}$. This ratio is

$$\frac{256^{28}[1 + 32(256 - 1)]}{256^{32}} = \frac{8161}{256^4} \approx 1.9 \times 10^{-6}.$$

By Exercise 325, if C_2 were used to its full error-correcting capability by correcting all double errors, then the probability that three or more errors would go undetected is about 7.5×10^{-3}. The difference in these probabilities indicates why the full error-correcting capability of C_2 is not used since the likelihood of three or more errors going undetected (or being miscorrected) is much higher with full error-correction.

Exercise 325   Verify the following:
(a) A sphere of radius 2 centered at a codeword of a $[32, 28]$ code over $\mathbb{F}_{256}$ contains

$$1 + 32(256 - 1) + \binom{32}{2}(256 - 1)^2$$

vectors in $\mathbb{F}_{256}^{32}$.

---

[2] The encoding of a digital video disc (DVD) involves Reed–Solomon codes in a fashion similar to the CD. Tom Høholdt has created a simulation of DVD decoding at the following web site: http://www.mat.dtu.dk/persons/Hoeholdt_Tom/.

(b) The probability that three or more errors would go undetected using the double error-correcting capability of a $[32, 28, 5]$ code over $\mathbb{F}_{256}$ is about 7.5×10^{-3}.   ◆

Step II: Decoding with C_1

If the decoder for C_2 determines that no errors in a 32-byte string are found, the 28-byte message is extracted and passed on to the next stage. If the decoder for C_2 detects a single error, the error is corrected and the 28-byte message is passed on. If the decoder detects more than two errors, it passes on a 28-byte string with all components flagged as erasures. These 28-byte strings correspond to the columns of the array in Figure 5.2, possibly with erasures. The diagonals of slope $-1/4$ are passed on as 28-byte received vectors to the decoder for C_1. C_1 can be used in different ways. In one scheme it is used only to correct erasures. By Theorem 1.11.6, C_1 can correct four erasures. Due to the 4-frame delay interleaving and the ability of C_1 to correct four erasures, a burst error covering 16 consecutive 588-bit strings on the disc can be corrected. Such a burst is approximately 2.8 mm in length along the track! In another scheme, again applying Theorem 1.11.6, C_1 is used to correct one error (which may have escaped the decoding performed by C_2) and two erasures. A comparison of the two schemes can be found in [253].

Step III: Errors that still survive

It is possible that there are samples that cannot be corrected by the use of C_2 and C_1 but are detected as errors and hence remain erased. One technique used is to "conceal" the error. Recall that consecutive samples are separated by two frames before any encoding was performed. When the final decoding is completed and these samples are brought back to their correct order, it is likely that the neighboring samples were correct or had been corrected. If this is the case, then the erased sample is replaced by an approximation obtained by linear interpolation using the two reliable samples on either side of the sample in error. Listening tests have shown that this process is essentially undetectable. If the neighbors are unreliable, implying a burst is still present, so that interpolation is not possible, then "muting" is used. Starting 32 samples prior to the burst, the reliable samples are gradually weakened until the burst occurs, the burst is replaced by a zero-valued sample, and the next 32 reliable samples are gradually strengthened. As this muting process occurs over a few milliseconds, it is essentially inaudible. Both linear interpolation and muting mask "clicks" that may otherwise occur. More details on interpolation and muting can be found in [154, 253].

6 Duadic codes

In Chapter 5 we described the family of cyclic codes called BCH codes and its subfamily of RS codes. In this chapter we define and study another family of cyclic codes called duadic codes. They are generalizations of quadratic residue codes, which we discuss in Section 6.6. Binary duadic codes were initially defined in [190] and were later generalized to arbitrary finite fields in [266, 270, 301, 315].

6.1 Definition and basic properties

We will define duadic codes in two different ways and show that the definitions are equivalent. We need some preliminary notation and results before we begin. Throughout this chapter $\mathbb{Z}_n$ will denote the ring of integers modulo n. We will also let $\mathcal{E}_n$ denote the subcode of even-like vectors in $\mathcal{R}_n = \mathbb{F}_q[x]/(x^n - 1)$. The code $\mathcal{E}_n$ is an $[n, n-1]$ cyclic code whose dual code $\mathcal{E}_n^{\perp}$ is the repetition code of length n. By Exercise 221 the repetition code has generating idempotent

$$\overline{j}(x) = \frac{1}{n}(1 + x + x^2 + \cdots + x^{n-1}).$$

So by Theorem 4.4.9, $\mathcal{E}_n$ has generating idempotent $1 - \overline{j}(x)\mu_{-1} = 1 - \overline{j}(x)$. We summarize this information in the following lemma.

Lemma 6.1.1 *The code $\mathcal{E}_n$ has the following properties*:
(i) $\mathcal{E}_n$ *is an $[n, n-1]$ cyclic code.*
(ii) $\mathcal{E}_n^{\perp}$ *is the repetition code with generating idempotent $\overline{j}(x) = (1/n)(1 + x + x^2 + \cdots + x^{n-1})$.*
(iii) $\mathcal{E}_n$ *has generating idempotent $1 - \overline{j}(x)$.*

In defining the duadic codes, we will obtain two pairs of codes; one pair will be two even-like codes, which are thus subcodes of $\mathcal{E}_n$, and the other pair will be odd-like codes. It will be important to be able to tell when either a vector or a cyclic code in $\mathcal{R}_n$ is even-like or odd-like.

Lemma 6.1.2 *Let $a(x) = \sum_{i=0}^{n-1} a_i x^i \in \mathcal{R}_n$. Also let $\mathcal{C}$ be a cyclic code in $\mathcal{R}_n$ with generator polynomial $g(x)$. Then*:
(i) $a(x)$ *is even-like if and only if $a(1) = 0$ if and only if $a(x)\overline{j}(x) = 0$,*
(ii) $a(x)$ *is odd-like if and only if $a(1) \neq 0$ if and only if $a(x)\overline{j}(x) = \alpha \overline{j}(x)$ for some nonzero $\alpha \in \mathbb{F}_q$,*

(iii) C is even-like if and only if $g(1) = 0$ if and only if $\overline{j}(x) \notin C$, and

(iv) C is odd-like if and only if $g(1) \neq 0$ if and only if $\overline{j}(x) \in C$.

Proof: Parts (ii) and (iv) follow from (i) and (iii), respectively. By definition, $a(x)$ is even-like precisely when $\sum_{i=0}^{n-1} a_i = 0$. This is the same as saying $a(1) = 0$. That this is equivalent to $a(x)\overline{j}(x) = 0$ follows from Exercise 221. This verifies (i). In part (iii), C is even-like if and only if $g(1) = 0$ from Exercise 238. Note that $(x - 1)n\overline{j}(x) = x^n - 1$ in $\mathbb{F}_q[x]$. As $g(x) \mid (x^n - 1)$ and $x^n - 1$ has distinct roots, $g(x) \mid \overline{j}(x)$ if and only if $g(1) \neq 0$. Since $\overline{j}(x) \in C$ if and only if $g(x) \mid \overline{j}(x)$, part (iii) follows. $\qquad\square$

We first define duadic codes in terms of their idempotents. Duadic codes come in two pairs, one even-like pair, which we usually denote C_1 and C_2, and one odd-like pair, usually denoted D_1 and D_2. Let $e_1(x)$ and $e_2(x)$ be two even-like idempotents with $C_1 = \langle e_1(x)\rangle$ and $C_2 = \langle e_2(x)\rangle$. The codes C_1 and C_2 form a pair of *even-like duadic codes* provided the following two criteria are met:

I. The idempotents satisfy

$$e_1(x) + e_2(x) = 1 - \overline{j}(x), \quad \text{and} \tag{6.1}$$

II. there is a multiplier μ_a such that

$$C_1\mu_a = C_2 \quad \text{and} \quad C_2\mu_a = C_1. \tag{6.2}$$

If $c(x) \in C_i$, then $c(x)e_i(x) = c(x)$ implying that $c(1) = c(1)e_i(1) = 0$ by Lemma 6.1.2(i); thus both C_1 and C_2 are indeed even-like codes. We remark that $e_1(x)\mu_a = e_2(x)$ and $e_2(x)\mu_a = e_1(x)$ if and only if $C_1\mu_a = C_2$ and $C_2\mu_a = C_1$ by Theorem 4.3.13(i); thus we can replace (6.2) in part II by

$$e_1(x)\mu_a = e_2(x) \quad \text{and} \quad e_2(x)\mu_a = e_1(x). \tag{6.3}$$

Associated to C_1 and C_2 is the pair of *odd-like duadic codes*

$$D_1 = \langle 1 - e_2(x)\rangle \quad \text{and} \quad D_2 = \langle 1 - e_1(x)\rangle. \tag{6.4}$$

As $1 - e_i(1) = 1$, by Lemma 6.1.2(ii), D_1 and D_2 are odd-like codes. We say that μ_a gives a *splitting* for the even-like duadic codes C_1 and C_2 or for the odd-like duadic codes D_1 and D_2.

Exercise 326 Prove that if C_1 and C_2 form a pair of even-like duadic codes and that C_1 and C_2' are also a pair of even-like duadic codes, then $C_2 = C_2'$. (This exercise shows that if we begin with a code C_1 that is one code in a pair of even-like duadic codes, there is no ambiguity as to what code it is paired with.) $\qquad\blacklozenge$

The following theorem gives basic facts about these four codes.

Theorem 6.1.3 Let $C_1 = \langle e_1(x)\rangle$ and $C_2 = \langle e_2(x)\rangle$ be a pair of even-like duadic codes of length n over $\mathbb{F}_q$. Suppose μ_a gives the splitting for C_1 and C_2. Let D_1 and D_2 be the associated odd-like duadic codes. Then:

(i) $e_1(x)e_2(x) = 0$,

(ii) $C_1 \cap C_2 = \{0\}$ and $C_1 + C_2 = \mathcal{E}_n$,

(iii) n is odd and C_1 and C_2 each have dimension $(n - 1)/2$,

(iv) $\mathcal{D}_1$ is the cyclic complement of $\mathcal{C}_2$ and $\mathcal{D}_2$ is the cyclic complement of $\mathcal{C}_1$,

(v) $\mathcal{D}_1$ and $\mathcal{D}_2$ each have dimension $(n+1)/2$,

(vi) $\mathcal{C}_i$ is the even-like subcode of $\mathcal{D}_i$ for $i = 1, 2$,

(vii) $\mathcal{D}_1\mu_a = \mathcal{D}_2$ and $\mathcal{D}_2\mu_a = \mathcal{D}_1$,

(viii) $\mathcal{D}_1 \cap \mathcal{D}_2 = \langle \overline{j}(x)\rangle$ and $\mathcal{D}_1 + \mathcal{D}_2 = \mathcal{R}_n$, and

(ix) $\mathcal{D}_i = \mathcal{C}_i + \langle \overline{j}(x)\rangle = \langle \overline{j}(x) + e_i(x)\rangle$ for $i = 1, 2$.

Proof: Multiplying (6.1) by $e_1(x)$ gives $e_1(x)e_2(x) = 0$, by Lemma 6.1.2(i). So (i) holds. By Theorem 4.3.7, $\mathcal{C}_1 \cap \mathcal{C}_2$ and $\mathcal{C}_1 + \mathcal{C}_2$ have generating idempotents $e_1(x)e_2(x) = 0$ and $e_1(x) + e_2(x) - e_1(x)e_2(x) = e_1(x) + e_2(x) = 1 - \overline{j}(x)$, respectively. Thus part (ii) holds by Lemma 6.1.1(iii). By (6.2), $\mathcal{C}_1$ and $\mathcal{C}_2$ are equivalent, and hence have the same dimension. By (ii) and Lemma 6.1.1(i), this dimension is $(n-1)/2$, and hence n is odd giving (iii). The cyclic complement of $\mathcal{C}_i$ has generating idempotent $1 - e_i(x)$ by Theorem 4.4.6(i); thus part (iv) is immediate from the definition of $\mathcal{D}_i$. Part (v) follows from the definition of cyclic complement and parts (iii) and (iv). As $\mathcal{D}_1$ is odd-like with generating idempotent $1 - e_2(x)$, by Exercise 257, the generating idempotent of the even-like subcode of $\mathcal{D}_1$ is $1 - e_2(x) - \overline{j}(x) = e_1(x)$. Thus $\mathcal{C}_1$ is the even-like subcode of $\mathcal{D}_1$; analogously $\mathcal{C}_2$ is the even-like subcode of $\mathcal{D}_2$ yielding (vi). The generating idempotent of $\mathcal{D}_1\mu_a$ is $(1 - e_2(x))\mu_a = 1 - e_2(x)\mu_a = 1 - e_1(x)$ by Theorem 4.3.13(i) and (6.3). Thus $\mathcal{D}_1\mu_a = \mathcal{D}_2$; analogously $\mathcal{D}_2\mu_a = \mathcal{D}_1$ producing (vii). By Theorem 4.3.7, $\mathcal{D}_1 \cap \mathcal{D}_2$ and $\mathcal{D}_1 + \mathcal{D}_2$ have generating idempotents $(1 - e_2(x))(1 - e_1(x)) = 1 - e_1(x) - e_2(x) = \overline{j}(x)$ and $(1 - e_2(x)) + (1 - e_1(x)) - (1 - e_2(x))(1 - e_1(x)) = 1$, respectively, as $e_1(x)e_2(x) = 0$. Thus (viii) holds as the generating idempotent of $\mathcal{R}_n$ is 1. Finally by (iii), (v), and (vi), $\mathcal{C}_i$ is a subspace of $\mathcal{D}_i$ of codimension 1; as $\overline{j}(x) \in \mathcal{D}_i \setminus \mathcal{C}_i$, $\mathcal{D}_i = \mathcal{C}_i + \langle \overline{j}(x)\rangle$. Also $\mathcal{D}_i = \langle \overline{j}(x) + e_i(x)\rangle$ by (6.1) and (6.4). □

Example 6.1.4 We illustrate the definition of duadic codes by constructing the generating idempotents of the binary duadic codes of length 7. The 2-cyclotomic cosets modulo 7 are $C_0 = \{0\}$, $C_1 = \{1, 2, 4\}$, and $C_3 = \{3, 6, 5\}$. Recall from Corollary 4.3.15 that every binary idempotent in $\mathcal{R}_n$ is of the form $e(x) = \sum_{j \in J}\sum_{i \in C_j} x^i$ and all such polynomials are idempotents. Thus there are $2^3 = 8$ idempotents, with four being even-like. These are $e_0(x) = 0$, $e_1(x) = 1 + x + x^2 + x^4$, $e_2(x) = 1 + x^3 + x^5 + x^6$, and $e_3(x) = x + x^2 + x^3 + x^4 + x^5 + x^6$. But $e_0(x)$ generates $\{0\}$ and $e_3(x)$ generates $\mathcal{E}_7$; see Exercise 327. So the only possible generating idempotents for even-like duadic codes are $e_1(x)$ and $e_2(x)$. Note that $e_1(x) + e_2(x) = 1 - \overline{j}(x)$ giving (6.1); also $e_1(x)\mu_3 = e_2(x)$ and $e_2(x)\mu_3 = e_1(x)$ giving (6.3). Thus there is one even-like pair of duadic codes of length 7 with one associated odd-like pair having generating idempotents $1 - e_2(x) = x^3 + x^5 + x^6$ and $1 - e_1(x) = x + x^2 + x^4$; the latter are Hamming codes. ■

Exercise 327 Prove that $x + x^2 + \cdots + x^{n-1}$ is the idempotent of the even-like code $\mathcal{E}_n$ of length n over $\mathbb{F}_2$. ♦

Exercise 328 Find idempotent generators of all of the binary even-like and odd-like duadic codes of length $n = 17$ and $n = 23$. The odd-like duadic codes of length 23 are Golay codes. ♦

Exercise 329 We could have defined duadic codes by beginning with the generating idempotents of the odd-like duadic codes. Let $\mathcal{D}_1$ and $\mathcal{D}_2$ be odd-like cyclic codes of length n over $\mathbb{F}_q$ with generating idempotents $d_1(x)$ and $d_2(x)$. Show that $\mathcal{D}_1$ and $\mathcal{D}_2$ are odd-like duadic codes if and only if:

I″. the idempotents satisfy $d_1(x) + d_2(x) = 1 + \overline{j}(x)$, and

II″. there is a multiplier μ_a such that $\mathcal{D}_1 \mu_a = \mathcal{D}_2$ and $\mathcal{D}_2 \mu_a = \mathcal{D}_1$.   ◆

Duadic codes can also be defined in terms of their defining sets (and thus ultimately by their generator polynomials). Let $\mathcal{C}_1$ and $\mathcal{C}_2$ be a pair of even-like duadic codes defined by I and II above. As these are cyclic codes, $\mathcal{C}_1$ and $\mathcal{C}_2$ have defining sets $T_1 = \{0\} \cup S_1$ and $T_2 = \{0\} \cup S_2$, respectively, relative to some primitive nth root of unity. Each of the sets S_1 and S_2 is a union of nonzero q-cyclotomic cosets. By Theorem 6.1.3(iii) $\mathcal{C}_1$ and $\mathcal{C}_2$ each have dimension $(n-1)/2$; by Theorem 4.4.2, S_1 and S_2 each have size $(n-1)/2$. The defining set of $\mathcal{C}_1 \cap \mathcal{C}_2 = \{0\}$ is $T_1 \cup T_2$, which must then be $\{0, 1, \ldots, n-1\}$ by Exercise 239. Thus $S_1 \cup S_2 = \{1, 2, \ldots, n-1\}$; since each S_i has size $(n-1)/2$, $S_1 \cap S_2 = \emptyset$. By (6.2) and Corollary 4.4.5, $T_1 \mu_{a^{-1}} = T_2$ and $T_2 \mu_{a^{-1}} = T_1$. Therefore $S_1 \mu_{a^{-1}} = S_2$ and $S_2 \mu_{a^{-1}} = S_1$. This leads to half of the following theorem.

Theorem 6.1.5 *Let $\mathcal{C}_1$ and $\mathcal{C}_2$ be cyclic codes over $\mathbb{F}_q$ with defining sets $T_1 = \{0\} \cup S_1$ and $T_2 = \{0\} \cup S_2$, respectively, where $0 \notin S_1$ and $0 \notin S_2$. Then $\mathcal{C}_1$ and $\mathcal{C}_2$ are a pair of even-like duadic codes if and only if:*

I′. *S_1 and S_2 satisfy*

$$S_1 \cup S_2 = \{1, 2, \ldots, n-1\} \quad and \quad S_1 \cap S_2 = \emptyset, \quad and \tag{6.5}$$

II′. *there is a multiplier μ_b such that*

$$S_1 \mu_b = S_2 \quad and \quad S_2 \mu_b = S_1. \tag{6.6}$$

Proof: The previous discussion proved that if $\mathcal{C}_1$ and $\mathcal{C}_2$ are a pair of even-like duadic codes, then I′ and II′ hold. Suppose that I′ and II′ hold. Because $0 \in T_i$, $\mathcal{C}_i$ is even-like, by Exercise 238, for $i = 1$ and 2. Let $e_i(x)$ be the generating idempotent of $\mathcal{C}_i$. As $\mathcal{C}_1 \cap \mathcal{C}_2$ has defining set $T_1 \cup T_2 = \{0, 1, \ldots, n-1\}$ by Exercise 239 and (6.5), $\mathcal{C}_1 \cap \mathcal{C}_2 = \{0\}$. By Theorem 4.3.7, $\mathcal{C}_1 \cap \mathcal{C}_2$ has generating idempotent $e_1(x)e_2(x)$, which therefore must be 0. As $\mathcal{C}_1 + \mathcal{C}_2$ has defining set $T_1 \cap T_2 = \{0\}$ by Exercise 239 and (6.5), $\mathcal{C}_1 + \mathcal{C}_2 = \mathcal{E}_n$. By Theorem 4.3.7, $\mathcal{C}_1 + \mathcal{C}_2$ has generating idempotent $e_1(x) + e_2(x) - e_1(x)e_2(x) = e_1(x) + e_2(x)$, which therefore must be $1 - \overline{j}(x)$ by Lemma 6.1.1(iii). Thus (6.1) holds. By Corollary 4.4.5, $\mathcal{C}_i \mu_{b^{-1}}$ has defining set $T_i \mu_b$ for $i = 1$ and 2. But by (6.6), $T_1 \mu_b = T_2$ and $T_2 \mu_b = T_1$. Thus $\mathcal{C}_1 \mu_{b^{-1}} = \mathcal{C}_2$ and $\mathcal{C}_2 \mu_{b^{-1}} = \mathcal{C}_1$, giving (6.2) with $a = b^{-1}$. Therefore $\mathcal{C}_1$ and $\mathcal{C}_2$ are a pair of even-like duadic codes.   □

We can use either our original definitions I and II for duadic codes defined in terms of their idempotents or I′ and II′ from Theorem 6.1.5. We give a name to conditions I′ and II′: we say that a pair of sets S_1 and S_2, each of which is a union of nonzero q-cyclotomic cosets, forms a *splitting of n given by μ_b over $\mathbb{F}_q$* provided conditions I′ and II′ from Theorem 6.1.5 hold. Note that the proof of Theorem 6.1.5 shows that μ_a in (6.2) and μ_b in (6.6) are related by $a = b^{-1}$. In other words, if μ_a gives a splitting for the duadic codes, then $\mu_{a^{-1}}$

gives the associated splitting of n. However, $S_1 \mu_{a^{-1}} = S_2$ implies $S_1 = S_2(\mu_{a^{-1}})^{-1} = S_2 \mu_a$. Similarly, $S_2 = S_1 \mu_a$, and we can in fact use the same multiplier for the splittings in either definition.

This theorem has an immediate corollary.

Corollary 6.1.6 *Duadic codes of length n over $\mathbb{F}_q$ exist if and only if there is a multiplier which gives a splitting of n.*

Example 6.1.7 We construct the generating idempotents of the duadic codes of length 11 over $\mathbb{F}_3$. We first use the splittings of 11 over $\mathbb{F}_3$ to show that there is only one pair of even-like duadic codes. The 3-cyclotomic cosets modulo 11 are $C_0 = \{0\}$, $C_1 = \{1, 3, 9, 5, 4\}$, and $C_2 = \{2, 6, 7, 10, 8\}$. The only possible splitting of 11 is $S_1 = C_1$ and $S_2 = C_2$, since S_1 and S_2 must contain five elements. Thus there is only one pair of even-like duadic codes. We now construct their idempotents. Let $i_0(x) = 1$, $i_1(x) = x + x^3 + x^4 + x^5 + x^9$, and $i_2(x) = x^2 + x^6 + x^7 + x^8 + x^{10}$. By Corollary 4.3.15, all idempotents are of the form $a_0 i_0(x) + a_1 i_1(x) + a_2 i_2(x)$, where $a_0, a_1, a_2 \in \mathbb{F}_3$. By Exercise 330, $i_1(x)^2 = -i_1(x)$. Thus $(1 + i_1(x))^2 = 1 + 2 i_1(x) - i_1(x) = 1 + i_1(x)$ and $1 + i_1(x)$ is an even-like idempotent. As $i_1(x)\mu_2 = i_2(x)$, then $i_2(x)^2 = -i_2(x)$ and $1 + i_2(x)$ is another even-like idempotent. Letting $e_1(x) = 1 + i_1(x)$ and $e_2(x) = 1 + i_2(x)$, we see that $e_1(x) + e_2(x) = 1 - \bar{j}(x)$ as $\bar{j}(x) = 2(1 + x + x^2 + \cdots + x^{10})$, giving (6.1). Also $e_1(x)\mu_2 = e_2(x)$ and $e_2(x)\mu_2 = e_1(x)$ giving (6.3). Thus $e_1(x)$ and $e_2(x)$ are the idempotent generators of the unique pair of even-like duadic codes. The corresponding generating idempotents for the odd-like duadic codes are $1 - e_2(x) = -i_2(x)$ and $1 - e_1(x) = -i_1(x)$. These odd-like codes are the ternary Golay codes. ∎

Exercise 330 In $\mathcal{R}_{11}$ over $\mathbb{F}_3$ show that $(x + x^3 + x^4 + x^5 + x^9)^2 = -(x + x^3 + x^4 + x^5 + x^9)$. ◆

Exercise 331 Find the generating idempotents of the duadic codes of length $n = 23$ over $\mathbb{F}_3$. ◆

Exercise 332 Find the splittings of $n = 13$ over $\mathbb{F}_3$ to determine the number of pairs of even-like duadic codes. Using these splittings, find those codes which are permutation equivalent using Theorem 4.3.17 and Corollary 4.4.5. ◆

Example 6.1.8 We construct the generating idempotents of the duadic codes of length 5 over $\mathbb{F}_4$. The 4-cyclotomic cosets modulo 5 are $C_0 = \{0\}$, $C_1 = \{1, 4\}$, and $C_2 = \{2, 3\}$. The only possible splitting of 5 over $\mathbb{F}_4$ is $S_1 = C_1$ and $S_2 = C_2$, since S_1 and S_2 must contain two elements. Thus there is only one pair of even-like duadic codes. Let $e_1(x)$ and $e_2(x)$ be their idempotents; let $i_0(x) = 1$, $i_1(x) = x + x^4$, and $i_2(x) = x^2 + x^3$. By Corollary 4.3.15, all idempotents are of the form $e(x) = a_0 i_0(x) + a_1 i_1(x) + a_2 i_2(x)$, where $a_0, a_1, a_2 \in \mathbb{F}_4$. Since $e(x)^2 = a_0^2 i_0(x) + a_1^2 i_1(x) + a_2^2 i_2(x) = e(x)$, we must have $a_0^2 = a_0$ and $a_2 = a_1^2$; thus $a_0 = 0$ or 1. Since μ_4 fixes cyclic codes over $\mathbb{F}_4$ by Theorem 4.3.13, the only multipliers that could interchange two cyclic codes are μ_2 or μ_3. Since $\mu_3 = \mu_4\mu_2$, we can assume that μ_2 interchanges the two even-like duadic codes. Suppose that $e_1(x) = a_0 i_0(x) + a_1 i_1(x) + a_1^2 i_2(x)$. Then $e_2(x) = e_1(x)\mu_2 = a_0 i_0(x) + a_1^2 i_1(x) + a_1 i_2(x)$

and $e_1(x) + e_2(x) = 1 - \overline{j}(x) = i_1(x) + i_2(x)$; thus $a_1 + a_1^2 = 1$ implying that $a_1 = \omega$ or $\overline{\omega}$. To make $e_i(x)$ even-like, $a_0 = 0$. So we can take the idempotents of $\mathcal{C}_1$ and $\mathcal{C}_2$ to be $e_1(x) = \omega(x + x^4) + \overline{\omega}(x^2 + x^3)$ and $e_2(x) = \overline{\omega}(x + x^4) + \omega(x^2 + x^3)$. The associated odd-like duadic codes have idempotents $1 + \overline{\omega}(x + x^4) + \omega(x^2 + x^3)$ and $1 + \omega(x + x^4) + \overline{\omega}(x^2 + x^3)$; these codes are each the punctured hexacode (see Exercise 363). ∎

Example 6.1.9 For comparison with the previous example, we construct the generating idempotents of the duadic codes of length 7 over $\mathbb{F}_4$. The codes in this example all turn out to be quadratic residue codes. The 4-cyclotomic cosets modulo 7 are $C_0 = \{0\}$, $C_1 = \{1, 4, 2\}$, and $C_3 = \{3, 5, 6\}$. Again the only possible splitting of 7 over $\mathbb{F}_4$ is $S_1 = C_1$ and $S_2 = C_3$, and there is only one pair of even-like duadic codes with idempotents $e_1(x)$ and $e_2(x)$. Let $i_0(x) = 1$, $i_1(x) = x + x^2 + x^4$, and $i_2(x) = x^3 + x^5 + x^6$. As in the previous example all idempotents are of the form $e(x) = a_0 i_0(x) + a_1 i_1(x) + a_2 i_2(x)$, where $a_0, a_1, a_2 \in \mathbb{F}_4$. However, now $e(x)^2 = a_0^2 i_0(x) + a_1^2 i_1(x) + a_2^2 i_2(x) = e(x)$; we must have $a_j^2 = a_j$ for $0 \leq j \leq 2$ and hence $a_j = 0$ or 1. Similarly to Example 6.1.8 we can assume that μ_3 interchanges the two even-like duadic codes; see also Exercise 333. If $e_1(x) = a_0 i_0(x) + a_1 i_1(x) + a_2 i_2(x)$, then $e_2(x) = e_1(x)\mu_3 = a_0 i_0(x) + a_2 i_1(x) + a_1 i_2(x)$ and $e_1(x) + e_2(x) = 1 - \overline{j}(x) = i_1(x) + i_2(x)$. Thus $a_1 + a_2 = 1$ implying that $\{a_1, a_2\} = \{0, 1\}$. To make $e_i(x)$ even-like, $a_0 = 1$. So the idempotents of $\mathcal{C}_1$ and $\mathcal{C}_2$ are $1 + x + x^2 + x^4$ and $1 + x^3 + x^5 + x^6$. The associated odd-like duadic codes have idempotents $x^3 + x^5 + x^6$ and $x + x^2 + x^4$. These are all binary idempotents; the subfield subcodes $\mathcal{C}_i|_{\mathbb{F}_2}$ and $\mathcal{D}_i|_{\mathbb{F}_2}$ are precisely the codes from Example 6.1.4. This is an illustration of Theorem 6.6.4, which applies to quadratic residue codes. ∎

The definition of duadic codes in terms of generating idempotents has the advantage that these idempotents can be constructed with the knowledge gained from the splittings of n over the fields $\mathbb{F}_2$ and $\mathbb{F}_4$ without factoring $x^n - 1$ as we saw in the preceding examples. With more difficulty, the generating idempotents can also sometimes be constructed over the field $\mathbb{F}_3$ without factoring $x^n - 1$.

Exercise 333 In this exercise, we examine which multipliers need to be checked to either interchange duadic codes (as in II of the duadic code definition) or produce a splitting (as in II' of Theorem 6.1.5). Suppose our codes are of length n over $\mathbb{F}_q$, where of course, n and q are relatively prime. The multipliers μ_a that need to be considered are indexed by the elements of the (multiplicative) group $\mathbb{Z}_n^\# = \{a \in \mathbb{Z}_n \mid \gcd(a, n) = 1\}$. Let Q be the subgroup of $\mathbb{Z}_n^\#$ generated by q. Prove the following:
(a) If $\mathcal{C}_1\mu_a = \mathcal{C}_2$ and $\mathcal{C}_2\mu_a = \mathcal{C}_1$ (from II of the duadic code definition), then $\mathcal{C}_1\mu_c\mu_a = \mathcal{C}_2$ and $\mathcal{C}_2\mu_c\mu_a = \mathcal{C}_1$ for all $c \in Q$.
(b) If $S_1\mu_b = S_2$ and $S_2\mu_b = S_1$ (from II' of Theorem 6.1.5), then $S_1\mu_c\mu_b = S_2$ and $S_2\mu_c\mu_b = S_1$ for all $c \in Q$.
(c) Prove that when checking II of the duadic code definition or II' of Theorem 6.1.5, one only needs to check multipliers indexed by one representative from each coset of Q in $\mathbb{Z}_n^\#$, and, in fact, the representative from Q itself need not be checked.
(d) What are the only multipliers that need to be considered when constructing duadic codes of length n over $\mathbb{F}_q$ where:

(i) $n = 5, q = 4$,

(ii) $n = 7, q = 4$,

(iii) $n = 15, q = 4$,

(iv) $n = 13, q = 3$, and

(v) $n = 23, q = 2$?   ♦

Exercise 334 Find the generating idempotents of the duadic codes of length n over $\mathbb{F}_4$ where:

(a) $n = 3$,

(b) $n = 9$,

(c) $n = 11$,

(d) $n = 13$, and

(e) $n = 23$.   ♦

If n is the length of our duadic codes, a splitting of n leads directly to the defining sets of the even-like and odd-like duadic codes, and hence the generator polynomials, once the primitive nth root of unity has been fixed. From there one can construct the generating idempotents, by using, for example, the Euclidean Algorithm and the technique in the proof of Theorem 4.3.2. In the binary case, by examining the exponents of two of these four idempotents another splitting of n is obtained as the next theorem shows. This provides a way to use splittings of n to obtain generating idempotents of duadic codes directly in the binary case. It is important to note that the splitting of n used to construct the defining sets and generator polynomials is not necessarily the same splitting as the one arising from the exponents of the idempotents; however, the multiplier used to give these splittings is the same. In Theorem 6.3.3, we will see that binary duadic codes exist only if $n \equiv \pm 1$ (mod 8).

Theorem 6.1.10 *Let $n \equiv \pm 1$ (mod 8) and let $e_1(x)$ and $e_2(x)$ be generating idempotents of even-like binary duadic codes of length n given by the multiplier μ_a. The following hold:*

(i) *If $n \equiv 1$ (mod 8), then $e_i(x) = \sum_{j \in S_i} x^j$, where S_1 and S_2 form a splitting of n given by μ_a.*

(ii) *If $n \equiv -1$ (mod 8), then $1 + e_i(x) = \sum_{j \in S_i} x^j$, where S_1 and S_2 form a splitting of n given by μ_a.*

Proof: Since μ_a is the multiplier giving the splitting of (6.2), from (6.3) $e_1(x)\mu_a = e_2(x)$ and $e_2(x)\mu_a = e_1(x)$. Then by (6.1),

$$e_1(x) + e_2(x) = \sum_{j=1}^{n-1} x^j. \tag{6.7}$$

By Corollary 4.3.15

$$e_i(x) = \epsilon_i + \sum_{j \in S_i} x^j, \tag{6.8}$$

where ϵ_i is 0 or 1 and S_i is a union of nonzero 2-cyclotomic cosets modulo n. Combining (6.7) and (6.8), we have

$$\epsilon_1 + \epsilon_2 + \sum_{j \in S_1} x^j + \sum_{j \in S_2} x^j = \sum_{j=1}^{n-1} x^j,$$

which implies that $\epsilon_1 = \epsilon_2$, $S_1 \cap S_2 = \emptyset$, and $S_1 \cup S_2 = \{1, 2, \ldots, n-1\}$. Furthermore as $e_1(x)\mu_a = e_2(x)$, $(\epsilon_1 + \sum_{j \in S_1} x^j)\mu_a = \epsilon_1 + \sum_{j \in S_1} x^{ja} = \epsilon_2 + \sum_{j \in S_2} x^j$ implying that $S_1 \mu_a = S_2$. Analogously, $S_2 \mu_a = S_1$ using $e_2(x)\mu_a = e_1(x)$. Thus S_1 and S_2 is a splitting of n given by μ_a. Now $e_1(x)$ is even-like; which in the binary case, means that $e_1(x)$ has even weight. The weight of $e_1(x)$ is the size of S_1, if $\epsilon_1 = 0$, and is one plus the size of S_1, if $\epsilon_1 = 1$. The size of S_1 is $(n-1)/2$, which is even if $n \equiv 1 \pmod 8$ and odd if $n \equiv -1 \pmod 8$. So $\epsilon_1 = 0$ if $n \equiv 1 \pmod 8$ and $\epsilon_1 = 1$ if $n \equiv -1 \pmod 8$, leading to (i) and (ii). □

The converse also holds.

Theorem 6.1.11 *Let $n \equiv \pm 1 \pmod 8$ and let S_1 and S_2 be a splitting of n over $\mathbb{F}_2$ given by μ_a. The following hold:*
(i) *If $n \equiv 1 \pmod 8$, then $e_i(x) = \sum_{j \in S_i} x^j$ with $i = 1$ and 2 are generating idempotents of an even-like pair of binary duadic codes with splitting given by μ_a.*
(ii) *If $n \equiv -1 \pmod 8$, then $e_i(x) = 1 + \sum_{j \in S_i} x^j$ with $i = 1$ and 2 are generating idempotents of an even-like pair of binary duadic codes with splitting given by μ_a.*

Proof: Define $e_i(x) = \epsilon + \sum_{j \in S_i} x^j$ for $i = 1$ and 2, where $\epsilon = 0$ if $n \equiv 1 \pmod 8$ and $\epsilon = 1$ if $n \equiv -1 \pmod 8$. Then as S_i has size $(n-1)/2$, $e_i(x)$ has even weight and hence is even-like. Furthermore, $e_i(x)$ is an idempotent as $\{0\} \cup S_i$ is a union of 2-cyclotomic cosets, by Corollary 4.3.15. As $S_1 \mu_a = S_2$ and $S_2 \mu_a = S_1$, $e_1(x)\mu_a = e_2(x)$ and $e_2(x)\mu_a = e_1(x)$ showing (6.3), which is equivalent to (6.2). Because $S_1 \cup S_2 = \{1, 2, \ldots, n-1\}$ and $S_1 \cap S_2 = \emptyset$, $e_1(x) + e_2(x) = 2\epsilon + \sum_{j=1}^{n-1} x^j = \sum_{j=1}^{n-1} x^j$ giving (6.1). □

Exercise 335 Use Theorems 6.1.10 and 6.1.11 to find all the idempotents of the even-like binary duadic codes of length 73 where $\mu_a = \mu_{-1}$. ♦

The proof of the following theorem is straightforward and is left as an exercise.

Theorem 6.1.12 *Let μ_c be any multiplier. The pairs $\mathcal{C}_1$, $\mathcal{C}_2$, and $\mathcal{D}_1$, $\mathcal{D}_2$ are associated pairs of even-like and odd-like duadic codes with splitting S_1, S_2 if and only if $\mathcal{C}_1 \mu_c$, $\mathcal{C}_2 \mu_c$, and $\mathcal{D}_1 \mu_c$, $\mathcal{D}_2 \mu_c$ are associated pairs of even-like and odd-like duadic codes with splitting $S_1 \mu_{c^{-1}}$, $S_2 \mu_{c^{-1}}$.*

Exercise 336 Prove Theorem 6.1.12. ♦

As we will see in Section 6.6, quadratic residue codes are duadic codes of prime length $n = p$ for which S_1 is the set of nonzero quadratic residues (that is, squares) in $\mathbb{F}_q$, S_2 is the set of nonresidues (that is, nonsquares), and the multiplier interchanging the codes is μ_a, where a is any nonresidue. Quadratic residue codes exist only for prime lengths, but there are duadic codes of composite length. At prime lengths there may be duadic codes that are not quadratic residue codes. Duadic codes possess many of the properties of quadratic residue codes. For example, all the duadic codes of Examples 6.1.4, 6.1.7, 6.1.8, and 6.1.9 and Exercises 328 and 331 are quadratic residue codes. In Exercises 332 and 334, some of the codes are quadratic residue codes and some are not.

6.2   A bit of number theory

In this section we digress to present some results in number theory that we will need in the remainder of this chapter. We begin with a general result that will be used in this section and the next. Its proof can be found in [195, Theorem 4–11]. In a ring with unity, a *unit* is an element with a multiplicative inverse. The units in a ring with unity form a group.

Lemma 6.2.1 *Let $n > 1$ be an integer and let U_n be the group of units in $\mathbb{Z}_n$. Then:*
(i) *U_n has order $\phi(n)$, where ϕ is the Euler totient first described in Section 3.3.*
(ii) *U_n is cyclic if and only if $n = 2, 4, p^t,$ or $2p^t$, where p is an odd prime.*

Note that the units in U_n are often called *reduced residues modulo n*, and if U_n is cyclic, its generator is called a *primitive root*. We will need to know when certain numbers are squares modulo an odd prime p.

Lemma 6.2.2 *Let p be an odd prime and let a be in $\mathbb{Z}_p$ with $a \not\equiv 0 \pmod{p}$. The following are equivalent:*
(i)   *a is a square.*
(ii) *The (multiplicative) order of a is a divisor of $(p-1)/2$.*
(iii) *$a^{(p-1)/2} \equiv 1 \pmod{p}$.*
Furthermore, if a is not a square, then $a^{(p-1)/2} \equiv -1 \pmod{p}$.

Proof: Since $\mathbb{Z}_p$ is the field $\mathbb{F}_p$, the group $\mathbb{Z}_p^*$ of nonzero elements of $\mathbb{Z}_p$ is a cyclic group by Theorem 3.3.1 (or Lemma 6.2.1); let α be a generator of this cyclic group. The nonzero squares in $\mathbb{Z}_p$ form a multiplicative group $\mathcal{Q}$ with generator α^2, an element of order $(p-1)/2$. Thus the nonzero squares have orders a divisor of $(p-1)/2$ showing that (i) implies (ii). Suppose that a has order d a divisor of $(p-1)/2$. So $a^d \equiv 1 \pmod{p}$ and if $m = (p-1)/2$, then $1 \equiv (a^d)^{m/d} \equiv a^{(p-1)/2} \pmod{p}$; hence (ii) implies (iii). If (iii) holds, then a has order a divisor d of $(p-1)/2$. Therefore as there is a unique subgroup of order d in any cyclic group and $\mathcal{Q}$ contains such a subgroup, $a \in \mathcal{Q}$. Thus (iii) implies (i). Since $x^2 \equiv 1 \pmod{p}$ has only two solutions ± 1 in a field of odd characteristic and $(a^{(p-1)/2})^2 \equiv 1 \pmod{p}$, if a is not a square, then $a^{(p-1)/2} \equiv -1 \pmod{p}$ by (iii). $\square$

By this lemma, to check if a is a square modulo p, one only checks whether or not $a^{(p-1)/2}$ is 1 or -1. We now show that if a is a square modulo an odd prime p, then it is a square modulo any prime power p^t, where $t > 0$.

Lemma 6.2.3 *Let p be an odd prime and t a positive integer. Then a is a square modulo p if and only if a is a square modulo p^t.*

Proof: If $a \equiv b^2 \pmod{p^t}$, then $a \equiv b^2 \pmod p$. For the converse, suppose that a is a square modulo p. By Lemma 6.2.2, $a^{(p-1)/2} \equiv 1 \pmod p$. By Exercise 337

$$a^{(p-1)p^{t-1}/2} \equiv \left[a^{(p-1)/2}\right]^{p^{t-1}} \equiv 1 \pmod{p^t}. \tag{6.9}$$

Let U be the units in $\mathbb{Z}_{p^t}$. This group is cyclic of order $\phi(p^t) = (p-1)p^{t-1}$ by Lemma 6.2.1. In particular, U has even order as p is odd. Let β be a generator of U. The set of squares in U is a subgroup R of order $\phi(p^t)/2$ generated by β^2. This is the unique subgroup of that order and contains all elements of orders dividing $\phi(p^t)/2$. Therefore by (6.9), $a \in R$, completing the proof. $\qquad\square$

Exercise 337 Prove that if p is a prime, x is an integer, and t is a positive integer, then $(1+xp)^{p^{t-1}} = 1 + yp^t$ for some integer y. $\qquad\blacklozenge$

We will be interested in the odd primes for which -1, 2, and 3 are squares modulo p.

Lemma 6.2.4 Let p be an odd prime. Then -1 is a square modulo p if and only if $p \equiv 1 \pmod 4$.

Proof: As p is odd, $p \equiv \pm 1 \pmod 4$. Suppose that $p = 4r + 1$, where r is an integer. Then $(-1)^{(p-1)/2} = (-1)^{2r} = 1$; by Lemma 6.2.2, -1 is a square modulo p. Suppose now that $p = 4r - 1$, where r is an integer. Then $(-1)^{(p-1)/2} = (-1)^{2r-1} = -1$; by Lemma 6.2.2, -1 is not a square modulo p. $\qquad\square$

Lemma 6.2.5 Let p be an odd prime. Then 2 is a square modulo p if and only if $p \equiv \pm 1 \pmod 8$.

Proof: We prove only the case $p = 8r + 1$, where r is an integer, and leave the cases $p = 8r - 1$ and $p = 8r \pm 3$ as exercises. Let $b = 2 \cdot 4 \cdot 6 \cdots 8r = 2^{4r}(4r)!$. Then

$$b = 2 \cdot 4 \cdot 6 \cdots 4r \cdot [p - (4r - 1)][p - (4r - 3)] \cdots (p - 1).$$

Considering b modulo p, we obtain $2^{4r}(4r)! \equiv b \equiv (-1)^{2r}(4r)! \pmod p$. Thus $2^{(p-1)/2} = 2^{4r} \equiv (-1)^{2r} \equiv 1 \pmod p$. By Lemma 6.2.2, 2 is a square modulo p. $\qquad\square$

Exercise 338 Complete the proof of Lemma 6.2.5 by examining the cases $p = 8r - 1$ and $p = 8r \pm 3$, where r is an integer. $\qquad\blacklozenge$

We will need to know information about $\mathrm{ord}_p(2)$ for later work. Using Lemma 6.2.5, we obtain the following result.

Lemma 6.2.6 Let p be an odd prime. The following hold:
(i)   If $p \equiv -1 \pmod 8$, then $\mathrm{ord}_p(2)$ is odd.
(ii) If $p \equiv 3 \pmod 8$, then $2 \mid \mathrm{ord}_p(2)$ but $4 \nmid \mathrm{ord}_p(2)$.
(iii) If $p \equiv -3 \pmod 8$, then $4 \mid \mathrm{ord}_p(2)$ but $8 \nmid \mathrm{ord}_p(2)$.

Proof: For (i), let $p = 8r - 1$, where r is an integer. By Lemmas 6.2.2 and 6.2.5, $2^{(p-1)/2} = 2^{4r-1} \equiv 1 \pmod p$. So $\mathrm{ord}_p(2)$ is a divisor of $4r - 1$, which is odd. Part (i) follows. For (ii), let $p = 8r + 3$, where r is an integer. By Lemmas 6.2.2 and 6.2.5, $2^{(p-1)/2} = 2^{4r+1} \equiv -1 \pmod p$. So as $\mathrm{ord}_p(2)$ is always a divisor of $p - 1 = 8r + 2 = 2(4r + 1)$ but not a divisor of $4r + 1$, $2 \mid \mathrm{ord}_p(2)$; since $4 \nmid (p - 1)$, $4 \nmid \mathrm{ord}_p(2)$ yielding part (ii). Finally for (iii), let $p = 8r - 3$, where r is an integer. By Lemmas 6.2.2 and 6.2.5, $2^{(p-1)/2} = 2^{4r-2} \equiv -1 \pmod{p}$. So as $\mathrm{ord}_p(2)$ is always a divisor of $p - 1 = 8r - 4 = 4(2r - 1)$ but not a divisor of $4r - 2 = 2(2r - 1)$, $4 \mid \mathrm{ord}_p(2)$; since $8 \nmid (p - 1)$, $8 \nmid \mathrm{ord}_p(2)$, showing (iii) holds. $\qquad\square$

Note that Lemma 6.2.6 does not address the case $p \equiv 1 \pmod 8$ because $\mathrm{ord}_p(2)$ can have various powers of 2 as a factor depending on p.

Corollary 6.2.7 *Let p be an odd prime. The following hold*:
(i)   *If $p \equiv -1 \pmod 8$, then $\mathrm{ord}_p(4) = \mathrm{ord}_p(2)$ and hence is odd.*
(ii) *If $p \equiv 3 \pmod 8$, then $\mathrm{ord}_p(4)$ is odd.*
(iii) *If $p \equiv -3 \pmod 8$, then $2 \mid \mathrm{ord}_p(4)$ but $4 \nmid \mathrm{ord}_p(4)$.*

Exercise 339 Prove Corollary 6.2.7.   ♦

Exercise 340 Compute $\mathrm{ord}_p(2)$ and $\mathrm{ord}_p(4)$ when $3 \le p < 100$ and p is prime. Compare the results to those implied by Lemma 6.2.6 and Corollary 6.2.7.   ♦

In order to discover when 3 is a square modulo p, it is simplest to use quadratic reciprocity. To do that we define the *Legendre symbol*

$$\left(\frac{a}{p}\right) = \begin{cases} 1 & \text{if } a \text{ is a nonzero square modulo } p \\ -1 & \text{if } a \text{ is a nonsquare modulo } p. \end{cases}$$

By Lemma 6.2.2 $(a/p) \equiv a^{(p-1)/2} \pmod p$. The Law of Quadratic Reciprocity, a proof of which can be found in [195], allows someone to decide if q is a square modulo p by determining if p is a square modulo q whenever p and q are distinct odd primes.

Theorem 6.2.8 (Law of Quadratic Reciprocity) *Suppose p and q are distinct odd primes. Then*

$$\left(\frac{q}{p}\right)\left(\frac{p}{q}\right) = (-1)^{\frac{p-1}{2}\frac{q-1}{2}}.$$

Lemma 6.2.9 *Let $p \ne 3$ be an odd prime. Then 3 is a square modulo p if and only if $p \equiv \pm 1 \pmod{12}$.*

Proof: We wish to find $\left(\frac{3}{p}\right)$. As p is an odd prime, $p = 12r \pm 1$ or $p = 12r \pm 5$, where r is an integer. Suppose that $p = 12r + 1$. Then $p \equiv 1 \equiv 1^2 \pmod 3$ and so $\left(\frac{p}{3}\right) = 1$. Also,

$$(-1)^{\frac{p-1}{2}\frac{3-1}{2}} = (-1)^{6r} = 1.$$

By the Law of Quadratic Reciprocity, $\left(\frac{3}{p}\right) = 1$. The other cases are left as an exercise. $\qquad\square$

Exercise 341 Complete the proof of Lemma 6.2.9 by examining the cases $p = 12r - 1$ and $p = 12r \pm 5$, where r is an integer.   ♦

Exercise 342 Make a table indicating when -1, 2, and 3 are squares modulo p, where p is a prime with $3 < p < 100$.   ♦

Exercise 343 Show that if p is an odd prime, then

$$\left(\frac{2}{p}\right) = (-1)^{\frac{p^2-1}{8}}.$$

◆

6.3 Existence of duadic codes

The existence of duadic codes for a given length n depends on the existence of a splitting of n by Corollary 6.1.6. In this section we determine, for each q, the integers n for which splittings exist. The following lemma reduces the problem to the case where n is a prime power; we sketch the proof leaving the details to Exercise 344.

Lemma 6.3.1 *Let $n = n_1 n_2$ where $\gcd(n_1, n_2) = 1$. There is a splitting of n given by μ_a if and only if there are splittings of n_1 and n_2 given by $\mu_{a \bmod n_1}$ and $\mu_{a \bmod n_2}$, respectively. Furthermore, q is a square modulo n if and only if q is a square modulo n_1 and a square modulo n_2.*

Proof: Since $\gcd(n_1, n_2) = 1$, it follows from the Chinese Remainder Theorem [195] that $z\theta = (z \bmod n_1, z \bmod n_2)$ defines a ring isomorphism θ from $\mathbb{Z}_n$ onto $\mathbb{Z}_{n_1} \times \mathbb{Z}_{n_2}$. The second assertion follows from this observation.

For the first assertion, let $Z_1 = \{(z, 0) \mid z \in \mathbb{Z}_{n_1}\}$ and $Z_2 = \{(0, z) \mid z \in \mathbb{Z}_{n_2}\}$. For $i = 1$ and 2, the projections $\pi_i : Z_i \mapsto \mathbb{Z}_{n_i}$ are ring isomorphisms. If C is a q-cyclotomic coset modulo n and $C\theta \cap Z_i \neq \emptyset$, $(C\theta \cap Z_i)\pi_i$ is a q-cyclotomic coset modulo n_i. Let S_1 and S_2 form a splitting of n given by μ_b. Then $(S_1\theta \cap Z_i)\pi_i$ and $(S_2\theta \cap Z_i)\pi_i$ form a splitting of n_i given by μ_{b_i} where $b_i \equiv b \pmod{n_i}$.

Conversely, let $S_{1,i}$ and $S_{2,i}$ form a splitting of n_i given by μ_{b_i} for $i = 1$ and 2. Then $((S_{1,1} \times \mathbb{Z}_{n_2}) \cup (\{0\} \times S_{1,2}))\theta^{-1}$ and $((S_{2,1} \times \mathbb{Z}_{n_1}) \cup (\{0\} \times S_{2,2}))\theta^{-1}$ form a splitting of n given by μ_b where $b = (b_1, b_2)\theta^{-1}$. □

Exercise 344 In this exercise, fill in some of the details of Lemma 6.3.1 by doing the following (where n_1 and n_2 are relatively prime and $n = n_1 n_2$):

(a) Prove that $z\theta = (z \bmod n_1, z \bmod n_2)$ defines a ring isomorphism θ from $\mathbb{Z}_n$ onto $\mathbb{Z}_{n_1} \times \mathbb{Z}_{n_2}$.

(b) Prove that if C is a q-cyclotomic coset modulo n and $C\theta \cap Z_i \neq \emptyset$, then $(C\theta \cap Z_i)\pi_i$ is a q-cyclotomic coset modulo n_i.

(c) Prove that if S_1 and S_2 form a splitting of n given by μ_b, then $(S_1\theta \cap Z_i)\pi_i$ and $(S_2\theta \cap Z_i)\pi_i$ form a splitting of n_i given by μ_{b_i} where $b_i \equiv b \pmod{n_i}$.

(d) Let $S_{1,i}$ and $S_{2,i}$ form a splitting of n_i given by μ_{b_i} for $i = 1$ and 2. Prove that $((S_{1,1} \times \mathbb{Z}_{n_2}) \cup (\{0\} \times S_{1,2}))\theta^{-1}$ and $((S_{2,1} \times \mathbb{Z}_{n_1}) \cup (\{0\} \times S_{2,2}))\theta^{-1}$ form a splitting of n given by μ_b, where $b = (b_1, b_2)\theta^{-1}$.
◆

We are now ready to give the criteria for the existence of duadic codes of length n over $\mathbb{F}_q$.

Theorem 6.3.2 *Duadic codes of length n over $\mathbb{F}_q$ exist if and only if q is a square modulo n.*

Proof: By Lemma 6.3.1, we may assume that $n = p^m$, where p is an odd prime. We first show that if a splitting of $n = p^m$ exists, then q is a square modulo n. Let U be the group of units in $\mathbb{Z}_n$. By Lemma 6.2.1 this group is cyclic of order $\phi(p^m) = (p-1)p^{m-1}$, which is even as p is odd. Since q is relatively prime to n, $q \in U$. Let R be the subgroup of U consisting of the squares in U. We only need to show that $q \in R$. If u generates U, then u^2 generates R. As U has even order, R has index 2 in U. Since $q \in U$, define Q to be the subgroup of U generated by q. Notice that if $a \in U$, then $aq \in U$ and hence U is a union of q-cyclotomic cosets modulo n; in fact, the q-cyclotomic cosets contained in U are precisely the cosets of Q in U. The number of q-cyclotomic cosets in U is then the index $|U : Q|$ of Q in U. Let S_1 and S_2 form a splitting of n given by μ_b. Each q-cyclotomic coset of U is in precisely one S_i as $U \subseteq S_1 \cup S_2$ and $S_1 \cap S_2 = \emptyset$. Because b and n are relatively prime, $b \in U$ and so $U\mu_b = U$ implying that $(U \cap S_1)\mu_b = U \cap S_2$. In particular, this says that U has an even number of q-cyclotomic cosets. Thus $|U : Q|$ is even; as $|U : R| = 2$ and U is cyclic, $Q \subseteq R$. Thus $q \in R$ as desired.

Now assume that q is a square modulo n. We show how to construct a splitting of n. For $1 \le t \le m$, let U_t be the group of units in $\mathbb{Z}_{p^t}$. Let R_t be the subgroup of U_t consisting of the squares of elements in U_t and let Q_t be the subgroup of U_t generated by q. As in the previous paragraph, U_t is cyclic of even order, and R_t has index 2 in U_t. As q is a square modulo p^m, then q is a square modulo p^t implying that $Q_t \subseteq R_t$. Finally, U_t is a union of q-cyclotomic cosets modulo p^t and these are precisely the cosets of Q_t in U_t. The nonzero elements of $\mathbb{Z}_n$ are the set

$$\bigcup_{t=1}^{m} p^{m-t} U_t. \tag{6.10}$$

We are now ready to construct the splitting of n. Since U_t is cyclic and $Q_t \subseteq R_t \subseteq U_t$ with $|U_t : R_t| = 2$, there is a unique subgroup K_t of U_t containing Q_t such that $|K_t : Q_t| = 2$. Note that U_t, R_t, Q_t, and K_t can be obtained from U_m, R_m, Q_m, and K_m by reducing the latter modulo p^t. Let $b \in K_m \setminus Q_m$. Then $K_m = Q_m \cup bQ_m$, and hence, by reducing modulo p^t, $K_t = Q_t \cup bQ_t$ for $1 \le t \le m$. Also, $b^2 \in Q_t$ modulo p^t. Let $g_1^{(t)}, g_2^{(t)}, \ldots, g_{i_t}^{(t)}$ be distinct coset representatives of K_t in U_t. Then the q-cyclotomic cosets modulo p^t in U_t are precisely the cosets $g_1^{(t)}Q_t, g_2^{(t)}Q_t, \ldots, g_{i_t}^{(t)}Q_t, bg_1^{(t)}Q_t, bg_2^{(t)}Q_t, \ldots, bg_{i_t}^{(t)}Q_t$. Let $S_{1,t} = g_1^{(t)}Q_t \cup g_2^{(t)}Q_t \cup \cdots \cup g_{i_t}^{(t)}Q_t$ and $S_{2,t} = bg_1^{(t)}Q_t \cup bg_2^{(t)}Q_t \cup \cdots \cup bg_{i_t}^{(t)}Q_t$. Then $S_{1,t}\mu_b = S_{2,t}$ and $S_{2,t}\mu_b = S_{1,t}$ as $b^2 \in Q_t$ modulo p^t, $S_{1,t} \cap S_{2,t} = \emptyset$, and $S_{1,t} \cup S_{2,t} = U_t$. Note that $p^{m-t}g_j^{(t)}Q_t$ and $p^{m-t}bg_j^{(t)}Q_t$ are q-cyclotomic cosets modulo p^m. Thus by (6.10), $S_1 = \bigcup_{t=1}^{m} p^{m-t}S_{1,t}$ and $S_2 = \bigcup_{t=1}^{m} p^{m-t}S_{2,t}$ form a splitting of n given by μ_b. $\square$

Exercise 345 Show that 2 is a square modulo 49. Use the technique in the proof to construct a splitting of 49 over $\mathbb{F}_2$. ♦

We can now give necessary and sufficient conditions on the length n for the existence of binary, ternary, and quaternary duadic codes. Recall from Theorem 6.1.3(iii) that n must be odd.

Theorem 6.3.3 *Let $n = p_1^{a_1} p_2^{a_2} \cdots p_r^{a_r}$ where $p_1, p_2, \ldots, p_r$ are distinct odd primes. The following assertions hold:*

(i)   Duadic codes of length n over $\mathbb{F}_2$ exist if and only if $p_i \equiv \pm 1 \pmod 8$ for $1 \leq i \leq r$.

(ii) Duadic codes of length n over $\mathbb{F}_3$ exist if and only if $p_i \equiv \pm 1 \pmod{12}$ for $1 \leq i \leq r$.

(iii) Duadic codes of length n over $\mathbb{F}_4$ exist for all (odd) n.

Proof: Duadic codes of length n over $\mathbb{F}_q$ exist if and only if q is a square modulo n by Theorem 6.3.2. By Lemma 6.3.1, q is a square modulo n if and only if q is a square modulo $p_i^{a_i}$ for $1 \leq i \leq r$. By Lemma 6.2.3, q is a square modulo $p_i^{a_i}$ if and only if q is a square modulo p_i. Part (i) now follows from Lemma 6.2.5 and (ii) from Lemma 6.2.9. Finally, part (iii) follows from the simple fact that $4 = 2^2$ is always a square modulo n.   □

Exercise 346 Find the integers n, with $3 < n < 200$, for which binary and ternary duadic codes exist.   ◆

In the binary case, there are results that count the number of duadic codes of prime length $n = p$. By Theorem 6.3.3, $p \equiv \pm 1 \pmod 8$. Let $e = (p - 1)/(2\operatorname{ord}_p(2))$. Then it can be shown [67] that the number of duadic codes of length p depends only on e. Further, the number of inequivalent duadic codes also depends only on e. If e is odd, the number of pairs of odd-like (or even-like) binary duadic codes is 2^{e-1}. When e is even, there is a more complicated bound for this number. For example, if $p = 31$, then $\operatorname{ord}_p(2) = 5$ and so $e = 3$. There are four pairs of odd-like (or even-like) duadic codes. One pair is a pair of quadratic residue codes; the other three pairs consist of six equivalent codes (see Example 6.4.8). These facts hold for any length $p \equiv \pm 1 \pmod 8$ for which $e = 3$, such as $p = 223, 433, 439, 457,$ and 727.

6.4   Orthogonality of duadic codes

The multiplier μ_{-1} plays a special role in determining the duals of duadic codes just as it does for duals of general cyclic codes; see Theorem 4.4.9. We first consider self-orthogonal codes.

Theorem 6.4.1 Let C be any $[n, (n-1)/2]$ cyclic code over $\mathbb{F}_q$. Then C is self-orthogonal if and only if C is an even-like duadic code whose splitting is given by μ_{-1}.

Proof: Suppose $C_1 = C$ is self-orthogonal with idempotent generator $e_1(x)$. Let $C_2 = \langle e_2(x) \rangle$, where $e_2(x) = e_1(x)\mu_{-1}$. Since C_1 is self-orthogonal and $\overline{j}(x)$ is not orthogonal to itself, $\overline{j}(x) \notin C_1$. Thus by Lemma 6.1.2(iii), C_1 is even-like, which implies that $\overline{j}(x) \in C_1^\perp$. By Theorem 4.4.9, $\overline{j}(x) \in C_1^\perp = \langle 1 - e_1(x)\mu_{-1} \rangle$. As $C_1^\perp$ has dimension $(n+1)/2$ and $C_1 \subset C_1^\perp$, we have $C_1^\perp = C_1 + \langle \overline{j}(x) \rangle$. Since $e_1(x)\overline{j}(x) = 0$ by Lemma 6.1.2(i), it follows from Theorem 4.3.7 that $C_1^\perp$ has generating idempotent $e_1(x) + \overline{j}(x) = 1 - e_1(x)\mu_{-1} = 1 - e_2(x)$, giving (6.1). Since $e_2(x) = e_1(x)\mu_{-1}$, $e_1(x) = e_2(x)\mu_{-1}^{-1} = e_2(x)\mu_{-1}$ yielding (6.3). Thus C_1 and C_2 are even-like duadic codes with splitting by μ_{-1}.

Conversely, let $C = C_1 = \langle e_1(x) \rangle$ be an even-like duadic code with splitting given by μ_{-1}. Then $C_2 = \langle e_1(x)\mu_{-1} \rangle$ and $C_1 \subset D_1 = \langle 1 - e_1(x)\mu_{-1} \rangle = C_1^\perp$ by Theorems 6.1.3(vi) and 4.4.9 and (6.4).   □

By Theorem 6.1.3(iii) and (v), the dimensions of even-like and odd-like duadic codes C_i and D_i indicate it is possible (although not necessary) that the dual of C_1 is either D_1 or D_2. The next two theorems describe when these two possibilities occur.

Theorem 6.4.2 *If C_1 and C_2 are a pair of even-like duadic codes over $\mathbb{F}_q$, with D_1 and D_2 the associated pair of odd-like duadic codes, the following are equivalent:*
(i) $C_1^\perp = D_1$.
(ii) $C_2^\perp = D_2$.
(iii) $C_1 \mu_{-1} = C_2$.
(iv) $C_2 \mu_{-1} = C_1$.

Proof: Since $C_1 \mu_a = C_2$, $C_2 \mu_a = C_1$, $D_1 \mu_a = D_2$, and $D_2 \mu_a = D_1$ for some a (by definition of the even-like duadic codes and Theorem 6.1.3(vii)), (i) is equivalent to (ii) by Exercise 35. Parts (iii) and (iv) are equivalent as $\mu_{-1}^{-1} = \mu_{-1}$. If (i) holds, C_1 is self-orthogonal by Theorem 6.1.3(vi), implying that (iii) holds by Theorem 6.4.1. Conversely, if (iii) holds, letting $e_i(x)$ be the generating idempotent for C_i, we have $e_1(x) \mu_{-1} = e_2(x)$ by Theorem 4.3.13. By Theorem 4.4.9, $C_1^\perp = \langle 1 - e_1(x) \mu_{-1} \rangle = \langle 1 - e_2(x) \rangle = D_1$ yielding (i). $\qquad\square$

Theorem 6.4.3 *If C_1 and C_2 are a pair of even-like duadic codes over $\mathbb{F}_q$ with D_1 and D_2 the associated pair of odd-like duadic codes, the following are equivalent:*
(i) $C_1^\perp = D_2$.
(ii) $C_2^\perp = D_1$.
(iii) $C_1 \mu_{-1} = C_1$.
(iv) $C_2 \mu_{-1} = C_2$.

Proof: Since $C_1 \mu_a = C_2$, $C_2 \mu_a = C_1$, $D_1 \mu_a = D_2$, and $D_2 \mu_a = D_1$ for some a (by definition of the even-like duadic codes and Theorem 6.1.3(vii)), (i) is equivalent to (ii) by Exercise 35. Let $e_i(x)$ be the generating idempotent of C_i. By Theorem 4.4.9, $C_1^\perp = D_2$ if and only if $1 - e_1(x) \mu_{-1} = 1 - e_1(x)$ if and only if $e_1(x) \mu_{-1} = e_1(x)$. So $C_1^\perp = D_2$ if and only if $C_1 \mu_{-1} = C_1$ by Theorem 4.3.13. Thus (i) and (iii) are equivalent and, analogously, (ii) and (iv) are equivalent. The theorem now follows. $\qquad\square$

Exercise 347 Identify the duadic codes and their duals in Exercises 328, 331, 332, and 335 and in Example 6.1.7. ◆

Results analogous to those in the last three theorems hold with μ_{-2} in place of μ_{-1} for codes over $\mathbb{F}_4$ where the orthogonality is with respect to the Hermitian inner product defined in Section 1.3. These codes are of interest because if a code over $\mathbb{F}_4$ has only even weight codewords, it must be Hermitian self-orthogonal by Theorem 1.4.10. The key to the analogy is the result of Theorem 4.4.15 that if C is a cyclic code over $\mathbb{F}_4$ with generating idempotent $e(x)$, then $C^{\perp_H}$ has generating idempotent $1 - e(x) \mu_{-2}$. We leave the proofs as an exercise.

Theorem 6.4.4 *Let C be any $[n, (n-1)/2]$ cyclic code over $\mathbb{F}_4$. Then C is Hermitian self-orthogonal if and only if C is an even-like duadic code whose splitting is given by μ_{-2}.*

Theorem 6.4.5 *If C_1 and C_2 are a pair of even-like duadic codes over $\mathbb{F}_4$ with D_1 and D_2 the associated pair of odd-like duadic codes, the following are equivalent:*

(i) $C_1^{\perp_H} = \mathcal{D}_1$.
(ii) $C_2^{\perp_H} = \mathcal{D}_2$.
(iii) $C_1 \mu_{-2} = C_2$.
(iv) $C_2 \mu_{-2} = C_1$.

Theorem 6.4.6 *If C_1 and C_2 are a pair of even-like duadic codes over $\mathbb{F}_4$ with $\mathcal{D}_1$ and $\mathcal{D}_2$ the associated pair of odd-like duadic codes, the following are equivalent:*
(i) $C_1^{\perp_H} = \mathcal{D}_2$.
(ii) $C_2^{\perp_H} = \mathcal{D}_1$.
(iii) $C_1 \mu_{-2} = C_1$.
(iv) $C_2 \mu_{-2} = C_2$.

Exercise 348 Identify the duadic codes over $\mathbb{F}_4$ and both their ordinary duals and Hermitian duals in Examples 6.1.8 and 6.1.9 and in Exercise 334.   ♦

Exercise 349 Prove Theorems 6.4.4, 6.4.5, and 6.4.6.   ♦

In the binary case, μ_{-1} gives every splitting of duadic codes of prime length $p \equiv -1$ (mod 8), as the following result shows.

Theorem 6.4.7 *If p is a prime with $p \equiv -1$ (mod 8), then every splitting of p over $\mathbb{F}_2$ is given by μ_{-1}. Furthermore, every binary even-like duadic code of length $p \equiv -1$ (mod 8) is self-orthogonal.*

Proof: The second part of the theorem follows from the first part and Theorem 6.4.1. For the first part, suppose that μ_a gives a splitting S_1 and S_2 for p. As μ_a interchanges S_1 and S_2, μ_a, and hence a, cannot have odd (multiplicative) order. Suppose that a has order $2w$. Then a^w is a solution of $x^2 = 1$ in $\mathbb{Z}_p = \mathbb{F}_p$, which has only the solutions 1 and -1. Since $a^w \neq 1$ in $\mathbb{Z}_p$, $a^w \equiv -1$ (mod p). If $w = 2v$, then -1 is the square of a^v in $\mathbb{Z}_p$, a contradiction to Lemma 6.2.4. So w is odd. Since μ_a swaps S_1 and S_2, applying μ_a an odd number of times swaps them as well. Hence $(\mu_a)^w = \mu_{a^w} = \mu_{-1}$ gives the same splitting as μ_a.   □

We use this result to find all the splittings of 31 over $\mathbb{F}_2$ leading to all the binary duadic codes of length 31.

Example 6.4.8 There are seven 2-cyclotomic cosets modulo 31, namely: $C_0 = \{0\}$, $C_1 = \{1, 2, 4, 8, 16\}$, $C_3 = \{3, 6, 12, 17, 24\}$, $C_5 = \{5, 9, 10, 18, 20\}$, $C_7 = \{7, 14, 19, 25, 28\}$, $C_{11} = \{11, 13, 21, 22, 26\}$, $C_{15} = \{15, 23, 27, 29, 30\}$. By Theorem 6.4.7 all splittings S_1 and S_2 can be obtained using μ_{-1} as $31 \equiv -1$ (mod 8). Since $C_1 \mu_{-1} = C_{15}$, one of C_1 and C_{15} is in S_1 and the other is in S_2. By renumbering S_1 and S_2 if necessary, we may assume that C_1 is in S_1. Similarly as $C_3 \mu_{-1} = C_7$ and $C_5 \mu_{-1} = C_{11}$, precisely one of C_3 or C_7 and one of C_5 and C_{11} is in S_1. Thus there are four possible splittings and hence four possible pairs of even-like duadic codes (and four odd-like pairs as well). We give the splittings in the following table.

Splitting	S_1	S_2
1	$C_1 \cup C_5 \cup C_7$	$C_3 \cup C_{11} \cup C_{15}$
2	$C_1 \cup C_5 \cup C_3$	$C_7 \cup C_{11} \cup C_{15}$
3	$C_1 \cup C_{11} \cup C_7$	$C_3 \cup C_5 \cup C_{15}$
4	$C_1 \cup C_{11} \cup C_3$	$C_7 \cup C_5 \cup C_{15}$

By Theorems 6.1.10 and 6.1.11, we can use these splittings to determine generating idempotents directly. The generating idempotents for the even-like duadic codes are $e_i(x) = 1 + \sum_{j \in S_i} x^j$; for the odd-like duadic codes they are $e_i(x) = \sum_{j \in S_i} x^j$. The multiplier μ_3 sends splitting number 2 to splitting number 3; it also sends splitting number 3 to splitting number 4. Hence the corresponding generating idempotents are mapped by μ_3 in the same way, showing that the six even-like duadic codes (and six odd-like duadic codes) arising from these splittings are equivalent by Theorem 4.3.13. These turn out to be the punctured Reed–Muller codes $\mathcal{R}(2, 5)^*$; see Section 1.10. All multipliers map splitting number 1 to itself (although a multiplier may reverse S_1 and S_2). By Theorem 4.3.17, as 31 is a prime, a duadic code arising from splitting number 1 is not equivalent to any duadic code arising from splitting number 2, 3, or 4. The duadic codes arising from this first splitting are quadratic residue codes. See Section 6.6. ■

Theorem 6.4.7 shows when all splittings over $\mathbb{F}_2$ of an odd prime p are given by μ_{-1}. The following result for splittings of an odd prime p over $\mathbb{F}_4$ are analogous; the proof is in [266].

Theorem 6.4.9 *If p is an odd prime, then every splitting of p over $\mathbb{F}_4$ is given by:*
(i)   *both μ_{-1} and μ_{-2} when $p \equiv -1 \pmod 8$,*
(ii) *both μ_{-1} and μ_2 when $p \equiv 3 \pmod 8$, and*
(iii) *both μ_{-2} and μ_2 when $p \equiv -3 \pmod 8$.*

Theorems 6.4.7 and 6.4.9 do not examine the case $p \equiv 1 \pmod 8$. The next result shows what happens there; see Lemma 6.4.16 and [266]. Part of this result is explained in Exercise 351.

Theorem 6.4.10 *Let $p \equiv 1 \pmod 8$ be a prime. The following hold:*
(i)   *If $\mathrm{ord}_p(2)$ is odd, then some splittings for p over $\mathbb{F}_2$, but not all, can be given by μ_{-1}; splittings given by μ_{-1} are precisely those given by μ_{-2}.*
(ii) *If $\mathrm{ord}_p(2)$ is odd, then some splittings for p over $\mathbb{F}_4$, but not all, can be given by μ_{-1}; splittings given by μ_{-1} are precisely those given by μ_{-2}.*
(iii) *If $2 \mid \mathrm{ord}_p(2)$ but $4 \nmid \mathrm{ord}_p(2)$, then some splittings for p over $\mathbb{F}_4$, but not all, can be given by μ_{-1}; splittings given by μ_{-1} are precisely those given by μ_2.*
(iv) *If $4 \mid \mathrm{ord}_p(2)$, then some splittings for p over $\mathbb{F}_4$, but not all, can be given by μ_{-2}; splittings given by μ_{-2} are precisely those given by μ_2.*

Example 6.4.11 We look at various primes $p \equiv 1 \pmod 8$ in light of Theorem 6.4.10. If $p = 17$, $\mathrm{ord}_p(2) = 8$ and the theorem indicates that some, but not all, splittings over $\mathbb{F}_4$ are given by both μ_{-2} and μ_2. If $p = 41$, $\mathrm{ord}_p(2) = 20$ and there is some splitting of 41 over $\mathbb{F}_4$ given by both μ_{-2} and μ_2. If $p = 57$, $\mathrm{ord}_p(2) = 18$ and there is some splitting of 57 over

$\mathbb{F}_4$ given by both μ_{-1} and μ_2. Finally, if $p = 73$, $\mathrm{ord}_p(2) = 9$ and there is some splitting of 73 over $\mathbb{F}_2$ given by both μ_{-1} and μ_{-2}; there is also some splitting of 73 over $\mathbb{F}_4$ given by both μ_{-1} and μ_{-2}.   ∎

Exercise 350 Find the splittings alluded to in Example 6.4.11. Also, what are the other splittings and what multipliers give them?   ♦

Exercise 351 This exercise explains part of Theorem 6.4.10. In Section 6.6 we will discuss quadratic residue codes over $\mathbb{F}_q$, which are merely duadic codes of odd prime length p with splittings over $\mathbb{F}_q$ given by $S_1 = \mathcal{Q}_p$ and $S_2 = \mathcal{N}_p$, where $\mathcal{Q}_p$ are the nonzero squares and $\mathcal{N}_p$ the nonsquares in $\mathbb{F}_p$. Binary quadratic residue codes exist if $p \equiv \pm 1 \pmod 8$; quadratic residue codes over $\mathbb{F}_4$ exist for any odd prime p. Do the following:
(a) Show that in $\mathbb{F}_p$ the product of two squares or two nonsquares is a square.
(b) Show that in $\mathbb{F}_p$ the product of a square and a nonsquare is a nonsquare.
(c) Show that if $p \equiv 1 \pmod 8$, then -1, 2, and -2 are all squares in $\mathbb{F}_p$.
(d) Prove that if $p \equiv 1 \pmod 8$, the splitting $S_1 = \mathcal{Q}_p$ and $S_2 = \mathcal{N}_p$ over $\mathbb{F}_q$ cannot be given by either μ_{-1} or μ_{-2}.   ♦

We now consider extending odd-like duadic codes over $\mathbb{F}_q$. Recall that if $\mathcal{D}$ is an odd-like code of length n, the ordinary extension $\widehat{\mathcal{D}}$ of $\mathcal{D}$ is defined by $\widehat{\mathcal{D}} = \{\widehat{\mathbf{c}} \mid \mathbf{c} \in \mathcal{D}\}$ where $\widehat{\mathbf{c}} = \mathbf{c}c_\infty = c_0 \cdots c_{p-1}c_\infty$ and

$$c_\infty = -\sum_{j=0}^{p-1} c_j.$$

Since the odd-like duadic codes are $[n, (n+1)/2]$ codes by Theorem 6.1.3(v), the extended codes would be $[n+1, (n+1)/2]$ codes. Hence these codes could potentially be self-dual. If $\mathcal{D}$ is an odd-like duadic code, $\mathcal{D}$ is obtained from its even-like subcode $\mathcal{C}$ by adding $\overline{j}(x)$ to a basis of $\mathcal{C}$. But $\overline{j}(x)$ is merely a multiple of the all-one vector $\mathbf{1}$, and so if we want to extend $\mathcal{D}$ in such a way that its extension is either self-dual or dual to the other extended odd-like duadic code, then we must extend $\mathbf{1}$ in such a way that it is orthogonal to itself. The ordinary extension will not always yield this. However, the next result shows that we can modify the method of extension for $\mathcal{D}_1$ and $\mathcal{D}_2$ if μ_{-1} gives the splitting and the equation

$$1 + \gamma^2 n = 0 \tag{6.11}$$

has a solution γ in $\mathbb{F}_q$. Let $\mathcal{D}$ be an odd-like duadic code and suppose (6.11) has a solution. If $\mathbf{c} = c_0 c_1 \cdots c_{n-1} \in \mathcal{D}$, let $\widetilde{\mathbf{c}} = c_0 c_1 \cdots c_{n-1}c_\infty$, where

$$c_\infty = -\gamma \sum_{i=0}^{n-1} c_i.$$

The new extension of $\mathcal{D}$ is defined by $\widetilde{\mathcal{D}} = \{\widetilde{\mathbf{c}} \mid \mathbf{c} \in \mathcal{D}\}$. Notice that the codes $\widehat{\mathcal{D}}$ and $\widetilde{\mathcal{D}}$ are monomially equivalent because $\widetilde{\mathcal{D}} = \widehat{\mathcal{D}}D$, where D is the $(n+1) \times (n+1)$ diagonal matrix $D = \mathrm{diag}(1, 1, \ldots, 1, \gamma)$.

Theorem 6.4.12 *Let $\mathcal{D}_1$ and $\mathcal{D}_2$ be a pair of odd-like duadic codes of length n over $\mathbb{F}_q$. Assume that there is a solution $\gamma \in \mathbb{F}_q$ to (6.11). The following hold:*
(i) *If μ_{-1} gives the splitting for $\mathcal{D}_1$ and $\mathcal{D}_2$, then $\widetilde{\mathcal{D}}_1$ and $\widetilde{\mathcal{D}}_2$ are self-dual.*
(ii) *If $\mathcal{D}_1\mu_{-1} = \mathcal{D}_1$, then $\widetilde{\mathcal{D}}_1$ and $\widetilde{\mathcal{D}}_2$ are duals of each other.*

Proof: Let $\mathcal{C}_1$ and $\mathcal{C}_2$ be the pair of even-like duadic codes associated to $\mathcal{D}_1$ and $\mathcal{D}_2$. We have the following two observations:

(a) $\widetilde{\overline{j}}(x)$ is orthogonal to itself.
(b) $\widetilde{\overline{j}}(x)$ is orthogonal to $\widetilde{\mathcal{C}}_i$.

The first holds because as a vector $\widetilde{\overline{j}}(x)$ is $(1/n, 1/n, \dots, 1/n, -\gamma) \in \mathbb{F}_q^{n+1}$. By the choice of γ, $\widetilde{\overline{j}}(x)$ is orthogonal to itself. The second holds because $\mathcal{C}_i$ is even-like and the extended coordinate in $\widetilde{\mathcal{C}}_i$ is always 0.

Assume that μ_{-1} gives the splitting of $\mathcal{D}_1$ and $\mathcal{D}_2$. Then $\mathcal{C}_1\mu_{-1} = \mathcal{C}_2$. By Theorem 6.4.2, $\mathcal{C}_1 = \mathcal{D}_1^\perp$; also $\mathcal{C}_1$ is self-orthogonal by Theorem 6.4.1. As the extension of $\mathcal{C}_1$ is trivial, $\widetilde{\mathcal{C}}_1$ is self-orthogonal. From (a) and (b) the code spanned by $\widetilde{\mathcal{C}}_1$ and $\widetilde{\overline{j}}(x)$ is self-orthogonal. By Theorem 6.1.3(v) and (ix), this self-orthogonal code is $\widetilde{\mathcal{D}}_1$ and so is self-dual. Analogously, $\widetilde{\mathcal{D}}_2$ is self-dual proving (i).

Now assume that $\mathcal{D}_1\mu_{-1} = \mathcal{D}_1$. Then $\mathcal{C}_1\mu_{-1} = \mathcal{C}_1$. By Theorem 6.4.3, $\mathcal{C}_2 = \mathcal{D}_1^\perp$ and hence $\widetilde{\mathcal{C}}_2$ and $\widetilde{\mathcal{C}}_1$ are orthogonal to each other by Theorem 6.1.3(vi). From (a) and (b) the codes spanned by $\widetilde{\mathcal{C}}_1$ and $\widetilde{\overline{j}}(x)$ and by $\widetilde{\mathcal{C}}_2$ and $\widetilde{\overline{j}}(x)$ are orthogonal to each other. By Theorem 6.1.3(v) and (ix), these codes are $\widetilde{\mathcal{D}}_1$ and $\widetilde{\mathcal{D}}_2$ of dimension $(n+1)/2$; hence they are duals of each other.   □

Before proceeding, we make a few remarks.
- A generator matrix for $\widetilde{\mathcal{D}}_i$ is obtained by taking any generator matrix for $\mathcal{C}_i$, adjoining a column of zeros and adding the row $1, 1, \dots, 1, -\gamma n$ representing $n\widetilde{\overline{j}}(x)$.
- In general γ satisfying (6.11) exists in $\mathbb{F}_q$ if and only if n and -1 are both squares or both nonsquares in $\mathbb{F}_q$.
- If q is a square, then $\mathbb{F}_q$ contains γ such that (6.11) holds, as we now see. If $q = r^{2s}$, where r is a prime, then every polynomial of degree 2 with entries in $\mathbb{F}_r$ has roots in $\mathbb{F}_{r^2}$, which is a subfield of $\mathbb{F}_q$ by Theorem 3.5.3. Since γ is a root of the quadratic $nx^2 + 1$, $\mathbb{F}_{r^2} \subseteq \mathbb{F}_q$ contains γ.
- In (6.11), $\gamma = 1$ if $n \equiv -1 \pmod{r}$ where $\mathbb{F}_q$ has characteristic r. For such cases, $\widetilde{\mathcal{D}}_i = \widehat{\mathcal{D}}_i$. In particular, this is the case if $\mathbb{F}_q$ has characteristic 2. More particularly, $\widetilde{\mathcal{D}}_i = \widehat{\mathcal{D}}_i$ if the codes are either over $\mathbb{F}_2$ or $\mathbb{F}_4$.
- There are values of n and q where duadic codes exist but γ does not. For example, if $q = 3$, duadic codes exist only if $n \equiv \pm 1 \pmod{12}$ by Theorem 6.3.3. Suppose that $n \equiv 1 \pmod{12}$. If $\gamma \in \mathbb{F}_3$ satisfies (6.11), then $1 + \gamma^2 = 0$ in $\mathbb{F}_3$, which is not possible. If $n \equiv -1 \pmod{12}$ and if $\gamma \in \mathbb{F}_3$ satisfies (6.11), then $1 - \gamma^2 = 0$ in $\mathbb{F}_3$, which has solution $\gamma = \pm 1$. Thus if $n \equiv -1 \pmod{12}$, we may choose $\gamma = 1$, and then $\widetilde{\mathcal{D}}_i$ is the ordinary extension $\widehat{\mathcal{D}}_i$.

Example 6.4.13 The 5-cyclotomic cosets modulo 11 are $C_0 = \{0\}$, $C_1 = \{1, 3, 4, 5, 9\}$, and $C_2 = \{2, 6, 7, 8, 10\}$. There is one splitting $S_1 = C_1$ and $S_2 = C_2$ of 11 over $\mathbb{F}_5$; it is given by μ_{-1}. Let $i_1(x) = x + x^3 + x^4 + x^5 + x^9$ and $i_2(x) = x^2 + x^6 + x^7 + x^8 + x^{10}$. The generating idempotent of $\mathcal{E}_{11}$ over $\mathbb{F}_5$ is $-i_1(x) - i_2(x)$. The even-like duadic codes $\mathcal{C}_1$ and $\mathcal{C}_2$ of length 11 over $\mathbb{F}_5$ have generating idempotents $e_1(x) = i_1(x) - 2i_2(x)$ and $e_2(x) = -2i_1(x) + i_2(x)$. The odd-like duadic codes $\mathcal{D}_1$ and $\mathcal{D}_2$ have generating idempotents $1 + 2i_1(x) - i_2(x)$ and $1 - i_1(x) + 2i_2(x)$. The solutions of (6.11) are $\gamma = \pm 2$. So $\widehat{\mathcal{D}}_i \neq \widetilde{\mathcal{D}}_i$; however $\widetilde{\mathcal{D}}_i$ is self-dual by Theorem 6.4.12(i). ∎

Exercise 352 Verify the claims made in Example 6.4.13. ◆

In case there is no solution to (6.11) in $\mathbb{F}_q$, as long as we are willing to rescale the last coordinate differently for $\mathcal{D}_1$ and $\mathcal{D}_2$, we can obtain dual codes if $\mathcal{D}_1 \mu_{-1} = \mathcal{D}_1$. It is left as an exercise to show that if $\mathcal{D}_1 \mu_{-1} = \mathcal{D}_1$, then $\widehat{\mathcal{D}}_1$ and $\widehat{\mathcal{D}}_2 D'$ are duals where $D' = \text{diag}(1, 1, \ldots, 1, -1/n)$.

Exercise 353 Let $\mathcal{D}_1$ and $\mathcal{D}_2$ be odd-like duadic codes over $\mathbb{F}_q$ of length n with $\mathcal{D}_i \mu_{-1} = \mathcal{D}_i$. Show that $\widehat{\mathcal{D}}_1$ and $\widehat{\mathcal{D}}_2 D'$ are dual codes where $D' = \text{diag}(1, 1, \ldots, 1, -1/n)$. ◆

In a manner similar to Theorem 6.4.12, the following theorem explains the duality of the extended odd-like duadic codes over $\mathbb{F}_4$ under the Hermitian inner product. Recall that in this case that $\gamma = 1$ is a solution of (6.11) and $\widetilde{\mathcal{D}}_i = \widehat{\mathcal{D}}_i$.

Theorem 6.4.14 *Let $\mathcal{D}_1$ and $\mathcal{D}_2$ be a pair of odd-like duadic codes of length n over $\mathbb{F}_4$. The following hold:*
(i) If μ_{-2} gives the splitting for $\mathcal{D}_1$ and $\mathcal{D}_2$, then $\widehat{\mathcal{D}}_1$ and $\widehat{\mathcal{D}}_2$ are Hermitian self-dual.
(ii) If $\mathcal{D}_1 \mu_{-2} = \mathcal{D}_1$, then $\widehat{\mathcal{D}}_1$ and $\widehat{\mathcal{D}}_2$ are Hermitian duals of each other.

Exercise 354 Prove Theorem 6.4.14. ◆

We can characterize the lengths for which there exist self-dual extended cyclic binary codes. To do that, we need the following two lemmas.

Lemma 6.4.15 *For an odd prime p and a positive integer b, -1 is a power of 2 modulo p^b if and only if $\text{ord}_p(2)$ is even.*

Proof: Suppose that $\text{ord}_p(2)$ is even. If $w = \text{ord}_{p^b}(2)$, as $2^w \equiv 1 \pmod{p^b}$ implies that $2^w \equiv 1 \pmod{p}$, w cannot be odd. So $w = 2r$ and $(2^r)^2 \equiv 1 \pmod{p^b}$. Thus $2^r \not\equiv 1 \pmod{p^b}$ is a solution of $x^2 \equiv 1 \pmod{p^b}$. By Exercise 355, $2^r \equiv -1 \pmod{p^b}$ and -1 is a power of 2 modulo p^b. Conversely, suppose that -1 is a power of 2 modulo p^b. Then -1 is a power of 2 modulo p and so $\text{ord}_p(-1) \mid \text{ord}_p(2)$; as $\text{ord}_p(-1) = 2$, $\text{ord}_p(2)$ is even. □

Exercise 355 Let p be an odd prime and b a positive integer. Prove that the only solutions of $x^2 \equiv 1 \pmod{p^b}$ are $x \equiv \pm 1 \pmod{p^b}$. ◆

Lemma 6.4.16 *Let p be an odd prime and a be a positive integer. A splitting of p^a over $\mathbb{F}_2$ given by the multiplier μ_{-1} exists if and only if $\text{ord}_p(2)$ is odd.*

Proof: Let C_i be the 2-cyclotomic coset modulo p^a containing i. Suppose that a splitting S_1 and S_2 of p^a over $\mathbb{F}_2$ given by μ_{-1} exists. Then C_1 and $C_1 \mu_{-1}$ are not in the same S_i. Hence $-1 \notin C_1$ and so -1 is not a power of 2. By Lemma 6.4.15, $\mathrm{ord}_p(2)$ is odd.

Conversely, suppose $\mathrm{ord}_p(2)$ is odd. Suppose $C_i \mu_{-1} = C_i$ for some $i \not\equiv 0 \pmod{p^a}$. Then $i2^j \equiv -i \pmod{p^a}$ for some j. Hence $2^j \equiv -1 \pmod{p^b}$ for some $1 \le b \le a$. But this contradicts Lemma 6.4.15. Therefore, $C_i \mu_{-1} \ne C_i$ for $i \not\equiv 0 \pmod{p^a}$. One splitting of p^a is given by placing exactly one of C_i or $C_i \mu_{-1}$ in S_1 and the other in S_2. This is possible as C_i and $C_i \mu_{-1}$ are distinct if $i \not\equiv 0 \pmod{p^a}$.   □

Exercise 356 Find all splittings of $p = 89$ over $\mathbb{F}_2$ given by μ_{-1}. Find one splitting not given by μ_{-1}.   ◆

Before continuing, we remark that Theorem 6.4.10(i) is a special case of Lemma 6.4.16.

Theorem 6.4.17 *Self-dual extended cyclic binary codes of length $n + 1$ exist if and only if $n = p_1^{a_1} p_2^{a_2} \cdots p_r^{a_r}$, where $p_1, \ldots, p_r$ are distinct primes such that for each i either:*
(i) $p_i \equiv -1 \pmod 8$, *or*
(ii) $p_i \equiv 1 \pmod 8$ *and* $\mathrm{ord}_{p_i}(2)$ *is odd.*

Proof: We first show that n has a splitting given by μ_{-1} if and only if (i) or (ii) hold. By Lemma 6.3.1, this is equivalent to showing that $p_i^{a_i}$ has a splitting given by μ_{-1} if and only if p_i satisfies either (i) or (ii). By Theorem 6.3.3 and Corollary 6.1.6, the only primes that can occur in any splitting of $p_i^{a_i}$ satisfy $p_i \equiv \pm 1 \pmod 8$. If $p_i \equiv -1 \pmod 8$, then $\mathrm{ord}_{p_i}(2)$ is odd by Lemma 6.2.6. So by Lemma 6.4.16, n has a splitting over $\mathbb{F}_2$ given by μ_{-1} if and only if (i) or (ii) hold.

Suppose that n has a splitting over $\mathbb{F}_2$ given by μ_{-1}. Let $\mathcal{D}_1$ and $\mathcal{D}_2$ be odd-like duadic codes given by this splitting. As $\widetilde{\mathcal{D}}_i = \widehat{\mathcal{D}}_i$ in the binary case, $\widehat{\mathcal{D}}_i$ are self-dual by Theorem 6.4.12. Conversely, suppose that there is a self-dual extended cyclic binary code $\widehat{\mathcal{D}}$ of length $n + 1$, where $\mathcal{D}$ is an $[n, (n + 1)/2]$ cyclic code. Let $\mathcal{C}$ be the $[n, (n - 1)/2]$ even-like subcode of $\mathcal{D}$; as $\widehat{\mathcal{D}}$ is self-dual, $\mathcal{C}$ is self-orthogonal. Then μ_{-1} gives a splitting of n over $\mathbb{F}_2$ by Theorem 6.4.1.   □

Exercise 357 Find all lengths n with $2 < n < 200$ where there exist self-dual extended cyclic binary codes of length $n + 1$.   ◆

6.5 Weights in duadic codes

In this section, we present two results about the possible codeword weights in duadic codes. The first deals with weights of codewords in binary duadic codes when the splitting is given by μ_{-1}. The second deals with the minimum weight codewords in duadic codes over arbitrary fields. We conclude the section with data about binary duadic codes of lengths up to 241, including minimum weights of these codes.

Theorem 6.5.1 *Let $\mathcal{D}_1$ and $\mathcal{D}_2$ be odd-like binary duadic codes of length n with splitting given by μ_{-1}. Then for $i = 1$ and 2,*
(i)   *the weight of every even weight codeword of $\mathcal{D}_i$ is 0 mod 4, and the weight of every odd weight codeword of $\mathcal{D}_i$ is n mod 4, and moreover*

(ii) $\widehat{\mathcal{D}}_i$ is self-dual doubly-even if $n \equiv -1$ (mod 8) and $\widehat{\mathcal{D}}_i$ is self-dual singly-even if $n \equiv 1$ (mod 8).

Proof: Let $\mathcal{C}_1$ and $\mathcal{C}_2$ be the associated even-like duadic codes. By Theorem 6.4.2 and Theorem 6.1.3(vi), $\mathcal{C}_i$ is self-orthogonal and is the even-like subcode of $\mathcal{D}_i$. By Theorem 4.4.18, $\mathcal{C}_i$ is doubly-even and so every even weight codeword of $\mathcal{D}_i$ has weight congruent to 0 mod 4. By Theorem 6.1.3(ix), the odd weight codewords of $\mathcal{D}_i$ are precisely $\overline{j}(x) + c(x)$, where $c(x) \in \mathcal{C}_i$. As $\overline{j}(x)$ is the all-1 codeword, $\mathrm{wt}(\overline{j}(x) + c(x)) \equiv n - \mathrm{wt}(c(x)) \equiv n$ (mod 4). Thus (i) holds.

By Theorem 6.3.3, $n \equiv \pm 1$ (mod 8). As μ_{-1} gives the splitting, Theorem 6.4.12(i) shows that $\widehat{\mathcal{D}}_i = \widetilde{\mathcal{D}}_i$ is self-dual. By part (i), the codewords of $\widehat{\mathcal{D}}_i$ that are extensions of even weight codewords have weights congruent to 0 mod 4; those that are extensions of odd weight codewords have weights congruent to $(n + 1)$ mod 4. Part (ii) now follows.   □

In the next theorem, we present a lower bound on the minimum odd-like weight in odd-like duadic codes, called the Square Root Bound. If this bound is actually met and the splitting is given by μ_{-1}, an amazing combinatorial structure arises involving the supports of the minimum weight codewords. This is a precursor of the material in Chapter 8, where we will investigate other situations for which the set of supports of codewords of a fixed weight form interesting combinatorial structures. Recall that in Chapter 3, we defined the support of a vector $\mathbf{v}$ to be the set of coordinates where $\mathbf{v}$ is nonzero. In our next result, the combinatorial structure that arises is called a projective plane. A *projective plane* consists of a set $\mathcal{P}$ of *points* together with a set $\mathcal{L}$ of *lines* for which the following conditions are satisfied. A line $\ell \in \mathcal{L}$ is a subset of points. For any two distinct points, there is exactly one line containing these two points (that is, two distinct points determine a unique line that passes though these two points). Any two distinct lines have exactly one point in common. Finally, to prevent degeneracy, we must also have at least four points no three of which are on the same line. From this definition, it can be shown (see Theorem 8.6.2) that if $\mathcal{P}$ is finite, then every line has the same number $\mu + 1$ of points, every point lies on $\mu + 1$ lines, and there are $\mu^2 + \mu + 1$ points and $\mu^2 + \mu + 1$ lines. The number μ is called the *order* of the projective plane. Any permutation of the points which maps lines to lines is an automorphism of the projective plane. The projective plane is *cyclic* provided the plane has an automorphism that is a $(\mu^2 + \mu + 1)$-cycle. The projective plane we will obtain arises from an odd-like duadic code of length n as follows. The set of points in the projective plane is the set of coordinates of the code. The set of lines in the plane is the set of supports of all the codewords of minimum weight. Thus as our code is cyclic, the resulting supports will form a projective plane that is also cyclic. Exercise 358 illustrates this result.

Exercise 358 In Example 6.1.4, we constructed the (only) pair of odd-like binary duadic codes of length 7. Pick one of them and write down the supports of the weight 3 codewords. Show that these supports form a cyclic projective plane of order 2.   ◆

Theorem 6.5.2 (Square Root Bound) *Let $\mathcal{D}_1$ and $\mathcal{D}_2$ be a pair of odd-like duadic codes of length n over $\mathbb{F}_q$. Let d_o be their (common) minimum odd-like weight. Then the following hold:*

(i) $d_o^2 \geq n$.

(ii) *If the splitting defining the duadic codes is given by μ_{-1}, then $d_o^2 - d_o + 1 \geq n$.*

(iii) *Suppose $d_o^2 - d_o + 1 = n$, where $d_o > 2$, and assume the splitting is given by μ_{-1}. Then for $i = 1$ and 2:*

(a) d_o *is the minimum weight of $\mathcal{D}_i$,*

(b) *the supports of the minimum weight codewords of $\mathcal{D}_i$ form a cyclic projective plane of order $d_o - 1$,*

(c) *the minimum weight codewords of $\mathcal{D}_i$ are multiples of binary vectors, and*

(d) *there are exactly $n(q - 1)$ minimum weight codewords in $\mathcal{D}_i$.*

Proof: Suppose the splitting defining the duadic codes is given by μ_a. Let $c(x) \in \mathcal{D}_1$ be an odd-like vector of weight d_o. Then $c'(x) = c(x)\mu_a \in \mathcal{D}_2$ is also odd-like and $c(x)c'(x) \in \mathcal{D}_1 \cap \mathcal{D}_2$ as $\mathcal{D}_1$ and $\mathcal{D}_2$ are ideals in $\mathcal{R}_n$. But $\mathcal{D}_1 \cap \mathcal{D}_2 = \langle \overline{j}(x) \rangle$ by Theorem 6.1.3(viii). By Lemma 6.1.2(ii), $c(x)c'(x)$ is odd-like, and in particular nonzero. Therefore $c(x)c'(x)$ is a nonzero multiple of $\overline{j}(x)$, and so $\mathrm{wt}(c(x)c'(x)) = n$. The number of terms in the product $c(x)c'(x)$ is at most d_o^2, implying (i). If $\mu_a = \mu_{-1}$, then the number of terms in $c(x)c'(x)$ is at most $d_o^2 - d_o + 1$ because, if $c(x) = \sum_{j=0}^{n-1} c_j x^j$, the coefficient of x^0 in $c(x)c'(x)$ is $\sum_j c_j^2$, where the sum is over all subscripts with $c_j \neq 0$. This produces (ii).

We prove (iii) for $i = 1$. Suppose the splitting is given by μ_{-1} and $n = d_o^2 - d_o + 1$. Let $c(x) = c_0 + c_1 x + \cdots + c_{n-1}x^{n-1} \in \mathcal{D}_1$ be an odd-like vector of weight d_o, and let $c'(x) = c(x)\mu_{-1}$. As in the proof of part (i), $c(x)c'(x) \in \mathcal{D}_1 \cap \mathcal{D}_2 = \langle \overline{j}(x) \rangle$, implying $c(x)c'(x) = \delta \overline{j}(x)$ for some $\delta \neq 0$. Let $c_{i_1}, \ldots, c_{i_{d_o}}$ be the nonzero coefficients of $c(x)$. Let $\mathcal{M}$ be the $d_o \times n$ matrix whose jth row corresponds to $c_{i_j} x^{-i_j} c(x)$ obtained by multiplying $c(x)$ by the jth nonzero term of $c'(x) = c(x^{-1})$. Adding the rows of $\mathcal{M}$ gives a vector corresponding to $c(x)c'(x) = \delta \overline{j}(x)$. Then $\mathcal{M}$ has d_o^2 nonzero entries. We label the columns of $\mathcal{M}$ by $x^0, x^1, x^2, \ldots, x^{n-1}$. Column x^0 of $\mathcal{M}$ has d_o nonzero entries. This leaves $d_o^2 - d_o = n - 1$ nonzero entries for the remaining $n - 1$ columns. Since the sum of the entries in the column labeled x^i is the coefficient of x^i in $\delta \overline{j}(x)$, which is nonzero, each of the last $n - 1$ columns contains exactly one nonzero entry. Thus the supports of any two rows of $\mathcal{M}$ overlap in only the x^0 coordinate.

Suppose $m(x) = m_0 + m_1 x + \cdots + m_{n-1}x^{n-1} \in \mathcal{D}_1$ is a nonzero even-like vector of weight w. We show that $w \geq d_o + 1$. By shifting $m(x)$, we may assume that $m_0 \neq 0$. The supports of $m(x)$ and each row of $\mathcal{M}$ overlap in coordinate x^0. If $\mathcal{C}_1$ and $\mathcal{C}_2$ are the even-like duadic codes corresponding to $\mathcal{D}_1$ and $\mathcal{D}_2$, then $\mathcal{C}_1$ is self-orthogonal and $\mathcal{C}_1^{\perp} = \mathcal{D}_1$ by Theorem 6.4.2. As $m(x) \in \mathcal{C}_1$ by Theorem 6.1.3(vi) and $\mathcal{C}_1^{\perp} = \mathcal{D}_1$, $m(x)$ is orthogonal to every row of $\mathcal{M}$. Therefore the support of $m(x)$ and the support of each row of $\mathcal{M}$ overlap in at least one more coordinate. As columns $x^1, \ldots, x^{n-1}$ of $\mathcal{M}$ each have exactly one nonzero entry, $\mathrm{wt}(m(x)) = w \geq d_o + 1$. This proves (iii)(a) and shows that the codewords in $\mathcal{D}_1$ of weight d_o must be odd-like.

Let $\mathcal{P} = \{x^0, x^1, \ldots, x^{n-1}\}$ be called points; let $\mathcal{L}$, which we will call lines, be the distinct supports of all codewords in $\mathcal{D}_1$ of weight d_o, with supports considered as subsets of $\mathcal{P}$. Let ℓ_c and ℓ_m be distinct lines associated with $c(x), m(x) \in \mathcal{D}_1$, respectively. As $c(x)$ and $m(x)$ have weight d_o, they are odd-like. Hence by Theorem 6.1.3(ix), $c(x) = \alpha \overline{j}(x) + a(x)$ and $m(x) = \beta \overline{j}(x) + b(x)$, where $a(x), b(x) \in \mathcal{C}_1$ and α, β are nonzero elements of $\mathbb{F}_q$. As

$C_1^{\perp} = \mathcal{D}_1$, the inner product of $c(x)$ and $m(x)$ is a nonzero multiple of the inner product of $\overline{j}(x)$ with itself, which is nonzero. Thus any two lines of $\mathcal{L}$ intersect in at least one point. The size of the intersection $\ell_c \cap \ell_m$ is not changed if both ℓ_c and ℓ_m are shifted by the same amount. Hence by shifting $c(x)$ and $m(x)$ by x^i for some i, we may assume $x^0 \in \ell_c$ and $x^0 \notin \ell_m$ as $\ell_c \neq \ell_m$. We construct the $d_o \times n$ matrix $\mathcal{M}$ from $c(x)$ as above. If ℓ is the support of any row of $\mathcal{M}$, then the size of $\ell \cap \ell_m$ is at least one, since we just showed that lines of $\mathcal{L}$ intersect in at least one point. Recall that each column of $\mathcal{M}$ except the first has exactly one nonzero entry. As ℓ_m intersects each of the rows of $\mathcal{M}$ in at least one coordinate and that cannot be the first coordinate, the size of $\ell \cap \ell_m$ must be exactly one because wt$(m(x)) = d_o$ and $\mathcal{M}$ has d_o rows. One of the rows of $\mathcal{M}$ has support ℓ_c, implying that $\ell_c \cap \ell_m$ has size one for all distinct lines. In particular, two points determine at most one line and every pair of lines intersects in exactly one point. To complete (iii)(b), we only need to show that two points determine at least one line. The set $\mathcal{L}$ is invariant under cyclic shifts as $\mathcal{D}_1$ is cyclic, and so we only need to show that x^0 and x^i, for $i > 0$, determine at least one line. But that is clear, as all lines corresponding to rows of $\mathcal{M}$ have a nonzero entry in coordinate x^0, and some row has a nonzero entry in coordinate x^i. It follows that $\mathcal{L}$ is a cyclic projective plane of order $d_o - 1$ giving (iii)(b).

Let $c(x) = c_0 + c_1 x + \cdots + c_{n-1} x^{n-1} \in \mathcal{D}_1$ have weight d_o. Defining $\mathcal{M}$ as above, if $r > s$ and c_r, c_s are nonzero, the unique nonzero entry of column x^{r-s} of $\mathcal{M}$ is $c_r c_s$. The rows of $\mathcal{M}$ sum to a vector corresponding to $c(x)c(x^{-1}) = \delta \overline{j}(x)$, where $\delta \neq 0$. Thus each column of $\mathcal{M}$ sums to $(1/n)\delta$. So if $i < j < k$ and c_i, c_j, c_k are nonzero (three such values exist as $d_o > 2$), $c_i c_j = c_i c_k = c_j c_k = (1/n)\delta$ implying that $c_i = c_j = c_k$. Hence (iii)(c) holds. There are

$$\binom{n}{2} \Big/ \binom{d_o}{2} = n$$

lines in $\mathcal{L}$; since every minimum weight codeword is a multiple of the binary vector corresponding to a line, (iii)(d) follows.   □

In part (iii) of the previous theorem, the restriction that $d_o > 2$ is not significant because if $d_o \leq 2$ and $n = d_o^2 - d_o + 1$, then $n \leq 3$.

Exercise 359 In this exercise, you will construct certain duadic codes over $\mathbb{F}_3$ of length 13. The 3-cylotomic cosets modulo 13 are:

$$C_0 = \{0\}, \quad C_1 = \{1, 3, 9\}, \quad C_2 = \{2, 5, 6\}, \quad C_4 = \{4, 10, 12\}, \quad C_7 = \{7, 8, 11\}.$$

Let $c_i(x) = \sum_{j \in C_i} x^j$.
(a) Show that $c_1(x)^2 = c_2(x) - c_4(x)$. Then use μ_2 to compute $c_2(x)^2$, $c_4(x)^2$, and $c_7(x)^2$.
(b) Show that $c_1(x)c_2(x) = c_1(x) + c_2(x) + c_7(x)$ and $c_1(x)c_4(x) = c_2(x) + c_7(x)$. Then use μ_2 to compute $c_i(x)c_j(x)$ for $i < j$ and $i, j \in \{1, 2, 4, 7\}$.
(c) Construct generating idempotents for the two pairs of even-like duadic codes of length 13 over $\mathbb{F}_3$ whose splitting is given by μ_{-1}. Hint: The generating idempotents each have weight 9. If $e(x)$ is such an idempotent, then $e(x) + e(x)\mu_{-1} = 1 - \overline{j}(x)$.
(d) Construct the generating idempotents for the two pairs of odd-like duadic codes of length 13 over $\mathbb{F}_3$ whose splitting is given by μ_{-1}.

Table 6.1 *Binary duadic codes of length* $n \le 119$

n	Idempotent	d	d_o	a	Number of codes
7	1*	3	3	$-1\ddagger$	2
17	0, 1*	5	5	$3\bowtie$	2
23	1*	7	7	$-1\ddagger$	2
31	1, 5, 7*	7	7	$-1\ddagger$	2
31	1, 3, 5	7	7	$-1\ddagger$	6
41	0, 1*	9	9	$3\bowtie$	2
47	1*	11	11	$-1\ddagger$	2
49	0, 1, 7	4	9	$-1\dagger$	4
71	1*	11	11	$-1\ddagger$	2
73	0, 1, 3, 5, 11	9	9	$-1\dagger$	8
73	0, 1, 3, 5, 13	9	9	$-1\dagger$	8
73	0, 1, 5, 9, 17	12	13	$3\bowtie$	4
73	0, 1, 3, 9, 25*	13	13	$5\bowtie$	2
79	1*	15	15	$-1\ddagger$	2
89	0, 1, 3, 5, 13	12	17	$-1\dagger$	8
89	0, 1, 3, 5, 19	12	17	$-1\dagger$	8
89	0, 1, 3, 11, 33	15	15	$5\bowtie$	4
89	0, 1, 5, 9, 11*	17	17	$3\bowtie$	2
97	0, 1*	15	15	$5\bowtie$	2
103	1*	19	19	$-1\ddagger$	2
113	0, 1, 9*	15	15	$3\bowtie$	2
113	0, 1, 3	18	19	$9\bowtie$	4
119	0, 1, 13, 17, 21	4	15	3	4
119	0, 1, 7, 11, 51	6	15	3	4
119	0, 1, 7, 13, 17	8	15	3	4
119	0, 1, 7, 11, 17	12	15	3	4

(e) Show that $1 + c_i(x)$ for $i \in \{1, 2, 4, 7\}$ are each in some code in part (d). Hint: Do this for $i = 1$ and then use μ_2.

(f) Write down all 13 cyclic shifts of $1 + c_1(x)$. What is the resulting structure? ◆

Table 6.1 gives information about the odd-like binary duadic codes $\mathcal{D}$ of length $n \le 119$. The splittings given in this table are obtained in a manner similar to that of Example 6.4.8. The idempotent of $\mathcal{D}$ is of the form $\sum_{s \in \mathcal{I}} \sum_{i \in C_s} x^i$. The column "Idempotent" gives the index set $\mathcal{I}$ for each equivalence class of odd-like duadic codes; a * in this column means that the code is a quadratic residue code. The columns "d" and "d_o" are the minimum weight and minimum odd weight, respectively. The column "a" indicates the multiplier μ_a giving the splitting. A $\bowtie$ in column "a" indicates that the two extended odd-like duadic codes are duals of each other (i.e. μ_{-1} fixes each code by Theorem 6.4.12). If the splitting is given by μ_{-1}, then by Theorem 6.5.1, the extended code is self-dual doubly-even if $n \equiv -1$ (mod 8) and self-dual singly-even if $n \equiv 1$ (mod 8). We denote the codes whose extensions are self-dual singly-even by † and the codes whose extensions are self-dual doubly-even by ‡ in the column "a". Finally, the column "Number of codes" is the number of either even-like or odd-like duadic codes in an equivalence class of duadic codes constructed from the given splitting. (For example, a 6 in the "Number of codes" column means that there are six equivalent even-like codes and six equivalent odd-like codes in the class of codes represented by the given splitting.) Theorem 4.3.17 shows that equivalence is determined by multipliers except in the case $n = 49$; however, in that case using multipliers is enough to show that there is only one class. The table comes from [277]. (The values of d_o and d for the [113, 57, 18] code in the table were computed probabilistically in [277] and exactly in [315]. This code has higher minimum weight than the quadratic residue code of the same length.) The minimum weight codewords in the [7, 4, 3] and [73, 37, 9] codes support projective planes of orders 2 and 8, respectively; see Exercise 358. There are 16 duadic codes of length 73 containing projective planes of order 8; while the codes fall into two equivalence classes of eight codes each, the projective planes are all equivalent. The codes of length 31, which are not quadratic residue codes, are punctured Reed–Muller codes; see Example 6.4.8.

Information about the binary duadic odd-like codes of lengths $127 \leq n \leq 241$ has been computed in [277]. We summarize some of this information in Table 6.2. In this table we list in the column labeled "d" the minimum weights that occur. The minimum weights reported here were computed probabilistically. Some of these were computed exactly in [315]; wherever exact minimum weights were found, they agreed with those computed in [277]. See [189, 277]. There may be several codes at each length. For example, if $n = 217$, there are 1024 duadic codes that fall into 88 equivalence classes by Theorem 4.3.17. Further information on the idempotents and splittings can be found in [277]. In the column "Comments", we indicate if the extended codes are self-dual singly-even by † or self-dual doubly-even by ‡, or if the extended codes are duals of each other by ⋈. By examining Tables 6.1 and 6.2, the Square Root Bound is seen to be very useful for duadic codes of relatively small length but becomes weaker as the length gets longer.

Table 6.3, taken from [266], with certain values supplied by Philippe Gaborit, gives information about the odd-like duadic codes over $\mathbb{F}_4$ of odd lengths 3 through 41. No prime lengths $p \equiv -1 \pmod{8}$ are listed as these duadic codes have the same generating idempotents as those given in Table 6.1; because these codes have a binary generator matrix, their minimum weight is the same as the minimum weight of the corresponding binary code by Theorem 3.8.8. For the same reasons we do not list those codes of prime length $p \equiv 1 \pmod 8$ which have binary idempotents (also described in Table 6.1). The codes of length $p \equiv -1 \pmod 8$ have extensions that are self-dual under both the ordinary and the Hermitian inner products; see Theorems 6.4.9 and 6.4.12 and Exercise 360. In Table 6.3, the column labeled "n" gives the length, "d" the minimum weight, and "$\widehat{d}$" the minimum weight of the extended code. Each row of the table represents a family of equivalent odd-like duadic codes that are permutation equivalent. The column "Number of codes" is the number of odd-like (or even-like) duadic codes in the equivalence class. In the "Comments" column, the symbol ♠ means that the splitting is given by μ_{-1} and so the extended codes are self-dual with respect to the ordinary inner product. The symbol ♡ means that the splitting is given by μ_{-2} and so the extended codes are self-dual with respect to the

Table 6.2 *Binary duadic codes of length* $119 < n \le 241$

n	d	Comments
127	15	‡ includes punctured Reed–Muller codes
127	16	‡
127	19	‡ includes quadratic residue codes
137	21	⋈ quadratic residue codes only
151	19	‡ includes quadratic residue codes
151	23	‡
161	4	†
161	8	†
161	16	†
167	23	‡ quadratic residue codes only
191	27	‡ quadratic residue codes only
193	27	⋈ quadratic residue codes only
199	31	‡ quadratic residue codes only
217	4	†
217	8	†
217	12	†
217	16	†
217	20	†
217	24	†
223	31	‡ includes quadratic residue codes
233	25	⋈ quadratic residue codes only
233	29	†
233	32	⋈
239	31	‡ quadratic residue codes only
241	25	⋈
241	30	⋈
241	31	⋈ quadratic residue codes only

Hermitian inner product; see Theorem 6.4.14. At length 21, one of the splittings given by μ_{-1} yields odd-like duadic codes whose weight 5 codewords support a projective plane of order 4.

Exercise 360 Prove that if $\mathcal{C}$ is an $[n, n/2]$ code over $\mathbb{F}_4$ with a generator matrix consisting of binary vectors, which is self-dual under the ordinary inner product, then $\mathcal{C}$ is also self-dual under the Hermitian inner product.   ♦

It has been a long-standing open problem to find a better bound than the Square Root Bound, since this seems to be a very weak bound. Additionally, it is not known whether the family of duadic codes over $\mathbb{F}_q$ is asymptotically good or bad. (See Section 5.1 for the definition of asymptotically good.) Also there is no efficient decoding scheme known for duadic codes. Finding such a scheme would enhance their usefulness greatly. We pose these questions as research problems.

Table 6.3 *Duadic codes over* $\mathbb{F}_4$ *of length* $n \leq 41$
(excluding those with binary generating idempotents)

n	d	$\widehat{d}$	Number of codes	Comments
3	2	3	2	♠ quadratic residue
5	3	4	2	♡ quadratic residue
9	3	3	4	♠
11	5	6	2	♠ quadratic residue
13	5	6	2	♡ quadratic residue
15	6	7	4	
15	6	6	4	
15	4	4	4	
15	3	3	4	
17	7	8	4	♡
19	7	8	2	♠ quadratic residue
21	4	4	4	♠
21	5	6	4	♠ projective plane
21	6	6	4	♠
21	3	3	4	♠
25	3	4	4	♡
27	3	3	4	♠
27	3	3	4	♠
29	11	12	2	♡ quadratic residue
33	3	3	4	♠
33	10	10	4	♠
33	6	6	4	♠
33	6	6	4	♠
35	8	8	8	♡
35	7	8	4	♡
35	4	4	4	♡
37	11	12	2	♡ quadratic residue
39	6	6	4	
39	3	3	4	
39	3	3	4	
39	10	11	4	
41	11	12	4	♡

Research Problem 6.5.3 *Improve the Square Root Bound for either the entire family of duadic codes or the subfamily of quadratic residue codes.*

Research Problem 6.5.4 *Decide whether or not the family of duadic codes is asymptotically good or bad.*

Research Problem 6.5.5 *Find an efficient decoding scheme for either the entire family of duadic codes or the subfamily of quadratic residue codes.*

6.6   Quadratic residue codes

In this section we study more closely the family of quadratic residue codes, which, as we have seen, are special cases of duadic codes. These codes, or extensions of them, include the Golay codes and the hexacode. Quadratic residue codes are duadic codes over $\mathbb{F}_q$ of odd prime length $n = p$; by Theorem 6.3.2, q must be a square modulo n. Throughout this section, we will let $n = p$ be an odd prime not dividing q; we will assume that q is a prime power that is a square modulo p. Let $\mathcal{Q}_p$ denote the set of nonzero squares modulo p, and let $\mathcal{N}_p$ be the set of nonsquares modulo p. The sets $\mathcal{Q}_p$ and $\mathcal{N}_p$ are called the nonzero *quadratic residues* and the *quadratic nonresidues* modulo p, respectively.

We begin with the following elementary lemma.

Lemma 6.6.1 *Let p be an odd prime. The following hold:*
(i) $|\mathcal{Q}_p| = |\mathcal{N}_p| = (p-1)/2$.
(ii) *Modulo p, we have $\mathcal{Q}_p a = \mathcal{Q}_p$, $\mathcal{N}_p a = \mathcal{N}_p$, $\mathcal{Q}_p b = \mathcal{N}_p$, and $\mathcal{N}_p b = \mathcal{Q}_p$ when $a \in \mathcal{Q}_p$ and $b \in \mathcal{N}_p$.*

Proof: The nonzero elements of the field $\mathbb{F}_p$ form a cyclic group $\mathbb{F}_p^*$ of even order $p - 1$ with generator α. $\mathcal{Q}_p$ is the set of even order elements, that is, $\mathcal{Q}_p = \{\alpha^{2i} \mid 0 \leq i < (p-1)/2\}$; this set forms a subgroup of index 2 in $\mathbb{F}_p^*$. Furthermore $\mathcal{N}_p$ is the coset $\mathcal{Q}_p \alpha$. The results now follow easily.   □

This lemma implies that the product of two residues or two nonresidues is a residue, while the product of a residue and a nonresidue is a nonresidue; these are facts we use throughout this section (see also Exercise 351). As a consequence of this lemma and Exercise 361, the pair of sets $\mathcal{Q}_p$ and $\mathcal{N}_p$ is a splitting of p given by the multiplier μ_b for any $b \in \mathcal{N}_p$. This splitting determines the defining sets for a pair of even-like duadic codes and a pair of odd-like duadic codes, called the *quadratic residue codes* or *QR codes*, of length p over $\mathbb{F}_q$. The odd-like QR codes have defining sets $\mathcal{Q}_p$ and $\mathcal{N}_p$ and dimension $(p + 1)/2$, while the even-like QR codes have defining sets $\mathcal{Q}_p \cup \{0\}$ and $\mathcal{N}_p \cup \{0\}$ and dimension $(p - 1)/2$. This discussion proves the following theorem.

Theorem 6.6.2 *Quadratic residue codes of odd prime length p exist over $\mathbb{F}_q$ if and only if $q \in \mathcal{Q}_p$.*

Exercise 361 Prove that if p is an odd prime, $\mathcal{Q}_p$ and $\mathcal{N}_p$ are each unions of q-cyclotomic cosets if and only if $q \in \mathcal{Q}_p$.   ♦

In the next two subsections, we will present the generating idempotents for the QR codes over fields of characteristic 2 and 3 as described in [274]. The following two theorems will assist us with this classification. The first theorem provides, among other things, a form that an idempotent must have if it is the generating idempotent for a quadratic residue code. In that theorem, we have to distinguish between QR codes and trivial codes. The *trivial codes* of length p over $\mathbb{F}_q$ are: $\mathbf{0}$, $\mathbb{F}_q^p$, the even-like subcode $\mathcal{E}_p$ of $\mathbb{F}_q^p$, and the code $\langle \mathbf{1} \rangle$ generated by the all-one codeword. The second theorem will give the generating idempotents of the four

QR codes from one of the even-like generators and describe how the generating idempotent of a QR code over some field is related to the generating idempotent of a QR code over an extension field.

Theorem 6.6.3 *Let C be a cyclic code of odd prime length p over $\mathbb{F}_q$, where q is a square modulo p. Let $e(x)$ be the generating idempotent of C. The following hold:*

(i) *C is a quadratic residue code or one of the trivial codes if and only if $e(x)\mu_c = e(x)$ for all $c \in \mathcal{Q}_p$.*

(ii) *If C is a quadratic residue code with generating idempotent $e(x)$, then*

$$e(x) = a_0 + a_1 \sum_{i \in \mathcal{Q}_p} x^i + a_2 \sum_{i \in \mathcal{N}_p} x^i,$$

for some a_0, a_1, and a_2 in $\mathbb{F}_q$.

(iii) *If $c \in \mathcal{Q}_p$ and C is a quadratic residue code, then $\mu_c \in \mathrm{PAut}(C)$.*

Proof: The trivial codes $\mathbf{0}$, $\mathbb{F}_q^p$, $\mathcal{E}_p$, and $\langle 1 \rangle$ have defining sets $\{0\} \cup \mathcal{Q}_p \cup \mathcal{N}_p$, $\emptyset$, $\{0\}$, and $\mathcal{Q}_p \cup \mathcal{N}_p$, respectively. Let T be the defining set of C. By Theorem 4.3.13 and Corollary 4.4.5, the code $C\mu_c$ is cyclic with generating idempotent $e(x)\mu_c$ and defining set $c^{-1}T \bmod p$. So $e(x)\mu_c = e(x)$ if and only if $cT \equiv T \pmod p$. Using Lemma 6.6.1, $e(x)\mu_c = e(x)$ for all $c \in \mathcal{Q}_p$ if and only if T is a union of some of $\{0\}$, $\mathcal{Q}_p$, or $\mathcal{N}_p$. So part (i) follows. Part (ii) follows from (i) and Lemma 6.6.1 as $e(x^c) = e(x)$ for all $c \in \mathcal{Q}_p$. Part (iii) also follows from part (i) because C and $C\mu_c$ have the same defining set for $c \in \mathcal{Q}_p$ and hence are equal, implying that $\mu_c \in \mathrm{PAut}(C)$. $\qquad\square$

Theorem 6.6.4 *Let C be an even-like quadratic residue code of prime length p over $\mathbb{F}_q$ with idempotent $e(x)$. The following hold:*

(i) *The four quadratic residue codes over $\mathbb{F}_q$ or any extension field of $\mathbb{F}_q$ have generating idempotents $e(x)$, $e(x)\mu_b$, $e(x) + \overline{j}(x)$, and $e(x)\mu_b + \overline{j}(x)$ for any $b \in \mathcal{N}_p$.*

(ii) *$e(x) + e(x)\mu_b = 1 - \overline{j}(x)$ for $b \in \mathcal{N}_p$.*

(iii) *The four quadratic residue codes over $\mathbb{F}_q$ have the same minimum weight and the same minimum weight codewords, up to scalar multiplication, as they do over an extension field of $\mathbb{F}_q$.*

Proof: By (6.3) and Theorem 6.1.3, the generating idempotents for the four QR codes over $\mathbb{F}_q$ are as claimed in (i). Because these four idempotents remain idempotents over any extension field of $\mathbb{F}_q$ and are associated with the same splitting of p into residues and nonresidues, they remain generating idempotents of QR codes over any extension field of $\mathbb{F}_q$, completing (i). Part (ii) follows from (6.1), while (iii) follows from Theorem 3.8.8. $\qquad\square$

Part (i) of this theorem was already illustrated in Example 6.1.9.

6.6.1 QR codes over fields of characteristic 2

In this subsection, we will find the generating idempotents of all the QR codes over any field of characteristic 2. We will see that we only have to look at the generating idempotents of QR codes over $\mathbb{F}_2$ and $\mathbb{F}_4$.

Theorem 6.6.5 *Let p be an odd prime. The following hold:*
(i) *Binary quadratic residue codes of length p exist if and only if $p \equiv \pm 1 \pmod 8$.*
(ii) *The even-like binary quadratic residue codes have generating idempotents*

$$\delta + \sum_{j \in \mathcal{Q}_p} x^j \quad and \quad \delta + \sum_{j \in \mathcal{N}_p} x^j,$$

where $\delta = 1$ if $p \equiv -1 \pmod 8$ and $\delta = 0$ if $p \equiv 1 \pmod 8$.
(iii) *The odd-like binary quadratic residue codes have generating idempotents*

$$\epsilon + \sum_{j \in \mathcal{Q}_p} x^j \quad and \quad \epsilon + \sum_{j \in \mathcal{N}_p} x^j,$$

where $\epsilon = 0$ if $p \equiv -1 \pmod 8$ and $\epsilon = 1$ if $p \equiv 1 \pmod 8$.

Proof: Binary QR codes of length p exist if and only if $2 \in \mathcal{Q}_p$ by Theorem 6.6.2, which is equivalent to $p \equiv \pm 1 \pmod 8$ by Lemma 6.2.5, giving (i). Let $e(x)$ be a generating idempotent of one of the QR codes. By Theorem 6.6.3,

$$e(x) = \sum_{i \in S} x^i,$$

where S is a union of some of $\{0\}$, $\mathcal{Q}_p$, and $\mathcal{N}_p$. As the cases $S = \{0\}$, $\mathcal{Q}_p \cup \mathcal{N}_p$, and $\{0\} \cup \mathcal{Q}_p \cup \mathcal{N}_p$ yield trivial codes by Exercise 362, for QR codes, S equals $\{0\} \cup \mathcal{Q}_p$, $\{0\} \cup \mathcal{N}_p$, $\mathcal{Q}_p$, or $\mathcal{N}_p$. These yield the idempotents in (ii) and (iii); one only needs to check that their weights are even or odd as required. $\square$

Exercise 362 Prove that if $\mathcal{C}$ is a binary cyclic code of length p with generating idempotent $e(x) = \sum_{i \in S} x^i$ where $S = \{0\}$, $\mathcal{Q}_p \cup \mathcal{N}_p$, or $\{0\} \cup \mathcal{Q}_p \cup \mathcal{N}_p$, then $\mathcal{C}$ is the code $\mathbb{F}_2^p$, the even subcode $\mathcal{E}_p$ of $\mathbb{F}_2^p$, or the subcode generated by the all-one vector. ◆

Because $4 = 2^2$ is obviously a square modulo any odd prime p, by Theorem 6.6.2, QR codes over $\mathbb{F}_4$ exist for any odd prime length. The idempotents for the QR codes over $\mathbb{F}_4$ of length $p \equiv \pm 1 \pmod 8$ are the same as the binary idempotents given in Theorem 6.6.5 by Theorem 6.6.4. We now find the generating idempotents for the QR codes over $\mathbb{F}_4$ of length p where $p \equiv \pm 3 \pmod 8$. (Recall that $\mathbb{F}_4 = \{0, 1, \omega, \overline{\omega}\}$, where $\overline{\omega} = 1 + \omega = \omega^2$.)

Theorem 6.6.6 *Let p be an odd prime. The following hold:*
(i) *If $p \equiv \pm 1 \pmod 8$, the generating idempotents of the quadratic residue codes over $\mathbb{F}_4$ are the same as those over $\mathbb{F}_2$ given in Theorem 6.6.5.*
(ii) *The even-like quadratic residue codes over $\mathbb{F}_4$ have generating idempotents*

$$\delta + \omega \sum_{j \in \mathcal{Q}_p} x^j + \overline{\omega} \sum_{j \in \mathcal{N}_p} x^j \quad and \quad \delta + \overline{\omega} \sum_{j \in \mathcal{Q}_p} x^j + \omega \sum_{j \in \mathcal{N}_p} x^j,$$

where $\delta = 0$ if $p \equiv -3 \pmod 8$ and $\delta = 1$ if $p \equiv 3 \pmod 8$.
(iii) *The odd-like quadratic residue codes over $\mathbb{F}_4$ have generating idempotents*

$$\epsilon + \omega \sum_{j \in \mathcal{Q}_p} x^j + \overline{\omega} \sum_{j \in \mathcal{N}_p} x^j \quad and \quad \epsilon + \overline{\omega} \sum_{j \in \mathcal{Q}_p} x^j + \omega \sum_{j \in \mathcal{N}_p} x^j,$$

where $\epsilon = 1$ if $p \equiv -3 \pmod 8$ and $\epsilon = 0$ if $p \equiv 3 \pmod 8$.

Proof: Part (i) follows from Theorem 6.6.4. Let $e(x)$ be a generating idempotent for an even-like QR code $\mathcal{C}_1$ over $\mathbb{F}_4$ with $p \equiv \pm 3 \pmod 8$. By Theorem 6.6.3, $e(x) = a_0 + a_1 Q(x) + a_2 N(x)$, where $Q(x) = \sum_{j \in \mathcal{Q}_p} x^j$ and $N(x) = \sum_{j \in \mathcal{N}_p} x^j$. By Lemma 6.2.5, $2 \in \mathcal{N}_p$. This implies that

$$Q(x)^2 = Q(x^2) = N(x) \quad \text{and} \quad N(x)^2 = N(x^2) = Q(x)$$

by Lemma 6.6.1. Therefore as $e(x)^2 = a_0^2 + a_1^2 Q(x)^2 + a_2^2 N(x)^2 = a_0^2 + a_1^2 N(x) + a_2^2 Q(x) = e(x)$, $a_0 \in \{0, 1\}$ and $a_2 = a_1^2$. The other even-like QR code $\mathcal{C}_2$ paired with $\mathcal{C}_1$ has generating idempotent $e(x)\mu_2$ by Theorem 6.6.4 as $2 \in \mathcal{N}_p$. Again by Lemma 6.2.5,

$$Q(x)\mu_2 = Q(x^2) = N(x) \quad \text{and} \quad N(x)\mu_2 = N(x^2) = Q(x)$$

as $2 \in \mathcal{N}_p$. Therefore $e(x)\mu_2 = a_0 + a_1^2 Q(x) + a_1 N(x)$. By Theorem 6.6.4(ii)

$$e(x) + e(x)\mu_2 = x + x^2 + \cdots + x^{p-1},$$

implying that $a_1 + a_1^2 = 1$. The only possibility is $a_1 \in \{\omega, \overline{\omega}\}$. Notice that, for either choice of a_1, $a_1 Q(x) + a_1^2 N(x)$ is odd-like if $p \equiv 3 \pmod 8$ and even-like if $p \equiv -3 \pmod 8$, as $Q(x)$ and $N(x)$ each have $(p-1)/2$ terms. Therefore parts (ii) and (iii) follow.   □

Example 6.6.7 We consider the binary QR codes of length 23. In that case,

$$\mathcal{Q}_{23} = \{1, 2, 3, 4, 6, 8, 9, 12, 13, 16, 18\} \text{ and }$$
$$\mathcal{N}_{23} = \{5, 7, 10, 11, 14, 15, 17, 19, 20, 21, 22\}.$$

The generating idempotents of the odd-like QR codes are

$$\sum_{j \in \mathcal{Q}_{23}} x^j \quad \text{and} \quad \sum_{j \in \mathcal{N}_{23}} x^j,$$

and the generating idempotents of the even-like QR codes are

$$1 + \sum_{j \in \mathcal{Q}_{23}} x^j \quad \text{and} \quad 1 + \sum_{j \in \mathcal{N}_{23}} x^j.$$

Note that the 2-cyclotomic cosets modulo 23 are $\{0\}$, $\mathcal{Q}_{23}$, and $\mathcal{N}_{23}$ implying that these are in fact the only binary duadic codes of length 23. The odd-like codes are the $[23, 12, 7]$ binary Golay code.   ■

Example 6.6.8 Now consider the QR codes of length 5 over $\mathbb{F}_4$; note that

$$\mathcal{Q}_5 = \{1, 4\} \quad \text{and} \quad \mathcal{N}_5 = \{2, 3\}.$$

The generating idempotents of the odd-like QR codes are

$$1 + \omega(x + x^4) + \overline{\omega}(x^2 + x^3) \quad \text{and} \quad 1 + \overline{\omega}(x + x^4) + \omega(x^2 + x^3),$$

and the generating idempotents of the even-like QR codes are

$$\omega(x + x^4) + \overline{\omega}(x^2 + x^3) \quad \text{and} \quad \overline{\omega}(x + x^4) + \omega(x^2 + x^3).$$

The 2-cyclotomic cosets modulo 5 are $\{0\}$, $\mathcal{Q}_5$, and $\mathcal{N}_5$ showing that the QR codes are the only duadic codes of length 5 over $\mathbb{F}_4$. The odd-like codes are equivalent to the punctured hexacode; see Exercise 363. See also Example 6.1.8.   ■

Exercise 363 Show that the odd-like QR codes of length 5 over $\mathbb{F}_4$ given in Example 6.6.8 are equivalent to the punctured hexacode, using the generator matrix of the hexacode found in Example 1.3.4. ♦

The idempotents that arise in Theorems 6.6.5 and 6.6.6 are the generating idempotents for QR codes over any field of characteristic 2 as the next result shows.

Theorem 6.6.9 *Let p be an odd prime. The following hold:*
(i) *Quadratic residue codes of length p over $\mathbb{F}_{2^t}$, where t is odd, exist if and only if $p \equiv \pm 1$ (mod 8), and the generating idempotents are those given in Theorem 6.6.5.*
(ii) *Quadratic residue codes of length p over $\mathbb{F}_{2^t}$, where t is even, exist for all p, and the generating idempotents are those given in Theorems 6.6.5 and 6.6.6.*

Proof: By Theorem 6.6.4, the result follows as long as we show that no quadratic residue codes exist when t is odd and $p \equiv \pm 3$ (mod 8). By Theorem 6.6.5, these codes do not exist if $t = 1$. If QR codes exist with $t = 2s + 1$ for some integer $s \geq 1$, then 2^t is a square modulo p by Theorem 6.6.2. If 2^t is a square modulo p, then 2 is also a square modulo p as $2^t = 2 \cdot (2^s)^2$, contradicting Lemma 6.2.5. □

Exercise 364 Give the generating idempotents for all quadratic residue codes of prime length $p \leq 29$ over $\mathbb{F}_{2^t}$. Distinguish between the idempotents that generate even-like codes and those that generate odd-like codes. Also distinguish between those that arise when t is even and when t is odd. ♦

Exercise 365 Let $\mathcal{D}_1$ and $\mathcal{D}_2$ be odd-like quadratic residue codes of prime length p over $\mathbb{F}_{2^t}$ with even-like subcodes $\mathcal{C}_1$ and $\mathcal{C}_2$. Prove the following:
(a) If $p \equiv -1$ (mod 8), then $\mathcal{C}_1$ and $\mathcal{C}_2$ are self-orthogonal under the ordinary inner product.
(b) If $p \equiv 1$ (mod 8), then $\mathcal{C}_1^\perp = \mathcal{D}_2$ and $\mathcal{C}_2^\perp = \mathcal{D}_1$.
(c) If $p \equiv 3$ (mod 8) and t is even, then $\mathcal{C}_1$ and $\mathcal{C}_2$ are self-orthogonal under the ordinary inner product.
(d) If $p \equiv -3$ (mod 8) and t is even, then $\mathcal{C}_1^\perp = \mathcal{D}_2$ and $\mathcal{C}_2^\perp = \mathcal{D}_1$.
(e) If $t = 2$ and either $p \equiv -3$ (mod 8) or $p \equiv -1$ (mod 8), then $\mathcal{C}_1$ and $\mathcal{C}_2$ are self-orthogonal under the Hermitian inner product.
(f) If $t = 2$ and either $p \equiv 1$ (mod 8) or $p \equiv 3$ (mod 8), then $\mathcal{C}_1^{\perp_H} = \mathcal{D}_2$ and $\mathcal{C}_2^{\perp_H} = \mathcal{D}_1$. ♦

6.6.2   QR codes over fields of characteristic 3

Analogous results hold for fields of characteristic 3. As in the last section, we let $Q(x) = \sum_{j \in \mathcal{Q}_p} x^j$ and $N(x) = \sum_{j \in \mathcal{N}_p} x^j$. We assume our QR codes have length p an odd prime that cannot equal 3. We first examine quadratic residue codes over $\mathbb{F}_3$.

Theorem 6.6.10 *Let $p > 3$ be prime. The following hold:*
(i) *Quadratic residue codes over $\mathbb{F}_3$ of length p exist if and only if $p \equiv \pm 1$ (mod 12).*

(ii) *The even-like quadratic residue codes over $\mathbb{F}_3$ have generating idempotents*

$$-\sum_{j \in \mathcal{Q}_p} x^j \quad and \quad -\sum_{j \in \mathcal{N}_p} x^j,$$

if $p \equiv 1 \pmod{12}$, and

$$1 + \sum_{j \in \mathcal{Q}_p} x^j \quad and \quad 1 + \sum_{j \in \mathcal{N}_p} x^j,$$

if $p \equiv -1 \pmod{12}$.

(iii) *The odd-like quadratic residue codes over $\mathbb{F}_3$ have generating idempotents*

$$1 + \sum_{j \in \mathcal{Q}_p} x^j \quad and \quad 1 + \sum_{j \in \mathcal{N}_p} x^j,$$

if $p \equiv 1 \pmod{12}$, and

$$-\sum_{j \in \mathcal{Q}_p} x^j \quad and \quad -\sum_{j \in \mathcal{N}_p} x^j,$$

if $p \equiv -1 \pmod{12}$.

Proof: Part (i) follows from Theorem 6.6.2 and Lemma 6.2.9. Let $p \equiv \pm 1 \pmod{12}$. If $e(x)$ is a generating idempotent for an even-like QR code $\mathcal{C}_1$ over $\mathbb{F}_3$, then by Theorem 6.6.3, $e(x) = a_0 + a_1 Q(x) + a_2 N(x)$, where $a_i \in \mathbb{F}_3$ for $0 \le i \le 2$. The other even-like QR code $\mathcal{C}_2$ paired with $\mathcal{C}_1$ has generating idempotent $e(x)\mu_b$, where $b \in \mathcal{N}_p$ by Theorem 6.6.4. Lemma 6.6.1 implies that

$$Q(x)\mu_b = Q(x^b) = N(x) \quad and \quad N(x)\mu_b = N(x^b) = Q(x).$$

Therefore $e(x)\mu_b = a_0 + a_2 Q(x) + a_1 N(x)$.

We first consider the case $p \equiv 1 \pmod{12}$. By Theorem 6.6.4(ii),

$$e(x) + e(x)\mu_b = -x - x^2 - \cdots - x^{p-1},$$

implying that $2a_0 = 0$ and $a_1 + a_2 = -1$. Thus $a_0 = 0$ and either $a_1 = a_2 = 1$ or $\{a_1, a_2\} = \{0, -1\}$. If $a_1 = a_2 = 1$, then $e(x) = Q(x) + N(x)$; but $Q(x) + N(x) = -(1 - \overline{j}(x))$, which is the negative of the idempotent generator of the even-like code $\mathcal{E}_p$. Thus $a_1 = a_2 = 1$ is not possible. So the two possibilities remaining for $\{a_1, a_2\}$ must lead to generating idempotents for the two even-like QR codes that we know must exist. The generating idempotents for the odd-like codes follow from Theorem 6.6.4.

We now consider the case $p \equiv -1 \pmod{12}$. By Theorem 6.6.4(ii),

$$e(x) + e(x)\mu_a = -1 + x + x^2 + \cdots + x^{p-1},$$

implying that $2a_0 = -1$ and $a_1 + a_2 = 1$. Thus $a_0 = 1$ and either $a_1 = a_2 = -1$ or $\{a_1, a_2\} = \{0, 1\}$. Again $1 - Q(x) - N(x) = -(1 - \overline{j}(x))$ generates $\mathcal{E}_p$ and so $a_1 = a_2 = -1$ is impossible. The generating idempotents for the even-like and odd-like codes follow as above.   □

Table 6.4 *The field $\mathbb{F}_9$*

ρ^i	$a + b\rho$	ρ^i	$a + b\rho$	ρ^i	$a + b\rho$
0	0	ρ^2	$1 + \rho$	ρ^5	$-\rho$
1	1	ρ^3	$1 - \rho$	ρ^6	$-1 - \rho$
ρ	ρ	ρ^4	-1	ρ^7	$-1 + \rho$

Example 6.6.11 We find the generating idempotents of the QR codes of length 11 over $\mathbb{F}_3$. Here

$$\mathcal{Q}_{11} = \{1, 3, 4, 5, 9\} \quad \text{and} \quad \mathcal{N}_{11} = \{2, 6, 7, 8, 10\}.$$

The generating idempotents of the odd-like QR codes are

$$-(x + x^3 + x^4 + x^5 + x^9) \quad \text{and} \quad -(x^2 + x^6 + x^7 + x^8 + x^{10}),$$

and the generating idempotents of the even-like QR codes are

$$1 + x + x^3 + x^4 + x^5 + x^9 \quad \text{and} \quad 1 + x^2 + x^6 + x^7 + x^8 + x^{10}.$$

The 3-cyclotomic cosets modulo 11 are $\{0\}$, $\mathcal{Q}_{11}$, and $\mathcal{N}_{11}$, implying that the QR codes are the only duadic codes of length 11 over $\mathbb{F}_3$. The odd-like codes are the [11, 6, 5] ternary Golay code. Compare this example to Example 6.1.7. ∎

We now turn to QR codes over $\mathbb{F}_9$. Because $9 = 3^2$ is a square modulo any odd prime p, by Theorem 6.6.2, QR codes over $\mathbb{F}_9$ exist for any odd prime length greater than 3. The idempotents for the QR codes over $\mathbb{F}_9$ of length $p \equiv \pm 1 \pmod{12}$ are the same as the idempotents given in Theorem 6.6.10 by Theorem 6.6.4. We now only need consider lengths p where $p \equiv \pm 5 \pmod{12}$. The field $\mathbb{F}_9$ can be constructed by adjoining an element ρ to $\mathbb{F}_3$, where $\rho^2 = 1 + \rho$. So $\mathbb{F}_9 = \{a + b\rho \mid a, b \in \mathbb{F}_3\}$. Multiplication in $\mathbb{F}_9$ is described in Table 6.4; note that ρ is a primitive 8th root of unity.

Theorem 6.6.12 *Let p be an odd prime. The following hold*:
(i) *If $p \equiv \pm 1 \pmod{12}$, the generating idempotents of the quadratic residue codes over $\mathbb{F}_9$ are the same as those over $\mathbb{F}_3$ given in Theorem 6.6.10.*
(ii) *The even-like quadratic residue codes over $\mathbb{F}_9$ have generating idempotents*

$$1 + \rho \sum_{j \in \mathcal{Q}_p} x^j + \rho^3 \sum_{j \in \mathcal{N}_p} x^j \quad \text{and} \quad 1 + \rho^3 \sum_{j \in \mathcal{Q}_p} x^j + \rho \sum_{j \in \mathcal{N}_p} x^j,$$

if $p \equiv 5 \pmod{12}$, and

$$-\rho \sum_{j \in \mathcal{Q}_p} x^j - \rho^3 \sum_{j \in \mathcal{N}_p} x^j \quad \text{and} \quad -\rho^3 \sum_{j \in \mathcal{Q}_p} x^j - \rho \sum_{j \in \mathcal{N}_p} x^j,$$

if $p \equiv -5 \pmod{12}$.
(iii) *The odd-like quadratic residue codes over $\mathbb{F}_9$ have generating idempotents*

$$-\rho \sum_{j \in \mathcal{Q}_p} x^j - \rho^3 \sum_{j \in \mathcal{N}_p} x^j \quad \text{and} \quad -\rho^3 \sum_{j \in \mathcal{Q}_p} x^j - \rho \sum_{j \in \mathcal{N}_p} x^j,$$

if $p \equiv 5 \pmod{12}$, *and*

$$1 + \rho \sum_{j \in \mathcal{Q}_p} x^j + \rho^3 \sum_{j \in \mathcal{N}_p} x^j \quad \text{and} \quad 1 + \rho^3 \sum_{j \in \mathcal{Q}_p} x^j + \rho \sum_{j \in \mathcal{N}_p} x^j,$$

if $p \equiv -5 \pmod{12}$.

Proof: Part (i) follows from Theorem 6.6.4. Let $e(x)$ be a generating idempotent for an even-like QR code $\mathcal{C}_1$ over $\mathbb{F}_9$ of length p with $p \equiv \pm 5 \pmod{12}$. Then by Theorem 6.6.3, $e(x) = a_0 + a_1 Q(x) + a_2 N(x)$, where $a_i \in \mathbb{F}_9$ for $0 \le i \le 2$. Using Lemma 6.6.1, notice that $Q(x)^3 = Q(x^3) = N(x)$ and $N(x)^3 = N(x^3) = Q(x)$ as $3 \in \mathcal{N}_p$ by Lemma 6.2.9. As $e(x)^2 = e(x)$, we must have $e(x)^3 = e(x)$. Thus $e(x)^3 = a_0^3 + a_1^3 N(x) + a_2^3 Q(x) = e(x)$, implying that $a_0^3 = a_0$ and $a_2 = a_1^3$. The other even-like QR code $\mathcal{C}_2$ paired with $\mathcal{C}_1$ has generating idempotent $e(x)\mu_b$, where $b \in \mathcal{N}_p$ by Theorem 6.6.4. Again by Lemma 6.6.1, $Q(x)\mu_b = Q(x^b) = N(x)$ and $N(x)\mu_b = N(x^b) = Q(x)$. Therefore, $e(x) = a_0 + a_1 Q(x) + a_1^3 N(x)$ and $e(x)\mu_b = a_0 + a_1^3 Q(x) + a_1 N(x)$.

We first consider the case $p \equiv 5 \pmod{12}$. By Theorem 6.6.4(ii),

$$e(x) + e(x)\mu_b = -1 + x + x^2 + \cdots + x^{p-1},$$

implying that $2a_0 = -1$ and $a_1 + a_1^3 = 1$. Thus $a_0 = 1$; by examining Table 6.4, either $a_1 = -1$ or $a_1 \in \{\rho, \rho^3\}$. As $1 - Q(x) - N(x) = -(1 - \bar{j}(x))$ generates $\mathcal{E}_p$, $a_1 = -1$ is impossible. So the two possibilities remaining for a_1 must lead to generating idempotents for the two even-like QR codes that we know exist. The generating idempotents for the odd-like codes follow from Theorem 6.6.4.

We leave the case $p \equiv -5 \pmod{12}$ as an exercise.   □

Exercise 366 Prove Theorem 6.6.12 in the case $p \equiv -5 \pmod{12}$.   ♦

The following theorem is analogous to Theorem 6.6.9 and is proved in the same way.

Theorem 6.6.13 *Let p be an odd prime with $p \neq 3$. The following hold:*
(i) *Quadratic residue codes of length p over $\mathbb{F}_{3^t}$, where t is odd, exist if and only if $p \equiv \pm 1 \pmod{12}$, and the generating idempotents are those given in Theorem 6.6.10.*
(ii) *Quadratic residue codes of length p over $\mathbb{F}_{3^t}$, where t is even, exist for all p, and the generating idempotents are those given in Theorems 6.6.10 and 6.6.12.*

Exercise 367 Give the generating idempotents for all quadratic residue codes of prime length $p \le 29$ over $\mathbb{F}_{3^t}$. Distinguish between the idempotents that generate even-like codes and those that generate odd-like codes. Also distinguish between those that arise when t is even and when t is odd.   ♦

Exercise 368 Let $\mathcal{D}_1$ and $\mathcal{D}_2$ be odd-like quadratic residue codes of prime length $p \neq 3$ over $\mathbb{F}_{3^t}$ with even-like subcodes $\mathcal{C}_1$ and $\mathcal{C}_2$. Prove the following:
(a) If $p \equiv -1 \pmod{12}$, then $\mathcal{C}_1$ and $\mathcal{C}_2$ are self-orthogonal.
(b) If $p \equiv 1 \pmod{12}$, then $\mathcal{C}_1^{\perp} = \mathcal{D}_2$ and $\mathcal{C}_2^{\perp} = \mathcal{D}_1$.
(c) If $p \equiv -5 \pmod{12}$ and t is even, then $\mathcal{C}_1$ and $\mathcal{C}_2$ are self-orthogonal.
(d) If $p \equiv 5 \pmod{12}$ and t is even, then $\mathcal{C}_1^{\perp} = \mathcal{D}_2$ and $\mathcal{C}_2^{\perp} = \mathcal{D}_1$.   ♦

6.6.3   Extending QR codes

As with any of the duadic codes, we can consider extending odd-like quadratic residue codes in such a way that the extensions are self-dual or dual to each other. These extensions may not be the ordinary extensions obtained by adding an overall parity check, but all the extensions are equivalent to that obtained by adding an overall parity check. Before examining the general case, we look at QR codes over $\mathbb{F}_2$, $\mathbb{F}_3$, and $\mathbb{F}_4$. In these cases, it is sufficient to use the ordinary extension.

Theorem 6.6.14 *Let $\mathcal{D}_1$ and $\mathcal{D}_2$ be the odd-like QR codes over $\mathbb{F}_q$ of odd prime length p.*

(i) *When $q = 2$, the following hold:*

   (a) $\widehat{\mathcal{D}}_1$ *and* $\widehat{\mathcal{D}}_2$ *are duals of each other when $p \equiv 1 \pmod 8$.*

   (b) $\widehat{\mathcal{D}}_i$ *is self-dual and doubly-even for $i = 1$ and 2 when $p \equiv -1 \pmod 8$.*

(ii) *When $q = 3$, the following hold:*

   (a) $\widehat{\mathcal{D}}_i$ *is self-dual for $i = 1$ and 2 when $p \equiv -1 \pmod{12}$.*

   (b) *If $p \equiv 1 \pmod{12}$, then $\widehat{\mathcal{D}}_1$ and $\widehat{\mathcal{D}}_2 D$ are duals of each other where D is the diagonal matrix $\operatorname{diag}(1, 1, \ldots, 1, -1)$.*

(iii) *When $q = 4$, the following hold:*

   (a) *When $p \equiv 1 \pmod 8$, $\widehat{\mathcal{D}}_1$ and $\widehat{\mathcal{D}}_2$ are duals of each other under either the ordinary or the Hermitian inner product.*

   (b) *When $p \equiv 3 \pmod 8$, $\widehat{\mathcal{D}}_1$ and $\widehat{\mathcal{D}}_2$ are duals of each other under the Hermitian inner product; furthermore, $\widehat{\mathcal{D}}_i$ is self-dual under the ordinary inner product for $i = 1$ and 2.*

   (c) *When $p \equiv -3 \pmod 8$, $\widehat{\mathcal{D}}_1$ and $\widehat{\mathcal{D}}_2$ are duals of each other under the ordinary inner product; furthermore, $\widehat{\mathcal{D}}_i$ is self-dual under the Hermitian inner product for $i = 1$ and 2.*

   (d) *When $p \equiv -1 \pmod 8$, $\widehat{\mathcal{D}}_i$ is self-dual under either the ordinary or the Hermitian inner product for $i = 1$ and 2.*

Proof: Let $\mathcal{C}_i$ be the even-like subcode of $\mathcal{D}_i$.

We first consider the case $q = 2$ or $q = 4$. In either case, $\overline{j}(x)$ is the all-one vector and its extension is the all-one vector of length $p + 1$; this extended all-one vector is orthogonal to itself under either the ordinary or Hermitian inner product. Suppose first that $p \equiv 1 \pmod 8$. By Lemmas 6.2.4 and 6.2.5, -1 and -2 are both in $\mathcal{Q}_p$. Thus $\mathcal{C}_i \mu_{-1} = \mathcal{C}_i \mu_{-2} = \mathcal{C}_i$ for $i = 1$ and 2. Applying Theorems 6.4.3 and 6.4.6, we obtain (i)(a) and (iii)(a). Consider next the case $p \equiv -1 \pmod 8$. This time by Lemmas 6.2.4 and 6.2.5, -1 and -2 are both in $\mathcal{N}_p$. Thus $\mathcal{C}_1 \mu_{-1} = \mathcal{C}_1 \mu_{-2} = \mathcal{C}_2$. By Theorems 6.4.2 and 6.4.5, we obtain part of (i)(b) and all of (iii)(d). To complete (i)(b), we note that the generating idempotent for $\mathcal{D}_i$ has weight $(p - 1)/2$ by Theorem 6.6.5 and hence $\mathcal{D}_i$ has a generator matrix consisting of shifts of this idempotent by Theorem 4.3.6. Thus $\widehat{\mathcal{D}}_i$ has a generator matrix consisting of vectors of weight $((p - 1)/2) + 1 \equiv 0 \pmod 4$. Thus $\widehat{\mathcal{D}}_i$ is doubly-even. The cases $p \equiv \pm 3 \pmod 8$ arise only when $q = 4$. Using the same argument, if $p \equiv 3 \pmod 8$, $-1 \in \mathcal{N}_p$ and $-2 \in \mathcal{Q}_p$, yielding (iii)(b) by Theorems 6.4.2 and 6.4.6. If $p \equiv -3 \pmod 8$, $-1 \in \mathcal{Q}_p$ and $-2 \in \mathcal{N}_p$, yielding (iii)(c) by Theorems 6.4.3 and 6.4.5.

Now let $q = 3$. Consider first the case $p \equiv -1 \pmod{12}$. Here the all-one vector extends to the all-one vector of length $p + 1$ and it is orthogonal to itself. By Lemma 6.2.4, $-1 \in \mathcal{N}_p$ and (ii)(a) follows from Theorem 6.4.2. Now suppose that $p \equiv 1 \pmod{12}$. By Lemma 6.2.4, $-1 \in \mathcal{Q}_p$ and by Theorem 6.4.3, $\mathcal{C}_1^\perp = \mathcal{D}_2$. In this case the all-one vector in $\mathcal{D}_i$ extends to $\widehat{\mathbf{1}} = 11 \cdots 1(-1)$ in $\widehat{\mathcal{D}}_i$. This vector is not orthogonal to itself, but is orthogonal to $\widehat{\mathbf{1}}D = \mathbf{1}_{p+1}$, where $\mathbf{1}_{p+1}$ is the all-one vector of length $p + 1$ and D is the diagonal matrix $\mathrm{diag}(1, 1, \ldots, 1, -1)$. This proves (ii)(b). $\square$

Example 6.6.15 We describe the extensions of the QR codes discussed in Examples 6.6.7, 6.6.8, and 6.6.11. We give the extension of the generating idempotent of odd-like codes; from this one can form a basis of the extended codes using shifts of the generating idempotent extended in the same way (see Theorem 4.3.6).

- The extended coordinate of either generating idempotent of an odd-like binary QR code of length 23 is 1. These extended codes are each self-dual and doubly-even by Theorem 6.6.14. They are extended binary Golay codes.
- The extended coordinate of either generating idempotent of an odd-like QR code of length 11 over $\mathbb{F}_3$ is -1. The extended codes are each self-dual by Theorem 6.6.14 and are extended ternary Golay codes.
- The extended coordinate of either generating idempotent of an odd-like QR code of length 5 over $\mathbb{F}_4$ is 1. These extended codes are each Hermitian self-dual; they are also dual to each other under the ordinary inner product by Theorem 6.6.14. They are equivalent to the hexacode.   ∎

We are now ready to describe, in general, the extensions of the odd-like QR codes $\mathcal{D}_1$ and $\mathcal{D}_2$ of length p over $\mathbb{F}_q$. We want them both to be extended in the same way whenever possible. Recall that we defined $\widetilde{\mathcal{D}}$ for an arbitrary odd-like duadic code $\mathcal{D}$ of length n using a solution γ of (6.11). As an odd-like QR code is obtained from its even-like subcode by adjoining the all-one vector $\mathbf{1}$, in order for either $\widetilde{\mathcal{D}}_1$ to be self-dual or dual to $\widetilde{\mathcal{D}}_2$, the extended vector $\widetilde{\mathbf{1}}$, which is $\mathbf{1}$ extended by some $\gamma \in \mathbb{F}_q$, must be orthogonal to itself. This means that $p + \gamma^2 p^2 = 0$ or

$$1 + \gamma^2 p = 0, \tag{6.12}$$

which is (6.11) with $n = p$. Suppose that $\mathcal{C}_i$ is the even-like subcode of $\mathcal{D}_i$. We know that either $-1 \in \mathcal{N}_p$ or $-1 \in \mathcal{Q}_p$, implying $\mathcal{C}_1\mu_{-1} = \mathcal{C}_2$ or $\mathcal{C}_i\mu_{-1} = \mathcal{C}_i$, respectively. By Theorems 6.4.2 or 6.4.3, these yield $\mathcal{C}_i^\perp = \mathcal{D}_i$ or $\mathcal{C}_1^\perp = \mathcal{D}_2$, respectively. Therefore, whenever (6.12) is satisfied, if $-1 \in \mathcal{N}_p$, $\widetilde{\mathcal{D}}_i$ is self-dual for $i = 1$ and 2, and if $-1 \in \mathcal{Q}_p$, $\widetilde{\mathcal{D}}_1$ and $\widetilde{\mathcal{D}}_2$ are duals of each other. This proves the following, using Lemma 6.2.4.

Theorem 6.6.16 *Let $\mathcal{D}_1$ and $\mathcal{D}_2$ be the odd-like QR codes of length p over $\mathbb{F}_q$, p an odd prime. Suppose that (6.12) is satisfied. Then if $p \equiv -1 \pmod{4}$, $\widetilde{\mathcal{D}}_i$ is self-dual for $i = 1$ and 2, and if $p \equiv 1 \pmod{4}$, $\widetilde{\mathcal{D}}_1$ and $\widetilde{\mathcal{D}}_2$ are duals of each other.*

Thus we are interested in the cases where (6.12) has a solution γ in $\mathbb{F}_q$. This is answered by the following. We only need the case where q is a square modulo p since we are assuming QR codes of length p over $\mathbb{F}_q$ exist.

Lemma 6.6.17 *Let r be a prime so that $q = r^t$ for some positive integer t. Let p be an odd prime with $p \neq r$ and assume that q is a square modulo p. There is a solution γ of (6.12) in $\mathbb{F}_q$ except when t is odd, $p \equiv 1 \pmod 4$, and $r \equiv -1 \pmod 4$. In that case there is a solution γ_1 in $\mathbb{F}_q$ of $-1 + \gamma_1^2 p = 0$.*

Proof: If t is even, then every quadratic equation with coefficients in $\mathbb{F}_r$, such as (6.12), has a solution in $\mathbb{F}_{r^2} \subseteq \mathbb{F}_q$. Assume that t is odd. Then q is a square modulo p if and only if r is a square modulo p as $q = (r^2)^{(t-1)/2} r$. So $\left(\frac{r}{p}\right) = 1$. Solving (6.12) is equivalent to solving $x^2 = -p$ in $\mathbb{F}_{r^t}$. As t is odd, there is a solution in $\mathbb{F}_q$ if and only if the solution is in $\mathbb{F}_r$. This equation reduces to $x^2 = 1$ if $r = 2$, which obviously has a solution $x = 1$. Assume that r is odd. Thus we have a solution to $x^2 = -p$ in $\mathbb{F}_{r^t}$ if and only if $\left(\frac{-p}{r}\right) = \left(\frac{-1}{r}\right)\left(\frac{p}{r}\right) = 1$. By the Law of Quadratic Reciprocity,

$$\left(\frac{r}{p}\right)\left(\frac{p}{r}\right) = (-1)^{\frac{r-1}{2}\frac{p-1}{2}}.$$

Hence as $\left(\frac{r}{p}\right) = 1$,

$$\left(\frac{-1}{r}\right)\left(\frac{p}{r}\right) = (-1)^{\frac{r-1}{2}}(-1)^{\frac{r-1}{2}\frac{p-1}{2}}$$

using Lemma 6.2.2. The only time the right-hand side of this equation is not 1 is when $p \equiv 1 \pmod 4$ and $r \equiv -1 \pmod 4$. In this exceptional case

$$\left(\frac{p}{r}\right) = (-1)^{\frac{r-1}{2}\frac{p-1}{2}} = 1,$$

showing that $x^2 = p$ has a solution in $\mathbb{F}_q$; hence $-1 + \gamma_1^2 p = 0$ has a solution γ_1 in $\mathbb{F}_q$. $\square$

Combining Theorem 6.6.16 and Lemma 6.6.17, we obtain the following.

Theorem 6.6.18 *Let r be a prime so that $q = r^t$ for some positive integer t. Let p be an odd prime with $p \neq r$ and assume that q is a square modulo p. Assume that if t is odd, then either $p \not\equiv 1 \pmod 4$ or $r \not\equiv -1 \pmod 4$. Let $\mathcal{D}_1$ and $\mathcal{D}_2$ be the two odd-like QR code over $\mathbb{F}_q$ of length p.*
(i) If $p \equiv -1 \pmod 4$, then $\widetilde{\mathcal{D}}_i$ is self-dual for $i = 1$ and 2.
(ii) If $p \equiv 1 \pmod 4$, then $\widetilde{\mathcal{D}}_1$ and $\widetilde{\mathcal{D}}_2$ are duals of each other.

This theorem gives the extension except when $q = r^t$ where t is odd, $p \equiv 1 \pmod 4$, and $r \equiv -1 \pmod 4$. But in that case $-1 + \gamma_1^2 p = 0$ has a solution γ_1 in $\mathbb{F}_q$ by Lemma 6.6.17. If $\mathcal{D}$ is an odd-like QR code of length p, define $\check{\mathcal{D}} = \{\check{\mathbf{c}} \mid \mathbf{c} \in \mathcal{D}\}$, where $\check{\mathbf{c}} = \mathbf{c}c_\infty = c_0 \cdots c_{p-1}c_\infty$ and

$$c_\infty = -\gamma_1 \sum_{j=0}^{p-1} c_j.$$

Theorem 6.6.19 *Let r be a prime so that $q = r^t$ for some odd positive integer t and $r \equiv -1 \pmod 4$. Assume $p \equiv 1 \pmod 4$ is a prime such that q is a square modulo p. Let $\mathcal{D}_1$ and $\mathcal{D}_2$ be the two odd-like QR codes over $\mathbb{F}_q$ of length p. Then $\check{\mathcal{D}}_1$ and $\check{\mathcal{D}}_2 D$ are duals of each other, where D is the diagonal matrix $\mathrm{diag}(1, 1, \ldots, 1, -1)$.*

Proof: By our assumption and Lemma 6.6.17, $-1 + \gamma_1^2 p = 0$ has a solution γ_1 in $\mathbb{F}_q$. As $p \equiv 1 \pmod 4$, $-1 \in \mathcal{Q}_p$. Let $\mathcal{C}_i$ be the even-like subcode of $\mathcal{D}_i$. Then $\mathcal{C}_i \mu_{-1} = \mathcal{C}_i$, and, by Theorem 6.4.3, $\mathcal{C}_1^\perp = \mathcal{D}_2$ and $\mathcal{C}_2^\perp = \mathcal{D}_1$. The result follows as $\check{\mathbf{1}}$ is orthogonal to $\check{\mathbf{1}}D$ because $-1 + \gamma_1^2 p = 0$. $\square$

Note that if $r \neq 2$, there are two solutions γ of $1 + \gamma^2 p = 0$ over $\mathbb{F}_{r^t}$, one the negative of the other. Similarly $-1 + \gamma_1^2 p = 0$ has two solutions. We can use either solution to define the extensions. When $r = 2$, we can always choose $\gamma = 1$, and hence for the codes over fields of characteristic 2, $\widetilde{\mathcal{D}} = \widehat{\mathcal{D}}$. We see that when $p \equiv -1 \pmod{12}$, the solution of (6.12) in $\mathbb{F}_3$ is $\gamma = 1$ and so here also $\widetilde{\mathcal{D}} = \widehat{\mathcal{D}}$. When $p \equiv 1 \pmod{12}$, (6.12) has no solution in $\mathbb{F}_3$ but one solution of $-1 + \gamma_1^2 p = 0$ in $\mathbb{F}_3$ is $\gamma_1 = 1$; thus in this case $\check{\mathcal{D}} = \widehat{\mathcal{D}}$.

We now look at the extended codes over fields of characteristic 2 and 3.

Corollary 6.6.20 *Let $\mathcal{D}_1$ and $\mathcal{D}_2$ be odd-like QR codes over $\mathbb{F}_{r^t}$ of length p.*
(i) *Suppose that $r = 2$. The following hold:*
   (a) *$\widehat{\mathcal{D}}_i$ is self-dual for $i = 1$ and 2 when $p \equiv -1 \pmod 8$ or when $p \equiv 3 \pmod 8$ with t even.*
   (b) *$\widehat{\mathcal{D}}_1$ and $\widehat{\mathcal{D}}_2$ are duals of each other when $p \equiv 1 \pmod 8$ or when $p \equiv -3 \pmod 8$ with t even.*
(ii) *Suppose that $r = 3$. The following hold:*
   (a) *$\widehat{\mathcal{D}}_i$ is self-dual for $i = 1$ and 2 when $p \equiv -1 \pmod{12}$.*
   (b) *If $p \equiv 1 \pmod{12}$, then $\widehat{\mathcal{D}}_1$ and $\widehat{\mathcal{D}}_2 D$ are duals of each other, where D is the diagonal matrix $\mathrm{diag}(1, 1, \ldots, 1, -1)$.*
   (c) *If $p \equiv -5 \pmod{12}$ with t even, then $\widetilde{\mathcal{D}}_i$ is self-dual for $i = 1$ and 2, where $\gamma = \rho^2$ from Table 6.4.*
   (d) *If $p \equiv 5 \pmod{12}$ with t even, then $\widehat{\mathcal{D}}_1$ and $\widehat{\mathcal{D}}_2$ are duals of each other.*

Exercise 369 Explicitly find the solutions of $1 + \gamma^2 p = 0$ and $-1 + \gamma_1^2 p = 0$ over $\mathbb{F}_{2^t}$ and $\mathbb{F}_{3^t}$. Then use that information to prove Corollary 6.6.20.   ◆

Exercise 370 Do the following:
(a) Give the appropriate extended generating idempotents for all odd-like quadratic residue codes of prime length $p \leq 29$ over $\mathbb{F}_{2^t}$. Distinguish between those that arise when t is even and when t is odd. Also give the duality relationships between the extended codes. See Exercise 364.
(b) Give the appropriate extended generating idempotents for all odd-like quadratic residue codes of prime length $p \leq 29$ over $\mathbb{F}_{3^t}$. Distinguish between those that arise when t is even and when t is odd. Also give the duality relationships between the extended codes. See Exercise 367.   ◆

6.6.4   Automorphisms of extended QR codes

In this section, we briefly present information about the automorphism groups of the extended QR codes. Those interested in a complete description of the automorphism groups of the extended QR codes (and extended generalized QR codes) should consult either [147] or [149]. Let $\mathcal{D}^{\text{ext}}$ be one of the extensions $\widehat{\mathcal{D}}$, $\widetilde{\mathcal{D}}$, or $\check{\mathcal{D}}$ of one of the odd-like QR codes $\mathcal{D}$, whichever is appropriate. The coordinates of $\mathcal{D}^{\text{ext}}$ are labeled $\{0, 1, \ldots, p-1, \infty\} = \mathbb{F}_p \cup \{\infty\}$. Obviously, the maps T_g, for $g \in \mathbb{F}_p$, given by $iT_g \equiv i + g \pmod{p}$ for all $i \in \mathbb{F}_p$ and $\infty T_g = \infty$ are in $\text{PAut}(\mathcal{D}^{\text{ext}})$ as these act as cyclic shifts on $\{0, 1, \ldots, p-1\}$ and fix ∞. Also if $a \in \mathcal{Q}_p$, the multiplier μ_a can be extended by letting $\infty \mu_a = \infty$; by Theorem 6.6.3, this is also in $\text{PAut}(\mathcal{D}^{\text{ext}})$. So far we have not found any automorphisms that move the coordinate ∞; however, the Gleason–Prange Theorem produces such a map. We do not prove this result.

Theorem 6.6.21 (Gleason–Prange Theorem) *Let $\mathcal{D}^{\text{ext}}$ be $\widehat{\mathcal{D}}$, $\widetilde{\mathcal{D}}$, or $\check{\mathcal{D}}$, where $\mathcal{D}$ is an odd-like QR code of length p over $\mathbb{F}_q$. Let P be the permutation matrix given by the permutation that interchanges ∞ with 0, and also interchanges g with $-1/g$ for $g \in \mathbb{F}_p$, $g \neq 0$. Then there is a diagonal matrix D, all of whose diagonal entries are ± 1, such that $DP \in \text{MAut}(\mathcal{D}^{\text{ext}})$.*

We have left the exact form of the diagonal matrix D rather vague as it depends on exactly which generating idempotent is used for the unextended code. Note, however, that if the field $\mathbb{F}_q$ has characteristic 2, then D is the identity matrix. The permutation matrices given by T_g with $g \in \mathbb{F}_p$, μ_a with $a \in \mathcal{Q}_p$, and P generate a group denoted $\text{PSL}_2(p)$ called the *projective special linear group*. In all but three cases, the full automorphism group $\Gamma\text{Aut}(\mathcal{D}^{\text{ext}})$ is only slightly bigger (we can add automorphisms related to field automorphisms of $\mathbb{F}_q$). The three exceptions occur when $p = 23$ and $q = 2^t$ (where $\mathcal{D}^{\text{ext}}$ has the same basis as the extended binary Golay code), when $p = 11$ and $q = 3^t$ (where $\mathcal{D}^{\text{ext}}$ has the same basis as the extended ternary Golay code), and when $p = 5$ and $q = 4^t$ (where $\mathcal{D}^{\text{ext}}$ has the same basis as the hexacode).

Exercise 371 Find the permutation in cycle form corresponding to the permutation matrix P in the Gleason–Prange Theorem when $p = 5$, $p = 11$, and $p = 23$. Note that these values of p are those arising in Example 6.6.15.   ♦

The automorphism from the Gleason–Prange Theorem together with T_g, for $g \in \mathbb{F}_p$, and μ_a, for $a \in \mathcal{Q}_p$, show that $\text{MAut}(\mathcal{D}^{\text{ext}})$ is transitive. This implies by Theorem 1.7.13, that the minimum weight of $\mathcal{D}$ is its minimum odd-like weight. Because QR codes are duadic codes, the Square Root Bound applies. We summarize this in the following.

Theorem 6.6.22 *Let $\mathcal{D}$ be an odd-like QR code of length p over $\mathbb{F}_q$. The minimum weight of $\mathcal{D}$ is its minimum odd-like weight d_o. Furthermore, $d_o^2 \geq p$. If $p \equiv -1 \pmod 4$, then $d_o^2 - d_o + 1 \geq p$. Additionally, every minimum weight codeword is odd-like. If $\mathcal{D}$ is binary, its minimum weight $d = d_o$ is odd, and if, in addition, $p \equiv -1 \pmod 8$, then $d \equiv 3 \pmod 4$.*

Proof: All statements but the last follow from the Square Root Bound and Theorem 1.7.13, together with the observation that μ_{-1} gives the splitting for QR codes precisely when $-1 \in \mathcal{N}_p$, that is when $p \equiv -1 \pmod 4$ by Lemma 6.2.4. If $\mathcal{D}$ is binary, its minimum weight d is odd as odd-like binary vectors have odd weight. By Theorem 6.6.14, $\widehat{\mathcal{D}}$ is doubly-even if $p \equiv -1 \pmod 8$; as $\widehat{\mathcal{D}}$ has minimum weight $d + 1$, $d \equiv 3 \pmod 4$.   □

Example 6.6.23 Using Theorem 6.6.22, we can easily determine the minimum weight of the binary odd-like quadratic residue codes of lengths $p = 23$ and 47 and dimensions 12 and 24, respectively. Let $\mathcal{D}$ be such a code. As $p \equiv -1 \pmod 4$, each satisfies $d_o^2 - d_o + 1 \geq p$ and d_o is odd. When $p = 23$, this bound implies that $d_o \geq 6$ and hence $d_o \geq 7$ as it is odd. The Sphere Packing Bound precludes a higher value of d_o. By Theorem 6.6.14, $\widehat{\mathcal{D}}$ is a [24, 12, 8] self-dual doubly-even code (the extended binary Golay code). When $p = 47$, $d_o \geq 8$ by the bound and $d_o \equiv 3 \pmod 4$ by Theorem 6.6.22. In particular, $d_o \geq 11$. The Sphere Packing Bound shows that $d_o < 15$, and as $d_o \equiv 3 \pmod 4$, $d_o = 11$. Thus $\mathcal{D}$ is a [47, 24, 11] code. Using Theorem 6.6.14, $\widehat{\mathcal{D}}$ is a [48, 24, 12] self-dual doubly-even code. At the present time there is no other known [48, 24, d] binary code with $d \geq 12$. ∎

Exercise 372 Find the minimum distance of an odd-like binary QR code of length 31. ♦

Example 6.6.24 Let $\mathcal{D}$ be an odd-like QR code over $\mathbb{F}_3$ of length $p = 11$ and dimension 6. Let $\mathcal{D}$ have minimum weight d and $\widehat{\mathcal{D}}$ have minimum weight $\widehat{d}$. By Theorem 6.6.14, $\widehat{\mathcal{D}}$ is self-dual as $p \equiv -1 \pmod{12}$. Hence, $3 \mid \widehat{d}$. By the Singleton Bound, $\widehat{d} \leq 7$ implying that $\widehat{d} = 3$ or $\widehat{d} = 6$. By Theorem 6.6.22, $d = d_o \geq 4$. As $d \leq \widehat{d}$, $\widehat{d} = 6$. As $d \geq \widehat{d} - 1$, $d \geq 5$. The generating idempotent of $\mathcal{D}$ has weight 5 by Theorem 6.6.10. Thus $\mathcal{D}$ is an [11, 6, 5] code, and $\widehat{\mathcal{D}}$ is a [12, 6, 6] self-dual code (the extended ternary Golay code). ∎

Example 6.6.25 Let $\mathcal{D}$ be an odd-like QR code over $\mathbb{F}_4$ of length $p = 5$ and dimension 3. Suppose that $\mathcal{D}$ and $\widehat{\mathcal{D}}$ have minimum weight d and $\widehat{d}$, respectively. By Theorem 6.6.14, $\widehat{\mathcal{D}}$ is Hermitian self-dual implying that $2 \mid \widehat{d}$. By the Singleton Bound, $d \leq 3$ and $\widehat{d} \leq 4$. By Theorem 6.6.22, $d = d_o \geq 3$. Thus $d = 3$ and $\widehat{d} = 4$. Therefore $\mathcal{D}$ is a [5, 3, 3] code, and $\widehat{\mathcal{D}}$ is a [6, 3, 4] Hermitian self-dual code (the hexacode). ∎

Example 6.6.26 Consider the duadic codes of length $n = 17$ over $\mathbb{F}_4$. The 4-cyclotomic cosets modulo 17 are

$$C_0 = \{0\}, \quad C_1 = \{1, 4, 16, 13\}, \quad C_2 = \{2, 8, 15, 9\}, \quad C_3 = \{3, 12, 14, 5\},$$
$$\text{and } C_6 = \{6, 7, 11, 10\}.$$

The odd-like quadratic residue codes $\mathcal{D}_1$ and $\mathcal{D}_2$ have defining sets $\mathcal{Q}_{17} = C_1 \cup C_2$ and $\mathcal{N}_{17} = C_3 \cup C_6$; the extended codes $\widehat{\mathcal{D}}_1$ and $\widehat{\mathcal{D}}_2$ are duals of each other under either the ordinary or Hermitian inner product by Theorem 6.6.14. Both $\mathcal{D}_1$ and $\mathcal{D}_2$ have binary idempotents by Theorem 6.6.6. By Theorem 6.6.22, $\mathcal{D}_1$ and $\mathcal{D}_2$ have minimum weight at least 5; as all minimum weight codewords are odd-like, $\widehat{\mathcal{D}}_1$ and $\widehat{\mathcal{D}}_2$ have minimum weight at least 6. In fact, both are [18, 9, 6] codes. This information is contained in Table 6.1. There are two other splittings given by $\mathcal{S}_1 = C_1 \cup C_3$ with $\mathcal{S}_2 = C_2 \cup C_6$, and $\mathcal{S}_1' = C_1 \cup C_6$ with $\mathcal{S}_2' = C_2 \cup C_3$. The splittings are interchanged by μ_6, yielding equivalent pairs of codes. The odd-like codes $\mathcal{D}_1'$ and $\mathcal{D}_2'$ with defining sets $\mathcal{S}_1$ and $\mathcal{S}_2$ have a splitting given by μ_{-2}. By Theorem 6.4.14, $\widehat{\mathcal{D}}_1'$ and $\widehat{\mathcal{D}}_2'$ are Hermitian self-dual. It turns out that both are [18, 9, 8] codes; these extended duadic codes have minimum weight higher than that of the extended quadratic residue codes. These are the codes summarized in Table 6.3. It was shown in [148] that a Hermitian self-dual [18, 9, 8] code over $\mathbb{F}_4$ is unique. Later, in [249], it was shown that an [18, 9, 8] code over $\mathbb{F}_4$ is unique. ∎

We conclude this section by presenting the automorphism groups of the binary extended odd-like quadratic residue codes. For a proof, see [149]; we discuss (i) in Example 9.6.2 and (ii) in Section 10.1.2.

Theorem 6.6.27 *Let p be a prime such that $p \equiv \pm 1$ (mod 8). Let $\mathcal{D}$ be a binary odd-like quadratic residue code of length p with extended code $\widehat{\mathcal{D}}$ of length $p + 1$. The following hold*:

(i)   *When $p = 7$, $\Gamma \mathrm{Aut}(\widehat{\mathcal{D}}) = \mathrm{PAut}(\widehat{\mathcal{D}})$ is isomorphic to the affine group $\mathrm{GA}_3(2)$ of order 1344.*

(ii)   *When $p = 23$, $\Gamma \mathrm{Aut}(\widehat{\mathcal{D}}) = \mathrm{PAut}(\widehat{\mathcal{D}})$ is isomorphic to the Mathieu group $\mathcal{M}_{24}$ of order 244 823 040.*

(iii) *If $p \notin \{7, 23\}$, then $\Gamma \mathrm{Aut}(\widehat{\mathcal{D}}) = \mathrm{PAut}(\widehat{\mathcal{D}})$ is isomorphic to the group $\mathrm{PSL}_2(p)$ of order $p(p^2 - 1)/2$.*

7 Weight distributions

In Chapter 1 we encountered the notion of the weight distribution of a code. In this chapter we greatly expand on this concept.

The weight distribution (or weight spectrum) of a code of length n specifies the number of codewords of each possible weight $0, 1, \ldots, n$. We generally denote the weight distribution of a code $\mathcal{C}$ by $A_0(\mathcal{C}), A_1(\mathcal{C}), \ldots, A_n(\mathcal{C})$, or, if the code is understood, by $A_0, A_1, \ldots, A_n$, where $A_i = A_i(\mathcal{C})$ is the number of codewords of weight i. As a code often has many values where $A_i(\mathcal{C}) = 0$, these values are usually omitted from the list. While the weight distribution does not in general uniquely determine a code, it does give important information of both practical and theoretical significance. However, computing the weight distribution of a large code, even on a computer, can be a formidable problem.

7.1 The MacWilliams equations

A linear code $\mathcal{C}$ is uniquely determined by its dual $\mathcal{C}^\perp$. The most fundamental result about weight distributions is a set of linear relations between the weight distributions of $\mathcal{C}$ and $\mathcal{C}^\perp$ which imply, in particular, that the weight distribution of $\mathcal{C}$ is uniquely determined by the weight distribution of $\mathcal{C}^\perp$ and vice versa. In other words, if we know the weight distribution of $\mathcal{C}$ we can determine the weight distribution of $\mathcal{C}^\perp$ without knowing specifically the codewords of $\mathcal{C}^\perp$ or anything else about its structure. These linear relations have been the most significant tool available for investigating and calculating weight distributions. They were first developed by MacWilliams in [213], and consequently are called the MacWilliams equations or the MacWilliams identities. Since then there have been variations, most notably the Pless power moments, which we will also examine.

Let $\mathcal{C}$ be an $[n, k, d]$ code over $\mathbb{F}_q$ with weight distribution $A_i = A_i(\mathcal{C})$ for $0 \le i \le n$, and let the weight distribution of $\mathcal{C}^\perp$ be $A_i^\perp = A_i(\mathcal{C}^\perp)$ for $0 \le i \le n$. The key to developing the MacWilliams equations is to examine the $q^k \times n$ matrix $\mathcal{M}$ whose rows are the codewords of $\mathcal{C}$ listed in some order. As an illustration of how $\mathcal{M}$ can be used, consider the following. The number of rows of $\mathcal{M}$ (i.e. the number of codewords of $\mathcal{C}$) equals q^k, but it also equals $\sum_{i=0}^{n} A_i$. Using the fact that $A_0^\perp = 1$ we obtain the linear equation

$$\sum_{j=0}^{n} A_j = q^k A_0^\perp. \tag{7.1}$$

We next count the total number of zeros in $\mathcal{M}$ in two different ways. By counting first by rows, we see that there are

$$\sum_{j=0}^{n-1}(n-j)A_j$$

zeros in $\mathcal{M}$. By Exercise 373 a column of $\mathcal{M}$ either consists entirely of zeros or contains every element of $\mathbb{F}_q$ an equal number of times, and additionally $\mathcal{M}$ has $A_1^{\perp}/(q-1)$ zero columns. Therefore, counting the number of zeros of $\mathcal{M}$ by columns, we see that the number of zeros in $\mathcal{M}$ also equals

$$q^k \frac{A_1^{\perp}}{q-1} + q^{k-1}\left(n - \frac{A_1^{\perp}}{q-1}\right) = q^{k-1}\left(nA_0^{\perp} + A_1^{\perp}\right),$$

again using $A_0^{\perp} = 1$. Equating these two counts, we obtain

$$\sum_{j=0}^{n-1}(n-j)A_j = q^{k-1}\left(nA_0^{\perp} + A_1^{\perp}\right). \tag{7.2}$$

Equations (7.1) and (7.2) are the first two equations in the list of $n+1$ *MacWilliams equations* relating the weight distributions of $\mathcal{C}$ and $\mathcal{C}^{\perp}$:

$$\sum_{j=0}^{n-\nu}\binom{n-j}{\nu}A_j = q^{k-\nu}\sum_{j=0}^{\nu}\binom{n-j}{n-\nu}A_j^{\perp} \quad \text{for } 0 \le \nu \le n. \tag{7.3}$$

Exercise 373 Let $\mathcal{M}$ be a $q^k \times n$ matrix whose rows are the codewords of an $[n,k]$ code $\mathcal{C}$ over $\mathbb{F}_q$. Let $A_1^{\perp}$ be the number of codewords in $\mathcal{C}^{\perp}$ of weight 1. Prove that:
(a) a column of $\mathcal{M}$ either consists entirely of zeros or contains every element of $\mathbb{F}_q$ an equal number of times, and
(b) $\mathcal{M}$ has $A_1^{\perp}/(q-1)$ zero columns.   ◆

We now show that all the MacWilliams equations follow from a closer examination of the matrix $\mathcal{M}$. This proof follows [40]. Before presenting the main result, we need two lemmas. Let the coordinates of the code $\mathcal{C}$ be denoted $\{1, 2, \ldots, n\}$. If $I \subseteq \{1, 2, \ldots, n\}$, then $\overline{I}$ will denote the complementary set $\{1, 2, \ldots, n\} \setminus I$. Recall that in Section 1.5 we introduced the punctured code $\mathcal{C}^I$ and the shortened code $\mathcal{C}_{\overline{I}}$, each of length $n - |\overline{I}|$, where $|\overline{I}|$ is the size of the set $\overline{I}$. If $\mathbf{x} \in \mathcal{C}$, $\mathbf{x}^I$ denotes the codeword in $\mathcal{C}^I$ obtained from $\mathbf{x}$ by puncturing on $\overline{I}$. Finally, let $A_j^{\perp}(I)$, for $0 \le j \le |I|$, be the weight distribution of $(\mathcal{C}^I)^{\perp}$.

Lemma 7.1.1 *If $0 \le j \le \nu \le n$, then*

$$\sum_{|I|=\nu} A_j^{\perp}(I) = \binom{n-j}{n-\nu}A_j^{\perp}.$$

Proof: Let

$$\mathcal{X} = \{(\mathbf{x}, I) \mid \mathbf{x} \in \mathcal{C}^{\perp},\ \mathrm{wt}(\mathbf{x}) = j,\ |I| = \nu,\ \mathrm{supp}(\mathbf{x}) \subseteq I\},$$

where supp(**x**) denotes the support of **x**. We count the number of elements in $\mathcal{X}$ in two different ways. If wt(**x**) $= j$ with $\mathbf{x} \in \mathcal{C}^\perp$, there are $\binom{n-j}{\nu-j} = \binom{n-j}{n-\nu}$ sets I of size $\nu \geq j$ that contain supp(**x**). Thus $|\mathcal{X}| = \binom{n-j}{n-\nu} A_j^\perp$. There are $A_j^\perp(I)$ vectors $\mathbf{y} \in (\mathcal{C}^T)^\perp$ with wt(**y**) $= j$. By Theorem 1.5.7, $\mathbf{y} \in (\mathcal{C}^\perp)_{\overline{I}}$, and so $\mathbf{y} = \mathbf{x}^T$ for a unique $\mathbf{x} \in \mathcal{C}^\perp$ with wt(**x**) $= j$ and supp(**x**) $\subseteq I$ for all I of size ν. Thus $|\mathcal{X}| = \sum_{|I|=\nu} A_j^\perp(I)$, and the result follows by equating the two counts.   □

In our second lemma, we consider again the $q^k \times n$ matrix $\mathcal{M}$ whose rows are the codewords of $\mathcal{C}$, and we let $\mathcal{M}(I)$ denote the $q^k \times |I|$ submatrix of $\mathcal{M}$ consisting of the columns of $\mathcal{M}$ indexed by I. Note that each row of $\mathcal{M}(I)$ is a codeword of $\mathcal{C}^T$; however, each codeword of $\mathcal{C}^T$ may be repeated several times.

Lemma 7.1.2 *Each codeword of $\mathcal{C}^T$ occurs exactly q^{k-k_I} times as a row of $\mathcal{M}(I)$, where k_I is the dimension of $\mathcal{C}^T$.*

Proof: The map $f : \mathcal{C} \rightarrow \mathcal{C}^T$ given by $f(\mathbf{x}) = \mathbf{x}^T$ (i.e. the puncturing map) is linear and surjective. So its kernel has dimension $k - k_I$ and the result follows.   □

Exercise 374 Prove that $f : \mathcal{C} \rightarrow \mathcal{C}^T$ given by $f(\mathbf{x}) = \mathbf{x}^T$ used in the proof of Lemma 7.1.2 is indeed linear and surjective.   ♦

The verification of the $n + 1$ equations of (7.3) is completed by counting the ν-tuples of zeros in the rows of $\mathcal{M}$ for each $0 \leq \nu \leq n$ in two different ways.

Theorem 7.1.3 *The equations of (7.3) are satisfied by the weight distributions of an $[n, k]$ code $\mathcal{C}$ over $\mathbb{F}_q$ and its dual.*

Proof: Let N_ν be the number of ν-tuples of zeros (not necessarily consecutive) in the rows of $\mathcal{M}$ for $0 \leq \nu \leq n$. Clearly, a row of $\mathcal{M}$ of weight j has $\binom{n-j}{\nu}$ ν-tuples of zeros. Thus

$$N_\nu = \sum_{j=0}^{n-\nu} \binom{n - j}{\nu} A_j,$$

which is the left-hand side of (7.3). By Lemma 7.1.1, the right-hand side of (7.3) is

$$q^{k-\nu} \sum_{j=0}^{\nu} \binom{n - j}{n - \nu} A_j^\perp = q^{k-\nu} \sum_{j=0}^{\nu} \sum_{|I|=\nu} A_j^\perp(I) = q^{k-\nu} \sum_{|I|=\nu} \sum_{j=0}^{\nu} A_j^\perp(I).$$

But $\sum_{j=0}^{\nu} A_j^\perp(I) = q^{\nu-k_I}$, as $(\mathcal{C}^T)^\perp$ is a $[\nu, \nu - k_I]$ code. Therefore

$$q^{k-\nu} \sum_{j=0}^{\nu} \binom{n - j}{n - \nu} A_j^\perp = q^{k-\nu} \sum_{|I|=\nu} q^{\nu-k_I} = \sum_{|I|=\nu} q^{k-k_I}.$$

By Lemma 7.1.2, q^{k-k_I} is the number of zero rows of $\mathcal{M}(I)$, which is the number of $|I|$-tuples of zeros in the coordinate positions I in $\mathcal{M}$. Thus $\sum_{|I|=\nu} q^{k-k_I} = N_\nu$. Equating the two counts for N_ν completes the proof.   □

Corollary 7.1.4 *The weight distribution of $\mathcal{C}$ uniquely determines the weight distribution of $\mathcal{C}^\perp$.*

Proof: The $(n+1) \times (n+1)$ coefficient matrix of the $A_j^\perp$s on the right-hand side of (7.3) is triangular with nonzero entries on the diagonal. Hence the $A_j^\perp$s are uniquely determined by the A_js. □

7.2　Equivalent formulations

There are several equivalent sets of equations that can be used in place of the MacWilliams equations (7.3). We state five more of these after introducing some additional notation. One of these equivalent formulations is most easily expressed by considering the generating polynomial of the weight distribution of the code. This polynomial has one of two forms: it is either a polynomial in a single variable of degree at most n, the length of the code, or it is a homogeneous polynomial in two variables of degree n. Both are called the *weight enumerator of the code* C and are denoted $W_C(x)$ or $W_C(x, y)$. The single variable weight enumerator of C is

$$W_C(x) = \sum_{i=0}^{n} A_i(C) x^i.$$

By replacing x by x/y and then multiplying by y^n, $W_C(x)$ can be converted to the two variable weight enumerator

$$W_C(x, y) = \sum_{i=0}^{n} A_i(C) x^i y^{n-i}.$$

Example 7.2.1 Consider the two binary codes C_1 and C_2 of Example 1.4.4 and Exercise 17 with generator matrices G_1 and G_2, respectively, where

$$G_1 = \begin{bmatrix} 1 & 1 & 0 & 0 & 0 & 0 \\ 0 & 0 & 1 & 1 & 0 & 0 \\ 0 & 0 & 0 & 0 & 1 & 1 \end{bmatrix} \quad \text{and} \quad G_2 = \begin{bmatrix} 1 & 1 & 0 & 0 & 0 & 0 \\ 0 & 1 & 1 & 0 & 0 & 0 \\ 1 & 1 & 1 & 1 & 1 & 1 \end{bmatrix}.$$

Both codes have weight distribution $A_0 = 1$, $A_2 = 3$, $A_4 = 3$, and $A_6 = 1$. Hence

$$W_{C_1}(x, y) = W_{C_2}(x, y) = y^6 + 3x^2 y^4 + 3x^4 y^2 + x^6 = (y^2 + x^2)^3.$$

Notice that $C_1 = C \oplus C \oplus C$, where C is the $[2, 1, 2]$ binary repetition code, and that $W_C(x, y) = y^2 + x^2$. ■

The form of the weight enumerator of C_1 from the previous example illustrates a general fact about the weight enumerator of the direct sum of two codes.

Theorem 7.2.2 *The weight enumerator of the direct sum* $C_1 \oplus C_2$ *is*

$$W_{C_1 \oplus C_2}(x, y) = W_{C_1}(x, y) W_{C_2}(x, y).$$

Exercise 375 Prove Theorem 7.2.2. ◆

The simplicity of the form of the weight enumerator of a direct sum shows the power of the notation.

Shortly we will list six equivalent forms of the MacWilliams equations, with (7.3) denoted (M_1). This form, as well as the next, denoted (M_2), involves only the weight distributions and binomial coefficients. Both (M_1) and (M_2) consist of a set of $n + 1$ equations. The third form, denoted (M_3), is a single identity involving the weight enumerators of a code and its dual. Although we originally called (M_1) the MacWilliams equations, in fact, any of these three forms is generally referred to as the MacWilliams equations. The fourth form is the actual solution of the MacWilliams equations for the weight distribution of the dual code in terms of the weight distribution of the original code; by Corollary 7.1.4 this solution is unique. In order to give the solution in compact form, we recall the definition of the Krawtchouk polynomial

$$K_k^{n,q}(x) = \sum_{j=0}^{k}(-1)^j(q-1)^{k-j}\binom{x}{j}\binom{n-x}{k-j} \quad \text{for } 0 \le k \le n$$

of degree k in x given in Chapter 2. The $n + 1$ equations that arise will be denoted (K). The final two sets of equations equivalent to (7.3), which are sometimes more convenient for calculations, involve the *Stirling numbers* $S(r, v)$ *of the second kind*. These are defined for nonnegative integers r, v by the equation

$$S(r, v) = \frac{1}{v!}\sum_{i=0}^{v}(-1)^{v-i}\binom{v}{i}i^r;$$

in addition, they satisfy the recursion

$$S(r, v) = vS(r - 1, v) + S(r - 1, v - 1) \quad \text{for } 1 \le v < r.$$

The number $v!S(r, v)$ equals the number of ways to distribute r distinct objects into v distinct boxes with no box left empty; in particular,

$$S(r, v) = 0 \quad \text{if } r < v \tag{7.4}$$

and

$$S(r, r) = 1. \tag{7.5}$$

The following is a basic identity for the Stirling numbers of the second kind (see [204]):

$$j^r = \sum_{v=0}^{r}v!\binom{j}{v}S(r, v). \tag{7.6}$$

The two forms of the equations arising here involve the weight distributions, binomial coefficients, and the Stirling numbers. They will be called the *Pless power moments* and denoted (P_1) and (P_2); they are due to Pless [258]. Each of (P_1) and (P_2) is an infinite set of equations.

Exercise 376   Compute a table of Stirling numbers $S(r, v)$ of the second kind for $1 \le v \le r \le 6$. ♦

We now state the six families of equations relating the weight distribution of an $[n, k]$ code C over $\mathbb{F}_q$ to the weight distribution of its dual $C^\perp$. The theorem that follows asserts their equivalence.

$$\sum_{j=0}^{n-v} \binom{n-j}{v} A_j = q^{k-v} \sum_{j=0}^{v} \binom{n-j}{n-v} A_j^{\perp} \quad \text{for } 0 \le v \le n. \tag{M_1}$$

$$\sum_{j=v}^{n} \binom{j}{v} A_j = q^{k-v} \sum_{j=0}^{v} (-1)^j \binom{n-j}{n-v} (q-1)^{v-j} A_j^{\perp} \quad \text{for } 0 \le v \le n. \tag{M_2}$$

$$W_{\mathcal{C}^{\perp}}(x, y) = \frac{1}{|\mathcal{C}|} W_{\mathcal{C}}(y - x, y + (q-1)x). \tag{M_3}$$

$$A_j^{\perp} = \frac{1}{|\mathcal{C}|} \sum_{i=0}^{n} A_i K_j^{n,q}(i) \quad \text{for } 0 \le j \le n. \tag{K}$$

$$\sum_{j=0}^{n} j^r A_j = \sum_{j=0}^{\min\{n,r\}} (-1)^j A_j^{\perp} \left[\sum_{v=j}^{r} v! S(r, v) q^{k-v} (q-1)^{v-j} \binom{n-j}{n-v} \right] \quad \text{for } 0 \le r. \tag{P_1}$$

$$\sum_{j=0}^{n} (n-j)^r A_j = \sum_{j=0}^{\min\{n,r\}} A_j^{\perp} \left[\sum_{v=j}^{r} v! S(r, v) q^{k-v} \binom{n-j}{n-v} \right] \quad \text{for } 0 \le r. \tag{P_2}$$

In showing the equivalence of these six families of equations, we leave many details as exercises.

Theorem 7.2.3 *The sets of equations* (M_1), (M_2), (M_3), (K), (P_1), *and* (P_2) *are equivalent.*

Proof: Exercise 377 shows that by expanding the right-hand side of (M_3) and equating coefficients, equations (K) arise; reversing the steps will give (M_3) from (K). Replacing y by $x + z$ in (M_3), expanding, and equating coefficients gives (M_1) as Exercise 378 shows; again the steps can be reversed so that (M_1) will produce (M_3). Exercise 379 shows that by replacing $\mathcal{C}$ by $\mathcal{C}^{\perp}$ (and $\mathcal{C}^{\perp}$ by $\mathcal{C}$) and x by $y + z$ in (M_3), expanding, and equating coefficients gives (M_2); again the steps are reversible so that (M_2) will produce (M_3) with $\mathcal{C}$ and $\mathcal{C}^{\perp}$ interchanged. Thus the first four families are equivalent.

We next prove that (M_2) and (P_1) are equivalent. Notice that in (M_2), the summation on the left-hand side can begin with $j = 0$ as $\binom{j}{v} = 0$ if $j < v$. Using (M_2) and (7.6), we calculate that

$$\sum_{j=0}^{n} j^r A_j = \sum_{j=0}^{n} \sum_{v=0}^{r} v! \binom{j}{v} S(r, v) A_j$$

$$= \sum_{v=0}^{r} v! S(r, v) \left[\sum_{j=0}^{n} \binom{j}{v} A_j \right]$$

$$= \sum_{v=0}^{r} v! S(r, v) \left[q^{k-v} \sum_{j=0}^{n} (-1)^j (q-1)^{v-j} \binom{n-j}{n-v} A_j^{\perp} \right]$$

$$= \sum_{j=0}^{n} (-1)^j A_j^{\perp} \left[\sum_{v=0}^{r} v! S(r, v) q^{k-v} (q-1)^{v-j} \binom{n-j}{n-v} \right].$$

The last expression is the right-hand side of (P_1) if we note that $\binom{n-j}{n-v} = 0$ if $n - j < n - v$, or equivalently $v < j$. This allows us to start the inner sum at $v = j$ rather than $v = 0$. As

$v \leq r$, we have $v < j$ if $r < j$, which allows us to stop the outer sum at $\min\{n, r\}$. Thus (M_2) holding implies that (P_1) holds. The converse follows from the preceding equations if we know that the $n + 1$ equations

$$\sum_{v=0}^{r} v! S(r, v) x_v = 0 \quad \text{for } 0 \leq r \leq n$$

have only the solution $x_0 = x_1 = \cdots = x_n = 0$. This is clear as the $(n + 1) \times (n + 1)$ matrix $[a_{r,v}] = [v! S(r, v)]$ is lower triangular by (7.4) with nonzero diagonal entries by (7.5). So (M_2) and (P_1) are equivalent.

A similar argument, which you are asked to give in Exercise 380 using (7.6) with j replaced by $n - j$, gives the equivalence of (M_1) and (P_2). Thus all six sets of equations are equivalent. □

Exercise 377 Show that by expanding the right-hand side of (M_3) and equating coefficients, equations (K) arise. Do this in such a way that it is easy to see that reversing the steps will give (M_3) from (K). ◆

Exercise 378 Show that by replacing y by $x + z$ in (M_3), expanding, and equating coefficients, you obtain (M_1). Do this in such a way that the steps can be reversed so that (M_1) will produce (M_3). ◆

Exercise 379 Show that by reversing the roles of C and $C^\perp$ in (M_3) and then replacing x by $y + z$, expanding, and equating coefficients produces (M_2). Do this in such a way that the steps can be reversed so that (M_2) will yield (M_3) with C and $C^\perp$ interchanged. ◆

Exercise 380 Using (7.6) with j replaced by $n - j$, show that beginning with (M_1), you can obtain (P_2). Prove that beginning with (P_2), you can arrive at (M_1) in an analogous way to what was done in the proof of Theorem 7.2.3. ◆

In the proof of Theorem 7.2.3, we interchanged the roles of C and $C^\perp$ in (M_3), thus reversing the roles of the A_is and $A_i^\perp$s. You can obviously reverse the A_is and $A_i^\perp$s in any of the other equivalent forms as long as the dimension k of C is replaced by $n - k$, the dimension of $C^\perp$. You are asked in Exercise 381 to do precisely that.

Exercise 381 Write the five families of equations corresponding to (M_1), (M_2), (K), (P_1), and (P_2) with the roles of C and $C^\perp$ reversed. ◆

Exercise 382 Let C be the $[5, 2]$ binary code generated by

$$G = \begin{bmatrix} 1 & 1 & 0 & 0 & 0 \\ 0 & 1 & 1 & 1 & 1 \end{bmatrix}.$$

(a) Find the weight distribution of C.
(b) Use one of (M_1), (M_2), (M_3), (K), (P_1), or (P_2) to find the weight distribution of $C^\perp$.
(c) Verify your result in (b) by listing the vectors in $C^\perp$. ◆

Exercise 383 Let C be an $[n, k]$ code over $\mathbb{F}_4$.

(a) Show that the number of vectors of weight n in $C^\perp$ is $\sum_{i=0}^{n} A_i 3^{n-i}(-1)^i$. Hint: Use (M_3).

(b) Show that if C has only even weight codewords, then $C^\perp$ contains a vector of weight n. ◆

7.3   A uniqueness result

The Pless power moments are particularly useful in showing uniqueness of certain solutions to any of the families of identities from the last section. This is illustrated in the next result, important for showing the existence of block designs in Chapter 8.

Theorem 7.3.1 *Let $S \subseteq \{1, 2, \ldots, n\}$ with $|S| = s$. Then the weight distributions of C and $C^\perp$ are uniquely determined by $A_1^\perp, A_2^\perp, \ldots, A_{s-1}^\perp$ and the A_i with $i \notin S$. These values can be found from the first s equations in (P_1).*

Proof: Assume that $A_1^\perp, A_2^\perp, \ldots, A_{s-1}^\perp$ and A_i for $i \notin S$ are known. The right-hand side of the first s equations for $0 \le r < s$ of (P_1) depend only on $A_0^\perp = 1, A_1^\perp, \ldots, A_{s-1}^\perp$. If we move the terms $j^r A_j$ for $j \notin S$ from the left-hand side to the right-hand side, we are left with s linear equations in unknowns A_j with $j \in S$. The coefficient matrix of these s equations is an $s \times s$ Vandermonde matrix

$$\begin{bmatrix} 1 & 1 & 1 & \cdots & 1 \\ j_1 & j_2 & j_3 & \cdots & j_s \\ j_1^2 & j_2^2 & j_3^2 & \cdots & j_s^2 \\ & & \vdots & & \\ j_1^{s-1} & j_2^{s-1} & j_3^{s-1} & \cdots & j_s^{s-1} \end{bmatrix},$$

where $S = \{j_1, \ldots, j_s\}$. This matrix is nonsingular by Lemma 4.5.1 and hence A_i with $i \in S$ are determined. Thus all the A_is are known, implying that the remaining $A_i^\perp$s are now uniquely determined by (K). □

Theorem 7.3.1 is often applied when s is the minimum weight of $C^\perp$ and $S = \{1, 2, \ldots, s\}$ so that $A_1^\perp = A_2^\perp = \cdots = A_{s-1}^\perp = 0$.

For ease of use, we compute the first five power moments from (P_1):

$$\sum_{j=0}^{n} A_j = q^k,$$

$$\sum_{j=0}^{n} j A_j = q^{k-1} \left(qn - n - A_1^\perp \right),$$

$$\sum_{j=0}^{n} j^2 A_j = q^{k-2} \big[(q-1)n(qn - n + 1) - (2qn - q - 2n + 2)A_1^\perp + 2A_2^\perp \big],$$

$$\sum_{j=0}^{n} j^3 A_j = q^{k-3}\big[(q-1)n(q^2n^2 - 2qn^2 + 3qn - q + n^2 - 3n + 2)$$

$$- (3q^2n^2 - 3q^2n - 6qn^2 + 12qn + q^2 - 6q + 3n^2 - 9n + 6)A_1^\perp$$
$$+ 6(qn - q - n + 2)A_2^\perp - 6A_3^\perp\big], \tag{7.7}$$

$$\sum_{j=0}^{n} j^4 A_j = q^{k-4}\big[(q-1)n(q^3n^3 - 3q^2n^3 + 6q^2n^2 - 4q^2n + q^2 + 3qn^3 - 12qn^2$$

$$+ 15qn - 6q - n^3 + 6n^2 - 11n + 6)$$
$$- (4q^3n^3 - 6q^3n^2 + 4q^3n - q^3 - 12q^2n^3 + 36q^2n^2 - 38q^2n + 14q^2$$
$$+ 12qn^3 - 54qn^2 + 78qn - 36q - 4n^3 + 24n^2 - 44n + 24)A_1^\perp$$
$$+ (12q^2n^2 - 24q^2n + 14q^2 - 24qn^2 + 84qn - 72q + 12n^2 - 60n + 72)A_2^\perp$$
$$- (24qn - 36q - 24n + 72)A_3^\perp + 24A_4^\perp\big].$$

In the binary case these become:

$$\sum_{j=0}^{n} A_j = 2^k,$$

$$\sum_{j=0}^{n} j A_j = 2^{k-1}(n - A_1^\perp),$$

$$\sum_{j=0}^{n} j^2 A_j = 2^{k-2}\big[n(n+1) - 2nA_1^\perp + 2A_2^\perp\big], \tag{7.8}$$

$$\sum_{j=0}^{n} j^3 A_j = 2^{k-3}\big[n^2(n+3) - (3n^2 + 3n - 2)A_1^\perp + 6nA_2^\perp - 6A_3^\perp\big],$$

$$\sum_{j=0}^{n} j^4 A_j = 2^{k-4}\big[n(n+1)(n^2 + 5n - 2) - 4n(n^2 + 3n - 2)A_1^\perp$$

$$+ 4(3n^2 + 3n - 4)A_2^\perp - 24nA_3^\perp + 24A_4^\perp\big].$$

Example 7.3.2 Let C be any self-dual $[12, 6, 6]$ ternary code. As we will see in Section 10.4.1, the only such code turns out to be the extended ternary Golay code $\mathcal{G}_{12}$ from Section 1.9.2. Since the weight of each codeword is a multiple of 3 by Theorem 1.4.5, $A_i = 0$ for $i \neq 0, 6, 9, 12$. By self-duality, $A_i = A_i^\perp$ for all i. As $A_0 = 1$, only A_6, A_9, and A_{12} are unknown. Using Theorem 7.3.1 with $s = 3$ and $S = \{6, 9, 12\}$, as $A_1^\perp = A_2^\perp = 0$, we can find these from the first three equations of (7.7):

$$A_6 + A_9 + A_{12} = 728,$$
$$6A_6 + 9A_9 + 12A_{12} = 5832,$$
$$36A_6 + 81A_9 + 144A_{12} = 48\,600.$$

The unique solution is $A_6 = 264$, $A_9 = 440$, and $A_{12} = 24$. Thus the weight enumerator of C, the extended ternary Golay code, is

$$W_C(x, y) = y^{12} + 264x^6y^6 + 440x^9y^3 + 24x^{12}.$$

■

Exercise 384 Let C be any self-dual doubly-even binary $[24, 12, 8]$ code. (We will see in Section 10.1.1 that the only such code is the extended binary Golay code $\mathcal{G}_{24}$ from Section 1.9.1.)

(a) Show that $A_0 = A_{24} = 1$ and the only unknown A_is are A_8, A_{12}, and A_{16}.

(b) Using (7.8), show that

$$W_C(x, y) = y^{24} + 759x^8y^{16} + 2576x^{12}y^{12} + 759x^{16}y^8 + x^{24}.$$

(c) By Theorem 1.4.5(iii), $A_8 = A_{16}$. Show that if we only use the first two equations of (P_1) together with $A_8 = A_{16}$, we do not obtain a unique solution. Thus adding the condition $A_8 = A_{16}$ does not reduce the number of power moment equations required to find the weight distribution. See also Exercise 385. ◆

Exercise 385 Let C be a binary code of length n where $A_i = A_{n-i}$ for all $0 \le i \le n/2$.

(a) Show that the first two equations of (P_1) are equivalent under the conditions $A_i = A_{n-i}$.

(b) What can be said about the third and fourth equations of (P_1) under the conditions $A_i = A_{n-i}$? ◆

Example 7.3.3 Let C be a $[16, 8, 4]$ self-dual doubly-even binary code. By Theorem 1.4.5(iv), $A_0 = A_{16} = 1$. Thus in the weight distribution, only A_4, A_8, and A_{12} are unknown as $A_i = 0$ when $4 \nmid i$. In Exercise 386, you will be asked to find these values by solving the first three equations in (7.8). We present an alternate solution. Letting $s = 3$ and $S = \{4, 8, 12\}$, there is a unique solution by Theorem 7.3.1 since $A_1^\perp = A_2^\perp = 0$. Thus if we can find one $[16, 8, 4]$ self-dual doubly-even binary code, its weight enumerator will give this unique solution. Let $C_1 = \widehat{\mathcal{H}}_3$, which is the unique $[8, 4, 4]$ code (see Exercise 56) and has weight enumerator $W_{C_1}(x, y) = y^8 + 14x^4y^4 + x^8$ (see Exercise 19 and Example 1.11.7). By Section 1.5.4, $C_1 \oplus C_1$ is a $[16, 8, 4]$ code. By Theorem 7.2.2,

$$\begin{aligned}
W_{C_1 \oplus C_1}(x, y) &= (y^8 + 14x^4y^4 + x^8)^2 \\
&= y^{16} + 28x^4y^{12} + 198x^8y^8 + 28x^{12}y^4 + x^{16}.
\end{aligned}$$

As all weights in $C_1 \oplus C_1$ are divisible by 4, this code is self-dual doubly-even by Theorem 1.4.8; thus we have produced the weight enumerator of C. We will explore this example more extensively later in this chapter. ∎

Exercise 386 Find the weight enumerator of a $[16, 8, 4]$ self-dual doubly-even binary code by solving the first three equations in (7.8). ◆

Example 7.3.4 Let C be the $[n, k]$ simplex code over $\mathbb{F}_q$ where $n = (q^k - 1)/(q - 1)$. These codes were developed in Section 1.8; the dual code $C^\perp$ is the $[n, n - k, 3]$ Hamming code $\mathcal{H}_{q,k}$. By Theorem 2.7.5, the weight distribution of C is $A_0 = 1$, $A_{q^{k-1}} = q^k - 1$, and $A_i = 0$ for all other i. Therefore the weight distribution $A_0^\perp, A_1^\perp, \ldots, A_n^\perp$ of $\mathcal{H}_{q,k}$ can be determined. For example, using (K) we obtain

$$\begin{aligned}
A_j^\perp &= q^{-k}\left[K_j^{n,q}(0) + (q^k - 1)K_j^{n,q}(q^{k-1})\right] \\
&= q^{-k}\left[(q - 1)^j \binom{n}{j} + (q^k - 1)K_j^{n,q}(q^{k-1})\right]
\end{aligned}$$

for $0 \le j \le n$. ∎

Exercise 387 Find the weight enumerator of the $[15, 11, 3]$ binary Hamming code. ◆

To conclude this section, we use (7.7) to produce an interesting result about ternary codes, due to Kennedy [164].

Theorem 7.3.5 *Let $\mathcal{C}$ be an $[n, k]$ ternary code with $k \geq 3$ and weight distribution $A_0, A_1, \ldots, A_n$. For $0 \leq i \leq 2$, let*

$$N_i = \sum_{j \equiv i \,(\mathrm{mod}\ 3)} A_j.$$

Then $N_0 \equiv N_1 \equiv N_2 \equiv 0 \,(\mathrm{mod}\ 3)$.

Proof: Because $k \geq 3$, the first three equations of (7.7) with $q = 3$ yield

$$N_0 + N_1 + N_2 \equiv \sum_{i=0}^{n} A_i \equiv 0 \,(\mathrm{mod}\ 3),$$

$$N_1 - N_2 \equiv \sum_{i=0}^{n} i A_i \equiv 0 \,(\mathrm{mod}\ 3),$$

$$N_1 + N_2 \equiv \sum_{i=0}^{n} i^2 A_i \equiv 0 \,(\mathrm{mod}\ 3),$$

since $i \equiv 0, 1,$ or -1 and $i^2 \equiv 0$ or 1 modulo 3. The only solution modulo 3 is $N_0 \equiv N_1 \equiv N_2 \equiv 0 \,(\mathrm{mod}\ 3)$. ☐

Exercise 388 Show that the following weight distributions for ternary codes cannot occur:
(a) $A_0 = 1, A_5 = 112, A_6 = 152, A_8 = 330, A_9 = 110, A_{11} = 24$.
(b) $A_0 = 1, A_3 = 22, A_4 = 42, A_5 = 67, A_6 = 55, A_7 = 32, A_8 = 24$. ◆

7.4   MDS codes

In this section, we show that the weight distribution of an $[n, k]$ MDS code over $\mathbb{F}_q$ is determined by the parameters $n, k,$ and q. One proof of this uses the MacWilliams equations; see [275]. We present a proof here that uses the inclusion–exclusion principle.

Theorem 7.4.1 *Let $\mathcal{C}$ be an $[n, k, d]$ MDS code over $\mathbb{F}_q$. The weight distribution of $\mathcal{C}$ is given by $A_0 = 1, A_i = 0$ for $1 \leq i < d$, and*

$$A_i = \binom{n}{i} \sum_{j=0}^{i-d} (-1)^j \binom{i}{j} (q^{i+1-d-j} - 1)$$

for $d \leq i \leq n$, where $d = n - k + 1$.

Proof: Clearly, $A_0 = 1$ and $A_i = 0$ for $1 \leq i < d$. Let $\mathcal{C}_T$ be the code of length $n - |T|$ obtained from $\mathcal{C}$ by shortening on T. By repeated application of Exercise 317, $\mathcal{C}_T$ is an $[n - |T|, k - |T|]$ MDS code if $|T| < k$ and $\mathcal{C}_T$ is the zero code if $|T| \geq k$. Let $\mathcal{C}(T)$ be the subcode of C which is zero on T. Then C_T is $C(T)$ punctured on T. In particular

$$|C(T)| = \begin{cases} q^{k-t} & \text{if } t = |T| < k, \\ 1 & \text{if } t = |T| \geq k, \end{cases} \tag{7.9}$$

and $C(T)$ is the set of codewords of C whose support is disjoint from T. For $0 \leq t \leq n$, define

$$N_t = \sum_{|T|=t} |C(T)|.$$

By (7.9),

$$N_t = \begin{cases} \binom{n}{t} q^{k-t} & \text{if } t = |T| < k, \\ \binom{n}{t} & \text{if } t = |T| \geq k. \end{cases} \tag{7.10}$$

Note that N_t counts the number of codewords of C that have weight $n - t$ or less, with a codeword multiply counted, once for every $C(T)$ it is in. By the inclusion–exclusion principle [204], for $d \leq i \leq n$,

$$A_i = \sum_{j=0}^{i} (-1)^j \binom{n-i+j}{j} N_{n-i+j}.$$

By (7.10),

$$A_i = \sum_{j=0}^{k+i-n-1} (-1)^j \binom{n-i+j}{j} \binom{n}{n-i+j} q^{k-(n-i+j)}$$

$$+ \sum_{j=k+i-n}^{i} (-1)^j \binom{n-i+j}{j} \binom{n}{n-i+j}$$

$$= \binom{n}{i} \left[\sum_{j=0}^{i-d} (-1)^j \binom{i}{j} q^{i+1-d-j} + \sum_{j=i-d+1}^{i} (-1)^j \binom{i}{j} \right],$$

as $d = n - k + 1$ and $\binom{n-i+j}{j}\binom{n}{n-i+j} = \binom{n}{i}\binom{i}{j}$. As

$$-\sum_{j=0}^{i-d} (-1)^j \binom{i}{j} = \sum_{j=i-d+1}^{i} (-1)^j \binom{i}{j}, \tag{7.11}$$

the result follows. □

Exercise 389 In the notation of the proof of Theorem 7.4.1, show that

$$N_t = A_{n-t} + \binom{t+1}{1} A_{n-t-1} + \binom{t+2}{2} A_{n-t-2} + \cdots + \binom{n}{n-t} A_0.$$

Exercise 390 Verify (7.11). Hint: $0 = (1-1)^i$.

Exercise 391 Find the weight enumerator of the $[6, 3, 4]$ hexacode over $\mathbb{F}_4$.

Exercise 392   (Hard) Do the following:

(a) Prove that

$$\sum_{j=0}^{m}(-1)^j\binom{i}{j} = (-1)^m\binom{i-1}{m}.$$

(b) The weight distribution of an $[n, k]$ MDS code over $\mathbb{F}_q$ can be given by

$$A_i = \binom{n}{i}(q-1)\sum_{j=0}^{i-d}(-1)^j\binom{i-1}{j}q^{i-d-j}$$

for $d \le i \le n$, where $d = n - k + 1$. Verify the equivalence of this form with that given in Theorem 7.4.1. Hint: Factor $q - 1$ out of the expression for A_i from the theorem to create a double summation. Interchange the order of summation and use (a).   ◆

Corollary 7.4.2 *If C is an $[n, k, d]$ MDS code over $\mathbb{F}_q$, then $A_d = (q-1)\binom{n}{d}$.*

Exercise 393 Prove Corollary 7.4.2 and give a combinatorial interpretation of the result.   ◆

Corollary 7.4.3 *Assume that there exists an $[n, k, d]$ MDS code C over $\mathbb{F}_q$.*
(i) *If $2 \le k$, then $d = n - k + 1 \le q$.*
(ii) *If $k \le n - 2$, then $k + 1 \le q$.*

Proof: By Theorem 7.4.1, if $d < n$, then $A_{d+1} = \binom{n}{d+1}((q^2-1) - (d+1)(q-1)) = \binom{n}{d+1}(q-1)(q-d)$. As $0 \le A_{d+1}$, (i) holds. Because $C^\perp$ is an $[n, n-k, k+1]$ MDS code by Theorem 2.4.3, applying (i) to $C^\perp$ shows that if $2 \le n - k$, then $k + 1 \le q$, which is (ii).   □

The preceding corollary provides bounds on the length and dimension of $[n, k]$ MDS codes over $\mathbb{F}_q$. If $k = 1$, there are arbitrarily long MDS codes, and these codes are all monomially equivalent to repetition codes; if $k = n - 1$, there are again arbitrarily long MDS codes that are duals of the MDS codes of dimension 1. These $[n, 1]$ and $[n, n-1]$ MDS codes are among the codes we called trivial MDS codes in Section 2.4. Certainly, the zero code and the whole space $\mathbb{F}_q^n$ are MDS and can also be arbitrarily long. By Corollary 7.4.3(ii), if $k \le n - 2$, then $k \le q - 1$. So nontrivial $[n, k]$ MDS codes can exist only when $2 \le k \le \min\{n - 2, q - 1\}$. By Corollary 7.4.3(i), $n \le q + k - 1$ implying that nontrivial MDS codes over $\mathbb{F}_q$ cannot be arbitrarily long; in particular as $k \le \min\{n - 2, q - 1\} \le q - 1$, $n \le 2q - 2$. In summary we have the following.

Corollary 7.4.4 *Assume that there exists an $[n, k, d]$ MDS code over $\mathbb{F}_q$.*
(i) *If the code is trivial, it can be arbitrarily long.*
(ii) *If the code is nontrivial, then $2 \le k \le \min\{n - 2, q - 1\}$ and $n \le q + k - 1 \le 2q - 2$.*

Recall that the generalized Reed–Solomon codes and extended generalized Reed–Solomon codes over $\mathbb{F}_q$ of lengths $n \le q + 1$ are MDS by Theorem 5.3.4. When q is even, there are $[q + 2, 3]$ and $[q + 2, q - 1]$ MDS codes related to GRS codes. These are the longest known MDS codes, which leads to the following:

MDS Conjecture *If there is a nontrivial $[n, k]$ MDS code over $\mathbb{F}_q$, then $n \leq q + 1$, except when q is even and $k = 3$ or $k = q - 1$ in which case $n \leq q + 2$.*

The MDS conjecture is true for $k = 2$ by Corollary 7.4.4(ii). The conjecture has been proved in certain other cases, an instance of which is the following.

Theorem 7.4.5 *Let C be an $[n, k]$ MDS code over $\mathbb{F}_q$.*
(i) *[299] If q is odd and $2 \leq k < (\sqrt{q} + 13)/4$, then $n \leq q + 1$.*
(ii) *[321] If q is even and $5 \leq k < (2\sqrt{q} + 15)/4$, then $n \leq q + 1$.*

7.5   Coset weight distributions

In this section we consider the weight distribution of a coset of a code. We will find that some cosets will have uniquely determined distributions. Before we proceed with the main result, we state a lemma that is used several times.

Lemma 7.5.1 *Let C be an $[n, k]$ code over $\mathbb{F}_q$. Let $\mathbf{v}$ be a vector in $\mathbb{F}_q^n$ but not in C, and let $\mathcal{D}$ be the $[n, k + 1]$ code generated by C and $\mathbf{v}$. The following hold:*
(i) *The weight distributions of $\mathbf{v} + C$ and $\alpha\mathbf{v} + C$, when $\alpha \neq 0$, are identical.*
(ii) *$A_i(\mathbf{v} + C) = A_i(\mathcal{D} \setminus C)/(q - 1)$ for $0 \leq i \leq n$.*

Proof: Part (i) follows from the two observations that $\mathrm{wt}(\alpha(\mathbf{v} + \mathbf{c})) = \mathrm{wt}(\mathbf{v} + \mathbf{c})$ and $\alpha\mathbf{v} + C = \alpha(\mathbf{v} + C)$. Part (ii) follows from part (i) and the fact that $\mathcal{D}$ is the disjoint union of the cosets $\alpha\mathbf{v} + C$ for $\alpha \in \mathbb{F}_q$. $\square$

The following theorem is an application of Theorem 7.3.1. Part (i) is the linear case of a more general result of Delsarte [62] on the covering radius $\rho(C)$ of a code C. The remainder of the theorem is due to Assmus and Pless [9]. Recall that the weight of a coset of a code is the smallest weight of any vector in the coset; a coset leader is a vector in a coset of this smallest weight.

Theorem 7.5.2 *Let C be an $[n, k]$ code over $\mathbb{F}_q$. Let $S = \{i > 0 \mid A_i(C^\perp) \neq 0\}$ and $s = |S|$. Then:*
(i)   *$\rho(C) \leq s$,*
(ii) *each coset of C of weight s has the same weight distribution,*
(iii) *the weight distribution of any coset of weight less than s is determined once the number of vectors of weights $1, 2, \ldots, s - 1$ in the coset are known, and*
(iv) *if C is an even binary code, each coset of weight $s - 1$ has the same weight distribution.*

Proof: Let $\mathbf{v}$ be a coset leader of a coset of C with $\mathrm{wt}(\mathbf{v}) = w$. Let $\mathcal{D}$ be the $(k + 1)$-dimensional code generated by C and $\mathbf{v}$. As $\mathcal{D}^\perp \subset C^\perp$, $A_i(\mathcal{D}^\perp) = 0$ if $i \neq 0$ and $i \notin S$. As $\mathbf{v}$ is a coset leader, $\mathrm{wt}(\mathbf{x}) \geq w$ for all $\mathbf{x} \in \mathcal{D} \setminus C$ by Lemma 7.5.1(ii). In particular, $A_i(\mathcal{D}) = A_i(C)$ for $0 \leq i < w$, and w is the smallest weight among all the vectors in $\mathcal{D} \setminus C$. Also, if the weight distribution of C is known, then knowing the weight distribution of $\mathcal{D}$ is equivalent to knowing the weight distribution of $\mathbf{v} + C$ by Lemma 7.5.1(ii).

Assume first that $w \geq s$. Thus $A_1(\mathcal{D}), A_2(\mathcal{D}), \ldots, A_{s-1}(\mathcal{D})$ are known as they are $A_1(\mathcal{C}), A_2(\mathcal{C}), \ldots, A_{s-1}(\mathcal{C})$. Also $A_i(\mathcal{D}^\perp)$ is known for $i \notin S$. Note that none of these numbers depend on $\mathbf{v}$, only that $w \geq s$. By Theorem 7.3.1 the weight distribution of $\mathcal{D}$ and $\mathcal{D}^\perp$ is uniquely determined; thus this distribution is independent of the choice of $\mathbf{v}$. But then the weight distribution of $\mathcal{D} \setminus \mathcal{C}$ is uniquely determined, and so the coset leader, being the smallest weight of vectors in $\mathcal{D} \setminus \mathcal{C}$, has a uniquely determined weight. Since we know that there is some coset leader $\mathbf{v}$ with $\mathrm{wt}(\mathbf{v}) = \rho(\mathcal{C})$ and that there are cosets of all smaller weights by Theorem 1.12.6(v), this uniquely determined weight must be $\rho(\mathcal{C})$. Hence it is not possible for s to be smaller than $\rho(\mathcal{C})$, proving (i). When $\mathrm{wt}(\mathbf{v}) = s$, the weight distribution of $\mathcal{D} \setminus \mathcal{C}$ is uniquely determined and so, therefore, is the weight distribution of $\mathbf{v} + \mathcal{C}$ by Lemma 7.5.1(ii), proving (ii). If $w < s$, Theorem 7.3.1 shows the weight distributions of $\mathcal{D}$ and $\mathcal{D}^\perp$ are known as long as $A_1(\mathcal{D}), A_2(\mathcal{D}), \ldots, A_{s-1}(\mathcal{D})$ are known; but this is equivalent to knowing the number of vectors in $\mathbf{v} + \mathcal{C}$ of weights $1, 2, \ldots, s - 1$, yielding (iii).

Assume now that $\mathcal{C}$ is a binary code with only vectors of even weight. Let $\mathbf{v}$ be a coset leader of a coset of $\mathcal{C}$ with $\mathrm{wt}(\mathbf{v}) = s - 1$. Then $\mathcal{D}$ has only even weight vectors if $s - 1$ is even, in which case $A_n(\mathcal{D}^\perp) = 1$; if $s - 1$ is odd, $\mathcal{D}$ has odd weight vectors and so $A_n(\mathcal{D}^\perp) = 0$. Therefore $A_i(\mathcal{D}^\perp)$ is known except for values in $S \setminus \{n\}$, a set of size $s - 1$ as $A_n(\mathcal{C}^\perp) = 1$. Because $A_i(\mathcal{D}) = A_i(\mathcal{C})$ for $0 \leq i \leq s - 2$, by Theorem 7.3.1, the weight distribution of $\mathcal{D}$, and hence of $\mathcal{D} \setminus \mathcal{C} = \mathbf{v} + \mathcal{C}$, is uniquely determined. □

Example 7.5.3 The complete coset weight distribution of the $[8, 4, 4]$ extended binary Hamming code $\mathcal{C} = \widehat{\mathcal{H}}_3$ was given in Example 1.11.7. We redo this example in light of the previous theorem. This code is self-dual with $A_0(\mathcal{C}) = 1$, $A_4(\mathcal{C}) = 14$, and $A_8(\mathcal{C}) = 1$. Thus in Theorem 7.5.2, $S = \{4, 8\}$ and so $s = 2$. By parts (ii) and (iv) of this theorem, the distributions of the cosets of weights 1 and 2 are each uniquely determined, and by part (i) there are no other coset weights. Recall from Exercise 72 that odd weight cosets have only odd weight vectors and even weight cosets have only even weight vectors as $\mathcal{C}$ has only even weight vectors. In particular, all odd weight vectors in $\mathbb{F}_2^8$ are in the cosets of weight 1. If there are two weight 1 vectors in a coset of weight 1, their difference is a weight two vector in the code, which is a contradiction. Hence there are precisely eight cosets of weight 1. If $\mathbf{v} + \mathcal{C}$ has weight 1, then $A_i(\mathbf{v} + \mathcal{C}) = \binom{8}{i}/8$ for i odd, which agrees with the results in Example 1.11.7. There are $2^4 - 9 = 7$ remaining cosets; all must have weight 2. Thus if $\mathbf{v} + \mathcal{C}$ has weight 2, then $A_i(\mathbf{v} + \mathcal{C}) = (\binom{8}{i} - A_i(\mathcal{C}))/7$ for i even, again agreeing with the results in Example 1.11.7. ∎

Exercise 394 Find the weight distribution of the cosets of the $[7, 4, 3]$ binary Hamming code $\mathcal{H}_3$. ◆

In the next example, we use the technique in the proof of Theorem 7.5.2 to find the weight distribution of a coset of weight 4 in a $[16, 8, 4]$ self-dual doubly-even binary code. In the exercises, the weight distributions of the cosets of weights 1 and 3 will be determined in the same manner. Determining the weight distributions of the cosets of weight 2 is more complicated and requires some specific knowledge of the code; this will also be considered in the exercises and in Example 7.5.6.

Example 7.5.4 Let C be a $[16, 8, 4]$ self-dual doubly-even binary code. From Example 7.3.3, its weight distribution is given by $A_0(C) = A_{16}(C) = 1$, $A_4(C) = A_{12}(C) = 28$, and $A_8(C) = 198$. In the notation of Theorem 7.5.2, $S = \{4, 8, 12, 16\}$ and $s = 4$. Hence the covering radius of C is at most 4. Suppose that $\mathbf{v}$ is a coset leader of a coset of C of weight 4. (Note that we do not know if such a coset leader exists.) Let D be the $[16, 9, 4]$ linear code generated by C and $\mathbf{v}$; thus $D = C \cup (\mathbf{v} + C)$. D is an even code, and $D^\perp \subset C$. Let $A_i = A_i(D^\perp)$ and $A_i^\perp = A_i(D)$. Since D is even, $D^\perp$ contains the all-one vector. Thus we know $A_0 = A_{16} = 1$ and $A_i = 0$ for all other i except possibly $i = 4, 8$, or 12 as $D^\perp \subset C$. Thus using the first three equations from (7.8), we obtain

$$1 + A_4 + A_8 + A_{12} + 1 = 2^7,$$
$$4A_4 + 8A_8 + 12A_{12} + 16 = 2^6 \cdot 16,$$
$$16A_4 + 64A_8 + 144A_{12} + 256 = 2^5 \cdot 16 \cdot 17.$$

Solving produces $A_4 = A_{12} = 12$ and $A_8 = 102$. As D contains the all-one vector, $A_i^\perp = A_{16-i}^\perp$. As D is even with minimum weight 4, $A_i^\perp = 0$ for i odd, and hence we only need $A_4^\perp$, $A_6^\perp$, and $A_8^\perp$ to determine the weight distribution of D completely. Using (M_3), (K), or (P_1) we discover that $A_4^\perp = 44$, $A_6^\perp = 64$, and $A_8^\perp = 294$. The weight distribution of the coset $\mathbf{v} + C$ is $A_i(\mathbf{v} + C) = A_i(D) - A_i(C)$. So the weight distribution of a coset of weight 4 is uniquely determined to be:

Weight	4	6	8	10	12
Number of vectors	16	64	96	64	16

Notice that we still do not know if such a coset exists.   ■

Exercise 395 Let C be a $[16, 8, 4]$ self-dual doubly-even binary code. Analogous to Example 7.5.4, let $\mathbf{v}$ be a coset leader of a coset of C of weight 3 and let D be the $[16, 9, 3]$ code $D = C \cup (\mathbf{v} + C)$. Let $A_i = A_i(D^\perp)$ and $A_i^\perp = A_i(D)$. (As with the coset of weight 4, we do not know if such a coset exists.)
(a) Show that $A_{16} = 0$ and that the only unknown A_is are A_4, A_8, and A_{12}.
(b) Show that $A_i^\perp = A_{16-i}^\perp$ and $A_1^\perp = A_2^\perp = 0$.
(c) Use the first three equations of (7.8) to compute A_4, A_8, and A_{12}.
(d) By Exercise 72, $\mathbf{v} + C$ has only odd weight vectors. Show that all odd weight vectors in D are in $\mathbf{v} + C$.
(e) Use (M_3), (K), or (P_1) to find the weight distribution of $\mathbf{v} + C$.   ◆

Exercise 396 Let C be a $[16, 8, 4]$ self-dual doubly-even binary code. Analogous to Exercise 395, let $\mathbf{v}$ be a coset leader of a coset of C of weight 1 and let D be the $[16, 9, 1]$ code $D = C \cup (\mathbf{v} + C)$. Let $A_i = A_i(D^\perp)$ and $A_i^\perp = A_i(D)$.
(a) Why do we know that a coset of weight 1 must exist?
(b) Show that $A_{16} = 0$ and that the only unknown A_is are A_4, A_8, and A_{12}.
(c) Show that $A_i^\perp = A_{16-i}^\perp$, $A_1^\perp = 1$, and $A_2^\perp = 0$.
(d) Use the first three equations of (7.8) to compute A_4, A_8, and A_{12}.
(e) By Exercise 72, $\mathbf{v} + C$ has only odd weight vectors. Show that all odd weight vectors in D are in $\mathbf{v} + C$.
(f) Use (M_3), (K), or (P_1) to find the weight distribution of $\mathbf{v} + C$.   ◆

Exercise 397 Let C be a $[16, 8, 4]$ self-dual doubly-even binary code. After having done Exercises 395 and 396, do the following:

(a) Show that all odd weight vectors of $\mathbb{F}_2^{16}$ are in a coset of C of either weight 1 or weight 3.

(b) Show that there are exactly 16 cosets of C of weight 1 and exactly 112 cosets of C of weight 3.

Notice that this exercise shows that the cosets of weight 3 in fact exist.   ◆

Before proceeding with the computation of the weight distributions of the cosets of weight 2 in the $[16, 8, 4]$ codes, we present the following useful result.

Theorem 7.5.5 *Let C be a code of length n over $\mathbb{F}_q$ with $A \in \Gamma\mathrm{Aut}(C)$. Let $\mathbf{v} \in \mathbb{F}_q^n$. Then the weight distribution of the coset $\mathbf{v} + C$ is the same as the weight distribution of the coset $\mathbf{v}A + C$.*

Proof: If $\mathbf{x} \in \mathbb{F}_q^n$, then applying A to $\mathbf{x}$ rescales the components with nonzero scalars, permutes the components, and then maps them to other field elements under a field automorphism. Thus $\mathrm{wt}(\mathbf{x}) = \mathrm{wt}(\mathbf{x}A)$. As $A \in \Gamma\mathrm{Aut}(C)$, $\mathbf{v}A + C = \{\mathbf{v}A + \mathbf{c} \mid \mathbf{c} \in C\} = \{(\mathbf{v} + \mathbf{c})A \mid \mathbf{c} \in C\}$ implying that $\mathbf{v}A + C$ and $\mathbf{v} + C$ have the same weight distribution.   □

Example 7.5.6 Let C be a $[16, 8, 4]$ self-dual doubly-even binary code. By [262], up to equivalence there are two such codes. Each code has different weight distributions for its cosets of weight 2. These two codes, denoted C_1 and C_2, have generator matrices G_1 and G_2, respectively, where

$$G_1 = \begin{bmatrix} 1 & 1 & 1 & 1 & 0 & 0 & 0 & 0 & 0 & 0 & 0 & 0 & 0 & 0 & 0 & 0 \\ 1 & 1 & 0 & 0 & 1 & 1 & 0 & 0 & 0 & 0 & 0 & 0 & 0 & 0 & 0 & 0 \\ 1 & 1 & 0 & 0 & 0 & 0 & 1 & 1 & 0 & 0 & 0 & 0 & 0 & 0 & 0 & 0 \\ 1 & 0 & 1 & 0 & 1 & 0 & 1 & 0 & 0 & 0 & 0 & 0 & 0 & 0 & 0 & 0 \\ 0 & 0 & 0 & 0 & 0 & 0 & 0 & 0 & 1 & 1 & 1 & 1 & 0 & 0 & 0 & 0 \\ 0 & 0 & 0 & 0 & 0 & 0 & 0 & 0 & 1 & 1 & 0 & 0 & 1 & 1 & 0 & 0 \\ 0 & 0 & 0 & 0 & 0 & 0 & 0 & 0 & 1 & 1 & 0 & 0 & 0 & 0 & 1 & 1 \\ 0 & 0 & 0 & 0 & 0 & 0 & 0 & 0 & 1 & 0 & 1 & 0 & 1 & 0 & 1 & 0 \end{bmatrix}$$

and

$$G_2 = \begin{bmatrix} 1 & 1 & 1 & 1 & 0 & 0 & 0 & 0 & 0 & 0 & 0 & 0 & 0 & 0 & 0 & 0 \\ 1 & 1 & 0 & 0 & 1 & 1 & 0 & 0 & 0 & 0 & 0 & 0 & 0 & 0 & 0 & 0 \\ 1 & 1 & 0 & 0 & 0 & 0 & 1 & 1 & 0 & 0 & 0 & 0 & 0 & 0 & 0 & 0 \\ 1 & 1 & 0 & 0 & 0 & 0 & 0 & 0 & 1 & 1 & 0 & 0 & 0 & 0 & 0 & 0 \\ 1 & 1 & 0 & 0 & 0 & 0 & 0 & 0 & 0 & 0 & 1 & 1 & 0 & 0 & 0 & 0 \\ 1 & 1 & 0 & 0 & 0 & 0 & 0 & 0 & 0 & 0 & 0 & 0 & 1 & 1 & 0 & 0 \\ 1 & 1 & 0 & 0 & 0 & 0 & 0 & 0 & 0 & 0 & 0 & 0 & 0 & 0 & 1 & 1 \\ 1 & 0 & 1 & 0 & 1 & 0 & 1 & 0 & 1 & 0 & 1 & 0 & 1 & 0 & 1 & 0 \end{bmatrix}.$$

In this example, we compute the weight distribution of the cosets of C_1 of weight 2. Notice that C_1 is the direct sum of an $[8, 4, 4]$ code with itself; the only such code is $\widehat{\mathcal{H}}_3$ by Exercise 56. Let $\mathbf{v}$ be a weight 2 vector that is a coset leader of a coset of C_1. As we want to find the weight distribution of $\mathbf{v} + C_1$, by Theorem 7.5.5, we may replace $\mathbf{v}$ by $\mathbf{v}A$, where A

is a product of the automorphisms of C_1 given in Exercise 398. We use the notation of that exercise. Let supp($\mathbf{v}$) = $\{a, b\}$; when we replace $\mathbf{v}$ by $\mathbf{v}A$, we will still denote the support of this new vector by $\{a, b\}$, where a and b will have different values. If both a and b are at least 9, apply P_5 to $\mathbf{v}$. Thus we may assume that $1 \le a < b \le 8$ or $1 \le a \le 8 < b$. In the first case, if $a \ne 1$, apply powers of P_3 followed by P_1 so that we may assume that $a = 1$; now apply powers of P_3 again so that we may assume that $b = 2$. In the second case, if $a \ne 1$, apply powers of P_3 followed by P_1 so that we may assume that $a = 1$; if $b \ne 9$, apply powers of P_4 followed by P_2 so that we may assume that $b = 9$. Thus there are only two possible coset leaders that we need to consider:

(a) $\mathbf{v}_1 = 1100000000000000$, and

(b) $\mathbf{v}_2 = 1000000010000000$.

We first consider $\mathbf{v} = \mathbf{v}_1$. Let $\mathcal{D}$ be the $[16, 9, 2]$ code $\mathcal{D} = C_1 \cup (\mathbf{v}_1 + C_1)$. Let $A_i = A_i(\mathcal{D}^\perp)$ and $A_i^\perp = A_i(\mathcal{D})$. $\mathcal{D}$ is an even code and so $\mathcal{D}^\perp$ contains the all-one vector. Thus we know $A_0 = A_{16} = 1$ and $A_i = 0$ for all other i except possibly $i = 4, 8$, or 12 as $\mathcal{D}^\perp \subset C_1$. We also know $A_1^\perp = 0$ and need to determine $A_2^\perp$. The only weight 2 vectors in $\mathbf{v}_1 + C_1$ other than $\mathbf{v}_1$ are $\mathbf{v}_1 + \mathbf{c}$, where $\mathbf{c}$ is a vector of weight 4 whose support contains $\{1, 2\}$. By inspecting G_1, it is clear that the only choices for such a $\mathbf{c}$ are the first three rows of G_1. Therefore $A_2^\perp = 4$. Thus using the first three equations from (7.8), we obtain

$$1 + A_4 + A_8 + A_{12} + 1 = 2^7,$$
$$4A_4 + 8A_8 + 12A_{12} + 16 = 2^6 \cdot 16,$$
$$16A_4 + 64A_8 + 144A_{12} + 256 = 2^5(16 \cdot 17 + 8).$$

Solving yields $A_4 = A_{12} = 20$ and $A_8 = 86$. As $\mathcal{D}$ contains the all-one vector, $A_i^\perp = A_{16-i}^\perp$. Using (M_3), (K), or (P_1), we have $A_4^\perp = 36$, $A_6^\perp = 60$, and $A_8^\perp = 310$. As $A_i(\mathbf{v}_1 + C_1) = A_i(\mathcal{D}) - A_i(C_1)$, the weight distribution of the coset $\mathbf{v}_1 + C_1$ is:

Weight	2	4	6	8	10	12	14
Number of vectors	4	8	60	112	60	8	4

Now consider $\mathbf{v} = \mathbf{v}_2$. Let $\mathcal{D}$ be the $[16, 9, 2]$ code $\mathcal{D} = C_1 \cup (\mathbf{v}_2 + C_1)$, which is still even. As in the case $\mathbf{v} = \mathbf{v}_1$, $A_0 = A_{16} = 1$ and only A_4, A_8, and A_{12} need to be determined as the other A_is are 0. Also $A_1^\perp = 0$, but $A_2^\perp = 1$ because there are clearly no weight 4 vectors in C_1 whose support contains $\{1, 9\}$. Thus using the first three equations from (7.8), we have

$$1 + A_4 + A_8 + A_{12} + 1 = 2^7,$$
$$4A_4 + 8A_8 + 12A_{12} + 16 = 2^6 \cdot 16,$$
$$16A_4 + 64A_8 + 144A_{12} + 256 = 2^5(16 \cdot 17 + 2).$$

In the same manner as for $\mathbf{v} = \mathbf{v}_1$, we obtain $A_4 = A_{12} = 14$, $A_8 = 98$, $A_4^\perp = 42$, $A_6^\perp = 63$, and $A_8^\perp = 298$. The weight distribution of the coset $\mathbf{v}_2 + C_1$ is:

Weight	2	4	6	8	10	12	14
Number of vectors	1	14	63	100	63	14	1

Note that the code C_1 and its cosets of weight 1 do not contain vectors of weight 2. Hence all weight 2 vectors in $\mathbb{F}_2^{16}$ are coset leaders of cosets of weight 2. We want to count the number of cosets of weight 2 whose weight distributions are the two possibilities we have produced. The cosets whose coset leaders have their support entirely contained in either $\{1, \ldots, 8\}$ or $\{9, \ldots, 16\}$ have the same distribution as $\mathbf{v}_1 + C_1$ has. There are $2\binom{8}{2} = 56$ such coset leaders. But each coset contains four coset leaders and thus there are $56/4 = 14$ such cosets. The cosets whose coset leaders have a 1 in coordinate $\{1, \ldots, 8\}$ and a 1 in coordinate $\{9, \ldots, 16\}$ have the same distribution as $\mathbf{v}_2 + C_1$ has. There are $8^2 = 64$ such coset leaders and hence 64 such cosets, as each coset has a unique coset leader. ∎

Exercise 398 Show that the following permutations are automorphisms of the code C_1 defined in Example 7.5.6 where the coordinates are labeled $1, 2, \ldots, 16$:
(a) $P_1 = (1, 2)(3, 4)$,
(b) $P_2 = (9, 10)(11, 12)$,
(c) $P_3 = (2, 3, 5, 4, 7, 8, 6)$,
(d) $P_4 = (10, 11, 13, 12, 15, 16, 14)$, and
(e) $P_5 = (1, 9)(2, 10)(3, 11)(4, 12)(5, 13)(6, 14)(7, 15)(8, 16)$. ◆

Exercise 399 Combining Examples 7.5.4 and 7.5.6 and Exercise 397, we have the following table for the weight distributions of the cosets of the code C_1 defined in Example 7.5.6.

Coset weight	Number of vectors of given weight																	Number of cosets
	0	1	2	3	4	5	6	7	8	9	10	11	12	13	14	15	16	
0	1	0	0	0	28	0	0	0	198	0	0	0	28	0	0	0	1	1
1																		16
2	0	0	4	0	8	0	60	0	112	0	60	0	8	0	4	0	0	14
2	0	0	1	0	14	0	63	0	100	0	63	0	14	0	1	0	0	64
3																		
4	0	0	0	0	16	0	64	0	96	0	64	0	16	0	0	0	0	

Fill in the missing entries using results from Exercises 395 and 396. Then complete the right-hand column for the cosets of weights 3 and 4. ◆

Exercise 400 In this exercise you are to find the weight distributions of the cosets of the code C_2 defined in Example 7.5.6. The cosets of weights 1, 3, and 4 have been considered in Example 7.5.4 and Exercises 395 and 396.
(a) Show that the following permutations are automorphisms of C_2:
   (i) $P_1 = (1, 2)(3, 4)$,
   (ii) $P_2 = (1, 3, 5, 7, 9, 11, 13, 15)(2, 4, 6, 8, 10, 12, 14, 16)$, and
   (iii) $P_3 = (3, 5, 7, 9, 11, 13, 15)(4, 6, 8, 10, 12, 14, 16)$.
(b) Show that the only two coset leaders of weight 2 that need to be considered are:
   (i) $\mathbf{v}_1 = 1100000000000000$, and
   (ii) $\mathbf{v}_2 = 1010000000000000$.
(c) Complete an analogous argument to that of Example 7.5.6 to find the weight distributions of the cosets of weight 2.
(d) For this code, complete a similar table to that in Exercise 399. ◆

Exercise 401 Let C be a [24, 12, 8] self-dual doubly-even binary code. In Section 10.1.1 we will see that the only such code is the extended binary Golay code $\mathcal{G}_{24}$ from Section 1.9.1. Its weight enumerator is given in Exercise 384.

(a) Show that C has covering radius at most 4. (It turns out that the covering radius is exactly 4; see Theorem 11.1.7.)

(b) Find the weight distributions of the cosets of weight 4 and the cosets of weight 3.   ◆

7.6   Weight distributions of punctured and shortened codes

The weight distribution of a code does not in general determine the weight distribution of a code obtained from it by either puncturing or shortening, but when certain uniformity conditions hold, the weight distribution of a punctured or shortened code can be determined from the original code. Let C be an $[n, k]$ code over $\mathbb{F}_q$. Let $\mathcal{M}$ be the $q^k \times n$ matrix whose rows are all codewords in C, and let $\mathcal{M}_i$ be the submatrix of $\mathcal{M}$ consisting of the codewords of weight i. A code is *homogeneous* provided that for $0 \leq i \leq n$, each column of $\mathcal{M}_i$ has the same weight. Recall from Section 1.7 that $\Gamma\mathrm{Aut}(C)$ is transitive if $\Gamma\mathrm{Aut}_{\mathrm{Pr}}(C)$ is a transitive permutation group. As Exercise 402 shows, if C has a transitive automorphism group, C is homogeneous. This is about the only simple way to decide if a code is homogeneous.

Exercise 402 Prove that if C has a transitive automorphism group, then C is homogeneous.
                                                                                            ◆

Theorem 7.6.1 (Prange) *Let C be a homogeneous $[n, k, d]$ code over $\mathbb{F}_q$ with $d > 1$. Let C^* be the code obtained from C by puncturing on some coordinate, and let C_* be the code obtained from C by shortening on some coordinate. Then for $0 \leq i \leq n - 1$ we have:*

(i) $A_i(C^*) = \dfrac{n-i}{n} A_i(C) + \dfrac{i+1}{n} A_{i+1}(C),$ *and*

(ii) $A_i(C_*) = \dfrac{n-i}{n} A_i(C).$

Assume further that C is an even binary code. Then for $1 \leq j \leq \lfloor n/2 \rfloor$ we have:

(iii) $A_{2j}(C) = A_{2j}(C^*) + A_{2j-1}(C^*),$

(iv) $A_{2j-1}(C^*) = \dfrac{2j}{n} A_{2j}(C)$ *and* $A_{2j}(C^*) = \dfrac{n-2j}{n} A_{2j}(C),$ *and*

(v) $2j A_{2j}(C^*) = (n - 2j) A_{2j-1}(C^*).$

Proof: The number of nonzero entries in $\mathcal{M}_i$ is $i A_i(C) = n w_i$, where w_i is the weight of a column of $\mathcal{M}_i$, since this weight is independent of the column as C is homogeneous. Thus $w_i = (i/n)A_i(C)$, and so each column of $\mathcal{M}_i$ has $((n-i)/n)A_i(C)$ zeros. A vector of weight i in C^* arises either from a vector of weight i in C with a zero in the punctured coordinate, or a vector of weight $i + 1$ in C with a nonzero component in the punctured coordinate; as $d > 1$ no vector in C^* can arise in both ways. Thus (i) holds. A vector of weight i in C_* arises from a vector of weight i in C with a zero on the shortened coordinate, yielding (ii).

Now assume that C is an even binary code. Vectors of weight $2j$ in C give rise to vectors of weights $2j$ or $2j - 1$ in C^*; all vectors of weight $2j$ or $2j - 1$ in C^* must arise from codewords of weight $2j$ in C. Thus (iii) holds. Since $A_{2j-1}(C) = A_{2j+1}(C) = 0$, substituting $i = 2j - 1$ and $i = 2j$ into (i) gives the two equations in (iv). Part (v) follows by solving each equation in (iv) for $A_{2j}(C)$. $\qquad\square$

Example 7.6.2 Let C be the $[12, 6, 6]$ extended ternary Golay code. As we will show in Section 10.4.2 $\Gamma\mathrm{Aut}(C)$ is transitive. By Exercise 402, C is homogeneous. Let C^* be C punctured on any coordinate. In Example 7.3.2, the weight distribution of C is given. Using Prange's Theorem,

$$W_{C^*}(x, y) = y^{11} + 132x^5y^6 + 132x^6y^5 + 330x^8y^3 + 110x^9y^2 + 24x^{11}.$$

The code C^* is the $[11, 6, 5]$ ternary Golay code. $\qquad\blacksquare$

Exercise 403 Let C be the $[24, 12, 8]$ extended binary Golay code, whose weight distribution is given in Exercise 384. As we will show in Section 10.1.2, $\Gamma\mathrm{Aut}(C)$ is transitive; and so, by Exercise 402, C is homogeneous. Use Prange's Theorem to show that

$$\begin{aligned}
W_{C^*}(x, y) &= y^{23} + 253x^7y^{16} + 506x^8y^{15} + 1288x^{11}y^{12} \\
&\quad + 1288x^{12}y^{11} + 506x^{15}y^8 + 253x^{16}y^7 + x^{23}.
\end{aligned}$$

The code C^* is the $[23, 12, 7]$ binary Golay code. $\qquad\blacklozenge$

Exercise 404 Let C be the $[24, 12, 8]$ extended binary Golay code. Let $C^{(i)}$ be the code obtained from C by puncturing on i points for $1 \le i \le 4$. These are $[24 - i, 12, 8 - i]$ codes each of which has a transitive automorphism group by the results in Section 10.1.2. The weight enumerator of $C^{(1)}$ is given in Exercise 403. Find the weight enumerators of $C^{(i)}$ for $2 \le i \le 4$. $\qquad\blacklozenge$

Exercise 405 Let C be the $[24, 12, 8]$ extended binary Golay code. Let $C_{(i)}$ be the code obtained from C by shortening on i points for $1 \le i \le 4$. These are $[24 - i, 12 - i, 8]$ codes each of which has a transitive automorphism group by the results in Section 10.1.2. Find the weight enumerators of $C_{(i)}$ for $1 \le i \le 4$. $\qquad\blacklozenge$

Exercise 406 Let C be either of the two $[16, 8, 4]$ self-dual doubly-even binary codes with generator matrices given in Example 7.5.6. By Exercises 398 and 400, $\Gamma\mathrm{Aut}(C)$ is transitive and so C is homogeneous. Use Prange's Theorem to find the weight distribution of C^*. $\qquad\blacklozenge$

Prange's Theorem also holds for homogeneous nonlinear codes of minimum distance $d > 1$ as no part of the proof requires the use of linearity. For example, a coset of weight 2 of the $[8, 4, 4]$ extended binary Hamming code $\widehat{\mathcal{H}}_3$ is homogeneous. Puncturing this coset gives a coset of weight 1 in $\mathcal{H}_3$, and (i) of Prange's Theorem gives its distribution from the coset distribution of the original weight 2 coset of $\widehat{\mathcal{H}}_3$; see Exercise 407. A coset of $\widehat{\mathcal{H}}_3$ of weight 1 is not homogeneous even though $\Gamma\mathrm{Aut}(\widehat{\mathcal{H}}_3)$ is triply transitive. Verifying homogeneity of cosets is in general quite difficult. One can show that a coset of the $[24, 12, 8]$ extended binary Golay code of weight 4 is homogeneous and so obtain the distribution of a weight 3 coset of the $[23, 12, 7]$ binary Golay code; see Exercise 408.

Exercise 407   Do the following:

(a) Choose a generator matrix for $\widehat{\mathcal{H}}_3$ and any convenient weight 2 vector $\mathbf{v}$ in $\mathbb{F}_2^8$. Show that the nonlinear code $\mathcal{C} = \mathbf{v} + \widehat{\mathcal{H}}_3$ is homogeneous.

(b) Apply part (i) of Prange's Theorem to find the weight distribution of $\mathcal{C}^*$ and compare your answer to the weight distribution of the weight 1 cosets of $\mathcal{H}_3$ found in Exercise 394.

(c) Show that a coset of $\widehat{\mathcal{H}}_3$ of weight 1 is not homogeneous.   ◆

Exercise 408   Assuming that a weight 4 coset of the [24, 12, 8] extended binary Golay code is homogeneous, use Prange's Theorem and the results of Exercise 401 to find the weight distribution of a weight 3 coset of the [23, 12, 7] binary Golay code.   ◆

7.7   Other weight enumerators

There are other weight enumerators for a code that contain more detailed information about the codewords. If $\mathbf{c} \in \mathbb{F}_q^n$ and $\alpha \in \mathbb{F}_q$, let $s_\alpha(\mathbf{c})$ be the number of components of $\mathbf{c}$ that equal α. Let $\mathbb{F}_q = \{\alpha_0, \alpha_1, \dots, \alpha_{q-1}\}$. The polynomial

$$W_{\mathcal{C}}'(x_{\alpha_0}, x_{\alpha_1}, \dots, x_{\alpha_{q-1}}) = \sum_{\mathbf{c} \in \mathcal{C}} \prod_{\alpha \in \mathbb{F}_q} x_\alpha^{s_\alpha(\mathbf{c})}$$

is called the *complete weight enumerator of* $\mathcal{C}$. Notice that if P is a permutation matrix, then $\mathcal{C}P$ and $\mathcal{C}$ have the same complete weight enumerator; however, generally $\mathcal{C}M$ and $\mathcal{C}$ do not have the same complete weight enumerator if M is a monomial matrix. If $q = 2$, then $W_{\mathcal{C}}'(x_0, x_1)$ is the ordinary weight enumerator $W_{\mathcal{C}}(x_1, x_0)$. MacWilliams [213] also proved that there is an identity between the complete weight enumerator of $\mathcal{C}$ and the complete weight enumerator of $\mathcal{C}^\perp$. This identity is the same as (M_3) if $q = 2$. If $q = 3$, with $\mathbb{F}_3 = \{0, 1, 2\}$, the identity is

$$W_{\mathcal{C}^\perp}'(x_0, x_1, x_2) = \frac{1}{|\mathcal{C}|} W_{\mathcal{C}}'(x_0 + x_1 + x_2, x_0 + \xi x_1 + \xi^2 x_2, x_0 + \xi^2 x_1 + \xi x_2), \qquad (7.12)$$

where $\xi = e^{2\pi i/3}$ is a primitive complex cube root of unity. If $q = 4$, with $\mathbb{F}_4 = \{0, 1, \omega, \overline{\omega}\}$ as in Section 1.1, there are two MacWilliams identities depending on whether one is considering the Hermitian or the ordinary dual; see [291]. For the ordinary dual it is

$$W_{\mathcal{C}^\perp}'(x_0, x_1, x_\omega, x_{\overline{\omega}}) = \frac{1}{|\mathcal{C}|} W_{\mathcal{C}}'(x_0 + x_1 + x_\omega + x_{\overline{\omega}}, x_0 + x_1 - x_\omega - x_{\overline{\omega}},$$
$$x_0 - x_1 - x_\omega + x_{\overline{\omega}}, x_0 - x_1 + x_\omega - x_{\overline{\omega}}). \qquad (7.13)$$

And for the Hermitian dual, the MacWilliams identity is

$$W_{\mathcal{C}^{\perp_H}}'(x_0, x_1, x_\omega, x_{\overline{\omega}}) = \frac{1}{|\mathcal{C}|} W_{\mathcal{C}}'(x_0 + x_1 + x_\omega + x_{\overline{\omega}}, x_0 + x_1 - x_\omega - x_{\overline{\omega}},$$
$$x_0 - x_1 + x_\omega - x_{\overline{\omega}}, x_0 - x_1 - x_\omega + x_{\overline{\omega}}). \qquad (7.14)$$

Exercise 409 Recall that the [4, 2] ternary tetracode $\mathcal{H}_{3,2}$, introduced in Example 1.3.3, has generator matrix

$$G = \begin{bmatrix} 1 & 0 & 1 & 1 \\ 0 & 1 & 1 & 2 \end{bmatrix}.$$

Find the complete weight enumerator of $\mathcal{H}_{3,2}$ and show that the MacWilliams identity (7.12) is satisfied; use of a computer algebra system will be most helpful.   ♦

Exercise 410 Let C be the [4, 2, 3] extended Reed–Solomon code over $\mathbb{F}_4$ with generator matrix

$$G = \begin{bmatrix} 1 & 1 & 1 & 1 \\ 1 & \omega & \overline{\omega} & 0 \end{bmatrix}.$$

(a) Show that C is self-dual under the ordinary inner product.
(b) Find the complete weight enumerator of C and show that the MacWilliams identity (7.13) is satisfied; use of a computer algebra system will be most helpful.   ♦

Exercise 411 Let C be the [6, 3, 4] hexacode over $\mathbb{F}_4$ with generator matrix

$$G = \begin{bmatrix} 1 & 0 & 0 & 1 & \omega & \omega \\ 0 & 1 & 0 & \omega & 1 & \omega \\ 0 & 0 & 1 & \omega & \omega & 1 \end{bmatrix},$$

as presented in Example 1.3.4. It is Hermitian self-dual. Find the complete weight enumerator of C and show that the MacWilliams identity (7.14) is satisfied; use of a computer algebra system will be most helpful.   ♦

Exercise 412 Recall that $\overline{}$ denotes conjugation in $\mathbb{F}_4$ as described in Section 1.3. Do the following:
(a) Show that $W'_{\overline{C}}(x_0, x_1, x_\omega, x_{\overline{\omega}}) = W'_C(x_0, x_1, x_{\overline{\omega}}, x_\omega)$.
(b) Show that (7.13) holds if and only if (7.14) holds.   ♦

Exercise 413 Show that for codes over $\mathbb{F}_4$, $W_C(x, y) = W'_C(y, x, x, x)$.   ♦

The Lee weight enumerator is sometimes used for codes over fields of characteristic $p \neq 2$. For such fields, let $\mathbb{F}_q = \{\alpha_0 = 0, \alpha_1, \ldots, \alpha_r, -\alpha_1, \ldots, -\alpha_r\}$ where $r = (q-1)/2$. The *Lee weight enumerator of C* is the polynomial

$$W''_C(x_0, x_{\alpha_1}, \ldots, x_{\alpha_r}) = \sum_{\mathbf{c} \in C} x_0^{s_0(\mathbf{c})} \prod_{i=1}^{r} x_{\alpha_i}^{s_{\alpha_i}(\mathbf{c}) + s_{-\alpha_i}(\mathbf{c})}.$$

The Lee weight enumerator reduces the number of variables in the complete weight enumerator essentially by half. If $q = 3$, then $W_C(x_1, x_0) = W''_C(x_0, x_1)$. There is also a relationship between the Lee weight enumerator of C and the Lee weight enumerator of $C^\perp$, which is obtained by equating x_{α_i} and $x_{-\alpha_i}$ in the MacWilliams identity for the complete weight enumerator.

MacWilliams identities are established for other families of codes and other types of weight enumerators in [291].

7.8   Constraints on weights

In this section, we will present constraints on the weights of codewords in even binary codes, where we recall that a binary code is called even if all of its codewords have even weight. The following identity from Theorem 1.4.3 for two vectors $\mathbf{x}, \mathbf{y} \in \mathbb{F}_2^n$ will prove useful:

$$\text{wt}(\mathbf{x} + \mathbf{y}) = \text{wt}(\mathbf{x}) + \text{wt}(\mathbf{y}) - 2\text{wt}(\mathbf{x} \cap \mathbf{y}), \tag{7.15}$$

where $\mathbf{x} \cap \mathbf{y}$ is the vector in $\mathbb{F}_2^n$ which has 1s precisely in those positions where both $\mathbf{x}$ and $\mathbf{y}$ have 1s. When the ordinary inner product between two vectors $\mathbf{x}$ and $\mathbf{y}$ is 0, we say the vectors are *orthogonal* and denote this by $\mathbf{x} \perp \mathbf{y}$.

The following is an easy consequence of (7.15).

Lemma 7.8.1 *Let $\mathbf{x}$ and $\mathbf{y}$ be two vectors in $\mathbb{F}_2^n$.*
(i) *Suppose* $\text{wt}(\mathbf{x}) \equiv 0 \pmod 4$. *Then* $\text{wt}(\mathbf{x} + \mathbf{y}) \equiv \text{wt}(\mathbf{y}) \pmod 4$ *if and only if* $\mathbf{x} \perp \mathbf{y}$, *and* $\text{wt}(\mathbf{x} + \mathbf{y}) \equiv \text{wt}(\mathbf{y}) + 2 \pmod 4$ *if and only if* $\mathbf{x} \not\perp \mathbf{y}$.
(ii) *Suppose* $\text{wt}(\mathbf{x}) \equiv 2 \pmod 4$. *Then* $\text{wt}(\mathbf{x} + \mathbf{y}) \equiv \text{wt}(\mathbf{y}) \pmod 4$ *if and only if* $\mathbf{x} \not\perp \mathbf{y}$, *and* $\text{wt}(\mathbf{x} + \mathbf{y}) \equiv \text{wt}(\mathbf{y}) + 2 \pmod 4$ *if and only if* $\mathbf{x} \perp \mathbf{y}$.

Exercise 414   Prove Lemma 7.8.1.   ◆

We say that a binary vector is *doubly-even* if its weight is divisible by 4. A binary vector is *singly-even* if its weight is even but not divisible by 4. Recall from Theorem 1.4.8 that if a binary code has only doubly-even vectors, the code is self-orthogonal. We will be primarily interested in codes where not all codewords are doubly-even. We will present a decomposition of even binary codes into pieces that are called the hull of the code, H-planes, and A-planes. An H-plane or A-plane is a 2-dimensional space spanned by $\mathbf{x}$ and $\mathbf{y}$; such a space is denoted span$\{\mathbf{x}, \mathbf{y}\}$. The hull, H-plane, and A-plane are defined as follows:
- The *hull* of a code $\mathcal{C}$ is the subcode $\mathcal{H} = \mathcal{C} \cap \mathcal{C}^\perp$.
- If $\mathbf{x}$ and $\mathbf{y}$ are doubly-even but not orthogonal, then we call span$\{\mathbf{x}, \mathbf{y}\}$ an *H-plane*. Note that if span$\{\mathbf{x}, \mathbf{y}\}$ is an H-plane, then $\text{wt}(\mathbf{x} + \mathbf{y}) \equiv 2 \pmod 4$ by Lemma 7.8.1(i), and so an H-plane contains two nonzero doubly-even vectors and one singly-even vector.
- If $\mathbf{x}$ and $\mathbf{y}$ are singly-even but not orthogonal, then we call span$\{\mathbf{x}, \mathbf{y}\}$ an *A-plane*. If span$\{\mathbf{x}, \mathbf{y}\}$ is an A-plane, then $\text{wt}(\mathbf{x} + \mathbf{y}) \equiv 2 \pmod 4$ still holds by Lemma 7.8.1(ii); an A-plane therefore contains three singly-even vectors.

Exercise 415   Prove that if $\mathcal{C}$ is an $[n, k]$ even binary code, then either $\mathcal{C}$ is self-orthogonal or its hull has dimension at most $k - 2$.   ◆

Exercise 416   Let $\mathcal{C}$ be a 2-dimensional even binary code that is not self-orthogonal. Prove the following:
(a) The hull of $\mathcal{C}$ is $\{\mathbf{0}\}$.
(b) Either $\mathcal{C}$ contains three singly-even vectors, in which case it is an A-plane, or $\mathcal{C}$ contains exactly two doubly-even nonzero vectors, in which case it is an H-plane.   ◆

Before proceeding with our decomposition, we need a few preliminary results.

Lemma 7.8.2 *Let C be an even binary code of dimension $k \geq 2$ whose hull is $\{0\}$. The following hold:*

(i) *If C contains a nonzero doubly-even vector, it contains an H-plane.*

(ii) *If C contains no nonzero doubly-even vectors, it is 2-dimensional and is an A-plane.*

Proof: Suppose that C contains a nonzero doubly-even vector $\mathbf{x}$. Then as $\mathbf{x} \notin C^{\perp}$, there is a codeword $\mathbf{y} \in C$ with $\mathbf{x} \not\perp \mathbf{y}$. If $\mathbf{y}$ is doubly-even, then $\mathrm{span}\{\mathbf{x}, \mathbf{y}\}$ is an H-plane. If $\mathbf{y}$ is singly-even, then $\mathbf{x} + \mathbf{y}$ is doubly-even by Lemma 7.8.1 and $\mathrm{span}\{\mathbf{x}, \mathbf{x} + \mathbf{y}\} = \mathrm{span}\{\mathbf{x}, \mathbf{y}\}$ is still an H-plane. This proves (i). Part (ii) follows from Exercise 417. $\square$

Exercise 417 Let C be an even binary code with dimension $k \geq 2$ whose hull is $\{0\}$. Prove that if C has no nonzero doubly-even vectors, then $k = 2$ and C is an A-plane. ◆

To describe our decomposition we need one more piece of notation. Suppose that C contains two subcodes, $\mathcal{A}$ and $\mathcal{B}$, such that $C = \mathcal{A} + \mathcal{B} = \{\mathbf{a} + \mathbf{b} \mid \mathbf{a} \in \mathcal{A} \text{ and } \mathbf{b} \in \mathcal{B}\}$, where $\mathcal{A} \cap \mathcal{B} = \{0\}$ and $\mathbf{a} \cdot \mathbf{b} = 0$ for all $\mathbf{a} \in \mathcal{A}$ and $\mathbf{b} \in \mathcal{B}$. Then we say that C is the *orthogonal sum* of $\mathcal{A}$ and $\mathcal{B}$, and denote this by $C = \mathcal{A} \perp \mathcal{B}$. (In vector space terminology, C is the orthogonal direct sum of $\mathcal{A}$ and $\mathcal{B}$.)

Exercise 418 Let $C = \mathcal{A}_1 \perp \mathcal{A}_2$, where $\mathcal{A}_1$ and $\mathcal{A}_2$ are A-planes. Prove that $C = \mathcal{H}_1 \perp \mathcal{H}_2$, where $\mathcal{H}_1$ and $\mathcal{H}_2$ are H-planes. ◆

Lemma 7.8.3 *Let C be an even binary code whose hull is $\{0\}$. Then C has even dimension $2m$ and one of the following occurs:*

(i) $C = \mathcal{H}_1 \perp \mathcal{H}_2 \perp \cdots \perp \mathcal{H}_m$, *where $\mathcal{H}_1, \mathcal{H}_2, \ldots, \mathcal{H}_m$ are H-planes.*

(ii) $C = \mathcal{H}_1 \perp \mathcal{H}_2 \perp \cdots \perp \mathcal{H}_{m-1} \perp \mathcal{A}$, *where $\mathcal{H}_1, \mathcal{H}_2, \ldots, \mathcal{H}_{m-1}$ are H-planes and $\mathcal{A}$ is an A-plane.*

Proof: This is proved by induction on the dimension of C. C cannot have dimension 1 as, otherwise, its nonzero vector, being even, is orthogonal to itself and hence in the hull. So C has dimension at least 2. If C has dimension 2, it is either an H-plane or an A-plane by Exercise 416. So we may assume that C has dimension at least 3. By Lemma 7.8.2, C contains an H-plane $\mathcal{H}_1$. The proof is complete by induction if we can show that $C = \mathcal{H}_1 \perp C_1$, where the hull of C_1 is $\{0\}$. Let $\mathcal{H}_1 = \mathrm{span}\{\mathbf{x}, \mathbf{y}\}$ where $\mathbf{x}$ and $\mathbf{y}$ are nonzero doubly-even vectors with $\mathbf{x} \not\perp \mathbf{y}$. Define $f : C \to \mathbb{F}_2^2$, where $f(\mathbf{c}) = (\mathbf{c} \cdot \mathbf{x}, \mathbf{c} \cdot \mathbf{y})$. By Exercise 419, f is a surjective linear transformation. Let C_1 be its kernel; C_1 has dimension $k - 2$. As $\mathbf{x} \not\perp \mathbf{y}$ and $\mathbf{x} \not\perp (\mathbf{x} + \mathbf{y})$, none of $\mathbf{x}, \mathbf{y}$, or $\mathbf{x} + \mathbf{y}$ are in C_1. Thus $\mathcal{H}_1 \cap C_1 = \{0\}$. By the definition of f, $\mathcal{H}_1$ is orthogonal to C_1. So $C = \mathcal{H}_1 \perp C_1$. If $\mathbf{z} \neq \mathbf{0}$ is in the hull of C_1, it is orthogonal to C_1 and to $\mathcal{H}_1$; hence $\mathbf{z}$ is in the hull of C, a contradiction. $\square$

Exercise 419 Let C be a binary code with two nonzero even vectors $\mathbf{x}$ and $\mathbf{y}$, where $\mathbf{x} \not\perp \mathbf{y}$. Define $f : C \to \mathbb{F}_2^2$, where $f(\mathbf{c}) = (\mathbf{c} \cdot \mathbf{x}, \mathbf{c} \cdot \mathbf{y})$. Show that f is a surjective linear transformation. ◆

The following is the converse of Lemma 7.8.3; we will need this result in Chapter 9.

Lemma 7.8.4 *If $C = P_1 \perp \cdots \perp P_k$ is an orthogonal sum where each P_i is either an H-plane or an A-plane, then the hull of C is $\{0\}$.*

Proof: If $\mathbf{x} \in C \cap C^\perp$, where $\mathbf{x} = \mathbf{x}_1 + \cdots + \mathbf{x}_k$ with $\mathbf{x}_i \in P_i$ for $1 \le i \le k$, then $\mathbf{x} \cdot \mathbf{y} = 0$ for all $\mathbf{y} \in C$. As $P_i \cap P_i^\perp = \{0\}$, if $\mathbf{x}_i \ne \mathbf{0}$, there exists $\mathbf{y}_i \in P_i$ with $\mathbf{x}_i \cdot \mathbf{y}_i \ne 0$. Thus $0 = \mathbf{x} \cdot \mathbf{y}_i$; but $\mathbf{x} \cdot \mathbf{y}_i = \mathbf{x}_i \cdot \mathbf{y}_i \ne 0$ as $\mathbf{x}_j \perp \mathbf{y}_i$ if $j \ne i$. This contradiction shows $\mathbf{x}_i = \mathbf{0}$. So $C \cap C^\perp = \{0\}$. $\qquad\square$

We now state our main decomposition result, found in [3].

Theorem 7.8.5 *Let C be an $[n, k]$ even binary code whose hull $\mathcal{H}$ has dimension r. Then $k = 2m + r$ for some nonnegative integer m and C has one of the following forms:*
(H) $C = \mathcal{H} \perp \mathcal{H}_1 \perp \mathcal{H}_2 \perp \cdots \perp \mathcal{H}_m$, *where $\mathcal{H}_1, \mathcal{H}_2, \ldots, \mathcal{H}_m$ are H-planes and $\mathcal{H}$ is doubly-even.*
(O) $C = \mathcal{H} \perp \mathcal{H}_1 \perp \mathcal{H}_2 \perp \cdots \perp \mathcal{H}_m$, *where $\mathcal{H}_1, \mathcal{H}_2, \ldots, \mathcal{H}_m$ are H-planes and $\mathcal{H}$ is singly-even.*
(A) $C = \mathcal{H} \perp \mathcal{H}_1 \perp \mathcal{H}_2 \perp \cdots \perp \mathcal{H}_{m-1} \perp \mathcal{A}$, *where $\mathcal{H}_1, \mathcal{H}_2, \ldots, \mathcal{H}_{m-1}$ are H-planes, $\mathcal{A}$ is an A-plane, and $\mathcal{H}$ is doubly-even.*

Proof: If $C = C^\perp$, we have either form (H) or form (O) with $m = 0$. Assume $C \ne C^\perp$. Let $\mathbf{x}_1, \mathbf{x}_2, \ldots, \mathbf{x}_r$ be a basis of $\mathcal{H}$; extend this to a basis $\mathbf{x}_1, \mathbf{x}_2, \ldots, \mathbf{x}_n$ of C. Let $C_1 = \text{span}\{\mathbf{x}_{r+1}, \mathbf{x}_{r+2}, \ldots, \mathbf{x}_n\}$. Clearly, $C = \mathcal{H} \perp C_1$ as every vector in $\mathcal{H}$ is orthogonal to every vector in C. If $\mathbf{z}$ is in the hull of C_1, it is orthogonal to everything in C_1 and in $\mathcal{H}$, and hence to everything in C. Thus $\mathbf{z}$ is also in $\mathcal{H}$, implying that $\mathbf{z} = \mathbf{0}$. So the hull of C_1 is $\{0\}$. By Lemma 7.8.3, the result follows once we show that the decomposition given in (A) with $\mathcal{H}$ singly-even reduces to (O). In that case, let $\mathcal{A} = \text{span}\{\mathbf{x}, \mathbf{y}\}$ and let $\mathbf{z}$ be a singly-even vector in $\mathcal{H}$. By Lemma 7.8.1, $\mathbf{x}_1 = \mathbf{z} + \mathbf{x}$ and $\mathbf{y}_1 = \mathbf{z} + \mathbf{y}$ are both doubly-even. As $\mathbf{x} \not\perp \mathbf{y}, \mathbf{x}_1 \not\perp \mathbf{y}_1$; hence $\mathcal{H}_m = \text{span}\{\mathbf{x}_1, \mathbf{y}_1\}$ is an H-plane. Clearly, $\mathcal{H} \perp \mathcal{H}_1 \perp \cdots \perp \mathcal{H}_{m-1} \perp \mathcal{A} = \mathcal{H} \perp \mathcal{H}_1 \perp \cdots \perp \mathcal{H}_m$ and we have form (O). $\qquad\square$

Exercise 420 If C is a singly-even self-orthogonal binary code, what is its form in Theorem 7.8.5? $\qquad\qquad\qquad\qquad\qquad\qquad\qquad\qquad\qquad\qquad\qquad\qquad\qquad\qquad\blacklozenge$

As a consequence of this theorem, we can obtain constraints on the number of singly-even and doubly-even codewords in an even binary code; see [31, 164, 271].

Theorem 7.8.6 *Let C be an $[n, k]$ even binary code whose hull has dimension r. Then $k = 2m + r$. Let a be the number of doubly-even vectors in C and b the number of singly-even vectors in C.*
(i) *If C has form (H), then $a = 2^r(2^{2m-1} + 2^{m-1})$ and $b = 2^r(2^{2m-1} - 2^{m-1})$.*
(ii) *If C has form (O), then $a = b = 2^{k-1}$.*
(iii) *If C has form (A), then $a = 2^r(2^{2m-1} - 2^{m-1})$ and $b = 2^r(2^{2m-1} + 2^{m-1})$.*

Proof: We leave the proofs of (i) and (ii) as exercises. Suppose C has form (A). Let $C_1 = \mathcal{H}_1 \perp \mathcal{H}_2 \perp \cdots \perp \mathcal{H}_{m-1} \perp \mathcal{A}$. As $\mathcal{H}$ is doubly-even, by Lemma 7.8.1, a, respectively b, is 2^r times the number of doubly-even, respectively singly-even, vectors in C_1. Therefore we only need look at C_1 where we prove the result by induction on m. When $m = 1$,

$\mathcal{C}_1 = \mathcal{A}$ and the result follows as $\mathcal{A}$ contains three singly-even vectors and one doubly-even vector. Suppose that the number of doubly-even vectors in $\mathcal{H}_1 \bot \mathcal{H}_2 \bot \cdots \bot \mathcal{H}_{i-1} \bot \mathcal{A}$ for some $i \geq 1$ is $2^{2i-1} - 2^{i-1}$ and the number of singly-even vectors is $2^{2i-1} + 2^{i-1}$. As $\mathcal{H}_i$ has one singly-even and three doubly-even vectors, the number of doubly-even vectors in $\mathcal{H}_1 \bot \mathcal{H}_2 \bot \cdots \bot \mathcal{H}_i \bot \mathcal{A}$ is $2^{2i-1} + 2^{i-1} + 3(2^{2i-1} - 2^{i-1}) = 2^{2i+1} - 2^i$ by Lemma 7.8.1. Similarly, the number of singly-even vectors in $\mathcal{H}_1 \bot \mathcal{H}_2 \bot \cdots \bot \mathcal{H}_i \bot \mathcal{A}$ is $3(2^{2i-1} + 2^{i-1}) + 2^{2i-1} - 2^{i-1} = 2^{2i+1} + 2^i$. $\qquad \square$

Exercise 421 Prove that a and b are as claimed in forms (H) and (O) of Theorem 7.8.6. $\qquad \blacklozenge$

Exercise 422 Find the form ((H), (O), or (A)) from Theorem 7.8.5, the corresponding parameters a, b, r, and m from Theorems 7.8.5 and 7.8.6, and the hull of each of the $[8, 4]$ even binary codes with the following generator matrices:

(a) $G_1 = \begin{bmatrix} 1 & 0 & 0 & 0 & 1 & 0 & 0 & 0 \\ 0 & 1 & 0 & 0 & 0 & 1 & 0 & 0 \\ 0 & 0 & 1 & 0 & 1 & 1 & 0 & 1 \\ 0 & 0 & 0 & 1 & 1 & 1 & 1 & 0 \end{bmatrix}$,

(b) $G_2 = \begin{bmatrix} 1 & 0 & 0 & 0 & 1 & 0 & 1 & 1 \\ 0 & 1 & 0 & 0 & 0 & 1 & 0 & 0 \\ 0 & 0 & 1 & 0 & 1 & 0 & 0 & 0 \\ 0 & 0 & 0 & 1 & 1 & 0 & 0 & 0 \end{bmatrix}$,

(c) $G_3 = \begin{bmatrix} 1 & 0 & 0 & 0 & 0 & 1 & 1 & 1 \\ 0 & 1 & 0 & 0 & 1 & 0 & 0 & 0 \\ 0 & 0 & 1 & 0 & 1 & 0 & 0 & 0 \\ 0 & 0 & 0 & 1 & 1 & 0 & 0 & 0 \end{bmatrix}$. $\qquad \blacklozenge$

Example 7.8.7 The MacWilliams equations, or their equivalent forms, can be used to determine whether or not a set of nonnegative integers could arise as the weight distribution of a code. If a set of integers were to be the weight distribution of a code, then the distribution of the dual code can be computed from one of the forms of the MacWilliams equations. The resulting dual distribution must obviously be a set of nonnegative integers; if not the original set of integers could not be the weight distribution of a code. Theorem 7.8.6 can be used to eliminate weight distributions that the MacWilliams equations cannot. For example, suppose that $\mathcal{C}$ is an $[8, 4]$ binary code with weight distribution $A_0 = A_2 = A_6 = A_8 = 1$ and $A_4 = 12$. Computing $A_i^\perp$, one finds that $A_i^\perp = A_i$, and so the MacWilliams equations do not eliminate this possibility. However, by using Theorem 7.8.6, this possibility is eliminated as Exercise 423 shows. $\qquad \blacksquare$

Exercise 423 Referring to Example 7.8.7, show that there is no $[8, 4]$ binary code with weight distribution $A_0 = A_2 = A_6 = A_8 = 1$ and $A_4 = 12$. $\qquad \blacklozenge$

Corollary 7.8.8 *If $\mathcal{C}$ is an $[n, k]$ even binary code with 2^{k-1} doubly-even and 2^{k-1} singly-even codewords, then the hull of $\mathcal{C}$ is singly-even.*

Exercise 424 Prove Corollary 7.8.8. $\qquad \blacklozenge$

If C is a binary code with odd weight vectors, there are constraints on the number of vectors of any weight equivalent to a fixed number modulo 4; see [272].

7.9 Weight preserving transformations

As codes are vector spaces, it is natural to study the linear transformations that map one code to another. An arbitrary linear transformation may send a codeword of one weight to a codeword of a different weight. As we have seen throughout this book, the weights of codewords play a significant role in the study of the codes and in their error-correcting capability. Thus if we are to consider two codes to be "the same," we would want to find a linear transformation between the two codes that preserves weights; such a map is called a *weight preserving linear transformation*. In this section, we will prove that such linear transformations must be monomial maps. This result, first proved by MacWilliams in her Ph.D. thesis [212], is the basis for our definition of equivalent codes in Chapter 1. The proof we present is found in [24]; a proof using abelian group characters can be found in [347].

Let C be an $[n, k]$ code over $\mathbb{F}_q$. Define

$$\mu(k) = \frac{q^k - 1}{q - 1}.$$

By Exercise 425, the vector space $\mathbb{F}_q^k$ has $\mu(k)$ 1-dimensional subspaces. Let $V_1, V_2, \ldots, V_{\mu(k)}$ be these subspaces, and let $\mathbf{v}_i$ be a nonzero vector in V_i; so $\{\mathbf{v}_i\}$ is a basis of V_i. The nonzero columns of a generator matrix for C are scalar multiples of some of the $\mathbf{v}_i$s (of course, viewed as column vectors). Recall from Section 7.8 that V_i is orthogonal to V_j, denoted $V_i \perp V_j$, provided V_i and V_j are orthogonal under the ordinary inner product; this is clearly equivalent to $\mathbf{v}_i \cdot \mathbf{v}_j = 0$. Now define a $\mu(k) \times \mu(k)$ matrix $A = [a_{i,j}]$ that describes the orthogonality relationship between the 1-dimensional spaces:

$$a_{i,j} = \begin{cases} 0 & \text{if } V_i \perp V_j, \\ 1 & \text{if } V_i \not\perp V_j. \end{cases}$$

Exercise 425 Prove that there are $(q^k - 1)/(q - 1)$ 1-dimensional subspaces of $\mathbb{F}_q^k$. ◆

We will first show that A is invertible over the rational numbers $\mathbb{Q}$. In order to do this, we need the following lemma.

Lemma 7.9.1 *Let A be the matrix described above.*

(i) *The sum of the rows of A is the vector $\mathbf{x} = x_1 \cdots x_{\mu(k)}$, with $x_i = \mu(k) - \mu(k - 1)$ for all i.*

(ii) *For $1 \leq j \leq \mu(k)$, let $\mathbf{y}^{(j)} = y_1^{(j)} \cdots y_{\mu(k)}^{(j)}$ be the sum of all of the rows of A having 0 in column j. Then $y_j^{(j)} = 0$ and $y_i^{(j)} = \mu(k - 1) - \mu(k - 2)$ for all $i \neq j$.*

Proof: To prove (i) it suffices to show that each column of A has $\mu(k) - \mu(k - 1)$ entries that equal 1; this is equivalent to showing that, for $1 \leq i \leq \mu(k)$, there are $\mu(k - 1)$ subspaces V_j orthogonal to V_i. Let $f_i : \mathbb{F}_q^k \to \mathbb{F}_q$, where $f_i(\mathbf{u}) = \mathbf{v}_i \cdot \mathbf{u}$. By Exercise 426(a), f_i is a surjective linear transformation. Therefore the dimension of the kernel of f_i is $k - 1$. Thus by Exercise 425, there are precisely $\mu(k-1)$ of the V_j in the kernel of f_i proving that there are $\mu(k-1)$ subspaces V_j orthogonal to V_i.

By (i), there are $\mu(k-1)$ rows of A that have 0 in column j. For part (ii) it therefore suffices to show that, for each $i \neq j$, there are $\mu(k-2)$ 1-dimensional subspaces V_m with $V_m \perp V_i$ and $V_m \perp V_j$. Let $f_{i,j} : \mathbb{F}_q^k \to \mathbb{F}_q^2$, where $f_{i,j}(\mathbf{u}) = (\mathbf{v}_i \cdot \mathbf{u}, \mathbf{v}_j \cdot \mathbf{u})$. By Exercise 426(b), $f_{i,j}$ is a surjective linear transformation whose kernel has dimension $k-2$. By Exercise 425, there are precisely $\mu(k-2)$ of the V_m in the kernel of $f_{i,j}$, as required. $\square$

Exercise 426 Let $\mathbf{v}$ and $\mathbf{w}$ be (nonzero) independent vectors in $\mathbb{F}_q^k$.
(a) Let $f : \mathbb{F}_q^k \to \mathbb{F}_q$, where $f(\mathbf{u}) = \mathbf{v} \cdot \mathbf{u}$. Show that f is a surjective linear transformation.
(b) Let $g : \mathbb{F}_q^k \to \mathbb{F}_q^2$, where $g(\mathbf{u}) = (\mathbf{v} \cdot \mathbf{u}, \mathbf{w} \cdot \mathbf{u})$. Show that g is a surjective linear transformation.
(See Exercise 419 for a comparable result.)   ♦

Theorem 7.9.2 *The matrix A is invertible over $\mathbb{Q}$.*

Proof: It suffices to show that A has rank $\mu(k)$ over $\mathbb{Q}$. This is accomplished if we show that by taking some linear combination of rows of A we can obtain all of the standard basis vectors $\mathbf{e}^{(j)}$, where $\mathbf{e}^{(j)}$ has a 1 in position j and 0 elsewhere. But in the notation of Lemma 7.9.1,

$$\mathbf{e}^{(j)} = \frac{1}{\mu(k) - \mu(k-1)}\mathbf{x} - \frac{1}{\mu(k-1) - \mu(k-2)}\mathbf{y}^{(j)}.$$

$\square$

Fix a generator matrix G of $\mathcal{C}$. Denote the columns of G by $\mathrm{col}(G)$. Define the linear transformation $f : \mathbb{F}_q^k \to \mathcal{C}$ by $f(\mathbf{u}) = \mathbf{u}G$. (This is the standard way of encoding the message $\mathbf{u}$ using G as described in Section 1.11.) We define a vector $\mathbf{w} = w_1 \cdots w_{\mu(k)}$ with integer entries:

$$w_i = |\{\mathbf{c} \in \mathrm{col}(G) \mid \mathbf{c} \neq \mathbf{0}, \mathbf{c}^\mathsf{T} \in V_i\}|.$$

Thus w_i tells how many nonzero columns of G are in V_i, that is, how many columns of G are nonzero scalar multiples of $\mathbf{v}_i$. So there are $n - \sum_{i=1}^{\mu(k)} w_i$ zero columns of G. By knowing $\mathbf{w}$ we can reproduce G up to permutation and scaling of columns. Notice that as the vectors $\mathbf{v}_1, \ldots, \mathbf{v}_{\mu(k)}$ consist of all the basis vectors of the 1-dimensional subspaces of $\mathbb{F}_q^k$, then $f(\mathbf{v}_1), \ldots, f(\mathbf{v}_{\mu(k)})$ consist of all the basis vectors of the 1-dimensional subspaces of $\mathcal{C}$, as G has rank k and $\mathcal{C}$ is k-dimensional. The weights of these codewords $f(\mathbf{v}_i)$ are determined by $A\mathbf{w}^\mathsf{T}$.

Theorem 7.9.3 *For $1 \leq i \leq \mu(k)$, $(A\mathbf{w}^\mathsf{T})_i = \mathrm{wt}(f(\mathbf{v}_i))$.*

Proof: We have

$$(A\mathbf{w}^\mathsf{T})_i = \sum_{j=1}^{\mu(k)} a_{i,j}w_j = \sum_{j \in S_i} w_j,$$

where $S_i = \{j \mid V_j \not\perp V_i\}$. By definition of w_j,

$$(A\mathbf{w}^\mathsf{T})_i = \sum_{j \in S_i} |\{\mathbf{c} \in \mathrm{col}(G) \mid \mathbf{c} \neq \mathbf{0}, \mathbf{c}^\mathsf{T} \in V_j\}|$$

$$= \sum_{j=1}^{\mu(k)} |\{\mathbf{c} \in \mathrm{col}(G) \mid \mathbf{c}^\mathsf{T} \in V_j \text{ and } \mathbf{v}_i \cdot \mathbf{c}^\mathsf{T} \neq 0\}|,$$

because $\mathbf{c}^\mathsf{T} \in V_j$ and $\mathbf{v}_i \cdot \mathbf{c}^\mathsf{T} \neq 0$ if and only if $\mathbf{c} \neq \mathbf{0}$, $\mathbf{c}^\mathsf{T} \in V_j$, and $V_i \not\perp V_j$. But

$$\mathrm{wt}(f(\mathbf{v}_i)) = \mathrm{wt}(\mathbf{v}_i G) = |\{\mathbf{c} \in \mathrm{col}(G) \mid \mathbf{v}_i \cdot \mathbf{c}^\mathsf{T} \neq 0\}|$$

$$= \sum_{j=1}^{\mu(k)} |\{\mathbf{c} \in \mathrm{col}(G) \mid \mathbf{c}^\mathsf{T} \in V_j \text{ and } \mathbf{v}_i \cdot \mathbf{c}^\mathsf{T} \neq 0\}|$$

as the transpose of every column of G is in some V_j. The result follows. $\qquad\square$

This result shows that the weight of every nonzero codeword, up to scalar multiplication, is listed as some component of $A\mathbf{w}^\mathsf{T}$. Now let $\mathcal{C}'$ be another $[n, k]$ code over $\mathbb{F}_q$ such that there is a surjective linear transformation $\phi : \mathcal{C} \to \mathcal{C}'$ that preserves weights, that is, $\mathrm{wt}(\phi(\mathbf{c})) = \mathrm{wt}(\mathbf{c})$ for all $\mathbf{c} \in \mathcal{C}$. Let $g : \mathbb{F}_q^k \to \mathcal{C}'$ be the composition $\phi \circ f$ so that $g(\mathbf{u}) = \phi(\mathbf{u}G)$ for all $\mathbf{u} \in \mathbb{F}_q^k$. As ϕ and f are linear, so is g. If we let G' be the matrix for g relative to the standard bases for $\mathbb{F}_q^k$ and $\mathbb{F}_q^n$, we have $g(\mathbf{u}) = \mathbf{u}G'$ and G' is a generator matrix for $\mathcal{C}'$; also g is the standard encoding function for the message $\mathbf{u}$ using G'. We can now prove our main result.

Theorem 7.9.4 *There is a weight preserving linear transformation between $[n, k]$ codes $\mathcal{C}$ and $\mathcal{C}'$ over $\mathbb{F}_q$ if and only if $\mathcal{C}$ and $\mathcal{C}'$ are monomially equivalent. Furthermore, the linear transformation agrees with the associated monomial transformation on every codeword in $\mathcal{C}$.*

Proof: If $\mathcal{C}$ and $\mathcal{C}'$ are monomially equivalent, the linear transformation determined by the monomial map preserves weights. For the converse, we use the notation developed up to this point. We first note that for $1 \leq i \leq \mu(k)$,

$$\mathrm{wt}(g(\mathbf{v}_i)) = \mathrm{wt}(\phi(f(\mathbf{v}_i))) = \mathrm{wt}(f(\mathbf{v}_i)) \tag{7.16}$$

as ϕ is weight preserving. Let $\mathbf{w}' = w_1' \cdots w_{\mu(k)}'$ be defined analogously to $\mathbf{w}$ by

$$w_i' = |\{\mathbf{c} \in \mathrm{col}(G') \mid \mathbf{c} \neq \mathbf{0}, \mathbf{c}^\mathsf{T} \in V_i\}|.$$

By Theorem 7.9.3 used twice and (7.16),

$$(A\mathbf{w}^\mathsf{T})_i = \mathrm{wt}(f(\mathbf{v}_i)) = \mathrm{wt}(g(\mathbf{v}_i)) = (A\mathbf{w}'^\mathsf{T})_i.$$

Thus $A\mathbf{w}^\mathsf{T} = A\mathbf{w}'^\mathsf{T}$ and hence $\mathbf{w} = \mathbf{w}'$ by Theorem 7.9.2. Since $\mathbf{w}$ and $\mathbf{w}'$ list the number of nonzero columns of G and G', respectively, that are in each V_i, the columns of G' are simply rearrangements of the columns of G upon possible rescaling. Thus there is a monomial matrix M with $G' = GM$, and $\mathcal{C}$ and $\mathcal{C}'$ are therefore monomially equivalent. Every codeword of $\mathcal{C}$ can be written as $\mathbf{u}G$ for some message $\mathbf{u} \in \mathbb{F}_q^k$. But $\phi(\mathbf{u}G) = \mathbf{u}G'$. Since $G' = GM$, we have $\phi(\mathbf{u}G) = \mathbf{u}GM$ and so the linear transformation ϕ and the monomial map M agree on every codeword of $\mathcal{C}$. $\qquad\square$

A consequence of the proof of Theorem 7.9.4 is the following result on constant weight codes due to Bonisoli [25]; the idea of this proof comes from Ward [347]. A linear code is *constant weight* if all nonzero codewords have the same weight. Examples of such codes are the simplex codes over $\mathbb{F}_q$ by Theorem 2.7.5.

Theorem 7.9.5 *Let C be an $[n, k]$ linear code over $\mathbb{F}_q$ with all nonzero codewords of the same weight. Assume that C is nonzero and no column of a generator matrix is identically zero. Then C is equivalent to the r-fold replication of a simplex code.*

Proof: We use the notation of this section. Let G be a generator matrix for C. Choose a column $\mathbf{c}^{\mathrm{T}}$ of G such that this column or any scalar multiple of this column occur the maximum number of times r as columns of G. Let $\mathbf{x}$ be an arbitrary nonzero element of $\mathbb{F}_q^k$. Then there exists a nonsingular matrix B such that $B\mathbf{c}^{\mathrm{T}} = \mathbf{x}^{\mathrm{T}}$. Thus BG is a $k \times n$ matrix that contains the column $\mathbf{x}^{\mathrm{T}}$, or scalar multiples of this column, at least r times. Note that $BG = G'$ is another generator matrix for C, and so B induces a nonsingular linear transformation of C. As all nonzero codewords of C have the same weight, this linear transformation is weight preserving. By the proof of Theorem 7.9.4, there is a monomial matrix M such that $G' = GM$. As G' contains the column $\mathbf{x}^{\mathrm{T}}$ or nonzero scalar multiples of this column at least r times and $G' = GM$ is merely the same matrix as G with its columns rearranged and rescaled, G has r or more columns that are nonzero scalar multiples of $\mathbf{x}^{\mathrm{T}}$. By maximality of r and the fact that $\mathbf{x}$ is arbitrary, every nonzero vector in $\mathbb{F}_q^k$ together with its nonzero scalar multiples occurs exactly r times as columns of G. But this means that C is the r-fold replication of the simplex code.   □

Example 7.9.6 Let C be the constant weight binary code with generator matrix

$$G = \begin{bmatrix} 0 & 0 & 0 & 1 & 1 & 1 & 1 \\ 0 & 1 & 1 & 0 & 0 & 1 & 1 \\ 1 & 0 & 1 & 0 & 1 & 0 & 1 \end{bmatrix};$$

this is the simplex code $\mathcal{S}_3$ of Section 1.8. The subcode generated by the second and third vector in the above matrix is also constant weight. Puncturing on its zero coordinate, we obtain the constant weight code C_1 with generator matrix

$$G_1 = \begin{bmatrix} 0 & 1 & 1 & 0 & 1 & 1 \\ 1 & 0 & 1 & 1 & 0 & 1 \end{bmatrix}.$$

This is obviously the 2-fold replication of the simplex code $\mathcal{S}_2$, which has generator matrix

$$G_2 = \begin{bmatrix} 0 & 1 & 1 \\ 1 & 0 & 1 \end{bmatrix}.$$ ■

7.10   Generalized Hamming weights

There is an increasing sequence of positive integers associated with a code called generalized Hamming weights which in many ways behave both like minimum weights of codes and weight distributions. This sequence, which includes the minimum weight of the code, satisfies both Generalized Singleton and Generalized Griesmer Bounds. As with weight distributions, the generalized Hamming weights for the dual code can be determined from those of the original code; furthermore, the generalized Hamming weights satisfy MacWilliams type equations.

Let C be an $[n, k, d]$ code over $\mathbb{F}_q$. If D is a subcode of C, the *support* supp(D) of D is the set of coordinates where not all codewords of D are zero. So $|\text{supp}(D)|$ is the number of nonzero columns in a generator matrix for D. For $1 \le r \le k$, the *rth-generalized Hamming weight of* C, which is also called the *rth-minimum support weight*, is

$$d_r(C) = d_r = \min\{|\text{supp}(D)| \mid D \text{ is an } [n, r] \text{ subcode of } C\}.$$

The set $\{d_1(C), d_2(C), \ldots, d_k(C)\}$ is called the *weight hierarchy of* C. The generalized Hamming weights were introduced in 1977 by Helleseth, Kløve, and Mykkeltveit [125]. Victor Wei [348] studied them in connection with wire-tap channels of type II; see Section 5 of [329]. They are also important in the study of the trellis structure of codes as described in Section 5 of [336].

The following theorem contains basic information about generalized Hamming weights.

Theorem 7.10.1 *Let C be an $[n, k, d]$ code over $\mathbb{F}_q$. The following hold:*
(i) $d = d_1(C) < d_2(C) < \cdots < d_k(C) \le n$.
(ii) *If M is an $n \times n$ monomial map and γ is an automorphism of $\mathbb{F}_q$, then the weight hierarchy of C is the same as the weight hierarchy of $CM\gamma$.*

Proof: Suppose that D is an $[n, r]$ subcode of C with $d_r(C) = |\text{supp}(D)|$. Choose one of the $d_r(C)$ nonzero columns of a generator matrix G_r of D and perform row operations on G_r, obtaining a new generator matrix G'_r of D, so that this column has a single nonzero entry in the first row. The subcode of D generated by rows 2 through r of G'_r is an $[n, r - 1]$ code with support size strictly less than $d_r(C)$. Thus $d_{r-1}(C) < d_r(C)$. By choosing a vector in C of minimum weight d to generate an $[n, 1]$ subcode of C, we see that $d = d_1(C)$; thus (i) follows.

In (ii), since applying $M\gamma$ rescales columns, permutes coordinates, and acts componentwise by γ, $|\text{supp}(D)| = |\text{supp}(DM\gamma)|$. Part (ii) follows. $\qquad \square$

Example 7.10.2 The $[15, 4, 8]$ binary simplex code C has generator matrix

$$G = \begin{bmatrix} 0 & 0 & 0 & 0 & 0 & 0 & 0 & 1 & 1 & 1 & 1 & 1 & 1 & 1 & 1 \\ 0 & 0 & 0 & 1 & 1 & 1 & 1 & 0 & 0 & 0 & 0 & 1 & 1 & 1 & 1 \\ 0 & 1 & 1 & 0 & 0 & 1 & 1 & 0 & 0 & 1 & 1 & 0 & 0 & 1 & 1 \\ 1 & 0 & 1 & 0 & 1 & 0 & 1 & 0 & 1 & 0 & 1 & 0 & 1 & 0 & 1 \end{bmatrix}.$$

By Theorem 2.7.5, all nonzero codewords have weight 8. The generalized Hamming weights are $d_1 = d = 8$, $d_2 = 12$, $d_3 = 14$, and $d_4 = 15$. To compute these values, first observe that because the sum of two distinct weight 8 codewords has weight 8, two weight 8 codewords have exactly four 1s in common positions. Therefore the support of a 2-dimensional subcode must be of size at least 12, and the first two rows of G generate a 2-dimensional code with support size 12. As G has no zero columns, $d_4 = 15$, and so by Theorem 7.10.1 $d_3 = 13$ or 14. Suppose that D is a 3-dimensional subcode with support size 13. Let $\mathcal{M}$ be a matrix whose rows are the codewords of D. In any given nonzero column, by Exercise 373, exactly half the rows of $\mathcal{M}$ have a 1 in that column. So $\mathcal{M}$ contains $13 \cdot 4 = 52$ 1s; but $\mathcal{M}$ has seven rows of weight 8, implying that $\mathcal{M}$ has 56 1s, a contradiction. So $d_3 = 14$.   ∎

There is a relationship between the weight hierarchy of C and the weight hierarchy of $C^\perp$ due to Wei [348]. First we prove a lemma.

Lemma 7.10.3 *Let C be an $[n, k]$ code over $\mathbb{F}_q$. For a positive integer $s < n$, let r be the largest integer such that C has a generator matrix G of the form*

$$G = \begin{bmatrix} G_1 & O \\ G_2 & G_3 \end{bmatrix},$$

where G_1 is an $r \times s$ matrix of rank r and G_3 is a $(k - r) \times (n - s)$ matrix. Then C has a parity check matrix of the form

$$H = \begin{bmatrix} H_1 & H_2 \\ O & H_3 \end{bmatrix},$$

where H_1 is a $(s - r) \times s$ matrix of rank $s - r$ and H_3 is an $(n - k - s + r) \times (n - s)$ matrix of rank $n - k - s + r$. Furthermore, $n - k - s + r$ is the largest dimension of a subspace of $C^\perp$ with support contained in the last $n - s$ coordinates.

Proof: Let $H = [A \;\; B]$, where A is an $(n - k) \times s$ matrix and B is an $(n - k) \times (n - s)$ matrix. Since the rows of H are orthogonal to the rows of G, the rows of A are orthogonal to the rows of G_1. Hence A has rank at most $s - r$. Suppose that the rank of A is less than $s - r$. Then there is a vector of length s not in the code generated by G_1 that is orthogonal to the rows of A. Appending $n - s$ zeros to the end of this vector, we obtain a vector orthogonal to the rows of H, which therefore is in C but is not in the code generated by $[G_1 \;\; O]$, contradicting the maximality of r. Thus the rank of A equals $s - r$ and hence C has a parity check matrix H satisfying the conclusions of the lemma.   □

Theorem 7.10.4 *Let C be an $[n, k]$ code over $\mathbb{F}_q$. Then*

$$\{d_r(C) \mid 1 \leq r \leq k\} = \{1, 2, \ldots, n\} \setminus \{n + 1 - d_r(C^\perp) \mid 1 \leq r \leq n - k\}.$$

Proof: Let $s = d_r(C^\perp)$ for some r with $1 \leq r \leq n - k$. It suffices to show that there does not exist a t with $1 \leq t \leq k$ such that $d_t(C) = n + 1 - s$.

As $s = d_r(C^\perp)$, there is a set of s coordinates that supports an r-dimensional subcode of $C^\perp$. Reorder columns so that these are the first s coordinates. Thus $C^\perp$ has an $(n - k) \times n$ generator matrix of the form

$$H = \begin{bmatrix} H_1 & O \\ H_2 & H_3 \end{bmatrix},$$

where H_1 is an $r \times s$ matrix of rank r. As $d_r(C^\perp) < d_{r+1}(C^\perp)$ by Theorem 7.10.1, no larger subcode has support in these s coordinates. Applying Lemma 7.10.3 with $C^\perp$ in place of C and $n - k$ in place of k, there is a $(k - s + r)$-dimensional subcode of C which is zero on the first s coordinate positions. Hence

$$d_{k-s+r}(C) \leq n - s. \tag{7.17}$$

Assume to the contrary that there is a t with $d_t(C) = n + 1 - s$. It follows from (7.17) that

$$t > k - s + r. \tag{7.18}$$

Replacing C by a permutation equivalent code, if necessary, C has a $k \times n$ generator matrix

$$G = \begin{bmatrix} G_1 & O \\ G_2 & G_3 \end{bmatrix},$$

where G_1 is a $t \times (n + 1 - s)$ matrix of rank t. Again as $d_t(C) < d_{t+1}(C)$ Lemma 7.10.3 applies with t in place of r and $n + 1 - s$ in place of s. So there is an $(s - 1 - k + t)$-dimensional subcode of $C^\perp$ that is zero on the first $n + 1 - s$ positions. Since $s = d_r(C^\perp)$, we have $s - 1 - k + t < r$ and this contradicts (7.18). □

We now illustrate how Theorem 7.10.4 can be used to compute the weight hierarchy of self-dual codes.

Example 7.10.5 Let C be a $[24, 12, 8]$ self-dual binary code. (This must be the extended binary Golay code.) We compute $\{d_r(C) \mid 1 \le r \le 12\}$ using the self-duality of C and Theorem 7.10.4. Since $d_1(C) = 8$, we have $1, 2, \ldots, 7 \notin \{d_r(C) \mid 1 \le r \le 12\}$ by Theorem 7.10.1, and thus $24, 23, \ldots, 18 \in \{d_r(C^\perp) \mid 1 \le r \le 12\}$ and $17 \notin \{d_r(C^\perp) \mid 1 \le r \le 12\}$. It is easy to see that a 2-dimensional self-orthogonal binary code of minimum weight 8 cannot exist for length less than 12. Therefore $d_2(C) \ge 12$ and $9, 10, 11 \notin \{d_r(C) \mid 1 \le r \le 12\}$. Thus $16, 15, 14 \in \{d_r(C^\perp) \mid 1 \le r \le 12\}$. As $C = C^\perp$, we now conclude that $\{8, 14, 15, 16, 18, 19, 20, 21, 22, 23, 24\}$ is contained in $\{d_r(C) \mid 1 \le r \le 12\}$ and $12 \le d_2(C) \le 13$. An easy argument shows that a $[13, 2, 8]$ self-orthogonal binary code must have a column of zeros in its generator matrix. Hence $d_2(C) = 12$. So we have:

r	1	2	3	4	5	6	7	8	9	10	11	12
$d_r(C)$	8	12	14	15	16	18	19	20	21	22	23	24

■

Exercise 427 Do the following:
(a) Show that a 2-dimensional self-orthogonal binary code of minimum weight 8 cannot exist for length less than 12.
(b) Show that a $[13, 2, 8]$ self-orthogonal binary code must have a column of zeros in its generator matrix. ◆

Exercise 428 Find the weight hierarchy of the $[16, 8, 4]$ self-dual doubly-even binary code with generator matrix G_1 of Example 7.5.6. ◆

Exercise 429 Find the weight hierarchy of the $[16, 8, 4]$ self-dual doubly-even binary code with generator matrix G_2 of Example 7.5.6. ◆

Exercise 430 What are the weight hierarchies of the $[7, 3, 4]$ binary simplex code and the $[7, 4, 3]$ binary Hamming code? ◆

Wei [348] observed that the minimum support weights generalize the Singleton Bound.

Theorem 7.10.6 (Generalized Singleton Bound) *For an $[n, k, d]$ linear code over $\mathbb{F}_q$, $d_r \le n - k + r$ for $1 \le r \le k$.*

Proof: The proof follows by induction on $k - r$. When $k - r = 0$, $d_r = d_k \le n = n - k + r$ by Theorem 7.10.1. Assuming $d_r \le n - k + r$ for some $r \le k$, then by the same theorem, $d_{r-1} \le d_r - 1 \le n - k + (r - 1)$, yielding the result.   □

The Singleton Bound is the case $r = 1$ of the Generalized Singleton Bound. It follows from the proof of the Generalized Singleton Bound that if a code meets the Singleton Bound, then $d_{r-1} = d_r - 1$ for $1 < r \le k$ and $d_k = n$. Codes meeting the Singleton Bound are MDS codes by definition; so MDS codes also meet the Generalized Singleton Bound and thus their generalized Hamming weights are determined.

Theorem 7.10.7 *Let C be an MDS code over $\mathbb{F}_q$. Then:*
(i) *C meets the Generalized Singleton Bound for all r with $1 \le r \le k$, and*
(ii) *$d_r = d + r - 1$ for $1 \le r \le k$.*

We now investigate a generalization of the Griesmer Bound [124, 128] that the minimum support weights satisfy. Let G be a generator matrix for a code C. Let $\mathrm{col}(G)$ be the set of distinct columns of G. If $\mathbf{x} \in \mathrm{col}(G)$, define $m(\mathbf{x})$ to be the multiplicity of $\mathbf{x}$ in G, and if $U \subseteq \mathrm{col}(G)$, define

$$m(U) = \sum_{\mathbf{x} \in U} m(\mathbf{x}).$$

For $U \subseteq \mathrm{col}(G)$, let $\mathrm{span}\{U\}$ be the subspace of $\mathbb{F}_q^k$ spanned by U. Finally, for $1 \le r \le k$, let $\mathcal{F}_{k,r}$ be the set of all r-dimensional subspaces of $\mathbb{F}_q^k$ spanned by columns of G. Before stating the Generalized Griesmer Bound, we need two lemmas. The first lemma gives a formula for $d_r(C)$ in terms of the function $m(U)$.

Lemma 7.10.8 *Let C be an $[n, k]$ code with generator matrix G. For $1 \le r \le k$,*

$$d_r(C) = n - \max\{m(U) \mid U \subseteq \mathrm{col}(G) \text{ and } \mathrm{span}\{U\} \in \mathcal{F}_{k,k-r}\}.$$

Proof: Let $\mathcal{D}$ be an $[n, r]$ subcode of C. Then there exists an $r \times k$ matrix A of rank r such that AG is a generator matrix of $\mathcal{D}$. Moreover, for each such matrix A, AG is a generator matrix of an $[n, r]$ subcode of C. By definition, $|\mathrm{supp}(\mathcal{D})| = n - m$, where m is the number of zero columns of AG. Hence $|\mathrm{supp}(\mathcal{D})| = n - m(U)$, where

$$U = \{\mathbf{y} \in \mathrm{col}(G) \mid A\mathbf{y} = \mathbf{0}\}.$$

Since the rank of A is r and the rank of G is k, $\mathrm{span}\{U\}$ is in $\mathcal{F}_{k,k-r}$. Conversely, if $U \subseteq \mathrm{col}(G)$ where $\mathrm{span}\{U\}$ is in $\mathcal{F}_{k,k-r}$, there is an $r \times k$ matrix A of rank r such that $A\mathbf{y} = \mathbf{0}$ if and only if $\mathbf{y} \in \mathrm{span}\{U\}$. But AG is a generator matrix for an $[n, r]$ subcode $\mathcal{D}$ of C. Thus as $d_r(C)$ is the minimum support size of any such $\mathcal{D}$, the result follows.   □

Using this lemma and its proof, we can relate $d_r(C)$ and $d_{r-1}(C)$.

Lemma 7.10.9 *Let C be an $[n, k]$ code over $\mathbb{F}_q$. Then for $1 < r \le k$,*

$$(q^r - 1)d_{r-1}(C) \le (q^r - q)d_r(C).$$

Proof: Let G be a generator matrix for C. By Lemma 7.10.8 and its proof, there exists an $r \times k$ matrix A of rank r such that $U = \{\mathbf{y} \in \mathrm{col}(G) \mid A\mathbf{y} = \mathbf{0}\}$ where $\mathrm{span}\{U\} \in \mathcal{F}_{k,k-r}$ and $d_r(C) = n - m(U)$. Furthermore, AG generates an r-dimensional subcode $\mathcal{D}$ of C. By Exercise 431 there are $t = (q^r - 1)/(q - 1)\,(r - 1)$-dimensional subcodes $V_1, V_2, \ldots, V_t$ of $\mathcal{D}$. To each V_i is associated an $(r - 1) \times k$ matrix A_i of rank $r - 1$ such that $A_i G$ is a generator matrix of V_i. Letting $U_i = \{\mathbf{y} \in \mathrm{col}(G) \mid A_i\mathbf{y} = \mathbf{0}\}$, we see that $\mathrm{span}\{U_i\} \in \mathcal{F}_{k,k-r+1}$ as in the proof of Lemma 7.10.8. Also $U \subseteq U_i$, as a column of AG is $\mathbf{0}$ implying that the same column of $A_i G$ is $\mathbf{0}$ because V_i is a subcode of $\mathcal{D}$. Conversely, if $U' \subseteq \mathrm{col}(G)$ where $U \subseteq U'$ and $\mathrm{span}\{U'\} \in \mathcal{F}_{k,k-r+1}$, there is an $(r - 1) \times k$ matrix A' of rank $r - 1$ such that $U' = \{\mathbf{y} \in \mathrm{col}(G) \mid A'\mathbf{y} = \mathbf{0}\}$. As $U \subseteq U'$, $\mathcal{D} = \mathrm{span}\{U\} \subseteq \mathrm{span}\{U'\}$. So $A'G$ generates an $(r - 1)$-dimensional subspace, say V_i, of $\mathcal{D}$ and hence $U' = U_i$.

By Lemma 7.10.8,

$$d_{r-1}(C) \leq n - m(U_i) \quad \text{for } 1 \leq i \leq t,$$

and thus

$$d_r(C) - d_{r-1}(C) \geq n - m(U) - [n - m(U_i)] = m(U_i \setminus U) \qquad \text{for } 1 \leq i \leq t.$$

Every column of G not in U is in exactly one U_i because U together with this column, plus possibly other columns, spans a subspace of $\mathcal{F}_{k,k-r+1}$ by the previous paragraph. Therefore

$$t[d_r(C) - d_{r-1}(C)] \geq \sum_{i=1}^{t} m(U_i \setminus U) = n - m(U) = d_r(C),$$

and the lemma follows. $\qquad \square$

Exercise 431 Prove that there are $t = (q^r - 1)/(q - 1)\,(r - 1)$-dimensional subcodes of an r-dimensional code over $\mathbb{F}_q$. $\qquad \blacklozenge$

Theorem 7.10.10 (Generalized Griesmer Bound) *Let C be an $[n, k]$ code over $\mathbb{F}_q$. Then for $1 \leq r \leq k$,*

$$n \geq d_r(C) + \sum_{i=1}^{k-r} \left\lceil \frac{q-1}{q^i(q^r - 1)} d_r(C) \right\rceil.$$

Proof: If $r = k$, the inequality reduces to the obvious assertion $n \geq d_k(C)$; see Theorem 7.10.1. Now assume that $r < k$. Without loss of generality, using Theorem 7.10.1(ii), we may assume that C has a generator matrix G of the form

$$G = \begin{bmatrix} G_1 & O \\ G_2 & G_3 \end{bmatrix},$$

where G_1 is an $r \times d_r(C)$ matrix of rank r and G_3 is a $(k - r) \times (n - d_r(C))$ matrix. The matrix $[G_2 \ G_3]$ has rank $k - r$. If G_3 has rank less than $k - r$, there is a nonzero codeword $\mathbf{x}$ in the code generated by $[G_2 \ G_3]$, which is zero on the last $n - d_r(C)$ coordinates, contradicting $d_r(C) < d_{r+1}(C)$ from Theorem 7.10.1. Therefore G_3 has rank $k - r$ and generates an $[n - d_r(C), k - r]$ code C_3.

Let $\mathbf{c} = c_1 \cdots c_n \in C$ where $\mathrm{wt}(c_{d_r(C)+1} \cdots c_n) = a > 0$. The subcode of C generated by $\mathbf{c}$ and the first r rows of G is an $(r + 1)$-dimensional subcode of C with support size

$a + d_r(\mathcal{C})$. So $a \geq d_{r+1}(\mathcal{C}) - d_r(\mathcal{C})$ implying that the minimum weight of $\mathcal{C}_3$ is at least $d_{r+1}(\mathcal{C}) - d_r(\mathcal{C})$, by choosing the above $\mathbf{c}$ so that $\text{wt}(c_{d_r(\mathcal{C})+1} \cdots c_n)$ has minimum weight in $\mathcal{C}_3$. By the Griesmer Bound applied to $\mathcal{C}_3$,

$$n - d_r(\mathcal{C}) \geq \sum_{i=0}^{k-r-1} \left\lceil \frac{d_{r+1}(\mathcal{C}) - d_r(\mathcal{C})}{q^i} \right\rceil = \sum_{i=1}^{k-r} \left\lceil \frac{d_{r+1}(\mathcal{C}) - d_r(\mathcal{C})}{q^{i-1}} \right\rceil.$$

By Lemma 7.10.9, $(q^{r+1} - 1)d_r(\mathcal{C}) \leq (q^{r+1} - q)d_{r+1}(\mathcal{C})$. Using this,

$$d_{r+1}(\mathcal{C}) - d_r(\mathcal{C}) \geq \left(1 - \frac{q^{r+1} - q}{q^{r+1} - 1}\right) d_{r+1}(\mathcal{C}) = \frac{q-1}{q^{r+1} - 1} d_{r+1}(\mathcal{C}).$$

Therefore,

$$n - d_r(\mathcal{C}) \geq \sum_{i=1}^{k-r} \left\lceil \frac{q-1}{q^{i-1}(q^{r+1} - 1)} d_{r+1}(\mathcal{C}) \right\rceil.$$

But again using Lemma 7.10.9,

$$n - d_r(\mathcal{C}) \geq \sum_{i=1}^{k-r} \left\lceil \frac{q-1}{q^{i-1}(q^{r+1} - q)} d_r(\mathcal{C}) \right\rceil = \sum_{i=1}^{k-r} \left\lceil \frac{q-1}{q^i(q^r - 1)} d_r(\mathcal{C}) \right\rceil. \qquad \square$$

The Griesmer Bound is the case $r = 1$ of the Generalized Griesmer Bound. We now show that if a code meets the Griesmer Bound, it also meets the Generalized Griesmer Bound for all r, and its weight hierarchy is uniquely determined; the binary case of this result is found in [128]. To simplify the notation we let

$$b_r = d_r(\mathcal{C}) + \sum_{i=1}^{k-r} \left\lceil \frac{q-1}{q^i(q^r - 1)} d_r(\mathcal{C}) \right\rceil \quad \text{for } 1 \leq r \leq k.$$

The Generalized Griesmer Bound then asserts that $n \geq b_r$ for $1 \leq r \leq k$. We first show that $b_{r+1} \geq b_r$ and determine when $b_{r+1} = b_r$.

Lemma 7.10.11 *Let* $1 \leq r < k$. *Then* $b_{r+1} \geq b_r$. *Furthermore,* $b_{r+1} = b_r$ *if and only if both of the following hold*:
(i) $d_{r+1}(\mathcal{C}) = \lceil (q^{r+1} - 1)d_r(\mathcal{C})/(q^{r+1} - q) \rceil$, *and*
(ii) $\lceil (q-1)d_{r+1}(\mathcal{C})/(q^i(q^{r+1} - 1)) \rceil = \lceil (q-1)d_r(\mathcal{C})/(q^{i+1}(q^r - 1)) \rceil$ *for* $1 \leq i \leq k - r - 1$.

Proof: Lemma 7.10.9 implies that

$$d_{r+1}(\mathcal{C}) \geq \left\lceil \frac{q^{r+1} - 1}{q^{r+1} - q} d_r(\mathcal{C}) \right\rceil \tag{7.19}$$

and

$$\left\lceil \frac{q-1}{q^i(q^{r+1} - 1)} d_{r+1}(\mathcal{C}) \right\rceil \geq \left\lceil \frac{q-1}{q^i(q^{r+1} - q)} d_r(\mathcal{C}) \right\rceil = \left\lceil \frac{q-1}{q^{i+1}(q^r - 1)} d_r(\mathcal{C}) \right\rceil. \tag{7.20}$$

By (7.19) and (7.20),

$$
b_{r+1} = d_{r+1}(\mathcal{C}) + \sum_{i=1}^{k-r-1} \left\lceil \frac{q-1}{q^i(q^{r+1}-1)} d_{r+1}(\mathcal{C}) \right\rceil
$$

$$
\geq \left\lceil \frac{q^{r+1}-1}{q^{r+1}-q} d_r(\mathcal{C}) \right\rceil + \sum_{i=1}^{k-r-1} \left\lceil \frac{q-1}{q^{i+1}(q^r-1)} d_r(\mathcal{C}) \right\rceil
$$

$$
= \left\lceil d_r(\mathcal{C}) + \frac{q-1}{q(q^r-1)} d_r(\mathcal{C}) \right\rceil + \sum_{i=2}^{k-r} \left\lceil \frac{q-1}{q^i(q^r-1)} d_r(\mathcal{C}) \right\rceil
$$

$$
= d_r(\mathcal{C}) + \sum_{i=1}^{k-r} \left\lceil \frac{q-1}{q^i(q^r-1)} d_r(\mathcal{C}) \right\rceil = b_r.
$$

Clearly, $b_{r+1} = b_r$ if and only if equality holds in (7.19) and in (7.20) when $1 \leq i \leq k - r - 1$. $\qquad\square$

Theorem 7.10.12 *Let $\mathcal{C}$ be an $[n, k, d]$ code over $\mathbb{F}_q$ meeting the Griesmer Bound. Then:*
(i) *$\mathcal{C}$ meets the Generalized Griesmer Bound for all r with $1 \leq r \leq k$, and*
(ii) *$d_r(\mathcal{C}) = \sum_{i=0}^{r-1} \lceil d/q^i \rceil$ for $1 \leq r \leq k$.*

Proof: By the Generalized Griesmer Bound $n \geq b_k$; using Lemma 7.10.11,

$$
n \geq b_k \geq b_{k-1} \geq \cdots \geq b_1.
$$

As $\mathcal{C}$ meets the Griesmer Bound, $b_1 = n$. Therefore $b_r = n$ for $1 \leq r \leq k$, giving (i).
Note that (ii) holds when $r = 1$, since $d_1(\mathcal{C}) = d$. Assume $1 < r \leq k$. By Lemma 7.10.11,

$$
\left\lceil \frac{q-1}{q^i(q^s-1)} d_s(\mathcal{C}) \right\rceil = \left\lceil \frac{q-1}{q^{i+1}(q^{s-1}-1)} d_{s-1}(\mathcal{C}) \right\rceil,
$$

for $1 < s \leq k$ and $1 \leq i \leq k - s$. Applying this inductively for $s = r, r-1, \ldots, 2$, we have

$$
\left\lceil \frac{q-1}{q^i(q^r-1)} d_r(\mathcal{C}) \right\rceil = \left\lceil \frac{q-1}{q^{i+r-1}(q-1)} d_1(\mathcal{C}) \right\rceil = \left\lceil \frac{d}{q^{i+r-1}} \right\rceil,
$$

for $1 \leq i \leq k - r$, as $d_1(\mathcal{C}) = d$ by Theorem 7.10.1. Since $b_r = n$, as $\mathcal{C}$ meets the Griesmer Bound, we have

$$
\sum_{i=0}^{k-1} \left\lceil \frac{d}{q^i} \right\rceil = n = d_r(\mathcal{C}) + \sum_{i=1}^{k-r} \left\lceil \frac{q-1}{q^i(q^r-1)} d_r(\mathcal{C}) \right\rceil = d_r(\mathcal{C}) + \sum_{i=1}^{k-r} \left\lceil \frac{d}{q^{i+r-1}} \right\rceil.
$$

So

$$
d_r(\mathcal{C}) = \sum_{i=0}^{k-1} \left\lceil \frac{d}{q^i} \right\rceil - \sum_{i=1}^{k-r} \left\lceil \frac{d}{q^{i+r-1}} \right\rceil = \sum_{i=0}^{r-1} \left\lceil \frac{d}{q^i} \right\rceil,
$$

and thus (ii) holds. $\qquad\square$

Example 7.10.13 By Theorem 2.7.5, the $[(q^k - 1)/(q - 1), k, q^{k-1}]$ simplex code $\mathcal{C}$ over $\mathbb{F}_q$ meets the Griesmer Bound. So by Theorem 7.10.12, the weight hierarchy of $\mathcal{C}$ is

$$d_r(\mathcal{C}) = \sum_{i=0}^{r-1} \left\lceil \frac{q^{k-1}}{q^i} \right\rceil = \sum_{i=0}^{r-1} q^{k-1-i} = q^{k-r} \frac{q^r - 1}{q - 1}.$$

∎

Exercise 432 From Example 7.10.13, what is the weight hierarchy of:
(a) the [15, 4, 8] binary simplex code (see also Example 7.10.2), and
(b) the [121, 5, 81] ternary simplex code? ◆

Exercise 433 Using the results of Exercise 432 and Theorem 7.10.4, find the weight hierarchy of:
(a) the [15, 11, 3] binary Hamming code, and
(b) the [121, 116, 3] ternary Hamming code. ◆

Exercise 434 Using the results of Example 7.10.13 and Theorem 7.10.4, find the weight hierarchy of the $[(q^k - 1)/(q - 1), (q^k - 1)/(q - 1) - k, 3]$ Hamming code $\mathcal{H}_{q,k}$ over $\mathbb{F}_q$. ◆

As remarked earlier, generalized MacWilliams equations for generalized Hamming weights have been established by Barg [14], Kløve [175], and Simonis [310]. We refer the interested reader to those papers for details.

The generalized Hamming weights have been computed for other codes. Wei [348] computed the weight hierarchy of the binary Reed–Muller codes of all orders and the Hamming codes; see Exercise 434 and Section 5 of [329]. (Note that even though the weight hierarchies of the binary Reed–Muller codes are known, the weight distributions of these codes are unknown for orders larger than 2.) The weight hierarchies are known for codes that meet the Griesmer Bound by Theorem 7.10.12; for codes that have length one more than the Griesmer Bound, the weight hierarchies are also known [127].[1]

Research Problem 7.10.14 *For families of codes such as BCH, quadratic residue, duadic, or lexicodes, find further information, either exact or asymptotic, about some or all of the generalized Hamming weights.*

---

[1] The proof in [127] is for binary codes only, but generalizes to nonbinary codes as well.

8 Designs

In this chapter we discuss some basic properties of combinatorial designs and their relationship to codes. In Section 6.5, we showed how duadic codes can lead to projective planes. Projective planes are a special case of t-designs, also called block designs, which are the main focus of this chapter. As with duadic codes and projective planes, most designs we study arise as the supports of codewords of a given weight in a code.

8.1 t-designs

A t-(v, k, λ) *design*, or briefly a t-*design*, is a pair $(\mathcal{P}, \mathcal{B})$ where $\mathcal{P}$ is a set of v elements, called *points*, and $\mathcal{B}$ is a collection of distinct subsets of $\mathcal{P}$ of size k, called *blocks*, such that every subset of points of size t is contained in precisely λ blocks. (Sometimes one considers t-designs in which the collection of blocks is a multiset, that is, blocks may be repeated. In such a case, a t-design without repeated blocks is called *simple*. We will generally only consider simple t-designs and hence, unless otherwise stated, the expression "t-design" will mean "simple t-design.") The number of blocks in $\mathcal{B}$ is denoted by b, and, as we will see shortly, is determined by the parameters t, v, k, and λ. There are several special cases of t-designs that have their own terminology:

- If $\lambda = 1$, a t-design is called a *Steiner* S(t, k, v) *system* or a *Steiner t-design*.
- If $b = v$, the t-design is *symmetric* and $k - \lambda$ is called its *order*. Nontrivial symmetric t-designs exist only for $t \leq 2$.
- A symmetric 2-$(v, k, 1)$ design (or, in the alternate notation, a symmetric S$(2, k, v)$ design) turns out to be a *projective plane* of order $k - 1$. This is not obvious from the definition of projective plane in Section 6.5. In fact, we prove in Theorem 8.6.1 that a set of points and lines forms a projective plane if and only if the set of points and lines forms a symmetric 2-$(v, k, 1)$ design.

It is often convenient to describe a t-design by giving a matrix that indicates the points that are in each block. The *incidence matrix* for a t-design $(\mathcal{P}, \mathcal{B})$ is a matrix with entries 0 or 1 whose rows are indexed by the blocks of $\mathcal{B}$ and whose columns are indexed by the points of $\mathcal{P}$ where the (i, j)-entry is 1 if and only if the ith block contains the jth point. The incidence matrix as defined here is the transpose of the incidence matrix defined by some other authors. This definition is used because, in some applications, the rows of the incidence matrix will represent the supports of codewords of a code, as we now illustrate.

Example 8.1.1 The binary [7, 4, 3] Hamming code $\mathcal{H}_3$ is a cyclic code with generator polynomial $g(x) = 1 + x + x^3$. The following matrix lists the codewords of weight 3 in $\mathcal{H}_3$:

$$A = \begin{bmatrix} 1 & 1 & 0 & 1 & 0 & 0 & 0 \\ 0 & 1 & 1 & 0 & 1 & 0 & 0 \\ 0 & 0 & 1 & 1 & 0 & 1 & 0 \\ 0 & 0 & 0 & 1 & 1 & 0 & 1 \\ 1 & 0 & 0 & 0 & 1 & 1 & 0 \\ 0 & 1 & 0 & 0 & 0 & 1 & 1 \\ 1 & 0 & 1 & 0 & 0 & 0 & 1 \end{bmatrix}.$$

If we label the coordinates $\mathcal{P} = \{0, 1, 2, 3, 4, 5, 6\}$, then the supports of these seven codewords are:

$$\mathcal{B} = \{\{0, 1, 3\}, \{1, 2, 4\}, \{2, 3, 5\}, \{3, 4, 6\}, \{4, 5, 0\}, \{5, 6, 1\}, \{6, 0, 2\}\}.$$

It is easy to check that $(\mathcal{P}, \mathcal{B})$ is a 2-(7, 3, 1) design (or an S(2, 3, 7) Steiner system). As the number of points is the number of blocks, the design is symmetric and so the design is a projective plane of order $3 - 1 = 2$. The matrix A is the incidence matrix of this design. Compare this example to Exercise 358. ∎

Example 8.1.2 Extend the Hamming code of the previous example, where we denote the extended coordinate by ∞, to obtain the [8, 4, 4] binary code $\widehat{\mathcal{H}}_3$. Let $\mathcal{P}$ be the coordinates $\{0, 1, 2, 3, 4, 5, 6, \infty\}$. There are 14 codewords of weight 4 in $\widehat{\mathcal{H}}_3$, and their supports are the set

$$\mathcal{B} = \{\{0, 1, 3, \infty\}, \{2, 4, 5, 6\}, \{1, 2, 4, \infty\}, \{0, 3, 5, 6\}, \{2, 3, 5, \infty\},$$
$$\{0, 1, 4, 6\}, \{3, 4, 6, \infty\}, \{0, 1, 2, 5\}, \{4, 5, 0, \infty\}, \{1, 2, 3, 6\},$$
$$\{5, 6, 1, \infty\}, \{0, 2, 3, 4\}, \{6, 0, 2, \infty\}, \{1, 3, 4, 5\}\}.$$

Notice how the supports containing ∞ are related to the supports of the weight 3 codewords of $\mathcal{H}_3$, and how the supports containing ∞ are related to the supports not containing ∞. It is easy to check that every set of three coordinates is contained in precisely one block. Thus $(\mathcal{P}, \mathcal{B})$ is a 3-(8, 4, 1) design or an S(3, 4, 8) Steiner system. This design is not symmetric. ∎

Exercise 435 Verify the claims of Examples 8.1.1 and 8.1.2 that $(\mathcal{P}, \mathcal{B})$ is a block design. ♦

As with codes, there is the notion of equivalence of designs. Two designs $(\mathcal{P}_1, \mathcal{B}_1)$ and $(\mathcal{P}_2, \mathcal{B}_2)$ are *equivalent* provided there is a bijection from $\mathcal{P}_1$ onto $\mathcal{P}_2$ that induces a bijection from $\mathcal{B}_1$ onto $\mathcal{B}_2$. A permutation of $\mathcal{P}$ is an *automorphism of* $(\mathcal{P}, \mathcal{B})$ provided the permutation induces a bijection on $\mathcal{B}$. The *automorphism group of* $(\mathcal{P}, \mathcal{B})$, denoted Aut$(\mathcal{P}, \mathcal{B})$, is the group of all automorphisms of $(\mathcal{P}, \mathcal{B})$.

Exercise 436 Consider the 2-(7, 3, 1) and 3-(8, 4, 1) designs presented in Examples 8.1.1 and 8.1.2. Prove that the permutation (0, 1, 2, 3, 4, 5, 6) is an automorphism of each design. ♦

If, as in Examples 8.1.1 and 8.1.2, the supports of the codewords of a fixed weight of a code are the blocks of a t-design for some t, then we say that the code *holds a design*.[1] Conversely, the row space, over some field, of the incidence matrix of a t-(v, k, λ) design defines a code. One cannot in general expect that the blocks are precisely the supports of all the codewords of weight k, but under suitable circumstances this is the case.

Remarkably, if $(\mathcal{P}, \mathcal{B})$ is a t-(v, k, λ) design, it is also an i-(v, k, λ_i) design for $i < t$, where λ_i is given in the next theorem.

Theorem 8.1.3 *Let $(\mathcal{P}, \mathcal{B})$ be a t-(v, k, λ) design. Let $0 \leq i \leq t$. Then $(\mathcal{P}, \mathcal{B})$ is an i-(v, k, λ_i) design, where*

$$\lambda_i = \lambda \frac{\dbinom{v-i}{t-i}}{\dbinom{k-i}{t-i}} = \lambda \frac{(v-i)(v-i-1)\cdots(v-t+1)}{(k-i)(k-i-1)\cdots(k-t+1)}.$$

Proof: Let $I \subseteq \mathcal{P}$, where $|I| = i$. Let N be the number of blocks that contain I. Define $\mathcal{X} = \{(T, B) \mid I \subseteq T \subseteq B \text{ with } |T| = t \text{ and } B \in \mathcal{B}\}$. We determine $|\mathcal{X}|$ in two ways. There are $\binom{v-i}{t-i}$ subsets of $\mathcal{P} \setminus I$ of size $t - i$; when I is added to each of these subsets, we get a t-element set T. As $(\mathcal{P}, \mathcal{B})$ is a t-(v, k, λ) design, each of these sets T containing I is in λ blocks. Therefore,

$$|\mathcal{X}| = \binom{v-i}{t-i}\lambda.$$

There are N blocks containing I. For each block B containing I, we can choose $t - i$ elements of $B \setminus I$ in $\binom{k-i}{t-i}$ ways so that when added to I form a t-element set T contained in B and containing I. So,

$$|\mathcal{X}| = N\binom{k-i}{t-i}.$$

Equating these two counts for $|\mathcal{X}|$, we see that N depends only on the size of I and the result follows by solving for N. $\qquad\qquad\square$

Example 8.1.4 Let $(\mathcal{P}, \mathcal{B})$ be the 2-$(7, 3, 1)$ symmetric design of Example 8.1.1. In the notation of Theorem 8.1.3, $\lambda_0 = 7, \lambda_1 = 3$, and $\lambda_2 = 1$. The interpretation of these numbers is as follows: The empty set is in all seven blocks, each point is in three blocks ($(\mathcal{P}, \mathcal{B})$ is a 1-$(7, 3, 3)$ design), and each pair of points is in one block ($(\mathcal{P}, \mathcal{B})$ is a 2-$(7, 3, 1)$ design). Each of these can be confirmed by direct verification from the blocks listed in the example. ∎

Example 8.1.5 Let $(\mathcal{P}, \mathcal{B})$ be the 3-$(8, 4, 1)$ design of Example 8.1.2. In the notation of Theorem 8.1.3, $\lambda_0 = 14, \lambda_1 = 7, \lambda_2 = 3$, and $\lambda_3 = 1$. The interpretation of these numbers is as follows, each of which can again be verified directly by looking at the blocks: The empty set is in all 14 blocks, each point is in seven blocks ($(\mathcal{P}, \mathcal{B})$ is a 1-$(8, 4, 7)$ design),

---

[1] If several distinct codewords have the same support, these codewords determine a unique block.

each pair of points is in three blocks (($\mathcal{P}$, $\mathcal{B}$) is a 2-(8, 4, 3) design), and each triple of points is in one block (($\mathcal{P}$, $\mathcal{B}$) is a 3-(8, 4, 1) design).   ∎

In the preceding examples we see that λ_0 is the number of blocks in the design; in Exercise 437, you are asked to prove that. By definition λ_1 is the number of blocks containing any given point. Thus we have the following theorem.

Theorem 8.1.6 *In a t-(v, k, λ) design, the number of blocks is*

$$b = \lambda_0 = \lambda \frac{\binom{v}{t}}{\binom{k}{t}},$$

and every point is in exactly

$$\lambda_1 = \frac{\lambda_0 k}{v} = \frac{bk}{v}$$

blocks.

Exercise 437 Let ($\mathcal{P}$, $\mathcal{B}$) be a t-(v, k, λ) design.
(a) Give a direct proof that there are

$$b = \lambda_0 = \lambda \frac{\binom{v}{t}}{\binom{k}{t}}$$

blocks in the design.
(b) Verify that $\lambda_1 = \lambda_0 k/v = bk/v$.   ◆

The number λ_1 of blocks containing a given point is called the *replication number* and is sometimes denoted r. A 2-(v, k, λ) design is sometimes called a *balanced incomplete block design* or a *(b, v, r, k, λ) design*. The values for b and r are determined from v, k, and λ using Theorem 8.1.6 with $t = 2$.

The fact that each λ_i is an integer implies certain constraints on the parameters t, v, k, and λ in order for a t-(v, k, λ) design to exist. A main problem in the study of designs is to determine whether numbers t, v, k, and λ, for which the λ_i are all integers, are actually the parameters of a t-(v, k, λ) design. A secondary problem is to classify all designs with given parameters up to equivalence when such designs exist.

Exercise 438 Show that designs with the following parameters cannot exist:
(a) $t = 2$, $v = 90$, $k = 5$, and $\lambda = 2$,
(b) $t = 4$, $v = 10$, $k = 5$, and $\lambda = 5$.   ◆

8.2   Intersection numbers

There are other integers associated with a t-design $(\mathcal{P}, \mathcal{B})$ that describe certain intersection properties of the design. Let I and J be subsets of $\mathcal{P}$ where $I \cap J = \emptyset$. Suppose that $|I| = i$ and $|J| = j$. Denote the number of blocks in $\mathcal{B}$ that contain I and are disjoint from J by λ_i^j. The next theorem shows that λ_i^j is independent of the choice of I and J provided $i + j \leq t$. We will also see in this theorem that these numbers satisfy a recursion reminiscent of the one satisfied by binomial coefficients. These numbers are called *intersection numbers* for the design. Certain of the intersection numbers have specific meaning:

- λ_0^0 is the number b of blocks,
- $\lambda_i^0 = \lambda_i$, and
- λ_0^j is the number of blocks not intersecting a given set of points of size j.

Theorem 8.2.1 *Let $(\mathcal{P}, \mathcal{B})$ be a t-(v, k, λ) design. Let I and J be disjoint subsets of $\mathcal{P}$ of size i and j, respectively. If $i + j \leq t$, λ_i^j is independent of I and J, and for $j \geq 1$,*

$$\lambda_i^j = \lambda_i^{j-1} - \lambda_{i+1}^{j-1}. \tag{8.1}$$

Also

$$\lambda_i^j = \lambda \frac{\binom{v - i - j}{k - i}}{\binom{v - t}{k - t}}. \tag{8.2}$$

Proof: We give an indication of the proof and ask the reader to give a formal proof in Exercise 440. Referring to Figure 8.1, the right-hand "edge" gives the entries λ_i^0, which are λ_i by definition. By Theorem 8.1.3, λ_i is independent of I for all $i \leq t$ and so the right-hand

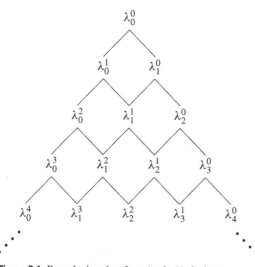

Figure 8.1 Pascal triangle of a t-(v, k, λ) design.

"edge" of the figure is uniquely determined. Let $I = \{p_1\}$. All λ_0^0 blocks either contain p_1 or they do not. The number containing p_1 is λ_1^0; thus the number not containing p_1 is $\lambda_0^0 - \lambda_1^0$. Hence $\lambda_0^1 = \lambda_0^0 - \lambda_1^0$ is independent of I. So the second row of the triangle in Figure 8.1 is uniquely determined. Now consider the third row. Let $I = \{p_1\}$ and $J = \{q_1\}$ with $p_1 \neq q_1$. The intersection number λ_1^0 is the number of blocks containing p_1. These blocks fall into two categories: those that contain q_1 and those that do not. The number containing q_1 is λ_2^0; hence the number not containing q_1 is $\lambda_1^0 - \lambda_2^0$. Thus the entry λ_1^1 is $\lambda_1^0 - \lambda_2^0$, which is therefore independent of both I and J. Now let $J = \{q_1, q_2\}$. There are λ_0^1 blocks that do not contain q_2; as this entry is in row 2, it is independent of q_2. Again these blocks either contain q_1 or they do not. There are λ_1^1 that do, independent of q_1, and again λ_0^2 must be $\lambda_0^1 - \lambda_1^1$ independent of J. Thus row three of Figure 8.1 is uniquely determined and has the values claimed. The formal proof of (8.1) follows inductively along similar lines.

Using the recurrence relation (8.1), we can determine all λ_i^j with $i + j \leq t$ from the λ_i^0 with $0 \leq i \leq t$. For example,

$$\lambda_0^2 = \lambda_0^1 - \lambda_1^1 = \lambda_0^0 - \lambda_1^0 - (\lambda_1^0 - \lambda_2^0) = \lambda_0^0 - 2\lambda_1^0 + \lambda_2^0 = \lambda_0 - 2\lambda_1 + \lambda_2.$$

Since the value for λ_i^0 given in (8.2) agrees with the value for $\lambda_i = \lambda_i^0$ given in Theorem 8.1.3, and since the values of λ_i^j given in (8.2) satisfy the recurrence relation (8.1), as Exercise 439 shows, the theorem follows.   □

Exercise 439 Show that the values of λ_i^j given in (8.2) satisfy the recurrence relation (8.1). ◆

Exercise 440 Give a formal proof of Theorem 8.2.1. ◆

As mentioned previously, (8.1) is reminiscent of the recurrence relation that defines the Pascal triangle of binomial coefficients; we call the family of intersection numbers $\{\lambda_i^j \mid i + j \leq t\}$ the *Pascal triangle* of $(\mathcal{P}, \mathcal{B})$. Figure 8.1 gives a visual representation of this triangle. Notice that the triangle is determined completely from either "edge," that is, from the values $\lambda_i^0 = \lambda_i$ for $0 \leq i \leq t$ or from the values λ_0^j for $0 \leq j \leq t$. By iterating (8.1), we obtain an explicit formula for the entries in the triangle from the values along the "right edge."

Corollary 8.2.2 *If* $0 \leq i + j \leq t$,

$$\lambda_i^j = \sum_{m=i}^{i+j} (-1)^{m-i} \binom{j}{m-i} \lambda_m.$$

Exercise 441 Prove Corollary 8.2.2. Hint: Show that if

$$\Lambda_i^j = \sum_{m=i}^{i+j} (-1)^{m-i} \binom{j}{m-i} \lambda_m,$$

then $\Lambda_i^0 = \lambda_i$ and Λ_i^j satisfies the recursion of (8.1). ◆

Exercise 442 Find an analogous formula to that of Corollary 8.2.2 involving the "left" edge λ_0^j of the Pascal triangle. Prove your formula is correct. ◆

Exercise 443 There is a 5-(18, 8, 6) design whose blocks are the supports of the minimum weight vectors in an [18, 9, 8] Hermitian self-dual code over $\mathbb{F}_4$.

(a) Construct the Pascal triangle for this design.

(b) How many minimum weight vectors are there in the code?   ◆

If the design is a Steiner design (that is $\lambda = 1$) it is possible to add $k - t$ new rows to the Pascal triangle provided we restrict the subsets I and J in our definition of λ_i^j. Let $(\mathcal{P}, \mathcal{B})$ be a t-$(v, k, 1)$ design and suppose $B \in \mathcal{B}$. Let I and J be subsets of $\mathcal{P}$ where $|I| = i$, $|J| = j$, and $I \cap J = \emptyset$, and assume that there is a block B such that $I \cup J \subseteq B$. If $t < i + j \leq k$, we define λ_i^j to be the number of blocks in $\mathcal{B}$ containing I and disjoint from J.

Theorem 8.2.3 *Let $(\mathcal{P}, \mathcal{B})$ be a t-$(v, k, 1)$ design. Let I and J be subsets of $\mathcal{P}$ where $|I| = i$, $|J| = j$, and $I \cap J = \emptyset$, and assume that $I \cup J$ is a subset of some block in $\mathcal{B}$. If $t < i + j \leq k$, then λ_i^j is independent of I and J. Also $\lambda_i^j = \lambda_i^{j-1} - \lambda_{i+1}^{j-1}$.*

Proof: As $\lambda = 1$, $\lambda_i^0 = 1$ for $t < i \leq k$. Thus λ_i^0 is independent of I. The remainder of the argument follows as in the proof of Theorem 8.2.1.   □

Notice that λ_i^j does not satisfy (8.2) if $t < i + j \leq k$ because λ_i^0 does not satisfy that equation for $t < i \leq k$. However, Corollary 8.2.2 does hold whenever $i + j \leq k$ as Exercise 441 shows. If $\lambda = 1$, the triangle formed by $\{\lambda_i^j \mid 0 \leq i + j \leq k\}$ is called the *extended Pascal triangle* of $(\mathcal{P}, \mathcal{B})$. Figures 8.2 and 8.3 give the extended Pascal triangles of a 3-(8, 4, 1) design and a 5-(24, 8, 1) design, respectively. The values λ_i^j for $t \leq i + j \leq k$ are connected by double lines to indicate the extended part of the triangle.

A 3-(8, 4, 1) design was given in Example 8.1.2; the design arose as the set of supports of the weight 4 vectors in $\widehat{\mathcal{H}}_3$. This code is the only [8, 4, 4] binary code by Exercise 56, but this does not imply that the design is unique. In the next example, we show that in fact any 3-(8, 4, 1) design is the set of supports of the weight 4 codewords in an [8, 4, 4] binary code and hence is unique.

Exercise 444 Verify the entries in Figures 8.2 and 8.3 given that the 3-(8, 4, 1) design has 14 blocks and the 5-(24, 8, 1) design has 759 blocks.   ◆

Example 8.2.4 We show that the 3-(8, 4, 1) design is unique up to equivalence. From the bottom row of the extended Pascal triangle of Figure 8.2, we see that any two blocks meet in an even number of points. Hence if we associate to each block of the design a vector in

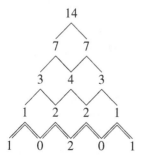

Figure 8.2 Extended Pascal triangle of a 3-(8, 4, 1) design.

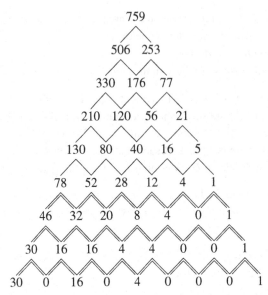

Figure 8.3 Extended Pascal triangle of a 5-(24, 8, 1) design.

$\mathbb{F}_2^8$ whose support is that block, these vectors generate a self-orthogonal doubly-even binary code $\widehat{\mathcal{C}}$ of length 8 by Theorem 1.4.8(i). By self-orthogonality, the dimension of $\widehat{\mathcal{C}}$ is at most 4; as there are 14 blocks, the dimension is 4. Thus $\widehat{\mathcal{C}}$ must be one of the equivalent forms of the [8, 4, 4] extended Hamming code $\widehat{\mathcal{H}}_3$. Thus an 3-(8, 4, 1) design is equivalent to the design of Example 8.1.2.   ■

8.3   Complementary, derived, and residual designs

If we begin with a t-(v, k, λ) design $(\mathcal{P}, \mathcal{B})$, there are three other natural designs that arise from this design:

- The *complementary design for* $(\mathcal{P}, \mathcal{B})$ is the design $(\mathcal{P}, \mathcal{B}')$ where $\mathcal{B}'$ consists of the complements of the blocks in $\mathcal{B}$. Theorem 8.2.1 shows that $(\mathcal{P}, \mathcal{B}')$ is in fact a t-$(v, v - k, \lambda_0')$ design whose Pascal triangle is the reflection, across the vertical line through the apex, of the Pascal triangle of $(\mathcal{P}, \mathcal{B})$. If the complement of every block is already in the design, the design is called *self-complementary*; in this case of course, k must be $v/2$. If $\lambda = 1$ and $k \leq v/2$, the Pascal triangle of the complementary design of $(\mathcal{P}, \mathcal{B})$ can also be extended by $k - t$ rows by reflecting the extended Pascal triangle of $(\mathcal{P}, \mathcal{B})$, provided the extended entries are interpreted correctly; if $t < i + j \leq k$, λ_i^j in the complementary design represents the number of blocks containing I and disjoint from J where $|I| = i$, $|J| = j$, and $I \cup J$ is disjoint from some block in the complementary design.
- Let $x \in \mathcal{P}$ be fixed. The *derived design for* $(\mathcal{P}, \mathcal{B})$ *with respect to* x is the design with points $\mathcal{P} \setminus \{x\}$ and blocks $\{B \setminus \{x\} \mid x \in B \in \mathcal{B}\}$. Theorem 8.2.1 shows that the derived design is a $(t - 1)$-$(v - 1, k - 1, \lambda)$ design whose Pascal triangle is the subtriangle of $(\mathcal{P}, \mathcal{B})$ consisting of λ_1^0 and the nodes below. If $\lambda = 1$, the extended Pascal triangle of the

derived design is the subtriangle of the extended triangle of $(\mathcal{P}, \mathcal{B})$ consisting of λ_1^0 and the nodes below.

- Again let $x \in \mathcal{P}$ be fixed. The *residual design for* $(\mathcal{P}, \mathcal{B})$ *with respect to* x is the design with points $\mathcal{P} \setminus \{x\}$ whose blocks are $B \in \mathcal{B}$ not containing x. Again Theorem 8.2.1 shows that the residual design is a $(t-1)$-$(v-1, k, \lambda_{t-1} - \lambda)$ design whose Pascal triangle is the subtriangle of $(\mathcal{P}, \mathcal{B})$ consisting of λ_0^1 and the nodes below.

Exercise 445 Verify all the claims made in connection with the definition of complementary, derived, and residual designs. ◆

Exercise 446 Let $\mathcal{D} = (\mathcal{P}, \mathcal{B})$ be a t-design with $t > 1$ and choose $x \in \mathcal{P}$. Let $\mathcal{D}_c$ be the complementary design of $\mathcal{D}$, $\mathcal{D}_d$ the derived design of $\mathcal{D}$ with respect to x, and $\mathcal{D}_r$ the residual design of $\mathcal{D}$ with respect to x.

(a) Let $\mathcal{D}_1$ be the complementary design of $\mathcal{D}_r$ and $\mathcal{D}_2$ the derived design of $\mathcal{D}_c$ with respect to x. Prove that $\mathcal{D}_1 = \mathcal{D}_2$.

(b) Let $\mathcal{D}_3$ be the complementary design of $\mathcal{D}_d$ and $\mathcal{D}_4$ the residual design of $\mathcal{D}_c$ with respect to x. Prove that $\mathcal{D}_3 = \mathcal{D}_4$. ◆

Example 8.3.1 In Example 8.2.4 we showed that a 3-(8, 4, 1) design $(\mathcal{P}, \mathcal{B})$ is equivalent to the design obtained from the weight 4 vectors of $\widehat{\mathcal{H}}_3$, the [8, 4, 4] extended binary Hamming code. We describe the complementary, derived, and residual designs obtained from $(\mathcal{P}, \mathcal{B})$:

- Since the complement of every block in $(\mathcal{P}, \mathcal{B})$ is already in $(\mathcal{P}, \mathcal{B})$, the design is self-complementary. Notice that this is related to the fact that $\widehat{\mathcal{H}}_3$ contains the all-one codeword and the blocks have size half the length of the code.
- There are $\lambda_1^0 = 7$ blocks of $(\mathcal{P}, \mathcal{B})$ that contain any given point x. The 2-(7, 3, 1) derived design, obtained by deleting x from these seven blocks, gives the supports of the weight 3 vectors in the code punctured on the coordinate represented by x. This design is the projective plane of order 2 and is equivalent to the design obtained in Example 8.1.1.
- There are $\lambda_0^1 = 7$ blocks of $(\mathcal{P}, \mathcal{B})$ that do not contain any given point x. The 2-(7, 4, 2) residual design with respect to x consists of the supports of the weight 4 vectors in the code punctured on the coordinate represented by x. ∎

Exercise 447 Find the parameters of the complementary, residual, and derived designs obtained from the 5-(18, 8, 6) design of Exercise 443. Also give the Pascal triangle of each of these designs. ◆

Example 8.3.2 In this example, we show that any 5-(24, 8, 1) design is held by the codewords of weight 8 in a [24, 12, 8] self-dual binary code. We, however, do not know that this design even exists. We will show in Example 8.4.3 that the design indeed does exist. In Section 10.1, we will show that both the design and the code are unique up to equivalence; the code is the extended binary Golay code.

Let $(\mathcal{P}, \mathcal{B})$ be such a 5-(24, 8, 1) design. A subset of $\mathcal{P}$ of size 4 will be called a *tetrad*. If T is a tetrad and p_1 is a point not in T, there is a unique block $B \in \mathcal{B}$ containing $T \cup \{p_1\}$ because $(\mathcal{P}, \mathcal{B})$ is a Steiner 5-design. Let T_1 be $B \setminus T$. Thus T_1 is the unique tetrad disjoint from T containing p_1 such that $T \cup T_1 \in \mathcal{B}$. Letting p_2 be a point not in $T \cup T_1$, we can similarly find a unique tetrad T_2 containing p_2 such that $T \cap T_2 = \emptyset$ and $T \cup T_2 \in \mathcal{B}$. As $(\mathcal{P}, \mathcal{B})$ is a Steiner 5-design, if $T \cup T_1$ and $T \cup T_2$ have five points in common, they must be equal implying $T_1 = T_2$, which is a contradiction as $p_2 \in T_2 \setminus (T \cup T_1)$; so $T_1 \cap T_2 = \emptyset$. Continuing in this manner we construct tetrads T_3, T_4, and T_5 such that T, $T_1, \ldots, T_5$ are pairwise disjoint and $T \cup T_i \in \mathcal{B}$ for $1 \leq i \leq 5$; the tetrads T, $T_1, \ldots, T_5$ are called the *sextet determined by T*.

Let $\mathcal{C}$ be the binary linear code of length 24 spanned by vectors whose supports are the blocks in $\mathcal{B}$. By examining the last line of Figure 8.3, we see that two distinct blocks intersect in either 0, 2, or 4 points because λ_i^{8-i} is nonzero only for $i = 0, 2, 4$, or 8; $\lambda_8^0 = 1$ simply says that only one block of size 8 contains the eight points of that block. In particular, $\mathcal{C}$ is self-orthogonal and therefore doubly-even, by Theorem 1.4.8(i). We prove that $\mathcal{C}$ has minimum weight 8. If it does not, it has minimum weight 4. Suppose $\mathbf{c} \in \mathcal{C}$ has weight 4 and support S. Let T be a tetrad intersecting S in exactly three places. Constructing the sextet determined by T, we obtain codewords $\mathbf{c}_i$ with support $T \cup T_i$ for $1 \leq i \leq 5$. As $|S| = 4$ and the tetrads are pairwise disjoint, we have $S \cap T_i = \emptyset$ for some i; therefore $\text{wt}(\mathbf{c} + \mathbf{c}_i) = 6$, which is a contradiction as $\mathcal{C}$ is doubly-even. Hence $\mathcal{C}$ is a self-orthogonal code with minimum weight $d = 8$.

The support of any codeword in $\mathcal{C}$ of weight 8 is called an *octad*. We next show that the octads are precisely the blocks. Clearly, the blocks of $\mathcal{B}$ are octads. Let $\mathbf{c}$ be a codeword in $\mathcal{C}$ of weight 8 with support S. We show S is a block of $\mathcal{B}$. Fix five points of S. There is a unique block $B \in \mathcal{B}$ containing these five points; as blocks are octads, there is a codeword $\mathbf{b} \in \mathcal{C}$ with support B. Since $\text{wt}(\mathbf{c} + \mathbf{b}) \leq 6$, $\mathbf{c} = \mathbf{b}$ as $\mathbf{c} + \mathbf{b} \in \mathcal{C}$ and $\mathcal{C}$ has minimum weight 8. So $S = B$ and hence every octad is a block of $\mathcal{B}$.

Since $\mathcal{C}$ is self-orthogonal, it has dimension at most 12. We show that it has dimension exactly 12. We do this by counting the number of cosets of $\mathcal{C}$ in $\mathbb{F}_2^{24}$. There is a unique coset leader in cosets of weight i, $0 \leq i \leq \lfloor (d-1)/2 \rfloor = 3$ by Exercise 66. Thus for $0 \leq i \leq 3$, there are $\binom{24}{i}$ cosets of weight i, accounting for 2325 cosets. Let $\mathbf{v}$ be a coset leader in a coset of weight 4; suppose its support is the tetrad T. If $\mathbf{w}$ is another coset leader of this coset, then $\mathbf{v} + \mathbf{w}$ must be a codeword and hence must be of weight 8. Thus the other coset leaders are determined once we find all weight 8 codewords whose supports contain T. Consider the sextet determined by T. If $\mathbf{c}_i \in \mathcal{C}$ has support $T \cup T_i$, the only codewords of weight 8 whose support contains T are $\mathbf{c}_1, \ldots, \mathbf{c}_5$ because all octads are blocks and the sextet determined by T is unique. So there are exactly six coset leaders in the coset, namely $\mathbf{v}$, $\mathbf{v} + \mathbf{c}_1, \ldots, \mathbf{v} + \mathbf{c}_5$. Therefore there are $\frac{1}{6}\binom{24}{4} = 1771$ cosets of weight 4, and hence, 2^{12} cosets of weight 4 or less. There are no cosets of weight 5 because, if $\mathbf{v}$ has weight 5, there is a codeword $\mathbf{c}$ of weight 8 such that $\text{wt}(\mathbf{v} + \mathbf{c}) = 3$ as $(\mathcal{P}, \mathcal{B})$ is a 5-design, which is a contradiction. By Theorem 1.12.6(v), the covering radius of $\mathcal{C}$ is 4. Therefore $\mathcal{C}$ is a [24, 12, 8] self-dual code whose codewords of weight 8 support the blocks $\mathcal{B}$. ∎

Example 8.3.3 Let $\mathcal{C}$ be the [24, 12, 8] extended binary Golay code, and let $\mathcal{D} = (\mathcal{P}, \mathcal{B})$ be the 5-(24, 8, 1) design held by the weight 8 codewords of $\mathcal{C}$. (See Example 8.4.3 where we show that the weight 8 codewords of $\mathcal{C}$ indeed hold a 5-design.) We produce a number of designs related to $\mathcal{D}$.

- The complementary design $\mathcal{D}_0$ for $\mathcal{D}$ is a 5-(24, 16, 78) design, since $\lambda_0^5 = 78$. As the all-one vector is in $\mathcal{C}$, this design must be held by the weight 16 codewords of $\mathcal{C}$. The extended Pascal triangle for $\mathcal{D}_0$ (with the appropriate interpretation for the extended nodes) is the extended Pascal triangle for $\mathcal{D}$ reflected across the vertical line through the apex.
- Puncturing $\mathcal{C}$ with respect to any coordinate p_1 gives the [23, 12, 7] binary Golay code $\mathcal{C}_1$. The weight 7 codewords of $\mathcal{C}_1$ must come from the weight 8 codewords of $\mathcal{C}$ with p_1 in their support. Thus the weight 7 codewords of $\mathcal{C}_1$ hold the 4-(23, 7, 1) design $\mathcal{D}_1$ derived from $\mathcal{D}$ with respect to p_1. The extended Pascal triangle of $\mathcal{D}_1$ is the subtriangle with apex $\lambda_1^0 = 253$ (the number of blocks of $\mathcal{D}_1$) obtained from Figure 8.3. Notice that by Exercise 403, $\mathcal{C}_1$ indeed has 253 weight 7 codewords.
- The weight 8 codewords of $\mathcal{C}_1$ must come from the weight 8 codewords of $\mathcal{C}$ that do not have p_1 in their support. Thus the weight 8 codewords of $\mathcal{C}_1$ hold the 4-(23, 8, 4) design $\mathcal{D}_2$, which is the residual design for $\mathcal{D}$ with respect to p_1. The Pascal triangle of $\mathcal{D}_2$ is the (non-extended) subtriangle with apex $\lambda_0^1 = 506$ (the number of blocks of $\mathcal{D}_2$) obtained from Figure 8.3. Notice again that by Exercise 403, $\mathcal{C}_1$ indeed has 506 weight 8 codewords.
- The complementary design for $\mathcal{D}_1$ is the 4-(23, 16, 52) design $\mathcal{D}_3$ held by the weight 16 codewords of $\mathcal{C}_1$ again because $\mathcal{C}_1$ contains the all-one codeword.
- The complementary design for $\mathcal{D}_2$ is the 4-(23, 15, 78) design $\mathcal{D}_4$ held by the weight 15 codewords of $\mathcal{C}_1$. The design $\mathcal{D}_4$ is also the derived design of $\mathcal{D}_0$ with respect to p_1; see Exercise 446.
- Puncturing $\mathcal{C}_1$ with respect to any coordinate p_2 gives the [22, 12, 6] code $\mathcal{C}_2$. The weight 6 codewords of $\mathcal{C}_2$ hold the 3-(22, 6, 1) design $\mathcal{D}_5$ derived from $\mathcal{D}_1$ with respect to p_2; the extended Pascal triangle of $\mathcal{D}_5$ is the subtriangle with apex $\lambda_2^0 = 77$ obtained from Figure 8.3.
- The weight 16 codewords of $\mathcal{C}_2$ hold the 3-(22, 16, 28) design $\mathcal{D}_6$, which is both the complementary design for $\mathcal{D}_5$ and the residual design for $\mathcal{D}_3$ with respect to p_2; again see Exercise 446.
- The weight 8 codewords of $\mathcal{C}_2$ hold the 3-(22, 8, 12) design $\mathcal{D}_7$, which is the residual design for $\mathcal{D}_2$ with respect to p_2, whose Pascal triangle is the (nonextended) subtriangle with apex $\lambda_0^2 = 330$ obtained from Figure 8.3.
- The weight 14 codewords of $\mathcal{C}_2$ hold the 3-(22, 14, 78) design $\mathcal{D}_8$, which is the complementary design for $\mathcal{D}_7$ and the derived design of $\mathcal{D}_4$ with respect to p_2.
- The supports of the weight 7 codewords of $\mathcal{C}_2$ are the union of the blocks of two different 3-designs. The weight 7 codewords of $\mathcal{C}_2$ arise from the codewords of weight 8 in $\mathcal{C}$ whose supports contain exactly one of p_1 or p_2. Thus the supports of the weight 7 codewords of $\mathcal{C}_2$ are either blocks in the 3-(22, 7, 4) residual design obtained from $\mathcal{D}_1$ with respect to p_2 or blocks in the 3-(22, 7, 4) design derived from $\mathcal{D}_2$ with respect to p_2. Hence there are 352 weight 7 codewords in $\mathcal{C}_2$ that hold a 3-(22, 7, 8) design $\mathcal{D}_9$ whose Pascal triangle is the (non-extended) subtriangle with apex $\lambda_1^1 = 176$ obtained from Figure 8.3 with all entries doubled.   ∎

The next two examples show how to use designs to find the weight distributions of the punctured codes $\mathcal{C}_1$ and $\mathcal{C}_2$ of Example 8.3.3 obtained from the [24, 12, 8] extended binary Golay code.

Example 8.3.4 In Exercise 403, we found the weight distribution of C_1 using Prange's Theorem. As an alternate approach, by examining the Pascal triangles for the designs that arise here, the weight distribution of C_1 can be obtained. We know that $A_i(C_1) = A_{23-i}(C_1)$, as C_1 contains the all-one codeword. Obviously, $A_0(C_1) = A_{23}(C_1) = 1$. The number of weight 7 codewords in C_1 is the number of blocks in D_1, which is the apex of its Pascal triangle (and the entry λ_1^0 of Figure 8.3); thus $A_7(C_1) = A_{16}(C_1) = 253$. The number of weight 8 codewords is the number of blocks in D_2, which is the apex of its Pascal triangle (and the entry λ_0^1 of Figure 8.3); thus $A_8(C_1) = A_{15}(C_1) = 506$. We complete the weight distribution by noting that $A_{11}(C_1) = A_{12}(C_1) = 2^{11} - 506 - 253 - 1 = 1288$. ∎

Example 8.3.5 A similar computation can be done to compute the weight distribution of C_2. Again since C_2 contains the all-one codeword, $A_i(C_2) = A_{22-i}(C_2)$ and $A_0(C_2) = A_{22}(C_2) = 1$. The number of weight 6 codewords in C_2 is the number of blocks in D_5, which is the entry λ_2^0 of Figure 8.3, implying $A_6(C_2) = A_{16}(C_2) = 77$. The number of weight 7 codewords is the number of blocks in D_9, which is two times the entry λ_1^1 of Figure 8.3; thus $A_7(C_2) = A_{16}(C_2) = 352$. The number of weight 8 codewords is the number of blocks in D_7, which is the entry λ_0^2 of Figure 8.3; hence $A_8(C_2) = A_{14}(C_2) = 330$. We will see later that the weight 12 codewords of C hold a 5-design. The Pascal triangle for that design is given in Figure 8.4. The number of weight 10 codewords in C_2 must be λ_2^0 from that figure. So $A_{10}(C_2) = A_{12}(C_2) = 616$. Finally, $A_{11}(C_2) = 2^{12} - 2 \cdot (1 + 77 + 352 + 330 + 616) = 1344$. ∎

Exercise 448 Find the Pascal triangle or the extended Pascal triangle, whichever is appropriate, for the designs $D_0, D_1, \ldots, D_9$ described in Example 8.3.3. ◆

Examples 8.3.1 and 8.3.3 illustrate the following result, whose proof is left as an exercise.

Theorem 8.3.6 *Let C be a binary linear code of length n such that the vectors of weight w hold a t-design $D = (P, B)$. Let $x \in P$ and let C^* be C punctured on coordinate x.*
(i) *If the all-one vector is in C, then the vectors of weight $n - w$ hold a t-design, and this design is the complementary design of D.*
(ii) *If C has no vectors of weight $w + 1$, then the residual $(t - 1)$ design of D with respect to x is the design held by the weight w vectors in the code C^*.*
(iii) *If C has no vectors of weight $w - 1$, then the $(t - 1)$ design derived from D with respect to x is the design held by the weight $w - 1$ vectors in the code C^*.*

Exercise 449 Prove Theorem 8.3.6. ◆

There is no known general result that characterizes all codes whose codewords of a given weight, such as minimum weight, hold a t-design. There are, however, results that show that codes satisfying certain conditions have weights where codewords of that weight hold a design. The best known of these is the Assmus–Mattson Theorem presented in the next section. It is this result that actually guarantees (without directly showing that supports of codewords satisfy the conditions required in the definition of a t-design) that the weight 4 codewords of the [8, 4, 4] extended binary Hamming code hold a 3-(8, 4, 1) design and the weight 8 codewords of the [24, 12, 8] extended binary Golay code hold a 5-(24, 8, 1) design.

8.4   The Assmus–Mattson Theorem

If the weight distribution of a code and its dual are of a particular form, a powerful result due to Assmus and Mattson [6] guarantees that t-designs are held by codewords in both the code and its dual. In fact, the Assmus–Mattson Theorem has been the main tool in discovering designs in codes.

For convenience, we first state the Assmus–Mattson Theorem for binary codes. We prove the general result, as the proof for binary codes is not significantly simpler than the proof for codes over an arbitrary field.

Theorem 8.4.1 (Assmus–Mattson) *Let C be a binary $[n, k, d]$ code. Suppose $C^\perp$ has minimum weight $d^\perp$. Suppose that $A_i = A_i(C)$ and $A_i^\perp = A_i(C^\perp)$, for $0 \leq i \leq n$, are the weight distributions of C and $C^\perp$, respectively. Fix a positive integer t with $t < d$, and let s be the number of i with $A_i^\perp \neq 0$ for $0 < i \leq n - t$. Suppose $s \leq d - t$. Then:*
(i) *the vectors of weight i in C hold a t-design provided $A_i \neq 0$ and $d \leq i \leq n$, and*
(ii) *the vectors of weight i in $C^\perp$ hold a t-design provided $A_i^\perp \neq 0$ and $d^\perp \leq i \leq n - t$.*

Theorem 8.4.2 (Assmus–Mattson) *Let C be an $[n, k, d]$ code over $\mathbb{F}_q$. Suppose $C^\perp$ has minimum weight $d^\perp$. Let w be the largest integer with $w \leq n$ satisfying*

$$w - \left\lfloor \frac{w + q - 2}{q - 1} \right\rfloor < d.$$

(So $w = n$ when $q = 2$.) Define $w^\perp$ analogously using $d^\perp$. Suppose that $A_i = A_i(C)$ and $A_i^\perp = A_i(C^\perp)$, for $0 \leq i \leq n$, are the weight distributions of C and $C^\perp$, respectively. Fix a positive integer t with $t < d$, and let s be the number of i with $A_i^\perp \neq 0$ for $0 < i \leq n - t$. Suppose $s \leq d - t$. Then:
(i) *the vectors of weight i in C hold a t-design provided $A_i \neq 0$ and $d \leq i \leq w$, and*
(ii) *the vectors of weight i in $C^\perp$ hold a t-design provided $A_i^\perp \neq 0$ and $d^\perp \leq i \leq \min\{n - t, w^\perp\}$.*

Proof: Let T be any set of t coordinate positions, and let C^T be the code of length $n - t$ obtained from C by puncturing on T. Let $C^\perp(T)$ be the subcode of $C^\perp$ that is zero on T, and let $(C^\perp)_T$ be the code $C^\perp$ shortened on T. Since $t < d$, it follows from Theorem 1.5.7 that C^T is an $[n - t, k, d^T]$ code with $d^T \geq d - t$ and $(C^T)^\perp = (C^\perp)_T$.

Let $A_i' = A_i(C^T)$ and $A_i'^\perp = A_i((C^T)^\perp) = A_i((C^\perp)_T)$, for $0 \leq i \leq n - t$, be the weight distributions of C^T and $(C^T)^\perp$, respectively. As $s \leq d - t \leq d^T$, $A_i' = 0$ for $1 \leq i \leq s - 1$. If $S = \{i \mid A_i^\perp \neq 0, 0 < i \leq n - t\}$, then, as $A_i'^\perp \leq A_i^\perp$ and $|S| = s$, the $A_i'^\perp$ are unknown only for $i \in S$. These facts about A_i' and $A_i'^\perp$ are independent of the choice of T. By Theorem 7.3.1, there is a unique solution for all A_i' and $A_i'^\perp$, which must therefore be the same for each set T of size t. The weight distribution of $C^\perp(T)$ is the same as the weight distribution of $(C^\perp)_T$; hence the weight distribution of $C^\perp(T)$ is the same for all T of size t.

In a code over any field, two codewords of minimum weight with the same support must be scalar multiples of each other by Exercise 451. Let $\mathcal{B}$ be the set of supports of the vectors in C of weight d. Let T be a set of size t. The codewords in C of weight d, whose support contains T, are in one-to-one correspondence with the vectors in C^T of weight $d - t$. There are A'_{d-t} such vectors in C^T and hence $A'_{d-t}/(q - 1)$ blocks in $\mathcal{B}$ containing T. Thus the codewords of weight d in C hold a t-design.

We prove the rest of (i) by induction. Assume that the codewords of weight x in C with $A_x \neq 0$ and $d \leq x \leq z - 1 < w$, for some integer z, hold t-designs. Suppose the intersection numbers of these designs are $\lambda_i^j(x)$. If $d \leq x \leq z - 1 < w$ but $A_x = 0$, set $\lambda_i^j(x) = 0$. By Exercise 451 the value w has been chosen to be the largest possible weight so that a codeword of weight w or less in C is determined uniquely up to scalar multiplication by its support. If $A_z \neq 0$, we show that the codewords of weight z in C hold a t-design. Suppose that there are $N(T)$ codewords in C of weight z whose support contains T. Every vector in C of weight z with support containing T is associated with a vector of weight $z - t$ in C^T. However, every vector of weight $z - t$ in C^T is associated with a vector of weight $z - \ell$ in C whose support intersects T in a set of size $t - \ell$ for $0 \leq \ell \leq z - d$. A calculation completed in Exercise 452 shows that

$$A'_{z-t} = N(T) + (q - 1) \sum_{\ell=1}^{z-d} \binom{t}{\ell} \lambda^\ell_{t-\ell}(z - \ell).$$

Therefore, $N(T)$ is independent of T, and hence the codewords of weight z in C hold a t-design. Thus (i) holds by induction.

Let $d^\perp \leq i \leq \min\{n - t, w^\perp\}$. Codewords in $C^\perp$ of weight $w^\perp$ or less are determined uniquely up to scalar multiplication by their supports by Exercise 451. Let $\mathcal{B}$ be the set of all supports of codewords in $C^\perp$ of weight i, and let $\mathcal{B}'$ be their complements. Let $\mathcal{B}'_T$ be the set of blocks in $\mathcal{B}'$ that contain T. These blocks are in one-to-one correspondence with the supports of codewords of weight i which are zero on T, that is, codewords of weight i in $C^\perp(T)$. The number of blocks in $\mathcal{B}'_T$ is independent of T as the weight distribution of $C^\perp(T)$ is independent of T. Therefore $|\mathcal{B}'_T|$ is independent of T, and $\mathcal{B}'$ is the set of blocks in a t-design. Hence $\mathcal{B}$ is the set of blocks in a t-design. This proves (ii).   □

Exercise 450 Show that in the $[12, 6, 6]$ extended ternary Golay code, two codewords, both of weight 6 or both of weight 9, with the same support must be scalar multiples of each other.   ◆

Exercise 451 Let C be a code over $\mathbb{F}_q$ of minimum weight d.
(a) Let $\mathbf{c}$ and $\mathbf{c}'$ be two codewords of weight d with $\text{supp}(\mathbf{c}) = \text{supp}(\mathbf{c}')$. Show that $\mathbf{c} = \alpha \mathbf{c}'$ for some nonzero α in $\mathbb{F}_q$.
(b) Let w be the largest integer with $w \leq n$ satisfying

$$w - \left\lfloor \frac{w + q - 2}{q - 1} \right\rfloor < d.$$

Show that if $\mathbf{c}$ and $\mathbf{c}'$ are two codewords of weight i with $d \leq i \leq w$ and $\text{supp}(\mathbf{c}) = \text{supp}(\mathbf{c}')$, then $\mathbf{c} = \alpha \mathbf{c}'$ for some nonzero α in $\mathbb{F}_q$.
(c) Let w be defined as in part (b). You are to show that w is the largest integer such that a codeword of weight w or less in an $[n, k, d]$ code over $\mathbb{F}_q$ is determined uniquely up to scalar multiplication by its support. Do this by finding two vectors in $\mathbb{F}_q^{w+1}$ of weight $w + 1$ that generate a $[w + 1, 2, d]$ code over $\mathbb{F}_q$.   ◆

Exercise 452 Show that in the notation of the proof of the Assmus–Mattson Theorem,

$$A'_{z-t} = N(T) + (q-1) \sum_{\ell=1}^{z-d} \binom{t}{\ell} \lambda^\ell_{t-\ell}(z-\ell).$$ ◆

Example 8.4.3 Let $\mathcal{C}$ be the [24, 12, 8] self-dual extended binary Golay code. $\mathcal{C}$ has codewords of weight 0, 8, 12, 16, and 24 only. In the notation of the Assmus–Mattson Theorem, $n = 24$ and $d = d^\perp = 8$. As $\mathcal{C}$ is self-dual, $A_i = A_i^\perp$. The values of s and t must satisfy $t < 8$ and $s \le 8 - t$, where $s = |\{i \mid 0 < i \le 24 - t$ and $A_i^\perp \ne 0\}|$; the highest value of t that satisfies these conditions is $t = 5$, in which case $s = 3$. Therefore the vectors of weight i in $\mathcal{C}^\perp = \mathcal{C}$ hold 5-designs for $i = 8, 12$, and 16. From the weight distribution of the Golay code given in Exercise 384, we know that $A_{12} = 2576$. Hence the value of λ for the 5-design held by the codewords of weight 12 is $\lambda = 2576\binom{12}{5}/\binom{24}{5} = 48$. In Figure 8.4, we give the Pascal triangle for the 5-(24, 12, 48) design held by the weight 12 vectors in $\mathcal{C}$. ■

Exercise 453 Verify the entries in Figure 8.4 using the fact that there are 2576 blocks in the 5-(24, 12, 48) design. ◆

Example 8.4.4 Let $\mathcal{C}$ be the [12, 6, 6] self-dual extended ternary Golay code presented in Section 1.9.2; its weight enumerator is given in Example 7.3.2. $\mathcal{C}$ has codewords of weight 0, 6, 9, and 12 only. In the notation of the Assmus–Mattson Theorem, $w = 11$, and the highest value of t that satisfies the hypothesis is $t = 5$; hence $s = 1$. Thus the codewords in $\mathcal{C}$ of weights 6 and 9 hold 5-designs. These designs have parameters 5-(12, 6, 1) and 5-(12, 9, 35), respectively. ■

Exercise 454 Verify that the parameters of the 5-designs held by weight 6 and weight 9 codewords of the [12, 6, 6] extended ternary Golay code are 5-(12, 6, 1) and 5-(12, 9, 35). ◆

Exercise 455 Find the Pascal triangles for designs with the following parameters:
(a) 5-(12, 6, 1),
(b) 5-(12, 9, 35). ◆

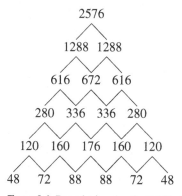

Figure 8.4 Pascal triangle of a 5-(24, 12, 48) design.

Exercise 456 In Section 1.10 we introduced the $[2^m, m + 1]$ Reed–Muller codes $\mathcal{R}(1, m)$. In Exercise 62 you showed that a generator matrix for $\mathcal{R}(1, m)$ can be obtained from a generator matrix of the $[2^m - 1, m]$ simplex code $\mathcal{S}_m$ by adjoining a column of 0s and then adding the all-one row to the matrix. The weight distribution of $\mathcal{S}_m$ is $A_0(\mathcal{S}_m) = 1$ and $A_{2^{m-1}}(\mathcal{S}_m) = 2^m - 1$ by Theorem 2.7.5. Therefore $\mathcal{R}(1, m)$ has weight distribution $A_0(\mathcal{R}(1, m)) = 1$, $A_{2^{m-1}}(\mathcal{R}(1, m)) = 2^{m+1} - 2$, and $A_{2^m}(\mathcal{R}(1, m)) = 1$. In addition, by Theorem 1.10.1 and Exercise 62 $\mathcal{R}(1, m)^\perp = \mathcal{R}(m - 2, m) = \widehat{\mathcal{H}}_m$, which is a $[2^m, 2^m - m - 1, 4]$ code.

(a) Prove that the codewords of weight 2^{m-1} in $\mathcal{R}(1, m)$ hold a 3-$(2^m, 2^{m-1}, \lambda)$ design.
(b) Prove that $\lambda = 2^{m-2} - 1$ in the 3-design of part (a). Hint: Use Theorem 8.1.6.
(c) Prove that the codewords of weight i with $4 \le i \le 2^m - 4$ in $\widehat{\mathcal{H}}_m$ hold 3-designs if $A_i(\widehat{\mathcal{H}}_m) \ne 0$.
(d) Find the parameter λ of the 3-$(8, 4, \lambda)$ design held by the words of weight 4 in $\widehat{\mathcal{H}}_3$.
(e) Find the weight distribution of the $[16, 11, 4]$ code $\widehat{\mathcal{H}}_4$. Hint: See Exercise 387.
(f) Find the parameters (block size, number of blocks, and λ) of all the 3-designs held by codewords of fixed weight in $\widehat{\mathcal{H}}_4$.   ◆

In Examples 8.3.4 and 8.3.5 we saw how to use designs to compute weight distributions of codes. We next use designs to compute coset weight distributions.

Example 8.4.5 In Example 1.11.7 we gave the complete coset distribution of the $[8, 4, 4]$ self-dual doubly-even binary code $\widehat{\mathcal{H}}_3$. This distribution is easily obtained by using the intersection numbers for the 3-$(8, 4, 1)$ design given in Figure 8.2. For example, in a coset of weight 1, the number of weight 3 vectors is the number of blocks of the 3-$(8, 4, 1)$ design containing a specific point; this value is $\lambda_1^0 = 7$. The number of weight 5 vectors is the number of blocks of the 3-$(8, 4, 1)$ design not containing a specific point; this value is $\lambda_0^1 = 7$. Note that because $\widehat{\mathcal{H}}_3$ contains the all-one vector, the number of weight 3 and weight 5 vectors must indeed agree. The remaining two vectors in the weight 1 coset must of course be a single weight 1 vector and a single weight 7 vector. Now consider a coset of weight 2. The weight 2 vectors in this coset come from adding the coset leader to $\mathbf{0}$ or adding the coset leader to a weight 4 vector in $\widehat{\mathcal{H}}_3$ whose support contains the support of the coset leader. Thus a weight 2 coset has $1 + \lambda_2^0 = 4$ vectors of weight 2. The weight 6 vectors in this coset come from adding the coset leader to the all-one codeword or adding the coset leader to a weight 4 vector in $\widehat{\mathcal{H}}_3$ whose support is disjoint from the support of the coset leader. Thus a weight 2 coset has $1 + \lambda_0^2 = 4$ vectors of weight 6. This leaves eight vectors of weight 4 in the coset. These weight 4 vectors come from a codeword of weight 4 whose support contains precisely one of the coordinates in the support of the coset leader. As there are two choices for this coordinate, the number of weight 4 vectors in a coset of weight 2 is $2\lambda_1^1 = 8$, as claimed.   ∎

Example 8.4.6 In this example, we compute the weight distribution of a weight 2 coset of the $[24, 12, 8]$ extended binary Golay code. Let $\mathcal{S}$ be a coset of weight 2. This coset has only even weight vectors. As the code contains the all-one codeword, we know that $A_i(\mathcal{S}) = A_{24-i}(\mathcal{S})$; so we only compute $A_i(\mathcal{S})$ for $i \le 12$. We use Figures 8.3 and 8.4 to do this computation. Let $\lambda_i^j(8)$ and $\lambda_i^j(12)$ denote the intersection numbers of the 5-$(24, 8, 1)$

Table 8.1 *Coset distribution of the* [24, 12, 8] *binary Golay code*

Coset	Number of vectors of given weight												Number
weight	0	1	2	3	4	5	6	7	8	9	10	11	12	of cosets
0	1	0	0	0	0	0	0	0	759	0	0	0	2576	1
1	0	1	0	0	0	0	0	253	0	506	0	1288	0	24
2	0	0	1	0	0	0	77	0	352	0	946	0	1344	276
3	0	0	0	1	0	21	0	168	0	640	0	1218	0	2024
4	0	0	0	0	6	0	64	0	360	0	960	0	1316	1771

design and the 5-(24, 12, 48) design, respectively. A weight 2 codeword in $\mathcal{S}$ is obtained only by adding the coset leader to **0**. So $A_2(\mathcal{S}) = 1$. There can be no weight 4 vectors in the coset. The weight 6 vectors come from the weight 8 codewords whose supports contain the support of the coset leader; thus $A_6(\mathcal{S}) = \lambda_2^0(8) = 77$. The weight 8 vectors come from the weight 8 codewords whose supports contain exactly one of the two coordinates of the support of the coset leader; thus $A_8(\mathcal{S}) = 2\lambda_1^1(8) = 352$. The weight 10 vectors come from the weight 8 codewords whose supports are disjoint from the support of the coset leader and the weight 12 codewords whose supports contain the support of the coset leader; thus $A_{10}(\mathcal{S}) = \lambda_0^2(8) + \lambda_2^0(12) = 946$. Finally, $A_{12}(\mathcal{S}) = 2^{12} - 2(1 + 77 + 352 + 946) = 1344$. The weight distributions of the cosets of weights 1, 3, and 4 are calculated in Exercise 457 with the final results given in Table 8.1. The number of cosets of each weight was determined in Example 8.3.2. ∎

Exercise 457 Verify that the weight distributions of the weight 1, 3, and 4 cosets of the [24, 12, 8] extended binary Golay code are as presented in Table 8.1. ◆

The Assmus–Mattson Theorem applied to $\mathcal{C}$ is most useful when $\mathcal{C}^\perp$ has only a few nonzero weights. This occurs for certain self-dual codes over $\mathbb{F}_2$, $\mathbb{F}_3$, and $\mathbb{F}_4$, as we will examine in more depth in Chapter 9. These are the cases to which the Assmus–Mattson Theorem has been most often applied. Curiously, the highest value of t for which t-designs have been produced from this theorem is $t = 5$ even though t-designs for higher values of t exist. A great many 5-designs have, in fact, been discovered as a consequence of this theorem. Designs held by codes may be stronger than predicted by the Assmus–Mattson Theorem. For example, there is a [22, 12, 8] Hermitian self-dual code over $\mathbb{F}_4$ whose weight 8 vectors hold a 2-design when the Assmus–Mattson Theorem says that the weight 8 vectors hold a 1-design [145].

Table 6.1 of Chapter 6 lists the binary odd-like duadic codes of length $n \leq 119$. The extended codes have the property that they have the same weight distribution as their duals; such codes are called *formally self-dual*. From the table, the number of nonzero weights in the extended codes can be evaluated. Using this information with the Assmus–Mattson Theorem all codewords of any weight (except 0) in the extended code hold t-designs for the following cases: $n + 1 = 18$ and $t = 2$; $n + 1 = 8, 32, 80,$ or 104 and $t = 3$; $n + 1 = 24$ or 48 and $t = 5$.

We conclude this section with a result that shows that t-designs are held by codewords of a code C of a fixed weight if the automorphism group $\Gamma\mathrm{Aut}(C)$ is t-transitive. Recall that $\Gamma\mathrm{Aut}(C)$ is *t-transitive* if for every pair of t-element ordered sets of coordinates, there is an element of the permutation group $\Gamma\mathrm{Aut}_{\mathrm{Pr}}(C)$ that sends the first set to the second set.

Theorem 8.4.7 *Let C be a code of length n over $\mathbb{F}_q$ where $\Gamma\mathrm{Aut}(C)$ is t-transitive. Then the codewords of any weight $i \geq t$ of C hold a t-design.*

Proof: Let $\mathcal{P}$ be the set of coordinates of the code and $\mathcal{B}$ the set of supports of the codewords of weight i. Of all the t-element subsets of $\mathcal{P}$, let $T_1 = \{i_1, \ldots, i_t\} \subseteq \mathcal{P}$ be one that is contained in the maximum number λ of blocks. Suppose these distinct blocks are $B_1, \ldots, B_\lambda$ whose supports are codewords $\mathbf{c}_1, \ldots, \mathbf{c}_\lambda$, respectively. Let T_2 be any other t-element subset of $\mathcal{P}$. Then there exists an automorphism g of C whose permutation part maps T_1 to T_2. The codewords $\mathbf{c}_1 g, \ldots, \mathbf{c}_\lambda g$ have distinct supports; these are blocks of $\mathcal{B}$. The maximality of λ shows that T_2 is in no more than λ blocks. Hence $(\mathcal{P}, \mathcal{B})$ is a t-(n, i, λ) design. $\square$

8.5   Codes from symmetric 2-designs

In the previous sections we focused on constructing designs from codes. In this section we reverse the concept: we investigate what can be said about a code over $\mathbb{F}_q$ that is generated by the rows of an incidence matrix of a design. We will be primarily interested in the minimum weight and dimension of the code. This is one situation where it is often more difficult to find the dimension of the code than the minimum distance. As an example of what can be said about the dimension of a code generated by the rows of the incidence matrix of a design, if the design is a Steiner t-design and the last row of its extended Pascal triangle indicates that the binary code generated is self-orthogonal, then the code has dimension at most half the length. For example, the last row of the extended Pascal triangle of the 5-(24, 8, 1) design in Figure 8.3 indicates that the blocks overlap in an even number of points and hence that the binary code generated by the incidence matrix of the design, which is the [24, 12, 8] extended binary Golay code, is self-orthogonal, a fact that we of course already knew. In the most important cases, the minimum weight is the block size and the minimum weight codewords have supports that are precisely the blocks.

In this section we will examine the dimension of codes arising from designs and also look at codewords whose weights are equal to the block size. If A is the incidence matrix of a t-design $(\mathcal{P}, \mathcal{B})$, let $C_q(A)$ be the linear code over $\mathbb{F}_q$ spanned by the rows of A. $C_q(A)$ is called the *code over $\mathbb{F}_q$ of the design* $(\mathcal{P}, \mathcal{B})$. We will focus on t-designs that are symmetric, and thus the number of blocks b equals the number of points v. In all other cases of nontrivial designs, $b > v$, a fact known as Fisher's inequality; see [204]. By [4], a nontrivial symmetric t-design does not exist for $t > 2$. So we will only consider symmetric 2-designs. A great deal can be said about the dimension of $C_q(A)$ in this case, information that has played a significant role in the study of projective planes as we will see in Section 8.6. In that section we will also consider the question of the minimum weight of $C_q(A)$ and what codewords have that minimum weight.

Before proceeding with our results, a few special properties of 2-(v, k, λ) designs are required. Let J_v be the $v \times v$ matrix all of whose entries are 1 and I_v the $v \times v$ identity matrix.

Lemma 8.5.1 *Let A be the incidence matrix of a 2-(v, k, λ) design. The following hold:*
(i) $A^{\mathrm{T}}A = (\lambda_1 - \lambda)I_v + \lambda J_v$ *and* $\lambda(v - 1) = \lambda_1(k - 1)$.
(ii) *If the design is symmetric, then:*
    (a) $\lambda_1 = k$ *and* $\lambda(v - 1) = k(k - 1)$,
    (b) $A^{\mathrm{T}}A = AA^{\mathrm{T}} = (k - \lambda)I_v + \lambda J_v$,
    (c) *every pair of blocks intersect in exactly λ points,*
    (d) $\det(A) = \pm k(k - \lambda)^{\frac{1}{2}(v-1)}$, *and*
    (e) *if v is even, $k - \lambda$ is a perfect square.*

Proof: The ith diagonal entry of $A^{\mathrm{T}}A$ counts the number of blocks containing the ith point; this is λ_1. The entry in row i and column $j \neq i$ of $A^{\mathrm{T}}A$ is the number of blocks containing both the ith and jth points; this is λ, proving the first equation in (i). The second equation follows from Theorem 8.1.3 with $t = 2$.

Now assume the design is symmetric so that $b = \lambda_0 = v$. Theorem 8.1.6 and part (i) yield (a). With $t = 2$ in Theorem 8.1.3,

$$\lambda = \lambda_2 = \lambda_1 \frac{k - 1}{v - 1}. \tag{8.3}$$

By Exercise 458,

$$\det(A^{\mathrm{T}}A) = (k - \lambda)^{v-1}(v\lambda + k - \lambda). \tag{8.4}$$

In particular, $A^{\mathrm{T}}A$ is nonsingular as $\lambda < k = \lambda_1$ by (8.3). Since A is a square matrix, it must then be nonsingular. As each row of A has k 1s, $AJ_v = kJ_v$. As each column of A has λ_1 1s, $J_vA = \lambda_1 J_v$. Since $\lambda_1 = k$, $J_vA = AJ_v$. Hence A commutes with J_v. This implies that $AA^{\mathrm{T}} = AA^{\mathrm{T}}AA^{-1} = A((k - \lambda)I_v + \lambda J_v)A^{-1} = ((k - \lambda)I_v + \lambda J_v)AA^{-1}$ and so $AA^{\mathrm{T}} = A^{\mathrm{T}}A$ giving (b). The entry in row i and column $j \neq i$ of AA^{T} is λ; this indicates that distinct blocks intersect in exactly λ points, which is (c). By (a), (8.3), and (8.4),

$$\det(A^{\mathrm{T}}A) = (k - \lambda)^{v-1}k^2.$$

This yields (d), which implies (e) as A is an integer matrix.   □

Exercise 458 Prove that

$$\det[(k - \lambda)I_v + \lambda J_v] = (k - \lambda)^{v-1}(v\lambda + k - \lambda). \qquad \blacklozenge$$

We are now ready to prove the two main theorems of this section that give the dimension of $\mathcal{C}_q(A)$ over $\mathbb{F}_q$ when A is the incidence matrix of a symmetric 2-design. We shall find it helpful to consider another code $\mathcal{D}_q(A)$ over $\mathbb{F}_q$ defined to be the code spanned by the rows of $A^\triangle$, where $A^\triangle$ is the matrix whose rows are the differences of all pairs of rows of A. Recall that $k - \lambda$ is the order of the design. If p is the characteristic of $\mathbb{F}_q$, the dimension can be bounded above or found exactly depending upon the divisibility of both $k - \lambda$ and k by p.

Theorem 8.5.2 *Let A be the incidence matrix of the symmetric 2-(v, k, λ) design $(\mathcal{P}, \mathcal{B})$. If p is the characteristic of $\mathbb{F}_q$, then the following hold:*

(i) $\mathcal{D}_q(A)$ is a subcode of $\mathcal{C}_q(A)$ with codimension at most 1 in $\mathcal{C}_q(A)$.

(ii) If $p \mid (k - \lambda)$, then $\mathcal{C}_q(A) \subseteq \mathcal{D}_q(A)^{\perp}$, and $\mathcal{D}_q(A)$ is self-orthogonal of dimension at most $v/2$.

(iii) If $p \mid (k - \lambda)$ and $p \mid k$, then $\mathcal{C}_q(A)$ is self-orthogonal of dimension at most $v/2$.

(iv) If $p \mid (k - \lambda)$ and $p \nmid k$, then $\mathcal{D}_q(A)$ is of codimension 1 in $\mathcal{C}_q(A)$, and $\mathcal{D}_q(A)$ has dimension less than $v/2$.

(v) If $p \nmid (k - \lambda)$ and $p \mid k$, then $\mathcal{C}_q(A)$ has dimension $v - 1$.

(vi) If $p \nmid (k - \lambda)$ and $p \nmid k$, then $\mathcal{C}_q(A)$ has dimension v.

Proof: Let $\mathbf{r}_1, \ldots, \mathbf{r}_v$ be the rows of A associated to the blocks $B_1, \ldots, B_v$ of $\mathcal{B}$. The supports of the rows of A^{Δ} are the symmetric differences of all pairs of blocks. Clearly, $\mathcal{C}_q(A) = \mathrm{span}\{\mathbf{r}_1\} + \mathcal{D}_q(A)$, and $\mathcal{D}_q(A)$ is of codimension at most 1 in $\mathcal{C}_q(A)$, giving (i).

We first consider the case $p \mid (k - \lambda)$. We have $\mathbf{r}_i \cdot \mathbf{r}_i \equiv k \equiv \lambda \pmod{p}$, and for $i \neq j$, $\mathbf{r}_i \cdot \mathbf{r}_j \equiv \lambda \pmod{p}$ by Lemma 8.5.1(ii)(c). Therefore $\mathbf{r}_i \cdot \mathbf{r}_j \equiv \lambda \pmod{p}$ for all i and j. So $(\mathbf{r}_i - \mathbf{r}_j) \cdot \mathbf{r}_m \equiv 0 \pmod{p}$ for all i, j, and m. Thus $\mathcal{D}_q(A)$ is self-orthogonal of dimension at most $v/2$, and $\mathcal{C}_q(A) \subseteq \mathcal{D}_q(A)^{\perp}$, proving (ii). If in addition $p \mid k$, then $p \mid \lambda$ and $\mathcal{C}_q(A)$ is self-orthogonal giving (iii) as $\mathbf{r}_i \cdot \mathbf{r}_j \equiv \lambda \equiv 0 \pmod{p}$. If $p \nmid k$, then $p \nmid \lambda$ and $\mathbf{r}_1 \cdot \mathbf{r}_1 \not\equiv 0 \pmod{p}$ implying that $\mathcal{D}_q(A)$ is of codimension 1 in $\mathcal{C}_q(A)$. As $\mathcal{D}_q(A)$ is properly contained in its dual by (ii), $\mathcal{D}_q(A)$ cannot be of dimension $v/2$, giving (iv).

Now assume that $p \nmid (k - \lambda)$. If $p \mid k$, then $\mathbf{r}_i \cdot \mathbf{1} \equiv k \equiv 0 \pmod{p}$. Thus $\mathcal{C}_q(A)$ is of dimension at most $v - 1$. Associate column i of A with the point $p_i \in \mathcal{P}$. Let $p_i, p_j \in \mathcal{P}$ with $i \neq j$. There are λ_1^1 blocks in $\mathcal{B}$ that contain p_i and not p_j. Thus the sum $\mathbf{s}_j$ of all rows of A associated with the blocks not containing p_j has 0 in column j and λ_1^1 in all other columns. By Corollary 8.2.2, $\lambda_1^1 = \lambda_1 - \lambda_2 = k - \lambda \not\equiv 0 \pmod{p}$. As $\mathrm{span}\{\mathbf{s}_j \mid 1 \leq j \leq v\}$ is at least $(v - 1)$-dimensional, $\mathcal{C}_q(A)$ has dimension $v - 1$ giving (v). If $p \nmid k$, by Lemma 8.5.1, $\det(A) \not\equiv 0 \pmod{p}$, yielding (vi). $\square$

From the coding theory standpoint, if $p \nmid (k - \lambda)$, the code $\mathcal{C}_q(A)$ is uninteresting. When $p \mid (k - \lambda)$, the code can potentially be worth examining. If in addition $p^2 \nmid (k - \lambda)$, the dimension of $\mathcal{C}_q(A)$ can be exactly determined.

Theorem 8.5.3 Let A be the incidence matrix of the symmetric 2-(v, k, λ) design $(\mathcal{P}, \mathcal{B})$. Let p be the characteristic of $\mathbb{F}_q$ and assume that $p \mid (k - \lambda)$, but $p^2 \nmid (k - \lambda)$. Then v is odd, and the following hold:

(i) If $p \mid k$, then $\mathcal{C}_q(A)$ is self-orthogonal and has dimension $(v-1)/2$.

(ii) If $p \nmid k$, then $\mathcal{D}_q(A) = \mathcal{C}_q(A)^{\perp} \subset \mathcal{C}_q(A) = \mathcal{D}_q(A)^{\perp}$. Furthermore, $\mathcal{C}_q(A)$ has dimension $(v + 1)/2$ and $\mathcal{D}_q(A)$ is self-orthogonal.

Proof: By Lemma 8.5.1(ii)(e), if v is even, $k - \lambda$ must be a perfect square contradicting $p \mid (k - \lambda)$ but $p^2 \nmid (k - \lambda)$. So v is odd.

View A as an integer matrix. Adding columns $1, 2, \ldots, v - 1$ of A to column v gives a matrix A_1 with ks in column v because each row of A has k 1s. Subtracting row v of A_1

from each of the previous rows gives a matrix A_2 with

$$A_2 = \begin{bmatrix} & & & 0 \\ & & & 0 \\ & H & & \vdots \\ & & & 0 \\ a_1 & \cdots & a_{v-1} & k \end{bmatrix},$$

where H is a $(v-1) \times (v-1)$ integer matrix and each a_i is an integer. Furthermore adding a row, or column, of a matrix to another row, or column, does not affect the determinant. Hence $\det(A) = k \det(H)$. By the theory of invariant factors of integer matrices (see for example, [187, Appendix C]), there are $(v-1) \times (v-1)$ integer matrices U and V of determinant ± 1 such that $UHV = \mathrm{diag}(h_1, \ldots, h_{v-1})$, where $h_1, \ldots, h_{v-1}$ are integers satisfying $h_i \mid h_{i+1}$ for $1 \le i < v - 1$. Hence,

$$\det(A) = k \det(H) = k h_1 h_2 \cdots h_{v-1} = \pm k(k - \lambda)^{\frac{1}{2}(v-1)}.$$

Furthermore, the rank r of A over $\mathbb{F}_q$, which is the dimension of $C_q(A)$, is the number of values among $\{k, h_1, \ldots, h_{v-1}\}$ that are not zero modulo p. As $p \mid (k - \lambda)$ but $p^2 \nmid (k - \lambda)$, at most $(v-1)/2$ of the h_is are 0 modulo p. Hence the dimension of $C_q(A)$ is at least $(v-1)/2$ if $p \mid k$ and at least $(v-1)/2 + 1 = (v+1)/2$ if $p \nmid k$. Thus if $p \mid k$, by Theorem 8.5.2(iii), $C_q(A)$ has dimension $(v-1)/2$ as v is odd, giving (i). If $p \nmid k$, $D_q(A)$ has codimension 1 in $C_q(A)$ and dimension less than $v/2$ by Theorem 8.5.2(iv), implying that $C_q(A)$ must have dimension $(v+1)/2$. This gives part of (ii), with the rest now following from Theorem 8.5.2. □

The condition that $p \mid (k - \lambda)$ but $p^2 \nmid (k - \lambda)$ is crucial to the theorem, as can be seen in the proof. When $p^2 \mid (k - \lambda)$, the code $C_q(A)$ may have smaller dimension, as we will discover in the next section.

Exercise 459 This exercise provides an alternate proof of Theorem 8.5.3(ii) that does not rely on invariant factors. The dimension r of $C_q(A)$ is the rank of A over $\mathbb{F}_q$, which is the rank of A over $\mathbb{F}_p$ as A has integer entries. Hence r is also the dimension of $C_p(A)$. Assume that the points of $\mathcal{P}$ have been ordered so that the right-most r coordinates form an information set for $C_p(A)$.

(a) Explain why there exists a $(v - r) \times r$ matrix B with entries in $\mathbb{F}_p$ such that $[I_{v-r} \ B]$ is a parity check matrix for $C_p(A)$.

(b) Let M be the $v \times v$ matrix

$$M = \begin{bmatrix} I_{v-r} & B \\ O & I_r \end{bmatrix}$$

viewed as an integer matrix. Explain why the integer matrix $A_1 = MA^{\mathsf{T}}$ satisfies $\det(A_1) = \det(A)$ and why each entry in the first $v - r$ rows of A_1 is a multiple of p.

(c) Prove that $p^{v-r} \mid \det(A)$.

(d) Prove that as $p \mid (k - \lambda)$, $p^2 \nmid (k - \lambda)$, and $p \nmid k$, then $v - r \le (v-1)/2$ and the dimension of $C_q(A)$ is $(v+1)/2$. ◆

We know that codewords in $C_q(A)$ whose supports are blocks have weight k. A natural question is to ask whether codewords of weight k exist whose supports are not blocks; this becomes even more interesting when k is the minimum weight of the code. We begin our investigation of this question with a lemma found in [187].

Lemma 8.5.4 *Let* $(\mathcal{P}, \mathcal{B})$ *be a symmetric* 2-(v, k, λ) *design. Let* $S \subset \mathcal{P}$ *with* $|S| = k$. *Suppose that* S *meets all blocks of* $\mathcal{B}$ *in at least* λ *points. Then* S *is a block.*

Proof: For all $B \in \mathcal{B}$, let $n_B = |S \cap B| - \lambda$; by assumption n_B is a nonnegative integer. Also,

$$\sum_{B \in \mathcal{B}} n_B = \sum_{B \in \mathcal{B}} |S \cap B| - v\lambda. \tag{8.5}$$

Let $\mathcal{X} = \{(x, B) \mid B \in \mathcal{B}, x \in S \cap B\}$. If $x \in S$, x is in k blocks B by Lemma 8.5.1. Thus $|\mathcal{X}| = k^2$. But for every block $B \in \mathcal{B}$, there are $|S \cap B|$ points $x \in S \cap B$. So $|\mathcal{X}| = \sum_{B \in \mathcal{B}} |S \cap B|$. Thus $\sum_{B \in \mathcal{B}} |S \cap B| = k^2$ implying from (8.5) and Lemma 8.5.1 that

$$\sum_{B \in \mathcal{B}} n_B = k - \lambda. \tag{8.6}$$

Let $\mathcal{Y} = \{(x, y, B) \mid B \in \mathcal{B}, x, y \in S \cap B, x \neq y\}$. There are $k(k-1)$ ordered pairs of elements in S each in exactly λ blocks. Thus $|\mathcal{Y}| = k(k-1)\lambda$. But for every block $B \in \mathcal{B}$, there are $|S \cap B|(|S \cap B| - 1) = (n_B + \lambda)(n_B + \lambda - 1)$ pairs of points in $S \cap B$. Thus

$$\sum_{B \in \mathcal{B}} (n_B + \lambda)(n_B + \lambda - 1) = k(k-1)\lambda.$$

Expanding the left-hand side and using (8.6), we obtain (see Exercise 460)

$$\sum_{B \in \mathcal{B}} n_B^2 = (k - \lambda)^2. \tag{8.7}$$

As the n_B are nonnegative integers, by (8.6) and (8.7), the only possibility is $n_B = 0$ for all but one B and $n_B = k - \lambda$ for this one B; but then $|S \cap B| = k$, implying $S = B$. $\square$

Exercise 460 In the proof of Lemma 8.5.4, verify (8.7). ◆

We now show that certain codewords of weight k in the code $C_q(A)$, where A is the incidence matrix of a symmetric 2-(v, k, λ) design $(\mathcal{P}, \mathcal{B})$, have supports that are blocks of the design. If $S \subset \mathcal{P}$, let $\mathbf{c}_S$ denote the vector in $\mathbb{F}_q^v$ whose nonzero components are all 1 and whose support is S.

Theorem 8.5.5 *Let* $(\mathcal{P}, \mathcal{B})$ *be a symmetric* 2-(v, k, λ) *design with incidence matrix* A. *Let* $p > \lambda$ *be a prime dividing* $k - \lambda$. *Suppose* $\mathbb{F}_q$ *has characteristic* p. *Finally, let* $S \subset \mathcal{P}$ *have size* k. *If* $\mathbf{c}_S \in C_q(A)$, *then* S *is a block in* $\mathcal{B}$.

Proof: As $\mathbf{c}_S \in C_q(A)$ and $C_q(A)$ is generated by $\{\mathbf{c}_B \mid B \in \mathcal{B}\}$, $\mathbf{c}_S = \sum_{B \in \mathcal{B}} a_B \mathbf{c}_B$ for $a_B \in \mathbb{F}_q$. Let B' be a block in $\mathcal{B}$. Then

$$\mathbf{c}_S \cdot \mathbf{c}_{B'} = \sum_{B \in \mathcal{B}} a_B \mathbf{c}_B \cdot \mathbf{c}_{B'}. \tag{8.8}$$

For $B \in \mathcal{B}$, as the only entries in $\mathbf{c}_B$ and $\mathbf{c}_{B'}$ are 0s and 1s, $\mathbf{c}_B \cdot \mathbf{c}_{B'}$ is actually in $\mathbb{F}_p \subseteq \mathbb{F}_q$ and clearly equals $|B \cap B'|$ modulo p. If $B \neq B'$, $|B \cap B'| = \lambda$ by Lemma 8.5.1; if $B = B'$, $|B \cap B'| = k$. As $p \mid (k - \lambda)$, $k \equiv \lambda \pmod{p}$. Thus $|B \cap B'| \equiv k \pmod{p}$ for all $B \in \mathcal{B}$ and the right-hand side of (8.8) becomes

$$\sum_{B \in \mathcal{B}} a_B \mathbf{c}_B \cdot \mathbf{c}_{B'} = \sum_{B \in \mathcal{B}} k a_B,$$

implying

$$\mathbf{c}_S \cdot \mathbf{c}_{B'} = \sum_{B \in \mathcal{B}} k a_B. \tag{8.9}$$

The codeword $\mathbf{c}_S$ is obtained by multiplying each row of A corresponding to B by a_B thus creating a matrix A' and adding the rows of A' in $\mathbb{F}_q$. The sum of all the entries in A' is $\sum_{B \in \mathcal{B}} k a_B$ as the row corresponding to B is the sum of a_B taken k times. This must be the sum of the entries in $\mathbf{c}_S$ in $\mathbb{F}_q$. But the sum of the entries in $\mathbf{c}_S$ is the sum of k 1s in $\mathbb{F}_q$. As above, $\mathbf{c}_S \cdot \mathbf{c}_{B'}$ is again in $\mathbb{F}_p \subseteq \mathbb{F}_q$ and equals $|S \cap B'|$ modulo p. Hence (8.9) becomes

$$|S \cap B'| \equiv k \equiv \lambda \pmod{p}.$$

As $p > \lambda$, we must have $|S \cap B'| \geq \lambda$ for all $B' \in \mathcal{B}$. The result follows from Lemma 8.5.4. $\qquad\qquad\square$

This theorem shows that if $p > \lambda$, the codewords in $\mathcal{C}_q(A)$ of weight k that are multiples of binary vectors must have supports that are blocks. There are examples of designs showing that you cannot drop either the condition $p > \lambda$ or the condition that the codeword of weight k must be constant on its support.

Example 8.5.6 There are three inequivalent symmetric 2-(16, 6, 2) designs. If A is the incidence matrix of one of these designs, then $\mathcal{C}_2(A)$ has dimension 6, 7, or 8. If we let A_i be the incidence matrix where $\mathcal{C}_i = \mathcal{C}_2(A_i)$ has dimension i, the weight enumerators are given in [302]:

$$W_{\mathcal{C}_6}(x, y) = y^{16} + 16x^6 y^{10} + 30x^8 y^8 + 16x^{10} y^6 + x^{16},$$
$$W_{\mathcal{C}_7}(x, y) = y^{16} + 4x^4 y^{12} + 32x^6 y^{10} + 54x^8 y^8 + 32x^{10} y^6 + 4x^{12} y^4 + x^{16},$$
$$W_{\mathcal{C}_8}(x, y) = y^{16} + 12x^4 y^{12} + 64x^6 y^{10} + 102x^8 y^8 + 64x^{10} y^6 + 12x^{12} y^4 + x^{16}.$$

Notice that the 16 blocks are precisely the supports of the weight 6 codewords only in the first code. In the other cases there are codewords of weight 6 whose supports are not blocks. Thus the condition $p > \lambda$ in Theorem 8.5.5 is necessary. Notice that in the last two cases the minimum weight is less than the block size. $\qquad\qquad\blacksquare$

Example 8.5.7 In Exercise 471, you are asked to construct an (11, 5, 2) cyclic difference set that will yield a symmetric 2-(11, 5, 2) design $\mathcal{D}$. Let A be its incidence matrix. By Theorem 8.5.3(ii), $\mathcal{C}_3(A)$ is an [11, 6] code. It turns out that the code has minimum weight 5 and that there are codewords of weight 5 whose supports are not blocks of the original design; these are codewords with nonzero components a mixture of both 1 and 2. Thus the condition that the codeword of weight k must be constant on its support is also necessary in Theorem 8.5.5. $\qquad\qquad\blacksquare$

There are other designs related to symmetric 2-designs that have a strong relationship to codes. For example, a *quasi-symmetric design* is a 2-design in which every pair of distinct blocks intersects in either α or β points. Whether or not there exists a quasi-symmetric 2-(49, 9, 6) design with distinct blocks intersecting in 1 or 3 points was an open question until 1995. If such a design exists, it has 196 blocks by Theorem 8.1.6. Its existence is demonstrated as follows. One possible weight enumerator of a binary self-dual [50, 25, 10] code $\mathcal{C}$ is $W_\mathcal{C}(x, y) = y^{50} + 196x^{10}y^{40} + 11368x^{12}y^{38} + \cdots$. If a code with this weight enumerator exists, the supports of all 196 minimum weight codewords amazingly must have one coordinate in common. Deleting this common coordinate from these supports produces a quasi-symmetric 2-(49, 9, 6) design with distinct blocks intersecting in 1 or 3 points. In [151], four inequivalent codes of this type are given, establishing the existence of at least four quasi-symmetric designs. Further discussion on quasi-symmetric designs and other connections between codes and designs can be found in [332].

We conclude this section with a result found in [13] that applies to codes holding 2-designs that may not be symmetric. We include it here because its proof is reminiscent of other proofs in this section.

Theorem 8.5.8 *Let $\mathcal{C}$ be a self-orthogonal $[n, k, d]$ code over $\mathbb{F}_q$ with $d \geq 2$. Suppose that the minimum weight codewords of $\mathcal{C}$ hold a 2-design. Then*

$$d \geq 1 + \sqrt{n - 1}.$$

Proof: Let B be a block in the 2-(n, d, λ) design $(\mathcal{P}, \mathcal{B})$ held by the minimum weight codewords of $\mathcal{C}$. Suppose $\mathcal{B}$ has b blocks. Let n_i be the number of blocks in $\mathcal{B}$, excluding B, that meet B in i points. Form a $(b - 1) \times d$ matrix M with columns indexed by the d points of B and rows indexed by the $b - 1$ blocks of $\mathcal{B}$ excluding B. The entry in row j and column k is 1 provided the kth point of B is in the block corresponding to the jth row of M. All other entries of M are 0.

We verify the following two equations:

$$\sum_{i=0}^{d-1} in_i = d(\lambda_1 - 1) \quad \text{and} \tag{8.10}$$

$$\sum_{i=0}^{d-1} i(i - 1)n_i = d(d - 1)(\lambda - 1). \tag{8.11}$$

In M there are n_i rows with i 1s. Thus the total number of 1s in M is $\sum_{i=0}^{d-1} in_i$. As every point of the design is in λ_1 blocks, there are $\lambda_1 - 1$ 1s in every column of M and hence a total of $d(\lambda_1 - 1)$ 1s in M. Equating these two counts gives (8.10). In a row of M with i 1s, there are $\binom{i}{2}$ pairs of 1s. Thus in the rows of M there are a total of $\sum_{i=0}^{d-1} \binom{i}{2}n_i$ pairs of 1s. As every pair of points is in λ blocks, fixing any pair of columns of M, there are $\lambda - 1$ rows of M that contain 1s in these two columns. Thus there are a total of $\binom{d}{2}(\lambda - 1)$ pairs of 1s in the rows of M. Equating these two counts and doubling the result gives (8.11).

As $\mathcal{C}$ is self-orthogonal, no pair of blocks of $\mathcal{B}$ can intersect in exactly one point. Hence $n_1 = 0$. Thus $i(i - 1)n_i \geq in_i$ for all i. Using this with (8.10) and (8.11) yields

$(d-1)(\lambda-1) \geq (\lambda_1-1)$. As $\lambda_1(d-1) = (n-1)\lambda$ by Theorem 8.5.1(i),

$$(d-1)^2(\lambda-1) \geq (\lambda_1-1)(d-1) = (n-1)\lambda - (d-1).$$

So $(d^2-2d+2-n)\lambda \geq (d-1)(d-2) \geq 0$, implying $d^2-2d+2-n \geq 0$. Solving this inequality yields the result. □

Example 8.5.9 We use this result to find a lower bound on the minimum weight of certain codes.

- Let $\mathcal{G}_{24}$ be the extended binary Golay code. This code is self-dual and doubly-even. In Section 10.1.2 we will see that PAut($\mathcal{G}_{24}$) is 5-transitive. In particular, PAut($\mathcal{G}_{24}$) is 2-transitive and, by Theorem 8.4.7, the minimum weight codewords of $\mathcal{G}_{24}$ hold a 2-design. Theorem 8.5.8 shows the minimum weight is at least $1 + \sqrt{23}$, that is, at least 6. Since $\mathcal{G}_{24}$ is doubly-even, the minimum weight is at least 8.
- Let $\mathcal{G}_{12}$ be the extended self-dual ternary Golay code. In Section 10.4.2 we show that MAut($\mathcal{G}_{12}$) is 5-transitive and so 2-transitive. By Theorem 8.4.7, the minimum weight codewords of $\mathcal{G}_{12}$ hold a 2-design, and Theorem 8.5.8 shows the minimum weight is at least $1 + \sqrt{11}$. Thus the minimum weight is at least 5. As all codewords of ternary self-dual codes have weights a multiple of 3, the minimum weight is at least 6.
- Let $\mathcal{C}$ be a self-orthogonal extended primitive BCH code of length q^m, where q is a prime power. By Theorem 5.1.9, $\mathcal{C}$ is affine-invariant and hence PAut($\mathcal{C}$) contains $GA_1(q^m)$, which is 2-transitive. Hence Theorem 8.4.7 shows the minimum weight codewords hold a 2-design, and Theorem 8.5.8 shows the minimum weight is at least $1 + \sqrt{q^m-1}$. ■

8.6 Projective planes

We now apply the results of the previous section to the specific case of projective planes.

In Section 8.1, we indicated that a symmetric 2-$(v, k, 1)$ design is precisely a projective plane as defined in Section 6.5; the lines of the plane are the blocks of the design. The order μ of a symmetric 2-$(v, k, 1)$ design is $k - 1$. (In many books the order of a plane is denoted by n; however, because we so often use n to denote the length of a code, we choose the symbol μ to denote the order of a plane.) By Theorem 8.1.6, $v = \mu^2 + \mu + 1$. Thus a projective plane is a symmetric 2-$(\mu^2 + \mu + 1, \mu + 1, 1)$ design. However, a 2-$(\mu^2 + \mu + 1, \mu + 1, 1)$ design is automatically symmetric because Theorem 8.1.6 shows that such a design has $\mu^2 + \mu + 1$ blocks. We now prove this basic equivalence.

Theorem 8.6.1 *A set of points and lines is a finite projective plane if and only if it is a symmetric 2-$(v, k, 1)$ design with $v \geq 4$.*

Proof: Let $\mathcal{P}$ and $\mathcal{L}$ denote the points and lines of a finite projective plane. We show that $(\mathcal{P}, \mathcal{B})$ is a symmetric 2-$(v, k, 1)$ design where $\mathcal{B} = \mathcal{L}$ and v is the number of points $\mathcal{P}$. We first must show that every line has the same number of points. Let ℓ be a line with k points where k is as large as possible. Let m be any other line. Pick any point P not on either ℓ

or m; such a point exists because the plane contains at least four points no three of which are on the same line. There are k distinct lines through P and the k points on ℓ. Each of these k lines intersects m in distinct points, since two points determine a unique line. Thus m contains at least k points. By maximality of k, m had exactly k points. So all lines have k points. As every pair of distinct points determines a unique line, $(\mathcal{P}, \mathcal{B})$ is a 2-$(v, k, 1)$ design. We need to show that it is symmetric. Let b be the number of lines and let

$$\mathcal{X} = \{(P, \ell) \mid P \in \mathcal{P}, \ell \in \mathcal{L}, P \in \ell\}.$$

As every line contains k points, $|\mathcal{X}| = bk$. If P is a point, then there are k lines through P as follows. Let m be a line not containing P. Each of its k points determines a unique line through P and hence there are at least k lines through P. Any line through P must intersect m in a unique point, with different lines intersecting at different points, and so there are at most k lines through P. Thus there are k lines through P. Hence $|\mathcal{X}| = vk$; equating the two counts for $|\mathcal{X}|$ gives $b = v$. Thus the design is symmetric; as a projective plane has at least four points, $v \geq 4$.

Now assume $(\mathcal{P}, \mathcal{B})$ is a symmetric 2-$(v, k, 1)$ design. Then every pair of points is in a unique block, and by Lemma 8.5.1, two distinct blocks intersect in a unique point. As $b = v \geq 4$, there are at least four points, no three of which are in the same block. Hence $(\mathcal{P}, \mathcal{B})$ is a projective plane.   □

From this proof, we now have some basic properties of projective planes.

Theorem 8.6.2 *A projective plane of order μ satisfies the following properties*:
(i)   *The plane has $\mu^2 + \mu + 1$ points and $\mu^2 + \mu + 1$ lines.*
(ii) *Distinct points determine a unique line and distinct lines intersect in a unique point.*
(iii) *Every line contains exactly $\mu + 1$ points and every point is on exactly $\mu + 1$ lines.*

We can determine the minimum distance, find the minimum weight codewords, and either bound or find exactly the dimension of $\mathcal{C}_2(A)$ when A is the incidence matrix of a projective plane of order μ.

Theorem 8.6.3 *Let A be the incidence matrix of a projective plane of order μ arising from the symmetric 2-$(v, k, 1)$ design $(\mathcal{P}, \mathcal{B})$.*
(i) *If $\mu = k - 1$ is odd, then $\mathcal{C}_2(A)$ is the $[v, v - 1, 2]$ code consisting of all even weight codewords in $\mathbb{F}_2^v$.*
(ii) *If $\mu = k - 1$ is even, then the extended code $\widehat{\mathcal{C}}_2(A)$ of $\mathcal{C}_2(A)$ is self-orthogonal of dimension at most $(v + 1)/2$. The minimum weight of $\mathcal{C}_2(A)$ is $\mu + 1$ and the lines of $(\mathcal{P}, \mathcal{B})$ are precisely the supports of the vectors of this weight.*

Proof: Suppose that $\mu = k - 1$ is odd. Then $2 \nmid \mu$ and $2 \mid k$. By Theorem 8.5.2, $\mathcal{C}_2(A)$ has dimension $v - 1$. As all blocks have an even number k of points, $\mathcal{C}_2(A)$ must be an even code. Therefore (i) holds.

Now suppose that $\mu = k - 1$ is even. Then as the blocks have odd weight and distinct blocks meet in one point, the extended code $\widehat{\mathcal{C}}_2(A)$ is generated by a set of codewords of even weight whose supports agree on two coordinates, one being the extended coordinate. Thus $\widehat{\mathcal{C}}_2(A)$ is self-orthogonal and hence has dimension at most $(v + 1)/2$. Also as $\widehat{\mathcal{C}}_2(A)$ is self-orthogonal, two even weight vectors in $C_2(A)$ meet in an even number of coordinates while two odd weight vectors in $C_2(A)$ meet in an odd number of coordinates. We only need to find the minimum weight codewords of $C_2(A)$.

Suppose $\mathbf{x}$ is a nonzero vector in $C_2(A)$ of even weight where supp($\mathbf{x}$) contains the point P. Let ℓ be any of the $\mu + 1$ lines through P. As $\mathbf{x}$ has even weight, supp($\mathbf{x}$) must intersect ℓ in an even number of points. Thus ℓ meets supp($\mathbf{x}$) at P and another point Q_ℓ. Since two points determine a unique line, the $\mu + 1$ lines ℓ determine $\mu + 1$ different points Q_ℓ showing that wt($\mathbf{x}$) $\geq \mu + 2$.

Suppose now that $\mathbf{x}$ is a vector in $C_2(A)$ of odd weight. If wt($\mathbf{x}$) $= 1$, then there is a codeword $\mathbf{c}$ whose support is a line containing supp($\mathbf{x}$); hence $\mathbf{x} + \mathbf{c}$ is a codeword of even weight μ, a contradiction to the above. So assume supp($\mathbf{x}$) contains points P and Q. Now supp($\mathbf{x}$) must meet each line in an odd number of points. Assume supp($\mathbf{x}$) is not the unique line ℓ containing P and Q. Then supp($\mathbf{x}$) must meet ℓ in at least one additional point R. If $\ell \subseteq$ supp($\mathbf{x}$), then wt($\mathbf{x}$) $\geq (\mu + 1) + 2$ as both $\mu + 1$ and wt($\mathbf{x}$) are odd. If $\ell \not\subseteq$ supp($\mathbf{x}$), then there is a point S on ℓ with S not in supp($\mathbf{x}$). The μ lines through S, excluding ℓ, meet supp($\mathbf{x}$) in at least μ distinct points none of which are P, Q, or R. Thus if supp($\mathbf{x}$) $\neq \ell$, wt($\mathbf{x}$) $\geq \mu + 3$ completing part (ii).   □

We remark that part (ii) of this theorem is consistent with Theorem 8.5.5; since $p = 2 > \lambda = 1$, the codewords of weight k in $C_2(A)$ are precisely those whose supports are the blocks. Theorem 8.6.3 indicates that the most interesting codes arise when μ is even. If $2 \mid \mu$ but $4 \nmid \mu$, we can give even more information about the code $C_2(A)$.

Theorem 8.6.4 *Let* $(\mathcal{P}, \mathcal{B})$ *be a projective plane of order* $\mu \equiv 2 \pmod{4}$ *with incidence matrix* A. *Then the following hold:*
(i) $C_2(A)$ *is a* $[v, (v + 1)/2, \mu + 1]$ *code, where* $v = \mu^2 + \mu + 1$. *Furthermore,* $C_2(A)$ *contains the all-one vector.*
(ii) *The code* $\mathcal{D}_2(A)$ *is the subcode of* $C_2(A)$ *consisting of all its even weight vectors,* $\mathcal{D}_2(A) = C_2(A)^\perp$, $\mathcal{D}_2(A)$ *is doubly-even, and* $C_2(A)$ *is generated by* $\mathcal{D}_2(A)$ *and the all-one vector.*
(iii) *The extended code* $\widehat{C}_2(A)$ *of* $C_2(A)$ *is a self-dual doubly-even* $[v + 1, (v + 1)/2, \mu + 2]$ *code, and all codewords of* $C_2(A)$ *have weights congruent to 0 or 3 modulo 4.*

Proof: By Theorem 8.6.2(i), $v = \mu^2 + \mu + 1$. As $2 \mid \mu$ but $2^2 \nmid \mu$, by Theorem 8.5.3(ii), the dimension of $C_2(A)$ is $(v + 1)/2$. The minimum weight is $\mu + 1$ by Theorem 8.6.3. The sum of all the rows of A is the all-one vector because every point is on $\mu + 1$ lines by Theorem 8.6.2(iii), completing the proof of (i).

The code $\mathcal{D}_2(A)$ is generated by vectors that are the difference of rows of A and hence have weight $2(\mu + 1) - 2 \equiv 0 \pmod{4}$ as distinct lines have $\mu + 1$ points with exactly one point in common. As $\mathcal{D}_2(A) = C_2(A)^\perp \subset C_2(A) = \mathcal{D}_2(A)^\perp$ by Theorem 8.5.3(ii), $\mathcal{D}_2(A)$ is self-orthogonal and hence doubly-even. The code $\mathcal{D}_2(A)$ is contained in the even weight subcode of $C_2(A)$. But both $\mathcal{D}_2(A)$ and the even weight subcode of $C_2(A)$ have codimension 1 in $C_2(A)$ by Theorem 8.5.2(iv) using the fact that the rows of A have odd weight $\mu + 1$. This verifies (ii).

Part (iii) follows from parts (i) and (ii) with the observation that the all-one vector in $C_2(A)$ extends to the all-one vector of length $\mu^2 + \mu + 2 \equiv 0 \pmod 4$ and codewords in $C_2(A)$ of minimum weight extend to codewords of weight $\mu + 2 \equiv 0 \pmod 4$ in $\widehat{C}_2(A)$. $\qquad\square$

When $4 \mid \mu$, the code $\widehat{C}_2(A)$ is still self-orthogonal of minimum weight $\mu + 2$ by Theorem 8.6.3; however, it may not be self-dual. We consider this case later.

We now examine sets of points in a projective plane no three of which are collinear. If $\mathcal{O}$ is such a set of points, fix a point P on $\mathcal{O}$. Then P together with every other point in $\mathcal{O}$ determines different lines through P. Therefore $\mathcal{O}$ can have at most $\mu + 2$ points as Theorem 8.6.2 shows that every point has $\mu + 1$ lines through it. Now assume $\mathcal{O}$ has exactly $\mu + 2$ points. Then every line through P must intersect $\mathcal{O}$ at exactly one other point. Any line that intersects $\mathcal{O}$ does so at some point P and hence meets $\mathcal{O}$ at another point. Thus every line either meets $\mathcal{O}$ in zero or two points. Now choose a point Q not on $\mathcal{O}$. Then any line through Q meeting $\mathcal{O}$ meets it in another point. All lines through Q meeting $\mathcal{O}$ partition $\mathcal{O}$ into pairs of points that determine these lines implying that μ is even. So if $\mathcal{O}$ has $\mu + 2$ points, then μ is even.

An *oval* of a projective plane of order μ is a set of $\mu + 2$ points, if μ is even, or a set of $\mu + 1$ points, if μ is odd, which does not contain three collinear points. So ovals are the largest sets possible in a projective plane with no three points collinear. There is no guarantee that ovals exist. The interesting case for us is when μ is even. Our above discussion proves the following.

Theorem 8.6.5 *Any line intersects an oval in a projective plane of even order in either zero or two points.*

If ovals exist in a projective plane of order $\mu \equiv 2 \pmod 4$, then the ovals play a special role in $C_2(A)$.

Theorem 8.6.6 *Let $(\mathcal{P}, \mathcal{B})$ be a projective plane of even order μ with incidence matrix A. Then the ovals of $(\mathcal{P}, \mathcal{B})$ are precisely the supports of the codewords of weight $\mu + 2$ in $C_2(A)^\perp$. Furthermore, the minimum weight of $C_2(A)^\perp$ is at least $\mu + 2$.*

Proof: Suppose first that $\mathbf{x} \in C_2(A)^\perp$ has weight $\mu + 2$. Then every line must be either disjoint from $\mathrm{supp}(\mathbf{x})$ or meet it in an even number of points as every line is the support of a codeword in $C_2(A)$. Suppose P is a point in $\mathrm{supp}(\mathbf{x})$. Every one of the $\mu + 1$ lines through P must intersect $\mathrm{supp}(\mathbf{x})$ in at least one other point; P and these $\mu + 1$ points account for all of the points of $\mathrm{supp}(\mathbf{x})$. Hence no line meets $\mathrm{supp}(\mathbf{x})$ in more than two points and so $\mathrm{supp}(\mathbf{x})$ is an oval.

For the converse, suppose that x is an oval in $(\mathcal{P}, \mathcal{B})$. Let $\mathbf{x}$ be the binary vector with $x = \mathrm{supp}(\mathbf{x})$. As x meets any line in an even number of points by Theorem 8.6.5, $\mathbf{x}$ is orthogonal to any codeword whose support is a line. As such codewords generate $C_2(A)$, $\mathbf{x} \in C_2(A)^\perp$.

The fact that the minimum weight of $C_2(A)^\perp$ is at least $\mu + 2$ is left as Exercise 461. $\qquad\square$

Exercise 461 Let $(\mathcal{P}, \mathcal{B})$ be a projective plane of even order μ with incidence matrix A. Prove that the minimum weight of $C_2(A)^\perp$ is at least $\mu + 2$. $\qquad\blacklozenge$

Corollary 8.6.7 *Let* $(\mathcal{P}, \mathcal{B})$ *be a projective plane of order* $\mu \equiv 2 \pmod 4$ *with incidence matrix* A. *Then the ovals of* $(\mathcal{P}, \mathcal{B})$ *are precisely the supports of the codewords of weight* $\mu + 2$ *in* $\mathcal{C}_2(A)$.

Proof: By Theorem 8.6.4, $\mathcal{C}_2(A)^\perp$ is the subcode of even weight vectors in $\mathcal{C}_2(A)$. The result follows from Theorem 8.6.6. ☐

We have settled the question of the dimension of $\mathcal{C}_2(A)$ for a projective plane of order μ except when $4 \mid \mu$. If $4 \mid \mu$, $\widehat{\mathcal{C}}_2(A)$ is self-orthogonal and $\mathcal{C}_2(A)$ has minimum weight $\mu + 1$ by Theorem 8.6.3; however, we do not know the dimension of $\mathcal{C}_2(A)$. A special case of this was considered by K. J. C. Smith [316], where he examined designs arising from certain projective geometries. The next theorem, which we present without proof, gives a part of Smith's result. The projective plane denoted PG(2, 2^s) is the plane whose points are the 1-dimensional subspaces of $\mathbb{F}_q^3$, where $q = 2^s$, and whose lines are the 2-dimensional subspaces of $\mathbb{F}_q^3$.

Theorem 8.6.8 *If* $(\mathcal{P}, \mathcal{B})$ *is the projective plane* PG(2, 2^s), *then* $\mathcal{C}_2(A)$ *has dimension* $3^s + 1$.

If μ is a power of any prime, then PG(2, μ) yields a projective plane in the same fashion that PG(2, 2^s) did above. The next theorem presents a special case of a result by F. J. MacWilliams and H. B. Mann [216] that gives the dimension of certain $\mathcal{C}_p(A)$ when p is an odd prime.

Theorem 8.6.9 *If* $(\mathcal{P}, \mathcal{B})$ *is the projective plane* PG(2, p^s) *where* p *is an odd prime, then* $\mathcal{C}_p(A)$ *has dimension*

$$\binom{p+1}{2}^s + 1.$$

Combinatorists have long been fascinated with the problem of finding values μ for which there is a projective plane of that order or proving that no such plane can exist. Since PG(2, μ) is a projective plane of order μ when μ is a power of a prime, projective planes exist for all prime power orders. There are three projective planes of order 9 that are not equivalent to PG(2, 9) or each other; see [182]. However, at the present time, there is no known projective plane of any order other than a prime power. In fact it is conjectured that projective planes of nonprime power orders do not exist.

Research Problem 8.6.10 *Find a projective plane of order that is not a prime power or show that no such plane exists.*

The most useful result that allows one to show that a symmetric 2-(v, k, λ) design cannot exist is the Bruck–Ryser–Chowla Theorem.

Theorem 8.6.11 (Bruck–Ryser–Chowla) *If there exists a symmetric* 2-(v, k, λ) *design where* $\mu = k - \lambda$, *then either:*
(i) *v is even and μ is a square, or*
(ii) *v is odd and $z^2 = \mu x^2 + (-1)^{(v-1)/2} \lambda y^2$ has an integer solution with not all of x, y, and z equal to 0.*

Table 8.2 *Parameters of binary codes from projective planes*

μ	$\widehat{C}_2(A)$	μ	$\widehat{C}_2(A)$
2	[8, 4, 4]	16*	[274, 82, 18]
4	[22, 10, 6]	18†	[344, 172, 20]
8	[74, 28, 10]	20†	[422, $\leq$ 211, 22]
12†	[158, $\leq$ 79, 14]	24†	[602, $\leq$ 301, 26]

Note that (i) of the Bruck–Ryser–Chowla Theorem is Lemma 8.5.1(ii)(e). The following is a consequence of the Bruck–Ryser–Chowla Theorem [187]. The notation in this theorem is as follows. If i is a positive integer, let i^* be defined as the *square-free part* of i; that is, $i = i^*j$, where j is a perfect square and i^* is 1 or the product of distinct primes.

Theorem 8.6.12 *Suppose that there exists a symmetric* 2-(v, k, λ) *design with order* $\mu = k - \lambda$. *Let* $p \mid \mu$ *where* p *is an odd prime.*
(i) *If* $p \nmid \mu^*$ *and* $p \mid \lambda^*$, *then* μ *is a square modulo* p.
(ii) *If* $p \mid \mu^*$ *and* $p \nmid \lambda^*$, *then* $(-1)^{(v-1)/2}\lambda^*$ *is a square modulo* p.
(iii) *If* $p \mid \mu^*$ *and* $p \mid \lambda^*$, *then* $(-1)^{(v+1)/2}(\lambda^*/p)(\mu^*/p)$ *is a square modulo* p.

As a corollary, we can prove the following result about the existence of projective planes.

Corollary 8.6.13 *Suppose that there exists a projective plane of order* μ *where* $\mu \equiv 1 \pmod 4$ *or* $\mu \equiv 2 \pmod 4$. *If* p *is an odd prime with* $p \mid \mu^*$, *then* $p \equiv 1 \pmod 4$.

Proof: The number of points v of the plane is $v = \mu^2 + \mu + 1$. Hence $(v - 1)/2$ is odd when μ is as stated. Thus by Theorem 8.6.12(ii), -1 is a square modulo p. By Lemma 6.2.4, $p \equiv 1 \pmod 4$. $\qquad\square$

Exercise 462 Do the following:
(a) Show that a projective plane of order 6 does not exist.
(b) Show that a projective plane of order 14 does not exist.
(c) Show that Corollary 8.6.13 does not settle the question of the existence of a projective plane of order 10.
(d) For what values of μ, with $\mu < 100$, does Corollary 8.6.13 show that a projective plane of order μ does not exist? $\qquad\blacklozenge$

Table 8.2 gives the length, dimension, and minimum distance of $\widehat{C}_2(A)$, where A is the incidence matrix of a projective plane of even order $\mu \leq 24$. If μ is odd, the code $C_2(A)$ is the $[\mu^2 + \mu + 1, \mu^2 + \mu, 2]$ code of all even weight vectors by Theorem 8.6.3. As mentioned earlier, planes of prime power order exist. The planes of orders 6, 14, and 22 do not exist by the Bruck–Ryser–Chowla Theorem (see Exercise 462). The smallest order not settled by Corollary 8.6.13 is 10. A monumental effort was made to construct a plane of order 10, and eventually it was shown that no such plane exists; we discuss this in Section 8.8. Planes of orders 12, 18, 20, and 24 may or may not exist; in the table these entries are marked with a†. We note that the projective planes of order 8 or less are unique up to equivalence [4], and there are four inequivalent projective planes of order 9 [182]. For given μ in the table, the minimum distance of $\widehat{C}_2(A)$ is $\mu + 2$ and the dimension is at most $(\mu^2 + \mu + 2)/2$ by Theorem 8.6.3(ii). When $\mu = 2$ or 18, the dimension equals $(\mu^2 + \mu + 2)/2$ by Theorem 8.6.4(iii). For $\mu = 4$ or 8, the unique projective plane is PG(2, μ), and the dimension of $\widehat{C}_2(A)$ is given by Theorem 8.6.8. If $\mu = 16$, the only plane included in the table is PG(2, 16), and for that reason the value $\mu = 16$ is marked with a*; the dimension of the associated code is again given by Theorem 8.6.8.

8.7   Cyclic projective planes

In this section we discuss cyclic projective planes and explore the relationship of difference sets to cyclic planes. We will use the theory of duadic codes to show the nonexistence of certain cyclic projective planes.

Recall that a projective plane on v points is cyclic if there is a v-cycle in its automorphism group. For example, the projective plane PG(2, q) of order $\mu = q$, where q is a prime power, has an automorphism which is a v-cycle where $v = \mu^2 + \mu + 1$; this is called a *Singer cycle* (see [4]). In general, a cyclic design is one that has a v-cycle in its automorphism group. By renaming the points, we may assume that the points are the cyclic group $\mathbb{Z}_v$ under addition modulo v and that the cyclic automorphism is the map $i \mapsto i + 1 \pmod{v}$. We make this assumption whenever we speak of cyclic designs. We begin with a theorem relating cyclic designs to duadic codes that will prove quite useful when we look at cyclic projective planes.

Theorem 8.7.1 *Let A be the incidence matrix of a cyclic symmetric 2-(v, k, λ) design $(\mathcal{P}, \mathcal{B})$. Let p be the characteristic of $\mathbb{F}_q$ and assume that $p \mid (k - \lambda)$ but $p^2 \nmid (k - \lambda)$. Then the following hold:*

(i) *If $p \mid k$, then $\mathcal{C}_q(A)$ is a self-orthogonal even-like duadic code of dimension $(v - 1)/2$ whose splitting is given by μ_{-1}.*

(ii) *If $p \nmid k$, then $\mathcal{D}_q(A)$ is a self-orthogonal even-like duadic code of dimension $(v - 1)/2$ whose splitting is given by μ_{-1}. Also $\mathcal{D}_q(A)^\perp = \mathcal{C}_q(A)$.*

Proof: As the rows of A are permuted by the map $i \mapsto i + 1 \pmod{v}$, $\mathcal{C}_q(A)$ must be a cyclic code. As differences of rows of A are also permuted by this map, $\mathcal{D}_q(A)$ is also a cyclic code. By Theorem 8.5.3 the dimensions and self-orthogonality of $\mathcal{C}_q(A)$ in (i) and $\mathcal{D}_q(A)$ in (ii) are as claimed. The remainder of the result follows from Theorems 6.4.1 and 8.5.3.   □

In this section we will need the following theorem comparing the action of a design automorphism on blocks and on points.

Theorem 8.7.2 *An automorphism of a symmetric 2-design fixes the same number of blocks as it fixes points.*

Proof: Let A be the incidence matrix of a symmetric 2-design $\mathcal{D} = (\mathcal{P}, \mathcal{B})$ and let $\tau \in \mathrm{Aut}(\mathcal{D})$. Then there are permutation matrices P and Q such that $PAQ = A$, where P gives the action of τ on the blocks $\mathcal{B}$, and Q gives the action of τ on the points $\mathcal{P}$. By Lemma 8.5.1, A is nonsingular over the rational numbers. Therefore $AQA^{-1} = P^{-1} = P^{\mathrm{T}}$, and Q and P have the same trace. As the number of fixed points of a permutation matrix is the trace of the matrix, the result follows. $\qquad\square$

Corollary 8.7.3 *An automorphism of a symmetric 2-design has the same cycle structure on blocks as it does on points.*

Proof: If σ and τ represent the same automorphism acting on blocks and points, respectively, then so do σ^i and τ^i. Let n_i, respectively n'_i, be the number of i-cycles of σ, respectively τ. If i is a prime p, the number of fixed points of σ^i, respectively τ^i, is $p \cdot n_p + n_1$, respectively $p \cdot n'_p + n'_1$. As $n_1 = n'_1$ by Theorem 8.7.2 applied to σ and τ, and since $p \cdot n_p + n_1 = p \cdot n'_p + n'_1$ by the same theorem applied to σ^i and τ^i, $n_p = n'_p$. We can continue inductively according to the number of prime factors of i. $\qquad\square$

Exercise 463 Fill in the details of the proof of Corollary 8.7.3. $\qquad\blacklozenge$

We are now ready to introduce the concept of a difference set; this will be intimately related to cyclic designs. Let $\mathbb{G}$ be an abelian group of order v. A (v, k, λ) *difference set* in $\mathbb{G}$ is a set D of k distinct elements of $\mathbb{G}$ such that the multiset $\{x - y \mid x \in D, y \in D, x \neq y\}$ contains every nonzero element of $\mathbb{G}$ exactly λ times. There are $k(k-1)$ differences between distinct elements of D, and these differences make up all nonzero elements of $\mathbb{G}$ exactly λ times. Thus

$$k(k-1) = (v-1)\lambda. \tag{8.12}$$

We will be most interested in the case where $\mathbb{G}$ is the cyclic group $\mathbb{Z}_v$, in which case the difference set is a *cyclic difference set*. In Theorem 8.7.5 we will see the connection between difference sets and symmetric designs.

Example 8.7.4 In $\mathbb{Z}_7$, the set $D = \{1, 2, 4\}$ is a $(7, 3, 1)$ cyclic difference set as one can see by direct computation (Exercise 464). By adding each element of $i \in \mathbb{Z}_7$ to the set D we obtain the sets $D_i = \{i + 1, i + 2, i + 4\}$. These form the blocks of a projective plane $\mathcal{D}$ of order 2. The map $i \mapsto i + 1 \pmod 7$ is a 7-cycle in $\mathrm{Aut}(\mathcal{D})$. $\qquad\blacksquare$

Exercise 464 Prove that in Example 8.7.4:
(a) $D = \{1, 2, 4\}$ is a $(7, 3, 1)$ cyclic difference set,
(b) the sets $D_i = \{i + 1, i + 2, i + 4\}$ are the blocks of a projective plane $\mathcal{D}$ of order 2, and
(c) the map $i \mapsto i + 1 \pmod 7$ is a 7-cycle in $\mathrm{Aut}(\mathcal{D})$. $\qquad\blacklozenge$

Exercise 465 Prove that:
(a) $\{1, 3, 4, 5, 9\}$ and $\{2, 6, 7, 8, 10\}$ are $(11, 5, 2)$ cyclic difference sets, and
(b) $\{1, 5, 6, 8\}$ and $\{0, 1, 3, 9\}$ are $(13, 4, 1)$ cyclic difference sets. $\qquad\blacklozenge$

For $i \in \mathbb{G}$, define the shift $g_i : \mathbb{G} \to \mathbb{G}$ by $zg_i = i + z$. It is easy to see that applying g_i to a difference set in $\mathbb{G}$ yields a difference set with the same parameters. The shifts of a (v, k, λ) difference set in $\mathbb{G}$ form a symmetric 2-(v, k, λ) design with $\mathbb{G}$ in its automorphism group.

Theorem 8.7.5 *Let* $\mathbb{G}$ *be an abelian group of order* v, *and let* $\mathcal{G} = \{g_i \mid i \in \mathbb{G}\}$. *The following hold*:

(i) *Let* D *be a* (v, k, λ) *difference set in* $\mathbb{G}$. *Let* $D_i = \{i + \delta \mid \delta \in D\}$ *and* $\mathcal{B} = \{D_i \mid i \in \mathbb{G}\}$. *Then* $\mathcal{D} = (\mathbb{G}, \mathcal{B})$ *is a symmetric* 2-(v, k, λ) *design with* $\mathcal{G} \subseteq \mathrm{Aut}(\mathcal{D})$.

(ii) *If* $\mathcal{D} = (\mathbb{G}, \mathcal{B})$ *is a symmetric* 2-(v, k, λ) *design which has* $\mathcal{G} \subseteq \mathrm{Aut}(\mathcal{D})$, *then any block* D *of* $\mathcal{B}$ *is a* (v, k, λ) *difference set in* $\mathbb{G}$ *and the blocks of* $\mathcal{B}$ *are* $D_i = \{i + \delta \mid \delta \in D\}$.

Proof: Let D be a (v, k, λ) difference set in $\mathbb{G}$ with $\mathcal{D} = (\mathbb{G}, \mathcal{B})$ as described. Let i and j be distinct elements of $\mathbb{G}$. By definition $i - j = x - y$ for λ pairs (x, y) of distinct elements in $D \times D$. Then $i - x = j - y$; as $i = i - x + x \in D_{i-x}$ and $j = j - y + y = i - x + y \in D_{i-x}$, every pair of points is in at least λ blocks. By (8.12),

$$v\binom{k}{2} = \binom{v}{2}\lambda.$$

The left-hand side counts the number of pairs of distinct elements in all the D_is and the right-hand side counts the number of pairs of distinct elements in $\mathbb{G}$ repeated λ times each. Therefore no pair can appear more than λ times in all the D_is. Thus $\mathcal{D}$ is a 2-(v, k, λ) design; since $\mathbb{G}$ has order v, $\mathcal{D}$ has v blocks and so is symmetric. As $(D_j)g_i = D_{i+j}$, $\mathcal{G} \subseteq \mathrm{Aut}(\mathcal{D})$ proving (i).

For the converse, let D be any block in $\mathcal{B}$. Then $D_i = \{i + \delta \mid \delta \in D\}$ is the image of D under g_i and hence is also in $\mathcal{B}$. Suppose first that $D_i = D_j$ for some $i \neq j$. Then $D = D_m$, where $m = i - j \neq 0$, implying that g_m fixes the block D. By Theorem 8.7.2 g_m must fix a point. Thus $m + \ell = \ell$ for some $\ell \in \mathbb{G}$, which is a contradiction as $m \neq 0$. Therefore the D_is are distinct and hence must be all v blocks as $\mathbb{G}$ has order v.

Let w be a nonzero element of $\mathbb{G}$. Then 0 and w are distinct points and so must be in λ blocks D_ℓ. Thus $0 = \ell + x$ and $w = \ell + y$ for some x and y in D. But $y - x = w$ and hence every nonzero element of $\mathbb{G}$ occurs as a difference of elements of D at least λ times. There are $k(k - 1)$ differences of distinct elements of D. But,

$$b = v = \lambda \frac{v(v - 1)}{k(k - 1)},$$

by Theorem 8.1.3. Thus $k(k - 1) = \lambda(v - 1)$ and hence each difference occurs exactly λ times, proving (ii). $\qquad\square$

This theorem shows that there is a one-to-one correspondence between difference sets in $\mathbb{G}$ and symmetric designs with points $\mathbb{G}$ having a group of automorphisms isomorphic to $\mathbb{G}$ acting naturally on the points and blocks. The design $\mathcal{D}$ arising from the difference set D as described in Theorem 8.7.5 is called the *development of* D.

We can use quadratic residues to construct difference sets.

Theorem 8.7.6 ([204]) *Let* $q = 4m - 1 = p^t$ *for a prime* p. *The set* $D = \{x^2 \mid x \in \mathbb{F}_q, x \neq 0\}$ *is a* $(4m - 1, 2m - 1, m - 1)$ *difference set in the additive group of* $\mathbb{F}_q$.

Proof: As $q = 4m - 1 = p^t$, $p \equiv -1 \pmod 4$ and t is odd. By Lemma 6.2.4, -1 is not a square in $\mathbb{F}_p$. Hence -1 is a root of $x^2 + 1$ in $\mathbb{F}_{p^2}$ but not in $\mathbb{F}_p$. Thus -1 is not a square in $\mathbb{F}_q$ as $\mathbb{F}_{p^2} \not\subseteq \mathbb{F}_q$. Since the nonzero elements of $\mathbb{F}_q$ form a cyclic group $\mathbb{F}_q^*$ with generator α, D

consists of the $2m - 1$ elements $\{\alpha^{2i} \mid 1 \le i \le 2m - 1\}$, and $(-1)D$ makes up the remaining nonzero elements (the nonsquares) of $\mathbb{F}_q$. Let M be the multiset $\{x - y \mid x \in D, y \in D, x \ne y\}$. If $x - y$ is a fixed difference in M, then for $1 \le i \le 2m - 1$, $\alpha^{2i}(x - y) = (\alpha^i)^2 x - (\alpha^i)^2 y$ and $-\alpha^{2i}(x - y) = (\alpha^i)^2 y - (\alpha^i)^2 x$ are each in M and comprise all $4m - 2$ nonzero elements of $\mathbb{F}_q$. Thus every nonzero element of $\mathbb{F}_q$ occurs λ times for some λ, making D a $(4m - 1, 2m - 1, \lambda)$ difference set in the additive group of $\mathbb{F}_q$. By (8.12), $\lambda = m - 1$. $\square$

Applying Theorem 8.7.6 to the case $q = p = 4m - 1$ where p is a prime, we obtain a cyclic difference set in $\mathbb{Z}_p$.

Exercise 466 Use Theorem 8.7.6 to obtain (v, k, λ) cyclic difference sets in $\mathbb{Z}_7$, $\mathbb{Z}_{11}$, $\mathbb{Z}_{19}$, $\mathbb{Z}_{23}$, and $\mathbb{Z}_{31}$. What are their parameters v, k, and λ? ◆

When $\lambda = 1$ and $\mathbb{G} = \mathbb{Z}_v$, the symmetric designs are cyclic projective planes. Since it is so difficult to prove the nonexistence of projective planes with parameters not covered by the Bruck–Ryser–Chowla Theorem, it is reasonable to consider planes with additional conditions. One natural possibility is to examine criteria under which cyclic projective planes do not exist. Our next two theorems show that if a cyclic projective plane has order μ where either $\mu \equiv 2 \pmod{4}$ or $\mu \equiv \pm 3 \pmod{9}$, then $\mu = 2$ or $\mu = 3$. These results were proved in the more general setting of $(\mu^2 + \mu + 1, \mu + 1, 1)$ difference sets in an abelian group of order $\mu^2 + \mu + 1$ in [353]. Using duadic codes, we can prove these two theorems rather easily. The proof of the first is found in [267].

Theorem 8.7.7 *Let $\mathcal{D}$ be a cyclic projective plane of order $\mu \equiv 2 \pmod{4}$. Then $\mu = 2$.*

Proof: Let A be the incidence matrix of $\mathcal{D}$. By Theorem 8.7.1(ii), $\mathcal{D}_2(A)$ is a $[v, (v - 1)/2]$ self-orthogonal even-like duadic code with splitting μ_{-1}. By Theorems 8.5.3 and 6.4.2, $\mathcal{C}_2(A)$ is an odd-like duadic code. By Theorem 8.6.3, $\mathcal{C}_2(A)$ has minimum weight $\mu + 1$, with minimum weight codewords being those whose supports are the lines. By Theorem 4.3.13, $\mu_2 \in \mathrm{PAut}(\mathcal{C}_2(A))$ and thus μ_2 permutes the minimum weight codewords. Hence $\mu_2 \in \mathrm{Aut}(\mathcal{D})$. Because μ_2 fixes the coordinate 0, by Theorem 8.7.2, μ_2 fixes a line ℓ and thus μ_2 fixes a minimum weight codeword $\mathbf{c}$ of $\mathcal{C}_2(A)$. But any nonzero binary codeword fixed by μ_2 must be an idempotent. By Theorem 8.7.5, the lines of $\mathcal{D}$ can be obtained by cyclic shifts of ℓ; as the codewords whose supports are the lines generate $\mathcal{C}_2(A)$, $\mathbf{c}$ must be the generating idempotent. But the generating idempotent of a binary duadic code of length v has weight $(v - 1)/2$ or $(v + 1)/2$ by equations (6.1) and (6.3). Thus $\mu + 1 = (v \pm 1)/2 = (\mu^2 + \mu + 1 \pm 1)/2$, which implies $\mu = 2$. $\square$

A similar argument involving ternary duadic codes yields the following result.

Theorem 8.7.8 *Let $\mathcal{D}$ be a cyclic projective plane of order μ where $3 \mid \mu$ but $9 \nmid \mu$. Then $\mu = 3$.*

Proof: Let A be the incidence matrix of the 2-$(v, k, 1)$ design $\mathcal{D}$ of order $\mu = k - 1$. By Theorem 8.5.3, as $3 \nmid (\mu + 1)$, $\mathcal{D}_3(A)$ has dimension $(v - 1)/2$ and $\mathcal{D}_3(A) = \mathcal{C}_3(A)^\perp \subset \mathcal{C}_3(A) = \mathcal{D}_3(A)^\perp$. Hence by Theorem 6.4.1, $\mathcal{D}_3(A)$ is an even-like duadic code with splitting given by μ_{-1}. By Theorem 6.4.2, $\mathcal{C}_3(A)$ is an odd-like duadic code. By Theorem 4.3.13,

$\mu_3 \in \mathrm{PAut}(C_3(A))$ and thus μ_3 permutes the minimum weight codewords; these codewords are precisely those with supports the lines of the plane by Theorem 6.5.2 and are multiples of the binary vectors corresponding to these supports. Hence $\mu_3 \in \mathrm{Aut}(\mathcal{D})$. Because μ_3 fixes the coordinate 0 by Theorem 8.7.2, μ_3 fixes a line ℓ and thus μ_3 fixes a minimum weight codeword $\mathbf{c}$ of $C_3(A)$ with $\mathrm{supp}(\mathbf{c}) = \ell$. Writing the codeword $\mathbf{c}$ as the polynomial $c(x)$ in $\mathcal{R}_v = \mathbb{F}_3[x]/(x^v - 1)$, we have $c(x) = c(x)\mu_3 = c(x^3) = c(x)^3$ as $\mathcal{R}_v$ has characteristic 3. Thus $c(x)^4 = c(x)^2$ implying that $e(x) = c(x)^2$ is an idempotent. Let C' be the cyclic code generated by $e(x)$ and all of its cyclic shifts. As $c(x)e(x) = c(x)^3 = c(x)$, C' contains $c(x)$, all of its cyclic shifts, and their scalar multiples. But these are all the minimum weight codewords of $C_3(A)$ by Theorem 6.5.2; hence $C' = C_3(A)$ implying that $e(x)$ is the generating idempotent of $C_3(A)$.

As $e(x)$ and $e(x)\mu_{-1}$ are the generating idempotents of an odd-like pair of duadic codes, by Exercise 329 $e(x) + e(x)\mu_{-1} = 1 + \overline{j}(x)$. Since $v \equiv 1 \pmod 3$ as $3 \mid \mu$, $\overline{j}(x) = 1 + x + x^2 + \cdots + x^{v-1}$ and so

$$e(x) + e(x^{-1}) = 2 + x + x^2 + \cdots + x^{\mu^2 + \mu}. \tag{8.13}$$

We examine how each of the terms in $e(x) = c(x)^2 = \sum_{m=0}^{v-1} e_m x^m$ can arise in the expansion of $c(x)^2$. The term $e_m x^m$ is the sum of terms of the form $x^i x^j$ where i and j are elements of ℓ. Suppose that $x^i x^j = x^{i'} x^{j'}$ with $i \neq i'$. Then $i - i' \equiv j' - j \pmod v$, implying that $i = j'$ and $j = i'$ since the difference $i - i'$ modulo v can be obtained only once because ℓ is a $(v, k, 1)$ difference set by Theorem 8.7.5. Therefore $e(x)$ has exactly $\mu + 1$ terms x^{2i} coming from the product $x^i x^i$ and $\binom{\mu+1}{2}$ terms $2x^{i+j}$ coming from $x^i x^j + x^j x^i$, where $i \neq j$. Thus $e(x)$ has exactly $\mu + 1$ terms with coefficient 1 and $\binom{\mu+1}{2}$ terms with coefficient 2. The same holds for $e(x)\mu_{-1}$. In order for (8.13) to hold, since the coefficient of x^0 in both $e(x)$ and $e(x)\mu_{-1}$ must agree, the coefficient of x^0 in both $e(x)$ and $e(x)\mu_{-1}$ must be 1. Furthermore, $e(x)$ and $e(x)\mu_{-1}$ must have exactly the same set of terms x^k with coefficients equal to 2; these terms when added in $e(x) + e(x)\mu_{-1}$ have coefficient equal to 1. Finally, whenever the coefficient of x^k, with $k \neq 0$, in $e(x)$ (or $e(x)\mu_{-1}$) is 1, then the coefficient of x^k in $e(x)\mu_{-1}$ (respectively, $e(x)$) is 0. This accounts for all terms of $e(x) + e(x)\mu_{-1}$, and so there is a total of $\binom{\mu+1}{2} + 2(\mu + 1) - 1$ terms, which must therefore equal $v = \mu^2 + \mu + 1$. Solving gives $\mu^2 - 3\mu = 0$; thus $\mu = 3$. $\square$

Another result, found in [266], eliminates certain orders for the existence of cyclic projective planes. To prove it, we need the following two lemmas.

Lemma 8.7.9 *Let v be an odd positive integer relatively prime to the integer b. Suppose v has the prime factorization $v = \prod_i p_i^{a_i}$, where p_i are distinct odd primes. We have $b^e \equiv -1 \pmod v$ for some odd positive integer e if and only if, for all i, $b^{e_i} \equiv -1 \pmod{p_i}$ for some odd positive integer e_i.*

Proof: If $b^e \equiv -1 \pmod v$ for some odd positive integer e, then clearly we may choose $e_i = e$.

The converse requires more work. By Exercise 467, as $b^{e_i} \equiv -1 \pmod{p_i}$, there is an odd positive integer f_i such that $b^{f_i} \equiv -1 \pmod{p_i^{a_i}}$. Therefore the converse is proved if we show that whenever n_1 and n_2 are relatively prime odd integers such that, for $i = 1$ and

2, $b^{g_i} \equiv -1 \pmod{n_i}$ where g_i is odd, then $b^{g_1 g_2} \equiv -1 \pmod{n_1 n_2}$. As g_i is odd, $b^{g_1 g_2} \equiv (-1)^{g_2} \equiv -1 \pmod{n_1}$ and $b^{g_1 g_2} \equiv (-1)^{g_1} \equiv -1 \pmod{n_2}$ implying $n_i \mid (b^{g_1 g_2} + 1)$ and hence $n_1 n_2 \mid (b^{g_1 g_2} + 1)$ because n_1 and n_2 are relatively prime. $\square$

Exercise 467 Let p be an odd prime and t a positive integer.
(a) Prove that if x is an integer, then $(-1 + xp)^{p^{t-1}} = -1 + yp^t$ for some integer y. (Compare this to Exercise 337.)
(b) Prove that if $b^e \equiv -1 \pmod{p}$ for some odd integer e, there is an odd integer f such that $b^f \equiv -1 \pmod{p^t}$. $\blacklozenge$

Lemma 8.7.10 *Let p be an odd prime.*
(i) *If $p \equiv 3 \pmod 8$, then $2^e \equiv -1 \pmod{p}$ for some odd positive integer e.*
(ii) *If $p \equiv 1 \pmod 8$ and $\mathrm{ord}_p(2) = 2e$ for some odd positive integer e, then $2^e \equiv -1 \pmod{p}$.*
(iii) *If $p \equiv -3 \pmod 8$, then $4^e \equiv -1 \pmod{p}$ for some odd positive integer e.*
(iv) *If $p \equiv 1 \pmod 8$ and $\mathrm{ord}_p(2) = 4e$ for some odd positive integer e, then $4^e \equiv -1 \pmod{p}$.*

Proof: If $p \equiv 3 \pmod 8$, then by Lemma 6.2.6, $\mathrm{ord}_p(2) = 2e$ for some odd positive integer e. So under the conditions of (i) or (ii), 2^e is a solution of $x^2 - 1$ modulo p; as this polynomial has only two solutions ± 1 and $2^e \not\equiv 1 \pmod{p}$, we must have $2^e \equiv -1 \pmod{p}$ yielding (i) and (ii). Parts (iii) and (iv) follow analogously since if $p \equiv -3 \pmod 8$, then $\mathrm{ord}_p(2) = 4e$ for some odd positive integer e by Lemma 6.2.6 implying $\mathrm{ord}_p(4) = 2e$. $\square$

Let p be an odd prime. We say that p is of type A if either $p \equiv 3 \pmod 8$ or $p \equiv 1 \pmod 8$ together with $\mathrm{ord}_p(2) \equiv 2 \pmod 4$. We say that p is of type B if either $p \equiv -3 \pmod 8$ or $p \equiv 1 \pmod 8$ together with $\mathrm{ord}_p(2) \equiv 4 \pmod 8$.

Theorem 8.7.11 *Suppose that μ is even and $v = \mu^2 + \mu + 1 = \prod p_i^{a_i}$, where the p_i are either all of type A or all of type B. Then there is no cyclic projective plane of order μ.*

Proof: The proof is by contradiction. Suppose that there is a cyclic projective plane of order μ. Let A be the incidence matrix of the plane. By Theorem 8.5.2, the code $\mathcal{D}_4(A)$ is self-orthogonal under the ordinary inner product and has dimension at most $(v - 1)/2$. The code $\mathcal{C}_4(A)$ is odd-like as the blocks have odd size and, by Theorem 8.5.2(i), $\mathcal{D}_4(A)$ must have codimension 1 in $\mathcal{C}_4(A)$. The code $\mathcal{C}_4(A)$ is cyclic as the plane is cyclic. As the sum of the entries in a column of A is $\mu + 1$, adding the rows of A in $\mathbb{F}_4$ gives the all-one vector $\mathbf{1}$. Thus $\mathcal{C}_4(A) = \mathcal{D}_4(A) + \langle \mathbf{1} \rangle$. By Corollary 4.4.12,

$$\mathcal{C}_4(A) \cap \mathcal{C}_4(A)\mu_{-1} = \langle \mathbf{1} \rangle. \tag{8.14}$$

First, assume that all the primes p_i are of type A. By Lemmas 8.7.9 and 8.7.10, there is an odd positive integer e such that $2^e \equiv -1 \pmod{v}$. So $4^{(e-1)/2} 2 \equiv -1 \pmod{v}$, implying that if C is a 4-cyclotomic coset modulo v, then $C\mu_{-1} = C\mu_2$ as multiplying a 4-cyclotomic coset by 4 fixes the coset. Therefore $\mathcal{C}_4(A)\mu_{-1} = \mathcal{C}_4(A)\mu_2$ as both must have the same defining set. As $\mathcal{C}_2(A)$ and $\mathcal{C}_2(A)\mu_2$ have the same defining set, since μ_2 fixes 2-cyclotomic cosets, $C_2(A) = C_2(A)\mu_2$. Thus we obtain

$$C_2(A) = C_2(A) \cap C_2(A)\mu_2 \subseteq C_4(A) \cap C_4(A)\mu_2 = \langle \mathbf{1} \rangle$$

by (8.14), an obvious contradiction as $C_2(A)$ contains binary vectors whose supports are the lines of the plane.

Now assume that all the primes p_i are of type B. As $\mathcal{D}_4(A)$ has a generating matrix consisting of binary vectors, it is self-orthogonal under the Hermitian inner product as well as the ordinary inner product (see the related Exercise 360). As $C_4(A) = \mathcal{D}_4(A) + \langle \mathbf{1} \rangle$, by Corollary 4.4.17,

$$C_4(A) \cap C_4(A)\mu_{-2} = \langle \mathbf{1} \rangle. \tag{8.15}$$

By Lemmas 8.7.9 and 8.7.10, $4^e \equiv -1 \pmod{v}$ for some odd positive integer e. So $4^e 2 \equiv -2 \pmod{v}$ implying that if C is a 4-cyclotomic coset modulo v, then $C\mu_{-2} = C\mu_2$. Hence $C_4(A)\mu_{-2} = C_4(A)\mu_2$. Again we obtain

$$C_2(A) = C_2(A) \cap C_2(A)\mu_2 \subseteq C_4(A) \cap C_4(A)\mu_2 = \langle \mathbf{1} \rangle$$

by (8.15), which is a contradiction.   □

Exercise 468   Using Corollary 8.6.13 of the Bruck–Ryser–Chowla Theorem and Theorems 8.7.7, 8.7.8, and 8.7.11, prove that there do not exist cyclic projective planes of non-prime power order $\mu \leq 50$ except possibly for $\mu = 28, 35, 36, 44$, or 45.   ◆

We conclude this section with a brief discussion of multipliers for difference sets. A *multiplier* of a difference set D in an abelian group $\mathbb{G}$ is an automorphism τ of $\mathbb{G}$ such that $D\tau = \{i + \delta \mid \delta \in D\}$ for some $i \in \mathbb{G}$. Notice that a multiplier of a difference set is an automorphism of the development of the difference set. In the case where $\mathbb{G} = \mathbb{Z}_v$, the automorphisms of $\mathbb{Z}_v$ are the maps $\mu_a : \mathbb{Z}_v \to \mathbb{Z}_v$ given by $i\mu_a = ia$ for all $i \in \mathbb{Z}_v$; here a is relatively prime to v. These are precisely the maps we called multipliers in our study of cyclic codes in Section 4.3. We can prove a multiplier theorem for cyclic difference sets by appealing to the theory of cyclic codes.

Theorem 8.7.12 ([287]) *Let D be a (v, k, λ) cyclic difference set. Suppose that p is a prime with $p \nmid v$ and $p \mid \mu = k - \lambda$. Assume that $p > \lambda$. Then μ_p is a multiplier of D.*

Proof: Let $\mathcal{D}$ be the development of D. Then $\mathcal{D}$ is a 2-(v, k, λ) symmetric design by Theorem 8.7.5. If $\mathcal{D}$ has incidence matrix A, then $C_p(A)$ is a cyclic code. By Theorem 4.3.13, μ_p is an automorphism of $C_p(A)$. Let $\mathbf{c}_D$, respectively $\mathbf{c}_{D\mu_p}$, be the vector in $\mathbb{F}_p^v$ whose nonzero components are all 1 and whose support is D, respectively $D\mu_p$. By construction $\mathbf{c}_D \in C_p(A)$. As μ_p is a permutation automorphism of $C_p(A)$, $\mathbf{c}_{D\mu_p} = \mathbf{c}_D\mu_p \in C_p(A)$. By Theorem 8.5.5, as $p > \lambda$, $D\mu_p$ is a block of $\mathcal{D}$; by Theorem 8.7.5, $D\mu_p = \{i + \delta \mid \delta \in D\}$ for some i, and μ_p is therefore a multiplier of D.   □

In the proof of this theorem, we specifically needed $p > \lambda$ in order to invoke Theorem 8.5.5. However, for every known difference set and every prime divisor p of $k - \lambda$ relatively prime to v, μ_p is a multiplier of the difference set even if $p \leq \lambda$. A corollary of this theorem, presented shortly, will illustrate its significance. To state the corollary, we need a preliminary definition and lemma. A subset of an abelian group $\mathbb{G}$ is *normalized* provided its entries sum to 0.

Lemma 8.7.13 *Let D be a (v, k, λ) difference set in an abelian group $\mathbb{G}$ with $\gcd(v, k) = 1$. Then the development $\mathcal{D}$ of D contains a unique normalized block.*

Proof: Suppose that the elements of D sum to d. Let $D_i = \{i + \delta \mid \delta \in D\}$ be an arbitrary block of $\mathcal{D}$. Then the entries in D_i sum to $d + ki$. As $\gcd(v, k) = 1$, there is a unique solution $i \in \mathbb{G}$ of $d + ki = 0$. $\qquad\square$

Corollary 8.7.14 ([204]) *Let $\mathcal{D} = (\mathbb{Z}_v, \mathcal{B})$ be a cyclic symmetric 2-(v, k, λ) design of order $\mu = k - \lambda$ with $\gcd(v, k) = 1$. Then the normalized block in $\mathcal{B}$ is a union of p-cyclotomic cosets modulo v for any prime $p \mid \mu$ with $p > \lambda$ and $p \nmid v$.*

Proof: If D is any block in $\mathcal{B}$, it is a cyclic difference set in $\mathbb{Z}_v$ and $\mathcal{D}$ is the development of D by Theorem 8.7.5. By Lemma 8.7.13, we may assume D is normalized. By Theorem 8.7.12, $D\mu_p$ is a block in $\mathcal{B}$ for any prime $p \mid \mu$ with $p > \lambda$ and $p \nmid v$. Clearly, $D\mu_p$ is normalized; by uniqueness in Lemma 8.7.13, $D\mu_p = D$. Thus if i is in D, then ip modulo v is also in D implying that D is the union of p-cyclotomic cosets modulo v. $\qquad\square$

Example 8.7.15 Let $\mathcal{D} = (\mathbb{Z}_{13}, \mathcal{B})$ be a cyclic projective plane of order $\mu = 3$. Then $\mathcal{D}$ is a symmetric 2-(13, 4, 1) design, and its normalized block D is a union of 3-cyclotomic cosets modulo 13. These cosets are $C_0 = \{0\}$, $C_1 = \{1, 3, 9\}$, $C_2 = \{2, 5, 6\}$, $C_4 = \{4, 10, 12\}$, and $C_7 = \{7, 8, 11\}$. As blocks have size 4, D must be equal to one of $C_0 \cup C_i$ for some $i \in \{1, 2, 4, 7\}$. Note that each of $C_0 \cup C_i$ is indeed a cyclic difference set; this same fact is explored in Exercise 359 via duadic codes. By Exercise 469, $D\mu_a$ for a relatively prime to v is also a (13, 4, 1) difference set and $\mathcal{D}\mu_a$ is the development of $D\mu_a$. Hence as $C_1\mu_2 = C_2$, $C_1\mu_4 = C_4$, and $C_1\mu_7 = C_7$, the four possibilities for D all produce equivalent cyclic designs. Thus up to equivalence there is only one cyclic projective plane of order 3. $\qquad\blacksquare$

Exercise 469 Let a be relatively prime to v. Let D be a (v, k, λ) cyclic difference set in $\mathbb{Z}_v$.
(a) Prove that $D\mu_a$ is also a cyclic difference set in $\mathbb{Z}_v$.
(b) Prove that if $\mathcal{D}$ is the development of D, then $\mathcal{D}\mu_a$ is the development of $D\mu_a$. $\qquad\blacklozenge$

Exercise 470 Prove that up to equivalence there is a unique cyclic projective plane of order 4. Also give a (21, 5, 1) difference set for this plane. $\qquad\blacklozenge$

Exercise 471 Prove that up to equivalence there is a unique cyclic symmetric 2-(11, 5, 2) design. Give an (11, 5, 2) difference set for this design. $\qquad\blacklozenge$

Exercise 472 Prove that up to equivalence there is a unique cyclic projective plane of order 5. Also give a (31, 6, 1) difference set for this plane. $\qquad\blacklozenge$

Example 8.7.16 Using Corollary 8.7.14 we can show that there is no $(\mu^2 + \mu + 1, \mu + 1, 1)$ cyclic difference set and hence no cyclic projective plane of order μ with $10 \mid \mu$. Suppose D is such a difference set. By Corollary 8.7.14, we may assume that D is normalized with multipliers μ_2 and μ_5, since $\lambda = 1$ and neither 2 nor 5 divide $\mu^2 + \mu + 1$. Let $x \in D$

with $x \neq 0$. As μ_2 and μ_5 are multipliers of D, $2x$, $4x$, and $5x$ are elements of D. If $4x \neq x$, then the difference $x = 2x - x = 5x - 4x$ occurs twice as a difference of distinct elements of D, contradicting $\lambda = 1$. There is an $x \in D$ with $4x \neq x$ as follows. If $4x = x$, then $3x = 0$. This has at most three solutions modulo $\mu^2 + \mu + 1$; since D has $\mu + 1 \geq 11$ elements, at least eight of them satisfy $4x \neq x$. ∎

Exercise 473 Show that there does not exist a cyclic projective plane of order μ whenever μ is a divisible by any of 6, 14, 15, 21, 22, 26, 33, 34, 35, 38, 46, 51, 57, 62, 87, and 91. ◆

Exercise 474 Show that there does not exist a cyclic projective plane of order μ where μ is divisible by 55. Hint: You may encounter the equation $120x = 0$ in $\mathbb{Z}_{\mu^2+\mu+1}$. If so, show that it has at most 24 solutions. ◆

Exercise 475 As stated at the beginning of this section, if μ is a power of a prime, then PG(2, μ) is a cyclic projective plane of order μ. In this problem we eliminate nonprime power orders μ up to 100.
(a) What nonprime power orders μ with $2 \leq \mu \leq 100$ are not eliminated as orders of cyclic projective planes based on Example 8.7.16 and Exercises 473 and 474?
(b) Which orders from part (a) cannot be eliminated by Theorems 8.7.7 and 8.7.8?
(c) Of those nonprime power orders left in (b), which cannot be eliminated by either the Bruck–Ryser–Chowla Theorem or Theorem 8.7.11?
With the techniques of this chapter and others, it has been shown that there is no cyclic projective plane of order $\mu \leq 3600$ unless μ is a prime power. ◆

Exercise 476 Do the following:
(a) Find all normalized (19, 9, 4) difference sets.
(b) Find all normalized (37, 9, 2) difference sets.
(c) Show that there does not exist a cyclic symmetric 2-(31, 10, 3) design (even though (8.12) holds). ◆

8.8   The nonexistence of a projective plane of order 10

We now discuss the search for the projective plane of order 10 and give an indication of how coding theory played a major role in showing the nonexistence of this plane. (See [181] for a survey of this problem.)

Suppose that $\mathcal{D}$ is a projective plane of order 10 with incidence matrix A. By Theorem 8.6.4, $\mathcal{C}_2(A)$ is a [111, 56, 11] code where all codewords have weights congruent to either 0 or 3 modulo 4. Furthermore, all the codewords in $\mathcal{C}_2(A)$ of weight 11 have supports that are precisely the lines of the plane by Theorem 8.6.3. In addition, $\mathcal{C}_2(A)$ contains the all-one codeword, and $\widehat{\mathcal{C}}_2(A)$ is self-dual and doubly-even. From this we deduce that if $A_i = A_i(\mathcal{C}_2(A))$, then $A_0 = A_{111} = 1$ and $A_{11} = A_{100} = 111$. Also, $A_i = 0$ if $1 \leq i \leq 10$ and $101 \leq i \leq 110$. Furthermore, $A_i = 0$ unless $i \equiv 0 \pmod 4$ or $i \equiv 3 \pmod 4$. If three further A_i are known, then by Theorem 7.3.1, the weight distribution of $\mathcal{C}_2(A)$ would be uniquely determined. A succession of authors [220, 48, 184, 185] showed that $A_{15} = A_{12} = A_{16} = 0$. In each case when a vector of weight 15, 12, or 16 was assumed to be in $C_2(A)$, a contradiction was reached before a projective plane could be constructed (by exhaustive computer search). Once these values for A_i were determined to be 0, the unique weight distribution of $C_2(A)$ was computed and all the entries, many of astronomical size, turned out to be nonnegative integers. It also turned out that A_{19} was nonzero. Then C. W. H. Lam tried to construct $\mathcal{D}$ where $C_2(A)$ contained a codeword of weight 19 or show that $A_{19} = 0$. When he got the latter result, he concluded that there is no plane of order 10 [186].

We give a very short description of what was involved in showing that $C_2(A)$ could not contain any vectors of weights 15, 12, or 16. In the first paper on this topic, MacWilliams, Sloane, and Thompson [220] assume that $C_2(A)$ contains a vector $\mathbf{x}$ of weight 15. As $\widehat{C}_2(A)$ is doubly-even, each line must meet $supp(\mathbf{x})$ in an odd number of points s. The minimum weight of $C_2(A)$ excludes values of s higher than 7. The value 7 can be eliminated since a vector $\mathbf{y} \in C_2(A)$ of weight 11 that meets $supp(\mathbf{x})$ in seven points when added to $\mathbf{x}$ gives a vector of weight 12 (hence the support of an oval by Corollary 8.6.7) that meets a line in four points, something that cannot happen by Theorem 8.6.5. Let b_i denote the number of lines that meet $supp(\mathbf{x})$ in i points for $i = 1, 3$, and 5. Then we have:

$$b_1 + b_3 + b_5 = 111,$$
$$b_1 + 3b_3 + 5b_5 = 165,$$
$$3b_3 + 10b_5 = 105.$$

The first equation counts the 111 lines in $C_2(A)$. Each of the 15 points of $supp(\mathbf{x})$ lies on 11 lines, and the second equation counts the incidences of points of $supp(\mathbf{x})$ on lines. Each of the 105 pairs of points on $supp(\mathbf{x})$ determines a line and the last equation counts these. The solution to this system is $b_1 = 90$, $b_3 = 15$, and $b_5 = 6$. It is a short step [220] to show that the six lines that meet $supp(\mathbf{x})$ in five points are uniquely determined (where "uniquely" means up to permutation of coordinates). A bit more analysis after this gives (uniquely) the 15 lines that meet $supp(\mathbf{x})$ in three points. It requires much more analysis and the use of a computer to show that the remaining 90 lines cannot be constructed.

The analysis of how the lines must meet a vector of a presumed weight, the number of cases to be considered, and the amount of computer time needed increases as one assumes that $C_2(A)$ contains a codeword of weight 12 [48, 184], weight 16 [185], and, the final case, weight 19 [186]. The last calculation used the equivalent of 800 days on a VAX-11/780 and 3000 hours on a Cray-1S. This unusual amount of computing prompted the authors [186] to analyze where possible errors could occur and the checks used. Solution of the following problem would be very interesting.

Research Problem 8.8.1 *Find a proof of the nonexistence of the projective plane of order 10 without a computer or with an easily reproducible computer program.*

8.9   Hadamard matrices and designs

There are many combinatorial configurations that are related to designs. One of these, Hadamard matrices, appears in codes in various ways. An $n \times n$ matrix H all of whose entries are ± 1 which satisfies

$$HH^{\mathrm{T}} = nI_n, \tag{8.16}$$

where I_n is the $n \times n$ identity matrix, is called a *Hadamard matrix of order n*.

Exercise 477 Show that two different rows of a Hadamard matrix of order $n > 1$ are orthogonal to each other; this orthogonality can be considered over any field including the real numbers. ◆

Exercise 478 Show that a Hadamard matrix H of order n satisfies

$$H^{\mathrm{T}}H = nI_n.$$

Note that this shows that the transpose of a Hadamard matrix is also a Hadamard matrix. ◆

Exercise 479 Show that two different columns of a Hadamard matrix of order $n > 1$ are orthogonal to each other; again this orthogonality can be considered over any field including the real numbers. ◆

Example 8.9.1 The matrix $H_1 = [1]$ is a Hadamard matrix of order 1. The matrix

$$H_2 = \begin{bmatrix} 1 & 1 \\ 1 & -1 \end{bmatrix}$$

is a Hadamard matrix of order 2. ∎

Two Hadamard matrices are *equivalent* provided one can be obtained from the other by a combination of row permutations, column permutations, multiplication of some rows by -1, and multiplication of some columns by -1.

Exercise 480 Show that if H and H' are Hadamard matrices of order n, then they are equivalent if and only if there are $n \times n$ permutation matrices P and Q and $n \times n$ diagonal matrices D_1 and D_2 with diagonal entries ± 1 such that

$$H' = D_1 P H Q D_2.$$ ◆

Exercise 481 Show that, up to equivalence, H_1 and H_2 as given in Example 8.9.1 are the unique Hadamard matrices of orders 1 and 2, respectively. ◆

Note that any Hadamard matrix is equivalent to at least one Hadamard matrix whose first row and first column consists of 1s; such a Hadamard matrix is called *normalized*. As we see next, the order n of a Hadamard matrix is divisible by four except when $n = 1$ or $n = 2$.

Theorem 8.9.2 *If H is a Hadamard matrix of order n, then either $n = 1$, $n = 2$, or $n \equiv 0 \pmod 4$.*

Proof: Assume that $n \geq 3$ and choose H to be normalized. By (8.16), each of the other rows of H must have an equal number of 1s and -1s. In particular, n must be even. By permuting columns of H, we may assume that the first three rows of H are

$$
\begin{array}{c|c|c|c}
\underbrace{11\cdots 1}_{a} & \underbrace{1\ \ 1\cdots\ \ \ 1}_{b} & \underbrace{1\ \ 1\cdots\ \ \ 1}_{c} & \underbrace{1\ \ 1\cdots\ \ \ 1}_{d} \\
11\cdots 1 & 1\ \ 1\cdots\ \ \ 1 & -1-1\cdots-1 & -1-1\cdots-1 \\
11\cdots 1 & -1-1\cdots-1 & 1\ \ 1\cdots\ \ \ 1 & -1-1\cdots-1
\end{array}
$$

Hence

$$a+b = c+d = n/2, \quad a+c = b+d = n/2, \quad \text{and } a+d = b+c = n/2.$$

Solving gives $a = b = c = d = n/4$, completing the proof. $\qquad\square$

Even though Hadamard matrices are known for many values of $n \equiv 0 \pmod 4$, including some infinite families, the existence of Hadamard matrices for any such n is still an open question. Hadamard matrices have been classified for modest orders. They are unique, up to equivalence, for orders 4, 8, and 12; up to equivalence, there are five Hadamard matrices of order 16, three of order 20, 60 of order 24, and 487 of order 28; see [155, 169, 170, 171, 330].

Exercise 482 Show that, up to equivalence, there is a unique Hadamard matrix of order 4. $\qquad\blacklozenge$

One way to construct Hadamard matrices is by using the tensor product. Let $A = [a_{i,j}]$ be an $n \times n$ matrix and B an $m \times m$ matrix. Then the *tensor product* of A and B is the $nm \times nm$ matrix

$$
A \otimes B = \begin{bmatrix}
a_{1,1}B & a_{1,2}B & \cdots & a_{1,n}B \\
a_{2,1}B & a_{2,2}B & \cdots & a_{2,n}B \\
 & & \vdots & \\
a_{n,1}B & a_{n,2}B & \cdots & a_{n,n}B
\end{bmatrix}.
$$

Example 8.9.3 Let $H_4 = H_2 \otimes H_2$, where H_2 is given in Example 8.9.1. Then

$$
H_4 = \begin{bmatrix}
1 & 1 & 1 & 1 \\
1 & -1 & 1 & -1 \\
1 & 1 & -1 & -1 \\
1 & -1 & -1 & 1
\end{bmatrix}
$$

is easily seen to be a Hadamard matrix. Notice that by Exercise 482, this is the unique Hadamard matrix of order 4. $\qquad\blacksquare$

This example illustrates the following general result.

Theorem 8.9.4 *If L and M are Hadamard matrices of orders ℓ and m, respectively, then $L \otimes M$ is a Hadamard matrix of order ℓm.*

Exercise 483 Prove Theorem 8.9.4. $\qquad\blacklozenge$

Let H_2 be the Hadamard matrix of Example 8.9.1. For $m \geq 2$, define H_{2^m} inductively by

$$H_{2^m} = H_2 \otimes H_{2^{m-1}}.$$

By Theorem 8.9.4, H_{2^m} is a Hadamard matrix of order 2^m. In particular, we have the following corollary.

Corollary 8.9.5 *There exist Hadamard matrices of all orders $n = 2^m$ for $m \geq 1$.*

The Hadamard matrices H_{2^m} can be used to generate the first order Reed–Muller codes. In H_2, replace each entry equal to 1 with 0 and each entry equal to -1 with 1. Then we get the matrix

$$\begin{bmatrix} 0 & 0 \\ 0 & 1 \end{bmatrix}. \tag{8.17}$$

The rows of this matrix can be considered as binary vectors. The *complement* of a binary vector $\mathbf{c}$ is the binary vector $\mathbf{c} + \mathbf{1}$, where $\mathbf{1}$ is the all-one vector; so the complement of $\mathbf{c}$ is the vector obtained from $\mathbf{c}$ by replacing 1 by 0 and 0 by 1. The two rows of matrix (8.17) together with their complements are all the vectors in $\mathbb{F}_2^2$. Apply the same process to H_4 from Example 8.9.3. Then the four rows obtained plus their complements give all the even weight vectors in $\mathbb{F}_2^4$. Exercise 484 shows what happens when this process is applied to H_8.

Exercise 484 Show that the binary vectors of length 8 obtained from H_8 by replacing 1 with 0 and -1 with 1 together with their complements are the codewords of the Reed–Muller code $\mathcal{R}(1, 3)$. Note that $\mathcal{R}(1, 3)$ is also the [8, 4, 4] Hamming code $\widehat{\mathcal{H}}_3$.   ◆

Theorem 8.9.6 *The binary vectors of length 2^m obtained from H_{2^m} by replacing 1 with 0 and -1 with 1 together with their complements are the codewords of $\mathcal{R}(1, m)$.*

Proof: In Section 1.10, $\mathcal{R}(1, m)$ is defined using the $(\mathbf{u} \mid \mathbf{u} + \mathbf{v})$ construction involving $\mathcal{R}(1, m - 1)$ and $\mathcal{R}(0, m - 1)$. As $\mathcal{R}(0, m - 1)$ consists of the zero codeword and all-one codeword of length 2^{m-1}, we have, from (1.7), that

$$\mathcal{R}(1, m) = \{(\mathbf{u}, \mathbf{u}) \mid \mathbf{u} \in \mathcal{R}(1, m - 1)\} \cup \{(\mathbf{u}, \mathbf{u} + \mathbf{1}) \mid \mathbf{u} \in \mathcal{R}(1, m - 1)\}. \tag{8.18}$$

We prove the result by induction on m, the cases $m = 1$ and 2 being described previously, and the case $m = 3$ verified in Exercise 484. By definition, $H_{2^m} = H_2 \otimes H_{2^{m-1}}$, and so the binary vectors obtained by replacing 1 with 0 and -1 with 1 in H_{2^m} together with their complements are precisely the vectors $(\mathbf{u}, \mathbf{u})$ and $(\mathbf{u}, \mathbf{u} + \mathbf{1})$ where $\mathbf{u}$ is a vector obtained by replacing 1 with 0 and -1 with 1 in $H_{2^{m-1}}$ together with their complements. By (8.18) and our inductive hypothesis that the result is true for $\mathcal{R}(1, m - 1)$, we see that the result is true for $\mathcal{R}(1, m)$.   □

It is usually not true that changing the rows of an order n Hadamard matrix to binary vectors and their complements, as just described, produces a binary linear code. However, it does obviously produce a possibly nonlinear binary code of $2n$ vectors of length n such that the distance between distinct codewords is $n/2$ or more. Using this, Levenshtein [193] has shown that if Hadamard matrices of certain orders exist, then the binary codes obtained from them meet the Plotkin Bound.

Exercise 485   Suppose that H is a Hadamard matrix of order $n = 4s$.

(a) By examining the proof of Theorem 8.9.2, we see that three rows, including the all-one row, have common 1s in exactly s columns. Prove that if we take any three rows of H, then there are exactly s columns where all the entries in these rows are the same (some columns have three 1s and some columns have three -1s). Hint: Rescale the columns so that one of the rows is the all-one row.

(b) Prove that if we take any three columns of H, then there are exactly s rows where all the entries in these columns are the same (some rows have three 1s and some rows have three -1s).

(c) Prove that if H is normalized and if we take any three columns of H, then there are exactly $s - 1$ rows excluding the first row where all the entries in these columns are the same (some rows have three 1s and some rows have three -1s).   ◆

Now let H be a normalized Hadamard matrix of order $n = 4s$. Delete the first row. Each of the remaining $4s - 1$ rows has $2s$ 1s and $2s$ -1s. The $4s$ coordinates are considered to be points. To each of the remaining rows associate two blocks of size $2s$: one block is the set of points on which the row is 1, and the other block is the set of points on which the row is -1. Denote this set of points and blocks $\mathcal{D}(H)$. There are $8s - 2$ blocks and they turn out to form a 3-$(4s, 2s, s - 1)$ design, called a *Hadamard 3-design*, as proved in Theorem 8.9.7.

Exercise 486   Show that in the point-block structure $\mathcal{D}(H_{2^m})$, the blocks are precisely the supports of the minimum weight codewords of $\mathcal{R}(1, m)$. Then using Exercise 456, show that $\mathcal{D}(H_{2^m})$ is indeed a 3-design. Note that this exercise shows that the Hadamard structure obtained from H_{2^m} is a 3-design without appealing to Theorem 8.9.7.   ◆

Theorem 8.9.7   *If H is a normalized Hadamard matrix of order $n = 4s$, then $\mathcal{D}(H)$ is a 3-$(4s, 2s, s - 1)$ design.*

Proof: The proof follows from Exercise 485(c).   □

Exercise 487   Find the Pascal triangle for a 3-$(4s, 2s, s - 1)$ design.   ◆

Exercise 488   Prove that the derived design, with respect to any point, of any 3-$(4s, 2s, s - 1)$ design is a symmetric 2-$(4s - 1, 2s - 1, s - 1)$ design.   ◆

Exercise 489   Prove that up to equivalence the Hadamard matrix of order 8 is unique. Hint: Use Example 8.2.4.   ◆

The converse of Theorem 8.9.7 is also true.

Theorem 8.9.8   *Let $\mathcal{D} = (\mathcal{P}, \mathcal{B})$ be a 3-$(4s, 2s, s - 1)$ design. Then there is a normalized Hadamard matrix H such that $\mathcal{D} = \mathcal{D}(H)$.*

Proof: Suppose that B and B' are distinct blocks of $\mathcal{B}$ that intersect in a point x. By Exercise 488 the derived design $\mathcal{D}_x$ of $\mathcal{D}$ with respect to x is a symmetric 2-$(4s - 1, 2s - 1, s - 1)$ design. By Lemma 8.5.1, any two blocks of $\mathcal{D}_x$ intersect in $s - 1$ points.

This implies that any two blocks of $\mathcal{B}$ are either disjoint or intersect in s points. As $\mathcal{D}_x$ has $4s - 1$ blocks, every point is in $4s - 1$ blocks of $\mathcal{D}$.

Fix a block B in $\mathcal{B}$ and define

$$\mathcal{X} = \{(x, B') \mid x \in B, \, x \in B' \in \mathcal{B}, \text{ and } B \neq B'\}.$$

We count the size of $\mathcal{X}$ in two different ways. There are $2s$ points x in B. For each such x there are $4s - 1$ blocks, including B, containing x. Therefore $|\mathcal{X}| = 2s(4s - 2)$. Let N be the number of blocks $B' \neq B$ with $B' \cap B$ nonempty. Then as $B' \cap B$ contains s points, $|\mathcal{X}| = sN$. Equating the two counts gives $N = 8s - 4$. Thus there are $8s - 3$ blocks, including B, which have a point in common with B. Therefore there is exactly one block in $\mathcal{B}$ disjoint from B; as each block contains half the points, this disjoint block must be the complement. So the blocks of $\mathcal{B}$ come in complementary pairs.

Fix a point x and consider the set $\mathcal{S}$ of $4s - 1$ blocks containing x. The remaining blocks are the complements of the blocks in $\mathcal{S}$. Order the points with x listed first; these points label the columns of a $4s \times 4s$ matrix H, determined as follows. Place the all-one vector in the first row; subsequent rows are associated with the blocks of $\mathcal{S}$ in some order and each row is obtained by placing a 1 in a column if the point labeling the column is in the block and a -1 in a column if the point is not in the block. Clearly, $\mathcal{D} = \mathcal{D}(H)$ if we show H is a normalized Hadamard matrix as the -1 entries in a row determine one of the blocks complementary to a block of $\mathcal{S}$. As all entries in the first row and column of H are 1, we only need to show H is a Hadamard matrix. As every entry in each row is ± 1, the inner product of a row with itself is $4s$. As every block has $2s$ points (and hence the corresponding row has $2s$ 1s and $2s$ -1s) and distinct blocks of $\mathcal{S}$ meet in s points, the inner product of distinct rows must be 0. Thus $HH^{\mathrm{T}} = 4s I_{4s}$, implying that H is a Hadamard matrix.   □

Another construction of an infinite family of Hadamard matrices is given by quadratic residues in a finite field $\mathbb{F}_q$, where $q \equiv \pm 1 \pmod 4$. Define $\chi : \mathbb{F}_q \rightarrow \{0, -1, 1\}$ by

$$\chi(a) = \begin{cases} 0 & \text{if } a = 0, \\ 1 & \text{if } a = b^2 \text{ for some } b \in \mathbb{F}_q \text{ with } b \neq 0, \\ -1 & \text{otherwise.} \end{cases}$$

Note that $\chi(ab) = \chi(a)\chi(b)$. Notice also that if $q = p$, where p is a prime, and if $a \neq 0$, then $\chi(a) = \left(\frac{a}{p}\right)$, the Legendre symbol defined in Section 6.6. Let $a_0 = 0, a_1 = 1$, $a_2, \ldots, a_{q-1}$ be an ordering of the elements of $\mathbb{F}_q$. If q is a prime, we let $a_i = i$. Let S_q be the $(q+1) \times (q+1)$ matrix $[s_{i,j}]$, with indices $\infty, a_0, \ldots, a_{q-1}$, where $s_{i,j}$ is defined as

$$s_{i,j} = \begin{cases} 0 & \text{if } i = \infty \, j = \infty, \\ 1 & \text{if } i = \infty \, j \neq \infty, \\ \chi(-1) & \text{if } i \neq \infty \, j = \infty, \\ \chi(a_i - a_j) & \text{if } i \neq \infty \, j \neq \infty. \end{cases} \tag{8.19}$$

Example 8.9.9 When $q = 5$, the matrix S_5 is

$$
\begin{array}{c c}
 & \begin{array}{c c c c c c} \infty & 0 & 1 & 2 & 3 & 4 \end{array} \\
\begin{array}{c} \infty \\ 0 \\ 1 \\ 2 \\ 3 \\ 4 \end{array} &
\left[\begin{array}{c c c c c c}
0 & 1 & 1 & 1 & 1 & 1 \\
1 & 0 & 1 & -1 & -1 & 1 \\
1 & 1 & 0 & 1 & -1 & -1 \\
1 & -1 & 1 & 0 & 1 & -1 \\
1 & -1 & -1 & 1 & 0 & 1 \\
1 & 1 & -1 & -1 & 1 & 0
\end{array} \right].
\end{array}
$$
∎

When q is a prime, S_q is a bordered circulant matrix. To analyze S_q further, we need the following result that extends Lemma 6.2.4.

Lemma 8.9.10 *If* $q \equiv \pm 1 \pmod 4$, *then* $\chi(-1) = 1$ *if* $q \equiv 1 \pmod 4$, *and* $\chi(-1) = -1$ *if* $q \equiv -1 \pmod 4$.

Proof: Let p be the characteristic of $\mathbb{F}_q$ and let $q = p^t$. If $p \equiv 1 \pmod 4$, then -1 is a square in $\mathbb{F}_p \subseteq \mathbb{F}_q$ by Lemma 6.2.4 and $q \equiv 1 \pmod 4$. Suppose that $p \equiv -1 \pmod 4$. If t is even, then $x^2 + 1$ has a root in $\mathbb{F}_{p^2} \subseteq \mathbb{F}_q$ and again $q \equiv 1 \pmod 4$. If t is odd, then $x^2 + 1$ has a root in $\mathbb{F}_{p^2}$ but not $\mathbb{F}_p$ by Lemma 6.2.4. As $\mathbb{F}_{p^2}$ is not contained in $\mathbb{F}_q$, -1 is not a square in $\mathbb{F}_q$ and in this case $q \equiv -1 \pmod 4$. $\square$

Theorem 8.9.11 *The following hold where* q *is an odd prime power*:
(i)   *If* $q \equiv 1 \pmod 4$, *then* $S_q = S_q^{\mathrm{T}}$.
(ii) *If* $q \equiv -1 \pmod 4$, *then* $S_q = -S_q^{\mathrm{T}}$.
(iii) $S_q S_q^{\mathrm{T}} = q I_{q+1}$.

Proof: To prove (i) and (ii), it suffices to show that $s_{i,j} = \chi(-1) s_{j,i}$ by Lemma 8.9.10. This is clear if one or both of i or j is ∞ as $\chi(-1) = 1$ if $q \equiv 1 \pmod 4$, and $\chi(-1) = -1$ if $q \equiv -1 \pmod 4$ by Lemma 8.9.10. As $s_{i,j} = \chi(a_i - a_j) = \chi(-1)\chi(a_j - a_i) = \chi(-1) s_{j,i}$, (i) and (ii) are true.

To show (iii), we must compute the inner product (over the integers) of two rows of S_q, and show that if the rows are the same, the inner product is q, and if the rows are distinct, the inner product is 0. The former is clear as every row has q entries that are ± 1 and one entry 0. Since $\mathbb{F}_q$ has odd characteristic, the nonzero elements of $\mathbb{F}_q$, denoted $\mathbb{F}_q^*$, form a cyclic group of even order, implying that half these elements are squares and half are nonsquares (a generalization of Lemma 6.6.1). In particular, this implies that

$$\sum_{a \in \mathbb{F}_q} \chi(a) = 0, \tag{8.20}$$

which shows that the inner product of the row labeled ∞ with any other row is 0. The inner product of row i and row j with $i \neq j$ and neither equaling ∞ is

$$1 + \sum_{k=0}^{q-1} \chi(a_i - a_k)\chi(a_j - a_k).$$

However, this is

$$1 + \sum_{b \in \mathbb{F}_q} \chi(b)\chi(c+b) = 1 + \sum_{b \in \mathbb{F}_q^*} \chi(b)\chi(c+b),$$

where $b = a_i - a_k$ runs through $\mathbb{F}_q$ as a_k does and $c = a_j - a_i \in \mathbb{F}_q^*$. Since $\chi(c+b) = \chi(b)\chi(cb^{-1}+1)$ if $b \neq 0$, the inner product becomes

$$1 + \sum_{b \in \mathbb{F}_q^*} \chi(b)^2 \chi(cb^{-1}+1) = 1 + \sum_{b \in \mathbb{F}_q^*} \chi(cb^{-1}+1).$$

As $cb^{-1} + 1$ runs through all elements of $\mathbb{F}_q$ except 1 as b runs through all elements of $\mathbb{F}_q^*$, by (8.20) the inner product is 0.   □

Corollary 8.9.12 *We have the following Hadamard matrices:*
(i) *If $q \equiv -1 \pmod 4$, then $I_{q+1} + S_q$ is a Hadamard matrix of order $q + 1$.*
(ii) *If $q \equiv 1 \pmod 4$, then*

$$\begin{bmatrix} I_{q+1} + S_q & I_{q+1} - S_q \\ -I_{q+1} + S_q & I_{q+1} + S_q \end{bmatrix}$$

is a Hadamard matrix of order $2q + 2$.
(iii) *If $q \equiv -1 \pmod 4$, then*

$$\begin{bmatrix} I_{q+1} + S_q & I_{q+1} + S_q \\ I_{q+1} + S_q & -I_{q+1} - S_q \end{bmatrix} \quad and \quad \begin{bmatrix} I_{q+1} + S_q & I_{q+1} + S_q \\ I_{q+1} - S_q & -I_{q+1} + S_q \end{bmatrix}$$

are Hadamard matrices of order $2q + 2$.

Exercise 490   Prove Corollary 8.9.12.   ◆

Exercise 491   Show that Hadamard matrices exist of all orders $n \leq 100$ where $n \equiv 0 \pmod 4$ except possibly $n = 92$. (Note that a Hadamard matrix of order 92 exists, but we have not covered its construction.)   ◆

9 Self-dual codes

In this chapter, we study the family of self-dual codes. Of primary interest will be upper bounds on the minimum weight of these codes, their possible weight enumerators, and their enumeration and classification for small to moderate lengths. We restrict our examination of self-dual codes to those over the three smallest fields $\mathbb{F}_2$, $\mathbb{F}_3$, and $\mathbb{F}_4$. In the case of codes over $\mathbb{F}_4$, the codes will be self-dual under the Hermitian inner product. We emphasize self-dual codes over these three fields not only because they are the smallest fields but also because self-dual codes over these fields are divisible, implying that there are regular gaps in their weight distributions. This is unlike the situation over larger fields, a result known as the Gleason–Pierce–Ward Theorem with which we begin.

A self-dual code has the obvious property that its weight distribution is the same as that of its dual. Some of the results proved for self-dual codes require only this equality of weight distributions. This has led to a broader class of codes known as formally self-dual codes that include self-dual codes. Recall from Section 8.4 that a code C is formally self-dual provided C and $C^\perp$ have the same weight distribution. This implies of course that C is an $[n, n/2]$ code and hence that n is even. Although this chapter deals mainly with self-dual codes, we will point out situations where the results of this chapter apply to formally self-dual codes and devote a section of the chapter to these codes.

9.1 The Gleason–Pierce–Ward Theorem

The Gleason–Pierce–Ward Theorem provides the main motivation for studying self-dual codes over $\mathbb{F}_2$, $\mathbb{F}_3$, and $\mathbb{F}_4$. These codes have the property that they are divisible. Recall that a code C is divisible by Δ provided all codewords have weights divisible by an integer Δ, called a divisor of C; the code is called divisible if it has a divisor $\Delta > 1$.

When C is an $[n, n/2]$ code, Theorems 1.4.5, 1.4.8, and 1.4.10 imply:
- If C is a self-dual code over $\mathbb{F}_2$, then C is divisible by $\Delta = 2$. If C is a divisible code over $\mathbb{F}_2$ with divisor $\Delta = 4$, then C is self-dual.
- C is a self-dual code over $\mathbb{F}_3$ if and only if C is divisible by $\Delta = 3$.
- C is a Hermitian self-dual code over $\mathbb{F}_4$ if and only if C is divisible by $\Delta = 2$.

The Gleason–Pierce–Ward Theorem states that divisible $[n, n/2]$ codes exist only for the values of q and Δ given above, except in one trivial situation, and that the codes are always self-dual except possibly when $q = \Delta = 2$.

Theorem 9.1.1 (Gleason–Pierce–Ward) *Let C be an $[n, n/2]$ divisible code over $\mathbb{F}_q$ with divisor $\Delta > 1$. Then one (or more) of the following holds:*
(i) $q = 2$ and $\Delta = 2$, or
(ii) $q = 2$, $\Delta = 4$, and C is self-dual, or
(iii) $q = 3$, $\Delta = 3$, and C is self-dual, or
(iv) $q = 4$, $\Delta = 2$, and C is Hermitian self-dual, or
(v) $\Delta = 2$ and C is equivalent to the code over $\mathbb{F}_q$ with generator matrix $[I_{n/2} \; I_{n/2}]$.

The Gleason–Pierce–Ward Theorem generalizes the Gleason–Pierce Theorem, a proof of which can be found in [8, 314]. In the original Gleason–Pierce Theorem, there is a stronger hypothesis; namely, C is required to be formally self-dual. In the original theorem, conclusions (i)–(iv) are the same, but conclusion (v) states that $\Delta = 2$ and C is a code with weight enumerator $(y^2 + (q - 1)x^2)^{n/2}$. The generalization was proved by Ward in [345, 347]. We present the proof of the Gleason–Pierce–Ward Theorem in Section 9.11. Notice that the codes arising in cases (ii) through (v) are self-dual (Hermitian if $q = 4$); however, the codes in (i) are merely even and not necessarily formally self-dual.

The codes that arise in the conclusion of this theorem have been given specific names. Doubly-even self-dual binary codes are called *Type II* codes; these are the codes arising in part (ii) of the Gleason–Pierce–Ward Theorem. The self-dual ternary codes from part (iii) are called *Type III* codes; the Hermitian self-dual codes over $\mathbb{F}_4$ are called *Type IV*. The codes in part (ii) also satisfy the conditions in part (i). The binary self-dual codes that are not doubly-even (or Type II) are called *Type I*. In other words, Type I codes are singly-even self-dual binary codes.[1]

Example 9.1.2 Part (i) of the Gleason–Pierce–Ward Theorem suggests that there may be binary codes of dimension half their length that are divisible by $\Delta = 2$ but are not self-dual. This is indeed the case; for instance, the $[6, 3, 2]$ binary code with generator matrix

$$G = \begin{bmatrix} 1 & 0 & 0 & 1 & 1 & 1 \\ 0 & 1 & 0 & 1 & 1 & 1 \\ 0 & 0 & 1 & 1 & 1 & 1 \end{bmatrix}$$

is a formally self-dual code divisible by $\Delta = 2$ that is not self-dual. As additional examples, by Theorem 6.6.14, the binary odd-like quadratic residue codes when extended are all formally self-dual codes divisible by $\Delta = 2$; those of length $p + 1$, where $p \equiv 1 \pmod 8$, are not self-dual.   ■

Exercise 492 Let C be the binary code with generator matrix

$$\begin{bmatrix} 1 & 1 & 0 & 0 & 0 & 0 \\ 1 & 0 & 1 & 0 & 0 & 0 \\ 0 & 0 & 0 & 1 & 1 & 0 \end{bmatrix}.$$

Show that C is divisible by $\Delta = 2$ and is not formally self-dual.   ◆

---

[1] Unfortunately in the literature there is confusion over the term Type I; some authors will simply call self-dual binary codes Type I codes allowing the Type II codes to be a subset of Type I codes. To add to the confusion further, Type I and Type II have been used to refer to certain other nonbinary codes.

We conclude this section with a theorem specifying exactly the condition on the length n which guarantees the existence of a self-dual code of that length; its proof is found in [260]. In Corollary 9.2.2, we give necessary and sufficient conditions on the length for the existence of self-dual codes over $\mathbb{F}_2$ and $\mathbb{F}_3$, which agree with the following theorem; the corollary will also give necessary and sufficient conditions on the length for existence of Type II codes that are not covered in this theorem.

Theorem 9.1.3 *There exists a self-dual code over $\mathbb{F}_q$ of even length n if and only if $(-1)^{n/2}$ is a square in $\mathbb{F}_q$. Furthermore, if n is even and $(-1)^{n/2}$ is not a square in $\mathbb{F}_q$, then the dimension of a maximal self-orthogonal code of length n is $(n/2) - 1$. If n is odd, then the dimension of a maximal self-orthogonal code of length n is $(n - 1)/2$.*

Example 9.1.4 Theorem 9.1.3 shows that a self-dual code of length n over $\mathbb{F}_3$ exists if and only if $4 \mid n$. An example of such a code is the direct sum of $m = n/4$ copies of the [4, 2, 3] self-dual tetracode introduced in Example 1.3.3. If $n \equiv 2 \pmod 4$, then no self-dual code of length n over $\mathbb{F}_3$ exists; however, a maximal self-orthogonal code of dimension $(n/2) - 1$ does exist. An example is the direct sum of $m = (n - 2)/4$ copies of the tetracode and one copy of the zero code of length 2. The theorem also guarantees that a self-dual code of length n over $\mathbb{F}_9$ exists if and only if n is even. An example is the direct sum of $m = n/2$ copies of the [2, 1, 2] self-dual code with generator matrix $[1 \ \rho^2]$, where ρ is the primitive element of $\mathbb{F}_9$ given in Table 6.4. Notice that the codes we have constructed show one direction of Theorem 9.1.3 when the fields are $\mathbb{F}_3$ and $\mathbb{F}_9$. ∎

Exercise 493 Use the codes constructed in Example 9.1.4 to show that if $4 \mid n$, there exist self-dual codes of length n over $\mathbb{F}_{3^t}$, and if n is even, there exist self-dual codes of length n over $\mathbb{F}_{3^t}$ when t is even. ♦

Exercise 494 Prove Theorem 9.1.3 for q a power of 2. ♦

9.2   Gleason polynomials

Self-dual codes over $\mathbb{F}_2$, $\mathbb{F}_3$, and $\mathbb{F}_4$, the last under the Hermitian inner product, all have weight enumerators that can be expressed as combinations of special polynomials that are the weight enumerators of specific codes of small length. These polynomials, known as Gleason polynomials, provide a powerful tool in the study of all self-dual codes over $\mathbb{F}_2$, $\mathbb{F}_3$, and $\mathbb{F}_4$.

Before stating Gleason's Theorem, we discuss the theorem in the case of formally self-dual binary codes. Let

$$g(x, y) = y^2 + x^2 \quad \text{and} \quad h(x, y) = y^8 + 14x^4y^4 + x^8.$$

Notice that $g(x, y)$ is the weight enumerator of the binary repetition code of length 2, and $h(x, y)$ is the weight enumerator of the [8, 4, 4] extended binary Hamming code $\widehat{\mathcal{H}}_3$. Gleason's Theorem states that if $\mathcal{C}$ is any binary formally self-dual code that is divisible by $\Delta = 2$, then $W_\mathcal{C}(x, y) = P(g, h)$, where P is a polynomial with rational coefficients in the variables g and h. Because $W_C(x, y)$ is homogeneous of degree n,

$$P(g, h) = \sum_{i=0}^{\lfloor \frac{n}{8} \rfloor} a_i g(x, y)^{\frac{n}{2}-4i} h(x, y)^i,$$

where a_i is rational. Also because $A_0(C) = 1$, $\sum_{i=0}^{\lfloor \frac{n}{8} \rfloor} a_i = 1$. For example, if $n = 6$, then $W_C(x, y) = g(x, y)^3 = (y^2 + x^2)^3$; the code C that is the direct sum of three copies of the [2, 1, 2] repetition code has this weight enumerator, by Theorem 7.2.2.

Theorem 9.2.1 (Gleason) *Let C be an $[n, n/2]$ code over $\mathbb{F}_q$, where $q = 2, 3$, or 4. Let*

$g_1(x, y) = y^2 + x^2$,

$g_2(x, y) = y^8 + 14x^4 y^4 + x^8$,

$g_3(x, y) = y^{24} + 759x^8 y^{16} + 2576x^{12} y^{12} + 759x^{16} y^8 + x^{24}$,

$g_4(x, y) = y^4 + 8x^3 y$,

$g_5(x, y) = y^{12} + 264x^6 y^6 + 440x^9 y^3 + 24x^{12}$,

$g_6(x, y) = y^2 + 3x^2$, *and*

$g_7(x, y) = y^6 + 45x^4 y^2 + 18x^6$.

Then:

(i)   *if $q = 2$ and C is formally self-dual and even,*

$$W_C(x, y) = \sum_{i=0}^{\lfloor \frac{n}{8} \rfloor} a_i g_1(x, y)^{\frac{n}{2}-4i} g_2(x, y)^i,$$

(ii) *if $q = 2$ and C is self-dual and doubly-even,*

$$W_C(x, y) = \sum_{i=0}^{\lfloor \frac{n}{24} \rfloor} a_i g_2(x, y)^{\frac{n}{8}-3i} g_3(x, y)^i,$$

(iii) *if $q = 3$ and C is self-dual,*

$$W_C(x, y) = \sum_{i=0}^{\lfloor \frac{n}{12} \rfloor} a_i g_4(x, y)^{\frac{n}{4}-3i} g_5(x, y)^i, \text{ and}$$

(iv) *if $q = 4$ and C is Hermitian self-dual,*

$$W_C(x, y) = \sum_{i=0}^{\lfloor \frac{n}{6} \rfloor} a_i g_6(x, y)^{\frac{n}{2}-3i} g_7(x, y)^i.$$

In all cases, all a_is are rational and $\sum_i a_i = 1$.

Proof: We will prove part (i) of Gleason's Theorem and leave parts (ii), (iii), and (iv) as exercises. Assume that C is a formally self-dual binary code of length n with all weights even. If $A_i = A_i(C)$ and $A_i^\perp = A_i(C^\perp)$ are the weight distributions of C and $C^\perp$, respectively, then $A_i = A_i^\perp$ for $0 \le i \le n$. As C is an even code, $C^\perp$ must contain the all-one vector $\mathbf{1}$ implying that $1 = A_n^\perp = A_n$ and hence that $\mathbf{1} \in C$. Therefore $A_i = 0$ if i is odd and $A_{n-i} = A_i$ for all i. For an integer j we choose shortly, let $S = \{2j, 2j + 2, \ldots, n - 2j\}$; then $|S| = s = (n - 4j)/2 + 1$. If $A_1, A_2, \ldots, A_{2j-1}$ are known, then the only unknowns are A_i for $i \in S$. By Theorem 7.3.1, if $2j - 1 \geq s - 1$ (that is $j \geq (n + 2)/8$), then the A_is with $i \in S$ are uniquely determined by the Pless power moments. Set $j = \lceil (n + 2)/8 \rceil$. Since $A_i = 0$ if i is odd, once the $j - 1$ values $A_2, A_4, \ldots, A_{2j-2}$ are known, then all A_i are uniquely determined.

The power moments (P_1) of Section 7.2 with $A_i = A_i^\perp$ can be viewed (by moving the right-hand side to the left) as a homogeneous system of linear equations. We have the additional linear homogeneous equations $A_i - A_{n-i} = 0$ for all i and $A_i = 0$ for i odd. Consider all solutions of this combined system over the rational numbers $\mathbb{Q}$. The solutions to this combined system form a subspace of $\mathbb{Q}^{n+1}$, and by the above argument the dimension of this subspace is $j = \lceil (n + 2)/8 \rceil$ (we do not assume that $A_0 = 1$ but allow A_0 to be any rational number). By Exercise 495, $j = \lfloor n/8 \rfloor + 1$. We give another complete set of solutions of this same system. By Theorem 7.2.2, for $0 \leq i \leq \lfloor n/8 \rfloor$, $g_1(x, y)^{\frac{n}{2}-4i} g_2(x, y)^i$ is the weight enumerator of the direct sum of $(n/2) - 4i$ copies of the $[2, 1, 2]$ binary repetition code and i copies of $\widehat{\mathcal{H}}_3$; this direct sum is a self-dual code of length n. Thus each of these polynomials leads to a solution of our homogeneous system. Therefore the set of polynomials $\sum_{i=0}^{\lfloor \frac{n}{8} \rfloor} a_i g_1(x, y)^{\frac{n}{2}-4i} g_2(x, y)^i$ where $a_i \in \mathbb{Q}$ also leads to solutions of this system. By Exercise 496, $\{g_1(x, y)^{\frac{n}{2}-4i} g_2(x, y)^i \mid 0 \leq i \leq \lfloor n/8 \rfloor\}$ is a linearly independent set of $j = \lfloor n/8 \rfloor + 1$ solutions, and hence all solutions of our homogeneous system can be derived from rational combinations of this set, proving part (i). Note that $\sum_i a_i = 1$ comes from the requirement that $A_0 = 1$. □

Exercise 495 Show that if n is an even positive integer then

$$\left\lceil \frac{n+2}{8} \right\rceil - 1 = \left\lfloor \frac{n}{8} \right\rfloor.$$ ♦

Exercise 496 With the notation of Gleason's Theorem do the following:
(a) Prove that if we let $g_2'(x, y) = x^2 y^2 (x^2 - y^2)^2$, then we can use $g_2'(x, y)$ in place of $g_2(x, y)$ by showing that given $a_i \in \mathbb{Q}$ (or $b_i \in \mathbb{Q}$), there exist $b_i \in \mathbb{Q}$ (or $a_i \in \mathbb{Q}$) such that

$$\sum_{i=0}^{\lfloor \frac{n}{8} \rfloor} a_i g_1(x, y)^{\frac{n}{2}-4i} g_2(x, y)^i = \sum_{i=0}^{\lfloor \frac{n}{8} \rfloor} b_i g_1(x, y)^{\frac{n}{2}-4i} g_2'(x, y)^i.$$

(b) Prove that $\{g_1(x, y)^{\frac{n}{2}-4i} g_2'(x, y)^i \mid 0 \leq i \leq \lfloor n/8 \rfloor\}$ is linearly independent over $\mathbb{Q}$.
(c) Prove that $\{g_1(x, y)^{\frac{n}{2}-4i} g_2(x, y)^i \mid 0 \leq i \leq \lfloor n/8 \rfloor\}$ is linearly independent over $\mathbb{Q}$. ♦

The polynomials $g_1, \ldots, g_7$ presented in Gleason's Theorem are all weight enumerators of certain self-dual codes and are called *Gleason polynomials*. The polynomials g_1 and g_2 were discussed earlier. The weight enumerators of the extended binary and ternary Golay codes are g_3 and g_5, by Exercise 384 and Example 7.3.2, respectively. The tetracode of Example 1.3.3 has weight enumerator g_4; see Exercise 19. The repetition code of length 2 over $\mathbb{F}_4$ has weight enumerator g_6, and the hexacode, given in Example 1.3.4, has weight enumerator g_7; see Exercises 19 and 391.

Exercise 497 Prove parts (ii), (iii), and (iv) of Gleason's Theorem. Note that in parts (iii) and (iv), you cannot assume that $A_i = A_{n-i}$. In part (ii) it may be helpful, as in

Exercise 496, to replace $g_3(x, y)$ by $g_3'(x, y) = x^4 y^4 (x^4 - y^4)^4$; in part (iii), you can replace $g_5(x, y)$ by $g_5'(x, y) = x^3 (x^3 - y^3)^3$; and in part (iv), you can replace $g_7(x, y)$ by $g_7'(x, y) = x^2 (x^2 - y^2)^2$. ◆

The proof of parts (i), (ii), and (iii) of Gleason's Theorem is found in [19]. There are other proofs. One of the nicest involves the use of invariant theory. The proof of part (iv) using invariant theory is found in [215, 217]. This approach has been applied to many other types of weight enumerators, such as complete weight enumerators, and to weight enumerators of other families of codes. In the proof, one can show that the weight enumerator of the code under consideration is invariant under a finite group of matrices, the specific group depending on the family of codes. This group includes a matrix directly associated to the MacWilliams equations (M_3) of Section 7.2. Once this group is known, the theory of invariants allows one to write a power series associated to a polynomial ring determined by this group. After this power series is computed, one can look for polynomials of degrees indicated by this power series to use in a Gleason-type theorem. Unfortunately, the development of invariant theory required for this is too extensive for this book. Readers interested in this approach should consult [215, 218, 291, 313].

Exercise 498 Show that the possible weight distributions of an even formally self-dual binary code of length $n = 8$ are:

$A_0 = A_8$	1	1	1	1	1	1	1	1
$A_2 = A_6$	0	1	2	3	4	5	6	7
A_4	14	12	10	8	6	4	2	0

◆

Exercise 499 Show that only two of the weight distributions in Exercise 498 actually arise as distributions of self-dual binary codes of length 8. ◆

Exercise 500 Make a table as in Exercise 498 of the possible weight distributions of an even formally self-dual binary code of length $n = 10$. ◆

Exercise 501 Do the following:
(a) Using the table from Exercise 500 give the weight distribution of a $[10, 5, 4]$ even formally self-dual binary code.
(b) Find a $[10, 5, 4]$ even formally self-dual binary code. (By Example 9.4.2, this code will not be self-dual.) ◆

Exercises 498 and 500 illustrate how Gleason's Theorem can be used to find possible weight distributions of even formally self-dual binary codes. These exercises together with Exercises 499 and 501 indicate that a possible weight distribution arising from Gleason's Theorem may not actually occur as the weight distribution of any code, or it may arise as the weight distribution of an even formally self-dual code but not of a self-dual code.

The following is an immediate consequence of Gleason's Theorem.

Corollary 9.2.2 *Self-dual doubly-even binary codes of length n exist if and only if* $8 \mid n$; *self-dual ternary codes of length n exist if and only if* $4 \mid n$; *and Hermitian self-dual codes over* $\mathbb{F}_4$ *of length n exist if and only if n is even.*

Exercise 502 Prove Corollary 9.2.2.   ◆

Note that Theorem 9.1.3 also implies that self-dual codes over $\mathbb{F}_3$ of length n exist if and only if $4 \mid n$; see Example 9.1.4. Hermitian self-dual codes over $\mathbb{F}_4$ of length n have only even weight codewords by Theorem 1.4.10; therefore by Exercise 383 these codes, being self-dual, have a codeword of weight n, and hence n must be even giving an alternate verification of part of Corollary 9.2.2. This argument also shows that every Hermitian self-dual code over $\mathbb{F}_4$ is equivalent to one that contains the all-one codeword.

9.3   Upper bounds

We can use Gleason's Theorem to determine upper bounds on the minimum distance of the codes arising in that theorem. We present some of these bounds in this section and give improvements where possible.

As an illustration, consider the even formally self-dual binary codes of length $n = 12$ that arise in (i) of Gleason's Theorem. The weight enumerator of such a code C must have the form $W_C(x, y) = a_1 g_1(x, y)^6 + (1 - a_1)g_1(x, y)^2 g_2(x, y)$. By examining the coefficient of $x^2 y^{10}$, we have

$$W_C(x, y) = \left(\frac{1}{4}A_2 - \frac{1}{2}\right) g_1(x, y)^6 + \left(-\frac{1}{4}A_2 + \frac{3}{2}\right) g_1(x, y)^2 g_2(x, y), \tag{9.1}$$

and hence $W_C(x, y)$ is completely determined when A_2 is known. If $A_2 = 0$,

$$W_C(x, y) = y^{12} + 15x^4 y^8 + 32x^6 y^6 + 15x^8 y^4 + x^{12}. \tag{9.2}$$

This argument shows that a $[12, 6]$ self-dual binary code has minimum distance at most 4. A generalization of this argument gives a bound on the minimum distance of the codes in Gleason's Theorem.

Exercise 503 Verify (9.1) and (9.2).   ◆

Theorem 9.3.1 *Let C be an* $[n, n/2, d]$ *code over* $\mathbb{F}_q$, *for* $q = 2, 3,$ *or 4.*
(i)   *If* $q = 2$ *and C is formally self-dual and even, then* $d \le 2 \lfloor n/8 \rfloor + 2$.
(ii) *If* $q = 2$ *and C is self-dual doubly-even, then* $d \le 4 \lfloor n/24 \rfloor + 4$.
(iii) *If* $q = 3$ *and C is self-dual, then* $d \le 3 \lfloor n/12 \rfloor + 3$.
(iv) *If* $q = 4$ *and C is Hermitian self-dual, then* $d \le 2 \lfloor n/6 \rfloor + 2$.
In all cases, if equality holds in the bounds, the weight enumerator of C is unique.

These bounds were proved in [217, 223], and we omit the proofs. However, we give the major idea used in the proof. Consider a Type II code C and assume that $A_i(C) = 0$ for $1 \le i < 4 \lfloor n/24 \rfloor + 4$. In our proof of Gleason's Theorem (which is a model for the Type II case you were asked to prove in Exercise 497), it is shown that the weight enumerator $W_C(x, y)$ of C is uniquely determined. In this polynomial, the value of $A_{4\lfloor n/24 \rfloor + 4}$ can be determined explicitly; this value is positive and so no Type II code of higher minimum weight could exist. A similar argument works in the other cases.

We turn now to the question of when these bounds can be met. This question has intrigued researchers for many years. Suppose first that C is a Type II or Type IV code with minimum weight meeting the appropriate bound of Theorem 9.3.1(ii) or (iv). The weight enumerator is unique. If the length n is large enough, this weight enumerator has a negative coefficient, which is clearly not possible. When searching for negative coefficients, a natural coefficient to examine is the first nonzero one after A_d, where d is the minimum weight. The exact situation is described by the following theorem found in [291].

Theorem 9.3.2 *Let $W_C(x, y)$ be the weight enumerator of a self-dual code C of length n of Type II or IV meeting bounds (ii) or (iv), respectively. Assume that n is divisible by 8 or 2, respectively. The following hold, where A_m is the coefficient of $y^m x^{n-m}$ in $W_C(x, y)$:*

- *Type II: If $n = 24i$ with $i \geq 154$, $n = 24i + 8$ with $i \geq 159$, or $n = 24i + 16$ with $i \geq 164$, then $A_m < 0$ for $m = 4 \lfloor n/24 \rfloor + 8$. In particular, C cannot exist for $n > 3928$.*
- *Type IV: If $n = 6i$ with $i \geq 17$, $n = 6i + 2$ with $i \geq 20$, or $n = 6i + 4$ with $i \geq 22$, then $A_m < 0$ for $m = 2 \lfloor n/6 \rfloor + 4$. In particular, C cannot exist for $n > 130$.*

This theorem only shows where one particular coefficient of $W_C(x, y)$ is negative. Of course other coefficients besides A_m given above may be negative eliminating those values of n. The following is known about the Type III case; see [291].

Theorem 9.3.3 *Let C be a Type III code of length n and minimum distance $d = 3 \lfloor n/12 \rfloor + 3$. Then C does not exist for $n = 72, 96, 120$, and all $n \geq 144$.*

The proof of this result involves showing that the weight enumerator of a Type III code meeting the bound of Theorem 9.3.1(iii) has a negative coefficient. J. N. Pierce was the first to use arguments on weight enumerators to show the nonexistence of codes when he showed that a Type III [72, 36, 21] code, which meets the bound of Theorem 9.3.1(iii), does not exist.

We now turn to the bound of Theorem 9.3.1(i). In [223] it was discovered that the weight enumerator of an even formally self-dual binary code of length n meeting the bound of Theorem 9.3.1(i) has a negative coefficient for $n = 32, 40, 42, 48, 50, 52$, and $n \geq 56$. This information allowed Ward to show in [344] that the only values of n for which self-dual codes exist meeting the bound are $n = 2, 4, 6, 8, 12, 14, 22$, and 24. There are also even formally self-dual binary codes of lengths $n = 10, 18, 20, 28$, and 30 that meet the bound, but no higher values of n; see [82, 83, 156, 165, 309]. (We note that for even n with $32 \leq n \leq 54$, there are no $[n, n/2, 2 \lfloor n/8 \rfloor + 2]$ binary codes, self-dual or not, by [32].) We summarize this in the following theorem.

Theorem 9.3.4 *There exist self-dual binary codes of length n meeting the bound of Theorem 9.3.1(i) if and only if $n = 2, 4, 6, 8, 12, 14, 22$, and 24. There exist even formally self-dual binary codes of length n meeting the bound of Theorem 9.3.1(i) if and only if n is even with $n \leq 30$ and $n \neq 16$ and $n \neq 26$. For all these codes, the weight enumerator is uniquely determined by the length.*

The bound in Theorem 9.3.1(i) can be significantly improved for self-dual codes. The following theorem is due to Rains [290] and shows that the bound for Type II codes is almost the bound for Type I codes also; its proof makes use of the shadow introduced in the next section.

Theorem 9.3.5 *Let C be an $[n, n/2, d]$ self-dual binary code. Then $d \leq 4 \lfloor n/24 \rfloor + 4$ if $n \not\equiv 22 \pmod{24}$. If $n \equiv 22 \pmod{24}$, then $d \leq 4 \lfloor n/24 \rfloor + 6$. If $24 \mid n$ and $d = 4 \lfloor n/24 \rfloor + 4$, then C is Type II.*

In Theorem 9.4.14, we will see that if $n \equiv 22 \pmod{24}$ and $d = 4 \lfloor n/24 \rfloor + 6$, then C can be obtained from a Type II $[n + 2, (n/2) + 1, d + 2]$ code in a very specific way.

A Type II, III, or IV code meeting the bound of Theorem 9.3.1(ii), (iii), or (iv), respectively, is called an *extremal* Type II, III, or IV code. A Type I code meeting the bound of Theorem 9.3.5 is called an *extremal* Type I code.[2] Note that the minimum distance for an extremal Type I code of length a multiple of 24 is 2 less than the Type II bound; that is $d = (n/6) + 2$ is the minimum weight of an extremal Type I code of length n a multiple of 24. The formally self-dual codes of Theorem 9.3.4 are also called *extremal*. Extremal codes of Type II, III, and IV all have unique weight enumerators and exist only for finitely many lengths, as described above. Those of Type I of length exceeding 24 do not necessarily have unique weight enumerators; it has not been proved that these exist for only finitely many lengths, but that is likely the case.

We summarize what is known about the Type I and Type II codes in Table 9.1 for length n with $2 \leq n \leq 72$. In the table, "d_I" is the largest minimum weight for which a Type I code is known to exist while "d_{II}" is the largest minimum weight for which a Type II code is known to exist. The superscript "E" indicates that the code is extremal; the superscript "O" indicates the code is not extremal but optimal – that is, no code of the given type can exist with a larger minimum weight. The number of inequivalent Type I and II codes of the given minimum weight is listed under "num_I" and "num_{II}", respectively. When the number in that column is exact (without $\geq$), the classification of those codes is complete; some of the codes that arise are discussed in Section 9.7. An entry in the column "d_I" or "d_{II}" such as "10 (12^E)" followed by a question mark in the next column indicates that there is a code known of the smaller minimum weight but there is no code known for the higher minimum weight. This table is taken from [291, Table X], where references are also provided, with updates from [30, 168].

The [8, 4, 4] extended Hamming code is the unique Type II code of length 8 indicated in Table 9.1. Two of the five [32, 16, 8] Type II codes are extended quadratic residue and Reed–Muller codes. The [24, 12, 8] and [48, 24, 12] binary extended quadratic residue codes are extremal Type II as exhibited in Example 6.6.23. However, the binary extended quadratic residue code of length 72 has minimum weight 12 and hence is not extremal. The first two lengths for which the existence of an extremal Type II code is undecided are 72 and 96. For lengths beyond 96, extremal Type II codes are known to exist only for lengths 104 and 136.

Research Problem 9.3.6 *Either construct or establish the nonexistence of a* [56, 28, 12],

---

[2] Before the bound of Theorem 9.3.5 was proved, extremal codes of Type I were those that met the bound of Theorem 9.3.1(i). The older bound is actually better than the newer bound for $n \leq 8$. There was an intermediate bound found in Theorem 1 of [58] that is also occasionally better than the bound of Theorem 9.3.5 for some $n \leq 68$, but not for any larger n.

Table 9.1 *Type I and II codes of length $2 \le n \le 72$*

n	d_I	num$_I$	d_{II}	num$_{II}$	n	d_I	num$_I$	d_{II}	num$_{II}$
2	2^O	1			38	8^E	≥ 368		
4	2^O	1			40	8^E	≥ 22	8^E	≥ 1000
6	2^O	1			42	8^E	≥ 30		
8	2^O	1	4^E	1	44	8^E	≥ 108		
10	2^O	2			46	10^E	≥ 1		
12	4^E	1			48	10^E	≥ 7	12^E	≥ 1
14	4^E	1			50	10°	≥ 6		
16	4^E	1	4^E	2	52	10°	≥ 499		
18	4^E	2			54	10°	≥ 54		
20	4^E	7			56	$10 \, (12^E)$	?	12^E	≥ 166
22	6^E	1			58	10°	≥ 101		
24	6^E	1	8^E	1	60	12^E	≥ 5		
26	6°	1			62	12^E	≥ 1		
28	6°	3			64	12^E	≥ 5	12^E	≥ 3270
30	6°	13			66	12^E	≥ 3		
32	8^E	3	8^E	5	68	12^E	≥ 65		
34	6°	≥ 200			70	$12 \, (14^E)$	?		
36	8^E	≥ 14			72	$12 \, (14^E)$	?	$12 \, (16^E)$	?

a [70, 35, 14], *and a* [72, 36, 14] *Type I code; do the same for a* [72, 36, 16] *and a* [96, 48, 20] *Type II code.*

Self-dual codes with an automorphism of certain prime orders have either been classified or shown not to exist. For example, it was shown in [54, 152, 265, 284] that there are no [72, 36, 16] Type II codes with an automorphism of prime order $p > 7$. Along the same lines, it is known (see [142]) that the only [48, 24, 12] Type II code with an automorphism of odd order is an extended quadratic residue code. By a long computer search [140, 141], it was shown that no other Type II code of length 48 exists. This leads to the following problem.

Research Problem 9.3.7 *Prove without a lengthy computer search that the* [48, 24, 12] *Type II code is unique.*

By Theorem 9.3.3, extremal Type III codes of length n cannot exist for $n = 72, 96, 120$, and all $n \ge 144$. The smallest length for which the existence of an extremal Type III code is undecided is length $n = 52$. Extremal Type III codes of length n are known to exist for all $n \equiv 0 \pmod 4$ with $4 \le n \le 64$ except $n = 52$; all other lengths through 140 except 72, 96, and 120 remain undecided. Table 9.2 summarizes this information. In the table, codes have length "n"; "d_{known}" is the largest minimum weight for which a Type III code is known to exist. The superscripts "E" and "O" and the notation "12 (15^E)" followed by a question mark are as in Table 9.1. The number of inequivalent Type III codes of the given minimum weight is listed under "num"; again those values given exactly indicate that the classification is complete. This table is taken from [291, Table XII], where references are provided.

Table 9.2 *Type III codes of length* $4 \leq n \leq 72$

n	d_{known}	num	n	d_{known}	num	n	d_{known}	num
4	3^E	1	28	9^E	≥ 32	52	$12\,(15^E)$	?
8	3^E	1	32	9^E	≥ 239	56	15^E	≥ 1
12	6^E	1	36	12^E	≥ 1	60	18^E	≥ 2
16	6^E	1	40	12^E	≥ 20	64	18^E	≥ 1
20	6^E	6	44	12^E	≥ 8	68	$15\,(18^E)$	?
24	9^E	2	48	15^E	≥ 2	72	18°	≥ 1

Table 9.3 *Type IV codes of length* $2 \leq n \leq 30$

n	d_{known}	num	n	d_{known}	num	n	d_{known}	num
2	2^E	1	12	4°	5	22	8^E	≥ 46
4	2^E	1	14	6^E	1	24	8°	≥ 17
6	4^E	1	16	6^E	4	26	$8\,(10^E)$	?
8	4^E	1	18	8^E	1	28	10^E	≥ 3
10	4^E	2	20	8^E	2	30	12^E	≥ 1

The $[4, 2, 3]$ tetracode is the unique Type III code of length 4, while the $[12, 6, 6]$ extended ternary Golay code is the unique Type III code of length 12. The two extremal Type III codes of length 24 are an extended quadratic residue code and a symmetry code, described in Section 10.5. The only known extremal Type III code of length 36 is a symmetry code, the only two known of lengths 48 and 60 are extended quadratic residue and symmetry codes, and the only one known of length 72 is an extended quadratic residue code.

Research Problem 9.3.8 *Either construct or establish the nonexistence of any of the extremal Type III codes of lengths* $n = 52, 68,$ *or* $76 \leq n \leq 140$ *(except 96 and 120), where* $4 \mid n$.

By Theorem 9.3.2, extremal Type IV codes of even length n cannot exist if $n > 130$; they are also known not to exist for lengths 12, 24, 102, 108, 114, 120, 122, 126, 128, and 130. They are known to exist for lengths 2, 4, 6, 8, 10, 14, 16, 18, 20, 22, 28, and 30. Existence is undecided for all other lengths, the smallest being length 26. It is known [144, 145] that no $[26, 13, 10]$ Type IV code exists with a nontrivial automorphism of odd order. Table 9.3 summarizes some of this information. Again, this table is taken from [291, Table XIII], where references are also provided, with updates from [167]; the notation is as in Table 9.2.

In Table 9.3, the unique $[6, 3, 4]$ Type IV code is the hexacode, the unique $[14, 7, 6]$ code is an extended quadratic residue code, and the unique $[18, 9, 8]$ code is an extended duadic (nonquadratic residue) code. The only known extremal Type IV code of length 30 is an extended quadratic residue code.

Research Problem 9.3.9 *Either construct or establish the nonexistence of any of the extremal Type IV codes of even lengths* $n = 26$ *or* $32 \leq n \leq 124$ *(except 102, 108, 114, 120, and 122).*

The extremal codes of Types II, III, and IV hold t-designs as a consequence of the Assmus–Mattson Theorem.

Theorem 9.3.10 *The following results on t-designs hold in extremal codes of Types II, III, and IV.*

(i) *Let C be a $[24m + 8\mu, 12m + 4\mu, 4m + 4]$ extremal Type II code for $\mu = 0, 1,$ or 2. Then codewords of any fixed weight except 0 hold t-designs for the following parameters:*

  (a) $t = 5$ *if* $\mu = 0$ *and* $m \geq 1$,

  (b) $t = 3$ *if* $\mu = 1$ *and* $m \geq 0$*, and*

  (c) $t = 1$ *if* $\mu = 2$ *and* $m \geq 0$.

(ii) *Let C be a $[12m + 4\mu, 6m + 2\mu, 3m + 3]$ extremal Type III code for $\mu = 0, 1,$ or 2. Then codewords of any fixed weight i with $3m + 3 \leq i \leq 6m + 3$ hold t-designs for the following parameters:*

  (a) $t = 5$ *if* $\mu = 0$ *and* $m \geq 1$,

  (b) $t = 3$ *if* $\mu = 1$ *and* $m \geq 0$*, and*

  (c) $t = 1$ *if* $\mu = 2$ *and* $m \geq 0$.

(iii) *Let C be a $[6m + 2\mu, 3m + \mu, 2m + 2]$ extremal Type IV code for $\mu = 0, 1,$ or 2. Then codewords of any fixed weight i with $2m + 2 \leq i \leq 3m + 2$ hold t-designs for the following parameters:*

  (a) $t = 5$ *if* $\mu = 0$ *and* $m \geq 2$,

  (b) $t = 3$ *if* $\mu = 1$ *and* $m \geq 1$*, and*

  (c) $t = 1$ *if* $\mu = 2$ *and* $m \geq 0$.

Proof: We prove (i)(a) and (ii)(a), leaving the remainder as an exercise. Suppose C is a $[24m, 12m, 4m + 4]$ Type II code with weight distribution $A_i = A_i(C)$. As $A_{24m-i} = A_i$ and $A_i = 0$ if $4 \nmid i$, then $\{i \mid 0 < i < 24m, A_i \neq 0\}$ has size at most $s = 4m - 1$. As $5 < 4m + 4$ and $s \leq (4m + 4) - 5$, (i)(a) follows from the Assmus–Mattson Theorem.

Suppose now that C is a $[12m, 6m, 3m + 3]$ Type III code. The largest integer w satisfying

$$w - \left\lfloor \frac{w + 1}{2} \right\rfloor < 3m + 3$$

is $w = 6m + 5$. As $A_i = 0$ if $3 \nmid i$, then $\{i \mid 0 < i < 12m - 5, A_i \neq 0\}$ has size at most $s = 3m - 2$. As $5 < 3m + 3$ and $s \leq (3m + 3) - 5$, (ii)(a) follows from the Assmus–Mattson Theorem. $\square$

Exercise 504 Complete the remaining parts of the proof of Theorem 9.3.10. ◆

Janusz has generalized Theorem 9.3.10(i) in [157].

Theorem 9.3.11 *Let C be a $[24m + 8\mu, 12m + 4\mu, 4m + 4]$ extremal Type II code for $\mu = 0, 1,$ or 2, where $m \geq 1$ if $\mu = 0$. Then either:*

(i) *the codewords of any fixed weight $i \neq 0$ hold t-designs for $t = 7 - 2\mu$, or*

(ii) *the codewords of any fixed weight $i \neq 0$ hold t-designs for $t = 5 - 2\mu$ and there is no i with $0 < i < 24m + 8\mu$ such that codewords of weight i hold a $(6 - 2\mu)$-design.*

Currently there are no known extremal Type II codes where Theorem 9.3.11(ii) is false. Applying this result to extremal Type II codes of length $n = 24m$, we see that codewords of any fixed weight $i \neq 0$ hold 5-designs, and if the codewords of some fixed weight i with $0 < i < n$ hold a 6-design, then the codewords of any fixed weight $i \neq 0$ hold 7-designs. We comment further about this at the end of the section.

In Example 8.4.6 we found the coset weight distribution of the $[24, 12, 8]$ extended binary Golay code with the assistance of the designs. The same technique can be used for any extremal Type II code to compute the coset weight distribution of any coset of weight t or less, where t is given in Theorem 9.3.10. By Theorem 7.5.2, the weight distribution of a coset is often uniquely determined once a few weights are known. The values of the intersection numbers λ_i^j for the t-designs will enable one to do the calculations as in Example 8.4.6.

Exercise 505 We will see later that there are five inequivalent Type II $[32, 16, 8]$ codes. Let $\mathcal{C}$ be such a code. The weight enumerator of $\mathcal{C}$ is

$$W_{\mathcal{C}}(x, y) = y^{32} + 620x^8 y^{24} + 13\,888x^{12} y^{20} + 36\,518x^{16} y^{16}$$
$$+ 13\,888x^{20} y^{12} + 620x^{24} y^8 + x^{32}.$$

By Theorem 9.3.10(i)(b), the codewords of fixed nonzero weight hold 3-designs.
(a) Find the Pascal triangle for the 3-designs held by codewords of $\mathcal{C}$ of weights 8, 12, 16, 20, and 24. Note that the designs held by the weight 24 and weight 20 codewords are the complements of the designs held by the weight 8 and weight 12 codewords, respectively.
(b) Compute the coset weight distributions for the cosets of $\mathcal{C}$ of weights 1, 2, and 3.
(c) How many cosets of weights 1, 2, and 3 are there?   ◆

In [176] the following result is presented; it is due to Venkov. It shows that there is surprising additional design structure in extremal Type II codes.

Theorem 9.3.12 Let C be an extremal Type II code of length $n = 24m + 8\mu$, with $\mu = 0$, 1, or 2. Let M be the set of codewords of minimum weight $4m + 4$ in C. For an arbitrary vector $\mathbf{a} \in \mathbb{F}_2^n$ and any j, define

$$u_j(\mathbf{a}) = |\{\mathbf{x} \in M \mid |\mathrm{supp}(\mathbf{a}) \cap \mathrm{supp}(\mathbf{x})| = j\}|.$$

Also define $k_5 = \binom{5m-2}{m-1}$, $k_3 = \binom{5m}{m}$, and $k_1 = 3\binom{5m+2}{m}$. Let $t = 5 - 2\mu$. If $\mathrm{wt}(\mathbf{a}) = t + 2$, then $u_{t+1}(\mathbf{a}) + (t - 4)u_{t+2}(\mathbf{a}) = k_t$, independent of the choice of $\mathbf{a}$.

Example 9.3.13 Let C be the $[24, 12, 8]$ extended binary Golay code. Then in Theorem 9.3.12, $m = 1$, $\mu = 0$, $t = 5$, and $k_5 = 1$. The theorem states that any vector $\mathbf{a} \in \mathbb{F}_2^{24}$ of weight 7 satisfies precisely one of the following: there is exactly one minimum weight codeword $\mathbf{c}$ whose support contains the support of $\mathbf{a}$ or meets it in six coordinates. By Theorem 9.3.10, the codewords of weight 8 hold a 5-design; this was a design we discussed in Chapter 8, and its intersection numbers are given in Figure 8.3. While this design is not a 7-design (or even a 6-design), it does have additional design-like properties. Theorem 9.3.12

implies that any set of points of size 7 in this design is either contained in exactly one block or meets exactly one block in six points but not both. ∎

Extremal self-dual codes have provided many important designs, particularly 5-designs. As Theorem 9.3.12 established, these designs have additional structure that almost turns a t-design into a $(t + 2)$-design. Theorem 9.3.11 indicates that if you find a $(t + 1)$-design, you find $(t + 2)$-designs. However, no one has even found nontrivial 6-designs in codes, and for a long time it was conjectured that no 6-designs existed. That conjecture has been disproved and nontrivial t-designs exist for all t [326].

Exercise 506   Let C be a $[48, 24, 12]$ binary extended quadratic residue code; see Example 6.6.23. By Theorem 9.3.10, all codewords of fixed weights k hold 5-$(48, k, \lambda)$ designs. The weight distribution of C is $A_0 = A_{48} = 1$, $A_{12} = A_{36} = 17\,296$, $A_{16} = A_{32} = 535\,095$, $A_{20} = A_{28} = 3\,995\,376$, and $A_{24} = 7\,681\,680$.
(a) Give the parameter λ for the 5-$(48, k, \lambda)$ design, with $k = 12, 16, 20, 24, 28, 32$, and 36, held by the codewords in C.
(b) Prove that none of the designs in (a) can be 6-designs.
(c) What does Theorem 9.3.12 imply about the intersection of sets of points of size 7 with blocks of the 5-$(48, 12, \lambda)$ design from (a)?   ◆

9.4   The Balance Principle and the shadow

Because extremal codes have the highest possible minimum weight for codes of their type and those of Types II, III, and IV hold designs, a great deal of effort has gone into either constructing these codes or showing they do not exist. Various arguments, including those involving the Balance Principle and the shadow code, have been used. We begin with the *Balance Principle*, which is also useful in gluing together self-orthogonal codes as we will see in Section 9.7.

Theorem 9.4.1 ([177, 273])   *Let C be a self-dual $[n, n/2]$ code. Choose a set of coordinate positions $\mathcal{P}_{n_1}$ of size n_1 and let $\mathcal{P}_{n_2}$ be the complementary set of coordinate positions of size $n_2 = n - n_1$. Let C_i be the subcode of C all of whose codewords have support in $\mathcal{P}_{n_i}$. The following hold:*
(i) *(Balance Principle)*

$$\dim C_1 - \frac{n_1}{2} = \dim C_2 - \frac{n_2}{2}.$$

(ii) *If we reorder coordinates so that $\mathcal{P}_{n_1}$ is the left-most n_1 coordinates and $\mathcal{P}_{n_2}$ is the right-most n_2 coordinates, then C has a generator matrix of the form*

$$G = \begin{bmatrix} A & O \\ O & B \\ D & E \end{bmatrix}, \tag{9.3}$$

where $[A \ \ O]$ is a generator matrix of C_1 and $[O \ \ B]$ is a generator matrix of C_2, O being the appropriate size zero matrix. We also have:

(a) If $k_i = \dim C_i$, then D and E each have rank $(n/2) - k_1 - k_2$.

(b) Let $\mathcal{A}$ be the code of length n_1 generated by A, $\mathcal{A}_D$ the code of length n_1 generated by the rows of A and D, $\mathcal{B}$ the code of length n_2 generated by B, and $\mathcal{B}_E$ the code of length n_2 generated by the rows of B and E. Then $\mathcal{A}^\perp = \mathcal{A}_D$ and $\mathcal{B}^\perp = \mathcal{B}_E$.

Proof: We first prove (ii). By definition of C_i, a generator matrix for C can be found as given, where A and B have k_1 and k_2 rows, respectively. Also, as C_i is the maximum size code all of whose codewords have support in $\mathcal{P}_{n_i}$, and the rows of G are independent, (ii)(a) must hold; otherwise, the size of C_1 or C_2 could be increased. The $(n/2) - k_2$ rows of A and D must be independent as otherwise we can again increase the size of C_2. Thus $\dim \mathcal{A}_D = (n/2) - k_2$. Furthermore, as C is self-dual, $\mathcal{A} \subseteq \mathcal{A}_D^\perp$ implying

$$k_1 + \left(\frac{n}{2} - k_2\right) \leq n_1. \tag{9.4}$$

Similarly, $\dim \mathcal{B}_E = (n/2) - k_1$, $\mathcal{B} \subseteq \mathcal{B}_E^\perp$, and

$$k_2 + \left(\frac{n}{2} - k_1\right) \leq n_2. \tag{9.5}$$

Adding (9.4) and (9.5), we obtain $n \leq n_1 + n_2$. Since we have equality here, we must have equality in both (9.4) and (9.5) implying (ii)(b). Part (i) is merely the statement that equality holds in (9.4). □

Example 9.4.2 In this example, we show there is no $[10, 5, 4]$ self-dual binary code; such a code would meet the bound of Theorem 9.3.1(i). Suppose C is such a code, where we write a generator matrix for C in form (9.3) with the first n_1 coordinates being the support of a weight 4 codeword. Then A has one row, and by the Balance Principle, B has two rows. As C contains the all-one codeword of length 10, the matrix B generates a $[6, 2, 4]$ self-orthogonal code containing the all-one vector of length 6. This is clearly impossible and C does not exist. ∎

Exercise 507 Let C be an $[18, 9]$ Type I code.
(a) Show that the bound on the minimum weight of C in Theorem 9.3.1(i) is $d \leq 6$.
(b) Show that there is no $[18, 9, 6]$ Type I code. ◆

Example 9.4.3 In Research Problem 9.3.7, we pose the question of finding all $[48, 24, 12]$ codes Type II C without a lengthy computer search. By permuting coordinates, we can assume C has generator matrix G in the form (9.3), where $n_1 = n_2 = 24$ and the first n_1 coordinates are the support of a weight 24 codeword. Koch proved in [176] that $1 \leq k_1 \leq 4$. It can be shown that every weight 12 codeword has support disjoint from the support of another weight 12 codeword and so, using the sum of these codewords as our weight 24 codeword, we may assume that $2 \leq k_1 \leq 4$. In [140], a lengthy computer search showed that if $k_1 = 4$, then C is the extended quadratic residue code. The values $k_1 = 2$ and $k_1 = 3$ were eliminated by another long computer search in [141]. ∎

Let C be an $[n, n/2]$ Type I code; C has a unique $[n, (n/2) - 1]$ subcode C_0 consisting of all codewords in C whose weights are multiples of four; that is, C_0 is the doubly-even subcode of C. Thus the weight distribution of C_0 is completely determined by the weight distribution of C. Hence the weight distribution of $C_0^{\perp}$ is determined by the MacWilliams equations. It can happen that the latter has negative or fractional coefficients even though the weight distribution of C does not, implying that C does not exist. This occurs for both $[10, 5, 4]$ and $[18, 9, 6]$ Type I codes and other Type I codes [58].

We consider the general situation of an $[n, n/2]$ self-dual binary code C; choose a fixed $[n, (n/2) - 1]$ subcode C_0, which is not necessarily a doubly-even subcode. Thus the code $C_0^{\perp}$ contains both C_0 and C, as C is self-dual, and has dimension $(n/2) + 1$ implying that there are four cosets of C_0 in $C_0^{\perp}$. Let $C_0^{\perp} = C_0 \cup C_1 \cup C_2 \cup C_3$, where C_i are these cosets of C_0 in $C_0^{\perp}$. Two of these cosets, including C_0, must be the two cosets of C_0 in C. We number the cosets so that $C = C_0 \cup C_2$; so $C_1 \cup C_3$ is a coset of C in $C_0^{\perp}$. If C is Type I and C_0 is the doubly-even subcode of C, then $C_1 \cup C_3$ is called the *shadow of C* [58]. The following lemma will give us a great deal of information about the cosets C_i.

Lemma 9.4.4 *Let C be an $[n, n/2]$ self-dual binary code with C_0 an $[n, (n/2) - 1]$ subcode. The following hold:*

(i) *If $\{i, j, k\} = \{1, 2, 3\}$, then the sum of a vector in C_i plus a vector in C_j is a vector in C_k.*

(ii) *For $i, j \in \{0, 1, 2, 3\}$, either every vector in C_i is orthogonal to every vector in C_j or no vector in C_i is orthogonal to any vector in C_j.*

The proof is left as an exercise. We give notation to the two possibilities in Lemma 9.4.4(ii). If every vector in C_i is orthogonal to every vector in C_j, we denote this by $C_i \perp C_j$; if no vector in C_i is orthogonal to any vector in C_j, we denote this by C_i / C_j. Notice that by definition $C_0 \perp C_i$ for all i.

Exercise 508   Prove Lemma 9.4.4.   ◆

Example 9.4.5 Let C be the $[6, 3, 2]$ self-dual binary code with generator matrix

$$G = \begin{bmatrix} 1 & 1 & 0 & 0 & 0 & 0 \\ 0 & 0 & 1 & 1 & 0 & 0 \\ 0 & 0 & 0 & 0 & 1 & 1 \end{bmatrix}.$$

Let C_0 be the doubly-even subcode of C. Then a generator matrix for $C_0^{\perp}$ is

$$G_0^{\perp} = \begin{bmatrix} 1 & 1 & 1 & 1 & 0 & 0 \\ 0 & 0 & 1 & 1 & 1 & 1 \\ 1 & 1 & 0 & 0 & 0 & 0 \\ 1 & 0 & 1 & 0 & 1 & 0 \end{bmatrix},$$

where the first two rows generate C_0. Let $c_1 = 101010$, $c_2 = 110000$, and $c_3 = c_1 + c_2 = 011010$. Then the cosets C_i of C_0 in $C_0^{\perp}$ for $1 \le i \le 3$ are $C_i = c_i + C_0$. Notice that the shadow $C_1 \cup C_3$ is the coset $c_1 + C$ of C in $C_0^{\perp}$; it consists of all the odd weight vectors in $C_0^{\perp}$. Also $\mathbf{1} \in C_2$.   ■

Exercise 509   With C and C_i as in Example 9.4.5, decide whether $C_i \perp C_j$ or C_i / C_j for all i and j (including $i = j$).   ◆

Table 9.4 *Orthogonality for an even weight coset of a self-dual code*

	C_0	C_2	C_1	C_3
C_0	$\perp$	$\perp$	$\perp$	$\perp$
C_2	$\perp$	$\perp$	/	/
C_1	$\perp$	/	$\perp$	/
C_3	$\perp$	/	/	$\perp$

Exercise 510   Let C be an $[n, k]$ self-orthogonal code over $\mathbb{F}_q$. Let $\mathbf{v} \in \mathbb{F}_q^n$ with $\mathbf{v} \notin C^\perp$. Show that $C_0 = \{\mathbf{c} \in C \mid \mathbf{c} \cdot \mathbf{c}_1 = 0 \text{ for all } \mathbf{c}_1 \in \mathbf{v} + C\}$ is a subcode of C of dimension $k - 1$.   ♦

Exercise 511   Let C be a self-dual binary code of length n. Show that there is a one-to-one correspondence between the subcodes of C of dimension $(n/2) - 1$ and the cosets of C. Hint: See Exercise 510.   ♦

In general, we can choose C_0 to be any $[n, (n/2) - 1]$ subcode of C and obtain the four cosets C_i of C_0 in $C_0^\perp$. Alternately, we can choose a coset C' of C and let $C_0 = (C \cup C')^\perp$. Then $C_0^\perp = C_0 \cup C_1 \cup C_2 \cup C_3$ where $C = C_0 \cup C_2$ and $C' = C_1 \cup C_3$.

Two possibilities arise for the weights of the cosets C_i depending on whether or not $\mathbf{1} \in C_0$. Recall from Exercise 72 that a coset of C, such as $C_1 \cup C_3$, has all even weight vectors or all odd weight vectors. If $\mathbf{1} \in C_0$, then all codewords in $C_0^\perp$ must be even weight, implying that $C_1 \cup C_3$ has only even weight vectors, or equivalently, $C_1 \cup C_3$ is an even weight coset of C. Notice that, in this case, $C_0 \cup C_1$ and $C_0 \cup C_3$ are self-dual codes. If $\mathbf{1} \notin C_0$, then $C_0^\perp$ is not an even code and C must be the even subcode of $C_0^\perp$. Thus $C_1 \cup C_3$ has only odd weight vectors, or equivalently, $C_1 \cup C_3$ is an odd weight coset of C. We summarize this in the following lemma.

Lemma 9.4.6 *Let C be an $[n, n/2]$ self-dual binary code with C_0 an $[n, (n/2) - 1]$ subcode. The following are the only possibilities:*
(i) *If $\mathbf{1} \in C_0$, then $C_1 \cup C_3$ is an even weight coset of C and contains only even weight vectors. Also $C_0 \cup C_1$ and $C_0 \cup C_3$ are self-dual codes.*
(ii) *If $\mathbf{1} \notin C_0$, then $\mathbf{1} \in C_2$. Furthermore, $C_1 \cup C_3$ is an odd weight coset of C and contains only odd weight vectors.*

We investigate the two possibilities that arise in Lemma 9.4.6. In each case we obtain an orthogonality table showing whether or not C_i is orthogonal to C_j. We also obtain additional information about the weights of vectors in C_1 and C_3.

Suppose that $C_1 \cup C_3$ is an even weight coset of C. Then Table 9.4 gives the orthogonality relationships between the cosets [38]; these follow from Lemma 9.4.4, with verification left as an exercise.

Exercise 512   Verify the entries in Table 9.4.   ♦

We give additional information about the cosets that arise in Table 9.4.

Theorem 9.4.7 *Let C be an $[n, n/2]$ self-dual binary code with C_0 an $[n, (n/2) - 1]$ subcode. Suppose that $\mathbf{1} \in C_0$. The following hold:*

(i) *If C is Type II, C_1 and C_3 can be numbered so that all vectors in C_1 have weight 0 modulo 4, and all vectors in C_3 have weight 2 modulo 4.*

(ii) *If C and C_0 are singly-even, then half the vectors in C_i are singly-even and half are doubly-even for $i = 1$ and 3.*

(iii) *(Shadow case) If C is singly-even and C_0 is the doubly-even subcode of C and if $\mathbf{c} \in C_1 \cup C_3$, then $\mathrm{wt}(\mathbf{c}) \equiv n/2 \,(\mathrm{mod}\ 4)$. Also $n \equiv 0 \,(\mathrm{mod}\ 8)$ or $n \equiv 4 \,(\mathrm{mod}\ 8)$.*

Proof: We leave the proofs of (i) and (ii) as an exercise. Suppose that C is singly-even and C_0 is the doubly-even subcode of C. Let $\mathbf{c}_i \in C_i$. Then $\mathbf{c}_1 + \mathbf{c}_3 \in C_2$ by Lemma 9.4.4 implying that $\mathrm{wt}(\mathbf{c}_1 + \mathbf{c}_3) \equiv 2 \,(\mathrm{mod}\ 4)$. But $\mathrm{wt}(\mathbf{c}_1 + \mathbf{c}_3) = \mathrm{wt}(\mathbf{c}_1) + \mathrm{wt}(\mathbf{c}_3) - 2\mathrm{wt}(\mathbf{c}_1 \cap \mathbf{c}_3) \equiv \mathrm{wt}(\mathbf{c}_1) + \mathrm{wt}(\mathbf{c}_3) - 2 \,(\mathrm{mod}\ 4)$ as C_1/C_3. Since both $\mathrm{wt}(\mathbf{c}_1)$ and $\mathrm{wt}(\mathbf{c}_3)$ are even, $\mathrm{wt}(\mathbf{c}_1) \equiv \mathrm{wt}(\mathbf{c}_3) \,(\mathrm{mod}\ 4)$.

If $n \not\equiv 0 \,(\mathrm{mod}\ 8)$, then all weights in C_1 and C_3 must be 2 modulo 4, as otherwise $C_0 \cup C_1$ or $C_0 \cup C_3$ would be a Type II code, which is a contradiction. Because $\mathbf{1} \in C_0, n \equiv 4 \,(\mathrm{mod}\ 8)$, and (iii) follows in this case. If $n \equiv 0 \,(\mathrm{mod}\ 8)$, then every $[n, (n/2) - 1]$ doubly-even code that contains $\mathbf{1}$, such as C_0, is contained in one Type I code and two Type II codes; see Exercise 526. These three codes must be in $C_0^{\perp}$; since $C = C_0 \cup C_2$ is Type I, $C_0 \cup C_1$ and $C_0 \cup C_3$ must be the Type II codes. Thus $\mathrm{wt}(\mathbf{c}) \equiv 0 \equiv n/2 \,(\mathrm{mod}\ 4)$. $\square$

Exercise 513 Prove Theorem 9.4.7(i) and (ii). ♦

Example 9.4.8 It is known [262] that there is a unique $[12, 6, 4]$ Type I code C, up to equivalence. Let C_0 be the doubly-even subcode of C, and let $C_1 \cup C_3$ be the shadow of C. So $C_0^{\perp} = C_0 \cup C_1 \cup C_2 \cup C_3$ where $C = C_0 \cup C_2$. As can be seen from the generator matrix for C, the shadow $C_1 \cup C_3$ of C is a weight 2 coset of C with six weight 2 vectors in the coset. Using Table 9.4 and Theorem 9.4.7, the weight distribution of C_i is

	0	2	4	6	8	10	12
C_0	1		15		15		1
C_2				32			
C_1		6		20		6	
C_3				32			

where we assume that C_1 contains a weight 2 vector. $C_0 \cup C_1$ is a $[12, 6, 2]$ self-dual code with a coset of weight 6, while $C_0 \cup C_3$ is a $[12, 6, 4]$ self-dual code with a coset of weight 2. ∎

Exercise 514 Verify that the weight distribution of each C_i given in Example 9.4.8 is correct. Note that you must show that if C_1 contains one weight 2 vector, it contains all six weight 2 vectors in $C_1 \cup C_3$. ♦

Example 9.4.9 Continuing with Example 9.4.8, let C be a $[12, 6, 4]$ Type I code. It turns out that there are two types of cosets of C of weight 2. There is a unique coset of weight 2 described in Example 9.4.8. There are 30 other cosets of weight 2, each containing two vectors of weight 2. Let C' be one of these. Let $C_0 = (C \cup C')^{\perp}$. The code C_0 is singly-even

Table 9.5 *Orthogonality for an odd weight coset of a self-dual code*

	C_0	C_2	C_1	C_3
C_0	$\perp$	$\perp$	$\perp$	$\perp$
C_2	$\perp$	$\perp$	/	/
C_1	$\perp$	/	/	$\perp$
C_3	$\perp$	/	$\perp$	/

and contains **1**. Again using Table 9.4 and Theorem 9.4.7, the weight distribution of C_i is given by

	0	2	4	6	8	10	12
C_0	1		7	16	7		1
C_2			8	16	8		
C_1		2	8	12	8	2	
C_3			8	16	8		

where we assume that C_1 contains a weight 2 vector. $C_0 \cup C_1$ is a $[12, 6, 2]$ self-dual code with a coset of weight 4, while $C_0 \cup C_3$ is a $[12, 6, 4]$ self-dual code with a coset of weight 2. ∎

Exercise 515 Verify that the weight distribution of each C_i given in Example 9.4.9 is correct. As in Exercise 514 you must show that if C_1 contains one weight 2 vector, it contains both weight 2 vectors in $C_1 \cup C_3$. ◆

We now turn to the second possibility given in Lemma 9.4.6(ii). Table 9.5 gives the orthogonality relationships between the cosets [38] when $C_1 \cup C_3$ contains only odd weight vectors.

Exercise 516 Verify the entries in Table 9.5. ◆

We have a result analogous to that of Theorem 9.4.7.

Theorem 9.4.10 *Let C be an $[n, n/2]$ self-dual binary code with C_0 an $[n, (n/2) - 1]$ subcode. Suppose that $\mathbf{1} \notin C_0$. The following hold:*
(i) *If C is Type II, C_1 and C_3 can be numbered so that all vectors in C_1 have weight 1 modulo 4, and all vectors in C_3 have weight 3 modulo 4.*
(ii) *If C_0 is singly-even, then, for $i = 1$ and 3, half the vectors in C_i have weight 1 modulo 4 and half the vectors in C_i have weight 3 modulo 4.*
(iii) *(Shadow case) If C is singly-even and C_0 is the doubly-even subcode of C and if $\mathbf{c} \in C_1 \cup C_3$, then $\mathrm{wt}(\mathbf{c}) \equiv n/2 \pmod 4$. Also $n \equiv 2 \pmod 8$ or $n \equiv 6 \pmod 8$.*

Proof: We leave the proofs of (i) and (ii) as an exercise. Suppose that C is singly-even and C_0 is the doubly-even subcode of C. We note that since $\mathbf{1} \notin C_0$, then $\mathbf{1} \in C_2$. As C_2 can only contain singly-even vectors, either $n \equiv 2 \pmod 8$ or $n \equiv 6 \pmod 8$.

We construct a new code C'' of length $n + 2$ by adjoining two components to $C_0^\perp$ where

$$C'' = \{c00 \mid c \in C_0\} \cup \{c11 \mid c \in C_2\} \cup \{c10 \mid c \in C_1\} \cup \{c01 \mid c \in C_3\}. \tag{9.6}$$

Clearly, C'' is an $[n + 2, (n/2) + 1]$ code. By Exercise 517, C'' is self-dual.

Consider first the case $n \equiv 2 \pmod 8$. Let $c_i \in C_i$. Then $c_1 + c_3 \in C_2$ by Lemma 9.4.4 implying that $\mathrm{wt}(c_1 + c_3) \equiv 2 \pmod 4$. But $\mathrm{wt}(c_1 + c_3) = \mathrm{wt}(c_1) + \mathrm{wt}(c_3) - 2\mathrm{wt}(c_1 \cap c_3) \equiv \mathrm{wt}(c_1) + \mathrm{wt}(c_3) \pmod 4$ as $C_1 \perp C_3$. Since both $\mathrm{wt}(c_1)$ and $\mathrm{wt}(c_3)$ are odd, $\mathrm{wt}(c_1) \equiv \mathrm{wt}(c_3) \pmod 4$. As C'' cannot be Type II since $n + 2 \equiv 4 \pmod 8$, $\mathrm{wt}(c_1) \equiv \mathrm{wt}(c_3) \equiv 1 \equiv n/2 \pmod 4$, proving (iii) in this case.

Now suppose $n \equiv 6 \pmod 8$. Let $\mathcal{D} = \{c00 \mid c \in C_0\} \cup \{c11 \mid c \in C_2\}$, which is doubly-even. Clearly, if $\mathcal{D}' = \mathcal{D} \cup (d + \mathcal{D})$ where $d = 00 \cdots 011$, $\mathcal{D}'$ is a Type I code. Also as the all-one vector of length n is in C_2, the all-one vector of length $n + 2$ is in $\mathcal{D}$. By Exercise 526, $\mathcal{D}$ is contained in only one Type I and two Type II codes of length $n + 2$, implying that C'' must be Type II. Hence if $c \in C_1 \cup C_3$, then $\mathrm{wt}(c) \equiv 3 \equiv n/2 \pmod 4$, proving (iii) in this case also.   □

Exercise 517 Show that the code constructed in (9.6) is self-dual.   ◆

Exercise 518 Prove Theorem 9.4.10(i) and (ii).   ◆

Example 9.4.11 Let C be a $[6, 3, 2]$ Type I code with C_0 its doubly-even subcode. See Example 9.4.5. As $1 \in C_2$, we can apply Theorem 9.4.10 and use Table 9.5 to obtain the weight distribution of C_i:

	0	1	2	3	4	5	6
C_0	1			3			
C_2			3		1		
C_1			4				
C_3			4				

■

Exercise 519 Verify that the weight distribution of each C_i given in Example 9.4.11 is correct. Do this in two ways. First, use Theorem 9.4.10 along with Table 9.5, and second, use the construction presented in Example 9.4.5.   ◆

The upper bound on the minimum distance of Type I codes given in Theorem 9.3.1(i) was first improved in [58]; this bound was later improved to the current bound given in Theorem 9.3.5. In proving the intermediate bound, the weight enumerator of the shadow was considered; if the weight enumerator of the shadow had either negative or fractional coefficients, the original code could not exist. Using this idea, Conway and Sloane in [58] have given possible weight enumerators of Type I codes of lengths through 64 with the highest possible minimum weight. A number of researchers have found Type I codes with a weight enumerator given as a possibility in [58].

If C is an $[n, n/2, d]$ self-dual code with $d > 2$, pick two coordinate positions and consider the $((n/2) - 1)$-dimensional subcode C' of C with either two 0s or two 1s in these positions. (The requirement $d > 2$ guarantees that C' is $((n/2) - 1)$-dimensional.) If we puncture C' on these positions, we obtain a self-dual code C'^* of length $n - 2$; C'^* is called a *child* of C and C is called a *parent*[3] of C'^*. This process reverses the construction of (9.6).

Example 9.4.12 There is a $[22, 11, 6]$ Type I code whose shadow has weight 7 and the extension in (9.6) gives the $[24, 12, 8]$ extended binary Golay code. The length 22 code is a child of the Golay code, sometimes called the *odd Golay code*.   ∎

If $n \equiv 6 \pmod 8$ and C is a Type I code with C_0 the doubly-even subcode, then $\mathbf{1} \notin C_0$. Thus when $n \equiv 6 \pmod 8$ our proof of Theorem 9.4.10 shows that any Type I code of length n can be extended to a Type II code of length $n + 2$, denoted C'' in the proof. Example 9.4.12 illustrates this.

If $n \equiv 0 \pmod 8$ and C is a Type II code of length n, it is also possible to define children of C of lengths $n - 4$ and $n - 6$ that are Type I codes. It can be shown [53, 264] that any Type I code of length $n \not\equiv 0 \pmod 8$ is a child of a Type II code of the next larger length equivalent to 0 modulo 8. This approach was used to classify all self-dual codes of lengths 26, 28, and 30 once all Type II codes of length 32 had been classified. See [53, 56, 264].

We can determine the weight enumerator of a parent from that of a child.

Theorem 9.4.13 *Let C be a Type I code of length $n \equiv 6 \pmod 8$ with C_0 its doubly-even subcode. Suppose that $W_i(x, y)$ is the weight enumerator of C_i. Then the weight enumerator of the parent C'' in (9.6) is*

$$W_{C''}(x, y) = y^2 W_0(x, y) + x^2 W_2(x, y) + xy[W_1(x, y) + W_3(x, y)].$$

Exercise 520 Prove Theorem 9.4.13.   ♦

Exercise 521 Do the following:
(a) Using Theorem 9.4.13 and the table of Example 9.4.11, find the weight distribution of the parent of a $[6, 3, 2]$ Type I code.
(b) Find a generator matrix for a $[6, 3, 2]$ Type I code whose first two rows generate its doubly-even subcode. Using the method described by (9.6), find a generator matrix for the $[8, 4, 4]$ parent.   ♦

With the notion of parents and children we can say more about Type I codes of length $n \equiv 22 \pmod{24}$ meeting the bound of Theorem 9.3.5. This result is due to Rains [290].

Theorem 9.4.14 *Let C be an $[n, n/2, d]$ Type I code with $n \equiv 22 \pmod{24}$ and $d = 4\lfloor n/24 \rfloor + 6$. Then C is the child of a Type II $[n + 2, (n/2) + 1, d + 2]$ code.*

In proving Theorem 9.4.14, we note that if C_0 is the doubly-even subcode of C, then $\mathbf{1} \notin C_0$. Thus the shadow has odd weight. The difficult part of the proof is showing that the shadow is a coset of C of weight $d + 1$. Once we know that, the parent of C has minimum weight $d + 2$. Example 9.4.12 is an illustration of this result.

---

[3] In Section 11.7 we use the terms "parent" and "child" in relation to cosets of a code. The current use is unrelated.

9.5   Counting self-orthogonal codes

As we will see in Section 9.7, the classification of self-dual codes of a fixed length n over $\mathbb{F}_2$, $\mathbb{F}_3$, or $\mathbb{F}_4$ depends on knowledge of the number of such codes. We begin with the binary case, where we will count the number of self-dual codes and the number of Type II codes. We first determine the total number of self-dual codes.

Theorem 9.5.1 ([259, 260]) *The number of self-dual binary codes of length n is*

$$\prod_{i=1}^{\frac{n}{2}-1}(2^i+1).$$

Proof: A self-dual binary code must contain the all-one vector **1**. Let $\sigma_{n,k}$ denote the number of $[n,k]$ self-orthogonal binary codes containing **1**. We note first that $\sigma_{n,1}=1$. We now find a recurrence relation for $\sigma_{n,k}$. We illustrate this recursion process by computing $\sigma_{n,2}$. There are $2^{n-1}-2$ even weight vectors that are neither **0** nor **1**. Each of these vectors is in a unique $[n,2]$ self-orthogonal code containing **1**; each such code contains two of these vectors. So $\sigma_{n,2}=2^{n-2}-1$.

Every $[n,k+1]$ self-orthogonal code containing **1** contains an $[n,k]$ self-orthogonal code also containing **1**. Beginning with an $[n,k]$ self-orthogonal code $\mathcal{C}$ containing **1**, the only way to find an $[n,k+1]$ self-orthogonal code $\mathcal{C}'$ containing $\mathcal{C}$ is by adjoining a vector $\mathbf{c}'$ from one of the $2^{n-2k}-1$ cosets of $\mathcal{C}$ in $\mathcal{C}^\perp$ that is unequal to $\mathcal{C}$. Thus $\mathcal{C}$ can be extended to $2^{n-2k}-1$ different $[n,k+1]$ self-orthogonal codes $\mathcal{C}'$. However, every such $\mathcal{C}'$ has 2^k-1 subcodes of dimension k containing **1** by Exercise 522(a). This shows that

$$\sigma_{n,k+1}=\frac{2^{n-2k}-1}{2^k-1}\sigma_{n,k}.$$

Note that using this recurrence relation, we obtain $\sigma_{n,2}=2^{n-2}-1$ as earlier. Using this recurrence relation, the number of self-dual binary codes of length n is

$$\sigma_{n,n/2}=\frac{2^2-1}{2^{n/2-1}-1}\cdot\frac{2^4-1}{2^{n/2-2}-1}\cdot\frac{2^6-1}{2^{n/2-3}-1}\cdots\frac{2^{n-2}-1}{2^1-1}\cdot\sigma_{n,1}$$

$$=\frac{2^2-1}{2^1-1}\cdot\frac{2^4-1}{2^2-1}\cdot\frac{2^6-1}{2^3-1}\cdots\frac{2^{n-2}-1}{2^{n/2-1}-1}$$

$$=(2^1+1)(2^2+1)(2^3+1)\cdots(2^{n/2-1}+1),$$

which is the desired result.   □

Exercise 522 In Exercise 431, you were asked to show that there are

$$t=(q^r-1)/(q-1)$$

$(r-1)$-dimensional subcodes of an r-dimensional code over $\mathbb{F}_q$. Do the following:
(a) Prove that a binary code of dimension $k+1$ containing the all-one vector **1** has 2^k-1 subcodes of dimension k also containing **1**.
(b) Prove that a binary code of dimension $k+1$ containing an $[n,m]$ subcode $\mathcal{C}$ has $2^{k+1-m}-1$ subcodes of dimension k also containing $\mathcal{C}$.   ♦

Example 9.5.2 Theorem 9.5.1 states that there are $(2^1 + 1) = 3$ $[4, 2]$ self-dual binary codes. They have generator matrices

$$\begin{bmatrix} 1 & 1 & 0 & 0 \\ 1 & 1 & 1 & 1 \end{bmatrix}, \quad \begin{bmatrix} 1 & 0 & 1 & 0 \\ 1 & 1 & 1 & 1 \end{bmatrix}, \quad \text{and} \quad \begin{bmatrix} 1 & 0 & 0 & 1 \\ 1 & 1 & 1 & 1 \end{bmatrix}.$$

Notice that these are all equivalent and so up to equivalence there is only one $[4, 2]$ self-dual binary code. ∎

Exercise 523 Verify Theorem 9.5.1 directly for $n = 6$ by constructing generator matrices for all $[6, 3]$ self-dual binary codes. ♦

We can also count the number of self-orthogonal binary codes of maximum dimension when the code has odd length. The proof of this result is found in [259], where the number of $[n, k]$ self-orthogonal codes with $k \leq (n-1)/2$ is also given.

Theorem 9.5.3 *If n is odd, the number of $[n, (n-1)/2]$ self-orthogonal binary codes is*

$$\prod_{i=1}^{\frac{n-1}{2}} (2^i + 1).$$

The next theorem extends Theorem 9.5.1.

Theorem 9.5.4 *If C is an $[n, m]$ self-orthogonal binary code containing $\mathbf{1}$, then C is contained in*

$$\prod_{i=1}^{\frac{n}{2}-m} (2^i + 1)$$

self-dual codes.

Proof: For $k \geq m$, let $\tau_{n,k}$ be the number of $[n, k]$ self-orthogonal codes containing C. Note that $\tau_{n,m} = 1$. Beginning with an $[n, k]$ self-orthogonal code C_1 containing C and adjoining an element of a coset of C_1 in $C_1^{\perp}$ provided the element is not in C_1, we obtain $2^{n-2k} - 1$ possible $[n, k+1]$ self-orthogonal codes C_2 containing C_1. By Exercise 522(b), there are $2^{k+1-m} - 1$ subcodes of C_2 of dimension k containing C. Thus

$$\tau_{n,k+1} = \frac{2^{n-2k} - 1}{2^{k+1-m} - 1} \tau_{n,k}.$$

Using this recurrence relation, the number of self-dual binary codes of length n containing C is

$$\begin{aligned}
\tau_{n,n/2} &= \frac{2^2 - 1}{2^{n/2-m} - 1} \cdot \frac{2^4 - 1}{2^{n/2-m-1} - 1} \cdot \frac{2^6 - 1}{2^{n/2-m-2} - 1} \cdots \frac{2^{n-2m} - 1}{2^1 - 1} \cdot \tau_{n,m} \\
&= \frac{2^2 - 1}{2^1 - 1} \cdot \frac{2^4 - 1}{2^2 - 1} \cdot \frac{2^6 - 1}{2^3 - 1} \cdots \frac{2^{n-2m} - 1}{2^{n/2-m} - 1} \\
&= (2^1 + 1)(2^2 + 1)(2^3 + 1) \cdots (2^{n/2-m} + 1),
\end{aligned}$$

which is the desired result. □

Exercise 524 Prove Theorem 9.5.1 as a corollary of Theorem 9.5.4. ◆

Now we would like to count the number of Type II codes, which is more complicated than counting the number of self-dual codes. We will state the results here and will leave most proofs to Section 9.12. We have the following theorem counting the number of Type II codes.

Theorem 9.5.5 *Let* $n \equiv 0 \pmod 8$.
(i) *Then there are*

$$\prod_{i=0}^{\frac{n}{2}-2}(2^i + 1)$$

Type II codes of length n.
(ii) *Let* C *be an* $[n, k]$ *doubly-even code containing* **1**. *Then* C *is contained in*

$$\prod_{i=0}^{\frac{n}{2}-k-1}(2^i + 1)$$

Type II codes of length n.

Example 9.5.6 Applying Theorem 9.5.5 to $n = 8$, we see that there are $2(2 + 1)(2^2 + 1) = 30$ Type II $[8, 4, 4]$ codes. By Exercise 56, all these codes are equivalent. ∎

Exercise 525 The matrix

$$G = \begin{bmatrix} 1 & 1 & 1 & 1 & 0 & 0 & 0 & 0 \\ 1 & 1 & 0 & 0 & 1 & 1 & 0 & 0 \\ 1 & 1 & 1 & 1 & 1 & 1 & 1 & 1 \end{bmatrix}$$

is the generator matrix of an $[8, 3, 4]$ doubly-even binary code C. How many Type II codes are there that contain C? ◆

Exercise 526 Prove that if $n \equiv 0 \pmod 8$, then every $[n, (n/2) - 1]$ doubly-even code containing **1** is contained in one Type I code and two Type II codes. ◆

When n is a multiple of eight, the number of $[n, k]$ doubly-even binary codes with $k \le n/2$ has been counted; for example, see Exercise 567 for the number of such codes containing **1**.

We can also count the number of maximal doubly-even binary codes of length n in the case when $8 \nmid n$.

Theorem 9.5.7 *The number of* $[n, (n/2) - 1]$ *doubly-even binary codes is:*
(i) $\prod_{i=2}^{\frac{n}{2}-1}(2^i + 1)$ *if* $n \equiv 4 \pmod 8$, *and*
(ii) $\prod_{i=1}^{\frac{n}{2}-1}(2^i + 1)$ *if* $n \equiv 2 \pmod 4$.

The proof of Theorem 9.5.7(i) is left to Section 9.12, but the proof of (ii) is quite easy once we observe that the count in (ii) agrees with the count in Theorem 9.5.1. As the proof is instructive, we give it here. Let C be a self-dual code of length n where $n \equiv 2 \pmod 4$. Then, C is singly-even as C contains **1**. There is a unique $[n, (n/2) - 1]$

doubly-even subcode C_0 by Theorem 1.4.6. Conversely, if C_0 is an $[n, (n/2) - 1]$ doubly-even code, it does not contain $\mathbf{1}$ as $n \equiv 2 \pmod 4$. So the only self-dual code containing C_0 must be obtained by adjoining $\mathbf{1}$. Thus the number of doubly-even $[n, (n/2) - 1]$ binary codes equals the number of self-dual codes; (ii) now follows by Theorem 9.5.1.

Example 9.5.8 The $[6, 2]$ doubly-even code C_0 with generator matrix

$$\begin{bmatrix} 1 & 1 & 1 & 1 & 0 & 0 \\ 0 & 0 & 1 & 1 & 1 & 1 \end{bmatrix}$$

is contained in the self-dual code C with generator matrix either

$$\begin{bmatrix} 1 & 1 & 1 & 1 & 0 & 0 \\ 0 & 0 & 1 & 1 & 1 & 1 \\ 1 & 1 & 1 & 1 & 1 & 1 \end{bmatrix} \quad \text{or} \quad \begin{bmatrix} 1 & 1 & 0 & 0 & 0 & 0 \\ 0 & 0 & 1 & 1 & 0 & 0 \\ 0 & 0 & 0 & 0 & 1 & 1 \end{bmatrix}. \qquad \blacksquare$$

Example 9.5.9 Notice that if $n = 4$, the product in Theorem 9.5.7(i) is empty and so equals 1. The unique doubly-even $[4, 1]$ code is $C = \{\mathbf{0}, \mathbf{1}\}$. All three self-dual codes given in Example 9.5.2 contain C. $\qquad \blacksquare$

Exercise 527 Do the following:
(a) Prove that if $n \equiv 4 \pmod 8$, there are three times as many self-dual binary codes of length n as there are doubly-even $[n, (n/2) - 1]$ binary codes.
(b) Prove that if $n \equiv 4 \pmod 8$ and if C is a doubly-even $[n, (n/2) - 1]$ binary code, then C is contained in exactly three self-dual $[n, n/2]$ codes. Hint: Consider the four cosets of C in the $[n, (n/2) + 1]$ code $C^{\perp}$. $\qquad \blacklozenge$

The number of doubly-even binary codes of odd length n and dimension $k < n/2$ has also been computed in [92].

There are analogous formulas for the number of Type III and Type IV codes. We state them here.

Theorem 9.5.10 *The following hold:*
(i) [259, 260] *The number of Type III codes over* $\mathbb{F}_3$ *of length* $n \equiv 0 \pmod 4$ *is*

$$2 \prod_{i=1}^{\frac{n}{2}-1} (3^i + 1).$$

(ii) [217] *The number of Type IV codes over* $\mathbb{F}_4$ *of length* $n \equiv 0 \pmod 2$ *is*

$$\prod_{i=0}^{\frac{n}{2}-1} (2^{2i+1} + 1).$$

The number of self-orthogonal ternary codes of length n with $4 \nmid n$ are also known; see [260]. We state the number of maximal ones.

Theorem 9.5.11 *The following hold*:

(i) *If n is odd, the number of* $[n, (n-1)/2]$ *self-orthogonal ternary codes is*

$$\prod_{i=1}^{\frac{n-1}{2}}(3^i + 1).$$

(ii) *If* $n \equiv 2 \pmod{4}$, *the number of* $[n, (n-2)/2]$ *self-orthogonal ternary codes is*

$$\prod_{i=2}^{\frac{n}{2}}(3^i + 1).$$

Exercise 528 In a Hermitian self-dual code over $\mathbb{F}_4$, all codewords have even weight and any even weight vector in $\mathbb{F}_4^n$ is orthogonal to itself under the Hermitian inner product. Let v_n be the number of even weight vectors in $\mathbb{F}_4^n$.

(a) Show that v_n satisfies the recurrence relation $v_n = v_{n-1} + 3(4^{n-1} - v_{n-1})$ with $v_1 = 1$.

(b) Solve the recurrence relation in part (a) to show that

$$v_n = 2 \cdot 4^{n-1} - (-2)^{n-1}.$$

(c) Show that the number of $[n, 1]$ Hermitian self-orthogonal codes is

$$\frac{2 \cdot 4^{n-1} - (-2)^{n-1} - 1}{3}.$$

◆

We conclude this section with a bound, found in [219], analogous to the Gilbert and Varshamov Bounds of Chapter 2.

Theorem 9.5.12 *Let* $n \equiv 0 \pmod{8}$. *Let r be the largest integer such that*

$$\binom{n}{4} + \binom{n}{8} + \binom{n}{12} + \cdots + \binom{n}{4(r-1)} < 2^{\frac{n}{2}-2} + 1. \tag{9.7}$$

Then there exists a Type II code of length n and minimum weight at least 4r.

Proof: First, let $\mathbf{v}$ be a vector in $\mathbb{F}_2^n$ of weight $4i$ with $0 < 4i < 4r$. The number of Type II codes containing the subcode $\{\mathbf{0}, \mathbf{1}, \mathbf{v}, \mathbf{1}+\mathbf{v}\}$ is $\prod_{i=0}^{n/2-3}(2^i + 1)$ by Theorem 9.5.5(ii). Thus there are at most

$$\left\{ \binom{n}{4} + \binom{n}{8} + \binom{n}{12} + \cdots + \binom{n}{4(r-1)} \right\} \prod_{i=0}^{n/2-3}(2^i + 1)$$

Type II codes of minimum weight less than $4r$. If this total is less than the total number of Type II codes, which is $\prod_{i=0}^{n/2-2}(2^i + 1)$ by Theorem 9.5.5(i), then there is at least one Type II code of minimum weight at least $4r$. The latter statement is equivalent to (9.7). □

In Table 9.6, we illustrate the bound of Theorem 9.5.12 and compare it to the Varshamov Bound in Chapter 2. The length is denoted "n" in the table. The highest value of d in the Varshamov Bound for which there is any $[n, n/2, d]$ code is given in the table in the column "d_{var}". (Note that the values of d obtained from the Gilbert Bound were no larger than the values from the Varshamov Bound.) The largest d in the bound of Theorem 9.5.12 is

Table 9.6 *Type II codes of length $n \le 96$*

n	d_{var}	d_{sdII}	d_{known}	n	d_{var}	d_{sdII}	d_{known}
8	3	4	4^E	56	9	8	12^E
16	4	4	4^E	64	9	8	12^E
24	5	4	8^E	72	10	12	12
32	6	4	8^E	80	11	12	16^E
40	7	8	8^E	88	12	12	16^E
48	8	8	12^E	96	13	12	16

denoted "d_{sdII}". The largest minimum weight for a Type II code of length n known to exist is denoted "d_{known}"; an "E" on this entry indicates the code is extremal. Construction of the latter codes, with references, for $8 \le n \le 88$ is described in [291]; the [96, 48, 16] Type II code was first constructed by Feit [80]. The Varshamov Bound states that codes meeting the bound exist; where to search for them is not obvious. From the table, one sees that to find good codes with dimension half the length, one can look at Type II codes when the length is a multiple of eight and obtain codes close to meeting the Varshamov Bound. It is instructive to compare Table 9.6 where we are examining codes whose minimum weights are compared to a lower bound with Table 9.1 where we are dealing with codes whose minimum weights are compared to an upper bound.

Exercise 529 Verify the entries in the columns d_{var} and d_{sdII} of Table 9.6.   ◆

There is a result analogous to that of Theorem 9.5.12 for self-dual codes over $\mathbb{F}_q$ found in [278]. Example 9.5.13 illustrates this for $q = 2$.

Example 9.5.13 Let n be even. Let r be the largest integer such that

$$\binom{n}{2} + \binom{n}{4} + \binom{n}{6} + \cdots + \binom{n}{2(r-1)} < 2^{\frac{n}{2}-1} + 1. \tag{9.8}$$

In Exercise 530, you are asked to prove that there exists a self-dual binary code of length n and minimum weight at least $2r$. Table 9.7 is analogous to Table 9.6 for self-dual binary codes of even length n with $4 \le n \le 50$ using the bound (9.8). The column "d_{var}" has the same meaning as it does in Table 9.6. The largest d from bound (9.8) is denoted "d_{sd}". The largest minimum weight for a self-dual binary code (Type I or Type II) of length n known to exist is denoted "d_{known}". In all cases, it has been proved by various means that there is no self-dual binary code of length $n \le 50$ and minimum weight higher than that given in the table under "d_{known}". For length $n \le 32$ there are no codes with higher minimum weight by the classifications discussed in Section 9.7. For $34 < n < 50$, there is no self-dual code with higher minimum weight by Theorem 9.3.5. There are no [34, 17, 8] or [50, 25, 12] self-dual codes by [58]; if either code exists, its shadow code would have a weight enumerator with noninteger coefficients. Furthermore, if $n \equiv 0 \pmod 8$, the codes of highest minimum weight are all attained by Type II codes (and possibly Type I codes as well). Again it is worth comparing this table to Table 9.1.   ∎

Table 9.7 *Self-dual binary codes of length*
$4 \le n \le 50$

n	d_{var}	d_{sd}	d_{known}	n	d_{var}	d_{sd}	d_{known}
4	2	2	2	28	5	4	6
6	3	2	2	30	6	4	6
8	3	2	4	32	6	4	8
10	3	2	2	34	6	6	6
12	3	2	4	36	6	6	8
14	4	2	4	38	7	6	8
16	4	4	4	40	7	6	8
18	4	4	4	42	7	6	8
20	4	4	4	44	7	6	8
22	5	4	6	46	7	6	10
24	5	4	8	48	8	6	12
26	5	4	6	50	8	8	10

Exercise 530 Prove that there exists a self-dual binary code of length n and minimum weight at least $2r$ provided r is the largest integer satisfying (9.8). ◆

Exercise 531 Verify the entries in the columns d_{var} and d_{sd} of Table 9.7. ◆

9.6 Mass formulas

We now know, from Theorem 9.5.1, that the total number of self-dual binary codes of length n is $\prod_{i=1}^{n/2-1}(2^i + 1)$. Ultimately we would like to find a representative from each equivalence class of these codes; this is called a *classification problem*. It would be of great help to know the number of inequivalent codes. At this point there does not seem to be a direct formula for this. However, we can create a formula that can be used to tell us when we have a representative from every class of inequivalent self-dual codes. This formula is called a *mass formula*.

Let $\mathcal{C}_1, \ldots, \mathcal{C}_s$ be representatives from every equivalence class of self-dual binary codes of length n. The number of codes equivalent to $\mathcal{C}_j$ is

$$\frac{|\text{Sym}_n|}{|\text{PAut}(\mathcal{C}_j)|} = \frac{n!}{|\text{PAut}(\mathcal{C}_j)|}.$$

Therefore, summing over j, we must obtain the total number of self-dual codes, which yields the following mass formula for self-dual binary codes:

$$\sum_{j=1}^{s} \frac{n!}{|\text{PAut}(\mathcal{C}_j)|} = \prod_{i=1}^{\frac{n}{2}-1}(2^i + 1). \tag{9.9}$$

We can find mass formulas for Type II, III, and IV codes also, in the same fashion, using the results of Theorems 9.5.5(i) and 9.5.10. For Type III, we use the group of all monomial transformations, which has size $2^n n!$, in place of Sym_n. For Type IV, we use the group of all monomial transformations followed possibly by the Frobenius map, which has size $2 \cdot 3^n n!$, in place of Sym_n. This gives the following theorem.

Theorem 9.6.1 *We have the following mass formulas:*
(i) *For self-dual binary codes of length n,*

$$\sum_j \frac{n!}{|\mathrm{PAut}(\mathcal{C}_j)|} = \prod_{i=1}^{\frac{n}{2}-1} (2^i + 1).$$

(ii) *For Type II codes of length n,*

$$\sum_j \frac{n!}{|\mathrm{PAut}(\mathcal{C}_j)|} = \prod_{i=0}^{\frac{n}{2}-2} (2^i + 1).$$

(iii) *For Type III codes of length n,*

$$\sum_j \frac{2^n n!}{|\mathrm{MAut}(\mathcal{C}_j)|} = 2 \prod_{i=1}^{\frac{n}{2}-1} (3^i + 1).$$

(iv) *For Type IV codes of length n,*

$$\sum_j \frac{2 \cdot 3^n n!}{|\Gamma\mathrm{Aut}(\mathcal{C}_j)|} = \prod_{i=0}^{\frac{n}{2}-1} (2^{2i+1} + 1).$$

In each case, the summation is over all j, where $\{\mathcal{C}_j\}$ is a complete set of representatives of inequivalent codes of the given type.

Example 9.6.2 We continue with Example 9.5.6 and illustrate the mass formula for Type II codes in Theorem 9.6.1(ii) for $n = 8$. Any Type II code of length 8 must be an $[8, 4, 4]$ code. Such a code $\mathcal{C}$ is unique and the left-hand side of the mass formula is $8!/|\mathrm{PAut}(\mathcal{C})|$; the right-hand side is 30. Therefore, $|\mathrm{PAut}(\mathcal{C})| = 1344$; it is known that $\mathrm{PAut}(\mathcal{C})$ is isomorphic to the 3-dimensional general affine group over $\mathbb{F}_2$ which indeed has order 1344. ∎

9.7   Classification

Once we have the mass formula for a class of codes, we can attempt to classify them. Suppose we want to classify the self-dual codes of length n. The strategy is to classify these codes for the smallest lengths first. Once that is accomplished, we proceed to larger lengths using the results obtained for smaller lengths. We discuss the overall strategy for classification focusing primarily on self-dual binary codes.

9.7.1   The Classification Algorithm

There is a general procedure that has been used in most of the classifications attempted. The *Classification Algorithm*, which we state for self-dual binary codes, is as follows:

I.   Find a self-dual binary code C_1 of length n.

II.   Compute the size of the automorphism group $\text{PAut}(C_1)$ of the code found in Step I and calculate the contribution of $n!/|\text{PAut}(C_1)|$ to the left-hand side of (9.9).

III.   Find a self-dual binary code C_j not equivalent to any previously found. Compute the size of the automorphism group $\text{PAut}(C_j)$ and calculate $n!/|\text{PAut}(C_j)|$; add this to the contributions to the left-hand side of (9.9) from the inequivalent codes previously found.

IV.   Repeat Step III until the total computed on the left-hand side of (9.9) gives the amount on the right-hand side.

The classification for Type II, III, and IV codes is analogous where we use the appropriate mass formula from Theorem 9.6.1.

Example 9.7.1   We classify all the self-dual binary codes of length $n = 8$. The right-hand side of the mass formula is $(2 + 1)(2^2 + 1)(2^3 + 1) = 135$. From Example 9.6.2, the unique $[8, 4, 4]$ code C_1 has automorphism group of size 1344. It contributes 30 to the left-hand side of the mass formula. There is also the self-dual code $C_2 = C \oplus C \oplus C \oplus C$, where C is the $[2, 1, 2]$ binary repetition code. Any of the two permutations on the two coordinates of any direct summand is an automorphism; this gives a subgroup of $\text{PAut}(C_2)$ of order 2^4. Furthermore, any of the four summands can be permuted in 4! ways. All the automorphisms can be generated from these; the group has order $2^4 4!$ and C_2 contributes 105 to the left-hand side of the mass formula. The total contribution is 135, which equals the right-hand side. Hence there are only two inequivalent self-dual binary codes of length 8.   ∎

Exercise 532   Find a complete list of all inequivalent self-dual binary codes of lengths:
(a) $n = 2$,
(b) $n = 4$,
(c) $n = 6$, and
(d) $n = 10$.   ◆

As the lengths grow, the classification becomes more difficult. There are methods to simplify this search. When many of the original classifications were done, computing the automorphism groups was also problematic; now, with computer programs that include packages for codes, finding the automorphism groups is simplified.

We now go into more detail regarding classification of binary codes and the specifics of carrying out the Classification Algorithm. There are several self-orthogonal codes that will arise frequently. They are denoted i_2, e_7, e_8, and d_{2m}, with $m \geq 2$. The subscript on these codes indicates their length. The codes i_2, e_7, and e_8 have generator matrices

$$i_2: \begin{bmatrix} 1 & 1 \end{bmatrix}, \quad e_7: \begin{bmatrix} 1 & 1 & 1 & 1 & 0 & 0 & 0 \\ 0 & 0 & 1 & 1 & 1 & 1 & 0 \\ 1 & 0 & 1 & 0 & 1 & 0 & 1 \end{bmatrix}, \quad e_8: \begin{bmatrix} 1 & 1 & 1 & 1 & 0 & 0 & 0 & 0 \\ 0 & 0 & 1 & 1 & 1 & 1 & 0 & 0 \\ 0 & 0 & 0 & 0 & 1 & 1 & 1 & 1 \\ 1 & 0 & 1 & 0 & 1 & 0 & 1 & 0 \end{bmatrix}.$$

The $[2m, m-1, 4]$ code d_{2m} has generator matrix

$$
d_{2m}:
\begin{bmatrix}
1 & 1 & 1 & 1 & 0 & 0 & 0 & 0 & \cdots & 0 & 0 & 0 & 0 & 0 & 0 \\
0 & 0 & 1 & 1 & 1 & 1 & 0 & 0 & \cdots & 0 & 0 & 0 & 0 & 0 & 0 \\
0 & 0 & 0 & 0 & 1 & 1 & 1 & 1 & \cdots & 0 & 0 & 0 & 0 & 0 & 0 \\
 & & & & & & & & \vdots & & & & & & \\
0 & 0 & 0 & 0 & 0 & 0 & 0 & 0 & \cdots & 1 & 1 & 1 & 1 & 0 & 0 \\
0 & 0 & 0 & 0 & 0 & 0 & 0 & 0 & \cdots & 0 & 0 & 1 & 1 & 1 & 1
\end{bmatrix}.
$$

The notation for these codes is analogous to that used for lattices; see [246].

Notice that e_8 is merely the $[8, 4, 4]$ extended Hamming code $\widehat{\mathcal{H}}_3$, while e_7 is the $[7, 3, 4]$ simplex code. As we mentioned in Example 9.6.2, the automorphism group of e_8 is isomorphic to the 3-dimensional general affine group over $\mathbb{F}_2$, often denoted $\mathrm{GA}_3(2)$, of order 1344. This group is a triply-transitive permutation group on eight points. The subgroup stabilizing one point is a group of order 168, denoted either $\mathrm{PSL}_2(7)$ or $\mathrm{PSL}_3(2)$; this is the automorphism group of e_7. The codes i_2 and d_4 have automorphism groups Sym_2 and Sym_4 of orders 2 and 24, respectively. When $m > 2$, the generator matrix for d_{2m} has identical columns $\{2i - 1, 2i\}$ for $1 \le i \le m$. Any permutation interchanging columns $2i - 1$ and $2i$ is an automorphism of d_{2m}; any permutation sending all pairs $\{2i - 1, 2i\}$ to pairs is also an automorphism. These generate all automorphisms, and $\mathrm{PAut}(d_{2m})$ has order $2^m m!$ when $m > 2$.

We need some additional terminology and notation. If the code C_1 is equivalent to C_2, then we denote this by $C_1 \simeq C_2$. A code C is *decomposable* if $C \simeq C_1 \oplus C_2$, where C_1 and C_2 are nonzero codes. If C is not equivalent to the direct sum of two nonzero codes, it is *indecomposable*. The codes i_2, e_7, e_8, and d_{2m} are indecomposable. The symbol mC will denote the direct sum of m copies of the code C; its length is obviously m times the length of C. Finally, when we say that C' (for example, i_2, e_7, e_8, or d_{2m}) is a "subcode" of C where the length n of C is longer than the length n' of C', we imply that $n - n'$ additional zero coordinates are appended to C' to make it length n.

The following basic theorem is very useful in classifying self-dual binary codes.

Theorem 9.7.2 *Let C be a self-orthogonal binary code of length n and minimum weight d. The following hold:*

(i) *If $d = 2$, then $C \simeq mi_2 \oplus C_1$ for some integer $m \ge 1$, where C_1 is a self-orthogonal code of length $n - 2m$ with minimum weight 4 or more.*

(ii) *If e_8 is a subcode of C, then $C \simeq e_8 \oplus C_1$ for some self-dual code C_1 of length $n - 8$.*

(iii) *If $d = 4$, then the subcode of C spanned by the codewords of C of weight 4 is a direct sum of copies of $d_{2m}s$, e_7s, and e_8s.*

Proof: We leave the proofs of (i) and (ii) to Exercise 533 and sketch the proof of (iii).

Let C' be the self-orthogonal subcode of C spanned by the weight 4 vectors. We prove (iii) by induction on the size of $\mathrm{supp}(C')$. If C' is decomposable, then it is the direct sum of two self-orthogonal codes each spanned by weight 4 vectors. By induction, each of these is a direct sum of copies of $d_{2m}s$, e_7s, and e_8s. Therefore it suffices to show that any indecomposable self-orthogonal code, which we still denote C', spanned by weight 4 vectors is either d_{2m}, e_7, or e_8. In the rest of this proof, any vector denoted $\mathbf{c}_i$ will be a weight 4 vector in $\mathcal{C}'$. For every pair of vectors $\mathbf{c}_1 \neq \mathbf{c}_2$, either supp($\mathbf{c}_1$) and supp($\mathbf{c}_2$) are disjoint or they intersect in two coordinates as $\mathcal{C}'$ is self-orthogonal. Also as $\mathcal{C}'$ is indecomposable, for every vector $\mathbf{c}_1$, there is a vector $\mathbf{c}_2$ such that their supports overlap in exactly two coordinates.

Suppose the supports of $\mathbf{c}_1$ and $\mathbf{c}_2$ overlap in two coordinates. By permuting coordinates, we may assume that

$$\mathbf{c}_1 = 1111000000\cdots \quad \text{and} \quad \mathbf{c}_2 = 0011110000\cdots. \tag{9.10}$$

By Exercise 534, if $\mathbf{c}_3$ has support neither contained in the first six coordinates nor disjoint from the first six coordinates, by reordering coordinates we may assume that one of the following holds:
(a) $\mathbf{c}_3 = 0000111100\cdots$ or
(b) $\mathbf{c}_3 = 1010101000\cdots$.

Suppose that (b) holds. If all weight 4 vectors of $\mathcal{C}'$ are in span$\{\mathbf{c}_1, \mathbf{c}_2, \mathbf{c}_3\}$, then $\mathcal{C}' = e_7$. Otherwise as $\mathcal{C}'$ is self-orthogonal and indecomposable, there is a vector $\mathbf{c}_4$ with support overlapping the first seven coordinates but not contained in them. By Exercise 535, span$\{\mathbf{c}_1, \mathbf{c}_2, \mathbf{c}_3, \mathbf{c}_4\}$ is e_8 possibly with zero coordinates added. By part (ii), $\mathcal{C}' = e_8$ as $\mathcal{C}'$ is indecomposable. Therefore we may assume (a) holds.

By this discussion we can begin with $\mathbf{c}_1$ as in (9.10). If $\mathcal{C}'$ has length 4, then $\mathcal{C}' = d_4$. Otherwise we may assume that $\mathbf{c}_2$ exists as in (9.10) and these two vectors span a d_6 possibly with zero coordinates added. If this is not all of $\mathcal{C}'$, then we may assume that there is a vector $\mathbf{c}_3$ as in (a). These three vectors span a d_8 again possibly with zero coordinates added. If this is not all of $\mathcal{C}'$, we can add another vector $\mathbf{c}_4$, which must have a form like (a) but not (b) above, and hence we create a d_{10} with possible zero coordinates appended. Continuing in this manner we eventually form a d_{2m} where $\mathcal{C}'$ has length $2m$ with $m \geq 5$. If this is not all of $\mathcal{C}'$, then $\mathcal{C}'$ is contained in $d_{2m}^{\perp}$, which is spanned by the vectors in d_{2m} together with $\mathbf{a} = 0101010101\cdots$ and $\mathbf{b} = 1100000000\cdots$. Since $m \geq 5$, all of the weight 4 vectors in $\mathcal{C}'$ must actually be in d_{2m} because adding vectors $\mathbf{a}$ and $\mathbf{b}$ to d_{2m} does not produce any new weight 4 vectors. Thus $\mathcal{C}' = d_{2m}$. $\qquad\square$

Exercise 533 Prove Theorem 9.7.2(i) and (ii). ◆

Exercise 534 With the notation as in the proof of Theorem 9.7.2, prove that if $\mathbf{c}_1 = 1111000000\cdots$ and $\mathbf{c}_2 = 0011110000\cdots$ and if $\mathbf{c}_3$ has support neither contained in the first six coordinates nor disjoint from the first six coordinates, by reordering coordinates we may assume that one of the following holds:
(a) $\mathbf{c}_3 = 0000111100\cdots$ or
(b) $\mathbf{c}_3 = 1010101000\cdots$. ◆

Exercise 535 With the notation as in the proof of Theorem 9.7.2, prove that if $\mathbf{c}_1 = 1111000000\cdots$, $\mathbf{c}_2 = 0011110000\cdots$, and $\mathbf{c}_3 = 1010101000\cdots$ and if $\mathbf{c}_4$ has support overlapping the first seven coordinates but not contained in them, then span$\{\mathbf{c}_1, \mathbf{c}_2, \mathbf{c}_3, \mathbf{c}_4\}$ is e_8 possibly with zero coordinates added. ◆

If we want to classify all self-dual codes of length n, we first classify all self-dual codes of length less than n. To begin Step I of the Classification Algorithm, choose some

Table 9.8 *Glue elements and group orders for d_{2m} and e_7*

Component	Glue element	Symbol	$|\mathcal{G}_0|$	$|\mathcal{G}_1|$
d_4	0000	**0**	4	6
	1100	**x**		
	1010	**y**		
	1001	**z**		
d_{2m} $(m > 2)$	$00000000\cdots0000$	**0**	$2^{m-1}m!$	2
	$01010101\cdots0101$	**a**		
	$11000000\cdots0000$	**b**		
	$10010101\cdots0101$	**c**		
e_7	0000000	**0**	168	1
	0101010	**d**		

decomposable code (e.g. $(n/2)i_2$). For Step II, compute its automorphism group order (e.g. $2^{n/2}(n/2)!$ if the code is $(n/2)i_2$) and its contribution to the mass formula (e.g. $n!/(2^{n/2}(n/2)!)$). Repeat Step III by finding the remaining decomposable codes; this is possible as the classification of self-dual codes of smaller lengths has been completed. The task now becomes more difficult as we must find the indecomposable codes. The methods used at times are rather *ad hoc*; however, the process known as "gluing" can be very useful.

9.7.2 Gluing theory

Gluing is a way to construct self-dual codes C of length n from shorter self-orthogonal codes systematically. We build C by beginning with a direct sum $C_1 \oplus C_2 \oplus \cdots \oplus C_t$ of self-orthogonal codes, which forms a subcode of C. If C is decomposable, then C is this direct sum. If not, then we add additional vectors, called *glue vectors*, so that we eventually obtain a generator matrix for C from the generator matrix for $C_1 \oplus C_2 \oplus \cdots \oplus C_t$ and the glue vectors. The codes C_i are called the *components* of C. Component codes are chosen so that any automorphism of C will either send a component code onto itself or possibly permute some components among themselves. This technique has several additional advantages; namely, it can be used to compute the order of the automorphism group of the code, which is crucial in the Classification Algorithm, and it also allows us to describe these codes efficiently. Gluing has been most effective in constructing self-dual codes of minimum weight 4. In this situation, the direct sum of component codes will contain all the vectors of weight 4 in the resulting code; if this does not produce the entire code, we add glue vectors taking care not to introduce any new weight 4 vectors. These component codes are determined in Theorem 9.7.2 to be d_{2m}, e_7, and e_8. However, if e_8 is a component, Theorem 9.7.2(ii) shows that C is decomposable; in this case we do not need to use glue.

The glue vectors for d_{2m} and e_7 consist of shorter length *glue elements* that are juxtaposed to form the glue vectors. The glue elements are simply elements of cosets of C_i in $C_i^{\perp}$. Since we can form any glue element from a coset leader and an element in the component, we can choose our possible glue elements from the coset leaders. In Table 9.8, we give the glue elements for the codes d_{2m} and e_7; we use the basis for the codes as given in Section 9.7.1. The code d_{2m} is a $[2m, m-1]$ code and so $d_{2m}^{\perp}$ is a $[2m, m+1]$ code; thus there are four cosets of d_{2m} in $d_{2m}^{\perp}$ and hence four glue elements for d_{2m}. Similarly, there are two glue elements for e_7. Recall that coset leaders have minimum weight in their cosets.

Exercise 536   Verify that the glue elements listed in Table 9.8 for d_4 are indeed coset leaders of different cosets of d_4 in $d_4^{\perp}$. Do the same for each of the other codes in the table.   ◆

Example 9.7.3   We construct a $[14, 7, 4]$ self-dual code using glue. We choose components $2e_7$; this subcode is 6-dimensional and has minimum weight 4. We need one glue vector. Since the glue vector must have weight at least 6, we choose **dd** from Table 9.8 to be the glue vector. Notice that as $\mathbf{d} + e_7$ is a coset of e_7 of weight 3, $\mathbf{dd} + 2e_7$ is a coset of $2e_7$ of weight 6. If $\mathcal{C}$ is the resulting code, $\mathcal{C}$ is indeed a $[14, 7, 4]$ code. As e_7 is self-orthogonal, so is $2e_7$. As **d** is in $e_7^{\perp}$ and **dd** is orthogonal to itself, the coset $\mathbf{dd} + 2e_7$ is self-orthogonal and orthogonal to $2e_7$; therefore $\mathcal{C}$ is indeed self-dual. This is the only indecomposable self-dual $[14, 7, 4]$ code. We label this code $2e_7$ to indicate the components. As we see, this is enough to determine the glue. From this label, a generator matrix is easy to construct.   ■

In Example 9.7.3, we see that the components determine the glue. That is the typical situation. In the literature, the label arising from the components is usually used to identify the code. There are situations where the label does not uniquely determine the code (up to equivalence); in that case, glue is attached to the label. Glue is also attached to the label if the glue is particularly difficult to determine.

Example 9.7.4   We construct two $[16, 8, 4]$ self-dual codes. The first code $\mathcal{C}$ has components $2e_8$; since this is already self-dual, we do not need to add glue. We would simply label this code as $2e_8$; this code is Type II. The second code has only one component, d_{16}, which is a $[16, 7, 4]$ self-orthogonal code; we need to add one glue vector. Since we want our resulting code to have minimum weight 4, we can use either **a** or **c** but not **b** from Table 9.8. But the code using **c** as glue is equivalent to the code using **a** as glue, simply by interchanging the first two coordinates, which is an automorphism of d_{16}. Letting the glue be **a**, which is in $d_{16}^{\perp}$ and which is orthogonal to itself, the resulting code, which we label d_{16}, is a $[16, 8, 4]$ Type I code.   ■

Exercise 537   Construct generator matrices for the self-dual binary codes with labels:
(a) $2e_7$ of length 14,
(b) $2e_8$ of length 16, and
(c) d_{16} of length 16.   ◆

We now describe how to determine the order of the automorphism group of a code obtained by gluing. To simplify notation, let $\mathcal{G}(\mathcal{C})$ be the automorphism group $\Gamma\mathrm{Aut}(\mathcal{C}) = \mathrm{PAut}(\mathcal{C})$ of $\mathcal{C}$. Suppose that $\mathcal{C}$ is a self-dual code obtained by gluing. Then any element of $\mathcal{G}(\mathcal{C})$ permutes the component codes. Therefore $\mathcal{G}(\mathcal{C})$ has a normal subgroup $\mathcal{G}'(\mathcal{C})$ consisting of those automorphisms that fix all components. Thus the quotient group $\mathcal{G}_2(\mathcal{C}) = \mathcal{G}(\mathcal{C})/\mathcal{G}'(\mathcal{C})$ acts as a permutation group on the components. Hence $|\mathcal{G}(\mathcal{C})|$, which is what we wish to calculate, is $|\mathcal{G}_2(\mathcal{C})||\mathcal{G}'(\mathcal{C})|$. Let $\mathcal{G}_0(\mathcal{C})$ be the normal subgroup of $\mathcal{G}'(\mathcal{C})$ that fixes each glue element modulo its component. Then $\mathcal{G}_1(\mathcal{C}) = \mathcal{G}'(\mathcal{C})/\mathcal{G}_0(\mathcal{C})$ permutes the glue. Therefore

$$|\mathcal{G}(\mathcal{C})| = |\mathcal{G}_0(\mathcal{C})||\mathcal{G}_1(\mathcal{C})||\mathcal{G}_2(\mathcal{C})|.$$

If the components of $\mathcal{C}$ are $\mathcal{C}_i$, then $|\mathcal{G}_0(\mathcal{C})|$ is the product of all $|\mathcal{G}_0(\mathcal{C}_i)|$. Since e_7 is the even-like subcode of the [7, 4, 3] quadratic residue code $\mathcal{Q}$ and $\mathbf{d} + e_7$ is all the odd weight vectors in $\mathcal{Q}$, all automorphisms of $\mathcal{Q}$ fix both e_7 and $\mathbf{d} + e_7$. Since the automorphism group of $\mathcal{Q}$ is isomorphic to $\text{PSL}_2(7)$, as described in Section 9.7.1, $|\mathcal{G}_0(e_7)| = 168$. Consider d_{2m} with $m > 2$. In Section 9.7.1, we indicated that $|\mathcal{G}(d_{2m})| = 2^m m!$ Half of the automorphisms in $\mathcal{G}(d_{2m})$ interchange the cosets $\mathbf{a} + d_{2m}$ and $\mathbf{c} + d_{2m}$, and all automorphisms fix $\mathbf{b} + d_{2m}$. Thus $|\mathcal{G}_0(d_{2m})| = 2^{m-1} m!$ and $|\mathcal{G}_1(d_{2m})| = 2$. The situation is a bit different if $m = 2$. There are exactly four permutations (the identity, $(1, 2)(3, 4)$, $(1, 3)(2, 4)$, and $(1, 4)(2, 3)$) that fix d_4, $\mathbf{x} + d_4$, $\mathbf{y} + d_4$, and $\mathbf{z} + d_4$; hence $|\mathcal{G}_0(d_4)| = 4$. As $\mathcal{G}(d_4) = \text{Sym}_4$, $|\mathcal{G}_1(d_4)| = 6$.

If $\mathcal{C}$ has component codes of minimum weight 4, usually $|\mathcal{G}_1(\mathcal{C})| = 1$, but not always as the next example illustrates. The group $\mathcal{G}_1(\mathcal{C})$ projects onto each component $\mathcal{C}_i$ as a subgroup of $\mathcal{G}_1(\mathcal{C}_i)$. In technical terms, $\mathcal{G}_1(\mathcal{C})$ is a subdirect product of the product of $\mathcal{G}_1(\mathcal{C}_i)$; that is, pieces of $\mathcal{G}_1(\mathcal{C}_1)$ are attached to pieces of $\mathcal{G}_1(\mathcal{C}_2)$, which are attached to pieces of $\mathcal{G}_1(\mathcal{C}_3)$, etc. In particular, $|\mathcal{G}_1(\mathcal{C})|$ divides the product of $|\mathcal{G}_1(\mathcal{C}_i)|$. The values of $|\mathcal{G}_i(d_{2m})|$ and $|\mathcal{G}_i(e_7)|$ for $i = 0$ and 1 are summarized in Table 9.8.

Example 9.7.5 We construct a [24, 12, 4] Type II code $\mathcal{C}$ labeled $6d_4$. This code requires six glue vectors. A generator matrix is

$$G = \begin{bmatrix} d_4 & 0 & 0 & 0 & 0 & 0 \\ 0 & d_4 & 0 & 0 & 0 & 0 \\ 0 & 0 & d_4 & 0 & 0 & 0 \\ 0 & 0 & 0 & d_4 & 0 & 0 \\ 0 & 0 & 0 & 0 & d_4 & 0 \\ 0 & 0 & 0 & 0 & 0 & d_4 \\ \hline \mathbf{x} & 0 & \mathbf{y} & \mathbf{x} & \mathbf{y} & 0 \\ \mathbf{x} & 0 & 0 & \mathbf{y} & \mathbf{x} & \mathbf{y} \\ \mathbf{x} & \mathbf{y} & 0 & 0 & \mathbf{y} & \mathbf{x} \\ \mathbf{x} & \mathbf{x} & \mathbf{y} & 0 & 0 & \mathbf{y} \\ \mathbf{x} & \mathbf{y} & \mathbf{x} & \mathbf{y} & 0 & 0 \\ \mathbf{y} & 0 & \mathbf{z} & \mathbf{y} & \mathbf{z} & 0 \end{bmatrix},$$

where d_4 represents the basis vector $\mathbf{1}$ of d_4, and $\mathbf{x}$, $\mathbf{y}$, and $\mathbf{z}$ are from Table 9.8. The first six rows generate the component code $6d_4$. Note that all weight 4 codewords are precisely the six codewords in the first six rows of G. As $|\mathcal{G}_0(d_4)| = 4$, $|\mathcal{G}_0(\mathcal{C})| = 4^6$. By Exercise 538, $|\mathcal{G}_1(\mathcal{C})| = 3$. Since any permutation in Sym_6 permutes the six components, $|\mathcal{G}_2(\mathcal{C})| = 720$. Thus

$$|\mathcal{G}(\mathcal{C})| = |\mathcal{G}_0(\mathcal{C})||\mathcal{G}_1(\mathcal{C})||\mathcal{G}_2(\mathcal{C})| = 4^6 \cdot 3 \cdot 720. \qquad \blacksquare$$

Table 9.9 *Labels of indecomposable self-dual codes*

Length	Components
8	e_8
12	d_{12}
14	$2e_7$
16	d_{16}, $2d_8$
18	$d_{10} \oplus e_7 \oplus f_1$, $3d_6$
20	d_{20}, $d_{12} \oplus d_8$, $2d_8 \oplus d_4$, $2e_7 \oplus d_6$, $3d_6 \oplus f_2$, $5d_4$

Exercise 538 In the notation of Example 9.7.5, do the following:
(a) Verify that $\mathcal{G}_1(\mathcal{C})$ contains the permutation $(1, 2, 3)$ acting on each component (if we label the coordinates of each d_4 with $\{1, 2, 3, 4\}$).
(b) Show that $|\mathcal{G}_1(\mathcal{C})| = 3$. ◆

Example 9.7.6 It is easier to construct a $[24, 12, 4]$ Type II code $\mathcal{C}$ labeled $d_{10} \oplus 2e_7$ as we only need to add two glue vectors since $d_{10} \oplus 2e_7$ has dimension 10. We can do this using the generator matrix

$$
G = \begin{bmatrix}
d_{10} & O & O \\
O & e_7 & O \\
O & O & e_7 \\
\hline
\mathbf{a} & \mathbf{d} & \mathbf{0} \\
\mathbf{c} & \mathbf{0} & \mathbf{d}
\end{bmatrix},
$$

where d_{10} and e_7 represent a basis for these codes, O is the appropriate size zero matrix, and $\mathbf{a}$, $\mathbf{c}$, and $\mathbf{d}$ are found in Table 9.8. It is easy to verify that the two glue vectors have weight 8, are orthogonal to each other, and have sum also of weight 8. The order of $\mathcal{G}(\mathcal{C})$ is

$$
|\mathcal{G}(\mathcal{C})| = |\mathcal{G}_0(\mathcal{C})||\mathcal{G}_1(\mathcal{C})||\mathcal{G}_2(\mathcal{C})| = [(2^4 5!)168^2] \cdot 1 \cdot 2 = 2^{14} \cdot 3^3 \cdot 5 \cdot 7^2. \quad \blacksquare
$$

Exercise 539 Find the size of the automorphism groups of the following binary self-dual codes:
(a) $2e_7$ of length 14,
(b) $2e_8$ of length 16, and
(c) d_{16} of length 16. ◆

Not every conceivable component code will lead to a self-dual code. For example, there are no $[24, 12, 4]$ Type II codes with the component codes d_4 or $2d_4$, although there are Type II codes of length 32 with these component codes. Note also that it is possible to have inequivalent codes with the same label; in such cases it is important to specify the glue.

It may happen that when constructing the component codes some positions have no codewords; then those positions are regarded as containing the free code, denoted f_n, whose only codeword is the zero vector of length n. Then $|\mathcal{G}_0(f_n)| = 1$. In Table 9.9 we give the labels for the indecomposable self-dual codes of length $n \leq 20$; note that the self-dual codes of length 10 are all decomposable.

Table 9.10 *Number of inequivalent self-dual binary codes*

Length	2	4	6	8	10	12	14	16
Number self-dual	1	1	1	2	2	3	4	7
Length	18	20	22	24	26	28	30	
Number self-dual	9	16	25	55	103	261	731	

Length	8	16	24	32
Number Type II	1	2	9	85

Exercise 540 Find the glue vectors and order of the automorphism group of the self-dual code of length 18 with label $d_{10} \oplus e_7 \oplus f_1$. ♦

Example 9.7.7 There are seven indecomposable Type II codes of length 24. Their labels are $d_{24}, 2d_{12}, d_{10} \oplus 2e_7, 3d_8, 4d_6, 6d_4$, and g_{24}, where g_{24} is the [24, 12, 8] extended binary Golay code. ■

Exercise 541 Find the glue vectors and order of the automorphism group of the [24, 12, 4] self-dual code with labels:
(a) $2d_{12}$, and
(b) $3d_8$. ♦

The classification of self-dual binary codes began with the classification for lengths $2 \le n \le 24$ in [262, 281]. Then the Type II codes of length 32 were classified in [53], and, finally, the self-dual codes of lengths 26, 28, and 30 were found in [53, 56, 264]. Table 9.10 gives the number of inequivalent self-dual codes of lengths up through 30 and the number of inequivalent Type II codes up through length 32.

Exercise 542 In Table 9.9 we give the labels for the components of the two indecomposable self-dual codes of length 16. By Table 9.10 there are seven self-dual codes of length 16. Give the labels of the decomposable self-dual codes of length 16. ♦

Recall by Theorem 9.7.2 that if a self-dual code C of length n has minimum weight 2, it is decomposable and equivalent to $mi_2 \oplus C_1$, where C_1 is a self-dual code of length $n - 2m$ and minimum weight at least 4. So in classifying codes of length n, assuming the codes of length less than n have been classified, we can easily find those of minimum weight 2. Also by Theorem 9.7.2, if e_8 is a subcode of C, then C is decomposable and can be found. In the classification of the indecomposable self-dual codes of lengths through 24 with minimum weight at least 4, the components for all codes, except the [24, 12, 8] extended binary Golay code, were chosen from d_{2m} and e_7. However, in classifying the Type II codes of length 32, it was also necessary to have components with minimum weight 8. There are several of these and they are all related to the binary Golay codes, in one way or another [53, 56, 291]. One of these, denoted g_{16}, is isomorphic to the [16, 5, 8] first order Reed–Muller code. Let g_{24} be the [24, 12, 8] extended binary Golay code. If **e** is any weight 8 codeword in g_{24} with support $\mathcal{P}_1$, by the Balance Principle, the subcode with support under the complementary

Table 9.11 *Number of inequivalent Type III and IV codes*

Length		4	8	12	16	20
Number Type III		1	1	3	7	24

Length		2	4	6	8	10	12	14	16
Number Type IV		1	1	2	3	5	10	21	55

positions to $\mathcal{P}_1$ must be a $[16, 5, 8]$ code, which is unique and hence equivalent to g_{16}. Reordering coordinates so that $\mathcal{P}_1$ is the first eight and writing a generator matrix for g_{24} in the form (9.3), the last six rows can be considered as glue for **e** glued to g_{16}. Note that we already used parts of this construction in obtaining the Nordstrom–Robinson code in Section 2.3.4.

In classifying the Type II codes of length 32, five inequivalent extremal $[32, 16, 8]$ codes were found. One of these is the extended quadratic residue code and another is the second order Reed–Muller code. As all five are extremal, they have the same weight distribution; their generator matrices can be found in [53, 56, 291]. The situation was too complicated to check the progress of the classification using only the mass formula of Theorem 9.5.5(i), as had been done for lower lengths. In this case, subformulas were developed that counted the number of codes containing a specific code, such as d_8. In the end, the original mass formula was used to check that the classification was complete. The Type I codes of lengths 26, 28, and 30 could then be found once it was observed that they were all children of Type II codes of length 32 [53, 56, 264]. Understanding the structure of a length 32 Type II code helps determine the equivalence or inequivalence of its numerous children.

At length 40, there are $N = (1 + 1) \cdot (2 + 1) \cdot (2^2 + 1) \cdots (2^{18} + 1)$ Type II codes by Theorem 9.5.5(i). As $N/40! \geq 17\,000$, it does not make sense to attempt to classify the Type II codes of length 40 or larger; see Exercise 543.

Exercise 543 Explain, in more detail, why there are at least 17 000 inequivalent Type II codes of length 40. ◆

As the extremal codes are most interesting, it makes sense only to classify those. However, formulas for the number of these are not known. On occasion, different methods have been used to classify extremal codes successfully. The number of inequivalent extremal binary codes of any length greater than 32, is, as yet, unknown.

Ternary self-dual codes have been classified up to length $n = 20$. In the ternary case, it is known that any self-orthogonal code generated by vectors of weight 3 is a direct sum of the $[3, 1, 3]$ code e_3, with generator matrix $[1 \ 1 \ 1]$, and the $[4, 2, 3]$ tetracode, e_4. In these classifications, components of minimum weight 6 are also needed; many of these are derived from the $[12, 6, 6]$ extended ternary Golay code. Table 9.11 gives the number of inequivalent Type III $[55, 222, 260, 282]$ and Type IV codes [55]. The extremal $[24, 12, 9]$ Type III codes have been classified in [191] using the classification of the 24×24 Hadamard matrices. The $[24, 12, d]$ Type III codes with $d = 3$ or 6 have not been classified and are of little interest since the extremal ones are known.

While self-dual codes have a number of appealing properties, there is no known efficient decoding algorithm for such codes. In Chapter 10 we give two algorithms to decode the [24, 12, 8] extended binary Golay code; these algorithms can be generalized, but only in a rather restricted manner.

Research Problem 9.7.8 *Find an efficient decoding algorithm that can be used on all self-dual codes or on a large family of such codes.*

9.8   Circulant constructions

Many of the codes we have studied or will study have a particular construction using circulant matrices. An $m \times m$ matrix A is *circulant* provided

$$
A = \begin{bmatrix}
a_1 & a_2 & a_3 & \cdots & a_m \\
a_m & a_1 & a_2 & \cdots & a_{m-1} \\
a_{m-1} & a_m & a_1 & \cdots & a_{m-2} \\
& & \vdots & & \\
a_2 & a_3 & a_4 & \cdots & a_1
\end{bmatrix}.
$$

An $(m + 1) \times (m + 1)$ matrix B is *bordered circulant* if

$$
B = \begin{bmatrix}
\alpha & \beta & \cdots & \beta \\
\gamma & & & \\
\gamma & & A & \\
\vdots & & & \\
\gamma & & &
\end{bmatrix},
\tag{9.11}
$$

where A is circulant. Notice that the identity matrix I_n is both a circulant and a bordered circulant matrix.

We say that a code has a *double circulant generator matrix* or a *bordered double circulant generator matrix* provided it has a generator matrix of the form

$$
[I_m \quad A] \quad \text{or} \quad [I_{m+1} \quad B],
\tag{9.12}
$$

where A is an $m \times m$ circulant matrix and B is an $(m + 1) \times (m + 1)$ bordered circulant matrix, respectively. A code has a *double circulant construction* or *bordered double circulant construction* provided it has a double circulant or bordered double circulant generator matrix, respectively.

Example 9.8.1 Using the generator matrix for the [12, 6, 6] extended ternary Golay code given in Section 1.9.2 we see that this code has a bordered double circulant construction. ∎

In a related concept, an $m \times m$ matrix R is called *reverse circulant* provided

$$R = \begin{bmatrix} r_1 & r_2 & r_3 & \cdots & r_m \\ r_2 & r_3 & r_4 & \cdots & r_1 \\ r_3 & r_4 & r_5 & \cdots & r_2 \\ & & \vdots & & \\ r_m & r_1 & r_2 & \cdots & r_{m-1} \end{bmatrix}.$$

An $(m+1) \times (m+1)$ matrix B is *bordered reverse circulant* if it has form (9.11) where A is reverse circulant. Again a code has a *reverse circulant generator matrix* or a *bordered reverse circulant generator matrix* provided it has a generator matrix of the form (9.12) where A is an $m \times m$ reverse circulant matrix or B is an $(m+1) \times (m+1)$ bordered reverse circulant matrix, respectively. Note that we drop the term "double" because the identity matrix is not reverse circulant. A code has a *reverse circulant construction* or *bordered reverse circulant construction* provided it has a reverse circulant or bordered reverse circulant generator matrix, respectively.

Example 9.8.2 The construction of the $[24, 12, 8]$ extended binary Golay code in Section 1.9.1 is a bordered reverse circulant construction. ∎

Exercise 544 Prove that if R is a reverse circulant matrix, then $R = R^{\mathrm{T}}$. ◆

Example 9.8.3 Let A be a 5×5 circulant matrix with rows $\mathbf{a}_1, \ldots, \mathbf{a}_5$. Form a matrix R with rows $\mathbf{a}_1, \mathbf{a}_5, \mathbf{a}_4, \mathbf{a}_3, \mathbf{a}_2$. Now do the same thing with the double circulant matrix $[I_5 \ A]$. So we obtain

$$[I_5 \quad A] = \begin{bmatrix} 1 & 0 & 0 & 0 & 0 & a_1 & a_2 & a_3 & a_4 & a_5 \\ 0 & 1 & 0 & 0 & 0 & a_5 & a_1 & a_2 & a_3 & a_4 \\ 0 & 0 & 1 & 0 & 0 & a_4 & a_5 & a_1 & a_2 & a_3 \\ 0 & 0 & 0 & 1 & 0 & a_3 & a_4 & a_5 & a_1 & a_2 \\ 0 & 0 & 0 & 0 & 1 & a_2 & a_3 & a_4 & a_5 & a_1 \end{bmatrix}$$

$$\rightarrow \begin{bmatrix} 1 & 0 & 0 & 0 & 0 & a_1 & a_2 & a_3 & a_4 & a_5 \\ 0 & 0 & 0 & 0 & 1 & a_2 & a_3 & a_4 & a_5 & a_1 \\ 0 & 0 & 0 & 1 & 0 & a_3 & a_4 & a_5 & a_1 & a_2 \\ 0 & 0 & 1 & 0 & 0 & a_4 & a_5 & a_1 & a_2 & a_3 \\ 0 & 1 & 0 & 0 & 0 & a_5 & a_1 & a_2 & a_3 & a_4 \end{bmatrix} = [P_5 \quad R],$$

where P_5 is a matrix obtained from I_5 by permuting its columns. Notice that R is reverse circulant and that by permuting the columns of P_5 we can obtain I_5. So the code generated by the double circulant generator matrix $[I_5 \ A]$ equals the code generated by the matrix $[P_5 \ R]$, which is equivalent to the code generated by the reverse circulant generator matrix $[I_5 \ R]$. You are asked to generalize this in Exercise 545. ∎

Exercise 545 Prove the following:

(a) A code has a double circulant construction if and only if it is equivalent to a code with a reverse circulant construction.

(b) A code has a bordered double circulant construction if and only if it is equivalent to a code with a bordered reverse circulant construction. ◆

A subfamily of formally self-dual codes is the class of isodual codes. A code is *isodual* if it is equivalent to its dual. Clearly, any isodual code is formally self-dual; however, a formally self-dual code need not be isodual as seen in [84]. Exercise 546 shows how to construct isodual codes and hence formally self-dual codes.

Exercise 546 Prove the following:

(a) A code of even length n with generator matrix $[I_{n/2} \ A]$, where $A = A^T$ is isodual.

(b) A code with a double circulant or reverse circulant construction is isodual. Hint: See Exercises 544 and 545.

(c) A code with a bordered double circulant or bordered reverse circulant construction is isodual provided the bordered matrix (9.11) used in the construction satisfies either $\beta = \gamma = 0$ or both β and γ are nonzero. ◆

9.9   Formally self-dual codes

In this section we examine general properties of formally self-dual codes and describe the current state of their classification. Recall that a code is formally self-dual if the code and its dual have the same weight enumerator. When considering codes over fields $\mathbb{F}_2$, $\mathbb{F}_3$, and $\mathbb{F}_4$, the only formally self-dual codes that are not already self-dual and whose weight enumerators are combinations of Gleason polynomials, are the even formally self-dual binary codes. Therefore in this section we will only consider even formally self-dual binary codes; relatively little work has been done on formally self-dual codes over nonbinary fields.

We begin by summarizing results about self-dual binary codes that either apply or do not apply to formally self-dual binary codes.

- While self-dual codes contain only even weight codewords, formally self-dual codes may contain odd weight codewords as well. See Exercise 547.
- A formally self-dual binary code $\mathcal{C}$ is even if and only if $\mathbf{1} \in \mathcal{C}$; see Exercise 548.
- If $\mathcal{C}$ is an even formally self-dual binary code of length n with weight distribution $A_i(\mathcal{C})$, then $A_i(\mathcal{C}) = A_{n-i}(\mathcal{C})$; again see Exercise 548.
- Gleason's Theorem applies to even formally self-dual binary codes. If $\mathcal{C}$ is such a code of length n, then

$$W_{\mathcal{C}}(x, y) = \sum_{i=0}^{\lfloor \frac{n}{8} \rfloor} a_i g_1(x, y)^{\frac{n}{2}-4i} g_2(x, y)^i,$$

where $g_1(x, y) = y^2 + x^2$ and $g_2(x, y) = y^8 + 14x^4y^4 + x^8$.

- In Theorem 9.3.4, a bound on the minimum weight of an even formally self-dual binary code is presented. If C is an $[n, n/2, d]$ even formally self-dual binary code, then

$$
d \le
\begin{cases}
2 \left\lfloor \dfrac{n}{8} \right\rfloor + 2 & \text{if } n \le 30, \\[2mm]
2 \left\lfloor \dfrac{n}{8} \right\rfloor & \text{if } n \ge 32.
\end{cases}
\tag{9.13}
$$

Furthermore, an even formally self-dual binary code meeting this bound with $n \le 30$ has a unique weight enumerator. For $n \le 30$, by Theorem 9.3.4, the bound is met precisely when $n \ne 16$ and $n \ne 26$. It is natural to ask if this bound can be improved. For example, if $n = 72$, the bound is $d \le 18$; however, by [32] there is no $[72, 36, 18]$ code of any type. The bound in Theorem 9.3.5 is certainly not going to hold without some modification on the values of n for which it might hold. For example, the bound in Theorem 9.3.5 when $n = 42$ is $d \le 8$; however, there is a $[42, 21, 10]$ even formally self-dual code.

Research Problem 9.9.1 *Improve the bound* (9.13) *for even formally self-dual binary codes.*

Exercise 547 Let C be the binary code with generator matrix

$$
\begin{bmatrix}
1 & 0 & 0 & 0 & 1 & 1 \\
0 & 1 & 0 & 1 & 0 & 1 \\
0 & 0 & 1 & 1 & 1 & 0
\end{bmatrix}.
$$

Show that C is formally self-dual with odd weight codewords and give its weight enumerator. ◆

Exercise 548 Let C be a formally self-dual binary code of length n with weight distribution $A_i(C)$.
(a) Prove that C is even if and only if $\mathbf{1} \in C$.
(b) Prove that if C is even, then $A_i(C) = A_{n-i}(C)$. ◆

The Balance Principle can be extended to formally self-dual codes. In fact, it extends to any code of dimension half its length. We state it in this general form but will apply it to even formally self-dual binary codes.

Theorem 9.9.2 ([83]) *Let C be an $[n, n/2]$ code over $\mathbb{F}_q$. Choose a set of coordinate positions $\mathcal{P}_{n_1}$ of size n_1 and let $\mathcal{P}_{n_2}$ be the complementary set of coordinate positions of size $n_2 = n - n_1$. Let C_i be the subcode of C all of whose codewords have support in $\mathcal{P}_{n_i}$, and let $\mathcal{Z}_i$ be the subcode of $C^\perp$ all of whose codewords have support in $\mathcal{P}_{n_i}$. The following hold:*
(i) *(Balance Principle)*

$$
\dim C_1 - \frac{n_1}{2} = \dim \mathcal{Z}_2 - \frac{n_2}{2} \quad \text{and} \quad \dim \mathcal{Z}_1 - \frac{n_1}{2} = \dim C_2 - \frac{n_2}{2}.
$$

(ii) *If we reorder coordinates so that $\mathcal{P}_{n_1}$ is the left-most n_1 coordinates and $\mathcal{P}_{n_2}$ is the right-most n_2 coordinates, then $\mathcal{C}$ and $\mathcal{C}^{\perp}$ have generator matrices of the form*

$$\mathrm{gen}(\mathcal{C}) = \begin{bmatrix} A & O \\ O & B \\ D & E \end{bmatrix} \quad and \quad \mathrm{gen}(\mathcal{C}^{\perp}) = \begin{bmatrix} F & O \\ O & J \\ L & M \end{bmatrix},$$

where $[A\ O]$ is a generator matrix of $\mathcal{C}_1$, $[O\ B]$ is a generator matrix of $\mathcal{C}_2$, $[F\ O]$ is a generator matrix of $\mathcal{Z}_1$, $[O\ J]$ is a generator matrix of $\mathcal{Z}_2$, and O is the appropriate size zero matrix. We also have the following:

(a) $\mathrm{rank}(D) = \mathrm{rank}(E) = \mathrm{rank}(L) = \mathrm{rank}(M)$.

(b) *Let $\mathcal{A}$ be the code of length n_1 generated by A, $\mathcal{A}_D$ the code of length n_1 generated by the rows of A and D, $\mathcal{B}$ the code of length n_2 generated by B, and $\mathcal{B}_E$ the code of length n_2 generated by the rows of B and E. Define $\mathcal{F}$, $\mathcal{F}_L$, $\mathcal{J}$, and $\mathcal{J}_M$ analogously. Then $\mathcal{A}^{\perp} = \mathcal{F}_L$, $\mathcal{B}^{\perp} = \mathcal{J}_M$, $\mathcal{F}^{\perp} = \mathcal{A}_D$, and $\mathcal{J}^{\perp} = \mathcal{B}_E$.*

The proof is left as an exercise.

Exercise 549 Prove Theorem 9.9.2. ◆

Example 9.9.3 In Example 9.4.2 we showed that there is no $[10, 5, 4]$ self-dual binary code. In this example, we show that, up to equivalence, there is a unique $[10, 5, 4]$ even formally self-dual binary code $\mathcal{C}$, a code we first encountered in Exercise 501. We use the notation of Theorem 9.9.2. Order the coordinates so that a codeword in $\mathcal{C}$ of weight 4 has its support in the first four coordinate positions and let $\mathcal{P}_{n_1}$ be these four coordinates. Thus the code $\mathcal{C}_1$ has dimension 1 and $A = [1\ 1\ 1\ 1]$. As $\mathcal{C}$ is even, $\mathbf{1} \in \mathcal{C}$, and so the all-one vector of length 6 is in $\mathcal{B}$. As there is no $[6, 2, 4]$ binary code containing the all-one vector, $\dim \mathcal{C}_2 = \dim \mathcal{B} = 1$. Therefore $B = [1\ 1\ 1\ 1\ 1\ 1]$. By the Balance Principle, $\dim \mathcal{Z}_1 + 1 = \dim \mathcal{C}_2 = 1$ implying that $\dim \mathcal{Z}_1 = 0$ and hence that $\mathcal{F}$ is the zero code. Since $\mathcal{F}^{\perp} = \mathcal{A}_D$, we may assume that

$$D = \begin{bmatrix} 1 & 0 & 0 & 0 \\ 0 & 1 & 0 & 0 \\ 0 & 0 & 1 & 0 \end{bmatrix}.$$

By Exercise 550, up to equivalence, we may choose the first two rows of E to be 111000 and 110100, respectively. Then, by the same exercise, again up to equivalence, we may choose the third row of E to be one of 110010, 101100, or 101010; but only the last leads to a $[10, 5, 4]$ code. Thus we have the generator matrix

$$\mathrm{gen}(\mathcal{C}) = \left[\begin{array}{cccc|cccccc} 1 & 1 & 1 & 1 & 0 & 0 & 0 & 0 & 0 & 0 \\ 0 & 0 & 0 & 0 & 1 & 1 & 1 & 1 & 1 & 1 \\ 1 & 0 & 0 & 0 & 1 & 1 & 1 & 0 & 0 & 0 \\ 0 & 1 & 0 & 0 & 1 & 1 & 0 & 1 & 0 & 0 \\ 0 & 0 & 1 & 0 & 1 & 0 & 1 & 0 & 1 & 0 \end{array} \right]. \tag{9.14}$$

By Exercise 551, the code generated by this matrix is formally self-dual. ∎

Exercise 550 In the notation of Example 9.9.3, show that, once A, B, and D are as determined in that example, up to equivalence we may choose the first two rows of E to

Table 9.12 *Even formally self-dual binary codes*

n	d	# s.d.	# e.f.s.d. (not s.d.)	n	d	# s.d.	# e.f.s.d. (not s.d.)
2	2	1	0	26	6*	1	≥ 30
4	2	1	0	28	8	0	1
6	2	1	1	30	8	0	≥ 6
8	4	1	0	32	8	8	≥ 10
10	4	0	1	34	8	0	≥ 7
12	4	1	2	36	8	≥ 4	≥ 23
14	4	1	9	38	8	≥ 7	≥ 24
16	4*	2	≥ 6	40	10	0	?
18	6	0	1	42	10	0	≥ 6
20	6	0	7	44	10	0	≥ 12
22	6	1	≥ 1000	46	10	≥ 1	≥ 41
24	8	1	0	48	12	≥ 1	?

be 111000 and 110100. Then show that the third row is, up to equivalence, either 110010, 101100, or 101010. Furthermore, show that the first two choices for the third row both lead to codewords in $\mathcal{C}$ of weight 2.   ◆

Exercise 551 Show that the binary code generated by the matrix in (9.14) is formally self-dual.   ◆

Exercise 552 Show that up to equivalence there are two $[6, 3, 2]$ even formally self-dual codes and give generator matrices for these codes.   ◆

The even formally self-dual binary codes of highest possible minimum weight have been classified completely through length 28 except for lengths 16, 22, and 26. We give the number of inequivalent codes in Table 9.12, taken from [82]. In this table, the length is denoted "n". The column "d" denotes the highest minimum distance possible as given in (9.13) except when $n = 16$ and $n = 26$ where there are no codes of any type meeting the bound by [32]; in these two cases, the value of d is reduced by 2 in the table (and marked with "*"). The column "# s.d." gives the number of inequivalent $[n, n/2, d]$ self-dual codes; the column "# e.f.s.d. (not s.d.)" gives the number of inequivalent $[n, n/2, d]$ even formally self-dual codes that are not self-dual. As we saw in Examples 9.4.2 and 9.9.3, there is exactly one $[10, 5, 4]$ even formally self-dual code and that code is not self-dual; the original proof of this, different from that presented in these examples, is found in [165]. The lengths 12, 14, and 20 cases were presented in [12], [83], and [84], respectively. Also in [84] more than 1000 inequivalent codes of length 22 were shown to exist. Binary $[18, 9, 6]$ and $[28, 14, 8]$ codes (not necessarily formally self-dual) were shown to be unique up to equivalence in [309] and [156], respectively; in each case these codes in fact turned out to be even formally self-dual. There are no self-dual codes of length 34 and minimum distance 8 (which would meet the bound of Theorem 9.3.5) by [58]. The complete classification of self-dual codes of lengths 18 and 20 [262] shows that there are no such codes of minimum distance 6.

Similarly, the complete classification of self-dual codes of lengths 28 and 30 [53, 56, 264] shows that there are no such codes of minimum distance 8. At lengths 40, 42, and 44, the bound on the minimum distance of self-dual codes is 8, by Theorem 9.3.5. A number of the even formally self-dual codes counted in Table 9.12 have double circulant constructions. All those of lengths through 12 have a double circulant construction. It can be shown that the [14, 7, 4] self-dual code does not have a double circulant construction but exactly two of the nine even formally self-dual codes of length 14 that are not self-dual do have such constructions. It can also be shown that the length 18 code does also, as do two of the seven even formally self-dual codes of length 20. All the codes of length through 20, except possibly those of length 16, are isodual. For the minimum weights given in the table, length 22 is the smallest length for which a non-isodual even formally self-dual code is known to exist.

In Table 9.12, we see that there exist even formally self-dual codes with higher minimum weight than self-dual codes at lengths $n = 10, 18, 20, 28, 30, 34, 42$, and 44. Relatively little is known about even formally self-dual codes at lengths greater than 48.

Research Problem 9.9.4 *The bound in* (9.13) *on the minimum distance for even formally self-dual binary codes of length $n = 40$ is $d \le 10$, while the bound on self-dual codes of that length is $d \le 8$ by Theorem 9.3.5. No* [40, 20, 10] *even formally self-dual binary code, which necessarily must be non-self-dual, is known. Find one or show it does not exist.*

By Theorem 9.3.5, if C is an $[n, n/2, d]$ self-dual binary code with $n \equiv 0 \pmod{24}$, then $d \le 4 \lfloor n/24 \rfloor + 4$, and when this bound is met, C is Type II. When $n \equiv 0 \pmod{24}$, there are no known even formally self-dual codes with $d = 4 \lfloor n/24 \rfloor + 4$ that are not self-dual. Note that this value of d may not be the highest value possible; by [32] it is the highest for $n = 24, 48$, and 72.

Research Problem 9.9.5 *Find a* [24m, 12m, 4m + 4] *even formally self-dual binary code that is not self-dual or prove that none exists.*

When codes are extremal Type II, the codewords support designs by Theorem 9.3.10. The following theorem, found in [165], illustrates the presence of designs in certain even formally self-dual codes.

Theorem 9.9.6 *Let C be an $[n, n/2, d]$ even formally self-dual binary code with $d = 2 \lfloor n/8 \rfloor + 2$ and $n > 2$. Let S_w be the set of vectors of weight w in C together with those of weight w in $C^\perp$; a vector in $C \cap C^\perp$ is included twice in S_w. Then the vectors in S_w hold a 3-design if $n \equiv 2 \pmod 8$ and a 1-design if $n \equiv 6 \pmod 8$.*

We note that this result holds only for the codes of lengths 6, 10, 14, 18, 22, and 30 from (9.13) and Table 9.12.

Exercise 553 Let C be the [10, 5, 4] even formally self-dual code C from Example 9.9.3.
(a) Show that $C \cap C^\perp = \{0, 1\}$.
(b) By part (a) and Theorem 9.9.6, the codewords of weight 4 in $C \cup C^\perp$ hold a 3-(10, 4, λ) design. Find λ.
(c) Show that the weight 4 codewords of C do not support a 3-design.   ♦

9.10 Additive codes over $\mathbb{F}_4$

We now turn our attention to a class of codes of interest because of their connection to quantum error-correction and quantum computing [44]. These codes over $\mathbb{F}_4$ may not be linear over $\mathbb{F}_4$ but are linear over $\mathbb{F}_2$. With an appropriate definition of inner product, the codes used in quantum error-correction are self-orthogonal. Many of the ideas already presented in this chapter carry over to this family of codes.

An *additive code* $\mathcal{C}$ over $\mathbb{F}_4$ of length n is a subset of $\mathbb{F}_4^n$ closed under vector addition. So $\mathcal{C}$ is an additive subgroup of $\mathbb{F}_4^n$. Because $\mathbf{x} + \mathbf{x} = \mathbf{0}$ for $\mathbf{x} \in \mathcal{C}$, $\mathcal{C}$ is a binary vector space. Thus $\mathcal{C}$ has 2^k codewords for some k with $0 \le k \le 2n$; $\mathcal{C}$ will be referred to as an $(n, 2^k)$ code or an $(n, 2^k, d)$ code if the minimum weight d is known. Note that an additive code over $\mathbb{F}_4$ differs from a linear code over $\mathbb{F}_4$ in that it may not be closed under scalar multiplication and hence may be nonlinear over $\mathbb{F}_4$. An $(n, 2^k)$ additive code $\mathcal{C}$ has a binary basis of k codewords. A *generator matrix* for $\mathcal{C}$ is any $k \times n$ matrix whose rows form a binary basis for $\mathcal{C}$.

Example 9.10.1 As a linear code over $\mathbb{F}_4$, the $[6, 3, 4]$ hexacode has generator matrix

$$\begin{bmatrix} 1 & 0 & 0 & 1 & \omega & \omega \\ 0 & 1 & 0 & \omega & 1 & \omega \\ 0 & 0 & 1 & \omega & \omega & 1 \end{bmatrix}.$$

However, thinking of the hexacode as an additive $(6, 2^6, 4)$ code, it has generator matrix

$$\begin{bmatrix} 1 & 0 & 0 & 1 & \omega & \omega \\ \omega & 0 & 0 & \omega & \overline{\omega} & \overline{\omega} \\ 0 & 1 & 0 & \omega & 1 & \omega \\ 0 & \omega & 0 & \overline{\omega} & \omega & \overline{\omega} \\ 0 & 0 & 1 & \omega & \omega & 1 \\ 0 & 0 & \omega & \overline{\omega} & \overline{\omega} & \omega \end{bmatrix}.$$

The second, fourth, and sixth rows of the latter matrix are obtained by multiplying the three rows of the former matrix by ω. ∎

In the connection between quantum codes and additive codes over $\mathbb{F}_4$ [44], a natural inner product, called the trace inner product, arises for the additive codes. Recall from Section 3.8 that the trace $\mathrm{Tr}_2 : \mathbb{F}_4 \to \mathbb{F}_2$ is defined by $\mathrm{Tr}_2(\alpha) = \alpha + \alpha^2$. Alternately, since conjugation $^-$ in $\mathbb{F}_4$ is given by $\overline{\alpha} = \alpha^2$, $\mathrm{Tr}_2(\alpha) = \alpha + \overline{\alpha}$. Thus $\mathrm{Tr}_2(0) = \mathrm{Tr}_2(1) = 0$ and $\mathrm{Tr}_2(\omega) = \mathrm{Tr}_2(\overline{\omega}) = 1$. By Lemma 3.8.5, $\mathrm{Tr}_2(\alpha + \beta) = \mathrm{Tr}_2(\alpha) + \mathrm{Tr}_2(\beta)$ indicating that Tr_2 is $\mathbb{F}_2$-linear. The *trace inner product* $\langle \cdot, \cdot \rangle_T$ on $\mathbb{F}_4^n$ is defined as follows. For $\mathbf{x} = x_1 x_2 \cdots x_n$ and $\mathbf{y} = y_1 y_2 \cdots y_n$ in $\mathbb{F}_4^n$, let

$$\langle \mathbf{x}, \mathbf{y} \rangle_T = \mathrm{Tr}_2(\langle \mathbf{x}, \mathbf{y} \rangle) = \sum_{i=1}^{n} (x_i \overline{y_i} + \overline{x_i} y_i), \tag{9.15}$$

where $\langle \cdot, \cdot \rangle$ is the Hermitian inner product. The right-most equality in (9.15) follows by the $\mathbb{F}_2$-linearity of Tr_2 and the fact that $\overline{\overline{\alpha}} = \alpha$. In working with the trace inner product it is helpful to observe that $\text{Tr}_2(\alpha\overline{\beta}) = 0$ if $\alpha = \beta$ or $\alpha = 0$ or $\beta = 0$ and $\text{Tr}_2(\alpha\overline{\beta}) = 1$ otherwise. The proof that the following basic properties of the trace inner product hold is left as an exercise.

Lemma 9.10.2 *If* $\mathbf{x}$, $\mathbf{y}$, *and* $\mathbf{z}$ *are in* $\mathbb{F}_4^n$, *then*:
(a) $\langle \mathbf{x}, \mathbf{x} \rangle_T = 0$,
(b) $\langle \mathbf{x}, \mathbf{y} \rangle_T = \langle \mathbf{y}, \mathbf{x} \rangle_T$,
(c) $\langle \mathbf{x} + \mathbf{y}, \mathbf{z} \rangle_T = \langle \mathbf{x}, \mathbf{z} \rangle_T + \langle \mathbf{y}, \mathbf{z} \rangle_T$, *and*
(d) $\langle \mathbf{x}, \omega\mathbf{x} \rangle_T = 0$ *if* $\text{wt}(\mathbf{x})$ *is even while* $\langle \mathbf{x}, \omega\mathbf{x} \rangle_T = 1$ *if* $\text{wt}(\mathbf{x})$ *is odd.*

Exercise 554 Compute $\langle 1\omega 1\overline{\omega}\omega\omega 1, 1\overline{\omega}\overline{\omega}\omega\omega 1\omega \rangle_T$. ◆

Exercise 555 Prove Lemma 9.10.2. ◆

As usual, define the *trace dual* $\mathcal{C}^{\perp_T}$ of an additive code $\mathcal{C}$ to be

$$\mathcal{C}^{\perp_T} = \left\{ \mathbf{v} \in \mathbb{F}_4^n \mid \langle \mathbf{u}, \mathbf{v} \rangle_T = 0 \text{ for all } \mathbf{u} \in \mathcal{C} \right\}.$$

Exercise 556 shows that if $\mathcal{C}$ is an additive $(n, 2^k)$ code, then $\mathcal{C}^{\perp_T}$ is an additive $(n, 2^{2n-k})$ code. The additive code $\mathcal{C}$ is *trace self-orthogonal* if $\mathcal{C} \subseteq \mathcal{C}^{\perp_T}$ and *trace self-dual* if $\mathcal{C} = \mathcal{C}^{\perp_T}$. Thus a trace self-dual additive code is an $(n, 2^n)$ code. Höhn [134] proved that the same MacWilliams equation that holds for linear codes over $\mathbb{F}_4$ also holds for additive codes over $\mathbb{F}_4$; namely (see (M_3))

$$W_{\mathcal{C}^{\perp_T}}(x, y) = \frac{1}{|\mathcal{C}|} W_{\mathcal{C}}(y - x, y + 3x).$$

Exercise 556 Prove that if $\mathcal{C}$ is an additive $(n, 2^k)$ code, then $\mathcal{C}^{\perp_T}$ is an additive $(n, 2^{2n-k})$ code. ◆

To obtain an equivalent additive code $\mathcal{C}_2$ from the additive code $\mathcal{C}_1$, as with linear codes over $\mathbb{F}_4$, we are allowed to permute and scale coordinates of $\mathcal{C}_1$. With linear codes we are then permitted to conjugate the resulting vectors; with additive codes we can in fact conjugate more generally as follows. Two additive codes $\mathcal{C}_1$ and $\mathcal{C}_2$ are *equivalent* provided there is a map sending the codewords of $\mathcal{C}_1$ onto the codewords of $\mathcal{C}_2$, where the map consists of a permutation of coordinates followed by a scaling of coordinates by elements of $\mathbb{F}_4$ followed by conjugation of some of the coordinates. In terms of generator matrices, if G_1 is a generator matrix of $\mathcal{C}_1$, then there is a generator matrix of $\mathcal{C}_2$ obtained from G_1 by permuting columns, scaling columns, and then conjugating some of the resulting columns. In the natural way we now define the *automorphism group* of the additive code $\mathcal{C}$, denoted $\Gamma\text{Aut}(\mathcal{C})$, to consist of all maps that permute coordinates, scale coordinates, and conjugate coordinates that send codewords of $\mathcal{C}$ to codewords of $\mathcal{C}$. By Exercise 557 permuting coordinates, scaling coordinates, and conjugating some coordinates of a trace self-orthogonal (or trace self-dual) code does not change trace self-orthogonality (or trace self-duality). However, Exercise 558 shows that if $\mathcal{C}_1$ is linear over $\mathbb{F}_4$ and if $\mathcal{C}_2$ is obtained from $\mathcal{C}_1$ by conjugating only some of the coordinates, then $\mathcal{C}_2$ may not be linear!

Exercise 557 Verify that if $\mathbf{x}'$ and $\mathbf{y}'$ are obtained from the vectors $\mathbf{x}$ and $\mathbf{y}$ in $\mathbb{F}_4^n$ by permuting coordinates, scaling coordinates, and conjugating some of the coordinates, then $\langle \mathbf{x}', \mathbf{y}' \rangle_T = \langle \mathbf{x}, \mathbf{y} \rangle_T$. ◆

Exercise 558 Find an example of a code $\mathcal{C}$ that is linear over $\mathbb{F}_4$ such that when the first coordinate only is conjugated, the resulting code is no longer linear over $\mathbb{F}_4$. ♦

We now focus on trace self-dual additive codes. For linear codes over $\mathbb{F}_4$, there is no distinction between Hermitian self-dual and trace self-dual codes.

Theorem 9.10.3 *Let $\mathcal{C}$ be a linear code over $\mathbb{F}_4$. Then $\mathcal{C}$ is Hermitian self-dual if and only if $\mathcal{C}$ is trace self-dual.*

Proof: First assume that $\mathcal{C}$ is an $[n, n/2]$ Hermitian self-dual code. Then $\langle \mathbf{x}, \mathbf{y} \rangle = 0$ for all $\mathbf{x}$ and $\mathbf{y}$ in $\mathcal{C}$. Thus $\langle \mathbf{x}, \mathbf{y} \rangle_T = \mathrm{Tr}_2(\langle \mathbf{x}, \mathbf{y} \rangle) = 0$ implying that $\mathcal{C} \subseteq \mathcal{C}^{\perp_T}$. As an additive code, $\mathcal{C}$ is an $(n, 2^n)$ code and hence, by Exercise 556, $\mathcal{C}^{\perp_T}$ is also an $(n, 2^n)$ code. Thus $\mathcal{C} = \mathcal{C}^{\perp_T}$ and $\mathcal{C}$ is trace self-dual.

Now assume that $\mathcal{C}$ is an $(n, 2^n)$ trace self-dual code. Because $\mathcal{C}$ is a linear code over $\mathbb{F}_4$, if $\mathbf{x} \in \mathcal{C}$, then $\omega \mathbf{x} \in \mathcal{C}$ and so $\langle \mathbf{x}, \omega \mathbf{x} \rangle_T = 0$. By Lemma 9.10.2(d), $\mathrm{wt}(\mathbf{x})$ is even for all $\mathbf{x} \in \mathcal{C}$. By Theorem 1.4.10, $\mathcal{C} \subseteq \mathcal{C}^{\perp_H}$. Because $\mathcal{C}$ has dimension $n/2$ as a linear code over $\mathbb{F}_4$, so does $\mathcal{C}^{\perp_H}$ implying that $\mathcal{C}$ is Hermitian self-dual. □

Unlike Hermitian self-dual codes that contain only even weight codewords, trace self-dual additive codes may contain odd weight codewords; such codes cannot be linear over $\mathbb{F}_4$ by Theorem 9.10.3. This leads to the following definitions. An additive code $\mathcal{C}$ is *Type II* if $\mathcal{C}$ is trace self-dual and all codewords have even weight. Exercise 383 applies to additive codes over $\mathbb{F}_4$ as the MacWilliams equation (M_3) holds. In particular, if $\mathcal{C}$ is Type II of length n, then $\mathcal{C}^{\perp_T} = \mathcal{C}$ contains a codeword of weight n implying that n is even. By Corollary 9.2.2, $\mathbb{F}_4$-linear Hermitian self-dual codes exist if and only if n is even. Hence Type II additive codes exist if and only if n is even. If $\mathcal{C}$ is trace self-dual but some codeword has odd weight (in which case the code cannot be $\mathbb{F}_4$-linear), the code is *Type I*. The $n \times n$ identity matrix generates an $(n, 2^n)$ Type I additive code implying Type I codes exist for all n.

There is an analog to Gleason's Theorem for Type II additive codes, which is identical to Gleason's Theorem for Hermitian self-dual linear codes; see [291, Section 7.7].

Theorem 9.10.4 *If $\mathcal{C}$ is a Type II additive code, then*

$$W_{\mathcal{C}}(x, y) = \sum_{i=0}^{\lfloor \frac{n}{6} \rfloor} a_i g_6(x, y)^{\frac{n}{2} - 3i} g_7(x, y)^i,$$

where $g_6(x, y) = y^2 + 3x^2$ and $g_7(x, y) = y^6 + 45x^4 y^2 + 18x^6$. All a_is are rational and $\sum_i a_i = 1$.

There is also a bound on the minimum distance of a trace self-dual additive code [291, Theorem 33]. If d_I and d_{II} are the minimum distances of Type I and Type II additive codes,

Table 9.13 *Number of inequivalent additive trace self-dual codes*

Length	1	2	3	4	5	6	7	8
Number Type I	1	1	3	4	11	20	59	?
Number Type II	0	1	0	2	0	6	0	21

respectively, of length $n > 1$, then

$$
d_I \leq \begin{cases} 2\left\lfloor\dfrac{n}{6}\right\rfloor + 1 & \text{if } n \equiv 0 \,(\text{mod } 6), \\[2mm] 2\left\lfloor\dfrac{n}{6}\right\rfloor + 3 & \text{if } n \equiv 5 \,(\text{mod } 6), \\[2mm] 2\left\lfloor\dfrac{n}{6}\right\rfloor + 2 & \text{otherwise,} \end{cases}
$$

$$
d_{II} \leq 2\left\lfloor\frac{n}{6}\right\rfloor + 2.
$$

A code that meets the appropriate bound is called *extremal*. Extremal Type II codes have a unique weight enumerator. This property is not true for extremal Type I codes.

As with our other self-dual codes, there is a mass formula [134] for counting the number of trace self-dual additive codes of a given length.

Theorem 9.10.5 *We have the following mass formulas*:
(i) *For trace self-dual additive codes of length n,*

$$
\sum_j \frac{6^n n!}{|\Gamma\mathrm{Aut}(\mathcal{C}_j)|} = \prod_{i=1}^{n}(2^i + 1).
$$

(ii) *For Type II additive codes of even length n,*

$$
\sum_j \frac{6^n n!}{|\Gamma\mathrm{Aut}(\mathcal{C}_j)|} = \prod_{i=0}^{n-1}(2^i + 1).
$$

In both cases, the summation is over all j where $\{\mathcal{C}_j\}$ is a complete set of representatives of inequivalent codes of the given type.

With the assistance of this mass formula, Höhn [134] has classified all Type I additive codes of lengths $n \leq 7$ and all Type II additive codes of lengths $n \leq 8$. Table 9.13 gives the number of inequivalent codes for these lengths.

The Balance Principle and the rest of Theorem 9.4.1 also hold for trace self-dual additive codes with minor modifications. Exercises 560, 561, and 562 use this theorem to find all inequivalent trace self-dual additive codes of lengths 2 and 3 and all inequivalent extremal ones of length 4.

Theorem 9.10.6 ([93]) *Let $\mathcal{C}$ be an $(n, 2^n)$ trace self-dual additive code. Choose a set of coordinate positions $\mathcal{P}_{n_1}$ of size n_1 and let $\mathcal{P}_{n_2}$ be the complementary set of coordinate*

Table 9.14 *Extremal Type I and II additive codes of length $2 \le n \le 16$*

n	d_I	num_I	d_{II}	num_{II}	n	d_I	num_I	d_{II}	num_{II}
2	1^O	1	2^E	1	10	4^E	≥ 51	4^E	≥ 5
3	2^E	1	–	–	11	5^E	1	–	–
4	2^E	1	2^E	2	12	5^E	≥ 7	6^E	1
5	3^E	1	–	–	13	5°	≥ 5	–	–
6	3^E	1	4^E	1	14	6^E	?	6^E	≥ 490
7	3°	3	–	–	15	6^E	≥ 2	–	–
8	4^E	2	4^E	3	16	6^E	≥ 3	6^E	≥ 4
9	4^E	8	–	–					

positions of size $n_2 = n - n_1$. Let $\mathcal{C}_i$ be the $(n, 2^{k_i})$ subcode of $\mathcal{C}$ all of whose codewords have support in $\mathcal{P}_{n_i}$. The following hold:

(i) *(Balance Principle)*

$$k_1 - n_1 = k_2 - n_2.$$

(ii) *If we reorder coordinates so that $\mathcal{P}_{n_1}$ is the left-most n_1 coordinates and $\mathcal{P}_{n_2}$ is the right-most n_2 coordinates, then $\mathcal{C}$ has a generator matrix of the form*

$$G = \begin{bmatrix} A & O \\ O & B \\ D & E \end{bmatrix},$$

where $[A \ \ O]$ is a generator matrix of $\mathcal{C}_1$ and $[O \ \ B]$ is a generator matrix of $\mathcal{C}_2$, O being the appropriate size zero matrix. We also have the following:

(a) *D and E each have $n - k_1 - k_2$ independent rows over $\mathbb{F}_2$.*

(b) *Let $\mathcal{A}$ be the code of length n_1 generated by A, $\mathcal{A}_D$ the code of length n_1 generated by the rows of A and D, $\mathcal{B}$ the code of length n_2 generated by B, and $\mathcal{B}_E$ the code of length n_2 generated by the rows of B and E. Then $\mathcal{A}^{\perp_T} = \mathcal{A}_D$ and $\mathcal{B}^{\perp_T} = \mathcal{B}_E$.*

Exercise 559 Prove Theorem 9.10.6.   ◆

Table 9.14, from [93], summarizes what is known about the number of inequivalent extremal Type I and Type II additive codes through length 16. In the table, codes have length "n"; "d_I" is the largest minimum weight for which a Type I code is known to exist, while "d_{II}" is the largest minimum weight for which a Type II code is known to exist. The superscript "E" indicates that the code is extremal; the superscript "O" indicates the code is not extremal, but no code of the given type can exist with a larger minimum weight, and so the code is optimal. The number of inequivalent Type I and II codes of the given minimum weight is listed under "num_I" and "num_{II}", respectively; when the number in the column is exact (without $\ge$), the classification of those codes is complete. There exist extremal

Type II additive codes for all even lengths from 2 to 22 inclusive. In this range, except for length 12, extremal Hermitian self-dual linear codes provide examples of extremal Type II additive codes; see Table 9.3 and Example 9.10.8. An extremal Type II additive code of length 24 has minimum weight 10; no such $\mathbb{F}_4$-linear Hermitian self-dual code exists [183]. The existence of a $(24, 2^{24}, 10)$ Type II code is still an open question, however.

Research Problem 9.10.7 *Either construct or establish the nonexistence of a $(24, 2^{24}, 10)$ Type II additive code.*

Example 9.10.8 By Table 9.3 there is no $[12, 6, 6]$ Hermitian self-dual code. However, Table 9.14 indicates that there is a unique $(12, 2^{12}, 6)$ Type II additive code. This code $\mathcal{C}$ is called the *dodecacode* and has generator matrix

$$
\begin{bmatrix}
0 & 0 & 0 & 0 & 0 & 0 & 1 & 1 & 1 & 1 & 1 & 1 \\
0 & 0 & 0 & 0 & 0 & 0 & \omega & \omega & \omega & \omega & \omega & \omega \\
1 & 1 & 1 & 1 & 1 & 1 & 0 & 0 & 0 & 0 & 0 & 0 \\
\omega & \omega & \omega & \omega & \omega & \omega & 0 & 0 & 0 & 0 & 0 & 0 \\
0 & 0 & 0 & 1 & \omega & \overline{\omega} & 0 & 0 & 0 & 1 & \omega & \overline{\omega} \\
0 & 0 & 0 & \omega & \overline{\omega} & 1 & 0 & 0 & 0 & \omega & \overline{\omega} & 1 \\
1 & \overline{\omega} & \omega & 0 & 0 & 0 & 1 & \overline{\omega} & \omega & 0 & 0 & 0 \\
\omega & 1 & \overline{\omega} & 0 & 0 & 0 & \omega & 1 & \overline{\omega} & 0 & 0 & 0 \\
0 & 0 & 0 & 1 & \overline{\omega} & \omega & \omega & \overline{\omega} & 1 & 0 & 0 & 0 \\
0 & 0 & 0 & \omega & 1 & \overline{\omega} & 1 & \omega & \overline{\omega} & 0 & 0 & 0 \\
1 & \omega & \overline{\omega} & 0 & 0 & 0 & 0 & 0 & 0 & \overline{\omega} & \omega & 1 \\
\overline{\omega} & 1 & \omega & 0 & 0 & 0 & 0 & 0 & 0 & 1 & \overline{\omega} & \omega
\end{bmatrix}.
$$

The dodecacode has weight enumerator

$$W_{\mathcal{C}}(x, y) = y^{12} + 396x^6y^6 + 1485x^8y^4 + 1980x^{10}y^2 + 234x^{12}.$$

The existence of the dodecacode, despite the nonexistence of a $[12, 6, 6]$ Hermitian self-dual code, leads one to hope that the code considered in Research Problem 9.10.7 exists. ∎

Exercise 560 Use the generator matrix given in Theorem 9.10.6 to construct generator matrices for the inequivalent trace self-dual additive codes of length $n = 2$. (Hint: Let n_1 be the minimum weight and place a minimum weight codeword in the first row of the generator matrix.) In the process of this construction verify that there are only two inequivalent codes, consistent with Table 9.13. Also give the weight distribution of each and identify which is Type I and which is Type II. Which are extremal?   ◆

Exercise 561 Repeat Exercise 560 with $n = 3$. This time there will be three inequivalent codes, verifying the entry in Table 9.13.   ◆

Exercise 562 Use the generator matrix given in Theorem 9.10.6 to construct generator matrices for all the inequivalent extremal trace self-dual additive codes of length $n = 4$. (Hint: Let $n_1 = 2$ and place a minimum weight codeword in the first row of the generator matrix.) In the process of this construction verify that there are only three inequivalent extremal codes, consistent with Table 9.14. Also give the weight distribution of each and identify which is Type I and which is Type II.   ♦

9.11 Proof of the Gleason–Pierce–Ward Theorem

In this section we prove the Gleason–Pierce–Ward Theorem stated at the beginning of the chapter. Our proof comes from [347].

We begin with some notation. Let C be a linear code of length n over $\mathbb{F}_q$, where $q = p^m$ with p a prime. As usual $\mathbb{F}_q^*$ denotes the nonzero elements of $\mathbb{F}_q$. Define the map $\tau : \mathbb{F}_q \to \mathbb{F}_q$ by $\tau(0) = 0$ and $\tau(\alpha) = \alpha^{-1}$ for all $\alpha \in \mathbb{F}_q^*$. This map will be crucial to the proof of the Gleason–Pierce–Ward Theorem; its connection with the theorem is hinted at in the following lemma.

Lemma 9.11.1 *We have the following*:
(i) $\tau(\alpha\beta) = \tau(\alpha)\tau(\beta)$ *for all q and all $\alpha, \beta \in \mathbb{F}_q$.*
(ii) *When $q = 2$ or 3, τ is the identity map, and when $q = 4$, τ is the map $\tau(\alpha) = \alpha^2$.*
(iii) *τ is also additive (that is, τ is a field automorphism) if and only if $q = 2, 3,$ or 4.*

Proof: Part (i) was proved in Exercise 176(a), while part (ii) is a straightforward exercise.

One direction of part (iii) is proved in Exercise 176(c). We give another proof different from the one suggested in the hint in that exercise. Assume that τ is a field automorphism of $\mathbb{F}_q$. Then $\tau(\alpha) = \alpha^{p^r}$ for some r with $0 \leq r < m$ and all $\alpha \in \mathbb{F}_q$ by Theorem 3.6.1. So if $\alpha \in \mathbb{F}_q^*$, then $\alpha^{p^r} = \alpha^{-1}$ and hence $\alpha^{p^r + 1} = 1$. This holds when α is primitive, yielding $(q - 1) \mid (p^r + 1)$ implying $p^m - 1 = p^r + 1$ if $q > 2$ by the restriction $0 \leq r < m$. The only solutions of $p^m - 1 = p^r + 1$ with $q > 2$ are $p = 3$ with $m = 1$ and $r = 0$ and $p = 2$ with $m = 2$ and $r = 1$. So $q = 2, 3,$ or 4. For the converse we assume $q = 2, 3,$ or 4, in which case τ is given in (ii). The identity map is always an automorphism and the map $\tau(\alpha) = \alpha^2$ in the case $q = 4$ is the Frobenius automorphism.   □

Exercise 563   Verify part (ii) of Lemma 9.11.1.   ♦

We now introduce notation that allows us to compare components of two different vectors. Let $\mathbf{u} = u_1 \cdots u_n$ and $\mathbf{v} = v_1 \cdots v_n$ be vectors in $\mathbb{F}_q^n$. For $\alpha \in \mathbb{F}_q$ define

$$\alpha_{\mathbf{u},\mathbf{v}} = |\{i \mid u_i \neq 0 \text{ and } v_i = \alpha u_i\}|, \quad \text{and}$$

$$\infty_{\mathbf{u},\mathbf{v}} = |\{i \mid u_i = 0 \text{ and } v_i \neq 0\}|.$$

We have the following elementary properties of $\alpha_{\mathbf{u},\mathbf{v}}$ and $\infty_{\mathbf{u},\mathbf{v}}$.

Lemma 9.11.2 *Let C be a divisible linear code over $\mathbb{F}_q$ with divisor Δ. Let $\mathbf{u}$ and $\mathbf{v}$ be vectors in $\mathbb{F}_q^n$ and α an element of $\mathbb{F}_q$. Then*:
(i) $\mathrm{wt}(\mathbf{u}) = \sum_{\alpha \in \mathbb{F}_q} \alpha_{\mathbf{u},\mathbf{v}}$,
(ii) $\mathrm{wt}(\mathbf{v} - \alpha\mathbf{u}) = \infty_{\mathbf{u},\mathbf{v}} + \mathrm{wt}(\mathbf{u}) - \alpha_{\mathbf{u},\mathbf{v}}$,
(iii) $\alpha_{\mathbf{u},\mathbf{v}} \equiv \infty_{\mathbf{u},\mathbf{v}} \pmod{\Delta}$ *if $\mathbf{u}$ and $\mathbf{v}$ are in C, and*
(iv) $q\infty_{\mathbf{u},\mathbf{v}} \equiv 0 \pmod{\Delta}$ *if $\mathbf{u}$ and $\mathbf{v}$ are in C.*

Proof: Part (i) is clear because wt(**u**) is the count of the number of nonzero coordinates of **u**, and for all such coordinates i, $v_i = \alpha u_i$ for $\alpha = v_i u_i^{-1} \in \mathbb{F}_q$. To prove (ii) we note that the coordinates are divided into four disjoint sets for each α:

(a) the coordinates i where $u_i = v_i = 0$,

(b) the coordinates i where $u_i = 0$ but $v_i \neq 0$,

(c) the coordinates i where $u_i \neq 0$ and $v_i = \alpha u_i$, and

(d) the coordinates i where $u_i \neq 0$ and $v_i \neq \alpha u_i$.

Clearly, wt(**v** − α**u**) is the sum of the number of coordinates in (b) and (d). By definition, there are $\infty_{\mathbf{u},\mathbf{v}}$ coordinates in (b). As there are wt(**u**) total coordinates in (c) and (d), and (c) has $\alpha_{\mathbf{u},\mathbf{v}}$ coordinates, then (d) has wt(**u**) − $\alpha_{\mathbf{u},\mathbf{v}}$ coordinates, proving (ii).

Part (iii) follows from part (ii) as wt(**v** − α**u**) $\equiv$ wt(**u**) $\equiv 0 \,(\mathrm{mod}\ \Delta)$ because $\mathcal{C}$ is linear and divisible by Δ. Finally, combining (i) and (iii) gives

$$0 \equiv \mathrm{wt}(\mathbf{u}) \equiv \sum_{\alpha \in \mathbb{F}_q} \alpha_{\mathbf{u},\mathbf{v}} \equiv \sum_{\alpha \in \mathbb{F}_q} \infty_{\mathbf{u},\mathbf{v}} \equiv q \infty_{\mathbf{u},\mathbf{v}} \,(\mathrm{mod}\ \Delta),$$

verifying (iv). □

We next examine the situation that will lead to part (v) of the Gleason–Pierce–Ward Theorem. To state this result, we need the concept of a coordinate functional of a code $\mathcal{C}$ over $\mathbb{F}_q$ of length n. For $1 \leq i \leq n$, define the *coordinate functional* $f_i : \mathcal{C} \to \mathbb{F}_q$ by $f_i(\mathbf{c}) = c_i$ where $\mathbf{c} = c_1 \cdots c_n$. Let $\mathcal{N}$ be the set of nonzero coordinate functionals of $\mathcal{C}$. We define the binary relation $\sim$ on $\mathcal{N}$ by $f_i \sim f_j$ provided there exists an element $\alpha \in \mathbb{F}_q^*$ such that $f_i(\mathbf{c}) = \alpha f_j(\mathbf{c})$ for all $\mathbf{c} \in \mathcal{C}$. We make two simple observations, which you are asked to prove in Exercise 564. First, f_i is a linear function and hence its values are completely determined by its values on the codewords in a generator matrix. Second, $\sim$ is an equivalence relation on $\mathcal{N}$. In the next lemma, we will see that $\mathcal{C}$ is equivalent to a replicated code whenever $\mathcal{C}$ has a divisor relatively prime to the order of the field. We say that a code $\mathcal{C}$ is a Δ-*fold replicated code* provided the set of coordinates of $\mathcal{C}$ is the disjoint union of subsets of coordinates of size Δ such that if $\{i_j \mid 1 \leq j \leq \Delta\}$ is any of these subsets and $\mathbf{c} = c_1 \cdots c_n \in \mathcal{C}$, then $c_{i_j} = c_{i_1}$ for $1 \leq j \leq \Delta$. In terms of coordinate functionals, these functionals are equal on the subsets of coordinates. The code with generator matrix $[I_{n/2} \mid I_{n/2}]$ in (v) of the Gleason–Pierce–Ward Theorem is a two-fold replicated code.

Exercise 564 Show the following:

(a) The coordinate functional f_i is a linear function from $\mathcal{C}$ to $\mathbb{F}_q$, and its values are completely determined by its values on the codewords in a generator matrix of $\mathcal{C}$.

(b) The relation $\sim$ on the set $\mathcal{N}$ of nonzero coordinate functionals is an equivalence relation on $\mathcal{N}$.   ◆

Lemma 9.11.3 *Let $\mathcal{C}$ be an $[n, k]$ code over $\mathbb{F}_q$ with divisor Δ relatively prime to p. Then the following hold:*

(i) *Each equivalence class of nonzero functionals has size a multiple of Δ.*

(ii) *$\mathcal{C}$ is monomially equivalent to a Δ-fold replicated code possibly with some coordinates that are always 0.*

Proof: The proof of (i) is by induction on n. For the initial case, we assume that all codewords have weight n, in which case $k = 1$ and C is an $[n, 1, n]$ code by the Singleton Bound. As a generator matrix for C is one codeword $\mathbf{u}$ with each entry nonzero, the n coordinate functionals are clearly equivalent. As $0 \equiv \text{wt}(\mathbf{u}) \equiv n \pmod{\Delta}$, the result holds in this case.

Now assume that $\mathbf{u}$ is a nonzero codeword of C with weight $w < n$. By assumption $\Delta \mid w$. By permuting the coordinates (which permutes the coordinate functionals but does not change the sizes of the equivalence classes), we may assume that $\mathbf{u} = u_1 \cdots u_w 0 \cdots 0$, where each u_i is nonzero. If $\mathbf{c} = c_1 \cdots c_n$ is a codeword in C, write $\mathbf{c} = (\mathbf{c}'|\mathbf{c}'')$, where $\mathbf{c}' = c_1 \cdots c_w$ and $\mathbf{c}'' = c_{w+1} \cdots c_n$. Define $C' = \{\mathbf{c}' \mid \mathbf{c} = (\mathbf{c}'|\mathbf{c}'') \in C\}$ and $C'' = \{\mathbf{c}'' \mid \mathbf{c} = (\mathbf{c}'|\mathbf{c}'') \in C\}$. So C' and C'' are both codes of lengths less than n. Notice that no coordinate functional f_i of C with $1 \le i \le w$ is equivalent to any coordinate functional f_j with $w + 1 \le j \le n$. Also the coordinate functionals for C are determined by the coordinate functionals for C' and C''. Therefore if we show that both C' and C'' are divisible by Δ, then by induction their coordinate functionals split into classes with sizes that are multiples of Δ, and thus the coordinate functionals of C split into classes of exactly the same sizes. To show both C' and C'' are divisible by Δ, let $\mathbf{c} = (\mathbf{c}'|\mathbf{c}'') \in C$. We need to show that $\text{wt}(\mathbf{c}')$ and $\text{wt}(\mathbf{c}'')$ are divisible by Δ. Since $\text{wt}(\mathbf{c})$ is divisible by Δ, it suffices to show that $\text{wt}(\mathbf{c}'')$ is divisible by Δ. Clearly, $\text{wt}(\mathbf{c}'') = \infty_{\mathbf{u},\mathbf{c}}$; as $p \nmid \Delta$, by Lemma 9.11.2(iv), $\infty_{\mathbf{u},\mathbf{c}} \equiv 0 \pmod{\Delta}$. Part (i) now follows.

Part (ii) is obvious from part (i) as you can rescale the coordinate functionals in an equivalence class so that the new functionals are identical. $\qquad\square$

This lemma shows what happens when we have a divisor relatively prime to p. The next two lemmas show what happens when we have a divisor that is a power of p. We extend the map τ to codewords of a code by acting with τ componentwise. Note that τ is injective. The first lemma gives its image.

Lemma 9.11.4 *Let C have divisor Δ, where Δ is a power of p. If $\Delta > 2$ or $q > 2$, then $\tau(C) \subseteq C^\perp$.*

Proof: It suffices to show that if $\mathbf{u}, \mathbf{v} \in C$, then $\tau(\mathbf{u}) \cdot \mathbf{v} = 0$. As

$$\alpha_{\mathbf{u},\mathbf{v}} = |\{i \mid u_i \ne 0 \text{ and } v_i = \alpha u_i\}| = |\{i \mid u_i \ne 0 \text{ and } \tau(u_i)v_i = \alpha\}|,$$

we have

$$\tau(\mathbf{u}) \cdot \mathbf{v} = \sum_{\alpha \in \mathbb{F}_q} \alpha(\alpha_{\mathbf{u},\mathbf{v}}),$$

noting that no component i contributes to the inner product when $u_i = 0$ or $v_i = 0$. By Lemma 9.11.2(iii), $\alpha_{\mathbf{u},\mathbf{v}} \equiv \infty_{\mathbf{u},\mathbf{v}} \pmod{\Delta}$, and so $\alpha_{\mathbf{u},\mathbf{v}} \equiv \infty_{\mathbf{u},\mathbf{v}} \pmod{p}$ as $p \mid \Delta$. Thus

$$\tau(\mathbf{u}) \cdot \mathbf{v} = \infty_{\mathbf{u},\mathbf{v}} \sum_{\alpha \in \mathbb{F}_q} \alpha.$$

If $q > 2$, $\sum_{\alpha \in \mathbb{F}_q} \alpha = 0$. If $q = 2$, then $4 \mid \Delta$ implying $\infty_{\mathbf{u},\mathbf{v}} \equiv 0 \pmod 2$ by Lemma 9.11.2(iv). Thus in either case $\tau(\mathbf{u}) \cdot \mathbf{v} = 0$. $\qquad\square$

Note that if the hypothesis of Lemma 9.11.4 holds and C has dimension $n/2$, then $\tau(C) = C^{\perp}$. We examine the possible values of Δ that could occur when $q \in \{2, 3, 4\}$.

Lemma 9.11.5 *Assume $\tau(C) = C^{\perp}$ and $q \in \{2, 3, 4\}$. Then:*
(i) *p is a divisor of C, and*
(ii) *the highest power of p dividing C is p if $q \in \{3, 4\}$ and is 4 if $q = 2$.*

Proof: As τ is the identity when $q \in \{2, 3\}$, C is self-dual. If $q = 4$, C is Hermitian self-dual by Lemma 9.11.1(ii). Thus (i) follows from Theorem 1.4.5.

To prove (ii), we permute the coordinates so that C has a generator matrix $G_1 = [I_{n/2} \ A]$ in standard form; note that this does not affect the hypothesis of the lemma. As $\tau(C) = C^{\perp}$, $[I_{n/2} \ \tau(A)]$ is a generator matrix for $C^{\perp}$; hence $G_2 = [-\tau(A)^{\mathrm{T}} \ I_{n/2}]$ is a generator matrix of C by Theorem 1.2.1. Let $\mathbf{u} = 10 \cdots 0a_1 \cdots a_{n/2}$ be the first row of G_1. Choose i so that $a_i \neq 0$; such an entry exists as all weights are divisible by at least 2. Let $\mathbf{v} = b_1 \cdots b_{n/2} 0 \cdots 1 \cdots 0$ be a row of G_2 where entry 1 is in coordinate i. Both $\mathbf{u}$ and $\mathbf{v}$ are in C, and so $0 = \mathbf{u} \cdot \tau(\mathbf{v}) = \tau(b_1) + a_i$ implying $\tau(b_1) = -a_i$. When $q = 2$, wt$(\mathbf{u} + \mathbf{v}) = $ wt$(\mathbf{u}) + $ wt$(\mathbf{v}) - 4$ as $a_i = b_1 = 1$ showing that the highest power of 2 dividing C is 4. When $q = 3$, wt$(\mathbf{u} + \mathbf{v}) = $ wt$(\mathbf{u}) + $ wt$(\mathbf{v}) - 3$ as $\{a_i, b_1\} = \{-1, 1\}$ implying that the highest power of 3 dividing C is 3. When $q = 4$, choose $\alpha \in \mathbb{F}_4$ with $\alpha \neq 0$ and $\alpha \neq a_i$. Then wt$(\mathbf{u} + \alpha\mathbf{v}) = $ wt$(\mathbf{u}) + $ wt$(\mathbf{v}) - 2$ proving that the highest power of 2 dividing C is 2. $\square$

We need one final lemma.

Lemma 9.11.6 *Let C be an $[n, k]$ code over $\mathbb{F}_q$ divisible by $\Delta > 1$ where either $\Delta > 2$ or $q > 2$. Then:*
(i) *$k \leq n/2$, and*
(ii) *if $k = n/2$ and f is some integer dividing Δ with $f > 1$ and $p \nmid f$, then $f = 2$ and C is equivalent to the code over $\mathbb{F}_q$ with generator matrix $[I_{n/2} \ I_{n/2}]$.*

Proof: Suppose $f \mid \Delta$ with $f > 1$ and $p \nmid f$. Then by Lemma 9.11.3(ii), C is monomially equivalent to an f-fold replicated code possibly with some coordinates that are always 0. As $f > 1$, $k \leq n/f \leq n/2$, which gives (i) in this case. This also shows that if $k = n/2$, then $f = 2$ and there can be no coordinates that are always 0 producing (ii). So assume that Δ is a power of p; part (i) follows in this case from Lemma 9.11.4 as $\tau(C) \subseteq C^{\perp}$. $\square$

We now complete the proof of the Gleason–Pierce–Ward Theorem. If Δ has a factor greater than 1 and relatively prime to p, we have (v) by Lemma 9.11.6. So we may assume that Δ is a power of p. We may also assume that $\Delta > 2$ or $q > 2$, as otherwise we have (i). We need to show that we have one (or more) of (ii), (iii), (iv), or (v). As C has dimension $n/2$, by Lemma 9.11.4 $\tau(C) = C^{\perp}$. Permuting columns as previously, we may assume that $G_1 = [I_{n/2} \ A]$ is a generator matrix for C and $G_2 = [I_{n/2} \ \tau(A)]$ is a generator matrix for $C^{\perp}$. Each row of A has at least one nonzero entry as C is divisible by at least 2. If every row and column of A has exactly one nonzero entry, then C satisfies (v). Hence, permuting columns again, we may assume the first two rows of G_1 are

$$\mathbf{u} = (1, 0, 0, \ldots, 0, a, \ldots) \quad \text{and}$$
$$\mathbf{v} = (0, 1, 0, \ldots, 0, b, \ldots),$$

where a and b are nonzero and located in coordinate $(n/2) + 1$. Let α and β be arbitrary elements of $\mathbb{F}_q$. Then:

$$\alpha \mathbf{u} + \beta \mathbf{v} = (\alpha, \beta, 0, \ldots, 0, \alpha a + \beta b, \ldots) \quad \text{and}$$
$$\tau(\alpha \mathbf{u} + \beta \mathbf{v}) = (\tau(\alpha), \tau(\beta), 0, \ldots, 0, \tau(\alpha a + \beta b), \ldots).$$

As $\tau(\alpha \mathbf{u} + \beta \mathbf{v}) \in \tau(\mathcal{C}) = \mathcal{C}^{\perp}$, by examining the generator matrix G_2, the only possibility is $\tau(\alpha \mathbf{u} + \beta \mathbf{v}) = \tau(\alpha)\tau(\mathbf{u}) + \tau(\beta)\tau(\mathbf{v})$ implying that $\tau(\alpha a + \beta b) = \tau(\alpha)\tau(a) + \tau(\beta)\tau(b) = \tau(\alpha a) + \tau(\beta b)$ by Lemma 9.11.1(i). Therefore as a and b are nonzero, τ is additive. By Lemma 9.11.1(iii), q is 2, 3, or 4. Lemma 9.11.5 completes the proof.

9.12 Proofs of some counting formulas

In this section we will present the proof of the counting formulas given in Section 9.5 for the number of doubly-even codes of length n. The basis for the proof is found in the orthogonal decomposition of even binary codes given in Section 7.8. Recall from Theorem 7.8.5 that an even binary code $\mathcal{C}$ can be decomposed as an orthogonal sum of its hull of dimension r and either m H-planes or $m - 1$ H-planes and one A-plane; $\mathcal{C}$ has dimension $2m + r$. This decomposition partially determines the form (H, O, or A) of $\mathcal{C}$: if one plane is an A-plane, the code has form A; if there are m H-planes, $\mathcal{C}$ has form H if the hull of $\mathcal{C}$ is doubly-even and form O if the hull of $\mathcal{C}$ is singly-even.

Let $\mathcal{E}_n$ be the $[n, n - 1]$ binary code of all even weight vectors where n is even. We can determine the form of $\mathcal{E}_n$ merely by knowing the value of n. In order to show this, we need the following lemma.

Lemma 9.12.1 *If $\mathcal{C}$ is a doubly-even binary code of dimension m that does not contain the all-one vector $\mathbf{1}$, then $\mathcal{C}$ is contained in an orthogonal sum of m H-planes. Furthermore $\mathbf{1}$ is orthogonal to and disjoint from this orthogonal sum.*

Proof: Let $\{\mathbf{x}_1, \mathbf{x}_2, \ldots, \mathbf{x}_m\}$ be a basis of $\mathcal{C}$. To prove the result we will construct $\{\mathbf{y}_1, \mathbf{y}_2, \ldots, \mathbf{y}_m\}$ so that $\mathcal{H}_i = \text{span}\{\mathbf{x}_i, \mathbf{y}_i\}$ is an H-plane for $1 \le i \le m$ and $\text{span}\{\mathbf{x}_1, \ldots, \mathbf{x}_m, \mathbf{y}_1, \ldots, \mathbf{y}_m\}$ is the orthogonal sum $\mathcal{H}_1 \perp \mathcal{H}_2 \perp \cdots \perp \mathcal{H}_m$.

To construct $\mathbf{y}_1$, first note that $\mathbf{x}_1 \notin \text{span}\{\mathbf{x}_2, \ldots, \mathbf{x}_m, \mathbf{1}\}$ since $\mathbf{1} \notin \mathcal{C}$. Thus $\text{span}\{\mathbf{x}_2, \ldots, \mathbf{x}_m, \mathbf{1}\}^{\perp} \setminus \text{span}\{\mathbf{x}_1\}^{\perp}$ is nonempty. Let $\mathbf{y}_1$ be in this set. As $\mathbf{y}_1 \perp \mathbf{1}$, $\mathbf{y}_1$ has even weight. As $\mathbf{y}_1 \not\perp \mathbf{x}_1$ and $\mathbf{x}_1$ is doubly-even, $\mathcal{H}_1 = \text{span}\{\mathbf{x}_1, \mathbf{y}_1\}$ is an H-plane by Exercise 416. Clearly, $\mathbf{x}_1 \notin \text{span}\{\mathbf{x}_2, \ldots, \mathbf{x}_m\}$. Neither $\mathbf{y}_1$ nor $\mathbf{x}_1 + \mathbf{y}_1$ is in $\text{span}\{\mathbf{x}_2, \ldots, \mathbf{x}_m\}$ as otherwise $\mathbf{y}_1 \in \mathcal{C}$, which is a contradiction. Therefore, $\mathcal{H}_1 \cap \text{span}\{\mathbf{x}_2, \ldots, \mathbf{x}_m\} = \{\mathbf{0}\}$. Clearly, $\mathbf{x}_1$ and $\mathbf{y}_1$ are orthogonal to $\text{span}\{\mathbf{x}_2, \ldots, \mathbf{x}_m\}$. Thus $\text{span}\{\mathbf{x}_1, \mathbf{x}_2, \ldots, \mathbf{x}_m, \mathbf{y}_1\} = \mathcal{H}_1 \perp \text{span}\{\mathbf{x}_2, \ldots, \mathbf{x}_m\}$.

We now want to choose $\mathbf{y}_2 \in \text{span}\{\mathbf{x}_1, \mathbf{x}_3, \ldots, \mathbf{x}_m, \mathbf{y}_1, \mathbf{1}\}^{\perp}$ with $\mathbf{y}_2 \notin \text{span}\{\mathbf{x}_2\}^{\perp}$. If no such $\mathbf{y}_2$ exists, then $\mathbf{x}_2 \in \text{span}\{\mathbf{x}_1, \mathbf{x}_3, \ldots, \mathbf{x}_m, \mathbf{y}_1, \mathbf{1}\}$, implying $\mathbf{x}_2 = \sum_{i \ne 2} \alpha_i \mathbf{x}_i + \beta \mathbf{y}_1 + \gamma \mathbf{1}$. As $\mathbf{x}_1 \cdot \mathbf{x}_2 = 0$, we obtain $\beta = 0$ and hence $\mathbf{x}_2 \in \text{span}\{\mathbf{x}_1, \mathbf{x}_3, \ldots, \mathbf{x}_m, \mathbf{1}\}$, which is a contradiction. So $\mathbf{y}_2$ exists; as with $\mathbf{y}_1$, it is of even weight and $\mathcal{H}_2 = \text{span}\{\mathbf{x}_2, \mathbf{y}_2\}$ is an H-plane. By our choices, $\mathcal{H}_2$ is orthogonal to both $\mathcal{H}_1$ and $\text{span}\{\mathbf{x}_3, \ldots, \mathbf{x}_m\}$. If $\{\mathbf{x}_1, \mathbf{x}_2, \ldots, \mathbf{x}_m, \mathbf{y}_1, \mathbf{y}_2\}$

is not linearly independent, then $\mathbf{y}_2 \in \text{span}\{\mathbf{x}_1, \mathbf{x}_2, \ldots, \mathbf{x}_m, \mathbf{y}_1\}$; this is a contradiction as $\mathbf{x}_2$ is orthogonal to $\text{span}\{\mathbf{x}_1, \mathbf{x}_2, \ldots, \mathbf{x}_m, \mathbf{y}_1\}$ but not $\mathbf{y}_2$. Thus $\text{span}\{\mathbf{x}_1, \mathbf{x}_2, \ldots, \mathbf{x}_m, \mathbf{y}_1, \mathbf{y}_2\} = \mathcal{H}_1 \perp \mathcal{H}_2 \perp \text{span}\{\mathbf{x}_3, \ldots, \mathbf{x}_m\}$.

The rest of the proof follows inductively with Exercise 566 proving the last statement.   □

Exercise 565   In the proof of Lemma 9.12.1, show how to construct the vector $\mathbf{y}_3$ and the H-plane $\mathcal{H}_3$ as the next step in the induction process and show that $\text{span}\{\mathbf{x}_1, \mathbf{x}_2, \ldots, \mathbf{x}_m, \mathbf{y}_1, \mathbf{y}_2, \mathbf{y}_3\} = \mathcal{H}_1 \perp \mathcal{H}_2 \perp \mathcal{H}_3 \perp \text{span}\{\mathbf{x}_4, \ldots, \mathbf{x}_m\}$.   ♦

Exercise 566   Prove the last statement in Lemma 9.12.1. Hint: Suppose $\mathbf{1} \in \mathcal{H}_1 \perp \cdots \perp \mathcal{H}_m$. Write $\mathbf{1} = \sum_{i=1}^m a_i \mathbf{x}_i + \sum_{i=1}^m b_i \mathbf{y}_i$. Then examine $\mathbf{x}_i \cdot \mathbf{1}$ for each i.   ♦

Theorem 9.12.2 [(270)] *The following hold*:
(i)   *If* $n \equiv 0 \pmod 8$, *then* $\mathcal{E}_n$ *has form* H.
(ii)   *If* $n \equiv 4 \pmod 8$, *then* $\mathcal{E}_n$ *has form* A.
(iii)   *If* $n \equiv 2 \pmod 4$, *then* $\mathcal{E}_n$ *has form* O.

Proof: By Theorem 7.8.5, decompose $\mathcal{E}_n$ as an orthogonal sum of its hull and either m H-planes or $m - 1$ H-planes and one A-plane. The hull of $\mathcal{E}_n$ is $\mathcal{H} = \mathcal{E}_n \cap \mathcal{E}_n^\perp = \{\mathbf{0}, \mathbf{1}\}$. Therefore the dimension of $\mathcal{E}_n$ is $n - 1 = 2m + 1$ as $r = 1$ is the dimension of $\mathcal{H}$; in particular $m = (n - 2)/2$. The hull of $\mathcal{E}_n$ is singly-even if $n \equiv 2 \pmod 4$, implying by Theorem 7.8.5 that $\mathcal{E}_n$ has form O giving (iii).

Suppose that $n \equiv 4 \pmod 8$. As $\mathcal{H}$ is doubly-even, $\mathcal{E}_n$ has either form H or A. If it has form H, then $\mathcal{E}_n = \text{span}\{\mathbf{1}\} \perp \mathcal{H}_1 \perp \cdots \perp \mathcal{H}_m$. Letting $\mathbf{x}_i$ be a doubly-even vector in $\mathcal{H}_i$, we have the set $\{\mathbf{1}, \mathbf{x}_1, \ldots, \mathbf{x}_m\}$ generating a doubly-even subcode of dimension $m + 1 = n/2$; such a subcode is Type II, contradicting Corollary 9.2.2, proving (ii).

Finally, suppose that $n \equiv 0 \pmod 8$. Then $\mathcal{E}_n$ contains the Type II subcode equal to the direct sum of $n/8$ copies of the $[8, 4, 4]$ extended Hamming code $\widehat{\mathcal{H}}_3$. Thus $\mathcal{E}_n$ contains a doubly-even subcode $\mathcal{C}'$ of dimension $n/2$. Let $\mathcal{C}$ be any $[n, n/2 - 1]$ vector space complement of $\mathbf{1}$ in $\mathcal{C}'$. By Lemma 9.12.1, $\mathcal{C}$ is contained in the orthogonal sum $\mathcal{H}_1 \perp \cdots \perp \mathcal{H}_{n/2-1}$ of $n/2 - 1$ H-planes. This orthogonal sum is orthogonal to, and disjoint from, the hull $\mathcal{H} = \text{span}\{\mathbf{1}\}$ of $\mathcal{E}_n$. Thus $\text{span}\{\mathbf{1}\} \perp \mathcal{H}_1 \perp \cdots \perp \mathcal{H}_{n/2-1} \subseteq \mathcal{E}_n$; as both codes have dimension $n - 1$, they are equal proving (i).   □

Once we know the form of $\mathcal{E}_n$, we can count the number of singly- and doubly-even vectors in $\mathbb{F}_2^n$.

Corollary 9.12.3 *Let n be even. Let a be the number of doubly-even vectors in $\mathbb{F}_2^n$ and let b be the number of singly-even vectors in $\mathbb{F}_2^n$. The values of a and b are as follows*:
(i)   *If* $n \equiv 0 \pmod 8$, *then* $a = 2^{n-2} + 2^{n/2-1}$ *and* $b = 2^{n-2} - 2^{n/2-1}$.
(ii)   *If* $n \equiv 4 \pmod 8$, *then* $a = 2^{n-2} - 2^{n/2-1}$ *and* $b = 2^{n-2} + 2^{n/2-1}$.
(iii)   *If* $n \equiv 2 \pmod 4$, *then* $a = b = 2^{n-2}$.

Proof: This follows directly from Theorems 7.8.6 and 9.12.2.   □

We need one more result before we can prove the counting formula for the number of Type II codes of length n where $n \equiv 0 \pmod 8$.

Lemma 9.12.4 *For $i = 1$ and 2, let C_i be an $[n, k_i]$ even binary code with $C_i \cap C_i^\perp = \{0\}$ and $k_1 + 2 \le k_2$. Suppose that $C_1 \subset C_2$. Then there exists either an H-plane or A-plane P in C_2 orthogonal to C_1 with $C_1 \cap P = \{0\}$.*

Proof: Let $\{\mathbf{x}_1, \ldots, \mathbf{x}_{k_1}\}$ be a basis of C_1. Define the linear function $f : C_2 \to \mathbb{F}_2^{k_1}$ by $f(\mathbf{x}) = (\mathbf{x} \cdot \mathbf{x}_1, \ldots, \mathbf{x} \cdot \mathbf{x}_{k_1})$. If $f|_{C_1}$ is the restriction of f to C_1, the kernel of $f|_{C_1}$ is $C_1 \cap C_1^\perp = \{0\}$. Thus $f|_{C_1}$, and hence f, is surjective. The kernel $\mathcal{K}$ of f is an $[n, k_2 - k_1]$ code orthogonal to C_1. As $C_1 \cap C_1^\perp = \{0\}$, $\mathcal{K} \cap C_1 = \{0\}$ and so $C_2 = C_1 \perp \mathcal{K}$. If $\mathbf{x} \in \mathcal{K} \cap \mathcal{K}^\perp$, then $\mathbf{x} \in C_2 \cap C_2^\perp$; hence $\mathcal{K} \cap \mathcal{K}^\perp = \{0\}$. The result follows by Lemma 7.8.2 applied to $\mathcal{K}$. $\qquad\square$

We now prove the counting formula given in Theorem 9.5.5(i). While the result is found in [219], the proof is different.

Theorem 9.12.5 *If $n \equiv 0 \pmod{8}$, then there are*

$$\prod_{i=0}^{\frac{n}{2}-2} (2^i + 1)$$

Type II codes of length n.

Proof: Let $\mu_{n,k}$ be the number of $[n, k]$ doubly-even codes containing $\mathbf{1}$. Clearly, $\mu_{n,1} = 1$. As in the proof of Theorem 9.5.1, we find a recurrence relation for $\mu_{n,k}$. We illustrate the recursion process by computing $\mu_{n,2}$. By Corollary 9.12.3, there are $2^{n-2} + 2^{n/2-1} - 2$ doubly-even vectors that are neither $\mathbf{0}$ nor $\mathbf{1}$. Each of these vectors is in a unique doubly-even $[n, 2]$ code containing $\mathbf{1}$; each such code contains two of these vectors. So $\mu_{n,2} = 2^{n-3} + 2^{n/2-2} - 1$.

Let C be any $[n, k]$ doubly-even code containing $\mathbf{1}$. As C is self-orthogonal, if C' is any vector space complement of span$\{\mathbf{1}\}$ in C, then $C = \text{span}\{\mathbf{1}\} \perp C'$.

By Lemma 9.12.1, applied to the $[n, k-1]$ code C', $C' \subseteq \mathcal{H}_1 \perp \cdots \perp \mathcal{H}_{k-1}$ where $\mathcal{H}_i$ are H-planes. Let $\mathcal{E}$ be any vector space complement of $\mathbf{1}$ in $\mathcal{E}_n$ containing $\mathcal{H}_1 \perp \cdots \perp \mathcal{H}_{k-1}$. Clearly, $\mathcal{E}_n = \text{span}\{\mathbf{1}\} \perp \mathcal{E}$ and $\mathcal{E} \cap \mathcal{E}^\perp = \{0\}$. Note that $\mathcal{E}$ has even dimension $n - 2$. Now apply Lemmas 7.8.4 and 9.12.4 inductively, initially beginning with $C_1 = \mathcal{H}_1 \perp \cdots \perp \mathcal{H}_{k-1}$ and always using $C_2 = \mathcal{E}$ in each step of the induction. Then there exist $\mathcal{H}_k, \ldots, \mathcal{H}_{n/2-1}$ which are either H-planes or A-planes such that $\mathcal{E} = \mathcal{H}_1 \perp \cdots \perp \mathcal{H}_{n/2-1}$. Thus $\mathcal{E}_n = \text{span}\{\mathbf{1}\} \perp \mathcal{H}_1 \perp \cdots \perp \mathcal{H}_{n/2-1}$. By Exercise 418, we may assume that at most one $\mathcal{H}_i$ is an A-plane. But, if one is an A-plane, then $\mathcal{E}_n$ has form A, contradicting Theorem 9.12.2.

If we want to extend C to an $[n, k+1]$ doubly-even code C'', we must add a nonzero doubly-even vector $\mathbf{c}''$ from $\mathcal{D} = \mathcal{H}_k \perp \cdots \perp \mathcal{H}_{n/2-1}$. By Lemma 7.8.4, $\mathcal{D} \cap \mathcal{D}^\perp = \{0\}$. By Theorem 7.8.6, there are $2^{n-2k-1} + 2^{n/2-k-1}$ doubly-even vectors in $\mathcal{D}$, implying that there are $2^{n-2k-1} + 2^{n/2-k-1} - 1$ choices for $\mathbf{c}''$. However, every such C'' has $2^k - 1$ doubly-even subcodes of dimension k containing $\mathbf{1}$ by Exercise 522(a). This shows that

$$\mu_{n,k+1} = \frac{2^{n-2k-1} + 2^{n/2-k-1} - 1}{2^k - 1} \mu_{n,k}.$$

When $k = 1$, we obtain $\mu_{n,2} = 2^{n-3} + 2^{n/2-2} - 1$, in agreement with the value we obtained above. Using this recurrence relation, the number of Type II codes of length n is

$$
\mu_{n,n/2} = \frac{2^1 + 2^0 - 1}{2^{n/2-1} - 1} \cdot \frac{2^3 + 2^1 - 1}{2^{n/2-2} - 1} \cdot \frac{2^5 + 2^2 - 1}{2^{n/2-3} - 1} \cdots \frac{2^{n-3} + 2^{n/2-2} - 1}{2^1 - 1} \cdot \mu_{n,1}
$$

$$
= \frac{2^1 + 2^0 - 1}{2^1 - 1} \cdot \frac{2^3 + 2^1 - 1}{2^2 - 1} \cdot \frac{2^5 + 2^2 - 1}{2^3 - 1} \cdots \frac{2^{n-3} + 2^{n/2-2} - 1}{2^{n/2-1} - 1}
$$

$$
= 2(2^1 + 1)(2^2 + 1) \cdots (2^{n/2-2} + 1),
$$

which is the desired result. $\square$

Exercise 567 Prove that if $n \equiv 0 \,(\mathrm{mod}\ 8)$, then there are

$$
\prod_{i=1}^{k-1} \frac{2^{n-2i-1} + 2^{n/2-i-1} - 1}{2^i - 1}
$$

$[n, k]$ doubly-even binary codes containing $\mathbf{1}$ for any $k \le n/2$. ◆

In Theorem 9.5.5(ii), there is also a counting formula for the number of Type II codes containing a given doubly-even code. We leave its proof, which follows along the lines of the proof of Theorem 9.12.5, to Exercise 568.

Exercise 568 Prove Theorem 9.5.5(ii). ◆

The proof of the counting formula for the number of $[n, (n/2) - 1]$ doubly-even codes when $n \equiv 4 \,(\mathrm{mod}\ 8)$, found in Theorem 9.5.7(i), is analogous to the proof of Theorem 9.12.5. (The counting formula for the number of $[n, (n/2) - 1]$ doubly-even codes when $n \equiv 2 \,(\mathrm{mod}\ 4)$ is found in Theorem 9.5.7(ii) and proved in Section 9.5.) For the proofs of the counting formulas for the number of Type III and Type IV codes, we refer the reader to [217, 259, 260].

Exercise 569 Prove Theorem 9.5.7(i). ◆

10 Some favorite self-dual codes

In this chapter we examine the properties of the binary and ternary Golay codes, the hexacode, and the Pless symmetry codes. The Golay codes and the hexacode have similar properties while the Pless symmetry codes generalize the [12, 6, 6] extended ternary Golay code. We conclude the chapter with a section showing some of the connections between these codes and lattices.

10.1 The binary Golay codes

In this section we examine in more detail the binary Golay codes of lengths 23 and 24. We have established the existence of a [23, 12, 7] and a [24, 12, 8] binary code in Section 1.9.1. Recall that our original construction of the [24, 12, 8] extended binary Golay code used a bordered reverse circulant generator matrix, and the [23, 12, 7] code was obtained by puncturing. Since then we have given different constructions of these codes, both of which were claimed to be unique codes of their length, dimension, and minimum distance. We first establish this uniqueness.

10.1.1 Uniqueness of the binary Golay codes

Throughout this section let C be a (possibly nonlinear) binary code of length 23 and minimum distance 7 containing $M \geq 2^{12}$ codewords, one of which is $\mathbf{0}$. In order to prove the uniqueness of C, we first show it has exactly 2^{12} codewords and is perfect. We then show it has a uniquely determined weight distribution and is in fact linear. This proof of linearity follows along the lines indicated by [244].

Lemma 10.1.1 C is a perfect code with 2^{12} codewords.

Proof: By the Sphere Packing Bound,

$$M \left[\binom{23}{0} + \binom{23}{1} + \binom{23}{2} + \binom{23}{3} \right] \leq 2^{23}, \tag{10.1}$$

implying $M \leq 2^{12}$. Hence $M = 2^{12}$, and we have equality in (10.1), which indicates that C is perfect. $\qquad \square$

Lemma 10.1.2 Let $A_i = A_i(C)$ be the weight distribution of C. Then

$$A_0 = A_{23} = 1, \quad A_7 = A_{16} = 253, \quad A_8 = A_{15} = 506, \quad A_{11} = A_{12} = 1288.$$

Proof: As C is perfect, the spheres of radius 3 centered at codewords are disjoint and cover $\mathbb{F}_2^{23}$. In this proof, let $N(w, c)$ denote the number of vectors in $\mathbb{F}_2^{23}$ of weight w in a sphere of radius 3 centered at a codeword of weight c; note that $N(w, c)$ is independent of the codeword chosen. Note also that $N(w, c) = 0$ when w and c differ by more than 3. Since the spheres of radius 3 pack $\mathbb{F}_2^{23}$ disjointly, we have

$$\sum_{c=0}^{23} N(w, c) A_c = \binom{23}{w}, \tag{10.2}$$

as every vector in $\mathbb{F}_2^{23}$ of weight w is in exactly one sphere of radius 3 centered at a codeword. The verification of the weight distribution merely exploits (10.2) repeatedly.

Clearly, $A_0 = 1$ and $A_i = 0$ for $1 \le i \le 6$. We now compute $A_7, A_8, \ldots, A_{23}$ in order. The strategy when computing A_i is to count the number of vectors in $\mathbb{F}_2^{23}$ of weight $w = i - 3$ that are in spheres of radius 3 centered at codewords and then use (10.2).

Computation of A_7: Any vector $\mathbf{x} \in \mathbb{F}_2^{23}$ with $\text{wt}(\mathbf{x}) = 4$ must be in a unique sphere centered at some codeword $\mathbf{c}$ with $\text{wt}(\mathbf{c}) = 7$. Clearly, $\text{supp}(\mathbf{x}) \subset \text{supp}(\mathbf{c})$ showing that $N(4, 7) = \binom{7}{4}$. Hence, (10.2) becomes $N(4, 7)A_7 = \binom{7}{4}A_7 = \binom{23}{4}$, implying $A_7 = 253$.

Computation of A_8: Any vector $\mathbf{x} \in \mathbb{F}_2^{23}$ with $\text{wt}(\mathbf{x}) = 5$ must be in a unique sphere centered at some codeword $\mathbf{c}$ with $\text{wt}(\mathbf{c}) = 7$ or 8. In either case, $\text{supp}(\mathbf{x}) \subset \text{supp}(\mathbf{c})$, showing that $N(5, 7) = \binom{7}{5}$ and $N(5, 8) = \binom{8}{5}$. Equation (10.2) becomes $N(5, 7)A_7 + N(5, 8)A_8 = \binom{7}{5}A_7 + \binom{8}{5}A_8 = \binom{23}{5}$ yielding $A_8 = 506$.

Computation of A_9: Any vector $\mathbf{x} \in \mathbb{F}_2^{23}$ with $\text{wt}(\mathbf{x}) = 6$ must be in a unique sphere centered at some codeword $\mathbf{c}$ with $\text{wt}(\mathbf{c}) = 7$, 8, or 9. If $\text{wt}(\mathbf{c}) = 7$, $\text{supp}(\mathbf{x}) \subset \text{supp}(\mathbf{c})$ or $\text{supp}(\mathbf{x})$ overlaps $\text{supp}(\mathbf{c})$ in five coordinates. Thus $N(6, 7) = \binom{7}{6} + \binom{7}{5}\binom{16}{1} = 343$. If $\text{wt}(\mathbf{c}) = 8$ or 9, $\text{supp}(\mathbf{x}) \subset \text{supp}(\mathbf{c})$. Hence, $N(6, 8) = \binom{8}{6} = 28$ and $N(6, 9) = \binom{9}{6} = 84$. Equation (10.2) is $N(6, 7)A_7 + N(6, 8)A_8 + N(6, 9)A_9 = 343 \cdot 253 + 28 \cdot 506 + 84A_9 = \binom{23}{6}$. Thus $A_9 = 0$.

The remaining values of A_i can be computed in a similar way. We only give the equations that arise when $10 \le i \le 12$, and leave their derivation and solution as exercises.

Computation of A_{10}: $\left(1 + \binom{7}{6}\binom{16}{1}\right)A_7 + \left(\binom{8}{7} + \binom{8}{6}\binom{15}{1}\right)A_8 + \binom{10}{7}A_{10} = \binom{23}{7}$.

Computation of A_{11}: $\left(\binom{16}{1} + \binom{7}{6}\binom{16}{2}\right)A_7 + \left(1 + \binom{8}{7}\binom{15}{1}\right)A_8 + \binom{11}{8}A_{11} = \binom{23}{8}$.

Computation of A_{12}: $\binom{16}{2}A_7 + \left(\binom{15}{1} + \binom{8}{7}\binom{15}{2}\right)A_8 + \binom{11}{9}A_{11} + \binom{12}{9}A_{12} = \binom{23}{9}$.

The reader is invited to complete the proof. $\square$

We remark that we do not actually have to calculate the weight distribution of C; examining the proof, we see that each A_i is unique. Since we know that the [23, 12, 7] binary Golay code $\mathcal{G}_{23}$ exists, C must have the same weight distribution as $\mathcal{G}_{23}$, which was presented in Exercise 403.

Exercise 570 In the proof of Lemma 10.1.2,
(a) verify and solve the equations for A_i with $10 \le i \le 12$, and
(b) find and solve the equations for A_i with $13 \le i \le 23$.   ◆

Exercise 571 Use the technique presented in the proof of Lemma 10.1.2 to compute the weight distribution of the perfect $[15, 11, 3]$ binary Hamming code $\mathcal{H}_4$. Compare this answer to that found in Exercise 387. ♦

Lemma 10.1.3 *Let $\widehat{C}$ be obtained from C by adding an overall parity check. Then $\widehat{C}$ is a $[24, 12, 8]$ linear Type II code, and C is also linear.*

Proof: By Lemma 10.1.2 every codeword of $\widehat{C}$ is doubly-even. We show that distinct codewords are orthogonal. For $\mathbf{c} \in C$, let $\widehat{\mathbf{c}}$ be its extension. Suppose that $\mathbf{c}_1$ and $\mathbf{c}_2$ are in C. Let $w = \mathrm{wt}(\widehat{\mathbf{c}}_1 + \widehat{\mathbf{c}}_2)$ and $C_1 = \mathbf{c}_1 + C$. As C has minimum distance 7, so does C_1. As $\mathbf{0} \in C_1$, Lemmas 10.1.1 and 10.1.2 can be applied to C_1. Thus $\widehat{C}_1 = \widehat{\mathbf{c}}_1 + \widehat{C}$ is doubly-even implying that $4 \mid w$. So $\widehat{\mathbf{c}}_1$ and $\widehat{\mathbf{c}}_2$ are orthogonal.

Let $\mathcal{L}$ be the linear code generated by the codewords of $\widehat{C}$; since all codewords of $\widehat{C}$ are orthogonal to each other (and to themselves as each is even), $\mathcal{L}$ has dimension at most 12. But $\widehat{C}$ has 2^{12} codewords, implying $\mathcal{L} = \widehat{C}$. Hence $\widehat{C}$ and its punctured code C are linear. The minimum weight of $\widehat{C}$ follows from Lemma 10.1.2. □

Corollary 10.1.4 *Let $\mathcal{A}$ be a binary, possibly nonlinear, code of length 24 and minimum distance 8 containing $\mathbf{0}$. If $\mathcal{A}$ has M codewords with $M \geq 2^{12}$, then $\mathcal{A}$ is a $[24, 12, 8]$ linear Type II code.*

Proof: Let C be the length 23 code obtained from $\mathcal{A}$ by puncturing on some coordinate i. Then C has M codewords, minimum distance at least 7, and contains $\mathbf{0}$. By Lemmas 10.1.1 and 10.1.3, $M = 2^{12}$ and C is linear. Note that $\mathcal{A}$ may not be the extension of C; however, we can puncture $\mathcal{A}$ on a different coordinate j to get a code C_1. The same argument applied to C_1 shows that C_1 is linear. Thus $\mathcal{A}$ is linear by Exercise 572; hence it is a $[24, 12, 8]$ code. By Lemma 10.1.3, $\mathcal{A}$ is Type II if we can show that $\mathcal{A} = \widehat{C}$. This is equivalent to showing that $\mathcal{A}$ is even. Suppose that it is not. Let $\mathbf{a} \in \mathcal{A}$ with $\mathrm{wt}(\mathbf{a})$ odd. By puncturing $\mathcal{A}$ on a coordinate where $\mathbf{a}$ is 0, we obtain a code with weight distribution given in Lemma 10.1.2. Thus $\mathrm{wt}(\mathbf{a})$ is 7, 11, 15, or 23. Now puncture $\mathcal{A}$ on a coordinate where $\mathbf{a}$ is 1. This punctured code still has the weight distribution in Lemma 10.1.2, but contains a vector of weight 6, 10, 14, or 22, which is a contradiction. □

Exercise 572 Let $\mathcal{A}$ be a possibly nonlinear binary code of minimum distance at least 2. Choose two coordinates i and j and puncture $\mathcal{A}$ on each to obtain two codes $\mathcal{A}_i$ and $\mathcal{A}_j$. Suppose that $\mathcal{A}_i$ and $\mathcal{A}_j$ are both linear. Prove that $\mathcal{A}$ is linear. ♦

We now know that C and $\widehat{C}$ are $[23, 12, 7]$ and $[24, 12, 8]$ linear codes. We show the uniqueness of $\widehat{C}$ by showing the uniqueness of a 5-$(24, 8, 1)$ design.

Theorem 10.1.5 *The 5-$(24, 8, 1)$ design is unique up to equivalence.*

Proof: Let $\mathcal{D} = (\mathcal{P}, \mathcal{B})$ be a 5-$(24, 8, 1)$ design. The intersection numbers of $\mathcal{D}$ are given in Figure 8.3; from these numbers we see that two distinct blocks intersect in either 0, 2, or 4 points. We will construct blocks B_i for $1 \leq i \leq 12$ and associated codewords $\mathbf{c}_i$ where

	1	2	3	4	5	6	7	8	9	1 0	1 1	1 2	1 3	1 4	1 5	1 6	1 7	1 8	1 9	2 0	2 1	2 2	2 3	2 4
c_1	1	1	1	1	1	1	1	1	0	0	0	0	0	0	0	0	0	0	0	0	0	0	0	0
c_2	1	1	1	1	0	0	0	0	1	1	1	1	0	0	0	0	0	0	0	0	0	0	0	0
c_3	1	1	1	0	1	0	0	0	1	0	0	0	1	1	1	0	0	0	0	0	0	0	0	0
c_4	1	1	0	1	1	0	0	0	1	0	0	0	0	0	0	1	1	1	0	0	0	0	0	0
c_5	1	0	1	1	1	0	0	0	1	0	0	0	0	0	0	0	0	0	1	1	1	0	0	0
c_6	0	1	1	1	1	0	0	0	1	0	0	0	0	0	0	0	0	0	0	0	0	1	1	1
c_7	1	1	1	0	0	1	0	0	1	0	0	0	0	0	0	1	0	0	1	0	0	1	0	0
c_8	1	1	0	1	0	1	0	0	1	0	0	0	1	0	0	0	0	0	0	1	0	0	1	0
c_9	1	0	1	1	0	1	0	0	1	0	0	0	0	1	0	0	1	0	0	0	0	0	0	1
c_{10}	1	1	0	0	1	1	0	0	1	1	0	0	0	0	0	0	0	0	0	0	1	0	0	1
c_{11}	1	0	1	0	1	1	0	0	1	0	1	0	0	0	0	0	0	1	0	0	0	0	1	0
c'_{12}	1	1	1	0	0	0	0	1	1	0	0	0	0	0	0	0	1	0	0	0	1	0	1	0
c''_{12}	1	1	1	0	0	0	0	1	1	0	0	0	0	0	0	0	0	1	0	1	0	0	0	1

Figure 10.1 Codewords from a 5-(24, 8, 1) design.

$\mathrm{supp}(c_i) = B_i$, which we give in Figure 10.1. Let $\mathcal{P} = \{1, 2, \ldots, 24\}$. Since we prove $\mathcal{D}$ is unique up to equivalence, we will be reordering points where possible.

By reordering points, we can assume $B_1 = \{1, 2, 3, 4, 5, 6, 7, 8\}$. We repeatedly use the fact that any five points determine a unique block. By reordering points, we may assume that the block containing $\{1, 2, 3, 4, 9\}$ is $B_2 = \{1, 2, 3, 4, 9, 10, 11, 12\}$. By similar reordering, the block containing $\{1, 2, 3, 5, 9\}$ is $B_3 = \{1, 2, 3, 5, 9, 13, 14, 15\}$, the block containing $\{1, 2, 4, 5, 9\}$ is $B_4 = \{1, 2, 4, 5, 9, 16, 17, 18\}$, the block containing $\{1, 3, 4, 5, 9\}$ is $B_5 = \{1, 3, 4, 5, 9, 19, 20, 21\}$, and finally the block containing $\{2, 3, 4, 5, 9\}$ is $B_6 = \{2, 3, 4, 5, 9, 22, 23, 24\}$. We repeatedly used the fact in constructing B_i, for $2 \le i \le 6$, that we began with five points, exactly four of which were in the previously constructed blocks. Let B_7 be the block containing $\{1, 2, 3, 6, 9\}$; B_7 intersects B_1, B_2, and B_3 in no other points and intersects B_4, B_5, and B_6 each in one more point. Thus we may assume $B_7 = \{1, 2, 3, 6, 9, 16, 19, 22\}$. Let B_8 be the block containing $\{1, 2, 4, 6, 9\}$; B_8 intersects B_1, B_2, B_4, and B_7 in no other points and intersects B_3, B_5, and B_6 each in one more point. Thus we may assume $B_8 = \{1, 2, 4, 6, 9, 13, 20, 23\}$. By Exercise 573, we may assume that $B_9 = \{1, 3, 4, 6, 9, 14, 17, 24\}$, $B_{10} = \{1, 2, 5, 6, 9, 10, 21, 24\}$, $B_{11} = \{1, 3, 5, 6, 9, 11, 18, 23\}$, and B_{12} is either $B'_{12} = \{1, 2, 3, 8, 9, 17, 21, 23\}$ or $B''_{12} = \{1, 2, 3, 8, 9, 18, 20, 24\}$.

For either B'_{12} or B''_{12}, the 12 weight 8 vectors whose supports are the blocks generate a 12-dimensional code in $\mathbb{F}_2^{24}$, as can be easily seen by row reduction. Call the resulting codes $\mathcal{C}'$ and $\mathcal{C}''$. By Example 8.3.2, the design $\mathcal{D}$ is held by the weight 8 codewords in a [24, 12, 8] binary code $\widehat{\mathcal{C}}$, which is Type II by Corollary 10.1.4. Thus there are at most two designs up to equivalence, one for each code $\mathcal{C}'$ and $\mathcal{C}''$. But using a computer algebra package, one can show that $\mathcal{C}' \simeq \mathcal{C}''$. Thus the design is unique up to equivalence. □

Exercise 573 In the proof of Theorem 10.1.5, show that after $B_1, \ldots, B_8$ have been determined, upon reordering points:

(a) the block B_9 containing $\{1, 3, 4, 6, 9\}$ is $B_9 = \{1, 3, 4, 6, 9, 14, 17, 24\}$,

(b) the block B_{10} containing $\{1, 2, 5, 6, 9\}$ is $B_{10} = \{1, 2, 5, 6, 9, 10, 21, 24\}$,

(c) the block B_{11} containing $\{1, 3, 5, 6, 9\}$ is $B_{11} = \{1, 3, 5, 6, 9, 11, 18, 23\}$, and

(d) the block B_{12} containing $\{1, 2, 3, 8, 9\}$ is either $\{1, 2, 3, 8, 9, 17, 21, 23\}$ or $\{1, 2, 3, 8, 9, 18, 20, 24\}$.   ◆

Theorem 10.1.6 *Both length* 23 *and length* 24, *possibly nonlinear, binary codes each containing* **0** *with* $M \geq 2^{12}$ *codewords and minimum distance* 7 *and* 8, *respectively, are unique up to equivalence. They are the* [23, 12, 7] *and* [24, 12, 8] *binary Golay codes.*

Proof: Let $\mathcal{A}$ be the code of length 24. By Corollary 10.1.4, $\mathcal{A}$ is a [24, 12, 8] Type II code. By Theorem 9.3.10, the codewords of weight 8 hold a 5-design. By Theorem 10.1.5 this design is unique, and by the argument in the proof, the weight 8 codewords span the code. Hence $\mathcal{A}$ is unique.

Let $\mathcal{C}$ be the code of length 23. By Lemma 10.1.3, $\widehat{\mathcal{C}}$ is a [24, 12, 8] code, which is unique. So $\mathcal{C}$ is $\widehat{\mathcal{C}}$ punctured on some coordinate. The proof is complete if we show all punctured codes are equivalent. By Example 6.6.23, the extended binary quadratic residue code is a [24, 12, 8] code. So it is equivalent to $\widehat{\mathcal{C}}$. The extended quadratic residue code is an extended cyclic code and by the Gleason–Prange Theorem, its automorphism group must be transitive; hence all punctured codes are equivalent by Theorem 1.6.6.   □

10.1.2 Properties of binary Golay codes

When we introduced the binary Golay codes in Section 1.9.1, we gave a generator matrix in standard form for the [24, 12, 8] extended binary Golay code $\mathcal{G}_{24}$. In Example 6.6.23, we showed that by extending either odd-like binary quadratic residue code of length 23, we obtain a [24, 12, 8] code, which by Theorem 10.1.6 is equivalent to $\mathcal{G}_{24}$. So one representation of $\mathcal{G}_{24}$ is as an extended quadratic residue code. Puncturing on the extended coordinate gives a representation of the [23, 12, 7] binary Golay code $\mathcal{G}_{23}$ as an odd-like quadratic residue code of length 23.

Viewing $\mathcal{G}_{23}$ as a quadratic residue code, there are two possible idempotents given in Theorem 6.6.5. One of them is

$$e_{23}(x) = x + x^2 + x^3 + x^4 + x^6 + x^8 + x^9 + x^{12} + x^{13} + x^{16} + x^{18}.$$

One can check that the generator polynomial for this code is

$$g_{23}(x) = 1 + x + x^5 + x^6 + x^7 + x^9 + x^{11}.$$

Label the coordinates of $\mathcal{G}_{23}$ as $\{0, 1, 2, \ldots, 22\}$ in the usual numbering for cyclic codes. To obtain $\mathcal{G}_{24}$, add an overall parity check in coordinate labeled ∞. In this form, we can describe $\Gamma\mathrm{Aut}(\mathcal{G}_{24}) = \mathrm{PAut}(\mathcal{G}_{24})$. $\mathrm{PAut}(\mathcal{G}_{24})$ contains the subgroup of cyclic shifts, generated by

$$s_{24} = (0, 1, 2, \ldots, 22).$$

$\mathrm{PAut}(\mathcal{G}_{24})$ also contains a group of multipliers generated by μ_2 by Theorem 4.3.13;

in cycle form

$$\mu_2 = (1, 2, 4, 8, 16, 9, 18, 13, 3, 6, 12)(5, 10, 20, 17, 11, 22, 21, 19, 15, 7, 14).$$

As $\mathcal{G}_{24}$ is an extended quadratic residue code, the Gleason–Prange Theorem applies; the automorphism arising from this theorem is

$$\nu_{24} = (0, \infty)(1, 22)(2, 11)(3, 15)(4, 17)(5, 9)(6, 19)(7, 13)(8, 20)(10, 16)(12, 21)(14, 18).$$

The permutations s_{24}, μ_2, and ν_{24} generate the projective special linear group denoted $\mathrm{PSL}_2(23)$. However, this is only a small subgroup of the entire group. The element

$$\eta_{24} = (1, 3, 13, 8, 6)(4, 9, 18, 12, 16)(7, 17, 10, 11, 22)(14, 19, 21, 20, 15)$$

is also in $\mathrm{PAut}(\mathcal{G}_{24})$; see [149, Theorem 6.7] for one proof of this. The elements s_{24}, μ_2, ν_{24}, and η_{24} generate $\mathrm{PAut}(\mathcal{G}_{24})$. The resulting group is the 5-fold transitive Mathieu group, often denoted $\mathcal{M}_{24}$, of order $244\,823\,040 = 24 \cdot 23 \cdot 22 \cdot 21 \cdot 20 \cdot 48$. $\mathcal{M}_{24}$ is the largest of the five sporadic simple groups known as Mathieu groups. The subgroup of $\mathcal{M}_{24}$ fixing the point ∞ is $\mathrm{PAut}(\mathcal{G}_{23})$. This is the 4-fold transitive Mathieu group $\mathcal{M}_{23}$, which is a simple group of order $10\,200\,960 = 23 \cdot 22 \cdot 21 \cdot 20 \cdot 48$.

Exercise 574 Verify that $g_{23}(x)$ is the generator polynomial for the cyclic code with generating idempotent $e_{23}(x)$.   ◆

Exercise 575 Verify that $\eta_{24} \in \mathrm{PAut}(\mathcal{G}_{24})$. One way to do this is to recognize that $\mathcal{G}_{24}$ is self-dual. To show that $\eta_{24} \in \mathrm{PAut}(\mathcal{G}_{24})$, it suffices to show that the image of a basis (such as one obtained by using cyclic shifts of $g_{23}(x)$ and adding a parity check) under η_{24} gives vectors orthogonal to this basis.   ◆

We also know the covering radius of each Golay code.

Theorem 10.1.7 *The covering radii of the* $[23, 12, 7]$ *and* $[24, 12, 8]$ *binary Golay codes are 3 and 4, respectively.*

Proof: As the $[23, 12, 7]$ code is perfect by Lemma 10.1.1, its covering radius is 3. The $[24, 12, 8]$ extended binary Golay code has covering radius 4 by Example 8.3.2.   □

10.2   Permutation decoding

Permutation decoding is a technique, developed in a 1964 paper by F. J. MacWilliams [214], which uses a fixed set of automorphisms of the code to assist in decoding a received vector. The idea of permutation decoding is to move all the errors in a received vector out of the information positions, using an automorphism of the code, so that the information symbols in the permuted vector are correct. Since the components of the permuted vector in the information positions are correct, apply the parity check equations to the now correct information symbols thus obtaining a vector that must be the correct codeword, only permuted. Finally, apply the inverse of the automorphism to this new codeword and obtain the codeword originally transmitted. This technique has been successfully used as a decoding method for the binary Golay codes.

To use permutation decoding on a t-error-correcting code we must settle two questions. First, how can you guarantee that all of the errors are moved out of the information positions? And second, how do you find a subset S of the automorphism group of the code that moves the nonzero entries in every possible error vector of weight t or less out of the information positions? Such a set is called a *PD-set*. The first question has a simple answer using the syndrome, as seen in the following theorem from [218]. The second question is not so easy to answer.

Theorem 10.2.1 *Let C be an $[n, k]$ t-error-correcting linear code with parity check matrix H having the identity matrix I_{n-k} in the redundancy positions. Suppose $\mathbf{y} = \mathbf{c} + \mathbf{e}$ is a vector with $\mathbf{c} \in C$ and $\mathrm{wt}(\mathbf{e}) \leq t$. Then the information symbols in $\mathbf{y}$ are correct if and only if $\mathrm{wt}(\mathrm{syn}(\mathbf{y})) \leq t$, where $\mathrm{syn}(\mathbf{y}) = H\mathbf{y}^{\mathrm{T}}$.*

Proof: As applying a permutation to the coordinates of C does not change weights of codewords, we may assume using Theorem 1.6.2 that the generator matrix G of C is in standard form $[I_k \ A]$ and that $H = [-A^{\mathrm{T}} \ I_{n-k}]$. Hence the first k coordinates are an information set. If the information symbols in $\mathbf{y}$ are correct, then $H\mathbf{y}^{\mathrm{T}} = H\mathbf{e}^{\mathrm{T}} = \mathbf{e}^{\mathrm{T}}$ as the first k coordinates of $\mathbf{e}$ are 0. Thus $\mathrm{wt}(\mathrm{syn}(\mathbf{y})) \leq t$. Conversely, suppose that some of the information symbols in $\mathbf{y}$ are incorrect. If $\mathbf{e} = e_1 \cdots e_n$, let $\mathbf{e}_1 = e_1 \cdots e_k$, which is nonzero, and $\mathbf{e}_2 = e_{k+1} \cdots e_n$. By Exercise 576, $\mathrm{wt}(\mathbf{x} + \mathbf{z}) \geq \mathrm{wt}(\mathbf{x}) - \mathrm{wt}(\mathbf{z})$; so we have

$$\mathrm{wt}(H\mathbf{y}^{\mathrm{T}}) = \mathrm{wt}(H\mathbf{e}^{\mathrm{T}}) = \mathrm{wt}\left(-A^{\mathrm{T}}\mathbf{e}_1^{\mathrm{T}} + \mathbf{e}_2^{\mathrm{T}}\right)$$
$$\geq \mathrm{wt}\left(-A^{\mathrm{T}}\mathbf{e}_1^{\mathrm{T}}\right) - \mathrm{wt}\left(\mathbf{e}_2^{\mathrm{T}}\right) = \mathrm{wt}(\mathbf{e}_1 A) - \mathrm{wt}(\mathbf{e}_2)$$
$$= \mathrm{wt}(\mathbf{e}_1 I_k) + \mathrm{wt}(\mathbf{e}_1 A) - [\mathrm{wt}(\mathbf{e}_1) + \mathrm{wt}(\mathbf{e}_2)] = \mathrm{wt}(\mathbf{e}_1 G) - \mathrm{wt}(\mathbf{e}).$$

As $\mathbf{e}_1 \neq \mathbf{0}$, $\mathbf{e}_1 G$ is a nonzero codeword and hence has weight at least $2t + 1$. Thus $\mathrm{wt}(H\mathbf{y}^{\mathrm{T}}) \geq 2t + 1 - t = t + 1$. $\qquad\square$

Exercise 576 Prove that if $\mathbf{x}$ and $\mathbf{z}$ are in $\mathbb{F}_q^n$, then $\mathrm{wt}(\mathbf{x} + \mathbf{z}) \geq \mathrm{wt}(\mathbf{x}) - \mathrm{wt}(\mathbf{z})$. ◆

Using this theorem, the Permutation Decoding Algorithm can be stated. Begin with an $[n, k]$ t-error-correcting code C having a parity check matrix H that has I_{n-k} in the redundancy positions. So the generator matrix G associated with H has I_k in the k information positions; the map $\mathbf{u} \mapsto \mathbf{u}G$, defined for any length k message $\mathbf{u}$, is a systematic encoder for C. Let $S = \{\sigma_1, \ldots, \sigma_P\}$ be a PD-set for C. The *Permutation Decoding Algorithm* is accomplished using the following algorithm:

I.   For a received vector $\mathbf{y}$, compute $\text{wt}(H(\mathbf{y}\sigma_i)^{\text{T}})$ for $i = 1, 2, \ldots$ until an i is found so that this weight is t or less.

II. Extract the information symbols from $\mathbf{y}\sigma_i$, and apply the parity check equations to the string of information symbols to obtain the redundancy symbols. Form a codeword $\mathbf{c}$ from these information and redundancy symbols.

III. $\mathbf{y}$ is decoded to the codeword $\mathbf{c}\sigma_i^{-1}$.

We now turn to the question of finding PD-sets. Clearly, the Permutation Decoding Algorithm is more efficient the smaller the size of the PD-set. In [107] a lower bound on this size is given. We will see shortly that a PD-set for the $[24, 12, 8]$ extended binary Golay code exists whose size equals this lower bound. Suppose $S = \{\sigma_1, \ldots, \sigma_P\}$ is a PD-set for a t-error-correcting $[n, k]$ code with redundancy locations $\mathcal{R}$ of size $r = n - k$. Let $\mathcal{R}_i$ be the image of $\mathcal{R}$ under the permutation part of σ_i^{-1} for $1 \le i \le P$. Because S is a PD-set, every subset of coordinates of size $e \le t$ is contained in some $\mathcal{R}_i$. In particular, the sets $\mathcal{R}_1, \ldots, \mathcal{R}_P$ cover all t-element subsets of the n-element set of coordinates. Let $N(t, r, n)$ be the minimum number of r-element subsets of an n-element set Ω that cover all t-element subsets of Ω. Our discussion shows that $P \ge N(t, r, n)$. If $R_1, \ldots, R_N$ with $N = N(t, r, n)$ is a collection of r-element subsets of Ω covering the t-element subsets, then form the set

$$\mathcal{X} = \{(R_i, \omega) \mid 1 \le i \le N \text{ and } \omega \in R_i\}.$$

Clearly, as $|R_i| = r$, $|\mathcal{X}| = rN(t, r, n)$. We compute a lower bound on $|\mathcal{X}|$ in another manner. If $\omega \in \Omega$, let $\mathcal{X}_\omega = \{(R_i, \omega) \in \mathcal{X}\}$. Notice that $|\mathcal{X}_\omega| \ge N(t - 1, r - 1, n - 1)$ because the sets obtained from R_i, where $(R_i, \omega) \in \mathcal{X}_\omega$, by deleting ω produce a collection of $(r - 1)$-element sets that cover all $(t - 1)$-element subsets of $\Omega \setminus \{\omega\}$. Hence there are at least $nN(t - 1, r - 1, n - 1)$ pairs in $\mathcal{X}$, and so

$$N(t, r, n) \ge \frac{n}{r} N(t - 1, r - 1, n - 1).$$

By repeated application of this inequality together with the observation that $N(1, r, n) = \lceil n/r \rceil$ we obtain the following.

Theorem 10.2.2 ([107]) *A PD-set of size P for a t-error-correcting $[n, k]$ code with redundancy $r = n - k$ satisfies*

$$P \ge \left\lceil \frac{n}{r} \left\lceil \frac{n-1}{r-1} \cdots \left\lceil \frac{n-t+1}{r-t+1} \right\rceil \cdots \right\rceil \right\rceil.$$

Example 10.2.3 By Theorem 10.2.2 we find that a PD-set for the $[24, 12, 8]$ extended binary Golay code has size at least 14. The permutations that follow form the PD-set found in [107] where the code is written as an extended quadratic residue code; here the generator polynomial of the (nonextended) cyclic quadratic residue code is $1 + x^2 + x^4 + x^5 + x^6 + x^{10} + x^{11}$, and the extended coordinate is labeled ∞.

(1)

(0, 14)(1, 11)(2, 19)(3, 17)(4, 16)(5, 15)(6, 12)(7, 18)(8, 20)(9, 21)(10, 13)(22, ∞)
(0, 16, 20, 13, 7, 4, 8)(1, 22, 11, 9, 18, ∞, 21, 12, 6, 17, 5, 15, 14, 19)(2, 10)
(0, 4, 15, 3, 14, 1, 9)(2, 17, 21, 6, 10, 11, 22)(5, 8, 18, 20, 7, 16, ∞)
(2, 20, 16, 21, 13, 6, 4, 12)(3, 19, 5, 8, 15, 14, 18, 9)(7, 17)(10, 22, 11, ∞)
(0, 11, 16, 6, 18, 19, 20, 9, 13, 22, 5, 17, 12, 14, 10, 8, 2, 7, 1, 21, ∞, 15, 4)
(0, 6, 20, 14, 8, 9, 13, 11, 17, 2, 19)(1, 5, 21, 12, 16, 3, 15, ∞, 4, 18, 10)
(0, 20, 4, 7, 12, 5, 6)(1, 15, 16, 17, 18, 21, 3)(2, 10, 22, ∞, 13, 8, 11)
(0, 6, 16, 1, 9, 17, 19, 11, 8, 22, 14, ∞, 5, 15, 2)(4, 10, 20, 21, 12)(7, 18, 13)
(0, 16, 4, 20, 19, 17, 18, 15, 3, 22, 11, 9)(1, 14, 10, ∞, 21, 12, 6, 7, 8, 5, 2, 13)
(0, 5, 11, 2, 21, 3, 12)(1, 8, 10, 6, 18, 22, 9, 13, 4, 14, 15, 20, 7, ∞)(16, 17)
(0, 15, 1, 12, 20, 19, 3, 9, 2, 10, 18, ∞, 21, 17)(4, 8, 16, 6, 7, 11, 14)(13, 22)
(0, 4, 5, 11, 15, 13, 19)(1, 7, 22, 2, 3, 8, 20)(6, 14, 18, 9, 17, 10, 21)
(0, 13, 18, 1, 21, 7, 6, ∞, 11, 16, 15, 17, 8, 4, 12, 20, 9, 3, 19, 10, 2, 14, 22)

These permutations were found by computer using orbits of $\mathcal{M}_{24}$ on certain sets.   ■

10.3   The hexacode

The hexacode, first presented in Example 1.3.4, is the unique [6, 3, 4] Hermitian self-dual code over $\mathbb{F}_4$. This MDS code possesses many of the same properties as the Golay codes including being extended perfect and having a representation as an extended quadratic residue code.

10.3.1   Uniqueness of the hexacode

Let C be a [6, 3, 4] code over $\mathbb{F}_4$ with generator matrix G, which we may assume is in standard form $[I_3 \; A]$ where A is a 3×3 matrix. As the minimum weight is 4, every entry in A is nonzero. By scaling the columns, we may assume that the first row of A is 111. By rescaling columns 2 and 3 of I_3, we may also assume that each entry in the first column of A is 1. If two entries in row two of A are equal to α, then α times row one of G added to row two of G gives a codeword of weight less than 4, which is a contradiction. Hence row two, and analogously row three, each have distinct entries. As rows two and three of A cannot be equal to each other, we may assume that

$$A = \begin{bmatrix} 1 & 1 & 1 \\ 1 & \omega & \overline{\omega} \\ 1 & \overline{\omega} & \omega \end{bmatrix}. \tag{10.3}$$

It is easy to see that G generates a Hermitian self-dual code and hence is equivalent to the hexacode. Notice that G is a bordered double circulant generator matrix (and bordered reverse generator matrix) for C. This proves the following result.

Theorem 10.3.1 *Let C be a [6, 3, 4] code over $\mathbb{F}_4$. Then C is unique and is equivalent to the hexacode.*

Exercise 577 Give an equivalence that maps the generator matrix for the hexacode as given in Example 1.3.4 to the generator matrix $[I_3 \ A]$, where A is given in (10.3). ◆

10.3.2 Properties of the hexacode

The extended binary and ternary Golay codes of lengths 24 and 12, respectively, were each extended quadratic residue codes. The same result holds for the hexacode by Exercise 363. In addition, the hexacode is an MDS code and by Exercise 578, the $[5, 3, 3]$ code obtained by puncturing the hexacode is a perfect code.

Exercise 578 Do the following:
(a) Show that a $[5, 3, 3]$ code over $\mathbb{F}_4$ is perfect.
(b) Show that a $[5, 3, 3]$ code over $\mathbb{F}_4$ is unique and that its extension is the hexacode.
(c) Show that the hexacode is an extended Hamming code over $\mathbb{F}_4$. ◆

By Theorem 6.6.6 (see Example 6.6.8 and Exercise 363), one $[5, 3, 3]$ odd-like quadratic residue code over $\mathbb{F}_4$ has generating idempotent

$$e_5(x) = 1 + \omega x + \overline{\omega} x^2 + \overline{\omega} x^3 + \omega x^4;$$

the associated generator polynomial is

$$g_5(x) = 1 + \omega x + x^2.$$

Extending this code we obtain the hexacode $\mathcal{G}_6$. Labeling the coordinates $0, 1, 2, 3, 4, \infty$, the permutation

$$s_6 = (0, 1, 2, 3, 4)$$

is in $\Gamma\text{Aut}(\mathcal{G}_6)$. Furthermore, the map

$$\eta_6 = \text{diag}(1, \overline{\omega}, \omega, \omega, \overline{\omega}, 1) \times (0, \infty) \times \sigma_2,$$

where σ_2 is the Frobenius map interchanging ω and $\overline{\omega}$, is in $\Gamma\text{Aut}(\mathcal{G}_6)$. These two elements generate $\Gamma\text{Aut}(\mathcal{G}_6)$, which is a group of order 2160. The group $\Gamma\text{Aut}_{\text{Pr}}(\mathcal{G}_6) = \{P \mid DP \in \text{MAut}(\widehat{G}_6)\}$ (the projection of $\Gamma\text{Aut}(\mathcal{G}_6)$ onto its permutation parts defined in Section 1.7) is the symmetric group Sym_6.

Exercise 579 Verify that $g_5(x)$ is the generator polynomial for the cyclic code with generating idempotent $e_5(x)$. ◆

Exercise 580 Verify that η_6 is in $\Gamma\text{Aut}(\mathcal{G}_6)$. ◆

The covering radius of the punctured hexacode is 1 as it is a perfect $[5, 3, 3]$ code. Hence by Theorem 1.12.6, the hexacode has covering radius at most 2. In fact we have the following.

Theorem 10.3.2 *The covering radius of the hexacode is 2.*

Exercise 581 Prove Theorem 10.3.2. ◆

10.3.3   Decoding the Golay code with the hexacode

In Section 10.2, we presented a method to decode the [24, 12, 8] extended binary Golay code. To be used effectively, this decoding must be done by computer. Here we present another method for decoding, found in [268], which can certainly be programmed on a computer but can also be done by hand. We use a version of the [24, 12, 8] extended binary Golay code defined in terms of the hexacode.

We first establish some notation. Let $\mathbf{v} = v_1 v_2 \cdots v_{24} \in \mathbb{F}_2^{24}$. We rearrange $\mathbf{v}$ into a 4×6 array

$$
\begin{bmatrix}
v_1 & v_5 & \cdots & v_{21} \\
v_2 & v_6 & \cdots & v_{22} \\
v_3 & v_7 & \cdots & v_{23} \\
v_4 & v_8 & \cdots & v_{24}
\end{bmatrix},
\tag{10.4}
$$

which we denote as $[\mathbf{v}]$. The *parity of a column* of $[\mathbf{v}]$ is even or odd if the binary sum of the entries in that column is 0 or 1, respectively. There is an analogous definition for the *parity of a row*. The set of vectors written as 4×6 matrices that satisfy two criteria will form the [24, 12, 8] extended binary Golay code. The second criterion involves the hexacode. The version of the hexacode $\mathcal{G}_6$ used throughout the remainder of this section has generator matrix

$$
G_6 =
\begin{bmatrix}
1 & 0 & 0 & 1 & \overline{\omega} & \omega \\
0 & 1 & 0 & 1 & \omega & \overline{\omega} \\
0 & 0 & 1 & 1 & 1 & 1
\end{bmatrix},
$$

which is obtained from the generator matrix $[I_3 \ A]$, where A is given in (10.3), by interchanging the first and third columns of $[I_3 \ A]$. This form is used because it is easy to identify the 64 codewords in the hexacode. To use the hexacode we send an element of $\mathbb{F}_2^{24}$ into an element of $\mathbb{F}_4^6$ by defining the *projection* $\mathrm{Pr}(\mathbf{v})$ *of* $\mathbf{v}$ *into* $\mathbb{F}_4^6$ as the matrix product

$$\mathrm{Pr}(\mathbf{v}) = [0 \ 1 \ \omega \ \overline{\omega}][\mathbf{v}].$$

Example 10.3.3 Suppose that $\mathbf{v} = 111111000000001101010110$. The projection operation can be visualized by writing

$$
\begin{matrix}
0 \\
1 \\
\omega \\
\overline{\omega}
\end{matrix}
\begin{bmatrix}
1 & 1 & 0 & 0 & 0 & 0 \\
1 & 1 & 0 & 0 & 1 & 1 \\
1 & 0 & 0 & 1 & 0 & 1 \\
1 & 0 & 0 & 1 & 1 & 0
\end{bmatrix},
$$
$$
 0 \ \ 1 \ \ 0 \ \ 1 \ \ \omega \ \ \overline{\omega}
$$

where the projection is accomplished by multiplying the first row of $[\mathbf{v}]$ by 0, multiplying the second row by 1, multiplying the third row by ω, multiplying the fourth row by $\overline{\omega}$, and then adding. ∎

Recall that the complement of a binary vector $\mathbf{v}$ is the binary vector $\mathbf{v} + \mathbf{1}$ obtained by replacing the 0s in $\mathbf{v}$ by 1s and the 1s by 0s. We will repeatedly use the following lemma without reference; its proof is left as an exercise.

Lemma 10.3.4 *The following hold for columns of* (10.4):
(i) *Given $\alpha \in \mathbb{F}_4$, there are exactly four possible choices for a column that projects to α. Exactly two choices have even parity and two have odd parity. The two of even parity are complements of each other, as are the two of odd parity.*
(ii) *The columns projecting to $0 \in \mathbb{F}_4$ are $[0000]^\mathrm{T}$, $[1111]^\mathrm{T}$, $[1000]^\mathrm{T}$, and $[0111]^\mathrm{T}$.*

Exercise 582 Prove Lemma 10.3.4.   ◆

Lemma 10.3.5 *Let C be the binary code consisting of all vectors $\mathbf{c}$ such that:*
(i) *all columns of $[\mathbf{c}]$ have the same parity and that parity equals the parity of the first row of $[\mathbf{c}]$, and*
(ii) *the vector $\mathrm{Pr}(\mathbf{c}) \in \mathcal{G}_6$.*
The code C is the $[24, 12, 8]$ extended binary Golay code.

Proof: If we fill in the first three rows of the matrix (10.4) arbitrarily, then the last row is uniquely determined if (i) is to be satisfied. The set of vectors satisfying (i) is clearly a linear space. Hence the set of vectors C_1 satisfying (i) is a $[24, 18]$ code. Similarly, if we fill in the first three columns of the matrix (10.4) arbitrarily, then these columns project onto three arbitrary elements of $\mathbb{F}_4$. Since the first three coordinates of the hexacode are the three information positions, the last three coordinates of the projection are uniquely determined if (ii) is to be satisfied. But there are precisely four choices for a column if it is to project to some fixed element of $\mathbb{F}_4$. Thus the set of vectors satisfying (ii) has size $2^{12} 4^3 = 2^{18}$. This set C_2 is clearly a linear space as Pr is a linear map and $\mathcal{G}_6$ is closed under addition. Thus C_2 is also a $[24, 18]$ code. Therefore $C = C_1 \cap C_2$ has dimension at least 12.

By Theorem 10.1.6, the proof is completed if we show that C has minimum distance 8. The vector $1111111100 \cdots 0$ satisfies (i) and projects onto the zero vector. Hence C has minimum distance at most 8. Suppose $\mathbf{c} \in C$ is a nonzero codeword. There are two cases: the top row of $[\mathbf{c}]$ has even parity or the top row has odd parity.

Suppose that the top row of $[\mathbf{c}]$ has even parity. If $\mathrm{Pr}(\mathbf{c}) \neq \mathbf{0}$, then as $\mathcal{G}_6$ has minimum weight 4, there are at least four nonzero columns in $[\mathbf{c}]$ all with a (nonzero) even number of 1s. Therefore $\mathrm{wt}(\mathbf{c}) \geq 8$. Suppose now that $\mathrm{Pr}(\mathbf{c}) = \mathbf{0}$. The only column vectors of even parity that project to $0 \in \mathbb{F}_4$ are the all-zero column and the all-one column. Since the top row of $[\mathbf{c}]$ has even parity, $[\mathbf{c}]$ must contain an even number of all-one columns. As $\mathbf{c} \neq \mathbf{0}$, we again have $\mathrm{wt}(\mathbf{c}) \geq 8$.

Suppose now that the top row of $[\mathbf{c}]$ has odd parity. Thus every column of $[\mathbf{c}]$ has an odd number of 1s and so there are at least six 1s in $[\mathbf{c}]$; we are done if one column has three 1s. So assume that every column of $[\mathbf{c}]$ has exactly one 1. In particular, the top row of $[\mathbf{c}]$ has a 1 in a given column precisely when that column projects to $0 \in \mathbb{F}_4$. Thus the number of 1s in the top row is $6 - \mathrm{wt}(\mathrm{Pr}(\mathbf{c}))$. However, the codewords of $\mathcal{G}_6$ have even weight and so the top row of $[\mathbf{c}]$ has even parity, which is a contradiction.   □

For the remainder of this section the code C in Lemma 10.3.5 will be denoted $\mathcal{G}_{24}$. In order to decode $\mathcal{G}_{24}$ by hand, we need to be able to recognize the codewords in $\mathcal{G}_6$. If we label the coordinates by $1, 2, 3, 4, 5, 6$, then the coordinates break into pairs $(1, 2)$, $(3, 4)$, and $(5, 6)$. There are permutation automorphisms that permute these three pairs in any manner and switch the order within any two pairs; see Exercise 583. We list the codewords

Table 10.1 *Nonzero codewords in the hexacode*

Group number	Hexacode codeword			Number in group
I	0 1	0 1	$\omega\overline{\omega}$	$3\times 12 = 36$
II	$\omega\overline{\omega}$	$\omega\overline{\omega}$	$\omega\overline{\omega}$	$3\times\ 4 = 12$
III	0 0	1 1	1 1	$3\times\ 3 = 9$
IV	1 1	$\omega\omega$	$\overline{\omega}\overline{\omega}$	$3\times\ 2 = 6$

in $\mathcal{G}_6$ in Table 10.1 in groups, telling how many are in each group. Each codeword in the table is separated into pairs. Additional codewords in the group are obtained by multiplying by ω or $\overline{\omega}$, permuting the pairs in any order, switching the order in any two pairs, and by any combination of these. In the table, the column "Number in group" tells how many codewords can be obtained by these operations. For example the group containing 00 11 11 also contains 11 00 11 and 11 11 00 together with all multiples by ω and $\overline{\omega}$.

Exercise 583 Do the following:
(a) Show that the permutations $p_1 = (1, 3)(2, 4)$, $p_2 = (1, 3, 5)(2, 4, 6)$, and $p_3 = (1, 2)(3, 4)$ are in $\mathrm{PAut}(\mathcal{G}_6)$.
(b) Find a product of p_1, p_2, and p_3 that:
  (i) interchanges the pairs $(3, 4)$ and $(5, 6)$,
  (ii) sends $(3, 4)$ to $(4, 3)$ and simultaneously $(5, 6)$ to $(6, 5)$. ◆

Exercise 584 Show that the number of codewords in each group of Table 10.1 is correct. ◆

Exercise 585 Decide if the following vectors are in the hexacode $\mathcal{G}_6$:
(a) $10\overline{\omega}\omega01$,
(b) $\omega\omega0011$,
(c) $0\overline{\omega}\overline{\omega}0\omega1$,
(d) $1\omega\omega1\omega1$. ◆

We now demonstrate how to decode a received vector $\mathbf{y} = \mathbf{c} + \mathbf{e}$ where $\mathbf{c} \in \mathcal{G}_{24}$ and $\mathrm{wt}(\mathbf{e}) \leq 3$. The decoding falls naturally into four cases depending on the parities of the columns of $[\mathbf{y}]$. The received vector uniquely determines which case we are in. In particular, if the received vector has odd weight, then one or three errors have occurred and we are in Case 1 or Case 3 below. (Note that the case number is **not** the number of errors.) If the received vector has even weight, then zero or two errors have occurred and we are in Case 2 or Case 4. If we cannot correct the received vector by following the procedure in the appropriate case, then more than three errors have been made. Because the covering radius of $\mathcal{G}_{24}$ is 4, if more than four errors are made, the nearest neighbor is not the correct codeword and the procedure will fail to produce $\mathbf{c}$. If four errors have been made, then Case 2 or Case 4 will arise; with more work, we can find six codewords within distance 4 of the received vector using these same techniques as in Cases 2 or 4. We will not describe the procedure for handling four errors; the interested reader can consult [268]. Please keep in mind that we are assuming at most three errors have been made.

Case 1: Three columns of [y] have even parity and three have odd parity

In this case, we know that three errors must have occurred and either there is one error in each odd parity column or there is one error in each even parity column. We look at each possibility separately. With each choice, we know three components of Pr(y) that are correct and hence there is only one way to fill in the coordinates in error to obtain a codeword in $\mathcal{G}_6$.

Example 10.3.6 Suppose $\mathbf{y} = 101000011100001000011001$. Then we obtain

$$
\text{Pr}(\mathbf{y}) =
\begin{matrix}
0 \\ 1 \\ \omega \\ \overline{\omega}
\end{matrix}
\begin{bmatrix}
1 & 0 & 1 & 0 & 0 & 1 \\
0 & 0 & 1 & 0 & 0 & 0 \\
1 & 0 & 0 & 1 & 0 & 0 \\
0 & 1 & 0 & 0 & 1 & 1
\end{bmatrix}
$$
$$
\qquad\quad \omega \;\; \overline{\omega} \;\; 1 \;\; \omega \;\; \overline{\omega} \;\; \overline{\omega}
$$

Notice that columns 1, 3, and 6 have even parity and columns 2, 4, and 5 have odd parity. Suppose the columns with odd parity are correct. Then $\text{Pr}(\mathbf{c}) = \star\overline{\omega}\star\omega\overline{\omega}\star$, where we must fill in the $\star$s. The only possibility for $\text{Pr}(\mathbf{c})$ is from group I or II in Table 10.1. If it is in group I, it must be $0\overline{\omega}\star\omega\overline{\omega}0$, which cannot be completed. So $\text{Pr}(\mathbf{c}) = \omega\overline{\omega}\overline{\omega}\omega\overline{\omega}\omega$. Since we make only one change in each of columns 1, 3, and 6, we obtain

$$
\begin{matrix}
0 \\ 1 \\ \omega \\ \overline{\omega}
\end{matrix}
\begin{bmatrix}
0 & 0 & 1 & 0 & 0 & 1 \\
0 & 0 & 1 & 0 & 0 & 1 \\
1 & 0 & 1 & 1 & 0 & 0 \\
0 & 1 & 0 & 0 & 1 & 1
\end{bmatrix}
$$
$$
\qquad\quad \omega \;\; \overline{\omega} \;\; \overline{\omega} \;\; \omega \;\; \overline{\omega} \;\; \omega
$$

However, the columns have odd parity while the first row has even parity, which is a contradiction. So the columns with even parity are correct and $\text{Pr}(\mathbf{c}) = \omega\star1\star\star\overline{\omega}$. The only possibility for $\text{Pr}(\mathbf{c})$ is from group IV; hence $\text{Pr}(\mathbf{c}) = \omega\omega11\overline{\omega}\overline{\omega}$. Since we make only one change in each of columns 2, 4, and 5, we obtain:

$$
\begin{matrix}
0 \\ 1 \\ \omega \\ \overline{\omega}
\end{matrix}
\begin{bmatrix}
1 & 0 & 1 & 0 & 1 & 1 \\
0 & 1 & 1 & 0 & 0 & 0 \\
1 & 0 & 0 & 1 & 0 & 0 \\
0 & 1 & 0 & 1 & 1 & 1
\end{bmatrix}
$$
$$
\qquad\quad \omega \;\; \omega \;\; 1 \;\; 1 \;\; \overline{\omega} \;\; \overline{\omega}
$$

As all columns and the first row have even parity, this is the codeword.   ∎

Case 2: Exactly four columns of [y] have the same parity

As we are assuming that at most three errors have been made, we know that the four columns with the same parity are correct and the other two columns each contain exactly one error.

Example 10.3.7 Suppose $\mathbf{y} = 100011110100000000101000$. Then:

$$
\Pr(\mathbf{y}) =
\begin{array}{c}
0 \\ 1 \\ \omega \\ \overline{\omega}
\end{array}
\left[
\begin{array}{cccccc}
1 & 1 & 0 & 0 & 0 & 1 \\
0 & 1 & 1 & 0 & 0 & 0 \\
0 & 1 & 0 & 0 & 1 & 0 \\
0 & 1 & 0 & 0 & 0 & 0
\end{array}
\right].
$$
$$
\ 0 \quad 0 \quad 1 \quad 0 \quad \omega \quad 0
$$

Thus columns 1, 3, 5, and 6 have odd parity and 2 and 4 have even parity. So $\Pr(\mathbf{c}) = 0\star1\star\omega0$. This must be in group I; hence $\Pr(\mathbf{c}) = 0\omega1\overline{\omega}\omega0$. We must make one change in each of columns 2 and 4 so that the top row has odd parity. This gives

$$
\begin{array}{c}
0 \\ 1 \\ \omega \\ \overline{\omega}
\end{array}
\left[
\begin{array}{cccccc}
1 & 1 & 0 & 0 & 0 & 1 \\
0 & 1 & 1 & 0 & 0 & 0 \\
0 & 0 & 0 & 0 & 1 & 0 \\
0 & 1 & 0 & 1 & 0 & 0
\end{array}
\right].
$$
$$
\ 0 \quad \omega \quad 1 \quad \overline{\omega} \quad \omega \quad 0
$$

Since the columns and first row all have odd parity, this is the codeword. ∎

Case 3: Exactly five columns of [y] have the same parity

As at most three errors have been made, either the five columns with the same parity are correct and the other column has either one or three errors, or four of the five columns are correct with two errors in the fifth and one error in the remaining column. It is easy using Table 10.1 to determine if all five columns of the same parity are correct and then to correct the remaining column. If four of the five columns are correct, a bit more analysis is required.

Example 10.3.8 Suppose $\mathbf{y} = 110010001001000010010110$. Then:

$$
\Pr(\mathbf{y}) =
\begin{array}{c}
0 \\ 1 \\ \omega \\ \overline{\omega}
\end{array}
\left[
\begin{array}{cccccc}
1 & 1 & 1 & 0 & 1 & 0 \\
1 & 0 & 0 & 0 & 0 & 1 \\
0 & 0 & 0 & 0 & 0 & 1 \\
0 & 0 & 1 & 0 & 1 & 0
\end{array}
\right].
$$
$$
\ 1 \quad 0 \quad \overline{\omega} \quad 0 \quad \overline{\omega} \quad \overline{\omega}
$$

Column 2 is the only column of odd parity, and so one or three errors have been made in this column. If all but column 2 of $\Pr(\mathbf{y})$ is correct, it is easy to see that no change in column 2 will yield a hexacode codeword from Table 10.1. So four of columns 1, 3, 4, 5, and 6 are correct. If columns 5 and 6 are both correct, then $\Pr(\mathbf{c})$ must be in either group III or IV in which case two of columns 1, 3, and 4 must be in error; this is an impossibility. In particular, columns 1, 3, and 4 are correct, and $\Pr(\mathbf{c}) = 1\star\overline{\omega}0\star\star$, which must be in group I.

The only possibility is $\Pr(\mathbf{c}) = 1\omega\overline{\omega}0\overline{\omega}0$. Thus the correct codeword must be:

$$
\begin{array}{c}
0 \\
1 \\
\omega \\
\overline{\omega}
\end{array}
\left[
\begin{array}{cccccc}
1 & 1 & 1 & 0 & 1 & 0 \\
1 & 0 & 0 & 0 & 0 & 0 \\
0 & 1 & 0 & 0 & 0 & 0 \\
0 & 0 & 1 & 0 & 1 & 0
\end{array}
\right].
$$
$$
1 \quad \omega \quad \overline{\omega} \quad 0 \quad \overline{\omega} \quad 0
$$

Note that the columns and first row all have even parity. ∎

Case 4: All six columns of [y] have the same parity

This is the easiest case to solve. If the top row has the same parity as the columns and $\Pr(\mathbf{y})$ is in $\mathcal{G}_6$, no errors were made. Otherwise, as we are assuming that at most three errors have been made and because the parity of the six columns is correct, any column that is incorrect must have two errors. Hence at most one column is incorrect. The correct vector $\Pr(\mathbf{c})$ is easy to find and hence the incorrect column, if there is one, is then clear.

Example 10.3.9 Suppose $\mathbf{y} = 011100101011000111011101$. Then:

$$
\Pr(\mathbf{y}) =
\begin{array}{c}
0 \\
1 \\
\omega \\
\overline{\omega}
\end{array}
\left[
\begin{array}{cccccc}
0 & 0 & 1 & 0 & 1 & 1 \\
1 & 0 & 0 & 0 & 1 & 1 \\
1 & 1 & 1 & 0 & 0 & 0 \\
1 & 0 & 1 & 1 & 1 & 1
\end{array}
\right].
$$
$$
0 \quad \omega \quad 1 \quad \overline{\omega} \quad \omega \quad \omega
$$

Clearly, all columns cannot be correct, and it is impossible for both columns 5 and 6 to be correct. Thus as columns 1, 2, 3, and 4 must be correct, the only possibility is for $\Pr(\mathbf{c})$ to be in group I and $\Pr(\mathbf{c}) = 0\omega1\overline{\omega}\omega0$. Thus two entries in column 6 must be changed and the correct codeword is

$$
\begin{array}{c}
0 \\
1 \\
\omega \\
\overline{\omega}
\end{array}
\left[
\begin{array}{cccccc}
0 & 0 & 1 & 0 & 1 & 1 \\
1 & 0 & 0 & 0 & 1 & 0 \\
1 & 1 & 1 & 0 & 0 & 0 \\
1 & 0 & 1 & 1 & 1 & 0
\end{array}
\right].
$$
$$
0 \quad \omega \quad 1 \quad \overline{\omega} \quad \omega \quad 0
$$

■

Exercise 586 Decode the following received vectors when $\mathcal{G}_{24}$ is used in transmission or show that more than three errors have been made:
(a) 001111001111101011000011,
(b) 110011010110111010010100,
(c) 000101001101111010111001,
(d) 101010111100100100100101,
(e) 101110111111010000100110. ◆

The tetracode can be used in a similar manner to decode the $[12, 6, 6]$ extended ternary Golay code; see [268]. This technique has been generalized in [143].

10.4   The ternary Golay codes

We now examine in more detail the $[11, 6, 5]$ and $[12, 6, 6]$ ternary Golay codes we first constructed in Section 1.9.2. Our original construction of the $[12, 6, 6]$ code used a bordered double circulant generator matrix. As in the binary case, the codes are unique and have relatively large automorphism groups.

10.4.1   Uniqueness of the ternary Golay codes

The uniqueness of the ternary Golay codes was proved in [65, 260, 269]; our proof follows these partially and is similar to the binary case. We first prove the uniqueness of the extended ternary Golay code through a series of exercises.

Exercise 587   Let C be a (possibly nonlinear) ternary code of length 11 and minimum distance 5 containing $M \geq 3^6$ codewords. Show that C has exactly 3^6 codewords and is perfect. Hint: Use the Sphere Packing Bound.   ◆

Exercise 588   Let C be a (possibly nonlinear) ternary code of length 11 and minimum distance 5 containing $M \geq 3^6$ codewords including the $\mathbf{0}$ codeword. By Exercise 587, C contains 3^6 codewords and is perfect. Show that the weight enumerator of C is

$$W_C(x, y) = y^{11} + 132x^5 y^6 + 132x^6 y^5 + 330x^8 y^3 + 110x^9 y^2 + 24x^{11}.$$

Hint: Mimic the proof of Lemma 10.1.2. Let $N(w, c)$ denote the number of vectors in $\mathbb{F}_3^{11}$ of weight w in a sphere of radius 2 centered at a codeword of weight c. The equation analogous to (10.2) is

$$\sum_{c=0}^{11} N(w, c)A_c = \binom{11}{w} 2^w.$$

The computations of $N(w, c)$ are more complicated than in the proof of Lemma 10.1.2.   ◆

Exercise 589   Let C' be a (possibly nonlinear) ternary code of length 12 and minimum distance 6 containing $M \geq 3^6$ codewords including the $\mathbf{0}$ codeword. Show that C' has exactly 3^6 codewords and weight enumerator

$$W_{C'}(x, y) = y^{12} + 264x^6 y^6 + 440x^9 y^3 + 24x^{12}.$$

Hint: Puncturing C' in any coordinate yields a code described in Exercises 587 and 588.   ◆

Exercise 590   Show that if $\mathbf{c}$ and $\mathbf{c}'$ are vectors in $\mathbb{F}_3^n$ such that wt($\mathbf{c}$), wt($\mathbf{c}'$), and wt($\mathbf{c} - \mathbf{c}'$) are all 0 modulo 3, then $\mathbf{c}$ and $\mathbf{c}'$ are orthogonal to each other. Hint: You may assume the nonzero entries of $\mathbf{c}$ all equal 1 as rescaling coordinates does not affect weight or orthogonality.   ◆

Exercise 591 Let $\mathcal{D}$ be a (possibly nonlinear) ternary code of length 12 and minimum distance 6 containing $M \geq 3^6$ codewords including the **0** codeword. Show that $\mathcal{D}$ is a $[12, 6, 6]$ self-orthogonal linear code. Hint: If $\mathbf{c} \in \mathcal{D}$, then $\mathcal{D}$ and $\mathbf{c} - \mathcal{D}$ satisfy the conditions of $\mathcal{C}'$ given in Exercise 589. Use Exercise 590.   ♦

Exercise 592 Let $\mathcal{C}$ be a $[12, 6, 6]$ ternary code. You will show that up to equivalence $\mathcal{C}$ has a unique generator matrix. Recall that given a generator matrix G for $\mathcal{C}$, an equivalent code is produced from the generator matrix obtained from G by permuting columns and/or multiplying columns by ± 1. You can also rearrange or rescale rows and obtain the same code. Without loss of generality we may assume that G is in standard form

$$G = \begin{bmatrix} I_6 & A \end{bmatrix},$$

where A is a 6×6 matrix, whose ith row is denoted $\mathbf{a}_i$. We label the columns of G by $1, 2, \ldots, 12$.

(a) Prove that for $1 \leq i \leq 6$, $\mathbf{a}_i$ has at most one component equal to 0.

(b) Prove that it is not possible for two different rows $\mathbf{a}_i$ and $\mathbf{a}_j$ of A to have components that agree in three or more coordinate positions.

(c) Prove that it is not possible for two different rows $\mathbf{a}_i$ and $\mathbf{a}_j$ of A to have components that are negatives of each other in three or more coordinate positions.

(d) Prove that every row of A has weight 5.

(e) Prove that no two rows of A have 0 in the same coordinate position.

(f) Prove that we may assume that

$$A = \begin{bmatrix} 0 & 1 & 1 & 1 & 1 & 1 \\ 1 & & & & & \\ 1 & & & & & \\ 1 & & & & & \\ 1 & & & & & \\ 1 & & & & & \end{bmatrix},$$

where the blank entries in each row are to be filled in with one 0, two 1s, and two -1s. Hint: You may have to rescale columns $2, \ldots, 6$ and rows $2, \ldots, 6$ of G to obtain the desired column 7 of G.

(g) Prove that we may assume that

$$A = \begin{bmatrix} 0 & 1 & 1 & 1 & 1 & 1 \\ 1 & 0 & 1 & -1 & -1 & 1 \\ 1 & & 0 & & & \\ 1 & & & 0 & & \\ 1 & & & & 0 & \\ 1 & & & & & 0 \end{bmatrix},$$

where the blank entries in each row are to be filled in with two 1s and two -1s.

(h) Show that, up to equivalence, $\mathbf{a}_3 = (1, 1, 0, 1, -1, -1)$.

(i) Prove that $\mathbf{a}_4$, $\mathbf{a}_5$, and $\mathbf{a}_6$ are uniquely determined and that

$$A = \begin{bmatrix} 0 & 1 & 1 & 1 & 1 & 1 \\ 1 & 0 & 1 & -1 & -1 & 1 \\ 1 & 1 & 0 & 1 & -1 & -1 \\ 1 & -1 & 1 & 0 & 1 & -1 \\ 1 & -1 & -1 & 1 & 0 & 1 \\ 1 & 1 & -1 & -1 & 1 & 0 \end{bmatrix}.$$

The generator matrix produced above is precisely the one given in Section 1.9.2 for the extended ternary Golay code which we showed in that section is indeed a $[12, 6, 6]$ code. ◆

Exercises 587–592 show that a, possibly nonlinear, ternary code of length 12 containing $\mathbf{0}$ with at least 3^6 codewords and minimum distance 6 is, up to equivalence, the $[12, 6, 6]$ extended ternary Golay code.

We now turn to a ternary code of length 11. Let C be a possibly nonlinear ternary code of length 11 and minimum distance 5 containing $M \geq 3^6$ codewords including the $\mathbf{0}$ codeword. We will show that this code is an $[11, 6, 5]$ linear code unique up to equivalence. By Exercises 587 and 588, C is a perfect code with 3^6 codewords including codewords of weight 11. By rescaling coordinates, we may assume that C contains the all-one codeword $\mathbf{1}$.

Note that if $\mathbf{c} \in C$, then $\mathbf{c} - C$ is a code of length 11 and minimum distance 5 containing 3^6 codewords including the $\mathbf{0}$ codeword. Thus its weight distribution is given in Exercise 588, implying that

$$d(\mathbf{c}, \mathbf{c}') \in \{5, 6, 8, 9, 11\} \tag{10.5}$$

for distinct codewords $\mathbf{c}$ and $\mathbf{c}'$ in C.

We wish to show that $-\mathbf{1}$ is in C. To do this we establish some notation. A vector in $\mathbb{F}_3^{11}$ has "type" $(-1)^a 1^b 0^c$ if it has a -1s, b 1s, and c 0s as components. Suppose that $\mathbf{x} \notin C$. As C is perfect, there is a unique codeword $\mathbf{c}$ at distance 2 or less from $\mathbf{x}$; we say that $\mathbf{c}$ "covers" $\mathbf{x}$. Assume $-\mathbf{1} \notin C$. Let $\mathbf{x}_i$ be the vector in $\mathbb{F}_3^{11}$ with 1 in coordinate i and -1 in the other ten coordinates. If some $\mathbf{x}_i$ is in C, then the distance between $\mathbf{x}_i$ and $\mathbf{1}$ is 10, contradicting (10.5). Let $\mathbf{c}_i$ cover $\mathbf{x}_i$. In particular $\mathrm{wt}(\mathbf{c}_i)$ is 9 or 11.

First assume that $\mathrm{wt}(\mathbf{c}_i) = 11$ for some i. No codeword can have exactly one 1 as its distance to $\mathbf{1}$ is 10, contradicting (10.5). Thus $\mathbf{c}_i$ must have at least two 1s. To cover $\mathbf{x}_i$, the only choices for $\mathbf{c}_i$ are type $(-1)^9 1^2$ or $(-1)^8 1^3$. In either case, a 1 is in coordinate i. Now assume that $\mathrm{wt}(\mathbf{c}_i) = 9$ for some i. Then $\mathbf{c}_i$ must have type $(-1)^9 0^2$ with a 0 in coordinate i. Consider the possibilities for a $\mathbf{c}_i$ and $\mathbf{c}_j$. If they are unequal, their distance apart must be at least 5. This implies that both cannot have type $(-1)^9 a^2$ and $(-1)^9 b^2$ where a and b are not -1. Also both cannot have type $(-1)^8 1^3$ with a 1 in a common coordinate. In particular it is impossible for all $\mathbf{c}_i$ to have type $(-1)^8 1^3$ since 11 is not a multiple of 3. Exercise 593 shows that by rearranging coordinates, the $\mathbf{c}_i$s are given in Figure 10.2.

Exercise 593 Fill in the details verifying that $\mathbf{c}_1, \ldots, \mathbf{c}_{11}$ are given in Figure 10.2. ◆

	1	2	3	4	5	6	7	8	9	10	11
$c_1 = c_2$	a	a	-1	-1	-1	-1	-1	-1	-1	-1	-1
$c_3 = c_4 = c_5$	-1	-1	1	1	1	-1	-1	-1	-1	-1	-1
$c_6 = c_7 = c_8$	-1	-1	-1	-1	-1	1	1	1	-1	-1	-1
$c_9 = c_{10} = c_{11}$	-1	-1	-1	-1	-1	-1	-1	-1	1	1	1

Figure 10.2 Configuration for $c_1, \ldots, c_{11}$, where $a = 0$ or $a = 1$.

Now let $\mathbf{x}$ have type $(-1)^9 1^2$ with 1s in coordinates 1 and 3. Since $d(\mathbf{x}, c_1) \leq 3$, $\mathbf{x} \notin C$. Suppose that c covers $\mathbf{x}$. Again c has weight 9 or 11. If $\mathrm{wt}(c) = 9$, then it has type $(-1)^9 0^2$ with 0s in coordinates 1 and 3. Then $d(c, c_1) \leq 3$, which is a contradiction. So $\mathrm{wt}(c) = 11$. If c has type $(-1)^a 1^b$, $b \leq 4$ since c covers $\mathbf{x}$. If $b = 1$ or $b = 4$, then $d(c, \mathbf{1})$ is 10 or 7, a contradiction. If $b = 2$, then $d(c, c_1) \leq 4$, a contradiction. So $b = 3$ and c must have 1s in both coordinates 1 and 3. Again $d(c, c_1) \leq 4$, a contradiction. Therefore $-\mathbf{1} \in C$.

So C has codewords $\mathbf{0}$, $\mathbf{1}$, and $-\mathbf{1}$. Furthermore, if $c \in C$ with $\mathrm{wt}(c) = 11$, $-c \in C$ as follows. Choose a diagonal matrix D so that $cD = \mathbf{1}$. Then CD has minimum distance 5 and codewords $\mathbf{0}$ and $\mathbf{1} = cD$. By the preceding argument, $-\mathbf{1} \in CD$ implying that $-cD \in CD$. So $-c \in C$.

In summary, we know the following. C is a perfect $(11, 3^6)$ code with minimum distance 5 containing $\mathbf{0}$ and $\mathbf{1}$. Furthermore, the negative of every weight 11 codeword of C is also in C. We next show that the same is true of all weight 9 codewords. Exercise 594 gives the possible types of codewords in C.

Exercise 594 Show that codewords of the given weight have only the listed types:
(a) weight 5: $1^5 0^6$, $(-1)^5 0^6$, $(-1)^2 1^3 0^6$, $(-1)^3 1^2 0^6$,
(b) weight 6: $(-1)^3 1^3 0^6$, $(-1)^6 0^6$, $1^6 0^6$,
(c) weight 8: $(-1)^2 1^6 0^3$, $(-1)^3 1^5 0^3$, $(-1)^5 1^3 0^3$, $(-1)^6 1^2 0^3$,
(d) weight 9: $(-1)^3 1^6 0^2$, $(-1)^6 1^3 0^2$,
(e) weight 11: $(-1)^{11}$, 1^{11}, $(-1)^5 1^6$, $(-1)^6 1^5$.
Hint: Consider each possible type for a given weight and compute its distance to each of $\mathbf{1}$ and $-\mathbf{1}$. Use (10.5). $\blacklozenge$

To show the negative of a weight 9 codeword c is also a codeword, rescaling all 11 coordinates by -1 if necessary, we may assume c has type $(-1)^3 1^6 0^2$ by Exercise 594. By permuting coordinates

$$c = (1, 1, 1, 1, 1, 1, -1, -1, -1, 0, 0).$$

Assume $-c \notin C$. Consider the weight 11 vector

$$\mathbf{x}_1 = (-1, -1, -1, -1, -1, -1, 1, 1, 1, 1, 1).$$

If $\mathbf{x}_1 \in C$, then $-\mathbf{x}_1 \in C$ and $d(-\mathbf{x}_1, c) = 2$, contradicting (10.5). Suppose c_1 covers $\mathbf{x}_1$. Then $\mathrm{wt}(c_1)$ is 9 or 11. By reordering coordinates and using Exercise 594, the only possibilities for c_1 are

$$(-1, -1, -1, -1, -1, 1, -1, 1, 1, 1, 1, 1),$$
$$(-1, -1, -1, -1, -1, 1, 1, 1, 1, 1, 1, 1),$$
$$(-1, -1, -1, -1, -1, 1, 1, 1, 1, 1, 1, -1),$$
$$(-1, -1, -1, -1, -1, -1, 1, 1, 0, 1, 0), \quad \text{or}$$
$$(-1, -1, -1, -1, -1, -1, 1, 0, 0, 1, 1).$$

In the first case, $d(-\mathbf{c}_1, \mathbf{c}) = 4$ and in the next three cases, $d(\mathbf{c}_1, \mathbf{c}) = 10$, contradicting (10.5). Only the last one is possible. Let

$$\mathbf{x}_2 = (-1, -1, -1, -1, -1, -1, 1, 1, 1, 1, -1),$$

which is not in C by Exercise 594. Suppose $\mathbf{c}_2$ covers $\mathbf{x}_2$. By testing possibilities as above and after reordering coordinates we may assume

$$\mathbf{c}_2 = (-1, -1, -1, -1, -1, 0, 0, 1, 1, 1, -1).$$

Finally, let

$$\mathbf{x}_3 = (-1, -1, -1, -1, -1, -1, 1, 1, 1, -1, 1),$$

which again is not in C by Exercise 594. No codeword $\mathbf{c}_3$ can be constructed covering $\mathbf{x}_3$ under the assumption that $\mathbf{c}_1$ and $\mathbf{c}_2$ are in C. This proves that the negative of a weight 9 codeword $\mathbf{c}$ is also a codeword.

Exercise 595 Fill in the details showing that $\mathbf{c}_2$ is as claimed and $\mathbf{c}_3$ does not exist.   ◆

Let $\mathcal{D}$ be the subcode of C consisting of the 110 codewords of weight 9. Let $\mathbf{x} \in C$ and $\mathbf{y} \in \mathcal{D}$. We show that $\mathbf{x}$ and $\mathbf{y}$ are orthogonal. By Exercise 596, since $\mathrm{wt}(\mathbf{y}) \equiv 0 \,(\mathrm{mod}\ 3)$,

$$\mathrm{wt}(\mathbf{x}) + \mathbf{x} \cdot \mathbf{y} \equiv d(\mathbf{x}, \mathbf{y}) \,(\mathrm{mod}\ 3) \text{ and} \tag{10.6}$$
$$\mathrm{wt}(\mathbf{x}) - \mathbf{x} \cdot \mathbf{y} \equiv d(\mathbf{x}, -\mathbf{y}) \,(\mathrm{mod}\ 3). \tag{10.7}$$

Since $\mathbf{y} \in \mathcal{D}$ and $-\mathbf{y} \in \mathcal{D}$, $d(\mathbf{x}, \mathbf{y})$ and $d(\mathbf{x}, -\mathbf{y})$ can each only be 0 or 2 modulo 3. If $\mathrm{wt}(\mathbf{x}) \equiv 0 \,(\mathrm{mod}\ 3)$ and either $d(\mathbf{x}, \mathbf{y})$ or $d(\mathbf{x}, -\mathbf{y})$ is 0 modulo 3, then $\mathbf{x} \cdot \mathbf{y} \equiv 0 \,(\mathrm{mod}\ 3)$ by (10.6) or (10.7), respectively. If $d(\mathbf{x}, \mathbf{y}) = d(\mathbf{x}, -\mathbf{y}) \equiv 2 \,(\mathrm{mod}\ 3)$, then $\mathbf{x} \cdot \mathbf{y} \equiv 2 \,(\mathrm{mod}\ 3)$ by (10.6) and $\mathbf{x} \cdot \mathbf{y} \equiv 1 \,(\mathrm{mod}\ 3)$ by (10.7), a conflict. If $\mathrm{wt}(\mathbf{x}) \equiv 2 \,(\mathrm{mod}\ 3)$ and either $d(\mathbf{x}, \mathbf{y})$ or $d(\mathbf{x}, -\mathbf{y})$ is 2 modulo 3, then $\mathbf{x} \cdot \mathbf{y} \equiv 0 \,(\mathrm{mod}\ 3)$ by (10.6) or (10.7), respectively. If $d(\mathbf{x}, \mathbf{y}) = d(\mathbf{x}, -\mathbf{y}) \equiv 0 \,(\mathrm{mod}\ 3)$, then $\mathbf{x} \cdot \mathbf{y} \equiv 1 \,(\mathrm{mod}\ 3)$ by (10.6) and $\mathbf{x} \cdot \mathbf{y} \equiv 2 \,(\mathrm{mod}\ 3)$ by (10.7), a conflict. In all cases $\mathbf{x}$ is orthogonal to $\mathbf{y}$.

Therefore the linear code $\mathcal{D}_1$ spanned by the codewords of $\mathcal{D}$ is orthogonal to the linear code C_1 spanned by C. As $\mathcal{D}$ has 110 codewords and $110 > 3^4$, $\mathcal{D}_1$ has dimension at least 5. As C has 3^6 codewords, C_1 has dimension at least 6. Since the sum of the dimensions of $\mathcal{D}_1$ and C_1 is at most 11, the dimensions of C_1 and $\mathcal{D}_1$ are exactly 6 and 5, respectively. Thus $C_1 = C$ implying C is linear. If we extend C, we obtain a $[12, 6]$ linear code $\widehat{C}$. By Exercise 594 the weight 5 codewords of C extend to weight 6 implying $\widehat{C}$ is a $[12, 6, 6]$ linear code. By Exercise 592, $\widehat{C}$ is the extended ternary Golay code. In the next subsection we will see that $\widehat{C}$ has a transitive automorphism group; by Theorem 1.7.13, all 12 codes obtained by puncturing $\widehat{C}$ in any of its coordinates are equivalent. Thus C is unique.

Exercise 596 Prove that if $\mathbf{x}$ and $\mathbf{y}$ are in $\mathbb{F}_3^n$, then

$$d(\mathbf{x}, \mathbf{y}) \equiv \mathrm{wt}(\mathbf{x}) + \mathrm{wt}(\mathbf{y}) + \mathbf{x} \cdot \mathbf{y} \ (\mathrm{mod}\ 3).$$

Hint: By rescaling, which does not change distance or inner product modulo 3, you may assume all nonzero components of $\mathbf{x}$ equal 1. ◆

This discussion and the accompanying exercises have proved the following theorem.

Theorem 10.4.1 *Both length* 11 *and length* 12, *possibly nonlinear, ternary codes each containing* $\mathbf{0}$ *with* $M \geq 3^6$ *codewords and minimum distance* 5 *and* 6, *respectively, are unique up to equivalence. They are the* $[11, 6, 5]$ *and* $[12, 6, 6]$ *ternary Golay codes.*

10.4.2 Properties of ternary Golay codes

In Example 6.6.11, we gave generating idempotents for the odd-like ternary quadratic residue codes of length 11. One such code has generating idempotent

$$e_{11}(x) = -(x^2 + x^6 + x^7 + x^8 + x^{10})$$

and generator polynomial

$$g_{11}(x) = -1 + x^2 - x^3 + x^4 + x^5.$$

This code has minimum distance $d_0 \geq 4$ by Theorem 6.6.22. By Theorem 6.6.14, the extended code is a $[12, 6]$ self-dual code with minimum distance a multiple of 3. Therefore $d_0 \geq 5$, and by Theorem 10.4.1 this is one representation of the $[11, 6, 5]$ ternary Golay code $\mathcal{G}_{11}$. Extending this odd-like quadratic residue code gives one representation of the $[12, 6, 6]$ extended ternary Golay code $\mathcal{G}_{12}$.

Exercise 597 Verify that $g_{11}(x)$ is the generator polynomial for the cyclic code with generating idempotent $e_{11}(x)$. ◆

Exercise 598 Create the generator matrix for the extended quadratic residue code extending the cyclic code with generator polynomial $g_{11}(x)$, adding the parity check coordinate on the right. Row reduce this generator matrix and show that columns can indeed be permuted and scaled to obtain the generator matrix in Exercise 592(i). ◆

As in Section 10.1.2, the following permutations are automorphisms of $\mathcal{G}_{12}$, viewed as the extended quadratic residue code obtained by extending the cyclic code with generating idempotent $e_{11}(x)$, where the coordinates are labeled $0, 1, \ldots, 10, \infty$:

$$s_{12} = (0, 1, 2, 3, 4, 5, 6, 7, 8, 9, 10), \quad \text{and}$$
$$\mu_3 = (1, 3, 9, 5, 4)(2, 6, 7, 10, 8).$$

The element from the Gleason–Prange Theorem is

$$v_{12} = \mathrm{diag}(1, 1, -1, 1, 1, 1, -1, -1, -1, 1, -1, -1)$$
$$\times (0, \infty)(1, 10)(2, 5)(3, 7)(4, 8)(6, 9).$$

These three elements along with

$$\eta_{12} = (1, 9)(4, 5)(6, 8)(7, 10)$$

generate $\Gamma\mathrm{Aut}(\mathcal{G}_{12}) = \mathrm{MAut}(\mathcal{G}_{12})$. This group has order $95040 \cdot 2$, and is isomorphic to a group sometimes denoted $\widetilde{\mathcal{M}}_{12}$.[1] This group has a center of order 2, and modulo this center, the quotient group is the 5-fold transitive Mathieu group $\mathcal{M}_{12}$ of order $95040 = 12 \cdot 11 \cdot 10 \cdot 9 \cdot 8$. This quotient group can be obtained by dropping the diagonal part of all elements of $\mathrm{MAut}(\mathcal{G}_{12})$; this is the group $\mathrm{MAut}_{\mathrm{Pr}}(\widehat{\mathcal{G}}_{12})$ defined in Section 1.7 to be the set $\{P \mid DP \in \mathrm{MAut}(\widehat{G}_{12})\}$.

Exercise 599 Verify that ν_{12} and η_{12} are in $\Gamma\mathrm{Aut}(\mathcal{G}_{12})$. See Exercise 575.   ◆

Again we can find the covering radius of each ternary Golay code.

Theorem 10.4.2 *The covering radii of the* $[11, 6, 5]$ *and* $[12, 6, 6]$ *ternary Golay codes are 2 and 3, respectively.*

Exercise 600 Prove Theorem 10.4.2.   ◆

The next example illustrates a construction of the extended ternary Golay code from the projective plane of order 3; it is due to Drápal [73].

Example 10.4.3 The projective plane PG(2, 3) has 13 points and 13 lines with four points per line and four lines through any point; see Theorem 8.6.2. Label the points $0, 1, 2, \ldots, 12$. Denote the four lines through 0 by $\ell_1, \ldots, \ell_4$ and the remaining lines by $\ell_5, \ldots, \ell_{13}$. Form vectors $\mathbf{v}_i$ in $\mathbb{F}_3^{13}$ of weight 4 with $\mathrm{supp}(\mathbf{v}_i) = \ell_i$ whose nonzero components are 1 if $1 \leq i \leq 4$ and 2 if $5 \leq i \leq 13$. For $1 \leq i < j \leq 13$, let $\mathbf{v}_{i,j} = \mathbf{v}_i - \mathbf{v}_j$ and let $\mathbf{v}_{i,j}^*$ be $\mathbf{v}_{i,j}$ punctured on coordinate 0. Let $\mathcal{C}$ be the ternary code of length 12 spanned by $\mathbf{v}_{i,j}^*$ for $1 \leq i < j \leq 13$. Then $\mathcal{C}$ is the $[12, 6, 6]$ extended ternary Golay code as Exercise 601 shows.   ∎

Exercise 601 Use the notation of Example 10.4.3.
(a) Show that $\mathrm{wt}(\mathbf{v}_{i,j}^*) = 6$.
(b) Show that $\mathbf{v}_{i,j}^*$ is orthogonal to $\mathbf{v}_{m,n}^*$ for $1 \leq i < j \leq 13$ and $1 \leq m < n \leq 13$. Hint: Consider the inner product of $\mathbf{v}_a^*$ with $\mathbf{v}_b^*$ where these vectors are $\mathbf{v}_a$ and $\mathbf{v}_b$ punctured on coordinate 0. It will be helpful to examine the cases $1 \leq a \leq 4, 1 \leq b \leq 4, 5 \leq a \leq 13$, and $5 \leq b \leq 13$ separately.
(c) Show that by labeling the coordinates and lines appropriately we can assume that $\ell_1 = \{0, 1, 2, 3\}$, $\ell_2 = \{0, 4, 5, 6\}$, $\ell_3 = \{0, 7, 8, 9\}$, $\ell_4 = \{0, 10, 11, 12\}$, $\ell_5 = \{1, 4, 7, 10\}$, $\ell_6 = \{1, 5, 8, 11\}$, $\ell_7 = \{1, 6, 9, 12\}$, and $\ell_8 = \{2, 4, 8, 12\}$.
(d) With the lines in part (c), show that $\{\mathbf{v}_{1,j}^* \mid 2 \leq j \leq 6, j = 8\}$ are six linearly independent vectors in $\mathcal{C}$.
(e) Show that $\mathcal{C}$ is a $[12, 6]$ self-dual code.
(f) Show that $\mathcal{C}$ has minimum weight 6. Hint: Assume $\mathcal{C}$ has minimum weight $d < 6$. By Theorem 1.4.10(i), $d = 3$. By part (e), $\mathcal{C}$ is self-dual. Show that there is no weight 3 vector in $\mathbb{F}_3^{12}$ that is orthogonal to all the vectors in part (d).   ◆

---

[1] $\widetilde{\mathcal{M}}_{12}$ is technically the double cover of $\mathcal{M}_{12}$, or the nonsplitting central extension of $\mathcal{M}_{12}$ by a center of order 2.

10.5 Symmetry codes

In Section 1.9.2 and Exercise 592 we gave a construction of the $[12, 6, 6]$ extended ternary Golay code using a double circulant generator matrix. This construction was generalized by Pless in [263]. The resulting codes are known as the *Pless symmetry codes*.

The symmetry codes are self-dual ternary codes of length $2q + 2$, where q is a power of an odd prime with $q \equiv 2 \pmod 3$; these codes will be denoted $\mathcal{S}(q)$. A generator matrix for $\mathcal{S}(q)$ is $[I_{q+1} \ S_q]$, where S_q is the $(q + 1) \times (q + 1)$ matrix defined by (8.19) in connection with our construction of Hadamard matrices in Section 8.9. When q is prime, S_q is a circulant matrix and $[I_{q+1} \ S_q]$ is a double circulant generator matrix for $\mathcal{S}(q)$. Example 8.9.9 gives S_5. This is the same as the matrix A in Exercise 592 and $[I_6 \ S_5]$ is the generator matrix G_{12} given in Section 1.9.2 for the $[12, 6, 6]$ extended ternary Golay code. Using the properties of S_q developed in Theorem 8.9.11 we can show that $\mathcal{S}(q)$ is a Type III code.

Theorem 10.5.1 *Let q be a power of an odd prime with $q \equiv 2 \pmod 3$. Then, modulo 3, the following matrix equalities hold:*
(i)   *If $q \equiv 1 \pmod 4$, then $S_q = S_q^{\mathrm{T}}$.*
(ii) *If $q \equiv -1 \pmod 4$, then $S_q = -S_q^{\mathrm{T}}$.*
(iii) $S_q S_q^{\mathrm{T}} = -I_{q+1}$.
Furthermore, $\mathcal{S}(q)$ is a Type III code.

Proof: Parts (i) and (ii) are the same as parts (i) and (ii) of Theorem 8.9.11. Since $q \equiv 2 \pmod 3$, part (iii) follows from Theorem 8.9.11(iii) upon reducing modulo 3.

The inner product of a row of S_q with itself is -1 by (iii). Hence the inner product of a row of $[I_{q+1} \ S_q]$ with itself is 0. The inner product of one row of S_q with a different row is 0 by (iii). Hence the inner product of one row of $[I_{q+1} \ S_q]$ with a different row is 0. Thus $[I_{q+1} \ S_q]$ generates a self-orthogonal code; as its dimension is clearly $q + 1$, the code is Type III.   □

There is another generator matrix for $\mathcal{S}(q)$ that, when used in combination with the original generator matrix, allows one to reduce the amount of computation required to calculate the minimum weight. The proof is left as an exercise.

Corollary 10.5.2 *Let q be a power of an odd prime with $q \equiv 2 \pmod 3$.*
(i) *If $q \equiv 1 \pmod 4$, then $[-S_q \ I_{q+1}]$ is a generator matrix of $\mathcal{S}(q)$.*
(ii) *If $q \equiv -1 \pmod 4$, then $[S_q \ I_{q+1}]$ is a generator matrix of $\mathcal{S}(q)$.*

Exercise 602 Prove Corollary 10.5.2.   ◆

Example 10.5.3 We give a quick proof that $\mathcal{S}(5)$ has minimum weight 6. By Theorem 10.5.1, $\mathcal{S}(5)$ is self-dual and by Corollary 10.5.2, $[I_6 \ S_5]$ and $[-S_5 \ I_6]$ both generate $\mathcal{S}(5)$. If $\mathcal{S}(5)$ has minimum weight 3, a minimum weight codeword has weight 0 or 1 on either the left-half or right-half of the coordinates. But as the left six and right six coordinates are each information positions, the only codeword that is zero on the left- or right-half is the zero codeword. The only codewords that have weight 1 on the left- or right-half are scalar multiples of the rows of $[I_6 \ S_5]$ and $[-S_5 \ I_6]$, all of which have weight 6. Hence $\mathcal{S}(5)$ has minimum weight 6 since its minimum weight is a multiple of 3. In a similar, but more complicated, fashion you can prove that the minimum weight of $\mathcal{S}(11)$ is 9; if it is not, then a minimum weight codeword must be a combination of at most three rows of $[I_{12} \ S_{11}]$ or $[S_{11} \ I_{12}]$. This requires knowing only the weights of combinations of at most three rows of S_{11}. $\blacksquare$

Corollary 10.5.2 leads directly to an automorphism of $\mathcal{S}(q)$ that interchanges the coordinates on the left-half with the coordinates on the right-half. The proof is again left as an exercise.

Corollary 10.5.4 *Let q be a power of an odd prime with $q \equiv 2 \,(\text{mod } 3)$, and let O_{q+1} be the $(q+1) \times (q+1)$ zero matrix.*
(i) *If $q \equiv 1 \,(\text{mod } 4)$, then*

$$\begin{bmatrix} O_{q+1} & I_{q+1} \\ -I_{q+1} & O_{q+1} \end{bmatrix} \tag{10.8}$$

is a monomial automorphism of $\mathcal{S}(q)$.
(ii) *If $q \equiv -1 \,(\text{mod } 4)$, then*

$$\begin{bmatrix} O_{q+1} & I_{q+1} \\ I_{q+1} & O_{q+1} \end{bmatrix} \tag{10.9}$$

is a permutation automorphism of $\mathcal{S}(q)$.

Exercise 603 Prove Corollary 10.5.4. ◆

The codes $\mathcal{S}(q)$ where $q = 5$, 11, 17, 23, and 29 are all extremal codes, as shown in [261, 263] using both theoretical and computer techniques. In each case, $2q + 2$ is a multiple of 12 and so by Theorem 9.3.10 codewords of weight k with $(q + 1)/2 + 3 \le k \le q + 4$ form 5-$(2q + 2, k, \lambda)$ designs. Table 10.2 gives the parameters of these designs where "b" denotes the number of blocks. Notice that the number of codewords of weight k in $\mathcal{S}(q)$ is twice the number of blocks in the design since the support of a codeword of weight k determines the codeword up to multiplication by ± 1; see Exercise 451.

The symmetry code $\mathcal{S}(5)$ is the $[12, 6, 6]$ extended ternary Golay code and the latter is unique. It can be shown that by extending the ternary quadratic residue code of length 23, a $[24, 12, 9]$ extremal Type III code is obtained. This code is not equivalent to the $[24, 12, 9]$ symmetry code $\mathcal{S}(11)$. (One way to show the two codes are inequivalent is to note that the extended quadratic residue code has an automorphism group of order 12 144, while the symmetry code has a subgroup of its automorphism group of order 5280 which is not a divisor of 12 144.) In [191], it was shown that these two codes are the only extremal Type III codes of length 24. The only extremal Type III code of length 36 that is known is the symmetry code $\mathcal{S}(17)$; it has been shown in [146] that this is the only extremal Type III code with an automorphism of prime order 5 or more. This suggests the following problem.

Research Problem 10.5.5 *Find all $[36, 18, 12]$ Type III codes.*

Table 10.2 *Parameters for 5-(v, k, λ) designs from $\mathcal{S}(q)$*

q	v	k	λ	b
5	12	6	1	132
5	12	9	35	220
11	24	9	6	2 024
11	24	12	576	30 912
11	24	15	8 580	121 440
17	36	12	45	21 420
17	36	15	5 577	700 128
17	36	18	209 685	9 226 140
17	36	21	2 438 973	45 185 184
23	48	15	364	207 552
23	48	18	50 456	10 083 568
23	48	21	2 957 388	248 854 848
23	48	24	71 307 600	2 872 677 600
23	48	27	749 999 640	15 907 684 672
29	60	18	3 060	1 950 540
29	60	21	449 820	120 728 160
29	60	24	34 337 160	4 412 121 480
29	60	27	1 271 766 600	86 037 019 040
29	60	30	24 140 500 956	925 179 540 912
29	60	33	239 329 029 060	5 507 375 047 020

In [263], a subgroup of $\Gamma\text{Aut}(\mathcal{S}(q)) = \text{MAut}(\mathcal{S}(q))$ has been found. There is evidence that this subgroup may be the entire automorphism group of $\mathcal{S}(q)$ for $q > 5$, at least when q is a prime. The monomial automorphism group contains $\mathcal{Z} = \{I_{2q+2}, -I_{2q+2}\}$. The group $\mathcal{Z}$ is the center of $\text{MAut}(\mathcal{S}(q))$. If $q \equiv 1 \pmod 4$, the automorphism (10.8) generates a subgroup $\mathcal{J}_4$ of $\text{MAut}(\mathcal{S}(q))$ of order 4 containing $\mathcal{Z}$. If $q \equiv -1 \pmod 4$, the automorphism (10.9) generates a subgroup $\mathcal{J}_2$ of $\text{MAut}(\mathcal{S}(q))$ of order 2 intersecting $\mathcal{Z}$ only in the identity I_{2q+2}. Finally, $\text{MAut}(\mathcal{S}(q))$ contains a subgroup $\mathcal{P}$ such that $\mathcal{P}/\mathcal{Z}$ is isomorphic to the *projective general linear group* $\text{PGL}_2(q)$, which is a group of order $q(q^2 - 1)$. If $q \equiv 1 \pmod 4$, $\text{MAut}(\mathcal{S}(q))$ contains a subgroup generated by $\mathcal{P}$ and $\mathcal{J}_4$ of order $4q(q^2 - 1)$; if $q \equiv -1 \pmod 4$, $\text{MAut}(\mathcal{S}(q))$ contains a subgroup generated by $\mathcal{P}$ and $\mathcal{J}_2$ also of order $4q(q^2 - 1)$.

Research Problem 10.5.6 *Find the complete automorphism group of $\mathcal{S}(q)$.*

10.6 Lattices and self-dual codes

In this section we give a brief introduction to lattices over the real numbers and their connection to codes, particularly self-dual codes. We will see that the Golay codes lead to important lattices. We will encounter lattices again in Section 12.5.3. The reader interested in further connections among codes, lattices, sphere packings, quantization, and groups should consult either [60] or [76].

We first define a lattice in $\mathbb{R}^n$. Let $\mathbf{v}_1, \ldots, \mathbf{v}_n$ be n linearly independent vectors in $\mathbb{R}^n$, where $\mathbf{v}_i = v_{i,1} v_{i,2} \cdots v_{i,n}$. The *lattice* Λ with *basis* $\{\mathbf{v}_1, \ldots, \mathbf{v}_n\}$ is the set of all integer combinations of $\mathbf{v}_1, \ldots, \mathbf{v}_n$; these integer combinations of basis vectors are the *points* of the lattice. Thus

$$\Lambda = \{z_1 \mathbf{v}_1 + z_2 \mathbf{v}_2 + \cdots + z_n \mathbf{v}_n \mid z_i \in \mathbb{Z}, 1 \leq i \leq n\}.$$

The matrix

$$M = \begin{bmatrix} v_{1,1} & v_{1,2} & \cdots & v_{1,n} \\ v_{2,1} & v_{2,2} & \cdots & v_{2,n} \\ & & \vdots & \\ v_{n,1} & v_{n,2} & \cdots & v_{n,n} \end{bmatrix}$$

is called the *generator matrix* of Λ; thus $\Lambda = \{\mathbf{z}M \mid \mathbf{z} \in \mathbb{Z}^n\}$. The $n \times n$ matrix

$$A = MM^{\mathsf{T}}$$

is called the *Gram matrix* for Λ; its (i, j)-entry is the (ordinary) inner product (over the real numbers) of $\mathbf{v}_i$ and $\mathbf{v}_j$. The *determinant* or *discriminant* of the lattice Λ is

$$\det \Lambda = \det A = (\det M)^2.$$

This determinant has a natural geometric meaning: $\det \Lambda$ is the volume of the *fundamental parallelotope of* Λ, which is the region

$$\{a_1 \mathbf{v}_1 + \cdots + a_n \mathbf{v}_n \mid 0 \leq a_i < 1, a_i \in \mathbb{R} \text{ for } 1 \leq i \leq n\}.$$

Example 10.6.1 The lattice $\mathbb{Z}^2$ with basis $\mathbf{v}_1 = (1, 0)$ and $\mathbf{v}_2 = (0, 1)$ has generator and Gram matrices both equal to the identity matrix I_2. Also $\det \mathbb{Z}^2 = 1$. This lattice is clearly the set of all points in the plane with integer coordinates. ∎

Example 10.6.2 Let $\mathbf{v}_1 = (\sqrt{2}, 0)$ and $\mathbf{v}_2 = (\sqrt{2}/2, \sqrt{6}/2)$. The lattice Λ_2, shown in Figure 10.3, with basis $\{\mathbf{v}_1, \mathbf{v}_2\}$ has generator and Gram matrices

$$M_2 = \begin{bmatrix} \sqrt{2} & 0 \\ \dfrac{\sqrt{2}}{2} & \dfrac{\sqrt{6}}{2} \end{bmatrix} \quad \text{and} \quad A_2 = \begin{bmatrix} 2 & 1 \\ 1 & 2 \end{bmatrix}.$$

This lattice is often called the *planar hexagonal lattice*; its determinant is 3. ∎

Exercise 604 Let Λ_2 be the lattice of Example 10.6.2. Do the following:
(a) Draw the fundamental region of Λ_2 in Figure 10.3.
(b) Verify geometrically that the fundamental region has area 3. ◆

If the Gram matrix of a lattice has integer entries, then the inner product between any two lattice points is always integer valued, and the lattice is called *integral*. The $\mathbb{Z}^2$ and planar hexagonal lattices defined in Examples 10.6.1 and 10.6.2 are integral. Note that in any lattice, we could rescale all the basis vectors by the same scalar; this does not change the "geometry" of the lattice, that is, the relative positions in $\mathbb{R}^n$ of the lattice points, but could change other properties, such as the integrality of the lattice or the size of the fundamental region ($\det \Lambda$). If we multiply each basis vector of Λ by the scalar c, then the resulting

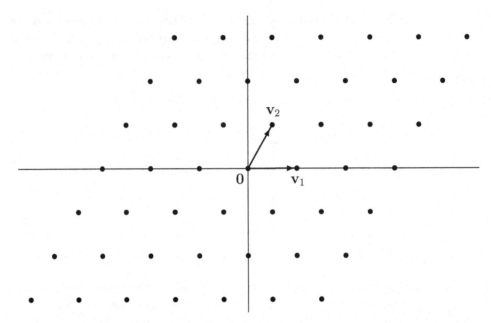

Figure 10.3 Planar hexagonal lattice in $\mathbb{R}^2$.

lattice is denoted $c\Lambda$. The generating matrix of $c\Lambda$ is cM and the Gram matrix is $c^2 A$. We also essentially obtain the same lattice if we rotate or reflect Λ. This leads to the concept of equivalence: the lattice Λ with generator matrix M is *equivalent* to the lattice Λ' if there is a scalar c, a matrix U with integer entries and $\det U = \pm 1$, and a matrix B with real entries and $BB^T = I_n$ such that $cUMB$ is a generator matrix of Λ'. (The matrix B acts as a series of rotations and the matrix U acts as a series of reflections.) Generally, when choosing a representative lattice in an equivalence class of lattices, one attempts to choose a representative that is integral and has determinant as small as possible.

Using the ordinary inner product in $\mathbb{R}^n$, we can define the *dual lattice* Λ^* of Λ to be the set of vectors in $\mathbb{R}^n$ whose inner product with every lattice point in Λ is an integer; that is,

$$\Lambda^* = \{ \mathbf{y} \in \mathbb{R}^n \mid \mathbf{x} \cdot \mathbf{y} \in \mathbb{Z} \text{ for all } \mathbf{x} \in \Lambda \}.$$

An integral lattice Λ is called *unimodular* or *self-dual* if $\Lambda = \Lambda^*$.

Theorem 10.6.3 *Let Λ be a lattice in $\mathbb{R}^n$ with basis $\mathbf{v}_1, \ldots, \mathbf{v}_n$, generator matrix M, and Gram matrix A. The following hold:*
(i) $\Lambda^* = \{ \mathbf{y} \in \mathbb{R}^n \mid \mathbf{v}_i \cdot \mathbf{y} \in \mathbb{Z} \text{ for all } 1 \leq i \leq n \} = \{ \mathbf{y} \in \mathbb{R}^n \mid \mathbf{y} M^T \in \mathbb{Z}^n \}$.
(ii) *The generator matrix of Λ^* is $(M^{-1})^T$.*
(iii) *The Gram matrix of Λ^* is A^{-1}.*
(iv) $\det \Lambda^* = 1/\det \Lambda$.
(v) Λ *is integral if and only if $\Lambda \subseteq \Lambda^*$.*

(vi) *If Λ is integral, then*

$$\Lambda \subseteq \Lambda^* \subseteq \frac{1}{\det \Lambda} \Lambda = (\det \Lambda^*)\Lambda.$$

(vii) *If Λ is integral, then Λ is unimodular if and only if $\det \Lambda = \pm 1$.*

Proof: We leave the proofs of (i) and (v) for Exercise 605. For (ii), let $\mathbf{w}_1, \dots, \mathbf{w}_n$ be the rows of $(M^{-1})^{\mathsf{T}}$. Then $\{\mathbf{w}_1, \dots, \mathbf{w}_n\}$ is a basis of $\mathbb{R}^n$ and, as $(M^{-1})^{\mathsf{T}} M^{\mathsf{T}} = (MM^{-1})^{\mathsf{T}} = I_n$,

$$\mathbf{w}_i \cdot \mathbf{v}_j = \begin{cases} 1 & \text{if } i = j, \\ 0 & \text{if } i \neq j. \end{cases} \tag{10.10}$$

In particular, $\{\mathbf{w}_1, \dots, \mathbf{w}_n\} \subseteq \Lambda^*$ by (i). Let $\mathbf{w} \in \Lambda^*$. Then $\mathbf{w} = a_1 \mathbf{w}_1 + \cdots + a_n \mathbf{w}_n$ as $\{\mathbf{w}_1, \dots, \mathbf{w}_n\}$ is a basis of $\mathbb{R}^n$. As $\mathbf{w} \in \Lambda^*$, $\mathbf{w} \cdot \mathbf{v}_j \in \mathbb{Z}$. But $\mathbf{w} \cdot \mathbf{v}_j = a_j$ for $1 \leq j \leq n$ by (10.10). Hence $a_j \in \mathbb{Z}$ and (ii) holds. The Gram matrix of Λ^* is $(M^{-1})^{\mathsf{T}} M^{-1} = (MM^{\mathsf{T}})^{-1} = A^{-1}$ yielding (iii). Thus

$$\det \Lambda^* = \det A^{-1} = \frac{1}{\det A} = \frac{1}{\det \Lambda},$$

producing (iv).

For (vi) we only need to prove the second containment by parts (iv) and (v). Let $\mathbf{y} \in \Lambda^*$. Then $\mathbf{y} M^{\mathsf{T}} \in \mathbb{Z}^n$ by (i), and hence there exists $\mathbf{z} \in \mathbb{Z}^n$ such that $\mathbf{y} = \mathbf{z}(M^{\mathsf{T}})^{-1} = \mathbf{z}(M^{\mathsf{T}})^{-1} M^{-1} M = \mathbf{z}(MM^{\mathsf{T}})^{-1} M = \mathbf{z} A^{-1} M$. But $A^{-1} = (\det A)^{-1} \text{adj}(A)$, where $\text{adj}(A)$ is the adjoint of A. Hence

$$\mathbf{y} = \mathbf{z}(\det A)^{-1} \text{adj}(A) M = \mathbf{z}'(\det A)^{-1} M,$$

where $\mathbf{z}' = \mathbf{z}\,\text{adj}(A) \in \mathbb{Z}^n$ as $\text{adj}(A)$ has integer entries since Λ is integral. Thus $\mathbf{y} \in (\det \Lambda)^{-1}\Lambda$ verifying (vi).

Finally by (iv) if Λ is unimodular, then $\det \Lambda = \det \Lambda^* = 1/(\det \Lambda)$ implying $\det \Lambda = \pm 1$. Conversely, if $\det \Lambda = \pm 1$ and Λ is integral, (vi) shows that $\Lambda \subseteq \Lambda^* \subseteq \pm\Lambda$ and so $\Lambda = \Lambda^*$ as clearly $\Lambda = \pm\Lambda$. □

Exercise 605 Prove Theorem 10.6.3(i) and (v). ◆

Exercise 606 Prove that if Λ is a lattice then $\Lambda = c\Lambda$ if and only if $c = \pm 1$. ◆

In an integral lattice Λ, $\mathbf{x} \cdot \mathbf{x}$ is an integer for all $\mathbf{x} \in \Lambda$. An integral lattice is called *even* if $\mathbf{x} \cdot \mathbf{x}$ is an even integer for all $\mathbf{x} \in \Lambda$. In particular, the Gram matrix has even entries on its main diagonal. An integral lattice that is not even is called *odd*. A lattice is *Type II* provided it is an even unimodular lattice and *Type I* if it is an odd unimodular lattice. Type I lattices exist for any dimension n; however, Type II lattices exist if and only if $n \equiv 0 \pmod 8$. The unimodular lattices have been classified up to equivalence through dimension 25; note that Type I codes have been classified through length 30 and Type II through length 32.

Exercise 607 Show that the lattice $\mathbb{Z}^2$ from Example 10.6.1 is a Type I lattice while Λ_2 from Example 10.6.2 is not unimodular. ◆

Analogous to measuring the Hamming distance between codewords, we measure the square of the Euclidean distance, or norm, between lattice points. The *norm* of a vector $\mathbf{v} = v_1 \cdots v_n$ in $\mathbb{R}^n$ is

$$N(\mathbf{v}) = \mathbf{v} \cdot \mathbf{v} = \sum_{i=1}^{n} v_i^2.$$

The minimum squared distance, or minimum norm, between two points in a lattice Λ, denoted μ, is

$$\mu = \min\{N(\mathbf{x} - \mathbf{y}) \mid \mathbf{x}, \mathbf{y} \in \Lambda, \ \mathbf{x} \neq \mathbf{y}\} = \min\{N(\mathbf{x}) \mid \mathbf{x} \in \Lambda, \ \mathbf{x} \neq \mathbf{0}\}.$$

The minimum norm is analogous to the minimum distance of a code. With lattices there is a power series, called the theta series, that corresponds to the weight enumerator of a code. The *theta series* $\Theta_\Lambda(q)$ of a lattice Λ is

$$\Theta_\Lambda(q) = \sum_{\mathbf{x} \in \Lambda} q^{\mathbf{x} \cdot \mathbf{x}}.$$

If Λ is integral and N_m is the number of lattice points of norm m, then

$$\Theta_\Lambda(q) = \sum_{m=0}^{\infty} N_m q^m.$$

For both Type I and Type II lattices, there is a result known as Hecke's Theorem, corresponding to Gleason's Theorem for self-dual codes, that states that the theta series of such a lattice is a complex power series in two specific power series which depend on whether the lattice is Type I or Type II. As with self-dual codes, Hecke's Theorem can also be used to produce an upper bound on the minimum norm of a Type I or II lattice: for Type I the bound is $\mu \leq \lfloor n/8 \rfloor + 1$, and for Type II the bound is $\mu \leq 2 \lfloor n/24 \rfloor + 2$. As Gleason's Theorem implies that Type II codes exist only for lengths a multiple of 8, Hecke's Theorem implies that Type II lattices exist in $\mathbb{R}^n$ only when $n \equiv 0 \pmod 8$.

Exercise 608 Let Λ_2 be the lattice of Example 10.6.2. Do the following:
(a) Verify that the minimum norm of the lattice is 2.
(b) Show that the norm of any lattice point of Λ_2 is an even integer.
(c) In Figure 10.3, find all lattice points whose norm is:
   (i)   2,
   (ii) 4,
   (iii) 6,
   and write these points as integer combinations of $\{\mathbf{v}_1, \mathbf{v}_2\}$.   ◆

We can tie the concept of minimum norm of a lattice to the notion of a lattice packing. If Λ is a lattice with norm μ, then the n-dimensional spheres of radius $\rho = \sqrt{\mu}/2$ centered at lattice points form a *lattice packing* in $\mathbb{R}^n$. These spheres do not overlap except at their boundary. The number of spheres touching the sphere centered at $\mathbf{0}$ is the number of lattice points of minimum norm μ; this number is called the *kissing number*. So the kissing number in lattices corresponds to the number of minimum weight codewords in a code. For a variety of reasons it is of interest to find lattices with large kissing numbers.

Exercise 609 Show that in a lattice packing, the number of spheres touching any sphere is the kissing number.   ♦

Exercise 610 Do the following:
(a) What is the radius of the spheres in a lattice packing using the lattices $\mathbb{Z}^2$ and Λ_2 from Examples 10.6.1 and 10.6.2?
(b) What is the kissing number of each of these lattices?   ♦

We now present a general construction, known as *Construction A*, of lattices directly from binary codes. Let $\mathcal{C}$ be an $[n, k, d]$ binary code. The lattice $\Lambda(\mathcal{C})$ is the set of all $\mathbf{x}$ in $\mathbb{R}^n$ obtained from a codeword in $\mathcal{C}$ by viewing the codeword as an integer vector with 0s and 1s, adding even integers to any components, and dividing the resulting vector by $\sqrt{2}$. In short

$$\Lambda(\mathcal{C}) = \{\mathbf{x} \in \mathbb{R}^n \mid \sqrt{2}\mathbf{x} \,(\mathrm{mod}\,2) \in \mathcal{C}\}.$$

The following can be found in [60, Chapter 7].

Theorem 10.6.4 *Let $\mathcal{C}$ be an $[n, k, d]$ binary code. The following hold:*
(i) *If $d \le 4$, the minimum norm μ of $\Lambda(\mathcal{C})$ is $\mu = d/2$; if $d > 4$, $\mu = 2$.*
(ii) *$\det \Lambda(\mathcal{C}) = 2^{n-2k}$.*
(iii) *$\Lambda(\mathcal{C}^{\perp}) = \Lambda(\mathcal{C})^*$.*
(iv) *$\Lambda(\mathcal{C})$ is integral if and only if $\mathcal{C}$ is self-orthogonal.*
(v) *$\Lambda(\mathcal{C})$ is Type I if and only if $\mathcal{C}$ is Type I.*
(vi) *$\Lambda(\mathcal{C})$ is Type II if and only if $\mathcal{C}$ is Type II.*

Proof: We leave the proofs of (i) and (vi) as an exercise. Without loss of generality, we may assume that $G = [I_k \ \ B]$ is a generator matrix for $\mathcal{C}$. Then, clearly,

$$M = \frac{1}{\sqrt{2}}\begin{bmatrix} I_k & B \\ O & 2I_{n-k} \end{bmatrix} \quad \text{and} \quad A = \frac{1}{2}\begin{bmatrix} I_k + BB^{\mathrm{T}} & 2B \\ 2B^{\mathrm{T}} & 4I_{n-k} \end{bmatrix}$$

are the generator and Gram matrices for $\Lambda(\mathcal{C})$, respectively. Hence

$$\det A = \frac{1}{2^n}\det\begin{bmatrix} I_k + BB^{\mathrm{T}} & 2B \\ 2B^{\mathrm{T}} & 4I_{n-k} \end{bmatrix} = \frac{1}{2^n}2^{n-k}\det\begin{bmatrix} I_k + BB^{\mathrm{T}} & 2B \\ B^{\mathrm{T}} & 2I_{n-k} \end{bmatrix}$$

$$= 2^{-k}\det\begin{bmatrix} I_k & O \\ B^{\mathrm{T}} & 2I_{n-k} \end{bmatrix} = 2^{-k}2^{n-k},$$

yielding (ii).

As $G^{\perp} = [B^{\mathrm{T}} \ \ I_{n-k}]$ is the generator matrix of $\mathcal{C}^{\perp}$, $\Lambda(\mathcal{C}^{\perp})$ has generator matrix

$$M^{\perp} = \frac{1}{\sqrt{2}}\begin{bmatrix} B^{\mathrm{T}} & I_{n-k} \\ 2I_k & O \end{bmatrix}.$$

The (binary) inner product of a row of G with a row of $G^{\perp}$ is 0; hence we see that the (real) inner product of a row of M and a row of $M^{\perp}$ is an integer. This proves that $M^{\perp}M^{\mathrm{T}}$ is an integer matrix and $\Lambda(\mathcal{C}^{\perp}) \subseteq \Lambda(\mathcal{C})^*$. To complete (iii) we must show $\Lambda(\mathcal{C})^* \subseteq \Lambda(\mathcal{C}^{\perp})$. Let $\mathbf{y} \in \Lambda(\mathcal{C})^*$. Then $\mathbf{y}M^{\mathrm{T}} \in \mathbb{Z}^n$ by Theorem 10.6.3(i). So there exists $\mathbf{z} \in \mathbb{Z}^n$ with

$\mathbf{y} = \mathbf{z}(M^T)^{-1} = \mathbf{z}(M^T)^{-1}(M^{\perp})^{-1}M^{\perp} = \mathbf{z}(M^{\perp}M^T)^{-1}M^{\perp}$. As $M^{\perp}M^T$ is an integer matrix and $\det(M^{\perp}M^T) = \det(M^{\perp})\det(M^T) = (\pm 2^{-n/2} \cdot 2^k) \cdot (2^{(n-2k)/2}) = \pm 1$,

$$(M^{\perp}M^T)^{-1} = \frac{1}{\det(M^{\perp}M^T)}\mathrm{adj}(M^{\perp}M^T)$$

is an integer matrix. Thus $\mathbf{y} = \mathbf{z}'M^{\perp}$ for some $\mathbf{z}' \in \mathbb{Z}^n$. Hence $\mathbf{y} \in \Lambda(\mathcal{C}^{\perp})$, completing (iii).

The (real) inner product of two rows of M is always an integer if and only if the (binary) inner product of two rows of G is 0, proving (iv).

Suppose $\Lambda(\mathcal{C})$ is Type I; then $\det \Lambda(\mathcal{C}) = \pm 1$, implying $k = n/2$ by (ii). As $\Lambda(\mathcal{C})$ is integral, $\mathcal{C} \subseteq \mathcal{C}^{\perp}$ by (iv). Hence $\mathcal{C}$ is self-dual. The (real) inner product of some lattice point with itself is an odd integer; the corresponding codeword must be singly-even implying that $\mathcal{C}$ is Type I. Conversely, suppose $\mathcal{C}$ is Type I. By (iv) $\Lambda(\mathcal{C})$ is integral and by (iii) $\Lambda(\mathcal{C}) = \Lambda(\mathcal{C}^{\perp}) = \Lambda(\mathcal{C})^*$. Hence $\Lambda(\mathcal{C})$ is unimodular. In addition, a singly-even codeword corresponds to a lattice point with odd integer norm. This proves (v). $\square$

Exercise 611 Prove Theorem 10.6.4(i) and (vi). ◆

Example 10.6.5 Let $\mathcal{C}$ be the $[8, 4, 4]$ extended Hamming code (denoted e_8 in Section 9.7). By Theorem 10.6.4, $\Lambda(\mathcal{C})$ is a Type II lattice with minimum norm 2. We ask you to verify this directly in Exercise 612. Up to equivalence, this lattice, which is usually denoted E_8, is known to be the unique Type II lattice in $\mathbb{R}^8$. We can determine precisely the lattice points of minimum norm 2. One can add ± 2 to any coordinate of the zero vector yielding 16 lattice points of norm 2 with "shape" $(1/\sqrt{2})(\pm 2, 0^7)$, indicating any vector with seven 0s and one $\pm 2/\sqrt{2}$. Beginning with any of the 14 weight 4 codewords, one can add -2 to any of the four nonzero coordinates obtaining $14 \cdot 16 = 224$ lattice points of norm 2 with shape $(1/\sqrt{2})((\pm 1)^4, 0^4)$. These are all of the 240 lattice points of norm 2. Hence, E_8 has kissing number 240. It is known that no other 8-dimensional lattice can have a higher kissing number. The theta series for E_8 is

$$\Theta_{E_8}(q) = 1 + 240q^2 + 2160q^4 + 6720q^6 + 17520q^8 + 30240q^{10} + \cdots$$

$$= 1 + 240 \sum_{m=1}^{\infty} \sigma_3(m)q^{2m},$$

where $\sigma_3(m) = \sum_{d|m} d^3$. It is also known that the lattice E_8 yields the densest possible lattice packing, with spheres of radius $1/\sqrt{2}$, where the *density* is the fraction of the space covered by the spheres; the density of E_8 is $\pi^4/384 \approx 0.253\,669\,51$. It seems that T. Gosset [110], in 1900, was the first to study E_8, along with 6- and 7-dimensional lattices denoted E_6 and E_7; hence these are sometimes called Gosset lattices. ∎

Exercise 612 Let e_8 be the $[8, 4, 4]$ extended binary Hamming code.
(a) Give a generator matrix for e_8 in the form $[I_4 \; B]$.
(b) Give the generator and Gram matrices for $\Lambda(e_8)$.
(c) Show directly, without quoting Theorem 10.6.4, that $\Lambda(e_8)$ is integral with determinant 1.
(d) Show directly, without quoting Theorem 10.6.4, that $\Lambda(e_8)$ is Type II with minimum norm 2. ◆

Exercise 613 Find the shapes of each of the 2160 lattice points in E_8 of norm 4. ◆

The disadvantage of Construction A is that the largest minimum norm of a resulting lattice is 2. There are other constructions of lattices from codes that produce lattices of great interest. To conclude this section, we present one construction, due to J. H. Conway [60, Chapter 10], of the 24-dimensional Leech lattice Λ_{24} using the [24, 12, 8] extended binary Golay code. This lattice was discovered in 1965 by J. Leech [188]. The lattice Λ_{24} is known to have the highest kissing number (196 560) of any 24-dimensional lattice. Furthermore, Λ_{24} is the unique Type II lattice of dimension 24 with minimum norm 4, the highest norm possible for Type II lattices in $\mathbb{R}^{24}$.

We need some notation. Label the coordinates $\Omega = \{1, 2, \ldots, 24\}$ and let $\{e_1, e_2, \ldots, e_{24}\}$ be the standard orthonormal basis of coordinate vectors in $\mathbb{R}^{24}$. If $S \subseteq \Omega$, let $e_S = \sum_{i \in S} e_i$. Fix some representation of the [24, 12, 8] extended binary Golay code $\mathcal{G}_{24}$. A subset S of Ω is called a $\mathcal{G}_{24}$-set if it is the support of a codeword in $\mathcal{G}_{24}$. Recall from Example 8.3.2 that the subsets of Ω of size 4 are called tetrads, and that each tetrad T uniquely determines five other tetrads, all six being pairwise disjoint, so that the union of any two tetrads is a $\mathcal{G}_{24}$-set of size 8; these six tetrads are called the sextet determined by T. We let $\mathcal{T}$ denote the set of all tetrads and $\mathcal{O}$ denote the set of 759 octads, that is, the set of all $\mathcal{G}_{24}$-sets of size 8. Recall from either Example 8.3.2 or the proof of the uniqueness of $\mathcal{G}_{24}$ in Section 10.1 that the codewords of weight 8 in $\mathcal{G}_{24}$ span $\mathcal{G}_{24}$.

We first construct an intermediate lattice Γ_0, then add one more spanning vector to form a lattice Γ_1, and finally rescale to obtain Λ_{24}. Define Γ_0 to be the 24-dimensional lattice spanned by $\{2e_S \mid S \in \mathcal{O}\}$.

Lemma 10.6.6 *The following hold in the lattice Γ_0:*
(i)   *Every lattice point has even integer components.*
(ii) *Γ_0 contains all vectors $4e_T$, where $T \in \mathcal{T}$.*
(iii) *Γ_0 contains $4e_i - 4e_j$ for all i and j in Ω.*
(iv) *A vector in $\mathbb{R}^{24}$ with all even integer components is in Γ_0 if and only if the sum of its components is a multiple of 16 and the set of coordinates in the vector where the components are not divisible by 4 forms a $\mathcal{G}_{24}$-set.*

Proof: Part (i) is immediate as $2e_S$ with $S \in \mathcal{O}$ has all even integer components. Let $T = T_0, T_1, \ldots, T_5$ be the sextet determined by the tetrad T. Since $4e_T = 2e_{T \cup T_1} + 2e_{T \cup T_2} - 2e_{T_1 \cup T_2}$, we have (ii). Now let $T = \{i, x, y, z\} \in \mathcal{T}$ and $T' = \{j, x, y, z\} \in \mathcal{T}$. Then $4e_T - 4e_{T'} = 4e_i - 4e_j$, showing (iii).

Let $\mathbf{v} \in \Gamma_0$. Since a spanning vector $2e_S$ for $S \in \mathcal{O}$ has its sum of components equaling 16, the sum of the components of $\mathbf{v}$ is a multiple of 16. If we take $\mathbf{v}$, divide its components by 2 and reduce modulo 2, we obtain a binary combination of some of the e_Ss where $S \in \mathcal{O}$, which can be viewed as codewords in $\mathcal{G}_{24}$ and hence must sum to e_C for some $\mathcal{G}_{24}$-set C; but C is precisely the set of coordinates where the components of $\mathbf{v}$ are not divisible by 4. Conversely, let $\mathbf{v}$ be a vector in $\mathbb{R}^{24}$ with all even integer components such that the sum of its components is a multiple of 16 and the set of coordinates in $\mathbf{v}$ where the components are not divisible by 4 forms a $\mathcal{G}_{24}$-set C. Let $\mathbf{c} \in \mathcal{G}_{24}$ have support C; then $\mathbf{c} = \mathbf{c}_1 + \cdots + \mathbf{c}_r$ where each $\mathbf{c}_i$ is a weight 8 codeword of $\mathcal{G}_{24}$, as the weight 8 codewords span $\mathcal{G}_{24}$. Let $S_i = \mathrm{supp}(\mathbf{c}_i)$. Define $\mathbf{x} = \sum_{i=1}^{r} 2e_{S_i} \in \Gamma_0$. Clearly, the coordinates of $\mathbf{x}$ that are 2 modulo 4 are precisely the coordinates of C and the rest are 0 modulo 4. Therefore the vector

$\mathbf{v} - \mathbf{x}$ has sum of components a multiple of 16 and all components are divisible by 4. By Exercise 614 and parts (ii) and (iii), $\mathbf{v} - \mathbf{x} \in \Gamma_0$ and hence $\mathbf{v} \in \Gamma_0$.   □

Exercise 614 Show that the set of vectors in $\mathbb{R}^{24}$ whose components are multiples of 4 and whose component sum is a multiple of 16 are precisely the vectors in the lattice spanned by $4\mathbf{e}_T$ for $T \in \mathcal{T}$ and $4\mathbf{e}_i - 4\mathbf{e}_j$ for i and j in Ω.   ◆

The lattice Γ_1 is defined to be the lattice spanned by Γ_0 and the vector $\mathbf{s} = (-3, 1, 1, \ldots, 1)$. Notice that we obtain a vector with -3 in coordinate i and 1s elsewhere by adding $4\mathbf{e}_1 - 4\mathbf{e}_i$ to $\mathbf{s}$.

Lemma 10.6.7 *The vector* $\mathbf{v} = v_1 v_2 \cdots v_{24}$ *is in* Γ_1 *if and only if the following three conditions all hold:*
(i)   *The components* v_i *are all congruent to the same value* m *modulo 2.*
(ii) *The set of* i *for which the components* v_i *are congruent to the same value modulo 4 form a* $\mathcal{G}_{24}$-set.
(iii) $\sum_{i=1}^{24} v_i \equiv 4m \pmod{8}$, *where* m *is defined in* (i).
Furthermore, if $\mathbf{x}$ *and* $\mathbf{y}$ *are in* Γ_1, *then* $\mathbf{x} \cdot \mathbf{y}$ *is a multiple of* 8 *and* $\mathbf{x} \cdot \mathbf{x}$ *is a multiple of* 16.

Proof: Statements (i), (ii), and (iii) hold for the vectors spanning Γ_1 and hence for any vector in Γ_1. The final statement also holds for vectors in Γ_1 as it too holds for the vectors spanning Γ_1, noting that if $\mathbf{x} = \sum_i \mathbf{x}_i$, then $\mathbf{x} \cdot \mathbf{x} = \sum_i \mathbf{x}_i \cdot \mathbf{x}_i + 2\sum_{i<j} \mathbf{x}_i \cdot \mathbf{x}_j$.

Suppose that $\mathbf{v}$ satisfies (i), (ii), and (iii). Then we can subtract $\alpha\mathbf{s}$ from $\mathbf{v}$ where α is one of 0, 1, 2, or 3, so that $\mathbf{v} - \alpha\mathbf{s}$ has only even coordinates and its component sum is a multiple of 16. By (ii), the set of coordinates in $\mathbf{v} - \alpha\mathbf{s}$ where the components are not divisible by 4 forms a $\mathcal{G}_{24}$-set. Hence $\mathbf{v} - \alpha\mathbf{s}$ is in Γ_0 by Lemma 10.6.6 and the proof is complete.   □

The lattice Γ_1 is not self-dual. We obtain Λ_{24} by multiplying all vectors in Γ_1 by $1/\sqrt{8}$. Thus

$$\Lambda_{24} = \frac{1}{\sqrt{8}}\Gamma_1.$$

By Lemma 10.6.7, the inner product of any two vectors in Λ_{24} is an integer, making Λ_{24} integral. By the same lemma, the norm of a vector in Λ_{24} is an even integer. It can be shown that $\det \Lambda_{24} = \pm 1$ implying that Λ_{24} is a Type II lattice. By Exercise 615, its minimum norm is 4 and its kissing number is 196 560.

Exercise 615 Upon scaling a vector in Γ_1 by multiplying by $1/\sqrt{8}$, the norm of the vector is reduced by a factor of eight. In this exercise you will show that no vector in Λ_{24} has norm 2 and also find the vectors in Λ_{24} of norm 4. This is equivalent to showing that no vector in Γ_1 has norm 16 and finding those vectors in Γ_1 of norm 32.
(a) Using the characterization of Γ_1 in Lemma 10.6.7, show that Γ_1 has no vectors of norm 16.
(b) Show that a vector of shape $((\pm 2)^8, 0^{16})$ with an even number of minus signs and the ± 2s in coordinates forming an octad is in Γ_1.
(c) Show that there are $2^7 \cdot 759$ vectors described in part (b).
(d) Show that a vector of shape $((\mp 3), (\pm 1)^{23})$ with the lower signs in a $\mathcal{G}_{24}$-set is in Γ_1.

(e) Show that there are $2^{12} \cdot 24$ vectors described in part (d).

(f) Show that a vector of shape $((\pm 4)^2, 0^{22})$ with the plus and minus signs assigned arbitrarily is in Γ_1.

(g) Show that there are $2 \cdot 24 \cdot 23$ vectors described in part (f).

(h) Show that a vector in Γ_1 of norm 32 is one of those found in part (b), (d), or (f).

(i) Show that the minimum norm of Λ_{24} is 4 and that the kissing number is 196 560. ◆

There are a number of other constructions for Λ_{24} including one from the $[24, 12, 9]$ symmetry code over $\mathbb{F}_3$; see Section 5.7 of Chapter 5 in [60]. Lattices over the complex numbers can also be defined; the $[12, 6, 6]$ extended ternary Golay code can be used to construct the 12-dimensional complex version of the Leech lattice; see Section 8 of Chapter 7 in [60].

11 Covering radius and cosets

In this chapter we examine in more detail the concept of covering radius first introduced in Section 1.12. In our study of codes, we have focused on codes with high minimum distance in order to have good error-correcting capabilities. An $[n, k, d]$ code C over $\mathbb{F}_q$ can correct $t = \lfloor (d-1)/2 \rfloor$ errors. Thus spheres in $\mathbb{F}_q^n$ of radius t centered at codewords are pairwise disjoint, a fact that fails for spheres of larger radius; t is called the packing radius of C. It is natural to explore the opposite situation of finding the smallest radius $\rho(C)$, called the covering radius, of spheres centered at codewords that completely cover the space $\mathbb{F}_q^n$; that is, every vector in $\mathbb{F}_q^n$ is in at least one of these spheres. Alternately, given a radius, it is mathematically interesting to find the centers of the fewest number of spheres of that radius which cover the space. When decoding C, if t or fewer errors are made, the received vector can be uniquely decoded. If the number of errors is more than t but no more than the covering radius, sometimes these errors can still be uniquely corrected. A number of applications have arisen for the coverings arising from codes, ranging from data compression to football pools. In general, if you are given a code, it is difficult to find its covering radius, and the complexity of this has been investigated; see [51].

In this chapter we will examine a portion of the basic theory and properties of the covering radius. We will also discuss what is known about the covering radius of specific codes or families of codes. Most of the work in this area has been done for binary codes, both linear and nonlinear. The reader interested in exploring covering radius in more depth should consult [34, 51].

11.1 Basics

We begin with some notation and recall some basic facts from Section 1.12. We also present a few new results that we will need in later sections.

Let C be a code over $\mathbb{F}_q$ of length n that is possibly nonlinear. We say that a vector in $\mathbb{F}_q^n$ is ρ-*covered* by C if it has distance ρ or less from at least one codeword in C. In this terminology the covering radius $\rho = \rho(C)$ of C is the smallest integer ρ such that every vector in $\mathbb{F}_q^n$ is ρ-covered by C. Equivalently,

$$\rho(C) = \max_{\mathbf{x} \in \mathbb{F}_q^n} \min_{\mathbf{c} \in C} d(\mathbf{x}, \mathbf{c}). \tag{11.1}$$

Exercise 616 Show that the definition of the covering radius of C given by (11.1) is equivalent to the definition that $\rho = \rho(C)$ is the smallest integer ρ such that every vector in $\mathbb{F}_q^n$ is ρ-covered by C. ♦

Exercise 617 Find the covering radius and the packing radius of the $[n, 1, n]$ repetition code over $\mathbb{F}_q$. ◆

Exercise 618 Prove that if $C_1 \subseteq C_2$, then $\rho(C_2) \le \rho(C_1)$. ◆

The definitions of covering radius and packing radius lead directly to a relationship between the two as we noted in Section 1.12.

Theorem 11.1.1 *Let C be a code of minimum distance d. Then*

$$\rho(C) \ge \left\lfloor \frac{d-1}{2} \right\rfloor.$$

Furthermore, we have equality if and only if C is perfect.

Another equivalent formulation of the covering radius is stated in Theorem 1.12.5; we restate this result but generalize slightly the first part to apply to nonlinear as well as linear codes. In order to simplify terminology, we will call the set $\mathbf{v} + C$ a *coset of C* with *coset representative* $\mathbf{v}$ even if C is nonlinear. Care must be taken when using cosets with nonlinear codes; for example, distinct cosets need not be disjoint, as Exercise 619 shows.

Theorem 11.1.2 *Let C be a code. The following hold*:
(i) $\rho(C)$ *is the largest value of the minimum weight of all cosets of C.*
(ii) *If C is an $[n, k]$ linear code over $\mathbb{F}_q$ with parity check matrix H, then $\rho(C)$ is the smallest number s such that every nonzero vector in $\mathbb{F}_q^{n-k}$ is a combination of s or fewer columns of H.*

Exercise 619 Find a nonlinear code and a pair of cosets of that code that are neither equal nor disjoint. ◆

Exercise 620 Let C be an $[n, k, d]$ linear code over $\mathbb{F}_q$ with covering radius $\rho(C) \ge d$. Explain why there is a linear code with more codewords than C but the same error-correcting capability. ◆

An obvious corollary of part (ii) of Theorem 11.1.2 is the following upper bound on the covering radius of a linear code, called the Redundancy Bound.

Corollary 11.1.3 (Redundancy Bound) *Let C be an $[n, k]$ code. Then $\rho(C) \le n - k$.*

Exercise 621 Prove Corollary 11.1.3. ◆

Recall from (1.11) that a sphere of radius r in $\mathbb{F}_q^n$ contains

$$\sum_{i=0}^{r} \binom{n}{i}(q-1)^i$$

vectors. The following is a lower bound on the covering radius, called the Sphere Covering Bound; its proof is left as an exercise.

Theorem 11.1.4 (Sphere Covering Bound) *If C is a code of length n over $\mathbb{F}_q$, then*

$$\rho(C) \geq \min \left\{ r \mid |C| \sum_{i=0}^{r} \binom{n}{i} (q-1)^i \geq q^n \right\}.$$

Exercise 622 Prove Theorem 11.1.4.   ◆

Exercise 623 Let C be a $[7, 3, 4]$ binary code. Give the upper and lower bounds for this code from the Redundancy Bound and the Sphere Covering Bound.   ◆

Another lower bound can be obtained from the following lemma, known as the Supercode Lemma.

Lemma 11.1.5 (Supercode Lemma) *If C and C' are linear codes with $C \subseteq C'$, then $\rho(C) \geq \min\{\text{wt}(\mathbf{x}) \mid \mathbf{x} \in C' \setminus C\}$.*

Proof: Let $\mathbf{x}$ be a vector in $C' \setminus C$ of minimum weight. Such a vector must be a coset leader of C as $C \subseteq C'$; therefore $\rho(C) \geq \text{wt}(\mathbf{x})$ and the result follows.   □

Theorem 1.12.6 will be used repeatedly, and we restate it here for convenience.

Theorem 11.1.6 *Let C be an $[n, k]$ code over $\mathbb{F}_q$. Let $\widehat{C}$ be the extension of C, and let C^* be a code obtained from C by puncturing on some coordinate. The following hold:*
(i)   *If $C = C_1 \oplus C_2$, then $\rho(C) = \rho(C_1) + \rho(C_2)$.*
(ii) *$\rho(C^*) = \rho(C)$ or $\rho(C^*) = \rho(C) - 1$.*
(iii) *$\rho(\widehat{C}) = \rho(C)$ or $\rho(\widehat{C}) = \rho(C) + 1$.*
(iv) *If $q = 2$, then $\rho(\widehat{C}) = \rho(C) + 1$.*
(v)   *Assume that $\mathbf{x}$ is a coset leader of C. If $\mathbf{x}' \in \mathbb{F}_q^n$ all of whose nonzero components agree with the same components of $\mathbf{x}$, then $\mathbf{x}'$ is also a coset leader of C. In particular, if there is a coset of weight s, there is also a coset of any weight less than s.*

Exercise 624 This exercise is designed to find the covering radius of a $[7, 3, 4]$ binary code C.
(a) Show that a $[7, 3, 4]$ binary code is unique up to equivalence.
(b) Apply the Supercode Lemma to obtain a lower bound on $\rho(C)$ using C' equal to the $[7, 4, 3]$ Hamming code; make sure you justify that C is indeed a subcode of C'.
(c) Show that there is no $[7, 4, 4]$ binary code.
(d) Find $\rho(C)$ and justify your answer.   ◆

Exercise 625 Let C be an even binary code. Show that if C^* is obtained from C by puncturing on one coordinate, then $\rho(C^*) = \rho(C) - 1$.   ◆

Recall from Theorem 11.1.1 that the covering radius of a code equals the packing radius of the code if and only if the code is perfect. We have the following result dealing with extended perfect codes and codes whose covering radius is one larger than the packing radius.

Theorem 11.1.7 *If C is an $[n, k, d]$ extended perfect code, then $\rho(C) = d/2$. Furthermore, if C is an $[n, k, d]$ binary code, then C is an extended perfect code if and only if $\rho(C) = d/2$.*

Exercise 626   Prove Theorem 11.1.7.   ◆

Exercise 627   Let C be an $[n, k]$ binary code. Define

$$C_1 = \{(c_1 + c_2 + \cdots + c_n, c_2, \ldots, c_n) \mid (c_1, c_2, \ldots, c_n) \in C\}.$$

(a) Prove that $\widehat{C}$ and $\widehat{C}_1$ are equivalent.
(b) Prove that C_1 is k-dimensional.
(c) Prove that $\rho(C) = \rho(C_1)$.   ◆

11.2   The Norse Bound and Reed–Muller codes

Another pair of upper bounds, presented in the next theorem, are called the Norse Bounds; they apply only to binary codes. The Norse Bounds are so named because they were discovered by the Norwegians Helleseth, Kløve, and Mykkeltveit [126]. They will be used to help determine the covering radius of some of the Reed–Muller codes. Before continuing, we need some additional terminology.

Let C be an (n, M) binary code. Let $\mathcal{M}$ be an $M \times n$ matrix whose rows are the codewords of C in some order. We say that C has *strength s*, where $1 \le s \le n$, provided that in each set of s columns of $\mathcal{M}$ each binary s-tuple occurs the same number $M/2^s$ of times. Exercise 628 gives a number of results related to the strength of a binary code. The code C is *self-complementary* provided for each codeword $\mathbf{c} \in C$, the complementary vector $\mathbf{1} + \mathbf{c}$, obtained from $\mathbf{c}$ by replacing 1 by 0 and 0 by 1, is also in C. Exercise 629 gives a few results about self-complementary codes.

Exercise 628   Let C be an (n, M) binary code whose codewords are the rows of the $M \times n$ matrix $\mathcal{M}$. Prove the following:
(a) If C has strength $s \ge 2$, then C also has strength $s - 1$.
(b) If C has strength s, then so does each coset $\mathbf{v} + C$.
(c) If C is linear, then C has strength s if and only if every s columns of a generator matrix of C are linearly independent if and only if the minimum weight of $C^{\perp}$ is at least $s + 1$. Hint: See Corollary 1.4.14.   ◆

Exercise 629   Let C be a self-complementary binary code. Prove the following:
(a) Each coset $\mathbf{v} + C$ is self-complementary.
(b) If C is also linear, $C^{\perp}$ is an even code.   ◆

We now state the Norse Bounds.

Theorem 11.2.1 (Norse Bounds)   *Let C be an (n, M) binary code.*
(i)   *If C has strength 1, then $\rho(C) \le \lfloor n/2 \rfloor$.*
(ii)   *If C has strength 2 and C is self-complementary, then $\rho(C) \le \lfloor (n - \sqrt{n})/2 \rfloor$.*

Proof: Let $\mathcal{M}$ be an $M \times n$ matrix whose rows are the codewords of C. First assume that C has strength 1. For $\mathbf{v} \in \mathbb{F}_2^n$ let $\mathcal{M}_{\mathbf{v}}$ be an $M \times n$ matrix whose rows are the codewords of $\mathbf{v} + C$. By Exercise 628(b), $\mathbf{v} + C$ also has strength 1. Therefore each column of $\mathcal{M}_{\mathbf{v}}$ has exactly half its entries equal to 0 and half equal to 1, implying that

$$\sum_{\mathbf{u} \in \mathbf{v}+C} \text{wt}(\mathbf{u}) = \frac{nM}{2},$$

or equivalently,

$$\sum_{i=0}^{n} i A_i(\mathbf{v}) = \frac{nM}{2}, \tag{11.2}$$

where $\{A_i(\mathbf{v}) \mid 0 \le i \le n\}$ is the weight distribution of $\mathbf{v} + C$. Equation (11.2) implies that the average distance between $\mathbf{v}$ and the codewords of C is $n/2$. Therefore the distance between $\mathbf{v}$ and some codeword $\mathbf{c} \in C$ is at most $n/2$. By Theorem 11.1.2(i) we have part (i).

Now assume that C, and hence $\mathbf{v} + C$, has strength 2. In each row of $\mathcal{M}_{\mathbf{v}}$ corresponding to the vector $\mathbf{u} \in \mathbf{v}+C$, there are $\binom{\text{wt}(\mathbf{u})}{2}$ ordered pairs of 1s. But each of the $\binom{n}{2}$ pairs of distinct columns of $\mathcal{M}_{\mathbf{v}}$ contains the binary 2-tuple $(1, 1)$ exactly $M/4$ times. Therefore

$$\sum_{\mathbf{u} \in \mathbf{v}+C} \binom{\text{wt}(\mathbf{u})}{2} = \sum_{\mathbf{u} \in \mathbf{v}+C} \frac{\text{wt}(\mathbf{u})(\text{wt}(\mathbf{u}) - 1)}{2} = \binom{n}{2}\frac{M}{4} = \frac{n(n-1)}{2}\frac{M}{4}.$$

Using (11.2) we obtain

$$\sum_{i=0}^{n} i^2 A_i(\mathbf{v}) = \frac{n(n+1)M}{4}. \tag{11.3}$$

Combining (11.2) and (11.3) produces

$$\sum_{i=0}^{n} (2i - n)^2 A_i(\mathbf{v}) = nM. \tag{11.4}$$

This implies that there is an i with $A_i(\mathbf{v}) \ne 0$ such that $(2i - n)^2 \ge n$. Taking the square root of both sides of this inequality shows that either

$$i \le \frac{n - \sqrt{n}}{2} \quad \text{or} \quad i \ge \frac{n + \sqrt{n}}{2}.$$

Suppose now that C is self-complementary. By Exercise 629, $\mathbf{v} + C$ is also self-complementary. Then there is an i satisfying the first of the above two inequalities with $A_i(\mathbf{v}) \ne 0$. Thus there is a vector $\mathbf{u} \in \mathbf{v} + C$ with $\text{wt}(\mathbf{u}) \le (n - \sqrt{n})/2$. Hence (ii) holds. $\square$

In the case of binary linear codes, the Norse Bounds can be restated as follows.

Corollary 11.2.2 *Let C be a binary linear code of length n.*
(i) *If $C^{\perp}$ has minimum distance at least 2, then $\rho(C) \le \lfloor n/2 \rfloor$.*
(ii) *If $C^{\perp}$ has minimum distance at least 4 and the all-one vector $\mathbf{1}$ is in C, then $\rho(C) \le \lfloor (n - \sqrt{n})/2 \rfloor$.*

Exercise 630 Prove Corollary 11.2.2. ◆

Exercise 631 Let C be a self-complementary binary code of strength 2 with $\rho(C) = (n - \sqrt{n})/2$, and let $\mathbf{v} + C$ have minimum weight $\rho(C)$. Show that half the vectors in $\mathbf{v} + C$

have weight $(n - \sqrt{n})/2$ and half the vectors have weight $(n + \sqrt{n})/2$. Hint: Examine (11.4). ◆

We are now ready to examine the covering radius of the Reed–Muller codes $\mathcal{R}(r, m)$ first defined in Section 1.10. Let $\rho_{RM}(r, m)$ denote the covering radius of $\mathcal{R}(r, m)$. Using Exercise 617, $\rho_{RM}(0, m) = 2^{m-1}$ as $\mathcal{R}(0, m)$ is the binary repetition code. Also $\rho_{RM}(m, m) = 0$ as $\mathcal{R}(m, m) = \mathbb{F}_2^{2^m}$. We consider the case $1 \leq r < m$. By Theorem 1.10.1, $\mathcal{R}(r, m)^\perp = \mathcal{R}(m - r - 1, m)$ has minimum weight $2^{r+1} \geq 4$; therefore $\mathcal{R}(r, m)$ has strength 2 by Exercise 628. In addition $\mathcal{R}(r, m)$ is self-complementary as $\mathbf{1} \in \mathcal{R}(0, m) \subset \mathcal{R}(r, m)$ also by Theorem 1.10.1. Therefore by the Norse Bound (ii) we have the following theorem.

Theorem 11.2.3 *Let* $1 \leq r < m$. *Then* $\rho_{RM}(r, m) \leq 2^{m-1} - 2^{(m-2)/2}$. *In particular,* $\rho_{RM}(1, m) \leq 2^{m-1} - 2^{(m-2)/2}$.

Notice that by Exercise 618 and the nesting properties of the Reed–Muller codes, any upper bound on $\rho_{RM}(1, m)$ is automatically an upper bound on $\rho_{RM}(r, m)$ for $1 < r$.

We now find a recurrence relation for the covering radius of first order Reed–Muller codes that, combined with Theorem 11.2.3, will allow us to get an exact value for $\rho_{RM}(1, m)$ when m is even.

Lemma 11.2.4 *For* $m \geq 2$,

$$\rho_{RM}(1, m) \geq 2\rho_{RM}(1, m - 2) + 2^{m-2}.$$

Proof: For a vector $\mathbf{x} \in \mathbb{F}_2^n$, let $\overline{\mathbf{x}} = \mathbf{1} + \mathbf{x}$ denote the complement of $\mathbf{x}$. From Exercise 632, every codeword in $\mathcal{R}(1, m)$ has one of the following forms:
(a) $(\mathbf{x}|\mathbf{x}|\mathbf{x}|\mathbf{x})$,
(b) $(\mathbf{x}|\overline{\mathbf{x}}|\overline{\mathbf{x}}|\mathbf{x})$,
(c) $(\mathbf{x}|\overline{\mathbf{x}}|\mathbf{x}|\overline{\mathbf{x}})$, or
(d) $(\mathbf{x}|\mathbf{x}|\overline{\mathbf{x}}|\overline{\mathbf{x}})$,
where $\mathbf{x} \in \mathcal{R}(1, m - 2)$. Choose a vector $\mathbf{v} \in \mathbb{F}_2^{2^{m-2}}$ such that the coset $\mathbf{v} + \mathcal{R}(1, m - 2)$ has weight equaling $\rho_{RM}(1, m - 2)$. Therefore,

$$d(\mathbf{v}, \mathbf{x}) \geq \rho_{RM}(1, m - 2) \tag{11.5}$$

for all $\mathbf{x} \in \mathcal{R}(1, m - 2)$. Notice that

$$d(\mathbf{v}, \overline{\mathbf{x}}) = d(\overline{\mathbf{v}}, \mathbf{x}) = 2^{m-2} - d(\mathbf{v}, \mathbf{x}) = 2^{m-2} - d(\overline{\mathbf{v}}, \overline{\mathbf{x}}). \tag{11.6}$$

We prove the lemma if we find a vector $\mathbf{y}$ in $\mathbb{F}_2^{2^m}$ which has distance at least $2\rho_{RM}(1, m - 2) + 2^{m-2}$ from all codewords $\mathbf{c} \in \mathcal{R}(1, m)$. Let $\mathbf{y} = (\mathbf{v}|\mathbf{v}|\mathbf{v}|\overline{\mathbf{v}})$. We examine each possibility (a) through (d) above for $\mathbf{c}$. For the corresponding case, using (11.5), (11.6), and the fact that $\overline{\mathbf{x}} \in \mathcal{R}(1, m - 2)$ if $\mathbf{x} \in \mathcal{R}(1, m - 2)$, we have:
(a) $d(\mathbf{y}, \mathbf{c}) = 3d(\mathbf{v}, \mathbf{x}) + d(\overline{\mathbf{v}}, \mathbf{x}) = 2^{m-2} + 2d(\mathbf{v}, \mathbf{x}) \geq 2^{m-2} + 2\rho_{RM}(1, m - 2)$,
(b) $d(\mathbf{y}, \mathbf{c}) = d(\mathbf{v}, \mathbf{x}) + 2d(\mathbf{v}, \overline{\mathbf{x}}) + d(\overline{\mathbf{v}}, \mathbf{x}) = d(\mathbf{v}, \mathbf{x}) + 3d(\mathbf{v}, \overline{\mathbf{x}}) = 2^{m-2} + 2d(\mathbf{v}, \overline{\mathbf{x}}) \geq 2^{m-2} + 2\rho_{RM}(1, m - 2)$,
(c) $d(\mathbf{y}, \mathbf{c}) = 2d(\mathbf{v}, \mathbf{x}) + d(\mathbf{v}, \overline{\mathbf{x}}) + d(\overline{\mathbf{v}}, \overline{\mathbf{x}}) = 3d(\mathbf{v}, \mathbf{x}) + d(\mathbf{v}, \overline{\mathbf{x}}) = 2d(\mathbf{v}, \mathbf{x}) + 2^{m-2} \geq 2\rho_{RM}(1, m - 2) + 2^{m-2}$, and

(d) $d(\mathbf{y}, \mathbf{c}) = 2d(\mathbf{v}, \mathbf{x}) + d(\mathbf{v}, \overline{\mathbf{x}}) + d(\overline{\mathbf{v}}, \overline{\mathbf{x}}) = 3d(\mathbf{v}, \mathbf{x}) + d(\mathbf{v}, \overline{\mathbf{x}}) = 2d(\mathbf{v}, \mathbf{x}) + 2^{m-2} \geq 2\rho_{RM}(1, m-2) + 2^{m-2}$.
The result follows. □

Exercise 632 Show that every codeword in $\mathcal{R}(1, m)$ has one of the following forms:
(a) $(\mathbf{x}|\mathbf{x}|\mathbf{x}|\mathbf{x})$,
(b) $(\mathbf{x}|\overline{\mathbf{x}}|\overline{\mathbf{x}}|\mathbf{x})$,
(c) $(\mathbf{x}|\overline{\mathbf{x}}|\mathbf{x}|\overline{\mathbf{x}})$, or
(d) $(\mathbf{x}|\mathbf{x}|\overline{\mathbf{x}}|\overline{\mathbf{x}})$,
where $\mathbf{x} \in \mathcal{R}(1, m-2)$. Hint: See Section 1.10.   ◆

Exercise 633 Do the following:
(a) Show that $\rho_{RM}(1, 1) = 0$.
(b) Show that if m is odd, then $\rho_{RM}(1, m) \geq 2^{m-1} - 2^{(m-1)/2}$. Hint: Use (a) and Lemma 11.2.4.
(c) Show that $\rho_{RM}(1, 2) = 1$.
(d) Show that if m is even, then $\rho_{RM}(1, m) \geq 2^{m-1} - 2^{(m-2)/2}$. Hint: Use (c) and Lemma 11.2.4.   ◆

Combining Theorem 11.2.3 and Exercise 633(d) yields the exact value for $\rho_{RM}(1, m)$ when m is even.

Theorem 11.2.5 *For m even,* $\rho_{RM}(1, m) = 2^{m-1} - 2^{(m-2)/2}$.

When m is odd, the inequality in Exercise 633(b) is an equality for the first four values of m. By Exercise 633(a), $\rho_{RM}(1, 1) = 0$. As $\mathcal{R}(1, 3)$ is the $[8, 4, 4]$ extended Hamming code, $\rho_{RM}(1, 3) = 2$ by Theorem 11.1.7. It is not difficult to show that $\rho_{RM}(1, 5) = 12$; one theoretical proof is given as a consequence of Lemma 9.4 in [34], or a computer package that calculates covering radius could also verify this. Mykkeltveit [242] showed that $\rho_{RM}(1, 7) = 56$. From these four cases one would conjecture that the lower bound in Exercise 633(b) for m odd is an equality. This, however, is false; when $m = 15$, the covering radius of $\mathcal{R}(1, 15)$ is at least $16\,276 = 2^{14} - 2^7 + 20$ [250, 251].

There are bounds on $\rho_{RM}(r, m)$ for $1 < r$ but relatively few results giving exact values for $\rho_{RM}(r, m)$. One such result is found in [237].

Theorem 11.2.6

$$\rho_{RM}(m - 3, m) = \begin{cases} m+2 & \text{for } m \text{ even,} \\ m+1 & \text{for } m \text{ odd.} \end{cases}$$

Table 11.1, found in [34], gives exact values and bounds on $\rho_{RM}(r, m)$ for $1 \leq r \leq m \leq 9$. References for the entries can be found in [34].

Exercise 634 A computer algebra package will be useful for this problem. Show that the Sphere Covering Bound gives the lower bounds on $\rho_{RM}(r, m)$ found in Table 11.1 for:
(a) $(r, m) = (3, 8)$ and
(b) $(r, m) = (r, 9)$ when $2 \leq r \leq 5$.   ◆

Table 11.1 *Bounds on* $\rho_{\text{RM}}(r, m)$ *for* $1 \le r \le m \le 9$

$r \setminus m$	1	2	3	4	5	6	7	8	9
1	0	1	2	6	12	28	56	120	240–244
2		0	1	2	6	18	40–44	84–100	171–220
3			0	1	2	8	20–23	43–67	111–167
4				0	1	2	8	22–31	58–98
5					0	1	2	10	23–41
6						0	1	2	10
7							0	1	2
8								0	1
9									0

Shortening a first order Reed–Muller code produces a binary simplex code; see Exercise 62. While we do not have exact values for the covering radius of all first order Reed–Muller codes, we do have exact values for the covering radius of all binary simplex codes.

Theorem 11.2.7 *The covering radius* ρ *of the* $[2^m - 1, m, 2^{m-1}]$ *binary simplex code* $\mathcal{S}_m$ *is* $2^{m-1} - 1$.

Proof: As no coordinate of $\mathcal{S}_m$ is identically 0, $\mathcal{S}_m$ has strength 1 by Exercise 628(c). Hence by the Norse Bound, $\rho \le 2^{m-1} - 1$. By Theorem 2.7.5, all nonzero codewords of $\mathcal{S}_m$ have weight 2^{m-1}. Hence the coset of $\mathcal{S}_m$ containing the all-one vector has weight $2^{m-1} - 1$. Therefore $2^{m-1} - 1 \le \rho$. □

11.3 Covering radius of BCH codes

As with Reed–Muller codes, the covering radii of codes from certain subfamilies of BCH codes have been computed. In particular, let $\mathcal{B}(t, m)$ be the primitive narrow-sense binary BCH code of length $n = 2^m - 1$ and designed distance $d = 2t + 1$. Exercise 635 gives the dimensions of these codes when $t \le 2^{\lceil m/2 \rceil - 1}$.

Exercise 635 Let C_i be the 2-cyclotomic coset modulo $n = 2^m - 1$ containing i.
(a) For $0 \le i \le n$, let $i = i_0 + i_1 2 + \cdots + i_{m-2} 2^{m-2} + i_{m-1} 2^{m-1}$ with $i_j \in \{0, 1\}$ be the binary expansion of i. Thus we can associate with i the binary string

$$i \leftrightarrow i_0 i_1 \cdots i_{m-2} i_{m-1} \tag{11.7}$$

of length m. Prove that C_i contains precisely the integers j whose associated binary string is a cyclic shift of (11.7).
(b) Prove that if j is an odd integer with $j \le 2^{\lceil m/2 \rceil} - 1$, then $C_1, C_3, \ldots, C_j$ are distinct 2-cyclotomic cosets each containing m elements.
(c) Prove that if $1 \le t \le 2^{\lceil m/2 \rceil - 1}$, $\mathcal{B}(t, m)$ has dimension $2^m - 1 - tm$.
(d) Demonstrate that the upper limit on j given in part (b) cannot be increased by showing that when $m = 5$, C_9 equals one of the C_i with $1 \le i \le 7$ and i odd.   ◆

Table 11.2 *Weight distribution of $\mathcal{B}(2, m)^{\perp}$*

	i	A_i
m odd:	0	1
	$2^{m-1} - 2^{(m-1)/2}$	$(2^m - 1)(2^{m-2} + 2^{(m-3)/2})$
	2^{m-1}	$(2^m - 1)(2^{m-1} + 1)$
	$2^{m-1} + 2^{(m-1)/2}$	$(2^m - 1)(2^{m-2} - 2^{(m-3)/2})$
m even:	0	1
	$2^{m-1} - 2^{m/2}$	$\frac{1}{3}2^{(m-2)/2-1}(2^m - 1)(2^{(m-2)/2} + 1)$
	$2^{m-1} - 2^{m/2-1}$	$\frac{1}{3}2^{(m+2)/2-1}(2^m - 1)(2^{m/2} + 1)$
	2^{m-1}	$(2^m - 1)(2^{m-2} + 1)$
	$2^{m-1} + 2^{m/2-1}$	$\frac{1}{3}2^{(m+2)/2-1}(2^m - 1)(2^{m/2} - 1)$
	$2^{m-1} + 2^{m/2}$	$\frac{1}{3}2^{(m-2)/2-1}(2^m - 1)(2^{(m-2)/2} - 1)$

Let $\rho_{\text{BCH}}(t, m)$ denote the covering radius $\rho(\mathcal{B}(t, m))$ of $\mathcal{B}(t, m)$. We can use the Supercode Lemma to obtain a lower bound on $\rho_{\text{BCH}}(t, m)$ when $t \leq 2^{\lceil m/2 \rceil - 1}$.

Theorem 11.3.1 *If* $t \leq 2^{\lceil m/2 \rceil - 1}$, *then* $\rho_{\text{BCH}}(t, m) \geq 2t - 1$.

Proof: By Exercise 635, $\mathcal{B}(t - 1, m)$ and $\mathcal{B}(t, m)$ have dimensions $2^m - 1 - (t - 1)m$ and $2^m - 1 - tm$, respectively. Therefore by the nesting property of BCH codes (Theorem 5.1.2), $\mathcal{B}(t, m)$ is a proper subcode of $\mathcal{B}(t - 1, m)$. Every vector $\mathbf{x}$ in $\mathcal{B}(t - 1, m) \setminus \mathcal{B}(t, m)$ has weight at least the minimum weight of $\mathcal{B}(t - 1, m)$, which is at least $2t - 1$ by the BCH Bound. The result now follows from the Supercode Lemma. $\square$

The case of the single error-correcting codes is quite easy to handle.

Theorem 11.3.2 *For* $m \geq 2$, $\rho_{\text{BCH}}(1, m) = 1$.

Proof: The code $\mathcal{B}(1, m)$ is the perfect single error-correcting Hamming code $\mathcal{H}_m$. As it is perfect, $\rho_{\text{BCH}}(1, m) = 1$. $\square$

We now turn our attention to the much more difficult case of the double error-correcting codes. When m is odd, we can use the lower bound in Theorem 11.3.1 and an upper bound due to Delsarte, which agrees with the lower bound. Because the Delsarte Bound, which we proved as part (i) of Theorem 7.5.2, will be used in this section and the next, we restate it here.

Theorem 11.3.3 (Delsarte Bound) *Let* $\mathcal{C}$ *be an* $[n, k]$ *code over* $\mathbb{F}_q$. *Let* $S = \{i > 0 \mid A_i(\mathcal{C}^{\perp}) \neq 0\}$ *and* $s = |S|$. *Then* $\rho(\mathcal{C}) \leq s$.

We remark that Delsarte actually proved this in [62] for nonlinear codes as well, with the appropriate interpretation of s.

We will show that $\rho_{\text{BCH}}(2, m) = 3$, a result due to Gorenstein, Peterson, and Zierler [108]. Theorem 11.3.1 provides a lower bound of 3 on the covering radius. The weight distribution is known for $\mathcal{B}(2, m)^{\perp}$. We present it without proof in Table 11.2; those interested in the proof are referred to [218, Chapter 15]. Using the Delsarte Bound we see from the table that an upper bound on the covering radius is 3 when m is odd and 5 when m is even. So when m is odd, $\rho_{\text{BCH}}(2, m) = 3$ as the upper and lower bounds agree; but this proof will not work if m is even. The case m even will require an additional lemma involving the trace function Tr_m from $\mathbb{F}_{2^m}$ to $\mathbb{F}_2$. Recall that

$$\text{Tr}_m(\alpha) = \sum_{i=0}^{m-1} \alpha^{2^i} \qquad \text{for all } \alpha \in \mathbb{F}_{2^m}.$$

Lemma 11.3.4 *The equation* $x^2 + x + \beta = 0$ *for* $\beta \in \mathbb{F}_{2^m}$ *has a solution in* $\mathbb{F}_{2^m}$ *if* $\text{Tr}_m(\beta) = 0$ *and no solutions if* $\text{Tr}_m(\beta) = 1$.

Proof: By the Normal Basis Theorem [196, Theorem 2.35], there is a basis

$$\left\{ \alpha, \alpha^2, \dots, \alpha^{2^{m-1}} \right\},$$

called a *normal basis*, of $\mathbb{F}_{2^m}$ as a vector space over $\mathbb{F}_2$. By Exercise 637, $\text{Tr}_m(\alpha) = \text{Tr}_m(\alpha^{2^i})$ for all i. As the trace function is not identically 0 by Lemma 3.8.5, this implies, using the $\mathbb{F}_2$-linearity of the trace function described in Lemma 3.8.5, that

$$\text{Tr}_m\left(\alpha^{2^i}\right) = 1 \qquad \text{for all } i. \tag{11.8}$$

Let

$$x = x_0\alpha + x_1\alpha^2 + x_2\alpha^{2^2} + \cdots + x_{m-1}\alpha^{2^{m-1}} \qquad \text{where } x_i \in \mathbb{F}_2, \quad \text{and}$$
$$\beta = b_0\alpha + b_1\alpha^2 + b_2\alpha^{2^2} + \cdots + b_{m-1}\alpha^{2^{m-1}} \qquad \text{where } b_i \in \mathbb{F}_2.$$

Then

$$x^2 = x_{m-1}\alpha + x_0\alpha^2 + x_1\alpha^{2^2} + \cdots + x_{m-2}\alpha^{2^{m-1}}.$$

If x is a solution of $x^2 + x + \beta = 0$, we must have

$$x_0 + x_{m-1} = b_0, \quad x_1 + x_0 = b_1, \dots, \quad x_{m-1} + x_{m-2} = b_{m-1}.$$

The sum of the left-hand sides of all these equations is 0, giving $\sum_{i=0}^{m-1} b_i = 0$. However, by (11.8), $\text{Tr}_m(\beta) = \sum_{i=0}^{m-1} b_i$ proving that $\text{Tr}_m(\beta) = 0$ is a necessary condition for the existence of a solution $x \in \mathbb{F}_{2^m}$ of $x^2 + x + \beta = 0$.

You are asked to prove the converse in Exercise 638. □

Exercise 636 Using Example 3.4.3, find a normal basis of $\mathbb{F}_8$ over $\mathbb{F}_2$. ♦

Exercise 637 If $\gamma \in \mathbb{F}_{2^m}$, prove that $\text{Tr}_m(\gamma) = \text{Tr}_m(\gamma^2)$. ♦

Exercise 638 With the notation as in Lemma 11.3.4, prove that if $\text{Tr}_m(\beta) = 0$, then $x^2 + x + \beta = 0$ has two solutions $x \in \mathbb{F}_{2^m}$ where we can take x_0 equal to either 0 or 1 and $x_i = x_0 + \sum_{j=1}^{i} b_j$ for $1 \le i \le m - 1$. ♦

We are now ready to show that $\rho_{\text{BCH}}(2, m) = 3$. The case m even uses some of the techniques we developed in Section 5.4 for decoding BCH codes.

Theorem 11.3.5 *If* $m \ge 3$, *then* $\rho_{\text{BCH}}(2, m) = 3$.

Proof: The case m odd was proved above. Now assume that m is even. By Theorem 11.3.1, we have $\rho_{\mathrm{BCH}}(2, m) \geq 3$. We are done if we show that every coset of $\mathcal{B}(2, m)$ contains a vector of weight at most 3.

Let α be a primitive element of $\mathbb{F}_{2^m}$. By Theorem 4.4.3 we can view

$$H = \begin{bmatrix} 1 & \alpha & \alpha^2 & \cdots & \alpha^{n-1} \\ 1 & \alpha^3 & \alpha^6 & \cdots & \alpha^{3(n-1)} \end{bmatrix}$$

as a parity check matrix for $\mathcal{B}(2, m)$. Let $y(x)$ be a polynomial in $\mathbb{F}_2[x]/(x^{2^m-1} - 1)$. In Section 5.4.1 we defined the syndrome S_i of $y(x)$ as $y(\alpha^i)$. By Exercise 295, we have

$$H\mathbf{y}^{\mathrm{T}} = \begin{bmatrix} S_1 \\ S_3 \end{bmatrix}, \tag{11.9}$$

where $\mathbf{y}$ is the vector in $\mathbb{F}_2^{2^m-1}$ associated to $y(x)$. By Theorem 1.11.5 and Exercise 295(b), there is a one-to-one correspondence between syndromes and cosets. Thus it suffices to show that for any pair of syndromes S_1 and S_3, there is a vector $\mathbf{y}$ with $\mathrm{wt}(\mathbf{y}) \leq 3$ that satisfies (11.9). Let X_i be the location numbers of $\mathbf{y}$; see Section 5.4. Thus, by (5.8), we only need to show that for any S_1 and S_3 in $\mathbb{F}_{2^m}$, there exists X_1, X_2, and X_3 in $\mathbb{F}_{2^m}$ such that

$$\begin{aligned} S_1 &= X_1 + X_2 + X_3, \\ S_3 &= X_1^3 + X_2^3 + X_3^3. \end{aligned} \tag{11.10}$$

If we let $z_i = X_i + S_1$, then by Exercise 639 we obtain the equivalent system

$$\begin{aligned} 0 &= z_1 + z_2 + z_3, \\ s &= z_1^3 + z_2^3 + z_3^3, \end{aligned} \tag{11.11}$$

where $s = S_1^3 + S_3$. If $s = 0$, $z_1 = z_2 = z_3 = 0$ is a solution of (11.11). So assume that $s \neq 0$. Substituting $z_3 = z_1 + z_2$ from the first equation in (11.11) into the second equation gives

$$z_1^2 z_2 + z_1 z_2^2 = s.$$

We will find a solution with $z_2 \neq 0$, in which case if we set $z = z_1/z_2$ we must solve

$$z^2 + z + \frac{s}{z_2^3} = 0, \tag{11.12}$$

which is of the form examined in Lemma 11.3.4. Thus by the lemma we must find a z_2 such that $\mathrm{Tr}_m(s/z_2^3) = 0$.

Since $s \neq 0$, we may assume that $s = \alpha^{3i+j}$, where $0 \leq j \leq 2$. There are two cases:

- Suppose $j = 0$. Let $z_2 = \alpha^i$. Then $\mathrm{Tr}_m(s/z_2^3) = \mathrm{Tr}_m(1) = 1 + 1 + \cdots + 1$ (m times). As m is even, $\mathrm{Tr}_m(1) = 0$ and so we have z_2 such that $\mathrm{Tr}_m(s/z_2^3) = 0$.
- Suppose $j = 1$ or 2. Recalling that Tr_m is a surjective $\mathbb{F}_2$-linear transformation of $\mathbb{F}_{2^m}$ onto $\mathbb{F}_2$, there are exactly 2^{m-1} elements of $\mathbb{F}_{2^m}$ whose trace is 0. Since m is even, $3 \mid (2^m - 1)$; thus there are exactly $(2^m - 1)/3$ elements of $\mathbb{F}_{2^m}$ that are nonzero cubes by Exercise 640. Hence these counts show that there must be some nonzero element of $\mathbb{F}_{2^m}$ which is not a cube but has trace 0; let $\theta = \alpha^{3k+r}$ be such an element where r then

Table 11.3 *Covering radius of binary*
$[n = 2^m - 1, k, d]$ *codes* $\mathcal{B}(t, m)$

m	n	k	t	d	$\rho_{BCH}(t, m)$
3	7	4	1	3	1
4	15	11	1	3	1
4	15	7	2	5	3
4	15	5	3	7	5
5	31	26	1	3	1
5	31	21	2	5	3
5	31	16	3	7	5
5	31	11	4	11	7
5	31	6	6	15	11
6	63	57	1	3	1
6	63	51	2	5	3
6	63	45	3	7	5
6	63	39	4	9	7
6	63	36	5	11	9

must be 1 or 2. Note that $0 = \text{Tr}_m(\theta) = \text{Tr}(\theta^2)$; hence if necessary we may replace θ by θ^2 and thereby assume that $j = r$. Now let $z_2 = \alpha^{i-k}$. Then $s/z_2^3 = \theta$, which has trace 0 as needed. Thus we again have z_2 such that $\text{Tr}_m(s/z_2^3) = 0$.

Therefore in each case we have produced a nonzero $z_2 \in \mathbb{F}_{2^m}$ such that there is a solution z of (11.12). Then $z_1 = z_2 z$ and $z_3 = z_1 + z_2$ give the solution of (11.11) as required. □

Exercise 639 Show that (11.10) and (11.11) are equivalent systems. Hint: Recall that $S_2 = S_1^2 = X_1^2 + X_2^2 + X_3^2$. ◆

Exercise 640 Show that if m is even, $3 \mid (2^m - 1)$ and there are exactly $(2^m - 1)/3$ elements of $\mathbb{F}_{2^m}$ that are nonzero cubes. Hint: $\mathbb{F}_{2^m}^* = \mathbb{F}_{2^m} \setminus \{0\}$ is a cyclic group of order divisible by 3. ◆

One would hope to try a similar method for the triple error-correcting BCH codes. Assmus and Mattson [7] showed that $\mathcal{B}(3, m)^\perp$ contains five nonzero weights when m is odd. So the Delsarte Bound and Theorem 11.3.1 combine to show that $\rho_{BCH}(3, m) = 5$ for m odd. When m is even, $\mathcal{B}(3, m)^\perp$ contains seven nonzero weights. The case m even was completed by van der Horst and Berger [138] and by Helleseth [121] yielding the following theorem.

Theorem 11.3.6 *If $m \geq 4$, then $\rho_{BCH}(3, m) = 5$.*

A computer calculation was done in [71, 72] that produced the covering radii of primitive narrow-sense binary BCH codes of lengths up to 63. The results are summarized in Table 11.3 and are also found in [34]. Note that each covering radius in the table satisfies $\rho_{BCH}(t, m) = 2t - 1$, which is the lower bound obtained in Theorem 11.3.1. The following theorem, proved by Vlăduţ and Skorobogatov [340], shows that for very long BCH codes the covering radius of $\mathcal{B}(t, m)$ is also equal to the lower bound $2t - 1$.

Theorem 11.3.7 *There is a constant M such that for all $m \geq M$, $\rho_{\mathrm{BCH}}(t, m) = 2t - 1$.*

11.4 Covering radius of self-dual codes

In Chapters 9 and 10 we examined some well-known binary self-dual codes of moderate lengths. In this section we want to compute the covering radius of some of these codes.

We begin by considering extremal Type II binary codes. In Example 1.11.7, we presented the weight distribution of each coset of the unique $[8, 4, 4]$ Type II binary code and conclude that it has covering radius 2. This same result comes from Theorem 11.1.7. In a series of examples and exercises in Section 7.5, we gave the complete coset weight distribution of each of the two $[16, 8, 4]$ Type II binary codes and concluded that the covering radius is 4 for each code. Exercise 641 gives another proof without resorting to specific knowledge of the generator matrix of the code.

Exercise 641 Let $\mathcal{C}$ be a $[16, 8, 4]$ Type II binary code.
(a) Use the Delsarte Bound to show that $\rho(\mathcal{C}) \leq 4$.
(b) Use Theorems 11.1.1 and 11.1.7 to show that $3 \leq \rho(\mathcal{C})$.
(c) Suppose that $\mathcal{C}$ has covering radius 3. By Exercise 625, if $\mathcal{C}^*$ is obtained from $\mathcal{C}$ by puncturing one coordinate, then $\rho(\mathcal{C}^*) = 2$. Show that this violates the Sphere Covering Bound. ◆

By Theorem 10.1.7 (or Example 8.3.2), we know that the unique $[24, 12, 8]$ Type II binary code has covering radius 4. As this code is extended perfect, Theorem 11.1.7 also shows this. We now turn to the $[32, 16, 8]$ Type II binary codes. In Section 9.7 we noted that there are exactly five such codes. Without knowing the specific structure of any of these except the number of weight 8 vectors, we can find their covering radii, a result found in [9].

Theorem 11.4.1 *Let $\mathcal{C}$ be a $[32, 16, 8]$ Type II code. Then $\rho(\mathcal{C}) = 6$.*

Proof: We remark that the weight distribution of $\mathcal{C}$ is determined uniquely; see Theorem 9.3.1. It turns out that $A_8(\mathcal{C}) = 620$. $\mathcal{C} = \mathcal{C}^\perp$ contains codewords of weights 8, 12, 16, 20, 24, and 32. So by the Delsarte Bound $\rho(\mathcal{C}) \leq 6$. By Theorems 11.1.1 and 11.1.7, $5 \leq \rho(\mathcal{C})$ as $\mathcal{C}$ is not extended perfect (Exercise 642). Suppose that $\rho(\mathcal{C}) = 5$. This implies that every weight 6 vector in $\mathbb{F}_2^{32}$ is in a coset of even weight less than 6. The only coset of weight 0 is the code itself, which does not contain weight 6 codewords. Let $\mathbf{x}$ be a vector of weight 6. Then there must be a codeword $\mathbf{c}$ such that $\mathrm{wt}(\mathbf{x} + \mathbf{c}) = 2$ or 4. The only possibility is that $\mathrm{wt}(\mathbf{c}) = 8$. Thus every weight 6 vector has support intersecting the support of one of the 620 codewords of weight 8 in $\mathcal{C}$ in either five or six coordinates. The number of weight 6 vectors whose support intersects the support of a weight 8 vector in this manner is $\binom{24}{1}\binom{8}{5} + \binom{8}{6} = 1372$. Thus there are at most $620 \cdot 1372 = 850\,640$ weight 6 vectors in $\mathbb{F}_2^{32}$. But as there are $\binom{32}{6} = 906\,192$ weight 6 vectors in $\mathbb{F}_2^{32}$, we have a contradiction. □

Exercise 642 Use the Sphere Packing Bound to show that a $[32, 16, 8]$ even code is not extended perfect. Remark: We know this if we use the fact that the only perfect

Table 11.4 *Coset distribution of* $C = \mathcal{R}(2, 5)$, *where* $A_i(\mathbf{v} + C) = A_{n-i}(\mathbf{v} + C)$

Coset weight	Number of vectors of given weight																	Number of cosets
	0	1	2	3	4	5	6	7	8	9	10	11	12	13	14	15	16	
0	1								620				13 888				36 518	1
1		1						153		467		5 208		8 680		18 251		32
2			1				35		240		2 193		6 720		14 155		18 848	496
3				1		7		84		892		3 949		10 507		17 328		4 960
4					1		28		322		1 964		6 895		14 392		18 322	11 160
4					2		24		324		1 976		6 878		14 384		18 360	6 200
4					3		20		326		1 988		6 861		14 376		18 388	2 480
4					4		16		328		2 000		6 844		14 368		18 416	930
4					5		12		330		2 012		6 827		14 360		18 444	248
5						6		106		850		3 934		10 620		17 252		27 776
6							32		320		1 952		6 912		14 400		18 304	11 253

multiple error-correcting codes are the Golay codes from Theorem 1.12.3, but we do not need to use this powerful theorem to verify that a [32, 16, 8] even code is not extended perfect. ◆

Exercise 643 In this exercise, you will find part of the coset weight distributions of cosets of a [32, 16, 8] Type II code C.

(a) The weight distribution of C is $A_0 = A_{32} = 1$, $A_8 = A_{24} = 620$, $A_{12} = A_{20} = 13\,888$, and $A_{16} = 36\,518$. Verify this by solving the first five Pless power moments (P_1); see also (7.8).

(b) By Theorem 9.3.10, the codewords of weights 8, 12, and 16 hold 3-designs. Find the Pascal triangle for each of these 3-designs.

(c) In Table 11.4, we give the complete coset weight distribution of the [32, 16, 8] Type II code $\mathcal{R}(2, 5)$, where blank entries are 0. For cosets of weights 1, 2, and 3, verify that any [32, 16, 8] Type II code has the same coset weight distribution and number of cosets as $\mathcal{R}(2, 5)$ does. Hint: See Example 8.4.6.

(d) Repeat part (c) for the cosets of weight 5.

(e) For cosets of weight 6, verify that any [32, 16, 8] Type II code has the same coset weight distribution as $\mathcal{R}(2, 5)$ does. Note that the number of cosets of weight 6 depends on the code.

By this exercise, the complete coset weight distribution of each of the five [32, 16, 8] Type II codes is known once the number and distribution of the weight 4 cosets is known, from which the number of weight 6 cosets can be computed. These have been computed in [45] and are different for each code. ◆

We summarize, in Table 11.5, what is known [9] about the covering radii of the extremal Type II binary codes of length $n \le 64$. An argument similar to that in the proof of Theorem 11.4.1 gives the upper and lower bounds for $n = 40, 56$, and 64. We remark that there exists an extremal Type II code of length 40 with covering radius 7 and another with covering radius 8; see [34]. The exact value for the covering radius of an extremal Type II code of length 48 is 8; this result, found in [9], uses the fact that the weight 12 codewords of such a code hold a 5-design by Theorem 9.3.10. The only [48, 24, 12] Type II code is

Table 11.5 *Covering radii ρ of extremal Type II binary codes of length $n \leq 64$*

n	8	16	24	32	40	48	56	64
k	4	8	12	16	20	24	28	32
d	4	4	8	8	8	12	12	12
ρ	2	4	4	6	6–8	8	8–10	8–12

the extended quadratic residue code; see Research Problem 9.3.7. Part of the coset weight distribution of the extended quadratic residue code is given in [62]; from this the complete coset distribution can be computed.

There are many similarities between the coset distributions of the length 32 Type II codes discussed in Exercise 643 and those of length 48. For example, the cosets of weights 1 through 5 and 7 in the length 48 codes have uniquely determined coset distributions and uniquely determined numbers of cosets. The cosets of weight 8 also have uniquely determined distributions but not number of cosets of weight 8.

Exercise 644 Verify the bounds given in Table 11.5 for $n = 40, 56$, and 64. When $n = 40$, the number of weight 8 codewords in an extremal Type II code is 285. When $n = 56$, the number of weight 12 codewords in an extremal Type II code is 8190. When $n = 64$, the number of weight 12 codewords in an extremal Type II code is 2976.   ♦

Near the end of Section 9.4 we defined the child of a self-dual code. Recall that if C is an $[n, n/2, d]$ self-dual binary code with $d \geq 3$, then a child C'^* of C is obtained by fixing two coordinates, taking the subcode C' that is 00 or 11 on those two coordinates, and then puncturing C' on those fixed coordinates. This produces an $[n - 2, (n - 2)/2, d'^*]$ self-dual code with $d'^* \geq d - 2$. The child of a self-dual code can have a larger covering radius than the parent as the following result shows.

Theorem 11.4.2 *Let C be an $[n, n/2, d]$ self-dual code with $d \geq 3$. Suppose that C'^* is a child of C. Then $\rho(C'^*) \geq d - 1$.*

Proof: Suppose that the two fixed coordinates which are used to produce C'^* are the two left-most coordinates. Let C' be as above. Choose $\mathbf{x} \in C$ to have minimum nonzero weight among all codewords of C which are 10 or 01 on the first two coordinates. So $\mathbf{x} = 10\mathbf{y}$ or $01\mathbf{y}$. As adding a codeword $\mathbf{c}'$ of C' to $\mathbf{x}$ produces a codeword $\mathbf{c} = \mathbf{c}' + \mathbf{x}$ of C with first two coordinates 01 or 10, wt($\mathbf{c}$) $\geq$ wt($\mathbf{x}$) by the choice of $\mathbf{x}$. This means that $\mathbf{y}$ must be a coset leader of the coset $\mathbf{y} + C'^*$. Since wt($\mathbf{y}$) $\geq d - 1$, the result follows.   □

A child of the $[24, 12, 8]$ extended binary Golay code is a $[22, 11, 6]$ extremal Type I code discussed in Example 9.4.12. Its covering radius is three more than that of its parent as Exercise 645 shows.

Exercise 645 Let C be the $[24, 12, 8]$ extended binary Golay code and C'^* its child. Prove that $\rho(C'^*) = 7$.   ♦

11.5   The length function

In this section we examine the *length function* $\ell_q(m, r)$ for $1 \le r \le m$, where $\ell_q(m, r)$ is the smallest length of any linear code over $\mathbb{F}_q$ of redundancy m and covering radius r. Closely related to the length function is the function $t_q(n, k)$, defined as the smallest covering radius of any $[n, k]$ code over $\mathbb{F}_q$. When $q = 2$, we will denote these functions by $\ell(m, r)$ and $t(n, k)$ respectively. Knowing one function will give the other, as we shall see shortly. Finding values of these functions and codes that realize these values is fundamental to finding codes with good covering properties. We first give some elementary facts about these two functions.

Theorem 11.5.1 *The following hold*:
(i) $t_q(n + 1, k + 1) \le t_q(n, k)$.
(ii) $\ell_q(m, r + 1) \le \ell_q(m, r)$.
(iii) $\ell_q(m, r) \le \ell_q(m + 1, r) - 1$.
(iv) $\ell_q(m + 1, r + 1) \le \ell_q(m, r) + 1$.
(v)   *If* $\ell_q(m, r) \le n$, *then* $t_q(n, n - m) \le r$.
(vi)   *If* $t_q(n, k) \le r$, *then* $\ell_q(n - k, r) \le n$.
(vii) $\ell_q(m, r)$ *equals the smallest integer* n *such that* $t_q(n, n - m) \le r$.
(viii) $t_q(n, k)$ *equals the smallest integer* r *such that* $\ell_q(n - k, r) \le n$.

Proof: We leave the proofs of (i), (vi), and (vii) as an exercise.

For (ii), let $n = \ell_q(m, r)$. Then there exists an $[n, n - m]$ code with covering radius r, which has parity check matrix $H = [I_m\ A]$. Among all such parity check matrices for $[n, n - m]$ codes with covering radius r, choose A to be the one with some column $\mathbf{w}^\mathrm{T}$ of smallest weight. This column cannot be the zero column as otherwise we can puncture the column and obtain a code of smaller length with redundancy m and covering radius r by Theorem 11.1.2(ii). Replace some nonzero entry in $\mathbf{w}^\mathrm{T}$ with 0 to form the matrix A_1 with $\mathbf{w}_1^\mathrm{T}$ in place of $\mathbf{w}^\mathrm{T}$; so $\mathbf{w} = \mathbf{w}_1 + \alpha_1 \mathbf{u}_1$ where $\mathbf{u}_1^\mathrm{T}$ is one of the columns of I_m. Then $H_1 = [I_m\ A_1]$ is the parity check matrix of an $[n, n - m]$ code $\mathcal{C}_1$ satisfying $\rho(\mathcal{C}_1) \ne r$ by the choice of A. Any syndrome $\mathbf{s}^\mathrm{T}$ can be written as a linear combination of r or fewer columns of H. If $\mathbf{w}^\mathrm{T}$ is involved in one of these linear combinations, we can replace $\mathbf{w}^\mathrm{T}$ by $\mathbf{w}_1^\mathrm{T} - \alpha_1 \mathbf{u}_1^\mathrm{T}$. This allows us to write $\mathbf{s}^\mathrm{T}$ as a linear combination of $r + 1$ or fewer columns of H_1. Thus $\rho(\mathcal{C}_1) \le r + 1$. Since $\rho(\mathcal{C}_1) \ne r$, $\rho(\mathcal{C}_1) \le r - 1$ or $\rho(\mathcal{C}_1) = r + 1$. We assume that $\rho(\mathcal{C}_1) \le r - 1$ and will obtain a contradiction. If $\mathbf{w}_1 = \mathbf{0}$, then every linear combination of columns of H_1 is in fact a linear combination of columns of H implying by Theorem 11.1.2(ii) that $\rho(\mathcal{C}) \le \rho(\mathcal{C}_1) \le r - 1$, a contradiction. Therefore replace some nonzero entry in $\mathbf{w}_1^\mathrm{T}$ with 0 to form the matrix A_2 with $\mathbf{w}_2^\mathrm{T}$ in place of $\mathbf{w}_1^\mathrm{T}$, where $\mathbf{w}_1 = \mathbf{w}_2 + \alpha_2 \mathbf{u}_2$ and $\mathbf{u}_2^\mathrm{T}$ is one of the columns of I_m. Then $H_2 = [I_m\ A_2]$ is the parity check matrix of an $[n, n - m]$ code $\mathcal{C}_2$. Arguing as above, $\rho(\mathcal{C}_2)$ is at most 1 more than $\rho(\mathcal{C}_1)$ implying $\rho(\mathcal{C}_2) \le r$. By the choice of A, $\rho(\mathcal{C}_2) \ne r$ and thus $\rho(\mathcal{C}_2) \le r - 1$. We can continue inductively constructing a sequence of codes $\mathcal{C}_i$ with parity check matrices $H_i = [I_m\ A_i]$, where A_i is obtained from A_{i-1} by replacing a nonzero entry in $\mathbf{w}_{i-1}$ by 0 to form $\mathbf{w}_i$. Each code satisfies $\rho(\mathcal{C}_i) \le r - 1$. This process continues until step $i = s$ where $\text{wt}(\mathbf{w}_s) = 1$. So this column $\mathbf{w}_s$ of H_s is a multiple of some column of I_m. But then every linear combination of the columns of H_s is a linear combination of the columns of H implying by Theorem 11.1.2(ii) that $\rho(C) \leq \rho(C_s) \leq r - 1$, which is a contradiction. Thus $\rho(C_1) = r + 1$ and therefore $\ell_q(m, r + 1) \leq n = \ell_q(m, r)$ proving (ii).

For part (iii), let H be a parity check matrix for a code of length $n = \ell_q(m + 1, r)$, redundancy $m + 1$, and covering radius r. By row reduction, we may assume that $[1\ 0\ 0 \cdots 0]^{\mathrm{T}}$ is the first column of H. Let H' be the $m \times (n - 1)$ matrix obtained from H by removing the first row and column of H. The code C', with parity check matrix H', has length $n - 1$ and redundancy m. Using the first column of H in a linear combination of the columns of H affects only the top entry; therefore $\rho(C') = r' \leq r$. Thus $\ell_q(m, r') \leq n - 1$. But $\ell_q(m, r) \leq \ell_q(m, r')$ by induction on (ii). Part (iii) follows.

To verify (iv), let C be a code of length $n = \ell_q(m, r)$, redundancy m, and covering radius r. Let Z be the zero code of length 1, which has covering radius 1. Then $C_1 = C \oplus Z$ is a code of length $n + 1$, redundancy $m + 1$, and covering radius $r + 1$ by Theorem 11.1.6(i). Therefore $\ell_q(m + 1, r + 1) \leq n + 1 = \ell_q(m, r) + 1$ and (iv) holds.

For part (v), suppose that $\ell_q(m, r) = n' \leq n$. Then there is an $[n', n' - m]$ code with covering radius r. So $t_q(n', n' - m) \leq r$. Applying (i) repeatedly, we obtain $t_q(n, n - m) \leq t_q(n', n' - m)$ completing (v).

Suppose that $t_q(n, k) = r$. Then there exists an $[n, k]$ code with covering radius r implying that $\ell_q(n - k, r) \leq n$. Let r' be the smallest value of ρ such that $\ell_q(n - k, \rho) \leq n$. Then $r' \leq r$, and there is an $[n', n' - n + k]$ code with $n' \leq n$ and covering radius r'. Thus $t_q(n', n' - n + k) \leq r'$. By repeated use of (i), $t_q(n, k) \leq t_q(n', n' - n + k) \leq r'$ implying $r \leq r'$. Hence $r = r'$ verifying (viii). $\qquad\square$

Exercise 646 Prove the remainder of Theorem 11.5.1 by doing the following:
(a) Prove Theorem 11.5.1(i). Hint: Suppose that $t_q(n, k) = r$. Thus there is an $[n, k]$ code C over $\mathbb{F}_q$ with covering radius r. Let H be a parity check matrix for C. Choose any column to adjoin to H. What happens to the covering radius of the code with this new parity check matrix?
(b) Prove Theorem 11.5.1(vi). Hint: See the proof of Theorem 11.5.1(v).
(c) Prove Theorem 11.5.1(vii). Hint: This follows directly from the definitions. ◆

Exercise 647 Suppose that C is a linear binary code of length $\ell(m, r)$ with $1 \leq r$ where C has redundancy m and covering radius r. Show that C has minimum distance at least 3 except when $m = r = 1$. ◆

There is an upper and lower bound on $\ell_q(m, r)$, as we now see.

Theorem 11.5.2 Let $r \leq m$. Then $m \leq \ell_q(m, r) \leq (q^{m-r+1} - 1)/(q - 1) + r - 1$. Furthermore:
(i) $m = \ell_q(m, r)$ if and only if $r = m$, and the only code satisfying this condition is the zero code, and
(ii) $\ell_q(m, 1) = (q^m - 1)/(q - 1)$, and the only code satisfying this condition is the Hamming code $\mathcal{H}_{q,m}$.

Proof: As the length of a code must equal or exceed its redundancy, $m \leq \ell_q(m, r)$, giving the lower bound on $\ell_q(m, r)$. The length of a code equals its redundancy if and only if the code is the zero code. But the zero code of length m has covering radius m. Any code of length m and covering radius m must have redundancy m by the Redundancy Bound. This completes (i).

We next prove the upper bound on $\ell_q(m, r)$. Let $\mathcal{C}$ be the direct sum of the Hamming code $\mathcal{H}_{q,m-r+1}$ of length $(q^{m-r+1} - 1)/(q - 1)$, which has covering radius 1, and the zero code of length $r - 1$, which has covering radius $r - 1$. By Theorem 11.1.6(i), $\rho(\mathcal{C}) = r$, verifying the upper bound.

This upper bound shows that $\ell_q(m, 1) \leq (q^m - 1)/(q - 1)$. Since the covering radius is 1, a code meeting the bound must have a nonzero scalar multiple of every nonzero syndrome in its parity check matrix. As the length is $(q^m - 1)/(q - 1)$, the parity check matrix cannot have two columns that are scalar multiples of each other, implying (ii). □

Notice that in the proof, the upper bound on $\ell_q(m, r)$ was obtained by constructing a code of redundancy m and covering radius r. This is analogous to the way upper bounds were obtained for $A_q(n, d)$ and $B_q(n, d)$ in Chapter 2.

Now suppose that we know all values of $t_q(n, k)$. From this we can determine all values of $\ell_q(m, r)$ as follows. Choose $r \leq m$. By Theorem 11.5.1(vii), we need to find the smallest n such that $t_q(n, n - m) \leq r$. By Theorem 11.5.2, there is an upper bound n' on $\ell_q(m, r)$; also by Theorem 11.5.1(i), $t_q(n, n - m) \leq t_q(n - 1, (n - 1) - m)$ implying that as n decreases, $t_q(n, n - m)$ increases. Therefore starting with $n = n'$ and decreasing n in steps of 1, find the first value of n where $t_q(n, n - m) > r$; then $\ell_q(m, r) = n + 1$.

Exercise 648 Do the following (where $q = 2$):
(a) Using Theorem 11.5.2, give upper bounds on $\ell(5, r)$ when $1 \leq r \leq 5$.
(b) What are $\ell(5, 1)$ and $\ell(5, 5)$?
(c) The following values of $t(n, n - 5)$ are known: $t(15, 10) = t(14, 9) = \cdots = t(9, 4) = 2$, $t(8, 3) = t(7, 2) = t(6, 1) = 3$, and $t(5, 0) = 5$. Use these values to compute $\ell(5, r)$ for $2 \leq r \leq 4$. You can check your answers in Table 11.6.
(d) For $2 \leq r \leq 4$, find parity check matrices for the codes that have length $\ell(5, r)$, redundancy 5, and covering radius r. ♦

Conversely, fix m and suppose we know $\ell_q(m, r)$ for $1 \leq r \leq m$, which by Theorem 11.5.1(ii) is a decreasing sequence as r increases. Then the smallest integer r such that $\ell_q(m, r) \leq m + j$ is the value of $t_q(m + j, j)$ for $1 \leq j$ by Theorem 11.5.1(viii).

Exercise 649 The following values of $\ell(7, r)$ are known:

r	1	2	3	4	5	6	7
$\ell(7, r)$	127	19	11	8	8	8	7

Using this information, compute $t(7 + j, j)$ for $1 \leq j \leq 20$. ♦

We now turn our focus to the length function for binary codes. We accumulate some results about $\ell(m, r) = \ell_2(m, r)$, most of which we do not prove.

Table 11.6 *Bounds on* $\ell(m, r)$

m/r	2	3	4	5	6	7	8	9	10
2	2								
3	4	3							
4	5	5	4						
5	9	6	6	5					
6	13	7	7	7	6				
7	19	11	8	8	8	7			
8	25–26	14	9	9	9	9	8		
9	34–39	17–18	13	10	10	10	10	9	
10	47–53	21–22	16	11	11	11	11	11	10
11	65–79	23	17–20	15	12	12	12	12	12
12	92–107	31–37	19–23	18	13	13	13	13	13
13	129–159	38–53	23–25	19	17	14	14	14	14
14	182–215	47–63	27–29	21–24	20	15	15	15	15
15	257–319	60–75	32–37	23–27	21	19	16	16	16
16	363–431	75–95	37–49	27–31	22–25	22	17	17	17
17	513–639	93–126	44–62	30–35	24–29	23	21	18	18
18	725–863	117–153	53–77	34–41	27–33	24–27	24	19	19
19	1025–1279	148–205	62–84	39–47	30–36	25–30	25	23	20
20	1449–1727	187–255	73–93	44–59	33–40	28–31	26–29	26	21
21	2049–2559	235–308	86–125	51–75	37–44	30–38	27–32	27	25
22	2897–3455	295–383	103–150	57–88	41–45	33–41	29–33	28–31	28
23	4097–5119	371–511	122–174	65–98	46–59	37–45	31–37	29–34	29
24	5794–6911	467–618	144–190	76–107	51–73	40–47	34–41	30–35	30–33

Theorem 11.5.3 *The following hold*:
(i) $\ell(m, r) = m + 1$ *if* $\lceil m/2 \rceil \le r < m$.
(ii) $\ell(2s + 1, s) = 2s + 5$ *if* $s \ge 1$.
(iii) $\ell(2s, s - 1) = 2s + 6$ *if* $s \ge 4$.
(iv) $\ell(2s + 1, s - 1) = 2s + 7$ *if* $s \ge 6$.
(v) $\ell(2s - 1, 2) \ge 2^s + 1$ *if* $s \ge 3$.

Proof: We only prove (i). Parts (ii), (iii), and (iv) are proved in [34]; they were first proved in [35, 41]. Part (v) is found in [322].

The binary repetition code of length $m + 1$ has covering radius $\lceil m/2 \rceil$ by Exercise 617. As it is of redundancy m, $\ell(m, \lceil m/2 \rceil) \le m + 1$. If $\lceil m/2 \rceil \le r < m$, then $\ell(m, r) \le \ell(m, \lceil m/2 \rceil)$ by Theorem 11.5.1(ii). By Theorem 11.5.2, $m + 1 \le \ell(m, r)$. Combining these inequalities yields (i). □

Table 11.6, found in [207], gives upper and lower bounds on $\ell(m, r)$ for $2 \le r \le \min\{m, 10\}$ with $m \le 24$. By Theorem 11.5.2, $\ell(m, 1) = 2^m - 1$ and these values are excluded from the table. References for the bounds can be found in [207]. The upper bounds are obtained by constructing codes with given covering radius and redundancy; the lower bounds are obtained by various results including those of this section.

Exercise 650   Using the theorems of this section, verify the entries in Table 11.6 for $m \leq 8$ excluding $\ell(6, 2)$, $\ell(7, 2)$, and $\ell(8, 2)$.   ◆

Exercise 651   Verify the lower bounds in Table 11.6 for $\ell(m, 2)$ when $m \geq 11$ and m is odd.   ◆

Exercise 652   Do the following:
(a) Use the Sphere Covering Bound to show that

$$\min_n \left\{ \sum_{i=0}^{r} \binom{n}{i} \geq 2^m \right\} \leq \ell(m, r).$$

(b) Use part (a) to show that $23 \leq \ell(11, 3)$.
(c) The existence of what code guarantees that $\ell(11, 3) = 23$?   ◆

We now give one construction, called the ADS construction, of binary codes that assists in providing many of the upper bounds in Table 11.6; for other constructions, see [34].

Before we give the construction, we must discuss the concept of normal codes and acceptable coordinates. Recall that if $S \subset \mathbb{F}_q^n$ and $\mathbf{x} \in \mathbb{F}_q^n$, then $d(\mathbf{x}, S)$ is the minimum distance from $\mathbf{x}$ to any vector in S; in the case that S is empty, set $d(\mathbf{x}, \emptyset) = n$. Let $\mathcal{C}$ be a binary linear code of length n. For $a \in \mathbb{F}_2$ and $1 \leq i \leq n$, let $\mathcal{C}^{a,(i)}$ denote the subset of $\mathcal{C}$ consisting of the codewords in $\mathcal{C}$ whose ith coordinate equals a. Note that $\mathcal{C}^{0,(i)}$ either equals $\mathcal{C}$, in which case $\mathcal{C}^{1,(i)} = \emptyset$, or $\mathcal{C}^{0,(i)}$ has codimension 1 in $\mathcal{C}$, in which case $\mathcal{C}^{1,(i)}$ is the coset $\mathcal{C} \setminus \mathcal{C}^{0,(i)}$. For $1 \leq i \leq n$, the *norm of $\mathcal{C}$ with respect to coordinate i* is defined by

$$N^{(i)}(\mathcal{C}) = \max_{\mathbf{x} \in \mathbb{F}_2^n} \left\{ d\big(\mathbf{x}, \mathcal{C}^{0,(i)}\big) + d\big(\mathbf{x}, \mathcal{C}^{1,(i)}\big) \right\}.$$

The *norm* of $\mathcal{C}$ is defined to be

$$N(\mathcal{C}) = \min_{1 \leq i \leq n} \left\{ N^{(i)}(\mathcal{C}) \right\}.$$

Coordinate i is called *acceptable* if $N^{(i)}(\mathcal{C}) = N(\mathcal{C})$. One relationship between the norm and covering radius of a code is given by the following lemma.

Lemma 11.5.4   *Let $\mathcal{C}$ be a binary linear code. Then*

$$\rho(\mathcal{C}) \leq \left\lfloor \frac{N(\mathcal{C})}{2} \right\rfloor.$$

Proof: Suppose that $\mathcal{C}$ has length n. For $\mathbf{x} \in \mathbb{F}_q^n$ and $1 \leq i \leq n$,

$$d(\mathbf{x}, \mathcal{C}) = \min \left\{ d\big(\mathbf{x}, \mathcal{C}^{0,(i)}\big), d\big(\mathbf{x}, \mathcal{C}^{1,(i)}\big) \right\} \leq \frac{1}{2} \big[d\big(\mathbf{x}, \mathcal{C}^{0,(i)}\big) + d\big(\mathbf{x}, \mathcal{C}^{1,(i)}\big) \big].$$

Therefore

$$\rho(\mathcal{C}) = \max_{\mathbf{x} \in \mathbb{F}_2^n} d(\mathbf{x}, \mathcal{C}) \leq \max_{\mathbf{x} \in \mathbb{F}_2^n} \left\{ \frac{1}{2} \big[d\big(\mathbf{x}, \mathcal{C}^{0,(i)}\big) + d\big(\mathbf{x}, \mathcal{C}^{1,(i)}\big) \big] \right\} = \frac{N^{(i)}(\mathcal{C})}{2}.$$

For some i, the latter equals $(N(\mathcal{C}))/2$. The result follows.   □

So, $2\rho(\mathcal{C}) \le N(\mathcal{C})$. We attach a specific name to codes with norms as close to this lower bound as possible. The code $\mathcal{C}$ is *normal* provided $N(\mathcal{C}) \le 2\rho(\mathcal{C}) + 1$. Thus for normal codes, $N(\mathcal{C}) = 2\rho(\mathcal{C})$ or $N(\mathcal{C}) = 2\rho(\mathcal{C}) + 1$. Interestingly to this point no one has found a binary linear code that is not normal. (There are binary nonlinear codes that are not normal [136].)

We are now ready to present the ADS construction of a binary code. As motivation, recall that the direct sum of two binary codes yields a code with covering radius the sum of the covering radii of the two component codes by Theorem 11.1.6(i). In certain circumstances, from these component codes, we can construct a code of length one less than that of the direct sum but with covering radius the sum of the covering radii of the two codes. Thus we "save" a coordinate. Let $\mathcal{C}_1$ and $\mathcal{C}_2$ be binary $[n_1, k_1]$ and $[n_2, k_2]$ codes, respectively. The *amalgamated direct sum*, or *ADS*, of $\mathcal{C}_1$ and $\mathcal{C}_2$ (with respect to the last coordinate of $\mathcal{C}_1$ and the first coordinate of $\mathcal{C}_2$) is the code $\mathcal{C}$ of length $n_1 + n_2 - 1$, where

$$\mathcal{C} = \left\{ (\mathbf{a}, 0, \mathbf{b}) \mid (\mathbf{a}, 0) \in \mathcal{C}_1^{0,(n_1)}, (0, \mathbf{b}) \in \mathcal{C}_2^{0,(1)} \right\}$$
$$\cup \left\{ (\mathbf{a}, 1, \mathbf{b}) \mid (\mathbf{a}, 1) \in \mathcal{C}_1^{1,(n_1)}, (1, \mathbf{b}) \in \mathcal{C}_2^{1,(1)} \right\}.$$

As long as either $\mathcal{C}_1$ is not identically zero in coordinate n_1 or $\mathcal{C}_2$ is not identically zero in coordinate 1, $\mathcal{C}$ is an $[n_1 + n_2 - 1, k_1 + k_2 - 1]$ code.

Exercise 653 Do the following:
(a) Let $\mathcal{C}_1$ and $\mathcal{C}_2$ be binary $[n_1, k_1]$ and $[n_2, k_2]$ codes, with parity check matrices $H_1 = [H_1' \ \mathbf{h}']$ and $H_2 = [\mathbf{h}'' \ H_2'']$, respectively, where $\mathbf{h}'$ and $\mathbf{h}''$ are column vectors. Let $\mathcal{C}$ be the ADS of $\mathcal{C}_1$ and $\mathcal{C}_2$, where we assume that $\mathcal{C}_1$ is not identically zero in coordinate n_1 or $\mathcal{C}_2$ is not identically zero in coordinate 1. Show that the parity check matrix of $\mathcal{C}$ is

$$\left[\begin{array}{c|c|c} H_1' & \mathbf{h}' & O \\ \hline O & \mathbf{h}'' & H_2'' \end{array} \right].$$

(b) What happens in the ADS of $\mathcal{C}_1$ and $\mathcal{C}_2$ when $\mathcal{C}_1$ is identically zero in coordinate n_1 and $\mathcal{C}_2$ is identically zero in coordinate 1?   ♦

We now show that the ADS construction yields a code with covering radius at most the sum of the covering radii of component codes provided these codes are normal. Notice that the redundancy of the ADS of $\mathcal{C}_1$ and $\mathcal{C}_2$ is the same as the redundancy of $\mathcal{C}_1 \oplus \mathcal{C}_2$. If they have the same covering radius, we have created a shorter code of the same redundancy and covering radius, which is precisely what the length function considers. The result is found in [52, 111].

Theorem 11.5.5 *Assume that $\mathcal{C}_1$ and $\mathcal{C}_2$ are normal codes of lengths n_1 and n_2, respectively, with the last coordinate of $\mathcal{C}_1$ and the first coordinate of $\mathcal{C}_2$ both acceptable. In addition assume that each of the sets $\mathcal{C}_1^{0,(n_1)}$, $\mathcal{C}_1^{1,(n_1)}$, $\mathcal{C}_2^{0,(1)}$, and $\mathcal{C}_2^{1,(1)}$ is nonempty. Let $\mathcal{C}$ be the ADS of $\mathcal{C}_1$ and $\mathcal{C}_2$ with respect to the last coordinate of $\mathcal{C}_1$ and the first coordinate of $\mathcal{C}_2$. Then $\rho(\mathcal{C}) \le \rho(\mathcal{C}_1) + \rho(\mathcal{C}_2)$.*

Proof: Let $z = (\mathbf{x}, 0, \mathbf{y}) \in \mathbb{F}_2^{n_1+n_2-1}$. Clearly,

$$d\big(\mathbf{z}, C^{0,(n_1)}\big) = d\big((\mathbf{x}, 0), C_1^{0,(n_1)}\big) + d\big((0, \mathbf{y}), C_2^{0,(1)}\big), \quad \text{and}$$
$$d\big(\mathbf{z}, C^{1,(n_1)}\big) = d\big((\mathbf{x}, 0), C_1^{1,(n_1)}\big) + d\big((0, \mathbf{y}), C_2^{1,(1)}\big) - 1.$$

Therefore adding these two equations and using the definitions

$$d\big(\mathbf{z}, C^{0,(n_1)}\big) + d\big(\mathbf{z}, C^{1,(n_1)}\big) \le N^{(n_1)}(C_1) + N^{(1)}(C_2) - 1. \tag{11.13}$$

This also holds for $\mathbf{z} = (\mathbf{x}, 1, \mathbf{y}) \in \mathbb{F}_2^{n_1+n_2-1}$. As (11.13) is true for all $\mathbf{z} \in \mathbb{F}_2^{n_1+n_2-1}$, $N^{(n_1)}(C) \le N^{(n_1)}(C_1) + N^{(1)}(C_2) - 1$.

Using Lemma 11.5.4 and the normality assumptions for C_1 and C_2, we see that

$$2\rho(C) \le N(C) \le N(C_1) + N(C_2) - 1 \le 2\rho(C_1) + 2\rho(C_2) + 1,$$

proving the theorem. □

Exercise 654   Let C be a binary code of length n. Prove the following:
(a) If $\text{PAut}(C)$ is transitive, then $N^{(i)}(C) = N^{(j)}(C)$ for $1 \le i < j \le n$. Furthermore all coordinates are acceptable.
(b) If $\mathbf{c} \in C$ and $\mathbf{x} \in \mathbb{F}_2^n$, then

$$d\big(\mathbf{x} + \mathbf{c}, C^{0,(i)}\big) + d\big(\mathbf{x} + \mathbf{c}, C^{1,(i)}\big) = d\big(\mathbf{x}, C^{0,(i)}\big) + d\big(\mathbf{x}, C^{1,(i)}\big).$$

(c) The $[7, 4, 3]$ binary Hamming code $\mathcal{H}_3$ is normal and every coordinate is acceptable. Hint: By part (a) you only have to compute $N^{(1)}(\mathcal{H}_3)$ as $\mathcal{H}_3$ is cyclic. By part (b) you only have to check coset representatives $\mathbf{x}$ of $\mathcal{H}_3$ in $\mathbb{F}_2^7$ when calculating $d(\mathbf{x}, \mathcal{H}_3^{0,(1)}) + d(\mathbf{x}, \mathcal{H}_3^{1,(1)})$.
(d) The $[3, 1, 3]$ binary repetition code is normal and every coordinate is acceptable. ◆

Example 11.5.6   Let $C_1 = C_2 = \mathcal{H}_3$ and let C be the ADS of C_1 and C_2. C is a $[13, 7]$ code. As $\mathcal{H}_3$ is perfect, $\rho(\mathcal{H}_3) = 1$. By Exercise 654, $\mathcal{H}_3$ is normal. Therefore by Theorem 11.5.5, $\rho(C) \le 2$. If $\rho(C) = 1$, then every syndrome in $\mathbb{F}_2^6$ must be a combination of one column of a parity check matrix of C, which is clearly impossible. Hence $\rho(C) = 2$, implying that $\ell(6, 2) \le 13$. Applying Exercise 652(a) shows that $11 \le \ell(6, 2)$. If there is an $[11, 5]$ binary code with covering radius 2, there is a $[12, 6]$ binary code also with covering radius 2 obtained by taking the direct sum of the $[11, 5]$ code with the repetition code of length 1. It can be shown that there is no $[12, 6]$ binary code with covering radius 2; see [34, Theorem 8.8]. This verifies that $\ell(6, 2) = 13$. ■

We conclude this section with a result, found in [41], that can be used to compute certain values of $\ell(m, 2)$; this result is important in showing that there is no $[12, 6]$ binary code with covering radius 2 as discussed in the preceding example.

Theorem 11.5.7   *If C is an $[n, n - m]$ binary code with covering radius 2 and w is the weight of a nonzero vector in $C^\perp$, then $w(n + 1 - w) \ge 2^{m-1}$.*

Proof: Let $\mathbf{x} \in C^{\perp}$ with $\text{wt}(\mathbf{x}) = w$. By permuting coordinates we may assume that C has the following parity check matrix, where $\mathbf{x}$ is the first row:

$$H = \begin{bmatrix} 1 & \cdots & 1 & 0 & \cdots & 0 \\ & H_1 & & & H_2 & \end{bmatrix}.$$

Consider a syndrome $\mathbf{s}^{\mathrm{T}}$ with 1 as the top entry. There are 2^{m-1} such syndromes. Either $\mathbf{s}^{\mathrm{T}}$ is one of the first w columns of H, or it is a sum of one of the first w columns of H and one of the last $n - w$ columns of H since $\rho(C) = 2$. Therefore $w + w(n - w) = w(n + 1 - w) \geq 2^{m-1}$. $\square$

Exercise 655 In this exercise you will verify that $\ell(5, 2) = 9$ as indicated in Table 11.6 in a different manner from Exercise 648.

(a) Show that $\ell(5, 2) \leq 9$ by showing that the ADS of the $[7, 4, 3]$ Hamming code and the $[3, 1, 3]$ repetition code is a $[9, 4]$ code with covering radius 2. Hint: See Exercise 654 and Example 11.5.6.

(b) Show that $8 \leq \ell(5, 2)$ using Exercise 652(a).

(c) Let C be an $[8, 3]$ binary code with covering radius 2.
  (i) Show that $C^{\perp}$ is an $[8, 5, d]$ code with $d \geq 3$. Hint: Use Theorem 11.5.7.
  (ii) Show that there does not exist an $[8, 5, d]$ binary code with $d \geq 3$. Note that this shows that $\ell(5, 2) = 9$.   ◆

11.6   Covering radius of subcodes

The relationship between the covering radius of a code and that of a subcode can be quite complex. Recall by Exercise 618 that if $C_0 \subseteq C$, then $\rho(C) \leq \rho(C_0)$. However, it is not clear how large the covering radius of $\rho(C_0)$ can be compared with $\rho(C)$. The next two examples give extremes for these possibilities.

Example 11.6.1 In this example we show how the covering radius of a subcode C_0 could be as small as the covering radius of the original code. Let H_1 be a parity check matrix of a binary code with covering radius ρ, and let C be the code with parity check matrix given by

$$H = [\, H_1 \quad H_1 \,].$$

Clearly, $\rho(C) = \rho$ as the list of columns of H is the same as the list of columns of H_1. Let C_0 be the subcode of codimension 1 in C with parity check matrix

$$\left[\begin{array}{ccc|ccc} & H_1 & & & H_1 & \\ 0 & \cdots & 0 & 1 & \cdots & 1 \end{array} \right].$$

We also have $\rho(C_0) = \rho$, as Exercise 656 shows.   ■

Exercise 656 Show that the codes with parity check matrices

$$[\, H_1 \quad H_1 \,] \quad \text{and} \quad \left[\begin{array}{ccc|ccc} & H_1 & & & H_1 & \\ 0 & \cdots & 0 & 1 & \cdots & 1 \end{array} \right]$$

have the same covering radius.   ◆

Example 11.6.2 In this example we find a subcode C_0 of codimension 1 of a code C where $\rho(C_0) = 2\rho(C) + 1$. In the next theorem we will see that this is as large as possible.

Let $m \geq 2$ be an integer and let H_m be the $m \times (2^m - 1)$ parity check matrix of the $[2^m - 1, 2^m - 1 - m, 3]$ binary Hamming code $\mathcal{H}_m$. Let H'_m be the $m \times (2^m - 2)$ matrix obtained from H_m by deleting its column of all 1s. Let r be a positive integer. Then the $rm \times (r(2^m - 2) + 1)$ matrix

$$
H = \begin{bmatrix}
1 & H'_m & O & \cdots & O \\
1 & O & H'_m & \cdots & O \\
\vdots & \vdots & \vdots & \ddots & \vdots \\
1 & O & O & \cdots & H'_m
\end{bmatrix}
$$

is a parity check matrix of a code C (which in fact is an ADS of r Hamming codes) of length $r(2^m - 2) + 1$ with $\rho(C) = r$ by Exercise 657. The matrix

$$
H_0 = \left[\begin{array}{c} H \\ \hline 1 \quad 0 \quad \cdots \quad 0 \end{array} \right]
$$

is a parity check matrix of a subcode C_0 of codimension 1 in C with covering radius $2r + 1$ also by Exercise 657. ∎

Exercise 657 With the notation of Example 11.6.2 do the following:
(a) Prove that $\rho(C) = r$.
(b) Prove that the syndrome of weight 1 with a 1 in its last position is not the sum of $2r$ or fewer columns of H_0.
(c) Prove that $\rho(C_0) = 2r + 1$. ♦

In these last two examples we have produced codes C and subcodes C_0 of codimension 1 where $\rho(C_0)$ is as small as possible (namely $\rho(C)$) and also as large as possible (namely $2\rho(C) + 1$) as the following theorem found in [1] shows.

Theorem 11.6.3 (Adams Bound) *Let C_0 be a subcode of codimension 1 in a binary linear code C. Then*

$$
\rho(C_0) \leq 2\rho(C) + 1. \tag{11.14}
$$

Proof: Suppose that C is an $[n, n - m]$ code. Let H be an $m \times n$ parity check matrix for C. We can add one row $\mathbf{v} = v_1 \cdots v_n$ to H to form an $(m + 1) \times n$ parity check matrix

$$
H_0 = \left[\frac{H}{\mathbf{v}} \right] \tag{11.15}
$$

for C_0. For $j = 1, 2, \ldots, n$, let the columns of H_0 and H be denoted h_0^j and h^j, respectively.

Let $\rho = \rho(C)$ and $\rho_0 = \rho(C_0)$. Assume (11.14) is false; so $\rho_0 \geq 2\rho + 2$. There is a syndrome $\mathbf{s}_0 \in \mathbb{F}_2^{m+1}$ of C_0 which is the sum of ρ_0 columns of H_0 but no fewer. Rearrange the columns of H_0 so that

$$
\mathbf{s}_0^{\mathrm{T}} = \sum_{j=1}^{\rho_0} h_0^j. \tag{11.16}
$$

Partition $\{1, 2, \ldots, \rho_0\}$ into two disjoint sets J_1 and J_2 of cardinality at least $\rho + 1$; this can be done since $\rho_0 \geq 2\rho + 2$. Hence (11.16) becomes

$$\mathbf{s}_0^{\mathrm{T}} = \sum_{j \in J_1} h_0^j + \sum_{j \in J_2} h_0^j. \tag{11.17}$$

As $\sum_{j \in J_1} h^j$ and $\sum_{j \in J_2} h^j$ are syndromes of $\mathcal{C}$ and $\rho(\mathcal{C}) = \rho$, there exist sets I_1 and I_2 of cardinality at most ρ such that

$$\sum_{j \in I_1} h^j = \sum_{j \in J_1} h^j \quad \text{and} \quad \sum_{j \in I_2} h^j = \sum_{j \in J_2} h^j. \tag{11.18}$$

Suppose that $\sum_{j \in I_1} v_j = \sum_{j \in J_1} v_j$; then $\sum_{j \in J_1} h_0^j = \sum_{j \in I_1} h_0^j$, and we can replace $\sum_{j \in J_1} h_0^j$ in (11.17) by $\sum_{j \in I_1} h_0^j$. Therefore we have expressed $\mathbf{s}_0^{\mathrm{T}}$ as a linear combination of fewer than ρ_0 columns of H_0, which is a contradiction. A similar contradiction arises if $\sum_{j \in I_2} v_j = \sum_{j \in J_2} v_j$. The remaining possibility is $\sum_{j \in I_1} v_j \neq \sum_{j \in J_1} v_j$ and $\sum_{j \in I_2} v_j \neq \sum_{j \in J_2} v_j$. However, then, $\sum_{j \in I_1} v_j + \sum_{j \in I_2} v_j = \sum_{j \in J_1} v_j + \sum_{j \in J_2} v_j$; hence by (11.18) $\sum_{j \in I_1} h_0^j + \sum_{j \in I_2} h_0^j = \sum_{j \in J_1} h_0^j + \sum_{j \in J_2} h_0^j$, and again we have expressed $\mathbf{s}_0^{\mathrm{T}}$ as a linear combination of fewer than ρ_0 columns of H_0, a contradiction. $\quad\square$

Exercise 658   Let $\mathcal{C}$ be the $[7, 4, 3]$ binary Hamming code $\mathcal{H}_3$. Let $\mathcal{C}_0$ be the $[7, 3, 4]$ subcode of even weight codewords of $\mathcal{C}$.
(a) Give a parity check matrix for $\mathcal{C}_0$ as in (11.15).
(b) Using the parity check matrix and Theorem 11.6.3, find the covering radius of $\mathcal{C}_0$.   ◆

This result is generalized in [37, 311]; the proof is similar but more technical.

Theorem 11.6.4 *Let $\mathcal{C}_0$ be a subcode of codimension i in a binary linear code $\mathcal{C}$. Then*

$$\rho(\mathcal{C}_0) \leq (i + 1)\rho(\mathcal{C}) + 1.$$

Furthermore, if $\mathcal{C}$ is even, then

$$\rho(\mathcal{C}_0) \leq (i + 1)\rho(\mathcal{C}).$$

Another upper bound on the covering radius of a subcode in terms of the covering radius of the original code, due to Calderbank [43], is sometimes an improvement of Theorem 11.6.4. Before stating this result, we need some additional terminology. A *graph* G is a pair consisting of a set V, called *vertices*, and a set E of unordered pairs of vertices, called *edges*. A *coloring* of G with a set C of *colors* is a map $f : V \to C$ such that $f(x) \neq f(y)$ whenever $\{x, y\}$ is an edge; that is, two vertices on the same edge do not have the same color, where the *color* of a vertex x is $f(x)$. The *chromatic number* of G is the smallest integer $\chi(G)$ for which G has a coloring using a set C containing $\chi(G)$ colors. We are interested in a specific graph called the *Kneser graph* $K(n, r + 1)$. The vertex set of $K(n, r + 1)$ is $\{\mathbf{v} \in \mathbb{F}_2^n \mid \mathrm{wt}(\mathbf{v}) = r + 1\}$, with vertices $\mathbf{u}$ and $\mathbf{v}$ forming an edge if and only if their supports are disjoint. The chromatic number of $K(n, r + 1)$ has been determined by Lovász [208]:

$$\chi(K(n, r + 1)) = n - 2r \quad \text{if } n \geq 2r + 2. \tag{11.19}$$

The Calderbank Bound is as follows.

Theorem 11.6.5 (Calderbank Bound) *Let C_0 be a subcode of codimension $i \geq 1$ in a binary linear code C of length n. Then*

$$\rho(C_0) \leq 2\rho(C) + 2^i - 1. \tag{11.20}$$

Proof: Let H be a parity check matrix for C. We can add i rows to H to form a parity check matrix

$$H_0 = \begin{bmatrix} H \\ H_1 \end{bmatrix}$$

for C_0. Let the jth column of H_0 be h_0^j for $1 \leq j \leq n$.

Let $\rho = \rho(C)$ and $\rho_0 = \rho(C_0)$. If $\rho_0 \leq 2\rho + 1$, then (11.20) clearly holds. So assume that $\rho_0 \geq 2\rho + 2$. There must be some syndrome s_0^T of C_0 which is a sum of ρ_0 columns of H_0 but no fewer. By rearranging the columns of H_0, we may assume that

$$s_0^T = \sum_{j=1}^{\rho_0} h_0^j. \tag{11.21}$$

Form the Kneser graph, $K(\rho_0, \rho + 1)$. This graph has as vertex set V the set of vectors in $\mathbb{F}_2^{\rho_0}$ of weight $\rho + 1$. We write each vector $\mathbf{v} \in \mathbb{F}_2^n$ as $(\mathbf{v}', \mathbf{v}'')$, where $\mathbf{v}' \in \mathbb{F}_2^{\rho_0}$. We wish to establish a coloring f of the vertices of $K(\rho_0, \rho + 1)$. For every $\mathbf{v}' \in V$, there is an $\mathbf{x}_{\mathbf{v}'} \in C$ of distance at most ρ from $(\mathbf{v}', \mathbf{0})$ as $\rho(C) = \rho$. For each $\mathbf{v}' \in V$ choose one such $\mathbf{x}_{\mathbf{v}'} \in C$ and define

$$f(\mathbf{v}') = H_1 \mathbf{x}_{\mathbf{v}'}^T.$$

Notice that $f(\mathbf{v}') \in \mathbb{F}_2^i$.

We first establish that $H_1 \mathbf{x}_{\mathbf{v}'}^T \neq \mathbf{0}$. Suppose that $H_1 \mathbf{x}_{\mathbf{v}'}^T = \mathbf{0}$. Then $H_0 \mathbf{x}_{\mathbf{v}'}^T = \mathbf{0}$ as $H \mathbf{x}_{\mathbf{v}'}^T = \mathbf{0}$ because $\mathbf{x}_{\mathbf{v}'} \in C$. Therefore,

$$H_0(\mathbf{v}', \mathbf{0})^T = H_0((\mathbf{v}', \mathbf{0}) - \mathbf{x}_{\mathbf{v}'})^T. \tag{11.22}$$

Because $\text{wt}(\mathbf{v}') = \rho + 1$, the left-hand side of (11.22) is a combination of $\rho + 1$ columns of H_0 from among the first ρ_0 columns of H_0, while the right-hand side of (11.22) is a combination of at most ρ columns of H_0 as $\text{wt}((\mathbf{v}', \mathbf{0}) - \mathbf{x}_{\mathbf{v}'}) \leq \rho$. Thus on the right-hand side of (11.21) we may replace $\rho + 1$ of the columns h_0^j by at most ρ columns of H_0 indicating that s_0^T is a linear combination of at most $\rho_0 - 1$ columns of H_0, which is a contradiction. This shows that $f(\mathbf{v}')$ is one of the $2^i - 1$ nonzero elements of $\mathbb{F}_2^i$.

We now establish that f is indeed a coloring of $K(\rho_0, \rho + 1)$. Let $\{\mathbf{v}', \mathbf{w}'\}$ be an edge of $K(\rho_0, \rho + 1)$. We must show that $H_1 \mathbf{x}_{\mathbf{v}'}^T \neq H_1 \mathbf{x}_{\mathbf{w}'}^T$. Suppose that $H_1 \mathbf{x}_{\mathbf{v}'}^T = H_1 \mathbf{x}_{\mathbf{w}'}^T$. Then as $\mathbf{v}'$ and $\mathbf{w}'$ are in V, where $\text{supp}(\mathbf{v}')$ and $\text{supp}(\mathbf{w}')$ are disjoint, $\text{wt}(\mathbf{v}' + \mathbf{w}') = 2\rho + 2$. Additionally, the distance between $(\mathbf{v}' + \mathbf{w}', \mathbf{0})$ and $\mathbf{x}_{\mathbf{v}'} + \mathbf{x}_{\mathbf{w}'}$ is at most 2ρ. Since $H_1(\mathbf{x}_{\mathbf{v}'} + \mathbf{x}_{\mathbf{w}'})^T = \mathbf{0}$ and $H(\mathbf{x}_{\mathbf{v}'} + \mathbf{x}_{\mathbf{w}'})^T = \mathbf{0}$ because $\mathbf{x}_{\mathbf{v}'} + \mathbf{x}_{\mathbf{w}'} \in C$, we have $H_0(\mathbf{x}_{\mathbf{v}'} + \mathbf{x}_{\mathbf{w}'})^T = \mathbf{0}$. As in (11.22),

$$H_0(\mathbf{v}' + \mathbf{w}', \mathbf{0})^T = H_0((\mathbf{v}' + \mathbf{w}', \mathbf{0}) - (\mathbf{x}_{\mathbf{v}'} + \mathbf{x}_{\mathbf{w}'}))^T. \tag{11.23}$$

Because $\mathrm{wt}(\mathbf{v}' + \mathbf{w}') = 2\rho + 2$, the left-hand side of (11.23) is a combination of $2\rho + 2$ columns of H_0 from among the first ρ_0 columns of H_0, while the right-hand side of (11.23) is a combination of at most 2ρ columns of H_0 as $\mathrm{wt}((\mathbf{v}' + \mathbf{w}', \mathbf{0}) - (\mathbf{x}_{\mathbf{v}'} + \mathbf{x}_{\mathbf{w}'})) \leq 2\rho$. As above, in the right-hand side of (11.21) we may replace $2\rho + 2$ of the columns h_0^j by at most 2ρ columns of H_0 indicating that $\mathbf{s}_0^{\mathsf{T}}$ is a linear combination of at most $\rho_0 - 2$ columns of H_0, which is a contradiction.

Therefore f is a coloring of $K(\rho_0, \rho + 1)$ with at most $2^i - 1$ colors. Since $\rho_0 \geq 2\rho + 2$, (11.19) implies that

$$\rho_0 - 2\rho = \chi(K(\rho_0, \rho + 1)) \leq 2^i - 1. \qquad \square$$

Notice that the Calderbank Bound with $i = 1$ is the bound of Theorem 11.6.3.

Exercise 659 Let $\mathcal{C}$ be a binary code with $\rho(\mathcal{C}) = 2$ and $\mathcal{C}_i$ a subcode of codimension i. For which values of i is the upper bound on $\rho(\mathcal{C}_i)$ better in Theorem 11.6.4 than in the Calderbank Bound?   ◆

Exercise 660 Let $\mathcal{C}$ be a binary code with $\rho(\mathcal{C}) = 10$ and $\mathcal{C}_i$ a subcode of codimension i. For which values of i is the upper bound on $\rho(\mathcal{C}_i)$ better in Theorem 11.6.4 than in the Calderbank Bound?   ◆

There is another upper bound on $\rho(\mathcal{C}_0)$ due to Hou [139] that in some cases improves the bound of Theorem 11.6.4 and the Calderbank Bound. Its proof relies on knowledge of the graph $G(n, s)$. The graph $G(n, s)$ has vertex set $\mathbb{F}_2^n$; two vertices $\mathbf{v}$ and $\mathbf{w}$ form an edge whenever $d(\mathbf{v}, \mathbf{w}) > s$.

The following theorem of Kleitman [172] gives a lower bound on the chromatic number of $G(n, 2r)$.

Theorem 11.6.6 *Let n and r be positive integers with $n \geq 2r + 1$. Then*

$$\chi(G(n, 2r)) \geq \frac{2^n}{\dbinom{n}{0} + \dbinom{n}{1} + \cdots + \dbinom{n}{r}}.$$

We now state the bound on $\rho(\mathcal{C}_0)$ due to Hou; we leave the details of the proof as an exercise.

Theorem 11.6.7 (Hou Bound) *Let $\mathcal{C}_0$ be a subcode of codimension $i \geq 1$ of a binary linear code $\mathcal{C}$. Let $\rho = \rho(\mathcal{C})$ and $\rho_0 = \rho(\mathcal{C}_0)$. Then*

$$\frac{2^{\rho_0}}{\dbinom{\rho_0}{0} + \dbinom{\rho_0}{1} + \cdots + \dbinom{\rho_0}{\rho}} \leq 2^i. \qquad (11.24)$$

Proof: If $\rho_0 \leq 2\rho$, then (11.24) is true by Exercise 661. Assume that $\rho_0 \geq 2\rho + 1$. Let H_0 be the parity check matrix of $\mathcal{C}_0$ as in the proof of Theorem 11.6.5. Analogously, define the same map f from the vertices of $G(\rho_0, 2\rho)$ (instead of $K(\rho_0, \rho + 1)$) to $\mathbb{F}_2^i$. As before it follows that f is a coloring of $G(\rho_0, 2\rho)$ with 2^i colors, and the result is a consequence of Theorem 11.6.6. $\square$

Exercise 661 Show that if $\rho_0 \leq 2\rho$, then (11.24) is true. ♦

Exercise 662 Fill in the details of the proof of Theorem 11.6.7. ♦

Exercise 663 Let C be a binary code with $\rho(C) = 1$ and C_i a subcode of codimension i. For each $i \geq 1$, which upper bound on $\rho(C_i)$ is better: the bound in Theorem 11.6.4, the Calderbank Bound, or the Hou Bound? ♦

11.7 Ancestors, descendants, and orphans

In this concluding section of the chapter we examine relationships among coset leaders of a code C. There is a natural relationship among cosets, and we will discover that all cosets of C of maximum weight (that is, those of weight $\rho(C)$) have a special place in this relationship. We will also introduce the concept of the Newton radius of a code and investigate it in light of this relationship among cosets.

There is a natural partial ordering $\leq$ on the vectors in $\mathbb{F}_2^n$ as follows. For $\mathbf{x}$ and $\mathbf{y}$ in $\mathbb{F}_2^n$, define $\mathbf{x} \leq \mathbf{y}$ provided that $\text{supp}(\mathbf{x}) \subseteq \text{supp}(\mathbf{y})$. If $\mathbf{x} \leq \mathbf{y}$, we will also say that $\mathbf{y}$ *covers* $\mathbf{x}$. We now use this partial order on $\mathbb{F}_2^n$ to define a partial order, also denoted $\leq$, on the set of cosets of a binary linear code C of length n. If C_1 and C_2 are two cosets of C, then $C_1 \leq C_2$ provided there are coset leaders $\mathbf{x}_1$ of C_1 and $\mathbf{x}_2$ of C_2 such that $\mathbf{x}_1 \leq \mathbf{x}_2$. As usual, $C_1 < C_2$ means that $C_1 \leq C_2$ but $C_1 \neq C_2$. Under this partial ordering the set of cosets of C has a unique minimal element, the code C itself.

Example 11.7.1 Let C be the $[5, 2, 3]$ binary code with generator matrix

$$\begin{bmatrix} 1 & 1 & 1 & 1 & 0 \\ 0 & 0 & 1 & 1 & 1 \end{bmatrix}.$$

Then the cosets of C with coset leaders are:

$$\begin{aligned}
C_0 &= 00000 + C, & C_4 &= 00010 + C, \\
C_1 &= 10000 + C, & C_5 &= 00001 + C, \\
C_2 &= 01000 + C, & C_6 &= 10100 + C = 01010 + C, \\
C_3 &= 00100 + C, & C_7 &= 10010 + C = 01100 + C.
\end{aligned}$$

The partial ordering of the cosets is depicted as

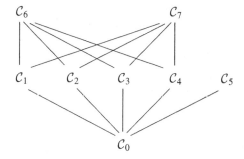

In this diagram notice that the coset weights go up in steps of one as you go up a chain starting with weight 0 at the very bottom.   ■

Exercise 664 List the cosets of the code C with generator matrix given below. Also give a pictorial representation of the partial ordering on these cosets as in Example 11.7.1.

(a) $\begin{bmatrix} 1 & 1 & 1 & 1 & 0 & 0 \\ 0 & 0 & 1 & 1 & 1 & 1 \end{bmatrix}$,

(b) $\begin{bmatrix} 1 & 0 & 0 & 0 & 1 & 1 \\ 0 & 1 & 0 & 1 & 0 & 1 \\ 0 & 0 & 1 & 1 & 1 & 0 \end{bmatrix}$.   ◆

If C_1 and C_2 are cosets of C with $C_1 < C_2$, then C_1 is called a *descendant* of C_2, and C_2 is an *ancestor* of C_1. If C_1 is a descendant of C_2, then clearly $\mathrm{wt}(C_1) \leq \mathrm{wt}(C_2) - 1$; if $C_1 < C_2$ with $\mathrm{wt}(C_1) = \mathrm{wt}(C_2) - 1$, then C_1 is a *child* of C_2, and C_2 is a *parent* of C_1.[1] By Theorem 11.1.6(v), if $C_1 < C_2$ with $\mathrm{wt}(C_1) < \mathrm{wt}(C_2) - 1$, then there is a coset C_3 such that $C_1 < C_3 < C_2$. A coset of C is called an *orphan* provided it has no parents, that is, provided it is a maximal element in the partial ordering of all cosets of C. Each coset of C of weight $\rho(C)$ is clearly an orphan but the converse does not hold in general. The presence of orphans with weight less than $\rho(C)$ contributes towards the difficulty in computing the covering radius of a code.

Example 11.7.2 In Example 11.7.1, $\rho(C) = 2$. The code C is a descendant of all other cosets and the child of C_i for $1 \leq i \leq 5$. C_6 and C_7 are parents of C_i for $1 \leq i \leq 4$. C_5, C_6, and C_7 are orphans; however, the weight of C_5 is not $\rho(C)$.   ■

Exercise 665 Give examples of parents and children in the partially ordered sets for the codes in Exercise 664. Which cosets have the most parents? Which cosets have the most children? Which cosets are orphans? Which orphans have weight equal to $\rho(C)$?   ◆

We now present two results dealing with the basic descendant/ancestor relationship among cosets. The first deals with weights of descendants illustrating a type of "balance" in the partial ordering.

Theorem 11.7.3 *Let C be a binary linear code and let C' be a coset of weight w of C. Let a be an integer with $0 < a < w$. Then the number of descendants of C' of weight a equals the number of descendants of C' of weight $w - a$.*

Proof: By Theorem 11.1.6(v), the coset leaders of the descendants of weight a (respectively, $w - a$) of C' are the vectors of weight a (respectively, of weight $w - a$) covered by some vector of C' of weight w, as the latter are precisely the coset leaders of C'. Let $\mathbf{x}$ be a coset leader of weight a of a descendant of C'. Then there exists a coset leader $\mathbf{x}'$ of C' such that $\mathbf{x} \leq \mathbf{x}'$; as $\mathbf{x}'$ covers $\mathbf{x}$, $\mathbf{x}' - \mathbf{x}$ is a coset leader of weight $w - a$ of a descendant of C'. Furthermore, every coset leader of weight $w - a$ of a descendant of C' arises in this manner, as seen by reversing the roles of a and $w - a$ in the above. Therefore there is a one-to-one

---

[1] We have used the terms "parent" and "child" in relation to self-dual codes in Chapter 9 and earlier in this chapter. The current use of "parent" and "child" is unrelated.

correspondence between the coset leaders of weight a of descendants of C' and those of weight $w - a$.

Suppose $\mathbf{x}$ and $\mathbf{y}$ are coset leaders of weight a of the same descendant of C'. Let $\mathbf{x}'$ and $\mathbf{y}'$ be vectors of weight w in C' that cover $\mathbf{x}$ and $\mathbf{y}$, respectively. Then $\mathbf{y} - \mathbf{x} \in C$, and since $\mathbf{x}' - \mathbf{y}'$ is also in C, so is

$$(\mathbf{x}' - \mathbf{x}) - (\mathbf{y}' - \mathbf{y}) = (\mathbf{x}' - \mathbf{y}') + (\mathbf{y} - \mathbf{x}).$$

Therefore $\mathbf{x}' - \mathbf{x}$ and $\mathbf{y}' - \mathbf{y}$ are coset leaders of the same coset. Thus there are at least as many descendants of C' of weight a as there are of weight $w - a$. Again by interchanging the roles of a and $w - a$, there are at least as many descendants of C' of weight $w - a$ as there are of weight a. $\qquad\square$

Example 11.7.4 We showed in Example 8.3.2 that the covering radius of the $[24, 12, 8]$ extended binary Golay code is 4. Furthermore, every coset of weight 4 contains exactly six coset leaders; in pairs their supports are disjoint. Each of these coset leaders covers exactly four coset leaders of weight 1 cosets. Because the cosets of weight 1 have a unique coset leader, there are 24 cosets of weight 1 that are descendants of a weight 4 coset. By Theorem 11.7.3 there are 24 cosets of weight 3 that are children of a given coset of weight 4. $\qquad\blacksquare$

Exercise 666 Describe the form of the coset leaders of the 24 cosets of weight 3 that are descendants of a fixed weight 4 coset in the $[24, 12, 8]$ extended binary Golay code as noted in Example 11.7.4. $\qquad\blacklozenge$

If C_1 and C_2 are two cosets of C with $C_1 < C_2$, then the following theorem shows how to obtain all of the coset leaders of C_1 from one coset leader of C_1 and all of the coset leaders of C_2 [36].

Theorem 11.7.5 *Let C_1 and C_2 be two cosets of a binary linear code C and assume that $C_1 < C_2$. Let $\mathbf{x}_1$ and $\mathbf{x}_2$ be coset leaders of C_1 and C_2, respectively, satisfying $\mathbf{x}_1 < \mathbf{x}_2$. Let $\mathbf{u} = \mathbf{x}_2 - \mathbf{x}_1$. Then a vector $\mathbf{x}$ is a coset leader of C_1 if and only if there is a coset leader $\mathbf{y}$ of C_2 such that $\mathbf{u} \le \mathbf{y}$ and $\mathbf{x} = \mathbf{y} - \mathbf{u}$.*

Proof: First suppose that $\mathbf{y}$ is a coset leader of C_2 with $\mathbf{u} \le \mathbf{y}$. Then

$$(\mathbf{y} - \mathbf{u}) - \mathbf{x}_1 = \mathbf{y} - (\mathbf{u} + \mathbf{x}_1) = \mathbf{y} - \mathbf{x}_2 \in C,$$

and thus $\mathbf{y} - \mathbf{u}$ is in C_1. Since

$$\mathrm{wt}(\mathbf{y} - \mathbf{u}) = \mathrm{wt}(\mathbf{y}) - \mathrm{wt}(\mathbf{u}) = \mathrm{wt}(\mathbf{x}_2) - \mathrm{wt}(\mathbf{u}) = \mathrm{wt}(\mathbf{x}_1),$$

$\mathbf{y} - \mathbf{u}$ is a coset leader of C_1.

Conversely, suppose that $\mathbf{x}$ is a coset leader of C_1. Then

$$(\mathbf{x} + \mathbf{u}) - \mathbf{x}_2 = (\mathbf{x} + \mathbf{x}_2 - \mathbf{x}_1) - \mathbf{x}_2 = \mathbf{x} - \mathbf{x}_1 \in C,$$

and thus $\mathbf{x} + \mathbf{u}$ is in C_2. As

$$\mathrm{wt}(\mathbf{x} + \mathbf{u}) \le \mathrm{wt}(\mathbf{x}) + \mathrm{wt}(\mathbf{u}) = \mathrm{wt}(\mathbf{x}_1) + \mathrm{wt}(\mathbf{u}) = \mathrm{wt}(\mathbf{x}_2),$$

$\mathbf{x} + \mathbf{u}$ is a coset leader of C_2. $\qquad\square$

Exercise 667 If $u \le y$ and $x = y - u$, show that $x \le x + u$ and $u \le x + u$.   ♦

Example 11.7.6 Let C be the [24, 12, 8] extended binary Golay code. Let $C_2 = x_2 + C$ be a coset of weight 4. Choose any vector x_1 of weight 3 with $x_1 \le x_2$. Let $C_1 = x_1 + C$. Then, in the notation of Theorem 11.7.5, u is a weight 1 vector with $\text{supp}(u) \subset \text{supp}(x_2)$. As all other coset leaders of C_2 have supports disjoint from $\text{supp}(x_2)$, x_1 is the only coset leader of C_1 by Theorem 11.7.5, a fact we already knew as all cosets of weight $t = \lfloor (d - 1)/2 \rfloor = 3$ or less have unique coset leaders.   ∎

The following corollaries are immediate consequences of Theorem 11.7.5.

Corollary 11.7.7 *Let C be a binary code and C' a coset of C with coset leader x. Let $i \in \text{supp}(x)$ and let u be the vector with 1 in coordinate i and 0 elsewhere. Then $x - u$ is a coset leader of a child of C', and no coset leader of $(x - u) + C$ has i in its support.*

Proof: By Theorem 11.1.6(v), $x - u$ is a coset leader of a child of C'. By Theorem 11.7.5, every coset leader of this child is of the form $y - u$, where $u \le y$; in particular, no coset leader of the child has i in its support.   □

Corollary 11.7.8 [(129)] *Let C_1 and C_2 be two cosets of a binary linear code C and assume that $C_1 < C_2$. Then for each coset leader x of C_1 there is a coset leader y of C_2 such that $x < y$.*

Exercise 668 Prove Corollary 11.7.8.   ♦

Example 11.7.9 It follows from Corollary 11.7.8 that to determine all the ancestors of a coset C' it suffices to know only one leader of C'. From that one coset leader x every coset leader of an ancestor has support containing $\text{supp}(x)$. However, to determine all the descendants of C', it does not suffice in general to know only one leader x of C' in that the coset leader of a descendent of C' may not have support contained in $\text{supp}(x)$. In Example 11.7.1, a coset leader of C_1 has support contained in the support of one of the two coset leaders of C_1's ancestor C_6. However, the coset leader 01010 of C_6 does not have support containing the support of the unique coset leader of C_6's descendant C_1.   ∎

The following theorem contains a characterization of orphans [33, 36].

Theorem 11.7.10 *Let C be an $[n, k, d]$ binary code. Let C' be a coset of C of weight w. The following hold:*
(i)   *C' is an orphan if and only if each coordinate position is covered by a vector in C' of weight w or $w + 1$.*
(ii) *If $d > 2$, then there is a one-to-one correspondence between the children of C' and the coordinate positions covered by the coset leaders of C'.*
(iii) *If $d > 2$ and C is an even code, then there is a one-to-one correspondence between the parents of C' and the coordinate positions not covered by the coset leaders of C'. In particular, C' is an orphan if and only if each coordinate position is covered by a coset leader of C'.*

Proof: For $1 \le i \le n$, let $e_i \in \mathbb{F}_2^n$ denote the vector of weight 1 with a 1 in coordinate i. By Corollary 11.7.8, each parent of C' equals $C' + e_i$ for some i. In order for C' to be an

Table 11.7 *Coset weight distribution of* $\mathcal{R}(1,4)$

Coset weight	Number of vectors of given weight																	Number of cosets
	0	1	2	3	4	5	6	7	8	9	10	11	12	13	14	15	16	
0	1								30								1	1
1		1						15		15						1		16
2			1				7		16		7				1			120
3				1		3		12		12		3		1				560
4					2		8		12		8		2					840
4					4				24				4					35
5						6		10		10		6						448
6							16				16							28

orphan, $C' + \mathbf{e}_i$ must contain vectors of weight w or less for all i. This is only possible if, for each i, there is a vector in C' of weight either w or $w + 1$ with a 1 in coordinate i. This proves (i).

Now assume that $d > 2$. Then for $i \neq j$, the cosets $C' + \mathbf{e}_i$ and $C' + \mathbf{e}_j$ are distinct. By definition, every child of C' equals $C' + \mathbf{e}_i$ for some i. Such a coset is a child if and only if some coset leader of C' has a 1 in coordinate i. This proves (ii).

Suppose now that C is even and $d > 2$. As in the proof of (i) each parent of C' equals $C' + \mathbf{e}_i$ for some i. In order for $C' + \mathbf{e}_i$ to be a parent of C', every vector in C' of weight w or $w + 1$ must not have a 1 in coordinate i. As C is even, a coset has only even weight vectors or only odd weight vectors; thus C' has no vectors of weight $w + 1$. The first part of (iii) now follows, noting that $C' + \mathbf{e}_i \neq C' + \mathbf{e}_j$ if $i \neq j$. The second part of (iii) follows directly from the first.   □

We leave the proof of the following corollary as an exercise.

Corollary 11.7.11 *Let C be an $[n, k, d]$ even binary code with $d > 2$. The following hold:*
(i) *An orphan of C has n descendants of weight 1 and n children.*
(ii) *A coset of weight w, which has exactly t coset leaders where $tw < n$, is not an orphan.*

Exercise 669 Prove Corollary 11.7.11.   ◆

Example 11.7.12 Table 11.7 presents the coset weight distribution of the $[16, 5, 8]$ Reed–Muller code $\mathcal{R}(1, 4)$. In a series of exercises, you will be asked to verify the entries in the table. By Theorem 11.2.5, $\rho(\mathcal{R}(1, 4)) = 6$. There are orphans of $\mathcal{R}(1, 4)$ of weights 4 and 6; see Exercise 676. The presence of these orphans and their weight distribution is part of a more general result found in [36].   ■

Exercise 670 By Theorem 1.10.1, $\mathcal{R}(1, 4)^\perp$ is the $[16, 11, 4]$ Reed–Muller code $\mathcal{R}(2, 4)$.
(a) By Theorem 1.10.1, $\mathcal{R}(1, 4)$ is a $[16, 5, 8]$ code containing the all-one vector. Show that the weight distribution of $\mathcal{R}(1, 4)$ is $A_0 = A_{16} = 1$ and $A_8 = 30$, verifying the entries in first row of Table 11.7.

(b) Show that the only possible weights of nonzero codewords in $\mathcal{R}(1, 4)^{\perp}$ are 4, 6, 8, 10, 12, and 16.

(c) Using the Assmus–Mattson Theorem, show that the 30 weight 8 codewords of $\mathcal{R}(1, 4)$ hold a 3-(16, 8, 3) design and draw its Pascal triangle.

(d) Verify the weight distribution of the cosets of weights 1, 2, and 3 in Table 11.7. Also verify that the number of such cosets is as indicated in the table. Hint: See Example 8.4.6.   ◆

Exercise 671 By Exercise 670 we have verified that the entries in Table 11.7 for the cosets of $\mathcal{R}(1, 4)$ of weights 1 and 3 are correct. In this exercise, you will verify that the information about weight 5 cosets is correct. By Exercise 670(b) and Theorem 7.5.2(iv), the weight distribution of a coset of weight 5 is uniquely determined.

(a) If there are x weight 5 vectors in a coset of weight 5, show that there are $16 - x$ vectors of weight 7 in the coset.

(b) How many vectors of weights 5 and 7 are there in $\mathbb{F}_2^{16}$?

(c) Let n be the number of cosets of weight 5. Since $\rho(\mathcal{R}(1, 4)) = 6$, each weight 5 and weight 7 vector in $\mathbb{F}_2^{16}$ is in a coset of weight 1, 3, or 5. Counting the total number of weight 5 and weight 7 vectors, find two equations relating x and n. Solve the equations to verify the information about weight 5 cosets in Table 11.7.   ◆

Exercise 672 Use Exercise 631 to verify that the weight distribution of a weight 6 coset of $\mathcal{R}(1, 4)$ is as in Table 11.7.   ◆

Exercise 673 In this exercise we show that any coset of weight 4 in $\mathcal{R}(1, 4)$ has either two or four coset leaders.

(a) Show that two different coset leaders in the same weight 4 coset of $\mathcal{R}(1, 4)$ have disjoint supports.

(b) Show that the maximum number of coset leaders in a weight 4 coset of $\mathcal{R}(1, 4)$ is 4.

(c) Show that if a weight 4 coset of $\mathcal{R}(1, 4)$ has at least three coset leaders, it actually has four coset leaders.

(d) Let $\mathbf{v} + \mathcal{R}(1, 4)$ be a coset of $\mathcal{R}(1, 4)$ of weight 4 whose only coset leader is $\mathbf{v}$. Suppose that $\text{supp}(\mathbf{v}) = \{i_1, i_2, i_3, i_4\}$.

   (i) Show that $\text{supp}(\mathbf{v})$ is not contained in the support of any weight 8 codeword of $\mathcal{R}(1, 4)$.

   (ii) Using Exercise 670(c), show that there are three weight 8 codewords of $\mathcal{R}(1, 4)$ with supports containing $\{i_2, i_3, i_4\}$.

   (iii) By (i), the supports of the three codewords of (ii) cannot contain i_1. Show that this is impossible.   ◆

Exercise 674 By Exercise 670(b) and Theorem 7.5.2(iii), once we know the number of coset leaders of a weight 4 coset of $\mathcal{R}(1, 4)$, the distribution of that coset is unique. By Exercise 673, there are either two or four coset leaders in a weight 4 coset. Verify the weight distributions of the weight 4 cosets given in Table 11.7. Hint: See Example 7.5.4. If $\mathcal{D} = \mathcal{C} \cup (\mathbf{v} + \mathcal{C})$, where $\mathbf{v}$ is a coset leader of weight 4, then $\mathcal{D}^{\perp} \subseteq \mathcal{C}^{\perp} = \mathcal{R}(2, 4)$, which has minimum weight 4.   ◆

Exercise 675 In this exercise, you will verify that there are 35 weight 4 cosets of $\mathcal{R}(1, 4)$ that have four coset leaders. From that you can compute the number of cosets of weight 4 with two coset leaders and the number of cosets of weight 6.

(a) Let $x_0 + \mathcal{R}(1, 4)$ be a coset of weight 4 with coset leaders x_0, x_1, x_2, and x_3. Show that $c_1 = x_0 + x_1$ and $c_2 = x_0 + x_2$ are weight 8 codewords in $\mathcal{R}(1, 4)$ whose supports intersect in the set supp(x_0) and in a natural way determine the supports of x_1, x_2, and x_3.

(b) Conversely, show how to obtain four coset leaders of a weight 4 coset by considering the intersection of the supports of two weight 8 codewords of $\mathcal{R}(1, 4)$ whose supports are not disjoint.

(c) There are 15 pairs of codewords of weight 8 in $\mathcal{R}(1, 4)$ where the weight 8 codewords in each pair have disjoint supports. Choose two distinct pairs. Show how these pairs determine the four coset leaders of a weight 4 coset of $\mathcal{R}(1, 4)$ and also determine a third pair of weight 8 codewords.

(d) Show that there are 35 weight 4 cosets of $\mathcal{R}(1, 4)$ with four coset leaders.

(e) Show that there are 840 weight 4 cosets of $\mathcal{R}(1, 4)$ with two coset leaders.

(f) Show that there are 28 weight 6 cosets of $\mathcal{R}(1, 4)$.   ◆

Exercise 676 Do the following:

(a) Show that every weight 6 coset of $\mathcal{R}(1, 4)$ has exactly 16 children and that every weight 5 coset has exactly one parent.

(b) Show that every weight 4 coset of $\mathcal{R}(1, 4)$ with four coset leaders is an orphan, but those with two coset leaders are not orphans.

(c) Show that every weight 4 coset of $\mathcal{R}(1, 4)$ with four coset leaders has exactly 16 children and that every weight 3 coset has precisely one parent that is an orphan and 12 parents that are not orphans.   ◆

We conclude this chapter with a brief introduction to the Newton radius first defined in [123]. The *Newton radius* $v(\mathcal{C})$ of a binary code $\mathcal{C}$ is the largest value v so that there exists a coset of weight v with only one coset leader. In particular, $v(\mathcal{C})$ is the largest weight of any error that can be uniquely corrected.

The first statement in the following lemma is from [123].

Lemma 11.7.13 *Let $\mathcal{C}$ be an $[n, k, d]$ binary code. The following hold*:

(i) $\lfloor (d - 1)/2 \rfloor \le v(\mathcal{C}) \le \rho(\mathcal{C})$.

(ii)   *If $\mathcal{C}$ is perfect, $(d - 1)/2 = v(\mathcal{C}) = \rho(\mathcal{C})$.*

(iii) *If $\mathcal{C}$ is even and $d > 2$, then $v(\mathcal{C}) < \rho(\mathcal{C})$.*

Proof: Proofs of (i) and (ii) are left as an exercise. Theorem 11.7.10(iii) implies that orphans in even codes have more than one coset leader. In particular, cosets of weight $\rho(\mathcal{C})$ cannot have unique coset leaders.   □

Exercise 677 Prove parts (i) and (ii) of Lemma 11.7.13.   ◆

Example 11.7.14 Let $\mathcal{C}$ be the $[15, 4, 8]$ simplex code. By Theorem 11.2.7, $\rho(\mathcal{C}) = 7$. It can be shown [123] that $v(\mathcal{C}) = 4$. So in this example, we have strict inequality in Lemma 11.7.13(i).   ■

We can use the ordering of binary vectors already developed in this section to simplify the proof of the following result from [123].

Theorem 11.7.15 *If C is an $[n, k]$ binary code, then*

$$0 \le \rho(C) - \nu(C) \le k.$$

Proof: That $0 \le \rho(C) - \nu(C)$ follows from Lemma 11.7.13. Let e_i be the binary vector of length n and weight 1 with 1 in coordinate i. Let x_1 be a weight $\rho = \rho(C)$ coset leader of a coset C_1 of C. By rearranging coordinates we may assume that $\operatorname{supp}(x_1) = \{1, 2, \ldots, \rho\}$. Then either $\nu(C) = \rho$, in which case we are done, or $\nu(C) \le \rho - 1$ by Lemma 11.7.13. Suppose the latter occurs. Then there exists another coset leader x_2 of C_1. Hence $x_1 + x_2 = c_1 \in C$. By rearranging coordinates we may assume that $1 \in \operatorname{supp}(c_1)$, since x_1 and x_2 must disagree on some coordinate. By Corollary 11.7.7, there is a child C_2 of C_1 with all coset leaders having first coordinate 0, one of which is $x_3 = e_1 + x_1$. If $\nu(C) \le \rho - 2$, there is a coset leader x_4 of C_2 that must disagree with x_3 on some coordinate in $\operatorname{supp}(x_3) = \{2, 3, \ldots, \rho\}$; by rearranging, we may assume that $c_2 = x_3 + x_4 \in C$ is 1 on the second coordinate. Since all coset leaders of C_2 are 0 on the first coordinate, so is c_2. Since c_1 begins $1 \cdots$ and c_2 begins $01 \cdots$, they are independent. By Corollary 11.7.7, there is a child C_3 of C_2 with all coset leaders having second coordinate 0, one of which is $x_5 = e_2 + x_3$. If C_3 has a coset leader that is 1 on the first coordinate, then C_2 must have a coset leader with a 1 in the coordinate by Corollary 11.7.8, which is a contradiction. So all coset leaders of C_3 are 0 on the first and second coordinates. If $\nu(C) \le \rho - 3$, there is a coset leader x_6 of C_3 which must disagree with x_5 on some coordinate in $\operatorname{supp}(x_5) = \{3, 4, \ldots, \rho\}$; by rearranging, we may assume that $c_3 = x_5 + x_6 \in C$ is 1 on the third coordinate. Since c_1 begins $1 \cdots$, c_2 begins $01 \cdots$, and c_3 begins $001 \cdots$, they are independent. We can continue inductively to find s independent codewords as long as $\nu(C) \le \rho - s$. Since the largest value of s possible is k, $\nu(C) \ge \rho - k$. $\qquad\square$

Exercise 678 In the proof of Theorem 11.7.15, carry out one more step in the induction. ◆

12 Codes over $\mathbb{Z}_4$

The study of codes over the ring $\mathbb{Z}_4$ attracted great interest through the work of Calderbank, Hammons, Kumar, Sloane, and Solé in the early 1990s which resulted in the publication of a paper [116] showing how several well-known families of nonlinear binary codes were intimately related to linear codes over $\mathbb{Z}_4$. In particular, for $m \geq 4$ an even integer, there is a family $\mathcal{K}(m)$ of $(2^m, 2^{2m}, 2^{m-1} - 2^{(m-2)/2})$ nonlinear binary codes, originally discovered by Kerdock [166], and a second family $\mathcal{P}(m)$ of $(2^m, 2^{2^m - 2m}, 6)$ nonlinear binary codes, due to Preparata [288], which possess a number of remarkable properties. First, the codes $\mathcal{K}(4)$ and $\mathcal{P}(4)$ are identical to one another and to the Nordstrom–Robinson code described in Section 2.3.4. Second, the Preparata codes have the same length as the extended double error-correcting BCH codes but have twice as many codewords. Third, and perhaps most intriguing, $\mathcal{K}(m)$ and $\mathcal{P}(m)$ behave as if they are duals of each other in the sense that the weight enumerator of $\mathcal{P}(m)$ is the MacWilliams transform of the weight enumerator of $\mathcal{K}(m)$; in other words, if $\mathcal{C} = \mathcal{K}(m)$, then the weight enumerator of $\mathcal{P}(m)$ is $\frac{1}{|C|}W_C(y - x, y + x)$ (see (M_3) of Section 7.2). A number of researchers had attempted to explain this curious relationship between weight enumerators without success until the connection with codes over $\mathbb{Z}_4$ was made in [116].

In this chapter we will present the basic theory of linear codes over $\mathbb{Z}_4$ including the connection between these codes and binary codes via the Gray map. We will also study cyclic, self-dual, and quadratic residue codes over $\mathbb{Z}_4$. The codes of Kerdock and Preparata will be described as the Gray image of certain extended cyclic codes over $\mathbb{Z}_4$. In order to present these codes we will need to examine the Galois ring $\mathrm{GR}(4^m)$, which plays the same role in the study of cyclic codes over $\mathbb{Z}_4$ as $\mathbb{F}_{2^m}$ (also denoted $\mathrm{GF}(2^m)$) does in the study of cyclic codes over $\mathbb{F}_2$.

12.1 Basic theory of $\mathbb{Z}_4$-linear codes

A $\mathbb{Z}_4$-*linear code*[1] $\mathcal{C}$ of length n is an additive subgroup of $\mathbb{Z}_4^n$. Such a subgroup is a $\mathbb{Z}_4$-module, which may or may not be free. (A $\mathbb{Z}_4$-module $\mathcal{M}$ is *free* if there exists a subset $\mathcal{B}$ of $\mathcal{M}$, called a *basis*, such that every element in $\mathcal{M}$ is uniquely expressible as a $\mathbb{Z}_4$-linear combination of the elements in $\mathcal{B}$.) We will still term elements of $\mathbb{Z}_4^n$ "vectors" even though

---

[1] Codes over $\mathbb{Z}_4$ are sometimes called quaternary codes in the literature. Unfortunately, codes over $\mathbb{F}_4$ are also called quaternary, a term we used in Chapter 1. In this book we reserve the term "quaternary" to refer to codes over $\mathbb{F}_4$.

$\mathbb{Z}_4^n$ is not a vector space. Note that if a vector $\mathbf{v}$ has components that are all 0s or 2s, then $2\mathbf{v} = \mathbf{0}$ implying that such a vector cannot be in a basis of a free submodule of $\mathbb{Z}_4^n$ since $2 \neq 0$. In describing codes over $\mathbb{Z}_4$, we denote the elements of $\mathbb{Z}_4$ in either of the natural forms $\{0, 1, 2, 3\}$ or $\{0, 1, 2, -1\}$, whichever is most convenient.

Example 12.1.1 Let C be the $\mathbb{Z}_4$-linear code of length 4 with the 16 codewords:

0000, 1113, 2222, 3331, 0202, 1311, 2020, 3133,
0022, 1131, 2200, 3313, 0220, 1333, 2002, 3111.

This is indeed an additive subgroup of $\mathbb{Z}_4^4$ as Exercise 679 shows. If this were a free $\mathbb{Z}_4$-module, it would have a basis of two vectors $\mathbf{b}_1$ and $\mathbf{b}_2$ such that every codeword would be a unique $\mathbb{Z}_4$-linear combination of $\mathbf{b}_1$ and $\mathbf{b}_2$. As described above, both $\mathbf{b}_1$ and $\mathbf{b}_2$ have at least one component equal to 1 or 3. However, if $\mathbf{b}_1$ and $\mathbf{b}_2$ are among the eight codewords with one component 1 or 3, then $2\mathbf{b}_1 = 2222 = 2\mathbf{b}_2$. Hence $\{\mathbf{b}_1, \mathbf{b}_2\}$ cannot be a basis of C implying C is not free. However, we can still produce a perfectly good generator matrix whose rows function for all intents and purposes as a "basis" for the code.   ■

Exercise 679 Let C be the code over $\mathbb{Z}_4$ of Example 12.1.1.
(a) Show that every codeword of C can be written uniquely in the form $x\mathbf{c}_1 + y\mathbf{c}_2 + z\mathbf{c}_3$, where $\mathbf{c}_1 = 1113$, $\mathbf{c}_2 = 0202$, $\mathbf{c}_3 = 0022$, $x \in \mathbb{Z}_4$, and y and z are in $\{0, 1\}$.
(b) Use part (a) to show that C is a $\mathbb{Z}_4$-linear code.   ◆

Example 12.1.2 Let C be the code over $\mathbb{Z}_4$ of Example 12.1.1. By Exercise 679(a), we can consider

$$G = \begin{bmatrix} 1 & 1 & 1 & 3 \\ 0 & 2 & 0 & 2 \\ 0 & 0 & 2 & 2 \end{bmatrix}$$

as a generator matrix for C in the following sense: every codeword of C is $(xyz)G$ for some x in $\mathbb{Z}_4$, and y and z in $\mathbb{Z}_2$.   ■

As Exercise 679 illustrates, every $\mathbb{Z}_4$-linear code C contains a set of $k_1 + k_2$ codewords $\mathbf{c}_1, \ldots, \mathbf{c}_{k_1}, \mathbf{c}_{k_1+1}, \ldots, \mathbf{c}_{k_1+k_2}$ such that every codeword in C is uniquely expressible in the form

$$\sum_{i=1}^{k_1} a_i \mathbf{c}_i + \sum_{i=k_1+1}^{k_1+k_2} a_i \mathbf{c}_i,$$

where $a_i \in \mathbb{Z}_4$ for $1 \leq i \leq k_1$ and $a_i \in \mathbb{Z}_2$ for $k_1 + 1 \leq i \leq k_1 + k_2$. Furthermore, each $\mathbf{c}_i$ has at least one component equal to 1 or 3 for $1 \leq i \leq k_1$ and each $\mathbf{c}_i$ has all components equal to 0 and 2 for $k_1 + 1 \leq i \leq k_1 + k_2$. If $k_2 = 0$, the code C is a free $\mathbb{Z}_4$-module. The matrix whose rows are $\mathbf{c}_i$ for $1 \leq i \leq k_1 + k_2$ is called a *generator matrix* for C. The code C has $4^{k_1} 2^{k_2}$ codewords and is said to be of *type* $4^{k_1} 2^{k_2}$. The code of Example 12.1.1 has type $4^1 2^2$.

As with codes over fields, we can describe equivalence of $\mathbb{Z}_4$-linear codes. Let C_1 and C_2 be two $\mathbb{Z}_4$-linear codes of length n. Then C_1 and C_2 are *permutation equivalent* provided there is an $n \times n$ permutation matrix P such that $C_2 = C_1 P$. We can also scale components of a code to obtain an equivalent code provided the scalars are invertible elements of $\mathbb{Z}_4$. To that end, we say that C_1 and C_2 are *monomially equivalent* provided there is an $n \times n$ monomial matrix M with all nonzero entries equal to 1 or 3 such that $C_2 = C_1 M$. The *permutation automorphism group* PAut(C) of C is the set of all permutation matrices P such that $CP = C$, while the *monomial automorphism group* MAut(C) of C is the set of all monomial matrices M (with nonzero entries 1 and 3) such that $CM = C$.

Exercise 680 Show that the $\mathbb{Z}_4$-linear codes with generator matrices

$$G_1 = \begin{bmatrix} 1 & 1 & 1 & 3 \\ 0 & 2 & 0 & 2 \\ 0 & 0 & 2 & 2 \end{bmatrix} \quad \text{and} \quad G_2 = \begin{bmatrix} 1 & 1 & 1 & 1 \\ 2 & 0 & 0 & 2 \\ 0 & 2 & 0 & 2 \end{bmatrix}$$

are monomially equivalent.   ◆

Exercise 681 Find PAut(C) and MAut(C), where C is the code in Example 12.1.1.   ◆

Every $[n, k]$ linear code over $\mathbb{F}_q$ is permutation equivalent to a code with generator matrix in standard form $[I_k \ A]$ by Theorem 1.6.2. In a similar way, there is a standard form for the generator matrix of a $\mathbb{Z}_4$-linear code. A generator matrix G of a $\mathbb{Z}_4$-linear code C is in *standard form* if

$$G = \begin{bmatrix} I_{k_1} & A & B_1 + 2B_2 \\ O & 2I_{k_2} & 2C \end{bmatrix}, \tag{12.1}$$

where A, B_1, B_2, and C are matrices with entries from $\mathbb{Z}_2$, and O is the $k_2 \times k_1$ zero matrix. The code C is of type $4^{k_1} 2^{k_2}$. Notice that the generator matrix for the code of Example 12.1.1 is in standard form. Although rather tedious, the proof of the following result is straightforward.

Theorem 12.1.3 *Every $\mathbb{Z}_4$-linear code is permutation equivalent to a code with generator matrix in standard form.*

Exercise 682 Prove Theorem 12.1.3.   ◆

There is a natural inner product, the ordinary dot product modulo 4, on $\mathbb{Z}_4^n$ defined by

$$\mathbf{x} \cdot \mathbf{y} = x_1 y_1 + x_2 y_2 + \cdots + x_n y_n \pmod{4},$$

where $\mathbf{x} = x_1 \cdots x_n$ and $\mathbf{y} = y_1 \cdots y_n$. As with linear codes over finite fields, we can define the *dual* code $C^{\perp}$ of a $\mathbb{Z}_4$-linear code C of length n by $C^{\perp} = \{\mathbf{x} \in \mathbb{Z}_4^n \mid \mathbf{x} \cdot \mathbf{c} = 0 \text{ for all } \mathbf{c} \in C\}$. By Exercise 683, $C^{\perp}$ is indeed a $\mathbb{Z}_4$-linear code. A $\mathbb{Z}_4$-linear code is *self-orthogonal* when $C \subseteq C^{\perp}$ and *self-dual* if $C = C^{\perp}$.

Exercise 683 Prove that the dual code of a $\mathbb{Z}_4$-linear code is a $\mathbb{Z}_4$-linear code.   ◆

Exercise 684 Prove that the code of Example 12.1.1 is self-dual.   ◆

Example 12.1.4 Unlike codes over finite fields, there are self-dual $\mathbb{Z}_4$-linear codes of odd length. For example, the code of length n and type 2^n with generator matrix $2I_n$ is self-dual.   ■

Exercise 685 Find generator matrices in standard form for all self-dual codes of lengths $n \leq 4$, up to monomial equivalence. Give the type of each code.   ◆

If C has generator matrix in standard form (12.1), then $C^{\perp}$ has generator matrix

$$G^{\perp} = \begin{bmatrix} -(B_1 + 2B_2)^{\mathrm{T}} - C^{\mathrm{T}} A^{\mathrm{T}} & C^{\mathrm{T}} & I_{n-k_1-k_2} \\ 2A^{\mathrm{T}} & 2I_{k_2} & O \end{bmatrix}, \tag{12.2}$$

where O is the $k_2 \times (n - k_1 - k_2)$ zero matrix. In particular, $C^{\perp}$ has type

$$4^{n-k_1-k_2} 2^{k_2}. \tag{12.3}$$

Exercise 686 Give the generator matrix (12.2) of $C^{\perp}$ for the code C of Example 12.1.1.   ◆

Exercise 687 Prove that $G^{\perp} G^{\mathrm{T}}$ is the zero matrix, where G is given in (12.1) and $G^{\perp}$ is given in (12.2).   ◆

Unlike codes over $\mathbb{F}_2$, the concept of weight of a vector in $\mathbb{Z}_4^n$ can have more than one meaning. In fact, three different weights, and hence three different distances, are used when dealing with codes over $\mathbb{Z}_4$. Let $\mathbf{x} \in \mathbb{Z}_4^n$; suppose that $n_a(\mathbf{x})$ denotes the number of components of $\mathbf{x}$ equal to a for all $a \in \mathbb{Z}_4$. The *Hamming weight* of $\mathbf{x}$ is $\mathrm{wt}_H(\mathbf{x}) = n_1(\mathbf{x}) + n_2(\mathbf{x}) + n_3(\mathbf{x})$, the *Lee weight* of $\mathbf{x}$ is $\mathrm{wt}_L(\mathbf{x}) = n_1(\mathbf{x}) + 2n_2(\mathbf{x}) + n_3(\mathbf{x})$, and the *Euclidean weight* of $\mathbf{x}$ is $\mathrm{wt}_E(\mathbf{x}) = n_1(\mathbf{x}) + 4n_2(\mathbf{x}) + n_3(\mathbf{x})$. Thus components equaling 1 or 3 contribute 1 to each weight while 2 contributes 2 to the Lee and 4 to the Euclidean weight. The *Hamming*, *Lee*, and *Euclidean distances* between $\mathbf{x}$ and $\mathbf{y}$ are $d_H(\mathbf{x}, \mathbf{y}) = \mathrm{wt}_H(\mathbf{x} - \mathbf{y})$, $d_L(\mathbf{x}, \mathbf{y}) = \mathrm{wt}_L(\mathbf{x} - \mathbf{y})$, and $d_E(\mathbf{x}, \mathbf{y}) = \mathrm{wt}_E(\mathbf{x} - \mathbf{y})$, respectively.

Exercise 688 Do the following:
(a) What are the Hamming, Lee, and Euclidean weights of the vector $120\,203\,303$?
(b) What are the Hamming, Lee, and Euclidean distances between the vectors $30\,012\,221$ and $20\,202\,213$?   ◆

Exercise 689 Find the Hamming, Lee, and Euclidean weight distributions of the code C of Example 12.1.1.   ◆

Exercise 690 Let C be a $\mathbb{Z}_4$-linear code that contains a codeword with only 1s and 3s. Prove that the Lee weight of all codewords in $C^{\perp}$ is even.   ◆

The following theorem is the $\mathbb{Z}_4$ analogue to Theorem 1.4.5(iv).

Theorem 12.1.5 Let C be a self-orthogonal $\mathbb{Z}_4$-linear code with $\mathbf{c} \in C$. Then:
(i) $\mathrm{wt}_L(\mathbf{c}) \equiv 0 \pmod{2}$, and
(ii) $\mathrm{wt}_E(\mathbf{c}) \equiv 0 \pmod{4}$.

Exercise 691 Prove Theorem 12.1.5.   ◆

There are also analogues to the MacWilliams equations for $\mathbb{Z}_4$-linear codes C of length n. Let

$$\mathrm{Ham}_C(x, y) = \sum_{\mathbf{c} \in C} x^{\mathrm{wt}_H(\mathbf{c})} y^{n - \mathrm{wt}_H(\mathbf{c})}$$

be the *Hamming weight enumerator* of C. Let

$$\text{Lee}_C(x, y) = \sum_{c \in C} x^{\text{wt}_L(c)} y^{2n - \text{wt}_L(c)} \tag{12.4}$$

be the *Lee weight enumerator* of C; note the presence of $2n$ in the exponent of y, which is needed because there may be codewords of Lee weight as high as $2n$. The analogs of (M_3) from Section 7.2 are [59, 116, 173, 174]:

$$\text{Ham}_{C^\perp}(x, y) = \frac{1}{|C|}\text{Ham}_C(y - x, y + 3x), \quad \text{and} \tag{12.5}$$

$$\text{Lee}_{C^\perp}(x, y) = \frac{1}{|C|}\text{Lee}_C(y - x, y + x). \tag{12.6}$$

Exercise 692   Give the Hamming and Lee weight enumerators for the self-dual code C of Example 12.1.1. Then verify (12.5) and (12.6) for this code.   ◆

There are two other weight enumerators of interest with $\mathbb{Z}_4$-linear codes. With $n_i(\mathbf{v})$ defined as before, let

$$\text{cwe}_C(a, b, c, d) = \sum_{\mathbf{v} \in C} a^{n_0(\mathbf{v})} b^{n_1(\mathbf{v})} c^{n_2(\mathbf{v})} d^{n_3(\mathbf{v})}$$

denote the *complete weight enumerator* of C, and let

$$\text{swe}_C(a, b, c) = \sum_{\mathbf{v} \in C} a^{n_0(\mathbf{v})} b^{n_1(\mathbf{v}) + n_3(\mathbf{v})} c^{n_2(\mathbf{v})} \tag{12.7}$$

denote the *symmetrized weight enumerator* of C. Notice that equivalent codes have the same symmetrized weight enumerators (making this one the more useful) while equivalent codes may not have the same complete weight enumerators. The corresponding MacWilliams equations are:

$$\text{cwe}_{C^\perp}(a, b, c, d) = \frac{1}{|C|}\text{cwe}_C(a + b + c + d, a + ib - c - id,$$

$$a - b + c - d, a - ib - c + id), \quad \text{and} \tag{12.8}$$

$$\text{swe}_{C^\perp}(a, b, c) = \frac{1}{|C|}\text{swe}_C(a + 2b + c, a - c, a - 2b + c), \tag{12.9}$$

where $i = \sqrt{-1}$.

Exercise 693   Give the complete and symmetrized weight enumerators for the self-dual code C of Example 12.1.1. Then verify (12.8) and (12.9) for this code.   ◆

Exercise 694   Do the following:
(a) Show that $\text{swe}_C(a, b, c) = \text{cwe}_C(a, b, c, b)$.
(b) Verify (12.9) assuming that (12.8) is valid.
(c) Show that $\text{Ham}_C(x, y) = \text{swe}_C(y, x, x)$.
(d) Verify (12.5) assuming that (12.9) is valid.   ◆

12.2   Binary codes from $\mathbb{Z}_4$-linear codes

The vehicle by which binary codes are obtained from $\mathbb{Z}_4$-linear codes is the *Gray map* $\mathfrak{G} : \mathbb{Z}_4 \to \mathbb{F}_2^2$ defined by

$$\mathfrak{G}(0) = 00, \quad \mathfrak{G}(1) = 01, \quad \mathfrak{G}(2) = 11, \quad \text{and } \mathfrak{G}(3) = 10.$$

This map is then extended componentwise to a map, also denoted $\mathfrak{G}$, from $\mathbb{Z}_4^n$ to $\mathbb{F}_2^{2n}$. If C is a $\mathbb{Z}_4$-linear code, its Gray image will be the binary code denoted $\mathfrak{G}(C)$. Note that C and $\mathfrak{G}(C)$ have the same size. In general, however, $\mathfrak{G}(C)$ will be nonlinear.

Exercise 695   Find all pairs of elements a and b of $\mathbb{Z}_4$ where $\mathfrak{G}(a + b) \neq \mathfrak{G}(a) + \mathfrak{G}(b)$.
    ◆

Exercise 696   Let C be the $\mathbb{Z}_4$-linear code of Example 12.1.1.
(a) List the 16 codewords of $\mathfrak{G}(C)$.
(b) Show that $\mathfrak{G}(C)$ is linear.
(c) What is the binary code obtained?   ◆

Exercise 697   Let C be the $\mathbb{Z}_4$-linear code of length 3 with generator matrix

$$G = \begin{bmatrix} 1 & 0 & 1 \\ 0 & 1 & 3 \end{bmatrix}.$$

(a) List the 16 codewords in C.
(b) List the 16 codewords in $\mathfrak{G}(C)$.
(c) Show that $\mathfrak{G}(C)$ is nonlinear.   ◆

In doing Exercises 696 and 697, you will note that the Lee weight of $\mathbf{v} \in \mathbb{Z}_4^n$ is the ordinary Hamming weight of its Gray image $\mathfrak{G}(\mathbf{v})$. This leads to an important property of the Gray images of $\mathbb{Z}_4$-linear codes. A code C (over $\mathbb{Z}_4$ or $\mathbb{F}_q$) is *distance invariant* provided the Hamming weight distribution of $\mathbf{c} + C$ is the same for all $\mathbf{c} \in C$. Note that all linear codes (over $\mathbb{Z}_4$ or $\mathbb{F}_q$) must be distance invariant simply because $\mathbf{c} + C = C$ for all $\mathbf{c} \in C$.

Example 12.2.1 When dealing with nonlinear binary codes the distance distribution of the code is more significant than the weight distribution because the former gives information about the error-correcting capability of the code. Recall from Chapter 2 that the Hamming distance distribution of a code C of length n is the set $\{B_i(C) \mid 0 \le i \le n\}$, where

$$B_i(C) = \frac{1}{|C|} \sum_{\mathbf{c} \in C} |\{\mathbf{v} \in C \mid d(\mathbf{v}, \mathbf{c}) = i\}|$$

and $d(\mathbf{v}, \mathbf{c})$ is the Hamming distance between $\mathbf{c}$ and $\mathbf{v}$; for codes over $\mathbb{Z}_4$, use the same definition with d_H in place of d. Thus $\{B_i(C) \mid 0 \le i \le n\}$ is the average of the weight distributions of $\mathbf{c} + C$ for all $\mathbf{c} \in C$. If the code is distance invariant, the distance distribution is the same as the weight distribution of any set $\mathbf{c} + C$; if in addition, the code contains the zero codeword, the distance distribution is the same as the weight distribution of C. Thus being distance invariant is a powerful property of a code. For instance, let C be the binary code of length 6 with codewords $\mathbf{c}_1 = 111100$, $\mathbf{c}_2 = 001111$, and $\mathbf{c}_3 = 101010$. The weight distribution of $\mathbf{c}_j + C$ is

$j\backslash i$	0	1	2	3	4	5	6
1	1	0	0	1	1	0	0
2	1	0	0	1	1	0	0 .
3	1	0	0	2	0	0	0

Thus C is not distance invariant, and its distance distribution is $B_0(C) = 1$, $B_3(C) = 4/3$, and $B_4(C) = 2/3$, with $B_i(C) = 0$ otherwise. The minimum weight 3 of this code happens to equal the minimum distance of the code as well; this does not hold in general. See Exercise 698. ∎

Exercise 698 Let C be the binary code of length 7 with codewords $\mathbf{c}_1 = 1111001$, $\mathbf{c}_2 = 0011111$, and $\mathbf{c}_3 = 1010101$.
(a) Find the weight distribution of $\mathbf{c}_j + C$ for $j = 1, 2$, and 3.
(b) Find the distance distribution of C.
(c) Find the minimum distance and the minimum weight of C. ♦

In what follows we will still denote the Hamming weight of a binary vector $\mathbf{v}$ by $\mathrm{wt}(\mathbf{v})$.

Theorem 12.2.2 *The following hold:*
(i) *The Gray map $\mathfrak{G}$ is a distance preserving map from $\mathbb{Z}_4^n$ with Lee distance to $\mathbb{F}_2^{2n}$ with Hamming distance.*
(ii) *If C is a $\mathbb{Z}_4$-linear code, then $\mathfrak{G}(C)$ is distance invariant.*
(iii) *If C is a $\mathbb{Z}_4$-linear code, then the Hamming weight distribution of $\mathfrak{G}(C)$ is the same as the Lee weight distribution of C.*

Proof: By Exercise 699, if a and b are in $\mathbb{Z}_4$, then $\mathrm{wt}_L(a - b) = \mathrm{wt}(\mathfrak{G}(a) - \mathfrak{G}(b))$. It follows that if $\mathbf{v} = v_1 \cdots v_n$ and $\mathbf{w} = w_1 \cdots w_n$ are in $\mathbb{Z}_4^n$, then

$$d_L(\mathbf{v}, \mathbf{w}) = \sum_{i=1}^{n} \mathrm{wt}_L(v_i - w_i) = \sum_{i=1}^{n} \mathrm{wt}(\mathfrak{G}(v_i) - \mathfrak{G}(w_i)) = d(\mathfrak{G}(\mathbf{v}), \mathfrak{G}(\mathbf{w})),$$

verifying (i).

By Exercise 699, if a and b are in $\mathbb{Z}_4$, then $\mathrm{wt}_L(a - b) = \mathrm{wt}(\mathfrak{G}(a) + \mathfrak{G}(b))$, implying that if $\mathbf{v} = v_1 \cdots v_n$ and $\mathbf{w} = w_1 \cdots w_n$ are in $\mathbb{Z}_4^n$, then

$$\mathrm{wt}_L(\mathbf{v} - \mathbf{w}) = \sum_{i=1}^{n} \mathrm{wt}_L(v_i - w_i)$$

$$= \sum_{i=1}^{n} \mathrm{wt}(\mathfrak{G}(v_i) + \mathfrak{G}(w_i)) = \mathrm{wt}(\mathfrak{G}(\mathbf{v}) + \mathfrak{G}(\mathbf{w})).$$

Therefore the Hamming weight distribution of $\mathfrak{G}(\mathbf{c}) + \mathfrak{G}(C)$ is the same as the Lee distribution of $\mathbf{c} - C = C$. Hence (ii) and (iii) follow. □

Exercise 699 Prove that if a and b are in $\mathbb{Z}_4$, then $\mathrm{wt}_L(a - b) = \mathrm{wt}(\mathfrak{G}(a) + \mathfrak{G}(b))$. ♦

In Exercises 696 and 697 we see that the Gray image of a code may be linear or nonlinear. It is natural to ask if we can tell when the Gray image will be linear. One criterion is the following, where $\mathbf{v} * \mathbf{w}$ is the componentwise product of the two vectors $\mathbf{v}$ and $\mathbf{w}$ in $\mathbb{Z}_4^n$.

Theorem 12.2.3 *Let C be a $\mathbb{Z}_4$-linear code. The binary code $\mathfrak{G}(C)$ is linear if and only if whenever $\mathbf{v}$ and $\mathbf{w}$ are in C, so is $2(\mathbf{v} * \mathbf{w})$.*

Proof: By Exercise 700, if a and b are in $\mathbb{Z}_4$, then $\mathfrak{G}(a) + \mathfrak{G}(b) = \mathfrak{G}(a + b + 2ab)$. Therefore if $\mathbf{v}$ and $\mathbf{w}$ are in $\mathbb{Z}_4^n$, then $\mathfrak{G}(\mathbf{v}) + \mathfrak{G}(\mathbf{w}) = \mathfrak{G}(\mathbf{v} + \mathbf{w} + 2(\mathbf{v} * \mathbf{w}))$. In particular, this shows that if $\mathbf{v}$ and $\mathbf{w}$ are in C, then $\mathfrak{G}(\mathbf{v}) + \mathfrak{G}(\mathbf{w})$ is in $\mathfrak{G}(C)$ if and only if $\mathbf{v} + \mathbf{w} + 2(\mathbf{v} * \mathbf{w}) \in C$ if and only if $2(\mathbf{v} * \mathbf{w}) \in C$, since $\mathbf{v} + \mathbf{w} \in C$. The result now follows.   □

Exercise 700 Prove that if a and b are in $\mathbb{Z}_4$, then

$$\mathfrak{G}(a) + \mathfrak{G}(b) = \mathfrak{G}(a + b + 2ab).$$

See the related Exercise 695.   ◆

Exercise 701 Verify that the code C of Example 12.1.1 (and Exercise 696) satisfies the criterion of Theorem 12.2.3, while the code C of Exercise 697 does not.   ◆

We will use the following theorem frequently; compare this with Theorems 1.4.3 and 1.4.8.

Theorem 12.2.4 *The following hold*:
(i) *Let $\mathbf{u}$ and $\mathbf{v}$ be in $\mathbb{Z}_4^n$. Then*

$$\text{wt}_E(\mathbf{u} + \mathbf{v}) \equiv \text{wt}_E(\mathbf{u}) + \text{wt}_E(\mathbf{v}) + 2(\mathbf{u} \cdot \mathbf{v}) \,(\text{mod } 8).$$

(ii) *Let C be a self-orthogonal $\mathbb{Z}_4$-linear code which has a generator matrix G such that all rows $\mathbf{r}$ of G satisfy $\text{wt}_E(\mathbf{r}) \equiv 0 \,(\text{mod } 8)$. Then $\text{wt}_E(\mathbf{c}) \equiv 0 \,(\text{mod } 8)$ for all $\mathbf{c} \in C$.*
(iii) *Let C be a $\mathbb{Z}_4$-linear code such that $\text{wt}_E(\mathbf{c}) \equiv 0 \,(\text{mod } 8)$ for all $\mathbf{c} \in C$. Then C is self-orthogonal.*

Exercise 702 Prove Theorem 12.2.4.   ◆

We now apply this theorem to show how to construct the Nordstrom–Robinson code as the Gray image of a $\mathbb{Z}_4$-linear code.

Example 12.2.5 Let o_8 be the $\mathbb{Z}_4$-linear code, called the *octacode*, with generator matrix

$$G = \begin{bmatrix} 1 & 0 & 0 & 0 & 3 & 1 & 2 & 1 \\ 0 & 1 & 0 & 0 & 1 & 2 & 3 & 1 \\ 0 & 0 & 1 & 0 & 3 & 3 & 3 & 2 \\ 0 & 0 & 0 & 1 & 2 & 3 & 1 & 1 \end{bmatrix}.$$

The rows of G are orthogonal to themselves and pairwise orthogonal to each other. Thus o_8 is self-orthogonal, and, as it has type 4^4, o_8 is self-dual. The Euclidean weight of each row of G equals 8. Hence by Theorem 12.2.4, every codeword of o_8 has Euclidean weight a multiple of 8. The minimum Euclidean weight is therefore 8 as each row of G has that Euclidean weight. By Exercise 703, o_8 has minimum Lee weight 6. The Gray image $\mathfrak{G}(o_8)$ therefore is a $(16, 256, 6)$ binary code by Theorem 12.2.2. By [317], $\mathfrak{G}(o_8)$ is the Nordstrom–Robinson code; see the discussion at the end of Section 2.3.4. We will examine this code more closely when we study the Kerdock codes in Section 12.7. In particular, Exercise 755 will show that the octacode is related to the Kerdock code $\mathcal{K}(4)$ and Exercise 759 will give its Lee weight distribution.   ■

Exercise 703   In this exercise, you will show that the octacode o_8 has minimum Lee weight 6. By Theorem 12.1.5, the codewords of o_8 all have even Lee weight; each row of the generator matrix has Lee weight 6.
(a) Show that there cannot be a vector in $\mathbb{Z}_4^n$ of Lee weight 2 and Euclidean weight 8.
(b) Show that the only vectors in $\mathbb{Z}_4^n$ of Lee weight 4 and Euclidean weight 8 have exactly two components equal to 2 and the remaining components equal to 0.
(c) By considering the generator matrix G for o_8 in Example 12.2.5, show that o_8 has no codewords of Lee weight 4.   ◆

Exercise 704   Using Theorem 12.2.3 show that $\mathfrak{G}(o_8)$ is nonlinear.   ◆

12.3   Cyclic codes over $\mathbb{Z}_4$

As with cyclic codes over a field, cyclic codes over $\mathbb{Z}_4$ form an important family of $\mathbb{Z}_4$-linear codes. A body of theory has been developed to handle these codes with obvious parallels to the theory of cyclic codes over fields.

As with usual cyclic codes over $\mathbb{F}_q$, we view codewords $\mathbf{c} = c_0 c_1 \cdots c_{n-1}$ in a cyclic $\mathbb{Z}_4$-linear code of length n as polynomials $c(x) = c_0 + c_1 x + \cdots + c_{n-1} x^{n-1} \in \mathbb{Z}_4[x]$. If we consider our polynomials as elements of the quotient ring

$$\mathfrak{R}_n = \mathbb{Z}_4[x]/(x^n - 1),$$

then $xc(x)$ modulo $x^n - 1$ represents the cyclic shift of $\mathbf{c}$. When studying cyclic codes over $\mathbb{F}_q$, we found generator polynomials and generating idempotents of these codes. It is natural to ask if such polynomials exist in the $\mathbb{Z}_4$ world. To do this we must study the factorization of $x^n - 1$ over $\mathbb{Z}_4$.

12.3.1   Factoring $x^n - 1$ over $\mathbb{Z}_4$

There are significant differences between the rings $\mathbb{Z}_4[x]$ and $\mathbb{F}_q[x]$. For example, an important obvious difference between the rings $\mathbb{Z}_4[x]$ and $\mathbb{F}_q[x]$ is that the degree of a product of two polynomials in $\mathbb{Z}_4[x]$ may be less than the sum of the degrees of the polynomials. Among other things this means that it is possible for a nonconstant polynomial in $\mathbb{Z}_4[x]$ to be invertible, that is, a unit. See Exercise 705.

Exercise 705   Compute $(1 + 2s(x))^2$ for any $s(x) \in \mathbb{Z}_4[x]$. Show how this illustrates the fact that the degree of a product of two polynomials in $\mathbb{Z}_4[x]$ may be less than the sum of the degrees of the polynomial factors. This computation shows that $1 + 2s(x)$ has a multiplicative inverse in $\mathbb{Z}_4[x]$; what is its inverse?   ◆

We still say that a polynomial $f(x) \in \mathbb{Z}_4[x]$ is *irreducible* if whenever $f(x) = g(x)h(x)$ for two polynomials $g(x)$ and $h(x)$ in $\mathbb{Z}_4[x]$, one of $g(x)$ or $h(x)$ is a unit. Since units do not have to be constant polynomials and the degree of $f(x)$ may be less than the sum of the degrees of $g(x)$ and $h(x)$, it is more difficult in $\mathbb{Z}_4[x]$ to check whether or not a polynomial is irreducible. Recall that in $\mathbb{F}_q[x]$ nonconstant polynomials can be factored into a product of irreducible polynomials which are unique up to scalar multiplication; see Theorem 3.4.1. This is not the case for polynomials in $\mathbb{Z}_4[x]$, even polynomials of the form we are most interested in, as the next example illustrates.

Example 12.3.1 The following are two factorizations of $x^4 - 1$ into irreducible polynomials in $\mathbb{Z}_4[x]$:

$$x^4 - 1 = (x - 1)(x + 1)(x^2 + 1) = (x + 1)^2(x^2 + 2x - 1).$$

The verification that these polynomials are in fact irreducible is tedious but straightforward. (If, for example, you assume that $x^2 + 2x - 1 = g(x)h(x)$, to show irreducibility then one of $g(x)$ or $h(x)$ must be shown to be a unit. Since the degree of a product is not necessarily the sum of the degrees, it cannot be assumed for instance that $g(x)$ and $h(x)$ are both of degree 1.)   ∎

The proper context for discussing factorization in $\mathbb{Z}_4[x]$ is not factoring polynomials into a product of irreducible polynomials but into a product of polynomials called *basic irreducible polynomials*. In order to discuss this concept, we need some notation and terminology. Define $\mu : \mathbb{Z}_4[x] \to \mathbb{F}_2[x]$ by $\mu(f(x)) = f(x) \pmod{2}$; that is, μ is determined by $\mu(0) = \mu(2) = 0$, $\mu(1) = \mu(3) = 1$, and $\mu(x) = x$. By Exercise 706, μ is a surjective ring homomorphism with kernel $(2) = \{2s(x) \mid s(x) \in \mathbb{Z}_4[x]\}$. In particular, this implies that if $f(x) \in \mathbb{F}_2[x]$, there is a $g(x) \in \mathbb{Z}_4[x]$ such that $\mu(g(x)) = f(x)$, and two such $g(x)$s differ by an element $2s(x)$ for some $s(x) \in \mathbb{Z}_4[x]$. The map μ is called the *reduction homomorphism*. We use these facts in what follows without reference.

Exercise 706 Prove that μ is a surjective ring homomorphism with kernel $(2) = \{2s(x) \mid s(x) \in \mathbb{Z}_4[x]\}$.   ◆

A polynomial $f(x) \in \mathbb{Z}_4[x]$ is *basic irreducible* if $\mu(f(x))$ is irreducible in $\mathbb{F}_2[x]$; it is *monic* if its leading coefficient is 1. An ideal $\mathcal{I}$ of a ring $\mathcal{R}$ is called a *primary ideal* provided $ab \in \mathcal{I}$ implies that either $a \in \mathcal{I}$ or $b^r \in \mathcal{I}$ for some positive integer r. A polynomial $f(x) \in \mathbb{Z}_4[x]$ is *primary* if the principal ideal $(f(x)) = \{f(x)g(x) \mid g(x) \in \mathbb{Z}_4[x]\}$ is a primary ideal.

Exercise 707 Examine the two factorizations of $x^4 - 1$ in Example 12.3.1. Which factors are basic irreducible polynomials and which are not?   ◆

Lemma 12.3.2 *If $f(x) \in \mathbb{Z}_4[x]$ is a basic irreducible polynomial, then $f(x)$ is a primary polynomial.*

Proof: Suppose that $g(x)h(x) \in (f(x))$. As $\mu(f(x))$ is irreducible, $d = \gcd(\mu(g(x)), \mu(f(x)))$ is either 1 or $\mu(f(x))$. If $d = 1$, then by the Euclidean Algorithm there exist polynomials $a(x)$ and $b(x)$ in $\mathbb{Z}_4[x]$ such that $\mu(a(x))\mu(g(x)) + \mu(b(x))\mu(f(x)) = 1$. Hence

$a(x)g(x) + b(x)f(x) = 1 + 2s(x)$ for some $s(x) \in \mathbb{Z}_4[x]$. Therefore $a(x)g(x)h(x)(1 + 2s(x)) + b(x)f(x)h(x)(1 + 2s(x)) = h(x)(1 + 2s(x))^2 = h(x)$, implying that $h(x) \in (f(x))$. Suppose now that $d = \mu(f(x))$. Then there exists $a(x) \in \mathbb{Z}_4[x]$ such that $\mu(g(x)) = \mu(f(x))\mu(a(x))$, implying that $g(x) = f(x)a(x) + 2s(x)$ for some $s(x) \in \mathbb{Z}_4[x]$. Hence $g(x)^2 = (f(x)a(x))^2 \in (f(x))$. Thus $f(x)$ is a primary polynomial. $\square$

If $\mathcal{R} = \mathbb{Z}_4[x]$ or $\mathbb{F}_2[x]$, two polynomials $f(x)$ and $g(x)$ in $\mathcal{R}$ are *coprime* or *relatively prime* provided $\mathcal{R} = (f(x)) + (g(x))$.

Exercise 708 Let $\mathcal{R} = \mathbb{Z}_4[x]$ or $\mathbb{F}_2[x]$ and suppose that $f(x)$ and $g(x)$ are polynomials in $\mathcal{R}$. Prove that $f(x)$ and $g(x)$ are coprime if and only if there exist $a(x)$ and $b(x)$ in $\mathcal{R}$ such that $a(x)f(x) + b(x)g(x) = 1$.   ♦

Exercise 709 Let $\mathcal{R} = \mathbb{Z}_4[x]$ or $\mathbb{F}_2[x]$. Let $k \geq 2$. Prove that if $f_i(x)$ are pairwise coprime polynomials in $\mathcal{R}$ for $1 \leq i \leq k$, then $f_1(x)$ and $f_2(x)f_3(x) \cdots f_k(x)$ are coprime.   ♦

Lemma 12.3.3 *Let $f(x)$ and $g(x)$ be polynomials in $\mathbb{Z}_4[x]$. Then $f(x)$ and $g(x)$ are coprime if and only if $\mu(f(x))$ and $\mu(g(x))$ are coprime polynomials in $\mathbb{F}_2[x]$.*

Proof: Suppose that $f(x)$ and $g(x)$ are coprime. By Exercise 708, $a(x)f(x) + b(x)g(x) = 1$ for some $a(x)$ and $b(x)$ in $\mathbb{Z}_4[x]$. Then $\mu(a(x))\mu(f(x)) + \mu(b(x))\mu(g(x)) = \mu(1) = 1$, implying that $\mu(f(x))$ and $\mu(g(x))$ are coprime. Conversely, suppose that $\mu(f(x))$ and $\mu(g(x))$ are coprime. Then there exist $a(x)$ and $b(x)$ in $\mathbb{Z}_4[x]$ such that $\mu(a(x))\mu(f(x)) + \mu(b(x))\mu(g(x)) = 1$. Thus $a(x)f(x) + b(x)g(x) = 1 + 2s(x)$ for some $s(x) \in \mathbb{Z}_4[x]$. But then $a(x)(1 + 2s(x))f(x) + b(x)(1 + 2s(x))g(x) = (1 + 2s(x))^2 = 1$ showing that $f(x)$ and $g(x)$ are coprime by Exercise 708. $\square$

The following, which is a special case of a result called Hensel's Lemma, shows how to get from a factorization of $\mu(f(x))$ to a factorization of $f(x)$.

Theorem 12.3.4 (Hensel's Lemma) *Let $f(x) \in \mathbb{Z}_4[x]$. Suppose $\mu(f(x)) = h_1(x)h_2(x) \cdots h_k(x)$, where $h_1(x), h_2(x), \ldots, h_k(x)$ are pairwise coprime polynomials in $\mathbb{F}_2[x]$. Then there exist $g_1(x), g_2(x), \ldots, g_k(x)$ in $\mathbb{Z}_4[x]$ such that:*
(i) *$\mu(g_i(x)) = h_i(x)$ for $1 \leq i \leq k$,*
(ii) *$g_1(x), g_2(x), \ldots, g_k(x)$ are pairwise coprime, and*
(iii) *$f(x) = g_1(x)g_2(x) \cdots g_k(x)$.*

Proof: The proof is by induction on k. Suppose $k = 2$. Choose $g_1'(x)$ and $g_2'(x)$ in $\mathbb{Z}_4[x]$ so that $\mu(g_1'(x)) = h_1(x)$ and $\mu(g_2'(x)) = h_2(x)$. So $f(x) = g_1'(x)g_2'(x) + 2s(x)$ for some $s(x) \in \mathbb{Z}_4[x]$. As $h_1(x)$ and $h_2(x)$ are coprime, so are $g_1'(x)$ and $g_2'(x)$ by Lemma 12.3.3. Thus by Exercise 708 there are polynomials $a_i(x) \in \mathbb{Z}_4[x]$ such that $a_1(x)g_1'(x) + a_2(x)g_2'(x) = 1$. Let $g_1(x) = g_1'(x) + 2a_2(x)s(x)$ and $g_2(x) = g_2'(x) + 2a_1(x)s(x)$. Then (i) and (ii) hold as $\mu(g_i(x)) = \mu(g_i'(x)) = h_i(x)$. Also $g_1(x)g_2(x) = g_1'(x)g_2'(x) + 2(a_1(x)g_1'(x) + a_2(x)g_2'(x))s(x) = g_1'(x)g_2'(x) + 2s(x) = f(x)$.

Now suppose $k = 3$. By Exercise 709, $h_1(x)$ and $h_2(x)h_3(x)$ are coprime. Thus using the case $k = 2$, there exist coprime polynomials $g_1(x)$ and $g_{23}(x)$ in $\mathbb{Z}_4[x]$ such that $\mu(g_1(x)) = h_1(x)$, $\mu(g_{23}(x)) = h_2(x)h_3(x)$, and $f(x) = g_1(x)g_{23}(x)$. Since $h_2(x)$ and $h_3(x)$ are coprime, again using the case $k = 2$, there are coprime polynomials $g_2(x)$ and $g_3(x)$

such that $\mu(g_2(x)) = h_2(x)$, $\mu(g_3(x)) = h_3(x)$, and $g_{23}(x) = g_2(x)g_3(x)$. This completes the case $k = 3$. Continuing inductively gives the result for all k.   □

Exercise 710 Let $a(x) \in \mathbb{Z}_4[x]$. Prove that the following are equivalent:
(a) $a(x)$ is invertible in $\mathbb{Z}_4[x]$.
(b) $a(x) = 1 + 2s(x)$ for some $s(x) \in \mathbb{Z}_4[x]$.
(c) $\mu(a(x)) = 1$. (Note that 1 is the only unit in $\mathbb{F}_2[x]$.)   ◆

Corollary 12.3.5 *Let $f(x) \in \mathbb{Z}_4[x]$. Suppose that all the roots of $\mu(f(x))$ are distinct. Then $f(x)$ is irreducible if and only if $\mu(f(x))$ is irreducible.*

Proof: If $\mu(f(x))$ is reducible, it factors into irreducible polynomials that are pairwise coprime as $\mu(f(x))$ has distinct roots. By Hensel's Lemma, $f(x)$ has a nontrivial factorization into basic irreducibles, which cannot be units by Exercise 710, implying that $f(x)$ is reducible. Suppose that $f(x) = g(x)h(x)$, where neither $g(x)$ nor $h(x)$ are units. Then $\mu(f(x)) = \mu(f(x))\mu(g(x))$ and neither $\mu(f(x))$ nor $\mu(g(x))$ are units by Exercise 710.   □

After presenting one final result without proof, we will be ready to discuss the factorization of $x^n - 1 \in \mathbb{Z}_4[x]$. This result, which is a special case of [230, Theorem XIII.6], applies to regular polynomials in $\mathbb{Z}_4[x]$. A *regular polynomial* in $\mathbb{Z}_4[x]$ is any nonzero polynomial that is not a divisor of zero.[2]

Lemma 12.3.6 *Let $f(x)$ be a regular polynomial in $\mathbb{Z}_4[x]$. Then there exist a monic polynomial $f'(x)$ and a unit $u(x)$ in $\mathbb{Z}_4[x]$ such that $\mu(f'(x)) = \mu(f(x))$ and $f'(x) = u(x)f(x)$.*

Since $\mathbb{F}_2[x]$ is a unique factorization domain, $\mu(x^n - 1) = x^n + 1 \in \mathbb{F}_2[x]$ has a factorization $h_1(x)h_2(x) \cdots h_k(x)$ into irreducible polynomials. These are pairwise coprime only if n is odd. So assuming that n is odd, by Hensel's Lemma, there is a factorization $x^n - 1 = g_1(x)g_2(x) \cdots g_k(x)$ into pairwise coprime basic irreducible polynomials $g_i(x) \in \mathbb{Z}_4[x]$ such that $\mu(g_i(x)) = h_i(x)$. By Lemma 12.3.2, each $g_i(x)$ is a primary polynomial. The polynomial $x^n - 1$ is not a divisor of zero in $\mathbb{Z}_4[x]$ and so is regular. Using the Factorization Theorem [230], which applies to regular polynomials that are factored into products of primary polynomials, this factorization of $x^n - 1$ into basic irreducible polynomials is unique up to multiplication by units. Furthermore, by Corollary 12.3.5, these basic irreducible polynomials are in fact irreducible. These $g_i(x)$ are also regular. Lemma 12.3.6 implies that since $g_i(x)$ is regular, there is a monic polynomial $g_i'(x) \in \mathbb{Z}_4[x]$ such that $\mu(g_i'(x)) = \mu(g_i(x))$ and $g_i'(x) = u_i(x)g_i(x)$ for some unit $u_i(x) \in \mathbb{Z}_4[x]$. Each $g_i'(x)$ is irreducible. Exercise 711 shows that $x^n - 1 = g_1'(x)g_2'(x) \cdots g_k'(x)$; hence we have shown that when n is odd, $x^n - 1$ can be factored into a unique product of monic irreducible polynomials in $\mathbb{Z}_4[x]$. Example 12.3.1 and Exercise 707 indicate why we cannot drop the condition that n be odd. This gives our starting point for discussing cyclic codes.

Theorem 12.3.7 *Let n be odd. Then $x^n - 1 = g_1(x)g_2(x) \cdots g_k(x)$ where $g_i(x) \in \mathbb{Z}_4[x]$ are unique monic irreducible (and basic irreducible) pairwise coprime polynomials in $\mathbb{Z}_4[x]$.*

---

[2] A nonzero element a in a ring $\mathcal{R}$ is a *divisor of zero* provided $ab = 0$ for some nonzero element $b \in \mathcal{R}$.

Furthermore, $x^n + 1 = \mu(g_1(x))\mu(g_2(x))\cdots\mu(g_k(x))$ *is a factorization into irreducible polynomials in* $\mathbb{F}_2[x]$.

Exercise 711 For n odd, $x^n - 1 = g_1(x)g_2(x)\cdots g_k(x)$ where $g_i(x)$ are irreducible, and hence regular. By Lemma 12.3.6, $x^n - 1 = a(x)g_1'(x)g_2'(x)\cdots g_k'(x)$, where $g_i'(x)$ are monic, $\mu(g_i'(x)) = \mu(g_i(x))$, and $a(x)$ is a unit. Show that $a(x) = 1$. Hint: See Exercise 710. ◆

Exercise 712 This illustrates Lemma 12.3.6. Let $g_i(x) = 2x + 1$.
(a) What is $\mu(g_i(x))$?
(b) Show that $g_i'(x) = 1$ satisfies both $\mu(g_i'(x)) = \mu(g_i(x))$ and $g_i'(x) = u_i(x)g_i(x)$ for the unit $u_i(x) = 2x + 1$. ◆

In order to factor $x^n - 1$ in $\mathbb{Z}_4[x]$, we first factor $x^n + 1$ in $\mathbb{F}_2[x]$ and then use Hensel's Lemma. The proof of Hensel's Lemma then gives a method for computing the factorization of $x^n - 1$ in $\mathbb{Z}_4[x]$. However, that method is too tedious. We introduce, without proof, another method that will produce the factorization; this method is due to Graeffe [334] but was adapted to $\mathbb{Z}_4$-linear codes in [318, Section 4 and Theorem 2]:

I. Let $h(x)$ be an irreducible factor of $x^n + 1$ in $\mathbb{F}_2[x]$. Write $h(x) = e(x) + o(x)$, where $e(x)$ is the sum of the terms of $h(x)$ with even exponents and $o(x)$ is the sum of the terms of $h(x)$ with odd exponents.
II. Then $g(x)$ is the irreducible factor of $x^n - 1$ in $\mathbb{Z}_4[x]$ with $\mu(g(x)) = h(x)$, where $g(x^2) = \pm(e(x)^2 - o(x)^2)$.

Example 12.3.8 In $\mathbb{F}_2[x]$, $x^7 + 1 = (x + 1)(x^3 + x + 1)(x^3 + x^2 + 1)$ is the factorization of $x^7 + 1$ into irreducible polynomials. We apply Graeffe's method to each factor to obtain the factorization of $x^7 - 1$ into monic irreducible polynomials of $\mathbb{Z}_4[x]$.
- If $h(x) = x + 1$, then $e(x) = 1$ and $o(x) = x$. So $g(x^2) = -(1 - x^2) = x^2 - 1$ and thus $g(x) = x - 1$.
- If $h(x) = x^3 + x + 1$, then $e(x) = 1$ and $o(x) = x^3 + x$. So $g(x^2) = -(1 - (x^3 + x)^2) = x^6 + 2x^4 + x^2 - 1$ and thus $g(x) = x^3 + 2x^2 + x - 1$.
- If $h(x) = x^3 + x^2 + 1$, then $e(x) = x^2 + 1$ and $o(x) = x^3$. So $g(x^2) = -((x^2 + 1)^2 - (x^3)^2) = x^6 - x^4 + 2x^2 - 1$ and thus $g(x) = x^3 - x^2 + 2x - 1$.
Therefore $x^7 - 1 = (x - 1)(x^3 + 2x^2 + x - 1)(x^3 - x^2 + 2x - 1)$ is the factorization of $x^7 - 1$ into monic irreducible polynomials in $\mathbb{Z}_4[x]$. ■

Exercise 713 Do the following:
(a) Verify that $x^9 + 1 = (x + 1)(x^2 + x + 1)(x^6 + x^3 + 1)$ is the factorization of $x^9 + 1$ into irreducible polynomials in $\mathbb{F}_2[x]$.
(b) Apply Graeffe's method to find the factorization of $x^9 - 1$ into monic irreducible polynomials in $\mathbb{Z}_4[x]$. ◆

Exercise 714 Do the following:
(a) Verify that $x^{15} + 1 = (x + 1)(x^2 + x + 1)(x^4 + x^3 + 1)(x^4 + x + 1)(x^4 + x^3 + x^2 + x + 1)$ is the factorization of $x^{15} + 1$ into irreducible polynomials in $\mathbb{F}_2[x]$.
(b) Apply Graeffe's method to find the factorization of $x^{15} - 1$ into monic irreducible polynomials in $\mathbb{Z}_4[x]$. ◆

Exercise 715 Do the following:

(a) Suppose that n is odd and the factorization of $x^n + 1$ into irreducible polynomials in $\mathbb{F}_2[x]$ is $x^n + 1 = (x + 1)(x^{n-1} + x^{n-2} + \cdots + x + 1)$. Show that the factorization of $x^n - 1$ into monic irreducible polynomials in $\mathbb{Z}_4[x]$ is $x^n - 1 = (x - 1)(x^{n-1} + x^{n-2} + \cdots + x + 1)$. Note that this does not require the use of Graeffe's method.

(b) Show that when $n = 3, 5, 11, 13$, and 19, the factorization of $x^n - 1$ in $\mathbb{Z}_4[x]$ is given by (a).   ♦

12.3.2 The ring $\mathfrak{R}_n = \mathbb{Z}_4[x]/(x^n - 1)$

To study cyclic codes over $\mathbb{F}_q$ we needed to find the ideals of $\mathbb{F}_q[x]/(x^n - 1)$. Similarly, we need to find the ideals of $\mathfrak{R}_n$ in order to study cyclic codes over $\mathbb{Z}_4$. We first need to know the ideal structure of $\mathbb{Z}_4[x]/(f(x))$, where $f(x)$ is a basic irreducible polynomial.

Lemma 12.3.9 *If $f(x) \in \mathbb{Z}_4[x]$ is a basic irreducible polynomial, then $\mathcal{R} = \mathbb{Z}_4[x]/(f(x))$ has only three ideals: (0), $(2) = \{2s(x) + (f(x)) \mid s(x) \in \mathbb{Z}_4[x]\}$, and $(1) = \mathcal{R}$.*

Proof: Suppose $\mathcal{I}$ is a nonzero ideal in $\mathcal{R}$. Let $g(x) + (f(x)) \in \mathcal{I}$ with $g(x) \notin (f(x))$. As $\mu(f(x))$ is irreducible, $\gcd(\mu(g(x)), \mu(f(x)))$ is either 1 or $\mu(f(x))$. Arguing as in the proof of Lemma 12.3.2, either $a(x)g(x) + b(x)f(x) = 1 + 2s(x)$ or $g(x) = f(x)a(x) + 2s(x)$ for some $a(x), b(x)$, and $s(x)$ in $\mathbb{Z}_4[x]$. In the former case, $a(x)g(x)(1 + 2s(x)) + b(x)f(x)(1 + 2s(x)) = (1 + 2s(x))^2 = 1$, implying that $g(x) + (f(x))$ has an inverse $a(x)(1 + 2s(x)) + (f(x))$ in $\mathcal{R}$. But this means that $\mathcal{I}$ contains an invertible element and hence equals $\mathcal{R}$ by Exercise 716. In the latter case, $g(x) + (f(x)) \in (2)$, and hence we may assume that $\mathcal{I} \subseteq (2)$. As $\mathcal{I} \neq \{0\}$, there exists $h(x) \in \mathbb{Z}_4[x]$ such that $2h(x) + (f(x)) \in \mathcal{I}$ with $2h(x) \notin (f(x))$. Assume that $\mu(h(x)) \in (\mu(f(x)))$; then $h(x) = f(x)a(x) + 2s(x)$ for some $a(x)$ and $s(x)$ in $\mathbb{Z}_4[x]$ implying $2h(x) = 2f(x)a(x) \in (f(x))$, a contradiction. Hence as $\mu(h(x)) \notin (\mu(f(x)))$ and $\mu(f(x))$ is irreducible, $\gcd(\mu(h(x)), \mu(f(x))) = 1$. Again, $a(x)h(x) + b(x)f(x) = 1 + 2s(x)$ for some $a(x)$, $b(x)$, and $s(x)$ in $\mathbb{Z}_4[x]$. Therefore $a(x)(2h(x)) + 2b(x)f(x) = 2$ showing that $2 + (f(x)) \in \mathcal{I}$ and so $(2) \subseteq \mathcal{I}$. Thus $\mathcal{I} = (2)$.   □

Exercise 716 Prove that if $\mathcal{I}$ is an ideal of a ring $\mathcal{R}$ and $\mathcal{I}$ contains an invertible element, then $\mathcal{I} = \mathcal{R}$.   ♦

We do not have the Division Algorithm in $\mathbb{Z}_4[x]$ and so the next result is not as obvious as the corresponding result in $\mathbb{F}_q[x]$.

Lemma 12.3.10 *Let $m(x)$ be a monic polynomial of degree r in $\mathbb{Z}_4[x]$. Then $\mathbb{Z}_4[x]/(m(x))$ has 4^r elements, and every element of $\mathbb{Z}_4[x]/(m(x))$ is uniquely expressible in the form $a(x) + (m(x))$, where $a(x)$ is the zero polynomial or has degree less than r.*

Proof: Let $a(x) = a_d x^d + a_{d-1} x^{d-1} + \cdots + a_0 \in \mathbb{Z}_4[x]$. If $d \geq r$, then $b(x) = a(x) - a_d x^{d-r} m(x)$ has degree less than d and $a(x) + (m(x)) = b(x) + (m(x))$. Hence every element $a(x) + (m(x))$ of $\mathbb{Z}_4[x]/(m(x))$ has a representative of degree less than r. Furthermore, if $b_1(x) + (m(x)) = b_2(x) + (m(x))$ with $b_i(x)$ each of degree less than r, then $b_1(x) - b_2(x)$ is a multiple of $m(x)$. By Exercise 717, $b_1(x) = b_2(x)$. The result follows.   □

Exercise 717 Let $m(x)$ be a monic polynomial of degree r and $h(x)$ a nonzero polynomial of degree s, where $m(x)$ and $h(x)$ are in $\mathbb{Z}_4[x]$. Prove that $h(x)m(x)$ has degree $r + s$.   ♦

After one more lemma we are ready to give the ideal structure of $\mathfrak{R}_n$. It is these ideals that produce cyclic codes. A principal ideal of $\mathfrak{R}_n$ is generated by an element $g(x) + (x^n - 1)$; to avoid confusion with the notation for ideals of $\mathbb{Z}_4[x]$, we denote this principal ideal by $\langle g(x) \rangle$. (Note that we should actually use $\langle g(x) + (x^n - 1) \rangle$, but we choose to drop this more cumbersome notation. Similarly, we will often say that a polynomial $a(x)$ is in $\mathfrak{R}_n$ rather than the more accurate expression that the coset $a(x) + (x^n - 1)$ is in $\mathfrak{R}_n$.)

Lemma 12.3.11 *Let n be odd. Suppose $m(x)$ is a monic polynomial of degree r which is a product of distinct irreducible factors of $x^n - 1$ in $\mathbb{Z}_4[x]$. Then the ideal $\langle m(x) \rangle$ of $\mathfrak{R}_n$ has 4^{n-r} elements and every element of $\langle m(x) \rangle$ is uniquely expressible in the form $m(x)a(x)$, where $a(x)$ is the zero polynomial or has degree less than $n - r$.*

Proof: By Theorem 12.3.7 and Exercise 717, there exists a monic polynomial $h(x)$ of degree $n - r$ such that $m(x)h(x) = x^n - 1$. Every element $f(x)$ of $\langle m(x) \rangle$ has the form $m(x)a(x)$. Let $a(x)$ be chosen to be of smallest degree such that $f(x) = m(x)a(x)$ in $\mathfrak{R}_n$. If $a(x) = a_d x^d + a_{d-1}x^{d-1} + \cdots + a_0$ with $a_d \neq 0$ and $d \geq n - r$, then $b(x) = a(x) - a_d x^{d-n+r}h(x)$ has degree less than d. So $m(x)b(x) = m(x)a(x) - a_d x^{d-n+r}(x^n - 1)$ in $\mathbb{Z}_4[x]$ implying that $f(x) = m(x)a(x) = m(x)b(x)$ in $\mathfrak{R}_n$, contradicting the choice of $a(x)$. Thus every element of $\langle m(x) \rangle$ has the form $m(x)a(x)$ where $a(x)$ has degree less than $n - r$. Furthermore, if $m(x)a_1(x) = m(x)a_2(x)$ in $\mathfrak{R}_n$ where each $a_i(x)$ has degree less than $n - r$, then $m(x)(a_1(x) - a_2(x))$ has degree less than n in $\mathbb{Z}_4[x]$ and is a multiple of $x^n - 1$, contradicting Exercise 717 unless $a_1(x) = a_2(x)$.   □

Theorem 12.3.12 *Let n be odd and let $x^n - 1 = g_1(x)g_2(x) \cdots g_k(x)$ be a factorization of $x^n - 1$ into pairwise coprime monic irreducible polynomials in $\mathbb{Z}_4[x]$. Let $\widehat{g}_i(x) = \prod_{j \neq i} g_j(x)$. Suppose that $g_i(x)$ has degree d_i. The following hold:*
(i) *$\mathfrak{R}_n$ has 4^n elements.*
(ii) *If $1 \leq i \leq k$, $\mathfrak{R}_n = \langle g_i(x) \rangle \oplus \langle \widehat{g}_i(x) \rangle$.*
(iii) *$\mathfrak{R}_n = \langle \widehat{g}_1(x) \rangle \oplus \langle \widehat{g}_2(x) \rangle \oplus \cdots \oplus \langle \widehat{g}_k(x) \rangle$.*
(iv) *If $1 \leq i \leq k$, $\langle \widehat{g}_i(x) \rangle = \langle \widehat{e}_i(x) \rangle$ where $\{\widehat{e}_i(x) \mid 1 \leq i \leq k\}$ are pairwise orthogonal idempotents of $\mathfrak{R}_n$, and $\sum_{i=1}^{k} \widehat{e}_i(x) = 1$.*
(v) *If $1 \leq i \leq k$, $\langle \widehat{g}_i(x) \rangle \simeq \mathbb{Z}_4[x]/(g_i(x))$ and $\langle \widehat{g}_i(x) \rangle$ has 4^{d_i} elements.*
(vi) *Every ideal of $\mathfrak{R}_n$ is a direct sum of $\langle \widehat{g}_i(x) \rangle$s and $\langle 2\widehat{g}_i(x) \rangle$s.*

Proof: Part (i) follows from Lemma 12.3.10. By Exercise 709, $(g_i(x)) + (\widehat{g}_i(x)) = \mathbb{Z}_4[x]$. Thus $\langle g_i(x) \rangle + \langle \widehat{g}_i(x) \rangle = \mathfrak{R}_n$. By Lemma 12.3.11, $\langle g_i(x) \rangle$ and $\langle \widehat{g}_i(x) \rangle$ have sizes 4^{n-d_i} and 4^{d_i}, respectively. Thus by (i) the sum $\langle g_i(x) \rangle + \langle \widehat{g}_i(x) \rangle = \mathfrak{R}_n$ must be direct, proving (ii).

To prove (iii) note that for $1 \leq j \leq k$

$$\sum_{i \neq j} \langle \widehat{g}_i(x) \rangle \bigcap \langle \widehat{g}_j(x) \rangle \subseteq \langle g_j(x) \rangle \cap \langle \widehat{g}_j(x) \rangle = \{0\};$$

the containment follows because each $\widehat{g}_i(x)$ with $i \neq j$ is a multiple of $g_j(x)$, and the equality follows from (ii). Thus the sum $\langle \widehat{g}_1(x) \rangle + \langle \widehat{g}_2(x) \rangle + \cdots + \langle \widehat{g}_k(x) \rangle$ is direct. Since

$\langle \widehat{g}_i(x) \rangle$ has 4^{d_i} elements by Lemma 12.3.11 and $d_1 + d_2 + \cdots + d_k = n$, (i) implies (iii). By part (iii)

$$1 = \sum_{i=1}^{k} \widehat{e}_i(x) \quad \text{in } \mathfrak{R}_n, \tag{12.10}$$

where $\widehat{e}_i(x) \in \langle \widehat{g}_i(x) \rangle$. As $\widehat{e}_i(x)\widehat{e}_j(x) \in \langle \widehat{g}_i(x) \rangle \cap \langle \widehat{g}_j(x) \rangle = \{0\}$ if $i \neq j$, multiplying (12.10) by $\widehat{e}_i(x)$ shows that $\{\widehat{e}_i(x) \mid 1 \leq i \leq k\}$ are pairwise orthogonal idempotents of $\mathfrak{R}_n$. Multiplying (12.10) by $\widehat{g}_i(x)$ shows that $\widehat{g}_i(x) = \widehat{g}_i(x)\widehat{e}_i(x)$, implying that $\langle \widehat{g}_i(x) \rangle = \langle \widehat{e}_i(x) \rangle$, verifying (iv). We leave part (v) as Exercise 718.

Let $\mathcal{I}$ be an ideal of $\mathfrak{R}_n$. If $a(x) \in \mathcal{I}$, multiplying (12.10) by $a(x)$ shows that $a(x) = \sum_{i=1}^{k} a(x)\widehat{e}_i(x)$. Since $a(x)\widehat{e}_i(x) \in \langle \widehat{g}_i(x) \rangle$, $\mathcal{I} = \sum_{i=1}^{k} \mathcal{I} \cap \langle \widehat{g}_i(x) \rangle$. By (v) and Lemma 12.3.9, part (vi) follows. $\square$

Exercise 718 Use the notation of Theorem 12.3.12.
(a) Prove that $\phi : \mathbb{Z}_4[x]/\langle g_i(x) \rangle \to \langle \widehat{g}_i(x) \rangle = \langle \widehat{e}_i(x) \rangle$ given by $\phi(a(x) + \langle g_i(x) \rangle) = a(x)\widehat{e}_i(x)$ is a well-defined ring homomorphism.
(b) Prove that ϕ is one-to-one and onto. Hint: Use the fact that $\langle \widehat{g}_i(x) \rangle = \langle \widehat{e}_i(x) \rangle$ along with Lemmas 12.3.10 and 12.3.11. ◆

12.3.3 Generating polynomials of cyclic codes over $\mathbb{Z}_4$

When n is odd we can give a pair of polynomials that "generate" any cyclic code of length n over $\mathbb{Z}_4$. The following result was proved in [279, 289].

Theorem 12.3.13 *Let C be a cyclic code over $\mathbb{Z}_4$ of odd length n. Then there exist unique monic polynomials $f(x)$, $g(x)$, and $h(x)$ such that $x^n - 1 = f(x)g(x)h(x)$ and $C = \langle f(x)g(x) \rangle \oplus \langle 2f(x)h(x) \rangle$. Furthermore, C has type $4^{\deg h} 2^{\deg g}$.*

Proof: Using the notation of Theorem 12.3.12, C is the direct sum of some $\langle \widehat{g}_i(x) \rangle$s and some $\langle 2\widehat{g}_i(x) \rangle$s where $x^n - 1 = g_1(x)g_2(x) \cdots g_k(x)$. Rearrange the $g_i(x)$ so that $C = \sum_{i=1}^{a} \langle \widehat{g}_i(x) \rangle \oplus \sum_{i=a+1}^{b} \langle 2\widehat{g}_i(x) \rangle$. Let $f(x) = \prod_{i=b+1}^{k} g_i(x)$, $g(x) = \prod_{i=a+1}^{b} g_i(x)$, and $h(x) = \prod_{i=1}^{a} g_i(x)$. Then $x^n - 1 = f(x)g(x)h(x)$. Since $\widehat{g}_i(x)$ is a multiple of $f(x)g(x)$ for $1 \leq i \leq a$, we have $\sum_{i=1}^{a} \langle \widehat{g}_i(x) \rangle \subseteq \langle f(x)g(x) \rangle$. By Lemma 12.3.11, $\langle f(x)g(x) \rangle$ has size $4^{n-\deg(fg)} = 4^{\deg h}$, which is also the size of $\sum_{i=1}^{a} \langle \widehat{g}_i(x) \rangle$ by Theorem 12.3.12(iii) and (v). Hence $\sum_{i=1}^{a} \langle \widehat{g}_i(x) \rangle = \langle f(x)g(x) \rangle$. Using the same argument and Exercise 719, $\sum_{i=a+1}^{b} \langle 2\widehat{g}_i(x) \rangle = \langle 2f(x)h(x) \rangle$. The uniqueness follows since any monic polynomial factor of $x^n - 1$ must factor into a product of basic irreducibles that must then be a product of some of the $g_i(x)$s that are unique by Theorem 12.3.7. The type of C follows from the fact that $\langle f(x)g(x) \rangle$ has size $4^{\deg h}$ and $\langle 2f(x)h(x) \rangle$ has size $2^{\deg g}$. $\square$

Exercise 719 Let n be odd and $m(x)$ a monic product of irreducible factors of $x^n - 1$ in $\mathbb{Z}_4[x]$ of degree r. Prove that the ideal $\langle 2m(x) \rangle$ of $\mathfrak{R}_n$ has 2^{n-r} elements and every element of $\langle 2m(x) \rangle$ is uniquely expressible in the form $2m(x)a(x)$, where $a(x)$ has degree less than $n - r$ with coefficients 0 and 1 only. ◆

Corollary 12.3.14 *With the notation as in Theorem 12.3.13, if $g(x) = 1$, then $C = \langle f(x) \rangle$, and C has type $4^{n-\deg f}$. If $h(x) = 1$, then $C = \langle 2f(x) \rangle$, and C has type $2^{n-\deg f}$.*

Exercise 720 Prove Corollary 12.3.14.   ♦

Corollary 12.3.15 *Let n be odd. Assume that $x^n - 1$ is a product of k irreducible polynomials in $\mathbb{Z}_4[x]$. Then there are 3^k cyclic codes over $\mathbb{Z}_4$ of length n.*

Proof: Let $x^n - 1 = g_1(x)g_2(x) \cdots g_k(x)$ be the factorization of $x^n - 1$ into monic irreducible polynomials. If C is a cyclic code, by Theorem 12.3.13, $C = \langle f(x)g(x) \rangle \oplus \langle 2f(x)h(x) \rangle$ where $x^n - 1 = f(x)g(x)h(x)$. Each $g_i(x)$ is a factor of exactly one of $f(x)$, $g(x)$, or $h(x)$. The result follows.   □

Example 12.3.16 By Example 12.3.8, $x^7 - 1 = g_1(x)g_2(x)g_3(x)$, where $g_1(x) = x - 1$, $g_2(x) = x^3 + 2x^2 + x - 1$, and $g_3(x) = x^3 - x^2 + 2x - 1$ are the monic irreducible factors of $x^7 - 1$. By Corollary 12.3.15, there are $3^3 = 27$ cyclic codes over $\mathbb{Z}_4$ of length 7. In Table 12.1 we give the generator polynomials of the 25 nontrivial cyclic codes of length 7 as described in Theorem 12.3.13. Each generator is given as a product of some of the $g_i(x)$s. If a single generator a is given, then the code is $\langle a(x) \rangle$; these are the codes that come from Corollary 12.3.14. If a pair $(a, 2b)$ of generators is given, then the code is $\langle a(x) \rangle \oplus \langle 2b(x) \rangle$. The information on idempotent generators and duality is discussed later.   ∎

Exercise 721 Verify that all the nontrivial cyclic codes of length 7 over $\mathbb{Z}_4$ have been accounted for in Table 12.1, and show that the codes with given generator polynomials have the indicated types.   ♦

If $C = \langle f(x)g(x) \rangle \oplus \langle 2f(x)h(x) \rangle$ as in Theorem 12.3.13, we can easily write down a generator matrix G for C. The first $\deg h$ rows of G correspond to $x^i f(x)g(x)$ for $0 \le i \le \deg h - 1$. The last $\deg g$ rows of G correspond to $2x^i f(x)h(x)$ for $0 \le i \le \deg g - 1$.

Example 12.3.17 Consider code No. 22 in Table 12.1. Since $g_1(x)g_3(x) = 1 + x + 3x^2 + 2x^3 + x^4$ and $2g_2(x) = 2 + 2x + 2x^3$, one generator matrix for this code is

$$
\begin{bmatrix}
1 & 1 & 3 & 2 & 1 & 0 & 0 \\
0 & 1 & 1 & 3 & 2 & 1 & 0 \\
0 & 0 & 1 & 1 & 3 & 2 & 1 \\
2 & 2 & 0 & 2 & 0 & 0 & 0 \\
0 & 2 & 2 & 0 & 2 & 0 & 0 \\
0 & 0 & 2 & 2 & 0 & 2 & 0 \\
0 & 0 & 0 & 2 & 2 & 0 & 2
\end{bmatrix}.
$$
    ∎

As with cyclic codes over $\mathbb{F}_q$, we can find generator polynomials of the dual codes. Let $f(x) = a_d x^d + a_{d-1} x^{d-1} + \cdots + a_0 \in \mathbb{Z}_4[x]$ with $a_d \ne 0$. Define the *reciprocal polynomial* $f^*(x)$ to be

$$f^*(x) = \pm x^d f(x^{-1}) = \pm(a_0 x^d + a_1 x^{d-1} + \cdots + a_d).$$

Table 12.1 *Generators of cyclic codes over $\mathbb{Z}_4$ of length 7*

Code number	Generator polynomials	"Generating idempotents"	Type	Dual code
1	$g_2 g_3$	$\widehat{e}_1$	4	6
2	$g_1 g_2$	$\widehat{e}_3$	4^3	4
3	$g_1 g_3$	$\widehat{e}_2$	4^3	5
4	g_2	$\widehat{e}_1 + \widehat{e}_3$	4^4	2
5	g_3	$\widehat{e}_1 + \widehat{e}_2$	4^4	3
6	g_1	$\widehat{e}_2 + \widehat{e}_3$	4^6	1
7	$2 g_2 g_3$	$2\widehat{e}_1$	2	25
8	$2 g_1 g_2$	$2\widehat{e}_3$	2^3	23
9	$2 g_1 g_3$	$2\widehat{e}_2$	2^3	24
10	$2 g_2$	$2\widehat{e}_1 + 2\widehat{e}_3$	2^4	21
11	$2 g_3$	$2\widehat{e}_1 + 2\widehat{e}_2$	2^4	22
12	$2 g_1$	$2\widehat{e}_2 + 2\widehat{e}_3$	2^6	16
13	2	2	2^7	self-dual
14	$(g_2 g_3, 2 g_2 g_1)$	$\widehat{e}_1 + 2\widehat{e}_3$	$4 \cdot 2^3$	19
15	$(g_3 g_2, 2 g_3 g_1)$	$\widehat{e}_1 + 2\widehat{e}_2$	$4 \cdot 2^3$	20
16	$(g_2 g_3, 2 g_1)$	$\widehat{e}_1 + 2\widehat{e}_2 + 2\widehat{e}_3$	$4 \cdot 2^6$	12
17	$(g_2 g_1, 2 g_2 g_3)$	$\widehat{e}_3 + 2\widehat{e}_1$	$4^3 2$	self-dual
18	$(g_3 g_1, 2 g_3 g_2)$	$\widehat{e}_2 + 2\widehat{e}_1$	$4^3 2$	self-dual
19	$(g_1 g_2, 2 g_1 g_3)$	$\widehat{e}_3 + 2\widehat{e}_2$	$4^3 2^3$	14
20	$(g_1 g_3, 2 g_1 g_2)$	$\widehat{e}_2 + 2\widehat{e}_3$	$4^3 2^3$	15
21	$(g_1 g_2, 2 g_3)$	$\widehat{e}_3 + 2\widehat{e}_1 + 2\widehat{e}_2$	$4^3 2^4$	10
22	$(g_1 g_3, 2 g_2)$	$\widehat{e}_2 + 2\widehat{e}_1 + 2\widehat{e}_3$	$4^3 2^4$	11
23	$(g_2, 2 g_1 g_3)$	$\widehat{e}_1 + \widehat{e}_3 + 2\widehat{e}_2$	$4^4 2^3$	8
24	$(g_3, 2 g_1 g_2)$	$\widehat{e}_1 + \widehat{e}_2 + 2\widehat{e}_3$	$4^4 2^3$	9
25	$(g_1, 2 g_2 g_3)$	$\widehat{e}_2 + \widehat{e}_3 + 2\widehat{e}_1$	$4^6 2$	7

We choose the $\pm$ sign so that the leading coefficient of $f^*(x)$ is 1 or 2. The proof of Lemma 4.4.8 also proves the following lemma.

Lemma 12.3.18 *Let $\mathbf{a} = a_0 a_1 \cdots a_{n-1}$ and $\mathbf{b} = b_0 b_1 \cdots b_{n-1}$ be vectors in $\mathbb{Z}_4^n$ with associated polynomials $a(x)$ and $b(x)$. Then $\mathbf{a}$ is orthogonal to $\mathbf{b}$ and all its shifts if and only if $a(x)b^*(x) = 0$ in $\mathfrak{R}_n$.*

We observe that if n is odd and $g_i(x)$ is a monic irreducible factor of $x^n - 1$ in $\mathbb{Z}_4[x]$, then $g_i^*(x)$ is also a monic irreducible factor of $x^n - 1$ as the constant term of $g_i(x)$ is ± 1. The reciprocal polynomial $\mu(g_i^*(x))$ of $\mu(g_i(x))$ in $\mathbb{F}_2[x]$ is an irreducible factor of $x^n + 1$ in $\mathbb{F}_2[x]$. Thus $g_i^*(x)$ is a basic irreducible factor of $x^n - 1$. By the unique factorization of $x^n - 1$ from Theorem 12.3.7, $g_i^*(x) = g_j(x)$ for some j since $g_i^*(x)$ is monic.

Example 12.3.19 In the notation of Example 12.3.16, $g_1^*(x) = g_1(x)$, $g_2^*(x) = g_3(x)$, and $g_3^*(x) = g_2(x)$. ∎

Theorem 12.3.20 *If $C = \langle f(x)g(x)\rangle \oplus \langle 2f(x)h(x)\rangle$ is a cyclic code of odd length n over $\mathbb{Z}_4$ with $f(x)g(x)h(x) = x^n - 1$, then $C^\perp = \langle h^*(x)g^*(x)\rangle \oplus \langle 2h^*(x)f^*(x)\rangle$. Furthermore, if $g(x) = 1$, then $C = \langle f(x)\rangle$ and $C^\perp = \langle h^*(x)\rangle$; if $h(x) = 1$, then $C = \langle 2f(x)\rangle$ and $C^\perp = \langle g^*(x)\rangle \oplus \langle 2f^*(x)\rangle$.*

Proof: Let $s(x) = h^*(x)g^*(x)$; then $s^*(x) = \pm h(x)g(x)$ and $f(x)g(x)s^*(x) = \pm(x^n - 1) \times g(x)$ in $\mathbb{Z}_4[x]$, implying $f(x)g(x)s^*(x) = 0$ in $\mathfrak{R}_n$. In the same way $2f(x)h(x)s^*(x) = 0$. Thus $h^*(x)g^*(x) \in C^\perp$ by Lemma 12.3.18. Likewise $2h^*(x)f^*(x)$ is orthogonal to $f(x)g(x)$ and its cyclic shifts. As $2a(x)2b^*(x) = 0$ in $\mathfrak{R}_n$ for all $a(x)$ and $b(x)$ in $\mathbb{Z}_4[x]$, $2h^*(x)f^*(x)$ is orthogonal to $2f(x)h(x)$ and its cyclic shifts. Thus $2h^*(x)f^*(x) \in C^\perp$. As $x^n - 1 = f^*(x)g^*(x)h^*(x)$, by Theorem 12.3.13, $\langle h^*(x)g^*(x)\rangle + \langle 2h^*(x)f^*(x)\rangle$ is a direct sum. Thus

$$\langle h^*(x)g^*(x)\rangle \oplus \langle 2h^*(x)f^*(x)\rangle \subseteq C^\perp. \tag{12.11}$$

Since C has type $4^{\deg h}2^{\deg g}$, $C^\perp$ has type $4^{n-\deg h-\deg g}2^{\deg g}$ by (12.3). But $\langle h^*(x)g^*(x)\rangle \oplus \langle 2h^*(x)f^*(x)\rangle$ has type $4^{\deg f^*}2^{\deg g^*} = 4^{\deg f}2^{\deg g}$. As these types agree, we have equality in (12.11) proving the first statement. The second statement follows from the first by direct computation as $g(x) = 1$ implies $g^*(x) = 1$ and $h^*(x)f^*(x) = x^n - 1$, while $h(x) = 1$ implies $h^*(x) = 1$. $\qquad\square$

Example 12.3.21 Code No. 14 in Table 12.1 has $f(x) = g_2(x)$, $g(x) = g_3(x)$, and $h(x) = g_1(x)$. Hence its dual is $\langle g_1^*(x)g_3^*(x)\rangle \oplus \langle 2g_1^*(x)g_2^*(x)\rangle = \langle g_1(x)g_2(x)\rangle \oplus \langle 2g_1(x)g_3(x)\rangle$, which is code No. 19. Code No. 22 has $f(x) = 1$, $g(x) = g_1(x)g_3(x)$, and $h(x) = g_2(x)$. So its dual is $\langle g_2^*(x)g_1^*(x)g_3^*(x)\rangle \oplus \langle 2g_2^*(x)\rangle = \langle 2g_3(x)\rangle$, which is code No. 11. Code No. 17 has $f(x) = g_2(x)$, $g(x) = g_1(x)$, and $h(x) = g_3(x)$. Its dual is $\langle g_3^*(x)g_1^*(x)\rangle \oplus \langle 2g_3^*(x)g_2^*(x)\rangle = \langle g_2(x)g_1(x)\rangle \oplus \langle 2g_2(x)g_3(x)\rangle$, making the code self-dual. There are three cyclic self-dual codes of length 7. $\blacksquare$

Exercise 722 Verify that the duals of the cyclic codes of length 7 over $\mathbb{Z}_4$ are as indicated in Table 12.1. $\qquad\blacklozenge$

12.3.4 Generating idempotents of cyclic codes over $\mathbb{Z}_4$

Let C be a nonzero cyclic code over $\mathbb{Z}_4$ of odd length n. By Theorem 12.3.13 there exist unique monic polynomials $f(x)$, $g(x)$, and $h(x)$ such that $x^n - 1 = f(x)g(x)h(x)$ and $C = \langle f(x)g(x)\rangle \oplus \langle 2f(x)h(x)\rangle$. By Theorem 12.3.12(vi), $\langle f(x)g(x)\rangle$ and $\langle 2f(x)h(x)\rangle$ can be expressed as a direct sum of ideals of the forms $\langle \widehat{g}_i(x)\rangle$ and $\langle 2\widehat{g}_i(x)\rangle$, respectively, where $x^n - 1 = g_1(x)g_2(x)\cdots g_k(x)$ is a factorization of $x^n - 1$ into irreducible polynomials in $\mathbb{Z}_4[x]$ and $\widehat{g}_i(x) = \prod_{j\neq i} g_j(x)$. Thus $\langle f(x)g(x)\rangle = \sum_{i\in\mathcal{I}} \langle \widehat{g}_i(x)\rangle$ and $\langle 2g(x)h(x)\rangle = \sum_{j\in\mathcal{J}} \langle 2\widehat{g}_j(x)\rangle$ for some subsets $\mathcal{I}$ and $\mathcal{J}$ of $\{1, 2, \ldots, k\}$. Theorem 12.3.12(iv) shows that $\langle \widehat{g}_i(x)\rangle = \langle \widehat{e}_i(x)\rangle$, where $\widehat{e}_i(x)$ is an idempotent of $\mathfrak{R}_n$. By Exercise 723(a) and (b), $\langle f(x)g(x)\rangle = \langle e(x)\rangle$ and $\langle g(x)h(x)\rangle = \langle E(x)\rangle$, where $e(x)$ and $E(x)$ are idempotents in $\mathfrak{R}_n$. This proves the following theorem.

Theorem 12.3.22 *Let $C = \langle f(x)g(x)\rangle \oplus \langle 2f(x)h(x)\rangle$ be a nonzero cyclic code over $\mathbb{Z}_4$ of odd length n where $x^n - 1 = f(x)g(x)h(x)$. Then:*

(i)   *if* $g(x) = 1, C = \langle f(x) \rangle = \langle e(x) \rangle$,
(ii) *if* $h(x) = 1, C = \langle 2f(x) \rangle = \langle 2E(x) \rangle$, *and*
(iii) *if* $g(x) \neq 1$ *and* $h(x) \neq 1, C = \langle f(x)g(x) \rangle \oplus \langle 2f(x)h(x) \rangle = \langle e(x) \rangle \oplus \langle 2E(x) \rangle$,
where $e(x)$ *and* $E(x)$ *are nonzero idempotents in* $\mathfrak{R}_n$.

Exercise 723 Let $I \subseteq \{1, 2, \ldots, k\}$ and let $I^c = \{1, 2, \ldots, k\} \setminus I$ be the complement of I.
Do the following:
(a) Prove that $\sum_{i \in I} \widehat{e}_i(x)$ is an idempotent in $\mathfrak{R}_n$. Hint: Recall that $\{\widehat{e}_i(x) \mid 1 \leq i \leq k\}$ are
    pairwise orthogonal idempotents.
(b) Prove that $\sum_{i \in I} \langle \widehat{g}_i(x) \rangle = \langle \sum_{i \in I} \widehat{e}_i(x) \rangle$.
(c) Prove that $\langle \prod_{j \in I^c} g_j(x) \rangle = \langle \sum_{i \in I} \widehat{e}_i(x) \rangle$. Hint: See the proof of Theorem 12.3.13.   ◆

By Theorem 12.3.22, $C = \langle e(x) \rangle \oplus \langle 2E(x) \rangle$, where $e(x)$ and $E(x)$ are idempotents. By
Exercise 724, $C = \langle e(x) + 2E(x) \rangle$. We call $e(x) + 2E(x)$ the "generating idempotent" of C.
We will use quote marks because $e(x) + 2E(x)$ is not really an idempotent (unless $E(x) =$
0); it is appropriate to use this term, however, because $e(x)$ and $E(x)$ are idempotents that
in combination generate C. We will use the term generating idempotent, without quotation
marks, when $E(x) = 0$ and hence C is actually generated by an idempotent (case (i) of
Theorem 12.3.22).

Exercise 724 Let $C = \langle e(x) \rangle \oplus \langle 2E(x) \rangle$, where $e(x)$ and $E(x)$ are idempotents. Show that
$C = \langle e(x) + 2E(x) \rangle$.   ◆

Recall that a multiplier μ_a defined on $\{0, 1, \ldots, n-1\}$ by $i\mu_a \equiv ia \pmod{n}$ is a permutation of the coordinate positions $\{0, 1, \ldots, n-1\}$ of a cyclic code of length n provided $\gcd(a, n) = 1$. We can apply multipliers to $\mathbb{Z}_4$-linear codes. As with cyclic codes
over $\mathbb{F}_q$, μ_a acts on $\mathfrak{R}_n$ by $f(x)\mu_a \equiv f(x^a) \pmod{x^n - 1}$ for $f(x) \in \mathfrak{R}_n$. See (4.4). Also
$(f(x)g(x))\mu_a = (f(x)\mu_a)(g(x)\mu_a)$ for $f(x)$ and $g(x)$ in $\mathfrak{R}_n$. This implies that if $e(x)$ is
an idempotent in $\mathfrak{R}_n$, so is $e(x)\mu_a$. In some circumstances we can compute the generating
idempotents of the dual of a cyclic code over $\mathbb{Z}_4$ and the intersection and sum of two such
codes; compare this to Theorems 4.3.7 and 4.4.9.

Lemma 12.3.23 *The following hold:*
(i) *Let* $C = \langle e(x) \rangle$ *be a* $\mathbb{Z}_4$-*linear cyclic code with generating idempotent* $e(x)$. *Then the*
     generating idempotent for $C^\perp$ *is* $1 - e(x)\mu_{-1}$.
(ii) *For* $i = 1$ *and* 2 *let* $C_i = \langle e_i(x) \rangle$ *be* $\mathbb{Z}_4$-*linear cyclic codes with generating idempotents*
     $e_i(x)$. *Then the generating idempotent for* $C_1 \cap C_2$ *is* $e_1(x)e_2(x)$, *and the generating*
     idempotent for $C_1 + C_2$ *is* $e_1(x) + e_2(x) - e_1(x)e_2(x)$.

Proof: Since $e(x)(1 - e(x)) = 0$ in $\mathfrak{R}_n$, $e(x)$ is orthogonal to the reciprocal polynomial of
$1 - e(x)$ and all of its shifts by Lemma 12.3.18; but these are merely scalar multiples of the
cyclic shifts of $1 - e(x)\mu_{-1}$. Thus $1 - e(x)\mu_{-1} \in C^\perp$. In the notation of Theorem 12.3.13,
C has generator polynomial $f(x)$ where $g(x) = 1$ and $f(x)h(x) = x^n - 1$ because C has
a generating idempotent (as opposed to being generated by an idempotent plus twice an
idempotent). By Theorem 12.3.20, $C^\perp$ has generator polynomial $h^*(x)$; however, $1 - e(x)$

is the generating idempotent of $\langle h(x) \rangle$ by Theorem 12.3.12. Hence $1 - e(x)\mu_{-1}$ is the generating idempotent of $\langle h^*(x) \rangle = \mathcal{C}^\perp$.

For (ii), clearly, $e_1(x)e_2(x) \in \mathcal{C}_1 \cap \mathcal{C}_2$. Thus $\langle e_1(x)e_2(x) \rangle \subseteq \mathcal{C}_1 \cap \mathcal{C}_2$. If $c(x) \in \mathcal{C}_1 \cap \mathcal{C}_2$, then $c(x) = s(x)e_1(x)$ and $c(x)e_2(x) = c(x)$, the latter by Exercise 725(a); thus $c(x) = c(x)e_2(x) = s(x)e_1(x)e_2(x) \in \langle e_1(x)e_2(x) \rangle$ and $\mathcal{C}_1 \cap \mathcal{C}_2 = \langle e_1(x)e_2(x) \rangle$. Next if $c(x) = c_1(x) + c_2(x)$ where $c_i(x) \in \mathcal{C}_i$ for $i = 1$ and 2, then by Exercise 725(a), $c(x)(e_1(x) + e_2(x) - e_1(x)e_2(x)) = c_1(x) + c_1(x)e_2(x) - c_1(x)e_2(x) + c_2(x)e_1(x) + c_2(x) - c_2(x)e_1(x) = c(x)$. Since $e_1(x) + e_2(x) - e_1(x)e_2(x)$ is clearly in $\mathcal{C}_1 + \mathcal{C}_2$, $\mathcal{C}_1 + \mathcal{C}_2 = \langle e_1(x) + e_2(x) - e_1(x)e_2(x) \rangle$ by Exercise 725(b). $\qquad\square$

Exercise 725 Let $\mathcal{C}$ be a cyclic code over either $\mathbb{F}_q$ or $\mathbb{Z}_4$. Do the following:
(a) Let $e(x)$ be a generating idempotent of $\mathcal{C}$. Prove that $c(x) \in \mathcal{C}$ if and only if $c(x)e(x) = c(x)$.
(b) Prove that if $e(x) \in \mathcal{C}$ and $c(x)e(x) = c(x)$ for all $c(x) \in \mathcal{C}$, then $e(x)$ is the generating idempotent of $\mathcal{C}$. ◆

One way to construct the "generating idempotent" of $\mathcal{C}$ is to begin with the primitive idempotents. We leave the proof of the following as an exercise.

Theorem 12.3.24 *Let n be odd and $g_1(x)g_2(x)\cdots g_k(x)$ be a factorization of $x^n - 1$ into irreducible polynomials in $\mathbb{Z}_4[x]$. Let $\widehat{g}_i(x) = \prod_{j \neq i} g_j(x)$ and $\langle \widehat{g}_i(x) \rangle = \langle \widehat{e}_i(x) \rangle$, where $\widehat{e}_i(x)$ is an idempotent of $\mathfrak{R}_n$. Let $\mathcal{C} = \langle f(x)g(x) \rangle \oplus \langle 2f(x)h(x) \rangle$ where $f(x)g(x)h(x) = x^n - 1$. Finally, suppose that $f(x)g(x) = \prod_{i \in I} g_i(x)$ and $f(x)h(x) = \prod_{j \in J} g_j(x)$ for subsets I and J of $\{1, 2, \dots, k\}$. Then:*
(i) $I^c \cap J^c = \emptyset$, *where I^c and J^c are the complements of I and J, respectively, in $\{1, 2, \dots, k\}$, and*
(ii) $\mathcal{C} = \langle e(x) + 2E(x) \rangle$, *where $e(x) = \sum_{i \in I^c} \widehat{e}_i(x)$ and $E(x) = \sum_{j \in J^c} \widehat{e}_j(x)$.*

Exercise 726 Fill in the details of the proof of Theorem 12.3.24. ◆

There is a method for constructing idempotents, including the primitive ones, in $\mathfrak{R}_n$ found in [27, 289]. The method begins with binary idempotents in $\mathcal{R}_n$; these can be obtained from generator polynomials using the Euclidean Algorithm as given in the proof of Theorem 4.3.2.

Theorem 12.3.25 *Let $\langle b(x) \rangle = \langle \mu(f(x)) \rangle$ where $f(x)g(x) = x^n - 1$ in $\mathbb{Z}_4[x]$, $b(x)$ is a binary idempotent in $\mathcal{R}_n = \mathbb{F}_2[x]/(x^n + 1)$, and n is odd. Let $e(x) = b^2(x)$ where $b^2(x)$ is computed in $\mathbb{Z}_4[x]$. Then $e(x)$ is the generating idempotent for $\langle f(x) \rangle$.*

Proof: As $b^2(x) = b(x)$ in $\mathcal{R}_n$, $\mu(e(x)) = \mu(b^2(x)) = b(x) + \mu(a(x)(x^n - 1))$ for some $a(x) \in \mathbb{Z}_4[x]$, implying that $e(x) = b(x) + a(x)(x^n - 1) + 2s(x)$ for some $s(x) \in \mathbb{Z}_4[x]$. Squaring and simplifying yields $e^2(x) = b^2(x) + d(x)(x^n - 1)$ in $\mathbb{Z}_4[x]$ or $e^2(x) = e(x)$ in $\mathfrak{R}_n$. Thus $e(x)$ is an idempotent in $\mathfrak{R}_n$.

By Theorem 4.4.6, $1 + b(x)$ is the generating idempotent in $\mathcal{R}_n$ of $\langle \mu(g(x)) \rangle$. So $1 + b(x) = r(x)g(x) + 2s(x)$ for some $r(x)$ and $s(x)$ in $\mathbb{Z}_4[x]$. Hence $b(x) = 1 + r(x)g(x) + 2(1 + s(x))$; squaring this, we obtain $e(x) = b^2(x) = 1 + t(x)g(x)$ for some $t(x)$ in $\mathbb{Z}_4[x]$. So $e(x)f(x) = f(x) + t(x)(x^n - 1)$ in $\mathbb{Z}_4[x]$ implying $f(x) \in \langle e(x) \rangle$ in $\mathfrak{R}_n$ or $\langle f(x) \rangle \subseteq \langle e(x) \rangle$.

Since $e(x) = b^2(x)$, $\mu(e(x)) = \mu(b^2(x)) \in \langle \mu(f(x)) \rangle$ in $\mathcal{R}_n$. So $e(x) = u(x)f(x) + 2v(x)$ in $\mathbb{Z}_4[x]$. Squaring yields $e^2(x) = u^2(x)f^2(x)$, implying that $e^2(x) \in \langle f(x) \rangle$ in $\mathcal{R}_n$. As $e(x)$ is an idempotent in $\mathcal{R}_n$, $e(x) \in \langle f(x) \rangle$ or $\langle e(x) \rangle \subseteq \langle f(x) \rangle$. Therefore $\langle e(x) \rangle = \langle f(x) \rangle$ and $e(x) = b^2(x)$ is the generating idempotent for $\langle f(x) \rangle$. $\qquad\square$

Example 12.3.26 In Example 12.3.8 we found the factorization of $x^7 - 1$ over $\mathbb{Z}_4[x]$ to be $x^7 - 1 = g_1(x)g_2(x)g_3(x)$, where $g_1(x) = x - 1$, $g_2(x) = x^3 + 2x^2 + x - 1$, and $g_3(x) = x^3 - x^2 + 2x - 1$. We wish to find the generating idempotents of $\langle \widehat{g}_i(x) \rangle$ in $\mathcal{R}_7$ where $\widehat{g}_i(x) = \prod_{j \neq i} g_j(x)$. We follow the notation of Theorem 12.3.24 and the method of Theorem 12.3.25. First, $\mu(\widehat{g}_1(x)) = x^6 + x^5 + x^4 + x^3 + x^2 + x + 1$. By Example 4.3.4, $b_1(x) = x^6 + x^5 + x^4 + x^3 + x^2 + x + 1$ is the generating idempotent of $\langle \mu(\widehat{g}_1(x)) \rangle$. With analogous notation, $\mu(\widehat{g}_2(x)) = x^4 + x^2 + x + 1$, $b_2(x) = x^4 + x^2 + x + 1$, $\mu(\widehat{g}_3(x)) = x^4 + x^3 + x^2 + 1$, and $b_3(x) = x^6 + x^5 + x^3 + 1$. Thus by Theorem 12.3.25:

$$\widehat{e}_1(x) = -x^6 - x^5 - x^4 - x^3 - x^2 - x - 1,$$
$$\widehat{e}_2(x) = 2x^6 + 2x^5 - x^4 + 2x^3 - x^2 - x + 1,$$
$$\widehat{e}_3(x) = -x^6 - x^5 + 2x^4 - x^3 + 2x^2 + 2x + 1.$$

By Theorem 12.3.22 and Exercise 724, each cyclic code has a "generating idempotent" $e(x) + 2E(x)$. In Table 12.1, we present the "generating idempotents" for the 25 nontrivial cyclic codes over $\mathbb{Z}_4$ of length 7. The idempotents $e(x)$ and $E(x)$ will each be a sum of some of $\widehat{e}_1(x), \widehat{e}_2(x)$, and $\widehat{e}_3(x)$. In the table we list $e(x) + 2E(x)$. Code No. 4 is $\langle g_2(x) \rangle$; so $f(x)g(x) = g_2(x)$ and $f(x)h(x) = x^n - 1$ implying that $e(x) = \widehat{e}_1(x) + \widehat{e}_3(x)$ and $2E(x) = 0$ by Theorem 12.3.24. Code No. 12 is $\langle 2g_1(x) \rangle$; so $f(x)g(x) = x^n - 1$ and $f(x)h(x) = g_1(x)$ implying that $e(x) = 0$ and $2E(x) = 2\widehat{e}_2(x) + 2\widehat{e}_3(x)$. Code No. 20 is $\langle g_1(x)g_3(x) \rangle + \langle 2g_1(x)g_2(x) \rangle$; so $f(x)g(x) = g_1(x)g_3(x)$ and $f(x)h(x) = g_1(x)g_2(x)$ implying that $e(x) = \widehat{e}_2(x)$ and $2E(x) = 2\widehat{e}_3(x)$. Thus code No. 20 is the ideal generated by $\widehat{e}_2(x) + 2\widehat{e}_3(x)$. Finally, code No. 21 is $\langle g_1(x)g_2(x) \rangle + \langle 2g_3(x) \rangle$; so $f(x)g(x) = g_1(x)g_2(x)$ and $f(x)h(x) = g_3(x)$ implying that $e(x) = \widehat{e}_3(x)$ and $2E(x) = 2\widehat{e}_1(x) + 2\widehat{e}_2(x)$. Therefore code No. 21 is the ideal generated by $\widehat{e}_3(x) + 2\widehat{e}_1(x) + 2\widehat{e}_2(x)$. $\qquad\blacksquare$

Exercise 727 Do the following:
(a) Verify that the method described in Theorem 12.3.25 actually leads to the idempotents $\widehat{e}_i(x)$ given in Example 12.3.26.
(b) Verify that the entries in the column "Generating idempotents" of Table 12.1 are correct as given. $\qquad\blacklozenge$

12.4 Quadratic residue codes over $\mathbb{Z}_4$

In Section 6.6 we examined the family of quadratic residue codes over $\mathbb{F}_q$. Among the cyclic codes over $\mathbb{Z}_4$ are codes that have properties similar to the ordinary quadratic residue codes and are thus termed $\mathbb{Z}_4$-quadratic residue codes [27, 279, 289]. We will give the generating idempotents and basic properties of these codes. Before doing so we establish our notation and present two preliminary lemmas.

Throughout this section, p will be a prime with $p \equiv \pm 1 \pmod 8$. Let $\mathcal{Q}_p$ denote the set of nonzero quadratic residues modulo p, and let $\mathcal{N}_p$ be the set of quadratic non-residues modulo p.

Lemma 12.4.1 *Define r by $p = 8r \pm 1$. Let $k \in \mathbb{F}_p$ with $k \neq 0$. Let $N_{\mathcal{Q}_p}(k)$ be the number of unordered pairs $\{\{i, j\} \mid i + j = k, \ i \neq j, i \text{ and } j \text{ in } \mathcal{Q}_p\}$. Define $N_{\mathcal{N}_p}(k)$ analogously. Then $N_{\mathcal{Q}_p}(k) = r - 1$ if $k \in \mathcal{Q}_p$, $N_{\mathcal{Q}_p}(k) = r$ if $k \in \mathcal{N}_p$, $N_{\mathcal{N}_p}(k) = r - 1$ if $k \in \mathcal{N}_p$, and $N_{\mathcal{N}_p}(k) = r$ if $k \in \mathcal{Q}_p$.*

Proof: We compute $N_{\mathcal{Q}_p}(k)$ first. By [66, p. 46] and Lemma 6.2.4, the number of pairs $(x, y) \in \mathbb{F}_p^2$ with

$$x^2 + y^2 = k \tag{12.12}$$

is $8r$. Solutions of (12.12) lead to solutions of $i + j = k$ with $i, j \in \mathcal{Q}_p$ by setting $\{i, j\} = \{x^2, y^2\}$. Three possibilities arise:
(a) $x = 0$ or $y = 0$,
(b) $x \neq 0$ and $y \neq 0$ with $x = \pm y$, and
(c) $x \neq 0$ and $y \neq 0$ with $x \neq \pm y$.
Solutions of (12.12) in either form (a) or (b) lead to solutions of $i + j = k$ where $i = 0$, $j = 0$, or $i = j$ and hence are not counted in $N_{\mathcal{Q}_p}(k)$. When $k \in \mathcal{N}_p$, (a) cannot occur; also (b) cannot occur because $x^2 + y^2 = k$ and $x = \pm y$ implies $2x^2 = k$, which means $2 \in \mathcal{N}_p$, contradicting Lemma 6.2.5. So if $k \in \mathcal{N}_p$, all $8r$ solutions of (12.12) have form (c). When $k \in \mathcal{Q}_p$, there are four solutions of form (a), namely $(\pm \gamma, 0)$ and $(0, \pm \gamma)$, where $\pm \gamma$ are the two solutions of $z^2 = k$. When $k \in \mathcal{Q}_p$, there are four solutions of (12.12) of form (b), namely $(\pm \gamma, \pm \gamma)$, where $\pm \gamma$ are the two solutions of $2z^2 = k$, noting that $2 \in \mathcal{Q}_p$ by Lemma 6.2.5. So if $k \in \mathcal{Q}_p$, $8r - 8$ solutions of (12.12) have form (c). In case (c) with $k \in \mathcal{Q}_p$ or $k \in \mathcal{N}_p$, any set of eight solutions of (12.12) in the group $(\pm x, \pm y)$ and $(\pm y, \pm x)$ lead to the same solution of $i + j = k$ counted by $N_{\mathcal{Q}_p}(k)$. Therefore, $N_{\mathcal{Q}_p}(k) = r - 1$ if $k \in \mathcal{Q}_p$ and $N_{\mathcal{Q}_p}(k) = r$ if $k \in \mathcal{N}_p$.

Let $\alpha \in \mathcal{N}_p$. Then $i + j = k$ if and only if $i\alpha + j\alpha = k\alpha$. Hence $N_{\mathcal{N}_p}(k) = N_{\mathcal{Q}_p}(k\alpha)$. Therefore, $N_{\mathcal{N}_p}(k) = r - 1$ if $k \in \mathcal{N}_p$ and $N_{\mathcal{N}_p}(k) = r$ if $k \in \mathcal{Q}_p$. $\square$

Let $Q(x) = \sum_{i \in \mathcal{Q}_p} x^i$ and $N(x) = \sum_{i \in \mathcal{N}_p} x^i$. Note that 1, $Q(x)$, and $N(x)$ are idempotents in $\mathcal{R}_p = \mathbb{F}_2[x]/(x^p + 1)$. As discovered in [279, 289], a combination of these will be idempotents in $\mathcal{R}_p$ that lead to the definition of $\mathbb{Z}_4$-quadratic residue codes. A multiple of the all-one vector is also an idempotent; let $\overline{j}(x) = p \sum_{i=0}^{p-1} x^i$. In particular, $\overline{j}(x) = 3 \sum_{i=0}^{p-1} x^i$ if $p \equiv -1 \pmod 8$ and $\overline{j}(x) = \sum_{i=0}^{p-1} x^i$ if $p \equiv 1 \pmod 8$.

Lemma 12.4.2 *Define r by $p = 8r \pm 1$. If r is odd, then $Q(x) + 2N(x)$, $N(x) + 2Q(x)$, $1 - Q(x) + 2N(x)$, and $1 - N(x) + 2Q(x)$ are idempotents in $\mathcal{R}_p$. If r is even, then $-Q(x)$, $-N(x)$, $1 + Q(x)$, and $1 + N(x)$ are idempotents in $\mathcal{R}_p$. Also, for r even or odd, $\overline{j}(x)$ is an idempotent in $\mathcal{R}_p$.*

Proof: We prove only the case when r is odd and leave the case when r is even to Exercise 728. Working in $\mathfrak{R}_p$, by Lemma 12.4.1,

$$Q(x)^2 = \left(\sum_{i \in \mathcal{Q}_p} x^i\right)^2 = \sum_{i \in \mathcal{Q}_p} x^{2i} + \sum_{i \neq j,\, i,j \in \mathcal{Q}_p} x^{i+j}$$
$$= Q(x) + 2[(r-1)Q(x) + rN(x)] = Q(x) + 2N(x)$$

since r is odd and $2 \in \mathcal{Q}_p$ by Lemma 6.2.5. Similarly,

$$N(x)^2 = \left(\sum_{i \in \mathcal{N}_p} x^i\right)^2 = \sum_{i \in \mathcal{N}_p} x^{2i} + \sum_{i \neq j,\, i,j \in \mathcal{N}_p} x^{i+j}$$
$$= N(x) + 2[(r-1)N(x) + rQ(x)] = N(x) + 2Q(x).$$

So $(Q(x) + 2N(x))^2 = Q(x)^2 = Q(x) + 2N(x)$, $(N(x) + 2Q(x))^2 = N(x)^2 = N(x) + 2Q(x)$, $(1 - Q(x) + 2N(x))^2 = (1 - Q(x))^2 = 1 - 2Q(x) + Q(x)^2 = 1 - Q(x) + 2N(x)$, and $(1 - N(x) + 2Q(x))^2 = (1 - N(x))^2 = 1 - 2N(x) + N(x)^2 = 1 - N(x) + 2Q(x)$.

Finally, $\overline{j}(x)^2 = p^2 \sum_{i=0}^{p-1} x^i \sum_{j=0}^{p-1} x^j = p^2\left(p \sum_{i=0}^{p-1} x^i\right) = \overline{j}(x)$ since $p^2 \equiv 1 \pmod 8$. $\qquad \square$

Exercise 728 Prove Lemma 12.4.2 when r is even. $\qquad\blacklozenge$

We now define the $\mathbb{Z}_4$-quadratic residue codes using the idempotents of Lemma 12.4.2. The definitions depend upon the value of p modulo 8.

12.4.1 $\mathbb{Z}_4$-quadratic residue codes: $p \equiv -1 \pmod 8$

We first look at the case where $p \equiv -1 \pmod 8$. Let $p + 1 = 8r$. If r is odd, define $\mathcal{D}_1 = \langle Q(x) + 2N(x)\rangle$, $\mathcal{D}_2 = \langle N(x) + 2Q(x)\rangle$, $\mathcal{C}_1 = \langle 1 - N(x) + 2Q(x)\rangle$, and $\mathcal{C}_2 = \langle 1 - Q(x) + 2N(x)\rangle$. If r is even, define $\mathcal{D}_1 = \langle -Q(x)\rangle$, $\mathcal{D}_2 = \langle -N(x)\rangle$, $\mathcal{C}_1 = \langle 1 + N(x)\rangle$, and $\mathcal{C}_2 = \langle 1 + Q(x)\rangle$. These codes are called $\mathbb{Z}_4$-*quadratic residue* codes when $p \equiv -1 \pmod 8$.

The next result, from [279, 289], shows why these codes are called quadratic residue codes; compare this result with those in Chapter 6.

Theorem 12.4.3 *Let* $p \equiv -1 \pmod 8$. *The $\mathbb{Z}_4$-quadratic residue codes satisfy the following:*

(i) $\mathcal{D}_i \mu_a = \mathcal{D}_i$ *and* $\mathcal{C}_i \mu_a = \mathcal{C}_i$ *for* $a \in \mathcal{Q}_p$; $\mathcal{D}_1 \mu_a = \mathcal{D}_2$ *and* $\mathcal{C}_1 \mu_a = \mathcal{C}_2$ *for* $a \in \mathcal{N}_p$; *in particular* $\mathcal{D}_1$ *and* $\mathcal{D}_2$ *are equivalent as are* $\mathcal{C}_1$ *and* $\mathcal{C}_2$.

(ii) $\mathcal{D}_1 \cap \mathcal{D}_2 = \langle \overline{j}(x)\rangle$ *and* $\mathcal{D}_1 + \mathcal{D}_2 = \mathfrak{R}_p$.

(iii) $\mathcal{C}_1 \cap \mathcal{C}_2 = \{0\}$ *and* $\mathcal{C}_1 + \mathcal{C}_2 = \langle \overline{j}(x)\rangle^{\perp}$.

(iv) $\mathcal{D}_1$ *and* $\mathcal{D}_2$ *have type* $4^{(p+1)/2}$; $\mathcal{C}_1$ *and* $\mathcal{C}_2$ *have type* $4^{(p-1)/2}$.

(v) $\mathcal{D}_i = \mathcal{C}_i + \langle \overline{j}(x)\rangle$ *for* $i = 1$ *and* 2.

(vi) $\mathcal{C}_1$ *and* $\mathcal{C}_2$ *are self-orthogonal and* $\mathcal{C}_i^{\perp} = \mathcal{D}_i$ *for* $i = 1$ *and* 2.

Proof: Suppose that $p + 1 = 8r$. We verify this result when r is odd and leave the proof for r even as an exercise.

For (i), if $a \in \mathcal{N}_p$, then $(Q(x) + 2N(x))\mu_a = N(x) + 2Q(x)$. Theorem 4.3.13(i) (whose proof is still valid for cyclic codes over $\mathbb{Z}_4$) implies that $\mathcal{D}_1\mu_a = \mathcal{D}_2$. Similarly, if $a \in \mathcal{Q}_p$, then $(Q(x) + 2N(x))\mu_a = Q(x) + 2N(x)$ and $(N(x) + 2Q(x))\mu_a = N(x) + 2Q(x)$, implying $\mathcal{D}_i\mu_a = \mathcal{D}_i$. The parts of (i) involving $\mathcal{C}_i$ are similar.

Since $p \equiv -1 \pmod 8$, $\overline{j}(x) = 3\sum_{i=0}^{p-1} x^i = 3 + 3Q(x) + 3N(x)$. Thus $(Q(x) + 2N(x))(N(x) + 2Q(x)) = (Q(x) + 2N(x))(\overline{j}(x) + 1 - (Q(x) + 2N(x))) = (Q(x) + 2N(x))\overline{j}(x) + Q(x) + 2N(x) - (Q(x) + 2N(x))^2 = (Q(x) + 2N(x))\overline{j}(x) = 3((p-1)/2) \times \sum_{i=0}^{p-1} x^i + 3(p-1)\sum_{i=0}^{p-1} x^i = (3/2)(p-1)\overline{j}(x) = (12r - 3)\overline{j}(x) = \overline{j}(x)$. By Lemma 12.3.23, $\mathcal{D}_1 \cap \mathcal{D}_2 = \langle \overline{j}(x) \rangle$. By the same result, $\mathcal{D}_1 + \mathcal{D}_2$ has generating idempotent $Q(x) + 2N(x) + N(x) + 2Q(x) - (Q(x) + 2N(x))(N(x) + 2Q(x)) = 3Q(x) + 3N(x) - \overline{j}(x) = 1$. Therefore $\mathcal{D}_1 + \mathcal{D}_2 = \mathfrak{R}_p$, proving part (ii).

Using $(Q(x) + 2N(x))(N(x) + 2Q(x)) = \overline{j}(x)$, for (iii) we have $(1 - N(x) + 2Q(x)) \times (1 - Q(x) + 2N(x)) = 1 - N(x) + 2Q(x) - Q(x) + 2N(x) + (N(x) + 2Q(x))(Q(x) + 2N(x)) = 1 + N(x) + Q(x) + \overline{j}(x) = 0$ proving that $\mathcal{C}_1 \cap \mathcal{C}_2 = \{0\}$. Since $(1 - N(x) + 2Q(x))(1 - Q(x) + 2N(x)) = 0$, $\mathcal{C}_1 + \mathcal{C}_2$ has generating idempotent $1 - N(x) + 2Q(x) + 1 - Q(x) + 2N(x) = 2 + N(x) + Q(x) = 1 - \overline{j}(x) = 1 - \overline{j}(x)\mu_{-1}$ as $\overline{j}(x)\mu_{-1} = \overline{j}(x)$. By Lemma 12.3.23, $\mathcal{C}_1 + \mathcal{C}_2 = \langle \overline{j}(x) \rangle^{\perp}$, completing (iii).

For (iv), we observe that $|\mathcal{D}_1 + \mathcal{D}_2| = |\mathcal{D}_1||\mathcal{D}_2|/|\mathcal{D}_1 \cap \mathcal{D}_2|$. By (i), $|\mathcal{D}_1| = |\mathcal{D}_2|$, and by (ii), $|\mathcal{D}_1 + \mathcal{D}_2| = 4^p$ and $|\mathcal{D}_1 \cap \mathcal{D}_2| = 4$. Thus $\mathcal{D}_1$ and $\mathcal{D}_2$ have size $4^{(p+1)/2}$; this is also their type as each has an idempotent generator (as opposed to a generator that is the sum of an idempotent and twice an idempotent; see Theorem 12.3.22 and Exercise 724). The remainder of (iv) is similar using (i) and (iii).

By (ii), $\overline{j}(x) \in \mathcal{D}_2$ implying that $(N(x) + 2Q(x))\overline{j}(x) = \overline{j}(x)$ as $N(x) + 2Q(x)$ is the multiplicative identity of $\mathcal{D}_2$. By Lemma 12.3.23, the generating idempotent for $\mathcal{C}_1 + \langle \overline{j}(x) \rangle$ is $1 - N(x) + 2Q(x) + \overline{j}(x) - (1 - N(x) + 2Q(x))\overline{j}(x) = 1 - N(x) + 2Q(x) + \overline{j}(x) - (\overline{j}(x) - \overline{j}(x)) = Q(x) + 2N(x)$, proving that $\mathcal{C}_1 + \langle \overline{j}(x) \rangle = \mathcal{D}_1$. Similarly, $\mathcal{C}_2 + \langle \overline{j}(x) \rangle = \mathcal{D}_2$, completing (v).

Finally, by Lemma 12.3.23, the generating idempotent for $\mathcal{C}_1^{\perp}$ is $1 - (1 - N(x) + 2Q(x))\mu_{-1} = N(x)\mu_{-1} + 2Q(x)\mu_{-1}$. By Lemma 6.2.4, $-1 \in \mathcal{N}_p$ as $p \equiv -1 \pmod 8$. Hence $N(x)\mu_{-1} = Q(x)$ and $Q(x)\mu_{-1} = N(x)$. Therefore the generating idempotent for $\mathcal{C}_1^{\perp}$ is $Q(x) + 2N(x)$, verifying that $\mathcal{C}_1^{\perp} = \mathcal{D}_1$. Similarly, $\mathcal{C}_2^{\perp} = \mathcal{D}_2$. This also implies that $\mathcal{C}_i$ is self-orthogonal as $\mathcal{C}_i \subseteq \mathcal{D}_i$ from (v), completing (vi).   □

Exercise 729 Prove Theorem 12.4.3 when r is even.   ◆

Example 12.4.4 Four of the cyclic codes of length 7 over $\mathbb{Z}_4$ given in Table 12.1 are quadratic residue codes. In this case $p = 7$ and $r = 1$. In the notation of that table (see Example 12.3.26), $\widehat{e}_1(x) = \overline{j}(x)$, $\widehat{e}_2(x) = 1 - Q(x) + 2N(x)$, and $\widehat{e}_3(x) = 1 - N(x) + 2Q(x)$. Therefore $\mathcal{D}_1$ is code No. 4, $\mathcal{D}_2$ is code No. 5, $\mathcal{C}_1$ is code No. 2, and $\mathcal{C}_2$ is code No. 3. Notice that the duality conditions given in Table 12.1 for these codes agree with Theorem 12.4.3(vi).   ■

Exercise 730 Do the following:
(a) Find the generating idempotents for the $\mathbb{Z}_4$-quadratic residue codes of length 23.
(b) Apply the reduction homomorphism μ to the idempotents found in part (a).
(c) What are the binary codes generated by the idempotents found in part (b)?   ◆

12.4.2 $\mathbb{Z}_4$-quadratic residue codes: $p \equiv 1 \pmod 8$

When $p \equiv 1 \pmod 8$, we simply reverse the C_is and D_is from the codes defined in the case $p \equiv -1 \pmod 8$. Again let $p - 1 = 8r$. If r is odd, define $D_1 = \langle 1 - N(x) + 2Q(x) \rangle$, $D_2 = \langle 1 - Q(x) + 2N(x) \rangle$, $C_1 = \langle Q(x) + 2N(x) \rangle$, and $C_2 = \langle N(x) + 2Q(x) \rangle$. If r is even, define $D_1 = \langle 1 + N(x) \rangle$, $D_2 = \langle 1 + Q(x) \rangle$, $C_1 = \langle -Q(x) \rangle$, and $C_2 = \langle -N(x) \rangle$. These codes are called $\mathbb{Z}_4$-*quadratic residue* codes when $p \equiv 1 \pmod 8$. We leave the proof of the next result as an exercise. Again see [279, 289].

Theorem 12.4.5   *Let* $p \equiv 1 \pmod 8$. *The* $\mathbb{Z}_4$-*quadratic residue codes satisfy the following*:
(i) $D_i \mu_a = D_i$ *and* $C_i \mu_a = C_i$ *for* $a \in Q_p$; $D_1 \mu_a = D_2$ *and* $C_1 \mu_a = C_2$ *for* $a \in N_p$; *in particular* D_1 *and* D_2 *are equivalent as are* C_1 *and* C_2.
(ii) $D_1 \cap D_2 = \langle \overline{j}(x) \rangle$ *and* $D_1 + D_2 = \Re_p$.
(iii) $C_1 \cap C_2 = \{0\}$ *and* $C_1 + C_2 = \langle \overline{j}(x) \rangle^{\perp}$.
(iv) D_1 *and* D_2 *have type* $4^{(p+1)/2}$; C_1 *and* C_2 *have type* $4^{(p-1)/2}$.
(v) $D_i = C_i + \langle \overline{j}(x) \rangle$ *for* $i = 1$ *and* 2.
(vi) $C_1^{\perp} = D_2$ *and* $C_2^{\perp} = D_1$.

Exercise 731   Prove Theorem 12.4.5.   ◆

Exercise 732   Do the following:
(a) Find the generating idempotents for the $\mathbb{Z}_4$-quadratic residue codes of length 17.
(b) Apply the reduction homomorphism μ to the idempotents found in part (a).
(c) What are the binary codes generated by the idempotents found in part (b)?   ◆

12.4.3   Extending $\mathbb{Z}_4$-quadratic residue codes

In Section 6.6.3, we described the extensions of quadratic residue codes over $\mathbb{F}_q$. We do the same for these codes over $\mathbb{Z}_4$. Let D_1 and D_2 be the quadratic residue codes of length $p \equiv \pm 1 \pmod 8$ as described in the two previous subsections. We will define two extensions of D_i. Define $\widehat{D}_i = \{c_\infty c_0 \cdots c_{p-1} \mid c_0 \cdots c_{p-1} \in D_i, c_\infty + c_0 + \cdots + c_{p-1} \equiv 0 \pmod 4)\}$ and $\widetilde{D}_i = \{c_\infty c_0 \cdots c_{p-1} \mid c_0 \cdots c_{p-1} \in D_i, \ -c_\infty + c_0 + \cdots + c_{p-1} \equiv 0 \pmod 4)\}$. $\widehat{D}_i$ and $\widetilde{D}_i$ are the *extended* $\mathbb{Z}_4$-*quadratic residue* codes of length $p + 1$. Note that $\widehat{D}_i$ and $\widetilde{D}_i$ are equivalent under the map that multiplies the extended coordinate by 3. If c_i is a codeword in D_i, let $\widehat{c}_i$ and $\widetilde{c}_i$ be the extended codewords in $\widehat{D}_i$ and $\widetilde{D}_i$, respectively.

Exercise 733   Do the following:
(a) Let C_i be the quadratic residue codes of length $p \equiv \pm 1 \pmod 8$ as described in the two previous subsections. Prove that the sum of the components of any codeword in C_i is 0 modulo 4. Hint: First show this for the generating idempotents and hence for the cyclic shifts of the idempotents.
(b) Prove that the sum of the components of $\overline{j}(x)$ is 1 modulo 4.
(c) Prove that the Euclidean weight of the generating idempotents of C_i is a multiple of 8 if $p \equiv -1 \pmod 8$.
(d) Prove that $\mathrm{wt}_E(\widehat{\overline{j}}(x)) = \mathrm{wt}_E(\widetilde{\overline{j}}(x)) \equiv 0 \pmod 8$ if $p \equiv -1 \pmod 8$.   ◆

Using Exercise 733 along with Theorems 12.4.3(v) and 12.4.5(v) we can find the generator matrices for $\widehat{\mathcal{D}}_i$ and $\widetilde{\mathcal{D}}_i$.

Theorem 12.4.6 *Let G_i be the generator matrix for the $\mathbb{Z}_4$-quadratic residue code C_i. Then generator matrices $\widehat{G}_i$ and $\widetilde{G}_i$ for $\widehat{\mathcal{D}}_i$ and $\widetilde{\mathcal{D}}_i$, respectively, are:*
(i) *If $p \equiv -1 \pmod 8$, then*

$$\widehat{G}_i = \begin{bmatrix} 3 & 3 & \cdots & 3 \\ 0 & & & \\ \vdots & & G_i & \\ 0 & & & \end{bmatrix} \quad and \quad \widetilde{G}_i = \begin{bmatrix} 1 & 3 & \cdots & 3 \\ 0 & & & \\ \vdots & & G_i & \\ 0 & & & \end{bmatrix}.$$

(ii) *If $p \equiv 1 \pmod 8$, then*

$$\widehat{G}_i = \begin{bmatrix} 3 & 1 & \cdots & 1 \\ 0 & & & \\ \vdots & & G_i & \\ 0 & & & \end{bmatrix} \quad and \quad \widetilde{G}_i = \begin{bmatrix} 1 & 1 & \cdots & 1 \\ 0 & & & \\ \vdots & & G_i & \\ 0 & & & \end{bmatrix}.$$

Exercise 734 Prove Theorem 12.4.6. ◆

The extended codes have the following properties; compare this result to Theorem 6.6.14.

Theorem 12.4.7 *Let $\mathcal{D}_i$ be the $\mathbb{Z}_4$-quadratic residue codes of length p. The following hold:*
(i) *If $p \equiv -1 \pmod 8$, then $\widehat{\mathcal{D}}_i$ and $\widetilde{\mathcal{D}}_i$ are self-dual. Furthermore, all codewords of $\widehat{\mathcal{D}}_i$ and $\widetilde{\mathcal{D}}_i$ have Euclidean weights a multiple of 8.*
(ii) *If $p \equiv 1 \pmod 8$, then $\widehat{\mathcal{D}}_1^{\perp} = \widetilde{\mathcal{D}}_2$ and $\widehat{\mathcal{D}}_2^{\perp} = \widetilde{\mathcal{D}}_1$.*

Proof: Let $p \equiv -1 \pmod 8$. Then by Theorem 12.4.3(vi), $C_i^{\perp} = \mathcal{D}_i$. Hence as the extended coordinate of any vector in C_i is 0 by Exercise 733, the extended codewords arising from C_i are orthogonal to all codewords in either $\widehat{\mathcal{D}}_i$ or $\widetilde{\mathcal{D}}_i$. Since the inner product of $\overline{\jmath}(x)$ with itself (and $\widetilde{\jmath}(x)$ with itself) is $3^2(p+1) \equiv 0 \pmod 4$ (and $1^2 + 3^2 p \equiv 0 \pmod 4$), $\widehat{\mathcal{D}}_i$ and $\widetilde{\mathcal{D}}_i$ are self-orthogonal using Theorem 12.4.3(v). By Theorem 12.4.3(iv), $\widehat{\mathcal{D}}_i$ and $\widetilde{\mathcal{D}}_i$ have $4^{(p+1)/2}$ codewords implying that $\widehat{\mathcal{D}}_i$ and $\widetilde{\mathcal{D}}_i$ are self-dual. To verify that the Euclidean weight of a codeword in $\widehat{\mathcal{D}}_i$ and $\widetilde{\mathcal{D}}_i$ is a multiple of 8, we only have to verify this for the rows of a generator matrix of $\widehat{\mathcal{D}}_i$ and $\widetilde{\mathcal{D}}_i$ by Theorem 12.2.4. The rows of the generator matrix for $\widehat{\mathcal{D}}_i$ and $\widetilde{\mathcal{D}}_i$ are the extensions of $\overline{\jmath}(x)$ and cyclic shifts of the generating idempotent of C_i; see Theorem 12.4.6. But the rows of the generator matrix for $\widehat{\mathcal{D}}_i$ and $\widetilde{\mathcal{D}}_i$ have Euclidean weight a multiple of 8 from Exercise 733.

Suppose now that $p \equiv 1 \pmod 8$. By Theorem 12.4.5(vi), $C_1^{\perp} = \mathcal{D}_2$ and $C_2^{\perp} = \mathcal{D}_1$. Hence as the extended coordinate of any vector in C_i is 0 by Exercise 733, the extended codewords arising from C_i are orthogonal to all codewords in either $\widehat{\mathcal{D}}_j$ or $\widetilde{\mathcal{D}}_j$ where $j \neq i$. Since the inner product of $\widetilde{\jmath}(x)$ with $\overline{\jmath}(x)$ is $3 + p \equiv 0 \pmod 4$, $\widehat{\mathcal{D}}_j^{\perp} \subseteq \widetilde{\mathcal{D}}_i$ where $j \neq i$. Part (ii) now follows from Theorem 12.4.5(iv). □

Example 12.4.8 By Theorem 12.4.7, the extended quadratic residue codes $\widehat{\mathcal{D}}_1$ and $\widehat{\mathcal{D}}_2$ of length $p + 1 = 24$ are self-dual codes where all codewords have Euclidean weights a multiple of 8. If $\mathcal{C}$ is either of these codes, the symmetrized weight enumerator (12.7) of $\mathcal{C}$, computed in [27], is:

$$
\begin{aligned}
\text{swe}_{\mathcal{C}}(a, b, c) = {} & a^{24} + c^{24} + 759(a^8c^{16} + a^{16}c^8) + 2576a^{12}c^{12} \\
& + 12\,144(a^2b^8c^{14} + a^{14}b^8c^2) + 170\,016(a^4b^8c^{12} + a^{12}b^8c^4) \\
& + 765\,072(a^6b^8c^{10} + a^{10}b^8c^6) + 1\,214\,400a^8b^8c^8 \\
& + 61\,824(ab^{12}c^{11} + a^{11}b^{12}c) + 1\,133\,440(a^3b^{12}c^9 + a^9b^{12}c^3) \\
& + 4\,080\,384(a^5b^{12}c^7 + a^7b^{12}c^5) + 24\,288(b^{16}c^8 + a^8b^{16}) \\
& + 680\,064(a^2b^{16}c^6 + a^6b^{16}c^2) + 1\,700\,160a^4b^{16}c^4 + 4096b^{24}.
\end{aligned}
$$

Thus we see that $\mathcal{C}$ has minimum Hamming weight 8, minimum Lee weight 12, and minimum Euclidean weight 16. The Gray image $\mathfrak{G}(\mathcal{C})$ is a nonlinear $(48, 2^{24}, 12)$ binary code. Recall that the extended binary quadratic residue code of length 48 is a $[48, 24, 12]$ self-dual doubly-even code as discussed in Example 6.6.23; see also the related Research Problem 9.3.7. ∎

Exercise 735 Let $\mathcal{C}$ be one of the extended quadratic residue codes of length 24 presented in Example 12.4.8.
(a) By examining $\text{swe}_{\mathcal{C}}(a, b, c)$, verify that $\mathcal{C}$ has minimum Hamming weight 8, minimum Lee weight 12, and minimum Euclidean weight 16.
(b) Give the Hamming weight distribution of $\mathfrak{G}(\mathcal{C})$ from $\text{swe}_{\mathcal{C}}(a, b, c)$.
(c) There is a symmetry in the entries in $\text{swe}_{\mathcal{C}}(a, b, c)$; for instance, the number of codewords in $\mathcal{C}$ contributing to $a^6b^8c^{10}$ is the same as the number of codewords contributing to $a^{10}b^8c^6$. The presence of what codeword in $\mathcal{C}$ explains this symmetry? ◆

Exercise 736 Do the following:
(a) Show that a generator matrix for $\widehat{\mathcal{D}}_1$ when $p = 7$ is

$$
\begin{bmatrix}
3 & 3 & 3 & 3 & 3 & 3 & 3 & 3 \\
0 & 1 & 2 & 2 & 3 & 2 & 3 & 3 \\
0 & 3 & 1 & 2 & 2 & 3 & 2 & 3 \\
0 & 3 & 3 & 1 & 2 & 2 & 3 & 2
\end{bmatrix}.
$$

(b) Row reduce the matrix found in (a). How is this matrix related to the generator matrix for o_8 found in Example 12.2.5? ◆

The automorphism groups of the extended $\mathbb{Z}_4$-quadratic residue codes of length $p + 1$ possess some of the same automorphisms as the ordinary quadratic residue codes. Denote the coordinates of the extended $\mathbb{Z}_4$-quadratic residue codes by $\{\infty, \mathbb{F}_p\}$. Then $\text{MAut}(\mathcal{D}^{\text{ext}})$, where $\mathcal{D}^{\text{ext}}$ is either $\widehat{\mathcal{D}}_i$ or $\widetilde{\mathcal{D}}_i$, contains the translation automorphisms T_g for $g \in \mathbb{F}_p$ given by $iT_g \equiv i + g \pmod{p}$ for all $g \in \mathbb{F}_p$, the multiplier automorphisms μ_a for $a \in \mathcal{Q}_p$, and an automorphism satisfying the Gleason–Prange Theorem of Section 6.6.4; see [27, 289]. In particular, $\text{MAut}(\mathcal{D}^{\text{ext}})$ is transitive. Call a vector with components in $\mathbb{Z}_4$ *even-like* if the sum of its components is 0 modulo 4 and *odd-like* otherwise. As in Theorem 1.7.13, the minimum Lee weight codewords in a $\mathbb{Z}_4$-quadratic residue $\mathcal{D}_i$ are all odd-like.

12.5   Self-dual codes over $\mathbb{Z}_4$

We have noticed that codes such as the octacode and some of the extended $\mathbb{Z}_4$-quadratic residue codes are self-dual. In this section we will study this family of codes. Again much of the study of self-dual codes over $\mathbb{Z}_4$ parallels that of self-dual codes over $\mathbb{F}_q$. We have already observed one important difference; namely, there are self-dual codes of odd length over $\mathbb{Z}_4$. For example, in Table 12.1 we found three self-dual cyclic codes of length 7.

By Theorem 12.1.5, the Euclidean weight of every codeword in a self-orthogonal code is a multiple of 4. By Theorem 12.2.4, if the Euclidean weight of every codeword is a multiple of 8, the code is self-orthogonal. This leads us to define Type I and II codes over $\mathbb{Z}_4$. A self-dual $\mathbb{Z}_4$-linear code is *Type II* if the Euclidean weight of every codeword is a multiple of 8. We will see later that Type II codes exist only for lengths $n \equiv 0 \pmod{8}$. These codes also contain a codeword with all coordinates ± 1. For example, by Theorem 12.4.7, the extended $\mathbb{Z}_4$-quadratic residue codes of length $p + 1$ with $p \equiv -1 \pmod{8}$ are Type II. A self-dual $\mathbb{Z}_4$-linear code is *Type I* if the Euclidean weight of some codeword is not a multiple of 8.

There are Gleason polynomials for self-dual codes over $\mathbb{Z}_4$ analogous to those that arise for self-dual codes over $\mathbb{F}_2$, $\mathbb{F}_3$, and $\mathbb{F}_4$ presented in Gleason's Theorem of Section 9.2. These polynomials can be found for the Hamming, symmetrized, and complete weight enumerators; see [291] for these polynomials.

There is also an upper bound on the Euclidean weight of a Type I or Type II code over $\mathbb{Z}_4$. The proof of the following can be found in [26, 292]; compare this to Theorems 9.3.1, 9.3.5, and 9.4.14.

Theorem 12.5.1 *Let C be a self-dual code over $\mathbb{Z}_4$ of length n. The following hold*:
(i) *If C is Type II, then the minimum Euclidean weight of C is at most $8 \lfloor n/24 \rfloor + 8$.*
(ii) *If C is Type I, then the minimum Euclidean weight of C is at most $8 \lfloor n/24 \rfloor + 8$ except when $n \equiv 23 \pmod{24}$, in which case the bound is $8 \lfloor n/24 \rfloor + 12$. If equality holds in this latter bound, then C is obtained by shortening[3] a Type II code of length $n + 1$.*

Codes meeting these bounds are called *Euclidean-extremal*.

Example 12.5.2 By Example 12.4.8 and Exercise 735, the extended $\mathbb{Z}_4$-quadratic residue codes of length 24 are Type II with minimum Euclidean weight 16. These are Euclidean-extremal codes.   ∎

The bounds of Theorem 12.5.1 are obviously bounds on the minimum Lee weight of self-dual codes, but highly unsatisfactory ones. No good bound on the minimum Lee weight is currently known.

---

[3] Shortening a $\mathbb{Z}_4$-linear code on a given coordinate is done as follows. If there are codewords that have every value in $\mathbb{Z}_4$ in the given coordinate position, choose those codewords with only 0 or 2 in that coordinate position and delete that coordinate to produce the codewords of the shortened code. If all codewords have only 0 or 2 in the given coordinate position (and some codeword has 2 in that position), choose those codewords with only 0 in that coordinate position and delete that coordinate to produce the codewords of the shortened code. In each case, the shortened code is linear with half as many codewords as the original. Furthermore, if the original code is self-dual, so is the shortened code.

Research Problem 12.5.3 *Give an improved upper bound on the minimum Lee weight of self-dual codes over $\mathbb{Z}_4$.*

Exercise 737 By Theorem 12.1.5, the Lee weight of every codeword in a self-orthogonal $\mathbb{Z}_4$-linear code is even. It is natural to ask what happens in a code over $\mathbb{Z}_4$ in which all codewords have Lee weight a multiple of 4.

(a) Give an example of a vector in $\mathbb{Z}_4^n$ of Lee weight 4 that is not orthogonal to itself.

(b) Unlike what occurs with codes where all Euclidean weights are multiples of 8, part (a) indicates that a code over $\mathbb{Z}_4$ in which all codewords have Lee weight a multiple of 4 is not necessarily self-orthogonal. However, if C is a self-dual $\mathbb{Z}_4$-linear code where all codewords have Lee weight a multiple of 4, then $\mathfrak{G}(C)$ is linear. Prove this. Hint: Let $\mathbf{u}, \mathbf{v} \in C$. By Theorem 12.2.2(i), $\mathrm{wt}_L(\mathbf{u} - \mathbf{v}) = d_L(\mathbf{u}, \mathbf{v}) = d(\mathfrak{G}(\mathbf{u}), \mathfrak{G}(\mathbf{v})) = \mathrm{wt}(\mathfrak{G}(\mathbf{u}) + \mathfrak{G}(\mathbf{v}))$. As C is linear, $\mathbf{u} - \mathbf{v} \in C$. Also $\mathrm{wt}_L(\mathbf{u}) = \mathrm{wt}(\mathfrak{G}(\mathbf{u}))$ and $\mathrm{wt}_L(\mathbf{v}) = \mathrm{wt}(\mathfrak{G}(\mathbf{v}))$. Use these facts and $\mathrm{wt}(\mathfrak{G}(\mathbf{u}) + \mathfrak{G}(\mathbf{v})) = \mathrm{wt}(\mathfrak{G}(\mathbf{u})) + \mathrm{wt}(\mathfrak{G}(\mathbf{v})) - 2\mathrm{wt}(\mathfrak{G}(\mathbf{u}) \cap \mathfrak{G}(\mathbf{v}))$ together with our assumption to show $\mathfrak{G}(\mathbf{u})$ and $\mathfrak{G}(\mathbf{v})$ are orthogonal. Assessing the size of $\mathfrak{G}(C)$, show that the self-orthogonal binary linear code spanned by $\mathfrak{G}(C)$ cannot be larger than $\mathfrak{G}(C)$.   ♦

Let C be a $\mathbb{Z}_4$-linear code of length n. There are two binary linear codes of length n associated with C. The *residue code* $\mathrm{Res}(C)$ is $\mu(C)$. The *torsion code* $\mathrm{Tor}(C)$ is $\{\mathbf{b} \in \mathbb{F}_2^n \mid 2\mathbf{b} \in C\}$. So the vectors in $\mathrm{Tor}(C)$ are obtained from the vectors in C with all components 0 or 2 by dividing these components in half. If C has generator matrix G in standard form (12.1), then $\mathrm{Res}(C)$ and $\mathrm{Tor}(C)$ have generator matrices

$$G_{\mathrm{Res}} = \begin{bmatrix} I_{k_1} & A & B_1 \end{bmatrix} \quad \text{and} \tag{12.13}$$

$$G_{\mathrm{Tor}} = \begin{bmatrix} I_{k_1} & A & B_1 \\ O & I_{k_2} & C \end{bmatrix}. \tag{12.14}$$

So $\mathrm{Res}(C) \subseteq \mathrm{Tor}(C)$. If C is self-dual, we have the following additional relationship between $\mathrm{Res}(C)$ and $\mathrm{Tor}(C)$.

Theorem 12.5.4 *If C is a self-dual $\mathbb{Z}_4$-linear code, then $\mathrm{Res}(C)$ is doubly-even and $\mathrm{Res}(C) = \mathrm{Tor}(C)^{\perp}$.*

Proof: Suppose C has generator matrix G given by (12.1). Denote one of the first k_1 rows of G by $\mathbf{r} = (\mathbf{i}, \mathbf{a}, \mathbf{b}_1 + 2\mathbf{b}_2)$, where $\mathbf{i}, \mathbf{a}, \mathbf{b}_1$, and $\mathbf{b}_2$ are rows of I_{k_1}, A, B_1, and B_2, respectively. Then $0 \equiv \mathbf{r} \cdot \mathbf{r} \equiv \mathbf{i} \cdot \mathbf{i} + \mathbf{a} \cdot \mathbf{a} + \mathbf{b}_1 \cdot \mathbf{b}_1 \pmod{4}$, implying that the rows of (12.13) are doubly-even and so are orthogonal to themselves. If $\mathbf{r}' = (\mathbf{i}', \mathbf{a}', \mathbf{b}_1' + 2\mathbf{b}_2')$ is another of the first k_1 rows of G, then

$$\mathbf{r} \cdot \mathbf{r}' \equiv \mathbf{i} \cdot \mathbf{i}' + \mathbf{a} \cdot \mathbf{a}' + \mathbf{b}_1 \cdot \mathbf{b}_1' + 2(\mathbf{b}_2 \cdot \mathbf{b}_1' + \mathbf{b}_1 \cdot \mathbf{b}_2') \pmod{4}. \tag{12.15}$$

If $\mathbf{s} = (\mathbf{0}, 2\mathbf{i}', 2\mathbf{c})$ is one of the bottom k_2 rows of G, using analogous notation, then

$$\mathbf{r} \cdot \mathbf{s} \equiv 2\mathbf{a} \cdot \mathbf{i}' + 2\mathbf{b}_1 \cdot \mathbf{c} \pmod{4}. \tag{12.16}$$

As $\mathbf{r} \cdot \mathbf{r}' \equiv \mathbf{r} \cdot \mathbf{s} \equiv 0 \pmod{4}$, (12.15) and (12.16) imply that the rows of (12.13) are orthogonal as binary vectors to the rows of (12.14). Thus $\mathrm{Res}(C) \subseteq \mathrm{Tor}(C)^{\perp}$; as C is self-dual,

$2k_1 + k_2 = n$, implying $\text{Res}(\mathcal{C}) = \text{Tor}(\mathcal{C})^\perp$. In particular, $\text{Res}(\mathcal{C})$ is self-orthogonal and as it has a generator matrix with doubly-even rows, $\text{Res}(\mathcal{C})$ is doubly-even. $\qquad\square$

One consequence of this result is that a Type II code must contain a codeword with all entries ± 1. This implies that Type II codes can exist only for lengths a multiple of 8. The simple proof we give of this was originally due to Gaborit.

Corollary 12.5.5 *Let $\mathcal{C}$ be a Type II code of length n. Then $\text{Tor}(\mathcal{C})$ is an even binary code, $\text{Res}(\mathcal{C})$ contains the all-one binary vector, $\mathcal{C}$ contains a codeword with all entries ± 1, and $n \equiv 0 \,(\text{mod } 8)$.*

Proof: Let $\mathbf{c} \in \text{Tor}(\mathcal{C})$. Then $2\mathbf{c} \in \mathcal{C}$; as $\text{wt}_E(2\mathbf{c}) \equiv 0 \,(\text{mod } 8)$, $\mathbf{c}$ is a binary vector of even weight. Hence $\text{Tor}(\mathcal{C})$ is an even binary code. So $\text{Res}(\mathcal{C})$ contains the all-one binary vector $\mathbf{1}$ as $\text{Res}(\mathcal{C}) = \text{Tor}(\mathcal{C})^\perp$ by Theorem 12.5.4. Any vector $\mathbf{v} \in \mathcal{C}$ with $\mu(\mathbf{v}) = \mathbf{1}$ has no entries 0 or 2; at least one such $\mathbf{v}$ exists. As $0 \equiv \text{wt}_E(\mathbf{v}) \equiv n \,(\text{mod } 8)$, the proof is complete. $\qquad\square$

If we start with an arbitrary doubly-even binary code, can we form a self-dual $\mathbb{Z}_4$-linear code with this binary code as its residue code? This is indeed possible, a fact we explore in the next section; but in order to do so, we need another form for the generator matrix.

Corollary 12.5.6 *Let $\mathcal{C}$ be a self-dual $\mathbb{Z}_4$-linear code with generator matrix in standard form. Then $\mathcal{C}$ has a generator matrix of the form*

$$G' = \begin{bmatrix} F & I_k + 2B \\ 2H & O \end{bmatrix},$$

where B, F, and H are binary matrices. Furthermore, generator matrices for $\text{Res}(\mathcal{C})$ and $\text{Tor}(\mathcal{C})$ are

$$G'_{\text{Res}} = [F \quad I_k] \quad and \quad G'_{\text{Tor}} = \begin{bmatrix} F & I_k \\ H & O \end{bmatrix},$$

respectively.

Proof: Let $k = k_1$ where $\mathcal{C}$ has type $4^{k_1} 2^{k_2}$. As $\mathcal{C}$ is self-dual, $k_2 = n - 2k$. We first show that $\mathcal{C}$ has a generator matrix of the form

$$G'' = \begin{bmatrix} D & E & I_k + 2B \\ O & 2I_{n-2k} & 2C \end{bmatrix},$$

where B, C, D, and E are binary matrices. To verify this, we only need to show that we can replace the first $k = k_1$ rows of G from (12.1) by $[D \ E \ I_k + 2B]$. By Theorems 12.5.4 and 1.6.2, the right-most k coordinates of $\text{Res}(\mathcal{C})$ are information positions, as the left-most $n - k$ coordinates of $\text{Tor}(\mathcal{C})$ are information positions. Thus by (12.13), B_1 has binary rank k and so has a binary inverse D. Hence the first k rows of G can be replaced by $D[I_k \ A \ B_1 + 2B_2] = [D \ E + 2E_1 \ I_k + 2B_3]$, where $DA = E + 2E_1$ and $D(B_1 + 2B_2) = I_k + 2B_3$. Adding $E_1[O \ 2I_{n-2k} \ 2C]$ to this gives G''. Now add $2C[D \ E \ I_k + 2B]$ to the bottom $n - 2k$ rows of G'' to obtain G'. That $\text{Res}(\mathcal{C})$ and $\text{Tor}(\mathcal{C})$ have the stated generator matrices follows from the definitions of the residue and torsion codes. $\qquad\square$

12.5.1   Mass formulas

We can use the form of the generator matrix for a self-dual $\mathbb{Z}_4$-linear code given in Corollary 12.5.6 to count the total number of such codes. Let $\mathcal{C}$ be any self-dual $\mathbb{Z}_4$-linear code of type $4^k 2^{n-2k}$ with $0 \le k \le \lfloor n/2 \rfloor$. By Theorem 12.5.4, $\mathrm{Res}(\mathcal{C})$ is an $[n, k]$ self-orthogonal doubly-even binary code, and we can apply a permutation matrix P to $\mathcal{C}$ so that the generator matrix G' of $\mathcal{C}P$ is given in Corollary 12.5.6. Conversely, if we begin with an $[n, k]$ self-orthogonal doubly-even binary code with generator matrix $[F\ I_k]$ that generates $\mathrm{Res}(\mathcal{C}P)$, we must be able to find the binary matrix B to produce the first k rows of G'. While the submatrix H in G' is not unique, its span is uniquely determined as $\mathrm{Res}(\mathcal{C})^{\perp} = \mathrm{Tor}(\mathcal{C})$. This produces a generator matrix for $\mathcal{C}P$ and from there we can produce $\mathcal{C}$. So to count the total number of codes $\mathcal{C}$ with a given residual code, we only need to count the number of choices for B.

Beginning with a generator matrix $[F\ I_k]$, which generates an $[n, k]$ self-orthogonal doubly-even binary code $\mathcal{C}_1$, choose H so that

$$\begin{bmatrix} F & I_k \\ H & O \end{bmatrix}$$

generates $\mathcal{C}_2 = \mathcal{C}_1^{\perp}$. We now show that there are $2^{k(k+1)/2}$ choices for B that yield self-dual $\mathbb{Z}_4$-linear codes with generator matrices of the form G' given in Corollary 12.5.6. As the inner product modulo 4 of vectors whose components are only 0s and 2s is always 0, the inner product of two of the bottom $n - 2k$ rows of G' is 0. Regardless of our choice of B, the inner product modulo 4 of one of the top k rows of G' with one of the bottom $n - 2k$ rows of G' is also always 0 as $\mathcal{C}_1^{\perp} = \mathcal{C}_2$. Furthermore, the inner product modulo 4 of one of the top k rows of G' with itself is 0 regardless of the choice of B because $\mathcal{C}_1$ is doubly-even. Let $B = [b_{i,j}]$ with $1 \le i \le k, 1 \le j \le k$. Choose the entries of B on or above the diagonal (that is, $b_{i,j}$ with $1 \le i \le j \le k$) arbitrarily; notice that we are freely choosing $k(k + 1)/2$ entries. We only need to make sure that the inner product of row i and row j of G', with $1 \le i < j \le k$, is 0 modulo 4. But this inner product modulo 4 is

$$\mathbf{f}_i \cdot \mathbf{f}_j + 2(b_{i,j} + b_{j,i}), \tag{12.17}$$

where $\mathbf{f}_i$ and $\mathbf{f}_j$ are rows i and j of F. As $\mathcal{C}_1$ is self-orthogonal, $\mathbf{f}_i \cdot \mathbf{f}_j \equiv 0 \,(\mathrm{mod}\ 2)$ implying that we can solve (12.17) for the binary value $b_{j,i}$ uniquely so that the desired inner product is 0 modulo 4. This proves the following result, first shown in [92].

Theorem 12.5.7 *For $0 \le k \le \lfloor n/2 \rfloor$, there are $v_{n,k} 2^{k(k+1)/2}$ self-dual codes over $\mathbb{Z}_4$ of length n and type $4^k 2^{n-2k}$, where $v_{n,k}$ is the number of $[n, k]$ self-orthogonal doubly-even binary codes. The total number of self-dual codes over $\mathbb{Z}_4$ of length n is*

$$\sum_{k=0}^{\lfloor \frac{n}{2} \rfloor} v_{n,k} 2^{k(k+1)/2}.$$

Exercise 738 Do the following:

(a) Fill in each entry $\star$ with 0 or 2 so that the resulting generator matrix yields a self-dual code:

$$G' = \begin{bmatrix} 0 & 1 & 1 & 1 & 3 & 2 & 0 & 2 \\ 1 & 0 & 1 & 1 & \star & 1 & 2 & 0 \\ 1 & 1 & 0 & 1 & \star & \star & 3 & 0 \\ 1 & 1 & 1 & 0 & \star & \star & \star & 1 \end{bmatrix}.$$

(b) Fill in a 4×6 matrix H with entries 0 or 1 so that

$$\begin{bmatrix} F & I_2 + 2B \\ 2H & O \end{bmatrix}$$

generates a self-dual code over $\mathbb{Z}_4$ of length 8 and type $4^2 2^4$, where

$$[F \quad I_2 + 2B] = \begin{bmatrix} 1 & 1 & 1 & 0 & 0 & 0 & 3 & 2 \\ 0 & 1 & 1 & 1 & 0 & 0 & 0 & 1 \end{bmatrix}. \qquad \blacklozenge$$

A similar argument allows us to count the number of Type II codes. Note that by Corollary 12.5.5, if $\mathcal{C}$ is Type II of length n, then $n \equiv 0 \pmod 8$ and $\mathrm{Res}(\mathcal{C})$ contains the all-one vector. For a proof of the following, see [92, 276] or Exercise 739.

Theorem 12.5.8 *If $n \equiv 0 \pmod 8$, there are $\mu_{n,k} 2^{1+k(k-1)/2}$ Type II codes over $\mathbb{Z}_4$ of length n and type $4^k 2^{n-2k}$ for $0 \le k \le n/2$, where $\mu_{n,k}$ is the number of $[n,k]$ self-orthogonal doubly-even binary codes containing $\mathbf{1}$. The total number of Type II codes over $\mathbb{Z}_4$ of length n is*

$$\sum_{k=0}^{\frac{n}{2}} \mu_{n,k} 2^{1+k(k-1)/2}.$$

Note that in the proof of Theorem 9.12.5, we gave a recurrence relation for $\mu_{n,k}$; namely $\mu_{n,1} = 1$ and

$$\mu_{n,k+1} = \frac{2^{n-2k-1} + 2^{n/2-k-1} - 1}{2^k - 1} \mu_{n,k}$$

for $k \ge 1$.

Exercise 739 Suppose F is a $k \times (n-k)$ binary matrix such that the binary code $\mathcal{R}$ of length $n \equiv 0 \pmod 8$ generated by $[F \ I_k]$ is doubly-even and contains the all-one vector. Suppose H is an $(n-2k) \times (n-k)$ binary matrix such that

$$\begin{bmatrix} F & I_k \\ H & O \end{bmatrix}$$

generates the binary dual of $\mathcal{R}$. By doing the following, show that there are exactly $2^{1+k(k-1)/2}$ $k \times k$ binary matrices $B = [b_{i,j}]$ such that

$$G' = \begin{bmatrix} F & I_k + 2B \\ 2H & O \end{bmatrix}$$

generates a Type II $\mathbb{Z}_4$-linear code. This will prove the main part of Theorem 12.5.8. Choose the entries of B in rows $2 \leq i \leq k$ on or above the main diagonal arbitrarily. Also choose the entry $b_{1,1}$ arbitrarily. Thus $1 + k(k-1)/2$ entries have been chosen arbitrarily. This exercise will show that a fixed choice for these entries determines all other entries uniquely so that G' generates a Type II $\mathbb{Z}_4$-linear code.

(a) Show that every row of H has an even number of ones and that the rows of $[2H \ \ O]$ have Euclidean weight a multiple of 8.

(b) Show that any row of $[2H \ \ O]$ is orthogonal to any row of G'.

(c) Let $2 \leq i \leq k$. Show that if all the entries in row i of B except $b_{i,1}$ are fixed, we can uniquely choose $b_{i,1}$ so that the Euclidean weight of row i of G' is a multiple of 8. Hint: First show that if all the entries of row i of B including $b_{i,1}$ are fixed, the Euclidean weight of row i of G' is either 0 or 4 modulo 8.

(d) The entries $b_{1,1}$ and $b_{i,j}$ for $2 \leq i \leq j \leq k$ have been fixed, and part (c) shows that $b_{i,1}$ is determined for $2 \leq i \leq k$. Show that the remaining entries of B are uniquely determined by the requirement that the rows of G' are to be orthogonal to each other. Hint: Use (12.17) noting that $\mathbf{f}_i \cdot \mathbf{f}_j \equiv 0 \pmod{2}$.

(e) We now show that G' generates a Type II $\mathbb{Z}_4$-linear code. At this point we know that G' generates a self-dual $\mathbb{Z}_4$-linear code where all rows, except possibly the first, have Euclidean weight a multiple of 8. Let G'' be a matrix identical to G' except with a different first row. Let the first row of G'' be the sum of rows $1, 2, \ldots, k$ of G'. Note that G' and G'' generate the same code and that the code is Type II if we show the first row of G'' has Euclidean weight a multiple of 8. Show that this is indeed true. Hint: The first row of G'' modulo 2 must be in the binary code $\mathcal{R}$, which contains the all-one vector. Modulo 2 what is this row? Show that this row has Euclidean weight $n \equiv 0 \pmod{8}$.

(f) Let

$$G' = \begin{bmatrix} 0 & 1 & 1 & 1 & 3 & \star & \star & \star \\ 1 & 0 & 1 & 1 & \star & 1 & 2 & 0 \\ 1 & 1 & 0 & 1 & \star & \star & 3 & 0 \\ 1 & 1 & 1 & 0 & \star & \star & \star & 1 \end{bmatrix}.$$

Fill in each entry $\star$ with 0 or 2 so that G' generates a Type II code.   ◆

In [59, 85], the self-dual $\mathbb{Z}_4$-linear codes of length n with $1 \leq n \leq 15$ are completely classified; in addition, the Type II codes of length 16 are classified in [276]. The mass formula together with the size of the automorphism groups of the codes is used, as was done in the classification of self-dual codes over $\mathbb{F}_q$, to insure that the classification is complete.

The heart of the classification is to compute the indecomposable codes. Recall that a code is decomposable if it is equivalent to a direct sum of two or more nonzero subcodes, called components, and indecomposable otherwise. If a self-dual code is decomposable, each of its component codes must be self-dual. Furthermore, the components of a decomposable Type II code must also be Type II codes. Recall that by Exercise 685, the only self-dual code over $\mathbb{Z}_4$ of length 1 is the code we denote A_1 with generator matrix $[\ 2 \ ]$. In that

Table 12.2 *Indecomposable self-dual codes over $\mathbb{Z}_4$ of length $1 \le n \le 16$*

n	Type I	Type II	n	Type I	Type II	n	Type I	Type II
1	1	0	7	1	0	13	7	0
2	0	0	8	2	4	14	92	0
3	0	0	9	0	0	15	111	0
4	1	0	10	4	0	16	?	123
5	0	0	11	2	0			
6	1	0	12	31	0			

exercise you deduced that, up to equivalence, the only self-dual codes of lengths 2 and 3 are $2A_1$ and $3A_1$, respectively, where $2A_1$ is the direct sum of two copies of A_1. You also discovered that, up to equivalence, there are two inequivalent self-dual codes of length 4: the decomposable code $4A_1$ and the indecomposable code D_4 with generator matrix

$$\begin{bmatrix} 1 & 1 & 1 & 1 \\ 0 & 2 & 0 & 2 \\ 0 & 0 & 2 & 2 \end{bmatrix}.$$

Table 12.2 contains the number of inequivalent indecomposable Type I and Type II codes of length n with $1 \le n \le 16$; this data is taken from [59, 85, 276]. While the real work in the classification is to find the indecomposable codes, using the mass formula requires a list of both the indecomposable and the decomposable codes.

Exercise 740 Let $\mathcal{C}$ be a self-orthogonal code of length n over $\mathbb{Z}_4$. Suppose that $\mathcal{C}$ contains a codeword with one component equal to 2 and the rest equal to 0. Prove that $\mathcal{C}$ is equivalent to a direct sum of A_1 and a self-dual code of length $n - 1$. ◆

Exercise 741 Let $\mathcal{C}$ be equivalent to a direct sum $\mathcal{C}_1 \oplus \mathcal{C}_2$ of the indecomposable codes $\mathcal{C}_1$ and $\mathcal{C}_2$. Also let $\mathcal{C}$ be equivalent to a direct sum $\mathcal{D}_1 \oplus \mathcal{D}_2$ of the indecomposable codes $\mathcal{D}_1$ and $\mathcal{D}_2$. Prove that $\mathcal{C}_1$ is equivalent to either $\mathcal{D}_1$ or $\mathcal{D}_2$, while $\mathcal{C}_2$ is equivalent to the other $\mathcal{D}_i$. ◆

Exercise 742 Construct one indecomposable self-dual code of length 6 over $\mathbb{Z}_4$. Also construct one indecomposable self-dual code of length 7. Hint: For length 7, check Table 12.1. ◆

Table 12.3 presents the total number of inequivalent Type I and Type II codes of length n with $1 \le n \le 16$. The Type II codes represented in this table all have minimum Euclidean weight 8 and hence are Euclidean-extremal by Theorem 12.5.1.

Example 12.5.9 From Table 12.2, there are four indecomposable Type II codes of length 8 and two indecomposable Type I codes. The decomposable self-dual codes of length 8 are $8A_1$, $4A_1 \oplus D_4$, $2A_1 \oplus E_6$, $A_1 \oplus E_7$, and $2D_4$, where E_6 and E_7 are the indecomposable codes of lengths 6 and 7, respectively, constructed in Exercise 742. All the decomposable

Table 12.3 *Self-dual codes over $\mathbb{Z}_4$ of length $1 \le n \le 16$*

n	Type I	Type II	n	Type I	Type II	n	Type I	Type II
1	1	0	7	4	0	13	66	0
2	1	0	8	7	4	14	170	0
3	1	0	9	11	0	15	290	0
4	2	0	10	16	0	16	?	133
5	2	0	11	19	0			
6	3	0	12	58	0			

codes are of Type I. The five decomposable codes are also all inequivalent to each other by a generalization of Exercise 741. This verifies the entries for length 8 in Table 12.3. ■

Exercise 743 Assuming the numbers in Table 12.2, verify that the numbers in Table 12.3 are correct for lengths 10, 11, 12, and 16. ◆

12.5.2 Self-dual cyclic codes

We have observed numerous times that self-dual cyclic codes over $\mathbb{Z}_4$ exist. Theorem 12.3.13 gives a pair of generating polynomials for cyclic codes. The next theorem, found in [283], gives conditions on these polynomials that lead to self-dual codes.

Theorem 12.5.10 *Let $C = \langle f(x)g(x) \rangle \oplus \langle 2f(x)h(x) \rangle$ be a cyclic code over $\mathbb{Z}_4$ of odd length n, where $f(x)$, $g(x)$, and $h(x)$ are monic polynomials such that $x^n - 1 = f(x)g(x)h(x)$. Then C is self-dual if and only if $f(x) = h^*(x)$ and $g(x) = g^*(x)$.*

Proof: We first remark that the constant term of any irreducible factor of $x^n - 1$ is not equal to either 0 or 2. In particular, by definition, $f^*(x)$, $g^*(x)$, and $h^*(x)$ are all monic and $f^*(x)g^*(x)h^*(x) = x^n - 1$.

Suppose that $f(x) = h^*(x)$ and $g(x) = g^*(x)$. By Theorem 12.3.20, $C^\perp = \langle h^*(x)g^*(x) \rangle \oplus \langle 2h^*(x)f^*(x) \rangle = \langle f(x)g(x) \rangle \oplus \langle 2f(x)h(x) \rangle = C$ and C is self-dual.

Now assume that C is self-dual. Since $C = \langle f(x)g(x) \rangle \oplus \langle 2f(x)h(x) \rangle$ and $C^\perp = \langle h^*(x)g^*(x) \rangle \oplus \langle 2h^*(x)f^*(x) \rangle$ and these decompositions are unique, we have

$$f(x)g(x) = h^*(x)g^*(x), \quad \text{and} \tag{12.18}$$

$$f(x)h(x) = h^*(x)f^*(x). \tag{12.19}$$

From (12.18), $x^n - 1 = f(x)g(x)h(x) = h^*(x)g^*(x)h(x)$. As $f^*(x)g^*(x)h^*(x) = x^n - 1$, we have $h^*(x)g^*(x)h(x) = h^*(x)g^*(x)f^*(x) = x^n - 1$. By the unique factorization of $x^n - 1$ into monic irreducible polynomials, $f^*(x) = h(x)$. Similarly, using (12.19), $x^n - 1 = f(x)h(x)g(x) = h^*(x)f^*(x)g(x) = h^*(x)f^*(x)g^*(x)$, implying that $g(x) = g^*(x)$. □

Example 12.5.11 In Exercise 714, we found that the factorization of $x^{15} - 1$ over $\mathbb{Z}_4$ is given by

$$x^{15} - 1 = g_1(x)g_2(x)g_3(x)g_4(x)g_4^*(x),$$

where $g_1(x) = x - 1$, $g_2(x) = x^4 + x^3 + x^2 + x + 1$, $g_3(x) = x^2 + x + 1$, and $g_4(x) = x^4 + 2x^2 + 3x + 1$. Let $\mathcal{C} = \langle f(x)g(x) \rangle \oplus \langle 2f(x)h(x) \rangle$ be a self-dual cyclic code of length 15. By Theorem 12.5.10, $f^*(x) = h(x)$ and $g^*(x) = g(x)$. This implies that if $g(x)$ contains a given factor, it must also contain the reciprocal polynomial of that factor. Also if an irreducible factor of $x^{15} - 1$ is its own reciprocal polynomial, it must be a factor of $g(x)$. Thus there are only the following possibilities:

(a) $f(x) = g_4(x)$, $g(x) = g_1(x)g_2(x)g_3(x)$, and $h(x) = g_4^*(x)$,
(b) $f(x) = g_4^*(x)$, $g(x) = g_1(x)g_2(x)g_3(x)$, and $h(x) = g_4(x)$, and
(c) $f(x) = 1$, $g(x) = x^{15} - 1$, and $h(x) = 1$.

The codes in (a) and (b) are of type $4^4 2^7$; they are equivalent under the multiplier μ_{-1}. The code in (c) is the trivial self-dual code with generator matrix $2I_{15}$.   ∎

Exercise 744 Consider the cyclic code $\mathcal{C}$ of length 15 with generator polynomials given by (a) in Example 12.5.11.
(a) Write down a 4×15 generator matrix for $\text{Res}(\mathcal{C})$. Note that $\text{Res}(\mathcal{C})$ is a cyclic code with generator polynomial $\mu(f(x)g(x))$.
(b) The generator matrix from part (a) is the parity check matrix for $\text{Tor}(\mathcal{C})$ by Theorem 12.5.4. What well-known code is $\text{Tor}(\mathcal{C})$? What well-known code is $\text{Res}(\mathcal{C})$?   ◆

Exercise 745 The factorization of $x^{21} - 1$ over $\mathbb{Z}_4$ is given by

$$x^{21} - 1 = g_1(x)g_2(x)g_3(x)g_3^*(x)g_4(x)g_4^*(x),$$

where $g_1(x) = x - 1$, $g_2(x) = x^2 + x + 1$, $g_3(x) = x^6 + 2x^5 + 3x^4 + 3x^2 + x + 1$, and $g_4(x) = x^3 + 2x^2 + x + 3$.
(a) As in Example 12.5.11, list all possible triples $(f(x), g(x), h(x))$ that lead to self-dual cyclic codes of length 21 over $\mathbb{Z}_4$. Note: There are nine such codes including the trivial self-dual code with generator matrix $2I_{21}$.
(b) Give the types of each code found in (a).   ◆

12.5.3 Lattices from self-dual codes over $\mathbb{Z}_4$

In Section 10.6 we introduced the concept of a lattice and described how to construct lattices from codes, particularly self-dual binary codes. One method to construct lattices from binary codes described in that section is called Construction A. There is a $\mathbb{Z}_4$-analogue called *Construction A$_4$*, beginning with a $\mathbb{Z}_4$-linear code $\mathcal{C}$ of length n. Construction A$_4$ produces a lattice $\Lambda_4(\mathcal{C})$, which is the set of all $\mathbf{x}$ in $\mathbb{R}^n$ obtained from a codeword in $\mathcal{C}$ by viewing the codeword as an integer vector with 0s, 1s, 2s, and 3s, adding integer multiples of 4 to any components, and dividing the resulting vector by 2. In particular,

$$\Lambda_4(\mathcal{C}) = \{\mathbf{x} \in \mathbb{R}^n \mid 2\mathbf{x} \,(\text{mod } 4) \in \mathcal{C}\}.$$

If the generator matrix G for C is written in standard form (12.1), then the generator matrix M for $\Lambda_4(C)$ is

$$M = \frac{1}{2} \begin{bmatrix} I_{k_1} & A & B_1 + 2B_2 \\ O & 2I_{k_2} & 2C \\ O & O & 4I_{n-k_1-k_2} \end{bmatrix}.$$ (12.20)

The following is the analogue of Theorem 10.6.4.

Theorem 12.5.12 *Let C be a $\mathbb{Z}_4$-linear code of length n and minimum Euclidean weight d_E. The following hold:*

(i) *If $d_E \leq 16$, the minimum norm μ of $\Lambda_4(C)$ is $\mu = d_E/4$; if $d_E > 16$, $\mu = 4$.*
(ii) $\det \Lambda_4(C) = 4^{n-2k_1-k_2}$.
(iii) $\Lambda_4(C^{\perp}) = \Lambda_4(C)^*$.
(iv) *$\Lambda_4(C)$ is integral if and only if C is self-orthogonal.*
(v) *$\Lambda_4(C)$ is Type I if and only if C is Type I.*
(vi) *$\Lambda_4(C)$ is Type II if and only if C is Type II.*

Proof: The proofs of (i) and (vi) are left as an exercise.

For part (ii), recall that $\det \Lambda_4(C) = \det A$, where A is the Gram matrix MM^{T}. Using M as in (12.20),

$$\det \Lambda_4(C) = (\det M)^2 = \left(\frac{1}{2^n} 2^{k_2} 4^{n-k_1-k_2} \right)^2 = 4^{n-2k_1-k_2},$$

verifying (ii).

The generator matrix $G^{\perp}$ for $C^{\perp}$ is given by (12.2). From this we see that the generator matrix $M^{\perp}$ of $\Lambda_4(C^{\perp})$ is

$$M^{\perp} = \frac{1}{2} \begin{bmatrix} -(B_1 + 2B_2)^{\mathsf{T}} - C^{\mathsf{T}} A^{\mathsf{T}} & C^{\mathsf{T}} & I_{n-k_1-k_2} \\ 2A^{\mathsf{T}} & 2I_{k_2} & O \\ 4I_{k_1} & O & O \end{bmatrix}.$$

The inner product (in $\mathbb{Z}_4$) of a row of G from (12.1) with a row of $G^{\perp}$ from (12.2) is 0; hence we see that the (real) inner product of a row of M and a row of $M^{\perp}$ is an integer. This proves that $M^{\perp} M^{\mathsf{T}}$ is an integer matrix and $\Lambda_4(C^{\perp}) \subseteq \Lambda_4(C)^*$. To complete (iii), we must show $\Lambda_4(C)^* \subseteq \Lambda_4(C^{\perp})$. Let $\mathbf{y} \in \Lambda_4(C)^*$. Then $\mathbf{y} M^{\mathsf{T}} \in \mathbb{Z}^n$ by Theorem 10.6.3(i). So there exists $\mathbf{z} \in \mathbb{Z}^n$ with $\mathbf{y} = \mathbf{z}(M^{\mathsf{T}})^{-1} = \mathbf{z}(M^{\mathsf{T}})^{-1}(M^{\perp})^{-1} M^{\perp} = \mathbf{z}(M^{\perp} M^{\mathsf{T}})^{-1} M^{\perp}$. As $M^{\perp} M^{\mathsf{T}}$ is an integer matrix and $\det(M^{\perp} M^{\mathsf{T}}) = \det(M^{\perp}) \det(M^{\mathsf{T}}) = (\pm 2^{-n} \cdot 4^{k_1} \cdot 2^{k_2}) \cdot (2^{-n} \cdot 2^{k_2} \cdot 4^{n-k_1-k_2}) = \pm 1$,

$$(M^{\perp} M^{\mathsf{T}})^{-1} = \frac{1}{\det(M^{\perp} M^{\mathsf{T}})} \mathrm{adj}(M^{\perp} M^{\mathsf{T}})$$

is an integer matrix. Thus $\mathbf{y} = \mathbf{z}' M^{\perp}$ for some $\mathbf{z}' \in \mathbb{Z}^n$. Hence $\mathbf{y} \in \Lambda_4(C^{\perp})$, completing (iii).

The (real) inner product of two rows of M is always an integer if and only if the inner product (in $\mathbb{Z}_4$) of two rows of G is 0, proving (iv).

Suppose $\Lambda_4(C)$ is Type I; then $\det \Lambda_4(C) = \pm 1$, implying $n = 2k_1 + k_2$ by (ii). As $\Lambda_4(C)$ is Type I, it is integral, and thus $C \subseteq C^{\perp}$ by (iv). Hence C is self-dual. The (real) inner product of some lattice point in $\Lambda_4(\mathcal{C})$ with itself is an odd integer; the corresponding codeword must have Euclidean weight 4 modulo 8, implying that $\mathcal{C}$ is Type I. Conversely, suppose $\mathcal{C}$ is Type I. By (iii), $\Lambda_4(\mathcal{C}) = \Lambda_4(\mathcal{C}^\perp) = \Lambda_4(\mathcal{C})^*$. Furthermore, $\Lambda_4(\mathcal{C})$ is integral by (iv). Thus by definition $\Lambda_4(\mathcal{C})$ is unimodular. In addition, a codeword in $\mathcal{C}$ with Euclidean weight 4 modulo 8 corresponds to a lattice point with odd integer norm. This verifies (v).   □

Exercise 746 Prove Theorem 12.5.12(i) and (vi).   ◆

Example 12.5.13 By Example 12.2.5, the octacode o_8 is a Type II code with minimum Euclidean weight 8. By Theorem 12.5.12, $\Lambda_4(o_8)$ is a Type II lattice in $\mathbb{R}^8$ with minimum norm 2. This is the Gosset lattice E_8 introduced in Example 10.6.5, which we recall is the unique Type II lattice in $\mathbb{R}^8$. From that example we saw that E_8 has precisely 240 lattice points of minimum norm 2. These can be found in $\Lambda_4(o_8)$ as follows. The symmetrized weight enumerator of o_8 is

$$\text{swe}_{o_8}(a, b, c) = a^8 + 16b^8 + c^8 + 14a^4c^4 + 112a^3b^4c + 112ab^4c^3.$$

The codewords of o_8 that are of Euclidean weight 8 are the 16 codewords which have all components ± 1 and the 112 codewords with four components ± 1 and one component equal to 2. The former yield 16 lattice points of "shape" $1/2(\pm 1^8)$, indicating a vector with eight entries equal to $\pm 1/2$. The latter yield 112 lattice points of shape $1/2(0^3, \pm 1^4, 2)$; to each of these 112 lattice points, four can be subtracted from the component that equaled 2, yielding 112 lattice points of shape $1/2(0^3, \pm 1^4, -2)$. (Note that this is not the same version of E_8 described in Example 10.6.5. However, the two forms are equivalent.)   ■

Exercise 747 Let $\mathcal{C}$ be one of the extended quadratic residue codes of length 24 presented in Example 12.4.8. It is a Type II code of minimum Euclidean weight 16. By Theorem 12.5.12, $\Lambda_4(\mathcal{C})$ is a Type II lattice in $\mathbb{R}^{24}$ with minimum norm 4. This lattice is equivalent to the Leech lattice Λ_{24} discussed in Section 10.6. Recall that Λ_{24} has 196 560 lattice points of minimum norm 4. Using $\text{swe}_{\mathcal{C}}(a, b, c)$, describe how these minimum norm lattice points arise from the codewords in $\mathcal{C}$.   ◆

12.6 Galois rings

When studying cyclic codes over $\mathbb{F}_q$, we often examine extension fields of $\mathbb{F}_q$ for a variety of reasons. For example the roots of $x^n - 1$, and hence the roots of the generator polynomial of a cyclic code, are in some extension field of $\mathbb{F}_q$. The extension field is often used to present a form of the parity check matrix; see Theorem 4.4.3. The analogous role for codes over $\mathbb{Z}_4$ is played by Galois rings. We will describe properties of these rings, mostly without proof; see [116, 230].

Let $f(x)$ be a monic basic irreducible polynomial of degree r. By Lemmas 12.3.9 and 12.3.10, $\mathbb{Z}_4[x]/(f(x))$ is a ring with 4^r elements and only one nontrivial ideal. (Recall that a finite field has no nontrivial ideals.) Such a ring is called a *Galois ring*. It turns out that all Galois rings of the same order are isomorphic (just as all finite fields of the same order are isomorphic), and we denote a Galois ring of order 4^r by $\text{GR}(4^r)$. Just as the field $\mathbb{F}_q$

contains the subfield $\mathbb{F}_p$, where p is the characteristic of $\mathbb{F}_q$, $\mathrm{GR}(4^r)$ is a ring of characteristic 4 containing the subring $\mathbb{Z}_4$.

The nonzero elements of a finite field form a cyclic group; a generator of this group is called a primitive element. The multiplicative structure of the field is most easily expressed using powers of a fixed primitive element. This is not the case with Galois rings, but an equally useful structure is present. We will use the following theorem, which we do not prove, along with its associated notation, throughout the remainder of this chapter without reference.

Theorem 12.6.1 *The Galois ring $\mathcal{R} = \mathrm{GR}(4^r)$ contains an element ξ of order $2^r - 1$. Every element $c \in \mathcal{R}$ can be uniquely expressed in the form $c = a + 2b$, where a and b are elements of $T(\mathcal{R}) = \{0, 1, \xi, \xi^2, \dots, \xi^{2^r - 2}\}$.*

The element ξ is called a *primitive element*; the expression $c = a + 2b$ with a and b in $T(\mathcal{R})$ is called the *2-adic representation* of c. The lone nontrivial ideal (2) is $2\mathcal{R} = \{2t \mid t \in T(\mathcal{R})\}$; the elements of $2\mathcal{R}$ consist of 0 together with all the divisors of zero in $\mathcal{R}$. The elements of $\mathcal{R} \setminus 2\mathcal{R}$ are precisely the invertible elements of $\mathcal{R}$.

Exercise 748 Let $\mathcal{R} = \mathrm{GR}(4^r)$ have primitive element ξ. Using the fact that every element $c \in \mathcal{R}$ is uniquely expressible in the form $c = a + 2b$, where a and b are in $T(\mathcal{R}) = \{0, 1, \xi, \xi^2, \dots, \xi^{2^r - 2}\}$, prove the following:
(a) The nontrivial ideal $2\mathcal{R} = \{2t \mid t \in T(\mathcal{R})\}$.
(b) The elements in $2\mathcal{R}$ are precisely 0 and the divisors of zero in $\mathcal{R}$.
(c) The invertible elements in $\mathcal{R}$ are exactly the elements in $\mathcal{R} \setminus 2\mathcal{R}$.   ◆

Since $f(x)$ is a monic basic irreducible polynomial of degree r in $\mathbb{Z}_4[x]$, $\mu(f(x))$ is an irreducible polynomial of degree r in $\mathbb{F}_2[x]$. The reduction homomorphism $\mu : \mathbb{Z}_4[x] \to \mathbb{F}_2[x]$ induces a homomorphism $\overline{\mu}$ from $\mathbb{Z}_4[x]/(f(x))$ onto $\mathbb{F}_2[x]/(\mu(f(x)))$ given by $\overline{\mu}(a(x) + (f(x))) = \mu(a(x)) + (\mu(f(x)))$ with kernel (2). Thus if $\mathcal{R}$ is the Galois ring $\mathbb{Z}_4[x]/(f(x))$, the quotient ring $\mathcal{R}/2\mathcal{R}$ is isomorphic to the field $\mathbb{F}_{2^r}$ with the primitive element $\xi \in \mathcal{R}$ mapped to a primitive element of $\mathbb{F}_{2^r}$. The map $\overline{\mu}$ is examined in Exercise 749.

Exercise 749 Prove that if $f(x)$ is a monic basic irreducible polynomial in $\mathbb{Z}_4[x]$, then $\overline{\mu} : \mathbb{Z}_4[x]/(f(x)) \to \mathbb{F}_2[x]/(\mu(f(x)))$ given by $\overline{\mu}(a(x) + (f(x))) = \mu(a(x)) + (\mu(f(x)))$ is a homomorphism onto $\mathbb{F}_2[x]/(\mu(f(x)))$ with kernel (2); be sure that you verify that $\overline{\mu}$ is well-defined.   ◆

Let p be a prime and $q = p^r$. Recall from Theorem 3.6.1 that the automorphism group of the finite field $\mathbb{F}_q$, called the Galois group $\mathrm{Gal}(\mathbb{F}_q)$ of $\mathbb{F}_q$, is a cyclic group of order r generated by the Frobenius automorphism $\sigma_p : \mathbb{F}_q \to \mathbb{F}_q$, where $\sigma_p(\alpha) = \alpha^p$. There is a similar structure for the automorphism group of $\mathrm{GR}(4^r)$, also denoted $\mathrm{Gal}(\mathrm{GR}(4^r))$, and called the *Galois group* of $\mathrm{GR}(4^r)$. This group is also cyclic of order r generated by the *Frobenius automorphism* $\nu_2 : \mathrm{GR}(4^r) \to \mathrm{GR}(4^r)$, with ν_2 defined by

$$\nu_2(c) = a^2 + 2b^2,$$

where $a + 2b$ is the 2-adic representation of c. The elements of $\mathbb{F}_q$ fixed by σ_p are precisely the elements of the prime subfield $\mathbb{F}_p$. Similarly, the elements of GR(4^r) fixed by v_2 are precisely the elements of the subring $\mathbb{Z}_4$.

Exercise 750 Let $\mathcal{R} = \mathrm{GR}(4^r)$ have primitive element ξ. The goal of this exercise is to show that v_2 is indeed an automorphism of $\mathcal{R}$ and that the only elements of $\mathcal{R}$ fixed by v_2 are the elements of $\mathbb{Z}_4$. You may use the fact that every element $c \in \mathcal{R}$ is uniquely expressible in the form $c = a + 2b$, where a and b are in $T(\mathcal{R}) = \{0, 1, \xi, \xi^2, \ldots, \xi^{2^r-2}\}$. Define $\tau(c) = c^{2^r}$ for all $c \in \mathcal{R}$.

(a) Show that if x and y are elements of $\mathcal{R}$, then $(x + 2y)^2 = x^2$.

(b) Show that if a and b are in $T(\mathcal{R})$, then $\tau(a + 2b) = a$. Note: Given $c \in \mathcal{R}$, this shows that $\tau(c) \in T(\mathcal{R})$ and also shows how to find a in order to write $c = a + 2b$ with a and b in $T(\mathcal{R})$.

(c) Show that if x and y are elements of $\mathcal{R}$, then $\tau(x + y) = \tau(x) + \tau(y) + 2(xy)^{2^{r-1}}$.

(d) Let a_1, a_2, b_1, and b_2 be in $T(\mathcal{R})$. Show that $(a_1 + 2b_1) + (a_2 + 2b_2) = (a_1 + a_2 + 2(a_1a_2)^{2^{r-1}}) + 2((a_1a_2)^{2^{r-1}} + b_1 + b_2)$ and that $a_1 + a_2 + 2(a_1a_2)^{2^{r-1}} \in T(\mathcal{R})$. (Note that $(a_1a_2)^{2^{r-1}} + b_1 + b_2$ may not be in $T(\mathcal{R})$ but $2((a_1a_2)^{2^{r-1}} + b_1 + b_2)$ is in $2T(\mathcal{R}) = 2\mathcal{R}$.)

(e) Show that $v_2(x + y) = v_2(x) + v_2(y)$ and $v_2(xy) = v_2(x)v_2(y)$ for all x and y in $\mathcal{R}$.

(f) Show that if a_1 and a_2 are elements of $T(\mathcal{R})$, then $a_1^2 = a_2^2$ implies that $a_1 = a_2$. Hint: Note that ξ has odd multiplicative order $2^r - 1$.

(g) Prove that v_2 is an automorphism of GR(4^r).

(h) Prove that the only elements $a + 2b \in \mathrm{GR}(4^r)$ with a and b in $T(\mathcal{R})$ such that $v_2(a + 2b) = a + 2b$ are those elements with $a \in \{0, 1\}$ and $b \in \{0, 1\}$. (These elements are the elements of $\mathbb{Z}_4$.)   ◆

By writing elements of GR(4^r) in the form $a + 2b$ where a and b are in $T(\mathcal{R})$, we can easily work with the multiplicative structure in GR(4^r). We need to describe the additive structure in GR(4^r). To make this structure easier to understand, we restrict our choice of the basic irreducible polynomial $f(x)$ used to define the Galois ring. Let $n = 2^r - 1$. There exists an irreducible polynomial $f_2(x) \in \mathbb{F}_2[x]$ of degree r that has a root in $\mathbb{F}_{2^r}$ of order n; recall that such a polynomial is called a *primitive polynomial* of $\mathbb{F}_2[x]$. Using Graeffe's method we can find a monic irreducible polynomial $f(x) \in \mathbb{Z}_4[x]$ such that $\mu(f(x)) = f_2(x)$; $f(x)$ is called a *primitive polynomial* of $\mathbb{Z}_4[x]$. In the Galois ring GR(4^r) $= \mathbb{Z}_4[x]/(f(x))$, let $\xi = x + (f(x))$. Then $f(\xi) = 0$ and ξ is in fact a primitive element of GR(4^r). Every element of $c \in \mathrm{GR}(4^r)$ can be expressed in its "multiplicative" form $c = a + 2b$ with a and b in $T(\mathcal{R})$ and in its additive form:

$$c = \sum_{i=0}^{r-1} c_i \xi^i, \quad \text{where } c_i \in \mathbb{Z}_4. \tag{12.21}$$

Exercise 751 Recall from Example 12.3.8 that $f(x) = x^3 + 2x^2 + x - 1$ is a basic irreducible polynomial in $\mathbb{Z}_4[x]$ and an irreducible factor of $x^7 - 1$. Also, $\mu(f(x)) = x^3 + x + 1$ is a primitive polynomial in $\mathbb{F}_2[x]$. In the Galois ring $\mathcal{R} = \mathrm{GR}(4^3) = \mathbb{Z}_4[x]/(f(x))$, let $\xi = x + (f(x))$. So $\xi^3 + 2\xi^2 + \xi - 1 = 0$ in GR(4^3). Do the following:

(a) Verify that the elements $c \in T(\mathcal{R})$ and $c \in 2T(\mathcal{R})$ when written in the additive form $c = b_0 + b_1\xi + b_2\xi^2$ are represented as follows:

Element	b_0	b_1	b_2	Element	b_0	b_1	b_2
0	0	0	0	0	0	0	0
1	1	0	0	2	2	0	0
ξ	0	1	0	2ξ	0	2	0
ξ^2	0	0	1	$2\xi^2$	0	0	2
ξ^3	1	3	2	$2\xi^3$	2	2	0
ξ^4	2	3	3	$2\xi^4$	0	2	2
ξ^5	3	3	1	$2\xi^5$	2	2	2
ξ^6	1	2	1	$2\xi^6$	2	0	2

(b) Compute ξ^7 and verify that it indeed equals 1.
(c) What is the additive form of $\xi^2 + 2\xi^6$?
(d) What is the multiplicative form of $3 + 2\xi + \xi^2$?
(e) Multiply $(\xi^3 + 2\xi^4)(\xi^2 + 2\xi^5)$ and write the answer in both multiplicative and additive forms.
(f) Add $(\xi^2 + 2\xi^4) + (\xi^5 + 2\xi^6)$ and write the answer in both additive and multiplicative forms. ◆

In [27], the $\mathbb{Z}_4$-quadratic residue codes of length n were developed using roots of $x^n - 1$ in a Galois ring. However, our approach in Section 12.4 was to avoid the use of these rings, and hence the roots, and instead focus on the idempotent generators.

Exercise 752 In Example 12.3.8 we factored $x^7 - 1$ over $\mathbb{Z}_4$.
(a) Using the table given in Exercise 751 find the roots of $g_2(x) = x^3 + 2x^2 + x - 1$ and $g_3(x) = x^3 - x^2 + 2x - 1$. Hint: The roots are also roots of $x^7 - 1$ and hence are seventh roots of unity implying that they are powers of ξ.
(b) What do you notice about the exponents of the roots of $g_2(x)$ and $g_3(x)$? ◆

In constructing the Kerdock codes, we will use the relative trace function from $\mathrm{GR}(4^r)$ into $\mathbb{Z}_4$, which is analogous to the trace function from $\mathbb{F}_{2^r}$ into $\mathbb{F}_2$. Recall that the trace function $\mathrm{Tr}_r : \mathbb{F}_{2^r} \to \mathbb{F}_2$ is given by

$$\mathrm{Tr}_r(\alpha) = \sum_{i=0}^{r-1} \alpha^{2^i} = \sum_{i=0}^{r-1} \sigma_2^i(\alpha) \qquad \text{for } \alpha \in \mathbb{F}_{2^r},$$

where σ_2 is the Frobenius automorphism of $\mathbb{F}_{2^r}$. The *relative trace function* $\mathrm{TR}_r : \mathrm{GR}(4^r) \to \mathbb{Z}_4$ is given by

$$\mathrm{TR}_r(\alpha) = \sum_{i=0}^{r-1} v_2^i(\alpha) \qquad \text{for } \alpha \in \mathrm{GR}(4^r).$$

By Exercise 753, $\mathrm{TR}_r(\alpha)$ is indeed an element of $\mathbb{Z}_4$. By Lemma 3.8.5, Tr_r is a surjective map; analogously, TR_r is a surjective map.

Exercise 753 Let ν_2 be the Frobenius automorphism of $GR(4^r)$ and TR_r the relative trace map. Do the following:

(a) Show that ν_2^r is the identity automorphism.

(b) Show that $\nu_2(TR_r(\alpha)) = TR_r(\alpha)$. Why does this show that $TR_r(\alpha)$ is an element of $\mathbb{Z}_4$?

(c) Show that $TR_r(\alpha + \beta) = TR_r(\alpha) + TR_r(\beta)$ for all α and β in $GR(4^r)$.

(d) Show that $TR_r(a\alpha) = aTR_r(\alpha)$ for all $a \in \mathbb{Z}_4$ and $\alpha \in GR(4^r)$. ◆

12.7 Kerdock codes

We now define the binary Kerdock code of length 2^{r+1} as the Gray image of the extended code of a certain cyclic $\mathbb{Z}_4$-linear code of length $n = 2^r - 1$. Let $H(x)$ be a primitive basic irreducible polynomial of degree r. Let $f(x)$ be the reciprocal polynomial of $(x^n - 1)/((x - 1)H(x))$. As in Section 12.3.3, $f(x)$ is a factor of $x^n - 1$. Define $K(r + 1)$ to be the cyclic code of length $2^r - 1$ over $\mathbb{Z}_4$ generated by $f(x)$. By Corollary 12.3.14, $K(r + 1)$ is a code of length $n = 2^r - 1$ and type $4^{n-\deg f} = 4^{r+1}$. Let $\widehat{K}(r + 1)$ be the extended cyclic code obtained by adding an overall parity check to $K(r + 1)$. This is a code of length 2^r and type 4^{r+1}. The *Kerdock code* $\mathcal{K}(r + 1)$ is defined to be the Gray image $\mathfrak{G}(\widehat{K}(r + 1))$ of $\widehat{K}(r + 1)$. So $\mathcal{K}(r + 1)$ is a code of length 2^{r+1} with 4^{r+1} codewords. In [116] it is shown that a simple rearrangement of the coordinates leads directly to the original definition given by Kerdock in [166]. The results of this section are taken primarily from [116]. Earlier Nechaev [243] discovered a related connection between $\mathbb{Z}_4$-sequences and Kerdock codes punctured on two coordinates.

Example 12.7.1 If $n = 2^3 - 1$, we can let $H(x) = x^3 + 2x^2 + x - 1$. In that case, $f^*(x) = (x^7 - 1)/((x - 1)H(x)) = x^3 - x^2 + 2x - 1$ and $f(x) = x^3 + 2x^2 + x - 1$. ■

In this section $GR(4^r)$ will denote the specific Galois ring $\mathbb{Z}_4[x]/(H(x))$. Let ξ be a primitive root of $H(x)$ in $GR(4^r)$. We will add the parity check position on the left of a codeword of $\widehat{K}(r + 1)$ and label the check position ∞. Hence if $c_\infty c_0 c_1 \cdots c_{n-1}$ is a codeword, the corresponding "extended" polynomial form would be $(c_\infty, c(x))$, where $c(x) = c_0 + c_1 x + \cdots + c_{n-1} x^{n-1}$.

Lemma 12.7.2 *Let $r \geq 2$. The following hold:*

(i) *A polynomial $c(x) \in \mathfrak{R}_n$ is in $K(r + 1)$ if and only if $(x - 1)H(x)c^*(x) = 0$ in $\mathfrak{R}_n$.*

(ii) *Generator matrices G_{r+1} and $\widehat{G}_{r+1}$ for $K(r + 1)$ and $\widehat{K}(r + 1)$, respectively, are*

$$G_{r+1} = \begin{bmatrix} 1 & 1 & 1 & \cdots & 1 \\ 1 & \xi & \xi^2 & \cdots & \xi^{n-1} \end{bmatrix}, \quad \widehat{G}_{r+1} = \begin{bmatrix} 1 & 1 & 1 & 1 & \cdots & 1 \\ 0 & 1 & \xi & \xi^2 & \cdots & \xi^{n-1} \end{bmatrix}.$$

We can replace each ξ^i by the column vector $[b_{i,0} \; b_{i,1} \; \cdots \; b_{i,r-1}]^{\mathrm{T}}$ obtained from (12.21), where $\xi^i = \sum_{j=0}^{r-1} b_{i,j} \xi^j$.

Proof: Since $\mathfrak{R}_n = \langle (x - 1)H(x) \rangle \oplus \langle f^*(x) \rangle$ by Theorem 12.3.12, $(x - 1)H(x)c^*(x) = 0$ in $\mathfrak{R}_n$ if and only if $c^*(x) \in \langle f^*(x) \rangle$ if and only if $c(x) = f(x)s(x)$ for some $s(x) \in \mathfrak{R}_n$, proving (i).

For (ii), we note that the matrix obtained from $[1 \ \xi \ \xi^2 \ \cdots \ \xi^{r-1}]$ by replacing each ξ^i with the coefficients from (12.21) is the $r \times r$ identity matrix. Hence the $r + 1$ rows of $\widehat{G}_{r+1}$ generate a code of type 4^{r+1}; the same will be true of the $r + 1$ rows of G_{r+1} once we show that the parity check of $\mathbf{r}_1 = [1 \ 1 \ 1 \ \cdots \ 1]$ is 1 and the parity check of $\mathbf{r}_2 = [1 \ \xi \ \xi^2 \ \cdots \ \xi^{n-1}]$ is 0. Since $\mathbf{r}_1$ has $2^r - 1$ 1s and $2^r - 1 \equiv 3 \pmod{4}$ as $r \geq 2$, the parity check for $\mathbf{r}_1$ is 1. The parity check for $\mathbf{r}_2$ is $-\sum_{i=0}^{n-1} \xi^i = -(\xi^n - 1)/(\xi - 1) = 0$ as $\xi^n = 1$.

To complete the proof of (ii), it suffices to show that $c(x) \in K(r + 1)$ where $c(x)$ is the polynomial associated to $\mathbf{r}_1$ and $\mathbf{r}_2$. For $\mathbf{r}_1$, let $c(x) = \sum_{i=0}^{n-1} x^i$. As $(x - 1)c^*(x) = (x - 1)\sum_{i=0}^{n-1} x^i = x^n - 1$ in $\mathbb{Z}_4[x]$, $(x - 1)H(x)c^*(x) = 0$ in $\mathfrak{R}_n$ showing $c(x) \in K(r + 1)$ by (i). For $\mathbf{r}_2$, we are actually working with r possible codewords. We do not wish to deal with them individually and so we let $c(x) = \sum_{i=0}^{n-1} \xi^i x^i$, thus allowing the coefficients of $c(x)$ to lie in the ring $\mathrm{GR}(4^r)$ rather than $\mathbb{Z}_4$. Clearly, by (i), we only need to show that $H(x)c^*(x) = 0$ even when we allow the coefficients of $c(x)$ to be elements of $\mathrm{GR}(4^r)$. So $c^*(x) = \xi \sum_{i=0}^{n-1} \xi^{n-1-i} x^i$ (where ξ in front of the summation makes $c^*(x)$ monic). Let $H(x) = \sum_{i=0}^{n-1} H_i x^i$. By Exercise 754, the coefficient of x^k for $0 \leq k \leq n - 1$ in the product $H(x)c^*(x)$ is

$$\xi \sum_{i+j\equiv k \,(\mathrm{mod}\ n)}^{n-1} H_i \xi^{n-1-j} = \xi \sum_{i=0}^{n-1} H_i \xi^{n-1-k+i} = \xi^{n-k} \sum_{i=0}^{n-1} H_i \xi^i, \tag{12.22}$$

since $\xi^n = 1$. The right-hand side of (12.22) is $\xi^{n-k} H(\xi) = 0$ as ξ is a root of $H(x)$. Therefore $(x - 1)H(x)c^*(x) = 0$. $\qquad\square$

We remark that $(x - 1)H(x)$ is called the *check polynomial* of $K(r + 1)$, analogous to the terminology used in Section 4.4 for cyclic codes over $\mathbb{F}_q$. Notice that the matrix G_{r+1} of Lemma 12.7.2 is reminiscent of the parity check matrix for the subcode of even weight vectors in the binary Hamming code, where we write this parity check matrix over an extension field of $\mathbb{F}_2$, rather than over $\mathbb{F}_2$.

Exercise 754 Let $\mathcal{R}$ be a commutative ring with $a(x) = \sum_{i=0}^{n-1} a_i x^i$ and $b(x) = \sum_{i=0}^{n-1} b_i x^i$ in $\mathcal{R}[x]/(x^n - 1)$. Prove that $a(x)b(x) = p(x) = \sum_{k=0}^{n-1} p_k x^k$ in $\mathcal{R}[x]/(x^n - 1)$, where

$$p_k = \sum_{i+j\equiv k \,(\mathrm{mod}\ n)} a_i b_j \quad \text{with } 0 \leq i \leq n - 1 \quad \text{and} \quad 0 \leq j \leq n - 1. \qquad \blacklozenge$$

Example 12.7.3 Using the basic irreducible polynomial from Exercise 751, a generator matrix for $\widehat{K}(4)$ from Theorem 12.7.2 is

$$\widehat{G}_4 = \begin{bmatrix} 1 & 1 & 1 & 1 & 1 & 1 & 1 & 1 \\ 0 & 1 & 0 & 0 & 1 & 2 & 3 & 1 \\ 0 & 0 & 1 & 0 & 3 & 3 & 3 & 2 \\ 0 & 0 & 0 & 1 & 2 & 3 & 1 & 1 \end{bmatrix}. \qquad \blacksquare$$

Exercise 755 Show that the generator matrices in Examples 12.2.5 and 12.7.3 generate the same $\mathbb{Z}_4$-linear code. ◆

We now use the relative trace to list all the codewords in $K(r+1)$ and $\widehat{K}(r+1)$. Let $\mathbf{1}_n$ denote the all-one vector of length n.

Lemma 12.7.4 Let $r \geq 2$ and $n = 2^r - 1$. Then $\mathbf{c} \in K(r+1)$ if and only if there exists $\lambda \in \mathcal{R} = \mathrm{GR}(4^r)$ and $\epsilon \in \mathbb{Z}_4$ such that

$$\mathbf{c} = (\mathrm{TR}_r(\lambda), \mathrm{TR}_r(\lambda\xi), \mathrm{TR}_r(\lambda\xi^2), \ldots, \mathrm{TR}_r(\lambda\xi^{n-1})) + \epsilon\mathbf{1}_n. \tag{12.23}$$

The parity check for $\mathbf{c}$ *is* ϵ.

Proof: Using Exercise 753, $\sum_{i=0}^{n-1}(\mathrm{TR}_r(\lambda\xi^i) + \epsilon) = \mathrm{TR}_r(\lambda\sum_{i=0}^{n-1}\xi^i) + (2^r - 1)\epsilon = \mathrm{TR}_r(\lambda(\xi^n - 1)/(\xi - 1)) + (2^r - 1)\epsilon = \mathrm{TR}_r(0) + (2^r - 1)\epsilon = (2^r - 1)\epsilon \equiv 3\epsilon \pmod{4}$. So the parity check for $\mathbf{c}$ is ϵ.

Let $\mathcal{C} = \{(\mathrm{TR}_r(\lambda), \mathrm{TR}_r(\lambda\xi), \mathrm{TR}_r(\lambda\xi^2), \ldots, \mathrm{TR}_r(\lambda\xi^{n-1})) + \epsilon\mathbf{1}_n \mid \lambda \in \mathrm{GR}(4^r), \epsilon \in \mathbb{Z}_4\}$. We first show that $\mathcal{C}$ has 4^{r+1} codewords; this is clear if we show that $\mathrm{TR}_r(\lambda\xi^i) + \epsilon = \mathrm{TR}_r(\lambda_1\xi^i) + \epsilon_1$ for $0 \leq i \leq n - 1$ implies $\lambda = \lambda_1$ and $\epsilon = \epsilon_1$. But $\epsilon = \epsilon_1$ since the parity check of $\mathbf{c}$ is ϵ. Therefore by Exercise 753, $\mathrm{TR}_r(\zeta\xi^i) = 0$ for $0 \leq i \leq n - 1$ where $\zeta = \lambda - \lambda_1$. We only need to show that $\zeta = 0$. Since $\mathrm{TR}_r(\zeta\xi^i) = 0$, $\mathrm{TR}_r(\zeta(2\xi^i)) = 0$ and hence $\mathrm{TR}_r(\zeta s) = 0$ for all $s \in \mathcal{R}$ again by Exercise 753. Thus TR_r is 0 on the ideal (ζ). Since TR_r is surjective, there is an $\alpha \in \mathcal{R}$ such that $\mathrm{TR}_r(\alpha) = 1$ and hence $\mathrm{TR}_r(2\alpha) = 2$. Therefore by Lemma 12.3.9, (ζ) can only be the zero ideal and hence $\zeta = 0$.

We now show that if $\mathbf{c}$ has the form in (12.23), then $\mathbf{c} \in K(r+1)$. Since $\mathbf{1}_n \in K(r+1)$, we only need to show that $\mathbf{c} \in K(r+1)$ when $\epsilon = 0$. If $\lambda = 0$, the result is clear. If $\lambda \neq 0$, then $c(x) \neq 0$ by the above argument. Now $c^*(x)$ is some cyclic shift of $p(x) = \alpha \sum_{i=0}^{n-1} \mathrm{TR}_r(\lambda\xi^{n-1-i})x^i$ for some $\alpha \in \mathbb{Z}_4$ chosen so that the leading coefficient is 1 or 2. ($c^*(x)$ may not equal $p(x)$ without the shift as we do not know the degree of $c(x)$.) By Lemma 12.7.2, $\mathbf{c} \in K(r+1)$ if $H(x)p(x) = 0$ in $\mathfrak{R}_n$. By the same argument as in the proof of Lemma 12.7.2, the coefficient of x^k for $0 \leq k \leq n - 1$ in the product $H(x)p(x)$ is

$$\alpha \sum_{i+j \equiv k \,(\mathrm{mod}\, n)} H_i \mathrm{TR}_r(\lambda\xi^{n-1-j}) = \alpha\mathrm{TR}_r\left(\lambda\sum_{i=0}^{n-1} H_i\xi^{n-1-k+i}\right)$$

$$= \alpha\mathrm{TR}_r\left(\lambda\xi^{n-1-k}\sum_{i=0}^{n-1} H_i\xi^i\right),$$

using Exercise 753, as $H_i \in \mathbb{Z}_4$. The latter trace is $\mathrm{TR}_r(\lambda\xi^{n-1-k}H(\xi)) = \mathrm{TR}_r(0) = 0$ and hence $H(x)p(x) = 0$ in $\mathfrak{R}_n$. So $\mathbf{c} \in K(r+1)$. Since $K(r+1)$ and $\mathcal{C}$ each have 4^{r+1} codewords and $\mathcal{C} \subseteq K(r+1)$, $\mathcal{C} = K(r+1)$. □

Exercise 756 Let $\mathcal{R} = \mathrm{GR}(4^r)$ with primitive element ξ; let $n = 2^r - 1$. Using the fact that $\mathrm{TR}_r : \mathcal{R} \to \mathbb{Z}_4$ is a surjective group homomorphism under addition (Exercise 753), show the following:

(a) For each $a \in \mathbb{Z}_4$, there are exactly 4^{r-1} elements of $\mathcal{R}$ with relative trace equal to a.

(b) For each $s \in \{0, 2, 2\xi, 2\xi^2, \ldots, 2\xi^{n-1}\} = 2\mathcal{R}$, $\mathrm{TR}_r(s) \in \{0, 2\}$ with at least one value equaling 2.

(c) For $a = 0$ or 2, there are exactly 2^{r-1} elements of $2\mathcal{R}$ with relative trace equal to a.

$\blacklozenge$

Before we give the Lee weight distribution of $\widehat{K}(r+1)$ we need one more computational result.

Lemma 12.7.5 *Let* $\mathcal{R} = \mathrm{GR}(4^r)$ *have primitive element* ξ. *Let* $n = 2^r - 1$. *Suppose that* $\lambda \in \mathcal{R}$ *but* $\lambda \notin 2\mathcal{R}$. *Then all the elements of* $\mathcal{S} = \mathcal{S}_1 \cup \mathcal{S}_2$, *where* $\mathcal{S}_1 = \{\lambda(\xi^j - \xi^k) \mid 0 \le j \le n-1, \ 0 \le k \le n-1, \ j \ne k\}$ *and* $\mathcal{S}_2 = \{\pm \lambda \xi^j \mid 0 \le j \le n-1\}$ *are distinct, and they are precisely the elements of* $\mathcal{R} \setminus 2\mathcal{R}$.

Proof: By Exercise 748 the set of invertible elements in $\mathcal{R}$ is $\mathcal{R} \setminus 2\mathcal{R}$. We only need to show that the elements of $\mathcal{S}$ are invertible and distinct because $\mathcal{R} \setminus 2\mathcal{R}$ has $4^r - 2^r$ elements and, if the elements of $\mathcal{S}$ are distinct, $\mathcal{S}$ has $(2^r - 1)(2^r - 2) + 2(2^r - 1) = 4^r - 2^r$ elements.

Since λ is invertible, the elements of $\mathcal{S}_2$ are invertible. Clearly, the elements of $\mathcal{S}_2$ are distinct as $-\xi^j = \xi^j + 2\xi^j$ cannot equal ξ^k for any j and k.

To show the elements of $\mathcal{S}_1$ are invertible it suffices to show that $\xi^j - \xi^k$ is invertible for $j \ne k$. If this element is not invertible, $\xi^j - \xi^k \in 2\mathcal{R}$. Recall that the reduction modulo 2 homomorphism $\overline{\mu} : \mathcal{R} \to \mathbb{F}_{2^r} = \mathcal{R}/2\mathcal{R}$, discussed before Exercise 749, maps the primitive element ξ of $\mathcal{R}$ to the primitive element $\overline{\mu}(\xi) = \theta$ of $\mathbb{F}_{2^r}$. Thus $\overline{\mu}(\xi)^j + \overline{\mu}(\xi)^k = 0$, implying that $\theta^j = \theta^k$. But $\theta^j \ne \theta^k$ for $j \ne k$ with j and k between 0 and $n-1$. Thus the elements of $\mathcal{S}_1$ are invertible. We now show that the elements of $\mathcal{S}_1$ are distinct. Suppose that $\xi^j - \xi^k = \xi^\ell - \xi^m$ with $j \ne k$ and $\ell \ne m$; then

$$1 + \xi^a = \xi^b + \xi^c, \tag{12.24}$$

where $m - j \equiv a \pmod{n}$, $\ell - j \equiv b \pmod{n}$, and $k - j \equiv c \pmod{n}$. So $(1 + \xi^a)^2 - v_2(1 + \xi^a) = (\xi^b + \xi^c)^2 - v_2(\xi^b + \xi^c)$, implying that $1 + \xi^{2a} + 2\xi^a - v_2(1) - v_2(\xi^a) = \xi^{2b} + \xi^{2c} + 2\xi^{b+c} - v_2(\xi^b) - v_2(\xi^c)$, where v_2 is the Frobenius automorphism. Thus $2\xi^a = 2\xi^{b+c}$, and hence $a \equiv b + c \pmod{n}$. Therefore if $x = \theta^a$, $y = \theta^b$, and $z = \theta^c$, then $x = yz$. Applying $\overline{\mu}$ to (12.24), we also have $1 + x = y + z$. Hence $0 = 1 + yz + y + z = (1 + y)(1 + z)$ holds in $\mathbb{F}_{2^r}$, implying $y = 1$ or $z = 1$. So $b \equiv 0 \pmod{n}$ or $c \equiv 0 \pmod{n}$. Hence $\ell = j$ or $k = j$; as the latter is impossible, we have $\ell = j$, which yields $k = m$ as well. So the elements of $\mathcal{S}_1$ are distinct.

Finally, we show that the elements of $\mathcal{S}_1 \cup \mathcal{S}_2$ are distinct. Assume $\xi^j - \xi^k = \pm \xi^m$. By rearranging and factoring out a power of ξ, $1 + \xi^a = \xi^b$. So $(1 + \xi^a)^2 - v_2(1 + \xi^a) = \xi^{2b} - v_2(\xi^b)$, giving $1 + 2\xi^a + \xi^{2a} - (1 + \xi^{2a}) = \xi^{2b} - \xi^{2b}$. Therefore $2\xi^a = 0$, which is a contradiction. $\square$

We now have the tools to find the Lee weight distribution of $\widehat{K}(r+1)$, which is also the Hamming weight distribution of the binary Kerdock code $\mathcal{K}(r+1) = \mathfrak{G}(\widehat{K}(r+1))$ by Theorem 12.2.2.

Theorem 12.7.6 *Let $r \geq 3$ with r odd. Let A_i be the number of vectors in $\widehat{K}(r+1)$ of Lee weight i. Then*

$$
A_i = \begin{cases}
1 & \text{if } i = 0, \\
2^{r+1}(2^r - 1) & \text{if } i = 2^r - 2^{(r-1)/2}, \\
2^{r+2} - 2 & \text{if } i = 2^r, \\
2^{r+1}(2^r - 1) & \text{if } i = 2^r + 2^{(r-1)/2}, \\
1 & \text{if } i = 2^{r+1}, \\
0 & \text{otherwise.}
\end{cases}
$$

This is also the Hamming weight distribution of the Kerdock code $\mathcal{K}(r+1)$.

Proof: Let $\mathcal{R} = \mathrm{GR}(4^r)$ and

$$\mathbf{v}_\lambda = (\mathrm{TR}_r(\lambda), \mathrm{TR}_r(\lambda\xi), \mathrm{TR}_r(\lambda\xi^2), \ldots, \mathrm{TR}_r(\lambda\xi^{n-1})),$$

where $n = 2^r - 1$. By Lemma 12.7.4, the vectors in $\widehat{K}(r+1)$ have the form $(0, \mathbf{v}_\lambda) + \epsilon \mathbf{1}_{n+1}$ for $\lambda \in \mathcal{R}$ and $\epsilon \in \mathbb{Z}_4$. We compute the Lee weight of each of these vectors by examining three cases. The first two cases give A_i for $i = 0$, 2^r, and 2^{r+1}. The last case gives the remaining A_i.

Case I: $\lambda = 0$
In this case $\mathbf{v}_\lambda$ is the zero vector, and $(0, \mathbf{v}_\lambda) + \epsilon \mathbf{1}_{n+1} = \epsilon \mathbf{1}_{n+1}$ has Lee weight 0 if $\epsilon = 0$, $n + 1 = 2^r$ if $\epsilon = 1$ or 3, and $2(n+1) = 2^{r+1}$ if $\epsilon = 2$. This case contributes one vector of Lee weight 0, two vectors of Lee weight 2^r, and one vector of Lee weight 2^{r+1}.

Case II: $\lambda \in 2\mathcal{R}$ with $\lambda \neq 0$
Thus $\lambda = 2\xi^i$ for some $0 \leq i \leq n - 1$. Hence the set of components of $\mathbf{v}_\lambda$ is always $\{\mathrm{TR}_r(2\xi^j) \mid 0 \leq j \leq n - 1\}$ (in some order) independent of i. But by Exercise 756(c), exactly 2^{r-1} of these values are 2 and the remaining $2^{r-1} - 1$ values are 0. Thus every vector $(0, \mathbf{v}_\lambda) + \epsilon \mathbf{1}_{n+1}$ has either 2^{r-1} 2s and 2^{r-1} 0s (when $\epsilon = 0$ or 2) or 2^{r-1} 1s and 2^{r-1} 3s (when $\epsilon = 1$ or 3). In all cases these vectors have Lee weight 2^r. This case contributes $4n = 2^{r+2} - 4$ vectors of Lee weight 2^r.

Case III: $\lambda \in \mathcal{R}$ with $\lambda \notin 2\mathcal{R}$
Let $\mathcal{R}^\# = \mathcal{R} \setminus 2\mathcal{R}$. For $j \in \mathbb{Z}_4$, let $n_j = n_j(\mathbf{v}_\lambda)$ be the number of components of $\mathbf{v}_\lambda$ equal to j. We set up a sequence of equations involving the n_j and then solve these equations. Let $i = \sqrt{-1}$. Define

$$S = \sum_{j=0}^{n-1} i^{\mathrm{TR}_r(\lambda\xi^j)} = n_0 - n_2 + i(n_1 - n_3). \tag{12.25}$$

If $\overline{S}$ is the complex conjugate of S, then

$$S\overline{S} = \sum_{j=0}^{n-1} i^{\mathrm{TR}_r(\lambda(\xi^j - \xi^j))} + \sum_{j \neq k} i^{\mathrm{TR}_r(\lambda(\xi^j - \xi^k))}$$

$$= 2^r - 1 + \sum_{a \in \mathcal{R}^\#} i^{\mathrm{TR}_r(a)} - \sum_{j=0}^{n-1} i^{\mathrm{TR}_r(\lambda\xi^j)} - \sum_{j=0}^{n-1} i^{\mathrm{TR}_r(-\lambda\xi^j)}$$

$$= 2^r - 1 + \sum_{a \in \mathcal{R}^\#} i^{\mathrm{TR}_r(a)} - S - \overline{S}$$

by Lemma 12.7.5. But $\sum_{a \in \mathcal{R}^\#} i^{\mathrm{TR}_r(a)} = \sum_{a \in \mathcal{R}} i^{\mathrm{TR}_r(a)} - \sum_{a \in 2\mathcal{R}} i^{\mathrm{TR}_r(a)} = 0 - 0 = 0$ by Exercise 756. Rearranging gives $(S + 1)(\overline{S} + 1) = 2^r$ and therefore

$$(n_0 - n_2 + 1)^2 + (n_1 - n_3)^2 = 2^r \tag{12.26}$$

by (12.25). Thus we need integer solutions to the equation $x^2 + y^2 = 2^r$. By Exercise 757, the solutions are $x = \delta_1 2^{(r-1)/2}$, and $y = \delta_2 2^{(r-1)/2}$, where δ_1 and δ_2 are ± 1.[4] By (12.26),

$$n_0 - n_2 = -1 + \delta_1 2^{(r-1)/2} \quad \text{and} \tag{12.27}$$
$$n_1 - n_3 = \delta_2 2^{(r-1)/2}. \tag{12.28}$$

However, $2\mathbf{v}_\lambda = \mathbf{v}_{2\lambda}$ is a vector from Case II, which means that

$$n_0 + n_2 = 2^{r-1} - 1 \quad \text{and} \tag{12.29}$$
$$n_1 + n_3 = 2^{r-1}. \tag{12.30}$$

Solving (12.27), (12.28), (12.29), and (12.30) produces

$$n_0 = 2^{r-2} + \delta_1 2^{(r-3)/2} - 1, \, n_1 = 2^{r-2} + \delta_2 2^{(r-3)/2}, \tag{12.31}$$
$$n_2 = 2^{r-2} - \delta_1 2^{(r-3)/2}, \text{ and } n_3 = 2^{r-2} - \delta_2 2^{(r-3)/2}. \tag{12.32}$$

We can easily obtain $n_j((0, \mathbf{v}_\lambda) + \epsilon\mathbf{1}_{n+1})$ in terms of $n_j = n_j(\mathbf{v}_\lambda)$ as the following table shows:

$\epsilon \setminus j$	0	1	2	3	Lee weight
0	$n_0 + 1$	n_1	n_2	n_3	$2^r - \delta_1 2^{(r-1)/2}$
1	n_3	$n_0 + 1$	n_1	n_2	$2^r + \delta_2 2^{(r-1)/2}$
2	n_2	n_3	$n_0 + 1$	n_1	$2^r + \delta_1 2^{(r-1)/2}$
3	n_1	n_2	n_3	$n_0 + 1$	$2^r - \delta_2 2^{(r-1)/2}$

The column "Lee weight" is obtained from the equation $\mathrm{wt}_L((0, \mathbf{v}_\lambda) + \epsilon\mathbf{1}_{n+1}) = n_1((0, \mathbf{v}_\lambda) + \epsilon\mathbf{1}_{n+1}) + 2n_2((0, \mathbf{v}_\lambda) + \epsilon\mathbf{1}_{n+1}) + n_3((0, \mathbf{v}_\lambda) + \epsilon\mathbf{1}_{n+1})$ by using (12.31) and (12.32). Thus two of the four codewords have Lee weight $2^r - 2^{(r-1)/2}$, while the other two have Lee weight $2^r + 2^{(r-1)/2}$. Therefore Case III contributes $2(4^r - 2^r) = 2^{r+1}(2^r - 1)$

---

[4] This is the only place in this proof where we require r to be odd.

codewords of Lee weight $2^r - 2^{(r-1)/2}$ and $2(4^r - 2^r) = 2^{r+1}(2^r - 1)$ codewords of Lee weight $2^r + 2^{(r-1)/2}$. ☐

Exercise 757 Let $r \geq 3$ be odd. Let x and y be integer solutions of $x^2 + y^2 = 2^r$. Do the following:
(a) Show that x and y are both even.
(b) Show that x and y are integer solutions of $x^2 + y^2 = 2^r$ if and only if x_1 and x_2 are integer solutions of $x_1^2 + y_1^2 = 2^{r-2}$, where $x_1 = x/2$ and $y_1 = y/2$.
(c) Prove that the only solutions of $x^2 + y^2 = 2^r$ are $x = \pm 2^{(r-1)/2}$ and $y = \pm 2^{(r-1)/2}$.
♦

Exercise 758 Verify the entries in the table in Case III of the proof of Theorem 12.7.6.
♦

Exercise 759 Find the Lee weight distribution of $\widehat{K}(4)$, which is the octacode o_8 of Example 12.2.5 by Exercise 755, and hence the weight distribution of $\mathcal{K}(4) = \mathfrak{G}(\widehat{K}(4))$. Show that this is the same as the weight distribution of the Nordstrom–Robinson code described in Section 2.3.4.
♦

12.8 Preparata codes

The binary Preparata codes were originally defined in [288]. They are nonlinear codes that are distance invariant of length 2^{r+1} and minimum distance 6. Also they are subsets of extended Hamming codes. The distance distribution of the Preparata code of length 2^{r+1} is related to the distance distribution of the Kerdock code of length 2^{r+1} by the MacWilliams equations. In [116], the authors observed that $\mathfrak{G}(\widehat{K}(r+1)^{\perp})$ has the same distance distribution as the original Preparata code of length 2^{r+1}. However, this code is not a subset of the extended Hamming code, as the original Preparata codes are, but rather of a nonlinear code with the same weight distribution as the extended Hamming code. So these codes are "Preparata-like."

For $r \geq 3$ and r odd, let $P(r+1) = \widehat{K}(r+1)^{\perp}$ and $\mathcal{P}(r+1) = \mathfrak{G}(P(r+1))$.

Theorem 12.8.1 Let $r \geq 3$ be odd. Then $P(r+1)$ and $\mathcal{P}(r+1)$ are distance invariant codes. The Lee weight enumerator of $P(r+1)$, which is also the Hamming weight enumerator of $\mathcal{P}(r+1)$, is

$$
\text{Lee}_{P(r+1)}(x, y) = \frac{1}{4^{r+1}}[(y-x)^{2^{r+1}} + (y+x)^{2^{r+1}}
$$
$$
+ 2^{r+1}(2^r - 1)(y-x)^{2^r - 2^{(r-1)/2}}(y+x)^{2^r + 2^{(r-1)/2}}
$$
$$
+ 2^{r+1}(2^r - 1)(y-x)^{2^r + 2^{(r-1)/2}}(y+x)^{2^r - 2^{(r-1)/2}}
$$
$$
+ (2^{r+2} - 2)(y-x)^{2^r}(y+x)^{2^r}].
$$

Alternately, if $B_j(r+1)$ is the number of codewords in $P(r+1)$ of Lee weight j, which is *also the number of codewords in $\mathcal{P}(r+1)$ of Hamming weight j, then*

$$B_j(r+1) = \frac{1}{4^{r+1}}\left[K_j^{2^{r+1},2}(0) + 2^{r+1}(2^r - 1)K_j^{2^{r+1},2}(2^r - 2^{(r-1)/2})\right.$$

$$+ (2^{r+2} - 2)K_j^{2^{r+1},2}(2^r) + 2^{r+1}(2^r - 1)K_j^{2^{r+1},2}(2^r + 2^{(r-1)/2})$$

$$\left. + K_j^{2^{r+1},2}(2^{r+1})\right],$$

where $K_j^{2^{r+1},2}$ is the Krawtchouck polynomial defined in Chapter 2. Furthermore, the minimum distance of $\mathcal{P}(r+1)$ is 6 and $B_j(r+1) = 0$ if j is odd.

Proof: The formula for $\mathrm{Lee}_{\mathcal{P}(r+1)}(x, y)$ follows from (12.6) and Theorem 12.7.6. The alternate form comes from (K) of Section 7.2. When applying (K) we must decide what values to use for q and n in $K_j^{n,q}$. We use $q = 2$ because the MacWilliams transform for the Lee weight enumerator satisfies (12.6), which is the same form used for binary codes (see (M_3) in Section 7.2). We use $n = 2 \cdot 2^r = 2^{r+1}$ because of the presence of "$2n$" in the exponent in (12.4). Another way to understand the choice of q and n is to realize that the Lee weight distribution of $\mathcal{P}(r+1)$ is the same as the Hamming weight distribution of $\mathcal{P}(r+1)$, which is a binary code of length 2^{r+1}. The last statement follows from Exercise 760.   □

Exercise 760 Do the following:
(a) If $r \geq 3$ and r is odd, show that the Lee weight of any codeword in $\mathcal{P}(r+1)$ is even. Hint: See Exercise 690.
(b) Using the formula for $B_j(r+1)$ in Theorem 12.8.1 (and a computer algebra package), show that $B_2(r+1) = B_4(r+1) = 0$ and $B_6(r+1) = (-2^{r+1} + 7 \cdot 2^{2r} - 7 \cdot 2^{3r} + 2^{4r+1})/45$.   ◆

13 Codes from algebraic geometry

Since the discovery of codes using algebraic geometry by V. D. Goppa in 1977 [105], there has been a great deal of research on these codes. Their importance was realized when in 1982 Tsfasman, Vlăduţ, and Zink [333] proved that certain algebraic geometry codes exceeded the Asymptotic Gilbert–Varshamov Bound, a feat many coding theorists felt could never be achieved. Algebraic geometry codes, now often called geometric Goppa codes, were originally developed using many extensive and deep results from algebraic geometry. These codes are defined using algebraic curves. They can also be defined using algebraic function fields as there is a one-to-one correspondence between "nice" algebraic curves and these function fields. The reader interested in the connection between these two theories can consult [320, Appendix B]. Another approach appeared in the 1998 publication by Høholdt, van Lint, and Pellikaan [135], where the theory of order and weight functions was used to describe a certain class of geometric Goppa codes.

In this chapter we choose to introduce a small portion of the theory of algebraic curves, enough to allow us to define algebraic geometry codes and present some simple examples. We will follow a very readable treatment of the subject by J. L. Walker [343]. Her monograph would make an excellent companion to this chapter. For those who want to learn more about the codes and their decoding but have a limited understanding of algebraic geometry, the Høholdt, van Lint, and Pellikaan chapter in the *Handbook of Coding Theory* [135] can be examined. For those who have either a strong background in algebraic geometry or a willingness to learn it, the texts by C. Moreno [240] or H. Stichtenoth [320] connect algebraic geometry to coding.

13.1 Affine space, projective space, and homogenization

Algebraic geometry codes are defined with respect to curves in either affine or projective space. In this section we introduce these concepts.

Let $\mathbb{F}$ be a field, possibly infinite. We define *n-dimensional affine space over* $\mathbb{F}$, denoted $\mathbb{A}^n(\mathbb{F})$, to be the ordinary n-dimensional vector space $\mathbb{F}^n$; the *points* in $\mathbb{A}^n(\mathbb{F})$ are $(x_1, x_2, \ldots, x_n)$ with $x_i \in \mathbb{F}$. Defining n-dimensional projective space is more complicated. First, let V_n be the nonzero vectors in $\mathbb{F}^{n+1}$. If $\mathbf{x} = (x_1, x_2, \ldots, x_{n+1})$ and $\mathbf{x}' = (x_1', x_2', \ldots, x_{n+1}')$ are in V_n, we say that $\mathbf{x}$ and $\mathbf{x}'$ are equivalent, denoted $\mathbf{x} \sim \mathbf{x}'$, if there exists a nonzero $\lambda \in \mathbb{F}$ such that $x_i' = \lambda x_i$ for $1 \le i \le n+1$. The relation $\sim$ is indeed an equivalence relation as Exercise 761 shows. The equivalence classes are denoted $(x_1 : x_2 : \cdots : x_{n+1})$ and consist of all nonzero scalar multiples of $(x_1, x_2, \ldots, x_{n+1})$. We define *n-dimensional projective space over* $\mathbb{F}$, denoted $\mathbb{P}^n(\mathbb{F})$, to be the set of all equivalence classes $(x_1 : x_2 : \cdots : x_{n+1})$, called *points*, with $x_i \in \mathbb{F}$. Note that the zero vector is excluded in projective space, and the points of $\mathbb{P}^n(\mathbb{F})$ can be identified with the 1-dimensional subspaces of $\mathbb{A}^{n+1}(\mathbb{F})$. The coordinates $x_1, x_2, \ldots, x_{n+1}$ are called *homogeneous coordinates* of $\mathbb{P}^n(\mathbb{F})$. If $P = (x_1 : x_2 : \cdots : x_{n+1}) \in \mathbb{P}^n(\mathbb{F})$ with $x_{n+1} = 0$, then P is called a *point at infinity*. Points in $\mathbb{P}^n(\mathbb{F})$ not at infinity are called *affine points*. By Exercise 761, each affine point in $\mathbb{P}^n(\mathbb{F})$ can be uniquely represented as $(x_1 : x_2 : \cdots : x_n : 1)$, and each point at infinity has a zero in its right-most coordinate and can be represented uniquely with a one in its right-most nonzero coordinate. $\mathbb{P}^1(\mathbb{F})$ is called the *projective line over* $\mathbb{F}$, and $\mathbb{P}^2(\mathbb{F})$ is called the *projective plane over* $\mathbb{F}$.

Exercise 761 Do the following:
(a) Prove that $\sim$ is an equivalence relation on V_n.
(b) Show that the affine points of $\mathbb{P}^n(\mathbb{F})$ can be represented uniquely by $(x_1 : x_2 : \cdots : x_n : 1)$ and hence correspond in a natural one-to-one manner with the points of $\mathbb{A}^n(\mathbb{F})$.
(c) Show that the points at infinity have a zero in the right-most coordinate and can be represented uniquely with a one in the right-most nonzero coordinate.
(d) Prove that if $\mathbb{F} = \mathbb{F}_q$, each equivalence class (point) $(x_1 : x_2 : \cdots : x_{n+1})$ contains $q - 1$ vectors from V_n.
(e) Prove that $\mathbb{P}^n(\mathbb{F}_q)$ contains $q^n + q^{n-1} + \cdots + q + 1$ points.
(f) Prove that $\mathbb{P}^n(\mathbb{F}_q)$ contains $q^{n-1} + q^{n-2} + \cdots + q + 1$ points at infinity.   ◆

Example 13.1.1 The projective line $\mathbb{P}^1(\mathbb{F}_q)$ has one point at infinity $(1 : 0)$ and q affine points $(x_1 : 1)$ where $x_1 \in \mathbb{F}_q$. The projective plane $\mathbb{P}^2(\mathbb{F}_q)$ has one point at infinity of the form $(1 : 0 : 0)$, q points at infinity of the form $(x_1 : 1 : 0)$ with $x_1 \in \mathbb{F}_q$, and q^2 affine points $(x_1 : x_2 : 1)$ where $x_1, x_2 \in \mathbb{F}_q$.   ■

Exercise 762 In Section 6.5 we defined a projective plane to be a set of points and a set of lines, consisting of points, such that any two distinct points determine a unique line that passes though these two points and any two distinct lines have exactly one point in common. In this exercise, you will show that $\mathbb{P}^2(\mathbb{F})$ is a projective plane in the sense of Section 6.5. To do this you need to know what the lines in $\mathbb{P}^2(\mathbb{F})$ are. Let a, b, and c be in $\mathbb{F}$, with at least one being nonzero. The set of points whose homogeneous coordinates satisfy $ax_1 + bx_2 + cx_3 = 0$ form a line.
(a) Show that if $ax_1 + bx_2 + cx_3 = 0$ and $(x_1 : x_2 : x_3) = (x_1' : x_2' : x_3')$, then $ax_1' + bx_2' + cx_3' = 0$. (This shows that it does not matter which representation you use for a point in order to decide if the point is on a given line or not.)
(b) Prove that two distinct points of $\mathbb{P}^2(\mathbb{F})$ determine a unique line that passes though these two points.
(c) Prove that any two distinct lines of $\mathbb{P}^2(\mathbb{F})$ have exactly one point in common.
(d) Parts (b) and (c) show that $\mathbb{P}^2(\mathbb{F})$ is a projective plane in the sense of Section 6.5. If $\mathbb{F} = \mathbb{F}_q$, what is the order of this plane?   ◆

Later we will deal with polynomials and rational functions (the ratios of polynomials). It will be useful to restrict to homogeneous polynomials or the ratios of homogeneous polynomials. In particular, we will be interested in the zeros of polynomials and in the zeros and poles of rational functions.

Let $x_1, \ldots, x_n$ be n independent indeterminates and let $\mathbb{F}[x_1, \ldots, x_n]$ denote the set of all polynomials in these indeterminates with coefficients from $\mathbb{F}$. A polynomial f in $\mathbb{F}[x_1, \ldots, x_n]$ is *homogeneous of degree d* if every term of f is of degree d. If a polynomial f of degree d is not homogeneous, we can "make" it be homogeneous by adding one more variable and appending powers of that variable to each term in f so that each modified term has degree d. This process is called the *homogenization* of f and the resulting polynomial is denoted f^H. More formally,

$$f^H(x_1, x_2, \ldots, x_n, x_{n+1}) = x_{n+1}^d f\left(\frac{x_1}{x_{n+1}}, \frac{x_2}{x_{n+1}}, \ldots, \frac{x_n}{x_{n+1}}\right), \tag{13.1}$$

where f has degree d.

Example 13.1.2 The polynomial $f(x_1, x_2) = x_1^3 + 4x_1^2 x_2 - 7x_1 x_2^2 + 8x_2^3$ is homogeneous of degree 3 over the real numbers. The polynomial $g(x_1, x_2) = 3x_1^5 - 6x_1 x_2^2 + x_2^4$ of degree 5 is not homogeneous; its homogenization is $g^H(x_1, x_2, x_3) = 3x_1^5 - 6x_1 x_2^2 x_3^2 + x_2^4 x_3$. ∎

Exercise 763 Give the homogenizations of the following polynomials over the real numbers:
(a) $f(x_1) = x_1^4 + 3x_1^2 - x_1 + 9$,
(b) $g(x_1, x_2) = 5x_1^8 - x_1 x_2^3 + 4x_1^2 x_2^5 - 7x_2^4$,
(c) $h(x_1, x_2, x_3) = x_1^2 x_2^3 x_3^5 - x_1 x_3^{12} + x_2^4 x_3$. ◆

Notice that $f^H(x_1, \ldots, x_n, 1) = f(x_1, \ldots, x_n)$ and so we can easily recover the original polynomial from its homogenization. Also, if we begin with a homogeneous polynomial $g(x_1, \ldots, x_{n+1})$ of degree d, then $f(x_1, \ldots, x_n) = g(x_1, \ldots, x_n, 1)$ is a polynomial of degree d or less; furthermore, if f has degree $k \le d$, then $g = x_{n+1}^{d-k} f^H$. In this manner we obtain a one-to-one correspondence between polynomials in n variables of degree d or less and homogeneous polynomials of degree d in $n+1$ variables. This proves the last statement of the following theorem; the proof of the remainder is left as an exercise.

Theorem 13.1.3 *Let $g(x_1, \ldots, x_{n+1})$ be a homogeneous polynomial of degree d over $\mathbb{F}$ and $f(x_1, \ldots, x_n)$ any polynomial of degree d over $\mathbb{F}$.*
(i) *If $\alpha \in \mathbb{F}$, then $g(\alpha x_1, \ldots, \alpha x_{n+1}) = \alpha^d g(x_1, \ldots, x_{n+1})$.*
(ii) *$f(x_1, \ldots, x_n) = 0$ if and only if $f^H(x_1, \ldots, x_n, 1) = 0$.*
(iii) *If $(x_1 : x_2 : \cdots : x_{n+1}) = (x_1' : x_2' : \cdots : x_{n+1}')$, then $g(x_1, \ldots, x_{n+1}) = 0$ if and only if $g(x_1', \ldots, x_{n+1}') = 0$. In particular, $f^H(x_1, \ldots, x_{n+1}) = 0$ if and only if $f^H(x_1', \ldots, x_{n+1}') = 0$.*
(iv) *There is a one-to-one correspondence between polynomials in n variables of degree d or less and homogeneous polynomials of degree d in $n+1$ variables.*

Exercise 764 Prove the first three parts of Theorem 13.1.3. ◆

The implications of Theorem 13.1.3 will prove quite important. Part (ii) implies that the zeros of f in $\mathbb{A}^n(\mathbb{F})$ correspond precisely to the affine points in $\mathbb{P}^n(\mathbb{F})$ that are zeros of f^H. f^H

may have other zeros that are points at infinity in $\mathbb{P}^n(\mathbb{F})$. Furthermore, part (iii) indicates that the concept of a point in $\mathbb{P}^n(\mathbb{F})$ being a zero of a homogeneous polynomial is well-defined.

Exercise 765 Give the one-to-one correspondence between the homogeneous polynomials of degree 2 in two variables over $\mathbb{F}_2$ and the polynomials of degree 2 or less (the zero polynomial is not included) in one variable over $\mathbb{F}_2$ as indicated in the proof of Theorem 13.1.3(iv). ◆

13.2 Some classical codes

Before studying algebraic geometry codes it is helpful to examine some classical codes that will motivate the definition of algebraic geometry codes.

13.2.1 Generalized Reed–Solomon codes revisited

In Section 5.2 we first defined Reed–Solomon codes as special cases of BCH codes. Theorem 5.2.3 gave an alternative formulation of narrow-sense RS codes, which we now recall. For $k \geq 0$, $\mathcal{P}_k$ denotes the set of polynomials of degree less than k, including the zero polynomial, in $\mathbb{F}_q[x]$. If α is a primitive nth root of unity in $\mathbb{F}_q$ where $n = q - 1$, then the code

$$\mathcal{C} = \{(f(1), f(\alpha), f(\alpha^2), \ldots, f(\alpha^{q-2})) \mid f \in \mathcal{P}_k\}$$

is the narrow-sense $[n, k, n - k + 1]$ RS code over $\mathbb{F}_q$. From Section 5.3, we saw that $\mathcal{C}$ could be extended to an $[n + 1, k, n - k + 2]$ code given by

$$\widehat{\mathcal{C}} = \{(f(1), f(\alpha), f(\alpha^2), \ldots, f(\alpha^{q-2}), f(0)) \mid f \in \mathcal{P}_k\}.$$

We now recall the definition of the generalized Reed–Solomon (GRS) codes. Let n be any integer with $1 \leq n \leq q$, $\boldsymbol{\gamma} = (\gamma_0, \ldots, \gamma_{n-1})$ an n-tuple of distinct elements of $\mathbb{F}_q$, and $\mathbf{v} = (v_0, \ldots, v_{n-1})$ an n-tuple of nonzero elements of $\mathbb{F}_q$. Let k be an integer with $1 \leq k \leq n$. Then the codes

$$\mathrm{GRS}_k(\boldsymbol{\gamma}, \mathbf{v}) = \{(v_0 f(\gamma_0), v_1 f(\gamma_1), \ldots, v_{n-1} f(\gamma_{n-1})) \mid f \in \mathcal{P}_k\}$$

are the GRS codes. The narrow-sense RS code $\mathcal{C}$ is the GRS code with $n = q - 1$, $\gamma_i = \alpha^i$, and $v_i = 1$; the extended code $\widehat{\mathcal{C}}$ is the GRS code with $n = q$, $\gamma_i = \alpha^i$ for $0 \leq i \leq q - 2$, $\gamma_{q-1} = 0$, and $v_i = 1$. By Theorem 13.1.3(iv), there is a one-to-one correspondence between the homogeneous polynomials of degree $k - 1$ in two variables over $\mathbb{F}_q$ and the nonzero polynomials of $\mathcal{P}_k$. We denote the homogeneous polynomials of degree $k - 1$ in two variables over $\mathbb{F}_q$ together with the zero polynomial by $\mathcal{L}_{k-1}$. If $f(x_1) \in \mathcal{P}_k$ with f nonzero of degree $d \leq k - 1$, then $g(x_1, x_2) = x_2^{k-1-d} f^H \in \mathcal{L}_{k-1}$ and $g(x_1, 1) = f^H(x_1, 1) = f(x_1)$ by (13.1). Therefore we can redefine the GRS codes as follows using homogeneous polynomials and projective points. Let $P_i = (\gamma_i, 1)$, which is a representative of the projective point $(\gamma_i : 1) \in \mathbb{P}^1(\mathbb{F}_q)$. Then

$$\mathrm{GRS}_k(\boldsymbol{\gamma}, \mathbf{v}) = \{(v_0 g(P_0), v_1 g(P_1), \ldots, v_{n-1} g(P_{n-1})) \mid g \in \mathcal{L}_{k-1}\}. \tag{13.2}$$

If we wish we can also include in our list a representative, say $(1, 0)$, of the point at infinity. Exercise 766 indicates that if the representatives of the projective points are changed, an equivalent code is obtained.

Exercise 766 Let $P_i' = (\beta_i \gamma_i, \beta_i)$ with $\beta_i \neq 0$. Prove that the code

$$\{(v_0 g(P_0'), v_1 g(P_1'), \ldots, v_{n-1} g(P_{n-1}')) \mid g \in \mathcal{L}_{k-1}\}$$

is equivalent to the code in (13.2).   ◆

13.2.2 Classical Goppa codes

Classical Goppa codes were introduced by V. D. Goppa in 1970 [103, 104]. These codes are generalizations of narrow-sense BCH codes and subfield subcodes of certain GRS codes.

We first give an alternate construction of narrow-sense BCH codes of length n over $\mathbb{F}_q$ to motivate the definition of Goppa codes. Let $t = \mathrm{ord}_q(n)$ and let β be a primitive nth root of unity in $\mathbb{F}_{q^t}$. Choose $\delta > 1$ and let $\mathcal{C}$ be the narrow-sense BCH code of length n and designed distance δ. Then $c(x) = c_0 + c_1 x + \cdots + c_{n-1} x^{n-1} \in \mathbb{F}_q[x]/(x^n - 1)$ is in $\mathcal{C}$ if and only if $c(\beta^j) = 0$ for $1 \leq j \leq \delta - 1$. We have

$$(x^n - 1) \sum_{i=0}^{n-1} \frac{c_i}{x - \beta^{-i}} = \sum_{i=0}^{n-1} c_i \sum_{\ell=0}^{n-1} x^\ell (\beta^{-i})^{n-1-\ell} = \sum_{\ell=0}^{n-1} x^\ell \sum_{i=0}^{n-1} c_i (\beta^{\ell+1})^i. \tag{13.3}$$

Because $c(\beta^{\ell+1}) = 0$ for $0 \leq \ell \leq \delta - 2$, the right-hand side of (13.3) is a polynomial whose lowest degree term has degree at least $\delta - 1$. Hence the right-hand side of (13.3) can be written as $x^{\delta-1} p(x)$, where $p(x)$ is a polynomial in $\mathbb{F}_{q^t}[x]$. Thus we can say that $c(x) \in \mathbb{F}_q[x]/(x^n - 1)$ is in $\mathcal{C}$ if and only if

$$\sum_{i=0}^{n-1} \frac{c_i}{x - \beta^{-i}} = \frac{x^{\delta-1} p(x)}{x^n - 1}$$

or equivalently

$$\sum_{i=0}^{n-1} \frac{c_i}{x - \beta^{-i}} \equiv 0 \pmod{x^{\delta-1}}.$$

This equivalence modulo $x^{\delta-1}$ means that if the left-hand side is written as a rational function $a(x)/b(x)$, then the numerator $a(x)$ will be a multiple of $x^{\delta-1}$ (noting that the denominator $b(x) = x^n - 1$).

This last equivalence is the basis of our definition of classical Goppa codes. To define a Goppa code of length n over $\mathbb{F}_q$, first fix an extension field $\mathbb{F}_{q^t}$ of $\mathbb{F}_q$; we do not require that $t = \mathrm{ord}_q(n)$. Let $L = \{\gamma_0, \gamma_1, \ldots, \gamma_{n-1}\}$ be n distinct elements of $\mathbb{F}_{q^t}$. Let $G(x) \in \mathbb{F}_{q^t}[x]$ with $G(\gamma_i) \neq 0$ for $0 \leq i \leq n - 1$. Then the *Goppa code* $\Gamma(L, G)$ is the set of vectors $c_0 c_1 \cdots c_{n-1} \in \mathbb{F}_q^n$ such that

$$\sum_{i=0}^{n-1} \frac{c_i}{x - \gamma_i} \equiv 0 \pmod{G(x)}. \tag{13.4}$$

Again this means that when the left-hand side is written as a rational function, the numerator is a multiple of $G(x)$. (Working modulo $G(x)$ is like working in the ring $\mathbb{F}_{q'}[x]/(G(x))$; requiring that $G(\gamma_i) \neq 0$ guarantees that $x - \gamma_i$ is invertible in this ring.) $G(x)$ is the *Goppa polynomial* of $\Gamma(L, G)$. The narrow-sense BCH code of length n and designed distance δ is the Goppa code $\Gamma(L, G)$ with $L = \{1, \beta^{-1}, \beta^{-2}, \ldots, \beta^{1-n}\}$ and $G(x) = x^{\delta-1}$.

Exercise 767 Construct a binary Goppa code of length 4 using (13.4) as follows. Use the field $\mathbb{F}_8$ given in Example 3.4.3 with primitive element α. Let $L = \{\gamma_0, \gamma_1, \gamma_2, \gamma_3\} = \{1, \alpha, \alpha^2, \alpha^4\}$ and $G(x) = x - \alpha^3$.

(a) When writing the left-hand side of (13.4) as a rational function $a(x)/b(x)$, the numerator $a(x)$ begins $c_0(x - \alpha)(x - \alpha^2)(x - \alpha^4) + \cdots$. Complete the rest of the numerator (but do not simplify).

(b) The numerator $a(x)$ found in part (a) is a multiple of $G(x) = x - \alpha^3$, which, in this case, is equivalent to saying $a(\alpha^3) = 0$. Simplify $a(\alpha^3)$ to obtain an $\mathbb{F}_8$-linear combination of c_0, c_1, c_2, and c_3. Hint: First compute $\alpha^3 - 1$, $\alpha^3 - \alpha$, $\alpha^3 - \alpha^2$, and $\alpha^3 - \alpha^4$.

(c) There are 16 choices for $c_0 c_1 c_2 c_3 \in \mathbb{F}_2^4$. Test each of these using the results of part (b) to find the codewords in $\Gamma(L, G)$. ◆

We now find a parity check matrix for $\Gamma(L, G)$. Notice that

$$\frac{1}{x - \gamma_i} \equiv -\frac{1}{G(\gamma_i)} \frac{G(x) - G(\gamma_i)}{x - \gamma_i} \pmod{G(x)}$$

since, by comparing numerators, $1 \equiv -G(\gamma_i)^{-1}(G(x) - G(\gamma_i)) \pmod{G(x)}$. So by (13.4) $\mathbf{c} = c_0 c_1 \cdots c_{n-1} \in \Gamma(L, G)$ if and only if

$$\sum_{i=0}^{n-1} c_i \frac{G(x) - G(\gamma_i)}{x - \gamma_i} G(\gamma_i)^{-1} \equiv 0 \pmod{G(x)}. \tag{13.5}$$

Suppose $G(x) = \sum_{j=0}^{w} g_j x^j$ with $g_j \in \mathbb{F}_{q'}$, where $w = \deg(G(x))$. Then

$$\frac{G(x) - G(\gamma_i)}{x - \gamma_i} G(\gamma_i)^{-1} = G(\gamma_i)^{-1} \sum_{j=1}^{w} g_j \sum_{k=0}^{j-1} x^k \gamma_i^{j-1-k}$$

$$= G(\gamma_i)^{-1} \sum_{k=0}^{w-1} x^k \left(\sum_{j=k+1}^{w} g_j \gamma_i^{j-1-k} \right).$$

Therefore, by (13.5), setting the coefficients of x^k equal to 0, in the order $k = w - 1$, $w - 2, \ldots, 0$, we have that $\mathbf{c} \in \Gamma(L, G)$ if and only if $H\mathbf{c}^{\mathsf{T}} = \mathbf{0}$, where

$$H = \begin{bmatrix} h_0 g_w & h_1 g_w & \cdots & h_{n-1} g_w \\ h_0(g_{w-1} + g_w \gamma_0) & h_1(g_{w-1} + g_w \gamma_1) & \cdots & h_{n-1}(g_{w-1} + g_w \gamma_{n-1}) \\ & & \vdots & \\ h_0 \sum_{j=1}^{w} g_j \gamma_0^{j-1} & h_1 \sum_{j=1}^{w} g_j \gamma_1^{j-1} & \cdots & h_{n-1} \sum_{j=1}^{w} g_j \gamma_{n-1}^{j-1} \end{bmatrix} \tag{13.6}$$

with $h_i = G(\gamma_i)^{-1}$. By Exercise 768, H can be row reduced to the $w \times n$ matrix H', where

$$H' = \begin{bmatrix} G(\gamma_0)^{-1} & G(\gamma_1)^{-1} & \cdots & G(\gamma_{n-1})^{-1} \\ G(\gamma_0)^{-1}\gamma_0 & G(\gamma_1)^{-1}\gamma_1 & \cdots & G(\gamma_{n-1})^{-1}\gamma_{n-1} \\ & & \vdots & \\ G(\gamma_0)^{-1}\gamma_0^{w-1} & G(\gamma_1)^{-1}\gamma_1^{w-1} & \cdots & G(\gamma_{n-1})^{-1}\gamma_{n-1}^{w-1} \end{bmatrix}. \tag{13.7}$$

Exercise 768 Show that the matrix H of (13.6) can be row reduced to H' of (13.7). ◆

By (5.4) we see that H' is the generator matrix for $\mathrm{GRS}_w(\gamma, \mathbf{v})$ where $\gamma = \{\gamma_0, \gamma_1, \ldots, \gamma_{n-1}\}$ and $v_i = G(\gamma_i)^{-1}$. This shows that $\Gamma(L, G)$ is the subfield subcode $\mathrm{GRS}_w(\gamma, \mathbf{v})^\perp|_{\mathbb{F}_q}$ of the dual of $\mathrm{GRS}_w(\gamma, \mathbf{v})$; see Section 3.8. Since $\mathrm{GRS}_w(\gamma, \mathbf{v})^\perp$ is also a GRS code by Theorem 5.3.3, a Goppa code is a subfield subcode of a GRS code.

The entries of H' are in $\mathbb{F}_{q^t}$. By choosing a basis of $\mathbb{F}_{q^t}$ over $\mathbb{F}_q$, each element of $\mathbb{F}_{q^t}$ can be represented as a $t \times 1$ column vector over $\mathbb{F}_q$. Replacing each entry of H' by its corresponding column vector, we obtain a $tw \times n$ matrix H'' over $\mathbb{F}_q$ which has the property that $\mathbf{c} \in \mathbb{F}_q^n$ is in $\Gamma(L, G)$ if and only if $H''\mathbf{c}^T = \mathbf{0}$. Compare this to Theorem 4.4.3.

We have the following bounds on the dimension and minimum distance of a Goppa code.

Theorem 13.2.1 *In the notation of this section, the Goppa code $\Gamma(L, G)$ with $\deg(G(x)) = w$ is an $[n, k, d]$ code, where $k \geq n - wt$ and $d \geq w + 1$.*

Proof: The rows of H'' may be dependent and hence this matrix has rank at most wt. Hence $\Gamma(L, G)$ has dimension at least $n - wt$. If a nonzero codeword $\mathbf{c} \in \Gamma(L, G)$ has weight w or less, then when the left-hand side of (13.4) is written as a rational function, the numerator has degree $w - 1$ or less; but this numerator must be a multiple of $G(x)$, which is impossible as $\deg(G(x)) = w$. □

Exercise 769 This is a continuation of Exercise 767. Do the following:
(a) Find the parity check matrix H' from (13.7) for the Goppa code found in Exercise 767.
(b) Give the parity check equation from the matrix in (a). How does this equation compare with the equation found in Exercise 767(b)?
(c) By writing each element of $\mathbb{F}_8$ as a binary 3-tuple from Example 3.4.3, construct the parity check matrix H'' as described above.
(d) Use the parity check matrix from part (c) to find the vectors in $\Gamma(L, G)$ directly. ◆

Exercise 770 Construct a binary Goppa code of length 7 from a parity check matrix as follows. Use the field $\mathbb{F}_8$ given in Example 3.4.3 with primitive element α. Let $L = \{0, 1, \alpha, \alpha^2, \alpha^4, \alpha^5, \alpha^6\}$ and $G(x) = x - \alpha^3$.
(a) Find the parity check matrix H' from (13.7) for $\Gamma(L, G)$.
(b) Find the binary parity check matrix H'' from H' as discussed above.
(c) What other code is $\Gamma(L, G)$ equivalent to? What are the dimension and minimum weight of $\Gamma(L, G)$, and how do these compare to the bounds of Theorem 13.2.1?
(d) Based on what you have discovered in part (c), state and prove a theorem about a binary Goppa code $\Gamma(L, G)$ of length $2^t - 1$ with $G(x) = x - \alpha^r$ for some r with $0 \leq r \leq 2^t - 2$ and $L = \{\alpha^i \mid 0 \leq i \leq 2^t - 2, i \neq r\} \cup \{0\}$ where α is a primitive element of $\mathbb{F}_{2^t}$. ◆

Another formulation for Goppa codes can be given as follows. Let $\mathcal{R}$ be the vector space of all rational functions $f(x) = a(x)/b(x)$ with coefficients in $\mathbb{F}_{q^t}$, where $a(x)$ and $b(x)$ are relatively prime and satisfy two requirements. First, the zeros of $a(x)$ include the zeros of $G(x)$ with at least the same multiplicity as in $G(x)$; second, the only possible zeros of $b(x)$, that is the poles of $f(x)$, are from $\gamma_0, \ldots, \gamma_{n-1}$ each with at most multiplicity one. Any rational function $f(x)$ has a Laurent series expansion about γ_i. The rational functions $f(x)$ in $\mathcal{R}$ have Laurent series expansion

$$\sum_{j=-1}^{\infty} f_j(x - \gamma_i)^j$$

about γ_i where $f_{-1} \neq 0$ if $f(x)$ has a pole at γ_i and $f_{-1} = 0$ otherwise. The *residue of* $f(x)$ *at* γ_i, denoted $\mathrm{Res}_{\gamma_i} f$, is the coefficient f_{-1}. Let

$$\mathcal{C} = \{(\mathrm{Res}_{\gamma_0} f, \mathrm{Res}_{\gamma_1} f, \ldots, \mathrm{Res}_{\gamma_{n-1}} f) \mid f(x) \in \mathcal{R}\}. \tag{13.8}$$

Exercise 771 shows that $\Gamma(L, G)$ is the subfield subcode $\mathcal{C}|_{\mathbb{F}_q}$.

Exercise 771 Prove that $\Gamma(L, G) = \mathcal{C}|_{\mathbb{F}_q}$, where $\mathcal{C}$ is given in (13.8). ◆

13.2.3   Generalized Reed–Muller codes

In Section 1.10 we introduced the binary Reed–Muller codes. We can construct similar codes, called generalized Reed–Muller codes, over other fields using a construction parallel to that of the GRS codes. Kasami, Lin, and Peterson [163] first introduced the primitive generalized Reed–Muller codes which we study here; the nonprimitive codes, which we do not investigate, were presented in [349]. Primitive and nonprimitive codes are also examined extensively in [4, 5].

Let m be a positive integer and let $P_1, P_2, \ldots, P_n$ be the $n = q^m$ points in the affine space $\mathbb{A}^m(\mathbb{F}_q)$. For any integer r with $0 \leq r \leq m(q-1)$, let $\mathbb{F}_q[x_1, x_2, \ldots, x_m]_r$ be the polynomials in $\mathbb{F}_q[x_1, x_2, \ldots, x_m]$ of degree r or less together with the zero polynomial. Define the *rth order generalized Reed–Muller*, or *GRM, code* of length $n = q^m$ to be

$$\mathcal{R}_q(r, m) = \{(f(P_1), f(P_2), \ldots, f(P_n)) \mid f \in \mathbb{F}_q[x_1, x_2, \ldots, x_m]_r\}.$$

If a term in f has a factor x_i^e where $e = q + d \geq q$, then the factor can be replaced by x_i^{1+d} without changing the value of f at any point P_j because $\beta^{q+d} = \beta^q \beta^d = \beta \beta^d$ as $\beta^q = \beta$ for all $\beta \in \mathbb{F}_q$. Thus we see that

$$\mathcal{R}_q(r, m) = \{(f(P_1), f(P_2), \ldots, f(P_n)) \mid f \in \mathbb{F}_q[x_1, x_2, \ldots, x_m]_r^*\},$$

where $\mathbb{F}_q[x_1, x_2, \ldots, x_m]_r^*$ is all polynomials in $\mathbb{F}_q[x_1, x_2, \ldots, x_m]_r$ with no term having an exponent q or higher on any variable. Clearly, this is a vector space over $\mathbb{F}_q$ with a basis

$$\mathcal{B}_q(r, m) = \left\{x_1^{e_1} \cdots x_m^{e_m} \mid 0 \leq e_i < q, \ e_1 + e_2 + \cdots + e_m \leq r\right\}.$$

Obviously, $\{(f(P_1), f(P_2), \ldots, f(P_n)) \mid f \in \mathcal{B}_q(r, m)\}$ spans $\mathcal{R}_q(r, m)$. These codewords are in fact independent; this fact is shown in the binary case in Exercise 772.

Example 13.2.2 We construct generator matrices for $\mathcal{R}_2(1, 2)$, $\mathcal{R}_2(2, 2)$, and $\mathcal{R}_2(2, 3)$; from these matrices we conclude that these codes are the original RM codes $\mathcal{R}(1, 2)$, $\mathcal{R}(2, 2)$, and $\mathcal{R}(2, 3)$. The basis $\mathcal{B}_2(1, 2)$ of $\mathbb{F}_2[x_1, x_2]_1^*$ is $\{f_1, f_2, f_3\}$, where $f_1(x_1, x_2) = 1$, $f_2(x_1, x_2) = x_1$, and $f_3(x_1, x_2) = x_2$. Let $P_1 = (0, 0)$, $P_2 = (1, 0)$, $P_3 = (0, 1)$, and $P_4 = (1, 1)$ be the points in $\mathbb{A}^2(\mathbb{F}_2)$. By evaluating f_1, f_2, f_3 in that order at P_1, P_2, P_3, P_4, we obtain the matrix

$$G'(1, 2) = \begin{bmatrix} 1 & 1 & 1 & 1 \\ 0 & 1 & 0 & 1 \\ 0 & 0 & 1 & 1 \end{bmatrix}.$$

This matrix clearly has independent rows and generates the same code $\mathcal{R}(1, 2)$ as $G(1, 2)$ did in Section 1.10. The basis $\mathcal{B}_2(2, 2)$ of $\mathbb{F}_2[x_1, x_2]_2^*$ is $\{f_1, f_2, f_3, f_4\}$, where $f_4(x_1, x_2) = x_1 x_2$. Using the same points and by evaluating f_1, f_2, f_3, f_4 in order, we obtain the matrix

$$G'(2, 2) = \begin{bmatrix} 1 & 1 & 1 & 1 \\ 0 & 1 & 0 & 1 \\ 0 & 0 & 1 & 1 \\ 0 & 0 & 0 & 1 \end{bmatrix}.$$

This matrix again clearly has independent rows and generates the same code as I_4 does, which is $\mathcal{R}(2, 2)$. The basis $\mathcal{B}_2(2, 3)$ of $\mathbb{F}_2[x_1, x_2, x_3]_2^*$ is $\{f_1, \ldots, f_7\}$, where $f_5(x_1, x_2, x_3) = x_3$, $f_6(x_1, x_2, x_3) = x_1 x_3$, and $f_7(x_1, x_2, x_3) = x_2 x_3$. Let $P_1 = (0, 0, 0)$, $P_2 = (1, 0, 0)$, $P_3 = (0, 1, 0)$, $P_4 = (1, 1, 0)$, $P_5 = (0, 0, 1)$, $P_6 = (1, 0, 1)$, $P_7 = (0, 1, 1)$, $P_8 = (1, 1, 1)$ be the points in $\mathbb{A}^3(\mathbb{F}_2)$. By evaluating $f_1, \ldots, f_7$ in that order we obtain the matrix

$$G'(2, 3) = \left[\begin{array}{cccc|cccc} 1 & 1 & 1 & 1 & 1 & 1 & 1 & 1 \\ 0 & 1 & 0 & 1 & 0 & 1 & 0 & 1 \\ 0 & 0 & 1 & 1 & 0 & 0 & 1 & 1 \\ 0 & 0 & 0 & 1 & 0 & 0 & 0 & 1 \\ 0 & 0 & 0 & 0 & 1 & 1 & 1 & 1 \\ 0 & 0 & 0 & 0 & 0 & 1 & 0 & 1 \\ 0 & 0 & 0 & 0 & 0 & 0 & 1 & 1 \end{array}\right].$$

Notice that

$$G'(2, 3) = \begin{bmatrix} G'(2, 2) & G'(2, 2) \\ O & G'(1, 2) \end{bmatrix},$$

which indicates that $G'(2, 3)$ generates $\mathcal{R}(2, 3)$ by (1.5) and (1.7). ∎

Exercise 772 By ordering the points $P_1, P_2, \ldots, P_n$ correctly with $n = 2^m$ and the basis $\mathcal{B}_2(r, m)$ of $\mathbb{F}_2[x_1, \ldots, x_m]_r^*$ appropriately, a generator matrix for $\mathcal{R}_2(r, m)$ decomposes to show that this code can be obtained from the $(\mathbf{u}|\mathbf{u} + \mathbf{v})$ construction.

(a) Study the construction in Example 13.2.2 and show how to generalize this inductively to obtain a generator matrix $G'(r, m)$ of $\mathcal{R}_2(r, m)$ from generator matrices $G'(r, m - 1)$ and $G'(r - 1, m - 1)$ for $\mathcal{R}_2(r, m - 1)$ and $\mathcal{R}_2(r - 1, m - 1)$ in the form

$$G'(r, m) = \begin{bmatrix} G'(r, m - 1) & G'(r, m - 1) \\ O & G'(r - 1, m - 1) \end{bmatrix}.$$

Hint: In Example 13.2.2 observe that $f_1, \ldots, f_4$ do not involve x_3, but $f_5 = f_1 x_3$, $f_6 = f_2 x_3$, $f_7 = f_3 x_3$ do.

(b) Explain why part (a) shows that $\mathcal{R}_2(r, m)$ is the ordinary binary RM code $\mathcal{R}(r, m)$.

(c) Explain why part (a) shows that $\{(f(P_1), f(P_2), \ldots, f(P_n)) \mid f \in \mathcal{B}_2(r, m)\}$ is independent.   ◆

13.3 Algebraic curves

The Reed–Solomon and generalized Reed–Muller codes have similar definitions, namely as the set of n-tuples $(f(P_1), \ldots, f(P_n))$ where $P_1, \ldots, P_n$ are fixed points in affine space and f runs through a specified set of functions. We can do the same for generalized Reed–Solomon codes with a minor modification to (13.2) where the points are in projective space. It is these constructions that were generalized by Goppa in [105, 106]. The codes that he constructed are now termed algebraic geometry codes or geometric Goppa codes.

In order to describe Goppa's construction, we need to study affine and projective curves with an emphasis on the points on these curves in given fields. We will limit our discussion to curves in the plane. These curves will be described in one of two ways. An *affine plane curve* $\mathcal{X}$ is the set of affine points $(x, y) \in \mathbb{A}^2(\mathbb{F})$, denoted $\mathcal{X}_f(\mathbb{F})$, where $f(x, y) = 0$ with $f \in \mathbb{F}[x, y]$. A *projective plane curve* $\mathcal{X}$ is the set of projective points $(x : y : z) \in \mathbb{P}^2(\mathbb{F})$, also denoted $\mathcal{X}_f(\mathbb{F})$, where $f(x, y, z) = 0$ with f a homogeneous polynomial in $\mathbb{F}[x, y, z]$.[1] Suppose that $f \in \mathbb{F}[x, y]$. If f^H is the homogenization of f, then $\mathcal{X}_{f^H}(\mathbb{F})$ is called the *projective closure of* $\mathcal{X}_f(\mathbb{F})$. In a sense, the only difference between $\mathcal{X}_f(\mathbb{F})$ and $\mathcal{X}_{f^H}(\mathbb{F})$ is that points at infinity have been added to $\mathcal{X}_f(\mathbb{F})$ to produce $\mathcal{X}_{f^H}(\mathbb{F})$; this follows from Theorem 13.1.3(ii). In many situations the function defining a curve will be defined over a field $\mathbb{F}$ but we will want the curve to be points in $\mathbb{A}^2(\mathbb{E})$ or $\mathbb{P}^2(\mathbb{E})$ where $\mathbb{E}$ is an extension field of $\mathbb{F}$. In that case we will attach the field to the notation for the curve, namely $\mathcal{X}_f(\mathbb{E})$. At other times we will define the curve simply by an equation without specifying the field and thus dropping the field from the notation.

We will need the concept of partial derivatives analogous to the same notion developed in calculus. If $f(x, y) = \sum a_{ij} x^i y^j \in \mathbb{F}[x, y]$, the *partial derivative f_x of f with respect to x* is

$$f_x(x, y) = \sum i a_{ij} x^{i-1} y^j.$$

The partial derivative with respect to y is defined analogously; if $f(x, y, z) \in \mathbb{F}[x, y, z]$, the partial derivative with respect to z can also be defined. A point (x_0, y_0) on an affine curve $\mathcal{X}_f(\mathbb{F})$ (where $f(x_0, y_0) = 0$) is *singular* if $f_x(x_0, y_0) = f_y(x_0, y_0) = 0$. A point on $\mathcal{X}_f(\mathbb{F})$ is *nonsingular* or *simple* if it is not singular. Analogous definitions hold for projective curves. A curve that has no singular points is called *nonsingular*, *regular*, or *smooth*.

Example 13.3.1 The *Fermat curve* $\mathcal{F}_m(\mathbb{F}_q)$ is a projective plane curve over $\mathbb{F}_q$ defined by $f(x, y, z) = x^m + y^m + z^m = 0$. As $f_x = mx^{m-1}$, $f_y = my^{m-1}$, and $f_z = mz^{m-1}$ and

---

[1] Note that when defining polynomials in two or three variables, we will not use subscripts; the variables will be denoted x and y, or x, y, and z.

since $(0 : 0 : 0)$ is not a projective point, the only time that singular points on $\mathcal{F}_m(\mathbb{F}_q)$ can exist are when $\gcd(m, q) \neq 1$; in that case every point on the curve is singular. So if m and q are relatively prime, $\mathcal{F}_m(\mathbb{F}_q)$ is nonsingular. ∎

Exercise 773 The Fermat curve $\mathcal{F}_m(\mathbb{F}_q)$ is defined in Example 13.3.1. Consider the curve $\mathcal{F}_3(\mathbb{F}_q)$ given by $x^3 + y^3 + z^3 = 0$.
(a) Find the three projective points $(x : y : z)$ of $\mathbb{P}^2(\mathbb{F}_2)$ on $\mathcal{F}_3(\mathbb{F}_2)$.
(b) Find the nine projective points $(x : y : z)$ of $\mathbb{P}^2(\mathbb{F}_4)$ on $\mathcal{F}_3(\mathbb{F}_4)$.
(c) Find the nine projective points $(x : y : z)$ of $\mathbb{P}^2(\mathbb{F}_8)$ on $\mathcal{F}_3(\mathbb{F}_8)$. The field $\mathbb{F}_8$ can be found in Example 3.4.3. ◆

Example 13.3.2 Let $q = r^2$ where r is a prime power. The *Hermitian curve* $\mathcal{H}_r(\mathbb{F}_q)$ is the projective plane curve over $\mathbb{F}_q$ defined by $f(x, y, z) = x^{r+1} - y^r z - y z^r = 0$. In Exercise 774, you will show that $\mathcal{H}_r(\mathbb{F}_q)$ has only one point at infinity, namely $(0 : 1 : 0)$. We show that there are r^3 affine points $(x : y : 1)$ of $\mathbb{P}^2(\mathbb{F}_q)$ on $\mathcal{H}_r(\mathbb{F}_q)$. As $z = 1$, $x^{r+1} = y^r + y$. But $y^r + y = \mathrm{Tr}_2(y)$, where $\mathrm{Tr}_2 : \mathbb{F}_{r^2} \to \mathbb{F}_r$ is the trace map from $\mathbb{F}_{r^2}$ to $\mathbb{F}_r$. By Lemma 3.8.5, Tr_2 is a nonzero $\mathbb{F}_r$-linear transformation from $\mathbb{F}_{r^2}$ to $\mathbb{F}_r$. As its image is a nonzero subspace of the 1-dimensional space $\mathbb{F}_r$, Tr_2 is surjective. So its kernel is a 1-dimensional $\mathbb{F}_r$-subspace of $\mathbb{F}_{r^2}$, implying there are r elements $y \in \mathbb{F}_{r^2}$ with $\mathrm{Tr}_2(y) = 0$. When $\mathrm{Tr}_2(y) = 0$, then $x^{r+1} = 0$ has one solution $x = 0$; this leads to r affine points on $\mathcal{H}_r(\mathbb{F}_q)$. If $x \in \mathbb{F}_{r^2}$, $x^{r+1} \in \mathbb{F}_r$ as $r^2 - 1 = (r + 1)(r - 1)$ and the nonzero elements of $\mathbb{F}_r$ in $\mathbb{F}_{r^2}$ are precisely those satisfying $\beta^{r-1} = 1$. When y is one of the $r^2 - r$ elements of $\mathbb{F}_{r^2}$ with $\mathrm{Tr}_2(y) \neq 0$, there are $r + 1$ solutions $x \in \mathbb{F}_{r^2}$ with $x^{r+1} = \mathrm{Tr}_2(y)$. This leads to $(r^2 - r)(r + 1) = r^3 - r$ more affine points. Hence, $\mathcal{H}_r(\mathbb{F}_q)$ has $r^3 - r + r = r^3$ affine points and a total of $r^3 + 1$ projective points. ∎

Exercise 774 Let $q = r^2$ where r is a prime power. The Hermitian curve $\mathcal{H}_r(\mathbb{F}_q)$ is defined in Example 13.3.2.
(a) Show that $\mathcal{H}_r(\mathbb{F}_q)$ is nonsingular. Hint: In $\mathbb{F}_q$, $r = 0$ as r is a multiple of the characteristic of $\mathbb{F}_q$.
(b) Show that $(0 : 1 : 0)$ is the only point at infinity on $\mathcal{H}_r(\mathbb{F}_q)$.
(c) Find the eight affine points $(x : y : 1)$ of $\mathbb{P}^2(\mathbb{F}_4)$ on $\mathcal{H}_2(\mathbb{F}_4)$.
(d) Find the 64 affine points $(x : y : 1)$ of $\mathbb{P}^2(\mathbb{F}_{16})$ on $\mathcal{H}_4(\mathbb{F}_{16})$. Table 5.1 gives the field $\mathbb{F}_{16}$. ◆

Exercise 775 The *Klein quartic* $\mathcal{K}_4(\mathbb{F}_q)$ is the projective curve over $\mathbb{F}_q$ defined by the fourth degree homogeneous polynomial equation $f(x, y, z) = x^3 y + y^3 z + z^3 x = 0$.
(a) Find the three partial derivatives of f.
(b) Show that if $\mathbb{F}_q$ has characteristic 3, $\mathcal{K}_4(\mathbb{F}_q)$ is nonsingular.
(c) If $(x : y : z)$ is a singular point of $\mathcal{K}_4(\mathbb{F}_q)$, show that $x^3 y = -3y^3 z$ using $f_y(x, y, z) = 0$, and that $z^3 x = 9y^3 z$ using $f_x(x, y, z) = 0$ and $f_y(x, y, z) = 0$.
(d) If $(x : y : z)$ is a singular point of $\mathcal{K}_4(\mathbb{F}_q)$, show that $7y^3 z = 0$ using $x^3 y + y^3 z + z^3 x = 0$ and part (c).
(e) Using part (d), show that if $\mathbb{F}_q$ does not have characteristic 7, then $\mathcal{K}_4(\mathbb{F}_q)$ is nonsingular.
(f) Find the three projective points in $\mathbb{P}^2(\mathbb{F}_2)$ on $\mathcal{K}_4(\mathbb{F}_2)$.
(g) Find the five projective points in $\mathbb{P}^2(\mathbb{F}_4)$ on $\mathcal{K}_4(\mathbb{F}_4)$.

(h) Find the 24 projective points in $\mathbb{P}^2(\mathbb{F}_8)$ on $\mathcal{K}_4(\mathbb{F}_8)$. The field $\mathbb{F}_8$ can be found in Example 3.4.3. ♦

When examining the points on a curve, we will need to know their degrees. The degree m of a point depends on the field under consideration. Let $q = p^r$ where p is a prime, and let $m \geq 1$ be an integer. The map $\sigma_q : \mathbb{F}_{q^m} \rightarrow \mathbb{F}_{q^m}$ given by $\sigma_q(\alpha) = \alpha^q$ is an automorphism of $\mathbb{F}_{q^m}$ that fixes $\mathbb{F}_q$; $\sigma_q = \sigma_p^r$ where σ_p is the Frobenius map defined in Section 3.6. If $P = (x, y)$ or $P = (x : y : z)$ have coordinates in $\mathbb{F}_{q^m}$, let $\sigma_q(P)$ denote $(\sigma_q(x), \sigma_q(y))$ or $(\sigma_q(x) : \sigma_q(y) : \sigma_q(z))$, respectively. By Exercise 776, in the projective case, it does not matter which representative is chosen for a projective point when applying σ_q to the point. Suppose f is in either $\mathbb{F}_q[x, y]$ or $\mathbb{F}_q[x, y, z]$, with f homogeneous in the latter case. So f defines either an affine or projective curve over $\mathbb{F}_q$ or over $\mathbb{F}_{q^m}$. Exercise 776 shows that if P is a point on $\mathcal{X}_f(\mathbb{F}_{q^m})$, so is $\sigma_q(P)$. Therefore $\{\sigma_q^i(P) \mid i \geq 0\}$ is a set of points on $\mathcal{X}_f(\mathbb{F}_{q^m})$; however, there are at most m distinct points in this set as σ_q^m is the identity automorphism of $\mathbb{F}_{q^m}$. This allows us to extend our notion of a point. A *point P on $\mathcal{X}_f(\mathbb{F}_q)$ of degree m over $\mathbb{F}_q$* is a set of m distinct points $P = \{P_0, \ldots, P_{m-1}\}$ with P_i on $\mathcal{X}_f(\mathbb{F}_{q^m})$, where $P_i = \sigma_q^i(P_0)$; the *degree of P* over $\mathbb{F}_q$ is denoted $\deg(P)$. The points of degree one on $\mathcal{X}_f(\mathbb{F}_q)$ are called *rational* or *$\mathbb{F}_q$-rational* points.[2] This definition is motivated by the well-known situation with polynomials over the real numbers. Conjugation is the automorphism of the complex numbers that fixes the real numbers. The polynomial $p(x) = x^2 - 4x + 5$ with real coefficients has no real roots but has two complex conjugate roots, $2 + i$ and $2 - i$. In our new terminology, the roots of $p(x)$ are lumped together as the pair $\{2 + i, 2 - i\}$ and is a point on $p(x) = 0$ of degree 2 over the real numbers (even though each individual root is not real). In the definition of a point $P = \{P_0, \ldots, P_{m-1}\}$ of degree m over $\mathbb{F}_q$, the P_is are required to be distinct. It is possible, beginning with a point P_0 in $\mathbb{A}^2(\mathbb{F}_{q^m})$ or $\mathbb{P}^2(\mathbb{F}_{q^m})$, that $P_i = \sigma_q^i(P_0)$ for $0 \leq i \leq m - 1$ are not distinct. In this case the distinct P_is form a point of smaller degree; see Exercise 777(c).

Exercise 776 Do the following:
(a) Let $(x : y : z) = (x' : y' : z') \in \mathbb{P}^2(\mathbb{F}_{q^m})$. Prove that $(\sigma_q(x) : \sigma_q(y) : \sigma_q(z)) = (\sigma_q(x') : \sigma_q(y') : \sigma_q(z'))$.
(b) Let f be either a polynomial in $\mathbb{F}_q[x, y]$ or a homogeneous polynomial in $\mathbb{F}_q[x, y, z]$. Show that if P is in either $\mathbb{A}^2(\mathbb{F}_{q^m})$ or in $\mathbb{P}^2(\mathbb{F}_{q^m})$, respectively, with $f(P) = 0$, then $f(\sigma_q(P)) = 0$. ♦

Example 13.3.3 Let $f(x, y, z) = x^3 + xz^2 + z^3 + y^2z + yz^2 \in \mathbb{F}_2[x, y, z]$. The curve determined by this equation is an example of an *elliptic curve*. A point at infinity on $\mathcal{X}_f(\mathbb{F})$ where $\mathbb{F}$ is any extension field of $\mathbb{F}_2$ satisfies $z = 0$; hence $x^3 = 0$, implying y can be any nonzero value. Thus there is only one point at infinity on $\mathcal{X}_f(\mathbb{F})$, namely $P_\infty = (0 : 1 : 0)$. The point at infinity is of degree 1 (that is, it is rational) over $\mathbb{F}_2$ as its coordinates are in $\mathbb{F}_2$.

---

[2] The objects that we previously called points are actually points of degree 1 over $\mathbb{F}_q$. Notice that points of degree m over $\mathbb{F}_q$ are fixed by σ_q, just as elements of $\mathbb{F}_q$ are; hence, this is why these points are considered to be on $\mathcal{X}_f(\mathbb{F}_q)$.

When considering the affine points, we can assume $z = 1$ and so

$$x^3 + x + 1 = y^2 + y. \tag{13.9}$$

What are the affine points of degree 1 over $\mathbb{F}_2$ on the curve? Since x and y are in $\mathbb{F}_2$, we see that $y^2 + y = 0$ but $x^3 + x + 1 = 1$. So the only point of degree 1 (that is, the only rational point) over $\mathbb{F}_2$ is the point at infinity. There are points of degree 2 over $\mathbb{F}_2$. Here x and y are in $\mathbb{F}_4$. If $y = 0$ or 1, then $y^2 + y = 0$. By (13.9), $x^3 + x + 1 = 0$, which has no solution in $\mathbb{F}_4$. If $y = \omega$ or $\overline{\omega}$, then $y^2 + y = 1$ and so $x^3 + x = x(x+1)^2 = 0$ by (13.9). Thus the points of degree 2 over $\mathbb{F}_2$ on the curve are $P_1 = \{p_1, p_1'\}$, where $p_1 = (0 : \omega : 1)$ and $p_1' = (0 : \overline{\omega} : 1)$, and $P_2 = \{p_2, p_2'\}$, where $p_2 = (1 : \omega : 1)$ and $p_2' = (1 : \overline{\omega} : 1)$, noting that σ_2 switches the two projective points in each pair. ∎

Exercise 777 Let $f(x, y, z) = x^3 + xz^2 + z^3 + y^2z + yz^2 \in \mathbb{F}_2[x, y, z]$ determine an elliptic curve as in Example 13.3.3.
(a) Prove that $\mathcal{X}_f(\mathbb{F})$ is nonsingular for any extension field $\mathbb{F}$ of $\mathbb{F}_2$.
(b) Find the four points of degree 3 on the curve over $\mathbb{F}_2$. The field $\mathbb{F}_8$ can be found in Example 3.4.3.
(c) Find the five points of degree 4 on the curve over $\mathbb{F}_2$. The field $\mathbb{F}_{16}$ can be found in Table 5.1. (Along the way you may rediscover the points of degree 2.) ◆

Exercise 778 This exercise uses the results of Exercise 773. Find the points on the Fermat curve defined by $x^3 + y^3 + z^3 = 0$ of degrees 1, 2, and 3 over $\mathbb{F}_2$. ◆

Exercise 779 This exercise uses the results of Exercise 774.
(a) There are eight affine points $(x : y : 1)$ of $\mathbb{P}^2(\mathbb{F}_4)$ on the Hermitian curve $\mathcal{H}_2(\mathbb{F}_4)$. Find the two affine points on this curve of degree 1 over $\mathbb{F}_2$ and the three affine points of degree 2 also over $\mathbb{F}_2$.
(b) There are 64 affine points $(x : y : 1)$ of $\mathbb{P}^2(\mathbb{F}_{16})$ on the Hermitian curve $\mathcal{H}_4(\mathbb{F}_{16})$. Find the two affine points on this curve of degree 1 over $\mathbb{F}_2$, the single affine point of degree 2 over $\mathbb{F}_2$, and the 15 affine points of degree 4 over $\mathbb{F}_2$. ◆

Exercise 780 This exercise uses the results of Exercise 775. Find the points on the Klein quartic defined by $x^3y + y^3z + z^3x = 0$ of degrees 1, 2, and 3 over $\mathbb{F}_2$. ◆

When defining algebraic geometry codes, we will need to be able to compute the points on the intersection of two curves. In addition to the degree of a point of intersection, we need to know the *intersection multiplicity*, which we shorten to *multiplicity*, at the point of intersection. We do not formally define multiplicity because the definition is too technical. As the following example illustrates, we can compute multiplicity similarly to the way multiplicity of zeros is computed for polynomials in one variable.

Example 13.3.4 In Example 13.3.3 and Exercise 777, we found some of the projective points from $\mathbb{P}^2(\mathbb{F}_q)$, where q is a power of 2, on the elliptic curve determined by $x^3 + xz^2 + z^3 + y^2z + yz^2 = 0$. We now explore how this curve intersects other curves.
• Intersection with $x = 0$. We either have $z = 0$ or can assume that $z = 1$. In the former case we obtain the point at infinity $P_\infty = (0 : 1 : 0)$; in the latter, the equation $y^2 + y + 1 = 0$ must be satisfied leading to the two points $p_1 = (0 : \omega : 1)$ and $p_1' = (0 : \overline{\omega} : 1)$ in $\mathbb{P}^2(\mathbb{F}_4)$.

We can view this in one of two ways. Over $\mathbb{F}_4$ or extension fields of $\mathbb{F}_4$, the curves $x^3 + xz^2 + z^3 + y^2z + yz^2 = 0$ and $x = 0$ intersect at the three points P_∞, p_1, and p'_1. Each of these points has degree 1, and the intersection multiplicity at each of these points is 1. Over $\mathbb{F}_2$ or extension fields of $\mathbb{F}_2$ not containing $\mathbb{F}_4$, the points p_1 and p'_1 combine to form a point of degree 2, and so in this case we have two points of intersection: P_∞ and $P_1 = \{p_1, p'_1\}$. The point P_∞ has degree 1 and P_1 has degree 2. Of course, the intersection multiplicity at each of these points is still 1.

- Intersection with $x^2 = 0$. We can view $x^2 = 0$ as the union of the line $x = 0$ with itself. Therefore every point on the elliptic curve and $x^2 = 0$ occurs twice as frequently as it did on the single line $x = 0$. Thus over $\mathbb{F}_4$ or extension fields of $\mathbb{F}_4$ there are three points of intersection, P_∞, p_1, and p'_1; each point continues to have degree 1 but now the multiplicity intersection at each point is 2. Similarly, over $\mathbb{F}_2$ or extension fields not containing $\mathbb{F}_4$, there are two points of intersection, P_∞ and $P_1 = \{p_1, p'_1\}$. They have degree 1 and degree 2, respectively, and the intersection multiplicity at each point is now 2.

- Intersection with $z = 0$. We saw from Example 13.3.3 that there was only one point P_∞ on the elliptic curve with $z = 0$. This point has degree 1 over $\mathbb{F}_2$ or any extension field of $\mathbb{F}_2$. Plugging $z = 0$ into the equation defining our elliptic curve, we obtain $x^3 = 0$. Hence P_∞ occurs on the intersection with multiplicity 3.

- Intersection with $z^2 = 0$. As in the case $x^2 = 0$, we double the multiplicities obtained in the case $z = 0$. Thus over any field of characteristic 2, the elliptic curve intersects the union of two lines each given by $z = 0$ at P_∞. This point has degree 1 over $\mathbb{F}_2$ or any extension field of $\mathbb{F}_2$. The intersection multiplicity at P_∞ doubles and therefore is 6.

- Intersection with $y = 0$. Here $z = 0$ is not possible as then x would have to be 0. So $z = 1$ and we must have $x^3 + x + 1 = 0$. The only solutions occur in $\mathbb{F}_8$ and lead to the points $p_3 = (\alpha : 0 : 1)$, $p'_3 = (\alpha^2 : 0 : 1)$, and $p''_3 = (\alpha^4 : 0 : 1)$ in $\mathbb{P}^2(\mathbb{F}_8)$. (Compare this to your answer in Exercise 777(b).) Thus over $\mathbb{F}_8$ or extension fields of $\mathbb{F}_8$ there are three points of intersection, p_3, p'_3, and p''_3; each has degree 1. The intersection multiplicity at each point is 1. Over $\mathbb{F}_2$ or extension fields not containing $\mathbb{F}_8$, the three points combine into a single point of intersection $P_3 = \{p_3, p'_3, p''_3\}$ of degree 3. Its intersection multiplicity remains 1. ∎

Exercise 781 As in Example 13.3.4, find the intersection of the elliptic curve $x^3 + xz^2 + z^3 + y^2z + yz^2 = 0$ with the curves given below over $\mathbb{F}_2$ and its extension fields. In addition give the degree of each point and the intersection multiplicity at each point of intersection.

(a) $x + z = 0$.
(b) $x^2 + z^2 = 0$. Hint: $x^2 + z^2 = (x + z)^2$.
(c) $y^2 = 0$.
(d) $xz = 0$. Hint: This is the union of two lines $x = 0$ and $z = 0$.
(e) $yz = 0$.
(f) $xy = 0$.
(g) $z^2 + y^2 + yz = 0$. ◆

Example 13.3.4 illustrates a certain uniformity in the number of points of intersection if counted properly. To count "properly" one must consider both multiplicity and degree; namely, an intersection of multiplicity m at a point of degree d counts md times. The total count of the points of intersection is computed by adding the products of all the multiplicities and degrees. From Example 13.3.4, the elliptic curve intersects $x = 0$ either in three points of degree 1 each with multiplicity 1, giving a total count of $1 \cdot 1 + 1 \cdot 1 + 1 \cdot 1 = 3$ points of intersection, or in one point of degree 1 intersecting with multiplicity 1 and one point of degree 2 intersecting with multiplicity 1, giving a total count of $1 \cdot 1 + 1 \cdot 2 = 3$ points of intersection. We also found that the elliptic curve intersects $x^2 = 0$ either in three points of degree 1 with intersection multiplicity 2, giving a total count of $2 \cdot 1 + 2 \cdot 1 + 2 \cdot 1 = 6$ points of intersection, or in one point of degree 1 with multiplicity 2 and one point of degree 2 also with multiplicity 2, giving a total count of $2 \cdot 1 + 2 \cdot 2 = 6$ points of intersection. Notice that these counts, 3 and 6, equal the product of the degree 3 of the homogeneous polynomial defining the elliptic curve and the degree, 1 or 2, of the homogeneous polynomial defining the curve intersecting the elliptic curve. This illustrates the following theorem due to Bézout.[3]

Theorem 13.3.5 (Bézout) *Let $f(x, y, z)$ and $g(x, y, z)$ be homogeneous polynomials over $\mathbb{F}$ of degrees d_f and d_g, respectively. Suppose that f and g have no common nonconstant polynomial factors. Then $\mathcal{X}_f$ and $\mathcal{X}_g$ intersect in $d_f d_g$ points when counted with multiplicity and degree.*

Exercise 782 Verify that Bézout's Theorem agrees with the results of Exercise 781. ◆

This discussion naturally leads to the concept of a divisor on a curve. Let $\mathcal{X}$ be a curve over the field $\mathbb{F}$. A *divisor D on $\mathcal{X}$ over $\mathbb{F}$* is a formal sum $D = \sum n_P P$, where n_P is an integer and P is a point of arbitrary degree on $\mathcal{X}$, with only a finite number of the n_P being nonzero. The divisor is *effective* if $n_P \geq 0$ for all P. The *support* $\mathrm{supp}(D)$ *of the divisor* $D = \sum n_P P$ is $\{P \mid n_P \neq 0\}$. The *degree of the divisor* $D = \sum n_P P$ is $\deg(D) = \sum n_P \deg(P)$.

Example 13.3.6 In Example 13.3.3 we found the points of degrees 1 and 2 over $\mathbb{F}_2$ on the elliptic curve $x^3 + xz^2 + z^3 + y^2 z + yz^2 = 0$. So one divisor on that curve over $\mathbb{F}_2$ is $D_1 = 7P_\infty + 4P_1 - 9P_2$; this is not effective but $D_2 = 7P_\infty + 4P_1 + 9P_2$ is an effective divisor. Both D_1 and D_2 have support $\{P_\infty, P_1, P_2\}$. Also $\deg(D_1) = 7 \cdot 1 + 4 \cdot 2 - 9 \cdot 2 = -3$ and $\deg(D_2) = 7 \cdot 1 + 4 \cdot 2 + 9 \cdot 2 = 33$. Note that $7P_\infty + 4p_1 - 9p_2'$ is not a divisor over $\mathbb{F}_2$ because p_1 and p_2' are not defined over $\mathbb{F}_2$; we are required to list the entire pair P_1 and P_2 in a divisor over $\mathbb{F}_2$. However, $7P_\infty + 4p_1 - 9p_2'$ is a divisor over $\mathbb{F}_4$ with support $\{P_\infty, p_1, p_2'\}$ and degree $7 \cdot 1 + 4 \cdot 1 - 9 \cdot 1 = 2$. ∎

We can use the notation of divisors to describe the intersection of two curves easily. If $\mathcal{X}_1$ and $\mathcal{X}_2$ are projective curves, then the *intersection divisor over $\mathbb{F}$ of $\mathcal{X}_1$ and $\mathcal{X}_2$*, denoted $\mathcal{X}_1 \cap \mathcal{X}_2$, is $\sum n_P P$, where the summation runs over all points over $\mathbb{F}$ on both $\mathcal{X}_1$ and $\mathcal{X}_2$ and n_P is the multiplicity of the point on the two curves. If these curves are defined by homogeneous polynomials, with no common nonconstant polynomial factors, of degrees d_f and d_g, then the intersection divisor has degree $d_f d_g$ by Bézout's Theorem.

---

[3] Bézout's Theorem dates back to 1779 and originated in remarks of Newton and MacLauren. A version of this theorem for the complex plane had been proved earlier by Euler in 1748 and Cramer in 1750.

Example 13.3.7 In Example 13.3.4 we found the intersection of the elliptic curve determined by $x^3 + xz^2 + z^3 + y^2z + yz^2 = 0$ with five other curves. Their intersection divisors are:

- Intersection with $x = 0$: $P_\infty + P_1$.
- Intersection with $x^2 = 0$: $2P_\infty + 2P_1$.
- Intersection with $z = 0$: $3P_\infty$.
- Intersection with $z^2 = 0$: $6P_\infty$.
- Intersection with $y = 0$: P_3.   ∎

Exercise 783 Find the intersection divisor of the elliptic curve in Exercise 781 with each of the other curves in that exercise.   ◆

We close this section with one final concept. In computing the minimum distance and dimension of algebraic geometry codes, the genus of a plane projective curve will play an important role. The *genus of a curve* is actually connected to a topological concept of the same name. Without giving this topological definition, we state a theorem, called Plücker's Formula[4] (see [135]), which gives the genus of a nonsingular projective plane curve; for our purposes, this can serve as the definition of genus.

Theorem 13.3.8 (Plücker's Formula) *The genus g of a nonsingular projective plane curve determined by a homogeneous polynomial of degree $d \geq 1$ is*

$$g = \frac{(d-1)(d-2)}{2}.$$

Exercise 784 Find the genus of the following curves which have already been shown to be nonsingular.
(a) The Fermat curve of Example 13.3.1.
(b) The Hermitian curve of Example 13.3.2.
(c) The Klein quartic of Exercise 775.
(d) The elliptic curve of Example 13.3.3.
Let $p(x) \in \mathbb{F}[x]$ be a polynomial of degree 3 that has no repeated roots. Show that the following curves are nonsingular and compute their genus.
(e) The curve $\mathcal{X}_{f^H}(\mathbb{F})$ where $f(x, y) = y^2 - p(x)$ and $\mathbb{F}$ does not have characteristic 2.
(f) The curve $\mathcal{X}_{f^H}(\mathbb{F})$ where $f(x, y) = y^2 + y + p(x)$ and $\mathbb{F}$ has characteristic 2.   ◆

13.4   Algebraic geometry codes

We are about ready to define the codes studied by Goppa that generalize the classical codes from Section 13.2. Recall that these codes are n-tuples consisting of functions evaluated at n fixed points where the functions run through a certain vector space. To define the more general codes we need to determine the n points, which will turn out to be points on a curve,

---

[4] This formula is actually a special case of a series of more general formulas developed by Plücker in 1834 and generalized by Max Noether in 1875 and 1883.

and the vector space of functions, which will actually be a vector space of equivalence classes of rational functions related to the function defining the curve.

We first concentrate on this latter vector space of functions. Let $p(x, y, z)$ be a homogeneous polynomial of positive degree that defines a projective curve $\mathcal{X}$ over a field $\mathbb{F}$. Define the *field of rational functions on $\mathcal{X}$ over $\mathbb{F}$* by

$$\mathbb{F}(\mathcal{X}) = \left(\left\{ \frac{g}{h} \,\middle|\, \begin{array}{l} g, h \in \mathbb{F}[x, y, z] \text{ homogeneous} \\ \text{of equal degree with } p \nmid h \end{array} \right\} \cup \{0\} \right) \Big/ \approx_{\mathcal{X}}.$$

This notation means that $\mathbb{F}(\mathcal{X})$ is actually a collection of equivalence classes of rational functions where the numerator and denominator are homogeneous of equal degree. By requiring that the denominator not be a multiple of $p(x, y, z)$, at least one of the points on $\mathcal{X}$ is not a zero of the denominator. To define the equivalence we say that $g/h \approx_{\mathcal{X}} g'/h'$ if and only if $gh' - g'h$ is a polynomial multiple of $p(x, y, z)$. Furthermore, we say $g/h \approx_{\mathcal{X}} 0$ precisely when $g(x, y, z)$ is a polynomial multiple of $p(x, y, z)$. By Exercise 785, $\mathbb{F}(\mathcal{X})$ is actually a field containing $\mathbb{F}$ as a subfield.

Exercise 785 Verify that $\mathbb{F}(\mathcal{X})$ is a field containing $\mathbb{F}$ as a subfield (where $\mathbb{F}$ is identified with the constant polynomials). ◆

Let $f = g/h \in \mathbb{F}(\mathcal{X})$ with $f \not\approx_{\mathcal{X}} 0$. Then the *divisor of f* is

$$\operatorname{div}(f) = (\mathcal{X} \cap \mathcal{X}_g) - (\mathcal{X} \cap \mathcal{X}_h), \tag{13.10}$$

that is, the difference in the intersection divisors $\mathcal{X} \cap \mathcal{X}_g$ and $\mathcal{X} \cap \mathcal{X}_h$. Essentially, $\operatorname{div}(f)$ is a mechanism to keep track easily of the zeros and poles of f that are on $\mathcal{X}$ together with their multiplicities and orders, respectively; in effect, $\operatorname{div}(f)$ is the "zeros of f on $\mathcal{X}$" minus the "poles of f on $\mathcal{X}$." There will be cancellation in (13.10) whenever a point on $\mathcal{X}$ is both a zero and a pole; ultimately, if P appears with positive coefficient in $\operatorname{div}(f)$, it is a zero of f and a pole if it appears with a negative coefficient. If g and h have degrees d_g and d_h, respectively, then by Bézout's Theorem, $\deg(\operatorname{div}(f)) = d_p d_g - d_p d_h = 0$ as $d_g = d_h$. Since f is only a representative of an equivalence class, we need to know that $\operatorname{div}(f)$ is independent of the representative chosen and therefore is well-defined. This is indeed true; we do not prove it but illustrate it in Exercise 786.

Exercise 786 Let $\mathcal{X}$ be the elliptic curve defined by $x^3 + xz^2 + z^3 + y^2z + yz^2 = 0$ over a field of characteristic 2. Let $f = g/h$ and $f' = g'/h'$, where $g(x, y, z) = x^2 + z^2$, $h(x, y, z) = z^2$, $g'(x, y, z) = z^2 + y^2 + yz$, and $h'(x, y, z) = xz$. As in Example 13.3.3, let $P_\infty = (0 : 1 : 0)$ and $P_2 = \{(1 : \omega : 1), (1 : \overline{\omega} : 1)\}$. Example 13.3.7 and Exercise 783 will help with parts (b) and (c).
(a) Show that $f \approx_{\mathcal{X}} f'$.
(b) Show that $\operatorname{div}(f) = 2P_2 - 4P_\infty$ using (13.10).
(c) Show that $\operatorname{div}(f') = 2P_2 - 4P_\infty$ using (13.10) with f', g', and h' in place of f, g, and h. ◆

There is a partial ordering of divisors on a curve given by $D = \sum n_P P \succeq D' = \sum n'_P P$ provided $n_P \geq n'_P$ for all P. Thus D is effective if $D \succeq 0$. We now define the space of functions that will help determine our algebraic geometry codes. Let D be a divisor on a projective curve $\mathcal{X}$ defined over $\mathbb{F}$. Let

$$L(D) = \{f \in \mathbb{F}(\mathcal{X}) \mid f \not\approx_{\mathcal{X}} 0, \operatorname{div}(f) + D \succeq 0\} \cup \{0\}.$$

In other words for $f \not\approx_{\mathcal{X}} 0$, $f \in L(D)$ if and only if $\operatorname{div}(f) + D$ is effective. This is a vector space over $\mathbb{F}$ as seen in Exercise 787. Suppose $D = \sum n_P P$. A rational function f is in $L(D)$ provided that any pole P of f has order not exceeding n_P and provided that any zero P of f has multiplicity at least $-n_P$.

Exercise 787 Prove that $L(D)$ is a vector space over $\mathbb{F}$.   ◆

The next theorem gives some simple facts about $L(D)$.

Theorem 13.4.1 *Let D be a divisor on a projective curve $\mathcal{X}$. The following hold:*
(i)   *If $\deg(D) < 0$, then $L(D) = \{0\}$.*
(ii) *The constant functions are in $L(D)$ if and only if $D \succeq 0$.*
(iii) *If P is a point on $\mathcal{X}$ with $P \notin \operatorname{supp}(D)$, then P is not a pole of any $f \in L(D)$.*

Proof: For (i), if $f \in L(D)$ with $f \not\approx_{\mathcal{X}} 0$, then $\operatorname{div}(f) + D \succeq 0$. In particular, $\deg(\operatorname{div}(f) + D) \geq 0$; but $\deg(\operatorname{div}(f) + D) = \deg(\operatorname{div}(f)) + \deg(D) = \deg(D)$, contradicting $\deg(D) < 0$.

Let $f \not\approx_{\mathcal{X}} 0$ be a constant function. If $f \in L(D)$ then $\operatorname{div}(f) + D \succeq 0$. But $\operatorname{div}(f) = 0$ when f is a constant function. So $D \succeq 0$. Conversely, if $D \succeq 0$, then $\operatorname{div}(f) + D \succeq 0$ provided $\operatorname{div}(f) = 0$, which is the case if f is a constant function. This proves (ii).

For (iii), if P is a pole of $f \in L(D)$ with $P \notin \operatorname{supp}(D)$, then the coefficient of P in the divisor $\operatorname{div}(f) + D$ of $\mathcal{X}$ is negative, contradicting the requirement that $\operatorname{div}(f) + D$ is effective for $f \in L(D)$.   □

With one more bit of notation we will finally be able to define algebraic geometry codes. Recall that the projective plane curve $\mathcal{X}$ is defined by $p(x, y, z) = 0$ where we now assume that the field $\mathbb{F}$ is $\mathbb{F}_q$. Let D be a divisor on $\mathcal{X}$; choose a set $\mathcal{P} = \{P_1, \ldots, P_n\}$ of n distinct $\mathbb{F}_q$-rational points on $\mathcal{X}$ such that $\mathcal{P} \cap \operatorname{supp}(D) = \emptyset$. Order the points in $\mathcal{P}$ and let $\operatorname{ev}_{\mathcal{P}} : L(D) \to \mathbb{F}_q^n$ be the *evaluation map* defined by

$$\operatorname{ev}_{\mathcal{P}}(f) = (f(P_1), f(P_2), \ldots, f(P_n)).$$

In this definition, we need to be careful. We must make sure $\operatorname{ev}_{\mathcal{P}}$ is well-defined, as the rational functions are actually representatives of equivalence classes. If $f \in L(D)$, then P_i is not a pole of f by Theorem 13.4.1(iii). However, if f is represented by g/h, h may still have a zero at P_i occurring a certain number of times in $\mathcal{X}_h \cap \mathcal{X}$. Since P_i is not a pole of f, P_i occurs at least this many times in $\mathcal{X}_g \cap \mathcal{X}$. If we choose g/h to represent f, $f(P_i)$ is really $0/0$, a situation we must avoid. It can be shown that for any $f \in L(D)$, we can choose a representative g/h where $h(P_i) \neq 0$; we will always make such a choice. Suppose now that f has two such representatives $g/h \approx_{\mathcal{X}} g'/h'$ where $h(P_i) \neq 0$ and $h'(P_i) \neq 0$. Then $gh' - g'h$ is a polynomial multiple of p and $p(P_i) = 0$, implying that $g(P_i)h'(P_i) = g'(P_i)h(P_i)$. Since $h(P_i)$ and $h'(P_i)$ are nonzero, $g(P_i)/h(P_i) =$

$g'(P_i)/h'(P_i)$. Thus $\mathrm{ev}_{\mathcal{P}}$ is well-defined on $L(D)$. Also $f(P_i) \in \mathbb{F}_q$ for any $f \in L(D)$ as the coefficients of the rational function and the coordinates of P_i are in $\mathbb{F}_q$. Hence the image $\mathrm{ev}_{\mathcal{P}}(f)$ of f is indeed contained in $\mathbb{F}_q^n$.

Exercise 788 Prove that the map $\mathrm{ev}_{\mathcal{P}}$ is linear. ◆

With this notation the *algebraic geometry code associated to $\mathcal{X}, \mathcal{P}$, and D* is defined to be

$$\mathcal{C}(\mathcal{X}, \mathcal{P}, D) = \{\mathrm{ev}_{\mathcal{P}}(f) \mid f \in L(D)\}.$$

Exercise 788 implies that $\mathcal{C}(\mathcal{X}, \mathcal{P}, D)$ is a linear code over $\mathbb{F}_q$. By Theorem 13.4.1(i), we will only be interested in codes where $\deg(D) \geq 0$ as otherwise $L(D) = \{0\}$ and $\mathcal{C}(\mathcal{X}, \mathcal{P}, D)$ is the zero code. We would like some information about the dimension and minimum distance of algebraic geometry codes. An upper bound on the dimension of the code is the dimension of $L(D)$. In the case that $\mathcal{X}$ is nonsingular, a major tool used to find the dimension of $L(D)$ is the Riemann–Roch Theorem, one version of which is the following.

Theorem 13.4.2 (Riemann–Roch) *Let D be a divisor on a nonsingular projective plane curve $\mathcal{X}$ over $\mathbb{F}_q$ of genus g. Then*

$$\dim(L(D)) \geq \deg(D) + 1 - g.$$

Furthermore, if $\deg(D) > 2g - 2$, then

$$\dim(L(D)) = \deg(D) + 1 - g.$$

The next theorem gives conditions under which we know the dimension of algebraic geometry codes exactly and have a lower bound on their minimum distance. The reader is encouraged to compare its proof to that of Theorem 5.3.1.

Theorem 13.4.3 *Let D be a divisor on a nonsingular projective plane curve $\mathcal{X}$ over $\mathbb{F}_q$ of genus g. Let $\mathcal{P}$ be a set of n distinct $\mathbb{F}_q$-rational points on $\mathcal{X}$ such that $\mathcal{P} \cap \mathrm{supp}(D) = \emptyset$. Furthermore, assume $2g - 2 < \deg(D) < n$. Then $\mathcal{C}(\mathcal{X}, \mathcal{P}, D)$ is an $[n, k, d]$ code over $\mathbb{F}_q$ where $k = \deg(D) + 1 - g$ and $d \geq n - \deg(D)$. If $\{f_1, \ldots, f_k\}$ is a basis of $L(D)$, then*

$$\begin{bmatrix} f_1(P_1) & f_1(P_2) & \cdots & f_1(P_n) \\ f_2(P_1) & f_2(P_2) & \cdots & f_2(P_n) \\ & & \vdots & \\ f_k(P_1) & f_k(P_2) & \cdots & f_k(P_n) \end{bmatrix}$$

is a generator matrix for $\mathcal{C}(\mathcal{X}, \mathcal{P}, D)$.

Proof: By the Riemann–Roch Theorem, $\dim(L(D)) = \deg(D) + 1 - g$ as $2g - 2 < \deg(D)$. Hence this value is also the dimension k of $\mathcal{C}(\mathcal{X}, \mathcal{P}, D)$ provided we show that the linear map $\mathrm{ev}_{\mathcal{P}}$ (see Exercise 788) has trivial kernel. Let $\mathcal{P} = \{P_1, \ldots, P_n\}$. Suppose that $\mathrm{ev}_{\mathcal{P}}(f) = \mathbf{0}$. Thus $f(P_i) = 0$ for all i, implying that P_i is a zero of f and the coefficient of P_i in $\mathrm{div}(f)$ is at least 1. But $P_i \notin \mathrm{supp}(D)$, and therefore $\mathrm{div}(f) + D - P_1 - \cdots - P_n \succeq 0$, implying that $f \in L(D - P_1 - \cdots - P_n)$. However, $\deg(D) < n$, which means that

$\deg(D - P_1 - \cdots - P_n) < 0$. By Theorem 13.4.1(i), $L(D - P_1 - \cdots - P_n) = \{0\}$, showing that $f = 0$ and hence that ev_P has trivial kernel. We conclude that $k = \deg(D) + 1 - g$. This also shows that the matrix given in the statement of the theorem is a generator matrix for $C(\mathcal{X}, \mathcal{P}, D)$.

Now suppose that $\mathrm{ev}_P(f)$ has minimum nonzero weight d. Thus $f(P_{i_j}) = 0$ for some set of $n - d$ distinct indices $\{i_j \mid 1 \le j \le n - d\}$. As above, $f \in L(D - P_{i_1} - \cdots - P_{i_{n-d}})$. Since $f \ne 0$, Theorem 13.4.1(i) shows that $\deg(D - P_{i_1} - \cdots - P_{i_{n-d}}) \ge 0$. Therefore, $\deg(D) - (n - d) \ge 0$ or $d \ge n - \deg(D)$. $\qquad\square$

Example 13.4.4 We show that the narrow-sense and the extended narrow-sense Reed–Solomon codes are in fact algebraic geometry codes. Consider the projective plane curve $\mathcal{X}$ over $\mathbb{F}_q$ given by $z = 0$. The points on the curve are $(x : y : 0)$, which essentially forms the projective line. Let $P_\infty = (1 : 0 : 0)$. There are q remaining $\mathbb{F}_q$-rational points on the line. Let P_0 be the $\mathbb{F}_q$-rational point represented by $(0 : 1 : 0)$. Let $P_1, \ldots, P_{q-1}$ be the remaining $\mathbb{F}_q$-rational points. For the narrow-sense Reed–Solomon codes we will let $n = q - 1$ and $\mathcal{P} = \{P_1, \ldots, P_{q-1}\}$. For the extended narrow-sense Reed–Solomon codes we will let $n = q$ and $\mathcal{P} = \{P_0, \ldots, P_{q-1}\}$. Fix k with $1 \le k \le n$ and let $D = (k - 1)P_\infty$; note that $D = 0$ when $k = 1$. Clearly, $\mathcal{P} \cap \mathrm{supp}(D) = \emptyset$, $\mathcal{X}$ is nonsingular of genus $g = 0$ by Plücker's Formula, and $\deg(D) = k - 1$. In particular, $\deg(D) > 2g - 2$ and, by the Riemann–Roch Theorem, $\dim(L(D)) = \deg(D) + 1 - g = k$.

We claim that

$$B = \left\{ 1, \frac{x}{y}, \frac{x^2}{y^2}, \ldots, \frac{x^{k-1}}{y^{k-1}} \right\}$$

is a basis of $L(D)$. First, $\mathrm{div}(x^j/y^j) = jP_0 - jP_\infty$. Thus $\mathrm{div}(x^j/y^j) + D = jP_0 + (k - 1 - j)P_\infty$, which is effective provided $0 \le j \le k - 1$. Hence, every function in B is in $L(D)$. We only need to show that B is independent over $\mathbb{F}_q$. Suppose that

$$f = \sum_{j=0}^{k-1} a_j \frac{x^j}{y^j} \approx_\mathcal{X} 0.$$

Then $f = g/h$, where $g(x, y, z) = \sum_{j=0}^{k-1} a_j x^j y^{k-1-j}$ and $h(x, y, z) = y^{k-1}$. By definition of $\approx_\mathcal{X}$, $g(x, y, z)$ is a polynomial multiple of z. Clearly, this polynomial multiple must be 0 as z does not occur in $g(x, y, z)$, which shows that $a_j = 0$ for $0 \le j \le k - 1$. Thus B is a basis of $L(D)$.

Using this basis we see that every nonzero polynomial in $L(D)$ can be written as $f(x, y, z) = g(x, y, z)/y^d$, where $g(x, y, z) = \sum_{j=0}^{d} c_j x^j y^{d-j}$ with $c_d \ne 0$ and $d \le k - 1$. But $g(x, y, z)$ is the homogenization (in $\mathbb{F}_q[x, y]$) of $m(x) = \sum_{j=0}^{d} c_j x^j$. So the association $f(x, y, z) \leftrightarrow m(x)$ is a one-to-one correspondence between $L(D)$ and the polynomials in $\mathbb{F}_q[x]$ of degree $k - 1$ or less, together with the zero polynomial, which we previously denoted $\mathcal{P}_k$. Furthermore, if $\beta \in \mathbb{F}_q$, then $m(\beta) = f(\beta, 1, 0)$; additionally, Theorem 13.1.3(i) implies that $f(\beta, 1, 0) = f(x_0, y_0, z_0)$, where $(x_0 : y_0 : z_0)$ is any representation of

$(\beta : 1 : 0)$ because f is the ratio of two homogeneous polynomials of equal degree d.[5] Let α be a primitive element of $\mathbb{F}_q$. Order the points so that P_i is represented by $(\alpha^{i-1} : 1 : 0)$ for $1 \leq i \leq q - 1$; then order $\mathcal{P}$ as $P_1, \ldots, P_{q-1}$ if $n = q - 1$ and as $P_1, \ldots, P_{q-1}, P_0$ if $n = q$. This discussion shows that

$$\{(m(1), m(\alpha), m(\alpha^2), \ldots, m(\alpha^{q-2})) \mid m \in \mathcal{P}_k\}$$
$$= \{(f(P_1), f(P_2), f(P_3), \ldots, f(P_{q-1})) \mid f \in L(D)\}$$

and

$$\{(m(1), m(\alpha), \ldots, m(\alpha^{q-2}), m(0)) \mid m \in \mathcal{P}_k\}$$
$$= \{(f(P_1), f(P_2), \ldots, f(P_{q-1}), f(P_0)) \mid f \in L(D)\}.$$

Hence from our presentation in Section 13.2.1, the narrow-sense and extended narrow-sense Reed–Solomon codes are algebraic geometry codes.

Theorem 13.4.3 shows that both the narrow-sense and extended narrow-sense Reed–Solomon codes have dimension $\deg(D) + 1 - g = k - 1 + 1 - 0 = k$ and minimum distance $d \geq n - \deg(D) = n - k + 1$. The Singleton Bound shows that $d \leq n - k + 1$ and hence $d = n - k + 1$; each code is MDS, as we already knew. ∎

Exercise 789 Let $\mathcal{X}$ be a projective curve of genus 0. Let D be a divisor on $\mathcal{X}$ such that $-2 < \deg(D) < n$. Prove that $\mathcal{C}(\mathcal{X}, \mathcal{P}, D)$ is MDS. ◆

Exercise 790 In Example 13.4.4 we show that Reed–Solomon codes are algebraic geometry codes. Here you will show that generalized Reed–Solomon codes are also algebraic geometry codes, as shown for example in [320].
(a) Let $\gamma_0, \gamma_1, \ldots, \gamma_{n-1}$ be n distinct elements of $\mathbb{F}_q$. Let $v_0, v_1, \ldots, v_{n-1}$ be n not necessarily distinct elements of $\mathbb{F}_q$. Prove the *Lagrange Interpolation Formula*, which states that there exists a polynomial $p(x) \in \mathbb{F}_q[x]$ of degree at most $n - 1$ such that $p(\gamma_i) = v_i$ given by

$$p(x) = \sum_{i=0}^{n-1} v_i \prod_{j \neq i} \frac{x - \gamma_j}{\gamma_i - \gamma_j}. \tag{13.11}$$

For the remainder of the problem we assume the notation of part (a), with the additional assumption that the v_is are nonzero and $\mathbf{v} = (v_0, \ldots, v_{n-1})$. Also, let $\gamma = (\gamma_0, \ldots, \gamma_{n-1})$. Let $\mathcal{X}$ be the projective plane curve over $\mathbb{F}_q$ given by $z = 0$. Let $h(x, y)$ be the homogenization in $\mathbb{F}_q[x, y]$ of the polynomial $p(x)$ of degree $d \leq n - 1$ given by (13.11); note that $h \neq 0$ as $v_i \neq 0$. Let $u(x, y, z) = h(x, y)/y^d$, which is an element of $\mathbb{F}_q(\mathcal{X})$. Let $\mathcal{P} = \{P_1, P_2, \ldots, P_n\}$ where P_i is a representative of the projective point $(\gamma_{i-1} : 1 : 0)$ in $\mathbb{P}^2(\mathbb{F}_q)$. Let $P_\infty = (1 : 0 : 0)$. Finally, for k an integer with $1 \leq k \leq n$, let $D = (k - 1) \times P_\infty - \operatorname{div}(u)$.
(b) Prove that $u(P_i) = v_{i-1}$ for $1 \leq i \leq n$.

---

[5] Recall that in our formulation (13.2) of GRS codes, the exact code depended on the representation of projective points chosen. The formulation of the narrow-sense and extended narrow-sense Reed–Solomon codes presented here has the nice feature that the exact code is independent of the representation chosen for the projective points.

(c) Prove that $\mathcal{P} \cap \mathrm{supp}(D) = \emptyset$.

(d) Prove that $\deg(D) = k - 1$. Hint: Recall that the degree of the divisor of any element in $\mathbb{F}_q(\mathcal{X})$ is 0.

(e) Prove that $\dim(L(D)) = k$ and that

$$B = \left\{ u, u\frac{x}{y}, u\frac{x^2}{y^2}, \ldots, u\frac{x^{k-1}}{y^{k-1}} \right\}$$

is a basis of $L(D)$.

(f) Prove that $\mathrm{GRS}_k(\gamma, \mathbf{v}) = \mathcal{C}(\mathcal{X}, \mathcal{P}, D)$.   ♦

Exercise 791 Let $\mathcal{X}$ be the elliptic curve over $\mathbb{F}_2$ defined by $x^3 + xz^2 + z^3 + y^2z + yz^2 = 0$. By Exercise 777, $\mathcal{X}$ is nonsingular. The genus of $\mathcal{X}$ is 1. Let $P_\infty = (0 : 1 : 0)$, which is the unique point at infinity on $\mathcal{X}$ by Example 13.3.3. Let $D = kP_\infty$ for some positive integer k.

(a) Prove that $\dim(L(D)) = k$.

(b) Compute $\mathrm{div}(x^i y^j / z^{i+j})$ where i and j are nonnegative integers (including the possibility that $i = j = 0$). Hint: Use Example 13.3.7.

(c) What condition must be satisfied for $x^i y^j / z^{i+j}$ to be in $L(D)$?

(d) Find a basis of $L(D)$ for $k \geq 1$ using the functions in part (c). Hint: Compute this first for $k = 1, 2, 3, 4, 5$ and then note that

$$\frac{x^3}{z^3} \approx_{\mathcal{X}} \frac{x}{z} + 1 + \frac{y^2}{z^2} + \frac{y}{z}.$$

(e) Let $\mathcal{P}$ be the four affine points on $\mathcal{X}$ in $\mathbb{P}^2(\mathbb{F}_4)$ (see Example 13.3.3 where these points are p_1, p_1', p_2, p_2'). For each of the codes $\mathcal{C}(\mathcal{X}, \mathcal{P}, D)$ over $\mathbb{F}_4$ with $1 \leq k \leq 3$, apply Theorem 13.4.3 to find the dimension of the code, a lower bound on its minimum distance, and a generator matrix. Also give an upper bound on the minimum distance from the Singleton Bound.

(f) Let $\mathcal{P}$ be the 12 affine points on $\mathcal{X}$ in $\mathbb{P}^2(\mathbb{F}_8)$ (see Exercise 777(b) where these points were found in the process of computing the points of degree 3 over $\mathbb{F}_2$ on the curve). For each of the codes $\mathcal{C}(\mathcal{X}, \mathcal{P}, D)$ over $\mathbb{F}_8$ with $1 \leq k \leq 11$, apply Theorem 13.4.3 to find the dimension of the code, a lower bound on its minimum distance, and a generator matrix. Also give an upper bound on the minimum distance from the Singleton Bound.   ♦

Exercise 791 illustrates some subtle concepts about the fields involved. The computation of $L(D)$ involves divisors over some ground field $\mathbb{F}_q$ ($\mathbb{F}_2$ in the exercise). The divisor D and the divisors of the functions in $L(D)$ involve points of various degrees over $\mathbb{F}_q$, that is, sets of points whose coordinates are in possibly many different extension fields of $\mathbb{F}_q$ ($\mathbb{F}_2$, $\mathbb{F}_4$, and $\mathbb{F}_8$ in the exercise). When constructing an algebraic geometry code, an extension field $\mathbb{F}_{q^m}$ of $\mathbb{F}_q$ is fixed, possibly $\mathbb{F}_q$ itself, and it is over the field $\mathbb{F}_{q^m}$ ($\mathbb{F}_4$ in part (e) of the exercise and $\mathbb{F}_8$ in part (f)) that the code is defined. All the points in $\mathcal{P}$ must be rational over $\mathbb{F}_{q^m}$.

Exercise 792 Let $\mathcal{X}$ be the Hermitian curve $\mathcal{H}_2(\mathbb{F}_4)$ over $\mathbb{F}_4$ of genus 1 defined by $x^3 + y^2z + yz^2 = 0$. By Exercise 774, $\mathcal{X}$ is nonsingular. In the same exercise the point at infinity

$P_\infty = (0 : 1 : 0)$ and the eight affine points on $\mathcal{X}$ in $\mathbb{P}^2(\mathbb{F}_4)$ are found; two of these affine points are $P_0 = (0 : 0 : 1)$ and $P_1 = (0 : 1 : 1)$. Let $D = kP_\infty$ for some positive integer k.

(a) Prove that $\dim(L(D)) = k$.

(b) Show that the intersection divisor of the Hermitian curve with the curve defined by $x = 0$ is $P_\infty + P_0 + P_1$.

(c) Show that the intersection divisor of the Hermitian curve with the curve defined by $y = 0$ is $3P_0$.

(d) Find the intersection divisor of the Hermitian curve with the curve defined by $z = 0$.

(e) Compute $\mathrm{div}(x^i y^j / z^{i+j})$ where i and j are nonnegative integers (including the possibility that $i = j = 0$).

(f) Show that $x^i y^j / z^{i+j}$ is in $L(D)$ if and only if $2i + 3j \le k$.

(g) Find a basis of $L(D)$ for $k \ge 1$ using the functions in part (f). Hint: Show that you can assume $0 \le i \le 2$ since

$$\frac{x^3}{z^3} \approx_\mathcal{X} \frac{y^2}{z^2} + \frac{y}{z}.$$

(h) Let $\mathcal{P}$ be the eight affine points on $\mathcal{X}$ in $\mathbb{P}^2(\mathbb{F}_4)$ (see Exercise 774). For the code $C(\mathcal{X}, \mathcal{P}, D)$ over $\mathbb{F}_4$ with $1 \le k \le 7$, apply Theorem 13.4.3 to find the dimension of the code, a lower bound on its minimum distance, and a generator matrix. Also give an upper bound on the minimum distance from the Singleton Bound.   ♦

Exercise 793 Let $\mathcal{X}$ be the Hermitian curve $\mathcal{H}_2(\mathbb{F}_{16})$ over $\mathbb{F}_{16}$ of genus 6 defined by $x^5 + y^4z + yz^4 = 0$. By Exercise 774, $\mathcal{X}$ is nonsingular. In the same exercise the point at infinity $P_\infty = (0 : 1 : 0)$ and the 64 affine points on $\mathcal{X}$ in $\mathbb{P}^2(\mathbb{F}_{16})$ are found; two of these affine points are $P_0 = (0 : 0 : 1)$ and $P_1 = (0 : 1 : 1)$. Let $D = kP_\infty$ for $k \ge 11$.

(a) Prove that $\dim(L(D)) = k - 5$.

(b) Show that the intersection divisor of the Hermitian curve with the curve defined by $x = 0$ is $P_\infty + P_0 + 3P_1$.

(c) Show that the intersection divisor of the Hermitian curve with the curve defined by $y = 0$ is $5P_0$.

(d) Find the intersection divisor of the Hermitian curve with the curve defined by $z = 0$.

(e) Compute $\mathrm{div}(x^i y^j / z^{i+j})$ where i and j are nonnegative integers (including the possibility that $i = j = 0$).

(f) Show that $x^i y^j / z^{i+j}$ is in $L(D)$ if and only if $4i + 5j \le k$.

(g) Find a basis of $L(D)$ for $k \ge 11$ using the functions in part (f). Hint: Show that you can assume $0 \le i \le 4$ since

$$\frac{x^5}{z^5} \approx_\mathcal{X} \frac{y^4}{z^4} + \frac{y}{z}.$$

(h) Let $\mathcal{P}$ be the 64 affine points on $\mathcal{X}$ in $\mathbb{P}^2(\mathbb{F}_{16})$ (see Exercise 774). For the code $C(\mathcal{X}, \mathcal{P}, D)$ over $\mathbb{F}_{16}$ with $11 \le k \le 63$, apply Theorem 13.4.3 to find the dimension of the code and a lower bound on its minimum distance. Also give an upper bound on the minimum distance from the Singleton Bound.   ♦

Exercise 794 Let $\mathcal{X}$ be the Fermat curve over $\mathbb{F}_2$ of genus 1 defined by $x^3 + y^3 + z^3 = 0$.

By Example 13.3.1, $\mathcal{X}$ is nonsingular. In Exercises 773 and 778, the points of degrees 1, 2, and 3 over $\mathbb{F}_2$ on $\mathcal{X}$ are found. Let $P_\infty = (1 : 1 : 0)$ and $P'_\infty = \{(\omega : 1 : 0), (\overline{\omega} : 1 : 0)\}$ be the points at infinity on $\mathcal{X}$ (see Exercise 773). Let $D = k(P_\infty + P'_\infty)$ for some positive integer k.

(a) Prove that $\dim(L(D)) = 3k$.

(b) Find the intersection divisor of the Fermat curve with the curve defined by $x = 0$.

(c) Find the intersection divisor of the Fermat curve with the curve defined by $y = 0$.

(d) Find the intersection divisor of the Fermat curve with the curve defined by $z = 0$.

(e) Compute $\mathrm{div}(x^i y^j / z^{i+j})$ where i and j are nonnegative integers (including the possibility that $i = j = 0$).

(f) What condition must be satisfied for $x^i y^j / z^{i+j}$ to be in $L(D)$?

(g) Find a basis of $L(D)$ for $k \geq 1$ using the functions in part (f). Hint: Show that you can assume $0 \leq i \leq 2$ since

$$\frac{x^3}{z^3} \approx_{\mathcal{X}} \frac{y^3}{z^3} + 1.$$

(h) Let $\mathcal{P}$ be the six affine points on $\mathcal{X}$ in $\mathbb{P}^2(\mathbb{F}_4)$ (see Exercise 773). For the code $\mathcal{C}(\mathcal{X}, \mathcal{P}, D)$ over $\mathbb{F}_4$ with $k = 1$, apply Theorem 13.4.3 to find the dimension of the code, a lower bound on its minimum distance, and a generator matrix. Also give an upper bound on the minimum distance from the Singleton Bound.

(i) Let $\mathcal{P}$ be the eight affine points on $\mathcal{X}$ in $\mathbb{P}^2(\mathbb{F}_8)$ (see Exercise 773). For the two codes $\mathcal{C}(\mathcal{X}, \mathcal{P}, D)$ over $\mathbb{F}_8$ with $1 \leq k \leq 2$, apply Theorem 13.4.3 to find the dimension of the code, a lower bound on its minimum distance, and a generator matrix. Also give an upper bound on the minimum distance from the Singleton Bound. ◆

Exercise 795 Let $\mathcal{X}$ be the Klein quartic over $\mathbb{F}_2$ of genus 3 defined by $x^3 y + y^3 z + z^3 x = 0$. By Exercise 775, $\mathcal{X}$ is nonsingular. In Exercises 775 and 780, the points on $\mathcal{X}$ of degrees 1, 2, and 3 are found. Let $P_\infty = (0 : 1 : 0)$ and $P'_\infty = (1 : 0 : 0)$ be the points at infinity on $\mathcal{X}$, and let $P_0 = (0 : 0 : 1)$ be the remaining $\mathbb{F}_2$-rational point on $\mathcal{X}$ (see Exercise 775). Let $D = kP_\infty$ for some integer $k \geq 5$.

(a) Prove that $\dim(L(D)) = k - 2$.

(b) Show that the intersection divisor of the Klein quartic with the curve defined by $x = 0$ is $3P_0 + P_\infty$.

(c) Show that the intersection divisor of the Klein quartic with the curve defined by $y = 0$ is $P_0 + 3P'_\infty$.

(d) Find the intersection divisor of the Klein quartic with the curve defined by $z = 0$.

(e) Compute $\mathrm{div}(x^i y^j / z^{i+j})$ where i and j are nonnegative integers (including the possibility that $i = j = 0$).

(f) Show that $x^i y^j / z^{i+j}$ is in $L(D)$ if and only if $2i + 3j \leq k$ and $i \leq 2j$.

(g) Find a basis of $L(D)$ for $k \geq 5$ using the functions in part (f). Hint: Compute this first for $k = 5, 6, \ldots, 12$ and then show that you can assume $0 \leq i \leq 2$ since

$$\frac{x^3 y}{z^4} \approx_{\mathcal{X}} \frac{y^3}{z^3} + \frac{x}{z}.$$

(h) Let $\mathcal{P}$ be the 22 affine points on $\mathcal{X}$ in $\mathbb{P}^2(\mathbb{F}_8)$ (see Exercise 775). For each of the codes

$C(\mathcal{X}, \mathcal{P}, D)$ over $\mathbb{F}_8$ with $5 \le k \le 21$, apply Theorem 13.4.3 to find the dimension of the code and a lower bound on its minimum distance. Give an upper bound on the minimum distance from the Singleton Bound. Finally, find a generator matrix for the code when $k = 21$.   ◆

Exercise 796 Prove that if $D \succeq 0$, then $C(\mathcal{X}, \mathcal{P}, D)$ contains the all-one codeword.   ◆

Recall that the dual of a generalized Reed–Solomon code is also a generalized Reed–Solomon code by Theorem 5.3.3. Since generalized Reed–Solomon codes are algebraic geometry codes by Exercise 790, it is natural to ask if duals of algebraic geometry codes are also algebraic geometry codes. The answer is yes. It can be shown (see [135, Theorem 2.72] or [320, Theorem II.2.10]) that $C(\mathcal{X}, \mathcal{P}, D)^{\perp} = C(\mathcal{X}, \mathcal{P}, E)$, where $E = P_1 + \cdots + P_n - D + (\eta)$ for a certain Weil differential η determined by $\mathcal{P}$; this differential is related to residues of functions mentioned at the conclusion of Section 13.2.2 and in Exercise 771.

13.5 The Gilbert–Varshamov Bound revisited

In Section 2.10 we derived upper and lower bounds on the largest possible rate

$$\alpha_q(\delta) = \limsup_{n \to \infty} n^{-1} \log_q A_q(n, \delta n)$$

for a family of codes over $\mathbb{F}_q$ of lengths going to infinity with relative distances approaching δ. The Asymptotic Gilbert–Varshamov Bound is the lower bound $\alpha_q(\delta) \ge 1 - H_q(\delta)$ guaranteeing that there exists a family of codes with relative distances approaching δ and rates approaching or exceeding $1 - H_q(\delta)$, where H_q is the Hilbert entropy function defined on $0 \le x \le r = 1 - q^{-1}$ by

$$H_q(x) = \begin{cases} 0 & \text{if } x = 0, \\ x \log_q(q-1) - x \log_q x - (1-x) \log_q(1-x) & \text{if } 0 < x \le r, \end{cases}$$

as we first described in Section 2.10.3. For 30 years after the publication of the Gilbert–Varshamov Bound, no family of codes had been demonstrated to exceed this bound until such a family of algebraic geometry codes was shown to exist in 1982. In this section we discuss this result; but before doing so we show that there is a family of Goppa codes that meet this bound.

13.5.1 Goppa codes meet the Gilbert–Varshamov Bound

Theorem 5.1.10 states that primitive BCH codes over $\mathbb{F}_q$ are asymptotically bad, where we recall that a set of codes of lengths going to infinity is asymptotically bad if either the rates go to 0 or the relative distances go to 0. This is true of other families of codes such as the family of cyclic codes whose extended codes are affine-invariant [161]. The family of binary codes obtained from Reed–Solomon codes over $\mathbb{F}_{2^m}$ by concatenation with the $[m, m, 1]$ binary code consisting of all binary m-tuples is also asymptotically bad [218, Chapter 10]. Recall that a set of codes of lengths approaching infinity is asymptotically good if both the rates and relative distances are bounded away from 0. Certainly, codes that meet the Asymptotic

Gilbert–Varshamov Bound are asymptotically good. For most families of codes, such as cyclic codes, it is not known if the family is asymptotically bad or if it contains a subset of asymptotically good codes. The first explicit construction of asymptotically good codes was due to Justesen [160] in 1972; his codes are a modification of the concatenated Reed–Solomon codes mentioned above. While the Justesen codes are asymptotically good, they do not meet the Asymptotic Gilbert–Varshamov Bound. However, Goppa codes include a set of codes that meet the Asymptotic Gilbert–Varshamov Bound. We prove this here. However, the reader will note that the construction of this set of Goppa codes is not explicit; that is, one cannot write down the family of specific codes that meet the bound.

In our construction we will need a lower bound on the number $I_{q^t}(e)$ of irreducible monic polynomials of degree e in $\mathbb{F}_{q^t}[x]$. The exact value of $I_{q^t}(e)$, computed in [18], is

$$I_{q^t}(e) = \frac{1}{e} \sum_{d|e} \mu(d) q^{te/d},$$

where μ is the Möbius function given by $\mu(d) = 1$ if $d = 1$, $\mu(d) = (-1)^r$ if d is a product of r distinct primes, and $\mu(d) = 0$ otherwise. The lower bound we will need is

$$I_{q^t}(e) \geq \frac{q^{te}}{e}\left(1 - q^{-te/2+1}\right); \tag{13.12}$$

the proof of this can also be found in [18].

Theorem 13.5.1 *There is a family of Goppa codes over $\mathbb{F}_q$ that meets the Asymptotic Gilbert–Varshamov Bound.*

Proof: Let t and d be positive integers. Consider the Goppa code $\Gamma(L, G)$ over $\mathbb{F}_q$ of length $n = q^t$ with $L = \{\gamma_0, \gamma_1, \ldots, \gamma_{n-1}\} = \mathbb{F}_{q^t}$ and $G \in \mathbb{F}_{q^t}[x]$. The polynomial G must not have roots in L, a requirement that is certainly satisfied if we choose G to be irreducible of degree $e > 1$ over $\mathbb{F}_{q^t}$. We need to decide what condition must be satisfied for the existence of such a G so that $\Gamma(L, G)$ has minimum distance at least d. To that end, let $\mathbf{c} = c_0 c_1 \cdots c_{n-1} \in \mathbb{F}_q^n$ where $0 < \text{wt}(\mathbf{c}) = w$. If $\mathbf{c} \in \Gamma(L, G)$, then $\sum_{i=0}^{n-1} c_i/(x - \gamma_i) = a(x)/b(x)$ where $b(x)$ is a product of w of the factors $(x - \gamma_i)$ and $\deg(a(x)) \leq w - 1$. Furthermore, $a(x)$ is a multiple of $G(x)$; in particular, $G(x)$ must be one of the irreducible factors of $a(x)$ of degree e. There are at most $\lfloor (w - 1)/e \rfloor$ such factors. So if we want to make sure that $\mathbf{c} \notin \Gamma(L, G)$, which we do when $w < d$, we must eliminate at most $\lfloor (w - 1)/e \rfloor$ choices for G from among all the $I_{q^t}(e)$ irreducible polynomials of degree e in $\mathbb{F}_{q^t}[x]$. For each $1 \leq w < d$ there are $\binom{n}{w}(q - 1)^w$ vectors $\mathbf{c}$ that are not to be in $\Gamma(L, G)$. Thus there are at most

$$\sum_{w=1}^{d-1} \left\lfloor \frac{w - 1}{e} \right\rfloor \binom{n}{w}(q - 1)^w < \frac{d}{e} V_q(n, d)$$

irreducible polynomials that cannot be our choice for G where $V_q(n, d) = \sum_{i=0}^{d} \binom{n}{i}(q - 1)^i$. Hence, we can find an irreducible G so that $\Gamma(L, G)$ has minimum distance at least d provided

$$\frac{d}{e} V_q(n, d) < \frac{q^{te}}{e}\left(1 - q^{-te/2+1}\right), \tag{13.13}$$

by (13.12). If $\delta = d/n$, taking logarithms base q and dividing by n, we obtain

$$n^{-1}[\log_q(\delta n) + \log_q V_q(n, \delta n)] < \frac{te}{n} + n^{-1}\left[\log_q\left(1 - q^{-te/2+1}\right)\right].$$

Taking limits as n approaches infinity, Lemma 2.10.3 yields $H_q(\delta) \leq \lim_{n\to\infty} te/n$ or $1 - H_q(\delta) \geq 1 - \lim_{n\to\infty} te/n$. Since $t = \log_q n$, we can choose an increasing sequence of es growing fast enough so that both the inequality in (13.13) is maintained (guaranteeing the existence of a sequence of Goppa codes of increasing lengths $n = q^t$ with relative minimum distances at least δn) and $1 - H_q(\delta) = 1 - \lim_{n\to\infty} te/n$. Theorem 13.2.1 implies that the codes in our sequence have rate at least $1 - te/n$; therefore this sequence meets the Asymptotic Gilbert–Varshamov Bound.   □

13.5.2 Algebraic geometry codes exceed the Gilbert–Varshamov Bound

The 1982 result of Tsfasman, Vlăduţ, and Zink [333] showed for the first time that there exists a sequence of codes whose relative distances approach δ with rates exceeding $1 - H_q(\delta)$ as their lengths go to infinity. We will only outline this result as the mathematics involved is beyond the scope of this book.

Let $\mathcal{X}$ be a curve of genus g over $\mathbb{F}_q$, $\mathcal{P}$ a set of size n consisting of $\mathbb{F}_q$-rational points on $\mathcal{X}$, and D a divisor on $\mathcal{X}$ with $\mathcal{P} \cap \mathrm{supp}(D) = \emptyset$ where $2g - 2 < \deg(D) < n$. By Theorem 13.4.3, $\mathcal{C}(\mathcal{X}, \mathcal{P}, D)$ is an $[n, k, d]$ code over $\mathbb{F}_q$ with rate $R = k/n = (\deg(D) + 1 - g)/n$ and relative distance $d/n \geq (n - \deg(D))/n$. Thus

$$R + \delta \geq \frac{\deg(D) + 1 - g}{n} + \frac{n - \deg(D)}{n} = 1 + \frac{1}{n} - \frac{g}{n}.$$

In particular,

$$R \geq -\delta + 1 + \frac{1}{n} - \frac{g}{n}. \tag{13.14}$$

Thus we wish to show that there exists a sequence of codes, defined from curves, of increasing length n with relative distances tending toward δ such that g/n tends to a number as close to 0 as possible. Such a construction is not possible using only curves in the projective plane $\mathbb{P}^2(\mathbb{F}_q)$ as the number of points on the curve is finite (at most $q^2 + q + 1$), implying that the length of a code defined on the curve cannot approach infinity since the length is bounded by $q^2 + q$. This limitation is overcome by allowing the curves we examine to come from higher dimensional projective space. Although individual curves still have only a finite number of points, that number can grow as the genus grows. Fortunately, (13.14) applies for curves in higher dimensional space, as does Theorem 13.4.3.

Since we wish g/n to be close to 0, we want n/g to be as large as possible. To explore the size of this quantity, define $N_q(g)$ to be the maximum number of $\mathbb{F}_q$-rational points on a nonsingular absolutely irreducible curve[6] $\mathcal{X}$ in $\mathbb{P}^m(\mathbb{F}_q)$ of genus g, and let

$$A(q) = \limsup_{g\to\infty} \frac{N_q(g)}{g}.$$

---

[6] A nonsingular plane curve is absolutely irreducible, but this is not necessarily true of curves in higher dimensions.

To construct the codes that we desire we first construct an appropriate family of curves and then define the codes over these curves. By our definitions, there exists a sequence of nonsingular absolutely irreducible curves $\mathcal{X}_i$ over $\mathbb{F}_q$ of genus g_i with $n_i + 1$ $\mathbb{F}_q$-rational points on $\mathcal{X}_i$ where n_i and g_i go to infinity and $\lim_{i\to\infty}(n_i + 1)/g_i = A(q)$. Choose an $\mathbb{F}_q$-rational point Q_i on $\mathcal{X}_i$ and let $\mathcal{P}_i$ be the n_i remaining $\mathbb{F}_q$-rational points on $\mathcal{X}_i$. Let r_i be an integer satisfying $2g_i - 2 < r_i < n_i$. By Theorem 13.4.3, the code $\mathcal{C}_i = \mathcal{C}(\mathcal{X}_i, \mathcal{P}_i, r_iQ_i)$ has length n_i, rate $R_i = (\deg(r_iQ_i) + 1 - g_i)/n_i = (r_i + 1 - g_i)/n_i$, and relative distance $\delta_i \geq 1 - \deg(r_iQ_i)/n_i = 1 - r_i/n_i$. Thus

$$R_i + \delta_i \geq \frac{r_i + 1 - g_i}{n_i} + 1 - \frac{r_i}{n_i} = 1 + \frac{1}{n_i} - \frac{g_i}{n_i},$$

implying

$$R \geq -\delta + 1 - \frac{1}{A(q)},$$

where $\lim_{i\to\infty} R_i = R$ and $\lim_{i\to\infty} \delta_i = \delta$. Hence,

$$\alpha_q(\delta) \geq -\delta + 1 - \frac{1}{A(q)}, \tag{13.15}$$

producing a lower bound on $\alpha_q(\delta)$. We must determine $A(q)$ in order to see if this bound is an improvement on the Asymptotic Gilbert–Varshamov Bound.

An upper bound on $A(q)$ was determined by Drinfeld and Vlăduţ [74].

Theorem 13.5.2 *For any prime power q, $A(q) \leq \sqrt{q} - 1$.*

There is actually equality in this bound by a result of Tsfasman, Vlăduţ, and Zink [333] for $q = p^2$ or $q = p^4$, p a prime, and a result of Ihara [153] for $q = p^{2m}$ with $m \geq 3$.

Theorem 13.5.3 *Let $q = p^{2m}$ where p is a prime. There is a sequence of nonsingular absolutely irreducible curves $\mathcal{X}_i$ over $\mathbb{F}_q$ with genus g_i and $n_i + 1$ rational points such that*

$$\lim_{i\to\infty} \frac{n_i}{g_i} = \sqrt{q} - 1.$$

This result and (13.15) give the following bound on $\alpha_q(\delta)$.

Theorem 13.5.4 (Asymptotic Tsfasman–Vlăduţ–Zink Bound) *If $q = p^{2m}$ where p is a prime, then*

$$\alpha_q(\delta) \geq -\delta + 1 - \frac{1}{\sqrt{q} - 1}.$$

This lower bound is determined by the straight line $R = -\delta + 1 - 1/(\sqrt{q} - 1)$ of negative slope that may or may not intersect the Gilbert–Varshamov curve $R = 1 - H_q(\delta)$, which is concave up. If the line intersects the curve in two points, we have shown that there is a sequence of codes exceeding the Asymptotic Gilbert–Varshamov Bound. This occurs when $q \geq 49$.

The curves in Theorem 13.5.3 are modular and unfortunately the proof does not give an explicit construction. In 1995, Garcia and Stichtenoth [96, 97] constructed a tower of function fields that potentially could be used to construct a sequence of algebraic geometry codes that would exceed the Asymptotic Gilbert–Varshamov Bound. A low complexity algorithm to accomplish this was indeed formulated in 2000 [307].

Exercise 797 For each of $q = 4$, $q = 49$, and $q = 64$ draw the graphs of the inequalities given by the Asymptotic Tsfasman–Vlăduţ–Zink and Asymptotic Gilbert–Varshamov Bounds. For $q = 49$ and $q = 64$, give a range of values of δ such that the Tsfasman–Vlăduţ–Zink Bound exceeds the Gilbert–Varshamov Bound.   ◆

14 Convolutional codes

The $[n, k]$ codes that we have studied to this point are called *block codes* because we encode a message of k information symbols into a block of length n. On the other hand convolutional codes use an encoding scheme that depends not only upon the current message being transmitted but upon a certain number of preceding messages. Thus "memory" is an important feature of an encoder of a convolutional code. For example, if $\mathbf{x}(1), \mathbf{x}(2), \ldots$ is a sequence of messages each from $\mathbb{F}_q^k$ to be transmitted at time $1, 2, \ldots$, then an (n, k) convolutional code with memory M will transmit codewords $\mathbf{c}(1), \mathbf{c}(2), \ldots$ where $\mathbf{c}(i) \in \mathbb{F}_q^n$ depends upon $\mathbf{x}(i), \mathbf{x}(i-1), \ldots, \mathbf{x}(i-M)$. In our study of linear block codes we have discovered that it is not unusual to consider codes of fairly high lengths n and dimensions k. In contrast, the study and application of convolutional codes has dealt primarily with (n, k) codes with n and k very small and a variety of values of M.

Convolutional codes were developed by Elias [78] in 1955. In this chapter we will only introduce the subject and restrict ourselves to binary codes. While there are a number of decoding algorithms for convolutional codes, the main one is due to Viterbi; we will examine his algorithm in Section 14.2. The early mathematical theory of these codes was developed extensively by Forney [88, 89, 90] and has been given a modern algebraic treatment by McEliece in [235]. We use the latter extensively in our presentation, particularly in Sections 14.3, 14.4, and 14.5. Those interested in a more in-depth treatment should consult the monograph [225] by Massey, or the books by Lin and Costello [197], McEliece [232], Piret [257], and Wicker [351]. One of the most promising applications of convolutional codes is to the construction of turbo codes, which we introduce in Section 15.7. Convolutional codes, along with several codes we have developed in previous chapters, have played a key role in deep space exploration, the topic of the final section of the book.

14.1 Generator matrices and encoding

There are a number of ways to define convolutional codes. We will present a simple definition that is reminiscent of the definition for a linear block code.

We begin with some terminology. Let D be an indeterminate; the symbol "D" is chosen because it will represent a delay operation. Then $\mathbb{F}_2[D]$ is the set of all binary polynomials in the variable D. This set is an integral domain and has a field of quotients that we denote $\mathbb{F}_2(D)$. Thus $\mathbb{F}_2(D)$ consists of all rational functions in D, that is, all ratios of polynomials in D. An (n, k) *convolutional code* $\mathcal{C}$ is a k-dimensional subspace of $\mathbb{F}_2(D)^n$. Notice that as $\mathbb{F}_2(D)$ is an infinite field, $\mathcal{C}$ contains an infinite number of codewords. The *rate* of $\mathcal{C}$

is $R = k/n$; for every k bits of information, $\mathcal{C}$ generates n codeword bits. A *generator matrix* G for $\mathcal{C}$ is a $k \times n$ matrix with entries in $\mathbb{F}_2(D)$ whose rows form a basis of $\mathcal{C}$. Any multiple of G by a nonzero element of $\mathbb{F}_2(D)$ will also be a generator matrix for $\mathcal{C}$. Thus if we multiply G by the least common multiple of all denominators occurring in entries of G, we obtain a generator matrix for $\mathcal{C}$ all of whose entries are polynomials. From this discussion we see that a convolutional code has a generator matrix all of whose entries are polynomials; such a matrix is called a *polynomial generator matrix* for $\mathcal{C}$. In Section 14.4 we will present a binary version of these polynomial generator matrices.

Before proceeding we make two observations. First, as with linear block codes, the choice of a generator matrix for a convolutional code determines the encoding. We will primarily consider polynomial generator matrices for our codes. Even among such matrices, some choices are better than others for a variety of reasons. Second, until the encoding procedure and the interpretation of "D" are explained, the connection between message and codeword will not be understood. We explore this after an example.

Example 14.1.1 Let $\mathcal{C}_1$ be the $(2, 1)$ convolutional code with generator matrix

$$G_1 = [1 + D + D^2 \quad 1 + D^2].$$

A second generator matrix for $\mathcal{C}_1$ is

$$G_1' = [1 + D^3 \quad 1 + D + D^2 + D^3],$$

which is obtained from G_1 by multiplying by $1 + D$, recalling that we are working in binary. Let $\mathcal{C}_2$ be the $(4, 2)$ convolutional code with generator matrix

$$G_2 = \begin{bmatrix} 1 & 1 + D + D^2 & 1 + D^2 & 1 + D \\ 0 & 1 + D & D & 1 \end{bmatrix}.$$

By Exercise 798, $\mathcal{C}_2$ is also generated by

$$G_2' = \begin{bmatrix} 1 & D & 1 + D & 0 \\ 0 & 1 + D & D & 1 \end{bmatrix}, \quad G_2'' = \begin{bmatrix} 1 & 1 & 1 & 1 \\ 0 & 1 + D & D & 1 \end{bmatrix}, \quad \text{and}$$

$$G_2''' = \begin{bmatrix} 1 + D & 0 & 1 & D \\ D & 1 + D + D^2 & D^2 & 1 \end{bmatrix}.$$

       ■

Exercise 798 By applying elementary row operations to G_2 given in Example 14.1.1, prove that $\mathcal{C}_2$ is also generated by G_2', G_2'', and G_2'''.   ◆

We now describe the encoding procedure for an (n, k) convolutional code $\mathcal{C}$ using a polynomial generator matrix G. To emphasize the intimate connection between G and the encoding process, we often refer to G as an *encoder*. To simplify matters, we first begin with the case $k = 1$. Because $k = 1$, at each time $i = 0, 1, \ldots$, a single bit (element of $\mathbb{F}_2$) $\mathbf{x}(i)$ is input to the generator matrix G. The input is thus a stream of bits "entering" the generator matrix one bit per time interval. Suppose that this input stream has L bits. We can express this input stream as the polynomial $\mathbf{x} = \sum_{i=0}^{L-1} \mathbf{x}(i)D^i$. The encoding is given by $\mathbf{x}G = \mathbf{c}$, analogous to the usual encoding of a linear block code. The codeword $\mathbf{c} = (\mathbf{c}_1, \mathbf{c}_2, \ldots, \mathbf{c}_n)$ has n components $\mathbf{c}_j$ for $1 \le j \le n$, each of which is a polynomial in D. At time 0, the generator matrix has produced the n bits $c_1(0), \ldots, c_n(0)$; at time 1, it has produced the n bits $c_1(1), \ldots, c_n(1)$, etc. Often the components $c_1, \ldots, c_n$ are interleaved so that the output stream looks like

$$c_1(0), \ldots, c_n(0), c_1(1), \ldots, c_n(1), c_1(2), \ldots, c_n(2), \ldots . \qquad (14.1)$$

Note that the left-most bits of the input stream $x(0), x(1), x(2), \ldots$ are actually input first and the left-most bits of (14.1) are output first.

Example 14.1.2 Suppose that in Example 14.1.1 G_1 is used to encode the bit stream 110101 using the $(2, 1)$ code $\mathcal{C}_1$. This bit stream corresponds to the polynomial $x = 1 + D + D^3 + D^5$; thus 1s enter the encoder at times 0, 1, 3, and 5, and 0s enter at times 2 and 4. This is encoded to $xG_1 = (c_1, c_2)$, where $c_1 = (1 + D + D^3 + D^5)(1 + D + D^2) = 1 + D^4 + D^6 + D^7$ and $c_2 = (1 + D + D^3 + D^5)(1 + D^2) = 1 + D + D^2 + D^7$. Thus c_1 is 10001011 and c_2 is 11100001. These are interleaved so that the output is 1101010010001011. If we look more closely at the multiplications and additions done to produce c_1 and c_2, we can see how memory plays a role. Consider the output of codeword c_1 at time $i = 5$, that is, $c_1(5)$. This value is the coefficient of D^5 in the product $x(1 + D + D^2)$, which is $x(5) + x(4) + x(3)$. So at time 5, the encoder must "remember" the two previous inputs. Similarly, we see that

$$c_1(i) = x(i) + x(i - 1) + x(i - 2), \qquad (14.2)$$
$$c_2(i) = x(i) + x(i - 2). \qquad (14.3)$$

For this reason, we say that the memory of the encoder is $M = 2$. We also see the role played by D as a *delay* operator. Since c_1 is $x(1 + D + D^2)$, at time i we have $c_1(i)$ is $x(i)(1 + D + D^2)$. Thus we obtain the above expression for $c_1(i)$ if we interpret $x(i)D$ as $x(i - 1)$ and $x(i)D^2$ as $x(i - 2)$. Each occurrence of D^j delays the input by j units of time. ∎

Exercise 799 Suppose that in Example 14.1.1 G_1' is used to encode messages.
(a) Give the resulting codeword (c_1, c_2) if 110101 is encoded. Also give the interleaved output.
(b) Give equations for $c_1(i)$ and $c_2(i)$ for the encoder G_1' corresponding to equations (14.2) and (14.3).
(c) What is the memory M for this encoder? ◆

We now examine the encoding process for arbitrary k. This time the input is k bit streams x_j for $j = 1, 2, \ldots, k$, which forms our message $x = (x_1, x_2, \ldots, x_k)$. Note that the k bits $x_j(0)$ enter the encoder at time 0 followed by the k bits $x_j(1)$ at time 1 and then $x_j(2)$ at time 2, etc. We can write each x_j as a polynomial in D as before. The codeword produced is again $xG = c = (c_1, c_2, \ldots, c_n)$ with n components each of which is a polynomial in D. These components are then interleaved as earlier. Notice that the resulting codeword is a polynomial combination of the rows of G, which we have given an interpretation as bit stream outputs. Recalling that the code $\mathcal{C}$ is the $\mathbb{F}_2(D)$-row span of the rows of a generator matrix, the codewords with polynomial components have a meaning as bit streams where the coefficient of D^i gives the output at time i. Codewords with some components equal to rational functions also have an interpretation, which we explore in Section 14.4.

Example 14.1.3 Suppose that in Example 14.1.1 G_2 is used to encode the message $(11010, 10111)$ using the $(4, 2)$ code C_2. This message corresponds to the polynomial pair $\mathbf{x} = (1 + D + D^3, 1 + D^2 + D^3 + D^4)$. Computing $\mathbf{x}G_2$ we obtain the codeword $\mathbf{c} = (1 + D + D^3, D + D^2 + D^4, 1 + D^2 + D^3 + D^4, 0)$, which corresponds to $(1101, 01101, 10111, 0)$. Because the resulting components do not have the same length, the question becomes how do you do the interleaving? It is simply done by padding the right with 0s. Note that components of the message each have length 5 and so correspond to polynomials of degree at most 4. As the maximum degree of any entry in G_2 is 2, the resulting codeword could have components of degree at most 6 and hence length 7. Thus it would be appropriate to pad each component with 0s up to length 7. So we would interleave $(1101000, 0110100, 1011100, 0000000)$ to obtain 10101100011010100110000000000. For this encoder the equations corresponding to (14.2) and (14.3) become

$$\mathbf{c}_1(i) = \mathbf{x}_1(i),$$
$$\mathbf{c}_2(i) = \mathbf{x}_1(i) + \mathbf{x}_1(i-1) + \mathbf{x}_1(i-2) + \mathbf{x}_2(i) + \mathbf{x}_2(i-1),$$
$$\mathbf{c}_3(i) = \mathbf{x}_1(i) + \mathbf{x}_1(i-2) + \mathbf{x}_2(i-1),$$
$$\mathbf{c}_4(i) = \mathbf{x}_1(i) + \mathbf{x}_1(i-1) + \mathbf{x}_2(i).$$

This encoder has memory $M = 2$. ∎

Exercise 800 Suppose that in Example 14.1.1 G_2' is used to encode messages.
(a) Give the resulting codeword $(\mathbf{c}_1, \mathbf{c}_2, \mathbf{c}_3, \mathbf{c}_4)$ if $(11010, 10111)$ is encoded. Also give the interleaved output.
(b) Give equations for $\mathbf{c}_1(i), \ldots, \mathbf{c}_4(i)$ for the encoder G_2' corresponding to the equations in Example 14.1.3.
(c) What is the memory M for this encoder? ◆

Exercise 801 Repeat Exercise 800 using the encoder G_2'' of Example 14.1.1. ◆

Exercise 802 Repeat Exercise 800 using the encoder G_2''' of Example 14.1.1. ◆

When using a convolutional code, one may be interested in constructing a physical encoder using shift-registers. Recall from Section 4.2 that we gave encoding schemes for cyclic codes and indicated how one of these schemes could be physically implemented using a linear feedback shift-register. For encoding convolutional codes we will use linear feedforward shift-registers. The main components of such a shift-register are again delay elements (also called flip-flops) and binary adders shown in Figure 4.1. As with encoders for cyclic codes, the encoder for a convolutional code is run by an external clock which generates a timing signal, or clock cycle, every t_0 seconds.

Example 14.1.4 Figure 14.1 gives the design of a physical encoder for the code C_1 of Example 14.1.1 using the encoder G_1. Table 14.1 shows the encoding of 110101 using this circuit. The column "Register before cycle" shows the contents of the shift-register before the start of clock cycle i. Notice that the shift-register initially contains $\mathbf{x}(0) = 1, \mathbf{x}(-1) = 0$,

Table 14.1 *Encoding* 110101 *using*
Figure 14.1

Register before cycle	Clock cycle i	$c_1(i)$	$c_2(i)$
100	0	1	1
110	1	0	1
011	2	0	1
101	3	0	0
010	4	1	0
101	5	0	0
010	6	1	0
001	7	1	1
?00	8		

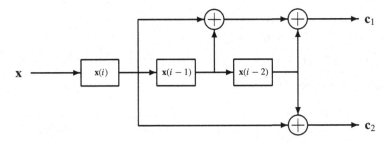

Figure 14.1 Physical encoder for $G_1 = [1 + D + D^2 \quad 1 + D^2]$.

and $x(-2) = 0$; this is the contents before clock cycle 0. At the start of clock cycle i, $x(i + 1)$ enters the shift-register from the left pushing $x(i)$ into the middle flip-flop and $x(i - 1)$ into the right-most flip-flop; simultaneously $c_1(i)$ and $c_2(i)$ are computed by the binary adders. Notice that the last digit $x(5) = 1$ enters before the start of clock cycle 5; at that point the shift-register contains 101. Because the encoder has memory $M = 2$, the clock cycles two more times to complete the encoding. Thus two more inputs must be given. These inputs are obviously 0 since $1 + D + D^3 + D^5$ is the same as $1 + D + D^3 + D^5 + 0D^6 + 0D^7$. Notice that the output c_1 and c_2 agrees with the computation in Example 14.1.2. Clock cycle 7 is completed with a new bit entering the shift-register on the left. This could be the first bit of the next message to be encoded; the two right-most registers both contain 0, preparing the circuit for transmission of the next message. This first bit of the next message is indicated with a "?" in Table 14.1. Note that the first column of "Register before cycle" is the input message. ∎

Exercise 803 Draw a physical encoder analogous to Figure 14.1 for the encoder G_1' from Example 14.1.1. Also construct a table analogous to Table 14.1 for the encoding of 110101 using the circuit you have drawn. Compare your answer with that obtained in Exercise 799. ♦

Exercise 804 Draw a physical encoder with two shift-registers analogous to Figure 14.1 for the encoder G_2 from Example 14.1.1. Then construct a table analogous to Table 14.1 for the encoding of $(11010, 10111)$ using the circuit you have drawn. Compare your answer with that obtained in Example 14.1.3. ◆

Exercise 805 Repeat Exercise 804 using the encoder G_2' from Example 14.1.1. Compare your answer with that obtained in Exercise 800. ◆

Exercise 806 Repeat Exercise 804 using the generator matrix G_2'' from Example 14.1.1. Compare your answer with that obtained in Exercise 801. ◆

Exercise 807 Repeat Exercise 804 using the generator matrix G_2''' from Example 14.1.1. Compare your answer with that obtained in Exercise 802. ◆

14.2   Viterbi decoding

In this section we present the Viterbi Decoding Algorithm for decoding convolutional codes. This algorithm was introduced by A. J. Viterbi [339] in 1967. The Viterbi Algorithm has also been applied in more general settings; see [234, 257, 336]. To understand this algorithm most easily, we first describe state diagrams and trellises.

14.2.1   State diagrams

To each polynomial generator matrix of a convolutional code we can associate a state diagram that allows us to do encoding. The state diagram is intimately related to the shift-register diagram and provides a visual way to find the output at any clock time.

Let G be a polynomial generator matrix for an (n, k) convolutional code C, from which a physical encoder has been produced. As earlier, it is easiest to define the state diagram first for the case $k = 1$. The state of an encoder at time i is essentially the contents of the shift-register at time i that entered the shift-register prior to time i. For example, the state of the encoder G_1 at time i with physical encoder given in Figure 14.1 is the contents $(\mathbf{x}(i - 1), \mathbf{x}(i - 2))$ of the two right-most delay elements of the shift-register. If we know the state of the encoder at time i and the input $\mathbf{x}(i)$ at time i, then we can compute the output $(\mathbf{c}_1(i), \mathbf{c}_2(i))$ at time i from this information. More generally, if the shift-register at time i contains $\mathbf{x}(i), \mathbf{x}(i - 1), \ldots, \mathbf{x}(i - M)$, then the encoder is in *the state* $(\mathbf{x}(i - 1), \mathbf{x}(i - 2), \ldots, \mathbf{x}(i - M))$. In the previous section we saw that the encoding equation $\mathbf{c} = \mathbf{x}G$ yields n equations for the n $\mathbf{c}_j(i)$s computed at time i in terms of $\mathbf{x}(i), \mathbf{x}(i - 1), \ldots, \mathbf{x}(i - M)$, where M is the memory of the encoder. Thus from these equations we see that if we have the state at time i and the input $\mathbf{x}(i)$, we can compute the output $\mathbf{c}(i)$.

Example 14.2.1 Consider Table 14.1. Before the start of clock cycle 3, the contents of the shift-register is 101 and hence the encoder is in state 01. Since $\mathbf{x}(3) = 1$ the encoder moves to the state 10, which is its state before clock cycle 4 begins, regardless of the input $\mathbf{x}(4)$. ∎

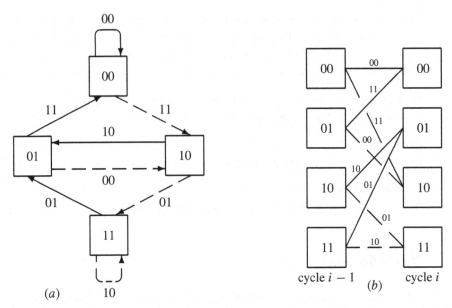

Figure 14.2 State diagrams for $G_1 = [1 + D + D^2 \ \ 1 + D^2]$.

The *set of states* of an encoder of an $(n, 1)$ convolutional code with memory M is the set of all ordered binary M-tuples. The set of states is of size 2^M and represents all possible states that an encoder can be in at any time.

The *state diagram* is a labeled directed graph determined as follows. The vertices of the graph are the set of states. There are two types of directed edges: solid and dashed. A directed edge from one vertex to another is solid if the input is 0 and dashed if the input is 1; the edge is then labeled with the output. In other words, a directed edge from the vertex $(\mathbf{x}(i-1), \ldots, \mathbf{x}(i-M))$ to the vertex $(\mathbf{x}(i), \mathbf{x}(i-1), \ldots, \mathbf{x}(i-M+1))$ is solid if $\mathbf{x}(i) = 0$ and dashed if $\mathbf{x}(i) = 1$ and is labeled $(\mathbf{c}_1(i), \ldots, \mathbf{c}_n(i))$. Notice that by looking at the vertices at the ends of the edge, we determine the contents of a shift-register at time i and can hence compute the edge label and whether the edge is solid or dashed.

Example 14.2.2 Figure 14.2 shows the state diagram for the generator matrix G_1 from Example 14.1.1. Two versions of the state diagram are given. One can encode any message by traversing through the diagram in Figure 14.2(a) always beginning at state 00 and ending at state 00. For example, if 101 is to be encoded, begin in state 00, traverse the dashed edge as the first input is 1; this edge ends in state 10 and outputs 11. Next traverse the solid edge, as the next input is 0, from state 10 to state 01 and output 10. Now traverse the dashed edge from state 01 to state 10 outputting 00. We have run out of inputs, but are not at state 00 – the shift-register is not all 0s yet. So we input 0 and hence traverse from state 10 to state 01 along the solid edge with output 10; finally, input another 0 and travel from state 01 to state 00 along the solid edge with output 11. The encoding of 101 is therefore 1110001011. Figure 14.2(b) is essentially the same diagram expanded to show the states at clock times $i - 1$ and i. ∎

Exercise 808 Draw two versions of the state diagram for the encoder G_1' from Example 14.1.1 as in Figure 14.2.   ◆

The same general idea holds for (n, k) convolutional codes with polynomial generator matrices $G = (g_{i,j}(D))$ when $k > 1$. For $1 \le i \le k$, let

$$m_i = \max_{1 \le j \le n} \deg(g_{i,j}(D)) \tag{14.4}$$

be the maximum degree of an entry in row i; m_i is called the *degree of the ith row* of G. We will adopt the convention that the zero polynomial has degree $-\infty$. Let $\mathbf{x} = (\mathbf{x}_1, \ldots, \mathbf{x}_k)$ be a message to be encoded with the jth input at time i being $\mathbf{x}_j(i)$. Let $\mathbf{x}G = \mathbf{c} = (\mathbf{c}_1, \ldots, \mathbf{c}_n)$ be the resulting codeword. Each $\mathbf{c}_j(i)$ is a combination of some of the $\mathbf{x}_J(I)$ with $I = i, i - 1, \ldots, i - m_J$ for $1 \le J \le k$. Hence if we know the inputs $\mathbf{x}_J(i)$ and all Jth inputs from times $i - 1$ back to time $i - m_J$, we can determine the output. So at time i, we say that the encoder is *in the state* $(\mathbf{x}_1(i - 1), \ldots, \mathbf{x}_1(i - m_1), \mathbf{x}_2(i - 1), \ldots, \mathbf{x}_2(i - m_2), \ldots, \mathbf{x}_k(i - 1), \ldots, \mathbf{x}_k(i - m_k))$. The *set of states* of G is the set of all ordered binary $(m_1 + m_2 + \cdots + m_k)$-tuples. The set of states is of size $2^{m_1 + m_2 + \cdots + m_k}$ and represents all possible states that an encoder can be in at any time. Note that the memory M of the encoder G is the maximum of the m_js; in particular when $k = 1$, $M = m_1$.

Example 14.2.3 For the encoder G_2 of the $(4, 2)$ convolutional code $\mathcal{C}_2$ of Example 14.1.1, $m_1 = 2$ and $m_2 = 1$. So there are eight states representing $(\mathbf{x}_1(i - 1), \mathbf{x}_1(i - 2), \mathbf{x}_2(i - 1))$ for G_2; notice that $\mathbf{c}_j(i)$ given in Example 14.1.3 is determined from the state $(\mathbf{x}_1(i - 1), \mathbf{x}_1(i - 2), \mathbf{x}_2(i - 1))$ and the input $(\mathbf{x}_1(i), \mathbf{x}_2(i))$. For the generator matrix G_2' of $\mathcal{C}_2$, $m_1 = m_2 = 1$, and the four states represent $(\mathbf{x}_1(i - 1), \mathbf{x}_2(i - 1))$. For the encoder G_2'' of $\mathcal{C}_2$, $m_1 = 0$, $m_2 = 1$, and the two states represent $\mathbf{x}_2(i - 1)$. Finally, for the encoder G_2''' of $\mathcal{C}_2$, $m_1 = 1$, $m_2 = 2$, and the eight states represent $(\mathbf{x}_1(i - 1), \mathbf{x}_2(i - 1), \mathbf{x}_2(i - 2))$.   ■

The state diagram for $k > 1$ is formed as in the case $k = 1$. The vertices of the labeled directed graph are the set of states. This time, however, there must be 2^k types of edges that represent each of the 2^k possible binary k-tuples coming from each of the possible inputs $(\mathbf{x}_1(i), \mathbf{x}_2(i), \ldots, \mathbf{x}_k(i))$. Again the edge is labeled with the output $(\mathbf{c}_1(i), \ldots, \mathbf{c}_n(i))$.

Example 14.2.4 Consider the state diagram analogous to that given in Figure 14.2(a) for the generator matrix G_2 of the $(4, 2)$ convolutional code $\mathcal{C}_2$ of Example 14.1.1. The state diagram has eight vertices and a total of 32 edges. The edges are of four types. There are four edges, one of each of the four types, leaving each vertex. For instance, suppose that all edges representing the input $(\mathbf{x}_1(i), \mathbf{x}_2(i)) = (0, 0)$ are solid red edges, those representing $(1, 0)$ are solid blue edges, those representing $(0, 1)$ are dashed blue edges, and those representing $(1, 1)$ are dashed red edges. Then, using the equations for $\mathbf{c}_j(i)$ given in Example 14.1.3, the solid red edge leaving $(\mathbf{x}_1(i - 1), \mathbf{x}_1(i - 2), \mathbf{x}_2(i - 1)) = (0, 1, 1)$ ends at vertex $(0, 0, 0)$ and is labeled $(0, 0, 0, 0)$; the solid blue edge leaving $(0, 1, 1)$ ends at vertex $(1, 0, 0)$ and is labeled $(1, 1, 1, 1)$; the dashed blue edge leaving $(0, 1, 1)$ ends at vertex $(0, 0, 1)$ and is labeled $(0, 1, 0, 1)$; and the dashed red edge leaving $(0, 1, 1)$ ends at vertex $(1, 0, 1)$ and is labeled $(1, 0, 1, 0)$.   ■

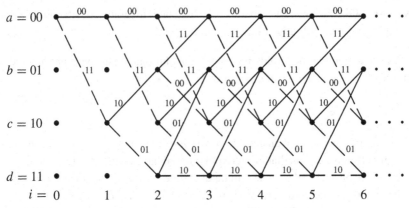

Figure 14.3 Trellis for the encoder $G_1 = [1 + D + D^2 \quad 1 + D^2]$.

Exercise 809 Verify that the edge labels for the four edges described in Example 14.2.4 are as claimed. ◆

Exercise 810 Give the state diagram for the encoder G_2' of the $(4, 2)$ convolutional code C_2 from Example 14.1.1. It may be easiest to draw the diagram in the form similar to that given in Figure 14.2(b). ◆

Exercise 811 Repeat Exercise 810 with the encoder G_2'' from Example 14.1.1. ◆

14.2.2 Trellis diagrams

The trellis diagram for an encoding of a convolutional code is merely an extension of the state diagram starting at time $i = 0$. We illustrate this in Figure 14.3 for the generator matrix G_1 of the $(2, 1)$ convolutional code C_1 from Example 14.1.1. The states are labeled $a = 00$, $b = 01$, $c = 10$, and $d = 11$. Notice that the trellis is a repetition of Figure 14.2(b) with the understanding that at time $i = 0$, the only state used is the zero state a; hence at time $i = 1$, the only states that can be reached at time $i = 1$ are a and c and thus they are the only ones drawn at that time. Encoding can be accomplished by tracing left-to-right through the trellis beginning at state a (and ending at the zero state a). For instance, if we let state $s \in \{a, b, c, d\}$ at time i be denoted by s_i, then the encoding of 1011 is accomplished by the path $a_0c_1b_2c_3d_4b_5a_6$ by following a dashed, then solid, then two dashed, and finally two solid edges (as we must input two additional 0s to reach state a); this yields the codeword 111000010111 by writing down the labels on the edges of the path.

Exercise 812 Use the trellis of Figure 14.3 to encode the following messages:
(a) 1001,
(b) 0011. ◆

Exercise 813 Draw the trellis diagram for the encoder G_1' of the $(2, 1)$ convolutional code C_1 presented in Example 14.1.1. The results of Exercise 808 will be useful. ◆

Exercise 814 Draw the trellis diagram for the encoder G_2' of the $(4, 2)$ convolutional code C_2 presented in Example 14.1.1. The results of Exercise 810 will be useful. ◆

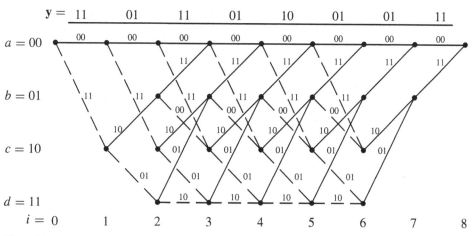

Figure 14.4 $L = 6$ truncated trellis for the encoder $G_1 = [1 + D + D^2 \; 1 + D^2]$.

14.2.3 The Viterbi Algorithm

The Viterbi Algorithm uses the trellis diagram for an encoder of a convolutional code to decode a received vector.[1] The version of the Viterbi Algorithm that we describe will accomplish nearest neighbor decoding. Suppose that a message to be encoded is input over L time periods using a generator matrix with memory M. The algorithm requires us to consider the portion of the trellis that starts at time $i = 0$ and ends at time $L + M$; this is called the L *truncated trellis*. We always move left-to-right through the trellis. Figure 14.4 shows the $L = 6$ truncated trellis for the generator matrix G_1 from Example 14.1.1 that we must examine to decode a message of length 6.

Suppose that the message $\mathbf{x}(i) = (\mathbf{x}_1(i), \dots, \mathbf{x}_k(i))$ for $i = 0, 1, \dots, L - 1$ is encoded using the generator matrix G to produce a codeword $\mathbf{c}(i) = (\mathbf{c}_1(i), \dots, \mathbf{c}_n(i))$ for $i = 0, 1, \dots, L + M - 1$. Assume $\mathbf{y}(i) = (\mathbf{y}_1(i), \dots, \mathbf{y}_n(i))$ for $i = 0, 1, \dots, L + M - 1$ is received. We define the weight of a path in the trellis. First, if e is an edge connecting state s at time $i - 1$ to state s' at time i, then the *weight of the edge e* is the Hamming distance between the edge label of e and the portion $\mathbf{y}(i - 1)$ of the received vector at time $i - 1$. The *weight of a path P* through the trellis is the sum of the weights of the edges of the path. The edge and path weights therefore depend upon the received vector. Denote the zero state a at time 0 by a_0. Suppose P is a path in the trellis starting at a_0 and ending at time i in state s; such a path P is a *survivor at time i* if its weight is smallest among all paths starting at a_0 and ending at time i in state s. The collection of all survivors starting at a_0 and ending in state s at time i will be denoted $\mathcal{S}(s, i)$. If P is a path in the trellis starting at a_0 and ending at time I, define $\mathbf{c}_P$ to be the *codeword associated with P* where $\mathbf{c}_P(i)$ is the label of the edge in P from the state at time i to the state at time $i + 1$ for $0 \le i < I$. Define $\mathbf{x}_P$ to be the *message associated with P* where $\mathbf{x}_P(i)$ is the input identified by the type of the edge in P from the state at time i to the state at time $i + 1$ for $0 \le i < \min\{I, L\}$.

---

[1] Trellises and the Viterbi Algorithm can also be used for decoding block codes. For a discussion of this, see [336].

We now describe the *Viterbi Decoding Algorithm* in four steps:

I.   Draw the L truncated trellis for G, replacing the edge labels by the edge weights. Let a denote the zero state.

II. Compute $\mathcal{S}(s, 1)$ for all states s using the trellis of Step I.

III. Repeat the following for $i = 2, 3, \ldots, L + M$ using the trellis of Step I. Assuming $\mathcal{S}(s, i - 1)$ has been computed for all states s, compute $\mathcal{S}(s, i)$ for all s as follows. For each state s' and each edge e from s' to s, form the path P made from P' followed by e where $P' \in \mathcal{S}(s', i - 1)$. Include P in $\mathcal{S}(s, i)$ if it has smallest weight among all such paths.

IV. A nearest neighbor to $\mathbf{y}$ is any $\mathbf{c}_P$ for $P \in \mathcal{S}(a, L + M)$ obtained from the message given by $\mathbf{x}_P$.

Why does this work? Clearly, a nearest neighbor $\mathbf{c}$ to $\mathbf{y}$ is a $\mathbf{c}_P$ obtained from any path P in $\mathcal{S}(a, L + M)$ since $\mathcal{S}(a, L + M)$ contains all the smallest weight paths from a_0 to a_{L+M} with $L + M$ edges, where a_{L+M} denotes the zero state at time $L + M$. The only thing we need to confirm is that Steps II and III produce $\mathcal{S}(a, L + M)$. This is certainly the case if we show these steps produce $\mathcal{S}(s, i)$. Step II by definition produces $\mathcal{S}(s, 1)$ for all s. If $P \in \mathcal{S}(s, i)$, then P is made up of a path P' ending at time $i - 1$ in state s' followed by an edge e from s' to s. If P' is not in $\mathcal{S}(s', i - 1)$, then there is a path P'' from a_0 ending in state s' at time $i - 1$ of lower weight than P'. But then the path P'' followed by e ends in state s at time i and has lower weight than P, a contradiction. So Step III correctly determines $\mathcal{S}(s, i)$.

Example 14.2.5 We illustrate the Viterbi Algorithm by decoding the received vector $\mathbf{y} = 1101110110010111$ using the trellis of Figure 14.4. Recall that for the code $\mathcal{C}_1$, two bits are received at each clock cycle; $\mathbf{y}$ is listed at the top of Figure 14.4 for convenience. The trellis for Step I is given in Figure 14.5. Again the states are $a = 00$, $b = 01$, $c = 10$, and $d = 11$; we denote state s at time i by s_i. So the zero state at time 0 is a_0.

For Step II, $\mathcal{S}(a, 1) = \{a_0 a_1\}$, $\mathcal{S}(b, 1) = \emptyset$, $\mathcal{S}(c, 1) = \{a_0 c_1\}$, and $\mathcal{S}(d, 1) = \emptyset$. From this we can compute $\mathcal{S}(s, 2)$ using the rules of Step III obtaining $\mathcal{S}(a, 2) = \{a_0 a_1 a_2\}$, $\mathcal{S}(b, 2) = \{a_0 c_1 b_2\}$, $\mathcal{S}(c, 2) = \{a_0 a_1 c_2\}$, and $\mathcal{S}(d, 2) = \{a_0 c_1 d_2\}$; notice that the paths listed in $\mathcal{S}(s, 2)$

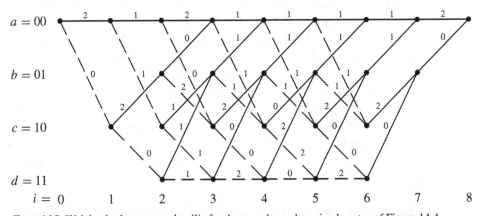

Figure 14.5 Weights in the truncated trellis for the encoder and received vector of Figure 14.4.

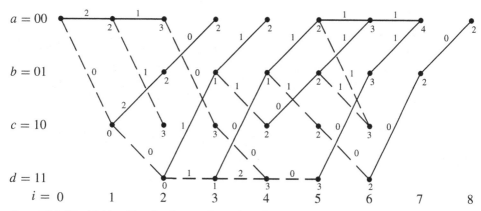

Figure 14.6 Viterbi Algorithm for the encoder and received vector of Figure 14.4.

begin with either a_0a_1 or a_0c_1, which are the paths in $S(s', 1)$. These paths are given in Figure 14.6. Consider how to construct $S(a, 3)$. Looking at the paths in each $S(s, 2)$, we see that there are only two possibilities: $a_0a_1a_2$ followed by the edge a_2a_3 giving the path $a_0a_1a_2a_3$ and $a_0c_1b_2$ followed by the edge b_2a_3 giving the path $a_0c_1b_2a_3$. However, the former has weight 5 while the latter has weight 2 and hence $S(a, 3) = \{a_0c_1b_2a_3\}$. Computing the other $S(s, 3)$ is similar. We continue inductively until $S(a, 8)$ is computed. All the paths are given in Figure 14.6; to make computations easier we have given the path weight under the end vertex of the path. Notice that some of the paths stop (e.g. $a_0a_1c_2$). That is because when one more edge is added to the path, the resulting path has higher weight than other paths ending at the same vertex.

To complete Step IV of the algorithm, we note that $S(a, 8)$ contains the lone path $P = a_0c_1d_2d_3b_4c_5d_6b_7a_8$ of weight 2. The codeword $\mathbf{c}_P$, obtained by tracing P in Figure 14.4, is 1101100100010111, which differs in two places (the weight of path P) from the received vector. All other paths through the trellis have weight 3 or more and so the corresponding codewords are distance at least 3 from $\mathbf{y}$. As solid edges come from input 0 and dashed edges come from input 1, the six inputs leading to $\mathbf{c}_P$, obtained by tracing the first six edges of P in Figure 14.4, give the message 111011. ∎

Exercise 815   Verify that the weights given in the truncated trellis of Figure 14.5 are correct.   ◆

Exercise 816   Verify that the Viterbi Algorithm described in Example 14.2.5 is correctly summarized by Figure 14.6.   ◆

Exercise 817   Use the Viterbi Algorithm to decode the following received vectors sent using the (2, 1) convolutional code C_1 with generator matrix G_1 from Example 14.1.1. Draw the trellis analogous to that of Figure 14.6. Give the path leading to the nearest codeword, the nearest codeword, the number of errors made in transmission (assuming the codeword you calculate is correct), and the associated message.
(a) 1100011111111101.
(b) 0110110110011001. (There are two answers.)   ◆

Convolutional codes came into more extensive use with the discovery of the Viterbi Algorithm in 1967. As can be seen from the examples and exercises in this section, the complexity of the algorithm for an (n, k) convolutional code depends heavily upon the memory M and upon k. For this reason, the algorithm is generally used only for M and k relatively small. There are other decoding algorithms, such as sequential decoding, for convolutional codes. However, each algorithm has its own drawbacks.

14.3   Canonical generator matrices

As we have seen, a convolutional code can have many different generator matrices, including ones whose entries are rational functions in D that are not polynomial. We have focused on polynomial generator matrices; among these polynomial generator matrices, the preferred ones are the canonical generator matrices.

We begin with some requisite terminology. Let $G = [g_{i,j}(D)]$ be a $k \times n$ polynomial matrix. Recall from (14.4) that the degree of the ith row of G is defined to be the maximum degree of the entries in row i. Define the *external degree* of G, denoted extdeg G, to be the sum of the degrees of the k rows of G.

Example 14.3.1 The generator matrices G_1 and G_1' for the $(2, 1)$ convolutional code C_1 of Example 14.1.1 have external degrees 2 and 3, respectively. The generator matrix G_2 for the code C_2 of the same example has external degree $2 + 1 = 3$. ∎

Exercise 818 Find the external degrees of the generator matrices G_2', G_2'', and G_2''' of the code C_2 from Example 14.1.1. ◆

A *canonical generator matrix* for a convolutional code C is any polynomial generator matrix whose external degree is minimal among all polynomial generator matrices for C. By definition, every convolutional code has a canonical generator matrix. This smallest external degree is called the *degree* of the code C.

Example 14.3.2 We can show that G_1 is a canonical generator matrix for the $(2, 1)$ code C_1 from Example 14.1.1 as follows. In this case the external degree of any polynomial generator matrix of C_1 is the maximum degree of its entries. Every other generator matrix G_1'' can be obtained from G_1 by multiplying G_1 by $p(D)/q(D)$, where $p(D)$ and $q(D)$ are relatively prime polynomials. Since we only need to consider polynomial generator matrices, $q(D)$ must divide both entries of G_1 and hence the greatest common divisor of both entries of G_1. As the latter is 1, $q(D) = 1$ and G_1'' is obtained from G_1 by multiplying the entries by the polynomial $p(D)$. If $p(D) \neq 1$, then the degrees of each entry in G_1'' are larger than the degrees of the corresponding entries in G_1. Thus G_1 has smallest external degree among all polynomial generator matrices for C_1 and therefore is canonical. ∎

Exercise 819 Let C be an $(n, 1)$ convolutional code. Prove that a polynomial generator matrix G of C is canonical if and only if the greatest common divisor of its entries is 1. Hint: Study Example 14.3.2. ◆

From Exercise 819, one can check relatively easily to see if a generator matrix for an $(n, 1)$ convolutional code is canonical or not. The process is not so clear for (n, k) codes with $k > 1$. It turns out (see Theorem 14.3.7) that a generator matrix is canonical if it is both basic and reduced, terms requiring more notation.

Let G be a $k \times n$ polynomial matrix with $k \le n$. A $k \times k$ *minor* of G is the determinant of a matrix consisting of k columns of G. G has $\binom{n}{k}$ minors. The *internal degree* of G, denoted intdeg G, is the maximum degree of all of the $k \times k$ minors of G. A *basic* polynomial generator matrix of a convolutional code is any polynomial generator matrix of minimal internal degree.

Exercise 820 Find the internal degree of each matrix G_1, G_1', G_2, G_2', G_2'', and G_2''' from Example 14.1.1. Notice that the internal degrees are always less than or equal to the external degrees found in Example 14.3.1 and Exercise 818.   ◆

A $k \times k$ matrix U with entries in $\mathbb{F}_2[D]$ (that is, polynomials in D) is *unimodular* if its determinant is 1. A polynomial matrix G is *reduced* if among all polynomial matrices of the form UG, where U is a $k \times k$ unimodular matrix, G has the smallest external degree. Since a unimodular matrix can be shown to be a product of elementary matrices, a matrix is reduced if and only if its external degree cannot be reduced by applying a sequence of elementary row operations to the matrix.

In [235], six equivalent formulations are given for a matrix to be basic and three are given for a matrix to be reduced. We present some of these equivalent statements here, omitting the proofs and referring the interested reader to [235, Appendix A].

Theorem 14.3.3 *Let G be a polynomial generator matrix of an (n, k) convolutional code. The matrix G is basic if and only if any of the following hold.*
(i)   *The greatest common divisor of the $k \times k$ minors of G is 1.*
(ii) *There exists an $n \times k$ matrix K with polynomial entries so that $GK = I_k$.*
(iii) *If $\mathbf{c} \in \mathbb{F}_2[D]^n$ and $\mathbf{c} = \mathbf{x}G$, then $\mathbf{x} \in \mathbb{F}_2[D]^k$.*

Recall that the matrix K in (ii) is a right inverse of G. Part (iii) of Theorem 14.3.3 states that if G is a basic encoder, then whenever the output is polynomial, so is the input. In Section 14.5, we will examine the situation where (iii) fails, namely where there is a polynomial output that comes from a non-polynomial input. Note that the proof that (ii) implies (iii) is simple. If (ii) holds and $\mathbf{c} = \mathbf{x}G$, then $\mathbf{x} = \mathbf{x}I_k = \mathbf{x}GK = \mathbf{c}K$; as $\mathbf{c}$ and K have only polynomial entries, so does $\mathbf{x}$.

The *degree* of a vector $\mathbf{v}(D) \in \mathbb{F}[D]^n$ is the largest degree of any of its components. Note that this definition is consistent with the definition of "degree" for a matrix given in (14.4).

Theorem 14.3.4 *Let G be a $k \times n$ polynomial matrix. Let $\mathbf{g}_i(D)$ be the ith row of G. The matrix G is reduced if and only if either of the following hold.*
(i)   intdeg G = extdeg G.
(ii) *For any $\mathbf{x}(D) = (x_1(D), \dots, x_k(D)) \in \mathbb{F}_2[D]^k$,*

$$\deg(\mathbf{x}(D)G) = \max_{1 \le i \le k} \{\deg x_i(D) + \deg \mathbf{g}_i(D)\}.$$

Part (ii) is called the *predictable degree property* of reduced matrices.

Example 14.3.5 Notice that Exercise 819 shows that a polynomial generator matrix of an $(n, 1)$ convolutional code is canonical if and only if it is basic. If G is any $1 \times n$ polynomial matrix, then the 1×1 minors of G are precisely the polynomial entries of the matrix. So G is reduced by Theorem 14.3.4. This shows that for an $(n, 1)$ code, a polynomial generator matrix is canonical if and only if it is basic and reduced. ■

We wish to show in general that a polynomial generator matrix for any (n, k) convolutional code is canonical if and only if it is basic and reduced. Exercise 820 hints at part of the following, which will be necessary for our result.

Lemma 14.3.6 *Let G be a $k \times n$ polynomial matrix over $\mathbb{F}_2(D)$ with $k \leq n$. Let N be a nonsingular $k \times k$ polynomial matrix with entries in $\mathbb{F}_2[D]$. The following hold:*
(i)   intdeg NG = intdeg G + deg det N.
(ii)   intdeg $G \leq$ intdeg NG. *Equality holds in this expression if and only if N is unimodular.*
(iii)   intdeg $G \leq$ extdeg G.

Proof: To prove (i), we observe that the $k \times k$ submatrices of NG are precisely the $k \times k$ submatrices of G each multiplied on the left by N. Thus the $k \times k$ minors of NG are exactly the $k \times k$ minors of G each multiplied by det N. This proves (i). Part (ii) follows from (i).
For (iii), suppose the degree of the ith row of G is m_i. Then extdeg $G = m_1 + m_2 + \cdots + m_k$. Any $k \times k$ minor of G is a sum of products of entries of G with one factor in each product from each row. Hence the highest degree product is at most $m_1 + m_2 + \cdots + m_k$, implying that this is the maximum degree of the minor. Thus intdeg $G \leq$ extdeg G. □

Theorem 14.3.7 *A polynomial generator matrix of an (n, k) convolutional code C is canonical if and only if it is basic and reduced.*

Proof: Let G be a canonical generator matrix for C. Since the basic generator matrices have a common internal degree, let this degree be d_0. Choose from among the basic generator matrices a matrix G_0 whose external degree is as small as possible. If U is any unimodular matrix, then intdeg UG_0 = intdeg $G_0 = d_0$ by Lemma 14.3.6(ii). By definition of G_0, extdeg $UG_0 \geq$ extdeg G_0 (as UG_0 generates C), implying that G_0 is reduced. Since G_0 has the smallest possible internal degree of any polynomial generator matrix of C, intdeg $G_0 \leq$ intdeg G. But intdeg $G \leq$ extdeg G by Lemma 14.3.6(iii) and extdeg $G \leq$ extdeg G_0 by definition as G is canonical. Thus

$$\text{intdeg } G_0 \leq \text{intdeg } G \leq \text{extdeg } G \leq \text{extdeg } G_0. \tag{14.5}$$

As G_0 is reduced, intdeg G_0 = extdeg G_0 by Theorem 14.3.4(i). Therefore we have equality everywhere in (14.5). Thus intdeg G = intdeg $G_0 = d_0$ showing that G has minimal internal degree among all polynomial generator matrices of C, implying that G is basic also. As intdeg G = extdeg G, G is reduced by Theorem 14.3.4(i).
Now suppose that G is both a basic and a reduced polynomial generator matrix of C. Let G_0 be any other polynomial generator matrix of C. By Lemma 14.3.6, extdeg $G_0 \geq$ intdeg G_0. Since G is basic, intdeg $G_0 \geq$ intdeg G. As G is reduced, intdeg G = extdeg G by Theorem 14.3.4(i). By combining these inequalities, extdeg $G_0 \geq$ extdeg G implying that G is canonical. □

Exercise 821 Which of the polynomial generator matrices G_2, G'_2, G''_2, and G'''_2 of the $(4, 2)$ convolutional code C_2 from Example 14.1.1 are basic? Which are reduced? Which are canonical? ◆

The proof of Theorem 14.3.7 shows that the minimal internal degree of any polynomial generator matrix of an (n, k) convolutional code C is also the minimal external degree of any polynomial generator matrix of C, that is, the sum of the row degrees of a canonical generator matrix. Amazingly, the set of row degrees is unique for a canonical generator matrix. To prove this, we need the following lemma.

Lemma 14.3.8 *Let G be a canonical generator matrix for an (n, k) convolutional code C. Reorder the rows of G so that the row degrees satisfy $m_1 \leq m_2 \leq \cdots \leq m_k$, where m_i is the degree of the ith row. Let G' be any polynomial generator matrix for C. Analogously reorder its rows so that its row degrees m'_i satisfy $m'_1 \leq m'_2 \leq \cdots \leq m'_k$. Then $m_i \leq m'_i$ for $1 \leq i \leq k$.*

Proof: Let $\mathbf{g}_i(D)$ be the ith row of G with $\deg \mathbf{g}_i(D) = m_i$. Analogously let $\mathbf{g}'_i(D)$ be the ith row of G' with $\deg \mathbf{g}'_i(D) = m'_i$. Suppose the result is false. Then there exists an integer j such that $m_1 \leq m'_1, \ldots, m_j \leq m'_j$ but $m_{j+1} > m'_{j+1}$. Let $1 \leq i \leq j+1$. Then $\deg \mathbf{g}'_i(D) < m_{j+1}$. Since $\mathbf{g}'_i(D)$ is a polynomial codeword, that is, a "polynomial output," and G is basic, there exists a "polynomial input" $\mathbf{x}_i(D) = (x_{i,1}(D), \ldots, x_{i,k}(D))$ such that $\mathbf{g}'_i(D) = \mathbf{x}_i(D)G$ by Theorem 14.3.3(iii). As G is also reduced, the "predictable degree property" of Theorem 14.3.4(ii) implies that:

$$m'_i = \deg \mathbf{g}'_i(D) = \deg (\mathbf{x}_i(D)G) = \max_{1 \leq \ell \leq k} \{\deg x_{i,\ell}(D) + \deg \mathbf{g}_\ell(D)\}$$
$$= \max_{1 \leq \ell \leq k} \{\deg x_{i,\ell}(D) + m_\ell\}.$$

As $\deg \mathbf{g}'_i(D) < m_{j+1}$, this implies that $\deg x_{i,\ell}(D) = -\infty$ for $\ell \geq j+1$. In other words $x_{i,\ell}(D) = 0$ for $\ell \geq j+1$. Since $\mathbf{g}'_i(D) = \mathbf{x}_i(D)G$, $\mathbf{g}'_i(D)$ must be a polynomial combination of the first j rows of G for $1 \leq i \leq j+1$. Hence the first $j+1$ rows of G' are dependent over $\mathbb{F}_2(D)$, a contradiction. □

This lemma immediately yields the following.

Theorem 14.3.9 *Let C be an (n, k) convolutional code. The set of row degrees is the same for all canonical generator matrices for C.*

This unique set of row degrees of a canonical generator matrix for C is called the set of *Forney indices* of C. The sum of the Forney indices is the external degree of a canonical generator matrix, which is the degree of C. (The degree of C is sometimes called the *constraint length* or the *overall constraint length*.) If m is the degree of the code, the code is often termed an (n, k, m) code. The Viterbi Algorithm requires 2^m states when a canonical encoder is used. If we construct a physical encoder for C from a canonical generator matrix, the memory of the encoder is the largest Forney index. Since this index is unique, it is called the *memory M* of the code. Note that by Lemma 14.3.8, the memory of a physical encoder derived from any polynomial generator matrix is at least the memory of the code. For example, the memory of the physical encoders G_2, G'_2, G''_2, and G'''_2 for C_2 of Example 14.1.1 are 2, 1, 1, and 2, respectively; C_2 has memory 1.

Example 14.3.10 The $(4, 2)$ code C_2 of Example 14.1.1 has canonical generator matrix G_2'' by Exercise 821. Thus the Forney indices for C_2 are 0 and 1. The degree and memory of C_2 are therefore each equal to 1, and C_2 is a $(4, 2, 1)$ code. ∎

Exercise 822 Let C_3 and C_4 be $(4, 3)$ convolutional codes with generator matrices

$$G_3 = \begin{bmatrix} 1 & 1 & 1 & 1 \\ 1+D & 1 & 0 & 0 \\ 0 & 1+D & 1 & 0 \end{bmatrix} \quad \text{and} \quad G_4 = \begin{bmatrix} 1 & 0 & 0 & 1 \\ 0 & 1 & 0 & 1 \\ 0 & 0 & 1 & 1+D^2 \end{bmatrix},$$

respectively.

(a) Verify that G_3 and G_4 are canonical generator matrices.
(b) Give the internal degree, external degree, degree, Forney indices, and memories of C_3 and C_4. ◆

Since every convolutional code has a canonical generator matrix, it is natural to ask how to construct such a generator matrix. A method for doing so, starting with any polynomial generator matrix, can be found in [235, Appendix B]. One reason for the interest in canonical generator matrices is that they lead directly to the construction of physical encoders for the code that have a minimum number of delay elements, as hinted at prior to Example 14.3.10. This minimum number of delay elements equals the degree of the code; see [235, Section 5].

14.4 Free distance

In this section we examine a parameter, called the free distance, of a convolutional code that gives an indication of the code's ability to correct errors. The free distance plays the role in a convolutional code that minimum distance plays in a block code, but its interpretation is not quite as straightforward as in the block code case. The free distance is a measure of the ability of the decoder to decode a received vector accurately: in general, the higher the free distance, the lower the probability that the decoder will make an error in decoding a received vector. The free distance is usually difficult to compute and so bounds on the free distance become important.

To define free distance, we first define the weight of an element of $\mathbb{F}_2(D)$. Every element $f(D) \in \mathbb{F}_2(D)$ is a quotient $p(D)/q(D)$ where $p(D)$ and $q(D)$ are in $\mathbb{F}_2[D]$. In addition, $f(D) = p(D)/q(D)$ has a one-sided Laurent series expansion $f(D) = \sum_{i \geq a} f_i D^i$, where a is an integer, possibly negative, and $f_i \in \mathbb{F}_2$ for $i \geq a$. The Laurent series is actually a power series $(a = 0)$ if $q(0) \neq 0$. We define the *weight* of $f(D)$ to be the number of nonzero coefficients in the Laurent series expansion of $f(D)$. We denote this value, which may be infinite, by $\text{wt}(f(D))$.

Example 14.4.1 To illustrate this weight, notice that $\text{wt}(1 + D + D^9) = 3$, whereas $\text{wt}((1 + D)/(1 + D^3)) = \text{wt}(1 + D + D^3 + D^4 + D^6 + D^7 + \cdots) = \infty$. ∎

When does an element of $\mathbb{F}_2(D)$ have finite weight? If $f(D) = \sum_{i \geq a} f_i D^i$ has finite weight, then multiplying by a high enough power of D produces a polynomial with the same weight. Hence $f(D)$ must have been a polynomial divided by some nonnegative integer power of D.

Lemma 14.4.2 *An element $f(D) \in \mathbb{F}_2(D)$ has finite weight if and only if either $f(D)$ is a polynomial or a polynomial divided by a positive integer power of D.*

We extend the definition of weight of an element of $\mathbb{F}_2(D)$ to an n-tuple $\mathbf{u}(D)$ of $\mathbb{F}_2(D)^n$ by saying that the weight of $\mathbf{u}(D) = (u_1(D), \ldots, u_n(D)) \in \mathbb{F}_2(D)^n$ is the sum $\sum_{i=1}^{n} \text{wt}(u_i(D))$ of the weights of each component of $\mathbf{u}(D)$. The *distance* between two elements $\mathbf{u}(D)$ and $\mathbf{v}(D)$ of $\mathbb{F}_2(D)^n$ is defined to be $\text{wt}(\mathbf{u}(D) - \mathbf{v}(D))$. The *free distance* d_{free} of an (n, k) convolutional code is defined to be the minimum distance between distinct codewords of $\mathcal{C}$. As $\mathcal{C}$ is linear over $\mathbb{F}_2(D)$, the free distance of $\mathcal{C}$ is the minimum weight of any nonzero codeword of $\mathcal{C}$. If the degree m and free distance d are known, then $\mathcal{C}$ is denoted an (n, k, m, d) convolutional code.

Example 14.4.3 We can show that the code $\mathcal{C}_1$ of Example 14.1.1 has free distance 5, as follows. Using the generator matrix G_1 for $\mathcal{C}_1$, we see that every codeword of $\mathcal{C}_1$ has the form $\mathbf{c}(D) = (f(D)(1 + D + D^2), f(D)(1 + D^2))$, where $f(D) = p(D)/q(D)$ is in $\mathbb{F}_2(D)$. Suppose $\mathbf{c}(D)$ has weight between 1 and 4, inclusive. As each component must have finite weight, by Lemma 14.4.2, we may multiply by some power of D, which does not change weight, and thereby assume that $\mathbf{c}(D)$ has polynomial components. Reducing $f(D)$ so that $p(D)$ and $q(D)$ are relatively prime, $q(D)$ must be a divisor of both $1 + D + D^2$ and $1 + D^2$. Thus $q(D) = 1$. Hence if the free distance of $\mathcal{C}_1$ is less than 5, there exists a codeword $\mathbf{c} = (f(D)(1 + D + D^2), f(D)(1 + D^2))$ of weight between 1 and 4 with $f(D)$ a polynomial. If $f(D)$ has only one term, $\text{wt}(\mathbf{c}(D)) = 5$. Suppose that $f(D)$ has at least two terms. Notice that both $\text{wt}(f(D)(1 + D + D^2)) \geq 2$ and $\text{wt}(f(D)(1 + D^2)) \geq 2$. Since $1 \leq \text{wt}(\mathbf{c}(D)) \leq 4$, $\text{wt}(f(D)(1 + D + D^2)) = \text{wt}(f(D)(1 + D^2)) = 2$. But $\text{wt}(f(D) \times (1 + D^2)) = 2$ only holds if $f(D) = D^a \sum_{i=0}^{b} D^{2i}$ for some nonnegative integer a and some positive integer b. In this case $\text{wt}(f(D)(1 + D + D^2)) > 2$, which is a contradiction. So $d_{free} > 4$. As $\text{wt}(1 + D + D^2, 1 + D^2) = 5$, $d_{free} = 5$. Thus $\mathcal{C}_1$ is a $(2, 1, 2, 5)$ code. ∎

Exercise 823 Let $\mathcal{C}_5$ be the $(2, 1)$ convolutional code with generator matrix

$$G_5 = [1 \quad 1 + D].$$

Show that the free distance of $\mathcal{C}_5$ is 3. ◆

Exercise 824 Show that the free distance of the code $\mathcal{C}_2$ from Example 14.1.1 is 4. Hint: Use the generator matrix G_2'. ◆

Example 14.4.3 and Exercises 823 and 824 illustrate the difficulty in finding the free distance of an (n, k) convolutional code even for very small n and k. There is, however, an upper bound on d_{free}, which we give in Theorem 14.4.10. For this it is helpful to give another way to generate a certain part of the code.

Let G be a fixed $k \times n$ generator matrix of an (n, k) convolutional code $\mathcal{C}$ where the maximum row degree of G is M, the memory of the encoder G. There are $M + 1$ $k \times n$ binary matrices $B_0, B_1, \ldots, B_M$ such that $G = B_0 + B_1 D + B_2 D^2 + \cdots + B_M D^M$. Form the binary matrix $B(G)$, with an infinite number of rows and columns, by

$$B(G) = \begin{bmatrix} B_0 & B_1 & B_2 & \cdots & B_M & & \\ & B_0 & B_1 & B_2 & \cdots & B_M & \\ & & B_0 & B_1 & B_2 & \cdots & B_M \\ & & & \ddots & & \ddots & & \ddots \end{bmatrix},$$

where blank entries are all 0. Similarly, for $\mathbf{x}$ a k-tuple of polynomials with maximum degree N, there are $(N+1)$ $k \times 1$ binary matrices $X_0, X_1, \ldots, X_N$ such that $\mathbf{x} = X_0 + X_1 D + \cdots + X_N D^N$. Let $b(\mathbf{x})$ be the binary vector of length $k(N+1)$ formed by interleaving $(X_0, X_1, \ldots, X_N)$. The codeword $\mathbf{c} = \mathbf{x}G$ when interleaved is precisely $b(\mathbf{x})B(G)$. Also the weight of $\mathbf{c}$ is the Hamming weight of $b(\mathbf{x})B(G)$.

Example 14.4.4 Using the encoder G_1 from Example 14.1.1 for the $(2, 1)$ code $\mathcal{C}_1$, we have $B_0 = [1\ 1]$, $B_1 = [1\ 0]$, and $B_2 = [1\ 1]$. So,

$$B(G_1) = \begin{bmatrix} 1 & 1 & 1 & 0 & 1 & 1 & & & & & & \\ & & 1 & 1 & 1 & 0 & 1 & 1 & & & & \\ & & & & 1 & 1 & 1 & 0 & 1 & 1 & & \\ & & & & & & 1 & 1 & 1 & 0 & 1 & 1 \\ & & & & & & & & 1 & 1 & 1 & 0 & 1 & 1 \\ & & & & & & & & & & 1 & 1 & 1 & 0 & 1 & 1 \\ & & & & & & & & & & & & & \ddots \end{bmatrix}.$$

Let $\mathbf{x} = 1 + D + D^3 + D^5$. Writing this in binary, interleaving being unnecessary as $k = 1$, gives $b(\mathbf{x}) = 110101$. Computing $b(\mathbf{x})B(G_1)$ yields the weight 8 vector 1101010010001011, which corresponds to $\mathbf{x}G_1 = (1 + D^4 + D^6 + D^7, 1 + D + D^2 + D^7)$ interleaved, having weight 8. This agrees with the results of Example 14.1.2. ∎

Exercise 825 Using the encoder G_2 from Example 14.1.1 for the $(4, 2)$ code $\mathcal{C}_2$, do the following:
(a) Find B_0, B_1, and B_2.
(b) Give the first ten rows of $B(G_2)$.
(c) Suppose $\mathbf{x} = (1 + D + D^3, 1 + D^2 + D^3 + D^4)$. What is $b(\mathbf{x})$?
(d) Compute $b(\mathbf{x})B(G_2)$ and check your result with Example 14.1.3. ◆

Exercise 826 Using the encoder G_3 from Exercise 822 for the $(4, 3)$ code $\mathcal{C}_3$, do the following:
(a) Find B_0 and B_1.
(b) Give the first nine rows of $B(G_3)$.
(c) Suppose $\mathbf{x} = (1 + D + D^2, 1 + D^2, D)$. What is $b(\mathbf{x})$?
(d) Compute $b(\mathbf{x})B(G_3)$. From this find $\mathbf{x}G_3$ written as a 4-tuple of polynomials. ◆

Exercise 827 Using the encoder G_4 from Exercise 822 for the $(4, 3)$ code $\mathcal{C}_4$, do the following:
(a) Find B_0, B_1, and B_2.
(b) Give the first nine rows of $B(G_4)$.

(c) Suppose $\mathbf{x} = (1 + D^2, D + D^2, 1 + D + D^2)$. What is $b(\mathbf{x})$?

(d) Compute $b(\mathbf{x})B(G_4)$. From this find $\mathbf{x}G_4$ written as a 4-tuple of polynomials.   ◆

Notice that the first block of $B(G)$, that is $[B_0\ B_1\ \cdots\ B_M\ \cdots]$ consisting of the first k rows of $B(G)$, corresponds to the k rows of G interleaved. The next block of k rows corresponds to the k rows of G multiplied by D and interleaved. In general, the ith block of k rows corresponds to the k rows of G multiplied by D^{i-1} and interleaved. Thus the rows of $B(G)$ span a binary code (of infinite dimension and length) whose vectors correspond to the codewords of C arising from polynomial inputs. If G is basic, then all polynomial outputs arise from polynomial inputs by Theorem 14.3.3. This proves the following result.

Theorem 14.4.5 *Let G be a basic generator matrix for an (n, k) convolutional code. Then every codeword with polynomial components (interleaved) corresponds to a finite linear combination of the rows of $B(G)$.*

The free distance of C is the weight of some finite weight codeword. Such codewords have components of the form $p(D)/D^i$ for some polynomial $p(D)$ and nonnegative integer i by Lemma 14.4.2. Multiplying such a codeword by a high enough power of D, which does not change its weight, gives a codeword with polynomial components having weight equal to the free distance. By Theorem 14.4.5, if G is a basic generator matrix of C, the free distance of C is the weight of some finite linear combination of the rows of $B(G)$. This observation leads to the following notation. Let L be a nonnegative integer; let $C^{(L)}$ be the set of polynomial codewords of the (n, k) convolutional code C with components of degree L or less. This set $C^{(L)}$ is in fact a binary linear code of length $n(L + 1)$; shortly we will find its dimension. We clearly have the following result as some codeword of minimum weight has polynomial components and hence is in some $C^{(L)}$.

Theorem 14.4.6 *Let C be an (n, k) convolutional code with free distance d_{free}. Let $C^{(L)}$ have minimum weight d_L. Then $d_{free} = \min_{L \geq 0} d_L$. Furthermore, some $C^{(L)}$ contain a codeword whose weight is d_{free}.*

The codewords of $C^{(L)}$, interleaved, correspond to the linear combinations of the rows of $B(G)$ whose support is entirely in the first $n(L + 1)$ columns of $B(G)$, where G is a basic generator matrix. By Theorem 14.4.6, if we look at the binary code generated by a big enough "upper left-hand corner" of $B(G)$ where G is basic, we will find a binary codeword whose minimum weight is the free distance of C. If G is also reduced, and hence is canonical, then we can in fact find how many rows of $B(G)$ generate the binary code $C^{(L)}$ (interleaved), as the proof of the next result, due to Forney [88], shows. Let k_L be the dimension of $C^{(L)}$ as a binary code.

Theorem 14.4.7 *Let C be an (n, k) convolutional code with Forney indices $m_1, \ldots, m_k$. Then the binary dimension k_L of $C^{(L)}$ satisfies*

$$k_L = \sum_{i=1}^{k} \max\{L + 1 - m_i, 0\}.$$

Proof: Let G be a canonical polynomial generator matrix for C with rows $\mathbf{g}_1(D), \ldots, \mathbf{g}_k(D)$. Order the rows so that $\deg \mathbf{g}_i(D) = m_i$ where $m_1 \leq m_2 \leq \cdots \leq m_k$. Let $\mathbf{c}(D)$ be any polynomial codeword with components of degree L or less. As G is basic, by Theorem 14.3.3(iii), there is a polynomial k-tuple $\mathbf{x}(D) = (x_1(D), \ldots, x_k(D))$ with $\mathbf{c}(D) = \mathbf{x}(D)G$. As G is reduced, by Theorem 14.3.4(ii) $\deg \mathbf{c}(D) = \max_{1 \leq i \leq k}(\deg x_i(D) + \deg \mathbf{g}_i(D))$. In particular, $\deg x_i(D) + \deg \mathbf{g}_i(D) = \deg x_i(D) + m_i \leq L$, implying that the set $\mathcal{B}_L = \{D^j \mathbf{g}_i(D) \mid 1 \leq i \leq k, \ j \geq 0, \ j + m_i \leq L\}$ spans the binary code $\mathcal{C}^{(L)}$. Furthermore, as $\{\mathbf{g}_i(D) \mid 1 \leq i \leq k\}$ is linearly independent over $\mathbb{F}_2(D)$, $\mathcal{B}_L$ is linearly independent over $\mathbb{F}_2$. Therefore $\mathcal{B}_L$ is a binary basis of $\mathcal{C}^{(L)}$. For each i, j ranges between 0 and $L - m_i$ as long as the latter is nonnegative. The result follows.   □

The result of this theorem shows how many rows of $B(G)$ we need to generate the binary code $\mathcal{C}^{(L)}$ (in interleaved form); the proof indicates which rows of $B(G)$ actually to take. This requires G to be canonical.

Example 14.4.8 Consider the $(2, 1)$ convolutional code $\mathcal{C}_5$ from Exercise 823 with generator matrix $G_5 = [1 \ 1 + D]$. Multiplying G_5 by $1 + D$, we obtain the generator matrix $G_5' = [1 + D \ 1 + D^2]$, which is not canonical by Exercise 819. Using G_5', we have $B_0 = [1 \ 1]$, $B_1 = [1 \ 0]$, and $B_2 = [0 \ 1]$. Thus

$$B(G_5') = \begin{bmatrix} 1 & 1 & 1 & 0 & 0 & 1 & & & \\ & & 1 & 1 & 1 & 0 & 0 & 1 & \\ & & & & 1 & 1 & 1 & 0 & 0 & 1 \\ & & & & & & 1 & 1 & 1 & 0 & 0 & 1 \\ & & & & & & & & & & \ddots \end{bmatrix}.$$

Notice that the codeword $(1/(1 + D))G_5' = (1, 1 + D)$ has weight 3 and is contained in $\mathcal{C}_5^{(1)}$. So we cannot use $B(G_5')$ to find all of $\mathcal{C}_5^{(1)}$. This is because G_5' is not canonical. Notice also that there is no polynomial codeword of weight 3 spanned by rows of $B(G_5')$, as the rows all have even weight. The free distance of $\mathcal{C}_5$ is 3, as Exercise 823 shows. So we cannot use $B(G_5')$ to discover the free distance of $\mathcal{C}_5$. However, the encoder G_5 is canonical by Exercise 819. If we form $B(G_5)$, we obtain

$$B(G_5) = \begin{bmatrix} 1 & 1 & 0 & 1 & & & \\ & 1 & 1 & 0 & 1 & & \\ & & 1 & 1 & 0 & 1 & \\ & & & 1 & 1 & 0 & 1 \\ & & & & & & \ddots \end{bmatrix}.$$

The Forney index of $\mathcal{C}_5$ is $m_1 = 1$. The code $\mathcal{C}_5^{(L)}$ has length $2(L + 1)$, binary dimension $k_L = L + 1 - m_1 = L$, and minimum distance 3 for all $L \geq 1$. The interleaved codewords of $\mathcal{C}_5^{(L)}$ have a binary basis consisting of the first L rows of $B(G_5)$.   ■

Example 14.4.9 The generator matrix G_2'' is a canonical generator matrix of $\mathcal{C}_2$ from Example 14.1.1. The Forney indices are $m_1 = 0$ and $m_2 = 1$; also

$$B_0 = \begin{bmatrix} 1 & 1 & 1 & 1 \\ 0 & 1 & 0 & 1 \end{bmatrix} \quad \text{and} \quad B_1 = \begin{bmatrix} 0 & 0 & 0 & 0 \\ 0 & 1 & 1 & 0 \end{bmatrix},$$

giving

$$
B(G_2'') = \begin{bmatrix}
1 & 1 & 1 & 1 & 0 & 0 & 0 & 0 & & & & \\
0 & 1 & 0 & 1 & 0 & 1 & 1 & 0 & & & & \\
 & & 1 & 1 & 1 & 1 & 0 & 0 & 0 & 0 & & \\
 & & 0 & 1 & 0 & 1 & 0 & 1 & 1 & 0 & & \\
 & & & & 1 & 1 & 1 & 1 & 0 & 0 & 0 & 0 \\
 & & & & 0 & 1 & 0 & 1 & 0 & 1 & 1 & 0 \\
 & & & & & & & & & & & \ddots
\end{bmatrix}.
$$

By Theorem 14.4.7, $k_L = \max\{L + 1 - m_1, 0\} + \max\{L + 1 - m_2, 0\} = \max\{L + 1, 0\} + \max\{L, 0\}$. So $k_L = 2L + 1$ for $L \geq 0$. Thus $C_2^{(L)}$, as a binary code, is a $[4(L + 1), 2L + 1]$ code; we can see that a basis of the interleaved version of $C_2^{(L)}$ is the first $2L + 1$ rows of $B(G_2'')$. As the rows of $B(G_2'')$ are of even weight, $C_2^{(L)}$, as a binary code, is an even code. It is not difficult to see that the span of the first $2L + 1$ rows of $B(G_2'')$ has no vector of weight 2. Thus, as a binary code, $C_2^{(L)}$ is a $[4(L + 1), 2L + 1, 4]$ code for all $L \geq 0$. By Theorem 14.4.6, C_2 has free distance 4; see also Exercise 824.   ∎

Exercise 828 In Example 14.4.9, a basis for $C_2^{(L)}$ was found for the code C_2 of Example 14.1.1 using the canonical generator matrix G_2''.
(a) By Example 14.4.9, $C_2^{(2)}$ has dimension 5, and the first five rows of $B(G_2'')$ form a binary basis of $C_2^{(2)}$. In Exercise 825(b), the first ten rows of $B(G_2)$ were found. Find a binary basis of $C_2^{(2)}$ using binary combinations of the first eight rows of $B(G_2)$. Note how much simpler it was to find a basis of $C_2^{(2)}$ using $B(G_2'')$.
(b) What property of G_2 allows you to find a basis for $C_2^{(2)}$ using $B(G_2)$? What property of G_2 is missing that causes the basis to be more difficult to find?
(c) Give the first ten rows of $B(G_2''')$. Give a maximum set of independent vectors found as binary combinations of rows of $B(G_2''')$ which are in $C_2^{(2)}$. What property is G_2''' missing that causes it not to contain a basis of $C_2^{(2)}$?   ◆

Exercise 829 Show that the free distance of the code C_3 from Exercise 822 is 3.   ◆

Exercise 830 Show that the free distance of the code C_4 from Exercise 822 is 2.   ◆

If we let $\Delta(n, k)$ be the largest minimum distance possible for a binary $[n, k]$ code, then an immediate consequence of Theorem 14.4.6 is the following bound.

Theorem 14.4.10 *Let C be an (n, k) convolutional code with Forney indices $m_1, \dots, m_k$. Let k_L be defined by Theorem 14.4.7. Then*

$$
d_{free} \leq \min_{L \geq 0} \Delta\left(n(L + 1), k_L\right).
$$

We can use bounds, such as those in Chapter 2, or tables, such as those found in [32], to give values of $\Delta(n(L + 1), k_L)$.

Example 14.4.11 A $(4, 3, 2)$ convolutional code has Forney indices that sum to $m = 2$. There are two possibilities: $m_1 = 0, m_2 = m_3 = 1$ or $m_1 = m_2 = 0, m_3 = 2$.

Consider first the case $m_1 = 0$, $m_2 = m_3 = 1$. By Theorem 14.4.7, $k_L = 3L + 1$ for all L. Hence Theorem 14.4.10 shows that

$$d_{free} \leq \min\{\Delta(4, 1), \Delta(8, 4), \Delta(12, 7), \ldots\}.$$

The values $\Delta(4, 1) = \Delta(8, 4) = \Delta(12, 7) = 4$ can be found in [32] or deduced from Table 2.1. Thus d_{free} is at most 4. In fact it can be shown that d_{free} is at most 3 for this code; see [235, Theorem 7.10]. C_3 from Exercise 822 is a $(4, 3, 2, 3)$ code with these Forney indices; see Exercise 829.

Next consider the case $m_1 = m_2 = 0$, $m_3 = 2$. By Theorem 14.4.7, $k_0 = 2$ and $k_L = 3L + 1$ for all $L \geq 1$. So,

$$d_{free} \leq \min\{\Delta(4, 2), \Delta(8, 4), \Delta(12, 7), \ldots\}.$$

Since $\Delta(4, 2)$ is easily seen to be 2, $d_{free} \leq 2$. The code C_4 from Exercise 822 is a $(4, 3, 2, 2)$ code with these Forney indices; see Exercise 830. ∎

Section 4 of [235] contains tables showing bounds on the free distance of (n, k, m) codes using a slightly weaker bound similar to that of Theorem 14.4.10 for $(n, k) = (2, 1)$, $(3, 1)$, $(3, 2)$, $(4, 1)$, and $(4, 3)$ with values of m up to 10. These tables also give codes that come closest to meeting these bounds. The bounds are most often met; when not met, except for one case, the codes have free distance one less than the value from the bound.

Exercise 831 Let C be a $(3, 2, 3)$ convolutional code.
(a) Find the two possible sets of Forney indices for C.
(b) Find the values of k_L for each set of Forney indices found in (a).
(c) Give an upper bound on the free distance of C for each set of Forney indices found in (a). ◆

14.5 Catastrophic encoders

When choosing an encoder for a convolutional code, there is one class of generator matrices that must be avoided – the so-called catastrophic encoders. In this section we see the problem that arises with their use.

We digress a bit before discussing catastrophic encoders. Let G be a $k \times n$ matrix over a field $\mathbb{F}$ with $k \leq n$. Recall that a right inverse of G is an $n \times k$ matrix K with entries in $\mathbb{F}$ such that $GK = I_k$. By a result of linear algebra G has a right inverse if and only if it has rank k. In general, if $k < n$, there are many possible right inverses. Now suppose that G is the generator matrix of either an $[n, k]$ block code or an (n, k) convolutional code. Then G has rank k and hence has a right inverse K. If $\mathbf{x}$ is a message of length k that was encoded to produce the codeword $\mathbf{c} = \mathbf{x}G$, then $\mathbf{x}$ can be recovered from $\mathbf{c}$ by observing that $\mathbf{x} = \mathbf{x}I_k = \mathbf{x}GK = \mathbf{c}K$; see also Section 1.11.1.

Exercise 832 Consider the binary matrix

$$G = \begin{bmatrix} 1 & 1 & 0 & 1 & 1 \\ 1 & 0 & 1 & 1 & 0 \end{bmatrix}.$$

(a) Find two 5×2 binary right inverses of G. Hint: Most of the entries can be chosen to be 0.

(b) Suppose that G is the generator matrix for a $[5, 2]$ binary code. Suppose that a message $\mathbf{x} \in \mathbb{F}_2^2$ is encoded by $\mathbf{x}G = \mathbf{c}$. Use each of the right inverses found in part (a) to find the message that was encoded to produce 01101.   ♦

Exercise 833   Do the following:

(a) Find a 2×1 right inverse for the generator matrix G_1 of the $(2, 1)$ convolutional code $\mathcal{C}_1$ from Example 14.1.1 where one of the entries of the right inverse is 0. Note: The other entry will be a rational function.

(b) Find a second right inverse for G_1 where both entries are polynomials in D. Note: This can be done with two polynomials of degree 1.

(c) Suppose that $\mathbf{x}$ is encoded so that $\mathbf{x}G_1 = (1 + D + D^5, 1 + D^3 + D^4 + D^5)$. Use the two right inverses of G_1 found in parts (a) and (b) to compute $\mathbf{x}$.   ♦

We now return to our discussion of catastrophic encoders. Let $\mathbf{x}$ be a message to be encoded using a generator matrix G for the (n, k) convolutional code $\mathcal{C}$. Let K be a right inverse for G with entries in $\mathbb{F}_2(D)$. The resulting codeword to be transmitted is $\mathbf{c} = \mathbf{x}G$. During transmission, $\mathbf{c}$ may be altered by noise with the result that $\mathbf{y}$ is received. A decoder finds a codeword $\mathbf{c}'$ that is close to $\mathbf{y}$, which hopefully actually equals $\mathbf{c}$. One is less interested in $\mathbf{c}$ and more interested in $\mathbf{x}$. Thus $\mathbf{x}' = \mathbf{c}'K$ can be computed with the hope that $\mathbf{x}'$ actually equals $\mathbf{x}$; this will certainly be the case if $\mathbf{c}' = \mathbf{c}$ but not otherwise. Let us investigate what might happen if $\mathbf{c}' \neq \mathbf{c}$. Let $\mathbf{e}_c = \mathbf{c}' - \mathbf{c}$ denote the *codeword error* and $\mathbf{e}_x = \mathbf{x}' - \mathbf{x}$ denote the *message error*. Note that $\mathbf{e}_c$, being the difference of two codewords of $\mathcal{C}$, must be a codeword of $\mathcal{C}$. The "message" obtained from the codeword $\mathbf{e}_c$ is $\mathbf{e}_c K = (\mathbf{c}' - \mathbf{c})K = \mathbf{x}' - \mathbf{x} = \mathbf{e}_x$. One expects that the number of erroneous symbols in the estimate $\mathbf{c}'$ would have a reasonable connection to the number of erroneous symbols in $\mathbf{x}'$. In particular, one would expect that if $\mathbf{c}'$ and $\mathbf{c}$ differ in a finite number of places, then $\mathbf{x}'$ and $\mathbf{x}$ should also differ in a finite number of places. In other words, if $\mathbf{e}_c$ has finite weight, $\mathbf{e}_x$ should also. If $\mathbf{e}_c$ were to have finite weight and $\mathbf{e}_x$ were to have infinite weight, that would be a "catastrophe." Recalling that $\mathbf{e}_c$ is a codeword, we give the following definition. A generator matrix G for an (n, k) convolutional code is called *catastrophic* if there is an infinite weight message $\mathbf{x} \in \mathbb{F}_2(D)^k$ such that $\mathbf{c} = \mathbf{x}G$ has finite weight. Otherwise, G is called *noncatastrophic*.

Example 14.5.1   The encoder G_1' of the $(2, 1)$ code $\mathcal{C}_1$ from Example 14.1.1 is catastrophic as we now show. One right inverse of G_1' is

$$K = \begin{bmatrix} \dfrac{D}{1 + D} \\ 1 \end{bmatrix}.$$

The code $\mathcal{C}_1$ contains the codeword $\mathbf{c} = (1 + D + D^2, 1 + D^2)$ (from the generator matrix G_1), whose interleaved binary form is 111011. This codeword has finite weight 5. Computing $\mathbf{c}K$ we obtain

$$\mathbf{x} = \mathbf{c}K = (1 + D + D^2)\frac{D}{1 + D} + 1 + D^2 = \frac{1}{1 + D} = \sum_{i=0}^{\infty} D^i.$$

Thus $\mathbf{x}$ has infinite weight, while $\mathbf{c}$ has finite weight. In other words, if an infinite string of 1s is input into the encoder G_1', the output is the codeword $1110110000\cdots$ of weight 5. ∎

Exercise 834 Confirm the results of Example 14.5.1 by showing that if an infinite string of 1s is input into the physical encoder for G_1' from Exercise 803, the output is the weight 5 codeword $1110110000\cdots$. ♦

Let us examine Example 14.5.1 more closely to see one reason why catastrophic encoders are to be avoided. The input $\mathbf{x}$ consisting of an infinite string of 1s has output $\mathbf{c} = 1110110000\cdots$ of weight 5. Clearly, the input $\mathbf{x}'$ consisting of an infinite string of 0s has output $\mathbf{c}'$ that is an infinite string of 0s. Suppose that five errors are made and that $\mathbf{c}'$ is received as $1110110000\cdots$. This would be decoded as if no errors were made and the message sent would be determined to be $\mathbf{x}$, the infinite string of 1s. This differs from the actual message $\mathbf{x}'$ in every position, certainly a very undesirable result!

Fortunately, a theorem of Massey and Sain [229] makes it possible to decide if an encoder is catastrophic. Their result gives two equivalent properties for a generator matrix G of an (n, k) convolutional code to be noncatastrophic. A matrix with entries in $\mathbb{F}_2(D)$ is a *finite weight matrix* provided all its entries have finite weight. Recall from Lemma 14.4.2 that each entry of a finite weight matrix is a polynomial or a polynomial divided by a positive integer power of D.

Theorem 14.5.2 (Massey–Sain) *Let G be a polynomial generator matrix for an (n, k) convolutional code C. The matrix G is a noncatastrophic encoder for C if and only if either of the following holds:*
(i) *The greatest common divisor of the $k \times k$ minors of G is a power of D.*
(ii) *G has a finite weight right inverse.*

We omit the proof of the Massey–Sain Theorem; see [229, 235]. Notice the clear connection between parts (i) and (ii) of this theorem and the parts (i) and (ii) of Theorem 14.3.3. In particular, a basic generator matrix and, more particularly, a canonical generator matrix, which we know exists for every code, is noncatastrophic.

Corollary 14.5.3 *Every basic generator matrix is a noncatastrophic generator matrix for a convolutional code. Every convolutional code has a noncatastrophic generator matrix.*

Example 14.5.4 In Example 14.1.1, we give two generator matrices, G_1 and G_1', of C_1. In Example 14.5.1, we exhibited an infinite weight input with a finite weight output using the generator matrix G_1'. So G_1' is catastrophic. We can obtain this same result by applying the Massey–Sain Theorem. The two minors of G_1' are $1 + D^3 = (1 + D)(1 + D + D^2)$ and $1 + D + D^2 + D^3 = (1 + D)(1 + D^2)$; the greatest common divisor of these minors is $1 + D$, which is not a power of D. (Notice the connection between the finite weight input that leads to an infinite weight codeword and the greatest common divisor $1 + D$ of the minors.) Alternately, suppose that $K = [a(D) \ b(D)]^\mathsf{T}$ is a finite weight right inverse of G_1'. Then there exist polynomials $p(D)$ and $q(D)$ together with nonnegative integers i and j such that $a(D) = p(D)/D^i$ and $b(D) = q(D)/D^j$. Then as $G_1' K = I_1$, $(1 + D^3)a(D) + (1 + D + D^2 + D^3)b(D) = 1$, or

$$D^j(1 + D^3)p(D) + D^i(1 + D + D^2 + D^3)q(D) = D^{i+j}.$$

But the left-hand side is a polynomial divisible by $1 + D$, while the right-hand side is not. Thus G_1' does not have a finite weight right inverse. In Example 14.5.1 we found a right inverse of G_1', but it was not of finite weight.

Turning to the encoder G_1, we see that the two minors of G_1 are $1 + D + D^2$ and $1 + D^2$, which have greatest common divisor 1. Thus G_1 is noncatastrophic. Also G_1 has the finite weight right inverse

$$K = \begin{bmatrix} D \\ 1 + D \end{bmatrix},$$

also confirming that G_1 is noncatastrophic.   ∎

Example 14.5.5 In Example 14.1.1, we give three generator matrices for the $(4, 2)$ convolutional code $\mathcal{C}_2$. All of these encoders are noncatastrophic. We leave the verification of this fact for G_2' and G_2'' as an exercise. The minor obtained from the first and second columns of G_2 is $1 + D$, while the minor from the first and third columns of G_2 is D. Hence the greatest common divisor of all six minors of G_2 must be 1.   ∎

Exercise 835 Verify that the encoders G_2' and G_2'' for the $(4, 2)$ convolutional code $\mathcal{C}_2$ from Example 14.1.1 are noncatastrophic by showing that the greatest common divisor of the minors of each encoder is 1.   ◆

Exercise 836 Find 4×2 finite weight right inverses for each of the encoders G_2, G_2', and G_2'' of the $(4, 2)$ convolutional code $\mathcal{C}_2$ from Example 14.1.1. Note: Many entries can be chosen to be 0.   ◆

Exercise 837 Let G_2''' be the generator matrix of the $(4, 2)$ convolutional code $\mathcal{C}_2$ in Example 14.1.1.
(a) Show that the greatest common divisor of the six minors of G_2''' is $1 + D + D^2$ and hence that G_2''' is a catastrophic encoder.
(b) Show that the input

$$\mathbf{x} = \left(\frac{D^2}{1 + D + D^2}, \frac{1}{1 + D + D^2} \right)$$

to the encoder G_2''' has infinite weight but that the encoded codeword $\mathbf{c} = \mathbf{x}G_2'''$ has finite weight. Also give $\mathbf{c}$.   ◆

Exercise 838 Let $\mathcal{C}_5$ be the $(2, 1)$ convolutional code from Exercise 823 and Example 14.4.8 with generator matrices

$$G_5 = \begin{bmatrix} 1 & 1 + D \end{bmatrix} \quad \text{and} \quad G_5' = \begin{bmatrix} 1 + D & 1 + D^2 \end{bmatrix}.$$

(a) Suppose that G'_5 is the encoder. Find a right inverse of G'_5 and use it to produce the input if the output is the codeword $(1, 1 + D)$.

(b) Draw the state diagram for the encoder G'_5.

(c) Draw the trellis diagram for the encoder G'_5.

(d) Apply the Viterbi Algorithm to decode the received output vector $1101000000 \cdots$ (of infinite length). Note that this is the interleaved codeword $(1, 1 + D)$. Note also that to apply the Viterbi Algorithm to a received vector of infinite length, follow the path of minimum weight through the trellis without trying to end in the zero state.

(e) Based on parts (a) or (d), why is G'_5 a catastrophic encoder?

(f) What is the greatest common divisor of the 1×1 minors of G'_5?

(g) Without resorting to the Massey–Sain Theorem, show that G'_5 does not have a finite weight inverse.

(h) Show that G_5 is a noncatastrophic generator matrix for $\mathcal{C}$.   ◆

15 Soft decision and iterative decoding

The decoding algorithms that we have considered to this point have all been hard decision algorithms. A *hard decision* decoder is one which accepts hard values (for example 0s or 1s if the data is binary) from the channel that are used to create what is hopefully the original codeword. Thus a hard decision decoder is characterized by "hard input" and "hard output." In contrast, a *soft decision* decoder will generally accept "soft input" from the channel while producing "hard output" estimates of the correct symbols. As we will see later, the "soft input" can be estimates, based on probabilities, of the received symbols. In our later discussion of turbo codes, we will see that turbo decoding uses two "soft input, soft output" decoders that pass "soft" information back and forth in an iterative manner between themselves. After a certain number of iterations, the turbo decoder produces a "hard estimate" of the correct transmitted symbols.

15.1 Additive white Gaussian noise

In order to understand soft decision decoding, it is helpful to take a closer look first at the communication channel presented in Figure 1.1. Our description relies heavily on the presentation in [158, Chapter 1]. The box in that figure labeled "Channel" is more accurately described as consisting of three components: a modulator, a waveform channel, and a demodulator; see Figure 15.1. For simplicity we restrict ourselves to binary data. Suppose that we transmit the binary codeword $\mathbf{c} = c_1 \cdots c_n$. The *modulator* converts each bit to a waveform. There are many modulation schemes used. One common scheme is called *binary phase-shift keying* (*BPSK*), which works as follows. Each bit is converted to a waveform whose duration is T seconds, the length of the clock cycle. If the bit 1 is to be transmitted beginning at time $t = 0$, the waveform is

$$
b_1(t) = \begin{cases} \sqrt{\dfrac{2E_s}{T}} \cos \omega t & \text{if } 0 \le t < T, \\ 0 & \text{otherwise,} \end{cases}
$$

where E_s is the energy of the signal and $\omega = 2\pi/T$ is the angular frequency. To transmit the bit 0 the waveform $b_0(t) = -b_1(t)$ is used. (Notice that $b_0(t)$ also equals $\sqrt{2E_s/T} \cos(\omega t + \pi)$, thus indicating why "phase-shift" is included in the name of the scheme.) Beginning at time $t = 0$, the codeword $\mathbf{c}$ is therefore transmitted as the

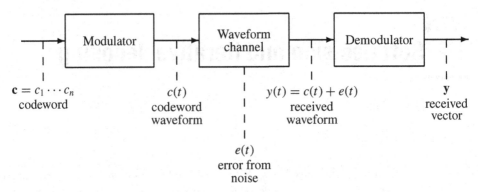

Figure 15.1 Modulated communication channel.

waveform

$$c(t) = \sum_{i=1}^{n} b_{c_i}(t - (i-1)T).$$

Exercise 839 Starting at time $t = 0$, graph the waveform for:
(a) the single bit 1,
(b) the single bit 0, and
(c) the codeword 00101.   ◆

The waveform $c(t)$ is transmitted over the waveform channel where noise $e(t)$ in the form of another wave may distort $c(t)$. By using a matched filter with impulse response

$$h(t) = \begin{cases} \sqrt{\dfrac{2}{T}} \cos \omega t & \text{if } 0 \le t < T, \\ 0 & \text{otherwise,} \end{cases}$$

the *demodulator* first converts the received waveform $y(t) = c(t) + e(t)$ into an n-tuple of real numbers $\mathbf{y'} = y'_1 y'_2 \cdots y'_n$ where

$$y'_i = \int_{(i-1)T}^{iT} y(t) h(iT - t)\, dt. \tag{15.1}$$

If no error occurs in the transmission of c_i, then $y'_i = \sqrt{E_s}$ if $c_i = 1$ and $y'_i = -\sqrt{E_s}$ if $c_i = 0$; see Exercise 840. We can produce a hard decision received binary vector $\mathbf{y} = \mathbf{c} + \mathbf{e} = y_1 y_2 \cdots y_n$ by choosing

$$y_i = \begin{cases} 1 & \text{if } y'_i > 0, \\ 0 & \text{if } y'_i \le 0. \end{cases} \tag{15.2}$$

These binary values y_i would be the final output of the demodulator if our decoder is a hard decision decoder.

Example 15.1.1 If the codeword $\mathbf{c} = 10110$ is sent, the transmitted waveform is $c(t) = b_1(t) - b_1(t - T) + b_1(t - 2T) + b_1(t - 3T) - b_1(t - 4T)$. Suppose $y(t) = c(t) + e(t)$ is received. Assume that the matched filter yields $y'_1 = 0.7$, $y'_2 = -1.1$, $y'_3 = 0.5$, $y'_4 = -0.1$, and $y_5' = -0.5$. The hard decision received vector is $\mathbf{y} = 10100$ and an error has occurred in the fourth coordinate. ∎

Exercise 840 Show that

$$\int_{(i-1)T}^{iT} \sqrt{\frac{2E_s}{T}} \cos \omega \, (t - (i-1)T) \sqrt{\frac{2}{T}} \cos \omega (iT - t) \, dt = \sqrt{E_s}.$$

Explain why this verifies the claim that if no error occurs during the transmission of a single bit c_i, then $y_i' = \pm\sqrt{E_s}$, where $+$ is chosen if $c_i = 1$ and $-$ is chosen if $c_i = 0$. ◆

The noise can often be modeled as *additive white Gaussian noise* (AWGN). This model describes noise in terms of the distribution of y_i' given by (15.1). If $c_i = 1$ is transmitted, the mean μ of y_i' is $\sqrt{E_s}$. If $c_i = 0$ is transmitted, the mean μ is $-\sqrt{E_s}$. Subtracting the appropriate value of μ, $y_i' - \mu$ is normally distributed with mean 0 and variance

$$\sigma^2 = \frac{N_0}{2}.$$

The value $N_0/2$ is called the *two-sided power spectral density*.

Using (15.1) and (15.2) to produce a binary vector from $y(t)$ is in fact a realization of the binary symmetric channel model described in Section 1.11.2. The crossover probability ϱ of this channel is computed as follows. Suppose, by symmetry, that the bit $c_i = 1$ and hence $b_1(t - (i-1)T)$ is transmitted. An error occurs if $y_i' \leq 0$ and the received bit would be declared to be 0 by (15.2). Since $y_i' - \mu$ is normally distributed with mean 0 and variance $\sigma^2 = N_0/2$,

$$\varrho = \text{prob}(y_i' \leq 0) = \frac{1}{\sqrt{2\pi}\sigma} \int_{-\infty}^{0} e^{\frac{-(y-\mu)^2}{2\sigma^2}} \, dy = \frac{1}{\sqrt{2\pi}} \int_{-\infty}^{-\sqrt{2E_s/N_0}} e^{-\frac{x^2}{2}} \, dx. \tag{15.3}$$

Exercise 841 Do the following:
(a) Verify the right-most equality in (15.3).
(b) Verify that

$$\frac{1}{\sqrt{2\pi}} \int_{-\infty}^{-\sqrt{2E_s/N_0}} e^{-\frac{x^2}{2}} \, dx = \frac{1}{2} - \frac{1}{2}\text{erf}(\sqrt{E_s/N_0}),$$

where

$$\text{erf}(z) = \frac{2}{\sqrt{\pi}} \int_0^z e^{-t^2} \, dt.$$

The function $\text{erf}(z)$ is called the *error function* and tabulated values of this function are available in some textbooks and computer packages. ◆

On the other hand the output of the demodulator does not have to be binary. It can be "soft" data that can be used by a soft decision decoder. Notice that the hard decision value y_i was chosen based on the threshold 0; that is, y_i is determined according to whether $y_i' > 0$ or $y_i' \leq 0$. This choice does not take into account how likely or unlikely the value of y_i' is, given whether 0 or 1 was the transmitted bit c_i. The determination of y_i from y_i' is called *quantization*. The fact that there are only two choices for y_i means that we have employed *binary quantization*.

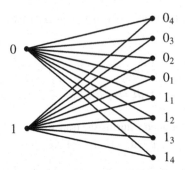

Figure 15.2 Binary input, 8-ary output discrete memoryless channel.

We could have other quantizations that determine the output of the demodulator. One example, which will become the basis of our soft decision decoding, is the following. Suppose we divide the real line into a disjoint union of eight intervals by choosing seven threshold values. Then we could assign y_i any one of eight values. We will use $0_1, 0_2, 0_3, 0_4, 1_1, 1_2, 1_3, 1_4$ for these eight values. For example, suppose that $(a, b]$ (where $a = -\infty$ or $b = \infty$ are possible) is the interval that determines 0_1; in other words, suppose that y_i is chosen to be 0_1 exactly when $y_i' \in (a, b]$. Then the probability that $y_i' \in (a, b]$, given that c_i is transmitted, is

$$\text{prob}(0_1 \mid c_i) = \text{prob}(a < y_i' \le b \mid c_i) = \frac{1}{\sqrt{2\pi}\sigma} \int_a^b e^{\frac{-(y-\mu)^2}{2\sigma^2}} \, dy, \tag{15.4}$$

where $\mu = \sqrt{E_s}$ if $c_i = 1$ and $\mu = -\sqrt{E_s}$ if $c_i = 0$. Therefore once the threshold levels are chosen, using equations similar to (15.4), we can calculate 16 probabilities $\text{prob}(y \mid c)$ where y ranges through the eight possible values $0_1, \ldots, 1_4$ and $c = 0$ or 1. These probabilities are characteristic of the channel under the assumption that the channel is subject to AWGN; they are the *channel statistics*. The probabilities can be computed from the knowledge of the signal energy E_s and the two-sided power spectral density $N_0/2$ recalling that $\sigma^2 = N_0/2$. The channel we have just described is a *binary input, 8-ary output discrete memoryless channel (DMC)* and is pictured in Figure 15.2. (The "binary input" is the input to the modulator, and the "8-ary output" is the output of the demodulator.) Along each edge we could write the probability $\text{prob}(y \mid c)$ as in the BSC of Figure 1.2. Of course the sum of the eight probabilities along the edges emanating from 0 add to 1 as do the eight probabilities along the edges emanating from 1. Presumably we would choose our notation so that the four highest probabilities emanating from 0 would terminate at $0_1, \ldots, 0_4$ and the four highest probabilities emanating from 1 would terminate at $1_1, \ldots, 1_4$. At some later point in time, the four values $0_1, \ldots, 0_4$ could be assigned 0 and the four values $1_1, \ldots, 1_4$ could be assigned 1.

By going to this 8-level quantization we are taking into account the likelihood of getting a value y_i' given that c_i is sent. It is not surprising this "soft" quantized data together with the probabilities $\text{prob}(y \mid c)$ can be used to create a more effective decoder. We can actually quantify how much "gain" there is when using coding with hard decision decoding compared with using no coding at all. Furthermore, we can measure the potential "gain" when soft decision decoding is used rather than hard decision decoding.

In Section 1.11.2 we presented Shannon's Theorem for a BSC. There is a corresponding theorem for the Gaussian channel if we give the appropriate channel capacity. Assume the Gaussian channel is subject to AWGN with two-sided spectral density $N_0/2$. Suppose, further, that the signaling power is S and bandwidth is W. Then Shannon showed that if the capacity $C_G(W)$ of the channel is defined to be

$$C_G(W) = W \log_2 \left(1 + \frac{S}{N_0 W} \right), \tag{15.5}$$

then arbitrarily reliable communication is possible provided the rate of information transmission is below capacity and impossible for rates above $C_G(W)$.

If we allow unlimited bandwidth, then the channel capacity approaches C_G^∞, where

$$C_G^\infty = \lim_{W \to \infty} W \log_2 \left(1 + \frac{S}{N_0 W} \right) = \frac{S}{N_0 \ln 2} \text{ bits/s.} \tag{15.6}$$

Earlier we let T be the time to transmit a single bit. So k information bits can be transmitted in $\tau = nT$ seconds. The *energy per bit*, denoted E_b, is then

$$E_b = \frac{S\tau}{k} = \frac{S}{R_t}, \tag{15.7}$$

where $R_t = k/\tau$ is the *rate of information transmission*. Combining this with (15.6), we see that

$$\frac{C_G^\infty}{R_t} = \frac{E_b}{N_0 \ln 2}. \tag{15.8}$$

Shannon's Theorem tells us that to have communication over this channel, then $R_t < C_G^\infty$.[1] Using (15.8), this implies that

$$\frac{E_b}{N_0} > \ln 2 \approx -1.6 \, \text{dB}. \tag{15.9}$$

Thus to have reliable communication over a Gaussian channel, the *signal-to-noise ratio* E_b/N_0, which measures the relative magnitude of the energy per bit and the noise energy, must exceed the value $-1.6 \, \text{dB}$, called the *Shannon limit*.[2] As long as the signal-to-noise ratio exceeds Shannon's limit, then Shannon's Theorem guarantees that there exists a communication system, possibly very complex, which can be used for reliable communication over the channel.

Exercise 842   Verify the right-most equality in (15.6).   ◆

If we require that hard decision decoding be used, then the channel has a different capacity. To have reliable communication with hard decision decoding, we must have

$$\frac{E_b}{N_0} > \frac{\pi}{2} \ln 2 \approx 0.4 \, \text{dB}.$$

---

[1] Capacity of a channel can be defined in more than one way. For the version defined here, Shannon's Theorem examines the connection between the information transmission rate R_t and the channel capacity. In the version of Shannon's Theorem for binary symmetric channels presented in Section 1.11.2, channel capacity was defined in the way that connects the information rate $R = k/n$ of an $[n, k]$ code to the channel capacity.

[2] A unitless quantity x, when converted to decibels, becomes $10 \log_{10}(x)$ dB.

Thus potentially there is a 2 dB gain if we use soft decision decoding rather than hard decision decoding over a Gaussian channel. Shortly we explain what such a gain will yield.

First consider what happens when we do not use coding to communicate over a channel. Let P_b be the *bit error rate* (BER), that is, the probability of a bit error in the decoding. If there is no coding, then $E_b = E_s$ since every signal is a single information bit. From (15.3), the BER of uncoded binary phase-shift keying is

$$P_b = Q(\sqrt{2E_b/N_0}),$$

where

$$Q(z) = \frac{1}{\sqrt{2\pi}} \int_z^\infty e^{-\frac{x^2}{2}} \, dx.$$

In Figure 15.3, we plot P_b for uncoded BPSK as a function of signal-to-noise ratio. (Note that the vertical scale is logarithmic, base 10.) We also indicate in the figure the Shannon limit of -1.6 dB and the hard decision limit of 0.4 dB.

As an illustration of what Figure 15.3 tells us, the acceptable bit error rate for image data transmitted by deep space satellites is typically in the range of 5×10^{-3} to 1×10^{-7} [352]. Suppose that we desire a BER of 1×10^{-5}. We see from Figure 15.3 that if we transmitted using BPSK without any coding, then the signal-to-noise ratio must be at least 9.6 dB. Thus with coding using hard decision decoding, there is a potential *coding gain* of 9.2 dB. With coding using soft decision decoding, there is a potential coding gain of 11.2 dB.

This improvement can be measured in several different ways. For instance, if we examine (15.7) and assume that N_0, k, and τ remain fixed, a 1 dB gain means that the signal power S can be reduced to about $10^{-0.1} S \approx 0.794S$. If the BER is 1×10^{-5} and if we could achieve a 9.2 dB coding gain by using a code C, then the signal power required by uncoded BPSK could be reduced to about $10^{-0.92} S \approx 0.120S$ with the use of C. Such a reduction in signal power is very significant. In deep space satellite communication this can mean smaller batteries (and thus less weight), longer operation of the spacecraft, or smaller transmitter size. In the late 1960s each dB of coding gain in satellite communications was estimated to be worth US\$1,000,000 in development and launch costs [227].

We can analyze potential coding gain when the channel is bandwidth limited using a code of information rate R. Assume communication uses an $[n, k]$ binary code, which therefore has information rate $R = k/n$. If we are sampling at the *Nyquist rate* of $2W$ samples per second and transmitting k information bits in τ seconds, then $n = 2W\tau$. Therefore the rate R_t of information transmission is

$$R_t = k/\tau = 2Wk/n = 2WR. \tag{15.10}$$

From (15.7) and (15.10),

$$\frac{S}{N_0 W} = \frac{2RE_b}{N_0}.$$

Using this equation, (15.5), and Shannon's Theorem, to have reliable communication

$$R_t = 2WR < W \log_2 \left(1 + \frac{2RE_b}{N_0} \right).$$

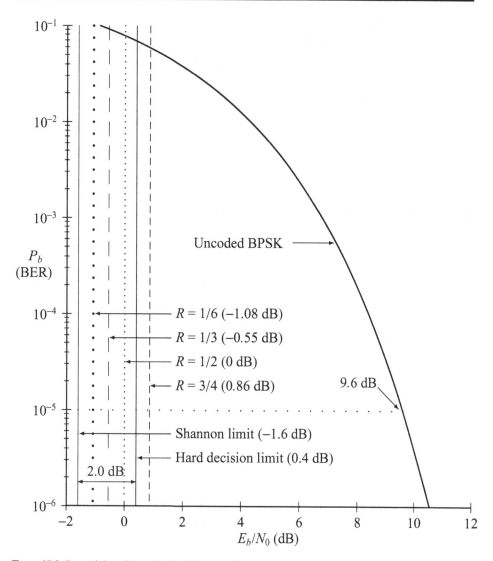

Figure 15.3 Potential coding gain in a Gaussian channel.

Solving for E_b/N_0 yields

$$\frac{E_b}{N_0} > \frac{2^{2R} - 1}{2R},$$ (15.11)

giving the potential coding gain for a code of information rate R.

Notice that if the information transmission rate R_t is held constant, then by (15.10), when $W \to \infty$, $R \to 0$. Letting $R \to 0$ in (15.11), we obtain $E_b/N_0 > \ln 2$, agreeing with (15.9) in the case where the channel has unlimited bandwidth. Suppose that we keep the bandwidth fixed as when the channel is bandwidth limited. The right-hand side of (15.11) is an increasing function of R. Hence to communicate near the Shannon limit, R must be close to 0; by (15.10), the information transmission rate R_t must also be close to 0. The following table illustrates the lower bound from (15.11), in dB, on signal-to-noise ratio for a variety of values of R.

R	3/4	1/2	1/3	1/6
$\dfrac{2^{2R}-1}{2R}$ (dB)	0.86	0	-0.55	-1.08

(15.12)

We include these values in Figure 15.3.[3]

Exercise 843 Verify the entries in (15.12).   ◆

15.2 A Soft Decision Viterbi Algorithm

In this section we examine how we might modify the Viterbi Algorithm (a hard decision decoding algorithm) of Section 14.2 to obtain a soft decision Viterbi decoding algorithm. The Soft Decision Viterbi Algorithm is carried out in precisely the same manner as the hard decision Viterbi Algorithm once the edge weights in the trellis are appropriately defined, except the survivor paths are the ones with highest path weights. The hard decision Viterbi Algorithm is nearest neighbor decoding that, over a binary symmetric channel, is also maximum likelihood decoding; see the discussion in Section 1.11.2. The Soft Decision Viterbi Algorithm will also be maximum likelihood decoding.

Suppose that an input message $\mathbf{x}(i) = (\mathbf{x}_1(i), \ldots, \mathbf{x}_k(i))$ for $i = 0, 1, \ldots, L-1$ is encoded using the generator matrix G of an (n, k) binary convolutional code to produce an output codeword $\mathbf{c}(i) = (\mathbf{c}_1(i), \ldots, \mathbf{c}_n(i))$ for $i = 0, 1, \ldots, L+M-1$. As previously, the state s of the encoder at time i is denoted s_i. The zero state will be state a, the initial and final state of the encoder. We will add M blocks of k zeros to the end of the message so that the encoder will terminate in the zero state a_{L+M} at time $L+M$. The bits of this codeword are interleaved as in (14.1) and this bit stream is modulated, transmitted, and received as a waveform $y(t)$ that is demodulated using (15.1). Although any demodulation scheme can be used, we will assume for concreteness that 8-level quantization is used as described in the previous section. After deinterleaving the quantized data, we have a received vector $\mathbf{y}(i) = (\mathbf{y}_1(i), \ldots, \mathbf{y}_n(i))$ for $i = 0, 1, \ldots, L+M-1$ where $\mathbf{y}_j(i) \in \{0_1, \ldots, 1_4\}$.

Since the Soft Decision Viterbi Algorithm is to perform maximum likelihood decoding, the algorithm must find the codeword $\mathbf{c}$ that maximizes $\text{prob}(\mathbf{y} \mid \mathbf{c})$. However, as the channel is memoryless,

$$\text{prob}(\mathbf{y} \mid \mathbf{c}) = \prod_{i=0}^{L+M-1} \prod_{j=1}^{n} \text{prob}(\mathbf{y}_j(i) \mid \mathbf{c}_j(i)).$$

---

[3] Equation (15.11) is valid if P_b is close to 0. If we can tolerate a BER of P_b not necessarily close to 0, then we can compress the data and obtain a modified version of (15.11). Taking this into account, the vertical lines in Figure 15.3 for the four values of R, for Shannon's limit, and for the hard decision limit actually curve slightly toward the left as they rise. So for BERs closer to 10^{-1}, we can obtain more coding gain than shown. See [158, Section 1.5] for a discussion and figure; see also [232].

Maximizing this probability is equivalent to maximizing its logarithm

$$\ln(\text{prob}(\mathbf{y} \mid \mathbf{c})) = \sum_{i=0}^{L+M-1} \sum_{j=1}^{n} \ln(\text{prob}(\mathbf{y}_j(i) \mid \mathbf{c}_j(i))),$$

which has the advantage of converting the product to a sum. (Recall that computing the weight of a path in the original Viterbi Algorithm was accomplished by summing edge weights.) We can employ a technique of Massey [226] that will allow us to make the edge weights into integers. Maximizing $\ln(\text{prob}(\mathbf{y} \mid \mathbf{c}))$ is equivalent to maximizing

$$\sum_{i=0}^{L+M-1} \sum_{j=1}^{n} \mu(\mathbf{y}_j(i), \mathbf{c}_j(i)), \tag{15.13}$$

with

$$\mu(\mathbf{y}_j(i), \mathbf{c}_j(i)) = A(\ln(\text{prob}(\mathbf{y}_j(i) \mid \mathbf{c}_j(i))) - f_{i,j}(\mathbf{y}_j(i))), \tag{15.14}$$

where A is a positive constant and $f_{i,j}(\mathbf{y}_j(i))$ is an arbitrary function. The value

$$\sum_{j=1}^{n} \mu(\mathbf{y}_j(i), \mathbf{c}_j(i)) \tag{15.15}$$

will be the edge weight in the trellis replacing the Hamming distance used in the hard decision Viterbi Algorithm. To make computations simpler first choose

$$f_{i,j}(\mathbf{y}_j(i)) = \min_{c \in \{0,1\}} \left\{ \ln(\text{prob}(\mathbf{y}_j(i) \mid c)) \right\}$$

and then choose A so that $\mu(\mathbf{y}_j(i), \mathbf{c}_j(i))$ is close to a positive integer. With these edge weights, the path weights can be determined. The Soft Decision Viterbi Algorithm now proceeds as in Section 14.2, except the surviving paths at a node are those of **maximum weight** rather than minimum weight.

Example 15.2.1 We repeat Example 14.2.5 using the Soft Decision Viterbi Algorithm. Assume the channel is a binary input, 8-ary output DMC subject to AWGN. Suppose thresholds have been chosen so that $\text{prob}(y \mid c)$ is given by

$c \backslash y$	0_4	0_3	0_2	0_1	1_1	1_2	1_3	1_4
0	0.368	0.207	0.169	0.097	0.065	0.051	0.028	0.015
1	0.015	0.028	0.051	0.065	0.097	0.169	0.207	0.368

Taking the natural logarithm of each probability yields

$c \backslash y$	0_4	0_3	0_2	0_1	1_1	1_2	1_3	1_4
0	−0.9997	−1.5750	−1.7779	−2.3330	−2.7334	−2.9759	−3.5756	−4.1997
1	−4.1997	−3.5756	−2.9759	−2.7334	−2.3330	−1.7779	−1.5750	−0.9997

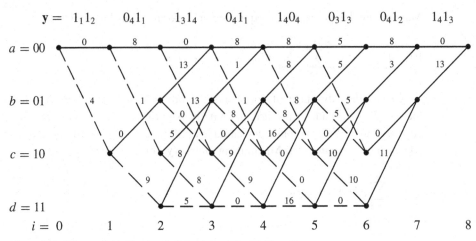

Figure 15.4 Edge weights for the Soft Decision Viterbi Algorithm.

By subtracting the smaller value in each column of the following table from each entry in the column, multiplying the result by $A = 2.5$, and rounding, we obtain $\mu(y, c)$:

$c \setminus y$	0_4	0_3	0_2	0_1	1_1	1_2	1_3	1_4
0	8	5	3	1	0	0	0	0
1	0	0	0	0	1	3	5	8

(We remark that the difference between the actual value and the rounded value in the above table is never more than 0.005. Also, other choices for A are possible, such as $A = 10000$.) Now assume that a six bit message followed by 00 has been encoded and

$$y = 1_1 1_2 \, 0_4 1_1 \, 1_3 1_4 \, 0_4 1_1 \, 1_4 0_4 \, 0_3 1_3 \, 0_4 1_2 \, 1_4 1_3$$

is the received demodulated vector. Note that when $0_1, \ldots, 0_4$ and $1_1, \ldots, 1_4$ are merged to the hard outputs 0 and 1, respectively, we have the same received vector as in Example 14.2.5. Figure 15.4 is the trellis of Figure 14.4 with the edge labels of that figure replaced by the weights from (15.15) analogous to the trellis of Figure 14.5. As an illustration, the dashed edge $a_0 c_1$ originally labeled 11 has weight $\mu(1_1, 1) + \mu(1_2, 1) = 1 + 3 = 4$. Figure 15.5 shows the survivor paths and their weights. The survivor path ending at state a when $i = 8$ is $a_0 a_1 a_2 c_3 d_5 d_6 b_7 a_8$ yielding the message 001111 (followed by two 0s) with encoding 0000110110100111. Recall that the message using hard decision Viterbi decoding was 111011 with encoding 1101100100010111. This example shows that hard decision and soft decision decoding can yield far different results. For comparison the path weight in the trellis of Figure 15.4 for the message 111011 from the hard decision decoding is 69, one less than the path weight of 70 for the message 001111. ∎

Exercise 844 Do the following:
(a) Beginning with the values for prob($y \mid c$), verify the values for $\mu(y, c)$ given in Example 15.2.1.
(b) Verify the edge weights shown in Figure 15.4.

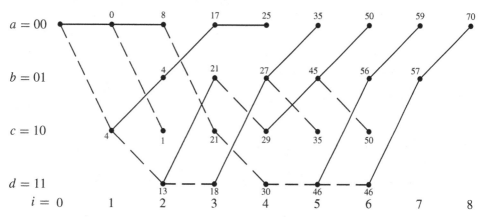

Figure 15.5 Survivor paths and their weights in the trellis of Figure 15.4.

(c) Verify that the survivor paths and weights for the Soft Decision Viterbi Algorithm of Example 15.2.1 are as shown in Figure 15.5. ◆

Exercise 845 Use the Soft Decision Viterbi Algorithm to decode the following received vectors sent using the code of Example 15.2.1 together with the edge weights determined by $\mu(y, c)$ in that example. (The trellis from Figure 14.4 will be useful.) Draw the trellis analogous to that of Figure 15.5. Give the most likely message and codeword. Finally, compare your results to those in Exercise 817.

(a) $1_4 1_3 0_3 0_2 0_4 1_3 1_2 1_3 1_4 1_2 1_3 1_1 1_2 1_4 0_3 1_4$

(b) $0_3 1_4 1_3 0_2 1_4 1_1 0_3 1_4 1_3 0_2 0_4 1_4 1_3 0_2 0_3 1_2$ ◆

It is interesting to apply the Soft Decision Viterbi Algorithm to a binary symmetric channel with crossover probability ϱ instead of the binary input 8-ary output channel. In the case of the BSC,

$$\text{prob}(\mathbf{y} \mid \mathbf{c}) = \prod_{i=0}^{L+M-1} \prod_{j=1}^{n} \text{prob}(\mathbf{y}_j(i) \mid \mathbf{c}_j(i))$$

$$= \prod_{i=0}^{L+M-1} \prod_{j=1}^{n} \varrho^{d(\mathbf{y}_j(i),\mathbf{c}_j(i))}(1 - \varrho)^{1-d(\mathbf{y}_j(i),\mathbf{c}_j(i))}.$$

Therefore from (15.13) and (15.14),

$$\mu(\mathbf{y}_j(i), \mathbf{c}_j(i)) = A(\ln(\varrho^{d(\mathbf{y}_j(i),\mathbf{c}_j(i))}(1 - \varrho)^{1-d(\mathbf{y}_j(i),\mathbf{c}_j(i))}) - f_{i,j}(\mathbf{y}_j(i)))$$

$$= d(\mathbf{y}_j(i), \mathbf{c}_j(i)) \left(A \ln \frac{\varrho}{1 - \varrho} \right) + A \ln(1 - \varrho) - A f_{i,j}(\mathbf{y}_j(i)).$$

Choosing $A = -(\ln(\varrho/(1 - \varrho)))^{-1}$, which is positive if $\varrho < 1/2$, and $f_{i,j}(\mathbf{y}_j(i)) = \ln(\varrho)$, we obtain

$$\mu(\mathbf{y}_j(i), \mathbf{c}_j(i)) = 1 - d(\mathbf{y}_j(i), \mathbf{c}_j(i)), \tag{15.16}$$

which implies, by (15.15), that the edge weights in the trellis equal

$$n - d(\mathbf{y}(i), \mathbf{c}(i)).$$

The path of maximum weight through the trellis with these edge weights is exactly the same path of minimum weight through the trellis with edge weights given by $d(\mathbf{y}(i), \mathbf{c}(i))$. The latter is the outcome of the hard decision Viterbi Algorithm of Section 14.2, and hence the hard decision and soft decision Viterbi Algorithms agree on a BSC.

Exercise 846   Verify (15.16).   ◆

15.3   The General Viterbi Algorithm

The Viterbi Algorithm can be placed in a more general setting that will make its use more transparent in a number of situations. This general setting, proposed by McEliece in [234], allows the labels (or weights) of the trellis edges to lie in any semiring.

A *semiring* is a set $\mathcal{A}$ with two binary operations: addition, usually denoted $+$, and multiplication, usually denoted $\cdot$. These operations must satisfy certain properties. First, $+$ is an associative and commutative operation; furthermore, there is an additive identity, denoted 0, such that $u + 0 = u$ for all $u \in \mathcal{A}$. Second, $\cdot$ is an associative operation; additionally, there is a multiplicative identity, denoted 1, such that $u \cdot 1 = 1 \cdot u = u$ for all $u \in \mathcal{A}$. Finally, the distributive law $(u + v) \cdot w = (u \cdot w) + (v \cdot w)$ holds for all u, v, and w in $\mathcal{A}$.

Let T be a truncated trellis of a convolutional code; we will use Figure 14.4 as the model for such a trellis.[4] The vertices of the trellis consist of states s at times i, which we continue to denote by s_i. An edge e in the trellis beginning in state s at time $i - 1$ and ending at state s' at time i is denoted $e = (s_{i-1}, s_i')$; the initial vertex s_{i-1} of e is denoted $*e$ and the final vertex is denoted $e*$. The label on the edge e is $\alpha(e)$. For example, the edge in Figure 14.4 from state d at time 4 to state b at time 5 is $e = (d_4, b_5)$, and so $*e = d_4$ and $e* = b_5$; also $\alpha(e) = 01$. The edge labels of T will come from a semiring $\mathcal{A}$ with operations $+$ and $\cdot$. The General Viterbi Algorithm, in essence, calculates the "flow" through the trellis. Suppose that $e_1 e_2 \cdots e_m$ is a path P in the trellis where $e_1, \ldots, e_m$ are edges satisfying $e_i* = *e_{i+1}$ for $1 \le i < m$. The *flow* along P is the product

$$v(P) = \alpha(e_1) \cdot \alpha(e_2) \cdots \alpha(e_m). \tag{15.17}$$

Note that the multiplication in $\mathcal{A}$ may not be commutative and hence the order in (15.17) is important. Now suppose that s_i and s_j' are two vertices in T with $i < j$. The *flow* from s_i to s_j' is

$$v(s_i, s_j') = \sum_P v(P), \tag{15.18}$$

where the summation runs through all paths P that start at vertex s_i and end at vertex s_j'. Of course the summation uses the addition $+$ of the semiring $\mathcal{A}$; since addition in $\mathcal{A}$ is commutative, the order of the summation is immaterial.

---

[4] A trellis can be defined for a block code that is similar to a truncated trellis of a convolutional code. Such a trellis will not necessarily have the same number of states at all times i ("time" is actually called "depth" in the trellis of a block code). However, as with convolutional codes, the trellis of a block code will have a single starting vertex and a single ending vertex. The results of this section will apply to such trellises as well; see [336] for an excellent exposition of this topic.

Example 15.3.1 Let $\mathcal{A}$ be the set of nonnegative integers together with the symbol ∞. $\mathcal{A}$ can be made into a semiring. Define the addition operation $+$ on $\mathcal{A}$ as follows: if u and v are in $\mathcal{A}$, then $u + v = \min\{u, v\}$ where $\min\{u, \infty\} = u$ for all $u \in \mathcal{A}$. Define the multiplication operation $\cdot$ on $\mathcal{A}$ to be ordinary addition where $u \cdot \infty = \infty \cdot u = \infty$ for all $u \in \mathcal{A}$. In Exercise 847, you are asked to verify that, under these operations, $\mathcal{A}$ is a semiring with additive identity ∞ and multiplicative identity 0. Consider the truncated trellis determined by a convolutional encoder and a received vector as in Figure 14.5. The edge labels of that trellis are in the set $\mathcal{A}$. Using the multiplication in $\mathcal{A}$ and (15.17), the flow along a path is the ordinary sum of the integer edge labels; this is exactly what we termed the "weight" of the path in Section 14.2.3. Consider the flow $v(a_0, s_i)$ from the initial state a at time 0 to a state s at time i. By definition of addition in $\mathcal{A}$ and (15.18), this flow is the minimum weight of all paths from a_0 to s_i. We called a path that gave this minimum weight a "survivor," and so the flow $v(a_0, s_i)$ is the weight of a survivor. The flow $v(a_0, a_8)$ in the truncated trellis of Figure 14.5 is therefore the weight of any survivor in $\mathcal{S}(a, 8)$ as determined by the Viterbi Algorithm presented in Section 14.2.3.   ∎

Exercise 847   Verify that the set $\mathcal{A}$ with operations defined in Example 15.3.1 is a semiring with additive identity ∞ and multiplicative identity 0.   ◆

Example 15.3.2 The set $\mathcal{A}$ of nonnegative integers can be made into a semiring under the following operations. Define $\cdot$ on $\mathcal{A}$ to be ordinary integer addition; if u and v are in $\mathcal{A}$, define $+$ to be $u + v = \max\{u, v\}$. Exercise 848 shows that, under these operations, $\mathcal{A}$ is a semiring with additive and multiplicative identity both equal to 0. The edge labels of the truncated trellis of Figure 15.4 are in the set $\mathcal{A}$. As in Example 15.3.1, the flow along a path in the trellis is the "weight" of the path. Using the definition of addition in $\mathcal{A}$ and (15.18), the flow $v(a_0, s_i)$ is the maximum weight of all paths from a_0 to s_i. A path giving this maximum weight is a survivor. The flow $v(a_0, a_8)$ in the truncated trellis of Figure 15.4 is again the weight of any survivor through the trellis as determined by the Soft Decision Viterbi Algorithm of Section 15.2.   ∎

Exercise 848   Verify that the set $\mathcal{A}$ with operations defined in Example 15.3.2 is a semiring with additive and multiplicative identity both equal to 0.   ◆

Exercise 849   Let $\mathcal{A}$ be the set of all polynomials in x with integer coefficients. $\mathcal{A}$ is a semiring under ordinary polynomial addition and multiplication. Consider a truncated trellis of an (n, k) binary convolutional code that is determined by an encoder with memory M used to encode messages of length L, followed by M blocks of k zeros; see Figure 14.4 for an example. The label of an edge is a binary n-tuple, which is the output from the encoder as determined by the state diagram described in Section 14.2. Relabel an edge of the trellis by x^w, where w is the weight of the binary n-tuple originally labeling that edge. For instance in Figure 14.4, the edge from a_0 to c_1 labeled 11 is relabeled x^2, and the edge from a_0 to a_1 labeled 00 is relabeled 1. Thus this new trellis has edges labeled by elements of $\mathcal{A}$.
(a) In the relabeled trellis, describe what the flow along a path represents.
(b) In the relabeled trellis, if the zero state a at time 0 is a_0 and the zero state at time $L + M$ is a_{L+M}, describe what the flow from a_0 to a_{L+M} represents.
(c) Compute the flow from a_0 to a_8 for the relabeled trellis in Figure 14.4.   ◆

Exercise 850 Let $\mathcal{B}$ be the set of all binary strings of any finite length including the empty string ϵ. Let $\mathcal{A}$ be the set of all finite subsets of $\mathcal{B}$ including the empty set $\emptyset$. If u and v are in $\mathcal{A}$, define $u + v$ to be the ordinary set union of u and v. If $u = \{u_1, \ldots, u_p\}$ and $v = \{v_1, \ldots, v_q\}$ where u_i and v_j are in $\mathcal{B}$, define $u \cdot v = \{u_i v_j \mid 1 \le i \le p, \ 1 \le j \le q\}$ where $u_i v_j$ is the string u_i concatenated (or juxtaposed) with the string v_j.
(a) If $u = \{01, 100, 1101\}$ and $v = \{001, 11\}$, what is $u \cdot v$?
(b) Show that $\mathcal{A}$ is a semiring with additive identity the empty set $\emptyset$ and multiplicative identity the single element set $\{\epsilon\}$ consisting of the empty string.
Now consider a truncated trellis of an (n, k) binary convolutional code that is determined by an encoder with memory M used to encode messages of length L, followed by M blocks of k zeros; see Figure 14.4 for an example. The label of an edge is a binary n-tuple, which is the output from the encoder as determined by the state diagram described in Section 14.2. Consider this label to be a single element set from the semiring $\mathcal{A}$.
(c) In this trellis, describe what the flow along a path represents.
(d) In this trellis, if the zero state a at time 0 is a_0 and the zero state at time $L + M$ is a_{L+M}, describe what the flow from a_0 to a_{L+M} represents.
(e) Compute the flow from a_0 to a_8 for the trellis in Figure 14.4.
(f) What is the connection between the answer in part (e) of this exercise and the answer in part (c) of Exercise 849?   ♦

We are now ready to state the generalization of the Viterbi Algorithm. We assume that the states of the trellis are in the set S where the zero state a at time 0 is a_0 and the final state at time $L + M$ is a_{L+M}. The *General Viterbi Algorithm* is as follows:

I. Set $v(a_0, a_0) = 1$.
II. Repeat the following in order for $i = 1, 2, \ldots, L + M$. Compute for each $s \in S$,

$$v(a_0, s_i) = \sum_e v(a_0, *e) \cdot \alpha(e), \tag{15.19}$$

where the summation ranges over all edges e of T such that $e* = s_i$.
We claim that equation (15.19) actually computes the flow as defined by equation (15.18). The two equations clearly agree for $i = 1$ by Step I as 1 is the multiplicative identity of $\mathcal{A}$. Inductively assume they agree for time $i = I$. From equation (15.18),

$$v(a_0, s_{I+1}) = \sum_P v(P),$$

where the summation ranges over all paths P from a_0 to s_{I+1} with $I + 1$ edges. But each such path consists of the first I edges making up a path $P_{s'}$ ending at some state s' at time I along with the final edge e where $*e = s'_I$ and $e* = s_{I+1}$. Thus

$$v(a_0, s_{I+1}) = \sum_{s' \in S} \sum_{P_{s'}} v(P_{s'}) \cdot \alpha(e) = \sum_e \left[\sum_{P_{s'}} v(P_{s'}) \right] \cdot \alpha(e),$$

where the last equality follows from the distributive property in $\mathcal{A}$. By induction the inner sum is $v(a_0, s'_I) = v(a_0, *e)$ and so $v(a_0, s_{I+1})$ is indeed the value computed in (15.19). Thus the General Viterbi Algorithm correctly computes the flow from a_0 to any vertex in the trellis. In particular, the flow $v(a_0, a_{L+M})$ is correctly computed by this algorithm.

15.4   Two-way APP decoding

In this section we present a soft decision decoding algorithm for binary convolutional codes that computes, at each instance of time, the probability that a message symbol is 0 based on knowledge of the received vector and the channel probabilities. This information can be used in two ways. First, knowing these probabilities, the decoder can decide whether or not the message symbol at any time is 0 or 1. Second, the decoder can pass these probabilities on for another decoder to use. From these probabilities, eventually hard decisions are made on the message symbols. As we present the algorithm, called Two-Way a Posteriori Probability (APP) Decoding, we will model it using a binary input, 8-ary output DMC subject to AWGN.

Adopting the notation of Section 15.2, a message $\mathbf{x}(i) = (\mathbf{x}_1(i), \ldots, \mathbf{x}_k(i))$ with $i = 0, 1, \ldots, L-1$ is encoded starting in the zero state a_0 using the generator matrix G of an (n, k) binary convolutional code to produce a codeword $\mathbf{c}$ where $\mathbf{c}(i) = (\mathbf{c}_1(i), \ldots, \mathbf{c}_n(i))$ for $i = 0, 1, \ldots, L+M-1$. Add M blocks of k zeros to the end of the message so that the encoder will also terminate in the zero state a_{L+M} at time $L + M$. After interleaving, modulating, transmitting, demodulating, and deinterleaving, we have the received vector $\mathbf{y}$, where $\mathbf{y}(i) = (\mathbf{y}_1(i), \ldots, \mathbf{y}_n(i))$ for $i = 0, 1, \ldots, L + M - 1$ with $\mathbf{y}_j(i) \in \{0_1, \ldots, 1_4\}$. We assume that, in addition to the truncated code trellis and the received vector, the decoder knows the channel statistics prob($y \mid c$), the probability that the bit y is received given that the bit c is transmitted.

The object of APP decoding is to compute the *a posteriori probability* prob($\mathbf{x}_j(i) = 0 \mid \mathbf{y}$). The decoder can either pass these probabilities on to another decoder or make hard decisions $\widehat{\mathbf{x}}_j(i)$ that estimate the message symbol $\mathbf{x}_j(i)$ according to

$$\widehat{\mathbf{x}}_j(i) = \begin{cases} 0 & \text{if prob}(\mathbf{x}_j(i) = 0 \mid \mathbf{y}) \geq \frac{1}{2}, \\ 1 & \text{otherwise.} \end{cases} \tag{15.20}$$

The computation of prob($\mathbf{x}_j(i) = 0 \mid \mathbf{y}$) relies on the equation

$$\text{prob}(\mathbf{x}_j(i) = 0 \mid \mathbf{y}) = \frac{\text{prob}(\mathbf{y} \text{ and } \mathbf{x}_j(i) = 0)}{\text{prob}(\mathbf{y})}. \tag{15.21}$$

Using the General Viterbi Algorithm described in Section 15.3 with the semiring $\mathcal{A}$ consisting of all nonnegative real numbers under ordinary addition and multiplication, the numerator and denominator of (15.21) can be determined. This makes up the Two-Way APP Decoding Algorithm.

As we describe APP decoding, we will reexamine Example 15.2.1 that was used to illustrate soft Viterbi decoding in Section 15.2; this example was initially studied in Example 14.2.5 to demonstrate hard Viterbi decoding. We will need the truncated trellis used in those examples. This trellis is found in both Figures 14.4 and 15.4; we recommend the reader refer to these figures for helpful insight into APP decoding.

To compute the numerator and denominator of (15.21), we will relabel each edge e of the truncated trellis using a label from $\mathcal{A}$. We will still need to refer to the original label $\mathbf{c}(i) = (\mathbf{c}_1(i), \ldots, \mathbf{c}_n(i))$ on the edge; we will call the original label the "output" of the edge. If the edge e starts at s_i and ends at s'_{i+1}, let $\alpha(e) = \text{prob}(\mathbf{y}(i) \text{ and } \mathbf{m}(i))$ be the new label on the edge. This label can be computed using the probability formula

$$\text{prob}(\mathbf{y}(i) \text{ and } \mathbf{m}(i)) = \text{prob}(\mathbf{y}(i) \mid \mathbf{m}(i))\text{prob}(\mathbf{m}(i)). \tag{15.22}$$

We illustrate this computation of the edge labels.

Example 15.4.1 Consider the code and the message of Example 15.2.1. The encoding uses a (2, 1) convolutional code originally considered in Example 14.1.1. The relevant trellis can be found in Figures 14.4 and 15.4. The set of states is $\{a, b, c, d\}$ and the received vector is

$$\mathbf{y} = 1_1 1_2\, 0_4 1_1\, 1_3 1_4\, 0_4 1_1\, 1_4 0_4\, 0_3 1_3\, 0_4 1_2\, 1_4 1_3.$$

We assume that the channel statistics $\text{prob}(y \mid c)$ are still given by the table

$c \setminus y$	0_4	0_3	0_2	0_1	1_1	1_2	1_3	1_4
0	0.368	0.207	0.169	0.097	0.065	0.051	0.028	0.015
1	0.015	0.028	0.051	0.065	0.097	0.169	0.207	0.368

as in Example 15.2.1. The message consists of six unknown bits followed by two 0s. We also assume that

$$\text{prob}(\mathbf{m}(i) = 0) = \begin{cases} 0.6 & \text{if } i = 0, 1, 2, 3, 4, 5, \\ 1 & \text{if } i = 6, 7, \end{cases} \quad \text{and}$$

$$\text{prob}(\mathbf{m}(i) = 1) = \begin{cases} 0.4 & \text{if } i = 0, 1, 2, 3, 4, 5, \\ 0 & \text{if } i = 6, 7, \end{cases}$$

where $\text{prob}(\mathbf{m}(i) = 0)$, respectively $\text{prob}(\mathbf{m}(i) = 1)$, is the probability that the ith message bit is 0, respectively 1. The values for these probabilities when $i = 6$ and 7 are clear because we know that the seventh and eighth message bits are 0. Normally we would expect the probabilities between times $i = 0$ and 5 to all equal 0.5. However, APP decoding can be applied to iterative decoding in which the probabilities $\text{prob}(\mathbf{m}(i) = 0)$ and $\text{prob}(\mathbf{m}(i) = 1)$ come from some other decoder and hence could be almost anything. Simply to illustrate this, we have chosen the values to be different from 0.5.

Consider the edge e from state c to state b between time $i = 1$ and $i = 2$. The input bit for this edge is 0, the edge output (original label) is 10, and the portion of the received vector between $i = 1$ and $i = 2$ is $0_4 1_1$. Thus by (15.22) and the probabilities in the previous paragraph,

$$\alpha(e) = \text{prob}(0_4 1_1 \mid 10)\text{prob}(\mathbf{m}(1) = 0)$$
$$= \text{prob}(0_4 \mid 1)\text{prob}(1_1 \mid 0)\text{prob}(\mathbf{m}(1) = 0)$$
$$= 0.015 \times 0.065 \times 0.6 = 5.850 \times 10^{-4}.$$

Similarly, if e is the edge between c_2 and d_3,

$$\alpha(e) = \text{prob}(1_3 1_4 \mid 01)\text{prob}(\mathbf{m}(2) = 1)$$
$$= \text{prob}(1_3 \mid 0)\text{prob}(1_4 \mid 1)\text{prob}(\mathbf{m}(2) = 1)$$
$$= 0.028 \times 0.368 \times 0.4 = 4.122 \times 10^{-3}.$$

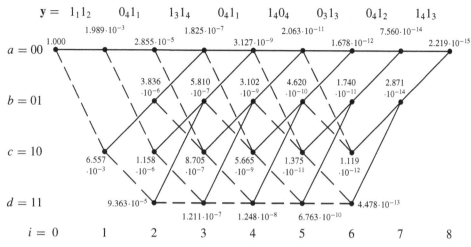

Figure 15.6 Flow $v(a_0, s_i)$ for Example 15.4.1.

If e is the edge between d_6 and b_7,

$$\alpha(e) = \text{prob}(0_4 1_2 \mid 01)\text{prob}(\mathbf{m}(6) = 0)$$
$$= \text{prob}(0_4 \mid 0)\text{prob}(1_2 \mid 1)\text{prob}(\mathbf{m}(6) = 0)$$
$$= 0.368 \times 0.169 \times 1 = 6.219 \times 10^{-2}.$$

The values of the new edge labels are given in Table 15.1. In the table, the label of the edge from s_i to s'_{i+1} is denoted $\alpha(s_i, s'_{i+1})$.   ■

Exercise 851 Verify the edge labels in Table 15.1.   ◆

With the General Viterbi Algorithm, it is amazingly simple to compute the numerator and the denominator of (15.21). To begin the process, first apply the General Viterbi Algorithm to the trellis with the new edge labels. The algorithm computes the flow $v(a_0, s_i)$ for all states s and $0 \le i \le L + M$. For the code and received vector considered in Example 15.4.1, the flow $v(a_0, s_i)$ is given in Figure 15.6; here the value of the flow is placed near the node s_i. Next, apply the General Viterbi Algorithm backwards. That is, find the flow from a_{L+M} to s_i beginning with $i = L + M$ and going down to $i = 0$; call this flow $v_b(a_{L+M}, s_i)$. Notice that $v(s_i, a_{L+M}) = v_b(a_{L+M}, s_i)$. For the code and received vector considered in Example 15.4.1, the backward flow $v_b(a_{L+M}, s_i)$ is given in Figure 15.7; again the value of the flow is placed near the node s_i.

Exercise 852 Verify the values of $v(a_0, s_i)$ in Figure 15.6 using the General Viterbi Algorithm.   ◆

Exercise 853 Verify the values of $v_b(a_{L+M}, s_i)$ in Figure 15.7 using the backwards version of the General Viterbi Algorithm.   ◆

To compute the denominator of (15.21), let $\mathcal{M}$ be the set of all possible message sequences $\mathbf{m} = (\mathbf{m}_1, \ldots, \mathbf{m}_k)$ where $\mathbf{m}(i) = (\mathbf{m}_1(i), \ldots, \mathbf{m}_k(i))$ is arbitrary for $0 \le i \le L - 1$ but is

Table 15.1 *Edge labels for Example 15.4.1*

	$i = 0$	$i = 1$	$i = 2$	$i = 3$	$i = 4$	$i = 5$	$i = 6$	$i = 7$
$\alpha(a_i, a_{i+1})$	$1.989 \cdot 10^{-3}$	$1.435 \cdot 10^{-2}$	$2.520 \cdot 10^{-4}$	$1.435 \cdot 10^{-2}$	$3.312 \cdot 10^{-3}$	$3.478 \cdot 10^{-3}$	$1.877 \cdot 10^{-2}$	$4.200 \cdot 10^{-4}$
$\alpha(a_i, c_{i+1})$	$6.557 \cdot 10^{-3}$	$5.820 \cdot 10^{-4}$	$3.047 \cdot 10^{-2}$	$5.820 \cdot 10^{-4}$	$2.208 \cdot 10^{-3}$	$2.318 \cdot 10^{-3}$		
$\alpha(b_i, a_{i+1})$			$4.571 \cdot 10^{-2}$	$8.730 \cdot 10^{-4}$	$3.312 \cdot 10^{-3}$	$3.478 \cdot 10^{-3}$	$2.535 \cdot 10^{-3}$	$7.618 \cdot 10^{-2}$
$\alpha(b_i, c_{i+1})$			$1.680 \cdot 10^{-4}$	$9.568 \cdot 10^{-3}$	$2.208 \cdot 10^{-3}$	$2.318 \cdot 10^{-3}$		
$\alpha(c_i, b_{i+1})$		$5.850 \cdot 10^{-4}$	$1.863 \cdot 10^{-3}$	$5.850 \cdot 10^{-4}$	$8.125 \cdot 10^{-2}$	$4.704 \cdot 10^{-4}$	$7.650 \cdot 10^{-4}$	
$\alpha(c_i, d_{i+1})$		$1.428 \cdot 10^{-2}$	$4.122 \cdot 10^{-3}$	$1.428 \cdot 10^{-2}$	$9.000 \cdot 10^{-5}$	$1.714 \cdot 10^{-2}$		
$\alpha(d_i, b_{i+1})$			$6.182 \cdot 10^{-3}$	$2.142 \cdot 10^{-2}$	$1.350 \cdot 10^{-4}$	$2.571 \cdot 10^{-2}$	$6.219 \cdot 10^{-2}$	
$\alpha(d_i, d_{i+1})$			$1.242 \cdot 10^{-3}$	$3.900 \cdot 10^{-4}$	$5.417 \cdot 10^{-2}$	$3.136 \cdot 10^{-4}$		

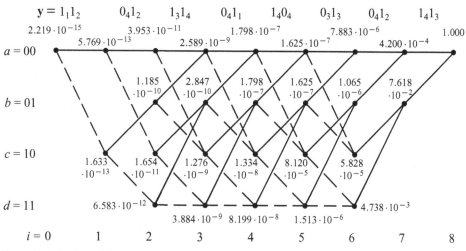

Figure 15.7 Backward flow $v_b(a_{L+M}, s_i)$ for Example 15.4.1.

a block of k zeros for $L \le i \le L + M - 1$. Thus each message corresponds to a single path through the truncated trellis. Notice that the denominator satisfies

$$\text{prob}(\mathbf{y}) = \sum_{\mathbf{m} \in \mathcal{M}} \text{prob}(\mathbf{y} \text{ and } \mathbf{m})$$

$$= \sum_{\mathbf{m} \in \mathcal{M}} \prod_{i=0}^{L+M-1} \text{prob}(\mathbf{y}(i) \text{ and } \mathbf{m}(i)).$$

The latter is precisely the flow $v(a_0, a_{L+M})$ through the trellis computed by the General Viterbi Algorithm using the semiring $\mathcal{A}$ of nonnegative real numbers under ordinary addition and multiplication.

To compute the numerator of (15.21), let $\mathcal{M}_j(i)$ denote the subset of $\mathcal{M}$ where $\mathbf{m}_j(i) = 0$. Each message in $\mathcal{M}_j(i)$ corresponds to a single path through the truncated trellis where, in going from time i to time $i + 1$, the edge chosen must correspond to one obtained only when the jth message bit at time i is 0. As an illustration, consider Figure 14.4. Since $k = 1$ for the code of this figure, we only need to consider $\mathcal{M}_1(i)$; this set of messages corresponds to all paths through the trellis that go along a solid edge between time i and time $i + 1$ as solid edges correspond to input 0. The numerator of (15.21) satisfies

$$\text{prob}(\mathbf{y} \text{ and } \mathbf{x}_j(i) = 0) = \sum_{\mathbf{m} \in \mathcal{M}_j(i)} \text{prob}(\mathbf{y} \text{ and } \mathbf{m})$$

$$= \sum_{\mathbf{m} \in \mathcal{M}_j(i)} \prod_{i=0}^{L+M-1} \text{prob}(\mathbf{y}(i) \text{ and } \mathbf{m}(i)).$$

Let $E(i, j)$ be all the edges from time i to time $i + 1$ that correspond to one obtained when the jth message bit at time i is 0. Then

$$\text{prob}(\mathbf{y} \text{ and } \mathbf{x}_j(i) = 0) = \sum_{e \in E(i,j)} v(a_0, *e)\alpha(*e, e*)v(e*, a_{L+M}),$$

where $\alpha(*e, e*)$ is the edge label $\alpha(e)$ of e. But $v(e*, a_{L+M}) = v_b(a_{L+M}, e*)$ and so

$$\text{prob}(\mathbf{y} \text{ and } \mathbf{x}_j(i) = 0) = \sum_{e \in E(i,j)} v(a_0, *e)\alpha(*e, e*)v_b(a_{L+M}, e*). \tag{15.23}$$

We now formally state the *Two-Way APP Decoding Algorithm* using an (n, k) binary convolutional code with received vector $\mathbf{y}$:

I.   Construct the truncated trellis for the code. Let a be the zero state.
II. Compute the edge labels of the truncated trellis using (15.22).
III. For $i = 0, 1, \ldots, L + M$, recursively compute the flow $v(a_0, s_i)$ for all states s using the General Viterbi Algorithm.
IV. For $i = L + M, L + M - 1, \ldots, 0$, recursively compute the backward flow $v_b(a_{L+M}, s_i)$ for all states s using the backward version of the General Viterbi Algorithm.
V.   For $0 \le i \le L + M - 1$ and $1 \le j \le k$, compute

$$\gamma_j(i) = \text{prob}(\mathbf{y} \text{ and } \mathbf{x}_j(i) = 0)$$

using (15.23).
VI. For $0 \le i \le L + M - 1$ and $1 \le j \le k$, compute

$$\text{prob}(\mathbf{x}_j(i) = 0 \mid \mathbf{y}) = \frac{\gamma_j(i)}{v(a_0, a_{L+M})}.$$

VII. For $0 \le i \le L + M - 1$ and $1 \le j \le k$, give an estimate $\widehat{\mathbf{x}}_j(i)$ of the message symbol $\mathbf{x}_j(i)$ according to (15.20).

Example 15.4.2 Continuing with Example 15.4.1, Steps I–IV of the Two-Way APP Decoding Algorithm have been completed in Table 15.1 and in Figures 15.6 and 15.7. The values for $\gamma_1(i)$ when $0 \le i \le 7$ computed in Step V are as follows. By (15.23),

$$\gamma_1(0) = v(a_0, a_0)\alpha(a_0, a_1)v_b(a_8, a_1) = 1.148 \times 10^{-15}$$

and

$$\gamma_1(1) = v(a_0, a_1)\alpha(a_1, a_2)v_b(a_8, a_2) + v(a_0, c_1)\alpha(c_1, b_2)v_b(a_8, b_2) = 1.583 \times 10^{-15}.$$

For $2 \le i \le 6$,

$$\gamma_1(i) = v(a_0, a_i)\alpha(a_i, a_{i+1})v_b(a_8, a_{i+1}) + v(a_0, b_i)\alpha(b_i, a_{i+1})v_b(a_8, a_{i+1})$$
$$+ v(a_0, c_i)\alpha(c_i, b_{i+1})v_b(a_8, b_{i+1}) + v(a_0, d_i)\alpha(d_i, b_{i+1})v_b(a_8, b_{i+1}),$$

yielding

$$\gamma_1(2) = 6.379 \times 10^{-16}, \quad \gamma_1(3) = 1.120 \times 10^{-15}, \quad \gamma_1(4) = 7.844 \times 10^{-17},$$
$$\gamma_1(5) = 3.175 \times 10^{-17}, \quad \text{and} \quad \gamma_1(6) = 2.219 \times 10^{-15}.$$

Also,

$$\gamma_1(7) = v(a_0, a_7)\alpha(a_7, a_8)v_b(a_8, a_8) + v(a_0, b_7)\alpha(b_7, a_8)v_b(a_8, a_8) = 2.219 \times 10^{-15}.$$

Since $\text{prob}(\mathbf{x}(i) = 0 \mid \mathbf{y}) = \gamma_1(i)/\nu(a_0, a_8)$, Step VI yields:

$\text{prob}(\mathbf{x}(0) = 0 \mid \mathbf{y}) = 0.5172,$ $\text{prob}(\mathbf{x}(1) = 0 \mid \mathbf{y}) = 0.7135,$

$\text{prob}(\mathbf{x}(2) = 0 \mid \mathbf{y}) = 0.2875,$ $\text{prob}(\mathbf{x}(3) = 0 \mid \mathbf{y}) = 0.5049,$

$\text{prob}(\mathbf{x}(4) = 0 \mid \mathbf{y}) = 0.0354,$ $\text{prob}(\mathbf{x}(5) = 0 \mid \mathbf{y}) = 0.0143,$ and

$\text{prob}(\mathbf{x}(6) = 0 \mid \mathbf{y}) = 1.0000,$ $\text{prob}(\mathbf{x}(7) = 0 \mid \mathbf{y}) = 1.0000.$

By (15.20), the APP decoder generates the message 00101100 in Step VII and therefore the codeword 0000111000010111. This message disagrees in component $\mathbf{x}(3)$ with the decoded message found using the Soft Decision Viterbi Algorithm in Example 15.2.1. Notice that the value of this bit $\mathbf{x}(3)$ (and also $\mathbf{x}(0)$) is highly uncertain by the above probability. Notice also that $\text{prob}(\mathbf{x}(6) = 0 \mid \mathbf{y}) = \text{prob}(\mathbf{x}(7) = 0 \mid \mathbf{y}) = 1.0000$, just as expected.   ∎

Exercise 854 Repeat Examples 15.4.1 and 15.4.2 using the same values for $\text{prob}(y \mid c)$, but

$$\text{prob}(\mathbf{m}(i) = 0) = \begin{cases} 0.5 & \text{if } i = 0, 1, 2, 3, 4, 5, \\ 1 & \text{if } i = 6, 7, \end{cases} \quad \text{and}$$

$$\text{prob}(\mathbf{m}(i) = 1) = \begin{cases} 0.5 & \text{if } i = 0, 1, 2, 3, 4, 5, \\ 0 & \text{if } i = 6, 7, \end{cases}$$

in place of the values of $\text{prob}(\mathbf{m}(i) = 0)$ and $\text{prob}(\mathbf{m}(i) = 1)$ of those examples.   ◆

Exercise 855 Use the Two-Way APP Decoding Algorithm to decode the following received vectors encoded by the code of Example 15.4.1 together with the probabilities of that example. Compare your results with those in Exercises 817 and 845.

(a) $1_4 1_3 0_3 0_2 0_4 1_3 1_2 1_3 1_4 1_2 1_3 1_1 1_2 1_4 0_3 1_4,$

(b) $0_3 1_4 1_3 0_2 1_4 1_1 0_3 1_4 1_3 0_2 0_4 1_4 1_3 0_2 0_3 1_2.$ ◆

There are other APP decoders for various types of codes; see [158, Chapter 7]. Also there are a number of connections between trellises and block codes; [336] gives an excellent survey. The February 2001 issue of *IEEE Transactions on Information Theory* is devoted to relationships among codes, graphs, and algorithms. The survey article [180] in that issue gives a nice introduction to these connections through a number of examples.

15.5 Message passing decoding

In this section we introduce the concept of message passing decoding; such decoding is used in several contexts, including some versions of turbo decoding. Message passing is our first example of iterative decoding. In the next section, we present two other types of iterative decoding applied to low density parity check codes.

To describe message passing it is helpful to begin with a bipartite graph, called a *Tanner graph*[5] [325], constructed from an $r \times n$ parity check matrix H for a binary code $\mathcal{C}$ of length

---

[5] Tanner graphs are actually more general than the definition we present. These graphs themselves were generalized by Wiberg, Loeliger, and Kötter [350].

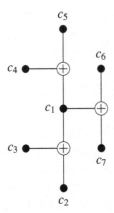

Figure 15.8 Tanner graph of the code in Example 15.5.1.

n. The Tanner graph has two types of vertices, called variable and check nodes. There are n *variable nodes*, one corresponding to each coordinate or column of H. There are r *check nodes*, one for each parity check equation or row of H. The Tanner graph has only edges between variable nodes and check nodes; a given check node is connected to precisely those variable nodes where there is 1 in the corresponding column of H.

Example 15.5.1 Let C be the $[7, 4, 2]$ binary code with parity check matrix

$$H = \begin{bmatrix} 1 & 1 & 1 & 0 & 0 & 0 & 0 \\ 1 & 0 & 0 & 1 & 1 & 0 & 0 \\ 1 & 0 & 0 & 0 & 0 & 1 & 1 \end{bmatrix}.$$

Figure 15.8 gives the Tanner graph for this code. The solid vertices are the variable nodes, while the vertices marked $\oplus$ represent the check nodes. ∎

Message passing sends information from variable to check nodes and from check to variable nodes along edges of the Tanner graph at discrete points of time. In the situation we examine later, the messages passed will represent pairs of probabilities. Initially every variable node has a message assigned to it; for example, these messages can be probabilities coming directly from the received vector. At time 1, some or all variable nodes send the message assigned to them to all attached check nodes via the edges of the graph. At time 2, some of those check nodes that received a message process the message and send along a message to some or all variable nodes attached to them. These two transmissions of messages make up one iteration. This process continues through several iterations. In the processing stage there is an important rule that must be followed. A message sent from a node along an adjacent edge must not depend on a message previously received along that edge.

We will illustrate message passing on a Tanner graph using an algorithm described in [91]. It has elements reminiscent of the Soft Decision Viterbi Algorithm presented in Section 15.2. First, we assume that T is the Tanner graph of a length n binary code C obtained from a parity check matrix H for C. We further assume that T has no cycles.[6] As in Section 15.2, for concreteness we assume that the demodulation scheme used for the received vector is 8-level quantization. Therefore suppose that $\mathbf{c} = c_1 c_2 \cdots c_n$ is the transmitted codeword, and $\mathbf{y} = y_1 y_2 \cdots y_n$ is received with $y_i \in \{0_1, \ldots, 1_4\}$. As with the Soft Decision Viterbi Algorithm, our goal is to perform maximum likelihood decoding and find the codeword $\mathbf{c}$ that maximizes $\text{prob}(\mathbf{y} \mid \mathbf{c})$. We assume the channel is memoryless and hence

$$\text{prob}(\mathbf{y} \mid \mathbf{c}) = \prod_{i=1}^{n} \text{prob}(y_i \mid c_i).$$

As in Section 15.2, take the logarithm turning the product into a sum and then rescale and shift the probabilities so that the value to maximize is

$$\sum_{i=1}^{n} \mu(y_i, c_i),$$

where the maximum is over all codewords $\mathbf{c} = c_1 c_2 \cdots c_n$, analogous to (15.15).

The Tanner graph T has no cycles and hence is a tree or forest. For simplicity, assume T is a tree; if T is a forest, carry out the algorithm on each tree in T separately. All the leaves are variable nodes provided no row of H has weight 1, which we also assume. At the end of the algorithm, messages will have been passed along all edges once in each direction. Each message is a pair of nonnegative integers. The first element in the pair is associated with the bit 0, and the second element in the pair is associated with the bit 1. Together the pair will represent relative probabilities of the occurrence of bit 0 or bit 1. The *Message Passing Decoding Algorithm* is as follows:

I.   To initialize, assign the pair $(\mu(y_i, 0), \mu(y_i, 1))$ to the ith variable node. No assignment is made to the check nodes. Also for every leaf, the same pair is to be the message, leaving that leaf node directed toward the attached check node.

II. To find a message along an edge e leaving a check node h, assume all messages directed toward h, except possibly along e, have been determined. Suppose the messages entering h except along e are

$$(a_{i_1}, b_{i_1}), (a_{i_2}, b_{i_2}), \ldots, (a_{i_t}, b_{i_t}).$$

Consider all sums

$$\sum_{j=1}^{t} x_{i_j}, \tag{15.24}$$

where x_{i_j} is one of a_{i_j} or b_{i_j}. The message along e leaving h is (m_0, m_1), where m_0 is the maximum of all such sums with an even number of b_{i_j}s, and m_1 is the maximum of all such sums with an odd number of b_{i_j}s.

III. To find a message along an edge e leaving a variable node v, assume all messages directed toward v, except possibly along e, have been determined. The message leaving

---

[6] Unfortunately, if the Tanner graph has no cycles, the code has low minimum weight. For example, such codes of rate 0.5 or greater have minimum weight 2 or less; see [79].

v along e toward a check node is assigned the pair at v (from Step I) added to the sum of all the incoming messages to v except along e.

IV. Once all messages in both directions along every edge have been determined by repetition of II and III, a new pair (p_0, p_1) is computed for every variable node v. This pair is the sum of the two messages along any edge attached to v.

V. Decode the received vector as follows. For the ith variable node, let $c_i = 0$ if $p_0 > p_1$ and $c_i = 1$ if $p_1 > p_0$.

The rationale for the message passed in II is that the summation in (15.24) is associated with bit 0, respectively bit 1, if there are an even, respectively odd, number of $b_{i,j}$s involved, as these sums correspond to adding an even, respectively odd, number of "probabilities" associated to 1s. We do not prove the validity of the algorithm; more discussion of the algorithm is found in [91] and the references therein. Note again how closely related this algorithm is to the Soft Decision Viterbi Algorithm.

Example 15.5.2 Suppose that $\mathbf{y} = 1_2 0_3 0_1 1_3 0_2 1_1 0_4$ is the received vector when a codeword is transmitted using the $[7, 4, 2]$ code $\mathcal{C}$ of Example 15.5.1. Even though this code has minimum distance 2, implying the code cannot correct any errors using nearest neighbor decoding, we can still decode using soft decision decoding to find the maximum likely codeword. We will use the Tanner graph for $\mathcal{C}$ in Figure 15.8 determined by the parity check matrix H given in Example 15.5.1. We will also use the values of $\mu(y, c)$ given in Example 15.2.1. The messages obtained using the Message Passing Decoding Algorithm are given in Figure 15.9. Step I of the algorithm is carried out at time 1 using the table of values for $\mu(y, c)$ in Example 15.2.1 applied to the received vector. At time 2, we pass messages from the three check nodes toward y_1 using II. For example, the message $(5, 8)$ directed toward y_1 is determined from the messages $(0, 5)$ and $(3, 0)$ entering the check node from y_4 and y_5 as follows:

$$(5, 8) = (\max\{0 + 3, 5 + 0\}, \max\{0 + 0, 5 + 3\}).$$

Similarly, the message $(6, 5)$ directed toward y_1 is

$$(6, 5) = (\max\{1 + 5, 0 + 0\}, \max\{0 + 5, 1 + 0\}).$$

After time 2, every edge has one message along it in some direction. At time 3, messages are sent along the three edges leaving y_1 using III. For example, the message $(14, 17)$ is the sum of the node pair $(0, 3)$ plus the sum $(8, 9) + (6, 5)$. At time 4, messages are sent to all leaf nodes using II. For instance, the message $(17, 20)$ sent to y_4 uses the edges $(3, 0)$ and $(14, 17)$ yielding

$$(17, 20) = (\max\{3 + 14, 0 + 17\}, \max\{3 + 17, 0 + 14\}).$$

After time 4, all edges have messages in each direction. By adding the edges attached to each variable node, Step IV gives:

	y_1	y_2	y_3	y_4	y_5	y_6	y_7
(p_0, p_1)	$(19, 25)$	$(25, 21)$	$(21, 25)$	$(17, 25)$	$(25, 19)$	$(19, 25)$	$(25, 16)$

Note that the pair for y_1 is obtained using any of the three edges attached to y_1,

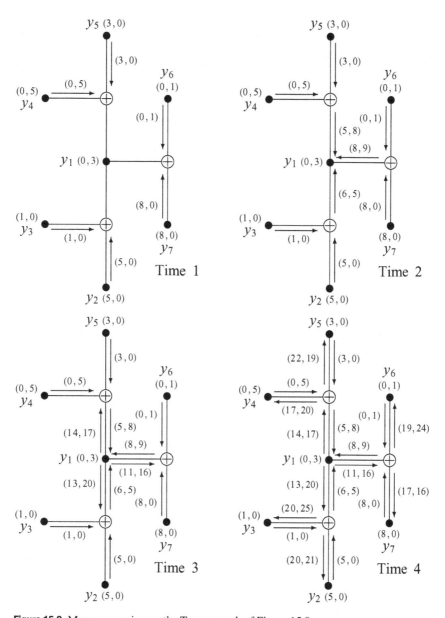

Figure 15.9 Message passing on the Tanner graph of Figure 15.8.

namely, $(19, 25) = (14, 17) + (5, 8) = (11, 16) + (8, 9) = (13, 20) + (6, 5)$. We decode $\mathbf{y}$ as $\mathbf{c} = 1011010$ from Step V. Notice that for our received vector $\mathbf{y}$ and this codeword $\mathbf{c}$, $\sum_{i=1}^{7} \mu(y_i, c_i) = 3 + 5 + 0 + 5 + 3 + 1 + 8 = 25$, which is the largest component of each of the seven (p_0, p_1)s.   ∎

Exercise 856 Verify the computations in Example 15.5.2.   ♦

Exercise 857 Let C be the code in Example 15.5.1 with Tanner graph in Figure 15.8. Use the Message Passing Decoding Algorithm to decode the following received vectors; use the values of $\mu(y, c)$ given in Example 15.2.1.

(a) $\mathbf{y} = 0_2 1_3 0_1 0_4 1_2 0_1 1_3$,

(b) $\mathbf{y} = 1_3 0_3 0_2 1_1 0_2 0_4 1_2$. ◆

Exercise 858 Construct the Tanner graph for the $[10, 6, 2]$ binary code with parity check matrix

$$H = \begin{bmatrix} 1 & 1 & 1 & 1 & 0 & 0 & 0 & 0 & 0 & 0 \\ 1 & 0 & 0 & 0 & 1 & 1 & 0 & 0 & 0 & 0 \\ 0 & 0 & 0 & 0 & 1 & 0 & 1 & 1 & 0 & 0 \\ 0 & 1 & 0 & 0 & 0 & 0 & 0 & 0 & 1 & 1 \end{bmatrix}.$$

Then use the Message Passing Decoding Algorithm to decode

$$\mathbf{y} = 1_1 0_2 1_3 0_1 1_1 1_2 1_3 1_4 0_3 1_1.$$

Use the values of $\mu(y, c)$ given in Example 15.2.1. ◆

15.6 Low density parity check codes

In the previous section we introduced our first example of iterative decoding through message passing. In the mid 1990s, iterative decoding became a significant technique for the decoding of turbo codes. Iterative decoding was introduced by Elias in 1954 [77]. Iterative decoding reappeared with the work of R. G. Gallager on low density parity check codes in 1963 but since then seems to have been largely neglected, with only a few exceptions, until researchers realized that the decoding of turbo codes was closely related to Gallager's techniques. In this section we will introduce low density parity check codes and examine Gallager's two versions of iterative decoding. Our presentation relies heavily on [94, 95, 211]. For the interested reader, the use of message passing in low density parity check codes is described in [295].

In short, a binary low density parity check code is a code having a parity check matrix with relatively few 1s which are dispersed with a certain regularity in the matrix. An $[n, k]$ binary linear code C is a *low density parity check (LDPC) code* provided it has an $m \times n$ parity check matrix H where every column has fixed weight c and every row of H has fixed weight r.[7] This type of code will be denoted an (n, c, r) LDPC code. Note that such a matrix may not have independent rows and so is technically not a parity check matrix in the sense we have used in this text. However, H can be row reduced and the zero rows removed to form a parity check matrix for C with independent rows. Thus the dimension of C is at least $n - m$. Counting the number of 1s in H by both rows and columns gives the relationship $nc = mr$ among the parameters. Generally, we want r and c to be relatively small compared to n and thus the density of 1s in H is low.

---

[7] A natural generalization for LDPC codes is to allow the row weights and column weights of H to vary, in some controlled manner. The resulting codes are sometimes termed "irregular" LDPC codes.

$$
H = \begin{bmatrix}
1 & 1 & 1 & 1 &   &   &   &   &   &   &   &   &   &   &   &   \\
  &   &   &   & 1 & 1 & 1 & 1 &   &   &   &   &   &   &   &   \\
  &   &   &   &   &   &   &   & 1 & 1 & 1 & 1 &   &   &   &   \\
  &   &   &   &   &   &   &   &   &   &   &   & 1 & 1 & 1 & 1 \\
1 &   &   &   & 1 &   &   &   & 1 &   &   &   & 1 &   &   &   \\
  & 1 &   &   &   & 1 &   &   &   & 1 &   &   &   & 1 &   &   \\
  &   & 1 &   &   &   & 1 &   &   &   & 1 &   &   &   & 1 &   \\
  &   &   & 1 &   &   &   & 1 &   &   &   & 1 &   &   &   & 1 \\
1 &   &   &   &   & 1 &   &   &   &   & 1 &   &   &   &   & 1 \\
  & 1 &   &   &   &   & 1 &   &   &   &   & 1 & 1 &   &   &   \\
  &   & 1 &   &   &   &   & 1 & 1 &   &   &   &   & 1 &   &   \\
  &   &   & 1 & 1 &   &   &   &   & 1 &   &   &   &   & 1 &   \\
\end{bmatrix}
$$

Figure 15.10 Parity check matrix for a $(16, 3, 4)$ LDPC code.

Figure 15.10 gives the parity check matrix of a $(16, 3, 4)$ code, which is a $[16, 6, 6]$ binary code.

Exercise 859 Show that the binary code C with parity check matrix given in Figure 15.10 has dimension 6 and minimum weight 6. Hint: Row reduce H to obtain the dimension. For the minimum weight, show that C is even and that every codeword has an even number of 1s in each of the four blocks of four coordinates, as presented in the figure. Then use Corollary 1.4.14 to show that C has no weight 4 codewords. ♦

Gallager developed two iterative decoding algorithms designed to decode LDPC codes of long length, several thousand bits for example. The first iterative decoding algorithm is a simple hard decision algorithm. Assume that the codeword $\mathbf{c}$ is transmitted using an (n, c, r) binary LDPC code C, and the vector $\mathbf{y}$ is received. In the computation of the syndrome $H\mathbf{y}^{\mathsf{T}}$, each received bit y_i affects at most c components of that syndrome, as the ith bit is in c parity checks. If among all the bits S involved in these c parity checks only the ith is in error, then these c components of $H\mathbf{y}^{\mathsf{T}}$ will equal 1 indicating the parity check equations are not satisfied. Even if there are some other errors among these bits S, one expects that several of the c components of $H\mathbf{y}^{\mathsf{T}}$ will equal 1. This is the basis of the *Gallager Hard Decision Decoding Algorithm*.

I.   Compute $H\mathbf{y}^{\mathsf{T}}$ and determine the unsatisfied parity checks, that is, the parity checks where the components of $H\mathbf{y}^{\mathsf{T}}$ equal 1.

II. For each of the n bits, compute the number of unsatisfied parity checks involving that bit.

III. Change the bits of $\mathbf{y}$ that are involved in the largest number of unsatisfied parity checks; call the resulting vector $\mathbf{y}$ again.

IV. Iteratively repeat I, II, and III until either $H\mathbf{y}^{\mathsf{T}} = \mathbf{0}$, in which case the received vector is decoded as this latest $\mathbf{y}$, or until a certain number of iterations is reached, in which case the received vector is not decoded.

Example 15.6.1 Let $\mathcal{C}$ be the $(16, 3, 4)$ LDPC code with parity check matrix H of Figure 15.10. Suppose that

$$\mathbf{y} = 0100101000100000$$

is received. Then $H\mathbf{y}^{\mathsf{T}} = 101011001001^{\mathsf{T}}$. There are six unsatisfied parity checks, corresponding to rows 1, 3, 5, 6, 9, and 12 of H. Bits 1 and 10 are each involved in three of these; all other bits are involved in two or fewer. Thus change bits 1 and 10 of $\mathbf{y}$ to obtain a new

$$\mathbf{y} = 1100101001100000.$$

Iterating, we see that $H\mathbf{y}^{\mathsf{T}} = \mathbf{0}$ and so this latest $\mathbf{y}$ is declared the transmitted codeword; since it has distance 2 from the received vector, we have accomplished nearest neighbor decoding.

Now suppose that

$$\mathbf{y} = 1100101000000000$$

is received. Then $H\mathbf{y}^{\mathsf{T}} = 000001101001^{\mathsf{T}}$. Parity checks 6, 7, 9, and 12 are unsatisfied; bits 6, 10, 11, and 15 are in two unsatisfied parity checks with the other bits in fewer. Changing these bits yields the new

$$\mathbf{y} = 1100111001100010.$$

Iterating, $H\mathbf{y}^{\mathsf{T}} = 010101101001^{\mathsf{T}}$. Parity checks 2, 4, 6, 7, 9, and 12 are unsatisfied; bits 6 and 15 are in three unsatisfied parity checks with the other bits in two or fewer. Changing bits 6 and 15 produces a new

$$\mathbf{y} = 1100101001100000.$$

Iterating again yields $H\mathbf{y}^{\mathsf{T}} = \mathbf{0}$ and so this latest $\mathbf{y}$ is declared the transmitted codeword, which is distance 2 from the received vector.

Finally, suppose that

$$\mathbf{y} = 0100101001001000$$

is received. Then $H\mathbf{y}^{\mathsf{T}} = 101100100100^{\mathsf{T}}$. There are five unsatisfied parity checks and bits 2, 3, 7, 11, 12, 13, and 15 are each involved in two of these; all other bits are involved in fewer than two. Changing these bits gives a new

$$\mathbf{y} = 0010100001110010. \tag{15.25}$$

Iterating, $H\mathbf{y}^{\mathsf{T}}$ is the all-one column vector, indicating that all parity checks are unsatisfied and all bits are in three unsatisfied checks. So all bits of $\mathbf{y}$ are changed and in the next iteration $H\mathbf{y}^{\mathsf{T}}$ is again the all-one column vector. Thus all bits are changed back giving (15.25) again. The iterations clearly are caught in a cycle and will never reach a codeword. So the original received vector is undecodable. ∎

Exercise 860 Verify the results illustrated in Example 15.6.1. ◆

Exercise 861 Let C be the $(16, 3, 4)$ LDPC code with parity check matrix H in Figure 15.10. Suppose the following vectors $\mathbf{y}$ are received. Decode them using the Gallager Hard Decision Decoding Algorithm when possible.
(a) $\mathbf{y} = 1000010110100100$,
(b) $\mathbf{y} = 1000010110100001$,
(c) $\mathbf{y} = 0110110110100101$. ◆

Exercise 862 The Gallager Hard Decision Decoding Algorithm is performed in "parallel;" all bits of $\mathbf{y}$ in the largest number of unsatisfied checks are changed at each iteration. There is a "sequential" version of the algorithm. In this version only one bit is changed at a time. Thus III is replaced by:

III.$'$ Among all bits of $\mathbf{y}$ that are involved in the largest number of unsatisfied parity checks, change only the bit y_i with smallest subscript i; call the resulting vector $\mathbf{y}$ again.

The sequential version is otherwise unchanged. Apply the sequential version of the algorithm to the same received vectors as in Example 15.6.1 and compare your answers with those in the example:
(a) $\mathbf{y} = 0100101000100000$,
(b) $\mathbf{y} = 1100101000000000$,
(c) $\mathbf{y} = 0100101001001000$. ◆

Gallager's Soft Decision Iterative Decoding Algorithm can be presented in a number of ways. We describe Mackay's version [211], which he calls the Sum-Product Decoding Algorithm. It applies more generally than we describe it here. We only present the algorithm and refer the reader to Mackay's paper for further discussion. Mackay has a visual animation of the algorithm in action at:

> `http://wol.ra.phy.cam.ac.uk/mackay/codes/gifs/`

Let C be an (n, c, r) binary low density parity check code with $m \times n$ parity check matrix H. A Tanner graph $\mathcal{T}$ for C can be formed. Number the variable nodes $1, 2, \ldots, n$ and the check nodes $1, 2, \ldots, m$. Every variable node is connected to c check nodes and every check node is connected to r variable nodes. Let $V(j)$ denote the r variable nodes connected to the jth check node; the set $V(j) \setminus s$ is $V(j)$ excluding the variable node s. Let $C(k)$ denote the c check nodes connected to the kth variable node; the set $C(k) \setminus t$ is $C(k)$ excluding the check node t.

Suppose the codeword $\mathbf{c}$ is transmitted and $\mathbf{y} = \mathbf{c} + \mathbf{e}$ is received, where $\mathbf{e}$ is the unknown error vector. Given the syndrome $H\mathbf{y}^{\mathsf{T}} = H\mathbf{e}^{\mathsf{T}} = \mathbf{z}^{\mathsf{T}}$, the object of the decoder is to compute

$$\text{prob}(e_k = 1 \mid \mathbf{z}) \quad \text{for } 1 \leq k \leq n. \tag{15.26}$$

From there, the most likely error vector and hence most likely codeword can be found. The algorithm computes, in an iterative manner, two probabilities associated to each edge of the Tanner graph and each $e \in \{0, 1\}$. The first probability q_{jk}^{e} for $j \in C(k)$ and $e \in \{0, 1\}$ is the probability that $e_k = e$ given information obtained from checks $C(k) \setminus j$. The second probability r_{jk}^{e} for $k \in V(j)$ and $e \in \{0, 1\}$ is the probability that the jth check is satisfied given $e_k = e$ and the other bits e_i for $i \in V(j) \setminus k$ have probability distribution given by $\{q_{ji}^{0}, q_{ji}^{1}\}$. There are initial probabilities $p_k^{e} = \text{prob}(e_k = e)$ for $1 \leq k \leq n$ and

$e \in \{0, 1\}$ used in the algorithm that come directly from the channel statistics. For example, if communication is over a binary symmetric channel with crossover probability ρ, then $p_k^0 = 1 - \rho$ and $p_k^1 = \rho$. The probabilities $\text{prob}(z_j \mid e_k = e, \{e_i \mid i \in V(j) \setminus k\})$ for $k \in V(j)$ and $e \in \{0, 1\}$ are also required by the algorithm. These probabilities can be computed efficiently with an algorithm similar to the Two-Way APP Algorithm of Section 15.4.

The *Sum-Product Decoding Algorithm* for binary LDPC codes is the following:

I.   Initially, set $q_{jk}^e = p_k^e$ for $j \in C(k)$ and $e \in \{0, 1\}$.

II. Update the values of r_{jk}^e for $k \in V(j)$ and $e \in \{0, 1\}$ according to the equation

$$r_{jk}^e = \sum_{e_i \in \{0,1\}, \, i \in V(j) \setminus k} \text{prob}(z_j \mid e_k = e, \{e_i \mid i \in V(j) \setminus k\}) \prod_{i \in V(j) \setminus k} q_{ji}^{e_i}.$$

III. Update the values of q_{jk}^e for $j \in C(k)$ and $e \in \{0, 1\}$ according to the equation

$$q_{jk}^e = \alpha_{jk} p_k^e \prod_{i \in C(k) \setminus j} r_{ik}^e,$$

where α_{jk} is chosen so that $q_{jk}^0 + q_{jk}^1 = 1$.

IV. For $j \in C(k)$ and $e \in \{0, 1\}$ compute

$$q_k^e = p_k^e \prod_{j \in C(k)} r_{jk}^e.$$

V.   For $1 \le k \le n$, set $\widehat{e}_k = 0$ if $q_k^0 > q_k^1$ and $\widehat{e}_k = 1$ if $q_k^1 > q_k^0$. Let $\widehat{\mathbf{e}} = \widehat{e}_1 \cdots \widehat{e}_n$. If $H\widehat{\mathbf{e}}^\mathsf{T} = \mathbf{z}^\mathsf{T}$, decode by setting $\mathbf{e} = \widehat{\mathbf{e}}$. Otherwise repeat II, III, and IV up to some maximum number of iterations. Declare a decoding failure if $H\widehat{\mathbf{e}}^\mathsf{T}$ never equals $\mathbf{z}^\mathsf{T}$.

If the Tanner graph $\mathcal{T}$ is without cycles and the algorithm successfully halts, then the probabilities in (15.26) are exactly $\alpha_k q_k^1$, where α_k is chosen so that $\alpha_k q_k^0 + \alpha_k q_k^1 = 1$. If the graph has cycles, these probabilities are approximations to (15.26). In the end, we do not care exactly what these probabilities are; we only care about obtaining a solution to $H\mathbf{e}^\mathsf{T} = \mathbf{z}^\mathsf{T}$. Thus the algorithm can be used successfully even when there are long cycles present. The algorithm is successful for codes of length a few thousand, say $n = 10\,000$, particularly with c small, say $c = 3$. Analysis of the algorithm can be found in [211].

15.7 Turbo codes

Turbo codes were introduced in 1993 by Berrou, Glavieux, and Thitimajshima [20]. A *turbo code* consists of a parallel concatenated encoder to which an iterative decoding scheme is applied. In this section we outline the ideas behind encoding while in the next section we will discuss turbo decoding; the reader interested in more detail can consult [118, 342, 352].

The importance of turbo coding has probably yet to be fully comprehended. On October 6, 1997, the *Cassini* spacecraft was launched bound for a rendezvous with Saturn in June, 2004; an experimental turbo code package was included as part of the mission. Turbo codes will have more earthly applications as well. According to Heegard and Wicker [118, p. 7], "It is simply not possible to overestimate the impact that the increase in range and/or data

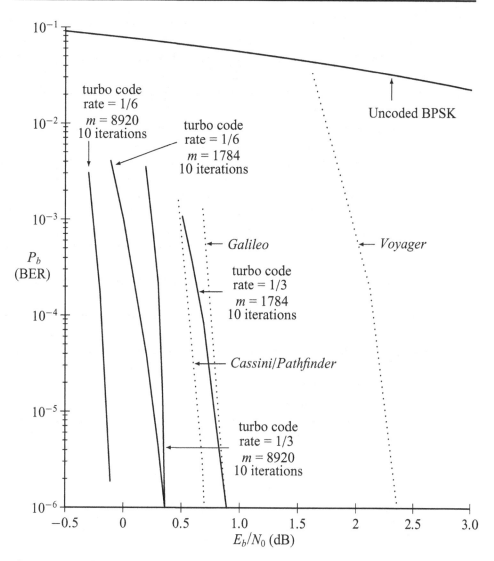

Figure 15.11 Coding gain in satellite communications.

rate resulting from turbo error control will have on the telecommunications industry. The effect will be particularly strong in wireless personal communications systems ...."

The excitement about turbo coding is illustrated by the results shown in Figure 15.11.[8] In Section 15.1, we indicated that one objective of coding is to communicate at signal-to-noise ratios close to the Shannon limit of -1.6 dB. Figure 15.11 shows the comparison between the signal-to-noise ratio E_b/N_0 and the bit error rate P_b of codes used in some satellite communication packages. This figure also compares four turbo codes to these satellite

---

[8] The data for this figure was provided by Bob McEliece of Caltech and Dariush Divsalar and Fabrizio Pollara of JPL.

packages. (The value of m is the message length; the number of iterations for the four turbo codes will be described in Section 15.8.) Figure 15.3 shows the Shannon limit and also the lowest signal-to-noise ratios that can be obtained for codes of various information rates. Comparing the two figures indicates the superiority of turbo codes; they achieve communication at low signal-to-noise ratios very close to the Shannon limit!

A *parallel concatenated encoder* consists of two or more component encoders which are usually either binary convolutional codes or block codes with a trellis structure that leads to efficient soft decision decoding.[9] The simplest situation, which we will concentrate on, involves two component codes, C_1 and C_2, each with systematic encoders. (Systematic convolutional encoders are described shortly.) In fact, the encoders for the two codes can be identical. To be even more concrete, assume that a component code C_i is either a $(2, 1)$ binary convolutional code with encoder $G_i = [1 \ g_i(D)]$, where $g_i(D)$ is a rational function or an $[n, n/2]$ binary block code with generator matrix (encoder) $G_i = [I_{n/2} \ A_i]$. (The encoders are not only in systematic form but in standard form.) If the component codes of the parallel concatenated code are convolutional, the resulting code is called a *parallel concatenated convolutional code* or PCCC. Notice that in either case, the component code is a rate 1/2 code as every message bit generates two codeword bits. We will assume the usual encoding where the encoding of the message x is xG_i. An encoder for the parallel concatenated code with component encoders G_1 and G_2 is formed as follows. The message x is encoded with the first code to produce (x, c_1), where $c_1 = x(D)g_1(D)$ if the code is convolutional or $c_1 = xA_1$ if the code is a block code. Next the message x is passed to a *permuter*, also called an *interleaver*.[10] The permuter applies a fixed permutation to the coordinates of x producing the permuted message $\tilde{x}$. The permuted message is encoded with G_2 to produce $(\tilde{x}, c_2)$, where $c_2 = \tilde{x}(D)g_2(D)$ or $c_2 = \tilde{x}A_2$. The codeword passed on to the channel is an interleaved version of the original message x and the two redundancy strings c_1 and c_2 that are generated by the codes C_1 and C_2. This process is pictured in Figure 15.12 and illustrated in Example 15.7.1. In the figure, SC_1 and SC_2 indicate that the encoders are in standard form. Notice that the resulting parallel concatenated code has rate 1/3.

Example 15.7.1 Let C_1 and C_2 each be the $[8, 4, 4]$ extended Hamming code with generator matrix

$$G_1 = G_2 = \begin{bmatrix} 1 & 0 & 0 & 0 & 0 & 1 & 1 & 1 \\ 0 & 1 & 0 & 0 & 1 & 0 & 1 & 1 \\ 0 & 0 & 1 & 0 & 1 & 1 & 0 & 1 \\ 0 & 0 & 0 & 1 & 1 & 1 & 1 & 0 \end{bmatrix}.$$

Assume the permuter is given by the permutation $(1, 3)(2, 4)$. Suppose that $x = 1011$ is the message to be encoded. Then $xG_1 = 10110100$ yielding $c_1 = 0100$. The permuted message is $\tilde{x} = 1110$, which is encoded as $\tilde{x}G_2 = 11100001$ to produce $c_2 = 0001$. Then

---

[9] We have not discussed how to form a trellis from a block code. This can be done in an efficient way that allows the use of either hard or soft decision Viterbi decoding presented in Sections 14.2 and 15.2. See [336].

[10] The term "interleaver" as used in parallel concatenated codes is somewhat different from the meaning of the term "interleaving" as used in Section 5.5. The term "interleaver" in connection with turbo codes is so prominent that we include it here, but will use the term "permuter" for the remainder of this chapter.

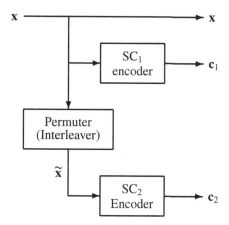

Figure 15.12 Parallel concatenated code with two standard form encoders.

$(\mathbf{x}, \mathbf{c}_1, \mathbf{c}_2) = (1011, 0100, 0001)$ is transmitted in interleaved form (the first bits of $\mathbf{x}, \mathbf{c}_1, \mathbf{c}_2$ followed by the second bits, third bits, and fourth bits) as $100\ 010\ 100\ 101$. ∎

Exercise 863 Find the interleaved form of all 16 codewords of the parallel concatenated code described in Example 15.7.1. What is the minimum distance of the resulting $[12, 4]$ binary code? ♦

In Chapter 14 the generator matrices for the convolutional codes that we presented were generally polynomial. However, the component convolutional codes in turbo codes use encoders that are most often nonpolynomial, because the encoders will be systematic, as the next example illustrates.

Example 15.7.2 We illustrate nonpolynomial encoders with the code $\mathcal{C}_1$ from Example 14.1.1. If we take the encoder G_1 and divide by $1 + D + D^2$, we obtain the generator matrix

$$G_1'' = \begin{bmatrix} 1 & \dfrac{1 + D^2}{1 + D + D^2} \end{bmatrix}.$$

A physical encoder for G_1'' is given in Figure 15.13. This is a feedback circuit. The state before clock cycle i is the two right-most delay elements which we denote by $\mathbf{s}(i)$ and $\mathbf{s}(i - 1)$. Notice these values are not $\mathbf{x}(i - 1)$ and $\mathbf{x}(i - 2)$ as they were in the feedforward circuit for the encoder G_1 given in Figure 14.1. The reader should compare Figures 14.1 and 15.13.

To understand why this circuit is the physical encoder for G_1'', we analyze the output $\mathbf{c}(i)$ at clock cycle i. Suppose $\mathbf{x}$ is the input message, and at time $i - 1$ the registers contain $\mathbf{x}(i), \mathbf{s}(i)$, and $\mathbf{s}(i - 1)$. As the next input enters from the left at time i, the output $\mathbf{c}_1(i)$ is clearly $\mathbf{x}(i)$, while the output $\mathbf{c}_2(i)$ is $\mathbf{x}(i) + (\mathbf{s}(i) + \mathbf{s}(i - 1)) + \mathbf{s}(i - 1) = \mathbf{x}(i) + \mathbf{s}(i)$. The contents of the registers are now $\mathbf{x}(i + 1), \mathbf{s}(i + 1) = \mathbf{x}(i) + \mathbf{s}(i) + \mathbf{s}(i - 1)$, and $\mathbf{s}(i)$. Thus

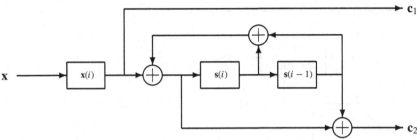

Figure 15.13 Physical encoder for $G_1'' = [1(1 + D^2)/(1 + D + D^2)]$.

we can solve for $c_2(i)$ using $s(i + 1) = x(i) + s(i) + s(i - 1)$ as follows:

$$
\begin{aligned}
c_2(i) &= x(i) + s(i) \\
&= x(i) + x(i - 1) + s(i - 1) + s(i - 2) \\
&= x(i) + x(i - 1) + x(i - 2) + s(i - 2) + s(i - 3) + s(i - 2) \\
&= x(i) + x(i - 1) + x(i - 2) + s(i - 3) \\
&= x(i) + x(i - 1) + x(i - 2) + x(i - 4) + s(i - 4) + s(i - 5).
\end{aligned}
$$

Continuing, we see that

$$
\begin{aligned}
c_2(i) &= x(i) + x(i - 1) + x(i - 2) + x(i - 4) + x(i - 5) \\
&\quad + x(i - 7) + x(i - 8) + x(i - 10) + x(i - 11) + \cdots .
\end{aligned}
\tag{15.27}
$$

By Exercise 864, we see that the Laurent series expansion for $(1 + D^2)/(1 + D + D^2)$ is

$$
1 + D + D^2 + D^4 + D^5 + D^7 + D^8 + D^{10} + D^{11} + D^{13} + D^{14} + \cdots .
\tag{15.28}
$$

Thus $c_2 = x(1 + D^2)/(1 + D + D^2) = x(1 + D + D^2 + D^4 + D^5 + D^7 + D^8 + D^{10} + D^{11} + \cdots)$, agreeing with (15.27). The physical encoder G_1'' for $\mathcal{C}_1$ has infinite memory by (15.27). (Recall that $\mathcal{C}_1$ has memory 2 since G_1 is a canonical encoder for $\mathcal{C}_1$ by Example 14.3.2.)   ∎

Exercise 864 Prove that the Laurent series for $(1 + D^2)/(1 + D + D^2)$ is given by (15.28). Hint: Multiply the series by $1 + D + D^2$.   ◆

Exercise 865 Consider the physical encoder G_1'' in Figure 15.13.
(a) Suppose that $x = 1000 \cdots$ is input to the encoder. So $x(0) = 1$ and $x(i) = 0$ for $i \geq 1$. Compute c_2 giving the contents of the shift-registers for several values of i until the pattern becomes clear. (The equations in Example 15.7.2 may prove helpful.) Check your answer by comparing it to the Laurent series for $(1 + D^2)/(1 + D + D^2)$ given by (15.28).
(b) Suppose that $x = 1010000 \cdots$ is input to the encoder. Compute c_2 giving the contents of the shift-registers for several values of i until the pattern becomes clear. Check your answer by comparing it to the product of $1 + D^2$ and the Laurent series for $(1 + D^2)/(1 + D + D^2)$ given by (15.28).   ◆

As with block codes, a *systematic generator matrix* (or *systematic encoder*) for an (n, k) convolutional code is any generator matrix with k columns which are, in some order, the k columns of the identity matrix I_k. For example, there are two systematic encoders for the code C_1 of Example 15.7.2: the matrix G_1'' of that example and

$$G_1''' = \left[\frac{1 + D + D^2}{1 + D^2} \quad 1 \right].$$

If the systematic encoder has form $[I_k \ A]$, the encoder is in *standard form*. The encoder G_1'' of Example 15.7.2 is in standard form. A *recursive systematic convolutional (RSC) encoder* for a convolutional code is a systematic encoder with some entry that is nonpolynomial of infinite weight. So both G_1'' and G_1''' are RSC encoders for C_1. A physical encoder representing an RSC encoder will involve feedback circuits as in Figure 15.13.

In most cases the component codes of a PCCC are encoded by RSC encoders in order to attempt to have high minimum weights. Suppose that $G = [1 \ f(D)/g(D)]$ is an RSC encoder for a $(2, 1)$ convolutional code where $f(D)$ and $g(D)$ are relatively prime polynomials and $g(D)$ is not a power of D. By Lemma 14.4.2, $\text{wt}(f(D)/g(D)) = \infty$. Suppose that G is the encoder for both component codes of a PCCC as pictured in Figure 15.12. If $x(D)$ is an input polynomial, then $x(D)G = (x(D), x(D)f(D)/g(D))$, which can have finite weight only if $x(D)$ is a multiple of $g(D)$. With convolutional codes, one wants all codewords to have high weight to make the free distance high. In particular, if a polynomial $x(D)$ is input to our PCCC, one hopes that at least one of the two outputs c_1 or c_2 is of high weight. Thus if c_1 has finite weight, which means $x(D)$ is a multiple of $g(D)$ (a somewhat rare occurrence), one hopes that the redundancy c_2 is of infinite weight. That is the purpose of the permuter; it permutes the input components of $x(D)$ so that if $x(D)$ is a multiple of $g(D)$, hopefully $\tilde{x}(D)$ is not a multiple of $g(D)$. Therefore if the permuter is designed properly, a very high percentage of outputs (c_1, c_2) will have high weight even if the component codes have small free distances individually. With only a relatively few codewords generated by polynomial input having small weight, the overall PCCC generally performs as if it has high free distance. The design and analysis of permuters can be found in [118, Chapter 3] and [342, Chapter 7].

Exercise 866 Suppose that G_1'' of Example 15.7.2 is used for both encoders in Figure 15.12.
(a) Let $\mathbf{x} = x(D) = 1 + D^3$ be the input to the first encoder. By multiplying $x(D)$ times $(1 + D^2)/(1 + D + D^2)$, compute c_1, giving the resulting polynomial. What is $\text{wt}(c_1)$?
(b) Suppose the interleaver permutes the first two components of $\mathbf{x}$ to produce $\tilde{\mathbf{x}} = D + D^3$. By multiplying $\tilde{\mathbf{x}}$ times the Laurent series (15.28) compute c_2, giving the resulting Laurent series. What is $\text{wt}(c_2)$? ◆

15.8 Turbo decoding

In this section we describe an iterative decoding algorithm that can be used to decode turbo codes. We rely heavily on the description given in [352]. Assume our turbo code involves two parallel concatenated codes C_1 and C_2 as presented in the previous section and summarized in Figure 15.12. The turbo decoding algorithm actually uses two decoders, one for each of the component codes. Suppose that the message $\mathbf{x} = (\mathbf{x}(1), \mathbf{x}(2), \ldots, \mathbf{x}(m))$ is transmitted interleaved with the two parity vectors $\mathbf{c}_1 = (\mathbf{c}_1(1), \mathbf{c}_1(2), \ldots, \mathbf{c}_1(m))$ and $\mathbf{c}_2 = (\mathbf{c}_2(1), \mathbf{c}_2(2), \ldots, \mathbf{c}_2(m))$. The received vectors $\mathbf{y} = (\mathbf{y}(1), \mathbf{y}(2), \ldots, \mathbf{y}(m))$, $\mathbf{y}_1 = (\mathbf{y}_1(1), \mathbf{y}_1(2), \ldots, \mathbf{y}_1(m))$, and $\mathbf{y}_2 = (\mathbf{y}_2(1), \mathbf{y}_2(2), \ldots, \mathbf{y}_2(m))$, obtained after deinterleaving, are noisy versions of $\mathbf{x}$, $\mathbf{c}_1$, and $\mathbf{c}_2$. Their components are received from the demodulator; if, for example, 8-level quantization was used, the components of the three received vectors would be in $\{0_1, \ldots, 1_4\}$. Ultimately, the job of the decoder is to find the a posteriori probability

$$\text{prob}(\mathbf{x}(i) = x \mid \mathbf{y}, \mathbf{y}_1, \mathbf{y}_2) \quad \text{for } x \in \{0, 1\} \text{ and } 1 \leq i \leq m. \tag{15.29}$$

Obtaining the exact value of these leads to a rather complex decoder. Berrou, Glavieux, and Thitimajshima [20] discovered a much less complex decoder that operates iteratively to obtain an estimate of these probabilities, thereby allowing reliable decoding. The iterative decoding begins with the code C_1 and uses a priori probabilities about $\mathbf{x}$, the channel statistics, and the received vectors $\mathbf{y}$ and $\mathbf{y}_1$. This first decoder passes on soft information to the decoder for C_2 that is to be used as a priori probabilities for this second decoder. This second decoder uses these a priori probabilities about $\mathbf{x}$, the channel statistics, and the received vectors $\mathbf{y}$ and $\mathbf{y}_2$ to compute soft information to be passed back to the first decoder. Passing soft information from the first decoder to the second and back to the first comprises a single iteration of the turbo decoder; the decoder runs for several iterations. The term "turbo" refers to the process of feeding information from one part of the decoder to the other and back again in order to gain improvement in decoding, much like "turbo-charging" an internal combustion engine.

We begin by examining the first half of the first iteration of the decoding process. This step uses the first code C_1 and the received vectors $\mathbf{y}$ and $\mathbf{y}_1$, with the goal of computing

$$\text{prob}(\mathbf{x}(i) = x \mid \mathbf{y}, \mathbf{y}_1) \quad \text{for } x \in \{0, 1\} \text{ and } 1 \leq i \leq m. \tag{15.30}$$

From these probabilities we can obtain the most likely message values using the partial information obtained from $\mathbf{y}$ and $\mathbf{y}_1$ but ignoring the information from $\mathbf{y}_2$. But

$$\text{prob}(\mathbf{x}(i) = x \mid \mathbf{y}, \mathbf{y}_1) = \frac{\text{prob}(\mathbf{x}(i) = x, \mathbf{y}, \mathbf{y}_1)}{\text{prob}(\mathbf{y}, \mathbf{y}_1)}$$
$$= \alpha \, \text{prob}(\mathbf{x}(i) = x, \mathbf{y}, \mathbf{y}_1), \tag{15.31}$$

where $\alpha = 1/\text{prob}(\mathbf{y}, \mathbf{y}_1)$. The exact value of α is unimportant because all we care about is which is greater: $\text{prob}(\mathbf{x}(i) = 0 \mid \mathbf{y}, \mathbf{y}_1)$ or $\text{prob}(\mathbf{x}(i) = 1 \mid \mathbf{y}, \mathbf{y}_1)$. This α *notation*, which is described more thoroughly in [252], allows us to remove conditional probabilities that do not affect the relative relationships of the probabilities we are computing; we use it freely throughout this section.

In order to determine the probabilities in (15.31), for $x \in \{0, 1\}$ and $1 \leq i \leq m$, let $\pi_i^{(0)}(x) = \text{prob}(\mathbf{x}(i) = x)$ be the a priori probability that the ith message bit is x. If, for example, 0 and 1 are equally likely to be transmitted, then $\pi_i^{(0)}(x) = 0.5$. With convolutional codes, we saw that certain bits at the end of a message are always 0 in order to send the encoder to the zero state; then $\pi_i^{(0)}(0) = 1$ and $\pi_i^{(0)}(1) = 0$ for those bits. We assume the channel statistics $\text{prob}(y \mid c)$ (see for instance Example 15.2.1) are known. Using these channel statistics, let $\lambda_i(x) = \text{prob}(\mathbf{y}(i) \mid \mathbf{x}(i) = x)$. We have

$$
\text{prob}(\mathbf{x}(i) = x, \mathbf{y}, \mathbf{y}_1) = \sum_{\mathbf{x}:\mathbf{x}(i)=x} \text{prob}(\mathbf{x}, \mathbf{y}, \mathbf{y}_1)
$$

$$
= \sum_{\mathbf{x}:\mathbf{x}(i)=x} \text{prob}(\mathbf{y}, \mathbf{y}_1 \mid \mathbf{x})\text{prob}(\mathbf{x})
$$

$$
= \sum_{\mathbf{x}:\mathbf{x}(i)=x} \text{prob}(\mathbf{y}_1 \mid \mathbf{x})\text{prob}(\mathbf{y} \mid \mathbf{x})\text{prob}(\mathbf{x}). \tag{15.32}
$$

But,

$$
\text{prob}(\mathbf{y} \mid \mathbf{x})\text{prob}(\mathbf{x}) = \prod_{j=1}^{m} \text{prob}(\mathbf{y}(j) \mid \mathbf{x}(j))\text{prob}(\mathbf{x}(j)) = \prod_{j=1}^{m} \lambda_j(\mathbf{x}(j))\pi_j^{(0)}(\mathbf{x}(j))
$$

under the assumption that the channel is memoryless. Also $\text{prob}(\mathbf{y}_1 \mid \mathbf{x}) = \text{prob}(\mathbf{y}_1 \mid \mathbf{c}_1)$ because $\mathbf{c}_1$ is determined directly from $\mathbf{x}$ by the encoding using $\mathcal{C}_1$. Combining these with (15.30), (15.31), and (15.32), we have

$$
\text{prob}(\mathbf{x}(i) = x \mid \mathbf{y}, \mathbf{y}_1) = \alpha \sum_{\mathbf{x}:\mathbf{x}(i)=x} \text{prob}(\mathbf{y}_1 \mid \mathbf{c}_1) \prod_{j=1}^{m} \lambda_j(\mathbf{x}(j))\pi_j^{(0)}(\mathbf{x}(j))
$$

$$
= \alpha\,\lambda_i(x)\pi_i^{(0)}(x) \sum_{\mathbf{x}:\mathbf{x}(i)=x} \text{prob}(\mathbf{y}_1 \mid \mathbf{c}_1) \prod_{\substack{j=1 \\ j\neq i}}^{m} \lambda_j(\mathbf{x}(j))\pi_j^{(0)}(\mathbf{x}(j)). \tag{15.33}
$$

Excluding α, (15.33) is a product of three terms. The first, $\lambda_i(x)$, is the *systematic term* containing information about $\mathbf{x}(i)$ derived from the channel statistics and the received bit $\mathbf{y}(i)$. The second, $\pi_i^{(0)}(x)$, is the *a priori term* determined only from the ith message bit. The third is called the *extrinsic term*, which contains information about $\mathbf{x}(i)$ derived from the received parity $\mathbf{y}_1$; we denote this term by

$$
\pi_i^{(1)}(x) = \sum_{\mathbf{x}:\mathbf{x}(i)=x} \text{prob}(\mathbf{y}_1 \mid \mathbf{c}_1) \prod_{\substack{j=1 \\ j\neq i}}^{m} \lambda_j(\mathbf{x}(j))\pi_j^{(0)}(\mathbf{x}(j)).
$$

Notice that the extrinsic information does not include the specific terms $\lambda_i(\mathbf{x}(i))\pi_i^{(0)}(\mathbf{x}(i))$, a phenomenon suggestive of the requirement placed on message passing that messages passed along an edge of a Tanner graph cannot involve information previously received along that edge. Additionally, the sum-product form of this extrinsic information is reminiscent of the sum-product forms that arise in both the Two-Way APP Decoding Algorithm and the Sum-Product Decoding Algorithm. This extrinsic information is passed on to the second decoder as the a priori probability used by that decoder.

The second decoder uses the a priori probabilities, channel statistics, and received vectors $\mathbf{y}$ and $\mathbf{y}_2$ to perform the second half of the first iteration. In a manner analogous to the first half,

$$\text{prob}(\mathbf{x}(i) = x \mid \mathbf{y}, \mathbf{y}_2) = \alpha \sum_{\mathbf{x}:\mathbf{x}(i)=x} \text{prob}(\mathbf{y}_2 \mid \mathbf{c}_2) \prod_{j=1}^{m} \lambda_j(\mathbf{x}(j)) \pi_j^{(1)}(\mathbf{x}(j))$$

$$= \alpha \lambda_i(x) \pi_i^{(1)}(x) \sum_{\mathbf{x}:\mathbf{x}(i)=x} \text{prob}(\mathbf{y}_2 \mid \mathbf{c}_2) \prod_{\substack{j=1 \\ j\neq i}}^{m} \lambda_j(\mathbf{x}(j)) \pi_j^{(1)}(\mathbf{x}(j)).$$

(The value of α above may be different from that in (15.31); again the exact value is immaterial.) The extrinsic information

$$\pi_i^{(2)}(x) = \sum_{\mathbf{x}:\mathbf{x}(i)=x} \text{prob}(\mathbf{y}_2 \mid \mathbf{c}_2) \prod_{\substack{j=1 \\ j\neq i}}^{m} \lambda_j(\mathbf{x}(j)) \pi_j^{(1)}(\mathbf{x}(j))$$

from this decoder is passed back to the first decoder to use as a priori probabilities to begin the first half of the second iteration.

Continuing in like manner, iteration I generates the following probabilities:

$$\pi_i^{(2I-1)}(x) = \sum_{\mathbf{x}:\mathbf{x}(i)=x} \text{prob}(\mathbf{y}_1 \mid \mathbf{c}_1) \prod_{\substack{j=1 \\ j\neq i}}^{m} \lambda_j(\mathbf{x}(j)) \pi_j^{(2I-2)}(\mathbf{x}(j)),$$

$$\pi_i^{(2I)}(x) = \sum_{\mathbf{x}:\mathbf{x}(i)=x} \text{prob}(\mathbf{y}_2 \mid \mathbf{c}_2) \prod_{\substack{j=1 \\ j\neq i}}^{m} \lambda_j(\mathbf{x}(j)) \pi_j^{(2I-1)}(\mathbf{x}(j)).$$

By (15.33) and analogs equations,

$$\text{prob}(\mathbf{x}(i) = x \mid \mathbf{y}, \mathbf{y}_1) = \alpha \lambda_i(x) \pi_i^{(2I-2)}(x) \pi_i^{(2I-1)}(x),$$
$$\text{prob}(\mathbf{x}(i) = x \mid \mathbf{y}, \mathbf{y}_2) = \alpha \lambda_i(x) \pi_i^{(2I-1)}(x) \pi_i^{(2I)}(x).$$

Solving these for $\pi_i^{(2I-1)}(x)$, and $\pi_i^{(2I)}(x)$, respectively, yields

$$\pi_i^{(n)}(x) = \begin{cases} \alpha \dfrac{\text{prob}(\mathbf{x}(i) = x \mid \mathbf{y}, \mathbf{y}_1)}{\lambda_i(x) \pi_i^{(n-1)}(x)} & \text{if } n = 2I - 1, \\[4mm] \alpha \dfrac{\text{prob}(\mathbf{x}(i) = x \mid \mathbf{y}, \mathbf{y}_2)}{\lambda_i(x) \pi_i^{(n-1)}(x)} & \text{if } n = 2I. \end{cases} \qquad (15.34)$$

(Note that the α values are not the same as before.) Our original goal was to compute (15.29). It turns out that this can be approximated after N complete iterations by

$$\alpha \lambda_i(x) \pi_i^{(2N)}(x) \pi_i^{(2N-1)}(x).$$

This leads to the *Turbo Decoding Algorithm*:

I.   Until some number of iterations N, compute

$$\pi_i^{(1)}(x), \ \pi_i^{(2)}(x), \ \pi_i^{(3)}(x), \dots, \pi_i^{(2N)}(x)$$

for $1 \leq i \leq m$ and $x \in \{0, 1\}$ using (15.34).

II. For $1 \leq i \leq m$ and $x \in \{0, 1\}$, compute

$$\gamma_i(x) = \alpha \lambda_i(x) \pi_i^{(2N)}(x) \pi_i^{(2N-1)}(x).$$

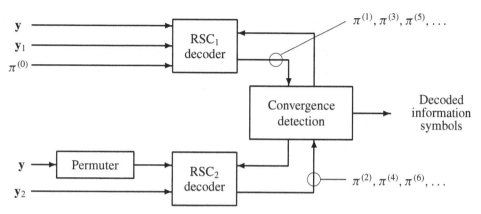

Figure 15.14 Turbo decoder.

III. The decoded message symbol $\widehat{\mathbf{x}}(i)$, estimating $\mathbf{x}(i)$, is

$$\widehat{\mathbf{x}}(i) = \begin{cases} 0 & \text{if } \gamma_i(0) > \gamma_i(1), \\ 1 & \text{if } \gamma_i(1) > \gamma_i(0), \end{cases}$$

for $1 \le i \le m$.

The value of N is determined in a number of different ways. For example, the value can simply be fixed, at say, $N = 10$ (see Figure 15.11). Or the sequence of probabilities $\pi_i^{(1)}(x)$, $\pi_i^{(2)}(x)$, $\pi_i^{(3)}(x)$, ... can be examined to see how much variability there is with an increasing number of iterations. The computation in Step I of $\pi_i^{(n)}(x)$ using (15.34) involves computing $\text{prob}(\mathbf{x}(i) = x \mid \mathbf{y}, \mathbf{y}_1)$ and $\text{prob}(\mathbf{x}(i) = x \mid \mathbf{y}, \mathbf{y}_2)$. These can often be computed using versions of Two-Way APP Decoding as in Section 15.4. The Turbo Decoding Algorithm is illustrated in Figure 15.14.

15.9 Some space history

We conclude this chapter with a brief summary of the use of error-correcting codes in the history of space exploration. A more thorough account can be found in [352].

The first error-correcting code developed for use in deep space was a nonlinear (32, 64, 16) code, a coset of the binary [32, 6, 16] Reed–Muller code $\mathcal{R}(1, 5)$, capable of correcting seven errors. This code was used on the *Mariner* 6 and 7 Mars missions launched, respectively, on February 24 and March 27, 1969. *Mariner* 6 flew within 2131 miles of the Martian equator on July 30, 1969, and *Mariner* 7 flew within 2130 miles of the southern hemisphere on August 4, 1969. The *Mariner* spacecrafts sent back a total of 201 gray-scale photographs with each pixel given one of $2^6 = 64$ shades of gray. The code was used on later *Mariner* missions and also on the *Viking* Mars landers. The coding gain using this code was 2.2 dB compared to uncoded BPSK at a BER of 5×10^{-3}. While the chosen code had a relatively low rate 6/32 among codes capable of correcting seven errors, it had the advantage of having an encoder and decoder that were easy to implement. The decoder, known as the *Green machine*, after its developer R. R. Green of NASA's Jet Propulsion

Laboratory (JPL), uses computations with Hadamard matrices to perform the decoding chores.

While the *Mariner* code was the first developed for space application, it was not actually the first to be launched. An encoding system based on a (2, 1, 20) convolutional code with generator matrix $G = [g_1(D) \ g_2(D)]$, where

$$g_1(D) = 1,$$
$$g_2(D) = 1 + D + D^2 + D^5 + D^6 + D^8 + D^9 + D^{12}$$
$$\quad + D^{13} + D^{14} + D^{16} + D^{17} + D^{18} + D^{19} + D^{20},$$

was launched aboard *Pioneer 9* on November 8, 1968, headed for solar orbit. Sequential decoding, described in [197, 232, 351], was used to decode received data. *Pioneer 9* continued to transmit information about solar winds, interplanetary electron density and magnetic fields, cosmic dust, and the results of other on-going experiments until 1983. This code with its sequential decoder provided a coding gain of 3.3 dB at a BER of 7.7×10^{-4}.

The *Pioneer 10* mission to Jupiter launched in 1972 and the *Pioneer 11* mission to Saturn launched a year later carried a (2, 1, 31, 21) quick-look-in convolutional code with encoder $[g_1(D) \ g_2(D)]$ developed by Massey and Costello [228], where

$$g_1(D) = 1 + D + D^2 + D^4 + D^5 + D^7 + D^8 + D^9 + D^{11}$$
$$\quad + D^{13} + D^{14} + D^{16} + D^{17} + D^{18} + D^{19} + D^{21} + D^{22}$$
$$\quad + D^{23} + D^{24} + D^{25} + D^{27} + D^{28} + D^{29} + D^{31},$$
$$g_2(D) = g_1(D) + D.$$

A *quick-look-in* convolutional code is a (2, 1) code with encoder $[g_1(D) \ g_2(D)]$, where $\deg g_1(D) = \deg g_2(D)$, the constant term of both $g_1(D)$ and $g_2(D)$ is 1, and $g_1(D) + g_2(D) = D$. Exercise 867 shows why this is called a quick-look-in code.

Exercise 867 Let C be a quick-look-in (2, 1) convolutional code with encoder $G = [g_1(D) \ g_2(D)]$.
(a) Prove that C is basic and noncatastrophic. Hint: Any divisor of $g_1(D)$ and $g_2(D)$ is a divisor of $g_1(D) + g_2(D)$.
(b) Let $x(D)$ be input to the code and $c(D) = (c_1(D), c_2(D)) = x(D)G$ be the output. Show that $c_1(D) + c_2(D) = x(D)D$.
Note that (b) shows the input can be obtained easily from the output by adding the two components of the corrected received vector; it is merely delayed by one unit of time. Hence the receiver can quickly look at the corrected vector to obtain the input. Note that (b) also shows that C is noncatastrophic because if the output has finite weight, the input must have finite weight as well.   ♦

The use of convolutional codes in deep space exploration became even more prevalent with the development of the Viterbi Algorithm. Sequential decoding has the disadvantage that it does not have a fixed decoding time. While the Viterbi Algorithm does not have this disadvantage, it cannot be used efficiently when the memory is too high. In the mid 1980s, the Consultative Committee for Space Data Systems (CCSDS) adopted as its standard a concatenated coding system with the inner encoder consisting of either a $(2, 1, 6, 10)$ or $(3, 1, 6, 15)$ convolutional code. The $(2, 1, 6, 10)$ code has generator matrix $[g_1(D) \ g_2(D)]$, where

$$g_1(D) = 1 + D + D^3 + D^4 + D^6,$$
$$g_2(D) = 1 + D^3 + D^4 + D^5 + D^6.$$

The $(3, 1, 6, 15)$ code has generator matrix $[g_1(D) \ g_2(D) \ g_3(D)]$, where $g_1(D)$ and $g_2(D)$ are as above and $g_3(D) = 1 + D^2 + D^4 + D^5 + D^6$. At a BER of 1×10^{-5}, the $(2, 1, 6, 10)$ code provides a coding gain of 5.4 dB, and the $(3, 1, 6, 15)$ code provides a coding gain of 5.7 dB. Both codes can be decoded efficiently using the Viterbi Algorithm but the $(2, 1, 6, 10)$ code has become the preferred choice.

Voyager 1 and 2 were launched in the summer of 1977 destined to explore Jupiter, Saturn, and their moons. These spacecraft transmitted two different types of data: imaging and GSE (general science and engineering). The imaging system required less reliability than the GSE system and used the $(2, 1, 6, 10)$ code for its encoding. The GSE data was transmitted using a concatenated code. The outer encoder was the $[24, 12, 8]$ extended binary Golay code. After encoding using the binary Golay code, codeword interleaving was performed with the result passed to the inner encoder, which was the $(2, 1, 6, 10)$ code. The process was reversed in decoding. (The imaging data merely bypassed the Golay code portion of the encoding and decoding process.) Partially because of the effective use of coding, NASA engineers were able to extend the mission of Voyager 2 to include flybys of Uranus and Neptune. To accomplish this, the GSE data was transmitted using a concatenated code with outer encoder a $[256, 224, 33]$ Reed–Solomon code over $\mathbb{F}_{256}$, in place of the Golay code, followed by the $(2, 1, 6, 10)$ inner encoder. Later NASA missions, including Galileo and some European Space Agency (ESA) missions such as the Giotto mission to Halley's Comet, used similar error-correction coding involving a Reed–Solomon code and a convolutional code with the added step of interleaving after encoding with the Reed–Solomon code. The coding gain for the Voyager coding system is shown in Figure 15.11.

The Galileo mission to study Jupiter and its atmosphere was fraught with problems from the outset. Originally scheduled for launch from the space shuttle in early 1986, the mission was delayed until late 1989 due to the shuttle Challenger in-flight explosion on January 28, 1986. Galileo was launched with two different concatenated encoders. The outer encoder of both systems was the $[256, 224, 33]$ Reed–Solomon code. The inner encoder for one system was the $(2, 1, 6, 10)$ CCSDS standard code, while the inner encoder for the other was a $(4, 1, 14)$ convolutional code with generator matrix $[g_1(D) \ g_2(D) \ g_3(D) \ g_4(D)]$, where

$$g_1(D) = 1 + D^3 + D^4 + D^7 + D^8 + D^{10} + D^{14},$$
$$g_2(D) = 1 + D^2 + D^5 + D^7 + D^9 + D^{10} + D^{11} + D^{14},$$
$$g_3(D) = 1 + D + D^4 + D^5 + D^6 + D^7 + D^9 + D^{10} + D^{12} + D^{13} + D^{14},$$
$$g_4(D) = 1 + D + D^2 + D^6 + D^8 + D^{10} + D^{11} + D^{12} + D^{14}.$$

At each time increment, the Viterbi decoder for the $(4, 1, 14)$ code had to examine $2^{14} = 16\,384$ states. To accomplish this daunting task, JPL completed the Big Viterbi Decoder in

1991 (over two years after *Galileo*'s launch!). This decoder could operate at one million bits per second. In route to Jupiter, the high gain X-band antenna, designed for transmission of *Galileo*'s data at a rate of 100 000 bits per second, failed to deploy. As a result a low gain antenna was forced to take over data transmission. However, the information transmission rate for the low gain antenna was only 10 bits per second. Matters were further complicated because the encoding by the (4, 1, 14) inner encoder was hard-wired into the high gain antenna. The JPL scientists were able to implement software changes to improve the data transmission and error-correction capability of *Galileo*, which in turn have improved data transmission in the CCSDS standard. In the end, the outer Reed–Solomon encoder originally on *Galileo* was replaced with another (2, 1) convolutional encoder; the inner encoder used was the original (2, 1, 6, 10) code already on board.

The *Cassini* spacecraft was launched on October 6, 1997, for a rendezvous with Saturn in June 2004. Aboard is the ESA probe *Huygens,* which will be dropped into the atmosphere of the moon Titan. Aboard *Cassini* are hardware encoders for the (2, 1, 6, 10) CCSDS standard convolutional code, the (4, 1, 14) code that is aboard *Galileo*, and a (6, 1, 14) convolutional code. A JPL team of Divsalar, Dolinar, and Pollara created a turbo code that has been placed aboard *Cassini* as an experimental package. This turbo code uses two (2, 1) convolutional codes in parallel. The turbo code is a modification of that given in Figure 15.12; the first code uses the RSC encoder $[1 \quad g_1(D)/g_2(D)]$, where $g_1(D) = 1 + D^2 + D^3 + D^5 + D^6$ and $g_2(D) = 1 + D + D^2 + D^3 + D^6$. The second encoder is $[g_1(D)/g_3(D) \quad g_2(D)/g_3(D)]$, where $g_3(D) = 1 + D$; this is not systematic but is recursive. On Earth, is a turbo decoder that is much less complex than the Big Viterbi Decoder to be used with the (6, 1, 14) convolutional code. Simulations have shown, as indicated in Figure 15.11, that the turbo code outperforms the concatenated code consisting of an outer Reed–Solomon code and the inner (6, 1, 14) code.

References

[1] M. J. Adams, "Subcodes and covering radius," *IEEE Trans. Inform. Theory* **IT–32** (1986), 700–701.

[2] E. Agrell, A. Vardy, and K. Zeger, "A table of upper bounds for binary codes," *IEEE Trans. Inform. Theory* **IT–47** (2001), 3004–3006.

[3] E. Artin, *Geometric Algebra*. Interscience Tracts in Pure and Applied Mathematics No. 3. New York: Interscience, 1957.

[4] E. F. Assmus, Jr. and J. D. Key, *Designs and Their Codes*. London: Cambridge University Press, 1993.

[5] E. F. Assmus, Jr. and J. D. Key, "Polynomial codes and finite geometries," in *Handbook of Coding Theory*, eds. V. S. Pless and W. C. Huffman. Amsterdam: Elsevier, 1998, pp. 1269–1343.

[6] E. F. Assmus, Jr. and H. F. Mattson, Jr., "New 5-designs," *J. Comb. Theory* **6** (1969), 122–151.

[7] E. F. Assmus, Jr. and H. F. Mattson, Jr. "Some 3-error correcting BCH codes have covering radius 5," *IEEE Trans. Inform. Theory* **IT–22** (1976), 348–349.

[8] E. F. Assmus, Jr., H. F. Mattson, Jr., and R. J. Turyn, "Research to develop the algebraic theory of codes," Report AFCRL-67-0365, Air Force Cambridge Res. Labs., Bedford, MA, June 1967.

[9] E. F. Assmus, Jr. and V. Pless, "On the covering radius of extremal self-dual codes," *IEEE Trans. Inform. Theory* **IT–29** (1983), 359–363.

[10] D. Augot, P. Charpin, and N. Sendrier, "Studying the locator polynomials of minimum weight codewords of BCH codes," *IEEE Trans. Inform. Theory* **IT–38** (1992), 960–973.

[11] D. Augot and L. Pecquet, "A Hensel lifting to replace factorization in list-decoding of algebraic-geometric and Reed–Solomon codes," *IEEE Trans. Inform. Theory* **IT–46** (2000), 2605–2614.

[12] C. Bachoc, "On harmonic weight enumerators of binary codes," *Designs, Codes and Crypt.* **18** (1999), 11–28.

[13] C. Bachoc and P. H. Tiep, "Appendix: two-designs and code minima," appendix to: W. Lempken, B. Schröder, and P. H. Tiep, "Symmetric squares, spherical designs, and lattice minima," *J. Algebra* **240** (2001), 185–208; appendix pp. 205–208.

[14] A. Barg, "The matroid of supports of a linear code," *Applicable Algebra in Engineering, Communication and Computing (AAECC Journal)* **8** (1997), 165–172.

[15] B. I. Belov, "A conjecture on the Griesmer bound," in *Proc. Optimization Methods and Their Applications, All Union Summer Sem., Lake Baikal* (1972), 100–106.

[16] T. P. Berger and P. Charpin, "The automorphism group of BCH codes and of some affine-invariant codes over extension fields," *Designs, Codes and Crypt.* **18** (1999), 29–53.

[17] E. R. Berlekamp, ed., *Key Papers in the Development of Coding Theory*. New York: IEEE Press, 1974.

[18] E. R. Berlekamp, *Algebraic Coding Theory*. Laguna Hills, CA: Aegean Park Press, 1984.

[19] E. R. Berlekamp, F. J. MacWilliams, and N. J. A. Sloane, "Gleason's theorem on self-dual codes," *IEEE Trans. Inform. Theory* **IT–18** (1972), 409–414.

[20] C. Berrou, A. Glavieux, and P. Thitimajshima, "Near Shannon limit error-correcting coding and decoding: turbo codes," *Proc. of the 1993 IEEE Internat. Communications Conf.*, Geneva, Switzerland (May 23–26, 1993), 1064–1070.

[21] R. E. Blahut, *Theory and Practice of Error Control Codes*. Reading, MA: Addison-Wesley, 1983.

[22] R. E. Blahut, "Decoding of cyclic codes and codes on curves," in *Handbook of Coding Theory*, eds. V. S. Pless and W. C. Huffman. Amsterdam: Elsevier, 1998, pp. 1569–1633.

[23] I. F. Blake, ed., *Algebraic Coding Theory: History and Development*. Stroudsburg, PA: Dowden, Hutchinson, & Ross, Inc., 1973.

[24] K. Bogart, D. Goldberg, and J. Gordon, "An elementary proof of the MacWilliams theorem on equivalence of codes," *Inform. and Control* **37** (1978), 19–22.

[25] A. Bonisoli, "Every equidistant linear code is a sequence of dual Hamming codes," *Ars Combin.* **18** (1984), 181–186.

[26] A. Bonnecaze, P. Solé, C. Bachoc, and B. Mourrain, "Type II codes over $\mathbb{Z}_4$," *IEEE Trans. Inform. Theory* **IT–43** (1997), 969–976.

[27] A. Bonnecaze, P. Solé, and A. R. Calderbank, "Quaternary quadratic residue codes and unimodular lattices," *IEEE Trans. Inform. Theory* **IT–41** (1995), 366–377.

[28] R. C. Bose and D. K. Ray-Chaudhuri, "On a class of error correcting binary group codes," *Inform. and Control* **3** (1960), 68–79. (Also reprinted in [17] pp. 75–78 and [23] pp. 165–176.)

[29] R. C. Bose and D. K. Ray-Chaudhuri, "Further results on error correcting binary group codes," *Inform. and Control* **3** (1960), 279–290. (Also reprinted in [17] pp. 78–81 and [23] pp. 177–188.)

[30] S. Bouyuklieva, "A method for constructing self-dual codes with an automorphism of order 2," *IEEE Trans. Inform. Theory* **IT–46** (2000), 496–504.

[31] A. E. Brouwer, "The linear programming bound for binary linear codes," *IEEE Trans. Inform. Theory* **IT–39** (1993), 677–688.

[32] A. E. Brouwer, "Bounds on the size of linear codes," in *Handbook of Coding Theory*, eds. V. S. Pless and W. C. Huffman. Amsterdam: Elsevier, 1998, pp. 295–461.

[33] R. A. Brualdi, N. Cai, and V. S. Pless, "Orphan structure of the first-order Reed–Muller codes," *Discrete Math.* **102** (1992), 239–247.

[34] R. A. Brualdi, S. Litsyn, and V. S. Pless, "Covering radius," in *Handbook of Coding Theory*, eds. V. S. Pless and W. C. Huffman. Amsterdam: Elsevier, 1998, pp. 755–826.

[35] R. A. Brualdi and V. Pless, "Subcodes of Hamming codes," *Congr. Numer.* **70** (1990), 153–158.

[36] R. A. Brualdi and V. S. Pless, "Orphans of the first order Reed–Muller codes," *IEEE Trans. Inform. Theory* **IT–36** (1990), 399–401.

[37] R. A. Brualdi and V. S. Pless, "On the covering radius of a code and its subcodes," *Discrete Math.* **83** (1990), 189–199.

[38] R. A. Brualdi and V. Pless, "Weight enumerators of self-dual codes," *IEEE Trans. Inform. Theory* **IT–37** (1991), 1222–1225.

[39] R. A. Brualdi and V. S. Pless, "Greedy Codes," *J. Comb. Theory* **64A** (1993), 10–30.

[40] R. A. Brualdi, V. Pless, and J. S. Beissinger, "On the MacWilliams identities for linear codes," *Linear Alg. and Its Applic.* **107** (1988), 181–189.

[41] R. A. Brualdi, V. Pless, and R. M. Wilson, "Short codes with a given covering radius," *IEEE Trans. Inform. Theory* **IT–35** (1989), 99–109.

[42] K. A. Bush, "Orthogonal arrays of index unity," *Ann. Math. Stat.* **23** (1952), 426–434.

[43] A. R. Calderbank, "Covering radius and the chromatic number of Kneser graphs," *J. Comb. Theory* **54A** (1990), 129–131.

[44] A. R. Calderbank, E. M. Rains, P. W. Shor, and N. J. A. Sloane, "Quantum error correction via codes over GF(4)," *IEEE Trans. Inform. Theory* **IT–44** (1998), 1369–1387.

[45] P. Camion, B. Courteau, and A. Monpetit, "Coset weight enumerators of the extremal self-dual binary codes of length 32," in *Proc. of Eurocode 1992, Udine, Italy, CISM Courses and Lectures No. 339*, eds. P. Camion, P. Charpin, and S. Harari. Vienna: Springer, 1993, pp. 17–29.

[46] A. Canteaut and F. Chabaud, "A new algorithm for finding minimum weight codewords in a linear code: application to primitive narrow-sense BCH codes of length 511," *IEEE Trans. Inform. Theory* **IT–44** (1998), 367–378.

[47] M. G. Carasso, J. B. H. Peek, and J. P. Sinjou, "The compact disc digital audio system," *Philips Technical Review* **40** No. 6 (1982), 151–155.

[48] J. L. Carter, "On the existence of a projective plane of order 10," Ph.D. Thesis, University of California, Berkeley, 1974.

[49] G. Castagnoli, J. L. Massey, P. A. Schoeller, and N. von Seeman, "On repeated-root cyclic codes," *IEEE Trans. Inform. Theory* **IT–37** (1991), 337–342.

[50] P. Charpin, "Open problems on cyclic codes," in *Handbook of Coding Theory*, eds. V. S. Pless and W. C. Huffman. Amsterdam: Elsevier, 1998, pp. 963–1063.

[51] G. D. Cohen, I. S. Honkala, S. Litsyn, and A. Lobstein, *Covering Codes*. Amsterdam: Elsevier, 1997.

[52] G. D. Cohen, A. C. Lobstein, and N. J. A. Sloane, "Further results on the covering radius of codes," *IEEE Trans. Inform. Theory* **IT–32** (1986), 680–694.

[53] J. H. Conway and V. Pless, "On the enumeration of self-dual codes," *J. Comb. Theory* **28A** (1980), 26–53.

[54] J. H. Conway and V. Pless, "On primes dividing the group order of a doubly-even (72, 36, 16) code and the group of a quaternary (24, 12, 10) code," *Discrete Math.* **38** (1982), 143–156.

[55] J. H. Conway, V. Pless, and N. J. A. Sloane, "Self-dual codes over GF(3) and GF(4) of length not exceeding 16," *IEEE Trans. Inform. Theory* **IT–25** (1979), 312–322.

[56] J. H. Conway, V. Pless, and N. J. A. Sloane, "The binary self-dual codes of length up to 32: a revised enumeration," *J. Comb. Theory* **60A** (1992), 183–195.

[57] J. H. Conway and N. J. A. Sloane, "Lexicographic codes: error-correcting codes from game theory," *IEEE Trans. Inform. Theory* **IT–32** (1986), 337–348.

[58] J. H. Conway and N. J. A. Sloane, "A new upper bound on the minimal distance of self-dual codes," *IEEE Trans. Inform. Theory* **IT–36** (1990), 1319–1333.

[59] J. H. Conway and N. J. A. Sloane, "Self-dual codes over the integers modulo 4," *J. Comb. Theory* **62A** (1993), 30–45.

[60] J. H. Conway and N. J. A. Sloane, *Sphere Packings, Lattices and Groups*, 3rd ed. New York: Springer-Verlag, 1999.

[61] P. Delsarte, "Bounds for unrestricted codes by linear programming," *Philips Research Report* **27** (1972), 272–289.

[62] P. Delsarte, "Four fundamental parameters of a code and their combinatorial significance," *Inform. and Control* **23** (1973), 407–438.

[63] P. Delsarte, "An algebraic approach to the association schemes of coding theory," *Philips Research Reports Supplements* No. 10 (1973).

[64] P. Delsarte, "On subfield subcodes of Reed–Solomon codes," *IEEE Trans. Inform. Theory* **IT–21** (1975), 575–576.

[65] P. Delsarte and J. M. Goethals, "Unrestricted codes with the Golay parameters are unique," *Discrete Math.* **12** (1975), 211–224.

[66] L. E. Dickson, *Linear Groups*. New York: Dover, 1958.

[67] C. Ding and V. Pless, "Cyclotomy and duadic codes of prime lengths," *IEEE Trans. Inform. Theory* **IT–45** (1999), 453–466.

[68] S. M. Dodunekov, "Zamechanie o vesovoy strukture porozhdayushchikh matrits lineĭnykh kodov," *Prob. peredach. inform.* **26** (1990), 101–104.

[69] S. M. Dodunekov and N. L. Manev, "Minimum possible block length of a linear code for some distance," *Problems of Inform. Trans.* **20** (1984), 8–14.

[70] S. M. Dodunekov and N. L. Manev, "An improvement of the Griesmer bound for some small minimum distances," *Discrete Applied Math.* **12** (1985), 103–114.

[71] R. Dougherty and H. Janwa, "Covering radius computations for binary cyclic codes," *Mathematics of Computation* **57** (1991), 415–434.

[72] D. E. Downie and N. J. A. Sloane, "The covering radius of cyclic codes of length up to 31," *IEEE Trans. Inform. Theory* **IT–31** (1985), 446–447.

[73] A. Drápal, "Yet another approach to ternary Golay codes," *Discrete Math.* **256** (2002), 459–464.

[74] V. G. Drinfeld and S. G. Vlăduţ, "The number of points on an algebraic curve," *Functional Anal.* **17** (1993), 53–54.

[75] I. Dumer, "Concatenated codes and their multilevel generalizations," in *Handbook of Coding Theory*, eds. V. S. Pless and W. C. Huffman. Amsterdam: Elsevier, 1998, pp. 1911–1988.

[76] W. Ebeling, *Lattices and Codes*. Wiesbaden, Germany: Friedr Vieweg & Sohn Verlagsgesellschaft, 1994.

[77] P. Elias, "Error-free coding," *IRE Trans. Inform. Theory* **IT–4** (1954), 29–37. (Also reprinted in [17] pp. 39–47.)

[78] P. Elias, "Coding for noisy channels," *1955 IRE International Convention Record* (part 4), 37–46. (Also reprinted in [17] pp. 48–55.)

[79] T. Etzion, A. Trachtenberg, and A. Vardy, "Which codes have cycle-free Tanner graphs?" *IEEE Trans. Inform. Theory* **IT–45** (1999), 2173–2181.

[80] W. Feit, "A self-dual even (96,48,16) code," *IEEE Trans. Inform. Theory* **IT–20** (1974), 136–138.

[81] L. Fejes Tóth, "Über einen geometrischen Satz," *Math. Zeit.* **46** (1940), 79–83.

[82] J. E. Fields, P. Gaborit, W. C. Huffman, and V. Pless, "On the classification of formally self-dual codes," *Proc. 36th Allerton Conf. on Commun. Control and Computing* (September 23–25, 1998), 566–575.

[83] J. E. Fields, P. Gaborit, W. C. Huffman, and V. Pless, "On the classification of extremal even formally self-dual codes," *Designs, Codes and Crypt.* **18** (1999), 125–148.

[84] J. E. Fields, P. Gaborit, W. C. Huffman, and V. Pless, "On the classification of extremal even formally self-dual codes of lengths 20 and 22," *Discrete Applied Math.* **111** (2001), 75–86.

[85] J. Fields, P. Gaborit, J. Leon, and V. Pless, "All self-dual $\mathbb{Z}_4$ codes of length 15 or less are known," *IEEE Trans. Inform. Theory* **IT–44** (1998), 311–322.

[86] G. D. Forney, Jr., "On decoding BCH codes," *IEEE Trans. Inform. Theory* **IT–11** (1965), 549–557. (Also reprinted in [17] pp. 136–144.)

[87] G. D. Forney, Jr., *Concatenated Codes*. Cambridge, MA: MIT Press, 1966.

[88] G. D. Forney, Jr., "Convolutional codes I: algebraic structure," *IEEE Trans. Inform. Theory* **IT–16** (1970), 720–738.

[89] G. D. Forney, Jr., "Structural analysis of convolutional codes via dual codes," *IEEE Trans. Inform. Theory* **IT–19** (1973), 512–518.

[90] G. D. Forney, Jr., "Minimal bases of rational vector spaces with applications to multivariable linear systems," *SIAM J. Control* **13** (1975), 493–502.

[91] G. D. Forney, Jr., "On iterative decoding and the two-way algorithm," preprint.

[92] P. Gaborit, "Mass formulas for self-dual codes over $\mathbb{Z}_4$ and $\mathbb{F}_q + u\mathbb{F}_q$ rings," *IEEE Trans. Inform. Theory* **IT-42** (1996), 1222–1228.

[93] P. Gaborit, W. C. Huffman, J.-L. Kim, and V. Pless, "On additive GF(4) codes," in *Codes and Association Schemes* (*DIMACS Workshop, November 9–12, 1999*), eds. A. Barg and S. Litsyn. Providence, RI: American Mathematical Society, 2001, pp. 135–149.

[94] R. G. Gallager, "Low-density parity-check codes," *IRE Trans. Inform. Theory* **IT-8** (1962), 21–28.

[95] R. G. Gallager, *Low-Density Parity-Check Codes*. Cambridge, MA: MIT Press, 1963.

[96] A. Garcia and H. Stichtenoth, "A tower of Artin–Schreier extensions of function fields attaining the Drinfeld–Vlăduţ bound," *Invent. Math.* **121** (1995), 211–222.

[97] A. Garcia and H. Stichtenoth, "On the asymptotic behavior of some towers of function fields over finite fields," *J. Number Theory* **61** (1996), 248–273.

[98] E. N. Gilbert, "A comparison of signaling alphabets," *Bell System Tech. J.* **31** (1952), 504–522. (Also reprinted in [17] pp. 14–19 and [23] pp. 24–42.)

[99] D. Goedhart, R. J. van de Plassche, and E. F. Stikvoort, "Digital-to-analog conversion in playing a compact disc," *Philips Technical Review* **40** No. 6 (1982), 174–179.

[100] J.-M. Goethals, "On the Golay perfect binary code," *J. Comb. Theory* **11** (1971), 178–186.

[101] J.-M. Goethals and S. L. Snover, "Nearly perfect codes," *Discrete Math.* **3** (1972), 65–88.

[102] M. J. E. Golay, "Notes on digital coding," *Proc. IEEE* **37** (1949), 657. (Also reprinted in [17] p. 13 and [23] p. 9.)

[103] V. D. Goppa, "A new class of linear error-correcting codes," *Problems of Inform. Trans.* **6** (1970), 207–212.

[104] V. D. Goppa, "Rational representation of codes and (L, g)-codes," *Problems of Inform. Trans.* **7** (1971), 223–229.

[105] V. D. Goppa, "Codes associated with divisors," *Problems of Inform. Trans.* **13** (1977), 22–26.

[106] V. D. Goppa, "Codes on algebraic curves," *Soviet Math. Dokl.* **24** (1981), 170–172.

[107] D. M. Gordon, "Minimal permutation sets for decoding the binary Golay codes," *IEEE Trans. Inform. Theory* **IT-28** (1982), 541–543.

[108] D. C. Gorenstein, W. W. Peterson, and N. Zierler, "Two error-correcting Bose–Chaudhury codes are quasi-perfect," *Inform. and Control* **3** (1960), 291–294.

[109] D. C. Gorenstein and N. Zierler, "A class of error-correcting codes in p^m symbols," *J. SIAM* **9** (1961), 207–214. (Also reprinted in [17] pp. 87–89 and [23] pp. 194–201.)

[110] T. Gosset, "On the regular and semi-regular figures in space of n dimensions," *Messenger Math.* **29** (1900), 43–48.

[111] R. L. Graham and N. J. A. Sloane, "On the covering radius of codes," *IEEE Trans. Inform. Theory* **IT-31** (1985), 385–401.

[112] J. H. Griesmer, "A bound for error-correcting codes," *IBM J. Research Develop.* **4** (1960), 532–542.

[113] V. Guruswami and M. Sudan, "Improved decoding of Reed–Solomon and algebraic-geometry codes," *IEEE Trans. Inform. Theory* **IT-45** (1999), 1757–1767.

[114] N. Hamada, "A characterization of some $[n, k, d; q]$-codes meeting the Griesmer bound using a minihyper in a finite projective geometry," *Discrete Math.* **116** (1993), 229–268.

[115] R. W. Hamming, *Coding and Information Theory*, 2nd ed. Englewood Cliffs, NJ: Prentice-Hall, 1986.

[116] A. R. Hammons, P. V. Kumar, A. R. Calderbank, N. J. A. Sloane, and P. Solé, "The $\mathbb{Z}_4$-linearity of Kerdock, Preparata, Goethals, and related codes," *IEEE Trans. Inform. Theory* **IT-40** (1994), 301–319.

[117] C. R. P. Hartmann and K. K. Tzeng, "Generalizations of the BCH bound," *Inform. and Control* **20** (1972), 489–498.

[118] C. Heegard and S. B. Wicker, *Turbo Coding*. Norwell, MA: Kluwer Academic Publishers, 1999.

[119] J. P. J. Heemskerk and K. A. S. Immink, "Compact disc: system aspects and modulation," *Philips Technical Review* **40** No. 6 (1982), 157–164.

[120] H. J. Helgert and R. D. Stinaff, "Minimum distance bounds for binary linear codes," *IEEE Trans. Inform. Theory* **IT–19** (1973), 344–356.

[121] T. Helleseth, "All binary 3-error-correcting BCH codes of length $2^m - 1$ have covering radius 5," *IEEE Trans. Inform. Theory* **IT–24** (1978), 257–258.

[122] T. Helleseth, "Projective codes meeting the Griesmer bound," *Discrete Math.* **107** (1992), 265–271.

[123] T. Helleseth and T. Kløve, "The Newton radius of codes," *IEEE Trans. Inform. Theory* **IT–43** (1997), 1820–1831.

[124] T. Helleseth, T. Kløve, V. I. Levenshtein, and O. Ytrehus, "Bounds on the minimum support weights," *IEEE Trans. Inform. Theory* **IT–41** (1995), 432–440.

[125] T. Helleseth, T. Kløve, and J. Mykkeltveit, "The weight distribution of irreducible cyclic codes with block length $n_1((q^\ell - 1)/N)$," *Discrete Math.* **18** (1977), 179–211.

[126] T. Helleseth, T. Kløve, and J. Mykkeltveit, "On the covering radius of binary codes," *IEEE Trans. Inform. Theory* **IT–24** (1978), 627–628.

[127] T. Helleseth, T. Kløve, and O. Ytrehus, "Codes and the chain condition," *Proc. Int. Workshop on Algebraic and Combinatorial Coding Theory* (Voneshta Voda, Bulgaria June 22–28, 1992), 88–91.

[128] T. Helleseth, T. Kløve, and O. Ytrehus, "Generalized Hamming weights of linear codes," *IEEE Trans. Inform. Theory* **IT–38** (1992), 1133–1140.

[129] T. Helleseth and H. F. Mattson, Jr., "On the cosets of the simplex code," *Discrete Math.* **56** (1985), 169–189.

[130] I. N. Herstein, *Abstract Algebra*. New York: Macmillan, 1990.

[131] R. Hill and D. E. Newton, "Optimal ternary linear codes," *Designs, Codes and Crypt.* **2** (1992), 137–157.

[132] A. Hocquenghem, "Codes correcteurs d'erreurs," *Chiffres* (Paris) **2** (1959), 147–156. (Also reprinted in [17] pp. 72–74 and [23] pp. 155–164.)

[133] H. Hoeve, J. Timmermans, and L. B. Vries, "Error correction and concealment in the compact disc system," *Philips Technical Review* **40** No. 6 (1982), 166–172.

[134] G. Höhn, "Self-dual codes over the Kleinian four group," preprint, 1996. See also http://xxx.lanl.gov/(math.CO/0005266).

[135] T. Høholdt, J. H. van Lint, and R. Pellikaan, "Algebraic geometry codes," in *Handbook of Coding Theory*, eds. V. S. Pless and W. C. Huffman. Amsterdam: Elsevier, 1998, pp. 871–961.

[136] I. S. Honkala and H. O. Hämäläinen, "Bounds for abnormal binary codes with covering radius one," *IEEE Trans. Inform. Theory* **IT–37** (1991), 372–375.

[137] I. Honkala and A. Tietäväinen, "Codes and number theory," in *Handbook of Coding Theory*, eds. V. S. Pless and W. C. Huffman. Amsterdam: Elsevier, 1998, pp. 1141–1194.

[138] J. A. van der Horst and T. Berger, "Complete decoding of triple-error-correcting binary BCH codes," *IEEE Trans. Inform. Theory* **IT–22** (1976), 138–147.

[139] X. D. Hou, "On the covering radius of subcodes of a code," *IEEE Trans. Inform. Theory* **IT–37** (1991), 1706–1707.

[140] S. Houghten, C. Lam, and L. Thiel, "Construction of (48, 24, 12) doubly-even self-dual codes," *Congr. Numer.* **103** (1994), 41–53.

[141] S. K. Houghten, C. W. H. Lam, L. H. Thiel, and R. A. Parker, "The extended quadratic residue code is the only (48, 24, 12) self-dual doubly-even code," preprint.

[142] W. C. Huffman, "Automorphisms of codes with applications to extremal doubly even codes of length 48," *IEEE Trans. Inform. Theory* **IT–28** (1982), 511–521.

[143] W. C. Huffman, "Decomposing and shortening codes using automorphisms," *IEEE Trans. Inform. Theory* **IT–32** (1986), 833–836.

[144] W. C. Huffman, "On extremal self-dual quaternary codes of lengths 18 to 28, I," *IEEE Trans. Inform. Theory* **IT–36** (1990), 651–660.

[145] W. C. Huffman, "On extremal self-dual quaternary codes of lengths 18 to 28, II," *IEEE Trans. Inform. Theory* **IT–37** (1991), 1206–1216.

[146] W. C. Huffman, "On extremal self-dual ternary codes of lengths 28 to 40," *IEEE Trans. Inform. Theory* **IT–38** (1992), 1395–1400.

[147] W. C. Huffman, "The automorphism groups of the generalized quadratic residue codes," *IEEE Trans. Inform. Theory* **IT–41** (1995), 378–386.

[148] W. C. Huffman, "Characterization of quaternary extremal codes of lengths 18 and 20," *IEEE Trans. Inform. Theory* **IT–43** (1997), 1613–1616.

[149] W. C. Huffman, "Codes and groups," in *Handbook of Coding Theory*, eds. V. S. Pless and W. C. Huffman. Amsterdam: Elsevier, 1998, pp. 1345–1440.

[150] W. C. Huffman, V. Job, and V. Pless, "Multipliers and generalized multipliers of cyclic codes and cyclic objects," *J. Comb. Theory* **62A** (1993), 183–215.

[151] W. C. Huffman and V. D. Tonchev, "The existence of extremal self-dual [50, 25, 10] codes and quasi-symmetric 2-(49, 9, 6) designs," *Designs, Codes and Crypt.* **6** (1995), 97–106.

[152] W. C. Huffman and V. Y. Yorgov, "A [72, 36, 16] doubly even code does not have an automorphism of order 11," *IEEE Trans. Inform. Theory* **IT–33** (1987), 749–752.

[153] Y. Ihara, "Some remarks on the number of rational points of algebraic curves over finite fields," *J. Fac. Sci. Univ. Tokyo Sect. IA Math.* **28** (1981), 721–724.

[154] K. A. S. Immink, "Reed–Solomon codes and the compact disc," in *Reed–Solomon Codes and Their Applications*, eds. S. B. Wicker and V. K. Bhargava. New York: IEEE Press, 1994, pp. 41–59.

[155] N. Ito, J. S. Leon, and J. Q. Longyear, "Classification of 3-(24, 12, 5) designs and 24-dimensional Hadamard matrices," *J. Comb. Theory* **31A** (1981), 66–93.

[156] D. B. Jaffe, "Optimal binary linear codes of length ≤ 30," *Discrete Math.* **223** (2000), 135–155.

[157] G. J. Janusz, "Overlap and covering polynomials with applications to designs and self-dual codes," *SIAM J. Discrete Math.* **13** (2000), 154–178.

[158] R. Johannesson and K. Sh. Zigangirov, *Fundamentals of Convolutional Coding*. New York: IEEE Press, 1999.

[159] S. M. Johnson, "A new upper bound for error-correcting codes," *IEEE Trans. Inform. Theory* **IT–8** (1962), 203–207.

[160] J. Justesen, "A class of constructive aymptotically good algebraic codes," *IEEE Trans. Inform. Theory* **IT–18** (1972), 652–656. (Also reprinted in [17] pp. 95–99 and [23] pp. 400–404.)

[161] T. Kasami, "An upper bound on k/n for affine-invariant codes with fixed d/n," *IEEE Trans. Inform. Theory* **IT–15** (1969), 174–176.

[162] T. Kasami, S. Lin, and W. Peterson, "Some results on cyclic codes which are invariant under the affine group and their applications," *Inform. and Control* **11** (1968), 475–496.

[163] T. Kasami, S. Lin, and W. Peterson, "New generalizations of the Reed–Muller codes. Part I:

Primitive codes," *IEEE Trans. Inform. Theory* **IT–14** (1968), 189–199. (Also reprinted in [23] pp. 323–333.)

[164] G. T. Kennedy, "Weight distributions of linear codes and the Gleason–Pierce theorem," *J. Comb. Theory* **67A** (1994), 72–88.

[165] G. T. Kennedy and V. Pless, "On designs and formally self-dual codes," *Designs, Codes and Crypt.* **4** (1994), 43–55.

[166] A. M. Kerdock, "A class of low-rate nonlinear binary codes," *Inform. and Control* **20** (1972), 182–187.

[167] J.-L. Kim, "New self-dual codes over GF(4) with the highest known minimum weights," *IEEE Trans. Inform. Theory* **IT–47** (2001), 1575–1580.

[168] J.-L. Kim, "New extremal self-dual codes of lengths 36, 38, and 58," *IEEE Trans. Inform. Theory* **IT–47** (2001), 386–393.

[169] H. Kimura, "New Hadamard matrix of order 24," *Graphs and Combin.* **5** (1989), 235–242.

[170] H. Kimura, "Classification of Hadamard matrices of order 28," *Discrete Math.* **133** (1994), 171–180.

[171] H. Kimura and H. Ohnmori, "Classification of Hadamard matrices of order 28," *Graphs and Combin.* **2** (1986), 247–257.

[172] D. J. Kleitman, "On a combinatorial conjecture of Erdös," *J. Comb. Theory* **1** (1966), 209–214.

[173] M. Klemm, "Über die Identität von MacWilliams für die Gewichtsfunktion von Codes," *Archiv Math.* **49** (1987), 400–406.

[174] M. Klemm, "Selbstduale Codes über dem Ring der ganzen Zahlen modulo 4," *Archiv Math.* **53** (1989), 201–207.

[175] T. Kløve, "Support weight distribution of linear codes," *Discrete Math.* **107** (1992), 311–316.

[176] H. Koch, "Unimodular lattices and self-dual codes," in *Proc. Intern. Congress Math., Berkeley 1986*, Vol. 1. Providence, RI: Amer. Math. Soc., 1987, pp. 457–465.

[177] H. Koch, "On self-dual, doubly-even codes of length 32," *J. Comb. Theory* **51A** (1989), 63–76.

[178] R. Kötter, "On algebraic decoding of algebraic-geometric and cyclic codes," Ph.D. Thesis, University of Linköping, 1996.

[179] R. Kötter and A. Vardy, "Algebraic soft-decision decoding of Reed–Solomon codes," preprint.

[180] F. R. Kschischang, B. J. Frey, and H.-A. Loeliger, "Factor graphs and the sum-product algorithm," *IEEE Trans. Inform. Theory* **IT–47** (2001), 498–519.

[181] C. W. H. Lam, "The search for a finite projective plane of order 10," *Amer. Math. Monthly* **98** (1991), 305–318.

[182] C. W. H. Lam, G. Kolesova, and L. Thiel, "A computer search for finite projective planes of order 9," *Discrete Math.* **92** (1991), 187–195.

[183] C. W. H. Lam and V. Pless, "There is no (24, 12, 10) self-dual quaternary code," *IEEE Trans. Inform. Theory* **IT–36** (1990), 1153–1156.

[184] C. W. H. Lam, L. Thiel, and S. Swiercz, "The nonexistence of ovals in a projective plane of order 10," *Discrete Math.* **45** (1983), 319–321.

[185] C. W. H. Lam, L. Thiel, and S. Swiercz, "The nonexistence of codewords of weight 16 in a projective plane of order 10," *J. Comb. Theory* **42A** (1986), 207–214.

[186] C. W. H. Lam, L. Thiel, and S. Swiercz, "The nonexistence of finite projective planes of order 10," *Canad. J. Math.* **41** (1989), 1117–1123.

[187] E. Lander, *Symmetric Designs: An Algebraic Approach*. London: Cambridge University Press, 1983.

[188] J. Leech, "Notes on sphere packings," *Canadian J. Math.* **19** (1967), 251–267.

[189] J. S. Leon, "A probabilistic algorithm for computing minimum weights of large error-correcting codes," *IEEE Trans. Inform. Theory* **IT–34** (1988), 1354–1359.

[190] J. S. Leon, J. M. Masley, and V. Pless, "Duadic codes," *IEEE Trans. Inform. Theory* **IT–30** (1984), 709–714.

[191] J. S. Leon, V. Pless, and N. J. A. Sloane, "On ternary self-dual codes of length 24," *IEEE Trans. Inform. Theory* **IT–27** (1981), 176–180.

[192] V. I. Levenshtein, "A class of systematic codes," *Soviet Math. Dokl.* **1**, No. 1 (1960), 368–371.

[193] V. I. Levenshtein, "The application of Hadamard matrices to a problem in coding," *Problems of Cybernetics* **5** (1964), 166–184.

[194] V. I. Levenshtein, "Universal bounds for codes and designs," in *Handbook of Coding Theory*, eds. V. S. Pless and W. C. Huffman. Amsterdam: Elsevier, 1998, pp. 499–648.

[195] W. J. LeVeque, *Topics in Number Theory*. Reading, MA: Addison-Wesley, 1956.

[196] R. Lidl and H. Niederreiter, *Finite Fields*. Reading, MA: Addison-Wesley, 1983.

[197] S. Lin and D. J. Costello, *Error-Control Coding – Fundamentals and Applications*. Englewood Cliffs, NJ: Prentice-Hall, 1983.

[198] S. Lin and E. J. Weldon, Jr., "Long BCH codes are bad," *Inform. and Control* **11** (1967), 445–451.

[199] K. Lindström, "The nonexistence of unknown nearly perfect binary codes," *Ann. Univ. Turku Ser. A. I* **169** (1975), 7–28.

[200] K. Lindström, "All nearly perfect codes are known," *Inform. and Control* **35** (1977), 40–47.

[201] J. H. van Lint, "Repeated root cyclic codes," *IEEE Trans. Inform. Theory* **IT–37** (1991), 343–345.

[202] J. H. van Lint, "The mathematics of the compact disc," *Mitteilungen der Deutschen Mathematiker-Vereinigung* **4** (1998), 25–29.

[203] J. H. van Lint and R. M. Wilson, "On the minimum distance of cyclic codes," *IEEE Trans. Inform. Theory* **IT–32** (1986), 23–40.

[204] J. H. van Lint and R. M. Wilson, *A Course in Combinatorics*. Cambridge: Cambridge University Press, 1992.

[205] S. Litsyn, "An updated table of the best binary codes known," in *Handbook of Coding Theory*, eds. V. S. Pless and W. C. Huffman. Amsterdam: Elsevier, 1998, pp. 463–498.

[206] S. P. Lloyd, "Binary block coding," *Bell System Tech. J.* **36** (1957), 517–535. (Also reprinted in [17] pp. 246–251.)

[207] A. C. Lobstein and V. S. Pless, "The length function, a revised table," in *Lecture Notes in Computer Science*, No. 781. New York: Springer-Verlag, 1994, pp. 51–55.

[208] L. Lovász, "Kneser's conjecture, chromatic number and homotopy," *J. Comb. Theory* **25A** (1978), 319–324.

[209] E. Lucas, "Sur les congruences des nombres euleriennes et des coefficients différentiels des fonctions trigonométriques, suivant un module premier," *Bull. Soc. Math. (France)* **6** (1878), 49–54.

[210] C. C. MacDuffee, *Theory of Equations*. New York: Wiley & Sons, 1954.

[211] D. J. C. Mackay, "Good error correcting codes based on very sparse matrices," *IEEE Trans. Inform. Theory* **IT–45** (1999), 399–431.

[212] F. J. MacWilliams, "Combinatorial problems of elementary abelian groups," Ph.D. Thesis, Harvard University, 1962.

[213] F. J. MacWilliams, "A theorem on the distribution of weights in a systematic code," *Bell System Tech. J.* **42** (1963), 79–94. (Also reprinted in [17] pp. 261–265 and [23] pp. 241–257.)

[214] F. J. MacWilliams, "Permutation decoding of systematic codes," *Bell System Tech. J.* **43** (1964), 485–505.

[215] F. J. MacWilliams, C. L. Mallows, and N. J. A. Sloane, "Generalizations of Gleason's theorem on weight enumerators of self-dual codes," *IEEE Trans. Inform. Theory* **IT–18** (1972), 794–805.

[216] F. J. MacWilliams and H. B. Mann, "On the p-rank of the design matrix of a difference set," *Inform. and Control* **12** (1968), 474–488.

[217] F. J. MacWilliams, A. M. Odlyzko, N. J. A. Sloane, and H. N. Ward, "Self-dual codes over GF(4)," *J. Comb. Theory* **25A** (1978), 288–318.

[218] F. J. MacWilliams and N. J. A. Sloane, *The Theory of Error-Correcting Codes*. New York: Elsevier/North Holland, 1977.

[219] F. J. MacWilliams, N. J. A. Sloane, and J. G. Thompson, "Good self-dual codes exist," *Discrete Math.* **3** (1972), 153–162.

[220] F. J. MacWilliams, N. J. A. Sloane, and J. G. Thompson, "On the nonexistence of a finite projective plane of order 10," *J. Comb. Theory* **14A** (1973), 66–78.

[221] C. L. Mallows, A. M. Odlyzko, and N. J. A. Sloane, "Upper bounds for modular forms, lattices, and codes," *J. Algebra* **36** (1975), 68–76.

[222] C. L. Mallows, V. Pless, and N. J. A. Sloane, "Self-dual codes over GF(3)," *SIAM J. Applied Mathematics* **31** (1976), 649–666.

[223] C. L. Mallows and N. J. A. Sloane, "An upper bound for self-dual codes," *Inform. and Control* **22** (1973), 188–200.

[224] J. L. Massey, "Shift-register synthesis and BCH decoding," *IEEE Trans. Inform. Theory* **IT–15** (1969), 122–127. (Also reprinted in [23] pp. 233–238.)

[225] J. L. Massey, "Error bounds for tree codes, trellis codes, and convolutional codes, with encoding and decoding procedures," *Coding and Complexity*, ed. G. Longo, CISM Courses and Lectures No. 216. New York: Springer, 1977.

[226] J. L. Massey, "The how and why of channel coding," *Proc. 1984 Int. Zurich Seminar on Digital Communications* (1984), 67–73.

[227] J. L. Massey, "Deep space communications and coding: a match made in heaven," *Advanced Methods for Satellite and Deep Space Communications*, ed. J. Hagenauer, Lecture Notes in Control and Inform. Sci. **182**. Berlin: Springer, 1992.

[228] J. L. Massey and D. J. Costello, Jr., "Nonsystematic convolutional codes for sequential decoding in space applications," *IEEE Trans. Commun. Technol.* **COM–19** (1971), 806–813.

[229] J. L. Massey and M. K. Sain, "Inverses of linear sequential circuits," *IEEE Trans. Comput.* **C–17** (1968), 330–337.

[230] B. R. McDonald, *Finite Rings with Identity*. New York: Marcel Dekker, 1974.

[231] R. J. McEliece, "Weight congruences for p-ary cyclic codes," *Discrete Math.* **3** (1972), 177–192.

[232] R. J. McEliece, *The Theory of Information and Coding*. Reading, MA: Addison-Wesley, 1977.

[233] R. J. McEliece, *Finite Fields for Computer Scientists and Engineers*. Boston: Kluwer Academic Publishers, 1987.

[234] R. J. McEliece, "On the BCJR trellis for linear block codes," *IEEE Trans. Inform. Theory* **IT–42** (1996), 1072–1092.

[235] R. J. McEliece, "The algebraic theory of convolutional codes," in *Handbook of Coding Theory*, eds. V. S. Pless and W. C. Huffman. Amsterdam: Elsevier, 1998, pp. 1065–1138.

[236] R. J. McEliece, E. R. Rodemich, H. Rumsey, Jr., and L. Welch, "New upper bounds on the rate of a code via the Delsarte–MacWilliams inequalities," *IEEE Trans. Inform. Theory* **IT–23** (1977), 157–166.

[237] A. McLoughlin, "The covering radius of the $(m - 3)$rd order Reed–Muller codes and a lower bound on the covering radius of the $(m - 4)$th order Reed–Muller codes," *SIAM J. Applied Mathematics* **37** (1979), 419–422.

[238] J. E. Meggitt, "Error-correcting codes for correcting bursts of errors," *IBM J. Res. Develop.* **4** (1960), 329–334.

[239] J. E. Meggitt, "Error-correcting codes and their implementation," *IRE Trans. Inform. Theory* **IT–6** (1960), 459–470.

[240] C. Moreno, *Algebraic Curves over Finite Fields*. Cambridge Tracts in Math. **97**. Cambridge: Cambridge University Press, 1991.

[241] D. E. Muller, "Application of Boolean algebra to switching circuit design and to error detection," *IEEE Trans. Computers* **3** (1954), 6–12. (Also reprinted in [17] pp. 20–26 and [23] pp. 43–49.)

[242] J. Mykkeltveit, "The covering radius of the (128, 8) Reed–Muller code is 56," *IEEE Trans. Inform. Theory* **IT–26** (1980), 359–362.

[243] A. A. Nechaev, "The Kerdock code in a cyclic form," *Diskret. Mat.* **1** (1989), 123–139. (English translation in *Discrete Math. Appl.* **1** (1991), 365–384.)

[244] A. Neumeier, private communication, 1990.

[245] R. R. Nielsen and T. Høholdt, "Decoding Reed–Solomon codes beyond half the minimum distance," in *Coding Theory, Cryptography, and Related Areas (Guanajunto, 1998)*, eds. J. Buchmann, T. Høholdt, H. Stichtenoth, and H. Tapia-Recillas. Berlin: Springer, 2000, pp. 221–236.

[246] H. V. Niemeier, "Definite Quadratische Formen der Dimension 24 und Diskriminante 1," *J. Number Theory* **5** (1973), 142–178.

[247] A. W. Nordstrom and J. P. Robinson, "An optimum nonlinear code," *Inform. and Control* **11** (1967), 613–616. (Also reprinted in [17] p. 101 and [23] pp. 358–361.)

[248] V. Olshevesky and A. Shokrollahi, "A displacement structure approach to efficient decoding of Reed–Solomon and algebraic-geometric codes," *Proc. 31st ACM Symp. Theory of Computing* (1999), 235–244.

[249] J. Olsson, "On the quaternary [18, 9, 8] code," *Proceedings of the Workshop on Coding and Cryptography, WCC99-INRIA* Jan. 10–14, 1999, pp. 65–73.

[250] N. J. Patterson and D. H. Wiedemann, "The covering radius of the $(2^{15}, 16)$ Reed–Muller code is at least 16 276," *IEEE Trans. Inform. Theory* **IT–29** (1983), 354–356.

[251] N. J. Patterson and D. H. Wiedemann, "Correction to 'The covering radius of the $(2^{15}, 16)$ Reed–Muller code is at least 16276'," *IEEE Trans. Inform. Theory* **IT–36** (1990), 443.

[252] J. Pearl, *Probabilistic Reasoning in Intelligent Systems: Networks of Plausible Inference*. San Mateo, CA: Morgan Kaufmann, 1988.

[253] J. B. H. Peek, "Communications aspects of the compact disc digital audio system," *IEEE Communications Mag.* **23** (1985), 7–15.

[254] W. W. Peterson, "Encoding and error-correction procedures for the Bose–Chaudhuri codes," *IRE Trans. Inform. Theory* **IT–6** (1960), 459–470. (Also reprinted in [17] pp. 109–120 and [23] pp. 221–232.)

[255] W. W. Peterson, *Error-Correcting Codes*. Cambridge, MA: MIT Press, 1961.

[256] W. W. Peterson and E. J. Weldon, Jr., *Error-Correcting Codes*, 2nd ed. Cambridge, MA.: MIT Press, 1972.

[257] P. Piret, *Convolutional Codes: An Algebraic Approach*. Cambridge, MA: MIT Press, 1988.

[258] V. Pless, "Power moment identities on weight distributions in error correcting codes," *Inform. and Control* **6** (1963), 147–152. (Also reprinted in [17] pp. 266–267 and [23] pp. 257–262.)

[259] V. Pless, "The number of isotropic subspaces in a finite geometry," *Rend. Cl. Scienze fisiche, matematiche e naturali,* Acc. Naz. Lincei **39** (1965), 418–421.

[260] V. Pless, "On the uniqueness of the Golay codes," *J. Comb. Theory* **5** (1968), 215–228.

[261] V. Pless, "The weight of the symmetry code for $p = 29$ and the 5-designs contained therein," *Annals N. Y. Acad. of Sciences* **175** (1970), 310–313.

[262] V. Pless, "A classification of self-orthogonal codes over GF(2)," *Discrete Math.* **3** (1972), 209–246.

[263] V. Pless, "Symmetry codes over GF(3) and new 5-designs," *J. Comb. Theory* **12** (1972), 119–142.

[264] V. Pless, "The children of the (32, 16) doubly even codes," *IEEE Trans. Inform. Theory* **IT–24** (1978), 738–746.

[265] V. Pless, "23 does not divide the order of the group of a (72, 36, 16) doubly-even code," *IEEE Trans. Inform. Theory* **IT–28** (1982), 112–117.

[266] V. Pless, "Q-Codes," *J. Comb. Theory* **43A** (1986), 258–276.

[267] V. Pless, "Cyclic projective planes and binary, extended cyclic self-dual codes," *J. Comb. Theory* **43A** (1986), 331–333.

[268] V. Pless, "Decoding the Golay codes," *IEEE Trans. Inform. Theory* **IT–32** (1986), 561–567.

[269] V. Pless, "More on the uniqueness of the Golay code," *Discrete Math.* **106/107** (1992), 391–398.

[270] V. Pless, "Duadic codes and generalizations," in *Proc. of Eurocode 1992, Udine, Italy, CISM Courses and Lectures* No. 339, eds. P. Camion, P. Charpin, and S. Harari. Vienna: Springer, 1993, pp. 3–16.

[271] V. Pless, "Parents, children, neighbors and the shadow," *Contemporary Math.* **168** (1994), 279–290.

[272] V. Pless, "Constraints on weights in binary codes," *Applicable Algebra in Engineering, Communication and Computing (AAECC Journal)* **8** (1997), 411–414.

[273] V. Pless, *Introduction to the Theory of Error-Correcting Codes,* 3rd ed. New York: J. Wiley & Sons, 1998.

[274] V. Pless, "Coding constructions," in *Handbook of Coding Theory,* eds. V. S. Pless and W. C. Huffman. Amsterdam: Elsevier, 1998, pp. 141–176.

[275] V. S. Pless, W. C. Huffman, and R. A. Brualdi, "An introduction to algebraic codes," in *Handbook of Coding Theory,* eds. V. S. Pless and W. C. Huffman. Amsterdam: Elsevier, 1998, pp. 3–139.

[276] V. Pless, J. Leon, and J. Fields, "All $\mathbb{Z}_4$ codes of Type II and length 16 are known," *J. Comb. Theory* **78A** (1997), 32–50.

[277] V. Pless, J. M. Masley, and J. S. Leon, "On weights in duadic codes," *J. Comb. Theory* **44A** (1987), 6–21.

[278] V. Pless and J. N. Pierce, "Self-dual codes over GF(q) satisfy a modified Varshamov–Gilbert bound," *Inform. and Control* **23** (1973), 35–40.

[279] V. Pless and Z. Qian, "Cyclic codes and quadratic residue codes over $\mathbb{Z}_4$," *IEEE Trans. Inform. Theory* **IT–42** (1996), 1594–1600.

[280] V. Pless and J. J. Rushanan, "Triadic Codes," *Lin. Alg. and Its Appl.* **98** (1988), 415–433.

[281] V. Pless and N. J. A. Sloane, "On the classification and enumeration of self-dual codes," *J. Comb. Theory* **18** (1975), 313–335.

[282] V. Pless, N. J. A. Sloane, and H. N. Ward, "Ternary codes of minimum weight 6 and the classification of self-dual codes of length 20," *IEEE Trans. Inform. Theory* **IT–26** (1980), 305–316.

[283] V. Pless, P. Solé, and Z. Qian, "Cyclic self-dual $\mathbb{Z}_4$-codes," *Finite Fields and Their Appl.* **3** (1997), 48–69.

[284] V. Pless and J. Thompson, "17 does not divide the order of the group of a (72, 36, 16) doubly-even code," *IEEE Trans. Inform. Theory* **IT–28** (1982), 537–544.

[285] M. Plotkin, "Binary codes with specified minimum distances," *IRE Trans. Inform. Theory* **IT–6** (1960), 445–450. (Also reprinted in [17] pp. 238–243.)

[286] K. C. Pohlmann, *Principles of Digital Audio*, 4th ed. New York: McGraw-Hill, 2000.

[287] A. Pott, "Applications of the DFT to abelian difference sets," *Archiv Math.* **51** (1988), 283–288.

[288] F. P. Preparata, "A class of optimum nonlinear double-error correcting codes," *Inform. and Control* **13** (1968), 378–400. (Also reprinted in [23] pp. 366–388.)

[289] Z. Qian, "Cyclic codes over $\mathbb{Z}_4$," Ph.D. Thesis, University of Illinois at Chicago, 1996.

[290] E. M. Rains, "Shadow bounds for self-dual codes," *IEEE Trans. Inform. Theory* **IT–44** (1998), 134–139.

[291] E. M. Rains and N. J. A. Sloane, "Self-dual codes," in *Handbook of Coding Theory*, eds. V. S. Pless and W. C. Huffman. Amsterdam: Elsevier, 1998, pp. 177–294.

[292] E. M. Rains and N. J. A. Sloane, "The shadow theory of modular and unimodular lattices," *J. Number Theory* **73** (1998), 359–389.

[293] I. S. Reed, "A class of multiple-error-correcting codes and the decoding scheme," *IRE Trans. Inform. Theory* **IT–4** (1954), 38–49. (Also reprinted in [17] pp. 27–38 and [23] pp. 50–61.)

[294] I. S. Reed and G. Solomon, "Polynomial codes over certain finite fields," *J. SIAM* **8** (1960), 300–304. (Also reprinted in [17] pp. 70–71 and [23] pp. 189–193.)

[295] T. J. Richardson and R. Urbanke, "The capacity of low-density parity-check codes under message-passing decoding," *IEEE Trans. Inform. Theory* **IT–47** (2001), 599–618.

[296] C. Roos, "A generalization of the BCH bound for cyclic codes, including the Hartmann–Tzeng bound," *J. Comb. Theory* **33A** (1982), 229–232.

[297] C. Roos, "A new lower bound on the minimum distance of a cyclic code," *IEEE Trans. Inform. Theory* **IT–29** (1983), 330–332.

[298] R. M. Roth, personal communication.

[299] R. M. Roth and A. Lempel, "On MDS codes via Cauchy matrices," *IEEE Trans. Inform. Theory* **IT–35** (1989), 1314–1319.

[300] R. M. Roth and G. Ruckenstein, "Efficient decoding of Reed–Solomon codes beyond half the minimum distance," *IEEE Trans. Inform. Theory* **IT–46** (2000), 246–257.

[301] J. J. Rushanan, "Generalized Q-codes," Ph.D. Thesis, California Institute of Technology, 1986.

[302] C. J. Salwach, "Planes, biplanes, and their codes," *American Math. Monthly* **88** (1981), 106–125.

[303] N. V. Semakov and V. A. Zinov'ev, "Complete and quasi-complete balanced codes," *Problems of Inform. Trans.* **5**(2) (1969), 11–13.

[304] N. V. Semakov and V. A. Zinov'ev, "Balanced codes and tactical configurations," *Problems of Inform. Trans.* **5**(3) (1969), 22–28.

[305] N. V. Semakov, V. A. Zinov'ev, and G. V. Zaitsev, "Uniformly packed codes," *Problems of Inform. Trans.* **7**(1) (1971), 30–39.

[306] C. Shannon, "A mathematical theory of communication," *Bell System Tech. J.* **27** (1948), 379–423 and 623–656.

[307] K. Shum, I. Aleshnikov, P. V. Kumar, and H. Stichtenoth, "A low complexity algorithm for the construction of algebraic geometry codes better than the Gilbert–Varshamov bound," *Proc. 38th Allerton Conf. on Commun. Control and Computing (October 4–6, 2000)*, 1031–1037.

[308] J. Simonis, "On generator matrices of codes," *IEEE Trans. Inform. Theory* **IT–38** (1992), 516.

[309] J. Simonis, "The [18, 9, 6] code is unique," *Discrete Math.* **106/107** (1992), 439–448.

[310] J. Simonis, "The effective length of subcodes," *Applicable Algebra in Engineering, Communication and Computing (AAECC Journal)* **5** (1994), 371–377.

[311] J. Simonis, "Subcodes and covering radius: a generalization of Adam's result," unpublished.

[312] R. C. Singleton, "Maximum distance q-ary codes," *IEEE Trans. Inform. Theory* **IT–10** (1964), 116–118.

[313] N. J. A. Sloane, "Weight enumerators of codes," in *Combinatorics*, eds. M. Hall, Jr. and J. H. van Lint. Dordrecht, Holland: Reidel Publishing, 1975, 115–142.

[314] N. J. A. Sloane, "Relations between combinatorics and other parts of mathematics," *Proc. Symposia Pure Math.* **34** (1979), 273–308.

[315] M. H. M. Smid, "Duadic codes," *IEEE Trans. Inform. Theory* **IT–33** (1987), 432–433.

[316] K. J. C. Smith, "On the p-rank of the incidence matrix of points and hyperplanes in a finite projective geometry," *J. Comb. Theory* **7** (1969), 122–129.

[317] S. L. Snover, "The uniqueness of the Nordstrom–Robinson and the Golay binary codes," Ph.D. Thesis, Michigan State University, 1973.

[318] P. Solé, "A quaternary cyclic code, and a family of quadriphase sequences with low correlation properties," *Lecture Notes in Computer Science* **388** (1989), 193–201.

[319] G. Solomon and J. J. Stiffler, "Algebraically punctured cyclic codes," *Inform. and Control* **8** (1965), 170–179.

[320] H. Stichtenoth, *Algebraic Function Fields and Codes*. New York: Springer-Verlag, 1993.

[321] L. Storme and J. A. Thas, "M.D.S. codes and arcs in PG(n, q) with q even: an improvement on the bounds of Bruen, Thas, and Blokhuis," *J. Comb. Theory* **62A** (1993), 139–154.

[322] R. Struik, "On the structure of linear codes with covering radius two and three," *IEEE Trans. Inform. Theory* **IT–40** (1994), 1406–1416.

[323] M. Sudan, "Decoding of Reed–Solomon codes beyond the error-correction bound," *J. Complexity* **13** (1997), 180–193.

[324] Y. Sugiyama, M. Kasahara, S. Hirasawa, and T. Namekawa, "A method for solving a key equation for decoding Goppa codes," *Inform. and Control* **27** (1975), 87–99.

[325] R. M. Tanner, "A recursive approach to low-complexity codes," *IEEE Trans. Inform. Theory* **IT–27** (1981), 533–547.

[326] L. Teirlinck, "Nontrivial t-designs without repeated blocks exist for all t," *Discrete Math.* **65** (1987), 301–311.

[327] A. Thue, "Über die dichteste Zusammenstellung von kongruenten Kreisen in einer Ebene," *Norske Vid. Selsk. Skr.* No. 1 (1910), 1–9.

[328] H. C. A. van Tilborg, "On the uniqueness resp. nonexistence of certain codes meeting the Griesmer bound," *Info. and Control* **44** (1980), 16–35.

[329] H. C. A. van Tilborg, "Coding theory at work in cryptology and vice versa," in *Handbook of Coding Theory*, eds. V. S. Pless and W. C. Huffman. Amsterdam: Elsevier, 1998, pp. 1195–1227.

[330] J. A. Todd, "A combinatorial problem," *J. Math. Phys.* **12** (1933), 321–333.

[331] V. D. Tonchev, *Combinatorial Configurations*. New York: Longman-Wiley, 1988.

[332] V. D. Tonchev, "Codes and designs," in *Handbook of Coding Theory*, eds. V. S. Pless and W. C. Huffman. Amsterdam: Elsevier, 1998, pp. 1229–1267.

[333] M. A. Tsfasman, S. G. Vlăduţ, and T. Zink, "Modular curves, Shimura curves and Goppa codes, better than Varshamov–Gilbert bound," *Math. Nachrichten* **109** (1982), 21–28.

[334] J. V. Uspensky, *Theory of Equations*. New York: McGraw-Hill, 1948.

[335] S. A. Vanstone and P. C. van Oorschot, *An Introduction to Error Correcting Codes with Applications*. Boston: Kluwer Academic Publishers, 1989.

[336] A. Vardy, "Trellis structure of codes," in *Handbook of Coding Theory*, eds. V. S. Pless and W. C. Huffman. Amsterdam: Elsevier, 1998, pp. 1989–2117.

[337] R. R. Varshamov, "Estimate of the number of signals in error correcting codes," *Dokl. Akad. Nauk SSSR* **117** (1957), 739–741. (English translation in [23] pp. 68–71.)

[338] J. L. Vasil'ev, "On nongroup close-packed codes," *Probl. Kibernet.* **8** (1962), 375–378. (English translation in [23] pp. 351–357.)

[339] A. J. Viterbi, "Error bounds for convolutional codes and an asymptotically optimum decoding algorithm," *IEEE Trans. Inform. Theory* **IT–13** (1967), 260–269. (Also reprinted in [17] pp. 195–204.)

[340] S. G. Vlăduţ and A. N. Skorobogatov, "Covering radius for long BCH codes," *Problemy Peredachi Informatsii* **25**(1) (1989) 38–45. (English translation in *Problems of Inform. Trans.* **25**(1) (1989), 28–34.)

[341] L. B. Vries and K. Odaka, "CIRC – the error correcting code for the compact disc," in *Digital Audio* (Collected papers from the AES Premier Conference, Rye, NY, June 3–6, 1982), eds. B. A. Blesser *et al.* Audio Engineering Society Inc., 1983, 178–186.

[342] B. Vucetic and J. Yuan, *Turbo Codes: Principles and Applications.* Norwell, MA: Kluwer Academic Publishers, 2000.

[343] J. L. Walker, *Codes and Curves.* Student Mathematical Library Series **7**. Providence, RI: American Mathematical Society, 2000.

[344] H. N. Ward, "A restriction on the weight enumerator of self-dual codes," *J. Comb. Theory* **21** (1976), 253–255.

[345] H. N. Ward, "Divisible codes," *Archiv Math. (Basel)* **36** (1981), 485–494.

[346] H. N. Ward, "Divisibility of codes meeting the Griesmer bound," *J. Comb. Theory* **83A** (1998), 79–93.

[347] H. N. Ward, "Quadratic residue codes and divisibility," in *Handbook of Coding Theory*, eds. V. S. Pless and W. C. Huffman. Amsterdam: Elsevier, 1998, pp. 827–870.

[348] V. K. Wei, "Generalized Hamming weights for linear codes," *IEEE Trans. Inform. Theory* **IT–37** (1991), 1412–1418.

[349] E. J. Weldon, Jr., "New generalizations of the Reed–Muller codes. Part II: Nonprimitive codes," *IEEE Trans. Inform. Theory* **IT–14** (1968), 199–205. (Also reprinted in [23] pp. 334–340.)

[350] N. Wiberg, H.-A. Loeliger, and R. Kötter, "Codes and iterative decoding on general graphs," *Euro. Trans. Telecommun.* **6** (1995), 513–526.

[351] S. B. Wicker, *Error Control Systems for Digital Communication and Storage.* Englewood Cliffs, NJ: Prentice-Hall, 1995.

[352] S. B. Wicker, "Deep space applications," in *Handbook of Coding Theory*, eds. V. S. Pless and W. C. Huffman. Amsterdam: Elsevier, 1998, pp. 2119–2169.

[353] H. A. Wilbrink, "A note on planar difference sets," *J. Comb. Theory* **38A** (1985), 94–95.

[354] X.-W. Wu and P. H. Siegel, "Efficient root-finding algorithm with applications to list decoding of algebraic-geometric codes," *IEEE Trans. Inform. Theory* **IT–47** (2001), 2579–2587.

Symbol index

Subject index